# Lecture Notes in Computer Science 13677

Founding Editors

Gerhard Goos
*Karlsruhe Institute of Technology, Karlsruhe, Germany*

Juris Hartmanis
*Cornell University, Ithaca, NY, USA*

Editorial Board Members

Elisa Bertino
*Purdue University, West Lafayette, IN, USA*

Wen Gao
*Peking University, Beijing, China*

Bernhard Steffen
*TU Dortmund University, Dortmund, Germany*

Moti Yung
*Columbia University, New York, NY, USA*

More information about this series at https://link.springer.com/bookseries/558

Shai Avidan · Gabriel Brostow ·
Moustapha Cissé · Giovanni Maria Farinella ·
Tal Hassner (Eds.)

# Computer Vision – ECCV 2022

## 17th European Conference
## Tel Aviv, Israel, October 23–27, 2022
## Proceedings, Part XVII

Springer

*Editors*
Shai Avidan
Tel Aviv University
Tel Aviv, Israel

Gabriel Brostow
University College London
London, UK

Moustapha Cissé
Google AI
Accra, Ghana

Giovanni Maria Farinella
University of Catania
Catania, Italy

Tal Hassner
Facebook (United States)
Menlo Park, CA, USA

ISSN 0302-9743 ISSN 1611-3349 (electronic)
Lecture Notes in Computer Science
ISBN 978-3-031-19789-5 ISBN 978-3-031-19790-1 (eBook)
https://doi.org/10.1007/978-3-031-19790-1

© The Editor(s) (if applicable) and The Author(s), under exclusive license
to Springer Nature Switzerland AG 2022
This work is subject to copyright. All rights are reserved by the Publisher, whether the whole or part of the material is concerned, specifically the rights of translation, reprinting, reuse of illustrations, recitation, broadcasting, reproduction on microfilms or in any other physical way, and transmission or information storage and retrieval, electronic adaptation, computer software, or by similar or dissimilar methodology now known or hereafter developed.
The use of general descriptive names, registered names, trademarks, service marks, etc. in this publication does not imply, even in the absence of a specific statement, that such names are exempt from the relevant protective laws and regulations and therefore free for general use.
The publisher, the authors, and the editors are safe to assume that the advice and information in this book are believed to be true and accurate at the date of publication. Neither the publisher nor the authors or the editors give a warranty, expressed or implied, with respect to the material contained herein or for any errors or omissions that may have been made. The publisher remains neutral with regard to jurisdictional claims in published maps and institutional affiliations.

This Springer imprint is published by the registered company Springer Nature Switzerland AG
The registered company address is: Gewerbestrasse 11, 6330 Cham, Switzerland

# Foreword

Organizing the European Conference on Computer Vision (ECCV 2022) in Tel-Aviv during a global pandemic was no easy feat. The uncertainty level was extremely high, and decisions had to be postponed to the last minute. Still, we managed to plan things just in time for ECCV 2022 to be held in person. Participation in physical events is crucial to stimulating collaborations and nurturing the culture of the Computer Vision community.

There were many people who worked hard to ensure attendees enjoyed the best science at the 16th edition of ECCV. We are grateful to the Program Chairs Gabriel Brostow and Tal Hassner, who went above and beyond to ensure the ECCV reviewing process ran smoothly. The scientific program includes dozens of workshops and tutorials in addition to the main conference and we would like to thank Leonid Karlinsky and Tomer Michaeli for their hard work. Finally, special thanks to the web chairs Lorenzo Baraldi and Kosta Derpanis, who put in extra hours to transfer information fast and efficiently to the ECCV community.

We would like to express gratitude to our generous sponsors and the Industry Chairs, Dimosthenis Karatzas and Chen Sagiv, who oversaw industry relations and proposed new ways for academia-industry collaboration and technology transfer. It's great to see so much industrial interest in what we're doing!

Authors' draft versions of the papers appeared online with open access on both the Computer Vision Foundation (CVF) and the European Computer Vision Association (ECVA) websites as with previous ECCVs. Springer, the publisher of the proceedings, has arranged for archival publication. The final version of the papers is hosted by SpringerLink, with active references and supplementary materials. It benefits all potential readers that we offer both a free and citeable version for all researchers, as well as an authoritative, citeable version for SpringerLink readers. Our thanks go to Ronan Nugent from Springer, who helped us negotiate this agreement. Last but not least, we wish to thank Eric Mortensen, our publication chair, whose expertise made the process smooth.

October 2022

Rita Cucchiara
Jiří Matas
Amnon Shashua
Lihi Zelnik-Manor

# Preface

Welcome to the proceedings of the European Conference on Computer Vision (ECCV 2022). This was a hybrid edition of ECCV as we made our way out of the COVID-19 pandemic. The conference received 5804 valid paper submissions, compared to 5150 submissions to ECCV 2020 (a 12.7% increase) and 2439 in ECCV 2018. 1645 submissions were accepted for publication (28%) and, of those, 157 (2.7% overall) as orals.

846 of the submissions were desk-rejected for various reasons. Many of them because they revealed author identity, thus violating the double-blind policy. This violation came in many forms: some had author names with the title, others added acknowledgments to specific grants, yet others had links to their github account where their name was visible. Tampering with the LaTeX template was another reason for automatic desk rejection.

ECCV 2022 used the traditional CMT system to manage the entire double-blind reviewing process. Authors did not know the names of the reviewers and vice versa. Each paper received at least 3 reviews (except 6 papers that received only 2 reviews), totalling more than 15,000 reviews.

Handling the review process at this scale was a significant challenge. To ensure that each submission received as fair and high-quality reviews as possible, we recruited more than 4719 reviewers (in the end, 4719 reviewers did at least one review). Similarly we recruited more than 276 area chairs (eventually, only 276 area chairs handled a batch of papers). The area chairs were selected based on their technical expertise and reputation, largely among people who served as area chairs in previous top computer vision and machine learning conferences (ECCV, ICCV, CVPR, NeurIPS, etc.).

Reviewers were similarly invited from previous conferences, and also from the pool of authors. We also encouraged experienced area chairs to suggest additional chairs and reviewers in the initial phase of recruiting. The median reviewer load was five papers per reviewer, while the average load was about four papers, because of the emergency reviewers. The area chair load was 35 papers, on average.

Conflicts of interest between authors, area chairs, and reviewers were handled largely automatically by the CMT platform, with some manual help from the Program Chairs. Reviewers were allowed to describe themselves as senior reviewer (load of 8 papers to review) or junior reviewers (load of 4 papers). Papers were matched to area chairs based on a subject-area affinity score computed in CMT and an affinity score computed by the Toronto Paper Matching System (TPMS). TPMS is based on the paper's full text. An area chair handling each submission would bid for preferred expert reviewers, and we balanced load and prevented conflicts.

The assignment of submissions to area chairs was relatively smooth, as was the assignment of submissions to reviewers. A small percentage of reviewers were not happy with their assignments in terms of subjects and self-reported expertise. This is an area for improvement, although it's interesting that many of these cases were reviewers hand-picked by AC's. We made a later round of reviewer recruiting, targeted at the list of authors of papers submitted to the conference, and had an excellent response which helped provide enough emergency reviewers. In the end, all but six papers received at least 3 reviews.

The challenges of the reviewing process are in line with past experiences at ECCV 2020. As the community grows, and the number of submissions increases, it becomes ever more challenging to recruit enough reviewers and ensure a high enough quality of reviews. Enlisting authors by default as reviewers might be one step to address this challenge.

Authors were given a week to rebut the initial reviews, and address reviewers' concerns. Each rebuttal was limited to a single pdf page with a fixed template.

The Area Chairs then led discussions with the reviewers on the merits of each submission. The goal was to reach consensus, but, ultimately, it was up to the Area Chair to make a decision. The decision was then discussed with a buddy Area Chair to make sure decisions were fair and informative. The entire process was conducted virtually with no in-person meetings taking place.

The Program Chairs were informed in cases where the Area Chairs overturned a decisive consensus reached by the reviewers, and pushed for the meta-reviews to contain details that explained the reasoning for such decisions. Obviously these were the most contentious cases, where reviewer inexperience was the most common reported factor.

Once the list of accepted papers was finalized and released, we went through the laborious process of plagiarism (including self-plagiarism) detection. A total of 4 accepted papers were rejected because of that.

Finally, we would like to thank our Technical Program Chair, Pavel Lifshits, who did tremendous work behind the scenes, and we thank the tireless CMT team.

October 2022

Gabriel Brostow
Giovanni Maria Farinella
Moustapha Cissé
Shai Avidan
Tal Hassner

# Organization

## General Chairs

| | |
|---|---|
| Rita Cucchiara | University of Modena and Reggio Emilia, Italy |
| Jiří Matas | Czech Technical University in Prague, Czech Republic |
| Amnon Shashua | Hebrew University of Jerusalem, Israel |
| Lihi Zelnik-Manor | Technion – Israel Institute of Technology, Israel |

## Program Chairs

| | |
|---|---|
| Shai Avidan | Tel-Aviv University, Israel |
| Gabriel Brostow | University College London, UK |
| Moustapha Cissé | Google AI, Ghana |
| Giovanni Maria Farinella | University of Catania, Italy |
| Tal Hassner | Facebook AI, USA |

## Program Technical Chair

| | |
|---|---|
| Pavel Lifshits | Technion – Israel Institute of Technology, Israel |

## Workshops Chairs

| | |
|---|---|
| Leonid Karlinsky | IBM Research, Israel |
| Tomer Michaeli | Technion – Israel Institute of Technology, Israel |
| Ko Nishino | Kyoto University, Japan |

## Tutorial Chairs

| | |
|---|---|
| Thomas Pock | Graz University of Technology, Austria |
| Natalia Neverova | Facebook AI Research, UK |

## Demo Chair

| | |
|---|---|
| Bohyung Han | Seoul National University, Korea |

## Social and Student Activities Chairs

Tatiana Tommasi — Italian Institute of Technology, Italy
Sagie Benaim — University of Copenhagen, Denmark

## Diversity and Inclusion Chairs

Xi Yin — Facebook AI Research, USA
Bryan Russell — Adobe, USA

## Communications Chairs

Lorenzo Baraldi — University of Modena and Reggio Emilia, Italy
Kosta Derpanis — York University & Samsung AI Centre Toronto, Canada

## Industrial Liaison Chairs

Dimosthenis Karatzas — Universitat Autònoma de Barcelona, Spain
Chen Sagiv — SagivTech, Israel

## Finance Chair

Gerard Medioni — University of Southern California & Amazon, USA

## Publication Chair

Eric Mortensen — MiCROTEC, USA

## Area Chairs

Lourdes Agapito — University College London, UK
Zeynep Akata — University of Tübingen, Germany
Naveed Akhtar — University of Western Australia, Australia
Karteek Alahari — Inria Grenoble Rhône-Alpes, France
Alexandre Alahi — École polytechnique fédérale de Lausanne, Switzerland
Pablo Arbelaez — Universidad de Los Andes, Columbia
Antonis A. Argyros — University of Crete & Foundation for Research and Technology-Hellas, Crete
Yuki M. Asano — University of Amsterdam, The Netherlands
Kalle Åström — Lund University, Sweden
Hadar Averbuch-Elor — Cornell University, USA

| | |
|---|---|
| Hossein Azizpour | KTH Royal Institute of Technology, Sweden |
| Vineeth N. Balasubramanian | Indian Institute of Technology, Hyderabad, India |
| Lamberto Ballan | University of Padova, Italy |
| Adrien Bartoli | Université Clermont Auvergne, France |
| Horst Bischof | Graz University of Technology, Austria |
| Matthew B. Blaschko | KU Leuven, Belgium |
| Federica Bogo | Meta Reality Labs Research, Switzerland |
| Katherine Bouman | California Institute of Technology, USA |
| Edmond Boyer | Inria Grenoble Rhône-Alpes, France |
| Michael S. Brown | York University, Canada |
| Vittorio Caggiano | Meta AI Research, USA |
| Neill Campbell | University of Bath, UK |
| Octavia Camps | Northeastern University, USA |
| Duygu Ceylan | Adobe Research, USA |
| Ayan Chakrabarti | Google Research, USA |
| Tat-Jen Cham | Nanyang Technological University, Singapore |
| Antoni Chan | City University of Hong Kong, Hong Kong, China |
| Manmohan Chandraker | NEC Labs America, USA |
| Xinlei Chen | Facebook AI Research, USA |
| Xilin Chen | Institute of Computing Technology, Chinese Academy of Sciences, China |
| Dongdong Chen | Microsoft Cloud AI, USA |
| Chen Chen | University of Central Florida, USA |
| Ondrej Chum | Vision Recognition Group, Czech Technical University in Prague, Czech Republic |
| John Collomosse | Adobe Research & University of Surrey, UK |
| Camille Couprie | Facebook, France |
| David Crandall | Indiana University, USA |
| Daniel Cremers | Technical University of Munich, Germany |
| Marco Cristani | University of Verona, Italy |
| Canton Cristian | Facebook AI Research, USA |
| Dengxin Dai | ETH Zurich, Switzerland |
| Dima Damen | University of Bristol, UK |
| Kostas Daniilidis | University of Pennsylvania, USA |
| Trevor Darrell | University of California, Berkeley, USA |
| Andrew Davison | Imperial College London, UK |
| Tali Dekel | Weizmann Institute of Science, Israel |
| Alessio Del Bue | Istituto Italiano di Tecnologia, Italy |
| Weihong Deng | Beijing University of Posts and Telecommunications, China |
| Konstantinos Derpanis | Ryerson University, Canada |
| Carl Doersch | DeepMind, UK |

| | |
|---|---|
| Matthijs Douze | Facebook AI Research, USA |
| Mohamed Elhoseiny | King Abdullah University of Science and Technology, Saudi Arabia |
| Sergio Escalera | University of Barcelona, Spain |
| Yi Fang | New York University, USA |
| Ryan Farrell | Brigham Young University, USA |
| Alireza Fathi | Google, USA |
| Christoph Feichtenhofer | Facebook AI Research, USA |
| Basura Fernando | Agency for Science, Technology and Research (A*STAR), Singapore |
| Vittorio Ferrari | Google Research, Switzerland |
| Andrew W. Fitzgibbon | Graphcore, UK |
| David J. Fleet | University of Toronto, Canada |
| David Forsyth | University of Illinois at Urbana-Champaign, USA |
| David Fouhey | University of Michigan, USA |
| Katerina Fragkiadaki | Carnegie Mellon University, USA |
| Friedrich Fraundorfer | Graz University of Technology, Austria |
| Oren Freifeld | Ben-Gurion University, Israel |
| Thomas Funkhouser | Google Research & Princeton University, USA |
| Yasutaka Furukawa | Simon Fraser University, Canada |
| Fabio Galasso | Sapienza University of Rome, Italy |
| Jürgen Gall | University of Bonn, Germany |
| Chuang Gan | Massachusetts Institute of Technology, USA |
| Zhe Gan | Microsoft, USA |
| Animesh Garg | University of Toronto, Vector Institute, Nvidia, Canada |
| Efstratios Gavves | University of Amsterdam, The Netherlands |
| Peter Gehler | Amazon, Germany |
| Theo Gevers | University of Amsterdam, The Netherlands |
| Bernard Ghanem | King Abdullah University of Science and Technology, Saudi Arabia |
| Ross B. Girshick | Facebook AI Research, USA |
| Georgia Gkioxari | Facebook AI Research, USA |
| Albert Gordo | Facebook, USA |
| Stephen Gould | Australian National University, Australia |
| Venu Madhav Govindu | Indian Institute of Science, India |
| Kristen Grauman | Facebook AI Research & UT Austin, USA |
| Abhinav Gupta | Carnegie Mellon University & Facebook AI Research, USA |
| Mohit Gupta | University of Wisconsin-Madison, USA |
| Hu Han | Institute of Computing Technology, Chinese Academy of Sciences, China |

| | |
|---|---|
| Bohyung Han | Seoul National University, Korea |
| Tian Han | Stevens Institute of Technology, USA |
| Emily Hand | University of Nevada, Reno, USA |
| Bharath Hariharan | Cornell University, USA |
| Ran He | Institute of Automation, Chinese Academy of Sciences, China |
| Otmar Hilliges | ETH Zurich, Switzerland |
| Adrian Hilton | University of Surrey, UK |
| Minh Hoai | Stony Brook University, USA |
| Yedid Hoshen | Hebrew University of Jerusalem, Israel |
| Timothy Hospedales | University of Edinburgh, UK |
| Gang Hua | Wormpex AI Research, USA |
| Di Huang | Beihang University, China |
| Jing Huang | Facebook, USA |
| Jia-Bin Huang | Facebook, USA |
| Nathan Jacobs | Washington University in St. Louis, USA |
| C.V. Jawahar | International Institute of Information Technology, Hyderabad, India |
| Herve Jegou | Facebook AI Research, France |
| Neel Joshi | Microsoft Research, USA |
| Armand Joulin | Facebook AI Research, France |
| Frederic Jurie | University of Caen Normandie, France |
| Fredrik Kahl | Chalmers University of Technology, Sweden |
| Yannis Kalantidis | NAVER LABS Europe, France |
| Evangelos Kalogerakis | University of Massachusetts, Amherst, USA |
| Sing Bing Kang | Zillow Group, USA |
| Yosi Keller | Bar Ilan University, Israel |
| Margret Keuper | University of Mannheim, Germany |
| Tae-Kyun Kim | Imperial College London, UK |
| Benjamin Kimia | Brown University, USA |
| Alexander Kirillov | Facebook AI Research, USA |
| Kris Kitani | Carnegie Mellon University, USA |
| Iasonas Kokkinos | Snap Inc. & University College London, UK |
| Vladlen Koltun | Apple, USA |
| Nikos Komodakis | University of Crete, Crete |
| Piotr Koniusz | Australian National University, Australia |
| Philipp Kraehenbuehl | University of Texas at Austin, USA |
| Dilip Krishnan | Google, USA |
| Ajay Kumar | Hong Kong Polytechnic University, Hong Kong, China |
| Junseok Kwon | Chung-Ang University, Korea |
| Jean-Francois Lalonde | Université Laval, Canada |

| | |
|---|---|
| Ivan Laptev | Inria Paris, France |
| Laura Leal-Taixé | Technical University of Munich, Germany |
| Erik Learned-Miller | University of Massachusetts, Amherst, USA |
| Gim Hee Lee | National University of Singapore, Singapore |
| Seungyong Lee | Pohang University of Science and Technology, Korea |
| Zhen Lei | Institute of Automation, Chinese Academy of Sciences, China |
| Bastian Leibe | RWTH Aachen University, Germany |
| Hongdong Li | Australian National University, Australia |
| Fuxin Li | Oregon State University, USA |
| Bo Li | University of Illinois at Urbana-Champaign, USA |
| Yin Li | University of Wisconsin-Madison, USA |
| Ser-Nam Lim | Meta AI Research, USA |
| Joseph Lim | University of Southern California, USA |
| Stephen Lin | Microsoft Research Asia, China |
| Dahua Lin | The Chinese University of Hong Kong, Hong Kong, China |
| Si Liu | Beihang University, China |
| Xiaoming Liu | Michigan State University, USA |
| Ce Liu | Microsoft, USA |
| Zicheng Liu | Microsoft, USA |
| Yanxi Liu | Pennsylvania State University, USA |
| Feng Liu | Portland State University, USA |
| Yebin Liu | Tsinghua University, China |
| Chen Change Loy | Nanyang Technological University, Singapore |
| Huchuan Lu | Dalian University of Technology, China |
| Cewu Lu | Shanghai Jiao Tong University, China |
| Oisin Mac Aodha | University of Edinburgh, UK |
| Dhruv Mahajan | Facebook, USA |
| Subhransu Maji | University of Massachusetts, Amherst, USA |
| Atsuto Maki | KTH Royal Institute of Technology, Sweden |
| Arun Mallya | NVIDIA, USA |
| R. Manmatha | Amazon, USA |
| Iacopo Masi | Sapienza University of Rome, Italy |
| Dimitris N. Metaxas | Rutgers University, USA |
| Ajmal Mian | University of Western Australia, Australia |
| Christian Micheloni | University of Udine, Italy |
| Krystian Mikolajczyk | Imperial College London, UK |
| Anurag Mittal | Indian Institute of Technology, Madras, India |
| Philippos Mordohai | Stevens Institute of Technology, USA |
| Greg Mori | Simon Fraser University & Borealis AI, Canada |

| | |
|---|---|
| Vittorio Murino | Istituto Italiano di Tecnologia, Italy |
| P. J. Narayanan | International Institute of Information Technology, Hyderabad, India |
| Ram Nevatia | University of Southern California, USA |
| Natalia Neverova | Facebook AI Research, UK |
| Richard Newcombe | Facebook, USA |
| Cuong V. Nguyen | Florida International University, USA |
| Bingbing Ni | Shanghai Jiao Tong University, China |
| Juan Carlos Niebles | Salesforce & Stanford University, USA |
| Ko Nishino | Kyoto University, Japan |
| Jean-Marc Odobez | Idiap Research Institute, École polytechnique fédérale de Lausanne, Switzerland |
| Francesca Odone | University of Genova, Italy |
| Takayuki Okatani | Tohoku University & RIKEN Center for Advanced Intelligence Project, Japan |
| Manohar Paluri | Facebook, USA |
| Guan Pang | Facebook, USA |
| Maja Pantic | Imperial College London, UK |
| Sylvain Paris | Adobe Research, USA |
| Jaesik Park | Pohang University of Science and Technology, Korea |
| Hyun Soo Park | The University of Minnesota, USA |
| Omkar M. Parkhi | Facebook, USA |
| Deepak Pathak | Carnegie Mellon University, USA |
| Georgios Pavlakos | University of California, Berkeley, USA |
| Marcello Pelillo | University of Venice, Italy |
| Marc Pollefeys | ETH Zurich & Microsoft, Switzerland |
| Jean Ponce | Inria, France |
| Gerard Pons-Moll | University of Tübingen, Germany |
| Fatih Porikli | Qualcomm, USA |
| Victor Adrian Prisacariu | University of Oxford, UK |
| Petia Radeva | University of Barcelona, Spain |
| Ravi Ramamoorthi | University of California, San Diego, USA |
| Deva Ramanan | Carnegie Mellon University, USA |
| Vignesh Ramanathan | Facebook, USA |
| Nalini Ratha | State University of New York at Buffalo, USA |
| Tammy Riklin Raviv | Ben-Gurion University, Israel |
| Tobias Ritschel | University College London, UK |
| Emanuele Rodola | Sapienza University of Rome, Italy |
| Amit K. Roy-Chowdhury | University of California, Riverside, USA |
| Michael Rubinstein | Google, USA |
| Olga Russakovsky | Princeton University, USA |

| | |
|---|---|
| Mathieu Salzmann | École polytechnique fédérale de Lausanne, Switzerland |
| Dimitris Samaras | Stony Brook University, USA |
| Aswin Sankaranarayanan | Carnegie Mellon University, USA |
| Imari Sato | National Institute of Informatics, Japan |
| Yoichi Sato | University of Tokyo, Japan |
| Shin'ichi Satoh | National Institute of Informatics, Japan |
| Walter Scheirer | University of Notre Dame, USA |
| Bernt Schiele | Max Planck Institute for Informatics, Germany |
| Konrad Schindler | ETH Zurich, Switzerland |
| Cordelia Schmid | Inria & Google, France |
| Alexander Schwing | University of Illinois at Urbana-Champaign, USA |
| Nicu Sebe | University of Trento, Italy |
| Greg Shakhnarovich | Toyota Technological Institute at Chicago, USA |
| Eli Shechtman | Adobe Research, USA |
| Humphrey Shi | University of Oregon & University of Illinois at Urbana-Champaign & Picsart AI Research, USA |
| Jianbo Shi | University of Pennsylvania, USA |
| Roy Shilkrot | Massachusetts Institute of Technology, USA |
| Mike Zheng Shou | National University of Singapore, Singapore |
| Kaleem Siddiqi | McGill University, Canada |
| Richa Singh | Indian Institute of Technology Jodhpur, India |
| Greg Slabaugh | Queen Mary University of London, UK |
| Cees Snoek | University of Amsterdam, The Netherlands |
| Yale Song | Facebook AI Research, USA |
| Yi-Zhe Song | University of Surrey, UK |
| Bjorn Stenger | Rakuten Institute of Technology |
| Abby Stylianou | Saint Louis University, USA |
| Akihiro Sugimoto | National Institute of Informatics, Japan |
| Chen Sun | Brown University, USA |
| Deqing Sun | Google, USA |
| Kalyan Sunkavalli | Adobe Research, USA |
| Ying Tai | Tencent YouTu Lab, China |
| Ayellet Tal | Technion – Israel Institute of Technology, Israel |
| Ping Tan | Simon Fraser University, Canada |
| Siyu Tang | ETH Zurich, Switzerland |
| Chi-Keung Tang | Hong Kong University of Science and Technology, Hong Kong, China |
| Radu Timofte | University of Würzburg, Germany & ETH Zurich, Switzerland |
| Federico Tombari | Google, Switzerland & Technical University of Munich, Germany |

| | |
|---|---|
| James Tompkin | Brown University, USA |
| Lorenzo Torresani | Dartmouth College, USA |
| Alexander Toshev | Apple, USA |
| Du Tran | Facebook AI Research, USA |
| Anh T. Tran | VinAI, Vietnam |
| Zhuowen Tu | University of California, San Diego, USA |
| Georgios Tzimiropoulos | Queen Mary University of London, UK |
| Jasper Uijlings | Google Research, Switzerland |
| Jan C. van Gemert | Delft University of Technology, The Netherlands |
| Gul Varol | Ecole des Ponts ParisTech, France |
| Nuno Vasconcelos | University of California, San Diego, USA |
| Mayank Vatsa | Indian Institute of Technology Jodhpur, India |
| Ashok Veeraraghavan | Rice University, USA |
| Jakob Verbeek | Facebook AI Research, France |
| Carl Vondrick | Columbia University, USA |
| Ruiping Wang | Institute of Computing Technology, Chinese Academy of Sciences, China |
| Xinchao Wang | National University of Singapore, Singapore |
| Liwei Wang | The Chinese University of Hong Kong, Hong Kong, China |
| Chaohui Wang | Université Paris-Est, France |
| Xiaolong Wang | University of California, San Diego, USA |
| Christian Wolf | NAVER LABS Europe, France |
| Tao Xiang | University of Surrey, UK |
| Saining Xie | Facebook AI Research, USA |
| Cihang Xie | University of California, Santa Cruz, USA |
| Zeki Yalniz | Facebook, USA |
| Ming-Hsuan Yang | University of California, Merced, USA |
| Angela Yao | National University of Singapore, Singapore |
| Shaodi You | University of Amsterdam, The Netherlands |
| Stella X. Yu | University of California, Berkeley, USA |
| Junsong Yuan | State University of New York at Buffalo, USA |
| Stefanos Zafeiriou | Imperial College London, UK |
| Amir Zamir | École polytechnique fédérale de Lausanne, Switzerland |
| Lei Zhang | Alibaba & Hong Kong Polytechnic University, Hong Kong, China |
| Lei Zhang | International Digital Economy Academy (IDEA), China |
| Pengchuan Zhang | Meta AI, USA |
| Bolei Zhou | University of California, Los Angeles, USA |
| Yuke Zhu | University of Texas at Austin, USA |

Todd Zickler — Harvard University, USA
Wangmeng Zuo — Harbin Institute of Technology, China

## Technical Program Committee

Davide Abati
Soroush Abbasi Koohpayegani
Amos L. Abbott
Rameen Abdal
Rabab Abdelfattah
Sahar Abdelnabi
Hassan Abu Alhaija
Abulikemu Abuduweili
Ron Abutbul
Hanno Ackermann
Aikaterini Adam
Kamil Adamczewski
Ehsan Adeli
Vida Adeli
Donald Adjeroh
Arman Afrasiyabi
Akshay Agarwal
Sameer Agarwal
Abhinav Agarwalla
Vaibhav Aggarwal
Sara Aghajanzadeh
Susmit Agrawal
Antonio Agudo
Touqeer Ahmad
Sk Miraj Ahmed
Chaitanya Ahuja
Nilesh A. Ahuja
Abhishek Aich
Shubhra Aich
Noam Aigerman
Arash Akbarinia
Peri Akiva
Derya Akkaynak
Emre Aksan
Arjun R. Akula
Yuval Alaluf
Stephan Alaniz
Paul Albert
Cenek Albl
Filippo Aleotti
Konstantinos P. Alexandridis
Motasem Alfarra
Mohsen Ali
Thiemo Alldieck
Hadi Alzayer
Liang An
Shan An
Yi An
Zhulin An
Dongsheng An
Jie An
Xiang An
Saket Anand
Cosmin Ancuti
Juan Andrade-Cetto
Alexander Andreopoulos
Bjoern Andres
Jerone T. A. Andrews
Shivangi Aneja
Anelia Angelova
Dragomir Anguelov
Rushil Anirudh
Oron Anschel
Rao Muhammad Anwer
Djamila Aouada
Evlampios Apostolidis
Srikar Appalaraju
Nikita Araslanov
Andre Araujo
Eric Arazo
Dawit Mureja Argaw
Anurag Arnab
Aditya Arora
Chetan Arora
Sunpreet S. Arora
Alexey Artemov
Muhammad Asad
Kumar Ashutosh
Sinem Aslan
Vishal Asnani
Mahmoud Assran
Amir Atapour-Abarghouei
Nikos Athanasiou
Ali Athar
ShahRukh Athar
Sara Atito
Souhaib Attaiki
Matan Atzmon
Mathieu Aubry
Nicolas Audebert
Tristan T. Aumentado-Armstrong
Melinos Averkiou
Yannis Avrithis
Stephane Ayache
Mehmet Aygün
Seyed Mehdi Ayyoubzadeh
Hossein Azizpour
George Azzopardi
Mallikarjun B. R.
Yunhao Ba
Abhishek Badki
Seung-Hwan Bae
Seung-Hwan Baek
Seungryul Baek
Piyush Nitin Bagad
Shai Bagon
Gaetan Bahl
Shikhar Bahl
Sherwin Bahmani
Haoran Bai
Lei Bai
Jiawang Bai
Haoyue Bai
Jinbin Bai
Xiang Bai
Xuyang Bai

Yang Bai
Yuanchao Bai
Ziqian Bai
Sungyong Baik
Kevin Bailly
Max Bain
Federico Baldassarre
Wele Gedara Chaminda Bandara
Biplab Banerjee
Pratyay Banerjee
Sandipan Banerjee
Jihwan Bang
Antyanta Bangunharcana
Aayush Bansal
Ankan Bansal
Siddhant Bansal
Wentao Bao
Zhipeng Bao
Amir Bar
Manel Baradad Jurjo
Lorenzo Baraldi
Danny Barash
Daniel Barath
Connelly Barnes
Ioan Andrei Bârsan
Steven Basart
Dina Bashkirova
Chaim Baskin
Peyman Bateni
Anil Batra
Sebastiano Battiato
Ardhendu Behera
Harkirat Behl
Jens Behley
Vasileios Belagiannis
Boulbaba Ben Amor
Emanuel Ben Baruch
Abdessamad Ben Hamza
Gil Ben-Artzi
Assia Benbihi
Fabian Benitez-Quiroz
Guy Ben-Yosef
Philipp Benz
Alexander W. Bergman
Urs Bergmann
Jesus Bermudez-Cameo
Stefano Berretti
Gedas Bertasius
Zachary Bessinger
Petra Bevandić
Matthew Beveridge
Lucas Beyer
Yash Bhalgat
Suvaansh Bhambri
Samarth Bharadwaj
Gaurav Bharaj
Aparna Bharati
Bharat Lal Bhatnagar
Uttaran Bhattacharya
Apratim Bhattacharyya
Brojeshwar Bhowmick
Ankan Kumar Bhunia
Ayan Kumar Bhunia
Qi Bi
Sai Bi
Michael Bi Mi
Gui-Bin Bian
Jia-Wang Bian
Shaojun Bian
Pia Bideau
Mario Bijelic
Hakan Bilen
Guillaume-Alexandre Bilodeau
Alexander Binder
Tolga Birdal
Vighnesh N. Birodkar
Sandika Biswas
Andreas Blattmann
Janusz Bobulski
Giuseppe Boccignone
Vishnu Boddeti
Navaneeth Bodla
Moritz Böhle
Aleksei Bokhovkin
Sam Bond-Taylor
Vivek Boominathan
Shubhankar Borse
Mark Boss
Andrea Bottino
Adnane Boukhayma
Fadi Boutros
Nicolas C. Boutry
Richard S. Bowen
Ivaylo Boyadzhiev
Aidan Boyd
Yuri Boykov
Aljaz Bozic
Behzad Bozorgtabar
Eric Brachmann
Samarth Brahmbhatt
Gustav Bredell
Francois Bremond
Joel Brogan
Andrew Brown
Thomas Brox
Marcus A. Brubaker
Robert-Jan Bruintjes
Yuqi Bu
Anders G. Buch
Himanshu Buckchash
Mateusz Buda
Ignas Budvytis
José M. Buenaposada
Marcel C. Bühler
Tu Bui
Adrian Bulat
Hannah Bull
Evgeny Burnaev
Andrei Bursuc
Benjamin Busam
Sergey N. Buzykanov
Wonmin Byeon
Fabian Caba
Martin Cadik
Guanyu Cai
Minjie Cai
Qing Cai
Zhongang Cai
Qi Cai
Yancheng Cai
Shen Cai
Han Cai
Jiarui Cai

Bowen Cai
Mu Cai
Qin Cai
Ruojin Cai
Weidong Cai
Weiwei Cai
Yi Cai
Yujun Cai
Zhiping Cai
Akin Caliskan
Lilian Calvet
Baris Can Cam
Necati Cihan Camgoz
Tommaso Campari
Dylan Campbell
Ziang Cao
Ang Cao
Xu Cao
Zhiwen Cao
Shengcao Cao
Song Cao
Weipeng Cao
Xiangyong Cao
Xiaochun Cao
Yue Cao
Yunhao Cao
Zhangjie Cao
Jiale Cao
Yang Cao
Jiajiong Cao
Jie Cao
Jinkun Cao
Lele Cao
Yulong Cao
Zhiguo Cao
Chen Cao
Razvan Caramalau
Marlène Careil
Gustavo Carneiro
Joao Carreira
Dan Casas
Paola Cascante-Bonilla
Angela Castillo
Francisco M. Castro
Pedro Castro
Luca Cavalli
George J. Cazenavette
Oya Celiktutan
Hakan Cevikalp
Sri Harsha C. H.
Sungmin Cha
Geonho Cha
Menglei Chai
Lucy Chai
Yuning Chai
Zenghao Chai
Anirban Chakraborty
Deep Chakraborty
Rudrasis Chakraborty
Souradeep Chakraborty
Kelvin C. K. Chan
Chee Seng Chan
Paramanand Chandramouli
Arjun Chandrasekaran
Kenneth Chaney
Dongliang Chang
Huiwen Chang
Peng Chang
Xiaojun Chang
Jia-Ren Chang
Hyung Jin Chang
Hyun Sung Chang
Ju Yong Chang
Li-Jen Chang
Qi Chang
Wei-Yi Chang
Yi Chang
Nadine Chang
Hanqing Chao
Pradyumna Chari
Dibyadip Chatterjee
Chiranjoy Chattopadhyay
Siddhartha Chaudhuri
Zhengping Che
Gal Chechik
Lianggangxu Chen
Qi Alfred Chen
Brian Chen
Bor-Chun Chen
Bo-Hao Chen
Bohong Chen
Bin Chen
Ziliang Chen
Cheng Chen
Chen Chen
Chaofeng Chen
Xi Chen
Haoyu Chen
Xuanhong Chen
Wei Chen
Qiang Chen
Shi Chen
Xianyu Chen
Chang Chen
Changhuai Chen
Hao Chen
Jie Chen
Jianbo Chen
Jingjing Chen
Jun Chen
Kejiang Chen
Mingcai Chen
Nenglun Chen
Qifeng Chen
Ruoyu Chen
Shu-Yu Chen
Weidong Chen
Weijie Chen
Weikai Chen
Xiang Chen
Xiuyi Chen
Xingyu Chen
Yaofo Chen
Yueting Chen
Yu Chen
Yunjin Chen
Yuntao Chen
Yun Chen
Zhenfang Chen
Zhuangzhuang Chen
Chu-Song Chen
Xiangyu Chen
Zhuo Chen
Chaoqi Chen
Shizhe Chen

Xiaotong Chen
Xiaozhi Chen
Dian Chen
Defang Chen
Dingfan Chen
Ding-Jie Chen
Ee Heng Chen
Tao Chen
Yixin Chen
Wei-Ting Chen
Lin Chen
Guang Chen
Guangyi Chen
Guanying Chen
Guangyao Chen
Hwann-Tzong Chen
Junwen Chen
Jiacheng Chen
Jianxu Chen
Hui Chen
Kai Chen
Kan Chen
Kevin Chen
Kuan-Wen Chen
Weihua Chen
Zhang Chen
Liang-Chieh Chen
Lele Chen
Liang Chen
Fanglin Chen
Zehui Chen
Minghui Chen
Minghao Chen
Xiaokang Chen
Qian Chen
Jun-Cheng Chen
Qi Chen
Qingcai Chen
Richard J. Chen
Runnan Chen
Rui Chen
Shuo Chen
Sentao Chen
Shaoyu Chen
Shixing Chen
Shuai Chen
Shuya Chen
Sizhe Chen
Simin Chen
Shaoxiang Chen
Zitian Chen
Tianlong Chen
Tianshui Chen
Min-Hung Chen
Xiangning Chen
Xin Chen
Xinghao Chen
Xuejin Chen
Xu Chen
Xuxi Chen
Yunlu Chen
Yanbei Chen
Yuxiao Chen
Yun-Chun Chen
Yi-Ting Chen
Yi-Wen Chen
Yinbo Chen
Yiran Chen
Yuanhong Chen
Yubei Chen
Yuefeng Chen
Yuhua Chen
Yukang Chen
Zerui Chen
Zhaoyu Chen
Zhen Chen
Zhenyu Chen
Zhi Chen
Zhiwei Chen
Zhixiang Chen
Long Chen
Bowen Cheng
Jun Cheng
Yi Cheng
Jingchun Cheng
Lechao Cheng
Xi Cheng
Yuan Cheng
Ho Kei Cheng
Kevin Ho Man Cheng
Jiacheng Cheng
Kelvin B. Cheng
Li Cheng
Mengjun Cheng
Zhen Cheng
Qingrong Cheng
Tianheng Cheng
Harry Cheng
Yihua Cheng
Yu Cheng
Ziheng Cheng
Soon Yau Cheong
Anoop Cherian
Manuela Chessa
Zhixiang Chi
Naoki Chiba
Julian Chibane
Kashyap Chitta
Tai-Yin Chiu
Hsu-kuang Chiu
Wei-Chen Chiu
Sungmin Cho
Donghyeon Cho
Hyeon Cho
Yooshin Cho
Gyusang Cho
Jang Hyun Cho
Seungju Cho
Nam Ik Cho
Sunghyun Cho
Hanbyel Cho
Jaesung Choe
Jooyoung Choi
Chiho Choi
Changwoon Choi
Jongwon Choi
Myungsub Choi
Dooseop Choi
Jonghyun Choi
Jinwoo Choi
Jun Won Choi
Min-Kook Choi
Hongsuk Choi
Janghoon Choi
Yoon-Ho Choi

Yukyung Choi
Jaegul Choo
Ayush Chopra
Siddharth Choudhary
Subhabrata Choudhury
Vasileios Choutas
Ka-Ho Chow
Pinaki Nath Chowdhury
Sammy Christen
Anders Christensen
Grigorios Chrysos
Hang Chu
Wen-Hsuan Chu
Peng Chu
Qi Chu
Ruihang Chu
Wei-Ta Chu
Yung-Yu Chuang
Sanghyuk Chun
Se Young Chun
Antonio Cinà
Ramazan Gokberk Cinbis
Javier Civera
Albert Clapés
Ronald Clark
Brian S. Clipp
Felipe Codevilla
Daniel Coelho de Castro
Niv Cohen
Forrester Cole
Maxwell D. Collins
Robert T. Collins
Marc Comino Trinidad
Runmin Cong
Wenyan Cong
Maxime Cordy
Marcella Cornia
Enric Corona
Huseyin Coskun
Luca Cosmo
Dragos Costea
Davide Cozzolino
Arun C. S. Kumar
Aiyu Cui
Qiongjie Cui
Quan Cui
Shuhao Cui
Yiming Cui
Ying Cui
Zijun Cui
Jiali Cui
Jiequan Cui
Yawen Cui
Zhen Cui
Zhaopeng Cui
Jack Culpepper
Xiaodong Cun
Ross Cutler
Adam Czajka
Ali Dabouei
Konstantinos M. Dafnis
Manuel Dahnert
Tao Dai
Yuchao Dai
Bo Dai
Mengyu Dai
Hang Dai
Haixing Dai
Peng Dai
Pingyang Dai
Qi Dai
Qiyu Dai
Yutong Dai
Naser Damer
Zhiyuan Dang
Mohamed Daoudi
Ayan Das
Abir Das
Debasmit Das
Deepayan Das
Partha Das
Sagnik Das
Soumi Das
Srijan Das
Swagatam Das
Avijit Dasgupta
Jim Davis
Adrian K. Davison
Homa Davoudi
Laura Daza
Matthias De Lange
Shalini De Mello
Marco De Nadai
Christophe De Vleeschouwer
Alp Dener
Boyang Deng
Congyue Deng
Bailin Deng
Yong Deng
Ye Deng
Zhuo Deng
Zhijie Deng
Xiaoming Deng
Jiankang Deng
Jinhong Deng
Jingjing Deng
Liang-Jian Deng
Siqi Deng
Xiang Deng
Xueqing Deng
Zhongying Deng
Karan Desai
Jean-Emmanuel Deschaud
Aniket Anand Deshmukh
Neel Dey
Helisa Dhamo
Prithviraj Dhar
Amaya Dharmasiri
Yan Di
Xing Di
Ousmane A. Dia
Haiwen Diao
Xiaolei Diao
Gonçalo José Dias Pais
Abdallah Dib
Anastasios Dimou
Changxing Ding
Henghui Ding
Guodong Ding
Yaqing Ding
Shuangrui Ding
Yuhang Ding
Yikang Ding
Shouhong Ding

Haisong Ding
Hui Ding
Jiahao Ding
Jian Ding
Jian-Jiun Ding
Shuxiao Ding
Tianyu Ding
Wenhao Ding
Yuqi Ding
Yi Ding
Yuzhen Ding
Zhengming Ding
Tan Minh Dinh
Vu Dinh
Christos Diou
Mandar Dixit
Bao Gia Doan
Khoa D. Doan
Dzung Anh Doan
Debi Prosad Dogra
Nehal Doiphode
Chengdong Dong
Bowen Dong
Zhenxing Dong
Hang Dong
Xiaoyi Dong
Haoye Dong
Jiangxin Dong
Shichao Dong
Xuan Dong
Zhen Dong
Shuting Dong
Jing Dong
Li Dong
Ming Dong
Nanqing Dong
Qiulei Dong
Runpei Dong
Siyan Dong
Tian Dong
Wei Dong
Xiaomeng Dong
Xin Dong
Xingbo Dong
Yuan Dong
Samuel Dooley
Gianfranco Doretto
Michael Dorkenwald
Keval Doshi
Zhaopeng Dou
Xiaotian Dou
Hazel Doughty
Ahmad Droby
Iddo Drori
Jie Du
Yong Du
Dawei Du
Dong Du
Ruoyi Du
Yuntao Du
Xuefeng Du
Yilun Du
Yuming Du
Radhika Dua
Haodong Duan
Jiafei Duan
Kaiwen Duan
Peiqi Duan
Ye Duan
Haoran Duan
Jiali Duan
Amanda Duarte
Abhimanyu Dubey
Shiv Ram Dubey
Florian Dubost
Lukasz Dudziak
Shivam Duggal
Justin M. Dulay
Matteo Dunnhofer
Chi Nhan Duong
Thibaut Durand
Mihai Dusmanu
Ujjal Kr Dutta
Debidatta Dwibedi
Isht Dwivedi
Sai Kumar Dwivedi
Takeharu Eda
Mark Edmonds
Alexei A. Efros
Thibaud Ehret
Max Ehrlich
Mahsa Ehsanpour
Iván Eichhardt
Farshad Einabadi
Marvin Eisenberger
Hazim Kemal Ekenel
Mohamed El Banani
Ismail Elezi
Moshe Eliasof
Alaa El-Nouby
Ian Endres
Francis Engelmann
Deniz Engin
Chanho Eom
Dave Epstein
Maria C. Escobar
Victor A. Escorcia
Carlos Esteves
Sungmin Eum
Bernard J. E. Evans
Ivan Evtimov
Fevziye Irem Eyiokur Yaman
Matteo Fabbri
Sébastien Fabbro
Gabriele Facciolo
Masud Fahim
Bin Fan
Hehe Fan
Deng-Ping Fan
Aoxiang Fan
Chen-Chen Fan
Qi Fan
Zhaoxin Fan
Haoqi Fan
Heng Fan
Hongyi Fan
Linxi Fan
Baojie Fan
Jiayuan Fan
Lei Fan
Quanfu Fan
Yonghui Fan
Yingruo Fan
Zhiwen Fan

Zicong Fan
Sean Fanello
Jiansheng Fang
Chaowei Fang
Yuming Fang
Jianwu Fang
Jin Fang
Qi Fang
Shancheng Fang
Tian Fang
Xianyong Fang
Gongfan Fang
Zhen Fang
Hui Fang
Jiemin Fang
Le Fang
Pengfei Fang
Xiaolin Fang
Yuxin Fang
Zhaoyuan Fang
Ammarah Farooq
Azade Farshad
Zhengcong Fei
Michael Felsberg
Wei Feng
Chen Feng
Fan Feng
Andrew Feng
Xin Feng
Zheyun Feng
Ruicheng Feng
Mingtao Feng
Qianyu Feng
Shangbin Feng
Chun-Mei Feng
Zunlei Feng
Zhiyong Feng
Martin Fergie
Mustansar Fiaz
Marco Fiorucci
Michael Firman
Hamed Firooz
Volker Fischer
Corneliu O. Florea
Georgios Floros
Wolfgang Foerstner
Gianni Franchi
Jean-Sebastien Franco
Simone Frintrop
Anna Fruehstueck
Changhong Fu
Chaoyou Fu
Cheng-Yang Fu
Chi-Wing Fu
Deqing Fu
Huan Fu
Jun Fu
Kexue Fu
Ying Fu
Jianlong Fu
Jingjing Fu
Qichen Fu
Tsu-Jui Fu
Xueyang Fu
Yang Fu
Yanwei Fu
Yonggan Fu
Wolfgang Fuhl
Yasuhisa Fujii
Kent Fujiwara
Marco Fumero
Takuya Funatomi
Isabel Funke
Dario Fuoli
Antonino Furnari
Matheus A. Gadelha
Akshay Gadi Patil
Adrian Galdran
Guillermo Gallego
Silvano Galliani
Orazio Gallo
Leonardo Galteri
Matteo Gamba
Yiming Gan
Sujoy Ganguly
Harald Ganster
Boyan Gao
Changxin Gao
Daiheng Gao
Difei Gao
Chen Gao
Fei Gao
Lin Gao
Wei Gao
Yiming Gao
Junyu Gao
Guangyu Ryan Gao
Haichang Gao
Hongchang Gao
Jialin Gao
Jin Gao
Jun Gao
Katelyn Gao
Mingchen Gao
Mingfei Gao
Pan Gao
Shangqian Gao
Shanghua Gao
Xitong Gao
Yunhe Gao
Zhanning Gao
Elena Garces
Nuno Cruz Garcia
Noa Garcia
Guillermo Garcia-Hernando
Isha Garg
Rahul Garg
Sourav Garg
Quentin Garrido
Stefano Gasperini
Kent Gauen
Chandan Gautam
Shivam Gautam
Paul Gay
Chunjiang Ge
Shiming Ge
Wenhang Ge
Yanhao Ge
Zheng Ge
Songwei Ge
Weifeng Ge
Yixiao Ge
Yuying Ge
Shijie Geng

Zhengyang Geng
Kyle A. Genova
Georgios Georgakis
Markos Georgopoulos
Marcel Geppert
Shabnam Ghadar
Mina Ghadimi Atigh
Deepti Ghadiyaram
Maani Ghaffari Jadidi
Sedigh Ghamari
Zahra Gharaee
Michaël Gharbi
Golnaz Ghiasi
Reza Ghoddoosian
Soumya Suvra Ghosal
Adhiraj Ghosh
Arthita Ghosh
Pallabi Ghosh
Soumyadeep Ghosh
Andrew Gilbert
Igor Gilitschenski
Jhony H. Giraldo
Andreu Girbau Xalabarder
Rohit Girdhar
Sharath Girish
Xavier Giro-i-Nieto
Raja Giryes
Thomas Gittings
Nikolaos Gkanatsios
Ioannis Gkioulekas
Abhiram Gnanasambandam
Aurele T. Gnanha
Clement L. J. C. Godard
Arushi Goel
Vidit Goel
Shubham Goel
Zan Gojcic
Aaron K. Gokaslan
Tejas Gokhale
S. Alireza Golestaneh
Thiago L. Gomes
Nuno Goncalves
Boqing Gong
Chen Gong
Yuanhao Gong
Guoqiang Gong
Jingyu Gong
Rui Gong
Yu Gong
Mingming Gong
Neil Zhenqiang Gong
Xun Gong
Yunye Gong
Yihong Gong
Cristina I. González
Nithin Gopalakrishnan Nair
Gaurav Goswami
Jianping Gou
Shreyank N. Gowda
Ankit Goyal
Helmut Grabner
Patrick L. Grady
Ben Graham
Eric Granger
Douglas R. Gray
Matej Grcić
David Griffiths
Jinjin Gu
Yun Gu
Shuyang Gu
Jianyang Gu
Fuqiang Gu
Jiatao Gu
Jindong Gu
Jiaqi Gu
Jinwei Gu
Jiaxin Gu
Geonmo Gu
Xiao Gu
Xinqian Gu
Xiuye Gu
Yuming Gu
Zhangxuan Gu
Dayan Guan
Junfeng Guan
Qingji Guan
Tianrui Guan
Shanyan Guan
Denis A. Gudovskiy
Ricardo Guerrero
Pierre-Louis Guhur
Jie Gui
Liangyan Gui
Liangke Gui
Benoit Guillard
Erhan Gundogdu
Manuel Günther
Jingcai Guo
Yuanfang Guo
Junfeng Guo
Chenqi Guo
Dan Guo
Hongji Guo
Jia Guo
Jie Guo
Minghao Guo
Shi Guo
Yanhui Guo
Yangyang Guo
Yuan-Chen Guo
Yilu Guo
Yiluan Guo
Yong Guo
Guangyu Guo
Haiyun Guo
Jinyang Guo
Jianyuan Guo
Pengsheng Guo
Pengfei Guo
Shuxuan Guo
Song Guo
Tianyu Guo
Qing Guo
Qiushan Guo
Wen Guo
Xiefan Guo
Xiaohu Guo
Xiaoqing Guo
Yufei Guo
Yuhui Guo
Yuliang Guo
Yunhui Guo
Yanwen Guo

Akshita Gupta
Ankush Gupta
Kamal Gupta
Kartik Gupta
Ritwik Gupta
Rohit Gupta
Siddharth Gururani
Fredrik K. Gustafsson
Abner Guzman Rivera
Vladimir Guzov
Matthew A. Gwilliam
Jung-Woo Ha
Marc Habermann
Isma Hadji
Christian Haene
Martin Hahner
Levente Hajder
Alexandros Haliassos
Emanuela Haller
Bumsub Ham
Abdullah J. Hamdi
Shreyas Hampali
Dongyoon Han
Chunrui Han
Dong-Jun Han
Dong-Sig Han
Guangxing Han
Zhizhong Han
Ruize Han
Jiaming Han
Jin Han
Ligong Han
Xian-Hua Han
Xiaoguang Han
Yizeng Han
Zhi Han
Zhenjun Han
Zhongyi Han
Jungong Han
Junlin Han
Kai Han
Kun Han
Sungwon Han
Songfang Han
Wei Han
Xiao Han
Xintong Han
Xinzhe Han
Yahong Han
Yan Han
Zongbo Han
Nicolai Hani
Rana Hanocka
Niklas Hanselmann
Nicklas A. Hansen
Hong Hanyu
Fusheng Hao
Yanbin Hao
Shijie Hao
Udith Haputhanthri
Mehrtash Harandi
Josh Harguess
Adam Harley
David M. Hart
Atsushi Hashimoto
Ali Hassani
Mohammed Hassanin
Yana Hasson
Joakim Bruslund Haurum
Bo He
Kun He
Chen He
Xin He
Fazhi He
Gaoqi He
Hao He
Haoyu He
Jiangpeng He
Hongliang He
Qian He
Xiangteng He
Xuming He
Yannan He
Yuhang He
Yang He
Xiangyu He
Nanjun He
Pan He
Sen He
Shengfeng He
Songtao He
Tao He
Tong He
Wei He
Xuehai He
Xiaoxiao He
Ying He
Yisheng He
Ziwen He
Peter Hedman
Felix Heide
Yacov Hel-Or
Paul Henderson
Philipp Henzler
Byeongho Heo
Jae-Pil Heo
Miran Heo
Sachini A. Herath
Stephane Herbin
Pedro Hermosilla Casajus
Monica Hernandez
Charles Herrmann
Roei Herzig
Mauricio Hess-Flores
Carlos Hinojosa
Tobias Hinz
Tsubasa Hirakawa
Chih-Hui Ho
Lam Si Tung Ho
Jennifer Hobbs
Derek Hoiem
Yannick Hold-Geoffroy
Aleksander Holynski
Cheeun Hong
Fa-Ting Hong
Hanbin Hong
Guan Zhe Hong
Danfeng Hong
Lanqing Hong
Xiaopeng Hong
Xin Hong
Jie Hong
Seungbum Hong
Cheng-Yao Hong
Seunghoon Hong

Yi Hong
Yuan Hong
Yuchen Hong
Anthony Hoogs
Maxwell C. Horton
Kazuhiro Hotta
Qibin Hou
Tingbo Hou
Junhui Hou
Ji Hou
Qiqi Hou
Rui Hou
Ruibing Hou
Zhi Hou
Henry Howard-Jenkins
Lukas Hoyer
Wei-Lin Hsiao
Chiou-Ting Hsu
Anthony Hu
Brian Hu
Yusong Hu
Hexiang Hu
Haoji Hu
Di Hu
Hengtong Hu
Haigen Hu
Lianyu Hu
Hanzhe Hu
Jie Hu
Junlin Hu
Shizhe Hu
Jian Hu
Zhiming Hu
Juhua Hu
Peng Hu
Ping Hu
Ronghang Hu
MengShun Hu
Tao Hu
Vincent Tao Hu
Xiaoling Hu
Xinting Hu
Xiaolin Hu
Xuefeng Hu
Xiaowei Hu
Yang Hu
Yueyu Hu
Zeyu Hu
Zhongyun Hu
Binh-Son Hua
Guoliang Hua
Yi Hua
Linzhi Huang
Qiusheng Huang
Bo Huang
Chen Huang
Hsin-Ping Huang
Ye Huang
Shuangping Huang
Zeng Huang
Buzhen Huang
Cong Huang
Heng Huang
Hao Huang
Qidong Huang
Huaibo Huang
Chaoqin Huang
Feihu Huang
Jiahui Huang
Jingjia Huang
Kun Huang
Lei Huang
Sheng Huang
Shuaiyi Huang
Siyu Huang
Xiaoshui Huang
Xiaoyang Huang
Yan Huang
Yihao Huang
Ying Huang
Ziling Huang
Xiaoke Huang
Yifei Huang
Haiyang Huang
Zhewei Huang
Jin Huang
Haibin Huang
Jiaxing Huang
Junjie Huang
Keli Huang
Lang Huang
Lin Huang
Luojie Huang
Mingzhen Huang
Shijia Huang
Shengyu Huang
Siyuan Huang
He Huang
Xiuyu Huang
Lianghua Huang
Yue Huang
Yaping Huang
Yuge Huang
Zehao Huang
Zeyi Huang
Zhiqi Huang
Zhongzhan Huang
Zilong Huang
Ziyuan Huang
Tianrui Hui
Zhuo Hui
Le Hui
Jing Huo
Junhwa Hur
Shehzeen S. Hussain
Chuong Minh Huynh
Seunghyun Hwang
Jaehui Hwang
Jyh-Jing Hwang
Sukjun Hwang
Soonmin Hwang
Wonjun Hwang
Rakib Hyder
Sangeek Hyun
Sarah Ibrahimi
Tomoki Ichikawa
Yerlan Idelbayev
A. S. M. Iftekhar
Masaaki Iiyama
Satoshi Ikehata
Sunghoon Im
Atul N. Ingle
Eldar Insafutdinov
Yani A. Ioannou
Radu Tudor Ionescu

Umar Iqbal
Go Irie
Muhammad Zubair Irshad
Ahmet Iscen
Berivan Isik
Ashraful Islam
Md Amirul Islam
Syed Islam
Mariko Isogawa
Vamsi Krishna K. Ithapu
Boris Ivanovic
Darshan Iyer
Sarah Jabbour
Ayush Jain
Nishant Jain
Samyak Jain
Vidit Jain
Vineet Jain
Priyank Jaini
Tomas Jakab
Mohammad A. A. K. Jalwana
Muhammad Abdullah Jamal
Hadi Jamali-Rad
Stuart James
Varun Jampani
Young Kyun Jang
YeongJun Jang
Yunseok Jang
Ronnachai Jaroensri
Bhavan Jasani
Krishna Murthy Jatavallabhula
Mojan Javaheripi
Syed A. Javed
Guillaume Jeanneret
Pranav Jeevan
Herve Jegou
Rohit Jena
Tomas Jenicek
Porter Jenkins
Simon Jenni
Hae-Gon Jeon
Sangryul Jeon
Boseung Jeong
Yoonwoo Jeong
Seong-Gyun Jeong
Jisoo Jeong
Allan D. Jepson
Ankit Jha
Sumit K. Jha
I-Hong Jhuo
Ge-Peng Ji
Chaonan Ji
Deyi Ji
Jingwei Ji
Wei Ji
Zhong Ji
Jiayi Ji
Pengliang Ji
Hui Ji
Mingi Ji
Xiaopeng Ji
Yuzhu Ji
Baoxiong Jia
Songhao Jia
Dan Jia
Shan Jia
Xiaojun Jia
Xiuyi Jia
Xu Jia
Menglin Jia
Wenqi Jia
Boyuan Jiang
Wenhao Jiang
Huaizu Jiang
Hanwen Jiang
Haiyong Jiang
Hao Jiang
Huajie Jiang
Huiqin Jiang
Haojun Jiang
Haobo Jiang
Junjun Jiang
Xingyu Jiang
Yangbangyan Jiang
Yu Jiang
Jianmin Jiang
Jiaxi Jiang
Jing Jiang
Kui Jiang
Li Jiang
Liming Jiang
Chiyu Jiang
Meirui Jiang
Chen Jiang
Peng Jiang
Tai-Xiang Jiang
Wen Jiang
Xinyang Jiang
Yifan Jiang
Yuming Jiang
Yingying Jiang
Zeren Jiang
ZhengKai Jiang
Zhenyu Jiang
Shuming Jiao
Jianbo Jiao
Licheng Jiao
Dongkwon Jin
Yeying Jin
Cheng Jin
Linyi Jin
Qing Jin
Taisong Jin
Xiao Jin
Xin Jin
Sheng Jin
Kyong Hwan Jin
Ruibing Jin
SouYoung Jin
Yueming Jin
Chenchen Jing
Longlong Jing
Taotao Jing
Yongcheng Jing
Younghyun Jo
Joakim Johnander
Jeff Johnson
Michael J. Jones
R. Kenny Jones
Rico Jonschkowski
Ameya Joshi
Sunghun Joung

Felix Juefei-Xu
Claudio R. Jung
Steffen Jung
Hari Chandana K.
Rahul Vigneswaran K.
Prajwal K. R.
Abhishek Kadian
Jhony Kaesemodel Pontes
Kumara Kahatapitiya
Anmol Kalia
Sinan Kalkan
Tarun Kalluri
Jaewon Kam
Sandesh Kamath
Meina Kan
Menelaos Kanakis
Takuhiro Kaneko
Di Kang
Guoliang Kang
Hao Kang
Jaeyeon Kang
Kyoungkook Kang
Li-Wei Kang
MinGuk Kang
Suk-Ju Kang
Zhao Kang
Yash Mukund Kant
Yueying Kao
Aupendu Kar
Konstantinos Karantzalos
Sezer Karaoglu
Navid Kardan
Sanjay Kariyappa
Leonid Karlinsky
Animesh Karnewar
Shyamgopal Karthik
Hirak J. Kashyap
Marc A. Kastner
Hirokatsu Kataoka
Angelos Katharopoulos
Hiroharu Kato
Kai Katsumata
Manuel Kaufmann
Chaitanya Kaul
Prakhar Kaushik
Yuki Kawana
Lei Ke
Lipeng Ke
Tsung-Wei Ke
Wei Ke
Petr Kellnhofer
Aniruddha Kembhavi
John Kender
Corentin Kervadec
Leonid Keselman
Daniel Keysers
Nima Khademi Kalantari
Taras Khakhulin
Samir Khaki
Muhammad Haris Khan
Qadeer Khan
Salman Khan
Subash Khanal
Vaishnavi M. Khindkar
Rawal Khirodkar
Saeed Khorram
Pirazh Khorramshahi
Kourosh Khoshelham
Ansh Khurana
Benjamin Kiefer
Jae Myung Kim
Junho Kim
Boah Kim
Hyeonseong Kim
Dong-Jin Kim
Dongwan Kim
Donghyun Kim
Doyeon Kim
Yonghyun Kim
Hyung-Il Kim
Hyunwoo Kim
Hyeongwoo Kim
Hyo Jin Kim
Hyunwoo J. Kim
Taehoon Kim
Jaeha Kim
Jiwon Kim
Jung Uk Kim
Kangyeol Kim
Eunji Kim
Daeha Kim
Dongwon Kim
Kunhee Kim
Kyungmin Kim
Junsik Kim
Min H. Kim
Namil Kim
Kookhoi Kim
Sanghyun Kim
Seongyeop Kim
Seungryong Kim
Saehoon Kim
Euyoung Kim
Guisik Kim
Sungyeon Kim
Sunnie S. Y. Kim
Taehun Kim
Tae Oh Kim
Won Hwa Kim
Seungwook Kim
YoungBin Kim
Youngeun Kim
Akisato Kimura
Furkan Osman Kınlı
Zsolt Kira
Hedvig Kjellström
Florian Kleber
Jan P. Klopp
Florian Kluger
Laurent Kneip
Byungsoo Ko
Muhammed Kocabas
A. Sophia Koepke
Kevin Koeser
Nick Kolkin
Nikos Kolotouros
Wai-Kin Adams Kong
Deying Kong
Caihua Kong
Youyong Kong
Shuyu Kong
Shu Kong
Tao Kong
Yajing Kong
Yu Kong

Zishang Kong
Theodora Kontogianni
Anton S. Konushin
Julian F. P. Kooij
Bruno Korbar
Giorgos Kordopatis-Zilos
Jari Korhonen
Adam Kortylewski
Denis Korzhenkov
Divya Kothandaraman
Suraj Kothawade
Iuliia Kotseruba
Satwik Kottur
Shashank Kotyan
Alexandros Kouris
Petros Koutras
Anna Kreshuk
Ranjay Krishna
Dilip Krishnan
Andrey Kuehlkamp
Hilde Kuehne
Jason Kuen
David Kügler
Arjan Kuijper
Anna Kukleva
Sumith Kulal
Viveka Kulharia
Akshay R. Kulkarni
Nilesh Kulkarni
Dominik Kulon
Abhinav Kumar
Akash Kumar
Suryansh Kumar
B. V. K. Vijaya Kumar
Pulkit Kumar
Ratnesh Kumar
Sateesh Kumar
Satish Kumar
Vijay Kumar B. G.
Nupur Kumari
Sudhakar Kumawat
Jogendra Nath Kundu
Hsien-Kai Kuo
Meng-Yu Jennifer Kuo
Vinod Kumar Kurmi
Yusuke Kurose
Keerthy Kusumam
Alina Kuznetsova
Henry Kvinge
Ho Man Kwan
Hyeokjun Kweon
Heeseung Kwon
Gihyun Kwon
Myung-Joon Kwon
Taesung Kwon
YoungJoong Kwon
Christos Kyrkou
Jorma Laaksonen
Yann Labbe
Zorah Laehner
Florent Lafarge
Hamid Laga
Manuel Lagunas
Shenqi Lai
Jian-Huang Lai
Zihang Lai
Mohamed I. Lakhal
Mohit Lamba
Meng Lan
Loic Landrieu
Zhiqiang Lang
Natalie Lang
Dong Lao
Yizhen Lao
Yingjie Lao
Issam Hadj Laradji
Gustav Larsson
Viktor Larsson
Zakaria Laskar
Stéphane Lathuilière
Chun Pong Lau
Rynson W. H. Lau
Hei Law
Justin Lazarow
Verica Lazova
Eric-Tuan Le
Hieu Le
Trung-Nghia Le
Mathias Lechner
Byeong-Uk Lee
Chen-Yu Lee
Che-Rung Lee
Chul Lee
Hong Joo Lee
Dongsoo Lee
Jiyoung Lee
Eugene Eu Tzuan Lee
Daeun Lee
Saehyung Lee
Jewook Lee
Hyungtae Lee
Hyunmin Lee
Jungbeom Lee
Joon-Young Lee
Jong-Seok Lee
Joonseok Lee
Junha Lee
Kibok Lee
Byung-Kwan Lee
Jangwon Lee
Jinho Lee
Jongmin Lee
Seunghyun Lee
Sohyun Lee
Minsik Lee
Dogyoon Lee
Seungmin Lee
Min Jun Lee
Sangho Lee
Sangmin Lee
Seungeun Lee
Seon-Ho Lee
Sungmin Lee
Sungho Lee
Sangyoun Lee
Vincent C. S. S. Lee
Jaeseong Lee
Yong Jae Lee
Chenyang Lei
Chenyi Lei
Jiahui Lei
Xinyu Lei
Yinjie Lei
Jiaxu Leng
Luziwei Leng

Jan E. Lenssen
Vincent Lepetit
Thomas Leung
María Leyva-Vallina
Xin Li
Yikang Li
Baoxin Li
Bin Li
Bing Li
Bowen Li
Changlin Li
Chao Li
Chongyi Li
Guanyue Li
Shuai Li
Jin Li
Dingquan Li
Dongxu Li
Yiting Li
Gang Li
Dian Li
Guohao Li
Haoang Li
Haoliang Li
Haoran Li
Hengduo Li
Huafeng Li
Xiaoming Li
Hanao Li
Hongwei Li
Ziqiang Li
Jisheng Li
Jiacheng Li
Jia Li
Jiachen Li
Jiahao Li
Jianwei Li
Jiazhi Li
Jie Li
Jing Li
Jingjing Li
Jingtao Li
Jun Li
Junxuan Li
Kai Li
Kailin Li
Kenneth Li
Kun Li
Kunpeng Li
Aoxue Li
Chenglong Li
Chenglin Li
Changsheng Li
Zhichao Li
Qiang Li
Yanyu Li
Zuoyue Li
Xiang Li
Xuelong Li
Fangda Li
Ailin Li
Liang Li
Chun-Guang Li
Daiqing Li
Dong Li
Guanbin Li
Guorong Li
Haifeng Li
Jianan Li
Jianing Li
Jiaxin Li
Ke Li
Lei Li
Lincheng Li
Liulei Li
Lujun Li
Linjie Li
Lin Li
Pengyu Li
Ping Li
Qiufu Li
Qingyong Li
Rui Li
Siyuan Li
Wei Li
Wenbin Li
Xiangyang Li
Xinyu Li
Xiujun Li
Xiu Li
Xu Li
Ya-Li Li
Yao Li
Yongjie Li
Yijun Li
Yiming Li
Yuezun Li
Yu Li
Yunheng Li
Yuqi Li
Zhe Li
Zeming Li
Zhen Li
Zhengqin Li
Zhimin Li
Jiefeng Li
Jinpeng Li
Chengze Li
Jianwu Li
Lerenhan Li
Shan Li
Suichan Li
Xiangtai Li
Yanjie Li
Yandong Li
Zhuoling Li
Zhenqiang Li
Manyi Li
Maosen Li
Ji Li
Minjun Li
Mingrui Li
Mengtian Li
Junyi Li
Nianyi Li
Bo Li
Xiao Li
Peihua Li
Peike Li
Peizhao Li
Peiliang Li
Qi Li
Ren Li
Runze Li
Shile Li

Sheng Li
Shigang Li
Shiyu Li
Shuang Li
Shasha Li
Shichao Li
Tianye Li
Yuexiang Li
Wei-Hong Li
Wanhua Li
Weihao Li
Weiming Li
Weixin Li
Wenbo Li
Wenshuo Li
Weijian Li
Yunan Li
Xirong Li
Xianhang Li
Xiaoyu Li
Xueqian Li
Xuanlin Li
Xianzhi Li
Yunqiang Li
Yanjing Li
Yansheng Li
Yawei Li
Yi Li
Yong Li
Yong-Lu Li
Yuhang Li
Yu-Jhe Li
Yuxi Li
Yunsheng Li
Yanwei Li
Zechao Li
Zejian Li
Zeju Li
Zekun Li
Zhaowen Li
Zheng Li
Zhenyu Li
Zhiheng Li
Zhi Li
Zhong Li
Zhuowei Li
Zhuowan Li
Zhuohang Li
Zizhang Li
Chen Li
Yuan-Fang Li
Dongze Lian
Xiaochen Lian
Zhouhui Lian
Long Lian
Qing Lian
Jin Lianbao
Jinxiu S. Liang
Dingkang Liang
Jiahao Liang
Jianming Liang
Jingyun Liang
Kevin J. Liang
Kaizhao Liang
Chen Liang
Jie Liang
Senwei Liang
Ding Liang
Jiajun Liang
Jian Liang
Kongming Liang
Siyuan Liang
Yuanzhi Liang
Zhengfa Liang
Mingfu Liang
Xiaodan Liang
Xuefeng Liang
Yuxuan Liang
Kang Liao
Liang Liao
Hong-Yuan Mark Liao
Wentong Liao
Haofu Liao
Yue Liao
Minghui Liao
Shengcai Liao
Ting-Hsuan Liao
Xin Liao
Yinghong Liao
Teck Yian Lim
Che-Tsung Lin
Chung-Ching Lin
Chen-Hsuan Lin
Cheng Lin
Chuming Lin
Chunyu Lin
Dahua Lin
Wei Lin
Zheng Lin
Huaijia Lin
Jason Lin
Jierui Lin
Jiaying Lin
Jie Lin
Kai-En Lin
Kevin Lin
Guangfeng Lin
Jiehong Lin
Feng Lin
Hang Lin
Kwan-Yee Lin
Ke Lin
Luojun Lin
Qinghong Lin
Xiangbo Lin
Yi Lin
Zudi Lin
Shijie Lin
Yiqun Lin
Tzu-Heng Lin
Ming Lin
Shaohui Lin
SongNan Lin
Ji Lin
Tsung-Yu Lin
Xudong Lin
Yancong Lin
Yen-Chen Lin
Yiming Lin
Yuewei Lin
Zhiqiu Lin
Zinan Lin
Zhe Lin
David B. Lindell
Zhixin Ling

Zhan Ling
Alexander Liniger
Venice Erin B. Liong
Joey Litalien
Or Litany
Roee Litman
Ron Litman
Jim Little
Dor Litvak
Shaoteng Liu
Shuaicheng Liu
Andrew Liu
Xian Liu
Shaohui Liu
Bei Liu
Bo Liu
Yong Liu
Ming Liu
Yanbin Liu
Chenxi Liu
Daqi Liu
Di Liu
Difan Liu
Dong Liu
Dongfang Liu
Daizong Liu
Xiao Liu
Fangyi Liu
Fengbei Liu
Fenglin Liu
Bin Liu
Yuang Liu
Ao Liu
Hong Liu
Hongfu Liu
Huidong Liu
Ziyi Liu
Feng Liu
Hao Liu
Jie Liu
Jialun Liu
Jiang Liu
Jing Liu
Jingya Liu
Jiaming Liu
Jun Liu
Juncheng Liu
Jiawei Liu
Hongyu Liu
Chuanbin Liu
Haotian Liu
Lingqiao Liu
Chang Liu
Han Liu
Liu Liu
Min Liu
Yingqi Liu
Aishan Liu
Bingyu Liu
Benlin Liu
Boxiao Liu
Chenchen Liu
Chuanjian Liu
Daqing Liu
Huan Liu
Haozhe Liu
Jiaheng Liu
Wei Liu
Jingzhou Liu
Jiyuan Liu
Lingbo Liu
Nian Liu
Peiye Liu
Qiankun Liu
Shenglan Liu
Shilong Liu
Wen Liu
Wenyu Liu
Weifeng Liu
Wu Liu
Xiaolong Liu
Yang Liu
Yanwei Liu
Yingcheng Liu
Yongfei Liu
Yihao Liu
Yu Liu
Yunze Liu
Ze Liu
Zhenhua Liu
Zhenguang Liu
Lin Liu
Lihao Liu
Pengju Liu
Xinhai Liu
Yunfei Liu
Meng Liu
Minghua Liu
Mingyuan Liu
Miao Liu
Peirong Liu
Ping Liu
Qingjie Liu
Ruoshi Liu
Risheng Liu
Songtao Liu
Xing Liu
Shikun Liu
Shuming Liu
Sheng Liu
Songhua Liu
Tongliang Liu
Weibo Liu
Weide Liu
Weizhe Liu
Wenxi Liu
Weiyang Liu
Xin Liu
Xiaobin Liu
Xudong Liu
Xiaoyi Liu
Xihui Liu
Xinchen Liu
Xingtong Liu
Xinpeng Liu
Xinyu Liu
Xianpeng Liu
Xu Liu
Xingyu Liu
Yongtuo Liu
Yahui Liu
Yangxin Liu
Yaoyao Liu
Yaojie Liu
Yuliang Liu

Yongcheng Liu
Yuan Liu
Yufan Liu
Yu-Lun Liu
Yun Liu
Yunfan Liu
Yuanzhong Liu
Zhuoran Liu
Zhen Liu
Zheng Liu
Zhijian Liu
Zhisong Liu
Ziquan Liu
Ziyu Liu
Zhihua Liu
Zechun Liu
Zhaoyang Liu
Zhengzhe Liu
Stephan Liwicki
Shao-Yuan Lo
Sylvain Lobry
Suhas Lohit
Vishnu Suresh Lokhande
Vincenzo Lomonaco
Chengjiang Long
Guodong Long
Fuchen Long
Shangbang Long
Yang Long
Zijun Long
Vasco Lopes
Antonio M. Lopez
Roberto Javier Lopez-Sastre
Tobias Lorenz
Javier Lorenzo-Navarro
Yujing Lou
Qian Lou
Xiankai Lu
Changsheng Lu
Huimin Lu
Yongxi Lu
Hao Lu
Hong Lu
Jiasen Lu
Juwei Lu
Fan Lu
Guangming Lu
Jiwen Lu
Shun Lu
Tao Lu
Xiaonan Lu
Yang Lu
Yao Lu
Yongchun Lu
Zhiwu Lu
Cheng Lu
Liying Lu
Guo Lu
Xuequan Lu
Yanye Lu
Yantao Lu
Yuhang Lu
Fujun Luan
Jonathon Luiten
Jovita Lukasik
Alan Lukezic
Jonathan Samuel Lumentut
Mayank Lunayach
Ao Luo
Canjie Luo
Chong Luo
Xu Luo
Grace Luo
Jun Luo
Katie Z. Luo
Tao Luo
Cheng Luo
Fangzhou Luo
Gen Luo
Lei Luo
Sihui Luo
Weixin Luo
Yan Luo
Xiaoyan Luo
Yong Luo
Yadan Luo
Hao Luo
Ruotian Luo
Mi Luo
Tiange Luo
Wenjie Luo
Wenhan Luo
Xiao Luo
Zhiming Luo
Zhipeng Luo
Zhengyi Luo
Diogo C. Luvizon
Zhaoyang Lv
Gengyu Lyu
Lingjuan Lyu
Jun Lyu
Yuanyuan Lyu
Youwei Lyu
Yueming Lyu
Bingpeng Ma
Chao Ma
Chongyang Ma
Congbo Ma
Chih-Yao Ma
Fan Ma
Lin Ma
Haoyu Ma
Hengbo Ma
Jianqi Ma
Jiawei Ma
Jiayi Ma
Kede Ma
Kai Ma
Lingni Ma
Lei Ma
Xu Ma
Ning Ma
Benteng Ma
Cheng Ma
Andy J. Ma
Long Ma
Zhanyu Ma
Zhiheng Ma
Qianli Ma
Shiqiang Ma
Sizhuo Ma
Shiqing Ma
Xiaolong Ma
Xinzhu Ma

Gautam B. Machiraju
Spandan Madan
Mathew Magimai-Doss
Luca Magri
Behrooz Mahasseni
Upal Mahbub
Siddharth Mahendran
Paridhi Maheshwari
Rishabh Maheshwary
Mohammed Mahmoud
Shishira R. R. Maiya
Sylwia Majchrowska
Arjun Majumdar
Puspita Majumdar
Orchid Majumder
Sagnik Majumder
Ilya Makarov
Farkhod F. Makhmudkhujaev
Yasushi Makihara
Ankur Mali
Mateusz Malinowski
Utkarsh Mall
Srikanth Malla
Clement Mallet
Dimitrios Mallis
Yunze Man
Dipu Manandhar
Massimiliano Mancini
Murari Mandal
Raunak Manekar
Karttikeya Mangalam
Puneet Mangla
Fabian Manhardt
Sivabalan Manivasagam
Fahim Mannan
Chengzhi Mao
Hanzi Mao
Jiayuan Mao
Junhua Mao
Zhiyuan Mao
Jiageng Mao
Yunyao Mao
Zhendong Mao
Alberto Marchisio
Diego Marcos
Riccardo Marin
Aram Markosyan
Renaud Marlet
Ricardo Marques
Miquel Martí i Rabadán
Diego Martin Arroyo
Niki Martinel
Brais Martinez
Julieta Martinez
Marc Masana
Tomohiro Mashita
Timothée Masquelier
Minesh Mathew
Tetsu Matsukawa
Marwan Mattar
Bruce A. Maxwell
Christoph Mayer
Mantas Mazeika
Pratik Mazumder
Scott McCloskey
Steven McDonagh
Ishit Mehta
Jie Mei
Kangfu Mei
Jieru Mei
Xiaoguang Mei
Givi Meishvili
Luke Melas-Kyriazi
Iaroslav Melekhov
Andres Mendez-Vazquez
Heydi Mendez-Vazquez
Matias Mendieta
Ricardo A. Mendoza-León
Chenlin Meng
Depu Meng
Rang Meng
Zibo Meng
Qingjie Meng
Qier Meng
Yanda Meng
Zihang Meng
Thomas Mensink
Fabian Mentzer
Christopher Metzler
Gregory P. Meyer
Vasileios Mezaris
Liang Mi
Lu Mi
Bo Miao
Changtao Miao
Zichen Miao
Qiguang Miao
Xin Miao
Zhongqi Miao
Frank Michel
Simone Milani
Ben Mildenhall
Roy V. Miles
Juhong Min
Kyle Min
Hyun-Seok Min
Weiqing Min
Yuecong Min
Zhixiang Min
Qi Ming
David Minnen
Aymen Mir
Deepak Mishra
Anand Mishra
Shlok K. Mishra
Niluthpol Mithun
Gaurav Mittal
Trisha Mittal
Daisuke Miyazaki
Kaichun Mo
Hong Mo
Zhipeng Mo
Davide Modolo
Abduallah A. Mohamed
Mohamed Afham Mohamed Aflal
Ron Mokady
Pavlo Molchanov
Davide Moltisanti
Liliane Momeni
Gianluca Monaci
Pascal Monasse
Ajoy Mondal
Tom Monnier

Aron Monszpart
Gyeongsik Moon
Suhong Moon
Taesup Moon
Sean Moran
Daniel Moreira
Pietro Morerio
Alexandre Morgand
Lia Morra
Ali Mosleh
Inbar Mosseri
Sayed Mohammad Mostafavi Isfahani
Saman Motamed
Ramy A. Mounir
Fangzhou Mu
Jiteng Mu
Norman Mu
Yasuhiro Mukaigawa
Ryan Mukherjee
Tanmoy Mukherjee
Yusuke Mukuta
Ravi Teja Mullapudi
Lea Müller
Matthias Müller
Martin Mundt
Nils Murrugarra-Llerena
Damien Muselet
Armin Mustafa
Muhammad Ferjad Naeem
Sauradip Nag
Hajime Nagahara
Pravin Nagar
Rajendra Nagar
Naveen Shankar Nagaraja
Varun Nagaraja
Tushar Nagarajan
Seungjun Nah
Gaku Nakano
Yuta Nakashima
Giljoo Nam
Seonghyeon Nam
Liangliang Nan
Yuesong Nan
Yeshwanth Napolean
Dinesh Reddy Narapureddy
Medhini Narasimhan
Supreeth Narasimhaswamy
Sriram Narayanan
Erickson R. Nascimento
Varun Nasery
K. L. Navaneet
Pablo Navarrete Michelini
Shant Navasardyan
Shah Nawaz
Nihal Nayak
Farhood Negin
Lukáš Neumann
Alejandro Newell
Evonne Ng
Kam Woh Ng
Tony Ng
Anh Nguyen
Tuan Anh Nguyen
Cuong Cao Nguyen
Ngoc Cuong Nguyen
Thanh Nguyen
Khoi Nguyen
Phi Le Nguyen
Phong Ha Nguyen
Tam Nguyen
Truong Nguyen
Anh Tuan Nguyen
Rang Nguyen
Thao Thi Phuong Nguyen
Van Nguyen Nguyen
Zhen-Liang Ni
Yao Ni
Shijie Nie
Xuecheng Nie
Yongwei Nie
Weizhi Nie
Ying Nie
Yinyu Nie
Kshitij N. Nikhal
Simon Niklaus
Xuefei Ning
Jifeng Ning
Yotam Nitzan
Di Niu
Shuaicheng Niu
Li Niu
Wei Niu
Yulei Niu
Zhenxing Niu
Albert No
Shohei Nobuhara
Nicoletta Noceti
Junhyug Noh
Sotiris Nousias
Slawomir Nowaczyk
Ewa M. Nowara
Valsamis Ntouskos
Gilberto Ochoa-Ruiz
Ferda Ofli
Jihyong Oh
Sangyun Oh
Youngtaek Oh
Hiroki Ohashi
Takahiro Okabe
Kemal Oksuz
Fumio Okura
Daniel Olmeda Reino
Matthew Olson
Carl Olsson
Roy Or-El
Alessandro Ortis
Guillermo Ortiz-Jimenez
Magnus Oskarsson
Ahmed A. A. Osman
Martin R. Oswald
Mayu Otani
Naima Otberdout
Cheng Ouyang
Jiahong Ouyang
Wanli Ouyang
Andrew Owens
Poojan B. Oza
Mete Ozay
A. Cengiz Oztireli
Gautam Pai
Tomas Pajdla
Umapada Pal

Simone Palazzo
Luca Palmieri
Bowen Pan
Hao Pan
Lili Pan
Tai-Yu Pan
Liang Pan
Chengwei Pan
Yingwei Pan
Xuran Pan
Jinshan Pan
Xinyu Pan
Liyuan Pan
Xingang Pan
Xingjia Pan
Zhihong Pan
Zizheng Pan
Priyadarshini Panda
Rameswar Panda
Rohit Pandey
Kaiyue Pang
Bo Pang
Guansong Pang
Jiangmiao Pang
Meng Pang
Tianyu Pang
Ziqi Pang
Omiros Pantazis
Andreas Panteli
Maja Pantic
Marina Paolanti
Joao P. Papa
Samuele Papa
Mike Papadakis
Dim P. Papadopoulos
George Papandreou
Constantin Pape
Toufiq Parag
Chethan Parameshwara
Shaifali Parashar
Alejandro Pardo
Rishubh Parihar
Sarah Parisot
JaeYoo Park
Gyeong-Moon Park
Hyojin Park
Hyoungseob Park
Jongchan Park
Jae Sung Park
Kiru Park
Chunghyun Park
Kwanyong Park
Sunghyun Park
Sungrae Park
Seongsik Park
Sanghyun Park
Sungjune Park
Taesung Park
Gaurav Parmar
Paritosh Parmar
Alvaro Parra
Despoina Paschalidou
Or Patashnik
Shivansh Patel
Pushpak Pati
Prashant W. Patil
Vaishakh Patil
Suvam Patra
Jay Patravali
Badri Narayana Patro
Angshuman Paul
Sudipta Paul
Rémi Pautrat
Nick E. Pears
Adithya Pediredla
Wenjie Pei
Shmuel Peleg
Latha Pemula
Bo Peng
Houwen Peng
Yue Peng
Liangzu Peng
Baoyun Peng
Jun Peng
Pai Peng
Sida Peng
Xi Peng
Yuxin Peng
Songyou Peng
Wei Peng
Weiqi Peng
Wen-Hsiao Peng
Pramuditha Perera
Juan C. Perez
Eduardo Pérez Pellitero
Juan-Manuel Perez-Rua
Federico Pernici
Marco Pesavento
Stavros Petridis
Ilya A. Petrov
Vladan Petrovic
Mathis Petrovich
Suzanne Petryk
Hieu Pham
Quang Pham
Khoi Pham
Tung Pham
Huy Phan
Stephen Phillips
Cheng Perng Phoo
David Picard
Marco Piccirilli
Georg Pichler
A. J. Piergiovanni
Vipin Pillai
Silvia L. Pintea
Giovanni Pintore
Robinson Piramuthu
Fiora Pirri
Theodoros Pissas
Fabio Pizzati
Benjamin Planche
Bryan Plummer
Matteo Poggi
Ashwini Pokle
Georgy E. Ponimatkin
Adrian Popescu
Stefan Popov
Nikola Popović
Ronald Poppe
Angelo Porrello
Michael Potter
Charalambos Poullis
Hadi Pouransari
Omid Poursaeed

Shraman Pramanick
Mantini Pranav
Dilip K. Prasad
Meghshyam Prasad
B. H. Pawan Prasad
Shitala Prasad
Prateek Prasanna
Ekta Prashnani
Derek S. Prijatelj
Luke Y. Prince
Véronique Prinet
Victor Adrian Prisacariu
James Pritts
Thomas Probst
Sergey Prokudin
Rita Pucci
Chi-Man Pun
Matthew Purri
Haozhi Qi
Lu Qi
Lei Qi
Xianbiao Qi
Yonggang Qi
Yuankai Qi
Siyuan Qi
Guocheng Qian
Hangwei Qian
Qi Qian
Deheng Qian
Shengsheng Qian
Wen Qian
Rui Qian
Yiming Qian
Shengju Qian
Shengyi Qian
Xuelin Qian
Zhenxing Qian
Nan Qiao
Xiaotian Qiao
Jing Qin
Can Qin
Siyang Qin
Hongwei Qin
Jie Qin
Minghai Qin
Yipeng Qin
Yongqiang Qin
Wenda Qin
Xuebin Qin
Yuzhe Qin
Yao Qin
Zhenyue Qin
Zhiwu Qing
Heqian Qiu
Jiayan Qiu
Jielin Qiu
Yue Qiu
Jiaxiong Qiu
Zhongxi Qiu
Shi Qiu
Zhaofan Qiu
Zhongnan Qu
Yanyun Qu
Kha Gia Quach
Yuhui Quan
Ruijie Quan
Mike Rabbat
Rahul Shekhar Rade
Filip Radenovic
Gorjan Radevski
Bogdan Raducanu
Francesco Ragusa
Shafin Rahman
Md Mahfuzur Rahman Siddiquee
Hossein Rahmani
Kiran Raja
Sivaramakrishnan Rajaraman
Jathushan Rajasegaran
Adnan Siraj Rakin
Michaël Ramamonjisoa
Chirag A. Raman
Shanmuganathan Raman
Vignesh Ramanathan
Vasili Ramanishka
Vikram V. Ramaswamy
Merey Ramazanova
Jason Rambach
Sai Saketh Rambhatla
Clément Rambour
Ashwin Ramesh Babu
Adín Ramírez Rivera
Arianna Rampini
Haoxi Ran
Aakanksha Rana
Aayush Jung Bahadur Rana
Kanchana N. Ranasinghe
Aneesh Rangnekar
Samrudhdhi B. Rangrej
Harsh Rangwani
Viresh Ranjan
Anyi Rao
Yongming Rao
Carolina Raposo
Michalis Raptis
Amir Rasouli
Vivek Rathod
Adepu Ravi Sankar
Avinash Ravichandran
Bharadwaj Ravichandran
Dripta S. Raychaudhuri
Adria Recasens
Simon Reiß
Davis Rempe
Daxuan Ren
Jiawei Ren
Jimmy Ren
Sucheng Ren
Dayong Ren
Zhile Ren
Dongwei Ren
Qibing Ren
Pengfei Ren
Zhenwen Ren
Xuqian Ren
Yixuan Ren
Zhongzheng Ren
Ambareesh Revanur
Hamed Rezazadegan Tavakoli
Rafael S. Rezende
Wonjong Rhee
Alexander Richard

Christian Richardt
Stephan R. Richter
Benjamin Riggan
Dominik Rivoir
Mamshad Nayeem Rizve
Joshua D. Robinson
Joseph Robinson
Chris Rockwell
Ranga Rodrigo
Andres C. Rodriguez
Carlos Rodriguez-Pardo
Marcus Rohrbach
Gemma Roig
Yu Rong
David A. Ross
Mohammad Rostami
Edward Rosten
Karsten Roth
Anirban Roy
Debaditya Roy
Shuvendu Roy
Ahana Roy Choudhury
Aruni Roy Chowdhury
Denys Rozumnyi
Shulan Ruan
Wenjie Ruan
Patrick Ruhkamp
Danila Rukhovich
Anian Ruoss
Chris Russell
Dan Ruta
Dawid Damian Rymarczyk
DongHun Ryu
Hyeonggon Ryu
Kwonyoung Ryu
Balasubramanian S.
Alexandre Sablayrolles
Mohammad Sabokrou
Arka Sadhu
Aniruddha Saha
Oindrila Saha
Pritish Sahu
Aneeshan Sain
Nirat Saini
Saurabh Saini
Takeshi Saitoh
Christos Sakaridis
Fumihiko Sakaue
Dimitrios Sakkos
Ken Sakurada
Parikshit V. Sakurikar
Rohit Saluja
Nermin Samet
Leo Sampaio Ferraz Ribeiro
Jorge Sanchez
Enrique Sanchez
Shengtian Sang
Anush Sankaran
Soubhik Sanyal
Nikolaos Sarafianos
Vishwanath Saragadam
István Sárándi
Saquib Sarfraz
Mert Bulent Sariyildiz
Anindya Sarkar
Pritam Sarkar
Paul-Edouard Sarlin
Hiroshi Sasaki
Takami Sato
Torsten Sattler
Ravi Kumar Satzoda
Axel Sauer
Stefano Savian
Artem Savkin
Manolis Savva
Gerald Schaefer
Simone Schaub-Meyer
Yoni Schirris
Samuel Schulter
Katja Schwarz
Jesse Scott
Sinisa Segvic
Constantin Marc Seibold
Lorenzo Seidenari
Matan Sela
Fadime Sener
Paul Hongsuck Seo
Kwanggyoon Seo
Hongje Seong
Dario Serez
Francesco Setti
Bryan Seybold
Mohamad Shahbazi
Shima Shahfar
Xinxin Shan
Caifeng Shan
Dandan Shan
Shawn Shan
Wei Shang
Jinghuan Shang
Jiaxiang Shang
Lei Shang
Sukrit Shankar
Ken Shao
Rui Shao
Jie Shao
Mingwen Shao
Aashish Sharma
Gaurav Sharma
Vivek Sharma
Abhishek Sharma
Yoli Shavit
Shashank Shekhar
Sumit Shekhar
Zhijie Shen
Fengyi Shen
Furao Shen
Jialie Shen
Jingjing Shen
Ziyi Shen
Linlin Shen
Guangyu Shen
Biluo Shen
Falong Shen
Jiajun Shen
Qiu Shen
Qiuhong Shen
Shuai Shen
Wang Shen
Yiqing Shen
Yunhang Shen
Siqi Shen
Bin Shen
Tianwei Shen

Xi Shen
Yilin Shen
Yuming Shen
Yucong Shen
Zhiqiang Shen
Lu Sheng
Yichen Sheng
Shivanand Venkanna Sheshappanavar
Shelly Sheynin
Baifeng Shi
Ruoxi Shi
Botian Shi
Hailin Shi
Jia Shi
Jing Shi
Shaoshuai Shi
Baoguang Shi
Boxin Shi
Hengcan Shi
Tianyang Shi
Xiaodan Shi
Yongjie Shi
Zhensheng Shi
Yinghuan Shi
Weiqi Shi
Wu Shi
Xuepeng Shi
Xiaoshuang Shi
Yujiao Shi
Zenglin Shi
Zhenmei Shi
Takashi Shibata
Meng-Li Shih
Yichang Shih
Hyunjung Shim
Dongseok Shim
Soshi Shimada
Inkyu Shin
Jinwoo Shin
Seungjoo Shin
Seungjae Shin
Koichi Shinoda
Suprosanna Shit
Palaiahnakote Shivakumara
Eli Shlizerman
Gaurav Shrivastava
Xiao Shu
Xiangbo Shu
Xiujun Shu
Yang Shu
Tianmin Shu
Jun Shu
Zhixin Shu
Bing Shuai
Maria Shugrina
Ivan Shugurov
Satya Narayan Shukla
Pranjay Shyam
Jianlou Si
Yawar Siddiqui
Alberto Signoroni
Pedro Silva
Jae-Young Sim
Oriane Siméoni
Martin Simon
Andrea Simonelli
Abhishek Singh
Ashish Singh
Dinesh Singh
Gurkirt Singh
Krishna Kumar Singh
Mannat Singh
Pravendra Singh
Rajat Vikram Singh
Utkarsh Singhal
Dipika Singhania
Vasu Singla
Harsh Sinha
Sudipta Sinha
Josef Sivic
Elena Sizikova
Geri Skenderi
Ivan Skorokhodov
Dmitriy Smirnov
Cameron Y. Smith
James S. Smith
Patrick Snape
Mattia Soldan
Hyeongseok Son
Sanghyun Son
Chuanbiao Song
Chen Song
Chunfeng Song
Dan Song
Dongjin Song
Hwanjun Song
Guoxian Song
Jiaming Song
Jie Song
Liangchen Song
Ran Song
Luchuan Song
Xibin Song
Li Song
Fenglong Song
Guoli Song
Guanglu Song
Zhenbo Song
Lin Song
Xinhang Song
Yang Song
Yibing Song
Rajiv Soundararajan
Hossein Souri
Cristovao Sousa
Riccardo Spezialetti
Leonidas Spinoulas
Michael W. Spratling
Deepak Sridhar
Srinath Sridhar
Gaurang Sriramanan
Vinkle Kumar Srivastav
Themos Stafylakis
Serban Stan
Anastasis Stathopoulos
Markus Steinberger
Jan Steinbrener
Sinisa Stekovic
Alexandros Stergiou
Gleb Sterkin
Rainer Stiefelhagen
Pierre Stock

Ombretta Strafforello
Julian Straub
Yannick Strümpler
Joerg Stueckler
Hang Su
Weijie Su
Jong-Chyi Su
Bing Su
Haisheng Su
Jinming Su
Yiyang Su
Yukun Su
Yuxin Su
Zhuo Su
Zhaoqi Su
Xiu Su
Yu-Chuan Su
Zhixun Su
Arulkumar Subramaniam
Akshayvarun Subramanya
A. Subramanyam
Swathikiran Sudhakaran
Yusuke Sugano
Masanori Suganuma
Yumin Suh
Yang Sui
Baochen Sun
Cheng Sun
Long Sun
Guolei Sun
Haoliang Sun
Haomiao Sun
He Sun
Hanqing Sun
Hao Sun
Lichao Sun
Jiachen Sun
Jiaming Sun
Jian Sun
Jin Sun
Jennifer J. Sun
Tiancheng Sun
Libo Sun
Peize Sun
Qianru Sun
Shanlin Sun
Yu Sun
Zhun Sun
Che Sun
Lin Sun
Tao Sun
Yiyou Sun
Chunyi Sun
Chong Sun
Weiwei Sun
Weixuan Sun
Xiuyu Sun
Yanan Sun
Zeren Sun
Zhaodong Sun
Zhiqing Sun
Minhyuk Sung
Jinli Suo
Simon Suo
Abhijit Suprem
Anshuman Suri
Saksham Suri
Joshua M. Susskind
Roman Suvorov
Gurumurthy Swaminathan
Robin Swanson
Paul Swoboda
Tabish A. Syed
Richard Szeliski
Fariborz Taherkhani
Yu-Wing Tai
Keita Takahashi
Walter Talbott
Gary Tam
Masato Tamura
Feitong Tan
Fuwen Tan
Shuhan Tan
Andong Tan
Bin Tan
Cheng Tan
Jianchao Tan
Lei Tan
Mingxing Tan
Xin Tan
Zichang Tan
Zhentao Tan
Kenichiro Tanaka
Masayuki Tanaka
Yushun Tang
Hao Tang
Jingqun Tang
Jinhui Tang
Kaihua Tang
Luming Tang
Lv Tang
Sheyang Tang
Shitao Tang
Siliang Tang
Shixiang Tang
Yansong Tang
Keke Tang
Chang Tang
Chenwei Tang
Jie Tang
Junshu Tang
Ming Tang
Peng Tang
Xu Tang
Yao Tang
Chen Tang
Fan Tang
Haoran Tang
Shengeng Tang
Yehui Tang
Zhipeng Tang
Ugo Tanielian
Chaofan Tao
Jiale Tao
Junli Tao
Renshuai Tao
An Tao
Guanhong Tao
Zhiqiang Tao
Makarand Tapaswi
Jean-Philippe G. Tarel
Juan J. Tarrio
Enzo Tartaglione
Keisuke Tateno
Zachary Teed

Ajinkya B. Tejankar
Bugra Tekin
Purva Tendulkar
Damien Teney
Minggui Teng
Chris Tensmeyer
Andrew Beng Jin Teoh
Philipp Terhörst
Kartik Thakral
Nupur Thakur
Kevin Thandiackal
Spyridon Thermos
Diego Thomas
William Thong
Yuesong Tian
Guanzhong Tian
Lin Tian
Shiqi Tian
Kai Tian
Meng Tian
Tai-Peng Tian
Zhuotao Tian
Shangxuan Tian
Tian Tian
Yapeng Tian
Yu Tian
Yuxin Tian
Leslie Ching Ow Tiong
Praveen Tirupattur
Garvita Tiwari
George Toderici
Antoine Toisoul
Aysim Toker
Tatiana Tommasi
Zhan Tong
Alessio Tonioni
Alessandro Torcinovich
Fabio Tosi
Matteo Toso
Hugo Touvron
Quan Hung Tran
Son Tran
Hung Tran
Ngoc-Trung Tran
Vinh Tran
Phong Tran
Giovanni Trappolini
Edith Tretschk
Subarna Tripathi
Shubhendu Trivedi
Eduard Trulls
Prune Truong
Thanh-Dat Truong
Tomasz Trzcinski
Sam Tsai
Yi-Hsuan Tsai
Ethan Tseng
Yu-Chee Tseng
Shahar Tsiper
Stavros Tsogkas
Shikui Tu
Zhigang Tu
Zhengzhong Tu
Richard Tucker
Sergey Tulyakov
Cigdem Turan
Daniyar Turmukhambetov
Victor G. Turrisi da Costa
Bartlomiej Twardowski
Christopher D. Twigg
Radim Tylecek
Mostofa Rafid Uddin
Md. Zasim Uddin
Kohei Uehara
Nicolas Ugrinovic
Youngjung Uh
Norimichi Ukita
Anwaar Ulhaq
Devesh Upadhyay
Paul Upchurch
Yoshitaka Ushiku
Yuzuko Utsumi
Mikaela Angelina Uy
Mohit Vaishnav
Pratik Vaishnavi
Jeya Maria Jose Valanarasu
Matias A. Valdenegro Toro
Diego Valsesia
Wouter Van Gansbeke
Nanne van Noord
Simon Vandenhende
Farshid Varno
Cristina Vasconcelos
Francisco Vasconcelos
Alex Vasilescu
Subeesh Vasu
Arun Balajee Vasudevan
Kanav Vats
Vaibhav S. Vavilala
Sagar Vaze
Javier Vazquez-Corral
Andrea Vedaldi
Olga Veksler
Andreas Velten
Sai H. Vemprala
Raviteja Vemulapalli
Shashanka Venkataramanan
Dor Verbin
Luisa Verdoliva
Manisha Verma
Yashaswi Verma
Constantin Vertan
Eli Verwimp
Deepak Vijaykeerthy
Pablo Villanueva
Ruben Villegas
Markus Vincze
Vibhav Vineet
Minh P. Vo
Huy V. Vo
Duc Minh Vo
Tomas Vojir
Igor Vozniak
Nicholas Vretos
Vibashan VS
Tuan-Anh Vu
Thang Vu
Mårten Wadenbäck
Neal Wadhwa
Aaron T. Walsman
Steven Walton
Jin Wan
Alvin Wan
Jia Wan

Jun Wan
Xiaoyue Wan
Fang Wan
Guowei Wan
Renjie Wan
Zhiqiang Wan
Ziyu Wan
Bastian Wandt
Dongdong Wang
Limin Wang
Haiyang Wang
Xiaobing Wang
Angtian Wang
Angelina Wang
Bing Wang
Bo Wang
Boyu Wang
Binghui Wang
Chen Wang
Chien-Yi Wang
Congli Wang
Qi Wang
Chengrui Wang
Rui Wang
Yiqun Wang
Cong Wang
Wenjing Wang
Dongkai Wang
Di Wang
Xiaogang Wang
Kai Wang
Zhizhong Wang
Fangjinhua Wang
Feng Wang
Hang Wang
Gaoang Wang
Guoqing Wang
Guangcong Wang
Guangzhi Wang
Hanqing Wang
Hao Wang
Haohan Wang
Haoran Wang
Hong Wang
Haotao Wang
Hu Wang
Huan Wang
Hua Wang
Hui-Po Wang
Hengli Wang
Hanyu Wang
Hongxing Wang
Jingwen Wang
Jialiang Wang
Jian Wang
Jianyi Wang
Jiashun Wang
Jiahao Wang
Tsun-Hsuan Wang
Xiaoqian Wang
Jinqiao Wang
Jun Wang
Jianzong Wang
Kaihong Wang
Ke Wang
Lei Wang
Lingjing Wang
Linnan Wang
Lin Wang
Liansheng Wang
Mengjiao Wang
Manning Wang
Nannan Wang
Peihao Wang
Jiayun Wang
Pu Wang
Qiang Wang
Qiufeng Wang
Qilong Wang
Qiangchang Wang
Qin Wang
Qing Wang
Ruocheng Wang
Ruibin Wang
Ruisheng Wang
Ruizhe Wang
Runqi Wang
Runzhong Wang
Wenxuan Wang
Sen Wang
Shangfei Wang
Shaofei Wang
Shijie Wang
Shiqi Wang
Zhibo Wang
Song Wang
Xinjiang Wang
Tai Wang
Tao Wang
Teng Wang
Xiang Wang
Tianren Wang
Tiantian Wang
Tianyi Wang
Fengjiao Wang
Wei Wang
Miaohui Wang
Suchen Wang
Siyue Wang
Yaoming Wang
Xiao Wang
Ze Wang
Biao Wang
Chaofei Wang
Dong Wang
Gu Wang
Guangrun Wang
Guangming Wang
Guo-Hua Wang
Haoqing Wang
Hesheng Wang
Huafeng Wang
Jinghua Wang
Jingdong Wang
Jingjing Wang
Jingya Wang
Jingkang Wang
Jiakai Wang
Junke Wang
Kuo Wang
Lichen Wang
Lizhi Wang
Longguang Wang
Mang Wang
Mei Wang

Min Wang
Peng-Shuai Wang
Run Wang
Shaoru Wang
Shuhui Wang
Tan Wang
Tiancai Wang
Tianqi Wang
Wenhai Wang
Wenzhe Wang
Xiaobo Wang
Xiudong Wang
Xu Wang
Yajie Wang
Yan Wang
Yuan-Gen Wang
Yingqian Wang
Yizhi Wang
Yulin Wang
Yu Wang
Yujie Wang
Yunhe Wang
Yuxi Wang
Yaowei Wang
Yiwei Wang
Zezheng Wang
Hongzhi Wang
Zhiqiang Wang
Ziteng Wang
Ziwei Wang
Zheng Wang
Zhenyu Wang
Binglu Wang
Zhongdao Wang
Ce Wang
Weining Wang
Weiyao Wang
Wenbin Wang
Wenguan Wang
Guangting Wang
Haolin Wang
Haiyan Wang
Huiyu Wang
Naiyan Wang
Jingbo Wang
Jinpeng Wang
Jiaqi Wang
Liyuan Wang
Lizhen Wang
Ning Wang
Wenqian Wang
Sheng-Yu Wang
Weimin Wang
Xiaohan Wang
Yifan Wang
Yi Wang
Yongtao Wang
Yizhou Wang
Zhuo Wang
Zhe Wang
Xudong Wang
Xiaofang Wang
Xinggang Wang
Xiaosen Wang
Xiaosong Wang
Xiaoyang Wang
Lijun Wang
Xinlong Wang
Xuan Wang
Xue Wang
Yangang Wang
Yaohui Wang
Yu-Chiang Frank Wang
Yida Wang
Yilin Wang
Yi Ru Wang
Yali Wang
Yinglong Wang
Yufu Wang
Yujiang Wang
Yuwang Wang
Yuting Wang
Yang Wang
Yu-Xiong Wang
Yixu Wang
Ziqi Wang
Zhicheng Wang
Zeyu Wang
Zhaowen Wang
Zhenyi Wang
Zhenzhi Wang
Zhijie Wang
Zhiyong Wang
Zhongling Wang
Zhuowei Wang
Zian Wang
Zifu Wang
Zihao Wang
Zirui Wang
Ziyan Wang
Wenxiao Wang
Zhen Wang
Zhepeng Wang
Zi Wang
Zihao W. Wang
Steven L. Waslander
Olivia Watkins
Daniel Watson
Silvan Weder
Dongyoon Wee
Dongming Wei
Tianyi Wei
Jia Wei
Dong Wei
Fangyun Wei
Longhui Wei
Mingqiang Wei
Xinyue Wei
Chen Wei
Donglai Wei
Pengxu Wei
Xing Wei
Xiu-Shen Wei
Wenqi Wei
Guoqiang Wei
Wei Wei
XingKui Wei
Xian Wei
Xingxing Wei
Yake Wei
Yuxiang Wei
Yi Wei
Luca Weihs
Michael Weinmann
Martin Weinmann

Congcong Wen
Chuan Wen
Jie Wen
Sijia Wen
Song Wen
Chao Wen
Xiang Wen
Zeyi Wen
Xin Wen
Yilin Wen
Yijia Weng
Shuchen Weng
Junwu Weng
Wenming Weng
Renliang Weng
Zhenyu Weng
Xinshuo Weng
Nicholas J. Westlake
Gordon Wetzstein
Lena M. Widin Klasén
Rick Wildes
Bryan M. Williams
Williem Williem
Ole Winther
Scott Wisdom
Alex Wong
Chau-Wai Wong
Kwan-Yee K. Wong
Yongkang Wong
Scott Workman
Marcel Worring
Michael Wray
Safwan Wshah
Xiang Wu
Aming Wu
Chongruo Wu
Cho-Ying Wu
Chunpeng Wu
Chenyan Wu
Ziyi Wu
Fuxiang Wu
Gang Wu
Haiping Wu
Huisi Wu
Jane Wu
Jialian Wu
Jing Wu
Jinjian Wu
Jianlong Wu
Xian Wu
Lifang Wu
Lifan Wu
Minye Wu
Qianyi Wu
Rongliang Wu
Rui Wu
Shiqian Wu
Shuzhe Wu
Shangzhe Wu
Tsung-Han Wu
Tz-Ying Wu
Ting-Wei Wu
Jiannan Wu
Zhiliang Wu
Yu Wu
Chenyun Wu
Dayan Wu
Dongxian Wu
Fei Wu
Hefeng Wu
Jianxin Wu
Weibin Wu
Wenxuan Wu
Wenhao Wu
Xiao Wu
Yicheng Wu
Yuanwei Wu
Yu-Huan Wu
Zhenxin Wu
Zhenyu Wu
Wei Wu
Peng Wu
Xiaohe Wu
Xindi Wu
Xinxing Wu
Xinyi Wu
Xingjiao Wu
Xiongwei Wu
Yangzheng Wu
Yanzhao Wu
Yawen Wu
Yong Wu
Yi Wu
Ying Nian Wu
Zhenyao Wu
Zhonghua Wu
Zongze Wu
Zuxuan Wu
Stefanie Wuhrer
Teng Xi
Jianing Xi
Fei Xia
Haifeng Xia
Menghan Xia
Yuanqing Xia
Zhihua Xia
Xiaobo Xia
Weihao Xia
Shihong Xia
Yan Xia
Yong Xia
Zhaoyang Xia
Zhihao Xia
Chuhua Xian
Yongqin Xian
Wangmeng Xiang
Fanbo Xiang
Tiange Xiang
Tao Xiang
Liuyu Xiang
Xiaoyu Xiang
Zhiyu Xiang
Aoran Xiao
Chunxia Xiao
Fanyi Xiao
Jimin Xiao
Jun Xiao
Taihong Xiao
Anqi Xiao
Junfei Xiao
Jing Xiao
Liang Xiao
Yang Xiao
Yuting Xiao
Yijun Xiao

Yao Xiao
Zeyu Xiao
Zhisheng Xiao
Zihao Xiao
Binhui Xie
Christopher Xie
Haozhe Xie
Jin Xie
Guo-Sen Xie
Hongtao Xie
Ming-Kun Xie
Tingting Xie
Chaohao Xie
Weicheng Xie
Xudong Xie
Jiyang Xie
Xiaohua Xie
Yuan Xie
Zhenyu Xie
Ning Xie
Xianghui Xie
Xiufeng Xie
You Xie
Yutong Xie
Fuyong Xing
Yifan Xing
Zhen Xing
Yuanjun Xiong
Jinhui Xiong
Weihua Xiong
Hongkai Xiong
Zhitong Xiong
Yuanhao Xiong
Yunyang Xiong
Yuwen Xiong
Zhiwei Xiong
Yuliang Xiu
An Xu
Chang Xu
Chenliang Xu
Chengming Xu
Chenshu Xu
Xiang Xu
Huijuan Xu
Zhe Xu
Jie Xu
Jingyi Xu
Jiarui Xu
Yinghao Xu
Kele Xu
Ke Xu
Li Xu
Linchuan Xu
Linning Xu
Mengde Xu
Mengmeng Frost Xu
Min Xu
Mingye Xu
Jun Xu
Ning Xu
Peng Xu
Runsheng Xu
Sheng Xu
Wenqiang Xu
Xiaogang Xu
Renzhe Xu
Kaidi Xu
Yi Xu
Chi Xu
Qiuling Xu
Baobei Xu
Feng Xu
Haohang Xu
Haofei Xu
Lan Xu
Mingze Xu
Songcen Xu
Weipeng Xu
Wenjia Xu
Wenju Xu
Xiangyu Xu
Xin Xu
Yinshuang Xu
Yixing Xu
Yuting Xu
Yanyu Xu
Zhenbo Xu
Zhiliang Xu
Zhiyuan Xu
Xiaohao Xu
Yanwu Xu
Yan Xu
Yiran Xu
Yifan Xu
Yufei Xu
Yong Xu
Zichuan Xu
Zenglin Xu
Zexiang Xu
Zhan Xu
Zheng Xu
Zhiwei Xu
Ziyue Xu
Shiyu Xuan
Hanyu Xuan
Fei Xue
Jianru Xue
Mingfu Xue
Qinghan Xue
Tianfan Xue
Chao Xue
Chuhui Xue
Nan Xue
Zhou Xue
Xiangyang Xue
Yuan Xue
Abhay Yadav
Ravindra Yadav
Kota Yamaguchi
Toshihiko Yamasaki
Kohei Yamashita
Chaochao Yan
Feng Yan
Kun Yan
Qingsen Yan
Qixin Yan
Rui Yan
Siming Yan
Xinchen Yan
Yaping Yan
Bin Yan
Qingan Yan
Shen Yan
Shipeng Yan
Xu Yan

Yan Yan
Yichao Yan
Zhaoyi Yan
Zike Yan
Zhiqiang Yan
Hongliang Yan
Zizheng Yan
Jiewen Yang
Anqi Joyce Yang
Shan Yang
Anqi Yang
Antoine Yang
Bo Yang
Baoyao Yang
Chenhongyi Yang
Dingkang Yang
De-Nian Yang
Dong Yang
David Yang
Fan Yang
Fengyu Yang
Fengting Yang
Fei Yang
Gengshan Yang
Heng Yang
Han Yang
Huan Yang
Yibo Yang
Jiancheng Yang
Jihan Yang
Jiawei Yang
Jiayu Yang
Jie Yang
Jinfa Yang
Jingkang Yang
Jinyu Yang
Cheng-Fu Yang
Ji Yang
Jianyu Yang
Kailun Yang
Tian Yang
Luyu Yang
Liang Yang
Li Yang
Michael Ying Yang
Yang Yang
Muli Yang
Le Yang
Qiushi Yang
Ren Yang
Ruihan Yang
Shuang Yang
Siyuan Yang
Su Yang
Shiqi Yang
Taojiannan Yang
Tianyu Yang
Lei Yang
Wanzhao Yang
Shuai Yang
William Yang
Wei Yang
Xiaofeng Yang
Xiaoshan Yang
Xin Yang
Xuan Yang
Xu Yang
Xingyi Yang
Xitong Yang
Jing Yang
Yanchao Yang
Wenming Yang
Yujiu Yang
Herb Yang
Jianfei Yang
Jinhui Yang
Chuanguang Yang
Guanglei Yang
Haitao Yang
Kewei Yang
Linlin Yang
Lijin Yang
Longrong Yang
Meng Yang
MingKun Yang
Sibei Yang
Shicai Yang
Tong Yang
Wen Yang
Xi Yang
Xiaolong Yang
Xue Yang
Yubin Yang
Ze Yang
Ziyi Yang
Yi Yang
Linjie Yang
Yuzhe Yang
Yiding Yang
Zhenpei Yang
Zhaohui Yang
Zhengyuan Yang
Zhibo Yang
Zongxin Yang
Hantao Yao
Mingde Yao
Rui Yao
Taiping Yao
Ting Yao
Cong Yao
Qingsong Yao
Quanming Yao
Xu Yao
Yuan Yao
Yao Yao
Yazhou Yao
Jiawen Yao
Shunyu Yao
Pew-Thian Yap
Sudhir Yarram
Rajeev Yasarla
Peng Ye
Botao Ye
Mao Ye
Fei Ye
Hanrong Ye
Jingwen Ye
Jinwei Ye
Jiarong Ye
Mang Ye
Meng Ye
Qi Ye
Qian Ye
Qixiang Ye
Junjie Ye

Sheng Ye
Nanyang Ye
Yufei Ye
Xiaoqing Ye
Ruolin Ye
Yousef Yeganeh
Chun-Hsiao Yeh
Raymond A. Yeh
Yu-Ying Yeh
Kai Yi
Chang Yi
Renjiao Yi
Xinping Yi
Peng Yi
Alper Yilmaz
Junho Yim
Hui Yin
Bangjie Yin
Jia-Li Yin
Miao Yin
Wenzhe Yin
Xuwang Yin
Ming Yin
Yu Yin
Aoxiong Yin
Kangxue Yin
Tianwei Yin
Wei Yin
Xianghua Ying
Rio Yokota
Tatsuya Yokota
Naoto Yokoya
Ryo Yonetani
Ki Yoon Yoo
Jinsu Yoo
Sunjae Yoon
Jae Shin Yoon
Jihun Yoon
Sung-Hoon Yoon
Ryota Yoshihashi
Yusuke Yoshiyasu
Chenyu You
Haoran You
Haoxuan You
Yang You
Quanzeng You
Tackgeun You
Kaichao You
Shan You
Xinge You
Yurong You
Baosheng Yu
Bei Yu
Haichao Yu
Hao Yu
Chaohui Yu
Fisher Yu
Jin-Gang Yu
Jiyang Yu
Jason J. Yu
Jiashuo Yu
Hong-Xing Yu
Lei Yu
Mulin Yu
Ning Yu
Peilin Yu
Qi Yu
Qian Yu
Rui Yu
Shuzhi Yu
Gang Yu
Tan Yu
Weijiang Yu
Xin Yu
Bingyao Yu
Ye Yu
Hanchao Yu
Yingchen Yu
Tao Yu
Xiaotian Yu
Qing Yu
Houjian Yu
Changqian Yu
Jing Yu
Jun Yu
Shujian Yu
Xiang Yu
Zhaofei Yu
Zhenbo Yu
Yinfeng Yu
Zhuoran Yu
Zitong Yu
Bo Yuan
Jiangbo Yuan
Liangzhe Yuan
Weihao Yuan
Jianbo Yuan
Xiaoyun Yuan
Ye Yuan
Li Yuan
Geng Yuan
Jialin Yuan
Maoxun Yuan
Peng Yuan
Xin Yuan
Yuan Yuan
Yuhui Yuan
Yixuan Yuan
Zheng Yuan
Mehmet Kerim Yücel
Kaiyu Yue
Haixiao Yue
Heeseung Yun
Sangdoo Yun
Tian Yun
Mahmut Yurt
Ekim Yurtsever
Ahmet Yüzügüler
Edouard Yvinec
Eloi Zablocki
Christopher Zach
Muhammad Zaigham Zaheer
Pierluigi Zama Ramirez
Yuhang Zang
Pietro Zanuttigh
Alexey Zaytsev
Bernhard Zeisl
Haitian Zeng
Pengpeng Zeng
Jiabei Zeng
Runhao Zeng
Wei Zeng
Yawen Zeng
Yi Zeng

Yiming Zeng
Tieyong Zeng
Huanqiang Zeng
Dan Zeng
Yu Zeng
Wei Zhai
Yuanhao Zhai
Fangneng Zhan
Kun Zhan
Xiong Zhang
Jingdong Zhang
Jiangning Zhang
Zhilu Zhang
Gengwei Zhang
Dongsu Zhang
Hui Zhang
Binjie Zhang
Bo Zhang
Tianhao Zhang
Cecilia Zhang
Jing Zhang
Chaoning Zhang
Chenxu Zhang
Chi Zhang
Chris Zhang
Yabin Zhang
Zhao Zhang
Rufeng Zhang
Chaoyi Zhang
Zheng Zhang
Da Zhang
Yi Zhang
Edward Zhang
Xin Zhang
Feifei Zhang
Feilong Zhang
Yuqi Zhang
GuiXuan Zhang
Hanlin Zhang
Hanwang Zhang
Hanzhen Zhang
Haotian Zhang
He Zhang
Haokui Zhang
Hongyuan Zhang
Hengrui Zhang
Hongming Zhang
Mingfang Zhang
Jianpeng Zhang
Jiaming Zhang
Jichao Zhang
Jie Zhang
Jingfeng Zhang
Jingyi Zhang
Jinnian Zhang
David Junhao Zhang
Junjie Zhang
Junzhe Zhang
Jiawan Zhang
Jingyang Zhang
Kai Zhang
Lei Zhang
Lihua Zhang
Lu Zhang
Miao Zhang
Minjia Zhang
Mingjin Zhang
Qi Zhang
Qian Zhang
Qilong Zhang
Qiming Zhang
Qiang Zhang
Richard Zhang
Ruimao Zhang
Ruisi Zhang
Ruixin Zhang
Runze Zhang
Qilin Zhang
Shan Zhang
Shanshan Zhang
Xi Sheryl Zhang
Song-Hai Zhang
Chongyang Zhang
Kaihao Zhang
Songyang Zhang
Shu Zhang
Siwei Zhang
Shujian Zhang
Tianyun Zhang
Tong Zhang
Tao Zhang
Wenwei Zhang
Wenqiang Zhang
Wen Zhang
Xiaolin Zhang
Xingchen Zhang
Xingxuan Zhang
Xiuming Zhang
Xiaoshuai Zhang
Xuanmeng Zhang
Xuanyang Zhang
Xucong Zhang
Xingxing Zhang
Xikun Zhang
Xiaohan Zhang
Yahui Zhang
Yunhua Zhang
Yan Zhang
Yanghao Zhang
Yifei Zhang
Yifan Zhang
Yi-Fan Zhang
Yihao Zhang
Yingliang Zhang
Youshan Zhang
Yulun Zhang
Yushu Zhang
Yixiao Zhang
Yide Zhang
Zhongwen Zhang
Bowen Zhang
Chen-Lin Zhang
Zehua Zhang
Zekun Zhang
Zeyu Zhang
Xiaowei Zhang
Yifeng Zhang
Cheng Zhang
Hongguang Zhang
Yuexi Zhang
Fa Zhang
Guofeng Zhang
Hao Zhang
Haofeng Zhang
Hongwen Zhang

Hua Zhang
Jiaxin Zhang
Zhenyu Zhang
Jian Zhang
Jianfeng Zhang
Jiao Zhang
Jiakai Zhang
Lefei Zhang
Le Zhang
Mi Zhang
Min Zhang
Ning Zhang
Pan Zhang
Pu Zhang
Qing Zhang
Renrui Zhang
Shifeng Zhang
Shuo Zhang
Shaoxiong Zhang
Weizhong Zhang
Xi Zhang
Xiaomei Zhang
Xinyu Zhang
Yin Zhang
Zicheng Zhang
Zihao Zhang
Ziqi Zhang
Zhaoxiang Zhang
Zhen Zhang
Zhipeng Zhang
Zhixing Zhang
Zhizheng Zhang
Jiawei Zhang
Zhong Zhang
Pingping Zhang
Yixin Zhang
Kui Zhang
Lingzhi Zhang
Huaiwen Zhang
Quanshi Zhang
Zhoutong Zhang
Yuhang Zhang
Yuting Zhang
Zhang Zhang
Ziming Zhang
Zhizhong Zhang
Qilong Zhangli
Bingyin Zhao
Bin Zhao
Chenglong Zhao
Lei Zhao
Feng Zhao
Gangming Zhao
Haiyan Zhao
Hao Zhao
Handong Zhao
Hengshuang Zhao
Yinan Zhao
Jiaojiao Zhao
Jiaqi Zhao
Jing Zhao
Kaili Zhao
Haojie Zhao
Yucheng Zhao
Longjiao Zhao
Long Zhao
Qingsong Zhao
Qingyu Zhao
Rui Zhao
Rui-Wei Zhao
Sicheng Zhao
Shuang Zhao
Siyan Zhao
Zelin Zhao
Shiyu Zhao
Wang Zhao
Tiesong Zhao
Qian Zhao
Wangbo Zhao
Xi-Le Zhao
Xu Zhao
Yajie Zhao
Yang Zhao
Ying Zhao
Yin Zhao
Yizhou Zhao
Yunhan Zhao
Yuyang Zhao
Yue Zhao
Yuzhi Zhao
Bowen Zhao
Pu Zhao
Bingchen Zhao
Borui Zhao
Fuqiang Zhao
Hanbin Zhao
Jian Zhao
Mingyang Zhao
Na Zhao
Rongchang Zhao
Ruiqi Zhao
Shuai Zhao
Wenda Zhao
Wenliang Zhao
Xiangyun Zhao
Yifan Zhao
Yaping Zhao
Zhou Zhao
He Zhao
Jie Zhao
Xibin Zhao
Xiaoqi Zhao
Zhengyu Zhao
Jin Zhe
Chuanxia Zheng
Huan Zheng
Hao Zheng
Jia Zheng
Jian-Qing Zheng
Shuai Zheng
Meng Zheng
Mingkai Zheng
Qian Zheng
Qi Zheng
Wu Zheng
Yinqiang Zheng
Yufeng Zheng
Yutong Zheng
Yalin Zheng
Yu Zheng
Feng Zheng
Zhaoheng Zheng
Haitian Zheng
Kang Zheng
Bolun Zheng

Haiyong Zheng
Mingwu Zheng
Sipeng Zheng
Tu Zheng
Wenzhao Zheng
Xiawu Zheng
Yinglin Zheng
Zhuo Zheng
Zilong Zheng
Kecheng Zheng
Zerong Zheng
Shuaifeng Zhi
Tiancheng Zhi
Jia-Xing Zhong
Yiwu Zhong
Fangwei Zhong
Zhihang Zhong
Yaoyao Zhong
Yiran Zhong
Zhun Zhong
Zichun Zhong
Bo Zhou
Boyao Zhou
Brady Zhou
Mo Zhou
Chunluan Zhou
Dingfu Zhou
Fan Zhou
Jingkai Zhou
Honglu Zhou
Jiaming Zhou
Jiahuan Zhou
Jun Zhou
Kaiyang Zhou
Keyang Zhou
Kuangqi Zhou
Lei Zhou
Lihua Zhou
Man Zhou
Mingyi Zhou
Mingyuan Zhou
Ning Zhou
Peng Zhou
Penghao Zhou
Qianyi Zhou
Shuigeng Zhou
Shangchen Zhou
Huayi Zhou
Zhize Zhou
Sanping Zhou
Qin Zhou
Tao Zhou
Wenbo Zhou
Xiangdong Zhou
Xiao-Yun Zhou
Xiao Zhou
Yang Zhou
Yipin Zhou
Zhenyu Zhou
Hao Zhou
Chu Zhou
Daquan Zhou
Da-Wei Zhou
Hang Zhou
Kang Zhou
Qianyu Zhou
Sheng Zhou
Wenhui Zhou
Xingyi Zhou
Yan-Jie Zhou
Yiyi Zhou
Yu Zhou
Yuan Zhou
Yuqian Zhou
Yuxuan Zhou
Zixiang Zhou
Wengang Zhou
Shuchang Zhou
Tianfei Zhou
Yichao Zhou
Alex Zhu
Chenchen Zhu
Deyao Zhu
Xiatian Zhu
Guibo Zhu
Haidong Zhu
Hao Zhu
Hongzi Zhu
Rui Zhu
Jing Zhu
Jianke Zhu
Junchen Zhu
Lei Zhu
Lingyu Zhu
Luyang Zhu
Menglong Zhu
Peihao Zhu
Hui Zhu
Xiaofeng Zhu
Tyler (Lixuan) Zhu
Wentao Zhu
Xiangyu Zhu
Xinqi Zhu
Xinxin Zhu
Xinliang Zhu
Yangguang Zhu
Yichen Zhu
Yixin Zhu
Yanjun Zhu
Yousong Zhu
Yuhao Zhu
Ye Zhu
Feng Zhu
Zhen Zhu
Fangrui Zhu
Jinjing Zhu
Linchao Zhu
Pengfei Zhu
Sijie Zhu
Xiaobin Zhu
Xiaoguang Zhu
Zezhou Zhu
Zhenyao Zhu
Kai Zhu
Pengkai Zhu
Bingbing Zhuang
Chengyuan Zhuang
Liansheng Zhuang
Peiye Zhuang
Yixin Zhuang
Yihong Zhuang
Junbao Zhuo
Andrea Ziani
Bartosz Zieliński
Primo Zingaretti

Nikolaos Zioulis
Andrew Zisserman
Yael Ziv
Liu Ziyin
Xingxing Zou
Danping Zou
Qi Zou
Shihao Zou
Xueyan Zou
Yang Zou
Yuliang Zou
Zihang Zou
Chuhang Zou
Dongqing Zou
Xu Zou
Zhiming Zou
Maria A. Zuluaga
Xinxin Zuo
Zhiwen Zuo
Reyer Zwiggelaar

# Contents – Part XVII

# Editing Out-of-Domain GAN Inversion via Differential Activations

Haorui Song[1], Yong Du[2], Tianyi Xiang[1], Junyu Dong[2], Jing Qin[3], and Shengfeng He[1(✉)]

[1] South China University of Technology, Guangzhou, China
hesfe@scut.edu.cn
[2] Ocean University of China, Qingdao, China
[3] The Hong Kong Polytechnic University, Hong Kong SAR, China

**Abstract.** Despite the demonstrated editing capacity in the latent space of a pretrained GAN model, inverting real-world images is stuck in a dilemma that the reconstruction cannot be faithful to the original input. The main reason for this is that the distributions between training and real-world data are misaligned, and because of that, it is unstable of GAN inversion for real image editing. In this paper, we propose a novel GAN prior based editing framework to tackle the out-of-domain inversion problem with a composition-decomposition paradigm. In particular, during the phase of composition, we introduce a differential activation module for detecting semantic changes from a global perspective, *i.e.*, the relative gap between the features of edited and unedited images. With the aid of the generated Diff-CAM mask, a coarse reconstruction can intuitively be composited by the paired original and edited images. In this way, the attribute-irrelevant regions can be survived in almost whole, while the quality of such an intermediate result is still limited by an unavoidable ghosting effect. Consequently, in the decomposition phase, we further present a GAN prior based deghosting network for separating the final fine edited image from the coarse reconstruction. Extensive experiments exhibit superiorities over the state-of-the-art methods, in terms of qualitative and quantitative evaluations. The robustness and flexibility of our method is also validated on both scenarios of single attribute and multi-attribute manipulations. Code is available at https://github.com/HaoruiSong622/Editing-Out-of-Domain.

## 1 Introduction

Generative Adversarial Networks (GANs) [5,12,17] have demonstrated impressive image editing capability. From a random noise input, GAN models can encode abundant semantic information and spontaneously excavate interpretable

H. Song and Y. Du—Contribute equally.

**Supplementary Information** The online version contains supplementary material available at https://doi.org/10.1007/978-3-031-19790-1_1.

© The Author(s), under exclusive license to Springer Nature Switzerland AG 2022
S. Avidan et al. (Eds.): ECCV 2022, LNCS 13677, pp. 1–17, 2022.
https://doi.org/10.1007/978-3-031-19790-1_1

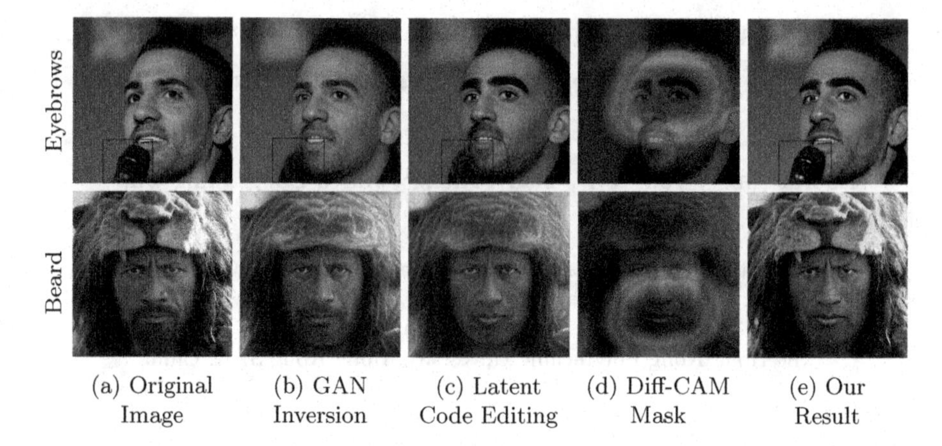


**Fig. 1.** We delve deep into the editing problem of out-of-domain GAN inversion. (b) shows that for out-of-domain real images, GAN inversion cannot obtain a faithful reconstruction and therefore produce unacceptable editing (c). Our framework localizes semantic changes with differential activations (d), enabling the preservation of out-of-domain image content (like the lion hat and microphone) while activating the editing ability of GAN priors.

directions in a latent space (*e.g.*, $\mathcal{W}$ space [18], $\mathcal{W}^+$ space [19] and etc.). By varying the latent codes along the controllable directions, highly realistic images with diverse attributes can be synthesized using GANs. However, such manipulations are applicable only in the latent space. For real images, a mapping function is required to transform the RGB input to a latent code.

GAN inversion [28,43] which aims at inverting a given image back into the latent space of a pretrained GAN model such as StyleGAN [18], can enable the corresponding semantic directions to be applicable for real image editing. As a consequence, numerous GAN inversion based image processing frameworks [2,13,34,38, 40,42] have emerged. However, existing inversion methods are stuck in a dilemma that they cannot faithfully invert those images that are not from the distribution of training data. For example, as shown in Fig. 1b, both real images are fed into the pSp encoder [28] for inversion, and then the codes are sent to a pretrained StyleGAN2 [19] for generation, but it turns out that the microphone and the lion hat are vanished or distorted. This is due to the misaligned data domains. Such an out-of-domain issue can undoubtedly lead to unstable editing performance and thus severely hinder the practicality of GAN inversion. On the other hand, the powerful attribute-aware manipulation capability of pretrained GANs is indispensable for image editing. These facts motivate us to raise a natural idea: it would be feasible if we can properly integrate the edited region from the corresponding inversion with its unedited counterpart from the original input.

To achieve this goal, we turn to consider how to detect the edited region in the inversion. This reminds us of class activation mapping (CAM) [29,41]. Such techniques focus on producing an attention map that highlights the regions that contribute to the classification decision, and have been widely used in visual expla-

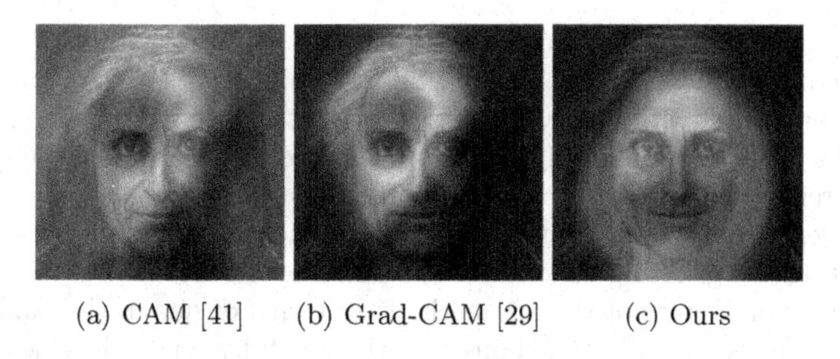

(a) CAM [41] (b) Grad-CAM [29] (c) Ours

**Fig. 2.** Activation maps generated by CAM, Grad-CAM and our Diff-CAM models. Our activation map has a broader and more comprehensive coverage.

nations, such as weakly-supervised localization [6] and visual question answering [26]. Also for image manipulation, Kim *et al.* [20] utilize Grad-CAM [29] to generate a mask for localizing the attribute-relevant regions. A critical problem of CAM-based methods is that, its principle is to locate the activation regions that make the final decision, but making such a decision does not require a comprehensive activation on the attribute-relevant regions (*e.g.*, locating the wrinkle instead of the whole face can classify the "old" attribute). As a result, they tend to produce localized activations (see Fig. 2a and 2b). Relying on CAM for editing is apparently not flexible, as the editing of some attributes like "sex" may change the entire face, but binarily classifying male or female typically concentrate on facial components. This contradicts with our objective to combine edited region and its unedited counterpart.

In this paper, we propose a novel GAN prior based editing framework to resolve the above problems. Specifically, our editing method is executed in a composition-decomposition manner. In the composition stage, our aim is to generate a coarse reconstruction via combining the edited inversion with the original input, weighting by a Diff-CAM mask which is used for indicating the edited region. In particular, we present a simple yet effective *differential activation* mechanism to track the semantic changes rather than locating classification-relevant regions. It is performed by capturing the variational features between the edited and the original inversions, and we shed light on the differential features that reveal the editing attributes. In this way, the produced mask can specify the range of the edited region more accurately (see in Fig. 2c), as the semantical differences are explicitly embedded in the hidden responses. While in the decomposition stage, we need to remove the ghosting effect occurred in the coarse result. To deal with it, we further design a deghosting network that reuses the GAN prior, which is used for separating the final fine edited inversion from the coarse reconstruction in a multi-scale aggregation manner. Extensive experiments show that our method is the first feasible real-image editing method that built upon GAN inversion.

In summary, our key contributions are as follows:

- We delve deep into the out-of-domain problem existed in GAN inversion, and propose a novel GAN prior based editing framework in a composition-

decomposition manner. Our method can use the original input to generate the unedited region, as well as maintaining a high quality of editing.
- We tailor a differential activation strategy to track semantic changes before and after editing. This design allows to embed more accurate range of the edited region with neglectable additional computational cost.
- We present a deghosting network with hierarchical GAN priors, for effectively alleviating the ghosting effect in the coarse reconstruction.
- We outperform state-of-the-art methods in terms of qualitative and quantitative evaluations, and we demonstrate the flexibility and robustness in both scenarios of single attribute and multi-attribute manipulations.

## 2 Related Work

**Non-GAN Prior Based Image Manipulation.** Non-GAN prior based methods [17,21,45] usually manipulate attributes of images via an adversarial training process. Kim *et al.* [21] propose a GAN based framework to discover cross-domain relations. Isola *et al.* [17] propose to use conditional GANs for image-to-image translation. And Zhu *et al.* [45] propose to translate images across different domains without paired training data. In general, these methods are designed to learn a model that corresponds to a specific translation, which leads to inflexibility in practical applications. To address this problem, StarGAN [9] is proposed to learn the mapping among multiple domains, using only a single generator and a discriminator. CMP [20] proposes to refine image-to-image translation results by introducing a cam-consistency loss to force the network to focus on attribute-relevant regions. Note that all these methods need to train models from scratch, and thus cannot capture GAN priors which is proven to be extremely effective for image manipulation [18,19]. Also, they are limited in synthesizing images at high resolution.

**GAN Prior Based Real Image Editing.** GAN prior based methods, *i.e.*, GAN inversion, are proposed to inference a latent code of a given image based on a pretrained GAN model such as StyleGAN [18,19]. These methods can be roughly divided into two categories, optimization-based [1,2,10,13,27,35] and learning-based [7,8,13,14,32,37,44]. The main advantage of the former techniques is that they can ensure superior image reconstruction, while the corresponding cost is a higher computational complexity. In contrast, learning-based methods have a fast inference speed. Richardson *et al.* [28] propose a pSp encoder that can embed real images into an extended $\mathcal{W}^+$ space. Xu *et al.* [36] propose to train a hierarchical encoder based on a feature pyramid network. Alaluf *et al.* [4] introduce an iterative refinement mechanism for learning the inversion of real images. However, these methods cannot faithfully reconstruct the image content, mainly due to the misalignment between training and test data. We aim to solve this problem in a novel composition-decomposition paradigm via differential activations.

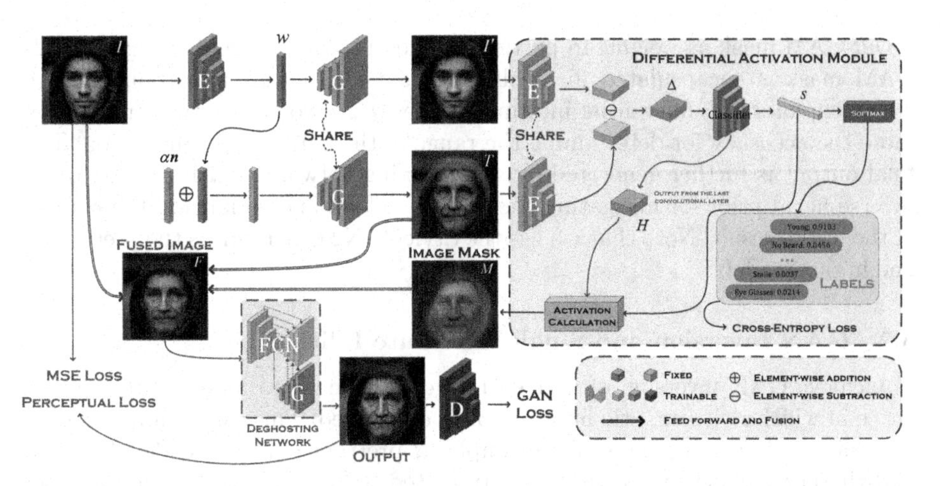


**Fig. 3.** Overall pipeline of our model. Given an input image, we first invert it to the latent space and perform user-desired editing. Then a differential activation module is applied to track the semantic changes of the manipulation. Edited region and unedited one are further combined using a Diff-CAM mask to produce a coarse reconstruction, on which the ghosting artifacts can be mitigated by the deghosting network with the aid of GAN priors.

**Interpreting CNN.** Recent interpreting CNN models [11,29,41] attempt to understand the behaviour of the networks. Zhou *et al.* [41] propose CAM which aims to highlight the model's attention regarding a specific class. Selvaraju *et al.* [29] propose Grad-CAM without relying on the global average pooling layer. Lee *et al.* [22] propose LFI-CAM which treats the feature maps as masks and learns the feature importance for generating the attention maps. Note that these methods are usually performed on responses themselves, while our strategy, with a clear aim of real image editing, explores to capture the variation between edit and unedited image features.

# 3 Approach

## 3.1 Overview

Due to the misaligned distributions between training and test data, existing GAN inversion methods cannot guarantee the fidelity of the reconstruction. And the quality of the subsequent edited image is therefore severely limited by such an out-of-domain problem. To remedy this, we propose a composition-decomposition paradigm for image editing and illustrate its overall pipeline in Fig. 3.

Specifically, given an image as input, our method firstly inverts it into the latent space. Semantic manipulation can then be produced by feeding and varying the latent code into a pretrained and fixed generator. Consequently, an initial result can be obtained by fusing the original input and the edited inversion with a Diff-CAM mask as weight. In particular, the procedure of generating the Diff-CAM mask is encapsulated in a self-contained differential activation module, which exploits the differential information between two reconstructions to promote the accuracy for determining the range of the editing-relevant region. The final output is further generated by a deghosting network, which resorts to the diverse facial prior for mitigating the ghosting effect and enhancing the realism of the initial result. Note that we use the StyleGAN2 generator as the pretrained one in our model.

### 3.2 GAN Inversion and Single Attribute Editing

To achieve GAN inversion of a given image $\boldsymbol{I}$, we need to map it into a latent space in which rich semantic information is embedded. This can be implemented via many existing methods, for example, a pretrained pSp encoder $E_{\text{fixed}}(\cdot)$, and the latent code $\boldsymbol{w}$ can thus be formulated by $\boldsymbol{w} = E_{\text{fixed}}(I)$. Then we can obtain the inverted image $\boldsymbol{I}'$ by a pretrained StyleGAN2 generator $G(\cdot)$, which is formulated by $\boldsymbol{I}' = G(\boldsymbol{w})$. Note that there most likely exists a bias between $\boldsymbol{I}'$ and $\boldsymbol{I}$, due to the out-of-domain problem.

And to manipulate the corresponding attributes, the latent code would be varied along various intepretable directions that are discovered in the latent space. The edited inversion $\boldsymbol{T}$ can then be produced based on the altered code by the same generator $G(\cdot)$. Given a specific direction $\boldsymbol{n}$ for single attribute editing, this process can be formulated as $\boldsymbol{T} = G(\boldsymbol{w} + \alpha\boldsymbol{n})$, where $\alpha$ is a scaling factor. Note that our method can also support multi-attribute editing, and we will discuss this in Sect. 3.6.

### 3.3 Differential Activation Module

Once we have the paired images $\{\boldsymbol{I}', \boldsymbol{T}\}$, we respectively feed them into a plain trainable encoder $E_{\text{trainable}}(\cdot)$, and can easily obtain the differential features $\boldsymbol{\Delta}$ via a simple subtraction operation:

$$\boldsymbol{\Delta} = E_{\text{trainable}}(\boldsymbol{I}') - E_{\text{trainable}}(\boldsymbol{T}). \tag{1}$$

Subsequently, these features are sent to a lightweight network which serves as a classifier and consists of convolutional and fully connected layers. We use the cross-entropy loss $\mathcal{L}_{\text{ce}}$ to train the encoder and the classifier together, which can be formulated as follows:

$$\mathcal{L}_{\text{ce}} = -\sum_{c=1}^{N} y_c log \frac{e^{s_c}}{\sum_{i=1}^{N} e^{s_i}}, \tag{2}$$

where $\boldsymbol{y} = \{y_1, y_2, \cdots, y_N\}$ is a one-hot vector that indicates which attribute has been edited, $\boldsymbol{s} = \{s_1, s_2, \cdots, s_N\}$ denotes the output vector of the classifier before softmax operation, $e$ denotes the natural constant, and $N$ is the total number of attributes.

Now we are ready for performing activation calculation. The first step is to define the weight $\beta_c^k$ that corresponded to the $k$th channel of $\boldsymbol{H}$ and the $c$th attribute, which is formulated as follows:

$$\beta_c^k = \overbrace{\frac{1}{Z}\sum_i\sum_j}^{\text{global average pooling}} \frac{\partial s_c}{\partial \boldsymbol{H}_{ij}^k}, \tag{3}$$

where $\boldsymbol{H}$ is the features generated by the last convolutional layer in the classifier, $i$ and $j$ respectively denotes the height and the width of the features.

Then our Diff-CAM mask $\boldsymbol{M}_{\text{Diff-CAM}}$ can be represented as a piecewise linear transformation of weighted differential features, that is

$$\boldsymbol{M}_{\text{Diff-CAM}} = \text{ReLU}(\sum_k \beta_c^k \boldsymbol{H}^k). \tag{4}$$

Finally, we normalize the above mask into the interval of $[0, 1]$ via $\boldsymbol{M}_{\text{Diff-CAM}} = \boldsymbol{M}_{\text{Diff-CAM}}/\max(\boldsymbol{M}_{\text{Diff-CAM}})$. Since the Diff-CAM mask is generated based on the differential features that describe semantically changes, the range of the editing-relevant region can be detected more accurately via a comprehensive activation. It is thus more suitable than other CAM-based masks for image editing.

### 3.4 Composition

After obtaining the Diff-CAM mask, it is time to composite the edited image with the original input for resolving the out-of-domain issue. We have the fused image $\boldsymbol{F}_{\text{fused}}$ by the following weighted average formula:

$$\boldsymbol{F}_{\text{fused}} = \boldsymbol{T} \odot \boldsymbol{M}_{\text{Diff-CAM}} + \boldsymbol{I} \odot (1 - \boldsymbol{M}_{\text{Diff-CAM}}), \tag{5}$$

where $\odot$ denotes the hadamard product. However, the quality of such an initial blending result is unsatisfactory due to an inevitable ghosting effect.

### 3.5 Deghosting Network

To cope with ghosting artifacts, we treat the coarse reconstruction $\boldsymbol{F}_{\text{fused}}$ as a combination of a target image and a ghost image. In order to decompose the target image out, we further perform a deghosting process on the coarse result via a deghosting network. As shown in Fig. 3, the architecture of the network includes a fully convolutional network which consists of an encoder (the orange part), a decoder (the pink part), a pretrained StyleGAN2 generator, and a discriminator $D(\cdot)$. Note that we denote the aggregation of the first three modules as $\phi(\cdot)$.

Our goal is to utilize the ghosting-free nature of the inherent facial prior in the pretrained GAN model, such that ghosting artifacts can be removed without destroying the original facial details. In particular, we first feed $\boldsymbol{F}_{\text{fused}}$ to an FCN-like [25] encoder-decoder architecture for two purposes: the encoder generates the latent code of $\boldsymbol{F}_{\text{fused}}$, and the decoder is trained to produce ghosting-free results. Meanwhile, with the predicted latent code, $\boldsymbol{F}_{\text{fused}}$ is inverted in the latent space and reconstructed by the StyleGAN2 generator without ghosting artifacts. We aggregate the corresponding features of the generator with the decoder hierarchically, yielding the final deghosting result.

Since the fused image $\boldsymbol{F}_{\text{fused}}$ has no ground-truth counterpart, we synthesize a set of paired data $\{\boldsymbol{F}_{\text{train}}, \boldsymbol{I}\}$ to train the deghosting network. The training image $\boldsymbol{F}_{\text{train}}$ is given by

$$\boldsymbol{F}_{\text{train}} = \boldsymbol{T} \odot \boldsymbol{M}_{\text{train}} + \boldsymbol{I} \odot (1 - \boldsymbol{M}_{\text{train}}), \tag{6}$$

and the corresponding Diff-CAM mask $\boldsymbol{M}_{\text{train}}$ is defined as follows:

$$\boldsymbol{M}_{\text{train}}(i,j) = \begin{cases} \boldsymbol{M}_{\text{Diff-CAM}}(i,j), & \text{if} \quad \boldsymbol{M}_{\text{Diff-CAM}}(i,j) \leq 0.5, \\ 1 - \boldsymbol{M}_{\text{Diff-CAM}}(i,j), & \text{if} \quad \boldsymbol{M}_{\text{Diff-CAM}}(i,j) > 0.5. \end{cases} \tag{7}$$

The rationale behind this setting is that: 1) the mask $\boldsymbol{M}_{\text{train}}$ is thus regularized into an interval of $[0, 0.5]$ so that the content of $\boldsymbol{I}$ dominates that of $\boldsymbol{F}_{\text{train}}$. We can then treat image $\boldsymbol{I}$ as the required ground truth. And 2) meanwhile, the ghosting effect still exists in $\boldsymbol{F}_{\text{train}}$. Note that the corresponding attribute in regard to generate the mask $\boldsymbol{M}_{\text{train}}$ is consistent with $\boldsymbol{T}$ and randomly selected. The total objective $\mathcal{L}_{\text{deghost}}$ for optimizing the deghosting network is defined as follows:

$$\mathcal{L}_{\text{deghost}} = \lambda_m \mathcal{L}_{\text{mse}} + \lambda_p \mathcal{L}_{\text{percep}} + \lambda_a \mathcal{L}_{\text{adv}}, \tag{8}$$

where $\mathcal{L}_{\text{mse}}$, $\mathcal{L}_{\text{percep}}$ and $\mathcal{L}_{\text{adv}}$ respectively denotes MSE loss, perceptual loss and adversarial loss, $\lambda_m$, $\lambda_p$, $\lambda_a$ are the balance factors. And the involved three losses are respectively defined as follows:

$$\mathcal{L}_{\text{mse}} = \frac{1}{Q} \|\boldsymbol{I} - \phi(\boldsymbol{F}_{\text{train}})\|_2, \tag{9}$$

$$\mathcal{L}_{\text{percep}} = \frac{1}{Q} \|V(\boldsymbol{I}) - V(\phi(\boldsymbol{F}_{\text{train}}))\|_2, \tag{10}$$

$$\mathcal{L}_{\text{adv}} = \underset{\boldsymbol{I} \sim P_r}{\mathbb{E}} log(D(\boldsymbol{I})) + \underset{\boldsymbol{F}_{\text{train}} \sim P_g}{\mathbb{E}} log(1 - D(\phi(\boldsymbol{F}_{\text{train}}))), \tag{11}$$

where $Q$ indicates the number of pixels, $P_r$ and $P_g$ respectively denotes the distribution of real data and generated data, $V(\cdot)$ denotes a pretrained VGG-16 network, and we select the features produced by the conv4_3 layer for modeling the loss.

### 3.6 Multi-attribute Editing

Our method also has a flexibility to handle multi-attribute editing. In fact, it can be decomposed into a sequence of single attribute editing tasks. Suppose the number of the attributes needed to be edited is $r$, three special points should be noted: 1) In the $i$th ($i \neq 1$) single attribute editing, the paired images $\{\boldsymbol{T}_i, \boldsymbol{T}_{i-1}\}$ are used for calculating the Diff-CAM mask. At last we will have a set of $r$ masks. 2) The final Diff-CAM mask is the result of performing element-wise maximization operation on the mask set. And 3) The final fused image is the composition of $\boldsymbol{T}_r$ and $\boldsymbol{I}$ with the final mask as weight.

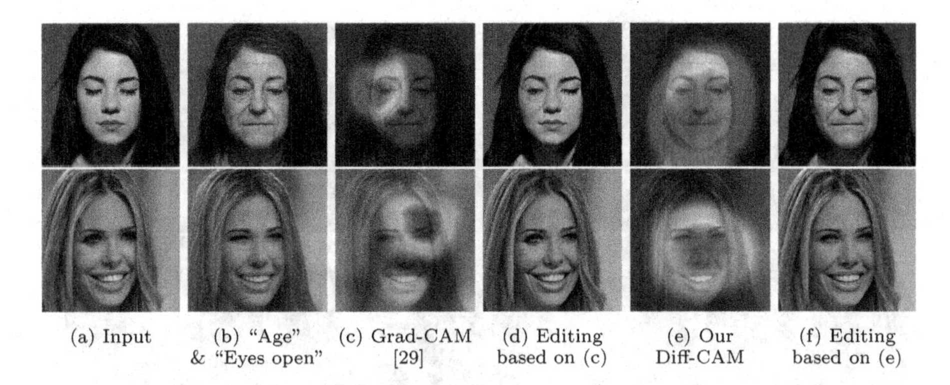


**Fig. 4.** Comparison with Grad-CAM in our editing framework. Grad-CAM exhibits localized attentions, while ours can correctly locate semantic changes during editing.

# 4 Experiments

## 4.1 Implementation Details

We implement our method in Pytorch on a PC with an Nvidia GeForce RTX 3090. During training, we use Adam as the optimizer with a learning rate of 0.0001, $\beta_1 = 0.9$, $\beta_2 = 0.99$. The hyperparameters in Eq. (8) are empirically set to $\lambda_m = 1$, $\lambda_p = 0.8$ and $\lambda_a = 0.01$. Before being sent to $E_{\text{train}}$, $\boldsymbol{I}'$ and $\boldsymbol{T}$ are downsampled from a resolution of $1024 \times 1024$ to that of $256 \times 256$. And the Diff-CAM mask computed by Eq. (4) will be upsampled to a resolution of $1024 \times 1024$ before being used in the composition process.

## 4.2 Experimental Data

FFHQ dataset [18] and Celeba-HQ dataset [24] both contain human face images of high quality and resolution, with 70000 and 30000 images respectively. We employ the FFHQ dataset for training the differential activation module and the deghosting network, while we utilize the Celeba-HQ dataset for testing. All the quantitative metrics are calculated on the Celeba-HQ dataset.

## 4.3 Component Analysis

**Effectiveness of DA Module.** First, in order to prove the effectiveness of our design of the DA module structure, we replace the DA module with the commonly used Grad-CAM [29] and check out how the masks differ and influence the editing.

The results are shown in Fig. 4. The results show the limitation of Grad-CAM that activates only in local areas. The resulted masks cannot suit for discovering semantic differences and therefore not suit for blending GAN inversion with the source image. On the contrary, the masks generated by our DA module successfully cover the editing-relevant regions, producing a global coverage for

**Fig. 5.** Effect of our deghosting network. Directly blending two images with a mask inevitably presents ghosting artifacts (teeth and face shape, zoom in for better view). Our deghosting network utilizes GAN priors to faithfully remove artifacts while retaining facial details.

"age" attribute while a local attention for "eyes open" attribute. This largely aids the blending of edited and unedited regions.

**Effectiveness of Deghosting Network.** Here we show the results before and after deghosting in Fig. 5. Even with a correctly detected mask, blending two images inevitably produces blurry and ghosting artifacts. Our deghosting network takes advantage of the rich facial priors from the pretrained GAN model, and effectively removes non-face artifacts as well as generating a clear blending of faces.

**Flexibility Analysis.** Our method is flexible and independent of the applied GAN inversion and interpretable directions. We choose several different inversion and interpretable direction models to work with our framework. Three combinations of inversion and direction methods are used, *i.e.*, the pSp encoder [28] together with directions found by StyleGAN2 distillation [33], the IdInvert encoder [43] with InterfaceGAN [30,31], and the e4e encoder [32] together with the directions obtained by StyleFlow [3].

The results are shown in Fig. 6. From the result we can see that all the encoding methods fail to retain the out-of-domain information. As for the first person wearing a blue hat and holding a fist, all encoders treat the hat as the hair and the fist as the background. Similar problem exists in the second person, in which his cap is inverted to hair and becomes white as he gets older. All these out-of-domain problems are addressed by our framework, regardless of their inversions and applied editings. The results show that our DA module and our deghosting network are encoder-independent and are robust enough to work with different types of inversion and editing methods.

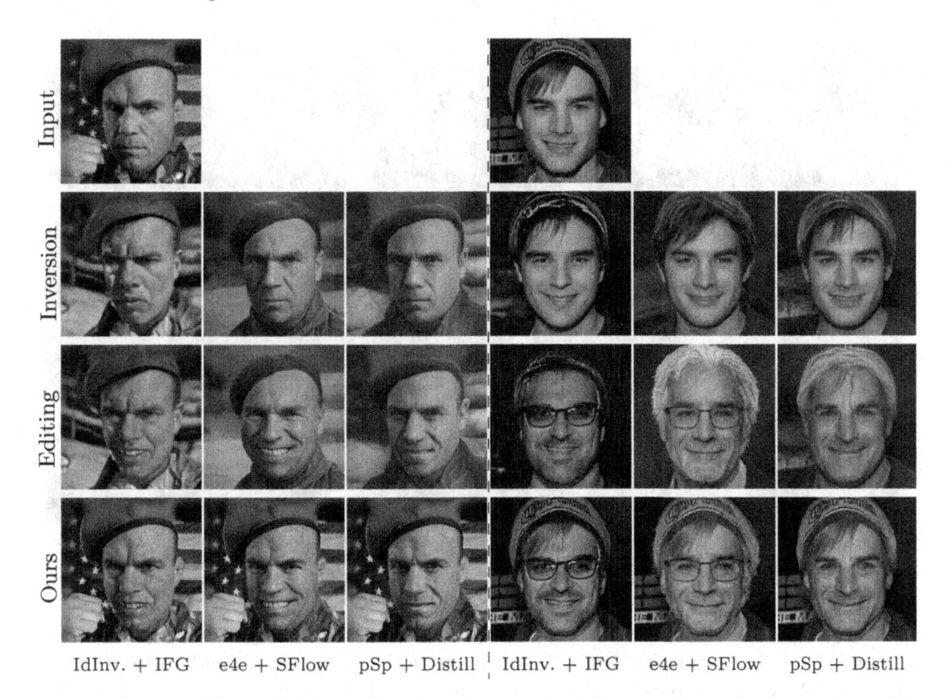


**Fig. 6.** Our framework is independent to GAN inversion and interpretable editing directions. We can work with arbitrary combinations of encoders and directions. (Zoom in for better view.)

### 4.4 Comparison with SOTAs

In order to prove the superiority of our model, we compare our model with other state-of-the-art facial attributes editing methods. Note that we do not make an additional comparison with StyleGAN inversion based editing method other than Fig. 6. This is due to that they would obviously fail on out-of-domain samples as the original GAN was not trained on them. To maintain fairness, here we mainly compare our results with those non-StyleGAN based image-to-image translation methods. In particular, we compare our model with StarGAN [9], AttGAN [15], and STGAN [23]. Kim *et al.* [20] propose to refine an image-to-image translation method by introducing a CAM-consistency loss to force the network to focus on attribute-relevant regions, and we also compare to the refined version of StarGAN and AttGAN (denoted as StarGAN* and AttGAN*). All the results are generated with their official codes, except the refinement method of Kim *et al.* [20] that is not publicly available and implemented by ourselves.

**Quantitative Evaluation.** To quantitatively compare our method with state-of-the-arts, we use the Fréchet inception distance (FID) [16] and learned perceptual image patch similarity (LPIPS) [39] metrics to measure the quality of the results. FID measures the distribution distance between the original image dataset and the manipulated dataset. We calculate FID metric for each model,

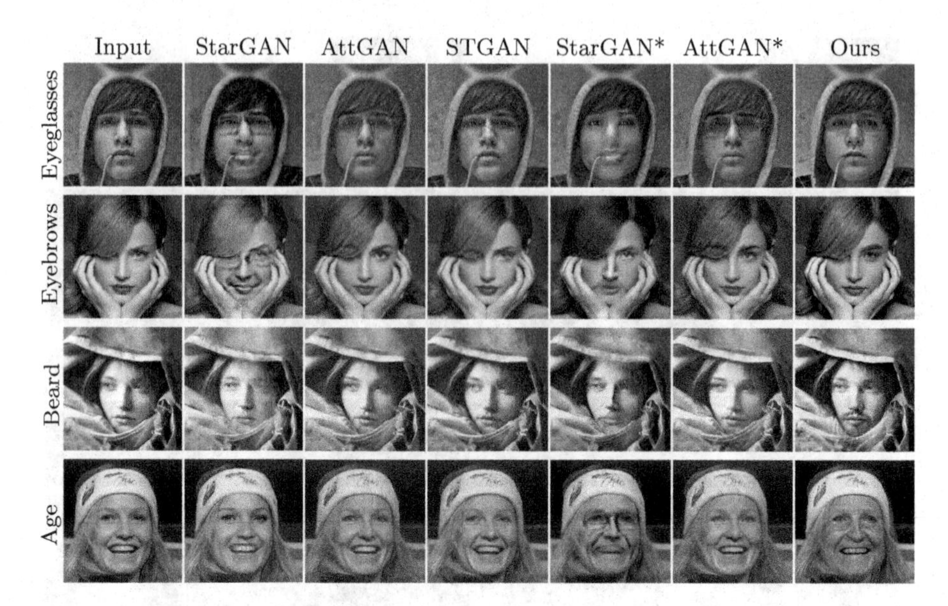


**Fig. 7.** Qualitative evaluation with respect to state-of-the-art image-to-image face editing methods on 4 attributes, "age", "bushy eyebrows", "eyeglasses", and "beard". StarGAN* and AttGAN* represent the refined version using the cam-consistency loss [20]. Our method can produce high-resolution and semantically correct editing on these challenging cases. (Zoom in for better view.)

and select 4 common attributes ("age", "bushy eyebrows", "eyeglasses", and "beard") that can be modified by all of the models to generate the manipulated dataset. The final FID value of each model is obtained by averaging the FID values corresponding to each attribute. LPIPS metric measures the perceptual similarity between the two images. The smaller the value of LPIPS, the greater the similarity. We use it to evaluate non-edited region consistency.

The numerical results are shown in Table 1. As can be seen, the proposed method shows the best FID and LPIPS scores among the competitors. This reveals that our model can better maintain the distribution of the original dataset, and a strong capability to preserve the image quality for non-edited regions.

**Qualitative Evaluation.** In order to demonstrate the superiority of our model, we conduct a qualitative study by contrasting the results generated from different models. Again, here we mainly compare with the image-to-image translation models. The results of editing 4 attributes by different models are shown in Fig. 7. We can see that, the outputs of our model are of the highest quality compared to all other methods. Our outputs best change the attributes while retaining other irrelevant information. StarGAN and its refined version suffer from checkerboard-like artifacts. AttGAN* can achieve better editing than the original version, but it still produces blurry details and semantically incorrect

**Table 1.** Quantitative comparison with state-of-the-art face editing methods. StarGAN* and AttGAN* represent the refined version using the cam-consistency loss [20]. Our method achieves the best numerical performance.

| | StarGAN | AttGAN | STGAN | StarGAN* | AttGAN* | Ours |
|---|---|---|---|---|---|---|
| FID↓ | 30.98 | 17.96 | 20.97 | 28.52 | 15.72 | **13.76** |
| LPIPS↓ | 0.208 | 0.107 | 0.178 | 0.138 | 0.099 | **0.094** |

**Fig. 8.** Evaluation on "car" domain and non-facial attributes "hairstyle", "hair color". (Zoom in for better view.)

editing (like the eyebrow on the left wrongly appears on the hair in the second example), indicating that using an additional mask-guided loss is not reliable for challenging cases. In contrast, our Diff-CAM mask driven framework obtains significantly preferable editing performance, not to mention the high-resolution features provided by StyleGAN-based editing.

### 4.5 Editing on Non-facial Attributes and Domains

We also verify the generalization ability of our model and display the qualitative results in Fig. 8. In particular, regarding non-facial attributes (like "hair color" or "hairstyle") and other domains (like "car"), the modifications may no longer happen in the center region like most of those occured in facial attributes editing, *e.g.*, logo, wheels, and hair. Nevertheless, our Diff-CAM can always precisely distinguish the edited and non-edited areas. And our framework can thus be robust to small regions and different attributes for editing.

### 4.6 Multi-attribute Editing

In addition to the single-attribute editing, our model shows its high degree of flexibility by also supporting editing multiple attributes. The process of modifying multiple attributes is completed by modifying the attributes one by one as is described in Sect. 3.6. Figure 9 shows two examples of editing two attributes, "eyes open" and "smile". The final outputs of our model successfully introduce the changes involved in the two editing steps and also manage to maintain the uninvertible information such as the hats and the fingers.

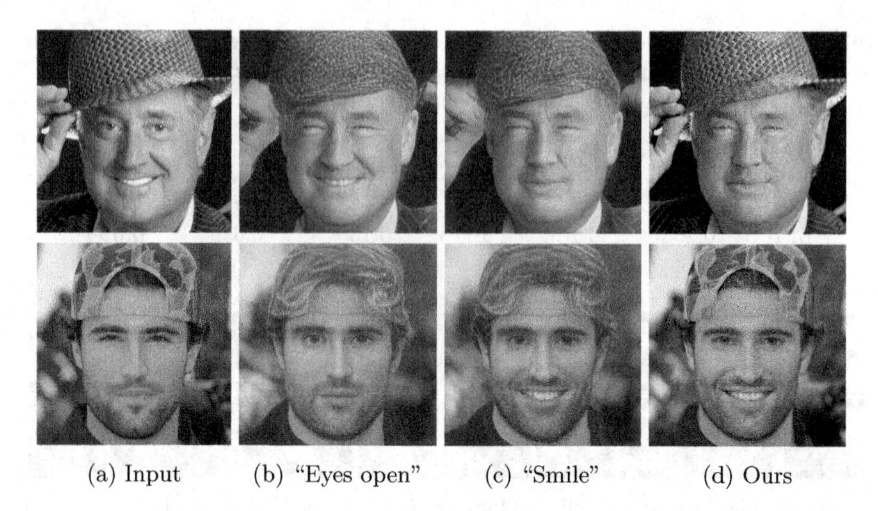


**Fig. 9.** Multi-attribute editing results. Our method can successfully edit multiple attributes one by one, while still retaining the out-of-domain regions.

**Fig. 10.** Challenging cases of our model. The two examples from different domains show the editing results when the inversion cannot faithfully reconstruct the original image.

### 4.7 Limitation

Although our model has achieved promising performance on the editing of facial or non-facial attributes and other domains, its ability to handle attribute changing is not unlimited. Figure 10 shows the examples of our limitation. Our model is heavily relied on the performance of GAN inversion methods and only introduces changes covered by the mask from the DA module. Therefore, if the inverted result cannot faithfully reconstruct the original image which is likely occurred in the case of non-human domains, serious distortion and ghosting artifacts will be existed in the final result.

## 5 Conclusion

In this paper, we propose a novel GAN prior based editing technique to tackle the out-of-domain inversion problem with a composition-decomposition paradigm. We introduce a differential activation mechanism to track semantic changes before and after editing. With the aid of the calculated Diff-CAM mask, a coarse reconstruction can be obtained by the composition of the edited image and the original input. We further present a deghosting network to mitigate the ghosting effect in the coarse result. Both qualitative and quantitative evaluations validate the superiority of our method.

**Acknowledgement.** This project is supported by the National Natural Science Foundation of China (62102381, U1706218, 41927805, 61972162); Shandong Natural Science Foundation (ZR2021QF035); Fundamental Research Funds for the Central Universities (202113035); the National Key R&D Program of China (2018AAA0100600); the China Postdoctoral Science Foundation (2020M682240, 2021T140631); Guangdong International Science and Technology Cooperation Project (No. 2021A0505030009); Guangdong Natural Science Foundation (2021A1515012625); Guangzhou Basic and Applied Research Project (202102021074); and CCF-Tencent Open Research fund (RAGR20210114).

## References

1. Abdal, R., Qin, Y., Wonka, P.: Image2stylegan: how to embed images into the stylegan latent space? In: ICCV, pp. 4432–4441 (2019)
2. Abdal, R., Qin, Y., Wonka, P.: Image2stylegan++: how to edit the embedded images? In: CVPR, pp. 8296–8305 (2020)
3. Abdal, R., Zhu, P., Mitra, N.J., Wonka, P.: Styleflow: attribute-conditioned exploration of stylegan-generated images using conditional continuous normalizing flows. ACM TOG **40**(3), 1–21 (2021)
4. Alaluf, Y., Patashnik, O., Cohen-Or, D.: Restyle: a residual-based StyleGAN encoder via iterative refinement. In: ICCV, pp. 6711–6720 (2021)
5. Arjovsky, M., Chintala, S., Bottou, L.: Wasserstein generative adversarial networks. In: ICML, pp. 214–223. PMLR (2017)
6. Bae, W., Noh, J., Kim, G.: Rethinking class activation mapping for weakly supervised object localization. In: Vedaldi, A., Bischof, H., Brox, T., Frahm, J.-M. (eds.) ECCV 2020. LNCS, vol. 12360, pp. 618–634. Springer, Cham (2020). https://doi.org/10.1007/978-3-030-58555-6_37
7. Bau, D., et al.: Inverting layers of a large generator. In: ICLR, vol. 2, p. 4 (2019)
8. Chai, L., Wulff, J., Isola, P.: Using latent space regression to analyze and leverage compositionality in GANS. arXiv preprint arXiv:2103.10426 (2021)
9. Choi, Y., Choi, M., Kim, M., Ha, J.W., Kim, S., Choo, J.: StarGAN: unified generative adversarial networks for multi-domain image-to-image translation. In: CVPR, pp. 8789–8797 (2018)
10. Collins, E., Bala, R., Price, B., Susstrunk, S.: Editing in style: uncovering the local semantics of GANS. In: CVPR, pp. 5771–5780 (2020)
11. Fukui, H., Hirakawa, T., Yamashita, T., Fujiyoshi, H.: Attention branch network: learning of attention mechanism for visual explanation. In: CVPR, pp. 10705–10714 (2019)
12. Goodfellow, I., et al.: Generative adversarial nets. Adv. Neural Inf. Process. Syst. **27** (2014)
13. Gu, J., Shen, Y., Zhou, B.: Image processing using multi-code GAN prior. In: CVPR, pp. 3012–3021 (2020)
14. Guan, S., Tai, Y., Ni, B., Zhu, F., Huang, F., Yang, X.: Collaborative learning for faster stylegan embedding. arXiv preprint arXiv:2007.01758 (2020)

15. He, Z., Zuo, W., Kan, M., Shan, S., Chen, X.: Attgan: facial attribute editing by only changing what you want. IEEE TIP **28**(11), 5464–5478 (2019)
16. Heusel, M., Ramsauer, H., Unterthiner, T., Nessler, B., Hochreiter, S.: GANS trained by a two time-scale update rule converge to a local nash equilibrium. In: NeurIPS (2017)
17. Isola, P., Zhu, J.Y., Zhou, T., Efros, A.A.: Image-to-image translation with conditional adversarial networks. In: CVPR, pp. 1125–1134 (2017)
18. Karras, T., Laine, S., Aila, T.: A style-based generator architecture for generative adversarial networks. In: CVPR, pp. 4401–4410 (2019)
19. Karras, T., Laine, S., Aittala, M., Hellsten, J., Lehtinen, J., Aila, T.: Analyzing and improving the image quality of StyleGAN. In: CVPR, pp. 8110–8119 (2020)
20. Kim, D., Khan, M.A., Choo, J.: Not just compete, but collaborate: local image-to-image translation via cooperative mask prediction. In: CVPR, pp. 6509–6518 (2021)
21. Kim, T., Cha, M., Kim, H., Lee, J.K., Kim, J.: Learning to discover cross-domain relations with generative adversarial networks. In: ICML, pp. 1857–1865. PMLR (2017)
22. Lee, K.H., Park, C., Oh, J., Kwak, N.: LFI-CAM: learning feature importance for better visual explanation. In: ICCV, pp. 1355–1363 (2021)
23. Liu, M., Ding, Y., Xia, M., Liu, X., Ding, E., Zuo, W., Wen, S.: STGAN: a unified selective transfer network for arbitrary image attribute editing. In: CVPR, pp. 3673–3682 (2019)
24. Liu, Z., Luo, P., Wang, X., Tang, X.: Deep learning face attributes in the wild. In: ICCV, pp. 3730–3738 (2015)
25. Long, J., Shelhamer, E., Darrell, T.: Fully convolutional networks for semantic segmentation. In: CVPR, pp. 3431–3440 (2015)
26. Patro, B.N., Lunayach, M., Patel, S., Namboodiri, V.P.: U-cam: visual explanation using uncertainty based class activation maps. In: ICCV, pp. 7444–7453 (2019)
27. Raj, A., Li, Y., Bresler, Y.: Gan-based projector for faster recovery with convergence guarantees in linear inverse problems. In: ICCV, pp. 5602–5611 (2019)
28. Richardson, E., et al.: Encoding in style: a StyleGAN encoder for image-to-image translation. In: CVPR (2021)
29. Selvaraju, R.R., Cogswell, M., Das, A., Vedantam, R., Parikh, D., Batra, D.: Grad-cam: visual explanations from deep networks via gradient-based localization. In: ICCV, pp. 618–626 (2017)
30. Shen, Y., Gu, J., Tang, X., Zhou, B.: Interpreting the latent space of GANS for semantic face editing. In: CVPR, pp. 9243–9252 (2020)
31. Shen, Y., Yang, C., Tang, X., Zhou, B.: InterFaceGAN: interpreting the disentangled face representation learned by GANS. IEEE TPAMI (2020)
32. Tov, O., Alaluf, Y., Nitzan, Y., Patashnik, O., Cohen-Or, D.: Designing an encoder for StyleGAN image manipulation. ACM TOG **40**(4), 1–14 (2021)
33. Viazovetskyi, Y., Ivashkin, V., Kashin, E.: StyleGAN2 distillation for feed-forward image manipulation. In: Vedaldi, A., Bischof, H., Brox, T., Frahm, J.-M. (eds.) ECCV 2020. LNCS, vol. 12367, pp. 170–186. Springer, Cham (2020). https://doi.org/10.1007/978-3-030-58542-6_11
34. Xu, Y., Deng, B., Wang, J., Jing, Y., Pan, J., He, S.: High-resolution face swapping via latent semantics disentanglement. In: CVPR, pp. 7642–7651 (2022)
35. Xu, Y., Du, Y., Xiao, W., Xu, X., He, S.: From continuity to editability: inverting GANS with consecutive images. In: ICCV, pp. 13910–13918 (2021)
36. Xu, Y., Shen, Y., Zhu, J., Yang, C., Zhou, B.: Generative hierarchical features from synthesizing images. In: CVPR, pp. 4432–4442 (2021)

37. Yang, H., Chai, L., Wen, Q., Zhao, S., Sun, Z., He, S.: Discovering interpretable latent space directions of GANS beyond binary attributes. In: CVPR, pp. 12177–12185 (2021)
38. Yang, T., Ren, P., Xie, X., Zhang, L.: Gan prior embedded network for blind face restoration in the wild. In: CVPR, pp. 672–681 (2021)
39. Zhang, R., Isola, P., Efros, A.A., Shechtman, E., Wang, O.: The unreasonable effectiveness of deep features as a perceptual metric. In: CVPR, pp. 586–595 (2018)
40. Zhong, Z., Chai, L., Zhou, Y., Deng, B., Pan, J., He, S.: Faithful extreme rescaling via generative prior reciprocated invertible representations. In: CVPR, pp. 5708–5717 (2022)
41. Zhou, B., Khosla, A., Lapedriza, A., Oliva, A., Torralba, A.: Learning deep features for discriminative localization. In: CVPR, pp. 2921–2929 (2016)
42. Zhou, Y., Xu, Y., Du, Y., Wen, Q., He, S.: Pro-pulse: Learning progressive encoders of latent semantics in GANS for photo upsampling. IEEE TIP **31**, 1230–1242 (2022)
43. Zhu, J., Shen, Y., Zhao, D., Zhou, B.: In-domain GAN inversion for real image editing. In: Vedaldi, A., Bischof, H., Brox, T., Frahm, J.-M. (eds.) ECCV 2020. LNCS, vol. 12362, pp. 592–608. Springer, Cham (2020). https://doi.org/10.1007/978-3-030-58520-4_35
44. Zhu, J.-Y., Krähenbühl, P., Shechtman, E., Efros, A.A.: Generative visual manipulation on the natural image manifold. In: Leibe, B., Matas, J., Sebe, N., Welling, M. (eds.) ECCV 2016. LNCS, vol. 9909, pp. 597–613. Springer, Cham (2016). https://doi.org/10.1007/978-3-319-46454-1_36
45. Zhu, J.Y., Park, T., Isola, P., Efros, A.A.: Unpaired image-to-image translation using cycle-consistent adversarial networks. In: ICCV, pp. 2223–2232 (2017)

# On the Robustness of Quality Measures for GANs

Motasem Alfarra[1(✉)], Juan C. Pérez[1], Anna Frühstück[1], Philip H. S. Torr[2], Peter Wonka[1], and Bernard Ghanem[1]

[1] King Abdullah University of Science and Tehchnology (KAUST), Thuwal, Saudi Arabia
motasem.alfarra@kaust.edu.sa

[2] University of Oxford, Oxford, UK

**Abstract.** This work evaluates the robustness of quality measures of generative models such as Inception Score (IS) and Fréchet Inception Distance (FID). Analogous to the vulnerability of deep models against a variety of adversarial attacks, we show that such metrics can also be manipulated by additive pixel perturbations. Our experiments indicate that one can generate a distribution of images with very high scores but low perceptual quality. Conversely, one can optimize for small imperceptible perturbations that, when added to real world images, deteriorate their scores. We further extend our evaluation to generative models themselves, including the state of the art network StyleGANv2. We show the vulnerability of both the generative model and the FID against additive perturbations in the latent space. Finally, we show that the FID can be robustified by simply replacing the standard Inception with a robust Inception. We validate the effectiveness of the robustified metric through extensive experiments, showing it is more robust against manipulation. Code: https://github.com/R-FID-Robustness-of-Quality-Measures-for-GANs

**Keywords:** Generative Adversarial Networks · Perceptual quality · Adversarial attacks · Network robustness

## 1 Introduction

Deep Neural Networks (DNNs) are vulnerable to small imperceptible perturbations known as adversarial attacks. For example, while two inputs $x$ and $(x + \delta)$ can be visually indistinguishable to humans, a classifier $f$ can output two different predictions. To address this deficiency in DNNs, adversarial attacks [7,11] and defenses [20,27] have prominently emerged as active areas of research. Starting from image classification [28], researchers also assessed the robustness of DNNs for other tasks, such as segmentation [1], object detection [30], and point

---

**Supplementary Information** The online version contains supplementary material available at https://doi.org/10.1007/978-3-031-19790-1_2.

© The Author(s), under exclusive license to Springer Nature Switzerland AG 2022
S. Avidan et al. (Eds.): ECCV 2022, LNCS 13677, pp. 18–33, 2022.
https://doi.org/10.1007/978-3-031-19790-1_2

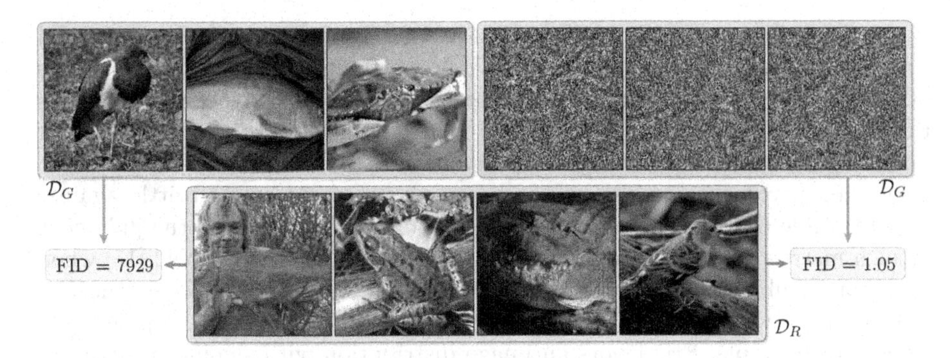


**Fig. 1. Does the Fréchet Inception Distance (FID) accurately measure the distances between image distributions?** We generate datasets that demonstrate the unreliability of FID in judging perceptual (dis)similarities between image distributions. The top left box shows a sample of a dataset constructed by introducing imperceptible noise to each ImageNet image. Despite the remarkable visual similarity between this dataset and ImageNet (bottom box), an extremely large FID (almost 8000) between these two datasets showcases FID's failure to capture perceptual similarities. On the other hand, a remarkably low FID (almost 1.0) between a dataset of random noise images (samples shown in the top right box) and ImageNet illustrates FID's failure to capture perceptual dissimilarities.

cloud classification [18]. While this lack of robustness questions the reliability of DNNs and hinders their deployment in the real world, DNNs are still widely used to evaluate performance in other computer vision tasks, such as that of generation.

Metrics in use for assessing generative models in general, and Generative Adversarial Networks (GANs) [10] in particular, are of utmost importance in the literature. This is because such metrics are widely used to establish the superiority of a generative model over others, hence guiding which GAN should be deployed in real world. Consequently, such metrics are expected to be not only useful in providing informative statistics about the distribution of generated images, but also reliable and robust. In this work, we investigate the robustness of metrics used to assess GANs. We first identify two interesting observations that are unique to this context. First, current GAN metrics are built on pre-trained classification DNNs that are nominally trained (*i.e.* trained on clean images only). A popular DNN of choice is the Inception model [25], on which the Inception Score (IS) [22] and Fréchet Inception Distance (FID) [12] rely. Since nominally trained DNNs are generally vulnerable to adversarial attacks [7], it is expected that DNN-based metrics for GANs also inherit these vulnerabilities. Second, current adversarial attacks proposed in the literature are mainly designed at the instance level (*e.g.* fooling a DNN into misclassifying a particular instance), while GAN metrics are distribution-based. Therefore, attacking these distribution-based metrics requires extending attack formulations from the paradigm of instances to that of distributions.

In this paper, we analyze the robustness of GAN metrics and recommend solutions to improve their robustness. We first attempt to assess the robustness of the quality measures used to evaluate GANs. We check whether such metrics are actually measuring the quality of image distributions by testing their vulnerability against additive pixel perturbations. While these metrics aim at measuring perceptual quality, we find that they are extremely brittle against imperceptible but carefully-crafted perturbations. We then assess the judgment of such metrics on the image distributions generated by StyleGANv2 [15] when its input is subjected to perturbations. While the output of GANs is generally well behaved, we still observe that such metrics provide inconsistent judgments where, for example, FID favors an image distribution with significant artifacts over more naturally-looking distributions. At last, we endeavor to reduce these metrics' vulnerability by incorporating robustly-trained models.

We summarize our contributions as follows:

- We are the first to provide an extensive experimental evaluation of the robustness of the Inception Score (IS) and the Fréchet Inception Distance (FID) against additive pixel perturbations. We propose two instance-based adversarial attacks that generate distributions of images that fool both IS and FID. For example, we show that perturbations $\delta$ with a small budget (*i.e.* $\|\delta\|_\infty \leq 0.01$) are sufficient to increase the FID between ImageNet [8] and a perturbed version of ImageNet to $\sim 7900$, while also being able to generate a distribution of random noise images whose FID to ImageNet is 1.05. We illustrate both cases in Fig. 1.
- We extend our evaluation to study the sensitivity of FID against perturbations in the latent space of state-of-the-art generative models. In this setup, we show the vulnerability of both StyleGANv2 and FID against perturbations in both its $z$- and $w$- spaces. We found that FID provides inconsistent evaluation of the distribution of generated images compared to their visual quality. Moreover, our attack in the latent space causes StyleGANv2 to generate images with significant artifacts, showcasing the vulnerability of StyleGANv2 to additive perturbations in the latent space.
- We propose to improve the reliability of FID by using adversarially-trained models in its computation. Specifically, we replace the traditional Inception model with its adversarially-trained counterpart to generate the embeddings on which the FID is computed. We show that our robust metric, dubbed R-FID, is more resistant against pixel perturbations than the regular FID.
- Finally, we study the properties of R-FID when evaluating different GANs. We show that R-FID is better than FID at distinguishing generated fake distributions from real ones. Moreover, R-FID provides more consistent evaluation under perturbations in the latent space of StyleGANv2.

## 2 Related Work

*GANs and Automated Assessment.* GANs [10] have shown remarkable generative capabilities, specially in the domain of images [4,14,15]. Since the advent of GANs, evaluating their generative capabilities has been challenging [10]. This challenge spurred research efforts into developing automated quantitative measures for GAN outputs. Metrics of particular importance for this purpose are the Inception Score (IS), introduced in [22], and the Fréchet Inception Distance (FID), introduced in [12]. Both metrics leverage the ImageNet-pretrained Inception architecture [25] as a rough proxy for human perception. The IS evaluates the generated images by computing conditional class distributions with Inception and measuring (1) each distribution's entropy—related to Inception's certainty of the image content, and (2) the marginal's entropy—related to diversity across generated images. Noting the IS does not compare the generated distribution to the (real world) target distribution, Heusel *et al.* [12] proposed the FID. The FID compares the generated and target distributions by (1) assuming the Inception features follow a Gaussian distribution and (2) using each distribution's first two moments to compute the Fréchet distance. Further, the FID was shown to be more consistent with human judgement [24].

Both the original works and later research criticized these quantitative assessments. On one hand, IS is sensitive to weight values, noisy estimation when splitting data, distribution shift from ImageNet, susceptibility to adversarial examples, image resolution, difficulty in discriminating GAN performance, and vulnerability to overfitting [2,3,22,29]. On the other hand, FID has been criticized for its over-simplistic assumptions ("Gaussianity" and its associated two-moment description), difficulty in discriminating GAN performance, and its inability to detect overfitting [3,19,29]. Moreover, both IS and FID were shown to be biased to both the number of samples used and the model to be evaluated [6]. In this work, we provide extensive empirical evidence showing that both IS and FID are not robust against perturbations that modify image quality. Furthermore, we also propose a new *robust* FID metric that enjoys superior robustness.

*Adversarial Robustness.* While DNNs became the *de facto* standard for image recognition, researchers found that such DNNs respond unexpectedly to small changes in their input [11,26]. In particular, various works [5,20] observed a widespread vulnerability of DNN models against input perturbations that did not modify image semantics. This observation spurred a line of research on adversarial attacks, aiming to develop procedures for finding input perturbations that fool DNNs [7]. This line of work found that these vulnerabilities are pervasive, casting doubt on the nature of the impressive performances of DNNs. Further research showed that training DNNs to be robust against these attacks [20] facilitated the learning of perceptually-correlated features [9,13]. Interestingly, a later work [23] even showed that such learnt features could be harnessed for image synthesis tasks. In this work, we show (1) that DNN-based scores for GANs are vulnerable against adversarial attacks, and (2) how these scores can be "robustified" by replacing nominally trained DNNs with robustly trained ones.

## 3 Robustness of IS and FID

To compare the output of generative models, two popular metrics are used: the *Inception Score* (IS) and the *Fréchet Inception Distance* (FID). These metrics depend only on the statistics of the distribution of generated images in an ImageNet-pretrained Inception's embedding space, raising the question:

*What do quality measures for generative models, such as IS and FID, tell us about image quality?*

We investigate this question from the robustness perspective. In particular, we analyze the sensitivity of these metrics to carefully crafted perturbations. We start with preliminary background about both metrics.

### 3.1 Preliminaries

We consider the standard image generation setup where a generator $G : \mathbb{R}^{d_z} \rightarrow \mathbb{R}^{d_x}$ receives a latent code $z \in \mathbb{R}^{d_z}$ and outputs an image $x \in \mathbb{R}^{d_x}$. Upon training, $G$ is evaluated based on the quality of the generated distribution of images $\mathcal{D}_G$ by computing either the IS [22] or the FID [12]. Both metrics leverage an ImageNet-pretrained [8] InceptionV3 [25]. Salimans *et al.* [22] proposed measuring the perceptual quality of the generated distribution $\mathcal{D}_G$ by computing the IS as:

$$\text{IS}(\mathcal{D}_G) = \exp\left(\mathbb{E}_{x \sim \mathcal{D}_G}\left(\text{KL}\left(p(y|x) \,||\, p(y)\right)\right)\right), \tag{1}$$

where $p(y|x)$ is the output probability distribution of the pretrained Inception model. While several works have argued about the effectiveness of the IS and its widely-used implementation [2], its main drawback is that it disregards the relation between the generated distribution, $\mathcal{D}_G$, and the real one, $\mathcal{D}_R$, used for training $G$ [12]. Consequently, Heusel *et al.* proposed the popular FID, which involves the statistics of the real distribution. In particular, FID assumes that the Inception features of an image distribution $\mathcal{D}$ follow a Gaussian distribution with mean $\mu_{\mathcal{D}}$ and covariance $\Sigma_{\mathcal{D}}$, and it measures the squared Wasserstein distance between the two Gaussian distributions of real and generated images. Hence, $\text{FID}(\mathcal{D}_R, \mathcal{D}_G)$, or FID for short, can be calculated as:

$$\text{FID} = \|\mu_R - \mu_G\|^2 + \text{Tr}\left(\Sigma_R + \Sigma_G - 2(\Sigma_R \Sigma_G)^{1/2}\right), \tag{2}$$

where $\cdot_R, \cdot_G$ are the statistics of the real and generated image distributions, respectively, and $\text{Tr}(\cdot)$ is the trace operator. Note that the statistics of both distributions are empirically estimated from their corresponding image samples. In principle, FID measures how close (realistic) the generated distribution $\mathcal{D}_G$ is to $\mathcal{D}_R$. We remark that the FID is the *de facto* metric for evaluating image generation-related tasks. Therefore, our study focuses mostly on FID.

We note here that both the IS and the FID are oblivious to $G$'s training process and can be computed to compare two arbitrary sets of images $\mathcal{D}_R$ and $\mathcal{D}_G$. In generative modeling, this is typically a set of real images (photographs) and a set of generated images. However, it is also possible to compare two sets of photographs, two sets of generated images, manipulated photographs with real photographs, *etc.* This flexibility allows us to study these metrics in a broader context next, where no generative model is involved.

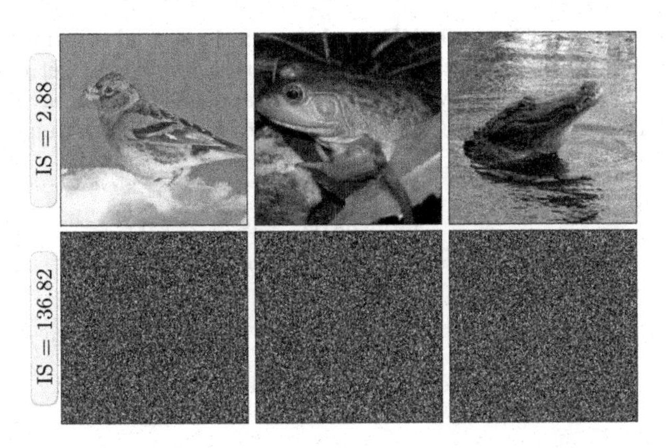


**Fig. 2. Sensitivity of Inception Score (IS) against pixel perturbations.** *First row:* real-looking images (sampled from $\mathcal{D}_G = \mathcal{D}_R + \delta$) with a low IS (below 3). *Second row:* random noise images with a high IS (over 135).

### 3.2 Robustness Under Pixel Perturbations

We first address the question presented earlier in Sect. 3 by analyzing the sensitivity of IS and FID to additive pixel perturbations. In particular, we assume $\mathcal{D}_R$ to be either CIFAR10 [17] or ImageNet [8] and ask: **(i)** can we generate a distribution of imperceptible additive perturbations $\delta$ that deteriorates the scores for $\mathcal{D}_G = \mathcal{D}_R + \delta$? Or, alternatively, **(ii)** can we generate a distribution of low visual quality images, *i.e.* noise images, that attain good quality scores? If the answer is yes to both questions, then FID and IS have limited capacity for providing information about image quality in the worst case.

**Good Images - Bad Scores.** We aim at constructing a distribution of real-looking images with *bad* quality measures, *i.e.* low IS or high FID. While both metrics are distribution-based, we design instance-wise proxy optimization problems to achieve our goal.

*Minimizing IS.* Based on Eq. (1), one could minimize the IS by having both the posterior $p(y|x)$ and the prior $p(y)$ be the same distribution. Assuming that $p(y)$ is a uniform distribution, we minimize the IS by maximizing the entropy of $p(y|x)$. Therefore, we can optimize a perturbation $\delta^*$ for each real image $x_r \sim \mathcal{D}_R$ by solving the following problem:

$$\begin{aligned} &\delta^* = \arg\max_{\|\delta\|_\infty \leq \epsilon} \mathcal{L}_{\text{ce}}\left(p(y|x_r + \delta), \hat{y}\right), \\ &\text{s.t. } \hat{y} = \arg\max_i \, p^i(y|x_r + \delta), \end{aligned} \tag{3}$$

where $\mathcal{L}_{\text{ce}}$ is the Cross Entropy loss. We solve the problem in Eq. (3) with 100 steps of Projected Gradient Descent (PGD) and zero initialization. We then compile the distribution $\mathcal{D}_G$, where each image $x_g = x_r + \delta^*$ is a perturbed version of the real dataset $\mathcal{D}_R$. Note that our objective aims to minimize the

**Table 1. Robustness of IS and FID against pixel perturbations.** We assess the robustness of IS and FID against perturbations with a limited budget $\epsilon$ on CIFAR10 and ImageNet. In the last row, we report the IS and FID of images with carefully-designed random noise having a resolution similar to CIFAR10 and ImageNet.

| $\epsilon$ | CIFAR10 | | ImageNet | |
|---|---|---|---|---|
| | IS | FID | IS | FID |
| 0.00 | 11.54 | 0.00 | 250.74 | 0.00 |
| $5 \times 10^{-3}$ | 2.62 | 142.45 | 3.08 | 3013.33 |
| 0.01 | 2.50 | 473.19 | 2.88 | 7929.01 |
| Random noise | 94.87 | 9.94 | 136.82 | 1.05 |

network's confidence in predicting all labels for each $x_g$. In doing so, both $p(y|x_g)$ and $p(y)$ tend to converge to a uniform distribution, thus, minimizing the KL divergence between them and effectively lowering the IS. Note how $\epsilon$ controls the allowed perturbation amount for each image $x_r$. Therefore, for small $\epsilon$ values, samples from $\mathcal{D}_G$ and $\mathcal{D}_R$ are perceptually indistinguishable.

*Maximizing FID.* Next, we extend our attack setup to the more challenging FID. Given an image $x$, we define $f(x) : \mathbb{R}^{d_x} \to \mathbb{R}^{d_e}$ to be the output embedding of an Inception model. We aim to maximize the FID by generating a perturbation $\delta$ that pushes the embedding of a real image away from its original position. In particular, for each $x_r \sim \mathcal{D}_R$, we aim to construct $x_g = x_r + \delta^*$ where:

$$\delta^* = \arg\max_{\|\delta\|_\infty \leq \epsilon} \ \|f(x_r) - f(x_r + \delta)\|_2 . \tag{4}$$

In our experiments, we solve the optimization problem in Eq. (4) with 100 PGD steps and a randomly initialized $\delta$ [20]. Maximizing this objective indirectly maximizes FID's first term (Eq. (2)), while resulting in a distribution of images $\mathcal{D}_G$ that is visually indistinguishable from the real $\mathcal{D}_R$ for small $\epsilon$ values.

*Experiments.* We report our results in Table 1. Our simple yet effective procedure illustrates how both metrics are very susceptible to attacks. In particular, solving the problem in Eq. (3) yields a distribution of noise that significantly decreases the IS from 11.5 to 2.5 in CIFAR10 and from 250.7 to 2.9 in ImageNet. We show a sample from $\mathcal{D}_G$ in Fig. 2, first row. Similarly, our optimization problem in Eq. (4) can create imperceptible perturbations that maximize the FID to $\approx$7900 between ImageNet and its perturbed version (examples shown in Fig. 1).

**Bad Images - Good Scores.** While the previous experiments illustrate the vulnerability of both the IS and FID against small perturbations (*i.e.* good images with bad scores), here we evaluate if the converse is also possible, *i.e.* bad images with good scores. In particular, we aim to construct a distribution of noise images (*e.g.* second row of Fig. 2) that enjoys good scores (high IS or low FID).

*Maximizing IS.* The IS has two terms: Inception's confidence on classifying a generated image, *i.e.* $p(y|x_g)$, and the diversity of the generated distribution of predicted labels, *i.e.* $p(y)$. One can maximize the IS by generating a distribution $\mathcal{D}_G$ such that: **(i)** each $x_g \sim \mathcal{D}_G$ is predicted with high confidence, and **(ii)** the distribution of predicted labels is uniform across Inception's output $\mathcal{Y}$. To that end, we propose the following procedure for constructing such $\mathcal{D}_G$. For each $x_g$, we sample a label $\hat{y} \sim \mathcal{Y}$ uniformly at random and solve the problem:

$$x_g = \arg\min_x \mathcal{L}_{ce}(p(y|x), \hat{y}). \tag{5}$$

In our experiments, we solve the problem in Eq. (5) with 100 gradient descent steps and random initialization for $x$.

*Minimizing FID.* Here, we analyze the robustness of FID against such a threat model. We follow a similar strategy to the objective in Eq. (4). For each image $x_r \sim \mathcal{D}_R$, we intend to construct $x_g$ such that:

$$x_g = \arg\min_x \|f(x) - f(x_r)\|_2 \tag{6}$$

with a randomly initialized $x$. In our experiments, we solve Eq. (6) with 100 gradient descent steps. As such, each $x_g$ will have a similar Inception representation to a real-world image, *i.e.* $f(x_g) \approx f(x_r)$, while being random noise.

*Experiments.* We report our results in the last row of Table 1. Both the objectives in Eqs. (5) and (6) are able to fool the IS and FID, respectively. In particular, we are able to generate distributions of noise images with resolutions $32 \times 32$ and $224 \times 224$ (*i.e.* CIFAR10 and ImageNet resolutions) but with IS of 94 and 136, respectively. We show a few qualitative samples in the second row of Fig. 2. Furthermore, we generate noise images that have embedding representations very similar to those of CIFAR10 and ImageNet images. This lowers the FID of both datasets to 9.94 and 1.05, respectively (examples are shown in Fig. 1).

### 3.3 Robustness Under Latent Perturbations

In the previous section, we established the vulnerability of both the IS and FID against pixel perturbations. Next, we investigate the vulnerability against perturbations in a GAN's latent space. Designing such an attack is more challenging in this case, since images can only be manipulated indirectly, and so there are fewer degrees of freedom for manipulating an image. To that end, we choose $G$ to be the state of the art generator StyleGANv2 [14] trained on the standard FFHQ dataset [14]. We limit the investigation to the FID metric, as IS is not commonly used in the context of unconditional generators, such as StyleGAN. Note that we always generate 70k samples from $G$ to compute the FID.

Recall that our generator $G$ accepts a random latent vector $z \sim \mathcal{N}(0, \mathrm{I})$[1] and maps it to the more expressive latent space $w$, which is then fed to the remaining layers of $G$. It is worthwhile to mention that "truncating" the latent $w$ with a

[1] The appendix presents results showing that sampling $z$ from different distributions still yields good looking StyleGANv2-generated images.

**Fig. 3. Effect of attacking truncated StyleGANv2's latent space on the Fréchet Inception Distance (FID).** We conduct attacks on the latent space of StyleGANv2 and record the effect on the FID. We display the resulting samples of these attacks for two truncation values, $\alpha = 0.7$ (top row) and $\alpha = 1.0$ (bottom row). Despite the stark differences in realism between the images in the top and bottom rows—*i.e.* the top row's remarkable quality and the bottom row's artifacts—the FID to FFHQ reverses this ranking, wherein the bottom row is judged as *farther* away from FFHQ than the top row.

pre-computed $\bar{w}^2$ and constant $\alpha \in \mathbb{R}$ (*i.e.* replacing $w$ with $\alpha w + (1-\alpha)\bar{w}$) controls both the quality and diversity of the generated images [14].

*Effect of Truncation on FID.* We first assess the effect of the truncation level $\alpha$ on both image quality and FID. We set $\alpha \in [0.7, 1.0, 1.3]$ and find FIDs to be $[21.81, 2.65, 9.31]$, respectively. Based on our results, we assert the following observation: while the visual quality of generated images at higher truncation levels, *e.g.* $\alpha = 0.7$, is better and has fewer artifacts than the other $\alpha$ values, the FID does not reflect this fact, showing lower (better) values for $\alpha \in \{1.0, 1.3\}$. We elaborate on this observation with qualitative experiments in the appendix.

*FID-Guided Sampling.* Next, we extend the optimization problem in Eq. (4) from image to latent perturbations. In particular, we aim at constructing a perturbation $\delta_z^*$ for each sampled latent $z$ by solving:

$$\delta_z^* = \arg\max_\delta \; \|f(G(z+\delta)) - f(x_r)\|_2 \,. \tag{7}$$

Thus, $\delta_z^*$ perturbs $z$ such that $G$ produces an image whose embedding differs from that of real image $x_r$. We solve the problem in Eq. (7) for $\alpha \in \{0.7, 1.0\}$.

*Experiments.* We visualize our results in Fig. 3 accompanied with their corresponding FID values (first and second rows correspond to $\alpha = 0.7$ and 1.0, respectively). While our attack in the latent space is indeed able to significantly increase the FID (from 2.65 to 31.68 for $\alpha = 1.0$ and 21.33 to 34.10 for $\alpha = 0.7$),

[2] $\bar{w}$ is referred to as the mean of the $w$-space. It is computed by sampling several latents $z$ and averaging their representations in the $w$-space.

we inspect the results and draw the following conclusions. **(i)** FID provides inconsistent evaluation of the generated distribution of images. For example, while both rows in Fig. 3 have comparable FID values, the visual quality is significantly different. This provides practical evidence of this metric's unreliability in measuring the performance of generative models. **(ii)** Adding crafted perturbations to the input of a state of the art GAN deteriorates the visual quality of its output space (second row in Fig. 3). This means that GANs are also vulnerable to adversarial attacks. This is confirmed in the literature for other generative models such as GLOW [16,21]. Moreover, we can formulate a problem similar to Eq. (7) but with the goal of perturbing the $w$-space instead of the $z$-space. We leave results of solving this formulation for different $\alpha$ values to the appendix.

***Section Summary.*** In this section, we presented an extensive experimental evaluation investigating if the quality measures (IS and FID) of generative models actually measure the perceptual quality of the output distributions. We found that such metrics are extremely vulnerable to pixel perturbations. We were able to construct images with very good scores but no visual content (Sect. 3.2), as well as images with realistic visual content but very bad scores (Sect. 3.2). We further studied the sensitivity of FID against perturbations in the latent space of StyleGANv2 (Sect. 3.3), allowing us to establish the inconsistency of FID under this setup as well. Therefore, we argue that such metrics, while measuring useful properties of the generated distribution, lead to questionable assessments of the visual quality of the generated images.

# 4 R-FID: Robustifying the FID

After establishing the vulnerability of IS and FID to perturbations, we analyze the cause of such behavior and propose a solution. We note that, while different metrics have different formulations, they rely on a pretrained Inception model that could potentially be a leading cause of such vulnerability. This observation suggests the following question:

*Can we robustify the FID by replacing its Inception component with a robustly trained counterpart?*

We first give a brief overview of adversarial training.

## 4.1 Leveraging Adversarially Trained Models

Adversarial training is arguably the *de facto* procedure for training robust models against adversarial attacks. Given input-label pairs $(x, y)$ sampled from a training set $\mathcal{D}_{tr}$, $\ell_2$-adversarial training solves the following min-max problem:

$$\min_{\theta} \mathbb{E}_{(x,y)\sim\mathcal{D}_{tr}} \left[ \max_{\|\delta\|_2 \leq \kappa} \mathcal{L}\left(x + \delta, y; \theta\right) \right] \tag{8}$$

**Fig. 4. Attacking R-FID with pixel perturbations.** We attack two variants of R-FID ($\kappa = 64$ and $\kappa = 128$) and visualize samples from the resulting datasets. Attempting to fool these R-FIDs at the pixel level yields perturbations that correlate with semantic patterns, in contrast to those obtained when attempting to fool the standard FID (as shown in Fig. 1).

**Table 2. R-FID against attacks in the pixel space.** We study the robustness of R-FID against the adversarial attacks in Eq. 4.

| $\epsilon$ | CIFAR10 | | ImageNet | |
|---|---|---|---|---|
| | $\kappa = 64$ | $\kappa = 128$ | $\kappa = 64$ | $\kappa = 128$ |
| 0.01 | 1.5 | 0.3 | 21.0 | 4.5 |
| 0.02 | 20.7 | 7.8 | 293.8 | 92.1 |
| 0.03 | 46.4 | 19.7 | 657.9 | 264.6 |

for a given loss function $\mathcal{L}$ to train a robust network with parameters $\theta$. We note that $\kappa$ controls the robustness-accuracy trade-off: models trained with larger $\kappa$ tend to have higher robust accuracy (accuracy under adversarial attacks) and lower clean accuracy (accuracy on clean images). Since robust models are expected to resist pixel perturbations, we expect such models to inherit robustness characteristics against the attacks constructed in Sect. 3.2. Moreover, earlier works showed that robustly-trained models tend to learn more semantically-aligned and invertible features [13]. Therefore, we hypothesize that replacing the pretrained Inception model with its robustly trained counterpart could increase FID's sensitivity to the visual quality of the generated distribution (*i.e.* robust against attacks in Sect. 3.3).

To that end, we propose the following modification to the FID computation. We replace the pretrained Inception model with a robustly trained version on ImageNet following Eq. (8) with $\kappa \in \{64, 128\}$. The training details are left to the appendix. We refer to this alternative as R-FID, and analyze its robustness against perturbations next.

**Table 3. Truncation's effect on R-FID.** We study how truncation affects R-FID against FFHQ (first two rows), and across different truncation levels (last two rows).

| $(\mathcal{D}_G(\alpha), \mathcal{D}_R)$ | 0.7 | 0.9 | 1.0 |
|---|---|---|---|
| $\kappa = 64$ | 98.3 | 90.0 | 88.1 |
| $\kappa = 128$ | 119.9 | 113.7 | 113.8 |
| $(\mathcal{D}_G(\alpha_i), \mathcal{D}_G(\alpha_j))$ | (0.7, 1.0) | (0.7, 0.9) | (0.9, 1.0) |
| $\kappa = 64$ | 10.5 | 4.9 | 0.48 |
| $\kappa = 128$ | 9.9 | 4.6 | 0.46 |

### 4.2 R-FID Against Pixel Perturbations

We first test the sensitivity of R-FID against additive pixel perturbations. For that purpose, we replace the Inception with a robust Inception, and repeat the experiments from Sect. 3.2 to construct real images with bad scores. We conduct experiments on CIFAR10 and ImageNet with $\epsilon \in \{0.01, 0.02, 0.03\}$ for the optimization problem in Eq. (4), and we report the results in Table 2. We observe that the use of a robustly-trained Inception significantly improves robustness against pixel perturbations. Our robustness improvement for the same value of $\epsilon = 0.01$ is of 3 orders of magnitude (an FID of 4 for $\kappa = 128$ compared to 7900 reported in Table 1). While both models consistently provide a notable increase in robustness against pixel perturbations, we find that the model most robust to adversarial attacks (*i.e.* $\kappa = 128$) is also the most robust to FID attacks. It is worthwhile to mention that this kind of robustness is expected since our models are trained not to alter their prediction under additive input perturbations. Hence, their feature space should enjoy robustness properties, as measured by our experiments. In Fig. 4 we visualize a sample from the adversarial distribution $\mathcal{D}_G$ (with $\epsilon = 0.08$) when $\mathcal{D}_R$ is ImageNet. We observe that our adversaries while aiming only at pushing the feature representation of samples of $\mathcal{D}_G$ away from those of $\mathcal{D}_R$, are also more correlated with human perception. This finding aligns with previous observations in the literature, which find robustly-trained models have a more interpretable (more semantically meaningful) feature space [9,13]. We leave the evaluation under larger values of $\epsilon$, along with experiments on unbounded perturbations, to the appendix.

### 4.3 R-FID Under Latent Perturbations

In Sect. 4.2, we tested R-FID's robustness against pixel-level perturbations. Next, we study R-FID for evaluating generative models. For this, we follow the setup in Sect. 3.3 using an FFHQ-trained StyleGANv2 as generator $G$.

*Effect of Truncation on R-FID.* Here, we analyze the R-FID when the generator is using different truncation levels. In particular, we choose $\alpha \in \{0.7, 0.9, 1.0\}$ and report results in Table 3. We observe that the robust Inception model clearly distinguishes the distribution generated by StyleGANv2 from the FFHQ dataset,

**Fig. 5. Robustness of R-FID against perturbations in StyleGANv2 latent space.** We conduct attacks on two variants of R-FID ($\kappa = 64$ on the left, and $\kappa = 128$ on the right) and two truncation values ($\alpha = 0.7$ on the top, and $\alpha = 1.0$ on the bottom) by perturbing the latent space. We also visualize samples from the generated distributions. For the pairs $(\kappa, \alpha) \in \{(64, 0.7), (64, 1.0), (128, 0.7), (128, 1.0)\}$, we find corresponding R-FID values of {128.1, 157.8, 126.6, 162,8}. In contrast to the minimal changes required to fool the standard FID (Fig. 3), fooling the R-FID leads to a dramatic degradation in visual quality of the generated images.

regardless of the truncation $\alpha$. In this case, we obtain an R-FID of 113.8, substantially larger than the 2.6 obtained when the nominally-trained Inception model is used. This result demonstrates that, while the visual quality of StyleGANv2's output is impressive, the generated image distribution is far from the FFHQ distribution. We further evaluate if the R-FID is generally large between any two distributions by measuring the R-FID between two distributions of images generated at two truncation levels $(\alpha_i, \alpha_j)$. Table 3 reports these results. We observe that **(i)** the R-FID between a distribution and itself is $\approx 0$, *e.g.* R-FID $= 10^{-3}$ at (1.0, 1.0). Please refer to the appendix for details. **(ii)** The R-FID gradually increases as the image distributions differ, *e.g.* R-FID at (0.9, 1.0) $<$ (0.7, 1.0). This observation validates that the large R-FID values found between FFHQ and various truncation levels are a result of the large separation in the embedding space that robust models induce between real and generated images.

*R-FID Guided Sampling.* Next, we assess the robustness of the R-FID against perturbations in the latent space of the generator $G$. For this purpose, we conduct the attack proposed in Eq. (7) with $f$ now being the robustly-trained Inception. We report results and visualize few samples in Fig. 5. We make the following observations. **(i)** While the R-FID indeed increases after the attack, the relative increment is far less than that of the non-robust FID. For example, R-FID increases by 44% at $\kappa = 64$ and $\alpha = 0.7$ compared to an FID increase of 1000% under the same setup. **(ii)** The increase in R-FID is associated with a significantly larger amount of artifacts introduced by the GAN in the generated images. This result further evidences the vulnerability of the generative model. However, it also highlights the changes in the image distribution that are

**Table 4. Sensitivity of R-FID against noise and blurring.** We measure R-FID ($\kappa = 128$) between ImageNet and a transformed version of it under Gaussian noise and blurring. As $\sigma$ increases, the image quality decreases and R-FID increases.

| $\sigma_N/\sigma_B$ | $0.1/1.0$ | $0.2/2.0$ | $0.3/3.0$ | $0.4/4.0$ |
|---|---|---|---|---|
| Gaussian (N)oise | 16.65 | 61.33 | 128.8 | 198.3 |
| Gaussian (B)lur | 15.54 | 54.07 | 78.67 | 89.11 |

required to increase the R-FID. We leave the $w$- space formulation for the attack on the R-FID, along with its experiments, to the appendix.

***Section Summary.*** In this section, we robustified the popular FID by replacing the pretrained Inception model with a robustly-trained version. We found this replacement results in a more robust metric (R-FID) against perturbations in both the pixel (Sect. 4.2) and latent (Sect. 4.3) spaces. Moreover, we found that pixel-based attacks yield much more perceptually-correlated perturbations when compared to the attacks that used the standard FID (Fig. 2). Finally, we observed that changing R-FID values requires a more significant and notable distribution shift in the generated images (Fig. 5).

### 4.4 R-FID Against Quality Degradation

At last, we analyze the effect of transformations that degrade image quality on R-FID. In particular, we apply Gaussian noise and Gaussian blurring on ImageNet and report the R-FID ($\kappa = 128$) between ImageNet and the degraded version in Table 4. Results show that as the quality of the images degrades (*i.e.* as $\sigma$ increases), the R-FID steadily increases. Thus, we find that R-FID is able to distinguish a distribution of images from its degraded version.

## 5 Discussion, Limitations, and Conclusions

In this work, we demonstrate several failure modes of popular GAN metrics, specifically IS and FID. We also propose a robust counterpart of FID (R-FID), which mitigates some of the robustness problems and yields significantly more robust behavior under the same threat models.

Measuring the visual quality for image distributions has two components: (1) the statistical measurement (*e.g.* Wasserstein distance) and (2) feature extraction using a pretrained model (*e.g.* InceptionV3). A limitation of our work is that we only focus on the second part (the pretrained model). As an interesting avenue for future work, we suggest a similar effort to assess the reliability of the statistical measurement as well, *i.e.* analyzing and finding better and more robust alternatives to the Wasserstein distance.

Current metrics mainly focus on comparing the distribution of features. In these cases, visual quality is only hoped to be a side effect and not directly optimized for nor tested by these metrics. Developing a metric that directly assesses visual quality remains an open problem that is not tackled by our work but is recommended for future work.

**Acknowledgments.** This work was supported by the King Abdullah University of Science and Technology (KAUST) Office of Sponsored Research (OSR) under Award No. OSR-CRG2019-4033.

## References

1. Arnab, A., Miksik, O., Torr, P.H.: On the robustness of semantic segmentation models to adversarial attacks. In: Proceedings of the IEEE Conference on Computer Vision and Pattern Recognition, pp. 888–897 (2018)
2. Barratt, S., Sharma, R.: A note on the inception score (2018)
3. Borji, A.: Pros and cons of GAN evaluation measures. Comput. Vis. Image Underst. **179**, 41–65 (2019)
4. Brock, A., Donahue, J., Simonyan, K.: Large scale GAN training for high fidelity natural image synthesis. In: International Conference on Learning Representations (ICLR) (2019)
5. Carlini, N., Wagner, D.: Towards evaluating the robustness of neural networks. In: 2017 IEEE Symposium on Security and Privacy (SP) (2017)
6. Chong, M.J., Forsyth, D.: Effectively unbiased fid and inception score and where to find them. In: Proceedings of the IEEE/CVF conference on computer vision and pattern recognition, pp. 6070–6079 (2020)
7. Croce, F., Hein, M.: Reliable evaluation of adversarial robustness with an ensemble of diverse parameter-free attacks. In: International Conference on Machine Learning (ICML) (2020)
8. Deng, J., Dong, W., Socher, R., Li, L.J., Li, K., Li, F.F.: Imagenet: a large-scale hierarchical image database. In: The IEEE Conference on Computer Vision and Pattern Recognition (CVPR) (2009)
9. Engstrom, L., Ilyas, A., Santurkar, S., Tsipras, D., Tran, B., Madry, A.: Adversarial robustness as a prior for learned representations (2020). https://openreview.net/forum?id=rygvFyrKwH
10. Goodfellow, I., et al.: Generative adversarial nets. In: Advances in Neural Information Processing Systems (2014)
11. Goodfellow, I., Shlens, J., Szegedy, C.: Explaining and harnessing adversarial examples. In: International Conference on Learning Representations (ICLR) (2015)
12. Heusel, M., Ramsauer, H., Unterthiner, T., Nessler, B., Hochreiter, S.: GANS trained by a two time-scale update rule converge to a local nash equilibrium. In: Advances in Neural Information Processing Systems (NeurIPS) (2017)
13. Ilyas, A., Santurkar, S., Tsipras, D., Engstrom, L., Tran, B., Madry, A.: Adversarial examples are not bugs, they are features. In: Advances in Neural Information Processing Systems (NeurIPS) (2019)
14. Karras, T., Laine, S., Aila, T.: A style-based generator architecture for generative adversarial networks. In: Proceedings of the IEEE/CVF Conference on Computer Vision and Pattern Recognition (CVPR) (2019)
15. Karras, T., Laine, S., Aittala, M., Hellsten, J., Lehtinen, J., Aila, T.: Analyzing and improving the image quality of StyleGAN. In: Proceedings of the IEEE/CVF Conference on Computer Vision and Pattern Recognition (CVPR) (2020)

16. Kingma, D.P., Dhariwal, P.: Glow: generative flow with invertible 1x1 convolutions. In: Bengio, S., Wallach, H., Larochelle, H., Grauman, K., Cesa-Bianchi, N., Garnett, R. (eds.) Advances in Neural Information Processing Systems, vol. 31. Curran Associates, Inc. (2018). https://proceedings.neurips.cc/paper/2018/file/d139db6a236200b21cc7f752979132d0-Paper.pdf
17. Krizhevsky, A., Hinton, G.: Learning multiple layers of features from tiny images. In: University of Toronto, Canada (2009)
18. Liu, H., Jia, J., Gong, N.Z.: PointGuard: provably robust 3D point cloud classification. In: Proceedings of the IEEE/CVF Conference on Computer Vision and Pattern Recognition, pp. 6186–6195 (2021)
19. Lucic, M., Kurach, K., Michalski, M., Gelly, S., Bousquet, O.: Are gans created equal? a large-scale study. Adv. Neural Inf. Process. Syst. (NeurIPS) **31** (2018)
20. Madry, A., Makelov, A., Schmidt, L., Tsipras, D., Vladu, A.: Towards deep learning models resistant to adversarial attacks. In: International Conference on Learning Representations (ICLR) (2018)
21. Pope, P., Balaji, Y., Feizi, S.: Adversarial robustness of flow-based generative models. In: Chiappa, S., Calandra, R. (eds.) Proceedings of the Twenty Third International Conference on Artificial Intelligence and Statistics. Proceedings of Machine Learning Research, vol. 108, pp. 3795–3805. PMLR (2020)
22. Salimans, T., et al.: Improved techniques for training GANS. In: Advances in Neural Information Processing Systems (NeurIPS) (2016)
23. Santurkar, S., Ilyas, A., Tsipras, D., Engstrom, L., Tran, B., Madry, A.: Image synthesis with a single (robust) classifier. In: Advances in Neural Information Processing Systems (NeurIPS) (2019)
24. Shmelkov, K., Schmid, C., Alahari, K.: How good is my GAN? In: Proceedings of the European Conference on Computer Vision (ECCV), pp. 213–229 (2018)
25. Szegedy, C., Vanhoucke, V., Ioffe, S., Shlens, J., Wojna, Z.: Rethinking the inception architecture for computer vision. In: Proceedings of the IEEE conference on computer vision and pattern recognition (CVPR) (2016)
26. Szegedy, C., et al.: Intriguing properties of neural networks. In: International Conference on Learning Representations (ICLR) (2014)
27. Wang, Y., Zou, D., Yi, J., Bailey, J., Ma, X., Gu, Q.: Improving adversarial robustness requires revisiting misclassified examples. In: International Conference on Learning Representations (ICLR) (2019)
28. Wu, D., Xia, S.T., Wang, Y.: Adversarial weight perturbation helps robust generalization. In: Advances in Neural Information Processing Systems (NeurIPS) (2020)
29. Xu, Q., et al.: An empirical study on evaluation metrics of generative adversarial networks. arXiv preprint arXiv:1806.07755 (2018)
30. Zhao, Y., Zhu, H., Liang, R., Shen, Q., Zhang, S., Chen, K.: Seeing isn't believing: towards more robust adversarial attack against real world object detectors. In: Proceedings of the 2019 ACM SIGSAC Conference on Computer and Communications Security, pp. 1989–2004 (2019)

# Sound-Guided Semantic Video Generation

Seung Hyun Lee[1], Gyeongrok Oh[1], Wonmin Byeon[2], Chanyoung Kim[1], Won Jeong Ryoo[1], Sang Ho Yoon[3], Hyunjun Cho[1], Jihyun Bae[1], Jinkyu Kim[4](✉), and Sangpil Kim[1](✉)

[1] Department of Artificial Intelligence, Korea University, Seongbuk, Korea
spk7@korea.ac.kr
[2] NVIDIA Research, NVIDIA Corporation, Santa Clara, USA
[3] Graduate School of Culture Technology, KAIST, Daejeon, South Korea
[4] Department of Computer Science and Engineering, Korea University, Seongbuk, Korea
jinkyukim@korea.ac.kr

**Abstract.** The recent success in StyleGAN demonstrates that pre-trained StyleGAN latent space is useful for realistic video generation. However, the generated motion in the video is usually not semantically meaningful due to the difficulty of determining the direction and magnitude in the StyleGAN latent space. In this paper, we propose a framework to generate realistic videos by leveraging multimodal (sound-image-text) embedding space. As sound provides the temporal contexts of the scene, our framework learns to generate a video that is semantically consistent with sound. First, our sound inversion module maps the audio directly into the StyleGAN latent space. We then incorporate the CLIP-based multimodal embedding space to further provide the audio-visual relationships. Finally, the proposed frame generator learns to find the trajectory in the latent space which is coherent with the corresponding sound and generates a video in a hierarchical manner. We provide the new high-resolution landscape video dataset (audio-visual pair) for the sound-guided video generation task. The experiments show that our model outperforms the state-of-the-art methods in terms of video quality. We further show several applications including image and video editing to verify the effectiveness of our method.

**Keywords:** Sound · Multi-modal representation · Video generation

## 1 Introduction

Existing video generation methods rely on motion generation from a noise vector given an initial frame [34,35,43]. As they create trajectory from noise vectors

Code and more diverse examples are available at https://kuai-lab.github.io/eccv2022sound/.

**Supplementary Information** The online version contains supplementary material available at https://doi.org/10.1007/978-3-031-19790-1_3.

© The Author(s), under exclusive license to Springer Nature Switzerland AG 2022
S. Avidan et al. (Eds.): ECCV 2022, LNCS 13677, pp. 34–50, 2022.
https://doi.org/10.1007/978-3-031-19790-1_3

without any guidance, the motion of the generated video is not semantically meaningful. On the other hand, sound provides the cue for motion and various context of the scene [23]. Specifically, sound may represent events of the scene such as 'viola playing' or 'birds singing'. It can also provide a tone of the scene such as 'Screaming' or 'Laughing'. This is an important cue for generating a video because the video's temporal component and motion are closely associated with sound. Generating high-fidelity video from sound is crucial for a variety of applications, such as multimedia content creation and filmmaking. However, most sound-based video generation works focus on generating a talking face matching from verbal sound using 2D facial landmarks [6,7,32] or 3D face representation [5,26,33,38,40,44,46].

Generating a realistic video from non-verbal sound is challenging. First, the spatial and temporal contexts of a generated video need to be semantically consistent with sound. Although some studies [3,20] have tried to synthesize video with sound as an independent variable, mapping from 1D semantic information from audio to visual signal is not straightforward. Furthermore, when a sound of waves enters as an input, these works can not generate a video with diverse waves. Second, the generated motion should be physically plausible and temporally coherent between frames. Recent works [1,12] use a latent mapper network with music as the input for navigating in StyleGAN [15] latent space. Since the direction in the latent space is randomly provided, however, the content semantics from the video are not realistic. Finally, there is a lack of high-fidelity video dataset for generating a realistic video from non-verbal sound.

To overcome these challenges, we propose a novel framework that can semantically invert sound to the StyleGAN latent space for video generation. The difficulty of inverting audio into the StyleGAN latent space is that the image regenerated after inverting the audio is visually irrelevant to the audio. To resolve this issue, we employ the joint embedding prior knowledge learned from large-scale multimodal data (image, text, sound) to sound inverting. A high fidelity video is generated by moving the latent vector in the $\mathcal{W}+$ space which is disentangled latent space of StyleGAN. Our sound-encoder learns to put the latent space $\mathcal{W}+$ from the sound input and find the trajectory from the initial latent code for video generation. As shown in Fig. 1, we generate a video whose meaning is consistent with the given sound-input.

Contrastive Language-Image Pre-Training (CLIP) [25] creates a very powerful joint multimodal embedding space with 400 million image-text pair data. We leverage a CLIP-based multimodal embedding space (sound-image, text-image) trained on a large scale at the same time. This multimodal embedding space provides guidance when walking in the $\mathcal{W}+$ space. StyleCLIP [24] and TediGAN [42] leverage the representational power of CLIP for text-driven image editing. Recently, the prior knowledge of CLIP has been transferred to the audio-visual relationship, making it possible to develop various applications using CLIP in sound modality [21,41]. In this paper, we exploit the prior knowledge of audio-visual to generate a complete video related to sound. Furthermore, we introduce a frame generator that finds the desired trajectory from the sound. The frame

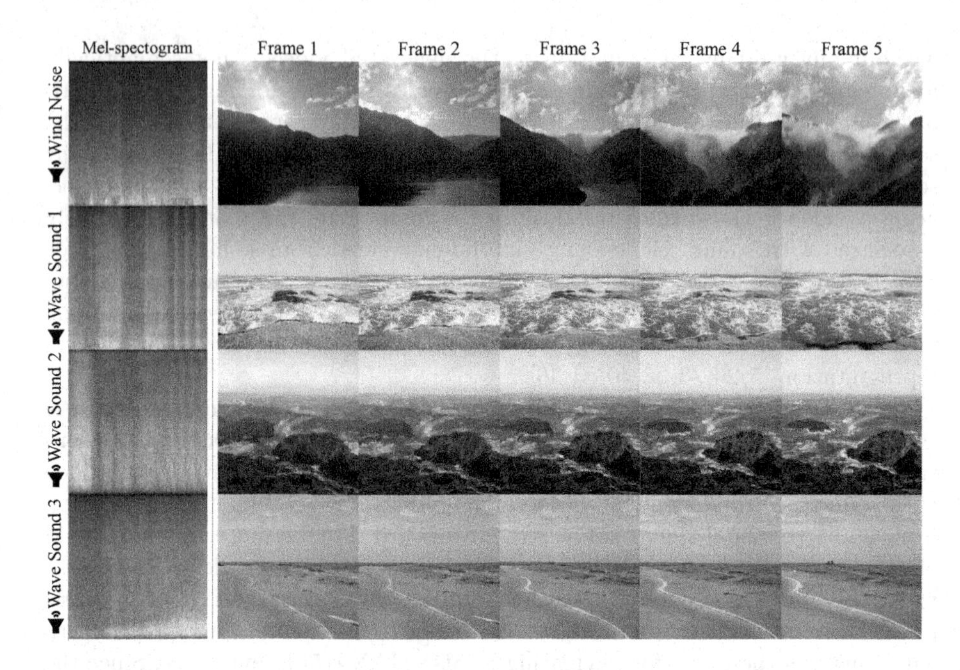


**Fig. 1.** An example of sound-guided semantic video generation. Our method generates semantic video reflecting temporal context using audio as an input feature.

generator predicts the latent code of the next time by using audio as a constraint. StyleGAN's style latent vector has a hierarchical representation, and our frame generator which is divided into three layers predicts coarse, mid, and fine style, respectively. Therefore, our proposed method provides nonlinear guidance for each time step beyond the linear interpolation of latent vectors (see Fig. 2). The latent code of our proposed method travels in the guidance of the audio in the latent space of StyleGAN with generating audio semantic meaningful and diverse output. In contrast, the general interpolation method travels in the latent space in one direction without any guide at regular intervals.

In this paper, our experimental results demonstrate that the proposed method supports a variety of sound sources and generates video that is a better reflection of given audio information. The existing audio-visual benchmark datasets [4,16,18] have limitations in high-fidelity video generation. We release a new high-resolution landscape video dataset, as the existing datasets are not designed for high-fidelity video generation [4] or do not support sound-video pairs [16,18]. Our dataset is effective for experimenting with high-resolution audio-visual generation models.

Our main contributions are listed as follows:

- We propose a novel method for mapping audio semantics into the StyleGAN latent space by matching multi-modal semantics with CLIP space.
- We introduce a framework for semantic-level video generation solely based on the given audio. We demonstrate the effectiveness of the proposed method by

outperforming the quality of generated videos from state-of-the-art methods in the sound-guided video generation task.

- Our framework not only determines the direction and size of movement of the latent code in the StyleGAN latent space, but also can generate rich style video given various sound attributes.
- We provide the new high-resolution landscape video dataset with audio-visual pairs for this task.

## 2 Related Work

**StyleGAN Latent Space Analysis.** StyleGAN [15] has a semantically rich latent space, so many studies [24,27,42] have analyzed the latent space of GAN. Most of the existing StyleGAN inversion modules minimize the difference between a latent vector and an embedding vector that encodes an image generated from the vector. Richardson *et al.* [27] have effectively inverted high-quality face images using prior knowledge related to the faces such as LPIPS [45], perceptual loss and pretrained ArcFace [8]. Patashnik *et al.* [24] defines a style latent mapper network that projects a given latent code to match the meaning of the text prompt. The latent mapper of the study learns given a pair of image and text prompts sampled from noise. However, there is a limitation in that randomly sampled image cannot have continuous motion like a real video. Our goal is to indicate the direction in which the style latent code will move in an audio-visual multimodal embedding space. Our study embeds the input audio directly into the StyleGAN latent space. The sound-guided latent code enables video generation in the expanded audio-visual domain. CLIP [25] represents the powerful joint embedding space between image and text modalities with 400 million image-text pairs, and this joint embedding space provides direction for traversing the StyleGAN latent space. Recent studies [10,21,41] extend the modalities of CLIP to audio. Lee *et al.* [21] especially focused on audio-visual representation learning for image editing, and we also leverage that audio-visual multimodal space embedding for navigating the latent code. Our works differ from previous works in that we consider temporal style variation guided by the CLIP embedding space. Sound semantics generate images at each time step using the latent vector optimized by the CLIP guidance.

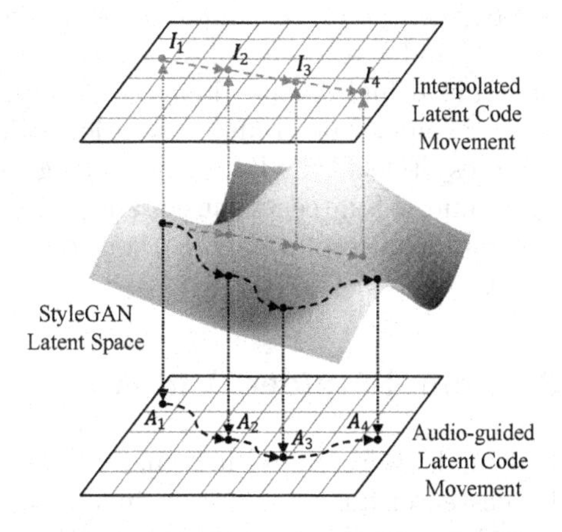


**Fig. 2.** Comparison of traversing method for the video generation. The top grid shows the conventional interpolated way, and the bottom grid shows our audio-guided latent code movement method on StyleGAN's latent space.

**Sound-Guided Video Generation.** Many studies use the temporal dynamics in sound as a source for vivid video generation [3,12,20]. There are mainly two approaches to generate video from sound as a source. The first is a conditional variational autoencoder method [17] that predicts the distribution of future video frames in the latent space. VAE-based method [3,20,43] mainly solve video prediction tasks that predict the next frame for a given frame. Among them, Chatterjee *et al.* [3] and Le *et al.* [20] consider the context of a scene with a non-verbal sound. Ji *et al.* [13] and Lahiri *et al.* [19] generate a video about facial expression with a verbal sound.

In this study, we enable the video generation task with a non-verbal sound condition that is more difficult than the verbal sound condition. GAN-based video generation is to sample the video from a noise vector [28,35,37,39]. Tian *et al.* [34] and Fox *et al.* [9] synthesize continuous videos with StyleGAN [15], a pre-trained high-fidelity image generator. Unlike previous studies, we consider a novel sound-guided high resolution video generation method. Jeong *et al.* [12] explores StyleGAN's latent space to generate a video. However, the domain of the sound is limited to music, and the guidance is not noticed by the user. We generate high-resolution video considering the semantics of sounds such as wind, raining, etc.

## 3 Sound-Guided Video Generation

Our model takes sound information as an input to generate a sequence of video frames accordingly, as shown in Fig. 3. For example, given a Laughing sound input, our model generates a video with a facial expression of laughing. To achieve this goal, our model needs two main capabilities. (i) The ability to understand the sound input and to condition it in the trained video generator. (ii) The ability to generate a video sequence that is perceptually realistic and is temporally consistent.

We propose that such capabilities can be learned via our two novel modules: (1) *Sound Inversion Encoder*, which learns a mapping from the sound input to a latent code in the (pre-trained) StyleGAN [15] latent space (Sect. 3.1). (2) *StyleGAN-based Video Generator*, which is conditioned on the sound-guided latent code and generates video frames accordingly (Sect. 3.2). Furthermore, we leverage the representation power of the CLIP [25]-based multimodal (image, text, and audio) joint embedding space, which regularizes perceptual consistency between the sound input and the generated video.

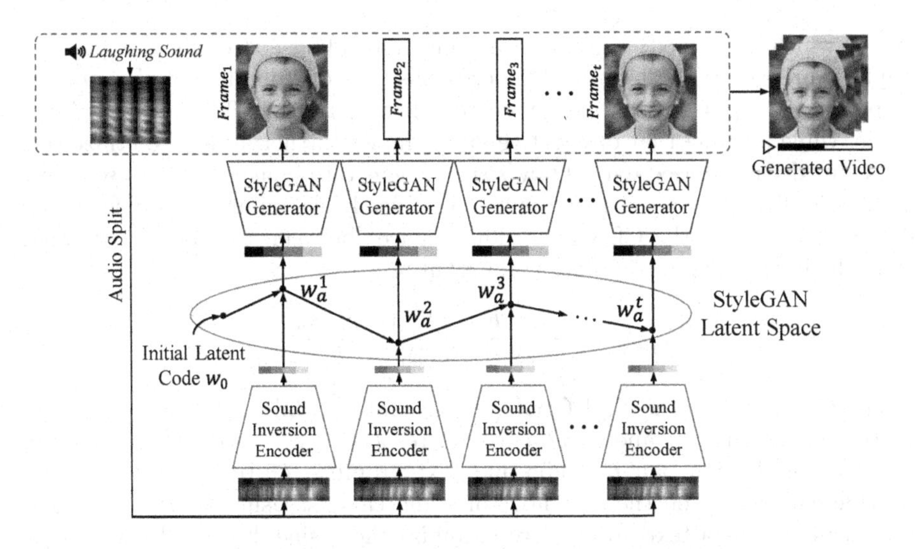


**Fig. 3.** An overview of our proposed sound-guided video generation model. Our model consists of two main modules: (i) Sound Inversion Encoder (Sect. 3.1), which takes a sequence of audio inputs as an input and outputs a latent code to generate video frames. (ii) StyleGAN-based Video Generator (Sect. 3.2), which generates temporally consistent and perceptually realistic video frames conditioned on the sound input.

## 3.1 Inverting Sound into the StyleGAN Latent Space

As shown in Fig. 4, our Sound Inversion Encoder takes as an input Mel-spectrogram Acoustic features and outputs a latent feature $\mathbf{w}_a \in \mathcal{W}+$ in the pre-trained StyleGAN feature space $\mathcal{W}+$. The sound-conditioned latent feature $\mathbf{w}_a$ is then augmented by the element-wise summation with the randomly sampled latent code $\mathbf{w} \in \mathcal{W}+$, yielding a sound-guided latent code $\hat{\mathbf{w}}_a \in \mathcal{W}+$. Conditioned on $\hat{\mathbf{w}}_a$, we generate an image, which maintains the content of the original image (generated with the random latent code $\mathbf{w}$) but its style is transferred according to the semantic of sound. Formally,

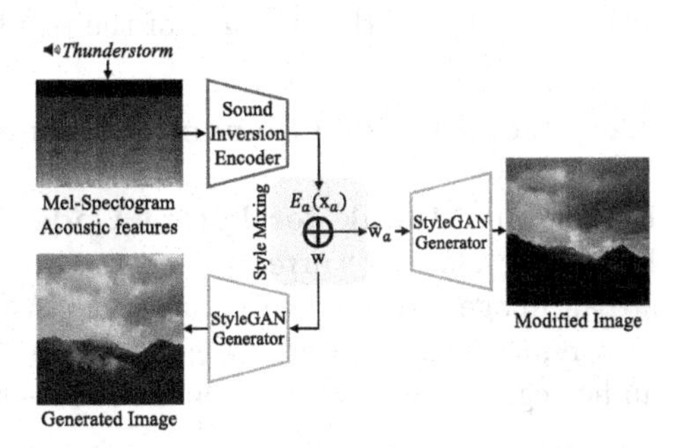


**Fig. 4.** Audio Inversion Encoder. From a given sound input, sound-guided latent code is obtained by the elementwise summation with the randomly sampled latent code.

$$\hat{\mathbf{w}}_a = E_a(\mathbf{x}_a) + \mathbf{w}. \tag{1}$$

where $E_a(\cdot)$ denotes our Sound Inversion Encoder given a sound input $\mathbf{x}_a$.

**Matching Multimodal Semantics via CLIP Space.** Lee *et al.* [21] introduced an extended CLIP-based multi-modal (image, text, and audio) feature space, which is trained to produce a joint embedding space where a positive triplet pair (e.g., audio input: "thunderstorms", text: "thunderstorm", and corresponding image) are mapped close together in the CLIP-based embedding space, while pushing that of negative pair samples further away. We utilize this pre-trained CLIP-based embedding space to generate images that are semantically well-aligned with the sound input. Specifically, we minimized the following cosine distance (in the CLIP embedding space) between the image generated from the latent code $\hat{\mathbf{w}}_a$ and an audio input $\mathbf{x}_a$.

$$\mathcal{L}_{\text{CLIP}}^{(a \leftrightarrow v)} = 1 - \frac{F_v(G(\hat{\mathbf{w}}_a)) \cdot F_a(\mathbf{x}_a)}{||F_v(G(\hat{\mathbf{w}}_a))||_2 \cdot ||F_a(\mathbf{x}_a)||_2}. \tag{2}$$

where $G(\cdot)$ denotes the StyleGAN [15] generator, while $F_v(\cdot)$ and $F_a(\cdot)$ are CLIP's image encoder and audio encoder, respectively. A similar loss function could be used for a pair of audio and text prompts. CLIP's text embedding provide constraints on the semantics of audio chunks. Using audio labels (e.g. thunderstorm) as a text prompt, we minimize the cosine distance between representations for image and text in the CLIP embedding space.

Lastly, we use $l_2$ distance between the latent code $\hat{\mathbf{w}}_a$ and their averaged latent code, i.e. $\bar{\mathbf{w}}_a = \sum_t \hat{\mathbf{w}}_a^t$, to explicitly constrain the generated sequence of images to share similar contents and semantics over time. Ultimately, we minimize the following loss to train our Sound Inversion Encoders.

$$\mathcal{L}_{\text{enc}} = \mathcal{L}_{\text{CLIP}}^{(a \leftrightarrow v)} + \mathcal{L}_{\text{CLIP}}^{(a \leftrightarrow t)} + \lambda_b||\hat{\mathbf{w}}_a - \bar{\mathbf{w}}_a||_2^2. \tag{3}$$

where $\lambda_b$ controls the strength of the regularization term.

### 3.2 Sound-Guided Semantic Video Generation

**Recurrent Module for Latent Code Sequence Generation.** As shown in Fig. 5, our model recurrently generates latent code at each timestep to generate an image sequence with the StyleGAN [15] generator. Formally, we train a recurrent neural network $E_{\text{RNN}}(\cdot)$ that outputs $\hat{\mathbf{w}}_a^t$ conditioned on the previous audio segment $\mathbf{x}_a^{t-1}$ and the hidden state $\mathbf{h}_t$ at timestep $t$.

$$\hat{\mathbf{w}}_a^t, \mathbf{h}_t = E_{\text{RNN}}(\mathbf{h}_{t-1}, E_a(\mathbf{x}_a^{t-1}) + \hat{\mathbf{w}}_a^{t-1}), \tag{4}$$

where $E_{\text{RNN}}(\cdot)$ comprises a fully-connected layers to recurrently yield the hidden state $\mathbf{h}_t$ and the current latent code $\hat{\mathbf{w}}_a^t$.

**Multiple Recurrent Blocks for Rich Style Information.** The input latent vectors are divided into three groups (0–2, 3–5, and 6–15 layers) and fed to different recurrent networks to generate different levels of detail. These three latent groups represent different styles (coarse, mid, and fine features) and control the attributes in StyleGAN [24]. Each recurrent block predicts the latent code of the next time step:

$$\hat{\mathbf{w}}_a^t = (E_{\text{RNN}}^{\text{coarse}}(\hat{\mathbf{w}}_a^{t-1}), E_{\text{RNN}}^{\text{mid}}(\hat{\mathbf{w}}_a^{t-1}), E_{\text{RNN}}^{\text{fine}}(\hat{\mathbf{w}}_a^{t-1})), \tag{5}$$

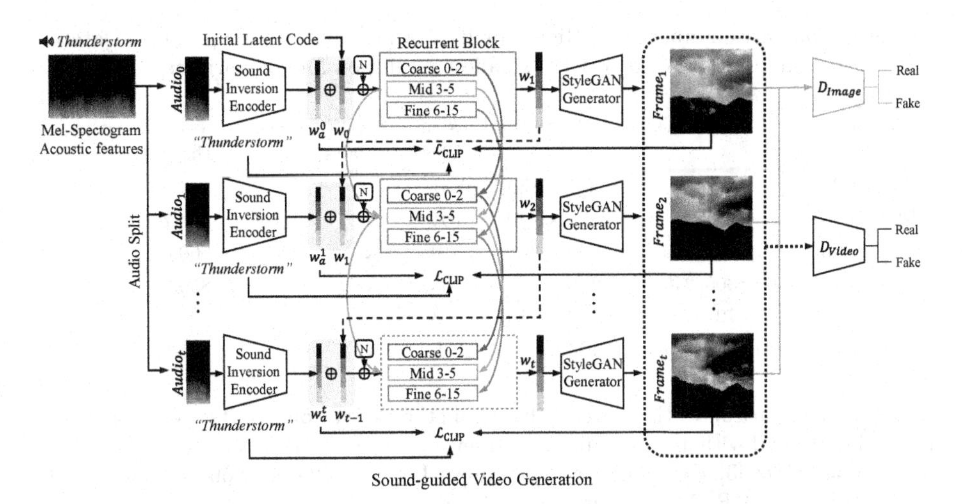


**Fig. 5.** An overview of our sound-guided video generation model, which consists of two main parts: (i) Sound Inversion Encoder, which iteratively generates sound-conditioned latent code $\mathbf{w}_a^t$ from corresponding audio time segments. (ii) StyleGAN-based Video Generator, which recurrently generates a video frame that is trained to be consistent with neighboring frames. Moreover, we train image and video discriminators adversarially to generate perceptually realistic video frames.

where $E_{\text{RNN}}^{\text{coarse}}(\cdot)$, $E_{\text{RNN}}^{\text{mid}}(\cdot)$, and $E_{\text{RNN}}^{\text{fine}}(\cdot)$ denote coarse, mid, and fine recurrent encoder-decoder network, respectively. Generated video $\tilde{\mathbf{v}}$ is a sequence of images synthesized from each latent code as follows:

$$\tilde{\mathbf{v}} = [G(\hat{\mathbf{w}}_a^1), G(\hat{\mathbf{w}}_a^2), ..., G(\hat{\mathbf{w}}_a^T)]. \tag{6}$$

where $T$ is the total sequence length of the video.

**Adversarial Image and Video Discriminators.** As shown in Fig. 5, we train a video discriminator $D_V$ adversarially by forwarding the generated video into the discriminator, which determines whether the input video is real or synthesized. Specifically, following MoCoGAN-HD [34], our video discriminator is based on the architecture of PatchGAN [11]. An input video (i.e. a real or fake example) is divided into small 3D patches, which are then classified as real or fake. The average response is used as the final output. We thus use the following adversarial loss $\mathcal{L}_{D_V}$:

$$\mathcal{L}_{D_V} = \mathbb{E}[\log D_V(\mathbf{v})] + \mathbb{E}[1 - \log D_V(\tilde{\mathbf{v}})], \tag{7}$$

where $\mathbf{v}$ and $\tilde{\mathbf{v}}$ are the real and fake example, respectively. Additionally, we also apply an image discriminator $D_I$, which similarly trains the model adversarially to determine whether the input image is real or fake on the time axis. Concretely, we optimize the following loss function:

$$\mathcal{L}_D = \mathcal{L}_{D_V} + \mathcal{L}_{D_I} = \mathcal{L}_{D_V} + \mathbb{E}[\log D_I(\mathbf{v})] + \mathbb{E}[1 - \log D_I(\tilde{\mathbf{v}})]. \tag{8}$$

**Table 1.** Comparison table of the High Fidelity Audio-visual Landscape Video Dataset and other datasets. (†): We report the smallest resolution for comparisons.

| Dataset | # of Videos | Resolution† | # of Classes | Audio-Video Pairs | Total Length (hours) |
|---|---|---|---|---|---|
| HMDB51 [18] | 6,766 | 340 × 256 | 51 | × | 4.9 |
| UCF-101 [31] | 13,320 | 320 × 240 | 101 | ✓ | 26.7 |
| VGG-Sound [4] | 200,000 | 640 × 360 | 310 | ✓ | 560.0 |
| Kinetics-400 [16] | 306,245 | 640 × 360 | 400 | ✓ | 850.7 |
| Sub-URMP [22] | 72 | 1920 × 1080 | 13 | ✓ | 1.0 |
| Landscape (ours) | 9,280 | 1280 × 720 | 9 | ✓ | 25.8 |

**Table 2.** Comparison to the state-of-the-art methods. We compare two version of our method (with and without sound inputs) to several state-of-the-art methods. We reproduce Sound2Sight [3], CCVS [20], and TräumerAI [12] as a baseline on two benchmark datasets: Sub-URMP [22] and our created Landscape.

| Method | *use* sound inputs | *use* the first frame | Sub-URMP [22] | | Landscape | |
|---|---|---|---|---|---|---|
| | | | IS (↑) | FVD (↓) | IS (↑) | FVD (↓) |
| Sound2Sight [3] | ✓ | ✓ | 1.64 ± 0.11 | 282.48 | 1.55 ± 0.48 | 311.55 |
| CCVS [20] | ✓ | ✓ | 2.06 ± 0.07 | 274.01 | 1.78 ± 0.64 | 305.40 |
| TräumerAI [12] | ✓ | - | 1.02 ± 0.14 | 350.80 | 1.07 ± 0.58 | 732.63 |
| Ours (*w/o sound inputs*) | - | - | 2.99 ± 0.48 | 272.27 | 1.68 ± 0.43 | 305.91 |
| Ours | ✓ | - | **3.05 ± 0.42** | **271.03** | **1.82 ± 0.26** | **291.88** |

**Loss Function.** Our model can be trained end-to-end by optimizing the following objective:

$$\min_{\theta_G} \max_{\theta_{D_V}} \mathcal{L}_{D_V} + \min_{\theta_G} \max_{\theta_{D_I}} \mathcal{L}_{D_I} + \min_{\theta_{E_a}} \lambda_{\text{enc}} \mathcal{L}_{\text{enc}}, \tag{9}$$

where $\lambda_{\text{enc}}$ controls the strength of the loss term.

# 4 Experiments

## 4.1 Datasets

There exist few publicly available Audio-Video paired datasets (e.g. UCF-101 [31], VGG-Sound [4], Kinetics-400 [16]), but they mostly support low-resolution video and contain a lot of noise in the audio. (see Table 1). To the best of our knowledge, there is the Sub-URMP (University of Rochester Musical Performance) dataset [22] that only provides pairs of high-fidelity audio-video. This dataset provides 72 video-audio pairs ($\approx$ 1 h in total) from recordings of 13 kinds of instruments played by different orchestra musicians. This dataset would be a good starting point to evaluate the performance of video generation models, but it only focuses on orchestra playing scenes with a limited number of video

**Table 3.** Quantitative evaluation. (a) Semantic Consistency between a given audio and generated video. (b) Ablation study of CLIP Loss. We evaluate the Inversion Score on the LHQ dataset.

(a) Semantic Consistency.

| Model | Similarity (t ↔ v)↑ | Similarity (a ↔ v)↑ |
|---|---|---|
| Sound2Sight [3] | 0.2491 | 0.1771 |
| CCVS [20] | 0.2514 | 0.1764 |
| TräumerAI [12] | 0.1932 | 0.1416 |
| Ours | **0.2556** | **0.1897** |

(b) The LHQ Dataset Inversion Score.

| Model | LPIPS ↓ | MSE ↓ |
|---|---|---|
| $\mathcal{L}_2 + \mathcal{L}_{reg}$ | 0.648 | 0.052 |
| $\mathcal{L}_2 + \mathcal{L}_{\text{LPIPS}} + \mathcal{L}_{reg}$ | 0.468 | 0.070 |
| $\mathcal{L}_2 + \mathcal{L}_{\text{LPIPS}} + \mathcal{L}_{reg} + \mathcal{L}_{\text{CLIP}}$ | **0.432** | **0.048** |

clips, limiting training efficiency. Thus, we create a new dataset that provides a high-fidelity audio-video paired dataset on landscape scenes.

**High Fidelity Audio-Video Landscape Dataset (Landscape).** We collect 928 high-resolution (at least 1280 × 720) video clips, where each clip is divided into 10 (non-overlapped) different clips of 10 s each. Overall, the total number of video clips available is 9,280, and the total length is approximately 26 h. Our Landscape dataset contains 9 different scenes, such as thunderstorms, waterfall burbling, volcano eruption, squishing water, wind noise, fire crackling, raining, underwater bubbling, and splashing water. We provide details of our created Landscape dataset in the supplemental material.

### 4.2 Comparison to Existing Sound-Guided Video Generation

**Quantitative Evaluation.** We use the latest VAE-based models as baselines, including Sound2Sight [3], CCVS [20] and StyleGAN-based model TräumerAI [12]. For our model and TräumerAI, we first pre-train StyleGAN on the high fidelity benchmark datasets (the Sub-URMP [22] and the LHQ datasets [30]) then train to navigate the latent space with the fixed image generator. The videos are generated with randomly sampled initial frames. In contrast, for Sound2Sight and CCVS, the first frames are provided. Table 2 shows that our approach produces the best quality results, and sound information is effective for video generation.

Table 3a demonstrates that our method produces more visually correlated videos from sound than other methods. To measure if the generated videos are semantically related to sound, we compare the cosine similarity between text-audio and video embedding. We obtain 512-dimensional video embeddings corresponding to the number of each frame with the CLIP [25] image encoder, and average them. Additionally, text embeddings are obtained from CLIP's text encoder, a 512-dimensional vector, and audio embeddings sharing CLIP space are obtained from Lee's [21] multi-modal embedding space. So leveraging the multi-modal embedding space helps to achieve semantic consistency in sound-guided video generation.

**Evaluation Metrics.** Evaluation of the quality of generated videos is known to be challenging. We first use the two widely-used quantitative metrics: Inception

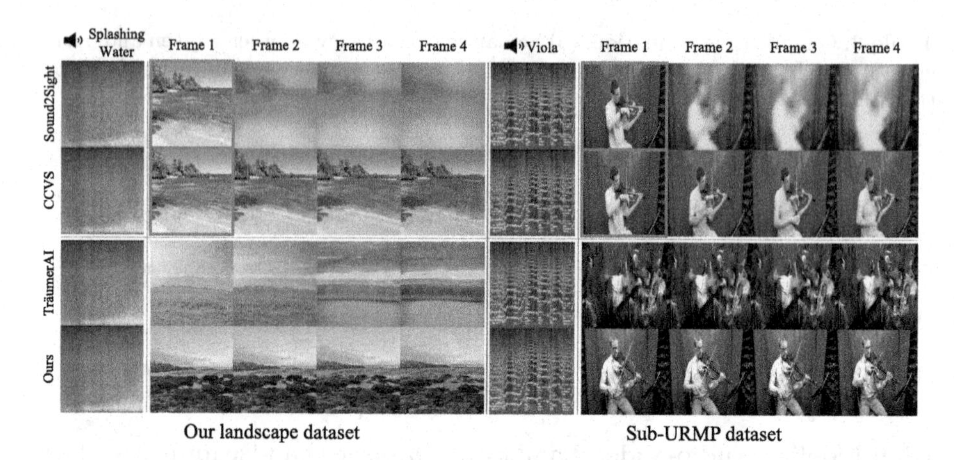


**Fig. 6.** Qualitative comparison between our sound-guided semantic video generation method and previous video generation results on our landscape dataset and Sub-URMP dataset [22]. Sound2Sight [3] and CCVS [20] (top) generate a video by conditioning on an image as the first frame highlighted in red, whereas TräumerAI [12], and our method (bottom) generate a video conditioned on audio input only. (Color figure online)

Score (IS) [29] and Fréchet Video Distance (FVD) [36]. The former is widely used to evaluate the outputs of GANs by measuring the KL-divergence between each image's label distribution and the marginal label distribution. Further, FVD quantifies the video quality by measuring the distribution gap between the real vs. synthesized videos in the latent space. We use an Inception3D [2] network, which is pre-trained on Kinetics-400 [16] and is fine-tuned on each benchmark accordingly.

**Qualitative Evaluation.** In Fig. 6, we visually compare the quality of the generated video with other models including Sound2Sight [3], CCVS [20] and TräumerAI [12] for the two benchmark datasets (Sub-URMP [22] and our landscape dataset). We found that TräumerAI is not able to produce any realistic video at all. Also, the generated video by Sound2Sight and CCVS is mostly distorted. Our method, on the other hand, produces high fidelity videos.

### 4.3 Ablation Studies

**CLIP Loss for StyleGAN Inversion.** In this study, we show that CLIP [25] prior knowledge is helpful for StyleGAN inversion. Table 3b compares the mean squared error (MSE) and LPIPS [45] between the original and the reconstruction image from the inversion module using the LHQ dataset. By minimizing the cosine distance between CLIP embeddings, the inversion reconstruction performance is improved in landscape images.

**Effect of Multiple Recurrent Blocks for Rich Style Information.** In sound-guided video generation, multiple recurrent blocks control the StyleGAN

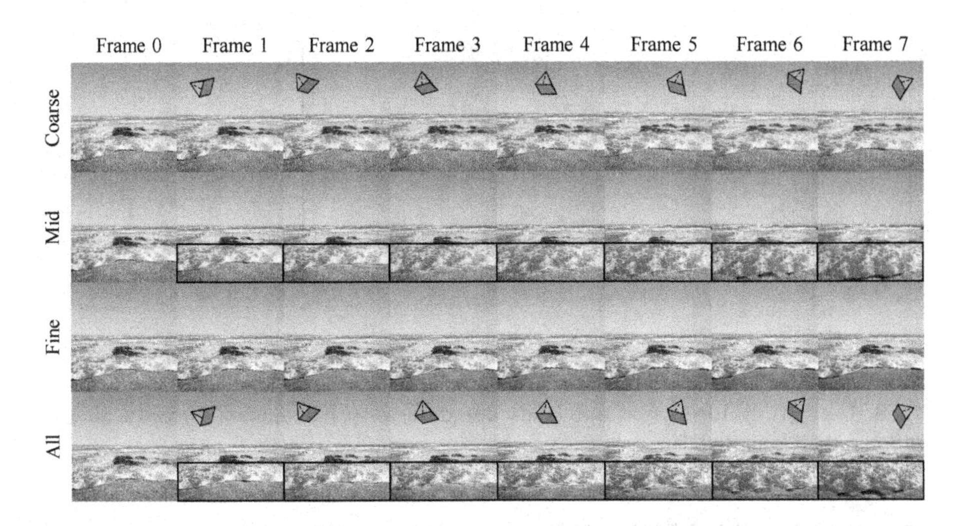


**Fig. 7.** Ablation of multiple recurrent blocks for rich style information. The first row shows the video generation result when only the coarse recurrent block is applied, the second is mid, the third is the fine recurrent block, and the last is the entire block. The red box is the area where the style has changed. (Color figure online)

attribute. Figure 7 compares video generation using only one coarse, mid, and fine recurrent block. Each block contains diverse style information. Specifically, the coarse and middle recurrent blocks control viewpoint and semantically meaningful motion changes (such as a wave strike) of the scene. Finally, the fine recurrent block handles fine texture changes.

### 4.4 User Study

In order to evaluate our proposed method, we request one hundred Amazon Mechanical Turk (AMT) participants. We show participants three types of generated videos that are generated by Sound2Sight [3], CCVS [20], TräumerAI [12], our model and ground truth. Participants answer the following questionnaire based on five-point Likert scale: (i) Realness - *Please evaluate the realness of the video* ("1 - very unrealistic" to "5 - very realistic") and (ii) Naturalness - *Which video generation result better expresses the target attribute?* (iii) Semantic Consistency - *Please evaluate the semantic consistency between video and audio* ("1 - very inconsistent" to "5 - very consistent").

The survey procedure is as follows. First, we have participants watch ground truth videos and measure realness and semantic consistency on a scale of 1 to 5. Then have participants measure realness against the results generated by other models. To evaluate naturalness, we ask participants to choose which of the videos generated by each model best expresses the target attribute. As shown in Fig. 8, our method significantly outperforms other state-of-the-art approaches (*Realness*, *Naturalness*, *Semantic Consistency*).

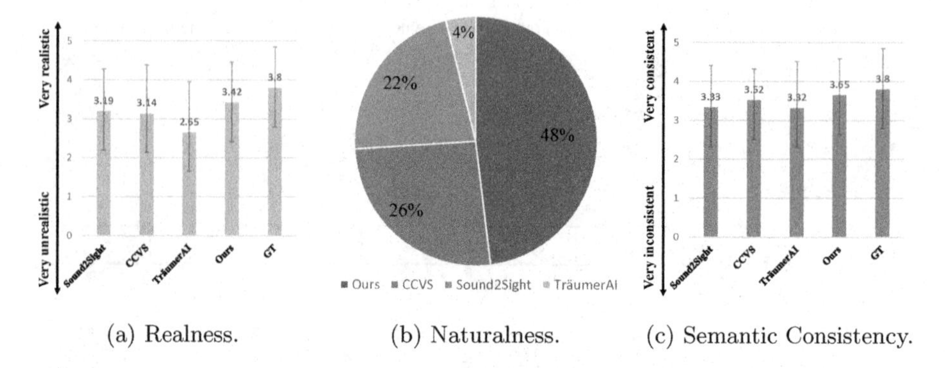


(a) Realness. (b) Naturalness. (c) Semantic Consistency.

**Fig. 8.** User Study Results on our landscape dataset. (a) Realness. (b) Naturalness. (c) Semantic Consistency.

### 4.5 Sound-Guided Video Editing with Text Constraints

The combination of text and sound makes it possible to generate various videos with rich information. Figure 9 demonstrates that the combination of text-query constraints and sound results in a much more versatile style of video. Because our model shares the CLIP [25] space trained with large-scale text and image pairs, video generation with text constraint is available. In addition, it is possible to apply styles of different levels with the style mixing technique. For video editing, we randomly sample multiple latent codes. The initial latent code for generating the first frame is the sample with the closest CLIP embedding cosine distance of the text prompt.

## 5 Discussion

**Limitations.** In our method, we sample the first frame from the StyleGAN [14, 15] latent space, and the space covers only the train domain. So, color change is rarely observed when the pre-trained StyleGAN and the video datasets have a domain gap. Still, we observe that our model is applicable to many cases such as sound-guided face video generation (see sup.).

In addition, in order for our model to learn to generate a video, the weights of the pre-trained image generator are required, which increases the training time. And maintaining the movement size and identity of the latent code predicted by the recurrent block has a tradeoff relationship (see sup.).

**Societal Impact.** Following this characteristic of our method, the user can generate a real video with the user's desire, and this usefulness allows the user to see the video that existed only in the user's imagination. However, although these features can provide a suitable output video to the user if the input video source is a work of art or is not ethically correct, the result produced may imply the contrary intentions of the original video creator or may raise ethical concerns.

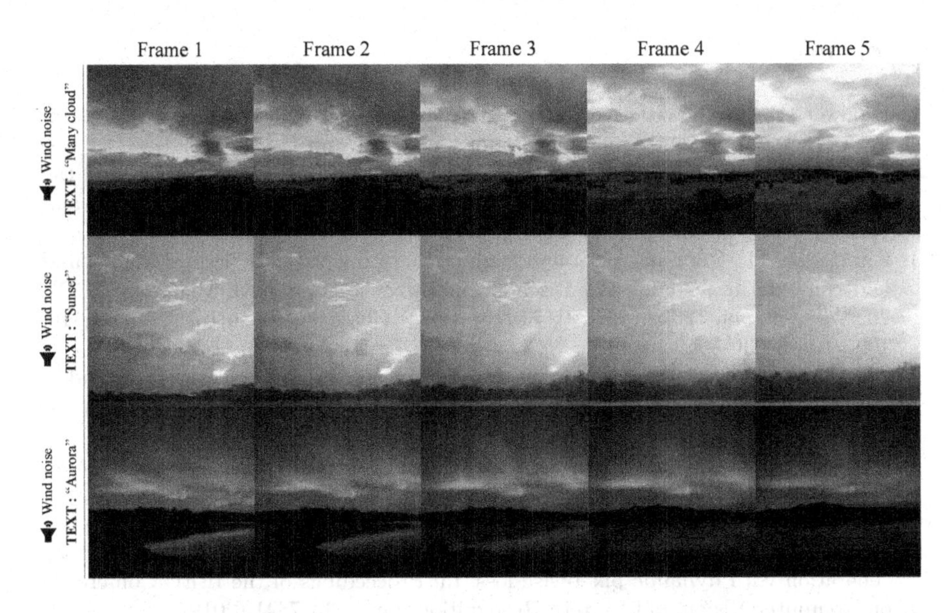


**Fig. 9.** Examples of sound-guided semantic video generation from sound with text prompts. Our model generates the videos by guiding sound from latent code containing text content. The time interval between each column is five frames.

## 6 Conclusion

This paper shows a new way to make sound-guided realistic videos by taking advantage of the multi-modal embedding space, which uses sound, images, and text. To be more specific, we use CLIP [25] space to encode the sound information into the StyleGAN [14,15] latent space, which allows generating videos that contain the corresponding source sound semantics. We design our model with the recurrent neural network since video consists of multiple frames and requires time-domain consistency. Multiple recurrent blocks temporarily represent rich style information (view point, sound-dependent frame's fine features). Additionally, we curate a new high-fidelity audio-video landscape dataset to validate our proposed method, which is superior to other methods. We demonstrate that our method qualitatively and quantitatively outperformed state-of-the-art methods for sound-guided video generation. The proposed model can be used in various applications such as video generation with text constraints.

**Acknowledgement.** This work is partially supported by Institute of Information & communications Technology Planning & Evaluation (IITP) grant funded by the Korea government(MSIT) (No. 2019-0-00079, Artificial Intelligence Graduate School Program(Korea University)). J. Kim is partially supported by the National Research Foundation of Korea grant (NRF-2021R1C1C1009608), Basic Science Research Program (NRF-2021R1A6A1A13044830), and ICT Creative Consilience program (IITP-2022-2022-0-01819). Any opinions, findings, and conclusions or recommendations expressed in this material are those of the authors and do not necessarily reflect the views of the funding agency.

## References


1. Brouwer, H.: Audio-reactive latent interpolations with StyleGAN. In: NeurIPS 2020 Workshop on Machine Learning for Creativity and Design (2020)
2. Carreira, J., Zisserman, A.: Quo Vadis, action recognition? a new model and the kinetics dataset. In: Proceedings of the IEEE Conference on Computer Vision and Pattern Recognition, pp. 6299–6308 (2017)
3. Chatterjee, M., Cherian, A.: Sound2Sight: generating visual dynamics from sound and context. In: Vedaldi, A., Bischof, H., Brox, T., Frahm, J.-M. (eds.) ECCV 2020. LNCS, vol. 12372, pp. 701–719. Springer, Cham (2020). https://doi.org/10.1007/978-3-030-58583-9_42
4. Chen, H., Xie, W., Vedaldi, A., Zisserman, A.: VGGSound: a large-scale audio-visual dataset. In: ICASSP 2020–2020 IEEE International Conference on Acoustics, Speech and Signal Processing (ICASSP), pp. 721–725. IEEE (2020)
5. Chen, L., et al.: Talking-head generation with rhythmic head motion. In: Vedaldi, A., Bischof, H., Brox, T., Frahm, J.-M. (eds.) ECCV 2020. LNCS, vol. 12354, pp. 35–51. Springer, Cham (2020). https://doi.org/10.1007/978-3-030-58545-7_3
6. Chen, L., Maddox, R.K., Duan, Z., Xu, C.: Hierarchical cross-modal talking face generation with dynamic pixel-wise loss. In: Proceedings of the IEEE Conference on Computer Vision and Pattern Recognition, pp. 7832–7841 (2019)
7. Das, D., Biswas, S., Sinha, S., Bhowmick, B.: Speech-driven facial animation using cascaded GANs for learning of motion and texture. In: Vedaldi, A., Bischof, H., Brox, T., Frahm, J.-M. (eds.) ECCV 2020. LNCS, vol. 12375, pp. 408–424. Springer, Cham (2020). https://doi.org/10.1007/978-3-030-58577-8_25
8. Deng, J., Guo, J., Xue, N., Zafeiriou, S.: Arcface: additive angular margin loss for deep face recognition. In: Proceedings of the IEEE/CVF Conference on Computer Vision and Pattern Recognition, pp. 4690–4699 (2019)
9. Fox, G., Tewari, A., Elgharib, M., Theobalt, C.: Stylevideogan: a temporal generative model using a pretrained stylegan. arXiv preprint arXiv:2107.07224 (2021)
10. Guzhov, A., Raue, F., Hees, J., Dengel, A.: AudioCLIP: extending clip to image, text and audio (2021)
11. Isola, P., Zhu, J.Y., Zhou, T., Efros, A.A.: Image-to-image translation with conditional adversarial networks. CVPR (2017)
12. Jeong, D., Doh, S., Kwon, T.: Träumerai: Dreaming music with stylegan. arXiv preprint arXiv:2102.04680 (2021)
13. Ji, X., et al.: Audio-driven emotional video portraits. In: Proceedings of the IEEE/CVF Conference on Computer Vision and Pattern Recognition, pp. 14080–14089 (2021)
14. Karras, T., et al.: Alias-free generative adversarial networks. Adv. Neural. Inf. Process. Syst. **34**, 852–863 (2021)
15. Karras, T., Laine, S., Aila, T.: A style-based generator architecture for generative adversarial networks. In: Proceedings of the IEEE/CVF Conference on Computer Vision and Pattern Recognition, pp. 4401–4410 (2019)
16. Kay, W., et al.: The kinetics human action video dataset. arXiv preprint arXiv:1705.06950 (2017)
17. Kingma, D.P., Welling, M.: Auto-encoding variational bayes. arXiv preprint arXiv:1312.6114 (2013)
18. Kuehne, H., Jhuang, H., Garrote, E., Poggio, T., Serre, T.: HMDB: a large video database for human motion recognition. In: Proceedings of the International Conference on Computer Vision (ICCV) (2011)

19. Lahiri, A., Kwatra, V., Frueh, C., Lewis, J., Bregler, C.: LipSync3D: data-efficient learning of personalized 3D talking faces from video using pose and lighting normalization. In: Proceedings of the IEEE/CVF Conference on Computer Vision and Pattern Recognition, pp. 2755–2764 (2021)
20. Le Moing, G., Ponce, J., Schmid, C.: Ccvs: context-aware controllable video synthesis. Adv. Neural. Inf. Process. Syst. **34**, 14042–14055 (2021)
21. Lee, S.H., et al.: Sound-guided semantic image manipulation. arXiv preprint arXiv:2112.00007 (2021)
22. Li, B., Liu, X., Dinesh, K., Duan, Z., Sharma, G.: Creating a multitrack classical music performance dataset for multimodal music analysis: challenges, insights, and applications. IEEE Trans. Multimedia **21**(2), 522–535 (2018)
23. Mesaros, A., Heittola, T., Virtanen, T., Plumbley, M.D.: Sound event detection: a tutorial. IEEE Signal Process. Mag. **38**(5), 67–83 (2021). https://doi.org/10.1109/MSP.2021.3090678
24. Patashnik, O., Wu, Z., Shechtman, E., Cohen-Or, D., Lischinski, D.: StyleCLIP: text-driven manipulation of stylegan imagery. In: Proceedings of the IEEE/CVF International Conference on Computer Vision (ICCV), pp. 2085–2094 (2021)
25. Radford, A., et al.: Learning transferable visual models from natural language supervision. Image, **2**, T2 (2021)
26. Richard, A., Lea, C., Ma, S., Gall, J., de la Torre, F., Sheikh, Y.: Audio- and gaze-driven facial animation of codec avatars. In: Proceedings of the IEEE/CVF Winter Conference on Applications of Computer Vision (WACV), pp. 41–50 (2021)
27. Richardson, E., et al.: Encoding in style: a StyleGAN encoder for image-to-image translation. In: Proceedings of the IEEE/CVF Conference on Computer Vision and Pattern Recognition, pp. 2287–2296 (2021)
28. Saito, M., Matsumoto, E., Saito, S.: Temporal generative adversarial nets with singular value clipping. In: Proceedings of the IEEE International Conference on Computer Vision, pp. 2830–2839 (2017)
29. Salimans, T., Goodfellow, I., Zaremba, W., Cheung, V., Radford, A., Chen, X.: Improved techniques for training gans. Adv. Neural Inf. Process. Syst. **29** (2016)
30. Skorokhodov, I., Sotnikov, G., Elhoseiny, M.: Aligning latent and image spaces to connect the unconnectable. In: Proceedings of the IEEE/CVF International Conference on Computer Vision (ICCV), pp. 14144–14153 (2021)
31. Soomro, K., Zamir, A.R., Shah, M.: Ucf101: a dataset of 101 human actions classes from videos in the wild. arXiv preprint arXiv:1212.0402 (2012)
32. Suwajanakorn, S., Seitz, S.M., Kemelmacher-Shlizerman, I.: Synthesizing Obama. ACM Trans. Graph. (TOG) **36**, 1–13 (2017)
33. Thies, J., Elgharib, M., Tewari, A., Theobalt, C., Nießner, M.: Neural voice puppetry: audio-driven facial reenactment. In: Vedaldi, A., Bischof, H., Brox, T., Frahm, J.-M. (eds.) ECCV 2020. LNCS, vol. 12361, pp. 716–731. Springer, Cham (2020). https://doi.org/10.1007/978-3-030-58517-4_42
34. Tian, Y., et al.: A good image generator is what you need for high-resolution video synthesis. In: International Conference on Learning Representations (2021). https://openreview.net/forum?id=6puCSjH3hwA
35. Tulyakov, S., Liu, M.Y., Yang, X., Kautz, J.: MoCoGAN: decomposing motion and content for video generation. In: Proceedings of the IEEE Conference on Computer Vision and Pattern Recognition, pp. 1526–1535 (2018)
36. Unterthiner, T., van Steenkiste, S., Kurach, K., Marinier, R., Michalski, M., Gelly, S.: Towards accurate generative models of video: A new metric & challenges. arXiv preprint arXiv:1812.01717 (2018)

37. Vondrick, C., Pirsiavash, H., Torralba, A.: Generating videos with scene dynamics. Adv. Neural. Inf. Process. Syst. **29**, 613–621 (2016)
38. Wang, T.C., Mallya, A., Liu, M.Y.: One-shot free-view neural talking-head synthesis for video conferencing. In: Proceedings of the IEEE Conference on Computer Vision and Pattern Recognition (2021)
39. Wang, Y., Bilinski, P., Bremond, F., Dantcheva, A.: G3AN: disentangling appearance and motion for video generation. In: Proceedings of the IEEE/CVF Conference on Computer Vision and Pattern Recognition, pp. 5264–5273 (2020)
40. Wu, H., Jia, J., Wang, H., Dou, Y., Duan, C., Deng, Q.: Imitating arbitrary talking style for realistic audio-driven talking face synthesis. In: Proceedings of the 29th ACM International Conference on Multimedia, pp. 1478–1486 (2021)
41. Wu, H.H., Seetharaman, P., Kumar, K., Bello, J.P.: Wav2clip: Learning robust audio representations from clip (2021)
42. Xia, W., Yang, Y., Xue, J.H., Wu, B.: TediGAN: text-guided diverse face image generation and manipulation. In: Proceedings of the IEEE/CVF Conference on Computer Vision and Pattern Recognition, pp. 2256–2265 (2021)
43. Yan, W., Zhang, Y., Abbeel, P., Srinivas, A.: VideoGPT: video generation using VQ-VAE and transformers. arXiv preprint arXiv:2104.10157 (2021)
44. Yi, R., Ye, Z., Zhang, J., Bao, H., Liu, Y.J.: Audio-driven talking face video generation with learning-based personalized head pose (2020)
45. Zhang, R., Isola, P., Efros, A.A., Shechtman, E., Wang, O.: The unreasonable effectiveness of deep features as a perceptual metric. In: CVPR (2018)
46. Zhou, Y., Han, X., Shechtman, E., Echevarria, J., Kalogerakis, E., Li, D.: MakeitTALK: speaker-aware talking-head animation. ACM Trans. Graph. **39**(6), 1–5 (2020)

# Inpainting at Modern Camera Resolution by Guided PatchMatch with Auto-curation

Lingzhi Zhang[1(✉)], Connelly Barnes[2], Kevin Wampler[2], Sohrab Amirghodsi[2], Eli Shechtman[2], Zhe Lin[2], and Jianbo Shi[1]

[1] University of Pennsylvania, Philadelphia, USA
zlz@seas.upenn.edu
[2] Adobe Research, Cambridge, USA

**Abstract.** Recently, deep models have established SOTA performance for low-resolution image inpainting, but they lack fidelity at resolutions associated with modern cameras such as 4K or more, and for large holes. We contribute an inpainting benchmark dataset of photos at 4K and above representative of modern sensors. We demonstrate a novel framework that combines deep learning and traditional methods. We use an existing deep inpainting model LaMa [27] to fill the hole plausibly, establish three guide images consisting of structure, segmentation, depth, and apply a multiply-guided PatchMatch [1] to produce eight candidate upsampled inpainted images. Next, we feed all candidate inpaintings through a novel curation module that chooses a good inpainting by column summation on an $8 \times 8$ antisymmetric pairwise preference matrix. Our framework's results are overwhelmingly preferred by users over 8 strong baselines, with improvements of quantitative metrics up to *7.4 times* over the best baseline LaMa, and our technique when paired with 4 different SOTA inpainting backbones improves each such that ours is overwhelmingly preferred by users over a strong super-res baseline.

**Keywords:** Inpainting · PatchMatch

## 1 Introduction

Image inpainting involves removing a region and replacing it with new pixels so the modified photo is visually plausible. We develop a method that can take an off-the-shelf low-res inpainting deep model and extend it to modern camera resolutions. Applied to LaMa [27], this yields a new a SOTA method for inpainting at modern camera resolutions that dramatically outperforms all existing models.

Traditional patch-based synthesis approaches, such as Wexler et al. [33], Barnes et al. [1], and Darabi et al. [6] were used for high-quality image inpainting at arbitrary resolutions. Recently, the state-of-the-art for *low resolution*

**Supplementary Information** The online version contains supplementary material available at https://doi.org/10.1007/978-3-031-19790-1_4.

© The Author(s), under exclusive license to Springer Nature Switzerland AG 2022
S. Avidan et al. (Eds.): ECCV 2022, LNCS 13677, pp. 51–67, 2022.
https://doi.org/10.1007/978-3-031-19790-1_4

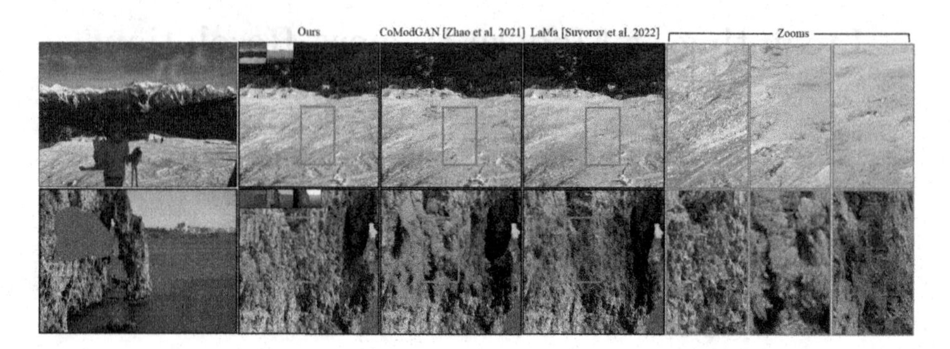


**Fig. 1.** Inpainting at modern camera resolutions via guided PatchMatch (guides shown in inset) with a novel automatic curation module. Photos are 12 and 30 MP. Our result has *significantly* higher fidelity high-res detail than CoModGAN [44] and LaMa [27], the strongest baselines according to our user study, which were upsampled by Real-ESRGAN [31]. The guides on the top left of the second columns indicate the chosen guides for the specific image.

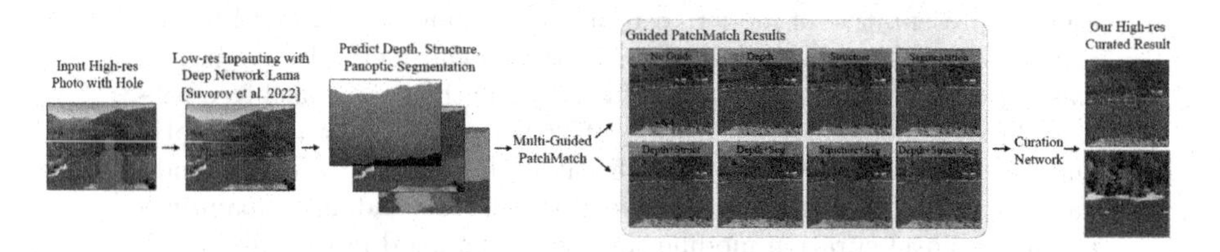


**Fig. 2.** Overview of our framework. See the intro for discussion of components.

image inpainting has been advanced by deep convolutional methods, such as Zeng et al. [40], Zhao et al. [44], and LaMa [27]. When we compared these methods, we noticed an interesting trade-off. Patch-based methods such as PatchMatch [1] can synthesize high-quality texture at arbitrarily high resolution but often make mistakes regarding structure and semantics. On the other hand, deep convolutional methods tend to generate inpaintings with plausible structure and semantics but less realistic textures. Worse still, most deep methods are limited to output resolutions such as 512 or 1024 pixels image along the max dimension[1], *much smaller* than modern camera resolutions which are typically 4K or above.

Based on our observations, we first collected a dataset of 1045 high-quality photos at 4K resolution or above representative of modern sensors and paired them with freeform and object-shaped hole masks. This dataset has approximately *2 orders of magnitude* more pixels per image than the dataset Places2 [45] that is commonly used for deep inpainting evaluation. Interestingly, photos in the Places2 dataset often used for evaluation have fewer pixels than the world's first digital camera [5], the FUJIX DS-1P, produced in 1988, so common evaluation practices are **more than 3 decades** behind sensor technology. We encourage the community to evaluate on our dataset, since datasets such as

[1] With the exception of HiFill [36] and LaMa [27], which we discuss in related work.

Places2 [45] are *not representative* of modern camera sensors, and which may therefore mislead researchers by giving improper guidance to the community.

After establishing the dataset, we next married the complementary advantages of patch-based synthesis in generating high texture quality at arbitrarily high resolution and deep networks, which can better predict structure and semantics. Our hybrid pipeline works as illustrated in Fig. 2. We first inpaint the hole region using an off-the-shelf deep network, such as LaMa [27], at 512 pixels along the long edge of the image. This method typically establishes a reasonable semantic layout and structure inside the hole. Still, as can be seen in Fig. 1, the texture quality is often poor if we zoom in. Next, we extract several guide images using existing methods: depth [37], structure [35], and panoptic segmentation [17]. We then use these guides in a multi-guided PatchMatch [1] implementation to perform patch-based image inpainting. Depending on the input photo, the best result may be obtained by different combinations of guides, so we produce a set of eight guided PatchMatch *candidate inpainting* results using multiple combinations of guides.

Besides the novelty of our overall framework, our key technical innovation is creating a **novel automatic curation module** whose architecture is shown in Fig. 3(a). This curation module automatically selects one good inpainted image out of the 8 candidates by an architecture suited for making subtle comparisons and contrasts between similar images. It does this by constructing an antisymmetric $8 \times 8$ matrix whose entries are populated in pairs by a network that learns to predict *antisymmetric pairwise human preferences* for every possible pair of candidate inpaintings. The preferred inpainted image is selected by taking the row with max column sum for the $8 \times 8$ matrix. Different from previous work such as RealismCNN [50] and LPIPS [43] and image quality assessment papers, our network estimates pairs of entries $M_{ij}$ and $M_{ji}$ at a time for an $8 \times 8$ matrix $M$ that is *antisymmetric* i.e. $M = -M^T$, and the pairwise structure is critical because it allows the network to differentiate between subtle feature differences. On pairs of images, our curation has near-human performance.

We conducted quantitative experiments with metrics commonly used to evaluate deep networks and multiple user studies for our evaluation. Our method dramatically outperforms state-of-the-art deep inpainting methods according to our quantitative metrics and user studies. User preferences are between 79% and 96% for our method, *every one of our 7* user study results in Tables 2, 3 and 4 has statistically significant preference for ours, and quantitative metrics improve by up to 7.4 times over strong baselines like LaMa [27] coupled with Real-ESRGAN [31]. Because our method uses patch-based synthesis, it can easily *scale to arbitrary resolution*: in the supplemental, we show results on images up to 62 megapixels. We show in our experiments that our method can be combined with four different deep inpainting baselines and improves **every** one of them for modern camera resolution inpainting according to quantitative metrics and user studies. Most convolutional models cannot scale to modern camera resolutions, except HiFill [36] and LaMa [27], which our evaluation shows we outperform.

Our paper contributes: (1) A benchmark dataset of 4K+ resolution images with holes appropriate for evaluating inpainting methods as they perform on modern camera sensors and evaluation of 9 methods on this dataset; (2) A novel high-resolution inpainting framework which shows that deep inpainting models need not give high quality results only on low resolutions. This framework has a choice of guides that perform well, which were nontrivial to choose and required extensive empirical investigations as discussed in Sect. 4.2, and is the first to explore combinations of multiple guides with a mechanism to automatically choose between them; (3) A curation module (seen for the first time in inpainting) with near-human performance that chooses a good inpainted image in a new way by populating an $8 \times 8$ antisymmetric preference matrix by comparing images one pair at a time and column-summing that matrix; (4) State-of-the-art results on the natural image inpainting task from both quantitative and user studies, with dramatic improvements over 8 strong baselines; (5) Our method can be combined with a variety of SOTA deep inpainting baselines and improves each of them for modern camera resolution inpainting. To enable reproducibility, we plan to release our full benchmark dataset including results of all methods and curation module test code and weights.

## 2 Related Work

**Patch-Based Synthesis and Inpainting.** Our approach uses the patch-based inpainting method of Wexler et al. [33] as implemented in the PatchMatch framework [1]. Inspired by image analogies [12], we added to this basic inpainting framework multiple guiding channels. Similar guided texture synthesis approaches have been used for stylizing renderings [9] and stylizing video by example [2,15], but the choice of guides is *nontrivial and application-dependent*: we performed many empirical investigations of alternative options to finally choose the guides used in this paper, as we discuss later in Sect. 4.2. Image Melding [6] improved patch-based synthesis results with geometric and photometric transformations and other means. We incorporated a gain and bias term from that paper to obtain better inpaintings for regions with smooth gradient changes in intensity. Kaspar et al. [16] performed texture synthesis using a patch-based optimization with guidance channels that preserve large-scale structures. Several papers explored initialization and search space constraints that use scene-level information that is different from ours [7,11,13].

Recently, some papers have explored learning good features by making PatchMatch differentiable [8,46]. Because of our modern camera resolution photos, multi-res and many iterations of filling used in PatchMatch [1], and because differentiable PatchMatch techniques track multiple candidate patches, the GPU memory requirements of applying a differentiable PatchMatch naively in our context are far beyond the capacity of today's GPUs. Thus, we use a non-differentiable PatchMatch with off-the-shelf guide features, and rely on the downstream curation module to pick good guide features.

**Neural Network Image Inpainting.** One significant advantage of convolutional neural network inpainting methods is that they can gain some understanding of semantic information such as global and local context, edges, and regions. Context encoders used a CNN to fill in a hole [24] by mapping an image with a square hole to a filled image. Iizuka et al. [14] used two discriminators to encourage the global and local appearance to both be reasonable. Yu et al. [38] proposed a contextual attention model which can effectively copy patches from outside the hole to inside the hole, however, it is limited in resolution because of the brute-force dense attention mechanism. Liu et al. [19] and Yu et al. [39] improved inpainting results by masking and gating convolutions, respectively. Zhao et al. [44] better address the situation of large holes by co-modulating the generative decoder using both a conditional and stochastic style representation.

A number of papers recently attempt to separate the scene in terms of edges [22,34], structure-texture separation, such as Ren et al. [26] which uses relative total variation (RTV) [35], or Liu et al. [20]. Our approach is inspired by these works and uses structure from RTV [35] and a panoptic segmentation map to guide PatchMatch.

A few recent works attempt to train neural networks that can output higher resolution results. ProFill [40] uses a guided upsampling network at a resolution of up to 1K on the long edge. HiFill [36] introduced a contextual residual aggregation module that weights residuals similar to those in a Laplacian pyramid [4], at up to 8K resolution, but according to our evaluation its results are worse than our method. The work LaMa [27] was trained on $256 \times 256$ patches but can be evaluated on images up to 2K.

**Visual Realism.** For pretraining our curation network, we use a similar idea as RealismCNN [50], which learns a visual realism score for composite images using a large dataset of images and a synthetic compositing pipeline. In our case, however, we generate millions of fake inpainted images using our pipeline, and pretrain a network to classify these images as fake and real images as real. Inspired by LPIPS [43] and image quality assessment (IQA) papers (e.g., [3, 28, 48]), we then fine-tune on real human preferences between pairs of synthetic inpainted images, but we use a different architecture and inference that involves predicting entries one pair at a time in an *antisymmetric matrix* and column-summing it. In contrast, LPIPS [43] learns a *symmetric* distance metric, full-reference IQA papers typically are also estimating some perceptual distance that is not antisymmetric, and no-reference IQA papers use a single image as input. See Fig. 3 for an illustration. Also related is Wang et al. [30], which trains a “universal” detector that can distinguish between CNN-generated images and real images.

## 3 Inpainting Dataset

To benchmark the performance of our method and 8 other methods, we collect 1045 high-quality images at modern sensor resolutions from two sources: DIV8K [10], and a test set portion of the dataset of photos taken by the authors and their collaborators mentioned earlier in Sect. 4.3. The photos span a diverse variety of scenes including indoors and outdoor urban photography including many architectural styles, nature and wildlife, macro photos, and human subject photos. The photos are all at least 4K pixels along the long edge. The mean megapixel count is 20 and the maximum is 45, excluding high-res panos that go up to 62 MP. DIV8K [10] has previously been used for super-resolution tasks and contains images with resolutions up to 8K. We chose all 583 images in DIV8K that were above 4K resolution on the long axis.

For each test image, with equal probability, we sample either a free-form mask or an instance mask using the same process and hole dataset as in ProFill [40]. Different from ProFill, because our photos are from modern sensors, we generate larger holes with a mean hole size of 2.3 MP: see the supplemental for details. No test images are seen during training. To enable reproducibility, we plan to release the testing dataset and results for all methods. Experimental results are shown later in Sect. 5.2.

## 4 Our Hybrid Synthesis Method

### 4.1 Multi-guided PatchMatch for Image Inpainting

PatchMatch [1] is an efficient randomized correspondence algorithm that is commonly used for patch-based synthesis for images, videos [2,15] and neural feature maps [18,47]. For image synthesis, PatchMatch repeatedly performs matching from the region being synthesized to a reference region: in our case, the matching is from the hole to the background region. We implement the method of Wexler et al. [33], with default hyperparameters for PatchMatch (e.g. $7 \times 7$ patches) and Wexler et al. One key advantage of PatchMatch is that it can efficiently scale to arbitrary resolutions while preserving high-texture fidelity. We extend this basic method in two ways: we allow multiple guides to be used (as in [12]), and we implement the gain and bias term from Image Melding [6].

For the multiple guides, we modify the sum of squared differences (SSD) used in Eq. (1) of PatchMatch. Instead of computing SSD over a 3 channel color image, we compute a weighted SSD over a $3+m$ channel image with channel weight $w_i$, where the first 3 channels are RGB and remaining channels are guides.

$$w_i = \begin{cases} w_c/3 & i \leq 3 \\ (1-w_c)/m & i > 3 \end{cases} \tag{1}$$

Using a separate validation set of images and manual inspection, we tried different settings of $w_c$ and empirically found the best setting is $w_c = 0.6$ if there is no structure guide, otherwise $w_c = 0.3$. Structure and RGB information are highly correlated so we decrease the RGB weight if there is a structure guide.

We implement the gain and bias term from Image Melding [6] because we find it helps improve inpainting quality in cases where there are subtle gradients, such as in the sky. We implement this term using RGB color space.

### 4.2 Guides Used for PatchMatch

We performed extensive investigations of many possible guides, and empirically settled on the three selected in this paper—structure, depth, and panoptic segmentation—because they worked the best on our validation set. Although our main novel low-level technical contribution is our curation module that we describe in the next section, the existence of appropriate guides, the empirical work needed to choose them and our overall novel multi-guided inpainting framework with curation also forms a high-level technical contribution.

Our pipeline works by first inpainting the hole using an off-the-shelf deep network LaMa [27] at 512 pixels on the long edge and then extracting three guide images: depth, segmentation, and structure. Depth is a useful cue as regions at a similar distance from the camera have usually more relevant content. This is especially important for slanted surfaces (e.g. the ground) where scale and focus properties of texture vary with depth. We also often want to sample semantically similar content, motivating the segmentation guide. However, the segmentation labeling might not be fine-grained enough to distinguish between different types of floor tiles or wall colors. Therefore we add the structure guide that captures the main edges and color regions in the image, and abstracts away the texture.

For depth prediction, we used a recent method by Yin et al. [37], which is retrained using a DPT [25] backbone to obtain better results. For panoptic segmentation prediction, we used PanopticFCN [17] retrained with a ResNeSt [41] backbone, which obtains a higher panoptic quality of 47.6 on the COCO val set. For structure prediction, we extract the structure using RTV [35], that was shown effective for inpainting by Ren et al. [26].

We performed extensive investigations of many alternative possible guides, and choose these three because they gave the best quality. We tried raw RGB color instead of structure, but this produced worse results due to patch matching becoming less flexible (see supplemental for details), Iterative Least Squares texture smoothing [21], but the latter had color bleeding artifacts that caused worse inpaintings, and Gaussian blurred RGB color and bilateral filtered RGB instead of structure, but these produced artifacts in the patch synthesis. We also tried different segmentation and depth modules and settled on the above ones because they gave better patch synthesis results.

Is there a universal 'best' guide for all images? We find that different choices of guides may give the best final inpainting result, depending on the input image and hole. Empirically, we find that the structure guide can help PatchMatch find patches with consistent edges and fine-scale structures, depth guides can be handy when the images have gradually changing depth such as outdoor photography, and segmentation guides can be useful when the segmentation map is accurate for preventing patches from leaking into the wrong semantic or object region. In general, we find no single guidance map by itself is sufficient to get the best quality of results, and multiple guidance maps are needed. In many applications it is desirable to have a fully automatic image inpainting process and we found that a fixed weighted combination of guides leads to sub-optimal results. Thus, as shown in Fig. 2, we generate a variety of results using different guides and use our curation module to automatically choose a good one. Specifically, we use a simple scheme where we can either enable or disable each of our 3 guides, so the total number of possibilities for guides are $2^3 = 8$, including the use of no guides, and then we select among those eight generated results. We show an example of how the guides influence the results in the supplemental.

### 4.3 Curation Module

Suppose you are given two inpainted candidate images and asked, *"Which of these do you prefer, or do you have no preference?"* We were inspired by how humans carry out this task: since the images usually look similar, we carefully compare and contrast between subtle differences in visual features to determine which image is slightly better overall. Our architecture thus is designed to enable such *subtle comparisons and contrasts* between features within inpainted images. In Fig. 3, we show an architecture diagram of our novel curation module and a comparison with other common architectures such as LPIPS [43] and no-reference IQA. Notably, our curation module has a different architecture and inference methodology that populates entries in a matrix $M$ that is *antisymmetric* due to the paired preference task and column-sums that matrix to determine the relative preference vector of one inpainting candidate as contrasted with others.

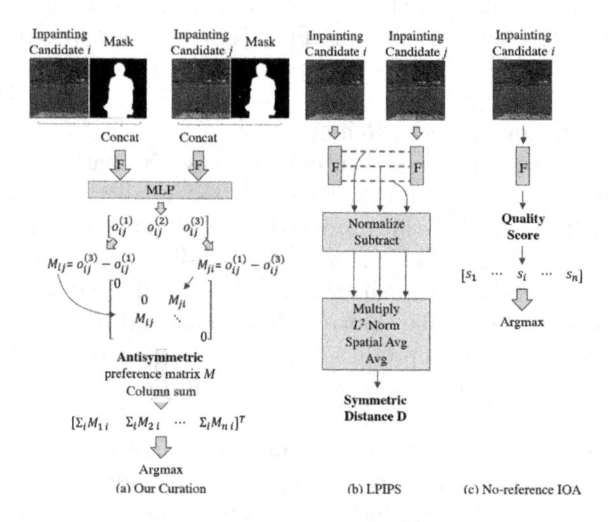


**Fig. 3.** Our curation architecture produces scores in an **antisymmetric** matrix $M$ that is column-summed. LPIPS [43] produces a **symmetric** distance score. Architectures used in RealismCNN [50] and no-reference IQA produce a score for each candidate image that is **independent** of the other candidates.

**Pretraining.** Inspired by RealismCNN [50], our curation network backbone $F$ is first pretrained to classify for a given image, whether it is a real image or a fake inpainted image as output by our pipeline. Our reasoning is that we observed that the initial pretrained network predictions have correlation with human perception of inpainting quality, and can allow the network to learn good features over a very large number of photos, but we need to later fine-tune on human preferences to obtain performance close to humans. We generated a dataset for pretraining the curation network by collecting 48229 diverse photos that are at least 2K in resolution, generating 10 synthetic holes for each, and then generating 8 guided

PatchMatch results. This results in more than 3 million inpainted images at 2K resolution. Our curation network backbone $F$ is EfficientNet-B0 [29]. We modify the input to take 4 input channels and pretrain the network from scratch. We trained for one epoch using a binary cross-entropy loss, after which we obtained train, test, val accuracies of 98.9%, 99.3%, 99.2%, respectively. Please see the supplemental for more details.

**Fine-Tuning on Human Preferences.** In nearly all cases, the pretrained curation network $F$ can easily distinguish between real and our inpainted results, but it was not specifically trained to predict human preference among different inpainted results. Therefore, we next fine-tune our network for a paired preference task. By subsampling the dataset described in the last subsection, and comparing sampled pairs of the 8 guided PatchMatch results against the others, we gather approximately 33000 synthetic inpainted image pairs for which we gather human preferences. We discuss in our supplemental lessons learned in gathering these preferences. For each image pair, our model works by featurizing each of the images in the pair through a shared-weight pretrained EfficientNet backbone $F$, and then using a small MLP to predict 3 classes that the human preference data contain: prefer left image, tie, prefer right image.

In contrast to perceptual distances such as LPIPS [43], we have a different task where our model predicts an antisymmetric preference. In particular, if one swaps the left and right image, one would expect the preference for left or right image to also swap. We thus impose this swapping as a data augmentation, by doubling each original batch to include a swapped copy of the batch: we found this accelerates and stabilizes training. We include a variety of standard augmentations that we list in the supplemental.

**Inference for Curation.** Our network is trained on paired preferences, but at inference time, we want to compare 8 guided PatchMatch inpaintings and establish a preference for each, and a preferred ordering. Moreover, in alternative implementations, one might wish to compare more or fewer images. Thus, to compare $n$ inpainted candidate images, we form an $n \times n$ matrix $M$, where $M_{ij}$ is the probability of preferring method $i$ over method $j$. We establish this probability for all pairs $i, j$ with $i < j$ by setting $o_{ij}^{(k)}$ for $k = 1, 2, 3$ as the 3 softmaxed outputs of the MLP for the pair, and then compute $M_{ij} = o_{ij}^{(3)} - o_{ij}^{(1)}$ and $M_{ji} = -M_{ij}$. The ground truth and prediction are antisymmetric matrices i.e. $M = -M^T$. The preference of a given inpainted image $i$ *in the context of the other images* is the average of row $i$ of $M$. In this way, we recover the same antisymmetric paired preferences in the special case of $n = 2$, but also generalize to establishing preferences among arbitrary numbers of images.

The input resolution for our curation network is $512 \times 512$, however, the photographs to be inpainted can have 1 to 2 orders of magnitude more pixels. We use an operation called "auto crop" to resize a crop region around the hole that contains approximately 25% hole pixels and surrounding context to the target resolution. Please see the supplemental for details about automatic cropping.

# 5 Experiments

## 5.1 Curation Module

**Table 1.** Curation network performance on human paired preference data, ablations, and comparisons. We report two test set accuracies: over the whole dataset, and over only easy cases. Our network is competitive with humans and outperforms all alternatives. Parenthetical numbers are relative to our network.

| Method | Accuracy | Accuracy for easy cases |
|---|---|---|
| Human performance | 57.1% (+0.7%) | 86.1% (+1.4%) |
| **Our curation network** | **56.4%** | **84.7%** |
| Ours no pretraining | 53.9% (−2.5%) | 78.8% (−5.9%) |
| Ours fewer augmentations | 53.8% (−2.7%) | 77.8% (−6.9%) |
| Ours no mask | 52.6% (−3.8%) | 72.2% (−12.5%) |
| Ours late fusion variant | 52.6% (−3.9%) | 82.1% (−2.6%) |
| Ours early fusion | 51.4% (−5.1%) | 77.8% (−6.9%) |
| Ours freeze backbone | 43.7% (−12.7%) | 53.0% (−33.1%) |
| NIMA [28] w/MLP | 41.4% (−15.0%) | 50.0% (−34.7%) |
| MetaIQA+ [49] w/MLP | 41.1% (−15.3%) | 50.4% (−34.3%) |
| Random Chance | 33.1% (−23.3%) | 30.6% (−54.2%) |

We show in Table 1 the performance on human paired preference data for our curation module, ablations, and comparisons. The table is computed from the human preference dataset previously discussed in Sect. 4.3. In our user study, if a human expresses a preference for one image, we ask if the preference is strong or weak. Because the task itself is challenging, we also report **easy cases** as those where mean human preference is strongly for one image.

We list our network and human performance in Table 1. Human preference is determined by collecting an additional opinion for a random subset of images. Our curation module outperforms all other alternatives and is only slightly worse than human performance by both metrics. Although the accuracy for both humans and ours is "only" 56–57%, this is already much better than random chance at 33%, and this is because of the difficulty of the task, where for many fills it is hard to tell whether they are tied in quality or one or the other is preferred. For easy or unambiguous cases, the accuracies for humans and ours are both much better, at 85–86%. Since models can overfit to the training set, we always report the checkpoint with highest validation accuracy.

We next discuss the ablations in Table 1. "Ours No Pretraining" skips the pretraining, which is necessary for best generalization. "Ours Fewer Augmentations" is an ablation where removing JPEG compression, rotation, and noise reduces accuracy. For "Ours No Mask," we do not input the hole mask. "Ours

Late Fusion Variant" modifies the pretraining so instead of using one classifier, both real and fake image are featurized with a shared-weight EfficientNet backbone and compared with an additional MLP. "Ours Early Fusion" modifies the network by concatenating both images with mask and feeding this through a single EfficientNet backbone followed by MLP.

We find that fine-tuning only the MLP for the human preference task and freezing the backbone network weights is insufficient. "Ours Freeze Backbone" freezes backbone weights after pretraining and only fine-tunes the MLP. Similarly, "NIMA [28] w/MLP" and "MetaIQA+ [49] w/MLP" use pretrained, frozen no-reference image quality assessment backbones, and fine-tune the MLP.

Our Table 4 shows that the guided inpainting chosen by our curation module out of all 8 options outperforms both a random choice of a guided fill and Photoshop's Content-Aware Fill: the outperformance in user preference is particularly strong. That table is for the inpainting dataset, which is described next.

## 5.2 Comparison with Other Methods

**Table 2.** A comparison study with the state-of-the-art inpainting methods. The top 3 methods are colored: 1, 2, 3.

| Methods | LPIPS ↓ | FID ↓ | | P-IDS ↑ | | U-IDS ↑ | | User Pref. ↑ | User Pref. ↑ |
|---|---|---|---|---|---|---|---|---|---|
| | | Full | Patch | Full | Patch | Full | Patch | Full image | Boundary patch |
| EdgeConnect [22] | 0.05017 | 35.06 | 41.05 | 0.04 | 4.56 | 0.00 | 0.55 | - | - |
| Deepfillv2 [39] | 0.05295 | 32.87 | 36.06 | 5.54 | 5.47 | 1.35 | 0.84 | - | - |
| MEDFE [20] | 0.05170 | 33.97 | 60.87 | 0.48 | 2.23 | 0.00 | 0.26 | - | - |
| HiFill [36] | 0.05213 | 34.39 | 31.74 | 4.15 | 5.20 | 0.75 | 0.97 | - | - |
| CoModGAN [44] | 0.05099 | 24.81 | 32.08 | 14.72 | 7.01 | 4.47 | 1.51 | 28 | 17 |
| MADF [51] | 0.04773 | 23.62 | 33.21 | 10.48 | 6.81 | 2.14 | 1.48 | 6 | 12 |
| ProFill [40] | 0.04783 | 24.25 | 31.26 | 11.35 | 6.89 | 2.26 | 1.31 | 10 | 16 |
| LaMa [27] | 0.04588 | 19.20 | 35.95 | 17.24 | 6.86 | 5.62 | 1.38 | 28 | 22 |
| SuperCAF (Ours) | 0.04156 | 18.74 | 15.63 | 22.46 | 19.77 | 10.70 | 10.22 | 128 | 133 |

We compare our algorithm with eight state-of-the-art image inpainting methods, quantitatively and qualitatively. Among all these methods, HiFill [36] can run on images up to 8K resolution like our method, the work LaMa [27] states that they can generalize to higher resolutions up to around 2K, ProFill [40] can run on images up to 1K resolution, and the rest of the methods can only run on images up to $512 \times 512$.

The main focus of our method is inpainting of high resolution images of size 4K and beyond. We thus generate full resolution images for all methods. We attempted to make the comparison as generous as possible for baseline methods by applying Real-ESRGAN [31] for super-resolution to increase all methods with limited output resolution back to the native image resolution. We chose Real-ESRGAN [31], since it is the state-of-the-art SR algorithm for real-world images

**Table 3.** A pairwise comparison study between high-resolution inpainting results for upsampled by Real-ESRGAN [31] and results upsampled by our framework, for four recent inpainting methods. Users are asked to choose the best image from a pair for each inpainting model. The best score is bold.

| Methods | LPIPS ↓ | FID ↓ | User Pref. ↑ |
|---|---|---|---|
| ProFill [40] + SR [31] | 0.0478 | 24.2589 | 21 |
| ProFill [40] + Ours | **0.0425** | **20.4362** | **179** |
| CoModGAN [44] + SR [31] | 0.0510 | 24.8189 | 36 |
| CoModGAN [44] + Ours | **0.0430** | **19.9224** | **164** |
| MADF [51] + SR [31] | 0.0477 | 23.6290 | 27 |
| MADF [51] + Ours | **0.0421** | **19.9022** | **173** |
| LaMa [27] + SR [31] | 0.04588 | 19.2022 | 42 |
| LaMa [27] + Ours | **0.04156** | **18.7414** | **158** |

**Table 4.** A comparison study with Photoshop's Content-Aware Fill, which is based on PatchMatch [1] and a randomly selected Guided PatchMatch baseline. Photos average 20 megapixels and holes average 2.3 megapixels. The user studies are performed on 100 photographs where all method outputs are different from each other. The best scores are bold text in the table.

| Methods | LPIPS ↓ | FID ↓ | User Pref. ↑ |
|---|---|---|---|
| Content-aware fill | 0.04675 | 23.0068 | 16 |
| Random guided PM | 0.04281 | 19.7704 | 24 |
| SuperCAF (ours) | **0.04156** | **18.7414** | **60** |

and is robust to visual artifacts in the input. We apply HiFill [36] at native resolution, and ProFill [40] at its maximum of 1K resolution. For LaMa [27], we found that although it can be applied at higher resolutions, we obtain best quality by applying it at a resolution where the maximum axis is 512 pixels. Additionally, we tried two scenarios for the baselines: running on the auto-crop region discussed earlier in Sect. 4.3 and running on the full image. In our evaluation we found no advantage for the baselines run on the auto-crop region so we used the full image scenario.

For the quantitative evaluation, we evaluated six popular metrics: Peak Signal-to-Noise Ratio (PSNR), SSIM [32], LPIPS [43], the recently improved version of FID [23], and P-IDS and U-IDS from CoModGAN [44], which were recently shown to correlate highly with human perception. Because the holes are very large (2.3 megapixel on average) and valid fills can have quite diverse contents that differ greatly from the original image, we feel the metrics *FID and P/U-IDS are most appropriate*, and we also show LPIPS, and we report PSNR and SSIM in the supplemental. The quantitative results are shown in Table 2. For P/U-IDS, because of the dataset size of 1045 images, we also apply vertical and horizontal augmentation for the full image to improve convergence of those metrics.

In Table 2, we report quantitative metrics for two scenarios: "full" indicates a square crop region around the entire inpainted region was used, as determined by auto-crop Sect. 4.3, and "patch" indicates 10 smaller randomly sampled $256 \times 256$ crop regions drawn at consistent position from locations where the patch center is a hole pixel. We note that ours outperforms all baselines by the metrics in the table. We particularly note that in the patch scenario, for the metrics FID, P-IDS, and U-IDS, we *dramatically* outperform the SOTA method LaMa [27] by factors *between 2.3 to 7.4 times*, and the outperformance can be even greater for other methods. This is because our method has much higher texture fidelity at the finest resolutions, since it can copy relevant background patches via PatchMatch. These textures form a coherent whole, as indicated by outperformance via other metrics.

As often mentioned in the inpainting literature (e.g., [38,39]), image inpainting lacks a good quantitative evaluation metric, and there is no single metric that can be used to gauge real image quality. Thus, we also conduct two user studies for randomly sampled 200 test images in comparison with the top four methods using Amazon Mechanical Turk. For the first user study, we use the "full" image scenario described earlier, and in the second user study, we use a "boundary patch" setup. The boundary patch is determined by randomly sampling a $512 \times 512$ crop region, centered on a boundary pixel of the hole region. This setup allows human participants to easily contrast the texture synthesized inside the hole with that in the background region, and thus assess the suitability of methods for inpainting at modern camera resolutions. In the user study, we give user output images from all methods with randomized order, and ask users to pick the best image for each case. For each of the two user studies, we recruited more than 150 users and asked each user to evaluate a randomly sampled batch of 20 images from the whole test set. The study results are shown in the last two columns of Table 2. In the "full" scenario, our automatically selected guided PatchMatch outperforms the other methods by a large margin of *4.6 times to 21 times*, and in the "boundary patch" scenario, our method outperforms alternatives even more strongly, by *6 times to 11 times*. Users in general very strongly prefer our method for inpainting at modern camera resolutions, but in the crop scenario where a user focuses on high-resolution detail—as might be important for large displays or large format prints—our method performs better still.

In Table 3, we show our method can be used in combination with 4 different baseline methods that perform the initial coarse-scale inpainting: ProFill [40], CoModGAN [44], MADF [51], and LaMa [27]. In every case, our method outperforms the baseline with Real-ESRGAN [31] super-resolution applied. We ran user study containing pairs of images, and found the user study preferences strongly prefer our method, with between *4.5 and 8.5 times* higher preference for ours over the baseline. Our method is suitable when combined with a variety of deep inpainting baselines, and greatly improves user preference over alternatives. We show similar results for 3 older inpainting models in the supplemental.

In Table 4, we compare our method to two other baselines: Photoshop's Content-Aware Fill (CAF), which is based on PatchMatch [1], and a baseline that randomly picks with equal probability one of our eight guided PatchMatch results. Our method again performs best for the quantitative metrics. We ran a user study on 100 images where all methods have different outputs, and again find our method is *strongly preferred 3.4 to 4 times* more than the other two baselines. This indicates that our method outperforms a strong commercial baseline of Photoshop's CAF, which is used by professionals to manipulate photos at modern camera resolutions, and shows that our curation outperforms a random guided PatchMatch result. We include many photographic results in our supplemental material, and show that the preference for ours is statistically significant for *all* 7 of the above user studies (Fig. 4).

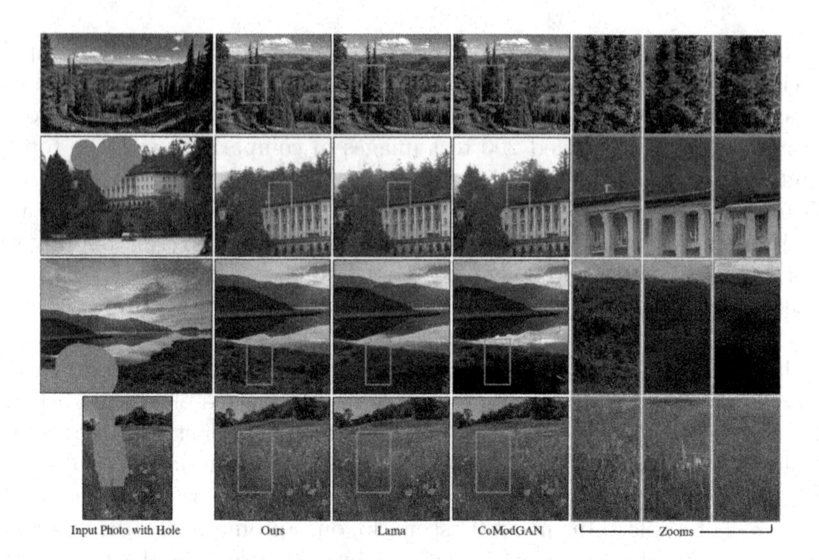


**Fig. 4.** Results for our method and two baselines with Real-ESRGAN, for (from top): a 26 MP nature panorama with real hole, a 20 MP photo with random scribble hole, a 12 MP lake photo with random scribble hole, a 12 MP field photo with an instance hole. Please check out supplemental for visual results.

### 5.3 Running Time

On a 3.6 GHz 8 core Intel i9-9900K with 11 GB NVidia RTX 2080 Ti, for a representative 12 MP image with 4K resolution, our naive implementation takes 23.0 s, and our optimized implementation takes 2.5 s by initially computing PatchMatch results at 1K, running curation, then using another PatchMatch to obtain the 4K result.

## 6 Discussion, Limitations, Future Work

Our method has some limitations, which could be mitigated through user interactions such as manually picking guides. Generally, PatchMatch is good at synthesizing texture and repetitive regular structures, but structures under perspective transformations in architecture can be broken. We occasionally observe small amounts of blur especially near the hole boundary: this might be addressed by using curation in a smarter way such as an iterative fill [40,42]. Occasionally, we observe repetitions of salient patches: these might be mitigated by incorporating patch usage budgets [9,16] combined with saliency. GANs may produce amazing results by hallucinating content not present in the input image, but they can also hallucinate bizarre artifacts. We use patch-based synthesis throughout the image, however, patch-based synthesis can remove unique features, so the result could be allowed to deviate from patch-based synthesis if we believe a hallucinated output is a good one.

## References

1. Barnes, C., Shechtman, E., Finkelstein, A., Goldman, D.B.: Patchmatch: a randomized correspondence algorithm for structural image editing. ACM Trans. Graph. **28**(3), 24 (2009)
2. Bénard, P., et al.: Stylizing animation by example. ACM Trans. Graph. (TOG) **32**(4), 1–12 (2013)
3. Bosse, S., Maniry, D., Müller, K.R., Wiegand, T., Samek, W.: Deep neural networks for no-reference and full-reference image quality assessment. IEEE Trans. Image Process. **27**(1), 206–219 (2017)
4. Burt, P.J., Adelson, E.H.: The Laplacian pyramid as a compact image code. In: Readings in Computer Vision, pp. 671–679. Elsevier (1987)
5. Cade, D.: The world's first 'fully' digital camera was created by Fuji (2016). https://petapixel.com/2016/06/09/photo-history-worlds-first-fully-digital-camera-invented-fuji/
6. Darabi, S., Shechtman, E., Barnes, C., Goldman, D.B., Sen, P.: Image melding: combining inconsistent images using patch-based synthesis. ACM Trans. Graph. (TOG) **31**(4), 1–10 (2012)
7. Diamanti, O., Barnes, C., Paris, S., Shechtman, E., Sorkine-Hornung, O.: Synthesis of complex image appearance from limited exemplars. ACM Trans. Graph. (TOG) **34**(2), 1–14 (2015)
8. Duggal, S., Wang, S., Ma, W.C., Hu, R., Urtasun, R.: DeepPruner: learning efficient stereo matching via differentiable PatchMatch. In: Proceedings of the IEEE/CVF International Conference on Computer Vision, pp. 4384–4393 (2019)
9. Fišer, J., et al.: StyLit: illumination-guided example-based stylization of 3D renderings. ACM Trans. Graph. (TOG) **35**(4), 1–11 (2016)
10. Gu, S., Lugmayr, A., Danelljan, M., Fritsche, M., Lamour, J., Timofte, R.: DIV8K: DIVerse 8K resolution image dataset. In: 2019 IEEE/CVF International Conference on Computer Vision Workshop (ICCVW), pp. 3512–3516. IEEE (2019)
11. He, K., Sun, J.: Statistics of patch offsets for image completion. In: Fitzgibbon, A., Lazebnik, S., Perona, P., Sato, Y., Schmid, C. (eds.) ECCV 2012. LNCS, vol. 7573, pp. 16–29. Springer, Heidelberg (2012). https://doi.org/10.1007/978-3-642-33709-3_2
12. Hertzmann, A., Jacobs, C.E., Oliver, N., Curless, B., Salesin, D.H.: Image analogies. In: Proceedings of the 28th Annual Conference on Computer Graphics and Interactive Techniques, pp. 327–340 (2001)
13. Huang, J.B., Kang, S.B., Ahuja, N., Kopf, J.: Image completion using planar structure guidance. ACM Trans. Graph. (TOG) **33**(4), 1–10 (2014)
14. Iizuka, S., Simo-Serra, E., Ishikawa, H.: Globally and locally consistent image completion. ACM Trans. Graph. (TOG) **36**(4), 1–14 (2017)
15. Jamriška, O., et al.: Stylizing video by example. ACM Trans. Graph. (TOG) **38**(4), 1–11 (2019)
16. Kaspar, A., Neubert, B., Lischinski, D., Pauly, M., Kopf, J.: Self tuning texture optimization. Comput. Graph. Forum **34**, 349–359 (2015)
17. Li, Y., et al.: Fully convolutional networks for panoptic segmentation. In: Proceedings of the IEEE/CVF Conference on Computer Vision and Pattern Recognition, pp. 214–223 (2021)
18. Liao, J., Yao, Y., Yuan, L., Hua, G., Kang, S.B.: Visual attribute transfer through deep image analogy. ACM Trans. Graph. **36**(4), 120:1–120:15 (2017). https://doi.org/10.1145/3072959.3073683. https://doi.acm.org/10.1145/3072959.3073683

19. Liu, G., Reda, F.A., Shih, K.J., Wang, T.-C., Tao, A., Catanzaro, B.: Image inpainting for irregular holes using partial convolutions. In: Ferrari, V., Hebert, M., Sminchisescu, C., Weiss, Y. (eds.) ECCV 2018. LNCS, vol. 11215, pp. 89–105. Springer, Cham (2018). https://doi.org/10.1007/978-3-030-01252-6_6
20. Liu, H., Jiang, B., Song, Y., Huang, W., Yang, C.: Rethinking image inpainting via a mutual encoder-decoder with feature equalizations. arXiv preprint arXiv:2007.06929 (2020)
21. Liu, W., Zhang, P., Huang, X., Yang, J., Shen, C., Reid, I.: Real-time image smoothing via iterative least squares. ACM Trans. Graph. (TOG) **39**(3), 1–24 (2020)
22. Nazeri, K., Ng, E., Joseph, T., Qureshi, F.Z., Ebrahimi, M.: EdgeConnect: generative image inpainting with adversarial edge learning. arXiv preprint arXiv:1901.00212 (2019)
23. Parmar, G., Zhang, R., Zhu, J.Y.: On buggy resizing libraries and surprising subtleties in FID calculation. arXiv preprint arXiv:2104.11222 (2021)
24. Pathak, D., Krahenbuhl, P., Donahue, J., Darrell, T., Efros, A.A.: Context encoders: feature learning by inpainting. In: Proceedings of the IEEE Conference on Computer Vision and Pattern Recognition, pp. 2536–2544 (2016)
25. Ranftl, R., Bochkovskiy, A., Koltun, V.: Vision transformers for dense prediction. In: Proceedings of the IEEE/CVF International Conference on Computer Vision, pp. 12179–12188 (2021)
26. Ren, Y., Yu, X., Zhang, R., Li, T.H., Liu, S., Li, G.: StructureFlow: image inpainting via structure-aware appearance flow. In: Proceedings of the IEEE/CVF International Conference on Computer Vision, pp. 181–190 (2019)
27. Suvorov, R., et al.: Resolution-robust large mask inpainting with Fourier convolutions. In: WACV: Winter Conference on Applications of Computer Vision (2022)
28. Talebi, H., Milanfar, P.: NIMA: neural image assessment. IEEE Trans. Image Process. **27**(8), 3998–4011 (2018)
29. Tan, M., Le, Q.: EfficientNet: rethinking model scaling for convolutional neural networks. In: International Conference on Machine Learning, pp. 6105–6114. PMLR (2019)
30. Wang, S.Y., Wang, O., Zhang, R., Owens, A., Efros, A.A.: CNN-generated images are surprisingly easy to spot... for now. In: Proceedings of the IEEE/CVF Conference on Computer Vision and Pattern Recognition, pp. 8695–8704 (2020)
31. Wang, X., Xie, L., Dong, C., Shan, Y.: Real-ESRGAN: training real-world blind super-resolution with pure synthetic data. In: Proceedings of the IEEE/CVF International Conference on Computer Vision, pp. 1905–1914 (2021)
32. Wang, Z., Bovik, A.C., Sheikh, H.R., Simoncelli, E.P.: Image quality assessment: from error visibility to structural similarity. IEEE Trans. Image Process. **13**(4), 600–612 (2004)
33. Wexler, Y., Shechtman, E., Irani, M.: Space-time completion of video. IEEE Trans. Pattern Anal. Mach. Intell. **29**(3), 463–476 (2007)
34. Xiong, W., et al.: Foreground-aware image inpainting. In: Proceedings of the IEEE/CVF Conference on Computer Vision and Pattern Recognition, pp. 5840–5848 (2019)
35. Xu, L., Yan, Q., Xia, Y., Jia, J.: Structure extraction from texture via relative total variation. ACM Trans. Graph. (TOG) **31**(6), 1–10 (2012)
36. Yi, Z., Tang, Q., Azizi, S., Jang, D., Xu, Z.: Contextual residual aggregation for ultra high-resolution image inpainting. In: Proceedings of the IEEE/CVF Conference on Computer Vision and Pattern Recognition, pp. 7508–7517 (2020)

37. Yin, W., et al.: Learning to recover 3D scene shape from a single image. In: Proceedings of IEEE Conference on Computer Vision and Pattern Recognition (CVPR) (2021)
38. Yu, J., Lin, Z., Yang, J., Shen, X., Lu, X., Huang, T.S.: Generative image inpainting with contextual attention. In: Proceedings of the IEEE Conference on Computer Vision and Pattern Recognition, pp. 5505–5514 (2018)
39. Yu, J., Lin, Z., Yang, J., Shen, X., Lu, X., Huang, T.S.: Free-form image inpainting with gated convolution. In: Proceedings of the IEEE/CVF International Conference on Computer Vision, pp. 4471–4480 (2019)
40. Zeng, Yu., Lin, Z., Yang, J., Zhang, J., Shechtman, E., Lu, H.: High-resolution image inpainting with iterative confidence feedback and guided upsampling. In: Vedaldi, A., Bischof, H., Brox, T., Frahm, J.-M. (eds.) ECCV 2020. LNCS, vol. 12364, pp. 1–17. Springer, Cham (2020). https://doi.org/10.1007/978-3-030-58529-7_1
41. Zhang, H., et al.: ResNeSt: split-attention networks. arXiv preprint arXiv:2004.08955 (2020)
42. Zhang, L., et al.: Perceptual artifacts localization for inpainting. In: Farinella, T. (ed.) ECCV 2022, LNCS 13689, pp. 146–164. Springer, Cham (2022)
43. Zhang, R., Isola, P., Efros, A.A., Shechtman, E., Wang, O.: The unreasonable effectiveness of deep features as a perceptual metric. In: Proceedings of the IEEE Conference on Computer Vision and Pattern Recognition (CVPR), pp. 586–595, June 2018
44. Zhao, S., et al.: Large scale image completion via co-modulated generative adversarial networks. In: International Conference on Learning Representations (ICLR) (2021)
45. Zhou, B., Lapedriza, A., Khosla, A., Oliva, A., Torralba, A.: Places: a 10 million image database for scene recognition. IEEE Trans. Pattern Anal. Mach. Intell. **40**(6), 1452–1464 (2017)
46. Zhou, X., et al.: CoCosNet v2: full-resolution correspondence learning for image translation. In: Proceedings of the IEEE/CVF Conference on Computer Vision and Pattern Recognition, pp. 11465–11475 (2021)
47. Zhou, X., et al.: Full-resolution correspondence learning for image translation. In: Proceedings of IEEE Conference on Computer Vision and Pattern Recognition (CVPR) (2021)
48. Zhu, H., Li, L., Wu, J., Dong, W., Shi, G.: MetaIQA: deep meta-learning for no-reference image quality assessment. In: Proceedings of the IEEE/CVF Conference on Computer Vision and Pattern Recognition, pp. 14143–14152 (2020)
49. Zhu, H., Li, L., Wu, J., Dong, W., Shi, G.: Generalizable no-reference image quality assessment via deep meta-learning. IEEE Trans. Circuits Syst. Video Technol. **32**(3), 1048–1060 (2021)
50. Zhu, J.Y., Krahenbuhl, P., Shechtman, E., Efros, A.A.: Learning a discriminative model for the perception of realism in composite images. In: Proceedings of the IEEE International Conference on Computer Vision, pp. 3943–3951 (2015)
51. Zhu, M., et al.: Image inpainting by end-to-end cascaded refinement with mask awareness. IEEE Trans. Image Process. **30**, 4855–4866 (2021)

# Controllable Video Generation Through Global and Local Motion Dynamics

Aram Davtyan(✉) and Paolo Favaro

Computer Vision Group, University of Bern, Bern, Switzerland
{aram.davtyan,paolo.favaro}@inf.unibe.ch

**Abstract.** We present GLASS, a method for Global and Local Action-driven Sequence Synthesis. GLASS is a generative model that is trained on video sequences in an unsupervised manner and that can animate an input image at test time. The method learns to segment frames into foreground-background layers and to generate transitions of the foregrounds over time through a global and local action representation. Global actions are explicitly related to 2D shifts, while local actions are instead related to (both geometric and photometric) local deformations. GLASS uses a recurrent neural network to transition between frames and is trained through a reconstruction loss. We also introduce W-Sprites (Walking Sprites), a novel synthetic dataset with a predefined action space. We evaluate our method on both W-Sprites and real datasets, and find that GLASS is able to generate realistic video sequences from a single input image and to successfully learn a more advanced action space than in prior work. Further details, the code and example videos are available at https://araachie.github.io/glass/.

**Keywords:** Video generation · Unsupervised action discovery · Controllable generation

## 1 Introduction

A long-standing objective in machine learning and computer vision is to build agents that can learn how to operate in an environment through visual data [12]. A successful approach to do so is to use supervised learning, *i.e.*, to train a model on a large, manually annotated dataset [25]. However, if we take inspiration from how infants learn to move, we are brought to conclude that they may not rely on extensive guidance. In fact, while supervision from adults might come through language [30], the signal is certainly not detailed enough to fully define the locomotion dynamics. One approach that does not require direct supervision is to learn just through direct scrutiny of other agents, *i.e.*, through passive imitation.

---

**Supplementary Information** The online version contains supplementary material available at https://doi.org/10.1007/978-3-031-19790-1_5.

© The Author(s), under exclusive license to Springer Nature Switzerland AG 2022
S. Avidan et al. (Eds.): ECCV 2022, LNCS 13677, pp. 68–84, 2022.
https://doi.org/10.1007/978-3-031-19790-1_5

**Fig. 1.** *W-Sprites* dataset sample videos. To play them use Acrobat Reader.

In fact, infants have an abundance of sensory exposure to the activities of adults before they themselves learn how to perform them [29].

The first step for an observing agent to learn how to operate in an environment through passive imitation and without explicit supervision is to build a model that: 1) separates an agent from its environment, 2) captures the appearance of the agent and its environment, and 3) builds a description of the agent's dynamics. The first requirement implies that the model incorporates some segmentation capability, and it allows to explain transitions over time more easily. The second requirement is dictated by the fact that we exploit the reconstruction of visual observations as our indirect supervision signal. Thus, our model also relates to the video generation literature. Finally, the third requirement is that the model includes an *action space*, which serves two purposes: i) it allows the model to decode a video into a sequence of actions (which is a representation of the agent's dynamics) and ii) it allows the model to control the generation of videos by editing the action sequence.

We introduce GLASS, a method for Global and Local Action-driven Sequence Synthesis. As shown in Fig. 2, GLASS first learns to segment each frame of a video into foreground and background layers. A basic principle to do that is to use motion as a cue, *i.e.*, the fact that agents exhibit, on average, a distinct motion flow compared to the environment. Motion-based segmentation could be achieved through background subtraction, which is however restricted to stationary backgrounds, or instead, more in general, via optical flow. For simplicity, we propose to use an explicit foreground-background motion segmentation based on 2D shifts. Then, GLASS regresses the relative shift between the foregrounds of two subsequent frames, which we call the *global action*, and between the backgrounds (see Fig. 3). The *local actions* are learned only from the foregrounds. The decomposition of the agent's motion into global and local components provides a computationally efficient representation of the video. In contrast, a global action space would require the much larger Cartesian product of the local and global action spaces. In practice, given an action space of finite size, global models tend to learn only global motions and to ignore local deformations. We train an RNN to predict, through a decoder, the next foreground by using an encoding of a foreground, the previous state, and an encoding of the local and global actions as input. All networks are trained via reconstruction losses.

We evaluate GLASS on both synthetic and real data. As synthetic data we introduce *W-Sprites* (Walking Sprites [1,2,23]) (see Fig. 1), a dataset with a pre-

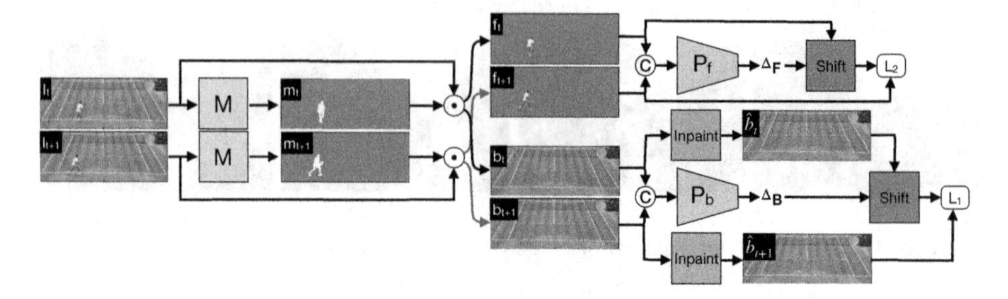


**Fig. 2.** GLASS Global Motion Analysis. Two input frames $I_t$ and $I_{t+1}$ are fed (separately) to a segmentation network $M$ to output the foreground masks $m_t$ and $m_{t+1}$ respectively. The masks are used to separate the foregrounds $f_t$ and $f_{t+1}$ from the backgrounds $b_t$ and $b_{t+1}$. The concatenated foregrounds are fed to the network $P_f$ to predict their relative shift $\Delta_F$. We use $\Delta_F$ to shift $f_t$ and match it to $f_{t+1}$ via an $L_2$ loss (foregrounds may not match exactly and this loss does not penalize small errors). In the case of the backgrounds we also train an inpainting network before shifting them with the predicted $\Delta_B$ and matching them with an $L_1$ loss (unlike foregrounds, we can expect backgrounds to match).

defined action space, and where the action labels between pairs of frames (as well as the agent segmentation and location, and the background shift) are known. We find that GLASS learns a robust representation of both global and local dynamics on W-Sprites. Moreover, GLASS is able to decode videos into sequences of actions that strongly correlate with the ground truth action sequences. Finally, users can generate novel sequences by controlling the input action sequences to GLASS. On real data, we find that GLASS can also generate realistic sequences by controlling the actions between frames.

**Contributions:** i) We introduce GLASS, a novel generative model with a global and local action space; the shifts estimated and generated through the global actions have an accuracy comparable to or higher than SotA; moreover, local actions allow a fine-grained modeling of dynamics that is not available in prior work; ii) We introduce W-Sprites, a novel dataset for the evaluation of action identification and generation; iii) We demonstrate GLASS on both synthetic and real datasets and show that it can: 1) segment an agent from its environment and estimate its global shift over time; 2) learn a disentangled action space that is consistent across agents; 3) decode videos into sequences of actions; 4) synthesize realistic videos under the guidance of a novel action policy.

## 2 Prior Work

**Video Generation.** Because GLASS is trained based on reconstruction losses, and it is built as a generative model, it relates to the generation of videos. Recent success in deep generative models for images [10,17,28] has aroused renewed interest in video generation. Several formulations tackling the problem of video

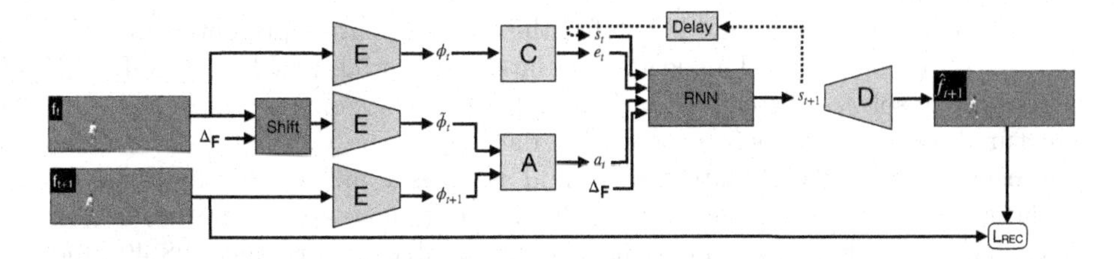


**Fig. 3.** GLASS Local Motion Analysis. We feed the segmented foreground $f_t$, its shifted version and $f_{t+1}$ separately as inputs to an encoder network $E$ to obtain features $\phi_t$, $\tilde{\phi}_t$ and $\phi_{t+1}$ respectively. The latter two features are then mapped to an action $a_t$ by the action network $A$. A further encoding of $\phi_t$ into $e_t$, the previous state $s_t$, and the local action $a_t$ and global action $\Delta_F$ are fed as input to the RNN to predict the next state $s_{t+1}$. Finally, a decoder maps the state $s_{t+1}$ to the next foreground $\hat{f}_{t+1}$, which is matched to the original foreground $f_{t+1}$ via the reconstruction loss.

generation exploit adversarial losses [3,4,9,22,32,35–37], autoregressive models [39] and use a wide range of architectures from RNNs [31] to transformers [40].

**Controllable Video Generation.** Video generation models can also differ in how they apply conditioning. While some prior work uses per-video class labels [19,38], *e.g.*, actions performed in a short sequence of frames, others, as in GLASS, use conditioning at each step [7,12,18,26,27]. For instance, in [12] the authors train a model to simulate the behavior of a robotic arm given the performed actions. Kim et al. [18] introduce GameGAN, a powerful generative model that can replace a game engine. It is trained to render the next frame given the current frame and the pressed keyboard action. One limitation of these methods is that they require knowledge of the ground truth actions and hence are restricted to synthetic data, such as video games. To become applicable to real data, several recent methods that learn an action space of the agent from raw videos without fine-grained annotations have been proposed. For instance, Rybkin et al. [29] propose a continuous latent space for the actions. They introduce arithmetical structure into their action space by exploiting the fact that two actions can be composed to get another action that would lead to the same result as when applying the original actions sequentially. [13] generates high-quality videos of moving agents with controllable actions. However, it builds an autoregressive model on top of pose maps obtained from supervised training, while our method is completely unsupervised. In [24] the continuous action space is replaced by a finite set. This allows a simpler control (playability) of the generated videos and favors interpretability of the learned actions. More recent work by Huang et al. [15] explicitly separates the foreground from the background and trains a network to predict the next frame given the current frame and the next segmentation mask. GLASS relates to this last family of methods as it also does not require any supervision signal.

**Unsupervised Learning of Structured Representations.** In GLASS we propose to learn the global and local actions from video frames. While the global ones are defined as foreground 2D shifts, the local ones are represented as a discrete set of action codes. This leads to a latent clustering problem. In GLASS, we propose to solve it through variational inference [21]. Some recent work learns structured representations from raw input data [5,6]. The VQ-VAE [34] formulation instead uses a discrete latent space and assumes a uniform distribution over the latent features. Recent advances in image and video generation has shown that such VQ-VAE based models have a remarkable performance [28,40] and this has encouraged us to adopt this approach.

## 3 Training GLASS

GLASS consists of two stages: One is the Global Motion Analysis (GMA) (shown in Fig. 2) and the other is the Local Motion Analysis (LMA) (shown in Fig. 3). GMA aims to separate the foreground agent from the background and to also regress the 2D shifts between foregrounds and backgrounds. LMA aims to learn a representation for local actions that can describe deformations other than 2D shifts. Towards this purpose it uses a Recurrent Neural Network (RNN) and a feature encoding of a frame and of the global and local actions as input. Both GMA and LMA stages are jointly trained in an unsupervised manner.

### 3.1 Global Motion Analysis

Let us denote a video as a sequence of $T$ frames $I_t \in \mathbb{R}^{3\times H\times W}$, where $t = 1,\dots,T$, and 3, $H$ and $W$ denote the number of color channels, the height and the width of the frame. Although GLASS is trained with video sequences, we can illustrate all the training losses with a single pair $(I_t, I_{t+1})$ of frames. Each frame is fed to a mask network M to output masks $m_t$ and $m_{t+1}$. The masks can take values between 0 and 1 (a sigmoid is used at the output), but are encouraged to take the extreme values through the following binarization loss

$$\mathcal{L}_{\text{BIN}} = \textstyle\sum_t \min\{m_t, 1-m_t\}. \tag{1}$$

We also discourage the mask from being empty or covering the whole frame by using a mask size loss

$$\mathcal{L}_{\text{SIZE}} = \textstyle\sum_t |\mathbb{E}[m_t] - \theta|, \tag{2}$$

where $\mathbb{E}[\cdot]$ denotes the average over all pixels and $\theta \in [0,1]$ is a tuning parameter (the percentage of image pixels covered by a mask on average). The masks are then used to extract the foregrounds $f_t = I_t \odot m_t$ and $f_{t+1} = I_{t+1} \odot m_{t+1}$ and the backgrounds $b_t = I_t \odot (1-m_t)$ and $b_{t+1} = I_{t+1} \odot (1-m_{t+1})$ ($\odot$ denotes the element-wise product). We assume that the foregrounds are approximately matching up to a relative shift $\bar{\Delta}_F$, *i.e.*, that $f_{t+1}[p] \simeq \left(f_t \circ \bar{\Delta}_F\right)[p] \doteq f_t[p+\bar{\Delta}_F]$, for all pixel coordinates $p \in \Omega \subset \mathbb{R}^2$. We then concatenate the foregrounds and feed them as input to the pose network $P_f$ to regress the relative shift $\Delta_F = P_f([f_t, f_{t+1}])$ between $f_t$ and $f_{t+1}$. Since we do not have the ground truth shift $\bar{\Delta}_F$, we cannot train $P_f$ via supervised learning. In alternative, we rely on the modeling assumption and define a reconstruction loss for the foreground by applying the estimated shift $\Delta_F$ to $f_t$ and by matching it to the frame $f_{t+1}$ in the $L_2$ norm (to allow for some error tolerance), *i.e.*,

$$\mathcal{L}_{\text{RECF}} = \sum_t \left\| f_{t+1} - f_t \circ \Delta_F \right\|_2^2. \tag{3}$$

A similar derivation pertains to the backgrounds. We concatenate the backgrounds and feed them as input to the pose network $P_b$ to regress the relative shift $\Delta_B = P_b([b_t, b_{t+1}])$ between $b_t$ and $b_{t+1}$. However, because of the holes left by the masks, learning the relative shift via a direct matching of the backgrounds would not work. Therefore, we also introduce an inpainting network N. To indicate the masked region to N we simply fill it with a value out of the image range (we use $[-1.1, -1.1, -1.1]$ as RGB values at the masked pixels). The inpainted regions are then copied to the corresponding backgrounds so that we obtain $\hat{b}_j = b_j \odot (1 - m_j) + \text{N}(b_j) \odot m_j$, with $j = \{t, t+1\}$. The background reconstructions are then matched with both an $L_1$ norm and a perceptual loss $\mathcal{L}_{\text{VGG}}$ based on VGG features[1] [16]

$$\mathcal{L}_{\text{RECB}} = \sum_t \left\| \hat{b}_{t+1} - \hat{b}_t \circ \Delta_B \right\|_1 + \lambda_{\text{VGG}} \mathcal{L}_{\text{VGG}} \left( \hat{b}_{t+1}, \hat{b}_t \circ \Delta_B \right). \tag{4}$$

Finally, we also have a joint reconstruction loss where we compose the foreground with the estimated foreground shift $\Delta_F$ and the inpainted background with the estimated background shift $\Delta_B$

$$\mathcal{L}_{\text{RECJ}} = \sum_t \left\| (f_t \odot m_t) \circ \Delta_F + (\hat{b}_t \circ \Delta_B) \odot (1 - m_t \circ \Delta_F) - I_{t+1} \right\|_1. \tag{5}$$

These losses are all we use to train the mask network M, the inpainting network N and the pose estimation networks $P_f$ and $P_b$. The inpainting network and the other networks could be further improved, but we find that the choices above are sufficient to obtain accurate segmentation masks and good shift estimates.

### 3.2 Local Motion Analysis

The LMA stage works directly on the foreground frames $f_t$ and $f_{t+1}$. It first shifts $f_t$ with $\Delta_F$. This is done to remove the global shift information from the input frames and to make the action network focus on the local variations. It further encodes the foreground frames with a convolutional neural network E and obtains $\phi_t = \text{E}(f_t)$, $\tilde{\phi}_t = \text{E}(f_t \circ \Delta_F)$ and similarly for $\phi_{t+1} = \text{E}(f_{t+1})$. The convolutional feature $\phi_t$ is then projected via C to give $e_t = \text{C}(\phi_t)$.

In the action network A there are a few pre-processing steps. First, both feature maps $\tilde{\phi}_t$ and $\phi_{t+1}$ are fed to a CNN and flat features $\psi_t$ and $\psi_{t+1}$ are obtained from the resulting feature maps through global average pooling. In

[1] VGG was obtained through supervised training, but GLASS can be trained equally well by replacing VGG with a similar network trained in a self-supervised manner.

CADDY [24] the actions are determined through a direct difference between Gaussian samples around $\psi_t$ and $\psi_{t+1}$. On average this means that the difference between features of images with the same action must align with the same direction. Although this works very well for CADDY, we find that this may be restrictive, especially if one wants to represent periodic motion (*e.g.*, in our case, an agent walking in place). Thus, we propose to learn a modified mapping of $\psi_{t+1}$ conditioned on $\psi_t$. We compute $\psi^i_{t+1} = \mathrm{T}^i(\psi_t, \psi^{i-1}_{t+1})$ with $i = 1, \ldots, P$, $\mathrm{T}^i$ are bilinear transformations, $\psi^0_{t+1} = \psi_{t+1}$, and we choose $P = 4$. We then compute the action direction $d_t = \psi^P_{t+1} - \psi_t$. Finally, the action $a_t$ is predicted through vector quantization after one additional MLP U to give $a_t = \mathrm{VQ}[\mathrm{U}(d_t)]$. The vector quantization VQ relies on $K$ learnable prototype vectors $c_k$, with $k = 1, \ldots, K$. The method identifies the prototype $c_q$ closest in $L_2$ norm to $\mathrm{U}(d_t)$, *i.e.*, $q = \arg\min_k \|c_k - \mathrm{U}(d_t)\|^2_2$, and uses that as the quantized action $\mathrm{VQ}[\mathrm{U}(d_t)] = c_q$. To train the VQ prototypes, we use the following loss [34]

$$\mathcal{L}_{VQ} = \|\mathrm{sg}[c_q] - U(d_t)\|^2_2 + \lambda_{\mathrm{VQ}}\|c_q - \mathrm{sg}[U(d_t)]\|^2_2, \tag{6}$$

where $\lambda_{\mathrm{VQ}} > 0$ is a tuning parameter and $\mathrm{sg}[\cdot]$ denotes stop-gradient. Vector quantization allows us to obtain latent space clustering in a simpler way compared to the Gaussian priors and the explicit tracking of cluster centroids as done in CADDY [24]. Moreover, our ablation studies show that VQ works better than the Gumbel-softmax trick (see Sect. 6 and Table 3).

Now, we have all the inputs needed for the RNN. We introduce an RNN state $s_t$ and feed it together with the encoding $e_t$ as input. Our RNN is split into 6 blocks as in CADDY [24]. Both the global action $\Delta_F$ and the local action $a_t$ are first mapped to embeddings of the same size and then fed to the modulated convolutional layers of the RNN similarly to StyleGAN [17]. To differentiate the roles of $\Delta_F$ and $a_t$ we feed the embeddings of $\Delta_F$ to the first two blocks of the RNN and that of $a_t$ to the remaining four blocks. The rationale is that early blocks correlate more with global changes, such as translations, and the later blocks correlate more with local deformations.

Finally, the decoder D takes the RNN prediction $s_{t+1}$ as input and outputs the frame $\hat{f}_{t+1} = D_f(s_{t+1})$ and the predicted mask $\hat{m}_{t+1} = D_m(s_{t+1})$. Moreover, the decoder predicts frames at 3 different scales (as also done in CADDY [24]). We introduce a reconstruction loss for each scale

$$\mathcal{L}_{\mathrm{RECU}} = \textstyle\sum_t \left\|\mathrm{sg}[\omega_{\mathrm{UNS}}] \odot \left(f_{t+1} - \hat{f}_{t+1}\right)\right\|_1, \tag{7}$$

where $\forall p \in \Omega$, $\omega_{\mathrm{UNS}}[p] = \|f_t[p] - f_{t+1}[p]\|_1 + \|\hat{f}_t[p] - \hat{f}_{t+1}[p]\|_1$ are weights that enhance the minimization at pixels where the input and predicted foregrounds differ, and also a perceptual loss

$$\mathcal{L}_{\text{LMA-VGG}} = \mathcal{L}_{\mathrm{VGG}}(f_{t+1}, \hat{f}_{t+1}). \tag{8}$$

To better learn local deformations, we also introduce a reconstruction loss that focuses on the differences between the foregrounds after aligning them with the estimated relative shifts, *i.e.*,

$$\mathcal{L}_{\text{RECS}} = \sum_t \left\| \text{sg}[\omega_{\text{ALIGN}}] \odot \left( f_{t+1} - \hat{f}_{t+1} \right) \right\|_1, \tag{9}$$

where $\omega_{\text{ALIGN}}[p] = \|f_t \circ \Delta_F[p] - f_{t+1}[p]\|_1 + \|\hat{f}_{t+1}[p] - f_{t+1}[p]\|_1$. To encourage the consistency between the predicted mask $\hat{m}_{t+1}$ and the mask $m_{t+1}$ obtained from $I_{t+1}$, we also minimize

$$\mathcal{L}_{\text{MSK}} = \sum_t \|\hat{m}_{t+1} - m_{t+1}\|_1. \tag{10}$$

Moreover, we encourage a cyclic consistency between the encoded features via

$$\mathcal{L}_{\text{CYC}} = \sum_t \|\text{sg}[\phi_{t+1}] - \text{E}(\hat{f}_{t+1})\|_1. \tag{11}$$

Our final loss consists of a linear combination of all the above losses (both from the GMA and LMA) through corresponding positive scalars $\lambda_{\text{VQ}}$, $\lambda_{\text{LMA-VGG}}$, $\lambda_{\text{RECU}}$, $\lambda_{\text{RECS}}$, $\lambda_{\text{MSK}}$, $\lambda_{\text{CYC}}$, $\lambda_{\text{RECF}}$, $\lambda_{\text{RECB}}$, $\lambda_{\text{RECJ}}$, $\lambda_{\text{BIN}}$, and $\lambda_{\text{SIZE}}$.

## 4 Implementation Details

At inference time, GLASS can generate a sequence of frames given only the first one. This setting is slightly different from training, where the model only predicts the next frame given the previous one. In order to prepare the model for test time, we adopt the mixed training procedure (Teacher Forcing) also used in [24]. That is, we select a video duration $T_f$, $0 < T_f < T$, and if $t \leq T_f$ we feed the encodings of the real frames to the RNN, otherwise if $t > T_f$ we use the encodings of the reconstructed frames. During the training we gradually decrease $T_f$ to 1 and increase $T$ to adapt the network to the generation of longer sequences. To speed up the convergence, we pretrain the GMA component for 3000 iterations. The coefficients before the loss terms are estimated on the training set. We found that the selected configuration works well across all datasets. The models are trained using the Adam optimizer [20] with a learning rate equal to 0.0004 and weight decay $10^{-6}$. For more details, see the supplementary material.

## 5 W-Sprites Dataset

In order to assess and ablate the components of GLASS, we build a synthetic video dataset of cartoon characters acting on a moving background. We call the dataset W-Sprites (for Walking Sprites). Each sequence is generated via the following procedure. First, one of 1296 different characters is sampled from the Sprites dataset [1,2,23]. This character is then animated in two stages. A random walk module produces a sequence of global coordinates of the sprite within a $96 \times 128$ resolution frame. We then sample one of 9 local actions conditioned on the shift induced by the global motion component. Those actions include: `walk front`, `walk left`, `walk right`, `spellcast front`, `spellcast`

**Table 1.** Global Motion Analysis (GMA). mIoU evaluations

| Configuration | $\not{\mathcal{L}}_{\text{RECB}}$ | $\not{\mathcal{L}}_{\text{RECF}}$ | $\not{\mathcal{L}}_{\text{SIZE}}$ | $\not{\mathcal{L}}_{\text{RECJ}}$ | $\not{\mathcal{L}}_{\text{BIN}}$ | $\not{\mathcal{L}}_{\text{VGG}}$ | GLASS |
|---|---|---|---|---|---|---|---|
| mIoU | 0.01 | 0.08 | 0.08 | 0.08 | 0.87 | **0.89** | 0.88 |

**Table 2.** Global Motion Analysis (GMA). Shift error estimation

| Configuration | Background shift error | | | | Foreground shift error | | | |
|---|---|---|---|---|---|---|---|---|
| | Mean | Min | Max | $\measuredangle$-ACC | Mean | Min | Max | $\measuredangle$-ACC |
| $\not{\mathcal{L}}_{\text{BIN}}$ | 0.55 | 0.01 | 1.16 | **1.00** | 4.46 | 0.05 | 8.50 | **1.00** |
| $\not{\mathcal{L}}_{\text{VGG}}$ | 0.52 | 0.02 | 0.90 | **1.00** | 4.38 | **0.01** | 8.51 | **1.00** |
| GLASS | **0.51** | **0.00** | **0.86** | **1.00** | **4.34** | 0.02 | **8.32** | **1.00** |

`left`, `spellcast right`, `slash front`, `slash left`, and `slash right`. The intuition under conditioning is that the global actions and the local ones should be correlated for more realism. For instance, when the global action module dictates a right shift, the only possible local action should be `walk right`. Analogously, the left shift induces the `walk left` action. The up and down shifts are animated with the `walk front` action. The remaining actions are used to animate the static sprite. To incorporate more generality and to reduce the gap with real data, we apply an independent random walk to the background image (this simulates camera motion). We use a single background image sampled from the "valleys" class of ImageNet [8]. Each video in the W-Sprites dataset is annotated with per frame actions (*i.e.*, global shifts and local action identifiers), background shifts and character masks. We show sequence samples from our dataset in Fig. 1 (to play the videos view the pdf with Acrobat Reader). The dataset contains videos with 10 to 90 frames per sprite. For testing purposes, we split the dataset into training (about 8/9 th) and validation sets (about 1/9 th). The validation set contains sprites never seen during training. For more details, see the supplementary material.

## 6 Ablations

In this section we separately ablate the global and local components of GLASS. We run the ablations on **W-Sprites**, which has been introduced in Sect. 5.

**GMA Ablations.** For the global motion analysis, we assess the impact of each loss function. Different loss terms are sequentially switched off and the performance of the model trained without those terms is reported. Given that W-Sprites is fully annotated, we propose several metrics to evaluate the training. First, we calculate the mean intersection over union (mIoU) between the ground truth and the predicted segmentation masks. Table 1 shows that the VGG loss seems to slightly hurt the segmentation performance. However, as shown in Table 2 the VGG loss benefits the shift estimation. Notice that in Table 2 we

**Table 3.** Local Motion Analysis (LMA). Component ablation results

| Configuration | LPIPS↓ | FID↓ | FVD↓ | $\text{mIoU}_{\text{RE}}$ ↑ | $\text{NMI}_{\text{G}}$ ↑ | $\text{NMI}_{\text{S}}$ ↑ | CON ↓ |
|---|---|---|---|---|---|---|---|
| Plain directions | 0.402 | 12.8 | 204 | 0.83 | 0.14 | 0.17 | 0.03 |
| Gumbel | 0.402 | 16.8 | 327 | 0.84 | 0.00 | 0.02 | 0.02 |
| No modulated convs | **0.398** | 10.4 | 172 | **0.89** | 0.35 | 0.38 | 0.03 |
| Joint input | 0.399 | **9.8** | **146** | **0.89** | 0.34 | 0.37 | 0.03 |
| $\not{\mathcal{L}}_{\text{RECS}}$ | 0.400 | 11.3 | 203 | **0.89** | 0.29 | 0.30 | **0.01** |
| GLASS 200K | 0.399 | 10.5 | 175 | 0.88 | **0.39** | **0.41** | 0.02 |
| CADDY [24] | 0.404 | **6.8** | 153 | – | – | – | – |
| GLASS 470K | **0.398** | 8.2 | **129** | 0.93 | 0.39 | 0.40 | 0.01 |

report only the cases where the masks are good enough (mIoU $> 0.8$). For the shift errors we show the $L_2$ norm of the difference between the ground truth foreground/background shift and the predicted one (in pixels). We also show the accuracy of the predicted foreground/background shift directions ($\measuredangle$-ACC). The direction is considered to be correctly predicted if the angle between the ground truth and the predicted shifts is less than 45°. Each model is trained for 60K iterations with a batch size of 4. The results are calculated on the validation set.

**LMA Ablations.** For the local motion analysis module we specifically design 5 cases that differ from GLASS in its essential components and show the results in Table 3. First, we evaluate swapping the modified mapping $T$ of the features $\psi_{t+1}$ for the direct difference between the features $\psi_{t+1} - \psi_t$ (as done in CADDY [24]). We refer to this configuration as "Plain directions". Second, we replace the vector quantization with an MLP that predicts the distribution over actions followed by the Gumbel-Softmax trick to sample a discrete action identifier. We name this model "Gumbel". We also ablate the impact of using modulated convolutional layers by feeding the action embeddings as normal inputs to common convolutional blocks. This cases is referred to as "No modulated convs". Also we consider the case where we feed the global and local action embeddings jointly to all RNN blocks instead of separate ones. We refer to this case as "Joint input". The last case that we evaluate for the ablations is the model trained without $\mathcal{L}_{\text{RECS}}$. All the models are trained for 200K iterations with a batch size of 4. Additionally we report the metrics of GLASS trained for 470K iterations. As a reference, we also show the performance of CADDY [24] trained on the W-Sprites dataset with the same configuration from the original paper for the Tennis dataset. For visual results, please, refer to the supplementary material.

Following CADDY [24], we generate the sequences from the first frames of the validation videos conditioned on the actions inferred from the remaining frames. We measure FID [14], FVD [33] and LPIPS [41] scores on the generated sequences to asses the quality of the generated videos. Additionally we segment the reconstructed sequences and report the mean IoU with the ground truth masks to asses the ability of the RNN to condition on the input global and

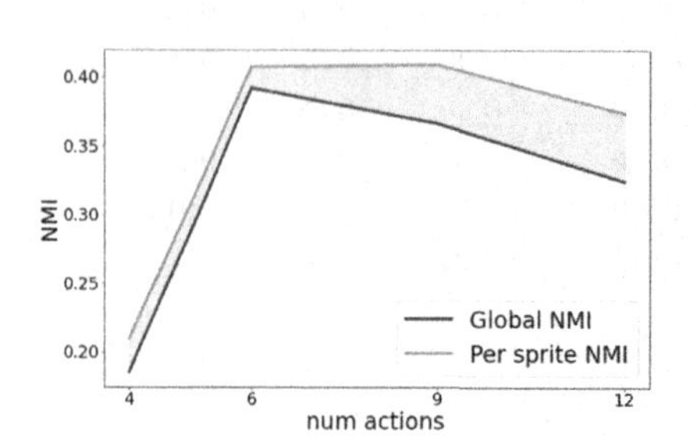


**Fig. 4.** Ablation of the number of actions fitted during the training of GLASS on the *W-Sprites* dataset.

**Fig. 5.** Example of transferring an action sequence decoded from the source video to the target. To play view the paper in Acrobat Reader.

local action embeddings. However, the most important aspect of our (and prior) work on controllable models is the identification of the action space. Thus, one needs a metric to asses the quality of the learned action space. For this purpose, we propose to use the normalized mutual information score (NMI) between the ground truth and inferred local actions

$$\mathrm{NMI}(X, Y) = \frac{2I(X,Y)}{H(X)+H(Y)}, \quad (12)$$

where $I(X, Y)$ is the mutual information between $X$ and $Y$ and $H(X)$ is the entropy of $X$. In our formulation, $X$ and $Y$ correspond to the predicted and ground truth actions respectively. One appealing advantage of NMI for GLASS is that NMI is invariant to permutations of the labels. Another advantage of using NMI is that NMI does not require the distributions to have the same number of actions. Thus, even with a given known number of ground truth actions, the model can be trained and assessed with a different number of actions. Indeed, the decomposition of a sequence into actions is not unique. For instance the `walk left` action can be decomposed into `turn left` and `walk`. We introduce two different protocols of measuring NMI. First, we use the trained model to map all the pairs of successive frames into actions. Then, the global $\mathrm{NMI_G}$ is computed between the ground truth actions and those predictions. Additionally, we average the per sprite NMI scores to obtain $\mathrm{NMI_S}$. Normally $\mathrm{NMI_S} > \mathrm{NMI_G}$. However, if the gap is large enough, this indicates the overfitting and the lack of consistency of the learned actions across different sprites. Therefore, we also report the consistency metric $\mathrm{CON} = \mathrm{NMI_S} - \mathrm{NMI_G}$. As a reference we use the $\mathrm{NMI_{RAND}}$, that is the NMI measured between the ground truth actions and random actions. The results are provided in Table 3. Given that $\mathrm{NMI_{RAND}} = 0.02$ on the W-Sprites test set, the full GLASS configuration with an NMI of 0.41

**Table 4.** *BAIR* dataset evaluation

| Method | LPIPS↓ | FID↓ | FVD↓ |
|---|---|---|---|
| MoCoGAN [32] | 0.466 | 198 | 1380 |
| MoCoGAN+ [24] | 0.201 | 66.1 | 849 |
| SAVP [22] | 0.433 | 220 | 1720 |
| SAVP+ [24] | 0.154 | 27.2 | 303 |
| Huang et al. [15] w/ *non-param* control | 0.176 | 29.3 | 293 |
| CADDY [24] | 0.202 | 35.9 | 423 |
| Huang et al. [15] w/ *positional* control | 0.202 | 28.5 | 333 |
| Huang et al. [15] w/ *affine* control | 0.201 | 30.1 | **292** |
| GLASS | **0.118** | **18.7** | 411 |

**Table 5.** *Tennis* dataset evaluation

| Method | LPIPS↓ | FID↓ | FVD↓ | ADD↓ | MDR↓ |
|---|---|---|---|---|---|
| MoCoGAN [32] | 0.266 | 132 | 3400 | 28.5 | 20.2 |
| MoCoGAN+ [24] | 0.166 | 56.8 | 1410 | 48.2 | 27.0 |
| SAVP [22] | 0.245 | 156 | 3270 | 10.7 | 19.7 |
| SAVP+ [24] | 0.104 | 25.2 | 223 | 13.4 | 19.2 |
| Huang et al. [15] w/ *non-param* control | 0.100 | 8.68 | 204 | 1.76 | 0.306 |
| CADDY [24] | 0.102 | 13.7 | 239 | 8.85 | 1.01 |
| Huang et al. [15] w/ *positional* control | 0.122 | 10.1 | 215 | 4.30 | 0.300 |
| Huang et al. [15] w/ *affine* control | 0.115 | 11.2 | **207** | 3.40 | 0.317 |
| GLASS | **0.046** | **7.37** | 257 | **2.00** | **0.214** |

shows that the model estimates action sequences with a high correlation to the ground truth actions. Note that since our main focus is learning the action space of the agent, the NMI metric is of higher importance than FID/FVD. However, since the model is based on reconstruction as a supervision signal, it is also important that the FID/FVD are within the high-quality range. Furthermore, we ablate the number of actions $K$ used to train GLASS. In Fig. 4 one can see that $K = 6$ is optimal in both NMI and CON.

## 7 Experiments on Real Data

We evaluate GLASS on 3 datasets. For synthetic data we use **W-Sprites**. For real data we use: 1) the **Tennis Dataset** and 2) the **BAIR Robot Pushing Dataset**. The Tennis Dataset was introduced in [24] and contains ~900 videos extracted from 2 Tennis matches from YouTube at $96 \times 256$ pixel resolution. The videos are cropped to contain only one half of the court, so that only one player

**Fig. 6.** A sequence generated with GLASS trained on the *W-Sprites* dataset. Note that the level of control provided by GLASS allows to generate unseen motion such as *jump*. Use Acrobat Reader to play the first frame.

is visible. The BAIR Robot Pushing Dataset [11] contains around 44K clips of a robot arm pushing toys on a flat square table at $256 \times 256$ pixel resolution.

**Baselines.** We compare to CADDY [24], because it allows frame-level playable control, and to Huang et al. [15]. However, the comparison to [15] is not fair, since it requires a prior knowledge of the future agent masks and also it lacks the ability to control the agent through discrete actions (playability). We also report the metrics on other conditional video generation models such as MoCoGAN [32], SAVP [22] and their large scale versions introduced in [24].

**Quantitative Analysis.** Following [24] we evaluate GLASS on the video reconstruction task. Given a test video, we use GMA to predict the global shifts and LMA to estimate the discrete actions performed along the video. Further, the agent is segmented using the masking network and the foreground is animated and pasted back to the shifted background using both global and local actions to reconstruct the whole sequence from the first frame. We report FID, FVD and LPIPS scores on the generated videos. On the Tennis dataset we additionally report the Average Detection Distance (ADD) and the Missing Detection Rate (MDR) suggested in [24]. Those metrics are supposed to assess the action space quality by detecting the tennis player with a pretrained human detector and by comparing the locations of the detected agents in the ground truth and generated sequences. On BAIR (see Table 4) our model performs almost 40% better in terms of frame-level quality, but lacks in FVD compared to [15]. However, it is still slightly better than CADDY. On the Tennis dataset (see Table 5) GLASS is around 50% better than the closest competitor in LPIPS, almost 30% better in FID, but loses in FVD. However, GLASS provides finer control over the agent according to ADD and MDR. It is worth noting, that the FVD is largely affected by the background modeling choices. We found that even different interpolation methods had a major impact on the FVD. However, since the main focus of this work is the action space of the foreground agent, we chose the simplest background model that yielded a performance on par (or better) with the SotA.

**Qualitative Analysis.** A trained GLASS allows a detailed control of the agent. On W-Sprites, we find that the LMA discovers such actions as `turn right`, `turn left`, `turn front`, `spellcast` and `slash`. Note that despite the difference between the discovered set of actions and the ground truth, all videos in the

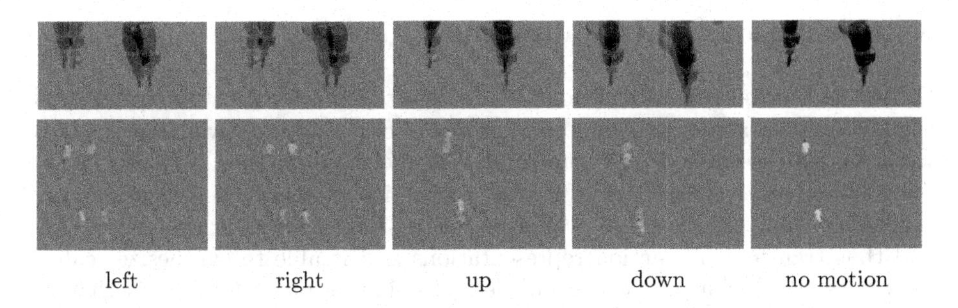


**Fig. 7.** Learned global actions on the *BAIR* and *Tennis* datasets (foreground-only generation). We use the green channel for the given initial foreground and the red channel for the foreground generated with the selected action. For both datasets we show 2 examples to demonstrate the action consistency. (Color figure online)

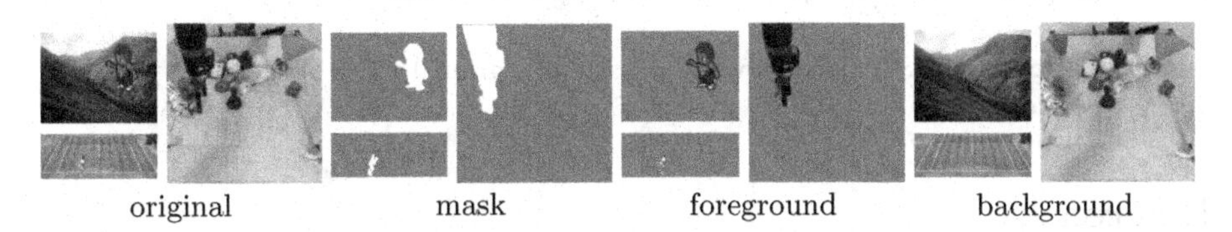


**Fig. 8.** Sample outputs from our GMA module. From left to right: original image, predicted segmentation, foreground, and inpainted background.

training set can be generated with this reduced set of actions (see Fig. 6). On Tennis we found that the local actions mostly correspond to some leg movements. On BAIR the LMA component discovers some small local deformations such as the state of the manipulator (closed or open).

In Fig. 7, we provide visual examples of the GLASS global action space. Given two different starting foregrounds from the BAIR and Tennis datasets (shown in the green channel), we show the generated foregrounds (in the red channel) after applying the `right`, `left`, `down`, `up` and `no motion` global shifts. We can also see that global actions apply consistently across different initial foregrounds. To show that the learned action space is consistent across different agents also in their fine-grained dynamics we use GLASS to transfer (both global and local) motion from one video to another. We first extract the sequence of actions in the first video using the GMA and LMA components of GLASS and then sequentially apply these actions to the first frame of the second video. In Fig. 5, we demonstrate it on the Tennis dataset.

Finally, in Fig. 8 we provide some sample outputs from our GMA module on test images from all three datasets. Given an input image, we can see that the segmentation network learns to extract accurate masks with which one can obtain high quality foreground images. These are necessary to model local dynamics. The inpainting of the background is sufficiently accurate to separate the two layers. For more visual examples, please see the supplementary material.

## 8 Conclusions and Limitations

GLASS is a novel generative model with a global and local action space that enables a fine-grained modeling and control of dynamics not available in prior work. GLASS is trained in a completely unsupervised manner. We also introduce W-Sprites, a novel dataset for the evaluation of action identification and generation. Our experimental evaluation shows that GLASS learns consistent, and thus transferrable, action representations and is able to synthesize realistic videos with arbitrary action policies. One limitation of GLASS is that it works only on data where the background dynamics are 2D shifts. This does not capture, for example, the motion of the background objects in the BAIR dataset. GLASS is capable of rendering multiple agents within the same frame, but does not learn the separate action spaces of multiple agents. This remains a challenging task, which we plan to tackle in future work.

**Acknowledgements.** This work was supported by grant 188690 of the Swiss National Science Foundation.

## References

1. Liberated pixel cup. https://lpc.opengameart.org. Accessed 07 Mar 2022
2. Universal LPC sprite sheet. https://github.com/makrohn/Universal-LPC-spritesheet/tree/7040e2fe85d2cb1e8154ec5fce382589d369bdb8. Accessed 07 Mar 2022
3. Acharya, D., Huang, Z., Paudel, D.P., Van Gool, L.: Towards high resolution video generation with progressive growing of sliced Wasserstein GANs. arXiv preprint arXiv:1810.02419 (2018)
4. Babaeizadeh, M., Finn, C., Erhan, D., Campbell, R.H., Levine, S.: Stochastic variational video prediction. arXiv preprint arXiv:1710.11252 (2017)
5. Burgess, C.P., et al.: Understanding disentangling in $\beta$-vae. arXiv preprint arXiv:1804.03599 (2018)
6. Caron, M., Bojanowski, P., Joulin, A., Douze, M.: Deep clustering for unsupervised learning of visual features. In: Proceedings of the European Conference on Computer Vision (ECCV), pp. 132–149 (2018)
7. Chiappa, S., Racaniere, S., Wierstra, D., Mohamed, S.: Recurrent environment simulators. arXiv preprint arXiv:1704.02254 (2017)
8. Deng, J., Dong, W., Socher, R., Li, L.J., Li, K., Fei-Fei, L.: ImageNet: a large-scale hierarchical image database. In: 2009 IEEE Conference on Computer Vision and Pattern Recognition, pp. 248–255. IEEE (2009)
9. Denton, E., Fergus, R.: Stochastic video generation with a learned prior. In: International Conference on Machine Learning, pp. 1174–1183. PMLR (2018)
10. Dhariwal, P., Nichol, A.: Diffusion models beat GANs on image synthesis. Adv. Neural Inf. Process. Syst. **34**, 8780–8794 (2021)
11. Ebert, F., Finn, C., Lee, A.X., Levine, S.: Self-supervised visual planning with temporal skip connections. In: CoRL, pp. 344–356 (2017)
12. Finn, C., Goodfellow, I., Levine, S.: Unsupervised learning for physical interaction through video prediction. Advances in Neural Inf. Process. Syst. **29** (2016)

13. Gafni, O., Wolf, L., Taigman, Y.: Vid2game: controllable characters extracted from real-world videos. arXiv preprint arXiv:1904.08379 (2019)
14. Heusel, M., Ramsauer, H., Unterthiner, T., Nessler, B., Hochreiter, S.: GANs trained by a two time-scale update rule converge to a local nash equilibrium. Adv. Neural Inf. Proces. Syst. **30** (2017)
15. Huang, J., Jin, Y., Yi, K.M., Sigal, L.: Layered controllable video generation. arXiv preprint arXiv:2111.12747 (2021)
16. Johnson, J., Alahi, A., Fei-Fei, L.: Perceptual losses for real-time style transfer and super-resolution. In: Leibe, B., Matas, J., Sebe, N., Welling, M. (eds.) ECCV 2016. LNCS, vol. 9906, pp. 694–711. Springer, Cham (2016). https://doi.org/10.1007/978-3-319-46475-6_43
17. Karras, T., et al.: Alias-free generative adversarial networks. Advances Neural Inf. Process. Syst. **34**, 852–863 (2021)
18. Kim, S.W., Zhou, Y., Philion, J., Torralba, A., Fidler, S.: Learning to simulate dynamic environments with GameGAN. In: Proceedings of the IEEE/CVF Conference on Computer Vision and Pattern Recognition, pp. 1231–1240 (2020)
19. Kim, Y., Nam, S., Cho, I., Kim, S.J.: Unsupervised keypoint learning for guiding class-conditional video prediction. Adv. Neural Inf. Process. Syst. **32** (2019)
20. Kingma, D.P., Ba, J.: Adam: a method for stochastic optimization. arXiv preprint arXiv:1412.6980 (2014)
21. Kingma, D.P., Welling, M.: Auto-encoding variational bayes. arXiv preprint arXiv:1312.6114 (2013)
22. Lee, A.X., Zhang, R., Ebert, F., Abbeel, P., Finn, C., Levine, S.: Stochastic adversarial video prediction. arXiv preprint arXiv:1804.01523 (2018)
23. Li, Y., Mandt, S.: Disentangled sequential autoencoder. In: International Conference on Machine Learning (2018)
24. Menapace, W., Lathuilière, S., Tulyakov, S., Siarohin, A., Ricci, E.: Playable video generation. In: Proceedings of the IEEE/CVF Conference on Computer Vision and Pattern Recognition, pp. 10061–10070 (2021)
25. Mottaghi, R., Bagherinezhad, H., Rastegari, M., Farhadi, A.: Newtonian scene understanding: unfolding the dynamics of objects in static images. In: Proceedings of the IEEE Conference on Computer Vision and Pattern Recognition, pp. 3521–3529 (2016)
26. Nunes, M.S., Dehban, A., Moreno, P., Santos-Victor, J.: Action-conditioned benchmarking of robotic video prediction models: a comparative study. In: 2020 IEEE International Conference on Robotics and Automation (ICRA), pp. 8316–8322. IEEE (2020)
27. Oh, J., Guo, X., Lee, H., Lewis, R.L., Singh, S.: Action-conditional video prediction using deep networks in Atari games. Adv. Neural Inf. Process. Syst. **28** (2015)
28. Razavi, A., Van den Oord, A., Vinyals, O.: Generating diverse high-fidelity images with VQ-VAE-2. Adv. Neural Inf. Process. Syst. **32** (2019)
29. Rybkin, O., Pertsch, K., Derpanis, K.G., Daniilidis, K., Jaegle, A.: Learning what you can do before doing anything. arXiv preprint arXiv:1806.09655 (2018)
30. Smith, L., Gasser, M.: The development of embodied cognition: six lessons from babies. Artif. Life **11**(1–2), 13–29 (2005)
31. Srivastava, N., Mansimov, E., Salakhudinov, R.: Unsupervised learning of video representations using LSTMs. In: International Conference on Machine Learning, pp. 843–852. PMLR (2015)
32. Tulyakov, S., Liu, M.Y., Yang, X., Kautz, J.: MocoGAN: decomposing motion and content for video generation. In: Proceedings of the IEEE Conference on Computer Vision and Pattern Recognition, pp. 1526–1535 (2018)

33. Unterthiner, T., van Steenkiste, S., Kurach, K., Marinier, R., Michalski, M., Gelly, S.: Towards accurate generative models of video: a new metric & challenges. arXiv preprint arXiv:1812.01717 (2018)
34. Van Den Oord, A., Vinyals, O., et al.: Neural discrete representation learning. Adv. Neural Inf. Process. Syst. **30** (2017)
35. Vondrick, C., Pirsiavash, H., Torralba, A.: Anticipating the future by watching unlabeled video. arXiv preprint arXiv:1504.08023 **2**, 2 (2015)
36. Vondrick, C., Pirsiavash, H., Torralba, A.: Generating videos with scene dynamics. Adv. Neural Inf. Process. Syst. **29** (2016)
37. Wang, Y., Bilinski, P., Bremond, F., Dantcheva, A.: G3AN: disentangling appearance and motion for video generation. In: Proceedings of the IEEE/CVF Conference on Computer Vision and Pattern Recognition, pp. 5264–5273 (2020)
38. Wang, Y., Bilinski, P., Bremond, F., Dantcheva, A.: Imaginator: conditional spatio-temporal GAN for video generation. In: Proceedings of the IEEE/CVF Winter Conference on Applications of Computer Vision, pp. 1160–1169 (2020)
39. Weissenborn, D., Täckström, O., Uszkoreit, J.: Scaling autoregressive video models. arXiv preprint arXiv:1906.02634 (2019)
40. Yan, W., Zhang, Y., Abbeel, P., Srinivas, A.: VideoGPT: video generation using VQ-VAE and transformers. arXiv preprint arXiv:2104.10157 (2021)
41. Zhang, R., Isola, P., Efros, A.A., Shechtman, E., Wang, O.: The unreasonable effectiveness of deep features as a perceptual metric. In: Proceedings of the IEEE Conference on Computer Vision and Pattern Recognition, pp. 586–595 (2018)

# StyleHEAT: One-Shot High-Resolution Editable Talking Face Generation via Pre-trained StyleGAN

Fei Yin[1], Yong Zhang[2(✉)], Xiaodong Cun[2], Mingdeng Cao[1], Yanbo Fan[2], Xuan Wang[2], Qingyan Bai[1], Baoyuan Wu[3], Jue Wang[2], and Yujiu Yang[1(✉)]

[1] Tsinghua Shenzhen International Graduate School, Tsinghua University, Beijing, China
yang.yujiu@sz.tsinghua.edu.cn
[2] Tencent AI Lab, Shenzhen, China
zhangyong201303@gmail.com
[3] School of Data Science, Secure Computing Lab of Big Data, The Chinese University of Hong Kong, Shenzhen, China

**Abstract.** One-shot talking face generation aims at synthesizing a high-quality talking face video from an arbitrary portrait image, driven by a video or an audio segment. In this work, we provide a solution from a novel perspective that differs from existing frameworks. We first investigate the latent feature space of a pre-trained StyleGAN and discover some excellent spatial transformation properties. Upon the observation, we propose a novel unified framework based on a pre-trained StyleGAN that enables a set of powerful functionalities, *i.e.*, high-resolution video generation, disentangled control by driving video or audio, and flexible face editing. Our framework elevates the resolution of the synthesized talking face to $1024 \times 1024$ for the first time, even though the training dataset has a lower resolution. Moreover, our framework allows two types of facial editing, *i.e.*, global editing via GAN inversion and intuitive editing via 3D morphable models. Comprehensive experiments show superior video quality and flexible controllability over state-of-the-art methods. Code is available at https://github.com/FeiiYin/StyleHEAT.

## 1 Introduction

One-shot talking face generation refers to the task of synthesizing a high-quality talking face video from a given portrait image, guided by a driving video or audio segment. The synthesized face inherits the identity information from the portrait image, while its pose and expression are transferred from the driving video or generated based on the driving audio. Talking face generation has a variety of important applications such as digital human animation, film production, *etc.*.

Work done during an internship at Tencent AI Lab.

**Supplementary Information** The online version contains supplementary material available at https://doi.org/10.1007/978-3-031-19790-1_6.

© The Author(s), under exclusive license to Springer Nature Switzerland AG 2022
S. Avidan et al. (Eds.): ECCV 2022, LNCS 13677, pp. 85–101, 2022.
https://doi.org/10.1007/978-3-031-19790-1_6

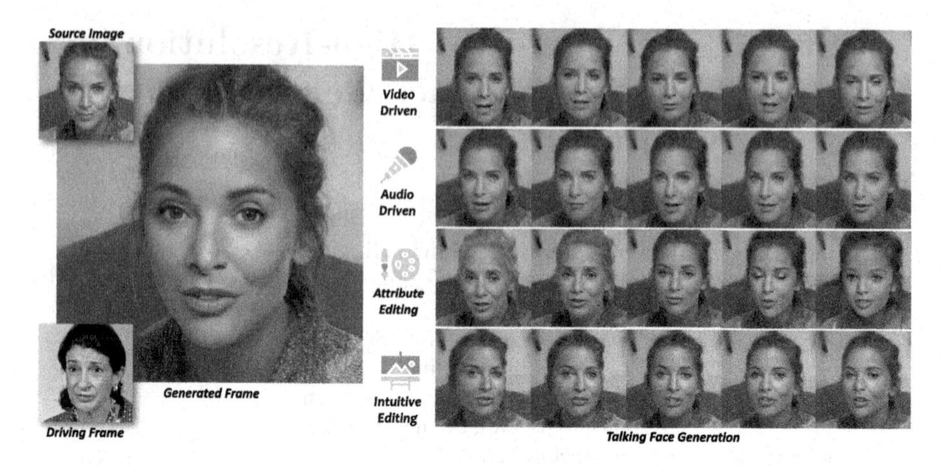


**Fig. 1.** Our unified framework enables high-resolution talking face generation, disentangled control by a driving video or audio, and flexible face editing.

Recent one-shot talking face generation methods [28,38,47] have made notable progress in driving expression and pose. However, they fail to generate high-resolution video frames. The video resolution of the ordinary methods still remains at 256 × 256. Few methods such as [38] and [47] have achieved the resolution of 512 × 512 by exploiting newly collected high-resolution datasets, but they are still bounded by the resolution of the training data. More importantly, improving the resolution requires properly designed network architectures and training strategies. Adding upsampling layers in a straightforward way into the network usually does not work well. Warping-based methods can be directly applied to higher-resolution images but will introduce inevitable artifacts. [28] and [14] utilize a post refining network to eliminate the artifacts, but limit the resolution of the finally synthesised results at the same time.

Face editing is an useful technique to enhance talking face videos, *e.g.*, the modification of facial appearance, pose, and expression. It has two categories, *i.e.*, intuitive editing (*e.g.*, pose and expression) and semantic facial attribute editing (*e.g.*, makeup, beard, and age). Only few talking face generation methods [14,28] enable intuitive editing via 3D morphable models (3DMMs). But there is no existing work that incorporates semantic attribute editing into the talking face generation framework. Besides, almost all previous methods provide frameworks for either the video-driven case or the audio-driven case, but few consider both except for [28]. Therefore, it is another challenge to integrate the driving and editing modules of different modalities into a unified framework.

We raise two ambitious questions: can we further improve the resolution of one-shot talking face to 1024 × 1024 even though the existing datasets have a lower resolution? Can we build a unified framework that enables different types of driving modalities as well as semantic and intuitive face editing? To achieve these goals, we resort to a powerful pre-trained generative model: StyleGAN [23]. StyleGAN has shown impressive results in various applications, *e.g.*, facial attribute editing [11], blind image restoration [39], portrait stylization [34], *etc.*. These methods utilize the learned image prior of StyleGAN to facilitate downstream tasks, removing the need of training a large model from scratch. The resolution is retained at 1024 × 1024 and visual details are also reserved. Despite these successes, to the best of our knowledge, there is no existing work that uses a pre-trained StyleGAN for one-shot talking face generation.

In this work, we first investigate the latent style space and the feature space of a pre-trained StyleGAN. The style space is also called $\mathcal{W}$ space. The style space is extensively explored by GAN inversion methods for face editing. The feature space is also called $\mathcal{F}$ space. In a talking face video, different facial expressions are achieved by deforming different facial regions in different ways. Given that style codes do not contain accurate spatial information, the style space might not be an appropriate choice for injecting facial motion. We then systematically study the feature space by applying a set of spatial transformations on the feature map of StyleGAN. Interestingly, we discover that the pre-trained model is robust to some geometric transformations as it can steadily generate high-quality images accordingly, indicating that the feature space has satisfying spatial properties.

Upon the above observation, we propose a novel unified framework for high-quality one-shot talking face generation based on a pre-trained StyleGAN. Specifically, we directly deform the StyleGAN spatial features using flow fields predicted by the video-based or audio-based motion generator, and then a calibration network is proposed to modulate the warped features. Such a design preserves the facial prior of the StyleGAN, enabling our model to generate high-resolution results while eliminating warping-induced flaws. Thanks to the pre-trained StyleGAN, our framework also allows two types of face editing, *i.e.*, global editing via GAN inversion and intuitive editing pose and expression based on 3DMM. Figure 1 illustrates the functionalities of the proposed framework.

Our main contributions are as follows:

- We propose a unified framework based on a pre-trained StyleGAN for one-shot talking face generation. It enables high-resolution video generation, disentangled control by driving video and audio, and flexible face editing.
- We conduct comprehensive experiments to illustrate the various capabilities of our framework and compare it with many state-of-the-art methods.

## 2 Related Work

**3D Structure-Based Talking-Face Generation.** Traditionally, 3D faces model priors (such as 3DMM [7]) provide a powerful tool for rendering and editing the portrait images by the parameters modulation. For example, DVP [25] modifies the parameters from source and target, then, a network is used to render the shading to video. Recent 3D model-based methods [10,14,16,25,28] can also do a good job for subject-agnostic face synthesis. HeadGAN [14] pre-processes the 3d mesh as input. PIRenderer [28] predicts a flow field for feature warping.

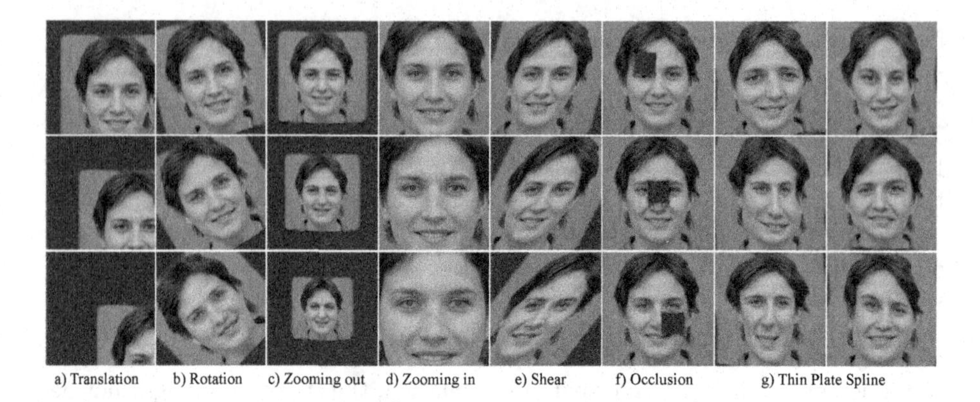


**Fig. 2.** Latent feature space investigation of a pre-trained StyleGAN. Different geometric transformations are applied to modify the feature maps.

**2D-Based Talking-Face Generation.** Instead of controlling the model parameters, mimicking the motions of another individual by the neural network is also a popular direction. Subject-agnostic approaches [4,9,30–32,38], which only need a single image of the target person are the most popular type. For the representative methods, Monkey-Net [31] propose a network to transfer the deformation from sparse to dense motion flow. FOMM [30] extends Monkey-Net via the first-order local affine transformations. Then, Face-vid2vid [38] improves FOMM via a learned 3D unsupervised key-points for free-view talking head generation. A concurrent work [8] also tries to explore the style space for face animation, but they need to fine-tune the pre-trained StyleGAN on the specific domain.

## 3 Investigating Feature Space of StyleGAN

StyleGAN2 [24] draw attention from the community since it can generate high-quality face images and the feature space is highly disentangled. To allow a pre-trained StyleGAN [24] for high-resolution talking-head video generation, one possible direction is StyleGAN based video generation [15,35], where they learn to generate videos via discovering an ideal trajectory in $\mathcal{W}^+$ latent space. However, the motion is randomly sampled without any control and the content is corrupted when the current pose differs from the initial one. This is because $\mathcal{W}^+$ is a highly semantic-condensed space and lacks explicit spatial prior [37]. Moreover, editing in $\mathcal{W}^+$ space [1–3,29,36,41] only allows changing high-level facial attributes, which cannot generate out-of-alignment images [20] since the StyleGAN is trained on aligned faces.

Thus, image editing in $\mathcal{F}$ feature space draws our close attention. Specifically, the latent code $\boldsymbol{f}$ in $\mathcal{F}$ feature space represents a spatial feature map in the generator. For StyleGAN [24], we define $\boldsymbol{f}$ as the feature map after a pair of upsampling and convolution layers at a certain scale. There are only a few previous methods [5,20,37,39,48] that edit the spatial features for GAN inversion [5,20,37], image composition [48], and blind face enhancement [39]. These approaches harvest the potential of spatial feature space editing and apply the spatial modulation (*e.g.* spatial feature transformation [40]) to the features. However, it has not been fully investigated whether the feature space of a pre-trained StyleGAN can still be used to generate realistic images after various geometric transformations.

We therefore conduct a detailed experiment to verify the spatial property of StyleGAN features and fully excavate its potential capability. We first randomly sample the style latent code $\boldsymbol{w}$ in $\mathcal{W}^+$ space to generate a random face image with the pre-trained StyleGAN. At the same time, various spatial features $[\boldsymbol{f}_{4\times4}, \boldsymbol{f}_{8\times8}, \dots, \boldsymbol{f}_{1024\times1024}]$ in $\mathcal{F}$ space can be obtained. We choose the feature resolution of $64 \times 64$ for a balanced trade-off between the inversion quality and editing capacity. To test the spatial property of the pre-trained StyleGAN features, several geometric transformations, including translation, rotation, zoom, shear, occlusion and Thin Plate Spline (TPS [42]), are used to manipulate $\boldsymbol{f}_{64\times64}$ directly. Finally, the transformed image can be generated by the forward pass with the edited feature map as input. Our experimental results are shown in Fig. 2. Either with simple affine transformations or complicated TPS deformations, we observe that the generated images maintain the same geometric changes as the deformations applied in the feature space. The generated images also share the same identity and appearance with a minor difference. This phenomenon demonstrates that the learned convolution kernels in the pre-trained generator perform in a transformation-invariant manner.

Based on the above observation, we can conclude that the features in a pre-trained StyleGAN preserve strong spatial prior and can be directly modified with geometric transformations. This spatial property makes it a promising direction to edit the feature space for talking face generation.

## 4 Methodology

We are interested in the task of controllable talking-head generation. Let $I$ be the source image and $\{d_1, d_2, \cdots, d_N\}$ be a talking-head video, where $d_i$ is the $i$-th video frame and $N$ is the total number of frames. An ideal framework is supposed to generate video $\{y_1, y_2, \cdots, y_N\}$ with the same identity as $I$ and the consistent motions derived from $\{d_1, d_2, \cdots, d_N\}$.

Inspired by our observation in Sect. 3, we propose a unified framework based on the $\mathcal{F}$ space excavation of the pre-trained StyleGAN $G$. As shown in Fig. 3, our approach contains several steps to achieve this goal. Given a single source image, we first use the GAN inversion method [37] to get the latent style code $\boldsymbol{w}$ and feature maps $\boldsymbol{f}$ of the source image. Then, to inject the accurate motion guidance, we predict a dense flow field by the motion generator $\Phi_{warp}$ from video (Sect. 4.1). Since the warping operation may introduce artifacts due to the occlusions and error mapping, a calibration network $\Phi_{cali}$ is introduced to renovate the edited spatial feature map (Sect. 4.2). Our framework can be extended

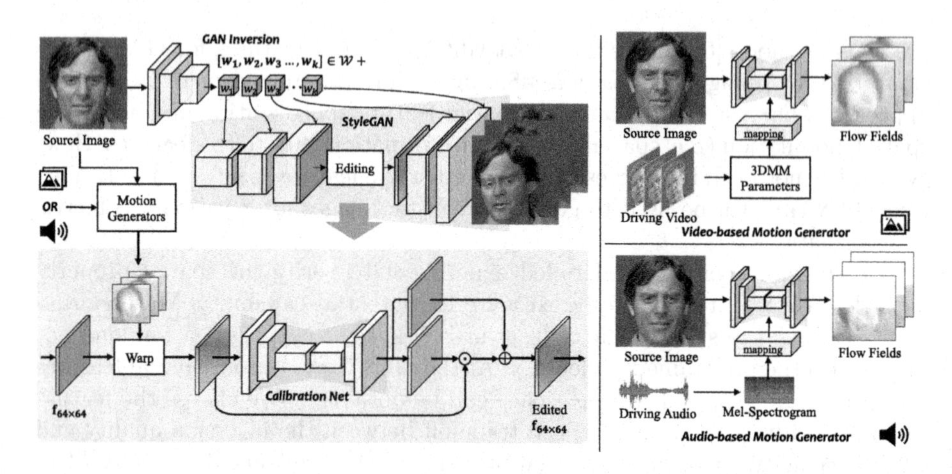


**Fig. 3.** The pipeline of our unified framework. The framework consists of four components, *i.e.*, a pre-trained StyleGAN, a video-driven motion generator, an audio-driven motion generator, and a calibration network. Given a source image, we can obtain the style codes and feature maps by the encoder of GAN inversion. The driven video or audio along with the source image are used to predict motion fields by the corresponding motion generator. The selected feature map is warped by the motion fields, followed by the calibration network for rectifying feature distortions. The refined feature map is then fed into the StyleGAN for the final face generation.

to audio-driven via similar flow prediction module (Sect. 4.3). The whole framework can be summarized as:

$$\hat{I}_i = G\left(\Phi_{cali}(\Phi_{warp}(I, d_i) \circ \boldsymbol{f}), \boldsymbol{w}\right), \tag{1}$$

where $\circ$ denotes the warping transformation.

### 4.1 Video-Driven Motion Generator

The goal of the video-driven motion generator is to generate dense flows with the driving video and the source image as inputs. Then, these flow fields will manipulate the feature map of the pre-trained StyleGAN for talking face generation. In this part, we first demonstrate the intermediate motion representation in our settings. Then, we give the details of the network structure and the training process for the dense motion field generation.

**Motion Representation.** To achieve accurate and intuitive motion control, semantic medium plays an important role in the generation process. Following previous works [14,28], we take advantage of the 3DMM [6] parameters for motion modeling. In 3DMM, the 3D shape $\boldsymbol{S}$ of a face can be decoupled as:

$$\boldsymbol{S} = \overline{\boldsymbol{S}} + \boldsymbol{\alpha}\boldsymbol{U}_{id} + \boldsymbol{\beta}\boldsymbol{U}_{exp}, \tag{2}$$

where $\overline{\boldsymbol{S}}$ is the average shape, $\boldsymbol{U}_{id}$ and $\boldsymbol{U}_{exp}$ are the orthonormal basis of identity and expression of LSFM morphable model [7]. Coeffcients $\boldsymbol{\alpha} \in \mathbb{R}^{80}$ and

$\boldsymbol{\beta} \in \mathbb{R}^{64}$ describe the person identity and expression, respectively. To preserve pose variance, coefficients $\boldsymbol{r} \in SO(3)$ and $\boldsymbol{t} \in \mathbb{R}^3$ denote the head rotation and translation. Then, we can model the motion of the driving face with a parameter set $\boldsymbol{p} = \{\boldsymbol{\beta}, \boldsymbol{r}, \boldsymbol{t}\}$ extracted by an existing 3D face reconstruction model [13].

Due to the inevitable prediction errors between consecutive frames in the same video, the parameters from a single input frame will cause jitter and instability in the finally generated video. Hence, we adopt a windowing strategy for better temporal consistency, where the parameters of the neighboring frames are also taken as the descriptor of the center frame to smooth the motion trajectory. Thus, the motion coefficient of the $i$-th driving frame is defined as:

$$\boldsymbol{p}_i \equiv \boldsymbol{p}_{i-k:i+k} \equiv \{\boldsymbol{\beta}_{i-k}, \boldsymbol{r}_{i-k}, \boldsymbol{t}_{i-k}, \ldots, \boldsymbol{\beta}_i, \boldsymbol{r}_i, \boldsymbol{t}_i, \ldots, \boldsymbol{\beta}_{i+k}, \boldsymbol{r}_{i+k}, \boldsymbol{t}_{i+k}\}, \tag{3}$$

where $k$ is the radius of the window.

**Network Structure.** Our network is built on a U-Net structure that requires the source image and the driving video as inputs, and the outputs are the desired flow fields for feature warping. It contains a 5-layer convolutional encoder and a 3-layer convolutional decoder for multi-scale feature extraction. We use the 3DMM parameters $\boldsymbol{p}_t$ from the driving frame $d_t$ as the motion representation. Specifically, these parameters are first mapped to a latent vector via a 3-layer MLP to aggregate the temporal information. Then, the motion parameters are injected into each convolutional layer via the adaptive instance normalization (AdaIN [18]). Next, the network can be trained by the source image $I$ and the motion condition $\boldsymbol{p}_t$ as inputs. Finally, the loss functions will be calculated between the target image $d_t$ and the generated image by the backward warping.

**Pre-training Strategy.** As we have investigated in Sect. 3, the feature warping shares the same geometry deformation with the final image. Thus, to simplify the learning of the whole framework, before joint training we first pre-train the video-based motion generator on the widely-used talking-face datasets (VoxCeleb [26]) to generate trustful flow fields in a self-supervised manner. Then, the low-resolution flow field are used to drive the spatial feature map of the pre-trained StyleGAN. Specifically, as the ground truth flow fields are not available, we predict the flow fields $\boldsymbol{n}$ using the network, and then the source frame $\boldsymbol{I}$ will be used to calculate the warped frame by $\hat{\boldsymbol{I}}_n = \boldsymbol{I} \circ \boldsymbol{n}$. Then, given the target frame $\boldsymbol{I}_t$, we use the perceptual loss [19] to calculate the $\mathcal{L}_1$ distance between the activation maps of the pre-trained VGG-19 network [33]:

$$\mathcal{L}^v = \sum_i \|\phi_i(\hat{\boldsymbol{I}}_n) - \phi_i(\boldsymbol{I}_t)\|_1, \tag{4}$$

where $\phi_i$ denotes the activation map of the $i$-th layer of the VGG-19 network. Similar to [30], we calculate the perceptual loss on a number of resolutions by applying pyramid down-sampling on $\boldsymbol{I}_t$ and $\hat{\boldsymbol{I}}_n$. After training, the generated flow field can be used to edit the feature map of StyleGAN.

### 4.2 Feature Calibration and Joint Training

The video-driven motion generators are pre-trained without considering any information about the pre-trained StyleGAN. Though the predicted motion fields can be used to warp the feature map of StyleGAN, it will inevitably introduce artifacts. For example, making a closed mouth open through 2D warping cannot fill correct teeth within the mouth. To alleviate the feature map distortion, we introduce a calibration network to rectify artifacts in the feature space.

**Calibration Network.** A calibration network is needed since the warped features still suffer from artifacts. As shown in Fig. 3, we adopt a U-Net architecture to extract multi-resolution spatial features. It consists of a 4-layer encoder and a 4-layer decoder. We feed the warped feature map $\boldsymbol{f}_w$ as the network's input. Then, the multi-scale conventional layers are used to refine the warped features. However, due to the high complexity of the intermediate features, instead of directly predicting the features, our calibration network performs the spatial feature transformation (SFT [38]) to the warped features, which is defined as:

$$\hat{\boldsymbol{f}}_c = SFT(\boldsymbol{f}_w|\boldsymbol{r}, \boldsymbol{t}) = \boldsymbol{r} \odot \boldsymbol{f}_w + \boldsymbol{t}, \tag{5}$$

where $\odot$ denotes element-wise multiplication. Then, the final high-quality and high-resolution result can be achieved as $\hat{I} = G(\hat{\boldsymbol{f}}_c, \boldsymbol{w})$.

**Overall End-to-End Training.** Directly applying the introduced calibration network is easy to encounter blur results (as shown in Fig. 12) since the quality of the frames in the video dataset is much lower than the high-resolution face dataset for training StyleGAN. Furthermore, inevitable detail lost of identity, attribute, texture, and background raised by the GAN inversion method will enlarge the gap between the generated images and the real images, which will further mislead the direction of the optimization.

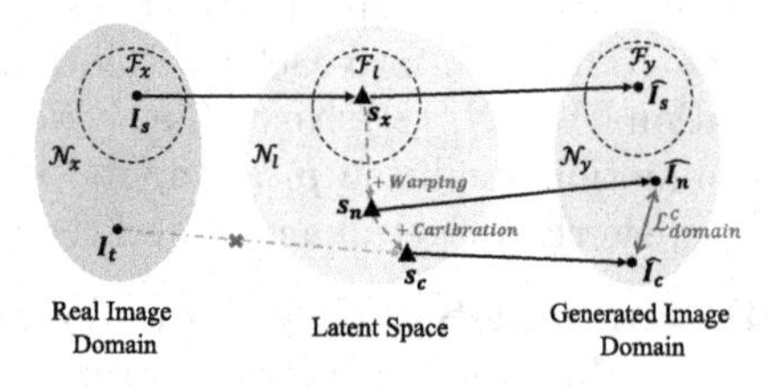


**Fig. 4.** Illustration of the domain loss.

Thus, we joint train the whole network except the pre-trained StyleGAN and design loss functions to solve the above problem. We first design a domain loss to restrict the differences between the reconstructed image of the warped feature map and that of the calibrated feature map in the generated image domain. As shown in Fig. 4, given a natural source image $\boldsymbol{I}_s$ in the aligned StyleGAN space $\mathcal{F}_x$, the GAN inversion method can invert and reconstruct the image in the latent space and the generated image domain, respectively. Differently, for the target image $\boldsymbol{I}_t$ which is out of the aligned domain, GAN inversion is hard to be applied. Thus, to obtain the desired latent space $\boldsymbol{s}_c$, the proposed method utilizes the flow fields to edit the images in the latent space. After editing, the warped feature $\boldsymbol{s}_n$ may not be in the aligned StyleGAN latent space anymore but it can still generate a high-quality image $\hat{\boldsymbol{I}}_n$ by forwarding pass as we have discussed in Sect. 3. Unfortunately, the warping artifacts may occur because of the low quality of the flow fields. Thus, we propose the calibration network to further edit the feature map as introduced previously. However, the results $\hat{\boldsymbol{I}}_c$ become blurry due to the feature shift. To preserve both advantages of $\hat{\boldsymbol{I}}_n$ and $\hat{\boldsymbol{I}}_c$, the domain loss is defined to measure their difference. Further, we take a masking strategy to enhance the weight of different areas. The calibration mask $M$ is comprised of the bounding boxes of the eyes and mouth because the artifacts often occur around them. Thus, the domain loss is:

$$\mathcal{L}^c_{domain} = \sum_i \|(1-M)\cdot\phi_i(\hat{\boldsymbol{I}}_n) - (1-M)\cdot\phi_i(\hat{\boldsymbol{I}}_c)\|_1. \tag{6}$$

Besides, for eliminating the artifacts of local facial features, the driving image $\boldsymbol{I}_t$ provides the most accurate high-frequency information. Hence, we calculate the $\mathcal{L}_1$ loss and the perceptual loss with the ground truth, which is weighted on the masked region:

$$\mathcal{L}^c_t = \sum_i \|M\cdot\phi_i(\boldsymbol{I}_t) - M\cdot\phi_i(\hat{\boldsymbol{I}}_c)\|_1 + \lambda^c_1 \cdot \|M\cdot\boldsymbol{I}_t - M\cdot\hat{\boldsymbol{I}}_c\|_1, \tag{7}$$

where $\lambda^c_1$ is the weight of the $\mathcal{L}_1$ loss.

Finally, to maintain the high fidelity of face generation, we also impose adversarial loss. Note that we freeze the parameters of the discriminator since the low-quality video frames may decline its performance. The adversarial loss is:

$$\mathcal{L}^c_{adv} = -\mathbb{E}[\log(D(\hat{\boldsymbol{I}}_c))], \tag{8}$$

where $D$ is a well-trained discriminator of StyleGAN2.

The framework is trained in an end-to-end manner together with the loss of the corresponding motion generators. Here, we calculate the perceptual loss between the intermediate results from the motion generator and the ground truth, which is the same as Eq. 4. The weight of other components (the StyleGAN generator and the inversion encoder) are frozen.

In summary, the overall loss is a weighted summation as follows:

$$\mathcal{L}^c = \mathcal{L}^c_t + \lambda^c_d \cdot \mathcal{L}^c_{domain} + \lambda^c_{adv} \cdot \mathcal{L}^c_{adv} + \beta^v \cdot \mathcal{L}^v, \tag{9}$$

where $\lambda^c_d$, $\lambda^c_{adv}$, and $\beta^v$ are the corresponding weights.

### 4.3 Extension on Audio-Driven Reenactment

We can further extend our framework to tackle the audio-driven facial reenactment task by extracting sequential motions from audio input. Audio-driven motion transfer is similar to video-driven motion transfer, but requires modeling the relationships between audio and

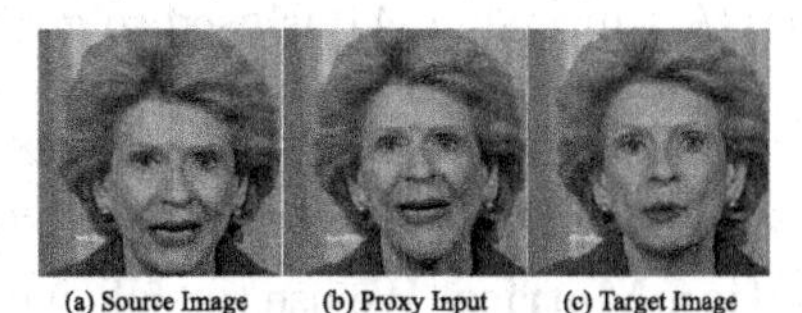


**Fig. 5.** Paired training data generation for audio-driven motion generator training.

face motions. Directly predicting the visual semantic parameters from audio information only is a difficult task and the two-stage converting procedure may accumulate errors. Consequently, we directly predict the motion from audio features as shown in the audio-based motion generator of Fig. 3.

In detail, we train the generator to predict the flow fields in the lower half face, since audio is closely related to lip movements. However, a major challenge, generating a video from audio lacks a paired dataset because the videos with the same pose but different lip shapes are hard to obtain. To address this issue, we construct the paired data with the same pose but different expressions under different audio conditions by utilizing the pre-trained video-driven motion generator in Sect. 4.1. Specifically, we generate the proxy input by mixing the 3DMM parameters extracted from the source and driving frame, *i.e.,* the proxy input has the same pose as the driving frame and the same expression as the source frame. We illustrate the main process in Fig. 5, where the head pose of the proxy input is high-aligned with the driving frame. By training on the paired dataset, our audio-driven generator will focus on the flow generation of expression.

After training the audio-driven motion generator, it can be added to the framework as a plugin to control the lip movement independently. The details of the network structure and the training procedure will be discussed in the supplementary materials.

## 5 Experiments

**Datasets.** We train the two motion generators on the VoxCeleb dataset [26] which consists of over 100K videos of 1,251 subjects. We joint train the whole framework on the HDTF dataset [47] which consists of 362 videos of over 300 subjects. HDTF is split into non-overlapping training and test sets. The test set contains 20 videos with around 10K frames. For cross-identity motion transfer evaluation, we select 1K high-resolution images from the CelebA-HQ dataset [21].

**Implementation Details.** We train the two motion generators and the calibration network in two stages. In the first stage, we pre-train the video-based motion generator on VoxCeleb. Then, we pre-train the audio-based generator with synthesized audio-motion pairs. As the motion from the pre-trained generators cannot be seamlessly applied to feature maps of StyleGAN, we need to finetune them along with the calibration network in the second stage. During inference, the two motion generators can be used individually or jointly.

The GAN inversion [44] is used to get the spatial feature maps in our framework. We exploit a learning-based inversion method [37] during training and an optimization-based inversion method [48] to optimize latent feature maps for more accurate reconstruction during inference.

**Evaluation Metrics.** We use the following metrics for evaluation: Learned Perceptual Image Patch Similarity (LPIPS) [46], Peak signal-to-noise ratio (PSNR), Structural Similarity (SSIM), Frechet Inception Distance (FID) [17], the cosine similarity (CSIM) of identity embeddings extracted from ArcFace [12], Average Expression Distance (AED) [28], and Average Pose Distance (APD) [28].

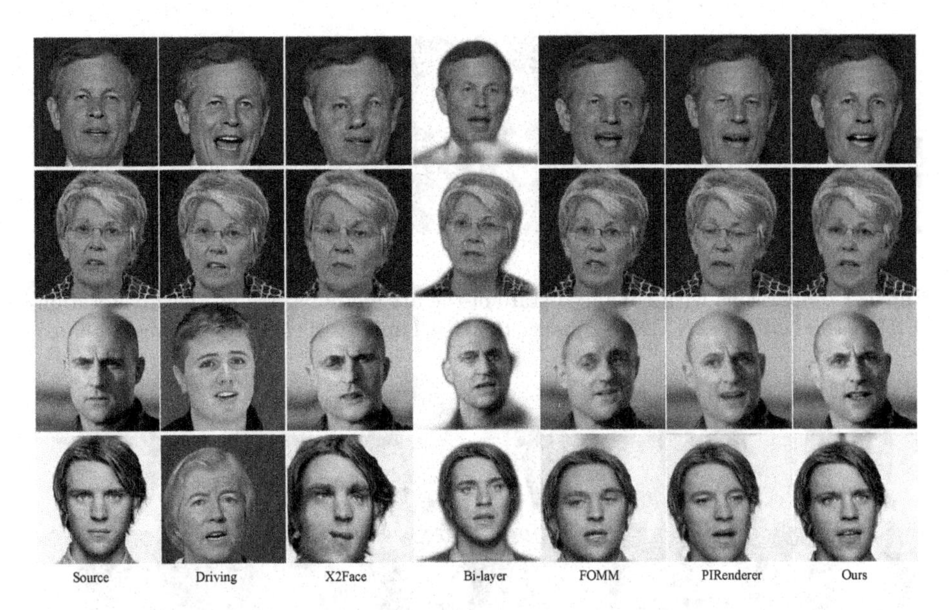


**Fig. 6.** Qualitative comparisons with state-of-the-art methods on the task of same-identity reenactment and cross-identity reenactment.

**Table 1.** Quantitative comparisons on talking face motion transfer.

| | Same-identity reenactment | | | | | | | Cross-identity reenactment | | | |
|---|---|---|---|---|---|---|---|---|---|---|---|
| | FID ↓ | LPIPS ↓ | PSNR ↑ | SSIM ↑ | CSIM ↑ | AED ↓ | APD ↓ | FID ↓ | CSIM ↑ | AED ↓ | APD ↓ |
| X2Face [43] | 44.32 | 0.2687 | 31.09 | 0.5926 | 0.6965 | 0.1680 | 0.03719 | 128.19 | 0.4449 | 0.3415 | 0.05156 |
| Bi-layer [45] | 118.46 | 0.5758 | 28.23 | 0.2906 | 0.3033 | 0.1219 | 0.01322 | 189.64 | 0.2252 | 0.2654 | **0.02054** |
| FOMM [30] | 29.17 | 0.2036 | 31.12 | **0.6353** | **0.8121** | **0.0946** | **0.00914** | 108.93 | 0.4517 | 0.2692 | 0.02576 |
| PIRenderer [28] | 27.14 | 0.2252 | 30.96 | 0.6028 | 0.7797 | 0.1073 | 0.01459 | 108.56 | 0.4812 | **0.2554** | 0.02962 |
| Ours | **18.02** | **0.1729** | **31.21** | 0.6019 | 0.7475 | 0.1151 | 0.01664 | **91.28** | **0.4890** | 0.2630 | 0.03484 |

### 5.1 Video-Driven Face Reenactment

To evaluate the performance of video-driven motion transfer, we conduct two facial reenactment tasks, *i.e.*, same-identity reenactment and cross-identity reenactment. For the same-identity reenactment, the identity of the source portrait is the same as that of the driving video. For cross-identity reenactment, the identity of the source portrait differs from that of the driving video.

**Qualitative Evaluation.** The visual results of the same-identity and cross-identity are shown in Fig. 6. Our method can achieve superior image resolution and quality over other methods. Here we focus on other aspects. In the same-identity case, all methods perform well in transfer pose except X2Face. For expression, our method outperforms other methods when there is a large expression difference between the source and driving images, especially when the mouth of the source is closed while that of the driving image is opened by

**Fig. 7.** Comparisons with enhanced state-of-the-art methods. We use face restoration (FR) method GFP-GAN [39] to enhance the visual quality of these competing methods.

**Fig. 8.** Global attribute editing via GAN inversion. The attribute is gradually modified in each generated talking video.

a large margin. In the cross-identity case, more issues occur for other methods while our method can work stably.

**Quantitative Evaluation.** Quantitative results of the two reenactment tasks are shown in Table 1. Our FID is the best in the cases, which indicates our synthesized faces are more realistic than those of other methods. Our better LPIPS and PSNR mean that we have better reconstruction performance. As our method uses a GAN inversion method to get the feature maps, it inevitably loses some identity information in the reconstruction. This might cause our lower CSIM in the same-identity case. On the contrary, our best CSIM in the cross-identity case indicates that our method can work stably in this more challenging setting and suffer from less distortion. Meanwhile, our AED and APD show comparable results to other methods although the StyleGAN inversion is imperfect and the extracted facial parameters will be further distorted.

**Comparisons with Enhanced Methods.** To eliminate the effect of resolution on our comparisons, we combine competing methods with a state-of-the-art blind face restoration method, *i.e.*, GFP-GAN [39], to improve the resolution and image quality. GFP-GAN improves the resolution of these methods to $1024 \times 1024$, which is shown in Fig. 7. We can observe that GFP-GAN greatly improves the visual quality for these low-resolution methods. However, it brings some side effects, including skin over-smoothing, details missing and color tone changing. Moreover, the face restoration cannot remedy the generated artifacts. The results

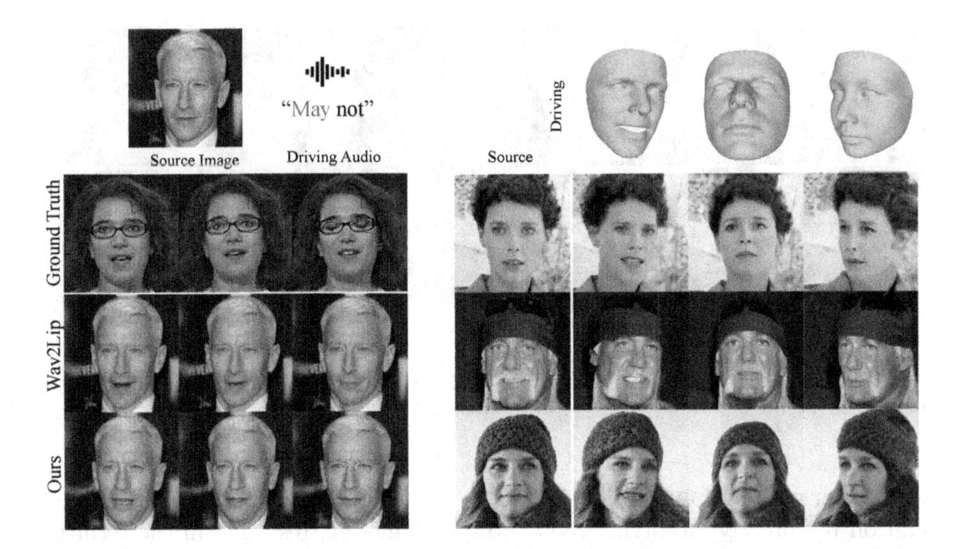


**Fig. 9.** Comparison with wav2lip [27].

**Fig. 10.** The results of intuitive face editing.

demonstrate that our method outperforms the competing methods in terms of image quality even though they are enhanced by the face restoration method.

### 5.2 Audio-Driven Talking Face Generation

In our framework, the audio-based motion generator can work either individually or jointly with the video-based motion generator. We compare with the state-of-the-art audio-driven method, wav2lip [27]. As shown in Fig. 9, our method have much better visual quality. The mouth of wav2lip is blurred and no teeth are synthesized. More results are shown in the supplementary material.

### 5.3 Talking Face Video Editing

**Global Attribute Editing.** As our model is built upon a pre-trained StyleGAN, it inherits a powerful property of StyleGAN, *i.e.,* facial attribute editing in the latent style space via existing GAN inversion methods. Our framework is convenient to edit attributes globally. We apply the GAN inversion method [37] to obtain the latent style codes for the first frame. Then, we can freely apply pre-defined style directions to change the style codes with a controllable extent in the video generation process at any timestamp. The results are shown in Fig. 8.

**Intuitive Editing.** Our video-based motion generator uses 3DMM parameters of the driving image to guide the motion generation for the source image. As 3DMM based talking face generation methods [14,28] always enable the intuitive

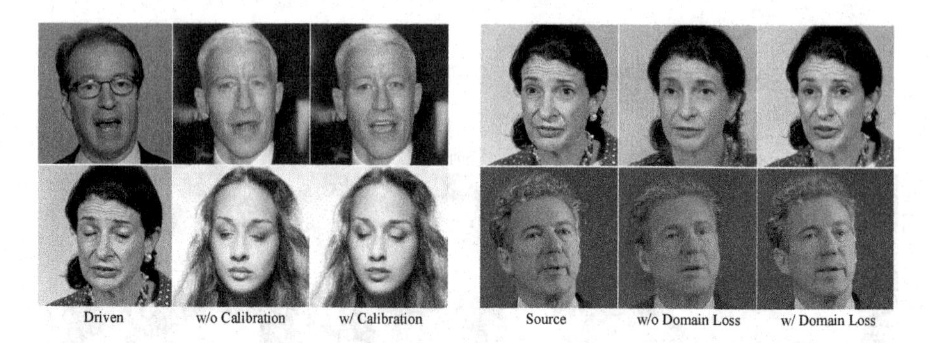


**Fig. 11.** Ablation study of calibration net

**Fig. 12.** Ablation study of the domain loss

editing on pose and expression, this also enables us to control the motion generation by directly modifying the 3DMM parameters, resulting in the intuitive editing on the final synthesis. The results are shown in Fig. 10.

### 5.4 Ablation Study

**Calibration Network.** Directly applying the flow fields to the feature map will lead to apparent artifacts around eyes and mouth, *e.g.,* 2D warping is unable to generate teeth for a closed mouth. Hence, we design the calibration network to rectify the artifacts. We compare the performance with or without the calibration network. The results are shown in Fig. 11. The calibration network greatly improves the shape and content around the eyes and mouth.

**Domain Loss.** The calibration network modifies the feature maps. To prevent the edited feature maps from going far away from the original feature maps, we design the domain loss. We compare the performance with or without it. As shown in Fig. 12, we can observe that dropping the loss makes the synthetic images blurry and lose facial details such as wrinkles and hair texture.

## 6 Conclusion

We propose a novel framework for one-shot talking face generation based on a pre-trained StyleGAN by exploring the properties of the latent feature space. Our framework supports video-driven and audio-driven reenactment. Besides, our framework allows two types of face editing, *i.e.*, global attribute editing via GAN inversion and intuitive editing based on 3DMM. We conduct comprehensive experiments to illustrate various capabilities of our unified framework.

**Limitation and Discussion.** As proposed in [22], there exist texture-sticking artefacts of images generated by StyleGAN2, which means the hair and face in synthesised videos typically do not move in unison. Alias-Free GAN [22] designs a specific architecture to overcome the problem. Our framework can be migrated to the new generator when high-quality GAN inversion methods are studied.

**Acknowledgement.** This work was supported in part by the National Natural Science Foundation of China under grant No. 61991450, the Shenzhen Key Laboratory of Marine IntelliSense and Computation under grant NO. ZDSYS20200811 142605016. Baoyuan Wu is supported by Shenzhen Science and Technology Program under grant No. ZDSYS20211021111415025.

## References

1. Abdal, R., Qin, Y., Wonka, P.: Image2StyleGAN: how to embed images into the StyleGAN latent space? In: CVPR (2019)
2. Abdal, R., Qin, Y., Wonka, P.: Image2StyleGAN++: how to edit the embedded images? In: CVPR (2020)
3. Alaluf, Y., Patashnik, O., Cohen-Or, D.: Restyle: a residual-based StyleGAN encoder via iterative refinement. In: ICCV (2021)
4. Anonymous: Latent image animator: learning to animate image via latent space navigation. In: ICLR (2022)
5. Bai, Q., Xu, Y., Zhu, J., Xia, W., Yang, Y., Shen, Y.: High-fidelity GAN inversion with padding space. arXiv preprint arXiv:2203.11105 (2022)
6. Blanz, V., Vetter, T.: A morphable model for the synthesis of 3D faces. In: SIGGRAPH (1999)
7. Booth, J., Roussos, A., Zafeiriou, S., Ponniah, A., Dunaway, D.: A 3D morphable model learnt from 10,000 faces. In: CVPR (2016)
8. Bounareli, S., Argyriou, V., Tzimiropoulos, G.: Finding directions in GAN's latent space for neural face reenactment. arXiv preprint arXiv:2202.00046 (2022)
9. Burkov, E., Pasechnik, I., Grigorev, A., Lempitsky, V.: Neural head reenactment with latent pose descriptors. In: CVPR (2020)
10. Cao, M., et al.: UniFaceGAN: a unified framework for temporally consistent facial video editing. IEEE TIP **30**, 6107–6116 (2021)
11. Chen, A., Liu, R., Xie, L., Chen, Z., Su, H., Yu, J.: SofGAN: a portrait image generator with dynamic styling. arXiv preprint arXiv:2007.03780 (2020)
12. Deng, J., Guo, J., Xue, N., Zafeiriou, S.: ArcFace: additive angular margin loss for deep face recognition. In: CVPR (2019)
13. Deng, Y., Yang, J., Xu, S., Chen, D., Jia, Y., Tong, X.: Accurate 3D face reconstruction with weakly-supervised learning: From single image to image set. In: CVPR Workshops (2019)
14. Doukas, M.C., Zafeiriou, S., Sharmanska, V.: HeadGAN: one-shot neural head synthesis and editing. In: ICCV (2021)
15. Fox, G., Tewari, A., Elgharib, M., Theobalt, C.: StyleVideoGAN: a temporal generative model using a pretrained StyleGAN. arXiv preprint arXiv:2107.07224 (2021)
16. Fried, O., et al.: Text-based editing of talking-head video. TOG **38**, 1–14 (2019)
17. Heusel, M., Ramsauer, H., Unterthiner, T., Nessler, B., Hochreiter, S.: GANs trained by a two time-scale update rule converge to a local nash equilibrium. In: NIPS (2017)
18. Huang, X., Belongie, S.: Arbitrary style transfer in real-time with adaptive instance normalization. In: ICCV (2017)

19. Johnson, J., Alahi, A., Fei-Fei, L.: Perceptual losses for real-time style transfer and super-resolution. In: Leibe, B., Matas, J., Sebe, N., Welling, M. (eds.) ECCV 2016. LNCS, vol. 9906, pp. 694–711. Springer, Cham (2016). https://doi.org/10.1007/978-3-319-46475-6_43
20. Kang, K., Kim, S., Cho, S.: GAN inversion for out-of-range images with geometric transformations. In: CVPR (2021)
21. Karras, T., Aila, T., Laine, S., Lehtinen, J.: Progressive growing of GANs for improved quality, stability, and variation. In: ICLR (2018)
22. Karras, T., et al.: Alias-free generative adversarial networks. In: NIPS (2021)
23. Karras, T., Laine, S., Aila, T.: A style-based generator architecture for generative adversarial networks. In: CVPR (2019)
24. Karras, T., Laine, S., Aittala, M., Hellsten, J., Lehtinen, J., Aila, T.: Analyzing and improving the image quality of StyleGAN. In: CVPR (2020)
25. Kim, H., et al.: Deep video portraits. TOG **37**, 1–14 (2018)
26. Nagrani, A., Chung, J.S., Zisserman, A.: VoxCeleb: a large-scale speaker identification dataset. In: INTERSPEECH (2017)
27. Prajwal, K., Mukhopadhyay, R., Namboodiri, V.P., Jawahar, C.: A lip sync expert is all you need for speech to lip generation in the wild. In: ACM Multimedia (2020)
28. Ren, Y., Li, G., Chen, Y., Li, T.H., Liu, S.: PIRenderer: controllable portrait image generation via semantic neural rendering. In: ICCV (2021)
29. Richardson, E., et al.: Encoding in style: a StyleGAN encoder for image-to-image translation. In: CVPR (2021)
30. Siarohin, A., Lathuilière, S., Tulyakov, S., Ricci, E., Sebe, N.: First order motion model for image animation. In: NIPS (2019)
31. Siarohin, A., Lathuilière, S., Tulyakov, S., Ricci, E., Sebe, N.: Animating arbitrary objects via deep motion transfer. In: CVPR (2019)
32. Siarohin, A., Woodford, O.J., Ren, J., Chai, M., Tulyakov, S.: Motion representations for articulated animation. In: CVPR (2021)
33. Simonyan, K., Zisserman, A.: Very deep convolutional networks for large-scale image recognition. arXiv preprint arXiv:1409.1556 (2014)
34. Song, G., et al.: AgileGAN: stylizing portraits by inversion-consistent transfer learning. TOG **40**, 1–13 (2021)
35. Tian, Y., et al.: A good image generator is what you need for high-resolution video synthesis. In: ICLR (2021)
36. Tzaban, R., Mokady, R., Gal, R., Bermano, A.H., Cohen-Or, D.: Stitch it in time: GAN-based facial editing of real videos. arXiv preprint arXiv:2201.08361 (2022)
37. Wang, T., Zhang, Y., Fan, Y., Wang, J., Chen, Q.: High-fidelity GAN inversion for image attribute editing. arXiv preprint arXiv:2109.06590 (2021)
38. Wang, T.C., Mallya, A., Liu, M.Y.: One-shot free-view neural talking-head synthesis for video conferencing. In: CVPR (2021)
39. Wang, X., Li, Y., Zhang, H., Shan, Y.: Towards real-world blind face restoration with generative facial prior. In: CVPR (2021)
40. Wang, X., Yu, K., Dong, C., Loy, C.C.: Recovering realistic texture in image super-resolution by deep spatial feature transform. In: CVPR (2018)
41. Wei, T., et al.: A simple baseline for StyleGAN inversion. arXiv preprint arXiv:2104.07661 (2021)
42. Wikipedia contributors: Thin plate spline—Wikipedia, the free encyclopedia (2020). https://en.wikipedia.org/wiki/Thin_plate_spline
43. Wiles, O., Koepke, A.S., Zisserman, A.: X2Face: a network for controlling face generation using images, audio, and pose codes. In: Ferrari, V., Hebert, M., Sminchisescu, C., Weiss, Y. (eds.) ECCV 2018. LNCS, vol. 11217, pp. 690–706. Springer, Cham (2018). https://doi.org/10.1007/978-3-030-01261-8_41

44. Xia, W., Zhang, Y., Yang, Y., Xue, J.H., Zhou, B., Yang, M.H.: GAN inversion: a survey. IEEE Trans. Pattern Anal. Mach. Intell. (2022)
45. Zakharov, E., Ivakhnenko, A., Shysheya, A., Lempitsky, V.: Fast bi-layer neural synthesis of one-shot realistic head avatars. In: Vedaldi, A., Bischof, H., Brox, T., Frahm, J.-M. (eds.) ECCV 2020. LNCS, vol. 12357, pp. 524–540. Springer, Cham (2020). https://doi.org/10.1007/978-3-030-58610-2_31
46. Zhang, R., Isola, P., Efros, A.A., Shechtman, E., Wang, O.: The unreasonable effectiveness of deep features as a perceptual metric. In: CVPR (2018)
47. Zhang, Z., Li, L., Ding, Y., Fan, C.: Flow-guided one-shot talking face generation with a high-resolution audio-visual dataset. In: CVPR (2021)
48. Zhu, P., Abdal, R., Femiani, J., Wonka, P.: Barbershop: GAN-based image compositing using segmentation masks. arXiv preprint arXiv:2106.01505 (2021)

# Long Video Generation with Time-Agnostic VQGAN and Time-Sensitive Transformer

Songwei Ge[1(✉)], Thomas Hayes[2], Harry Yang[2], Xi Yin[2], Guan Pang[2], David Jacobs[1], Jia-Bin Huang[1,2], and Devi Parikh[2,3]

[1] University of Maryland, College Park, USA
songweig@cs.umd.edu
[2] Meta AI, Menlo Park, USA
[3] Georgia Tech, Atlanta, USA

**Abstract.** Videos are created to express emotion, exchange information, and share experiences. Video synthesis has intrigued researchers for a long time. Despite the rapid progress driven by advances in visual synthesis, most existing studies focus on improving the frames' quality and the transitions between them, while little progress has been made in generating longer videos. In this paper, we present a method that builds on 3D-VQGAN and transformers to generate videos with thousands of frames. Our evaluation shows that our model trained on 16-frame video clips from standard benchmarks such as UCF-101, Sky Time-lapse, and Taichi-HD datasets can generate diverse, coherent, and high-quality long videos. We also showcase conditional extensions of our approach for generating meaningful long videos by incorporating temporal information with text and audio. Videos and code can be found at https://songweige.github.io/projects/tats.

**Keywords:** Video · Generation

## 1 Introduction

From conveying emotions that break language and cultural barriers to being the most popular medium on social platforms, videos are arguably the most informative, expressive, diverse, and entertaining visual form. Video synthesis has been an exciting yet long-standing problem. The challenges include not only achieving high visual quality in each frame and a natural transitions between frames, but a consistent theme, and even a meaningful storyline throughout

S. Ge—Work done primarily during an internship at Meta AI.

**Supplementary Information** The online version contains supplementary material available at https://doi.org/10.1007/978-3-031-19790-1_7.

© The Author(s), under exclusive license to Springer Nature Switzerland AG 2022
S. Avidan et al. (Eds.): ECCV 2022, LNCS 13677, pp. 102–118, 2022.
https://doi.org/10.1007/978-3-031-19790-1_7

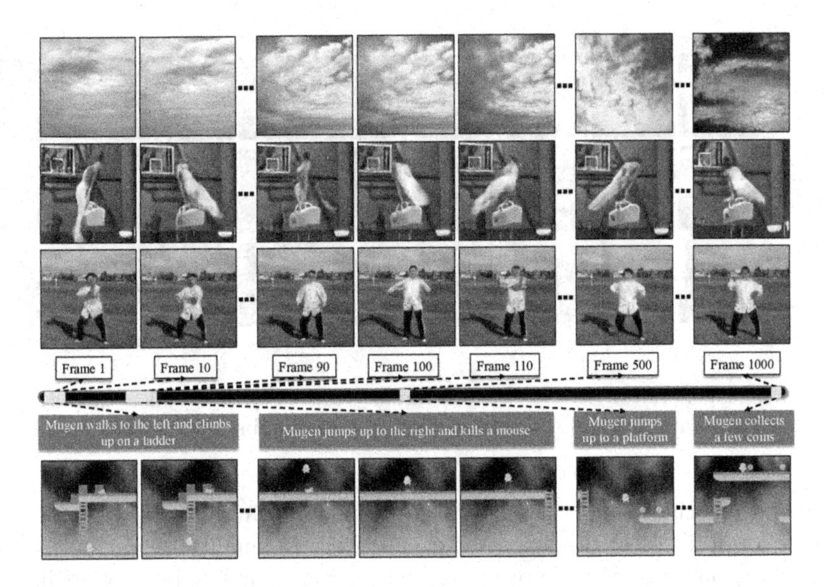


**Fig. 1.** Long videos generated by our model TATS with 1024 frames.

the video. The former has been the focus of many existing studies [9,36,45,49,57] working with tens of frames. The latter is largely unexplored and requires the ability to model *long-range temporal dependence* in videos with many more frames.

**Prior Works and Their Limitations.** *GAN-based methods* [9,26,36,49] can generate plausible short videos, but extending them to longer videos requires prohibitively high memory and time cost of training and inference.[1] *Autoregressive methods* alleviate the training cost constraints through *sequential prediction*. For example, RNNs and LSTMs generate temporal noise vectors for an image generator [9,27,36,37,45,47]; transformers either directly generate pixel values [20,52] or indirectly predict latent tokens [10,25,29,33,54,57]. These approaches circumvent training on the long videos directly by unrolling the RNN states or using a sliding window during inference. Recent works use *implicit neural representations* to reduce cost by directly mapping temporal positions to either pixels [58] or StyleGAN [23] feature map values [41]. However, the visual quality of the generated frames deteriorates quickly when performing generation beyond the training video length for all such methods as shown in Fig. 2.

**Our Work.** We tackle the problem of long video generation. Building upon the recent advances of VQGAN [10] for high-resolution *image generation*, we first develop a baseline by extending the 2D-VQGAN to 3D (2D space and 1D time) for modeling videos. This naively extended method, however, fails to produce high-quality, coherent long videos. Our work investigates the model

[1] Training DVD-GAN [9] or DVD-GAN-FP [26] on 16-frame videos requires 32–512 TPU replicas and 12–96 h.

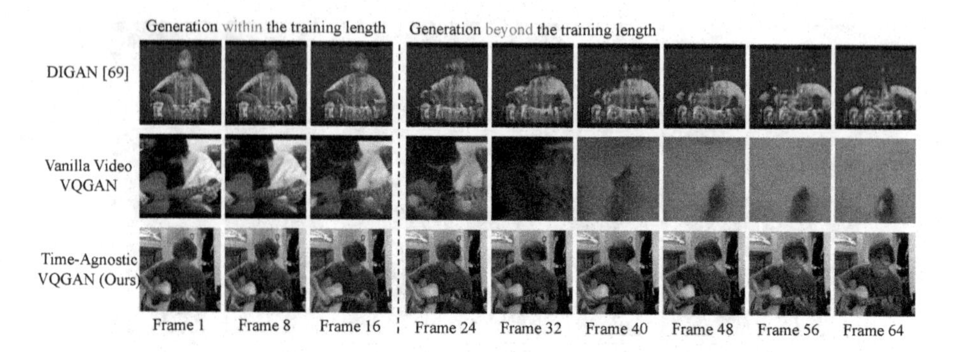


**Fig. 2.** Video generation results with a vanilla video VQGAN and a time-agnostic VQGAN, within and beyond the training length using sliding window attention.

design and identifies simple changes that significantly improve the capability to generate long videos of thousands of frames without quality degradation when conditioning on no or weak information. Our core insights lie in 1) removing the undesired dependence on time from VQGAN and 2) enabling the transformer to capture long-range temporal dependence. Below we outline these two key ideas.

**Time-Agnostic VQGAN.** Our model is trained on short video clips, e.g., 16 frames, like the previous methods [9,36,37,45,47,49]. At inference time, we use a sliding window approach [5] on the transformer to sample tokens for a longer length. The sliding window repeatedly appends the most recently generated tokens to the partial sequence and drops the earliest tokens to maintain a fixed sequence length. However, applying a sliding attention window to 3D-VQVAE (e.g. VideoGPT [57]) or 3D-VQGAN fails to preserve the video quality *beyond the training length*, as shown in Fig. 2. The reason turns out to be that the zero paddings used in these models *corrupts* the latent tokens and results in token sequences at the inference time that are drastically different from those observed during training when using sliding window. The amount of corruption depends on the temporal position of the token.[2] We address this issue by using *replicate padding* that mitigates the corruption by better approximating the real frames and brings no computational overhead. As a result, the transformer trained on the tokens encoded by our time-agnostic VQGAN effectively preserves the visual quality *beyond the training video length.*

**Time-Sensitive Transformer.** While removing the temporal dependence in VQGAN is desirable, long video generation certainly needs temporal information! This is necessary to model long-range dependence through the video and follow a sequence of events a storyline might suggest. While transformers can generate arbitrarily long sequences, errors tend to accumulate, leading to quality degradation for long video generation. To mitigate this, we introduce a hierarchical architecture where an *autoregressive transformer* first generates a set

[2] The large spatial span in image synthesis [10] disguises the issue. When applying sliding window to border tokens, the problem resurfaces in supp. mat. Figure 12.

of sparse latent frames, providing a more global structure. Then an *interpolation transformer* fills in the skipped frames autoregressively while attending to the generated sparse frames on both ends. With these modifications, our transformer models long videos more effectively and efficiently. Together with our proposed 3D-VQGAN, we call our model Time-Agnostic VQGAN and Time-Sensitive Transformer (TATS). We highlight the capability of TATS by showing generated video samples of 1024 frames in Fig. 1.

**Our Results.** We evaluate our model on several video generation benchmarks. We first consider a standard short video generation setting. Then, we carefully analyze its effectiveness for long video generation, comparing it against several recent models. Given that the evaluation of long video generation has not been well studied, we generalize several popular metrics for this task considering important evaluation axes for video generation models, including long term quality and coherence. Our model achieves state-of-the-art short and long video generation results on the UCF-101 [42], Sky Time-lapse [55], Taichi-HD [38], and AudioSet-Drum [12] datasets. We further demonstrate the effectiveness of our model by conditioning on temporal information such as text and audio.

**Our Contributions**

- We identify the undesired temporal dependence introduced by the zero paddings in VQGAN as a cause of the ineffectiveness of applying a sliding window for long video generation. We propose a simple yet effective fix.
- We propose a hierarchical transformer that can model longer time dependence and delay the quality degradation, and show that our model can generate meaningful videos according to the story flow provided by text or audio.
- To our knowledge, we are the first to generate long videos and analyze their quality. We do so by generalizing several popular metrics to a longer video span and showing that our model can generate more diverse, coherent, and higher-quality long videos.

## 2 Methodology

In this section, we first briefly recap the VQGAN framework and describe its extension to video generation. Next, we present our time-agnostic VQGAN and time-sensitive transformer models for long video generation (Fig. 3).

### 2.1 Extending the VQGAN Framework for Video Generation

Vector Quantised Variational AutoEncoder (VQVAE) [29] uses a discrete bottleneck as the latent space for reconstruction. An autoregressive model such as a transformer is then used to model the prior distribution of the latent space. VQGAN [10] is a variant of VQVAE that uses perceptual and GAN losses to achieve better reconstruction quality when increasing the bottleneck compression rates.

**Vanilla Video VQGAN.** We adapt the VQGAN architecture for video generation by replacing its 2D convolution operations with 3D convolutions. Given a video $\mathbf{x} \in \mathbb{R}^{T \times H \times W \times 3}$, the VQVAE consists of an encoder $f_{\mathcal{E}}$ and a decoder $f_{\mathcal{G}}$. The discrete latent tokens $\mathbf{z} = \mathbf{q}(f_{\mathcal{E}}(\mathbf{x})) \in \mathbb{Z}^{t \times h \times w}$ with embeddings $\mathbf{c}_z \in \mathbb{R}^{t \times h \times w \times c}$ are computed using a quantization operation $\mathbf{q}$ which applies nearest neighbor search using a trainable codebook $\mathcal{C} = \{\mathbf{c}_i\}_{i=1}^{K}$. The embeddings of the tokens are then fed into the decoder to reconstruct the input $\hat{\mathbf{x}} = f_{\mathcal{G}}(\mathbf{c}_z)$. The VQVAE is trained with the following loss:

$$\mathcal{L}_{\text{vqvae}} = \underbrace{\|\mathbf{x} - \hat{\mathbf{x}}\|_1}_{\mathcal{L}_{\text{rec}}} + \underbrace{\|\text{sg}[f_{\mathcal{E}}(\mathbf{x})] - \mathbf{c}_z\|_2^2}_{\mathcal{L}_{\text{codebook}}} + \underbrace{\beta \|\text{sg}\left[\mathbf{c}_z\right] - f_{\mathcal{E}}(\mathbf{x})\|_2^2}_{\mathcal{L}_{\text{commit}}},$$

where sg is a stop-gradient operation and we use $\beta = 0.25$ following the VQGAN paper [10]. We optimize $\mathcal{L}_{\text{codebook}}$ using an EMA update and circumvent the non-differentiable quantization step $\mathbf{q}$ with a straight-through gradient estimator [29]. VQGAN additionally adopts a perceptual loss [18,59] and a discriminator $f_{\mathcal{D}}$ to improve the reconstruction quality. Similar to other GAN-based video generation models [9,47], we use two types of discriminators in our model - a spatial discriminator $f_{\mathcal{D}_s}$ that takes in random reconstructed frames $\hat{\mathbf{x}}_i \in \mathbb{R}^{H \times W \times 3}$ to encourage frame quality and a temporal discriminator $f_{\mathcal{D}_t}$ that takes in the entire reconstruction $\hat{\mathbf{x}} \in \mathbb{R}^{T \times H \times W \times 3}$ to penalize implausible motions:

$$\mathcal{L}_{\text{disc}} = \log f_{\mathcal{D}_{s/t}}(\mathbf{x}) + \log(1 - f_{\mathcal{D}_{s/t}}(\hat{\mathbf{x}}))$$

We also use feature matching losses [50,51] to stabilize the GAN training:

$$\mathcal{L}_{\text{match}} = \sum_i p_i \left\| f^{(i)}_{\mathcal{D}_{s/t}/\text{VGG}}(\hat{\mathbf{x}}) - f^{(i)}_{\mathcal{D}_{s/t}/\text{VGG}}(\mathbf{x}) \right\|_1,$$

where $f^{(i)}_{\mathcal{D}_{s/t}/\text{VGG}}$ denotes the $i^{\text{th}}$ layer of either a trained VGG network [39] or discriminators with a scaling factor $p_i$. When using a VGG network, this loss is known as the perceptual loss [59]. $p_i$ is a learned constant for VGG and the reciprocal of the number of elements in the layer for discriminators. Our overall VQGAN training objective is as follows:

$$\begin{aligned} &\min_{f_{\mathcal{E}}, f_{\mathcal{G}}, \mathcal{C}} \left( \max_{f_{\mathcal{D}_s}, f_{\mathcal{D}_t}} (\lambda_{\text{disc}} \mathcal{L}_{\text{disc}}) \right) \\ &+ \min_{f_{\mathcal{E}}, f_{\mathcal{G}}, \mathcal{C}} (\lambda_{\text{match}} \mathcal{L}_{\text{match}} + \lambda_{\text{rec}} \mathcal{L}_{\text{rec}} + \mathcal{L}_{\text{codebook}} + \beta \mathcal{L}_{\text{commit}}) \end{aligned}$$

Directly applying GAN losses to VideoGPT [57] or 3D VQGAN [10] leads to training stability issues. In addition to the feature matching loss, we discuss other necessary architecture choices and training heuristics in supp. mat. A.1.

**Autoregressive Prior Model.** After training the video VQGAN, each video can be encoded into its discrete representation $\mathbf{z} = \mathbf{q}(f_{\mathcal{E}}(\mathbf{x}))$. Following VQGAN [10], we unroll these tokens into a 1D sequence using the row-major

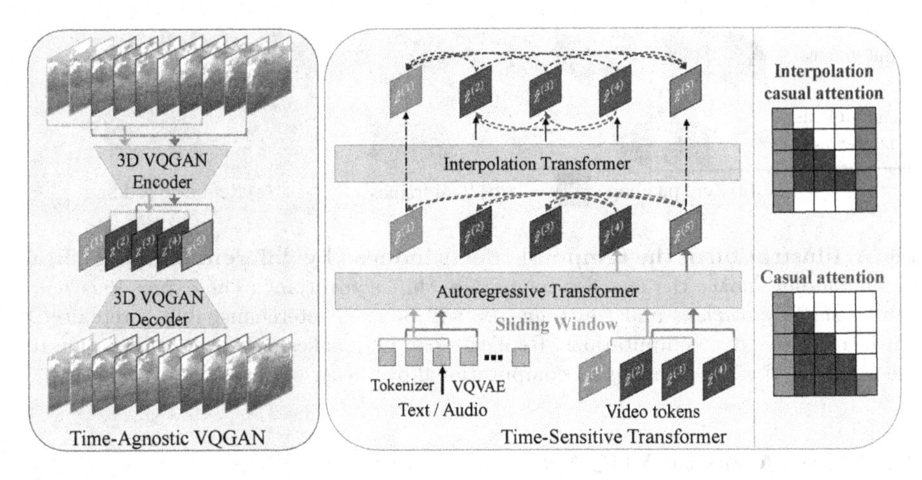


**Fig. 3. Overview of the proposed framework.** Our model contains two modules: time-agnostic VQGAN and time-sensitive transformer. The former compresses the videos both temporally and spatially into discrete tokens without injecting any dependence on the relative temporal position, which allows the usage of a sliding window during inference for longer video generation. The latter uses a hierarchical transformer for capturing longer temporal dependence.

order frame by frame. We then train a transformer $f_{\mathcal{T}}$ to model the prior categorical distribution of $\mathbf{z}$ in the dataset autoregressively:

$$p(\mathbf{z}) = p(\mathbf{z}_0) \prod_{i=0}^{t \times h \times w - 1} p(\mathbf{z}_{i+1} | \mathbf{z}_{0:i}),$$

where $p(\mathbf{z}_{i+1}|\mathbf{z}_{0:i}) = f_{\mathcal{T}}(\mathbf{z}_{0:i})$ and $\mathbf{z}_0$ is given as the start of sequence token. We train the transformer to minimize the negative log-likelihood over training samples:

$$\mathcal{L}_{\text{transformer}} = \mathbb{E}_{\mathbf{z} \sim p(\mathbf{z}_{\text{data}})}[-\log p(\mathbf{z})]$$

At inference time, we randomly sample video tokens from the predicted categorical distribution $p(\mathbf{z}_{i+1}|\mathbf{z}_{0:i})$ in sequence and feed them into the decoder to generate the videos $\hat{\mathbf{x}} = f_{\mathcal{G}}(\hat{\mathbf{c}}_z)$. To synthesize videos longer than the training length, we generalize sliding attention window for our use [5]. A similar idea has been used in 2D to generate images of higher resolution [10]. We denote the $j^{th}$ temporal slice of $\mathbf{z}$ to be $\mathbf{z}^{(j)} \in \mathbb{Z}^{h \times w}$, where $0 \leq j \leq t-1$. For instance, to generate $\mathbf{z}^{(t)}$ that is beyond the training length, we condition on the $t-1$ slices before it to match the transformer sequence length $p(\mathbf{z}^{(t)}|\mathbf{z}^{(1:t-1)}) = f_{\mathcal{T}}(\mathbf{z}^{(1:t-1)})$. However, as shown in Fig. 2, when paired with sliding window attention, the vanilla video VQGAN and transformer cannot generate longer videos without quality degradation. Next, we discuss the reason and a simple yet effective fix.

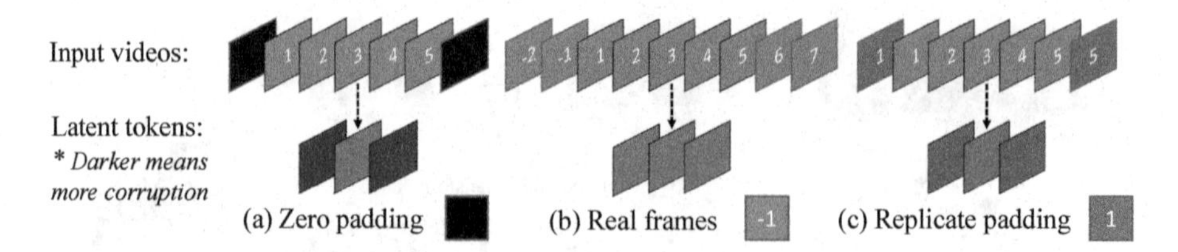


**Fig. 4. Illustration of the temporal effects induced by different paddings.** Real frame padding makes the encoder temporally shift-equivariant (The expressions *temporally shift-equivariant* and *time-agnostic* will be used interchangeably hereinafter.) but introduces extra computations. Replicate padding makes decent approximation to the real frames while bringing no computational overhead.

## 2.2 Time-Agnostic VQGAN

When the Markov property holds, a transformer with a sliding window can generate arbitrarily long sequences as demonstrated in long article generation [5]. However, a crucial premise that has been overlooked is that the transformer needs to see sequences that start with tokens similar to $\mathbf{z}^{(1:t-1)}$ during training to predict token $\mathbf{z}^t$. This premise breaks down in VQGAN. We provide some intuitions about the reason and defer a detailed discussion to supp. mat. A.2.

Different from natural language modeling where the tokens come from realistic data, VQGAN tokens are produced by an encoder $f_\mathcal{E}$ which by default adopts zero paddings for the desired output shape. When a short video clip is encoded, the zero paddings in the temporal dimension also get encoded and affect the output tokens, causing an unbalanced effects to tokens at different temporal position [2,17,24,56]. The tokens closer to the temporal boundary will be affected more significantly. As a result, for real data to match $\mathbf{z}^{(1:t-1)}$, they have to contain these zero-frames, which is not the case in practice. Therefore, removing those paddings in the temporal dimension is crucial for making the encoder *time-agnostic* and enabling sliding window.

After removing all the padding, one needs to pad real frames to both ends of the input videos to obtain the desired output size [21,41]. Note that the zeros are also padded in the intermediate layers when applied, and importantly, they need not be computed. But, if we want to pad with realistic values, the input needs to be padded long enough to cover the entire receptive field, and all these extra values in the intermediate layers need to be computed. The number of needed real frames can be as large as $\mathcal{O}(Ld)$, where $L$ is the number of layers and $d$ is the compression rate in the temporal direction of the video. Although padding with real frames makes the encoder fully time agnostic, it can be expensive with a large compression rate or a deep network. In addition, frames near both ends of the videos need to be discarded for not having enough real frames to pad, and consequently some short videos in the dataset may be completely dropped.

Instead of padding with real values, we propose to approximate real frames with the values close to them. An assumption, which is often referred to as the "boring videos" assumption in the previous literature [6], is to assume that the videos are frozen beyond the given length. Following the assumption, the last boundary slices can be used for padding without computation. More importantly, this can be readily implemented by the replicate padding mode using a standard machine learning library. An illustration of the effects induced by different paddings can be found in Fig. 4. Other reasonable assumptions can also be adopted and may correspond to different padding modes. For instance, the reflected padding mode can be used if videos are assumed to play in reverse beyond their length, which introduces more realistic motions than the frozen frames. In supp. mat. A.3, we provide a careful analysis of the time dependence when applying different padding types and numbers of padded real frames. We find that the replicate padding alone resolves the issue well in practice and inherits the low computational cost merit of zero paddings. Therefore, in the following experiments, we use the replicate paddings and no real frames padded.

### 2.3 Time-Sensitive Transformer

The time-agnostic property of the VQGAN makes it feasible to generate long videos using a transformer with sliding window attention. However, long video generation does need temporal information! To maintain a consistent theme running from the beginning of the video to the end requires the capacity to model long-range dependence. And besides, the spirit of a long video is in its underlying story, which requires both predicting the motions in the next few frames and the ability to plan on how the events in the video proceed. This section discusses the time-sensitive transformer for improved long video generation.

Due to the probabilistic nature of transformers, errors can be introduced when sampling from a categorical distribution, which then accumulate over time. One common strategy to improve the long-term capacity is to use a hierarchical model [7,11] to reduce the chance of drifting away from the target theme. Specifically, we propose to condition one interpolation transformer on the tokens generated by the other autoregressive transformer that outputs more sparsely sampled tokens. The autoregressive transformer is trained in a standard way but on the sampled frames with larger intervals. For the interpolation transformer, it fills in the missing frames between any two adjacent frames generated by the autoregressive transformer. To do so, we propose interpolation attention as shown in Fig. 3, where the predicted tokens attend to both the previously generated tokens in a causal way and the tokens from the sparsely sampled frames at both ends. A more detailed description can be found in supp. mat. A.4. We consider an autoregressive transformer that generates $4\times$ more sparse video tokens. Generalizing this model to even more extreme cases such as video generation based on key-frames would be an interesting future work.

Another simple yet effective way to improve the long period coherence is to provide the underlying "storyline" of the desired video directly. The VQVAE framework has been shown effective in conditional image generation [10,34]. We hence consider several kinds of conditional information that provide additional temporal information such as audio [8], text [53], and so on. To utilize conditional information for long video generation, we use either a tokenizer or an additional

VQVAE to discretize the text or audio and prepend the obtained tokens to the video tokens and remove the start of the sequence token $\mathbf{z}_0$. At inference time, we extend the sliding window by simultaneously applying it to the conditioned tokens and video tokens.

# 3 Experiments

In this section, we evaluate the proposed method on several benchmark datasets for video generation with an emphasis on long video generation.

## 3.1 Experimental Setups

**Datasets and Evaluation.** We show results on UCF-101 [42], Sky Time-lapse [55], and Taichi-HD [38] for unconditional or class-conditioned video generation with $128 \times 128$ resolution following [58], AudioSet-Drum [12] for audio-conditioned video generation with $64 \times 64$ resolution following [25], and MUGEN [14] with $256 \times 256$ resolution for text-conditioned video generation. We follow the previous methods [45,58] to use Fréchet Video Distance (FVD) [48] and Kernel Video Distance (KVD) [48] as the evaluation metrics on UCF-101, Sky Time-lapse, and Taichi-HD datasets. In addition, we follow the methods evaluated on UCF-101 [9,25] to report the Inception Score (IS) [37] calculated by a trained C3D model [46]. For audio-conditioned generation evaluation, we measure the SSIM and PSNR at the $45^{\text{th}}$ frame which is the longest videos considered in the previous methods [8,25]. See supp. mat. B.1 for more details about the datasets.

**Training Details.** To compare our methods with previous ones, we train our VQGAN on videos with 16 frames. We adopt a compression rate $d = 4$ in temporal dimension and $d = 8$ in spatial dimensions. For transformer models, we train a decoder-only transformer with size between GPT-1 [31] and GPT-2 [32] for a consideration of computational cost. We refer to our model with a single autoregressive transformer as **TATS-base** and the proposed hierarchical transformer as **TATS-hierarchical**. For audio-conditioned model, we train another VQGAN to compress the Short-Time Fourier Transform (STFT) data into a discrete space. For text-conditioned model, we use a BPE tokenizer pretrained by CLIP [30]. See supp. mat. B.2 for more details about the training and inference, and supp. mat. B.3. for the comparison of computational costs.

## 3.2 Quantitative Evaluation on Short Video Generation

In this section, we demonstrate the effectiveness of our TATS-base model under a standard short video generation setting, where only 16 frames are generated for each video. The quantitative results are shown in Table 1.

Our model achieves state-of-art FVD and KVD on the UCF-101 and Taichi-HD datasets, and state-of-art KVD on the Sky Time-lapse dataset for unconditional video generation, and improved the quality on the AudioSet-Drum for

**Table 1.** Quantitative results of standard video generation on different datasets. We report FVD and KVD on the Taichi-HD and Sky Time-lapse datasets, IS and FVD on the UCF-101 dataset, SSIM and PSNR at the 45$^{th}$ frame on the AudioSet-Drum dataset. * denotes training on the entire UCF-101 dataset instead of the train split. The class column indicates whether the class labels are used as conditional information.

(a) Sky Time-lapse

| Method | FVD (↓) | KVD (↓) |
|---|---|---|
| MoCoGAN-HD | 183.6±5.2 | 13.9±0.7 |
| DIGAN | **114.6**±4.9 | 6.8±0.5 |
| TATS-base | 132.56±2.6 | **5.7**±0.3 |

(b) TaiChi-HD

| Method | FVD (↓) | KVD (↓) |
|---|---|---|
| MoCoGAN-HD | 144.7±6.0 | 25.4±1.9 |
| DIGAN | 128.1±4.9 | 20.6±1.1 |
| TATS-base | **94.60**±2.7 | **9.8**±1.0 |

(c) AudioSet-Drum

| Method | SSIM (↑) | PSNR (↑) |
|---|---|---|
| SVG-LP | 0.510±0.008 | 13.5±0.1 |
| Vougioukas *et al.* | 0.896±0.015 | 23.3±0.3 |
| Sound2Sight | 0.947±0.007 | 27.0±0.3 |
| CCVS | 0.945±0.008 | 27.3±0.5 |
| TATS-base | **0.964**±0.005 | **27.7**±0.4 |

(d) UCF-101

| Method | Class | IS (↑) | FVD (↓) |
|---|---|---|---|
| VGAN | ✓ | 8.31±.09 | - |
| TGAN | ✗ | 11.85±.07 | - |
| TGAN | ✓ | 15.83±.18 | - |
| MoCoGAN | ✓ | 12.42±.07 | - |
| ProgressiveVGAN | ✓ | 14.56±.05 | - |
| LDVD-GAN | ✗ | 22.91±.19 | - |
| VideoGPT | ✗ | 24.69±.30 | - |
| TGANv2 | ✓ | 28.87±.67 | 1209±28 |
| DVD-GAN* | ✓ | 27.38±.53 | - |
| MoCoGAN-HD* | ✗ | 32.36 | 838 |
| DIGAN | ✗ | 29.71±.53 | 655±22 |
| DIGAN* | ✗ | 32.70±.35 | 577±21 |
| CCVS*+Real frame | ✗ | 41.37±.39 | 389±14 |
| CCVS*+StyleGAN | ✗ | 24.47±.13 | 386±15 |
| StyleGAN-V | ✗ | 23.94±.73 | - |
| CogVideo | ✓ | 50.46 | 626 |
| Video Diffusion | ✗ | 57.00±.62 | **295**±3 |
| Real data | - | 90.52 | - |
| TATS-base | ✗ | 57.63±.24 | 420±18 |
| TATS-base | ✓ | **79.28**±.38 | 332±18 |

audio-conditioned video generation. TATS-base improves on IS by 76.2% over previous method with synthetic initial frames [25] and is competitive against concurrent works [15,16,41]. On the UCF-101 dataset, we also explore generation with class labels as conditional information. Following the previous method [36], we sample labels from the prior distribution as the input to the generation. We find that conditioning on the labels significantly eases the transformer modeling task and boosts the generation quality, improving IS from 57.63 to 79.28, which is close to the upper bound shown in the "Real data" row [1,19]. This has been extensively observed in image generation [3,10] but not quite yet revealed in the video generation. TATS-base significantly advances the IS over the previous methods, demonstrating its power in modeling diverse video datasets.

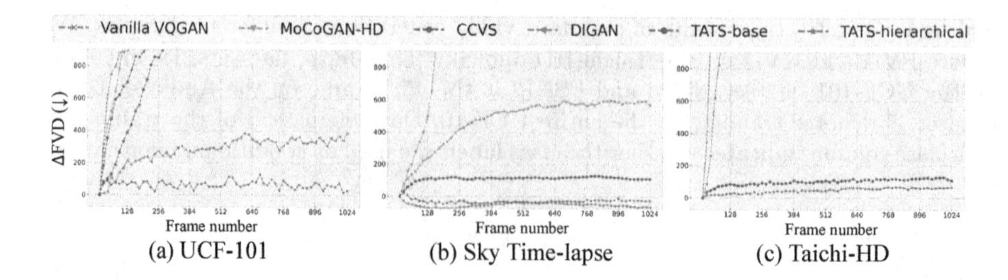


**Fig. 5. Quality degradation.** FVD changes of non-overlapping 16-frame clips from the long videos generated by models trained on the UCF-101, Sky Time-lapse, and Taichi-HD datasets. The lower value indicates the slower degradation.

### 3.3 Quantitative Evaluation on Long Video Generation

Quantitatively evaluating long video generation results has been under explored. In this section, we propose several metrics by generalizing existing metrics to evaluate the crucial aspects of the long videos. We generate 512 videos with 1024 frames on each of the Sky Time-lapse, Taichi-HD, and UCF-101 datasets. We compare our proposed methods, TATS-base and TATS-hierarchical, with the baseline Vanilla VQGAN and the state-of-the-art models including MoCoGAN-HD [45], DIGAN [58], and CCVS [25] by unrolling RNN states, directly sampling, and sliding attention window using their official model checkpoints.

**Quality.** We measure video quality with respect to the duration by evaluating every 16 frames extracted side-by-side from the generated videos. Ideally, every set of 16-frame videos should come from the same distribution as the training set. Therefore, we report the FVD changes of these generated clips compared with the first generated 16 frames in Fig. 5, to measure the degradation of the video quality. The figure shows that our TATS-base model successfully delays the quality degradation compared with the vanilla VQGAN baseline, MoCoGAN-HD, and CCVS models. In addition, the TATS-hierarchical model further improves the long-term quality of the TATS-base model. The concurrent work DIGAN [58] also claims the ability of extrapolation, while we show that the generation still degrades severely after certain number frames on the UCF-101 and Taichi-HD datasets. We conjecture the unusual decrease of the FVD w.r.t. the duration of DIGAN and TATS-hierarchical on Sky Time-lapse can be explained by that the I3D model [6] used to calculate FVD is trained on Kinetics-400 dataset, and the sky videos can be outliers of the training data and lead to weak activation in the logit layers and therefore such unusual behaviors. We further perform qualitative and human evaluations to compare our method with DIGAN in supp. mat. C. The results confirm that the sky videos generated by TATS have better quality.

**Coherence.** The generated long videos should follow a consistent topical theme. We evaluate the coherence of the videos on the UCF-101 dataset since it has multiple themes (classes). We expect the generated long videos to be classified as the same class all the time. We adopt the same trained C3D model for IS

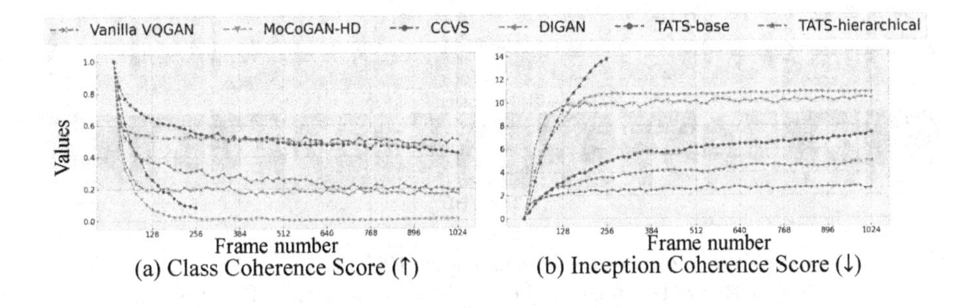

(a) Class Coherence Score (↑)

(b) Inception Coherence Score (↓)

**Fig. 6. Video coherency.** CCS and ICS values of every 16-frame clip extracted from long videos generated by different models trained on the UCF-101 dataset.

calculated [37], and propose two metrics, Class Coherence Score (CCS) and Inception Coherence Score (ICS) at time step $t$, measuring the theme similarity between the non-overlapped 16 frames w.r.t. the first 16 frames, defined as below:

$$\mathrm{CCS}_t = \sum_i \frac{\mathbf{1}(\arg\max p_{C3D}(y|x_i^{(0:15)}), \arg\max p_{C3D}(y|x_i^{(t:t+15)}))}{N}$$

$$\mathrm{ICS}_t = \sum_i p_{C3D}(y|x_i^{(t:t+15)}) \log \frac{p_{C3D}(y|x_i^{(t:t+15)})}{p_{C3D}(y|x_i^{(0:15)})}$$

The ICS captures class shift more accurately than the CCS, which only looks at the most probable class. On the other hand, CCS is more intuitive and allows us to define single metrics such as the area under the curve. CCS also doesn't have the asymmetry issue (unlike KL divergence used in ICS). We show the scores of TATS models and several baselines in Fig. 6. TATS-base achieves decent coherence as opposite to its quality degradation shown previously. Such difference can be explained by its failure on a small portion of generated videos as shown in supp. mat. Figure 20, which dominates the FVD. This also shows that CCS and ICS measure coherence on individual videos that complementary to FVD changes. Furthermore, TATS-hierarchical outperforms the baselines on both metrics. For example, more than half of the videos are still classified consistently in the last 16 frames of the entire 1024 frames.

### 3.4 Qualitative Evaluation on Long Video Generation

This section shows qualitative results of videos with 1024 frames and discusses their common properties. 1024 frames per video approaches the upper bound in the training data as shown in supp. mat. Figure 15, which means the generated videos are probably different from the training videos, at least in duration.

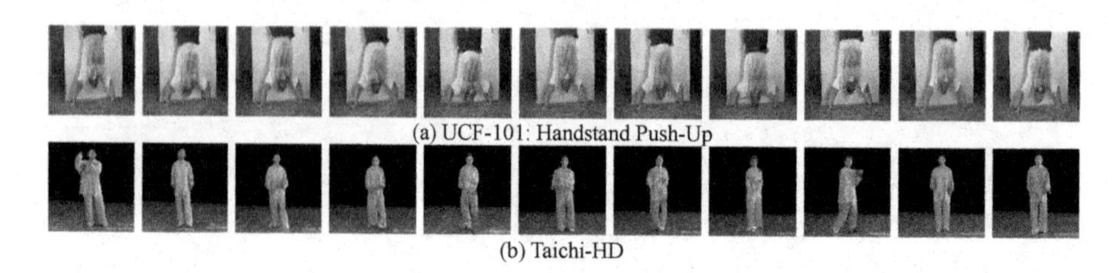


**Fig. 7. Videos with recurrent events.** Every 100$^{th}$ frame is extracted from the generated videos with 1024 frames on the UCF-101 and Taichi-HD datasets.

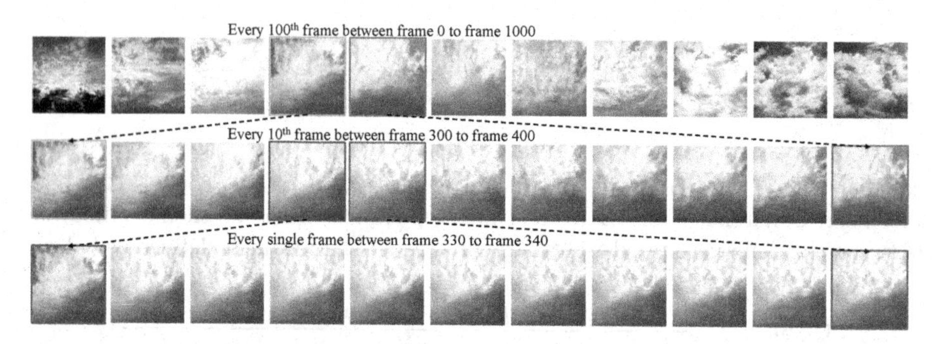


**Fig. 8. Videos with homomorphisms.** Every 100$^{th}$, 10$^{th}$, and consecutive frames are extracted from a generated sky video with 1024 frames.

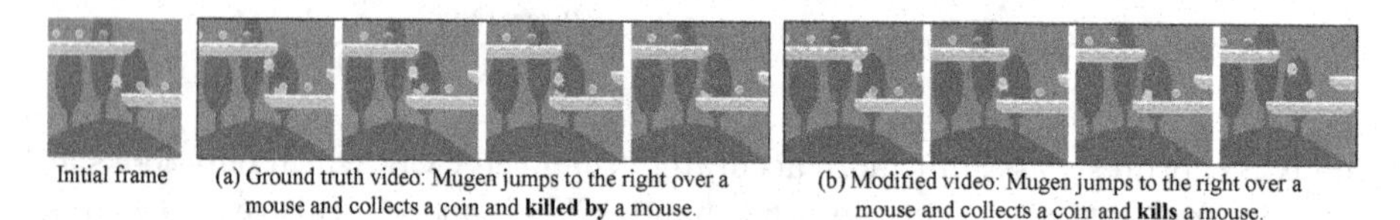


**Fig. 9. Videos with meanings.** Video manipulation by modifying the texts.

**Recurrent Actions.** We find that some video themes contain repeated events such as the *Handstand Push-Up* class in the UCF-101 dataset and videos in the Taichi-HD dataset. As shown in Fig. 7, our model generalizes to long video generation by producing realistic and recurrent actions. However, we find that all the 1024 generated frames in these videos are unique and motions are not identical, which shows that our model is not simply copying the short loops.

**Smooth Transitions.** Videos with the same theme often share visual features such as scenes and motions. We show that with enough training data available for a single theme, our model learns to "stitch" the long videos through generating smooth transitions between them. For

**Table 2.** LPIPS and color histogram correlation between the first and the last frames of the sky videos.

| Videos | LPIPS metric | Color Similarity |
|---|---|---|
| Real | $0.1839 \pm 0.0683$ | $0.7330 \pm 0.2401$ |
| Fake | $0.3461 \pm 0.1184$ | $0.0797 \pm 0.1445$ |

example in Fig. 8, we show that our model generates a long sky video containing different weather and timing while the transitions between these conditions are still natural and realistic. To show that this is not the case in the training data, we compute the LPIPS score [59] and color histrogram correlation between the $1^{st}$ and the $1024^{th}$ frames and report the mean and standard deviation on the 500 generated and 216 real long videos in Table 2. It shows that such transitions are much more prominent on the generated videos than the training videos. Similar examples of the UCF-101 and Taichi-HD videos can be found in supp. mat. Figure 19. A separation of content and motion latent features [45,47,58] may better leverage such transitions, which we leave as future work.

**Meaningful synthesis.** By conditioning on the temporal information, we can achieve more controllable generation, which allows us to directly create or modify videos based our own will. For example, in Fig. 9, we show that it is possible to manipulate the videos by changing the underlying storyline - by replacing "killed by" with "kills" we completely change the destiny of Mugen!

## 4 Related Work

In this section, we discuss the related work in video generation using different models. We focus on the different strategies adopted by these methods to handle temporal dynamics and their potential for and challenges associated with long video generation. Also see supp. mat. D for other relevant models and tasks.

**GAN-Based Video Generator.** Adapting GANs [13,22,23] to video synthesis requires modeling the temporal dimension. Both 3D deconvolutionals [49] and additional RNN or LSTM [9,27,36,37,45,47,47] have been used. By unrolling the steps taken by the RNNs, videos of longer duration can be generated. However, as shown in our experiments, the quality of these videos degrades quickly. In addition, without further modifications, the length of videos generated by these models is limited by the GPU memory (e.g., at most 140 frames can be generated on 32GB V100 GPU by HVG [7]).

**AR-Based Video Generator.** Autoregressive models have become a ubiquitous generative model for video synthesis [20,35,43,52]. A common challenge faced by AR models is their slow inference speed. This issue is mitigated by training on the compressed tokens with VQVAEs [25,33,54,57]. Our model falls into this line of video generators. We show that such a VQVAE-based AR model is promising to generate long videos with long-range dependence.

**INR-Based Video Generator.** Implicit Neural Representations [40,44] represent continuous signals such as videos by mapping the coordinate space to RGB space [41,58]. The advantage of these models is their ability to generate arbitrarily long videos non-autoregressively. However, their generated videos still suffer from the quality degradation or contain periodic artifacts due to the positional encoding and struggle at synthesizing new content.

**Concurrent Works on Long Video Generation.** In parallel to our work, [15] proposes a gradient conditioning method for sampling longer videos with a diffusion model, [28] and [16] explore a hierarchical model with frame-level VQVAE embeddings or RGB frames. [4] uses a low-resolution long video generator and short-video super-resolution network to generate videos of dynamic scenes.

## 5 Conclusion

We propose TATS, a time-agnostic VQGAN and time-sensitive transformer model, that is only trained on clips with tens of frames and can generate thousands of frames using a sliding window during inference time. Our model generates meaningful videos when conditioned on text and audio. Our paper is a small step but we hope it can encourage future works on more interesting forms of video synthesis with a realistic number of frames, perhaps even movies.

**Acknowledgements.** We thank Oran Gafni, Sasha Sheng, and Isabelle Hu for helpful discussion and feedback; Patrick Esser and Robin Rombach for sharing additional insights for training VQGAN models; Anoop Cherian and Moitreya Chatterjee for sharing the pre-processing code for the AudioSet dataset.

## References

1. Acharya, D., Huang, Z., Paudel, D.P., Van Gool, L.: Towards high resolution video generation with progressive growing of sliced Wasserstein GANs. arXiv preprint arXiv:1810.02419 (2018)
2. Alsallakh, B., Kokhlikyan, N., Miglani, V., Yuan, J., Reblitz-Richardson, O.: Mind the pad - CNNs can develop blind spots. In: ICLR (2021)
3. Brock, A., Donahue, J., Simonyan, K.: Large scale GAN training for high fidelity natural image synthesis. In: ICLR (2018)
4. Brooks, T., et al.: Generating long videos of dynamic scenes. arXiv preprint arXiv:2206.03429 (2022)
5. Brown, T., et al.: Language models are few-shot learners. In: NeurIPS (2020)
6. Carreira, J., Zisserman, A.: Quo Vadis, action recognition? A new model and the kinetics dataset. In: CVPR (2017)
7. Castrejon, L., Ballas, N., Courville, A.: Hierarchical video generation for complex data. arXiv preprint arXiv:2106.02719 (2021)
8. Chatterjee, M., Cherian, A.: Sound2Sight: generating visual dynamics from sound and context. In: Vedaldi, A., Bischof, H., Brox, T., Frahm, J.-M. (eds.) ECCV 2020. LNCS, vol. 12372, pp. 701–719. Springer, Cham (2020). https://doi.org/10.1007/978-3-030-58583-9_42
9. Clark, A., Donahue, J., Simonyan, K.: Adversarial video generation on complex datasets. arXiv preprint arXiv:1907.06571 (2019)
10. Esser, P., Rombach, R., Ommer, B.: Taming transformers for high-resolution image synthesis. In: CVPR (2021)
11. Fan, A., Lewis, M., Dauphin, Y.: Hierarchical neural story generation (2018)

12. Gemmeke, J.F., et al.: Audio set: an ontology and human-labeled dataset for audio events. In: ICASSP (2017)
13. Goodfellow, I., et al.: Generative adversarial nets. In: NeurIPS (2014)
14. Hayes, T., et al.: MUGEN: a playground for video-audio-text multimodal understanding and generation. arXiv preprint arXiv:2204.08058 (2022)
15. Ho, J., Salimans, T., Gritsenko, A., Chan, W., Norouzi, M., Fleet, D.J.: Video diffusion models. arXiv preprint arXiv:2204.03458 (2022)
16. Hong, W., Ding, M., Zheng, W., Liu, X., Tang, J.: CogVideo: large-scale pretraining for text-to-video generation via transformers. arXiv preprint arXiv:2205.15868 (2022)
17. Islam, M.A., Jia, S., Bruce, N.D.: How much position information do convolutional neural networks encode? In: ICLR (2019)
18. Johnson, J., Alahi, A., Fei-Fei, L.: Perceptual losses for real-time style transfer and super-resolution. In: Leibe, B., Matas, J., Sebe, N., Welling, M. (eds.) ECCV 2016. LNCS, vol. 9906, pp. 694–711. Springer, Cham (2016). https://doi.org/10.1007/978-3-319-46475-6_43
19. Kahembwe, E., Ramamoorthy, S.: Lower dimensional kernels for video discriminators. Neural Netw. **132**, 506–520 (2020)
20. Kalchbrenner, N., et al.: Video pixel networks. In: ICML (2017)
21. Karras, T., et al.: Alias-free generative adversarial networks. In: NeurIPS (2021)
22. Karras, T., Laine, S., Aila, T.: A style-based generator architecture for generative adversarial networks. In: CVPR (2019)
23. Karras, T., Laine, S., Aittala, M., Hellsten, J., Lehtinen, J., Aila, T.: Analyzing and improving the image quality of StyleGAN. In: CVPR (2020)
24. Kayhan, O.S., van Gemert, J.C.: On translation invariance in CNNs: convolutional layers can exploit absolute spatial location. In: CVPR (2020)
25. Le Moing, G., Ponce, J., Schmid, C.: CCVS: context-aware controllable video synthesis. In: NeurIPS (2021)
26. Luc, P., et al.: Transformation-based adversarial video prediction on large-scale data. arXiv preprint arXiv:2003.04035 (2020)
27. Munoz, A., Zolfaghari, M., Argus, M., Brox, T.: Temporal shift GAN for large scale video generation. In: WACV (2021)
28. Nash, C., et al.: Transframer: arbitrary frame prediction with generative models. arXiv preprint arXiv:2203.09494 (2022)
29. van den Oord, A., Vinyals, O., Kavukcuoglu, K.: Neural discrete representation learning. In: NeurIPS (2017)
30. Radford, A., et al.: Learning transferable visual models from natural language supervision. In: ICML (2021)
31. Radford, A., Narasimhan, K., Salimans, T., Sutskever, I.: Improving language understanding by generative pre-training (2018)
32. Radford, A., Wu, J., Child, R., Luan, D., Amodei, D., Sutskever, I.: Language models are unsupervised multitask learners (2019)
33. Rakhimov, R., Volkhonskiy, D., Artemov, A., Zorin, D., Burnaev, E.: Latent video transformer. arXiv preprint arXiv:2006.10704 (2020)
34. Ramesh, A., et al.: Zero-shot text-to-image generation. arXiv preprint arXiv:2102.12092 (2021)
35. Ranzato, M., Szlam, A., Bruna, J., Mathieu, M., Collobert, R., Chopra, S.: Video (language) modeling: a baseline for generative models of natural videos. arXiv preprint arXiv:1412.6604 (2014)
36. Saito, M., Matsumoto, E., Saito, S.: Temporal generative adversarial nets with singular value clipping. In: ICCV (2017)

37. Saito, M., Saito, S., Koyama, M., Kobayashi, S.: Train sparsely, generate densely: memory-efficient unsupervised training of high-resolution temporal GAN. Int. J. Comput. Vis. **128**(10), 2586–2606 (2020). https://doi.org/10.1007/s11263-020-01333-y
38. Siarohin, A., Lathuilière, S., Tulyakov, S., Ricci, E., Sebe, N.: First order motion model for image animation. In: NeurIPS (2019)
39. Simonyan, K., Zisserman, A.: Very deep convolutional networks for large-scale image recognition. arXiv preprint arXiv:1409.1556 (2014)
40. Sitzmann, V., Martel, J., Bergman, A., Lindell, D., Wetzstein, G.: Implicit neural representations with periodic activation functions. In: NeurIPS (2020)
41. Skorokhodov, I., Tulyakov, S., Elhoseiny, M.: StyleGAN-V: a continuous video generator with the price, image quality and perks of StyleGAN2. arXiv preprint arXiv:2112.14683 (2021)
42. Soomro, K., Zamir, A.R., Shah, M.: UCF101: a dataset of 101 human actions classes from videos in the wild. arXiv preprint arXiv:1212.0402 (2012)
43. Srivastava, N., Mansimov, E., Salakhudinov, R.: Unsupervised learning of video representations using LSTMs. In: ICML (2015)
44. Tancik, M., et al.: Fourier features let networks learn high frequency functions in low dimensional domains. In: NeurIPS (2020)
45. Tian, Y., et al.: A good image generator is what you need for high-resolution video synthesis. In: ICLR (2021)
46. Tran, D., Bourdev, L., Fergus, R., Torresani, L., Paluri, M.: Learning spatiotemporal features with 3D convolutional networks. In: ICCV (2015)
47. Tulyakov, S., Liu, M.Y., Yang, X., Kautz, J.: MoCoGAN: decomposing motion and content for video generation. In: CVPR (2018)
48. Unterthiner, T., van Steenkiste, S., Kurach, K., Marinier, R., Michalski, M., Gelly, S.: Towards accurate generative models of video: a new metric & challenges. In: ICLR (2019)
49. Vondrick, C., Pirsiavash, H., Torralba, A.: Generating videos with scene dynamics. In: NeurIPS (2016)
50. Wang, T.C., et al.: Video-to-video synthesis. In: NeurIPS (2018)
51. Wang, T.C., Liu, M.Y., Zhu, J.Y., Tao, A., Kautz, J., Catanzaro, B.: High-resolution image synthesis and semantic manipulation with conditional GANs. In: CVPR (2018)
52. Weissenborn, D., Täckström, O., Uszkoreit, J.: Scaling autoregressive video models. In: ICLR (2020)
53. Wu, C., et al.: GODIVA: generating open-domain videos from natural descriptions. arXiv preprint arXiv:2104.14806 (2021)
54. Wu, C., et al.: NÜWA: visual synthesis pre-training for neural visual world creation. arXiv preprint arXiv:2111.12417 (2021)
55. Xiong, W., Luo, W., Ma, L., Liu, W., Luo, J.: Learning to generate time-lapse videos using multi-stage dynamic generative adversarial networks. In: CVPR (2018)
56. Xu, R., Wang, X., Chen, K., Zhou, B., Loy, C.C.: Positional encoding as spatial inductive bias in GANs. In: CVPR (2021)
57. Yan, W., Zhang, Y., Abbeel, P., Srinivas, A.: VideoGPT: cideo generation using VQ-VAE and transformers. arXiv preprint arXiv:2104.10157 (2021)
58. Yu, S., et al.: Generating videos with dynamics-aware implicit generative adversarial networks. In: ICLR (2021)
59. Zhang, R., Isola, P., Efros, A.A., Shechtman, E., Wang, O.: The unreasonable effectiveness of deep features as a perceptual metric. In: CVPR (2018)

# Combining Internal and External Constraints for Unrolling Shutter in Videos

Eyal Naor, Itai Antebi, Shai Bagon(✉), and Michal Irani

Department of Computer Science and Applied Mathematics,
The Weizmann Institute of Science, Rehovot, Israel
shai.bagon@weizmann.ac.il
http://www.wisdom.weizmann.ac.il/~vision/VideoRS

**Abstract.** Videos obtained by rolling-shutter (RS) cameras result in spatially-distorted frames. These distortions become significant under fast camera/scene motions. Undoing effects of RS is sometimes addressed as a *spatial* problem, where objects need to be rectified/displaced in order to generate their correct global shutter (GS) frame. However, the cause of the RS effect is inherently temporal, not spatial. In this paper we propose a space-time solution to the RS problem. We observe that despite the severe differences between their $xy$ frames, a RS video and its corresponding GS video tend to share the exact same $xt$ slices – up to a known sub-frame temporal shift. Moreover, they share the same distribution of small 2D $xt$-patches, despite the strong temporal aliasing within each video. This allows to constrain the GS output video using *video-specific* constraints imposed by the RS input video. Our algorithm is composed of 3 main components: (i) Dense temporal upsampling between consecutive RS frames using an off-the-shelf method, (which was trained on regular video sequences), from which we extract GS "proposals". (ii) Learning to correctly merge an ensemble of such GS "proposals" using a dedicated MergeNet. (iii) A *video-specific* zero-shot optimization which imposes the similarity of $xt$-patches between the GS output video and the RS input video. Our method obtains state-of-the-art results on benchmark datasets, both numerically and visually, despite being trained on a small synthetic RS/GS dataset. Moreover, it generalizes well to new complex RS videos with motion types outside the distribution of the training set (e.g., complex non-rigid motions) – videos which competing methods trained on much more data cannot handle well. We attribute these generalization capabilities to the combination of external and internal constraints.

## 1 Introduction

Rolling shutter (RS) cameras are widely used in many consumer products. In contrast to global shutter (GS) cameras, which capture all pixels of a single frame simultaneously, RS cameras capture the image pixels row by row. Consequently,

© The Author(s), under exclusive license to Springer Nature Switzerland AG 2022
S. Avidan et al. (Eds.): ECCV 2022, LNCS 13677, pp. 119–134, 2022.
https://doi.org/10.1007/978-3-031-19790-1_8

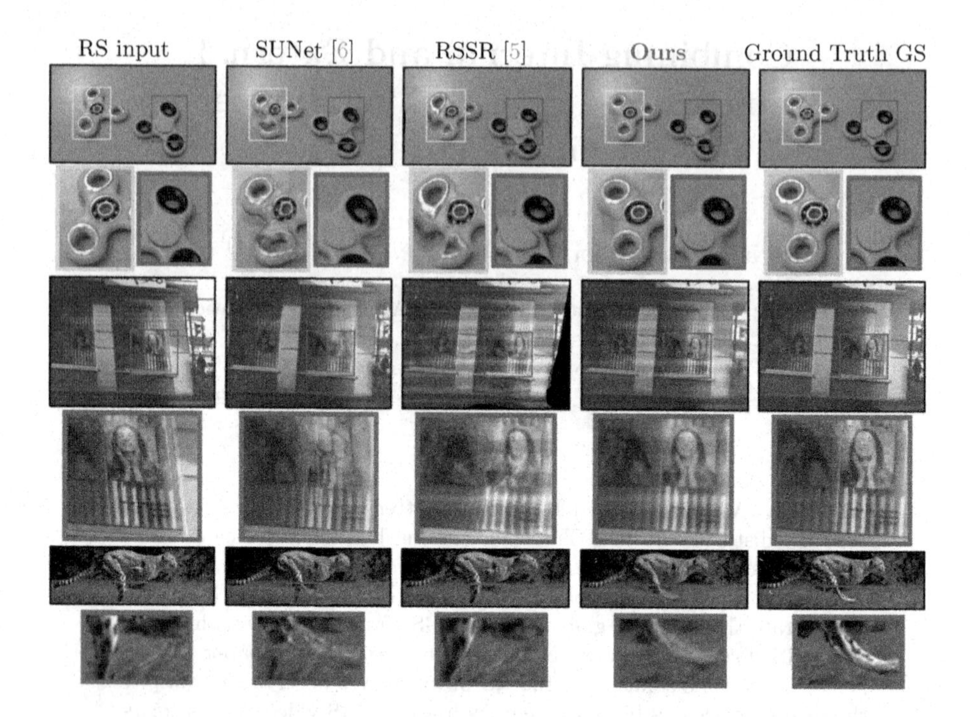


**Fig. 1. Examples of RS-induced distortions for various scene dynamics (and attempts to fix them).** (Top) Rotational motion: the round tip of the rotating pink spinner turns into an ellipse in the RS video, and its position is displaced within the frame. (Middle) Camera translation: straight vertical lines get tilted in RS. (Bottom) Non-rigid motion: the limbs of a fast running cheetah become completely distorted and dislocated (in a non-parametric way). SotA methods (SUNet [6], RSSR [5]) fail to generalize to RS distortion types outside their trainning set (especially non-rigid scenes), whereas our method does favorably.

a variety of spatial distortions (e.g., tilt, stretch, curve, wobble) appear under fast camera/scene motion. Examples of such distortions can be seen in Fig. 1 (e.g., the round hole at the tip of the rotating pink spinner turns from a circle into an ellipse, and its position is displaced within the frame). Figure 2 exhibits the degree of misalignment between temporally corresponding RS and GS frames in a video of a fast running cheetah.

RS correction methods can be broadly classified as either *single-frame* [7, 11,12,20] or *multi-frame* [1,5,6,8,13,15,18,19]. Attempting to reconstruct a GS frame from a single RS frame is highly ill-posed, as it does not exploit the inherent temporal aspect of the RS problem. Single-frame methods thus require significant assumptions on either the camera motion (pure translation, pure rotation, etc.) or on the scene (planar scene, straight lines, etc.) As a result, the current leading methods [5,6] are multi-frame ones. These are the methods we compare against.

In this paper we propose a space-time solution to the RS problem. The RS problem is fundamentally temporal, since it stems from different rows being

(a) Initial RS-GS Misalignment:

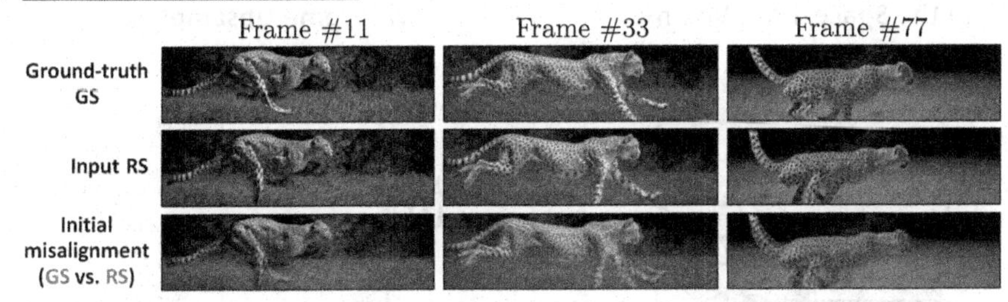


(b) Residual misalignemnts after fixing RS Effects: *(shown for Frame#11)*

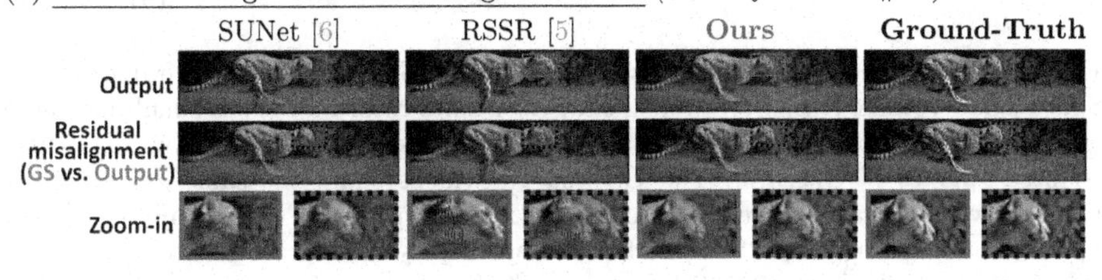


**Fig. 2. Visualizing the RS-GS Misalignments.** (a) Visualization of the initial distortion/misalignment between the RS frame (inserted into the G band) and its corresponding GS frame (inserted into the R&B bands). Notice the complexity of the artifacts, especially in non-rigid areas. Grayscale indicates good alignment, whereas Green and Magenta indicate misalignment. (b) Visualizing residual misalignment between reconstructed GS and ground-truth GS. While state-of-the-art competitors fail to properly correct the distortions and position of the running cheetah (e.g., see zoom-in on face), our method does so favorably.

captured at different times. In fact, the RS frame captures the "correct" (undistorted) image rows, but at the wrong times. Thus, despite the severe spatial distortions between RS and GS frames, we observe that a RS video and its corresponding GS video share the exact same $xt$ slices – up to a known sub-frame temporal shift. This observation is invariant to the types or complexity of the camera/scene motions. We thus repose the problem of rectifying RS videos as a problem of correctly shifting and interpolating $xt$ slices.

More specifically, our algorithm is composed of 3 steps: (i) We use an off-the-shelf frame interpolation algorithm [2] (pre-trained on regular videos) to densely fill the space-time volume between consecutive RS frames. Were the temporal-upsampling perfect, recovering the GS frame would then be trivial – simply sample the correct row from each interpolated video frame (see Fig. 3(b)). (ii) Since general-purpose frame interpolation methods are prone to errors (and more so in presence of RS effects), we apply the temporal-interpolation algorithm to multiple augmentations of the RS video, to generate multiple GS "proposals". We train a small RS-specific MergeNet to correctly merge an ensemble of such GS "proposals", while being sensitive to local RS idiosyncrasies. Due to its simplicity, it suffices to train MergeNet on a *small and synthetic* dataset of RS/GS video

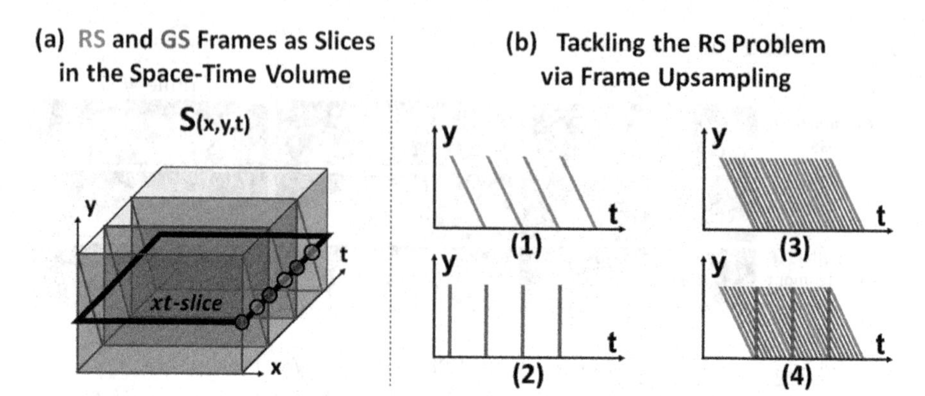


**Fig. 3. Space-time relations between RS and GS. (a)** RS and GS frames in the Space-Time Volume $S(x, y, t)$. In magenta: GS frames record all pixels simultaneously, thus capturing $xy$ planes in $S(x, y, t)$. In green: RS frames record row-by-row, capturing "slanted" planes in $S(x, y, t)$. The GS and RS videos share the same $xt$ slices, up to a known sub-frame shift. **(b)** Re-casting the RS problem as a temporal frame-upsampling problem. (1) RS input frames are displayed as green slanted lines in a side $yt$-view of $S(x, y, t)$. (2) GS frames are displayed as vertical magenta lines in the side view. (3) Temporal frame-upsampling "fills" the space time volume by generating intermediate RS frames. (4) Sampling the relevant row from each interpolated RS frame allows to reconstruct the GS frames.

pairs. This is in sharp contrast to competing SotA methods [5,6], which train a different model for each new dataset. (iii) Finally, we observe that a RS/GS video pair tends to share the same distribution of small 2D $xt$-patches. We use this observation to impose *test-time video-specific* constrains on the $xt$-patches of the GS output video (to match those of the RS input video).

Our method thus benefits from both Internal and External constraints: on the one hand, we utilize a frame interpolation method [2], *externally* trained on large datasets of videos, while on the other hand, our zero-shot test-time optimization takes advantage of *internal* video-specific distribution of $xt$-patches.

Our method outperforms existing RS video methods on a large variety of RS video types – evaluated both on existing RS benchmark datasets, as well as on a new dataset we collected of challenging videos with highly non-rigid motions (which are lacking in existing benchmark datasets). Particularly, our method outperforms prior methods ***by a large margin*** on RS videos of complex non-rigid scenes. We attribute the generalization capabilities of our algorithm to its combination of external and internal constraints.

Our contributions are thus several fold:

- We re-cast the RS correction problem as a temporal upsampling problem. As such, we can leverage advanced frame interpolation methods (which have been pre-trained on a large variety of real-world videos).
- We observe that a RS video and its corresponding GS video share the same small $xt$-patches, despite significant temporal aliasing exhibited in both

videos. This allows to impose *video-specific* constraints on the GS output, *at test-time*.
- We curated and released a new dataset of RS/GS video pairs, which pushes the envelope of RS benchmarks to include also complex non-rigid motions.
- We provide state-of-the-art results on RS video benchmarks.

## 2 Related Work

RSSR [5] and SUNet [6] are the SotA methods, and are also the most closely related to our approach. SUNet [6] introduced symmetric consistency constraints while warping consecutive RS frames to produce a single in-between GS frame. They use a constant velocity motion model, and train a network to convert the optical flow between the RS frames to produce "RS undistortion-flow". RSSR [5] inverts the mechanism of RS in order to recover a high framerate GS video from consecutive RS frames. Both methods have complex networks that require much training data. In fact, they have a different trained model for each dataset. This limits their performance and generalization capabilities on new out-of-distribution RS videos (see Figs. 1, 2). Furthermore, since [5] rectifies the frames using RS undistortion-flow, the results contain holes on occlusion boundaries. In contrast, our method uses a single light network (trained on a synthetic dataset) on all benchmarks, with leading performance, while exhibiting good generalization capabilities on new out-of-distribution RS videos.

A number of other previous Deep-Learning multi-frame methods were also proposed. In [9] an array of networks was trained to warp and predict a GS frame from two consecutive RS frames. They further provide 2 benchmark datasets of paired RS/GS videos for training and evaluation. These datasets were later used by [5,6], showing SotA results. We too experiment on these datasets, and compare to the leading methods [5,6]. Two other nice recent RS methods, but less closely related to us (as they require different input types) have been recently presented. The method of [17] solves for a GS output from a *blurry* RS input, whereas [1] proposes a method which uses two simultaneous RS cameras mounted on a single platform to produce one GS output.

## 3 Inherent Relations Between GS & RS Videos

**Claim:** The $xt$-slice of the RS video is a *shifted version* of the $xt$-slice of the GS video, at the same row $j$, at a "sub-pixel" shift of $j/N<1$ along the $t$-axis, where $N$ is the number of rows in each frame.

**Proof:** Let $S(x, y, t)$ denote the continuous space-time volume (Fig. 3a). It is defined in the camera coordinate system, hence any camera motion relative to the scene can be regarded as scene motion relative to a static camera. Let $GS(i, j, k)$ and $RS(i, j, k)$ denote the GS and RS videos, respectively, where $(i, j)$ are integer pixel coordinates, and $k$ is the frame number, taken at time gaps of $\Delta T$ along

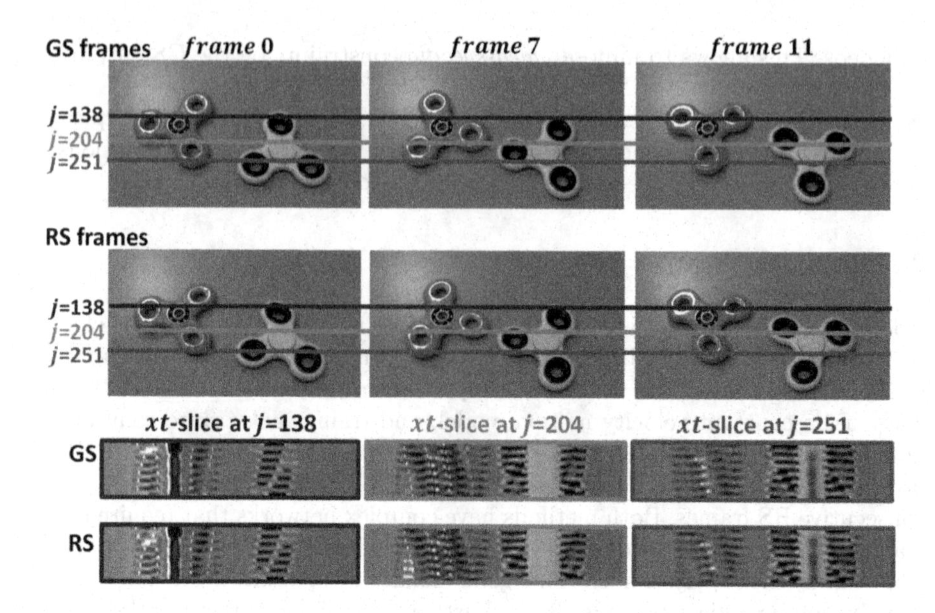


**Fig. 4. GS-RS video pairs share the same $xt$-slices.** Corresponding $xt$-slices from the RS and GS video, at a few rows $j$ (= y). Although the RS frames exhibit strong distortions compared to their corresponding GS frames (circles turn into ellipses; angles of the spinner arms appear different), corresponding $xt$-slices (at same row $j$) are very similar, as they sample the same $xt$-plane in the continuous space-time volume $S(x, y, t)$ at a 1D 'sub-pixel' shift in the $t$-direction (the vertical axis in those slices). See Sect. 3 and Fig. 3a for more details.

the $t$ axis (w.l.o.g. we define $\Delta T = 1$). *By definition*, RS and GS videos are just different space-time samplings of the continuous $S(x, y, t)$ as follows:

$$GS(i, j, k) = S(i, j, k) \quad \text{and} \quad RS(i, j, k) = S\left(i, j, k + \frac{j}{N}\right) \tag{1}$$

Note that the above 'sampling' equation explicitly entails that, for any row $j_0$ ($j_0 = 1,..N$), $GS(i, j_0, k)$ and $RS(i, j_0, k)$ are related by a fixed 1D shift of $j_0/N$ $< 1$ along the $t$-axis, $\forall i, k$. These are exactly the $xt$-slices of the GS and RS videos, at row $j_0$. This proves our claim. ■

Although unintuitive, note that this observation, which is simply derived from Eq.(1), is independent of the content of the space-time volume $S$, hence is *invariant to the type of scene/camera motions* (whether rotation, translation, etc.).

More intuitively: The *spatial frames* of the GS and RS videos are *different planes* within the space-time volume $S(x, y, t)$ (see purple and green slices in Fig. 3a). Hence they naturally have very different appearances under severe camera or scene motions. However, these RS distortions are manifested only in the $y$ direction (since every row $y = j$ is sampled at different time), but are not expressed in the $xt$ slices of $S$. The $xt$-slices of GS and RS videos at row $j$ are just different samples of the *same shared plane* (the black plane in Fig. 3a). Their samples are marked by purple and green points inside the black $xt$-plane. Fig. 4 exemplifies this phenomenon, displaying 3 corresponding $xt$ slices of a GS-RS video pair of a complex dynamic scene (fast rotating spinners). Our algorithm for undoing RS effects builds on top of this simple yet powerful observation, in 2 ways:

1. In principle, given this observation, the solution to the RS problem seems trivial: just back-warp each $xt$ slice of the RS video by its *known* sub-frame temporal shift $j/N$ (determined by its row index $j$). However, such sub-frame warping is far from being trivial in the presence of *temporal aliasing*, which is very characteristic of video data (due to low frame-rate compared to fast scene/camera motions). To address this issue, we resort to a pre-trained state-of-the-art temporal upsampling/interpolation network [2] (which was trained on a huge collection of regular videos). However, video temporal upsampling methods have their own inaccuracies, and even more so on RS videos (which are video types they were not trained on). In Sect. 4.2 we propose an approach to address this issue (with very few RS-GS training data).
2. We further employ in our algorithm another inherent relation of a RS-GS video pair: These 2 videos *share the same pool of small xt-patches*. It was shown in [14] that small 3D space-time patches (e.g., $7 \times 7 \times 3$) recur abundantly within a *single* natural video of a dynamic scene. This space-time recurrence is an inherent property of the continuous dynamic world. Moreover, such patches were shown to appear in different aliased forms at different locations within the video (which gave rise to temporal super-resolution from a single video [14,21]). The RS distortions affect the $y$ direction of small space-time patches; however, these distortions do not affect the $xt$ direction of these patches. Thus, while the GS and RS videos may not share the same small 3D space-time patches, they do share the same small 2D $xt$-patches (e.g., $7 \times 3$), despite the temporal aliasing. The continuous version of a 2D $xt$-patch will appear many times in the continuous dynamic scene, hence will appear at multiple $xt$ slices in each video, each time sampled at a different sub-frame ("sub-pixel") temporal shift. Therefore, despite the temporal aliasing, each small $xt$-patch in the GS video will likely have similar patches (with the appropriate sub-frame temporal sampling) within some $xt$ slices of the RS video.
   We empirically measured the strength of GS-RS *cross*-video recurrence of $xt$-patches, compared to their *internal*-recurrence within the GS video itself. This was estimated as follows: We randomly sampled a variety of GS-RS video pairs, which cover a variety of different motion types (rotation, translation, zoom, and non-rigid motions). For each $7 \times 3$ $xt$-patch $p(x,t)$ in each GS video, we computed 2 distances: (i) the distance to its nearest-neighbor (NN) $xt$-patch in the GS video $d_{GS}(p) = \|p - \mathrm{NN}(p, GS)\|$, and (ii) the distance to its nearest-neighbor $xt$-patch in the paired RS video $d_{RS}(p) = \|p - \mathrm{NN}(p, RS)\|$. We then measure for each patch the ratio $r = \frac{d_{RS}(p)}{d_{GS}(p)}$, which

tells us how worse is the patch similarity *across* the 2 videos compared its similarity *within* the GS video. Our empirical evaluations show that $mean(r)$ = 1.13, i.e., on average, the cross GS-RS patch distance $d_{RS}(p)$ is only $\times 1.13$ larger than the internal patch distance $d_{GS}(p)$. This holds not only for smooth patches, but also for patches with high gradient content (which correspond to sharp edges and high temporal changes). In fact, when measured only for the *top 25% of xt-patches with the highest gradient magnitude*, for 61% of them $r \leq 1.1$, and for 85% of them $r \leq 1.5$. This indicates high similarity of small $xt$-patches between the 2 videos.
We employ this observation to impose an additional *video-specific* prior on our GS output video, at test-time, constraining the output by the collection of $7 \times 3$ $xt$-patches of the RS input video (see Sect. 4.3 for more details).

## 4 Method

Our algorithm is composed of 3 main steps. First, we use an off-the-shelf general-purpose frame-upsampling algorithm, in order to extract the relevant rows from different temporally-interpolated RS frames, and compose them into GS frames (Sect. 4.1). To compensate for the fact that the frame-upsampling algorithm is general-purpose (not RS-specific), we apply this process repeatedly on several different augmentations of the RS input video, which result in several "GS proposals" per frame. In the second step, we train and use a RS-specific MergeNet, to merge the GS proposals into a coherent GS frame. Since this is a simple network with few layers and a very narrow receptive field, it suffices to train it with a small synthetic RS-GS dataset (Sect. 4.2). Lastly, we use a zero-shot approach to refine the resulting GS video to adhere to the patch statistics of the input RS video, thus reducing blurriness and other undesirable visual artifacts (Sect. 4.3).

We note that these 3 main steps are separate as they are trained at different times, *on different types of data: Stage1* - leverages off-the-shelf SotA frame-interpolation methods, pretrained on large datasets of *general* videos. *Stage2* - MergeNet is trained on RS videos, at train-time. *Stage3* - is applied to the specific test video, at test-time.

### 4.1 Generating GS Proposals via Temporal Frame-Upsampling

As we observe in Sect. 3, GS $xt$-slices can be recovered by shifting the RS $xt$-slices at each row $j$ by a "subpixel" (subframe) shift of size $j/N$. However, such sub-frame warping is challenging due to severe temporal-aliasing, which is very characteristic of video data (regardless of whether a RS or a GS camera was used). To address this issue, we resort to a state-of-the-art off-the-shelf temporal frame interpolation network, DAIN [2], which was trained on a large in-the-wild video dataset [16] depicting a wide range of motions and complex scene dynamics. DAIN works by estimating a robust depth-aware flow between consecutive frames and utilizes this flow to efficiently perform temporal frame upsampling to an arbitrary (user-defined) framerate. Using the same flow to interpolate all

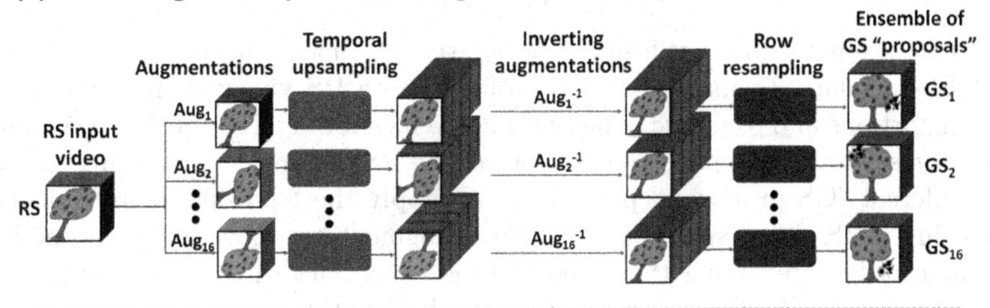


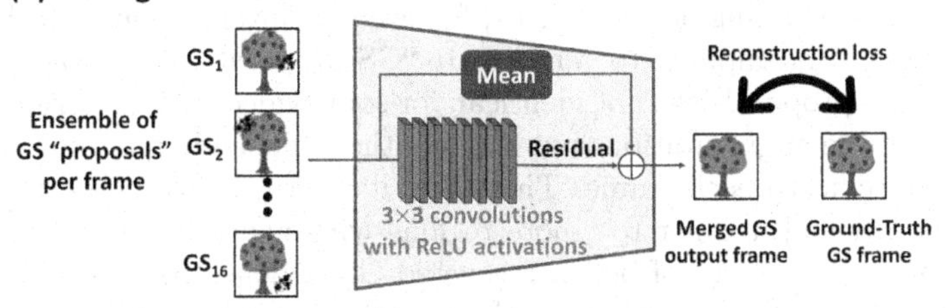


**Fig. 5. MergeNet.** (a) We adapt a general-purpose frame-upsampling method [2] (trained on regular videos) to temporally-upsample RS videos, by applying it to multiple augmentations of the RS video. This generates 16 GS "proposals" per frame (see Sect. 4.2). (b) We train a small RS-specific "MergeNet" to correctly merge an ensemble of GS frame "proposals" into a single coherent GS frame. MergeNet learns the residual correction w.r.t. the mean of the 16 frame proposals, while being sensitive to local RS idiosyncrasies. Being a rather shallow CNN (8 layers of $3 \times 3$ convolutions followed by ReLU activations), with a small ($17 \times 17$) receptive field, it suffices to train MergeNet on a small and synthetic dataset of RS/GS video pairs. All hidden layers are with 64 channels.

in-between frames makes DAIN's output temporally consistent, reducing flickers and other undesired artifacts in its predictions (regardless of the temporal interpolation rate).

Since each row in the RS frame comes from a different temporal offset, the number of interpolated frames between every two RS input frames is determined by the number of rows in each frame. That is, for a RS video with $N$ rows/frame we need to temporally-upsample $\times N$ the original frame-rate, producing $N-1$ additional frames between every 2 input RS frames (typical values of $N$ are on the order of hundreds of rows). Once interpolated, we compose a *GS proposal* frame by taking the relevant row from each temporally-interpolated RS frame, as illustrated in Fig. 3(b). However, since DAIN did not train on videos with noticeable RS distortions, the interpolated frames of DAIN on RS data are often imperfect, affecting the quality of the GS proposal. This problem is addressed using the second step of our algorithm, described next.

### 4.2 MergeNet: Merging Multiple GS Proposals

DAIN [2] is general-purpose frame-interpolation method, trained on many videos (not RS-specific). To make better use of DAIN on RS videos, which may contain distortions and dynamic behavior outside its training distribution, we apply DAIN on several different augmentations of the RS input video, resulting in several different "GS proposals" per frame. We apply the following augmentations to the input RS video (spatial and temporal augmentations, which ignore the RS scanning order): reversing the video in time (play backwards), spatially rotating it by $k \cdot 90°$ ($k = 1, 2, 3, 4$), and spatial horizontal flipping. This results in 16 pre-determined augmentations in total. DAIN is then applied to temporally upsample each of these 16 augmented videos, followed by inverse augmentation and appropriate row-subsampling, to generate 16 "GS proposals" (Fig. 5(a)).

The resulting GS proposals are not identical: for some videos DAIN performs better on several of the augmentations but not on others, depending on motions and distortions specific to each frame. Therefore, it is crucial to merge these proposals in a non-trivial manner to ignore regions with unwanted artifacts in proposals and taking advantage of better recovered GS regions. To that end we use a RS-specific "MergeNet" to combine these proposals into a coherent GS frame. Note that although motions and distortions in videos may be *globally* complex, they are still characterized by *locally* simple linear motions. Therefore, we deliberately design MergeNet with a small receptive field ($17 \times 17$ pixels), to learn how to merge and fix small patch-wise GS-proposals. Consequently, a limited synthetic video dataset with a variety of simple global affine motions provides enough diversity of locally-linear ones. These offer sufficient examples to train MergeNet to learn how to correctly merge and fix small patch-wise GS-proposals. We train MergeNet on a small available synthetic dataset of affine-induced RS/GS videos pairs (the "Carla-RS" train-set of [9] – see Sect. 5.1). Although trained on local patch-wise synthetic examples, MergeNet generalizes well to real complex RS videos of highly non-rigid scenes.

To conclude, we reduce the difficult task of correcting RS-distorted frames, to a much simpler task of adapting a generally-applicable frame interpolation algorithm to handle RS videos. The "heavy-lifting" global motions considerations are done by a general-purpose frame interpolation method, while the local adaptation to the idiosyncrasies of RS is done by our small MergeNet. Figure 5(b) shows the architecture of MergeNet.

### 4.3 Imposing *Video-Specific* Patch Statistics at test-time

The final step of our algorithm makes use of our observation that small $xt$-patches are shared by a RS and GS video of the same dynamic scene (see Sect. 3). We thus constrain the $xt$-patches in our GS output video, to be from the same $xt$-patch distribution as the RS input video. This is obtained via a short *test-time* optimization over the $7 \times 3$ $xt$-patches of the GS frames predicted from our previous MergeNet step. This process changes the predicted GS patches to have smaller distances to their nearest-neighbor (NN) patches in the input RS video, while not allowing them to deviate too far from their initial predicted value. We formulate this via the minimization of the loss function: $\mathcal{L}_{NN} + \lambda \cdot \mathcal{L}_{validity}$

where
$$\mathcal{L}_{NN} = \sum_{ijk} \alpha_{ijk} \left\| GS_{patch_{ijk}} - RS_{patch_{NN[ijk]}} \right\|^2$$
$$\mathcal{L}_{validity} = \sum_{ijk} (1 - \alpha_{ijk}) \left\| GS_{patch_{ijk}} - GS^{initial}_{patch_{ijk}} \right\|^2$$

$\mathcal{L}_{NN}$ incorporates the distance of each $xt$-patch in location $i, j, k$ to its NN patch in the input RS video. Minimizing $\mathcal{L}_{NN}$ brings the predicted GS patches closer to those of the RS, thus improving the recurrence of $xt$ patches between the input RS and the output GS. $\mathcal{L}_{validity}$ measures the distance between the current GS prediction and the initial output of MergeNet, $GS^{initial}$. Minimizing $\mathcal{L}_{validity}$ helps stabilize the optimization process. Following [10], we give higher weight $\alpha_{ijk}$ to patches with strong edges, as their NNs are more reliable and less prone to over-fitting noise (see discussion on PatchSNR in [10]). Accordingly, the values of $\alpha_{ijk} \in [0, 1]$ are determined by Canny edge responses [3] (computed on the $xt$-slices of the predicted GS output from MergeNet), to weigh the patches accordingly. We set $\lambda$ so that the 2 loss terms are of the same order of magnitude.

# 5 Experimental Results

We validate our approach both quantitatively and qualitatively on existing RS/GS benchmark datasets, as well as on a new challenging dataset we curated.

## 5.1 Datasets

RS benchmark datasets with ground-truth GS data are comprised of aligned RS/GS video pairs. Curating such datasets is technically challenging: one needs to capture or synthesize a very high framerate video, and then sub-sample it (vertically or diagonally – see Fig. 3) at $1/N$ of the framerate in order to generate aligned RS/GS videos (where $N$ is the number of rows per frame). Existing benchmark datasets are thus relatively small (~2K frames), with a very limited variety of dynamic motions.

- **Fastec-RS** [9]**:** This dataset was created using a high-speed camera mounted on a driving car. Consequently, the motions are mostly horizontal translation, and the RS distortions are mostly affine ones. *Fastec-RS* dataset comprises 76 sequences with at most ***34 frames*** per sequence – ***56 Train-set, 20 Test-set***.
- **Carla-RS** [9]**:** This dataset was synthesized using the Carla simulator [4]; a virtual 3D environment. The virtual environment allows to simulate more complex camera motions; thus, Carla-RS, albeit synthetic, contains a wider variety of RS artifacts. *Carla-RS* comprises 250 sequences with ***10 frames each – 210 Train-set, 40 Test-set***. Being synthetic, this dataset further

**Table 1. Numerical Evaluation:** Our method outperforms [5,6] on all three benchmarks. Note that SUNet and RSSR trained benchmark-specific models for Fastec-RS and Carla-RS, while our method uses a single trained model for all datasets. * In-the-wild-RS has no training set, hence for SUNet and RSSR we evaluated both their models, and reported their best result. As can be seen, existing methods struggle to generalize to RS video types which are beyond those explicitly represented in their training sets. In contrast, our method (trained only on Carla-RS), generalizes much better (with a significant margin of +3.39 dB) on the challenging In-the-wild-RS.

| Dataset | | Ours | SUNet [6] | RSSR [5] |
|---|---|---|---|---|
| Fastec-RS [9] | PSNR [dB], SSIM | **28.573, 0.8436** | 28.249, 0.8277 | 21.175, 0.7649 |
| Carla-RS [9] | PSNR [dB], SSIM | **31.430, 0.9187** | 29.170, 0.8499 | 24.776, 0.8661 |
| Carla-RS masked | PSNR [dB], SSIM | **31.840, 0.9187** | 29.269, 0.8499 | 30.137, 0.8661 |
| In-the-wild-RS* | PSNR [dB], SSIM | **27.919, 0.900** | 24.531, 0.8262 | 24.163, 0.8376 |

comes with occlusion maps between consecutive RS frames (regions which can potentially be ignored when evaluating the GS reconstruction results). Masks are used to calculate numerical results in *Carla-RS masked*, which are also reported in Table 1.

- **In-the-wild-RS [NEW]:** Neither *Fastec-RS* nor *Carla-RS* contain real complex non-rigid motions, and as such are quite limited. To mitigate this lacuna, we curated *In-the-wild-RS*, generated from 15 Youtube videos captured with high speed GS cameras, featuring complex non-rigid scene motions (running animals, flying birds, turbulent water, rotating spinners, etc.), captured with unrestricted camera motions. For a few of these videos, we further generated 2–3 versions of GS/RS pairs, with varying degree of RS complexity for the same scenes. On average, ***30 frames*** per sequence (some longer, some shorter) – ***NO Train-set, 15 Test-set***. This dataset can be accessed through our project page.

### 5.2 Quantitative Results

Table 1 shows PSNR and SSIM results of our reconstructed GS frames for the test sets of all three benchmarks. We used only the synthetic training set of *Carla-RS* to train our MergeNet, and used the same network for *all* 3 benchmarks.

We compare our results to the state-of-the-art methods SUNet [6] and RSSR [5]. These methods trained a *different instance of their network* for each dataset – *Fastec-RS* and *Carla-RS* (as opposed to our single MergeNet). Furthermore, since *In-the-wild-RS* has no training set, we ran both trained models of [5,6], and reported ***the best performing one*** on *In-the-wild-RS*'s test-set for each method (in both cases, it was the network trained on the synthetic Carla-RS dataset that performed best on *In-the-wild-RS*). Nevertheless, our method significantly outperforms SUNet and RSSR on all three benchmarks.

It is interesting to see that although we did not use the training set of Fastec-RS at all, we outperform the models trained specifically on that benchmark, by

***+0.32*** **dB** and ***+7.4*** **dB**. On Carla-RS, we outperform the competing models by ***+2.26*** **dB** and ***+6.7*** **dB**. More importantly, our method significantly outperforms SUNet and RSSR when evaluated on the challenging In-the-wild-RS dataset (with complex non-rigid motions, and no training set), by ***+3.39*** **dB** and ***+3.76*** **dB**. All in all, existing methods have more difficulty generalizing to new types of RS distortions that are outside the distribution of their training set.

Note that the numerical results of RSSR [5] are low due to the holes in their predicted GS frames, where no pixels were warped to by their undistortion-flow. The synthetic Carla-RS further comes with GT masks on occluded pixels, allowing RSSR to compare only on non-occluded pixels. These results are shown in the third row of Table 1. RSSR performs significantly better on *non-occluded* pixels, surpassing SUNet. However, our method still performs better than both (***+2.57*** **dB** and ***+1.7*** **dB**).

*Ablation:* We further used Fastec-RS benchmark to evaluate the contribution of each step in our method. We note that most of the "heavy lifting" comes from applying DAIN. This first step already yields good PSNR of 27.67 dB. Applying MergeNet further improves results by additional ~1 dB. Our last *video-specific* test-time optimization step improved 30% of the sequences by ~0.2 dB, while yielding a smaller improvement on the other sequences.

### 5.3 Qualitative Results

Figures 1, 2, 6 show visual results and comparisons. Rectifying RS frames requires not only reconstructing good visual quality, but no less important – achieving good alignment w.r.t. the ground-truth GS frames. This is a difficult non-trivial task, as shown in Fig. 2. To better highlight the degree of *residual misalignment* between the predicted GS frames and the ground-truth ones, we use the following visualization: We convert the ground-truth and predicted GS frames to grayscale images. We place the ground-truth RS in the red and blue channels, and the predicted GS in the green channel. Properly rectified areas are gray in the new visualization, while green or magenta highlighted areas indicate misalignments. Figures 2 and 6 use this visualization to highlight how our method better rectifies the scene in a variety of complex motion types. Compare, for example, our proper alignment of the fast-moving sign-pole in the middle of Fig. 6; the non-rigid motion of the water ripples, the foot of the flipping man at the bottom of Fig. 6, or the cheetah head in Fig. 2. Moreover, note our reconstruction of the round shape and position of the fast rotating spinner, as well as the hind leg of the cheetah, in Fig. 1. These are a few examples of complex dynamic scenes from our new In-the-wild-RS dataset.

## 6 Limitations

While current methods for temporal video-upsampling are quite advanced, this still forms the main bottleneck of our method. Our performance is bounded

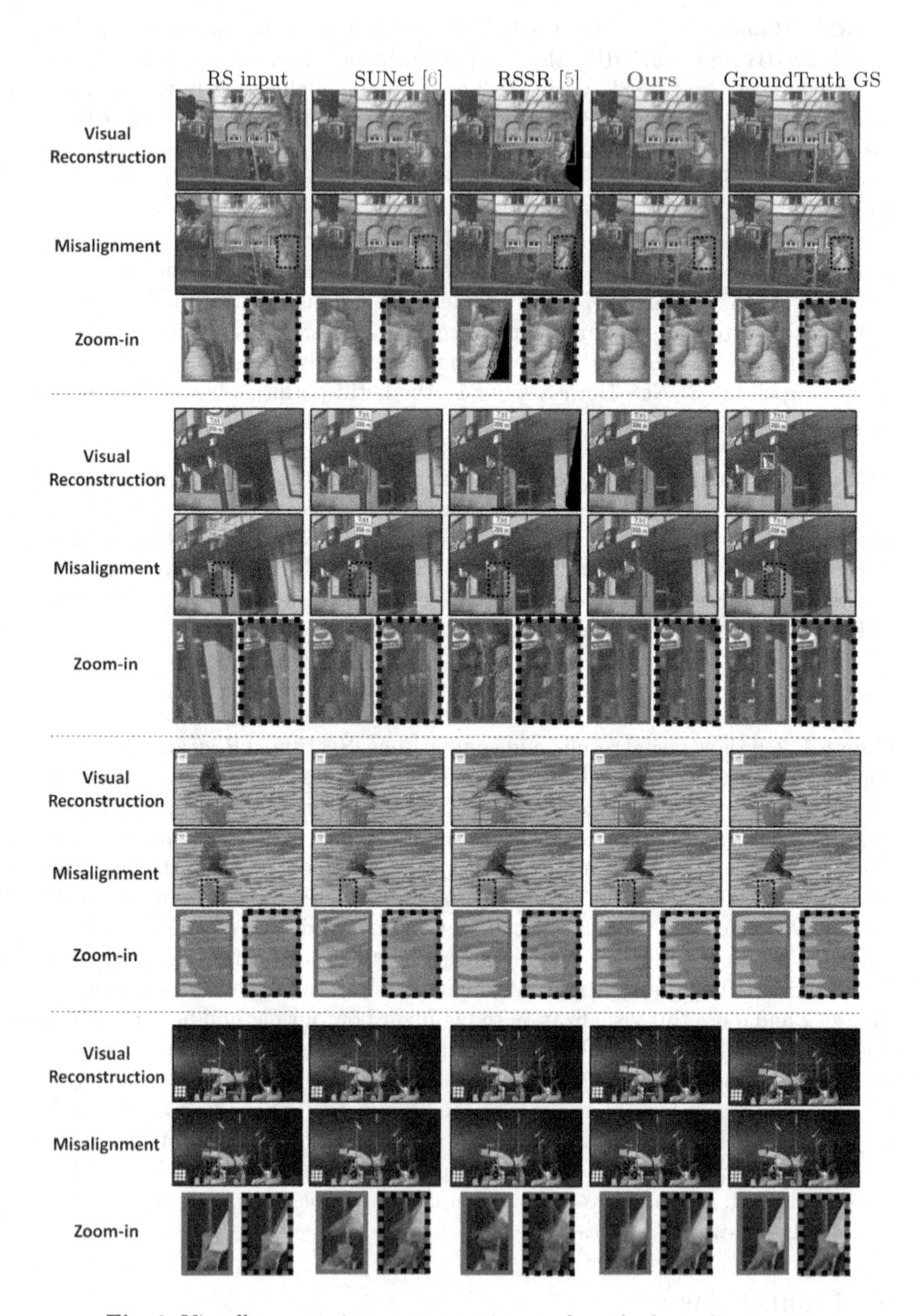


**Fig. 6.** Visually comparing reconstructions and residual misalignments.

by the limitations of SotA frame-interpolation methods, which currently cannot handle videos with severe motion aliasing. For example, a video recording of an extremely fast rotating propeller (much faster than the camera framerate), will appear in the video to be rotating in the reversed/wrong direction (even when recorded by a GS camera). Current temporal interpolation methods cannot undo such severe motion aliasing, thus fail to generate the correct intermediate frames. Our method fails when the frame-interpolation method breaks down. However, since the frame-upsampling is a standalone module in our method, it can be replaced as SotA frame-interpolation methods improve, leading to an immediate improvement in our algorithm, at no extra cost or effort.

## 7 Conclusion

We re-cast the RS problem as a temporal frame-upsampling problem. As such, we can leverage advanced frame interpolation methods (which have been pre-trained on a large variety of complex real-world videos). We bridge the gap between frame-interpolation of general videos to frame-interpolation of RS videos using a dedicated MergeNet. We further observe that a RS video and its corresponding GS video share the same small $xt$-patches, despite significant temporal aliasing exhibited in both videos. This allows to impose video-specific constraints on the GS output, at test-time. Our method obtains state-of-the-art results on a variety of benchmark datasets, both numerically and visually, despite being trained only on a small synthetic RS/GS dataset. Moreover, it generalizes well to new complex RS videos containing highly non-rigid motions – videos which competing methods trained on more data cannot handle well.

**Acknowledgments.** Project received funding from: the European Research Council (ERC grant No 788535), the Carolito Stiftung, and the D. Dan and Betty Kahn Foundation. Dr. Bagon is a Robin Chemers Neustein AI Fellow.

## References

1. Albl, C., Kukelova, Z., Larsson, V., Polic, M., Pajdla, T., Schindler, K.: From two rolling shutters to one global shutter. In: Proceedings of the IEEE/CVF Conference on Computer Vision and Pattern Recognition, pp. 2505–2513 (2020)
2. Bao, W., Lai, W.S., Ma, C., Zhang, X., Gao, Z., Yang, M.H.: Depth-aware video frame interpolation. In: Proceedings of the IEEE Conference on Computer Vision and Pattern Recognition (CVPR), pp. 3703–3712 (2019)
3. Canny, J.: A computational approach to edge detection. IEEE Trans. Pattern Anal. Mach. Intell. **PAMI–8**, 679–698 (1986)
4. Dosovitskiy, A., Ros, G., Codevilla, F., Lopez, A., Koltun, V.: CARLA: an open urban driving simulator. In: Conference on Robot Learning, pp. 1–16. PMLR (2017)
5. Fan, B., Dai, Y.: Inverting a rolling shutter camera: bring rolling shutter images to high framerate global shutter video. In: Proceedings of the IEEE/CVF International Conference on Computer Vision, pp. 4228–4237 (2021)

6. Fan, B., Dai, Y., He, M.: SUNet: symmetric undistortion network for rolling shutter correction. In: Proceedings of the IEEE/CVF International Conference on Computer Vision, pp. 4541–4550 (2021)
7. Lao, Y., Ait-Aider, O.: A robust method for strong rolling shutter effects correction using lines with automatic feature selection. In: Proceedings of the IEEE Conference on Computer Vision and Pattern Recognition, pp. 4795–4803 (2018)
8. Liu, P., Cui, Z., Larsson, V., Pollefeys, M.: Deep shutter unrolling network. In: Proceedings of the IEEE/CVF Conference on Computer Vision and Pattern Recognition, pp. 5941–5949 (2020)
9. Liu, P., Cui, Z., Larsson, V., Pollefeys, M.: Deep shutter unrolling network. In: Proceedings of the IEEE/CVF Conference on Computer Vision and Pattern Recognition (CVPR) (2020)
10. Mosseri, I., Zontak, M., Irani, M.: Combining the power of internal and external denoising. In: IEEE International Conference on Computational Photography (ICCP) (2013)
11. Rengarajan, V., Balaji, Y., Rajagopalan, A.: Unrolling the shutter: CNN to correct motion distortions. In: Proceedings of the IEEE Conference on computer Vision and Pattern Recognition, pp. 2291–2299 (2017)
12. Rengarajan, V., Rajagopalan, A.N., Aravind, R.: From bows to arrows: rolling shutter rectification of urban scenes. In: Proceedings of the IEEE Conference on Computer Vision and Pattern Recognition, pp. 2773–2781 (2016)
13. Ringaby, E., Forssén, P.E.: Efficient video rectification and stabilisation for cell-phones. Int. J. Comput. Vis. **96**(3), 335–352 (2012)
14. Shahar, O., Faktor, A., Irani, M.: Space-time super-resolution from a single video. IEEE (2011)
15. Vasu, S., Rajagopalan, A., et al.: Occlusion-aware rolling shutter rectification of 3D scenes. In: Proceedings of the IEEE Conference on Computer Vision and Pattern Recognition, pp. 636–645 (2018)
16. Xue, T., Chen, B., Wu, J., Wei, D., Freeman, W.T.: Video enhancement with task-oriented flow. Int. J. Comput. Vis. (IJCV) **127**(8), 1106–1125 (2019)
17. Zhong, Z., Zheng, Y., Sato, I.: Towards rolling shutter correction and deblurring in dynamic scenes. In: Proceedings of the IEEE/CVF Conference on Computer Vision and Pattern Recognition, pp. 9219–9228 (2021)
18. Zhuang, B., Cheong, L.F., Hee Lee, G.: Rolling-shutter-aware differential SFM and image rectification. In: Proceedings of the IEEE International Conference on Computer Vision, pp. 948–956 (2017)
19. Zhuang, B., Tran, Q.-H.: Image stitching and rectification for hand-held cameras. In: Vedaldi, A., Bischof, H., Brox, T., Frahm, J.-M. (eds.) ECCV 2020. LNCS, vol. 12352, pp. 243–260. Springer, Cham (2020). https://doi.org/10.1007/978-3-030-58571-6_15
20. Zhuang, B., Tran, Q.H., Ji, P., Cheong, L.F., Chandraker, M.: Learning structure-and-motion-aware rolling shutter correction. In: Proceedings of the IEEE/CVF Conference on Computer Vision and Pattern Recognition, pp. 4551–4560 (2019)
21. Zuckerman, L.P., Naor, E., Pisha, G., Bagon, S., Irani, M.: Across scales and across dimensions: temporal super-resolution using deep internal learning. In: Vedaldi, A., Bischof, H., Brox, T., Frahm, J.-M. (eds.) ECCV 2020. LNCS, vol. 12352, pp. 52–68. Springer, Cham (2020). https://doi.org/10.1007/978-3-030-58571-6_4

# WISE: Whitebox Image Stylization by Example-Based Learning

Winfried Lötzsch[1], Max Reimann[1(✉)], Martin Büssemeyer[1], Amir Semmo[2], Jürgen Döllner[1], and Matthias Trapp[1]

[1] Hasso Plattner Institute, University of Potsdam, Potsdam, Germany
max.reimann@hpi.uni-potsdam.de
[2] Digital Masterpieces GmbH, Potsdam, Germany

**Abstract.** Image-based artistic rendering can synthesize a variety of expressive styles using algorithmic image filtering. In contrast to deep learning-based methods, these heuristics-based filtering techniques can operate on high-resolution images, are interpretable, and can be parameterized according to various design aspects. However, adapting or extending these techniques to produce new styles is often a tedious and error-prone task that requires expert knowledge. We propose a new paradigm to alleviate this problem: implementing algorithmic image filtering techniques as differentiable operations that can learn parametrizations aligned to certain reference styles. To this end, we present WISE, an example-based image-processing system that can handle a multitude of stylization techniques, such as watercolor, oil or cartoon stylization, within a common framework. By training parameter prediction networks for global and local filter parameterizations, we can simultaneously adapt effects to reference styles and image content, e.g., to enhance facial features. Our method can be optimized in a style-transfer framework or learned in a generative-adversarial setting for image-to-image translation. We demonstrate that jointly training an XDoG filter and a CNN for postprocessing can achieve comparable results to a state-of-the-art GAN-based method. https://github.com/winfried-loetzsch/wise.

## 1 Introduction

Image stylization has become a major part of visual communication, with millions of edited and stylized photos shared every day. At this, a large body of research in Non-photorealistic Rendering (NPR) has been dedicated to imitating hand-drawn artistic styles [31,47]. Traditionally, such *heuristics-based algorithms* [51] for image-based artistic rendering emulate a certain artistic style using a series of specifically developed *algorithmic* image processing operations. Thus, creating new styles is often a time-consuming process that requires the knowledge of domain experts.

W. Lötzsch and M. Reimann—Both authors contributed equally to this work.

**Supplementary Information** The online version contains supplementary material available at https://doi.org/10.1007/978-3-031-19790-1_9.

© The Author(s), under exclusive license to Springer Nature Switzerland AG 2022
S. Avidan et al. (Eds.): ECCV 2022, LNCS 13677, pp. 135–152, 2022.
https://doi.org/10.1007/978-3-031-19790-1_9

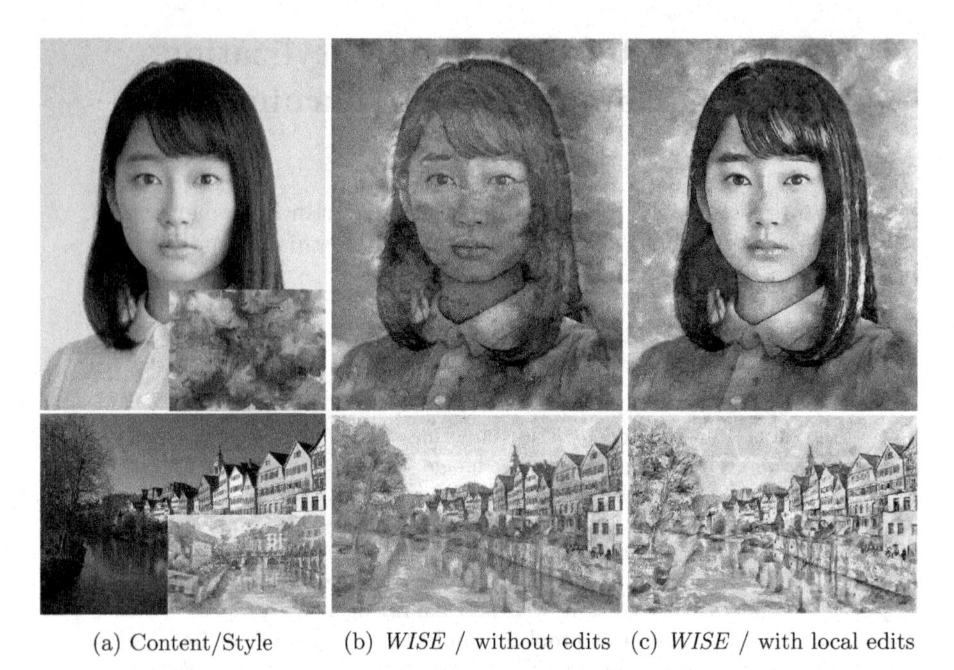

(a) Content/Style (b) *WISE* / without edits (c) *WISE* / with local edits

**Fig. 1. Example-based effect adjustment.** *WISE* optimizes effect parameters, e.g., of a watercolor stylization effect, to match stylized outputs to a reference style a. The results b can then be interactively adjusted by tuning the obtained parameters globally and locally for increased artistic control c (View examples of editing in our supplemental video: https://youtu.be/wIndN7cr0PE).

Recently, deep learning-based techniques for stylization and image-to-image translation have gained popularity by enabling the learning of stylistic abstractions from example data. In particular, Neural Style Transfer (NST) [21,22] that transfers the artistic style of a reference image and Generative Adversarial Network (GAN)-based [11,19] methods for fitting style distributions have achieved impressive results and are increasingly used in commercial applications [1].

Classical heuristics-based filters and filter-based image stylization pipelines, such as the eXtended difference-of-Gaussians (XDoG) filter [59], cartoon effect [60], or watercolor effect [2,57], expose a range of parameters to the user that enable fine-grained global and local control over artistic aspects of the stylized output. By contrast, learning-based techniques are commonly limited in their modes of control, i.e., NST [6] only offers control over a general content-style tradeoff. Furthermore, their learned representations are generally not interpretable as a set of design aspects and configurations. Thus, these approaches often do not meet the requirements of interactive image editing tasks that go beyond one-shot global stylizations towards editing with high-level and low-level artistic control [10,14,18]. Additionally, deep network-based methods are often computationally expensive in both training and inference on high image resolutions [6,24,25]. This further limits their applicability in interactive or mobile

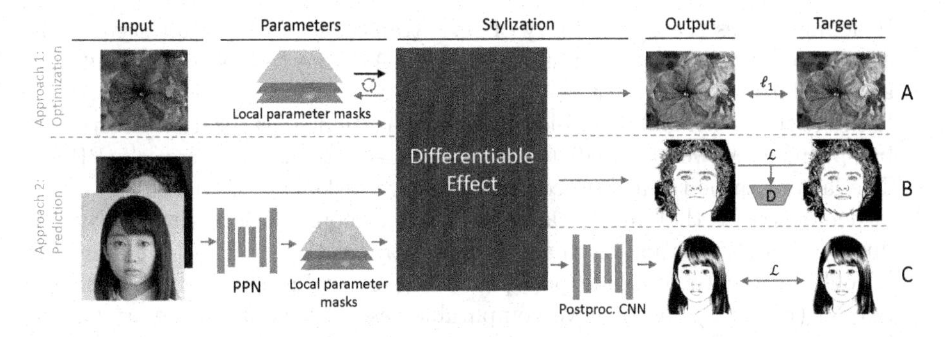


**Fig. 2. Overview of *WISE*.** Differentiable effects can be adapted to example data, demonstrated for three use-cases: *parametric style transfer* (**A**) optimizes parameter masks to match a hand-drawn or synthesized stylization target (e.g., from NST as in Fig. 1) which enables style transfer results to remain editable and resolution-independent. *Local parameter prediction* (**B**) trains PPNs to predict parameter masks to adapt the effect to the content (e.g., for facial structure enhancement as shown here). Combined with a postprocessing CNN, local parameter prediction can learn sophisticated *image-to-image translation* tasks such as learning hand-drawn sketch-styles (**C**).

applications [9] and their capability to simulate fine-grained (pigment-based) local effects and phenomena of artistic media such as watercolor and oil paint.

To counterbalance these limitations, this work aims to combine the strengths of *heuristics-based* and *learning-based* image stylization by implementing algorithmic effects as *differentiable* operations that can be trained to learn filter-based parameters aligning to certain reference styles. The goal is to enable (1) the creation of complex, example-based stylizations using lightweight algorithmic approaches that remain interpretable and can operate on very high image resolutions, and enable (2) the editing with artistic control on a fine-granular level according to design aspects. To this end, we present *WISE*, a *whitebox* system for example-based image processing that can handle a multitude of stylization techniques in a common framework. Our system integrates existing algorithmic effects such as XDoG-based stylization [59], cartoon stylization [60], watercolor effects [2,57], and oilpaint effects [50], by creating a library of differentiable image filters that match their shader kernel-based counterparts. We show that the majority of filters (e.g., bilateral filtering) can be transformed into auto-differentiable formulations, while for the remaining filters, gradients can be approximated (e.g., for color quantization). Using our framework, effects can be adapted to reference styles using popular, deep network-based image-to-image translation losses. We train exemplary effects using both NST and GAN-based losses and show qualitatively and quantitatively that the results are comparable to state-of-the-art deep networks while retaining the advantages of filter-based stylization. To summarize, this paper contributes the following:

1. It provides an end-to-end framework for example-based image stylization using differentiable algorithmic filters. Figure 2 shows an overview of the system.

2. It demonstrates the applicability of style transfer-losses to adapt stylization effects to a reference style (Fig. 1 and Fig. 2 A). The results remain editable and resolution-independent.
3. It shows that both global and local parametrizations of stylization effects can be optimized as well as predicted by a parameter prediction network (PPN). The latter can be trained on content-adaptive tasks (Fig. 2 B).
4. It demonstrates that filters can be trained in combination with CNNs for improved generalization on image-to-image translation tasks. Combining the XDoG effect with a simple post-processing Convolutional Neural Network (CNN) (Fig. 2 C) can achieve comparable results to state-of-the-art GAN-based image stylization for hand-drawn sketch styles, but at much lower system complexity.

## 2 Related Work

*Heuristics-Based Stylization.* In NPR, image-based artistic rendering deals with emulating traditional artistic styles, using a pipeline of rendering stages [31,47,51]. Commonly, edge detection and content abstraction are important parts of such pipelines. The XDoG filter [59,60] is an extended version of the Difference-of-Gaussians (DoG) band-pass filter, and can be used to create smooth edge stylizations. Furthermore, edge-aware smoothing filters such as bilateral filtering [56] or Kuwahara filtering [30] can abstract image contents, and can be combined with image flow to adapt the results to local image structures [32,34]. These techniques can be found in heuristics-based effects such as cartoon stylization [60], oil-paint abstraction [52] and image watercolorization [2,57], each consisting of a series of rendering stages such as image blending, wobbling, pigment dispersion and wet-in-wet stylization. For a comprehensive taxonomy of techniques, the interested reader is referred to the survey by Kyprianidis *et al.* [31].

These effects are typically parameterized globally, and can be further adjusted within pre-defined parameter ranges, or locally on a per-pixel level using parameter masks [52]. In this work, we implement variants of the XDoG, cartoon filtering, and watercolor pipeline in our framework using auto-differentiable formulations of each rendering stage. At runtime, users choose between one of these different effect pipelines; and results are generally achieved by optimizing the exact chain of filters as introduced in [2,50,57,59,60].

*Deep Learning-Based Methods.* With the advent of deep learning, CNNs for image generation and transformation have led to a range of impressive results.

In NST, first introduced by Gatys *et al.* [6], the stylistic characteristics of a reference image are transferred to a content image by matching deep feature statistics using an optimization process. Fast feed-forward networks have been trained to reproduce a single [22] or even arbitrarily many styles [12,17,41]. Furthermore, there have been efforts to increase controllability of NSTs, e.g., by control over color [7], sub-styles [45] or strokes [20,44], however, the representations are not interpretable. Recently, style transfer has been formulated as a neurally-guided stroke rendering optimization approach [29], that retains interpretability, however, is slow to optimize. We show that our method can obtain comparable results to a state-of-the-art NST [27], while retaining interactive editing control.

GANs, first introduced by Goodfellow *et al.* [11], learn powerful generative networks that model the input distribution. They have been widely used for conditional image generation tasks, with both paired [19] and unpaired [65] training data. In the stylization domain, it has found applications for collection style transfer [3,49], cartoon generation [4,55], and sketch styles [62]. However, domain-specific applications often require sophisticated losses and multiple networks to prevent artifacts [4,63]. We show that our differentiable implementation of the XDoG filter can be trained as a generator network in a GAN framework and can produce comparable sketches to state-of-the-art (CNN-based) GANs [62,63].

*Learnable Filters.* While the previous end-to-end CNNs deliver impressive results, they are limited in their output resolution. A few recent methods have proposed training fast algorithmic filters to operate efficiently at high resolutions. Getreuer *et al.* [8] introduce learnable approximations of algorithmic image filters, such as of the XDoG filter. At run-time, a linear filter is selected per image pixel according to the local structure tensor; filters can be combined in pipelines for image enhancement [5]. Gharbi *et al.* [9] train a CNN to predict affine transformations for bilateral image enhancement, e.g., to approximate edge-aware image filters or tone adjustments. The transform filters are predicted at a low resolution and then applied in full resolution to the image. “Exposure” framework [16] combines learning linear image filters with reinforcement learning, where an actor-critic model decides which filters to include to achieve a desired photo enhancement effect.

These methods have in common that they learn several simple, linear functions to approximate image processing operations [8,9,16,39,61]. Our framework consists of pipelines of differentiable filters as well. However, in contrast to previous work, we make a variety of heuristics-based stylization operators differentiable and learn to predict their parameterizations. Thereby, sophisticated stylization effects (e.g., those found in stylization applications) can be ported and directly used in our framework.

## 3 Differentiable Image Filters

Heuristics-based stylization effects consist of pipelines of image filtering operations. To compute gradients for effect input parameters, all image operations within the pipeline are required to be differentiable with respect to their parameters and the image input. Gradients throughout the pipeline can then be obtained by applying the chain rule. Figure 3 outlines the gradient flow from a loss function to the effect parameters by the cartoon pipeline example, gradients for parameters can be obtained both globally as well as locally using per-pixel parameter masks.

In previous works, individual operations in such pipelines are typically implemented as shader kernels for fast GPU-based processing. To achieve an end-to-end gradient flow, we implement these operations in an auto-grad enabled framework, the implemented filtering stages are listed in Table 1. Point-based

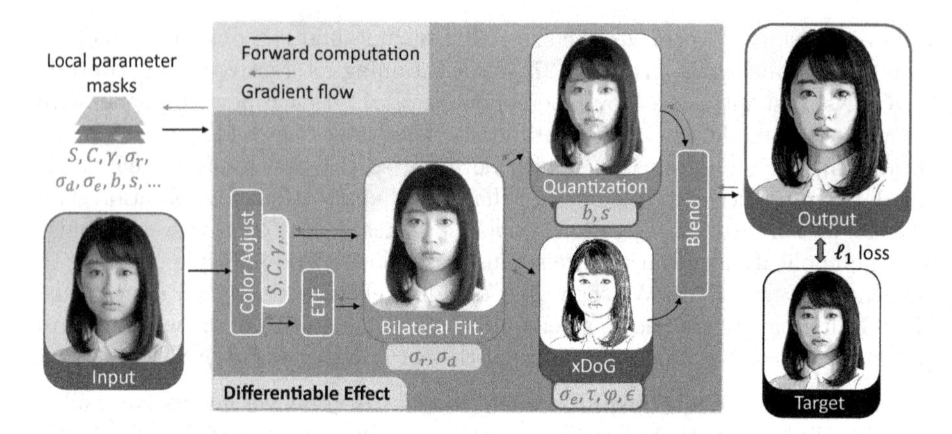


**Fig. 3. Exemplary differentiable effect pipeline.** Shown for the cartoon effect proposed by Winnemöller *et al.* [60]. Gradients are backpropagated from the loss to the filter parameter masks. In the effect, colors are first adjusted based on parameters such as saturation $S$, contrast $C$ or gamma $\gamma$, and then, using an Edge Tangent Flow (ETF) [23], orientation-aligned bilateral filtering [32], and XDoG [59] are computed. Additionally, the image is quantized with respect to the number of bins $b$ and softness $s$. (Color figure online)

and fixed-neighbourhood operations such as color space conversions, structure tensor computation [32], or DoG [59] can generally be converted into differentiable filters by transforming any kernelized function into a sequence of its constituent auto-differentiable transformations. The exception to this are functions which are inherently not differentiable, such as color quantization, and which require the approximation of a numeric gradient. An example for numeric gradient approximation of color quantization is shown in the supplemental material. Structure-adaptive neighborhood operations, such as the orientation-aligned bilateral filter and flow-based Gaussian smoothing filter, iteratively determine sampling locations based on the structure of the content (often oriented along a flow field). To preserve gradients and make use of the in-built fixed-neighborhood functions of auto-grad enabled frameworks, per-pixel iteration is transformed into a grid-sampling operation where neighborhood values are accumulated into a new dimension with size $D$ which represents the expected maximum kernel neighbourhood. A structure-adaptive filter transformation by example of the orientation-aligned bilateral filter is shown in the supplementary material.

*Implementation Aspects.* We implement differentiable filters in PyTorch and create reference implementations of the same effects using OpenGL shaders. The learnability of each effect parameter is validated in a functional benchmark (shown in supplemental material) by optimizing the differentiable effect to match reference effects with randomized parameters. At inference time, OpenGL shaders can be interchangeably used with the differentiable effect, and can be efficiently executed on high-resolution images using parameters predicted on low-resolution images. At training time, memory usage of differentiable filters can be reduced by controlling their kernel-size (shown in supplemental material).

**Table 1.** Differentiable filters by type and effect. Filters can be classified by their sampling approach, which is either point-based (PB) or in a fixed neighborhood (FN) or structure-adaptive neighborhood (SN). Some filters are non-differentiable and require numerical gradient (NG) approximations for training.

| Filtering operation | Differentiable filter type | Cartoon | Watercolor | Oil paint |
|---|---|---|---|---|
| Anisotropic Kuwahara [34] | SN | | ✓ | |
| Bilateral [56] | FN | | ✓ | ✓ |
| Bump Mapping/Phong Shading [42] | FN | | | ✓ |
| Color Adjustment | PB | ✓ | ✓ | ✓ |
| Color Quantization [60] | PB,NG | ✓ | | |
| Flow-based Gaussian Smoothing [32] | SN | ✓ | ✓ | ✓ |
| Gaps [38] | FN | | ✓ | |
| Joint Bilateral Upsampling [28] | SN | | | ✓ |
| Flow-based Laplacian of Gaussian [33] | SN | | | ✓ |
| Image Composition [43] | PB | | ✓ | |
| Orientation-aligned Bilateral [32] | SN | ✓ | | ✓ |
| Warping/Wobbling [2] | FN | | ✓ | |
| Wet-in-Wet [57] | SN | | ✓ | |
| XDoG [59] | SN | ✓ | ✓ | ✓ |

# 4 Parameter Prediction

With the introduced differentiable filter pipelines, parameters can be optimized using image-based losses. To generalize to unseen data, we explore Parameter Prediction Networks (PPNs) that are trained to predict global parameters or spatially varying (local) parameters.

## 4.1 Parameter Prediction Networks

*Global Parameter Prediction.* We construct a PPN that predicts the effect parameters of a stylized example image, given both the stylized and source image. Thereby, the network is trained to effectively reverse-engineer the stylization effect. During training, gradients are back-propagated through the effect, the parameters, and finally to the PPN. Formally, let $I$ denote the input image, $T$ the target image, $O(\cdot)$ the differentiable effect, $P_G(\cdot)$ the PPN network. The loss for the global PPN is computed using an $\ell_1$ image space-based loss as:

$$\mathcal{L}_{\text{global}} = \|O(I, P_G(I))) - T\|_1 \tag{1}$$

Our global PPN architecture consists of a VGG backbone [53] that extracts features of the input and stylized image and computes layer-wise Gram matrices to encode important style information [6]. The accumulated features are passed

**Table 2. Global PPN loss functions.** The PPNs are trained with different loss functions on the NPR benchmark [48]. Networks trained in parameter space use reference parameters as the loss signal directly. We measure the Structural Similarity Index Measure (SSIM) and Peak Signal-to-Noise Ratio (PSNR)

| Loss domain | SSIM | PSNR | Parameter loss |
|---|---|---|---|
| Parameter space $\ell_1/\ell_2$ | 0.738/0.737 | 12.530/12.927 | **0.158**/0.162 |
| Image space $\ell_1/\ell_2$ | **0.780**/0.764 | **13.875**/13.286 | 0.183/0.190 |

to a multi-head module to predict the final global parameters. We found this network architecture to perform superior against other common architecture variants, please refer to the supplementary material for an ablation study and details on the architecture.

*Local Parameter Prediction.* While the global PPN can predict settings of similar algorithmic effects, real-world, hand-drawn images often vary significantly based on the local content, which cannot be modeled by a global parameterization. Therefore, we construct a PPN to predict local *parameter masks.* We use a U-Net architecture [46] for mask prediction at input resolution, where each output channel represents a parameter. Gradients are back-propagated to the PPN through the differentiable effects for all parameter masks.

For training, a paired data GAN approach is used, where a patch-based Pix2Pix discriminator [19] matches the distribution of patches in the reference image and an additional weighted $\ell_1$ image space loss enforces a more strict pixel-wise similarity. Formally, let $D(\cdot)$ denote the discriminator, $\mathcal{L}_{TV}(\cdot)$ the total variation regularizer [22] to enforce smooth parameter masks, and $P_L(\cdot)$ the local PPN which acts as the generator network. The final loss $\mathcal{L}$ for the PPN generator is computed as:

$$\begin{aligned}\mathcal{L} = \mathbb{E}\big[&\alpha \log(1 - D(I, O(I, P_L(I)))) \\ &+ \beta \|O(I, P_L(I)) - T\|_1 + \gamma \mathcal{L}_{TV}(P_L(I))\big]\end{aligned} \tag{2}$$

### 4.2 PPN Experiments

We conduct several experiments to validate our approach for global and local parameter prediction.

*Global Parameter Prediction.* We compare the loss function and loss space of global PPNs in (Table 2). We find that while directly predicting in parameter space (without obtaining gradients from the effect) yields closer parameter values, the highest visual accuracy is achieved using an $\ell_1$ image space-based loss. This validates the usefulness of the differentiable effect being part of the training pipeline. Global PPNs can accurately match reference stylizations created by the same effect, as shown in Fig. 4b. Furthermore, they can approximate similar hand-drawn styles, albeit with significant local deviations, as shown in Fig. 4e.

(a) Source (b) Predicted (c) Reference (d) Source (e) Predicted (f) Reference

**Fig. 4. Predictions using the global PPN for XDoG.** The stylized reference c is synthetic (generated using the reference XDoG implementation), while the reference f is hand-drawn, taken from APDrawing [62].

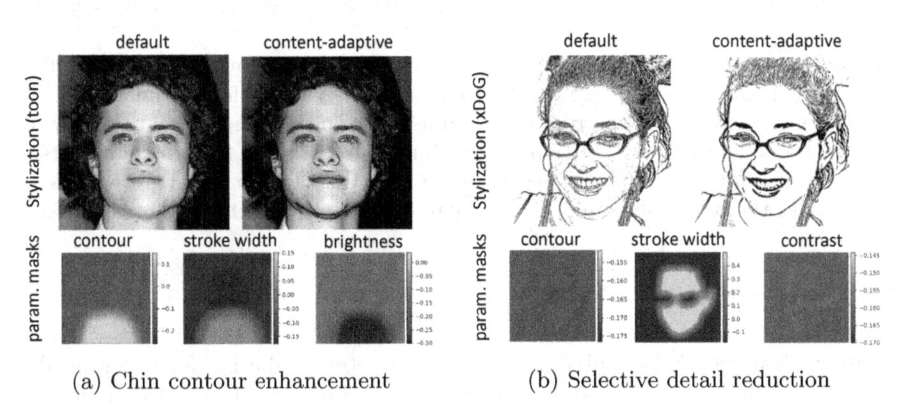


(a) Chin contour enhancement (b) Selective detail reduction

**Fig. 5. Local PPN results.** Networks are trained on CelebAMask-HQ [35] to generate selective enhancements by predicting parameter masks. The default result is created by a global parameter configuration, while the shown predicted parameter masks create the content-adaptive result.

*Content-Adaptive Effects.* Using the previously described approach for local PPN training, we demonstrate its applicability to several common problems that are often present in purely algorithmic image-stylization techniques. We consider three tasks to improve stylization quality: (1) highlighting facial features, for example by increasing contours at low-contrast edges such as the chin (Fig. 5a), (2) selectively reducing details such as small wrinkles in the face (Fig. 5b), and (3) background removal (refer to supplemental material). We use the CelebAMask-HQ dataset [35] for training, which consists of 30,000 high-resolution face images and segmentation masks for all parts of the face. For the above tasks, we each create a synthetic training dataset by stylizing images using a reference effect and adjusting its parameters for certain parts of the face (obtained from dataset annotations) according to the task, e.g., increasing the amount of contours in the chin area. In Fig. 5 trained PPN networks are evaluated by plotting the predicted local parameter masks together with the generated stylizations. It shows that the networks learn to predict parameter masks for the relevant regions accurately solely by observing pixels without additional supervision.

(a) Content image (b) Style image (c) Stylized result (d) Locally edited

**Fig. 6. Parametric style transfer.** Effect parameters are optimized (watercolor effect in top row, oilpaint effect in bottom row) to match the stylistic reference image b. Users can then interactively edit the result c by adjusting resulting parameter masks. In the top row, the saturation parameter mask is adjusted to highlight foreground objects d. In the bottom row, the oilpaint-specific bump scale and flow-smoothing are adjusted in d to make the background appear to be painted wet-in-wet with long brushstrokes.

# 5 Applications

Using our framework for global and local parameter prediction for differentiable algorithmic pipelines, example-based stylization with closely related references is made possible. To adapt to real-world, more diverse example data, our framework can be integrated with existing stylization paradigms. In the following, we demonstrate our approach for the task of (statistics-based) style transfer reconstruction and GAN-based image-to-image translation based on the APDrawing dataset [62].

## 5.1 Style Transfer

We investigate the combination of iterative style transfer and algorithmic effects. We use Style Transfer by Relaxed Optimal Transport and Self-Similarity (STROTSS) by Kolkin *et al.* [27] to create stylized references for our effect. We subsequently try to recreate the style transfer result with our algorithmic effects by optimizing parameter masks (Fig. 6). For this, a $\ell_1$ loss in image space is again used to match effect output and reference image.

*Optimization.* We run the style transfer algorithm [27] for 200 steps to create a stylized reference with a resolution of 1024 × 1024 pixels. The local parameter masks are then optimized using 1000 iterations of Adam [26] and a learning rate of 0.1. The learning rate is decreased by a factor of 0.98 every 5 iterations starting from iteration 50. To avoid the generation of artifacts in parameter masks, we smooth all masks at increasing iterations (10, 25, 50, 100, 250, 500) using a Gaussian filter. Alternatively to image-space matching, parameters can also be

**Table 3.** Comparison of our method to STROTSS [27] for style loss $\mathcal{L}_S$, content loss $\mathcal{L}_C$, and the $\ell_1$ difference between respective results. We test against in-domain style images and against a set of common (arbitrary domain) NST styles. In each case, results are averaged over 10 styles and NPRB [48] as content (Exemplary style images and results in suppl. material.)

| | | XDoG | Cartoon | Watercolor | | Oilpaint | |
|---|---|---|---|---|---|---|---|
| Style domain: | | Bw-drawing | Cartoon | Watercolor | Common | Oilpaint | Common |
| $\mathcal{L}_S$ | STROTSS | 0.340 | 0.289 | 0.351 | 0.39 | 0.209 | 0.39 |
| | Our results | 0.246 | 0.406 | 0.359 | 0.42 | 0.384 | 0.52 |
| $\mathcal{L}_C$ | STROTSS | 0.099 | 0.094 | 0.081 | 0.036 | 0.037 | 0.036 |
| | Our results | 0.172 | 0.148 | 0.092 | 0.034 | 0.023 | 0.033 |
| | $\ell_1$ Difference | 0.188 | 0.136 | 0.007 | 0.021 | 0.036 | 0.039 |

directly optimized using NST [6] losses $\mathcal{L}_S + \mathcal{L}_C$, however this often fails to transfer more complex stylistic elements (shown in the supplemental material).

*Results.* As Table 3 shows, parametric style transfer works better for effects that have a high expressivity and are closer to hand-drawn styles, such as the watercolor or oilpaint, compared to more restricted effects such as XDoG or cartoon. After the generation of local parameter masks, the parameters can be refined by the user as shown in Fig. 6d. By optimizing parameters, we obtain an interpretable "whitebox" representation of a style that, in contrast to current pixel-optimizing NSTs [6,22,27], retains controllability according to artistic design aspects. Furthermore, our method is resolution independent, i.e., parameter masks can be optimized at lower resolutions and then scaled up to high resolutions for editing. In Fig. 6d (bottom row) the effect is applied at $4096 \times 4096$ pixels, while current style transfers are mostly memory-limited to much lower resolutions. We further compare matching performance of in-domain styles with a set of common NST styles and observe that the $\ell_1$ difference (Table 3) is only marginally higher for the latter. Thus, highly parameterized effects such as watercolor or oilpaint can emulate any out-of-domain style by per-pixel optimization of parameter combinations reasonably well. However, as this creates highly fragmented parameter masks, a tradeoff between generalizability and interpretability of masks can be made using a weighted total variation-loss.

### 5.2 GAN-Based Image-to-Image Translation with PPNs

For learning a style distribution, i.e., the characteristics of an artistic style over a larger collection of artworks, GAN-based approaches have achieved impressive results [4,58,62]. We investigate training PPNs in a GAN-setting for image-to-image translation. While the global and local PPNs discussed in Sect. 4 can match in-domain styles very well (Sect. 4.2), they cannot produce local image structures that are not synthesizable by their constituent image filters. Artistic reference styles, however, often contain stylistic elements that have not been

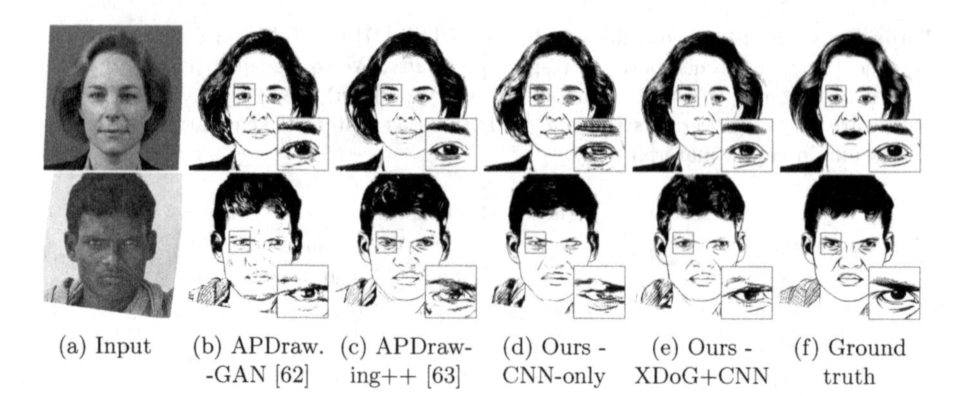


**Fig. 7. Results on APDrawing.** While APDrawing GAN b and APDrawing++ GAN c can produce inconsistent lines, our proposed method e generally produces flow-consistent lines. The differentiable filter in our approach is important for consistent quality, as solely using an image translation CNN [22] often produces local dithering artefacts (upper row) and patchy features (lower row) d.

modeled in the heuristics-based filter - this holds especially true for more simple effects such as xDoG. For reference styles that are stylistically close to such effects (e.g., xDoG), such as line-drawings, we hypothesize that combining our filter pipeline with a lightweight CNN-based post-processing operation and learning them end-to-end can close the domain gap while retaining the positive properties of the filter parametrization and beeing computationally efficient.

*Dataset.* For our experiment, we select the APDrawing [62] dataset which consists of closely matching photos and their hand-drawn stylistic counterparts. Its hand-drawn images are reasonably similar to the XDoG results, while still containing many stylistic abstractions that cannot be emulated solely by the XDoG. We hypothesize, that re-creating such an effect entails both edge detection and content abstraction, which could be performed by our differential XDoG pipeline combined with a separate convolution network for content abstraction.

APDrawing contains a set of 140 portrait photographs along with paired drawings of these portraits. The GAN-based local PPN approach introduced in Sect. 4 is used to train on this paired dataset, where solely the generator is extended using a CNN $N(\cdot)$ to post-process the XDoG output, i.e., following Eq. 2 our generator now combines PPN, effect and CNN: $N(O(I, P_L(I)))$.

*Architecture.* We investigate the efficacy of each component in our proposed approach for APDrawing in Table 4. Following Yi *et al.* [63], we measure the Fréchet Inception Distance (FID) score [15] and Learned Perceptual Image Patch Similarity (LPIPS) [64] to the test set. We train for 200 epochs and otherwise use the same hyperparameters as Pix2Pix [19]. We observed that using the ResNet-based architecture for image translation introduced by Johnson *et al.* [22] works best for the post-processing CNN. Furthermore integrating the XDoG in the pipeline improves the results vs. a convolutional-only pipeline. Note that this combination of algorithmic effects and CNNs in a training pipeline is only made possible by our introduced approach for end-to-end differentiable filter pipelines and PPNs. Omitting the PPN and using fixed parameters for XDoG significantly degrades the results, which validates the integrated training of filter and CNN. Further, we observe that training with XDoG as a postprocessing instead of as a preprocessing step does not converge. All architecture choices are extensively evaluated in an ablation study, please refer to the supplemental material.

*Results.* While the CNN alone (without XDoG) already achieves good FID and LPIPS scores, we show in Fig. 7 that it creates major artifacts especially around eyebrows and eyes, which are not detected by those metrics. Compared to the APDrawing GAN approach by Yi *et al.* [62], our model improves the FID score (Table 4). The state-of-the-art APDrawing++ [63] improves on these metrics and quantitatively performs better than our model, however qualitatively it can suffer from artifacts in small structures such as the eyes (Fig. 7c) whereas our approach leads to more consistent lines. We note that their approach consists of a sophisticated combination of several losses and task-specific discriminators that require facial landmarks to train multiple local generator networks for facial features such as eyes, nose, and mouth separately. This limits their generalizability to other datasets, while our approach, on the other hand, represents a general setup for image-to-image translation consisting of a globally trained CNN and a simple effect, making it applicable to any paired training data without further annotation requirements.

**Table 4.** Our results on APDrawing [62]

| PPN | XDoG | CNN | FID | LPIPS |
|---|---|---|---|---|
| ✗ | ✗ | U-Net | 71.26 | 0.322 |
| ✗ | ✗ | ResNet | 62.44 | **0.275** |
| ✗[a] | ✓ | U-Net | 75.40 | 0.329 |
| ✗[a] | ✓ | ResNet | 71.56 | 0.305 |
| ✓ | ✓ | U-Net | 89.93 | 0.366 |
| ✓ | ✓ | ResNet | **60.55** | 0.285 |
| APDrawing GAN | | | 62.14[b] | 0.291[b] |
| APDrawing++ | | | **54.40** | **0.258**[b] |
| Train vs. Test | | | 49.72 | - |

[a] a fixed parameter preset is used
[b] results obtained from [62] [63]

## 6 Discussion

*Applicability.* In the previous sections, we have demonstrated the applicability of differentiable filters to several example-based stylization tasks using four established heuristics-based filter pipelines. Their constituent image filters (Table 1) form a common basis of many image-based artistic rendering approaches [31]. We expect that other filtering-based effects, such as pencil-hatching [37], or stippling [54], can be transferred to our framework with relative ease due to their pipeline-based, GPU-optimized formulations. Stroke-based rendering approaches, on the other hand, are typically optimized globally [40] or locally [13], and are thus challenging to transform into differentiable formulations. However, a recent approach by Liu *et al.* [36] has shown that strokes can be predicted in a single feedforward pass of a CNN, which could be regarded as a complementary approach.

*Limitations.* Our PPN-based approaches make use of a paired data training regime. While paired data can be synthetically generated for content-adaptive effects aiming at solving filter-specific problems, datasets with paired real-world paintings are subject to limited availability. As our training approach follows Pix2Pix GAN [19], future work extending the method to train with unpaired training losses, such as cycle-consistency losses [65], could alleviate this limitation. An inherent limitation of predicting parameters in comparison to directly predicting pixels (as with convolutional GANs), remains the constraint of only being able to produce styles that lie in the manifold of achievable effects of the underlying image filters. While this can be mitigated using a post-processing CNN, this represents a trade-off with respect to interpretability and range of low-level control (we examine this aspect in the supplemental material). On the other hand, our parametric style transfer is able to match arbitrary styles when optimizing highly parameterized effects such as watercolor. Training a PPN with such an effect on a large dataset, e.g., using unpaired training, could similarly already have sufficient representation capability without postprocessing CNNs.

## 7 Conclusions

In this work, we propose the combination of algorithmic stylization effects and example-based learning by implementing heuristics-based stylization effects as differentiable operations and learning their parametrizations. The results show that both optimization of parameters, e.g., to achieve style transfers, and their global and local prediction, e.g., for content-adaptive effects, are viable approaches for example-based algorithmic stylizations. Our experiments demonstrate that our approach is especially suitable for applications that require fast adaptation to new styles while retaining full artistic control and low computation times for high image resolutions. Furthermore, stylizations beyond the filters' abstraction capabilities are achieved by adding convolutional post-processing. This approach can generate results on-par with state-of-the-art CNN-based methods. For future research, learning the composition of filters as building blocks of a generic algorithmic effect pipeline would allow for seamless integration of user control and example-based stylization without the limitation to a specific stylization technique.

**Acknowledgments.** We thank Sumit Shekhar and Sebastian Pasewaldt for comments and discussions. This work was partially funded by the BMBF through grants 01IS18092 ("mdViPro") and 01IS19006 ("KI-LAB-ITSE").

## References

1. Adobe. Photoshop neural filters overview. https://helpx.adobe.com/photoshop/using/neural-filters.html (2021)
2. Bousseau, A., Kaplan, M., Thollot, J., Sillion, FX.: Interactive watercolor rendering with temporal coherence and abstraction. In: Proceedings of the 4th International Symposium on Non-Photorealistic Animation and Rendering (NPAR), pp. 141–149 (2006)

3. Chen, X., Chang, X., Yang, X., Song, L., Tao, D.: Gated-GAN: adversarial gated networks for multi-collection style transfer. IEEE Trans. Image Process. **28**(2), 546–560 (2018)
4. Chen, Y., Lai, Y-K., Liu, Y-J.: CartoonGAN: generative adversarial networks for photo cartoonization. In: Proceedings of the IEEE Conference on Computer Vision and Pattern Recognition (CVPR), pp. 9465–9474 (2018)
5. Garcia-Dorado, I., Getreuer, P., Wronski, B., Milanfar, P.: Image stylisation: from predefined to personalised. IET Comput. Vis. **14**(6), 291–303 (2020)
6. Gatys, L.A. Ecker, A.S., Bethge, M.: Image style transfer using convolutional neural networks. In: Proceedings of the IEEE Conference on Computer Vision and Pattern Recognition (CVPR), pp. 2414–2423 (2016)
7. Gatys, L.A., Ecker, A.S., Bethge, M., Hertzmann, A., Shechtman, E.: Controlling perceptual factors in neural style transfer. In Proceedings of the IEEE Conference on Computer Vision and Pattern Recognition (CVPR), pp. 3730–3738 (2017)
8. Getreuer, P., Garcia-Dorado, I., Isidoro, J., Choi, S., Ong, F., Milanfar, P.: BLADE: filter learning for general purpose computational photography. In Proceedings of the IEEE International Conference on Computational Photography (ICCP), pp. 1–11 (2018)
9. Gharbi, M., Chen, J., Barron, J.T., Hasinoff, S.W., Durand, F.: Deep bilateral learning for real-time image enhancement. ACM Trans. Graph. **36**(4), 1–12 (2017)
10. Gooch, A.A., Long, J., Ji, L., Estey, A., Gooch, B.S.: Viewing progress in non-photorealistic rendering through Heinlein's lens. In: Proceedings of the International Symposium on Non-Photorealistic Animation and Rendering (NPAR), pp. 165–171 (2010)
11. Goodfellow, I., et al. Generative Adversarial Nets. In Advances in Neural Information Processing Systems (NIPS) (2014)
12. Gu, S., Chen, C., Liao, J., Yuan, L.: Arbitrary style transfer with deep feature reshuffle. In: Proceedings of the IEEE Conference on Computer Vision and Pattern Recognition (CVPR), pp. 8222–8231 (2018)
13. Hertzmann, A.: Painterly rendering with curved brush strokes of multiple sizes. In: Proceedings of the 25th Annual Conference on Computer Graphics and Interactive Techniques (SIGGRAPH), pp. 453–460 (1998)
14. Hertzmann, A.: Computers do not make art, people do. Commun. ACM **63**(5), 45–48 (2020)
15. Martin, H., Hubert, R., Thomas, U., Bernhard, N., Sepp, H.: GANs trained by a two time-scale update rule converge to a local Nash equilibrium. In: Advances in Neural Information Processing Systems (NIPS) (2017)
16. Yuanming, H., He, H., Chenxi, X., Wang, B., Lin, S.: Exposure: a white-box photo post-processing framework. ACM Trans. Graph. **37**(2), 1–17 (2018)
17. Huang, X., Belongie, S.: Arbitrary style transfer in real-time with adaptive instance normalization. In: Proceedings of the IEEE International Conference on Computer Vision (ICCV), pp. 1501–1510 (2017)
18. Isenberg, T.: Interactive NPAR: what type of tools should we create? In: Proceedings of the Non-Photorealistic Animation and Rendering (NPAR) (2016)
19. Isola, P., Zhu, J.Y., Zhou, T., Efros, A.A.: Image-to-image translation with conditional adversarial networks. In: Proceedings of the IEEE Conference on Computer Vision and Pattern Recognition (CVPR), pp. 1125–1134 (2017)
20. Jing, Y., et al.: Stroke controllable fast style transfer with adaptive receptive fields. In: Proceedings of the European Conference on Computer Vision (ECCV) (2018)
21. Jing, Y., Yang, Y., Feng, Z., Ye, J., Yizhou, Yu., Song, M.: Neural style transfer: a review. IEEE Trans. Visual. Comput. Graph. **26**(11), 3365–3385 (2019)

22. Johnson, J., Alahi, A., Fei-Fei, L.: Perceptual losses for real-time style transfer and super-resolution. In: Leibe, B., Matas, J., Sebe, N., Welling, M. (eds.) ECCV 2016. LNCS, vol. 9906, pp. 694–711. Springer, Cham (2016). https://doi.org/10.1007/978-3-319-46475-6_43
23. Kang, H., Lee, S., Chui, C.K.: Flow-based image abstraction. IEEE Trans. Visual. Comput. Graph. **15**(1), 62–76 (2009)
24. Karras, T., Aila, T., Laine, S., Lehtinen, J.: Progressive growing of GANs for improved quality, stability, and variation. In: Proceedings of the International Conference on Learning Representations (ICLR) (2018)
25. Karras, T., Laine, S., Aila, T.: A style-based generator architecture for generative adversarial networks. In: Proceedings of the IEEE Conference on Computer Vision and Pattern Recognition (CVPR), pp. 4401–4410 (2019)
26. Kingma, D.P., Ba, J.: Adam: a method for stochastic optimization. In: Proceedings of the International Conference on Learning Representations (ICLR) (2015)
27. Kolkin, N., Salavon, J., Shakhnarovich, G.: Style transfer by relaxed optimal transport and self-similarity. In: Proceedings of the IEEE Conference on Computer Vision and Pattern Recognition (CVPR), pp. 10051–10060 (2019)
28. Kopf, J., Cohen, M.F., Lischinski, D., Uyttendaele, M.: Joint bilateral upsampling. ACM Trans. Gaph. (ToG) **26**(3), 96-es (2007)
29. Kotovenko, D., Wright, M., Heimbrecht, A., Ommer, B.: Rethinking style transfer: from pixels to parameterized brushstrokes. In: Proceedings of the IEEE Conference on Computer Vision and Pattern Recognition (CVPR), pp. 12196–12205 (2021)
30. Kuwahara, M., Hachimura, K., Eiho, S., Kinoshita, M.: Processing of RI-angiocardiographic images. In: Digital Processing of Biomedical Images, pp. 187–202. Springer, Cham (1976). https://doi.org/10.1007/978-1-4684-0769-3_13
31. Kyprianidis, J.E., Collomosse, J., Wang, T., Isenberg, T.: State of the "Art": a taxonomy of artistic stylization techniques for images and video. IEEE Trans. Visual. Comput. Graph. **19**(5), 866–885 (2012)
32. Kyprianidis, J.E., Döllner, J.: Image abstraction by structure adaptive filtering. In: Proceedings of the EG UK Theory and Practice of Computer Graphics (TPCG), pp. 51–58 (2008)
33. Kyprianidis, J.E., Kang, H.: Image and video abstraction by coherence-enhancing filtering. In: Computer Graphics Forum, vol. 30, pp. 593–602. Wiley Online Library (2011)
34. Kyprianidis, J.E., Kang, H., Döllner, J.: Image and video abstraction by anisotropic kuwahara filtering. Comput. Graph. Forum **28**(7), 1955–1963 (2009)
35. Lee, C.H., Liu, Z., Wu, L., Luo, P.: MaskGAN: towards diverse and interactive facial image manipulation. In: Proceedings of the IEEE Conference on Computer Vision and Pattern Recognition (CVPR), pp. 5549–5558 (2020)
36. Liu, S., et al.: Paint transformer: feed forward neural painting with stroke prediction. In: Proceedings of the IEEE International Conference on Computer Vision (ICCV), pp. 6598–6607 (2021)
37. Cewu, L., Li, X., Jia, J.: Combining sketch and tone for pencil drawing production. In: Proceedings of the International Symposium on Non-Photorealistic Animation and Rendering (NPAR) (2012)
38. Montesdeoca, S.E., et al.: Edge- and substrate-based effects for watercolor stylization. In: Proceedings of the International Symposium on Non-Photorealistic Animation and Rendering (NPAR), pp. 1–10 (2017)
39. Moran, S., Marza, P., McDonagh, S., Parisot, S., Slabaugh, G.: DeepLPF: deep local parametric filters for image enhancement. In: Proceeings of the IEEE Con-

ference on Computer Vision and Pattern Recognition (CVPR), pp. 12826–12835 (2020)
40. Nakano, R.: Neural painters: a learned differentiable constraint for generating brushstroke paintings. arXiv preprint arXiv:1904.08410 (2019)
41. Park, D.Y, Lee, K.H.: Arbitrary style transfer in real-time with adaptive instance normalization. In: Proceedings of the IEEE Conference on Computer Vision and Pattern Recognition (CVPR), pp. 5880–5888 (2019)
42. Phong, B.T.: Illumination for computer generated pictures. Commun. ACM **18**(6), 311–317 (1975)
43. Porter, T., Duff, T.: Compositing digital images. SIGGRAPH. Comput. Graph. **18**(3), 253–259 (1984)
44. Reimann, M., Buchheim, B., Semmo, A., Döllner, J., Trapp, M.: Controlling strokes in fast neural style transfer using content transforms. Vis. Comput. 1–15 (2022)
45. Reimann, M., Klingbeil, M., Pasewaldt, S., Semmo, A., Trapp, M., Döllner, J.: Locally controllable neural style transfer on mobile devices. Vis. Comput. **35**(11), 1531–1547 (2019). https://doi.org/10.1007/s00371-019-01654-1
46. Ronneberger, O., Fischer, P., Brox, T.: U-Net: convolutional networks for biomedical image segmentation. In: Navab, N., Hornegger, J., Wells, W.M., Frangi, A.F. (eds.) MICCAI 2015. LNCS, vol. 9351, pp. 234–241. Springer, Cham (2015). https://doi.org/10.1007/978-3-319-24574-4_28
47. Rosin, P., Collomosse, J.:. Image and Video-based Artistic Stylisation, vol. 42 (2012)
48. Rosin, P.L., et al.: NPRportrait 1.0: a three-level benchmark for non-photorealistic rendering of portraits. Comput. Vis. Med. **8**(3), 445–465 (2022). https://doi.org/10.1007/s41095-021-0255-3
49. Sanakoyeu, A., Kotovenko, D., Lang, S., Ommer, B.: A style-aware content loss for real-time HD style transfer. In: Proceedings of the European Conference on Computer Vision (ECCV) (2018)
50. Semmo, A., Dürschmid, T., Trapp, M., Klingbeil, M., Döllner, J., Pasewaldt, S.: Interactive image filtering with multiple levels-of-control on mobile devices. In: Proceedings of the SIGGRAPH ASIA Mobile Graphics and Interactive Applications (MGIA), pp. 1–8 (2016)
51. Semmo, A., Isenberg, T., Döllner, J.: Neural style transfer: a paradigm shift for image-based artistic rendering? In Proceedings of International Symposium on Non-Photorealistic Animation and Rendering (NPAR), pp. 1–13 (2017)
52. Semmo, A., Limberger, D., Kyprianidis, J.E., Döllner, J.: Image stylization by interactive oil paint filtering. Comput. Graph. **55**, 171 (2016)
53. Simonyan, K., Zisserman, A.: Very deep convolutional networks for large-scale image recognition. In: Proceedings of the International Conference on Learning Representations (ICLR) (2015)
54. Son, M., Lee, Y., Kang, H., Lee, S.: Structure grid for directional stippling. Graph. Models **73**(3), 74–87 (2011)
55. Su, H., Niu, J., Liu, X., Li, Q., Cui, J., Wan, J.: MangaGAN: unpaired photo-to-manga translation based on the methodology of manga drawing. In: Proceedings of the AAAI Conference on Artificial Intelligence, pp. 2611–2619 (2021)
56. Tomasi, C., Manduchi, Ro.: Bilateral filtering for gray and color images. In: Proceedings of the IEEE International Conference on Computer Vision (ICCV), pp. 839–846 (1998)
57. Wang, M., et al.: Towards photo watercolorization with artistic verisimilitude. IEEE Trans. Vis. Comput. Graph. **20**(10), 1451–1460 (2014)

58. Wang, X., Yu, J.: Learning to cartoonize using white-box cartoon representations. In: Proceedings of the IEEE/CVF Conference on Computer Vision and Pattern Recognition (CVPR), pp. 8090–8099 (2020)
59. Winnemöller, H.: XDoG: advanced image stylization with extended difference-of-gaussians. In: Proceedings of the ACM SIGGRAPH/Eurographics Symposium on Non-Photorealistic Animation and Rendering (NPAR), pp. 147–156 (2011)
60. Winnemöller, H., Olsen, S.C., Gooch, B.: Real-time video abstraction. ACM Trans. Graph. (TOG) **25**(3), 1221–1226 (2006)
61. Yan, Z., Zhang, H., Wang, B., Paris, S., Yizhou, Yu.: Automatic photo adjustment using deep neural networks. ACM Trans. Graph. (TOG) **35**(2), 1–15 (2016)
62. Yi, R., Liu, Y.-J., Lai, Y.-K., Rosin, P.L.: APDrawingGAN: generating artistic portrait drawings from face photos with hierarchical GANs. In: Proceedings of the IEEE Conference on Computer Vision and Pattern Recognition (CVPR), pp. 10743–10752 (2019)
63. Yi, R., Xia, M., Liu, Y.-J., Lai, Y.-K., Rosin, P.L.: Line drawings for face portraits from photos using global and local structure based GANs. In: Proceedings of the IEEE Transactions on Pattern Analysis and Machine Intelligence (TPAMI) (2020)
64. Zhang, R., Isola, P., Efros, A.A., Shechtman, E., Wang, O.: The unreasonable effectiveness of deep features as a perceptual metric. In: Proceedings of the IEEE Conference on Computer Vision and Pattern Recognition (CVPR), pp. 586–595 (2018)
65. Zhu, J.-Y., Park, T., Isola, P., Efros, A.A.: Unpaired image-to-image translation using cycle-consistent adversarial networks. In: Proceedings of the IEEE International Conference on Computer Vision (ICCV), pp. 2223–2232 (2017)

# Neural Radiance Transfer Fields for Relightable Novel-View Synthesis with Global Illumination

Linjie Lyu[1(✉)], Ayush Tewari[2], Thomas Leimkühler[1], Marc Habermann[1], and Christian Theobalt[1]

[1] Max Planck Institute for Informatics, Saarland Informatics Campus, Saarbrücken, Germany
llyu@mpi-inf.mpg.de
[2] MIT, Cambridge, USA

**Abstract.** Given a set of images of a scene, the re-rendering of this scene from novel views and lighting conditions is an important and challenging problem in Computer Vision and Graphics. On the one hand, most existing works in Computer Vision usually impose many assumptions regarding the image formation process, e.g. direct illumination and predefined materials, to make scene parameter estimation tractable. On the other hand, mature Computer Graphics tools allow modeling of complex photo-realistic light transport given all the scene parameters. Combining these approaches, we propose a method for scene relighting under novel views by learning a neural precomputed radiance transfer function, which implicitly handles global illumination effects using novel environment maps. Our method can be solely supervised on a set of real images of the scene under a single unknown lighting condition. To disambiguate the task during training, we tightly integrate a differentiable path tracer in the training process and propose a combination of a synthesized OLAT and a real image loss. Results show that the recovered disentanglement of scene parameters improves significantly over the current state of the art and, thus, also our re-rendering results are more realistic and accurate.

## 1 Introduction

The image formation process is influenced by many factors such as the scene geometry, the object materials, the lighting, and the properties of the recording camera. Recovering these properties solely from the final images of the scene is an important inverse problem in Computer Vision and enables several applications such as scene understanding, virtual reality, and controllable image synthesis. Since this is an ill-posed and challenging inverse problem, existing methods make several assumptions about the 3D scene. Common assumptions are that scenes are diffuse [47], or can be described by some predefined material models [51,54]. Importantly, most methods only consider the direct scene illumination [3,4,25,

**Supplementary Information** The online version contains supplementary material available at https://doi.org/10.1007/978-3-031-19790-1_10.

© The Author(s), under exclusive license to Springer Nature Switzerland AG 2022
S. Avidan et al. (Eds.): ECCV 2022, LNCS 13677, pp. 153–169, 2022.
https://doi.org/10.1007/978-3-031-19790-1_10

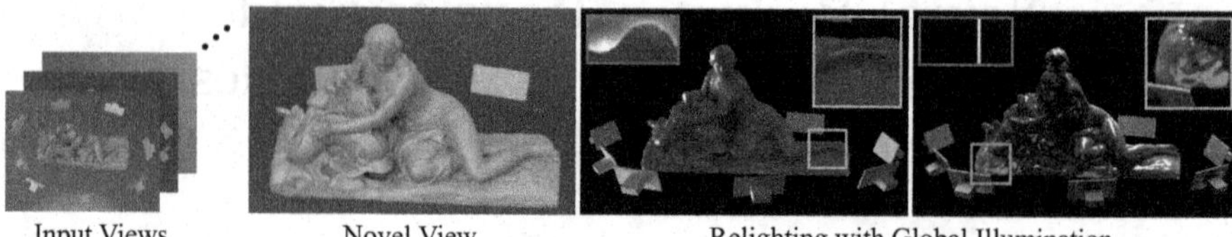


**Fig. 1.** Our method takes multiple views of a scene under one unknown illumination condition as input and allows novel-view synthesis and relighting (corresponding environment maps in green insets) with intricate multi-bounce illumination (orange insets). (Color figure online)

30,38,51,54]. These assumptions limit existing methods from recovering accurate and rich scene properties, resulting in limited re-rendering results as well, e.g. global illumination effects cannot be modeled.

In parallel, the field of Computer Graphics has extensively researched the problem of photorealistic image synthesis. These methods take a well-defined 3D scene and render a realistic image. Methods have explored different ways of modeling indirect illumination using path tracing. Since most path tracing methods are inefficient, precomputed radiance transfer (PRT) was introduced as an efficient approximation of global illumination [16,35–37,43,46]. However, these approaches usually do not consider *recovering the PRT solely from images.*

In this paper, we combine the learning of precomputed radiance transfer function with inverse rendering, thus combining the best of Computer Vision and Computer Graphics. The precomputed radiance transfer function is parameterized as a neural network. Thus, it does not require any predefined approximation function, e.g. spherical harmonics. As we model the material using a learned PRT, our method does not share common limitations with existing inverse rendering methods – our method is capable of dealing with complex light paths such as indirect reflections and shadows, and is also not limited to any predefined BRDF model. Our method is learned on multi-view observations of a scene under a single unknown light condition. In addition to the PRT, it also recovers the scene illumination as an environment map, and the scene geometry defined as a neural signed distance field. Thus, our method enables applications such as novel-view synthesis and global relighting using environment maps. Existing methods, which enable these applications while taking global illumination into account, rely on light-stage datasets, where the object is captured from multiple views under different light conditions. We show that such a setup is not essential. In contrast to real light-stage data, we generate synthetic light-stage data of the scene using a high-quality renderer. This enables correct disentanglement of the material and illumination properties in the scene, while the real multi-view data allows us to capture photorealistic details and to overcome the common assumptions made by the renderer. In summary, our contributions are:

- A method for recovering the radiance transfer field from images of objects under an unknown light condition, hence enabling free-viewpoint relighting with realistic global illumination.

- A *neural* precomputed radiance transfer (PRT) field for multi-bounce global illumination computation and neural implicit surface rendering.
- A novel supervision strategy leveraging a differentiable ray-tracer for physically based scene reconstruction, multiple light bounce rendering, and a new synthetic OLAT supervision.

Our qualitative and quantitative results demonstrate a clear step forward in terms of the recovery of scene parameters as well as the synthesis quality of our approach under novel views and lighting conditions when comparing to the previous state of the art. We will make the code and the new dataset publicly available.

## 2 Related Work

In the following, we focus on previous work concerning radiance transfer and inverse rendering. Although our method also recovers the scene geometry using an off-the-shelf implicit geometry reconstruction approach [44], it is not the main focus of this work and, thus, we do not review related work in this area.

*Precomputed Radiance Transfer.* Precomputed Radiance Transfer (PRT) [36] is a powerful approach for efficient rendering of global illumination [35]. Typically, static geometry and reflectance in combination with distant illumination are assumed, which allows to partially precompute light transport for free-viewpoint synthesis and dynamic lighting. Extensions to e.g., dynamic objects [37] or near-field illumination [43,46] exist. The generic formalization of PRT [16] enables the incorporation of arbitrarily complex light transport, including multi-bounce light paths. While these works improve the runtime of the forward rendering pipeline, they do not consider recovering the PRT solely from a set of real world images of the object. In contrast, we employ this concept to efficiently decompose illumination and reflectance for view synthesis and relighting, taking into account full global illumination, but apply these concepts in an inverse setting where we aim to recover the PRT from images by means of training a neural PRT network.

The versatility of PRT has encouraged the exploration of different angular basis functions, such as spherical harmonics [13,36], Haar wavelets [26], spherical isotropic [42] and anisotropic [49] radial basis functions. While the inherent prior of such basis functions can be beneficial for inverse problems, they also limit the range of illumination effects that can be explained by such a basis. As a remedy, we use the primal directional basis [9] in combination with a neural network, to overcome the limitations of classical basis functions. We encode the full radiance transfer into a neural field [40,48]. Recently, Rainer et al. [32] have explored PRT-inspired neural field-based forward rendering of synthetic scenes – with full knowledge of all scene parameters, as in most works discussed above. In contrast, our framework is concerned with global illumination-aware novel-view synthesis and relighting from multi-view data under one unknown illumination condition. Further, the transfer from distant illumination to local lighting is traditionally concerned with the *incoming* radiance at a surface point [12], i.e., the convolution with reflectance is excluded from precomputation to increase efficiency and to reduce storage requirements. In contrast, we follow ideas from PRT-based relighting [26,41] in that we directly predict *outgoing* radiance.

*Inverse Rendering and Relighting.* Inverse rendering [22,33] aims at estimating scene properties such as geometry, lighting, and materials from image observations. In this work, we are particularly interested in decoupling lighting using multi-view data, and therefore focus our literature review on corresponding related work in illumination decomposition and relighting.

Controllable illumination in a multi-view light stage [7] is conceptually the most straightforward way of obtaining a light-reflectance decomposition in the presence of global illumination, via one-light-at-a-time (OLAT) captures. Even though not trivial, novel-view synthesis and relighting boils down to clever interpolation [29,39,53]. In contrast, input to our method is casually-captured multi-view data under unknown illumination, while embedding *synthetic* OLAT data generation into the training process to aid disentanglement.

Techniques for inverse rendering from multi-view data typically impose strong assumptions on lighting and material, with shading models commonly only considering *direct* illumination [3,4,25,30,38,51,54]. Different scene representations have been explored in this context, including meshes [25,30], signed distance functions (SDFs) [51], or neural radiance or reflectance fields [3,4,23,38,54]. A common paradigm is the explicit reconstruction of a material representation, e.g., an albedo and roughness map, limiting them to recover appearance effects within the range of these predefined representations. In contrast, our approach seeks to decompose observed color into illumination and a radiance transfer function in a surface-based scene representation, enabling relighting with intricate *indirect* illumination, while reconstructing materials only for supervision.

Incorporating multiple light bounces into inverse rendering and relighting can be done by using heuristic lighting models [14,21], by assuming known illumination [8], or by employing physically-based rendering to approximate irradiance [31]. Chen et al. [5] approximate PRT in neural rendering, given geometry, without physically-based modeling of multiple light bounces. Thul et al. [41] utilize PRT in a custom optimization to perform global illumination-aware decomposition of lighting and materials, approximating the required gradients with a single-bounce model. In contrast, differentiable path tracing [2,18,28] can be used to obtain full gradients for global illumination-aware inverse rendering [1,27]. We also leverage the concept of differentiable path tracing [28] during training as a means for achieving disentanglement. Different from path tracing, our performance at inference time is independent of light transport complexity and by design produces noise-free renderings of multi-bounce illumination.

## 3 Method

Our method takes $m \approx 64$ posed multi-view images under an unknown illumination condition as input and allows efficient novel-view synthesis and relighting for the object depicted in these images. To this end, our approach leverages the concept of precomputed radiance transfer (PRT) to factor multi-view observations into illumination and reflectance. Thus, at inference, novel illumination

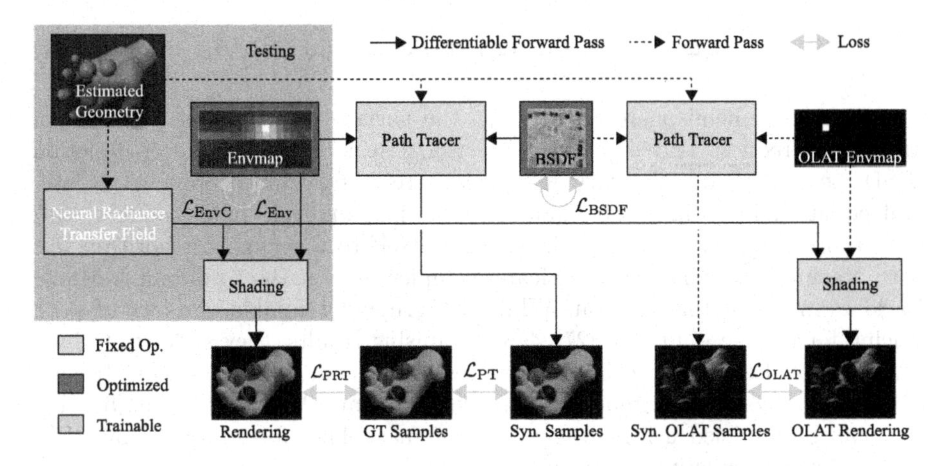


**Fig. 2.** Overview of our pipeline. The Shading blocks evaluate Eq. 2. The directional inputs to the radiance transfer field are omitted to avoid clutter. The blue area marks the parts run at test time, when any environment map can be used to light the scene with full global illumination.

conditions in the form of environment maps can be multiplied with our learned reflectance field given a user-defined camera viewpoint.

In more detail, the rendering equation and its equivalent formulation in the PRT framework forms the theoretical foundation of our approach, and we provide a brief overview of it (Sect. 3.1). Then, we introduce our neural radiance transfer field (NRTF), a neural network that takes as input a point on the surface and its normal, as well as the incoming and outgoing light directions and predicts the radiance transfer of the scene. This radiance can then be multiplied with an arbitrary environment map enabling global illumination relighting (Sect. 3.2). To achieve this, we first estimate scene geometry from the available multi-view observations using implicit surface reconstruction [44] (Sect. 3.2). In order to train the NRTF, an approximate disentanglement between the observed material and lighting in the multi-view training images is required. Our solution for this is to leverage a differentiable path tracer [28]. It allows the joint optimization of a spatially-varying BSDF and the environment map (Sect. 3.3). Once the BSDF is obtained, the path tracer can be used to synthesize one-light-at-a-time (OLAT) renderings of the scene (Sect. 3.4). The NRTF is trained using a combined loss, consisting of a real image loss that helps to recover photoreal material effects beyond the effects possible with the BSDF model, as well as a synthetic OLAT loss that acts as a prior improving generalization to novel lighting conditions (Sect. 3.4). An overview of our pipeline is shown in Fig. 2.

## 3.1 Background

We are interested in estimating radiance $L$ arriving from scene point $\mathbf{x} \in \mathbb{R}^3$ in direction $\boldsymbol{\omega}_o \in \Omega$, where $\Omega$ denotes the space of 3D directions, i.e., points on the unit sphere. The rendering equation [11] describing global light transport can be formulated as

$$L(\mathbf{x}, \boldsymbol{\omega}_o) = \int_{\Omega_+} L(x, \boldsymbol{\omega}_i)\, \rho(\mathbf{x}, \boldsymbol{\omega}_i, \boldsymbol{\omega}_o)\, (\boldsymbol{\omega}_i \cdot \mathbf{n})\, d\boldsymbol{\omega}_i, \tag{1}$$

where $\Omega_+$ is a hemisphere centered at the surface normal $\mathbf{n}$ of $\mathbf{x}$, $\boldsymbol{\omega}_i$ is an incoming direction, and $\rho$ is the bidirectional scattering distribution function (BSDF) encoding spatially-varying surface material reflectance. Solving this integral equation including global illumination, i.e., multiple light bounces with potentially complex inter-reflections, lends itself to a recursive algorithm like path tracing [11], which stochastically samples light paths to obtain a Monte Carlo estimate of the solution. While modern differentiable variants of path tracing for inverse rendering [27] show promising results, they suffer from high computational costs, especially in the presence of complex light paths. To gain efficiency, we consider distant but otherwise arbitrary illumination, i.e., lighting that can be modeled using an environment map. Therefore, inspired by PRT, we rewrite the rendering equations as

$$L(\mathbf{x}, \boldsymbol{\omega}_o) = \int_{\Omega_+} L_{\mathrm{e}}(\boldsymbol{\omega}_i)\, T(\mathbf{x}, \mathbf{n}, \boldsymbol{\omega}_i, \boldsymbol{\omega}_o)\, d\boldsymbol{\omega}_i, \tag{2}$$

where $L_e(\boldsymbol{\omega}_i)$ is the incoming environment light from direction $\boldsymbol{\omega}_i$, which is notably independent of $\mathbf{x}$. The crucial ingredient of this formulation is the collapsed radiance transfer function $T$, which transforms the global distant illumination $L_e$ from direction $\boldsymbol{\omega}_i$ into local reflected radiance at position $\mathbf{x}$ into direction $\boldsymbol{\omega}_o$. Given an environment map and the scene-specific transfer function $T$, all that is needed to compute global illumination for a pixel is to obtain the primary intersection point $\mathbf{x}$, evaluate $T$ for all environment map texels, multiply with the respective illumination, and sum all contributions. If $T$ is compact and easy to evaluate, arbitrarily complex global illumination can be efficiently computed on a GPU in a map-reduce fashion.

### 3.2 Neural Radiance Transfer Field (NRTF)

We model the radiance transfer function $T$ using our neural radiance transfer field

$$T_\theta(\mathcal{H}(\mathbf{x}), \mathbf{n}, \mathcal{F}(\boldsymbol{\omega}_i), \mathcal{F}(\boldsymbol{\omega}_o)) = \mathbf{c}_t, \tag{3}$$

where $\mathbf{c}_t \in \mathbb{R}^3$ denotes transferred RGB color and $\theta$ indicates the trainable parameters. We parameterize $T_\theta$ using a multi-layer perceptron (MLP). Here, we apply a multi-resolution hash encoding $\mathcal{H}(\cdot)$ [24] to the 3D position $\mathbf{x}$, and a spherical harmonics encoding $\mathcal{F}(\cdot)$ [50] to light directions $\boldsymbol{\omega}_i$ and $\boldsymbol{\omega}_o$. The hash encoding enables faster training and evaluation of our networks. Details about the network architecture and the encoding strategy can be found in the supplemental document. When rendering an image from an arbitrary camera view centered at $\mathbf{o}$, we shoot a ray $\mathbf{r}(t) = \mathbf{o} + t\boldsymbol{\omega}_o$ through a pixel with 2D coordinate $\mathbf{u}$, and compute the intersection point $\mathbf{x}$ with respect to the scene geometry. At $\mathbf{x}$, we now evaluate a discretized version of Eq. 2 using our learned $T_\theta$: With $\hat{\boldsymbol{\omega}}$ denoting discrete incoming directions, corresponding to the pixels of a discretized environment map $\hat{L}_{\mathrm{e}}$, we write

$$L_\theta(\mathbf{u}) = L_\theta(\mathbf{x}, \boldsymbol{\omega}_o) = \sum_{\hat{\boldsymbol{\omega}}_i} \hat{L}_e(\hat{\boldsymbol{\omega}}_i) \cdot T_\theta(\mathcal{H}(\mathbf{x}), \mathbf{n}, \mathcal{F}(\hat{\boldsymbol{\omega}}_i), \mathcal{F}(\boldsymbol{\omega}_o)) . \quad (4)$$

Note that this process is repeated for each pixel of the output image. It is worth emphasizing again that this formulation can capture multi-bounce lighting effects and complex material reflectance. Importantly, $\mathbf{x}$, $\boldsymbol{\omega}_o$, and $\hat{L}_e$, i.e., the camera and the environment map can be modified at test time, enabling free-viewpoint rendering and scene relighting. In the following, we explain how we first obtain the scene geometry from the set of multi-view images and then provide details on how the NRTF can be trained without ground truth scene lighting and material.

*Geometry Estimation.* In general, our approach is agnostic to the type of surface-based geometry representation. Recent neural rendering methods [40] have demonstrated state-of-the-art shape reconstruction results using implicit neural SDF representations. We leverage the recently proposed NeuS [44] for computing the SDF geometry of the object. NeuS takes multi-view images and camera poses as input and reconstructs the geometry, represented as a neural field. Since rendering an explicit mesh is significantly more efficient than rendering an SDF, we extract a mesh from the implicit surface using Marching Cubes [20] and use this mesh in our method. We utilize Blender's "Smart UV Project" operator [6] to automatically generate the texture map for the mesh extracted from the SDF.

### 3.3 Path Tracing for Initial Light and Material Estimation

As an initial step, we leverage the state-of-art differentiable path tracer Mitsuba 2 [28] to optimize material properties and scene illumination. We choose a blended BSDF type, where a rough conductor BSDF with roughness $\alpha$ and a diffuse BSDF with a $512 \times 512$ texture $A$ is combined using a convex combination with weight $w$. Illumination is represented as a $32 \times 16$ environment map $\hat{L}_e$. Jointly optimizing light and material properties is difficult due to the ambiguities in the image formation process. In order to make our optimization stable, we assume the object to have a specular material that does not vary spatially. However, we use a spatially-varying diffuse texture $A$ for capturing details. While these assumptions are often incorrect for many complex scenes, we show that our neural radiance transfer function is capable of reconstructions beyond these assumptions. Using the reconstructed geometry, it is straightforward to obtain foreground masks for each input view $I_i$, and we define the set of all foreground pixels as $\mathcal{M}_i$.

We jointly optimize $w, \alpha, A, \hat{L}_e$ from the input multi-view images and the precomputed geometry, using the following loss term:

$$\mathcal{L}(w, \alpha, A, \hat{L}_e) = \mathcal{L}_{\mathrm{PT}}(w, \alpha, A, \hat{L}_e) + \lambda_{\mathrm{reg}} \mathcal{L}_{\mathrm{reg}}(A, \hat{L}_e) . \quad (5)$$

It consists of a data term and a regularizer that is weighted by $\lambda_{\mathrm{reg}}$. The data term is defined as

$$\mathcal{L}_{\mathrm{PT}}(w, \alpha, A, \hat{L}_e) = \sum_{i=1}^{m} \sum_{\mathbf{u} \in \mathcal{M}_i} ||\hat{I}_i(\mathbf{u}) - I_i(\mathbf{u})||^2 . \quad (6)$$

Here, $\hat{I}_i$ is the path-traced reconstruction using up to five light bounces from the estimated scene parameters rendered from the $i$th viewpoint, $\mathbf{u}$ denotes 2D pixel coordinates, and $||\cdot||$ is the Euclidean norm. We use two regularizers to better constrain the problem:

$$\mathcal{L}_{\text{reg}}(A, \hat{L}_{\text{e}}) = \sum_{\hat{\omega}_i} |\nabla \hat{L}_{\text{e}}(\hat{\omega}_i)| + \lambda_{\text{BDSF}} \sum_{\mathbf{u}' \in \mathcal{M}_{\text{tex}}} |\nabla A(\mathbf{u}')| \tag{7}$$

where $\nabla(\cdot)$ denotes the image gradient, $\lambda_{\text{BDSF}}$ is a weighting factor, and $|\cdot|$ denotes the L1 norm. $\mathbf{u}'$ are 2D $uv$-coordinates in the texture map and $\mathcal{M}_{\text{tex}}$ is the set of texels that is covered by the unwrapped geometry. The first term, $\mathcal{L}_{\text{Env}}$, is a regularizer on the environment map reconstruction, while the second term, $\mathcal{L}_{\text{BSDF}}$, regularizes the texture reconstruction. Both encourage image gradient sparsity. We refer to the supplemental document for more details.

### 3.4 Training the Neural Radiance Transfer Field

*OLAT Synthesis.* Our goal is to train the neural transfer field for the input scene. If we train the neural network only with the input illumination condition, the network can easily overfit, thus, not being able to disentangle illumination and material. Traditionally, learning-based methods, which disentangle material and illumination properties, rely on light-stage capture setups [7,29,39,53]. In contrast to these approaches, we only rely on a single illumination condition. We show that it is possible to train for disentanglement even in this more challenging setup, by simulating a virtual light stage. Using the reconstruction obtained with the differentiable path tracer, we render synthetic images of the scene under novel one-light-at-a-time (OLAT) illumination conditions. Here, only one pixel on the environment map is active at a time. We sample OLAT images for training and novel camera views from every incoming light direction. We use Blender [6] to render the OLAT images with the reconstructed geometry and material as input. In total, we synthesize $N_{\text{c}} * N_{\text{e}}$ OLAT images as extra supervision, where $N_{\text{c}}$ is the number of sampled camera views and $N_{\text{e}}$ is the number of texels in the environment map. Note that the OLAT representation forms a complete basis for illumination conditions, i.e., any environment map can be computed as a linear combination of OLAT environment maps. Using these OLATs for our network supervision enables generalization to unseen illumination conditions and camera views.

*NRTF Training.* We train our NRTF in two stages. First, we train on the OLAT dataset using the following loss:

$$\mathcal{L}_{\text{OLAT}}(\theta) = \sum_{i=1}^{N_{\text{c}}} \sum_{\mathbf{u} \in \mathcal{M}_{\text{i}}} \left|\left| \frac{L_{\theta,i}(\mathbf{u}) - O_i(\mathbf{u})}{\text{sg}\left(L_{\theta,i}(\mathbf{u})\right) + \epsilon} \right|\right|^2, \tag{8}$$

Here, $O_i$ is the $i$th OLAT image from Blender and $L_{\theta,i}$ is the corresponding estimate from our NRTF using Eq. 4. Stop gradient is denoted by $\text{sg}(\cdot)$. We set

$\epsilon = 1e-3$ to avoid division by zero and optimize for the network parameters $\theta$. As shown in Noise2Noise [17], this loss works better for high-dymanic range images in the presence of path-tracing noise. Training on the OLAT images enables relighting and novel-view synthesis using the learned transfer function.

However, the method so far is heavily influenced by the lighting-reflectance ambiguity, and by the assumption of a global specularity parameter. Thus, in a second step, we further finetune the network on the input multi-view images. Here, we sample images from the real images as well as the synthetic OLAT images in a minibatch for training. The loss for this stage is defined as

$$\mathcal{L}(\theta, \tilde{L}_{\mathrm{e}}) = \mathcal{L}_{\mathrm{OLAT}}(\theta, \tilde{L}_{\mathrm{e}}) + \lambda_{\mathrm{PRT}}\mathcal{L}_{\mathrm{PRT}}(\theta, \tilde{L}_{\mathrm{e}}) + \lambda_{\mathrm{EnvC}}\mathcal{L}_{\mathrm{EnvC}}(\tilde{L}_{\mathrm{e}})\,. \qquad (9)$$

$\mathcal{L}_{\mathrm{OLAT}}$ is used for the OLAT images in the batch. It is defined as in Eq. 8, however, we also finetune the environment map $\tilde{L}_{\mathrm{e}}$ in this stage using the previously obtained environment map $\hat{L}_{\mathrm{e}}$ for initialization. We further use a masked L2 loss for real images as:

$$\mathcal{L}_{\mathrm{PRT}}(\theta, \tilde{L}_{\mathrm{e}}) = \sum_{i=1}^{m} \sum_{\mathbf{u} \in \mathcal{M}_i} ||L_{\theta,i}(\mathbf{u}) - I_i(\mathbf{u})||^2\,, \qquad (10)$$

Training on real images allows us to update the environment map. We add a regularizer, which penalizes the output to be too far from the initial environment map.

$$\mathcal{L}_{\mathrm{EnvC}}(\tilde{L}_{\mathrm{e}}) = \sum_{\hat{\omega}_i} ||\tilde{L}_{\mathrm{e}}(\hat{\omega}_i)) - \hat{L}_{\mathrm{e}}(\hat{\omega}_i)||^2\,, \qquad (11)$$

where $\hat{L}_e(\hat{\omega}_i)$ denotes the initial environment map estimate.

## 4 Results

Next, we report results of the experiments we conducted to evaluate our method. We construct five synthetic scenes to showcase global illumination effects and further utilize four real scenes from the DTU dataset [10]. For each scene, we take 32–64 input views with a resolution of $800 \times 600$ pixels. During training, all our environment maps have a resolution of $32 \times 16$ pixels in latlong format, but this resolution can be different at test time due to our continuous neural-field formulation. On a single Quadro RTX 8000 GPU, training takes half an hour for initial light and material estimation, eight hours for OLAT training, and an additional 16 h for the final joint optimization to reach highest-quality results. Factorizing lighting and reflectance is fundamentally ambiguous [15] and cannot be resolved from image observations in general [19,34], especially when allowing for spatially-varying materials [33]. To aid meaningful comparisons nevertheless, we follow the procedure of Zhang et al. [54] and other works for all qualitative and quantitative results on synthetic scenes: We compute the mean RGB value of the ground truth environment map and normalize our estimated lighting by its inverse.

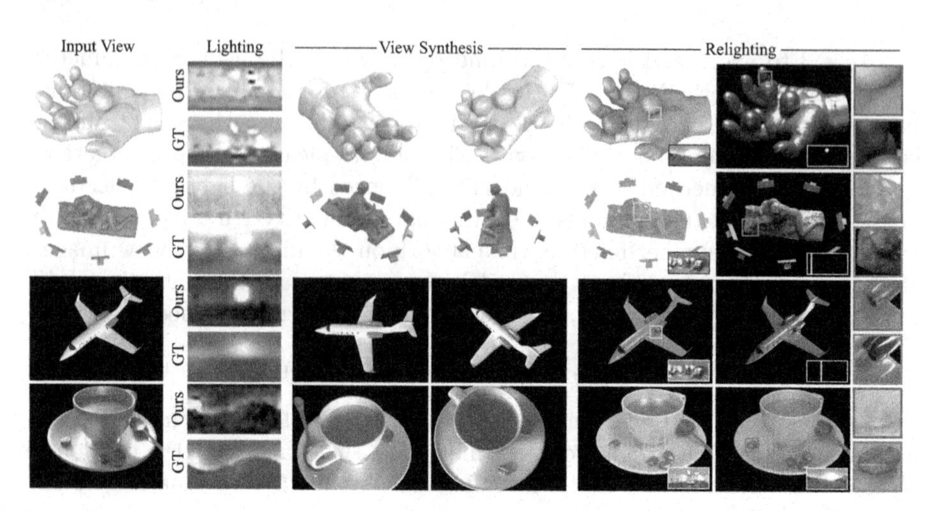


**Fig. 3.** Qualitative results on synthetic data. First column: example input view for training. Second column: our estimated environment map and the corresponding ground truth. Third and fourth column: novel view synthesis results of our approach. Note that our method achieves sharp and accurate novel views that are almost indistinguishable from the input views in terms of quality. Last two columns: relighting results of our method using novel environment maps (see insets). Also for novel lighting conditions our approach achieves convincing results with sharp specular reflections and secondary light bounce effects, e.g. indirect reflections on the wing of the airplane.

### 4.1 Qualitative Results

In Fig. 3, we demonstrate qualitative results of our method. Next to a representative input view (first column), we show the estimated and ground truth lighting (second column), followed by two exemplary novel views created with our method (third and fourth column). Finally, we show relighting results (remaining two columns) using different environment maps (insets). We see that our method produces high-quality relightable novel views, while successfully incorporating global-illumination effects like higher-order specular reflections and subtle color bleeding (see also Fig. 1). In our supplemental video, we further show that our method is also temporally stable when smoothly changing the camera view or rotating an environment map.

*Real Data.* In Fig. 4, we show results of our method on real scenes of the DTU dataset [10]. We can successfully synthesize high-quality novel views and plausible relighting. This shows that our method is robust to such real world captures, which are very challenging due to the lack of very precise camera calibration and foreground segmentation, camera noise, and other effects that are typically not present in synthetic datasets.

*Beyond the Model Assumptions.* In Fig. 5, we show that our approach can learn spatially-varying material effects beyond the ones that can be explained by the

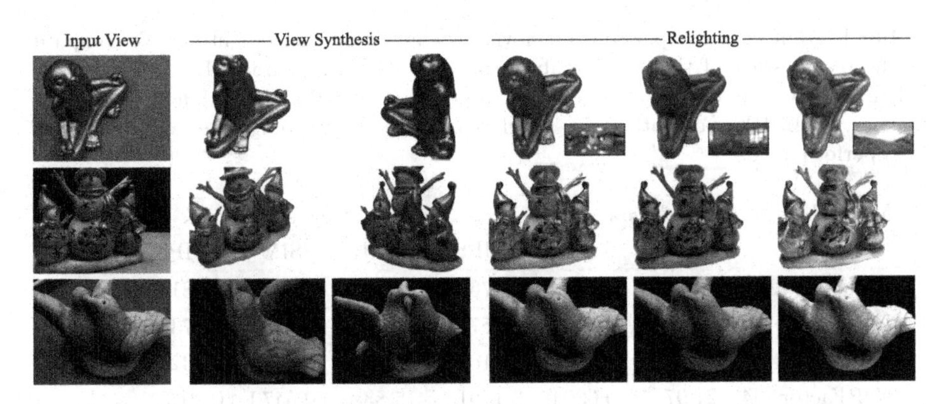


**Fig. 4.** Qualitative results on real data [10]. First column: example input views used for training our method. Second and third column: Novel view results. Note that even for real data our method achieves realistic novel view renderings. Last three columns: relighting results using the environment maps depicted in the insets. Also here, note that our method can achieve plausible relighting effects.

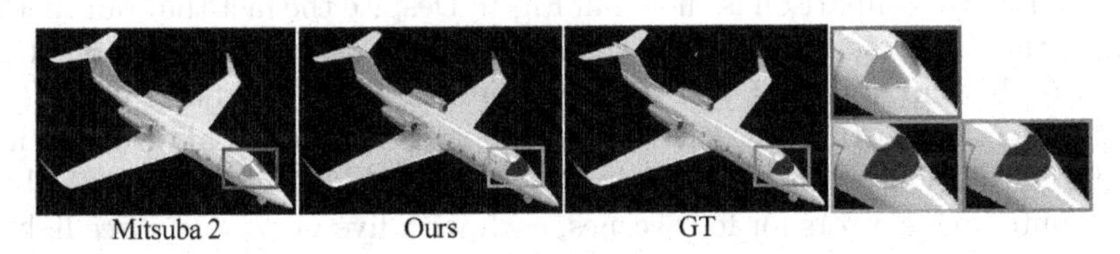


**Fig. 5.** Supervision with real input images lets our NRTF (center) learn appearance effects beyond the material model used during initialization with Mitsuba 2 (left), producing images close to the ground truth (right). Notice that our approach captures spatially-varying material roughness, see close-ups. Images show a relit novel view.

light and material models of the differentiable path tracer [28]. This is due to the real image loss, which lets the network learn appearance effects from real observations.

## 4.2 Comparisons

We compare our approach against several alternatives on the task of novel-view synthesis with relighting: On the one hand, we analyze the capabilities of stand-alone differentiable path-tracing using Mitsuba2 [28], which can be used to perform inverse rendering in the presence of global illumination. On the other hand, we consider three recent neural field-based inverse-rendering approaches PhySG [51], Neural-PIL [4], and NeRFactor [54], all employing only a direct illumination model. We omit a comparison to NeRD [3] as Neural-PIL can be considered as the follow-up. We also compare with RNR [5], providing it with the same geometry as ours. Further, we provide more results on NeRFactor [54] dataset in the supplemental document.

**Table 1.** Numerical comparisons for novel-view synthesis and relighting. We compare to the recent state of the art Mitsuba2 [28], PhySG [51], Neural-PIL [4], NeRFactor [54] and RNR [5] in terms of image-based metrics, i.e. PSNR and SSIM, and perceptual metrics, i.e. LPIPS. For both tasks, novel view synthesis and relighting, we achieve the best performance.

| Method | Novel view synthesis | | | Novel view synthesis & relighting | | |
|---|---|---|---|---|---|---|
| | PSNR ↑ | SSIM ↑ | LPIPS ↓ | PSNR ↑ | SSIM ↑ | LPIPS ↓ |
| Mitsuba2 [28] | 23.50 | 0.7567 | 0.0763 | 21.69 | 0.5722 | 0.0812 |
| PhySG [51] | 20.52 | 0.8563 | 0.2577 | 17.30 | 0.6252 | 0.2736 |
| Neural-PIL [4] | 17.07 | 0.5563 | 0.1159 | 14.76 | 0.4895 | 0.1328 |
| NeRFactor [54] | 21.97 | 0.6394 | 0.1691 | 15.83 | 0.6470 | 0.2033 |
| RNR [5] | 22.54 | 0.8122 | 0.0960 | 18.06 | 0.7009 | 0.1081 |
| Ours | **28.73** | **0.9151** | **0.0454** | **23.06** | **0.8247** | **0.0692** |

A qualitative comparison is shown in Fig. 6. Despite the fact that our method provides the sharpest and most realistic results, it is worth noting that our method is the only one that can recover accurate indirect lighting effects, e.g. the self-reflection on the wing of the airplane and the color spill of the squares onto the statue. This is further confirmed by the quantitative analysis in Table 1. We compute image errors for four scenes, each with five views and three lighting conditions according to three metrics on the tasks of novel-view synthesis and novel-view synthesis with relighting. In particular, we evaluate the Peak Signal-to-noise Ratio (PSNR), the Structural Similarity Index Measure (SSIM) [45], and the learned perceptual image patch similarity (LPIPS) [52]. PSNR and SSIM are purely image-based metrics and, thus, sometimes do not reflect the *perceived* image quality. For this reason, we also provide the perceptual LPIPS metric. We observe that our approach again delivers the highest-quality results for both tasks across all metrics.

### 4.3 Ablation and Extension

Here, we study ablations and extensions in order to gain further insights into our system. All results are compiled into Table 2, where the evaluation protocol is the same as in Sect. 4.2. First, we consider omitting the OLAT loss (Sect. 3.4). We observe that result quality reduces significantly for the relighting task compared to our full method. This is due to the poor disentanglement of lighting and reflectance and the fact that the network can overfit to the lighting condition in the training data, which also explains why the novel view synthesis without the OLAT loss is slightly more accurate than our method.

Second, we investigate the behavior of our approach when input views are captured under *multiple* unknown illumination conditions. In this experiment, we use three different environment maps. When reconstructing geometry (Sect. 3.2), we select only a subset of multi-view images with the same illumination condi-

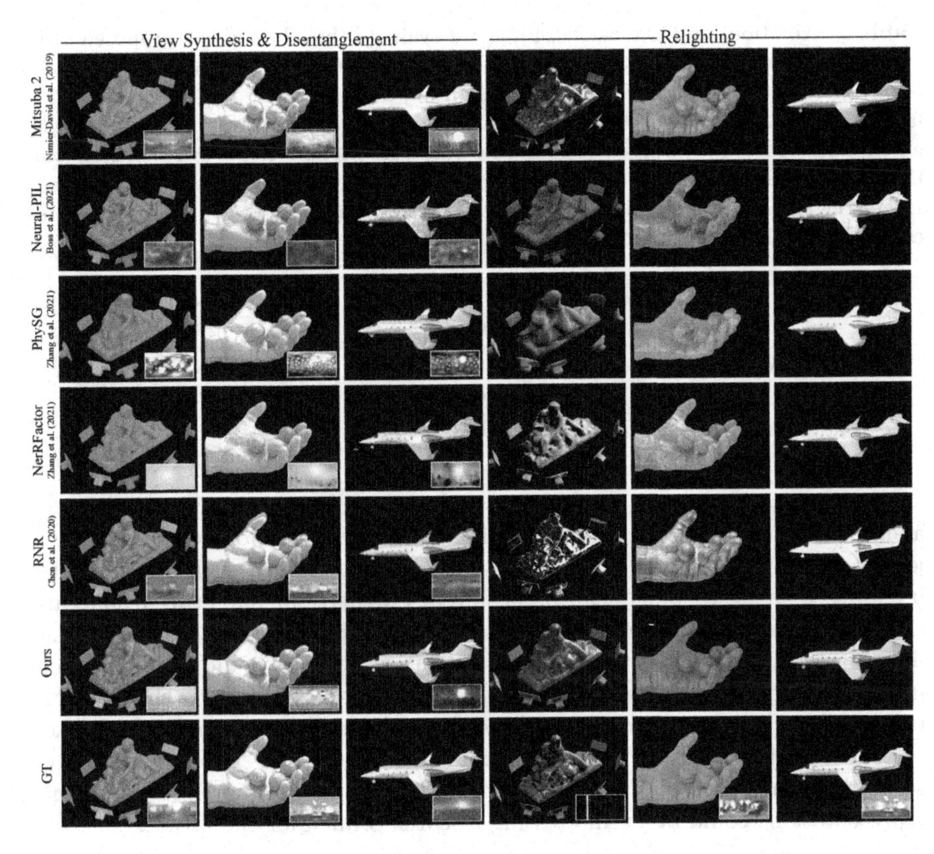


**Fig. 6.** Comparisons to related works [4,5,28,51,54] for novel-view synthesis and disentanglement of lighting (left three columns), and relighting (right three columns). Note that for both tasks we achieve the best results in terms of rendering quality. It is also worth noting that we are the only method, which can accurately reproduce the indirect illumination effects such as the self-reflection on the wing of the airplane.

tion, while during initial light and material estimation (Sect. 3.3), we optimize for three individual environment maps. Not surprisingly, we observe that this extended setup increases result quality even more compared to our full single-lighting approach. Yet, it has a significantly less pronounced effect compared to the omission of the OLAT training stage, indicating that our pipeline achieves a solid disentanglement for the single-illumination condition.

## 5 Limitations and Future Work

Although our method improves the state of the art in terms of image quality and global illumination handling, it still has some limitations, which open up future work in this direction. In particular, our method relies on an accurate geometry estimate of the scene and we are not jointly optimizing the scene geometry

**Table 2.** Ablations and extensions. First, we evaluate the effect of the proposed synthetic OLAT loss. One can clearly see that without the OLAT loss the performance of our method drastically drops for the relighting task. This can be explained by the fact that the OLAT loss acts as a regularizer and prevents overfitting to the single environment map that is recovered during training. Moreover, we evaluate how our method performs when the object was observed under *multiple* lighting conditions. Interestingly, with this additional input, our method can achieve even better results, especially for the relighting task.

| Method | Novel view synthesis | | | Novel view synthesis & relighting | | |
|---|---|---|---|---|---|---|
| | PSNR ↑ | SSIM ↑ | LPIPS ↓ | PSNR ↑ | SSIM ↑ | LPIPS ↓ |
| w/o OLAT Loss | 31.35 | 0.9668 | 0.063 | 10.03 | 0.4547 | 0.4487 |
| Full | 29.56 | 0.9418 | 0.066 | 24.76 | 0.8665 | 0.069 |
| Multiple envmaps | 30.62 | 0.9166 | 0.043 | 26.67 | 0.9071 | 0.047 |

along with the material and lighting of the scene. Future work could involve a joint reasoning of all these aspects in a differentiable manner such that optimizing all scene aspects jointly can be achieved. Further, our relighting results are only correct up to a global scale due to the inherent ambiguity between scene illumination and the object material. Here, future research could explore a minimal setup required to disentangle such ambiguities, e.g. it may be that a single measurement on the surface can resolve the ambiguity. Last, our method takes several seconds per frame. Ideally, it would run at real time enabling interactive scene relighting with global illumination. Thus, exploring more efficient scene representations could be an interesting research branch for the future.

## 6 Conclusion

We presented neural radiance transfer fields, which enable global illumination scene relighting and view synthesis given multi-view images of the object. At the technical core, our method implements the concept of precomputed radiance transfer that disentangles illumination from appearance. To this end, we propose a neural radiance transfer field represented as an MLP and show how at train time differentiable path tracing and a dedicated OLAT loss can be used to let the network accurately learn such a disentanglement. Once trained, our rendering formulation allows novel-view synthesis and relighting, which is aware of global-illumination effects. Our results demonstrate a clear improvement over the current state of the art while future work could involve further improving the runtime and a joint reasoning of geometry, material, and scene lighting.

**Acknowledgement.** We would like to thank Xiuming Zhang for his help with the NeRFactor comparisons. Authors from MPII were supported by the ERC Consolidator Grant 4DRepLy (770784).

## References


1. Azinović, D., Li, T.M., Kaplanyan, A., Nießner, M.: Inverse path tracing for joint material and lighting estimation. In: Proceedings of the Computer Vision and Pattern Recognition (CVPR). IEEE (2019)
2. Bangaru, S., Michel, J., Mu, K., Bernstein, G., Li, T.M., Ragan-Kelley, J.: Systematically differentiating parametric discontinuities. ACM Trans. Graph. **40**(107), 107:1–107:17 (2021)
3. Boss, M., Braun, R., Jampani, V., Barron, J.T., Liu, C., Lensch, H.P.: NeRD: neural reflectance decomposition from image collections. In: Proceedings of the IEEE/CVF International Conference on Computer Vision (ICCV), pp. 12684–12694 (2021)
4. Boss, M., Jampani, V., Braun, R., Liu, C., Barron, J., Lensch, H.: Neural-PIL: neural pre-integrated lighting for reflectance decomposition. Adv. Neural. Inf. Process. Syst. **34**, 10691–10704 (2021)
5. Chen, Z., et al.: A neural rendering framework for free-viewpoint relighting. In: CVPR (2020)
6. Community, B.O.: Blender - a 3D modelling and rendering package (2018). https://www.blender.org
7. Debevec, P., Hawkins, T., Tchou, C., Duiker, H.P., Sarokin, W., Sagar, M.: Acquiring the reflectance field of a human face. In: Proceedings of the 27th Annual Conference on Computer Graphics and Interactive Techniques, pp. 145–156 (2000)
8. Goel, P., Cohen, L., Guesman, J., Thamizharasan, V., Tompkin, J., Ritchie, D.: Shape from tracing: towards reconstructing 3D object geometry and svbrdf material from images via differentiable path tracing. In: 2020 International Conference on 3D Vision (3DV), pp. 1186–1195. IEEE (2020)
9. Hao, X., Baby, T., Varshney, A.: Interactive subsurface scattering for translucent meshes. In: Proceedings of the 2003 Symposium on Interactive 3D Graphics, pp. 75–82 (2003)
10. Jensen, R., Dahl, A., Vogiatzis, G., Tola, E., Aanæs, H.: Large scale multi-view stereopsis evaluation. In: 2014 IEEE Conference on Computer Vision and Pattern Recognition, pp. 406–413. IEEE (2014)
11. Kajiya, J.T.: The rendering equation. In: Proceedings of the 13th Annual Conference on Computer Graphics and Interactive Techniques, pp. 143–150 (1986)
12. Kautz, J., Sloan, P.P., Lehtinen, J.: Precomputed radiance transfer: theory and practice. In: ACM SIGGRAPH 2005 Courses, pp. 1–es (2005)
13. Kautz, J., Snyder, J., Sloan, P.P.J.: Fast arbitrary BRDF shading for low-frequency lighting using spherical harmonics. Rendering Tech. **2**(291–296), 1 (2002)
14. Laffont, P.Y., Bousseau, A., Drettakis, G.: Rich intrinsic image decomposition of outdoor scenes from multiple views. IEEE Trans. Visual. Comput. Graph. **19**(2), 210–224 (2012)
15. Land, E.H., McCann, J.J.: Lightness and retinex theory. Josa **61**(1), 1–11 (1971)
16. Lehtinen, J.: A framework for precomputed and captured light transport. ACM Trans. Graph. (TOG) **26**(4), 13-es (2007)
17. Lehtinen, J., et al.: Noise2noise: learning image restoration without clean data. arXiv preprint arXiv:1803.04189 (2018)
18. Li, T.M., Aittala, M., Durand, F., Lehtinen, J.: Differentiable monte carlo ray tracing through edge sampling. ACM Trans. Graph. (TOG) **37**(6), 1–11 (2018)
19. Lombardi, S., Nishino, K.: Reflectance and illumination recovery in the wild. IEEE Trans. Pattern Anal. Mach. Intell. **38**(1), 129–141 (2015)

20. Lorensen, W.E., Cline, H.E.: Marching cubes: a high resolution 3D surface construction algorithm. ACM siggraph comput. graph. **21**(4), 163–169 (1987)
21. Lyu, L., Habermann, M., Liu, L., Tewari, A., Theobalt, C., et al.: Efficient and differentiable shadow computation for inverse problems. In: ICCV, pp. 13107–13116 (2021)
22. Marschner, S.R.: Inverse Rendering for Computer Graphics. Cornell University, Ithaca (1998)
23. Mildenhall, B., Srinivasan, P.P., Tancik, M., Barron, J.T., Ramamoorthi, R., Ng, R.: NeRF: representing scenes as neural radiance fields for view synthesis. In: Vedaldi, A., Bischof, H., Brox, T., Frahm, J.-M. (eds.) ECCV 2020. LNCS, vol. 12346, pp. 405–421. Springer, Cham (2020). https://doi.org/10.1007/978-3-030-58452-8_24
24. Müller, T., Evans, A., Schied, C., Keller, A.: Instant neural graphics primitives with a multiresolution hash encoding. arXiv:2201.05989 (2022)
25. Munkberg, J., et al.: Extracting triangular 3d models, materials, and lighting from images. arXiv preprint arXiv:2111.12503 (2021)
26. Ng, R., Ramamoorthi, R., Hanrahan, P.: All-frequency shadows using non-linear wavelet lighting approximation. In: ACM SIGGRAPH 2003 Papers, pp. 376–381 (2003)
27. Nimier-David, M., Dong, Z., Jakob, W., Kaplanyan, A.: Material and lighting reconstruction for complex indoor scenes with texture-space differentiable rendering (2021)
28. Nimier-David, M., Vicini, D., Zeltner, T., Jakob, W.: Mitsuba 2: a retargetable forward and inverse renderer. ACM Trans. Graph. (TOG) **38**(6), 1–17 (2019)
29. Pandey, R., et al.: Total relighting: learning to relight portraits for background replacement, vol. 40 (2021). https://doi.org/10.1145/3450626.3459872
30. Philip, J., Gharbi, M., Zhou, T., Efros, A.A., Drettakis, G.: Multi-view relighting using a geometry-aware network. ACM Trans. Graph. **38**(4), 78–1 (2019)
31. Philip, J., Morgenthaler, S., Gharbi, M., Drettakis, G.: Free-viewpoint indoor neural relighting from multi-view stereo. ACM Trans. Graph. (TOG) **40**(5), 1–18 (2021)
32. Rainer, G., Bousseau, A., Ritschel, T., Drettakis, G.: Neural precomputed radiance transfer. In: Computer Graphics Forum (Proceedings of Eurographics), vol. 41, no. 2 (2022). https://www-sop.inria.fr/reves/Basilic/2022/RBRD22
33. Ramamoorthi, R., Hanrahan, P.: A signal-processing framework for inverse rendering. In: Proceedings of the 28th Annual Conference on Computer Graphics and Interactive Techniques, pp. 117–128 (2001)
34. Ramamoorthi, R., Hanrahan, P.: A signal-processing framework for reflection. ACM Trans. Graph. (TOG) **23**(4), 1004–1042 (2004)
35. Ritschel, T., Dachsbacher, C., Grosch, T., Kautz, J.: The state of the art in interactive global illumination. In: Computer Graphics Forum, vol. 31, pp. 160–188. Wiley Online Library (2012)
36. Sloan, P.P., Kautz, J., Snyder, J.: Precomputed radiance transfer for real-time rendering in dynamic, low-frequency lighting environments. In: Proceedings of the SIGGRAPH, pp. 527–536 (2002)
37. Sloan, P.P., Luna, B., Snyder, J.: Local, deformable precomputed radiance transfer. ACM Trans. Graph. (TOG) **24**(3), 1216–1224 (2005)
38. Srinivasan, P.P., Deng, B., Zhang, X., Tancik, M., Mildenhall, B., Barron, J.T.: NeRV: neural reflectance and visibility fields for relighting and view synthesis. In: Proceedings of the IEEE/CVF Conference on Computer Vision and Pattern Recognition, pp. 7495–7504 (2021)

39. Sun, T., et al.: Light stage super-resolution: continuous high-frequency relighting. ACM Trans. Graph. (TOG) **39**(6), 1–12 (2020)
40. Tewari, A., et al.: Advances in neural rendering. arXiv preprint arXiv:2111.05849 (2021)
41. Thul, D., Tsiminaki, V., Ladický, L., Pollefeys, M.: Precomputed radiance transfer for reflectance and lighting estimation. In: 2020 International Conference on 3D Vision (3DV), pp. 1147–1156. IEEE (2020)
42. Tsai, Y.T., Shih, Z.C.: All-frequency precomputed radiance transfer using spherical radial basis functions and clustered tensor approximation. ACM Trans. graph. (TOG) **25**(3), 967–976 (2006)
43. Wang, J., Ramamoorthi, R.: Analytic spherical harmonic coefficients for polygonal area lights. ACM Trans. Graph. (Proc. SIGGRAPH 2018) **37**(4), 1–11 (2018)
44. Wang, P., Liu, L., Liu, Y., Theobalt, C., Komura, T., Wang, W.: NeuS: learning neural implicit surfaces by volume rendering for multi-view reconstruction. arXiv preprint arXiv:2106.10689 (2021)
45. Wang, Z., Bovik, A., Sheikh, H., Simoncelli, E.: Image quality assessment: from error visibility to structural similarity. IEEE Trans. Image Process. **13**(4), 600–612 (2004). https://doi.org/10.1109/TIP.2003.819861
46. Wu, L., Cai, G., Zhao, S., Ramamoorthi, R.: Analytic spherical harmonic gradients for real-time rendering with many polygonal area lights. ACM Trans. Graph. (TOG) **39**(4), 134–1 (2020)
47. Wu, S., Rupprecht, C., Vedaldi, A.: Unsupervised learning of probably symmetric deformable 3D objects from images in the wild. In: Proceedings of the IEEE/CVF Conference on Computer Vision and Pattern Recognition, pp. 1–10 (2020)
48. Xie, Y., et al.: Neural fields in visual computing and beyond. arXiv preprint arXiv:2111.11426 (2021)
49. Xu, K., Sun, W.L., Dong, Z., Zhao, D.Y., Wu, R.D., Hu, S.M.: Anisotropic spherical gaussians. ACM Trans. Graph. (TOG) **32**(6), 1–11 (2013)
50. Yu, A., Fridovich-Keil, S., Tancik, M., Chen, Q., Recht, B., Kanazawa, A.: Plenoxels: radiance fields without neural networks. arXiv preprint arXiv:2112.05131 (2021)
51. Zhang, K., Luan, F., Wang, Q., Bala, K., Snavely, N.: PhySG: inverse rendering with spherical gaussians for physics-based material editing and relighting. In: CVPR (2021)
52. Zhang, R., Isola, P., Efros, A.A., Shechtman, E., Wang, O.: The unreasonable effectiveness of deep features as a perceptual metric. In: 2018 IEEE/CVF Conference on Computer Vision and Pattern Recognition (CVPR), pp. 586–595. IEEE Computer Society, Los Alamitos (2018). https://doi.org/10.1109/CVPR.2018.00068, https://doi.ieeecomputersociety.org/10.1109/CVPR.2018.00068
53. Zhang, X., et al.: Neural light transport for relighting and view synthesis. ACM Trans. Graph. (TOG) **40**(1), 1–17 (2021)
54. Zhang, X., Srinivasan, P.P., Deng, B., Debevec, P., Freeman, W.T., Barron, J.T.: NeRFactor: neural factorization of shape and reflectance under an unknown illumination. ACM Trans. Graph. (TOG) **40**(6), 1–18 (2021)

# Transformers as Meta-learners for Implicit Neural Representations

Yinbo Chen and Xiaolong Wang(✉)

UC San Diego, San Diego, USA
xiw012@ucsd.edu

**Abstract.** Implicit Neural Representations (INRs) have emerged and shown their benefits over discrete representations in recent years. However, fitting an INR to the given observations usually requires optimization with gradient descent from scratch, which is inefficient and does not generalize well with sparse observations. To address this problem, most of the prior works train a hypernetwork that generates a single vector to modulate the INR weights, where the single vector becomes an information bottleneck that limits the reconstruction precision of the output INR. Recent work shows that the whole set of weights in INR can be precisely inferred without the single-vector bottleneck by gradient-based meta-learning. Motivated by a generalized formulation of gradient-based meta-learning, we propose a formulation that uses Transformers as hypernetworks for INRs, where it can directly build the whole set of INR weights with Transformers specialized as set-to-set mapping. We demonstrate the effectiveness of our method for building INRs in different tasks and domains, including 2D image regression and view synthesis for 3D objects. Our work draws connections between the Transformer hypernetworks and gradient-based meta-learning algorithms and we provide further analysis for understanding the generated INRs. The project page with code is at https://yinboc.github.io/trans-inr/.

## 1 Introduction

In recent years, Implicit Neural Representations (INRs) have been proposed as continuous data representations for various tasks in computer vision. With INR, data is represented as a neural function that maps continuous coordinates to signals. For example, an image can be represented as a neural function that maps 2D coordinates to RGB values, a 3D scene can be represented as a neural radiance field (NeRF [34]) that maps 3D locations with view directions to densities and RGB values. Compared to discrete data representations such as pixels, voxels, and meshes, INRs do not require resolution-dependent quadratic or cubic storage. Their representation capacity does not depend on grid resolution but instead on the capacity of a neural network, which may capture the underlying data structure and reduce the redundancy in representation, therefore providing a compact yet powerful continuous data representation.

However, learning the neural functions of resolution-free INRs from given observations usually requires optimization with gradient descent steps, which

© The Author(s), under exclusive license to Springer Nature Switzerland AG 2022
S. Avidan et al. (Eds.): ECCV 2022, LNCS 13677, pp. 170–187, 2022.
https://doi.org/10.1007/978-3-031-19790-1_11

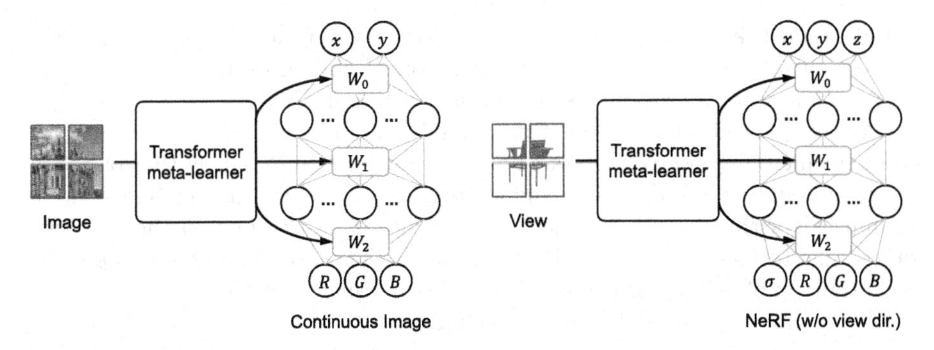


**Fig. 1.** Implicit Neural Representation (INR) is a function that maps coordinates to signals. We propose to use Transformers as meta-learners for directly building the whole weights in INRs from given observations. Our method supports various types of INRs, such as continuous images and neural radiance fields.

has several challenges: (i) Optimization can be slow if every INR is learned independently from a random initialization; (ii) The learned INR does not generalize well to unseen coordinates if the given observations are sparse and no strong prior is shared.

From the perspective of efficiently building INRs, previous works [47] proposed to learn a latent space where each INR can be decoded by a single vector with a hypernetwork [20]. However, a single vector may not have enough capacity to capture the fine details of a complex real-world image or 3D object, while these works show promising results in generative tasks [1,5,49], they do not have high precision in reconstruction tasks [5]. The single-vector modulated INRs are mostly used for representing local tiles [31] for reconstruction. Recent works [7,44,57] revisit the grid-based discrete representations and define INRs over deep feature maps, where the capacity and storage will be resolution-dependent and the decoding is bounded by feature maps as the INRs rely on local features. Going beyond the limitation of the resolution, our work is inspired by recent works [46,52] which explore a promising direction in the intersection between gradient-based meta-learning and INRs. Without grid-based representation, these works can efficiently and precisely infer the whole set of INR weights without the single-vector bottleneck. However, the computation of higher-order derivatives and a learned fixed initialization make these methods less flexible, and gradient descent that involves sequential forward and backward passing is still necessary for obtaining INRs from given observations in these works.

Motivated by a generalized formulation of the gradient-based meta-learning methods, we propose the formulation that uses Transformers [55] as effective hypernetworks for INRs (Fig. 1). Our general idea is to use Transformers to transfer the knowledge from image observations to INR weights. Specifically, we first convert the input observations to data tokens, then we view the weights in INR as the set of column vectors in weight matrices of different layers, for which we create initialization tokens each representing one column vector. These initialization tokens are passed together with data tokens into a Transformer, and the output tokens are mapped to their corresponding location (according to the location of initialization tokens) as the weights in INRs.

We verify the effectiveness of our method for building INRs in both 2D and 3D domains, including image regression and view synthesis. We show that our approach can efficiently build INRs and outperform previous gradient-based meta-learning algorithms on reconstruction and synthesis tasks. Our further analysis shows qualitatively that the INRs built by the Transformer meta-learner may potentially exploit the data structures without explicit supervision.

To summarize, our contributions include:

- We propose a Transformer hypernetwork to infer the whole weights in an INR, which removes the single-vector bottleneck and does not rely on grid-based representation or gradient computation.
- We draw connections between the Transformer hypernetwork and the gradient-based meta-learning for INRs.
- Our analysis sheds light on the structures of the generated INRs.

## 2 Related Work

**Implicit Neural Representation.** Implicit neural representations (INRs) have been demonstrated as flexible and compact continuous data representations in recent works. A main branch of these works use INRs for representing 3D objects or scenes, their wide applications include 3D reconstruction [10,18,19,33] and generation [5,13,45]. Typical examples of resolution-free INRs include DeepSDF [37] which represents 3D shapes as a field of signed distance function, Occupancy Networks [32] and IM-NET [8] which represents 3D shapes as binary classification neural network that classifies each 3D coordinate for being inside or outside the shape. NeRF and its follow-up works [27,30,34,38] are proposed to represent a scene as a neural radiance field that maps each position to a density and a view-dependent RGB value, with differentiable volumetric rendering that allows optimizing the representation from 2D views. The idea of INR has also been adapted for representing 2D images in recent works [1,7,25,49], which allows decoding for arbitrary output resolution. Several recent works observe that coordinate-based MLPs with ReLU activation may lack the capacity for representing fine details, solutions proposed to address this issue include replacing ReLU with sine activation function [47], and using Fourier features of input coordinates [53].

**Hypernetworks for INRs.** A hypernetwork [20] $g$ generates the weights $\theta$ for another network $f_\theta$ from some input $z$, i.e. $\theta = g(z)$. Directly building an INR from given observations will usually require performing gradient descent steps, which is inefficient and does not generalize well with sparse observations. A common way to tackle this short-come is learning a latent space for INRs [32,37,47,48], where each INR corresponds to a latent vector that can be decoded by a hypernetwork. Since a single vector may have limited capacity for representing the fine details (e.g. lack of details in reconstructing face image [5,47]), many recent works [4,7,9,18,24,31,39] address this issue by revisiting discrete representation and defining INRs with feature maps in a hybrid way, where the data still corresponds to a grid-based representation. Different from these hybrid methods, our goal is to obtain a hypernetwork that allows for building a resolution-free neural function (i.e. a global function instead of a grid-based representation).

**Meta-learning.** Learning to build a neural function from given observations is related to the topic of meta-learning, where a differentiable meta-learner is trained for inferring the weights in a neural network. Most previous works on meta-learning have been focus on few-shot learning [35,43,50,51,56] and reinforcement learning [16,17,23], where a meta-learner allows fast adaptation for new observations and better generalization with few samples. Gradient-based methods is a popular branch in meta-learning algorithms, typical examples include MAML [17], Reptile [36], and their extentions [2,15,42]. A recent paper provides a comprehensive survey on meta-learning algorithms [21].

While most previous works in meta-learning aim at building a neural function for processing the data, the recent rising topic of implicit neural representation connects neural functions and data representations, which extends the idea of meta-learning with new possibilities for building neural functions that represent the data. MetaSDF [46] adopts a gradient-based meta-learning algorithm for learning signed distance functions, which leads to much faster convergence than standard learning. Learned Init [52] generalizes this idea to wider classes of INRs and shows the effectiveness of using the meta-learned initialization as encoded prior. While these works have shown promising results, their methods only learn a fixed initialization and require test-time optimization. We show that it is possible to directly build the whole INR with a Transformer meta-learner and it is more flexible than a fixed initialization.

**Transformers.** Transformers [55] were initially proposed for machine translation, and has later been a state-of-the-art architecture used in various methods [3,12,40,41] in natural language processing. Recent works [14,28,54] also demonstrate the potential of Transformers for encoding visual data. In this work, we show promising results of using Transformers in meta-learning for directly inferring the whole weights in a neural function of INR.

## 3 Method

### 3.1 Problem Formulation

We are interested in the problem of recovering a signal $I$ from observations $O$. The signal is a function $I : X \to Y$ defined in a bounded domain that $X \subseteq \mathbb{R}^c, Y \subseteq \mathbb{R}^d$. For instance, an image can be represented as a function that maps 2D coordinates to 3D tuples of RGB values. A 3D object or scene can be represented as a neural radiance field (NeRF) [34], which maps 3D locations with view directions $v$ (normalized 3D vectors) to 4D tuples that describe the densities and RGB values.

In implicit neural representation, the signal $I$ is estimated and parameterized by a neural function $f_\theta$ with $\theta$ as its weights (learnable parameters). A typical example of $f_\theta$ is a multilayer perceptron (MLP). We consider a more general class of $f_\theta$ where its weights consist of a set of matrices

$$\theta = \{W_i \mid W_i \in \mathbb{R}^{\mathrm{in}_i \times \mathrm{out}_i}\}_{i=0}^{m-1},$$

the biases to add (if exist) are merged into these matrices. Given the observations $O$, our goal is now to obtain $\theta$ that fits the signal $I$ with the neural function $f_\theta$.

Observations $O$ is a set $O = \{T_i(I)\}_{i=0}^{|O|-1}$ with transform functions $T_i$. For example, to estimate a continuous image, each pixel $i$ in the given image can be approximately viewed as $T_i(I) = I(x_i)$, where $x_i$ is the center coordinate of pixel $i$ and $I(x_i)$ are the RGB values. To estimate a 3D object with NeRF, an input view provides each pixel $i$ with its corresponding rendering ray $r_i$, that can be represented as $T_i(I) = R(I, r_i)$, where $R$ is the function renders the RGB values from ray $r_i$ in the radience field $I$.

Given the observation set $O$, estimating $I$ with the INR $f_\theta$ can be addressed by minimizing the L2 loss

$$L(\theta; O) = \frac{1}{|O|} \sum_{T_i \in O} \|T_i(f_\theta) - T_i(I)\|_2^2. \tag{1}$$

If we assume $T_i(f_\theta)$ is differentiable to $\theta$, minimizing this loss with gradient descent steps is referred to as fitting an INR to given observations or learning the INR. The goal of a meta-learner is to efficiently find $\theta$ with given $O$ and improve the generalization of the neural function $f_\theta$.

### 3.2 Motivating from Gradient-Based Meta-learning

In meta-learning, the goal is to train a meta-learner that infers the weights $\theta$ of a target network $f_\theta$ from given observations. In MAML [17], the learnable component is an initialization $\theta_0$, $\theta = \theta_n$ is inferred by updating $\theta_0$ for $n$ steps

$$\theta_{i+1} = \theta_i + (-\nabla_\theta \mathcal{L}(\theta; O)|_{\theta=\theta_i}), \tag{2}$$

where $\mathcal{L}$ is the differentiable loss function computed with observations $O$. The update formula above defines a computation graph from $\theta_0$ to $\theta_n$, if the computation graph (with higher-order derivatives) is differentiable, the gradient for optimizing $\theta_n$ can be back-propagated to $\theta_0$ for training this meta-learner.

We consider a more general class of meta-learners, where its learnable components contain: (i) A learnable initialization $\theta_0$; (ii) A total number of update steps $n$; (iii) A step-specific learnable update rule $U_{\psi_i}$ (with $\psi_i$ as its parameters) that conditions on some provided data $D_i$:

$$\theta_{i+1} = \theta_i + U_{\psi_i}(\theta_i; D_i). \tag{3}$$

The meta-learning objective is applied to the final vector $\theta_n$, which is typically fitting the seen observations or generalizing to unseen observations.

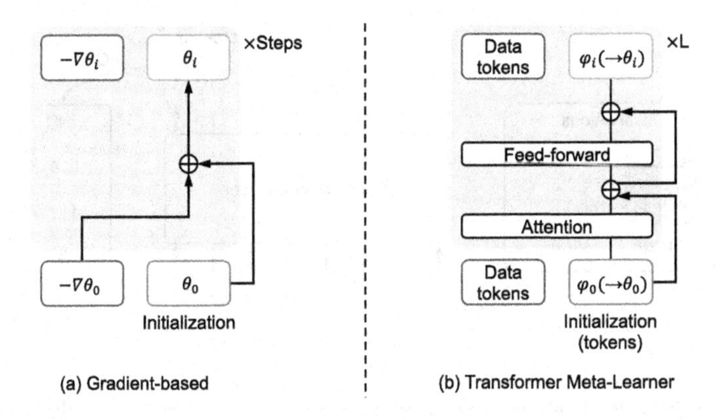


**Fig. 2. Motivating from gradient-based meta-learners.** The residual link in the Transformer meta-learner shares a similar formulation as subtracting the gradients in gradient descent for updating the weights.

We observe that this formulation can be naturally instantiated with a Transformer architecture. In general, we propose to represent observations as a set of data tokens, which are passed into a Transformer encoder with a set of initialization tokens that are learnable parameters defined in addition, as shown in Fig. 2(b). The computation graph with the residual link can be written as

$$\varphi_{i+1} = \varphi_i + U_{\psi_i}(\varphi_i; d_i), \tag{4}$$

where $d_i$ are the data tokens at layer $i$, $U_{\psi_i}$ is the function that describes how the output residual is conditioned on $i$-th layer's input, i.e. the function that is composed of the attention layer and the feed-forward layer, $\varphi_0$ is the learnable initialization tokens and tokens $\varphi_i$ corresponds to the target weights $\theta_i$.

### 3.3 Transformers as Meta-learners

In this section, we introduce the details of our Transformer hypernetwork. We use Transformers to directly build the whole weights $\theta$ by transferring the knowledge from encoded information of observations $O$. Our method is demonstrated in Fig. 3, in general, it represents the observations as data tokens and decodes them to weight tokens, that each weight token corresponds to some locations in the INR weights.

In practice, the observation set usually consists of images (or with given poses). We follow a similar strategy as in Vision Transformer [14], where the images are split into patches. The patches are flattened and then mapped by a fully connected (FC) layer to vectors in the dimension of the input to the Transformer. We denote these vectors as data tokens, i.e. the tokens that represent the observation data, which are the blue input tokens in Fig. 3.

To decode for the whole INR weights $\theta = \{W_i\}_{i=0}^{m-1}$, we view each weight matrix $W_i$ as a set of column vectors, and $\theta$ can be represented as the joint of the

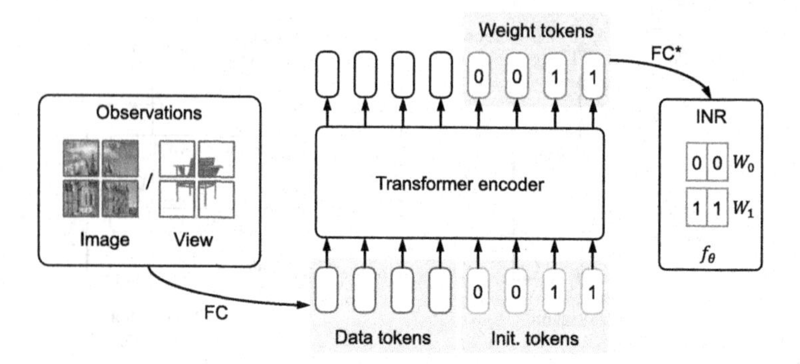


**Fig. 3. Transformers as meta-learners.** We propose to use a Transformer encoder as the meta-learner that directly builds the whole weights of an INR from given observations. The observations are split into patches and mapped to data tokens by a fully connected (FC) layer. The INR weights are viewed as the set of column vectors in weight matrices. For each column vector, we create a corresponding initialization token at the input. The data tokens and the initialization tokens are passed together into the Transformer encoder. The weight tokens are generated at the output and are mapped to the column vectors in INR weights with layerwise FCs (denoted by FC*).

column vector sets. For each of these column vectors, we create an initialization token (which is a learnable vector parameter) correspondingly at the input for the Transformer. In Fig. 3, they are illustrated as green tokens.

These initialization tokens and data tokens are passed together into the Transformer encoder, which jointly models: (i) building features of the observations through interactions in data tokens; (ii) transferring the knowledge of observations to the weights through interactions between data tokens and initialization tokens; (iii) the relation of different weights in INR through interactions in initialization tokens.

Finally, the output vectors at the positions of the input initialization tokens are denoted as weights tokens, which are shown in Fig. 3 as the tokens in orange color. To map them into the INR weights, since the dimensions of the column vectors in $W_i$ can be different for different $i$, we have $m$ independent FC layers for each $i \in \{0, \dots, m-1\}$ that maps the weight tokens to their corresponding column vectors in $W_i$, which gets whole INR weights $\theta = \{W_i\}_{i=0}^{m-1}$.

To train this Transformer meta-learner, a loss is computed with regard to the INR weights $\theta$. Let $O$ denotes the observations from which we generate $\theta$, for the optimization goal of the meta-learner, the loss can be defined as $L(\theta; O)$ in Eq. 1. In tasks that require improving the generalization of the INR $f_\theta$ (e.g. view synthesis from a single input image), we sample $O' \neq O$ from the training set and compute the loss $L(\theta; O')$ instead. $L(\theta; O')$ requires the estimated $f_\theta$ to generalize to unseen observations, which explicitly adds generalization of INR as an objective.

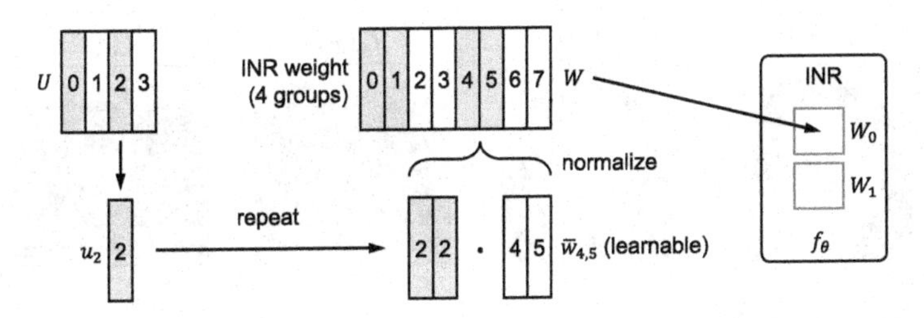


**Fig. 4. Weight Grouping.** Columns in weight matrix $W$ are divided into groups, each group can be generated by a single vector. $\bar{w}_i$ are learnable vectors assigned for every column in $W$, which are independent of the input observations.

### 3.4 Weight Grouping

Assigning tokens for each column vector in weight matrices might be inefficient when the size of $\theta$ is large. To improve the efficiency and scalability of our Transformer meta-learner, we present a weight grouping strategy that offers control for the trade-off of precision and cost.

The general idea is to divide the columns in a weight matrix into groups and assign a single token for each group, as illustrated in Fig. 4. Specifically, let $W \in \theta$ denotes a weight matrix that can be viewed as column vectors $W = [w_0 \dots w_{r-1}]$. For weight grouping with a group size of $k$, $W$ will be defined by a new set of column vectors $U = [u_0 \dots u_{r/k-1}]$ (assume $r$ is divisible by $k$). Specifically, $w_i$ is defined by $u_{\lfloor i/k \rfloor}$ with the formula

$$w_i = \text{normalize}(u_{\lfloor i/k \rfloor} \cdot \bar{w}_i),$$

where normalize refers to L2 normalization, $\bar{w}_i$ are the learnable parameters assigned for every weight $w_i$, note that they are independent of the given observations. With this formulation, $U$ will replace $W$ as the new weights for the Transformer meta-learner to build.

The weight grouping strategy will roughly reduce the number of weight tokens by a factor of $k$, which makes it more efficient for the Transformer meta-learner to build the weights while maintaining the representation capacity of the inferred INR $f_\theta$. $\bar{w}_i$ for every column vector $w_i \in W$ are learnable vectors that do not need to be generated by the Transformer, and they make the columns vector within the same group still different from each other.

## 4 Experiments

### 4.1 Image Regression

Image regression is a basic task commonly used for evaluating the representation capacity of INRs in recent works [47,52]. In image regression, a target image $J$ is sampled from an image distribution $J \sim \mathcal{J}$. An INR $f_\theta$ is a neural network that takes as input a 2D coordinate in the image and outputs the RGB value. The

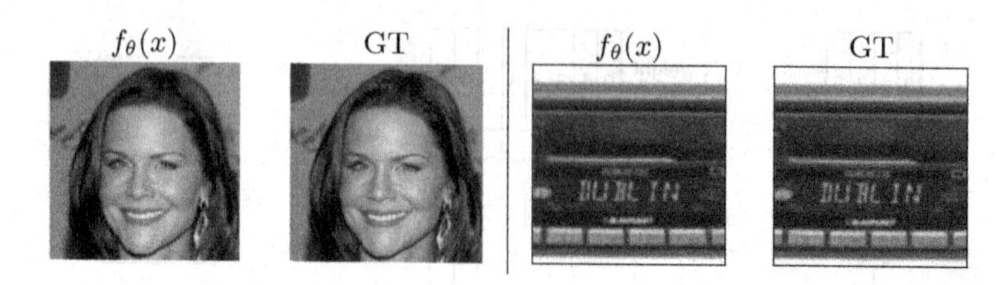


**Fig. 5. Qualitative results of image regression.** Our method builds the weights of $f_\theta$ that fit the observations of the target image and recovers the details of real-world images. Examples are shown on CelebA (left) and Imagenette (right), which are face images and natural images of general objects.

goal is to infer the weights $\theta$ in INR $f_\theta$ for a given target image $J$ so that $f_\theta$ can reconstruct $J$ by outputting the RGB values at the center coordinates of pixels in $J$. Unlike previous works that perform gradient descent steps to optimize the INR weights for given observations, our goal is to use a Transformer to directly generate the INR that can fit the pixel values in the target image without test-time optimization.

**Setup.** We follow the datasets of real-world images used in recent work [52]. *CelebA* [29] is a large-scale dataset of face images. It contains about 202K images of celebrities, which are split into 162K, 20K, and 20K images as training, validation, and test sets. *Imagenette* [22] is a dataset of common objects. It is a subset of 10 classes chosen from the 1K classes in ImageNet [11], which contains about 9K images for training and 4K images for testing.

**Input Encoding.** To apply the Transformer meta-learner to the task of image regression, we will need to encode the given input of the target image to a set of tokens as the Transformer's data tokens. To achieve this, we follow the practice in Vision Transformer (ViT) [14] that split the input image into patches. Specifically, the input image is represented by a set of patches of shape $P \times P$, which are converted to flattened vectors $\{p_i\}_{i=0}^{n_p-1}$ with dimension $P \times P \times 3$ for RGB images. For each patch $p_i$, it is assigned with a learnable positional embedding $e_i$. The $i$-th data token is obtained by $\text{FC}(p_i + e_i)$ with a FC layer.

**Implementation Details.** On the Imagenette dataset, we apply RandomCrop data augmentation for training our Transformer meta-learner. For both CelebA and Imagenette datasets, the resolution of target images is $178 \times 178$ which follows prior practice. We apply a zero-padding of 1 to get the input resolution $180 \times 180$, and split the image with patch size $P = 9$. For INR, we follow the same 5-layer MLP structure as in prior work [52], which has the hidden dimension of 256. The number of groups in weight grouping is 64 by default for a good balance in performance and efficiency. The Transformer follows a similar structure as ViT-Base [14], but we reduce the number of layers by half to 6 layers for efficiency. The networks are trained end-to-end with Adam [26] with a learning rate $1 \cdot 10^{-4}$ and the learning rate decays once by 10 when the loss plateaus.

**Table 1. Quantitative results of image regression (PSNR).** Learned Init is a gradient-based meta-learning algorithm that adapts to an image with a few gradient steps.

| | CelebA | Imagenette |
|---|---|---|
| Learned Init [52] | 30.37 | 27.07 |
| Ours | **31.96** | **29.01** |

**Table 2. Ablations on the number of weight groups.** The PSNR is evaluated on CelebA dataset. Having more groups ($G$) in weight grouping will make the output INR more flexible and help for representing the details (the yellow box in the shown example).

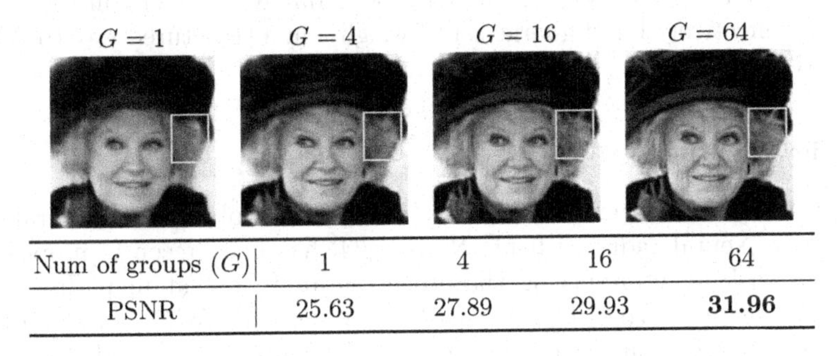


| Num of groups ($G$) | 1 | 4 | 16 | 64 |
|---|---|---|---|---|
| PSNR | 25.63 | 27.89 | 29.93 | **31.96** |

**Qualitative Results.** We first show qualitative results in Fig. 5. We observe that the Transformer meta-learners are surprisingly effective for building INRs of images in high precision, that can even recover the fine details in complex real-world images. For example, the left example from CelebA shows that our inferred INR $f_\theta$ can successfully reconstruct various details in a face image, including the teeth, lighting effect, and even the background patterns which is not a part of faces. From the right figure of Imagenette dataset, we observe that our inferred INR can recover the digital texts on the object with high fidelity. While it is observed in prior work [47] that learning a latent space of vectors and decoding INRs by hypernetwork can not recover the details in a face image, we show that an INR with precise information can be directly built by a Transformer without any gradient computation.

**Quantitative Results.** In Table 1, we show quantitative results of our Transformer meta-learner and compare our performance with the gradient-based meta-learning algorithm Learned Init proposed in prior work. Learned Init meta-learns an initialization that can be quickly adapted to target images within a few gradient steps. On both real-world image datasets, we observe that our method achieves the PSNR at around 30 for image regression, and our method without any gradient computation outperforms prior gradient-based meta-learning. The gradient steps involve the repeated process of forward and backward passing through the whole INR sequentially, while ours can directly build the INR by forwarding the information into a shallow Transformer. In summary, our method provides a precise yet efficient hypernetwork-based way of converting image pixels to a global neural function as their underlying INR.

**Ablations on the Number of Weight Groups.** To justify that the Transformer meta-learner learns about building a complex INR, we show by experiments that the number of groups in weight grouping is not redundant. The qualitative and quantitative results are both shown in Table 2. We observe that by increasing the number of groups $G$ from 1 to 64, the recovered details for the target image are significantly improved in vision, and the PSNR is consistently improving by large margins. The results demonstrate the effectiveness of the weight grouping strategy, and it indicates that the Transformer meta-learner can learn about the complex relations between different weights in the INR so that it can effectively build a large set of weights in a structured way to achieve high precision.

### 4.2 View Synthesis

View synthesis aims at generating a novel view of a 3D object with several given input views. Neural radiance field (NeRF) [34] has been recently proposed to tackle this task by representing the object as an INR that maps from a 3D coordinate and a viewing direction to the density and RGB value. With the volumetric rendering, the generated views of NeRF are differentiable to the INR weights. View synthesis can be then achieved by first fitting INR for the given input views, then rendering the INR from novel views. The goal of a meta-learner is to infer the INR from given input views efficiently, and improve its generalization so that view synthesis can be achieved with fewer input views.

**Setup.** We perform view synthesis on objects from ShapeNet [6] dataset. We follow prior work [52] which considers 3 categories: chairs, cars, and lamps. For each category, the objects are split into two sets for training and test, where for each object 25 views (with known camera pose) are provided for training. During testing, a random input view is sampled for evaluating the performance of novel view synthesis.

**Input Encoding.** For each input view image, given the known camera pose, we first compute the ray emitted from every pixel for rendering. The emitted ray at each pixel can be represented as a 3D starting point and a 3D direction vector (normalized as a unit vector). With the original image which has RGB channels, we concatenate all the information at every pixel, which gets an extended image with 9 channels. The extended image contains all the information about an input view, therefore, it can be then split into patches and mapped to data tokens in the Transformer meta-learner. This representation naturally generalizes to multiple input views. Since the information of a single view is represented by a set of patches, when multiple input views are available, their data tokens can be simply merged as a set for representing all the observation information for passing into the Transformer.

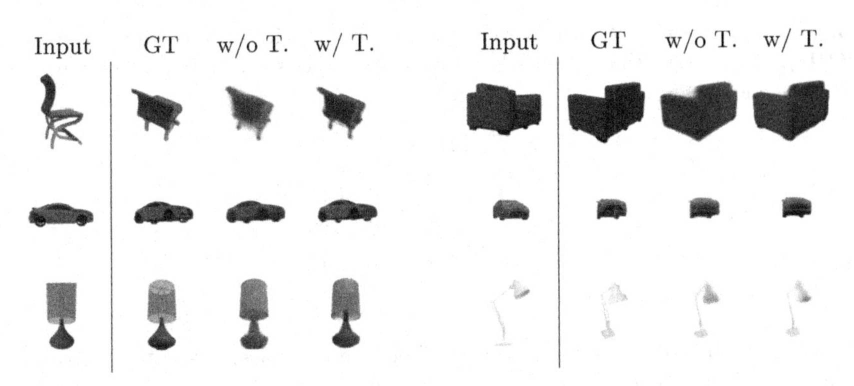


**Fig. 6. View synthesis with Transformer meta-learner on ShapeNet.** The rows show for chairs, cars, and lamps categories. "w/o T." denotes the results of using the Transformer to infer the INR weights without test-time optimization. "w/ T." performs a few test-time optimization steps on the generated INR for the sparse input views, which further helps to reconstruct the fine details in the input views. The corresponding quantitative results are shown in Table 4.

**Table 3. Comparison of building INR for single image view synthesis (PSNR).** The compared methods are baselines and the gradient-based meta-learning algorithm in prior work. Ours does not perform test-time optimization.

| | Chairs | Cars | Lamps |
|---|---|---|---|
| NeRF [34] (Standard [52]) | 12.49 | 11.45 | 15.47 |
| Matched [52] | 16.40 | 22.39 | 20.79 |
| Shuffled [52] | 10.76 | 11.30 | 13.88 |
| Learned Init [52] | 18.85 | 22.80 | 22.35 |
| Ours | **19.66** | **23.78** | **22.76** |

**Adaptive Sampling.** To improve the training stability, we propose an adaptive sampling strategy for the first training epoch. Specifically, when we sample the pixel locations for computing the reconstruction loss, we ensure that half of them are sampled from the foreground of the image. This is implemented by selecting the non-white pixels as the background in ShapeNet image is white. We found that the training process is stable after having this simple sampling strategy.

**Implementation Details.** In ShapeNet, input views are provided in resolution $128 \times 128$. We split input views with patch size 8 for the Transformer input. We use NeRF as the INR representation, it follows the architecture in [52] which consists of 6 layers with the hidden dimension of 256 and does not use "coarse" and "fine" models for simplicity. The architecture of the Transformer and the optimizer are the same as the experiments for image regression.

**Results.** We first compare our method to the prior gradient-based meta-learning algorithm of building INR for single image view synthesis, the results are shown

**Table 4. Effect of test-time optimization in single image view synthesis (PSNR).** We observe that our method for view synthesis can naturally take benefits from test-time optimization.

| | w/o T. | w/T. |
|---|---|---|
| Chairs | 19.66 | **20.56** |
| Cars | 23.78 | **24.73** |
| Lamps | 22.76 | **24.71** |

in Table 3. Standard, Matched, and Shuffled are the baselines trained from different initializations from the prior work [52]. Specifically, Standard denotes a random initialization, Matched is the initialization learned from scratch which matches the output of the meta-learned initialization, Shuffled is permuting the weights in the meta-learned initialization. We observe that our method outperforms the baselines and the gradient-based meta-learning algorithm for inferring the weights in an INR.

Our method can also naturally take benefits from test-time optimization. The qualitative and quantitative results are shown in Fig. 6 and Table 4. We observe that the Transformer meta-learner can effectively build the INR of a 3D object with sparse input views. Since our method builds the whole INR, we can perform further test-time optimization on the INR with given input views just as the original training in NeRF. For efficiency, our test-time optimization only contains 100 gradient steps, it further helps for constructing the fine details in input views.

## 5 Does the INR Exploit Data Structures?

A key potential advantage of INRs is that their representation capacity does not depend on grid resolution but instead on the capacity of the neural network, which allows it to exploit the underlying structures in data and reduce the representation redundancy. To explore whether the structure is modeled in INRs, we visualize the attention weights at the last layer from the weight tokens to the data tokens. Intuitively, since each data token corresponds to a patch in the original image, the attention weights may represent that, for weight columns in different layers, which part of the original image they mostly depend on.

We reshape the attention weights to the 2D grid of patches and bilinearly up-sample the grid to a mask with the same resolution as the input image. We mask the input image so that parts with higher attention will be shown, the visualization results on CelebA dataset are shown in Fig. 7. We observe that there exist weight columns in different layers that attend to structured parts. For example, there are weights roughly attending to the nose and mouth in layer 1, the forehead in layer 2, and the whole face in layer 3. Our observations suggest that the generated INRs may potentially capture the structure of data, which is different from the grid-based discrete representation, where every entry independently represents a pixel and the data structure is not well exploited.

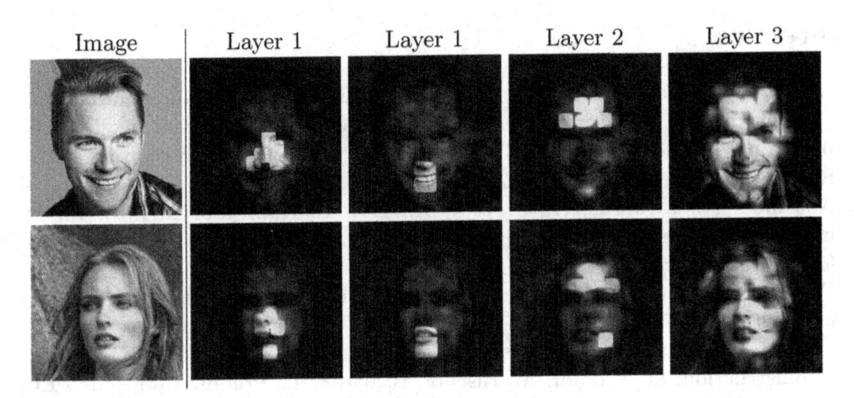


**Fig. 7. Attention masks from weight tokens to data tokens.** Representative examples are selected from tokens for different INR layers. The attention map shows where the corresponding INR weight is attending to.

## 6 Conclusion

In this work, we proposed a simple and general formulation that uses Transformers as meta-learners for building neural functions of INRs, which opens up a promising direction with new possibilities. While most of the prior works of hypernetworks for INRs are based on single-vector modulation and high precision reconstruction as a global INR function was mostly achieved by gradient-based meta-learning, our proposed Transformer hypernetwork can efficiently build an INR in one forward pass without any gradient steps, and we observed it can outperform the previous gradient-based meta-learning algorithms for building INRs in the tasks of image regression and view synthesis. While gradient information is not necessary for our model, our method simply builds the weights in a standard INR, therefore it is also flexible to be further combined with any INRs that involve test-time optimization.

Our method draws connections between the Transformer hypernetworks and the gradient-based meta-learning algorithms, and our further analysis sheds light on the generated INRs. We observed that the INR which represents data as a global function may potentially capture the underlying structures without any explicit supervision. Understanding and utilizing these encoded structures could be a promising direction for future works.

**Acknowledgement.** This work was supported, in part, by grants from DARPA LwLL, NSF CCF-2112665 (TILOS), NSF 1730158 CI-New: Cognitive Hardware and Software Ecosystem Community Infrastructure (CHASE-CI), NSF ACI-1541349 CC*DNI Pacific Research Platform, and gifts from Meta, Google, Qualcomm and Picsart.

## References


1. Anokhin, I., Demochkin, K., Khakhulin, T., Sterkin, G., Lempitsky, V., Korzhenkov, D.: Image generators with conditionally-independent pixel synthesis. In: Proceedings of the IEEE/CVF Conference on Computer Vision and Pattern Recognition, pp. 14278–14287 (2021)
2. Antoniou, A., Edwards, H., Storkey, A.: How to train your MAML. In: International Conference on Learning Representations (2019). http://openreview.net/forum?id=HJGven05Y7
3. Brown, T.B., et al.: Language models are few-shot learners. arXiv preprint arXiv:2005.14165 (2020)
4. Chabra, R., et al.: Deep local shapes: learning local SDF priors for detailed 3D recolnstruction. In: Vedaldi, A., Bischof, H., Brox, T., Frahm, J.-M. (eds.) ECCV 2020. LNCS, vol. 12374, pp. 608–625. Springer, Cham (2020). https://doi.org/10.1007/978-3-030-58526-6_36
5. Chan, E.R., Monteiro, M., Kellnhofer, P., Wu, J., Wetzstein, G.: Pi-GAN: periodic implicit generative adversarial networks for 3D-aware image synthesis. In: Proceedings of the IEEE/CVF Conference on Computer Vision and Pattern Recognition, pp. 5799–5809 (2021)
6. Chang, A.X., et al.: ShapeNet: an information-rich 3d model repository. arXiv preprint arXiv:1512.03012 (2015)
7. Chen, Y., Liu, S., Wang, X.: Learning continuous image representation with local implicit image function. In: Proceedings of the IEEE/CVF Conference on Computer Vision and Pattern Recognition, pp. 8628–8638 (2021)
8. Chen, Z., Zhang, H.: Learning implicit fields for generative shape modeling. In: Proceedings of the IEEE/CVF Conference on Computer Vision and Pattern Recognition (CVPR) (2019)
9. Chibane, J., Alldieck, T., Pons-Moll, G.: Implicit functions in feature space for 3D shape reconstruction and completion. In: Proceedings of the IEEE/CVF Conference on Computer Vision and Pattern Recognition, pp. 6970–6981 (2020)
10. Deng, B., et al.: NASA neural articulated shape approximation. In: Vedaldi, A., Bischof, H., Brox, T., Frahm, J.-M. (eds.) ECCV 2020. LNCS, vol. 12352, pp. 612–628. Springer, Cham (2020). https://doi.org/10.1007/978-3-030-58571-6_36
11. Deng, J., Dong, W., Socher, R., Li, L.J., Li, K., Fei-Fei, L.: ImageNet: a large-scale hierarchical image database. In: 2009 IEEE conference on computer vision and pattern recognition, pp. 248–255. IEEE (2009)
12. Devlin, J., Chang, M.W., Lee, K., Toutanova, K.: BERT: pre-training of deep bidirectional transformers for language understanding. In: Proceedings of the 2019 Conference of the North American Chapter of the Association for Computational Linguistics: Human Language Technologies, Volume 1 (Long and Short Papers), pp. 4171–4186. Association for Computational Linguistics, Minneapolis (2019). https://doi.org/10.18653/v1/N19-1423, http://aclanthology.org/N19-1423
13. DeVries, T., Bautista, M.A., Srivastava, N., Taylor, G.W., Susskind, J.M.: Unconstrained scene generation with locally conditioned radiance fields. arXiv preprint arXiv:2104.00670 (2021)
14. Dosovitskiy, A., et al.: An image is worth 16x16 words: transformers for image recognition at scale. In: International Conference on Learning Representations (2021). http://openreview.net/forum?id=YicbFdNTTy
15. Fallah, A., Mokhtari, A., Ozdaglar, A.: On the convergence theory of gradient-based model-agnostic meta-learning algorithms. In: International Conference on Artificial Intelligence and Statistics, pp. 1082–1092. PMLR (2020)

16. Fernando, C., et al.: Meta-learning by the baldwin effect. In: Proceedings of the Genetic and Evolutionary Computation Conference Companion, pp. 1313–1320 (2018)
17. Finn, C., Abbeel, P., Levine, S.: Model-agnostic meta-learning for fast adaptation of deep networks. In: International Conference on Machine Learning, pp. 1126–1135. PMLR (2017)
18. Genova, K., Cole, F., Sud, A., Sarna, A., Funkhouser, T.: Local deep implicit functions for 3D shape. In: Proceedings of the IEEE/CVF Conference on Computer Vision and Pattern Recognition, pp. 4857–4866 (2020)
19. Genova, K., Cole, F., Vlasic, D., Sarna, A., Freeman, W.T., Funkhouser, T.: Learning shape templates with structured implicit functions. In: Proceedings of the IEEE/CVF International Conference on Computer Vision, pp. 7154–7164 (2019)
20. Ha, D., Dai, A.M., Le, Q.V.: Hypernetworks. In: ICLR (2017)
21. Hospedales, T., Antoniou, A., Micaelli, P., Storkey, A.: Meta-learning in neural networks: a survey. arXiv preprint arXiv:2004.05439 (2020)
22. Howard, J.: Imagenette. http://github.com/fastai/imagenette (2020)
23. Jaderberg, M., et al.: Human-level performance in 3D multiplayer games with population-based reinforcement learning. Science **364**(6443), 859–865 (2019)
24. Jiang, C., et al.: Local implicit grid representations for 3D scenes. In: Proceedings of the IEEE/CVF Conference on Computer Vision and Pattern Recognition, pp. 6001–6010 (2020)
25. Karras, T., et al.: Alias-free generative adversarial networks. arXiv preprint arXiv:2106.12423 (2021)
26. Kingma, D.P., Ba, J.: Adam: a method for stochastic optimization. arXiv preprint arXiv:1412.6980 (2014)
27. Liu, L., Gu, J., Lin, K.Z., Chua, T.S., Theobalt, C.: Neural sparse voxel fields. arXiv preprint arXiv:2007.11571 (2020)
28. Liu, Z., et al.: Swin transformer: hierarchical vision transformer using shifted windows. In: International Conference on Computer Vision (ICCV) (2021)
29. Liu, Z., Luo, P., Wang, X., Tang, X.: Deep learning face attributes in the wild. In: Proceedings of the IEEE International Conference on Computer Vision, pp. 3730–3738 (2015)
30. Martin-Brualla, R., Radwan, N., Sajjadi, M.S., Barron, J.T., Dosovitskiy, A., Duckworth, D.: NeRF in the wild: neural radiance fields for unconstrained photo collections. In: Proceedings of the IEEE/CVF Conference on Computer Vision and Pattern Recognition, pp. 7210–7219 (2021)
31. Mehta, I., Gharbi, M., Barnes, C., Shechtman, E., Ramamoorthi, R., Chandraker, M.: Modulated periodic activations for generalizable local functional representations. In: Proceedings of the IEEE/CVF International Conference on Computer Vision, pp. 14214–14223 (2021)
32. Mescheder, L., Oechsle, M., Niemeyer, M., Nowozin, S., Geiger, A.: Occupancy networks: learning 3D reconstruction in function space. In: Proceedings of the IEEE/CVF Conference on Computer Vision and Pattern Recognition (CVPR) (2019)
33. Michalkiewicz, M., Pontes, J.K., Jack, D., Baktashmotlagh, M., Eriksson, A.: Implicit surface representations as layers in neural networks. In: Proceedings of the IEEE/CVF International Conference on Computer Vision, pp. 4743–4752 (2019)
34. Mildenhall, B., Srinivasan, P.P., Tancik, M., Barron, J.T., Ramamoorthi, R., Ng, R.: NeRF: representing scenes as neural radiance fields for view synthesis. In: Vedaldi, A., Bischof, H., Brox, T., Frahm, J.-M. (eds.) ECCV 2020. LNCS, vol.

12346, pp. 405–421. Springer, Cham (2020). https://doi.org/10.1007/978-3-030-58452-8_24
35. Mishra, N., Rohaninejad, M., Chen, X., Abbeel, P.: A simple neural attentive meta-learner. In: International Conference on Learning Representations (2018). http://openreview.net/forum?id=B1DmUzWAW
36. Nichol, A., Achiam, J., Schulman, J.: On first-order meta-learning algorithms. arXiv preprint arXiv:1803.02999 (2018)
37. Park, J.J., Florence, P., Straub, J., Newcombe, R., Lovegrove, S.: DeepSDF: learning continuous signed distance functions for shape representation. In: Proceedings of the IEEE/CVF Conference on Computer Vision and Pattern Recognition (CVPR) (2019)
38. Park, K., et al.: Deformable neural radiance fields. arXiv preprint arXiv:2011.12948 (2020)
39. Peng, S., Niemeyer, M., Mescheder, L., Pollefeys, M., Geiger, A.: Convolutional occupancy networks. In: Vedaldi, A., Bischof, H., Brox, T., Frahm, J.-M. (eds.) ECCV 2020, Part III. LNCS, vol. 12348, pp. 523–540. Springer, Cham (2020). https://doi.org/10.1007/978-3-030-58580-8_31
40. Radford, A., Narasimhan, K.: Improving language understanding by generative pre-training. Preprint (2018)
41. Radford, A., Wu, J., Child, R., Luan, D., Amodei, D., Sutskever, I.: Language models are unsupervised multitask learners. Preprint (2019)
42. Rajeswaran, A., Finn, C., Kakade, S.M., Levine, S.: Meta-learning with implicit gradients. In: NeurIPS (2019)
43. Ravi, S., Larochelle, H.: Optimization as a model for few-shot learning. In: In International Conference on Learning Representations (ICLR) (2017)
44. Saito, S., Huang, Z., Natsume, R., Morishima, S., Kanazawa, A., Li, H.: PiFu: pixel-aligned implicit function for high-resolution clothed human digitization. In: Proceedings of the IEEE/CVF International Conference on Computer Vision, pp. 2304–2314 (2019)
45. Schwarz, K., Liao, Y., Niemeyer, M., Geiger, A.: GRAF: generative radiance fields for 3d-aware image synthesis. arXiv preprint arXiv:2007.02442 (2020)
46. Sitzmann, V., Chan, E.R., Tucker, R., Snavely, N., Wetzstein, G.: MetaSDF: meta-learning signed distance functions. In: Proceedings of NeurIPS (2020)
47. Sitzmann, V., Martel, J.N., Bergman, A.W., Lindell, D.B., Wetzstein, G.: Implicit neural representations with periodic activation functions. In: Proceedings of NeurIPS (2020)
48. Sitzmann, V., Zollhöfer, M., Wetzstein, G.: Scene representation networks: continuous 3D-structure-aware neural scene representations. In: Advances in Neural Information Processing Systems (2019)
49. Skorokhodov, I., Ignatyev, S., Elhoseiny, M.: Adversarial generation of continuous images. In: Proceedings of the IEEE/CVF Conference on Computer Vision and Pattern Recognition, pp. 10753–10764 (2021)
50. Snell, J., Swersky, K., Zemel, R.: Prototypical networks for few-shot learning. In: Guyon, I., et al. (eds.) Advances in Neural Information Processing Systems, vol. 30. Curran Associates, Inc. (2017). http://proceedings.neurips.cc/paper/2017/file/cb8da6767461f2812ae4290eac7cbc42-Paper.pdf
51. Sung, F., Yang, Y., Zhang, L., Xiang, T., Torr, P.H., Hospedales, T.M.: Learning to compare: relation network for few-shot learning. In: Proceedings of the IEEE Conference on Computer Vision and Pattern Recognition, pp. 1199–1208 (2018)
52. Tancik, M., et al.: Learned initializations for optimizing coordinate-based neural representations. In: CVPR (2021)

53. Tancik, M., et al.: Fourier features let networks learn high frequency functions in low dimensional domains. In: NeurIPS (2020)
54. Touvron, H., Cord, M., Douze, M., Massa, F., Sablayrolles, A., Jégou, H.: Training data-efficient image transformers & distillation through attention. In: International Conference on Machine Learning, pp. 10347–10357. PMLR (2021)
55. Vaswani, A., et al.: Attention is all you need. In: Advances in Neural Information Processing Systems, pp. 5998–6008 (2017)
56. Vinyals, O., Blundell, C., Lillicrap, T., Wierstra, D., et al.: Matching networks for one shot learning. Adv. Neural. Inf. Process. Syst. **29**, 3630–3638 (2016)
57. Yu, A., Ye, V., Tancik, M., Kanazawa, A.: pixelNeRF: neural radiance fields from one or few images. In: Proceedings of the IEEE/CVF Conference on Computer Vision and Pattern Recognition, pp. 4578–4587 (2021)

# Style Your Hair: Latent Optimization for Pose-Invariant Hairstyle Transfer via Local-Style-Aware Hair Alignment

Taewoo Kim, Chaeyeon Chung, Yoonseo Kim, Sunghyun Park, Kangyeol Kim, and Jaegul Choo(✉)

Korea Advanced Institute of Science and Technology, Daejeon, South Korea
{specia1ktu,cy_chung,grandchasevs,psh01087,kangyeolk,jchoo}@kaist.ac.kr

**Abstract.** Editing hairstyle is unique and challenging due to the complexity and delicacy of hairstyle. Although recent approaches significantly improved the hair details, these models often produce undesirable outputs when a pose of a source image is considerably different from that of a target hair image, limiting their real-world applications. HairFIT, a pose-invariant hairstyle transfer model, alleviates this limitation yet still shows unsatisfactory quality in preserving delicate hair textures. To solve these limitations, we propose a high-performing pose-invariant hairstyle transfer model equipped with latent optimization and a newly presented local-style-matching loss. In the StyleGAN2 latent space, we first explore a pose-aligned latent code of a target hair with the detailed textures preserved based on local style matching. Then, our model inpaints the occlusions of the source considering the aligned target hair and blends both images to produce a final output. The experimental results demonstrate that our model has strengths in transferring a hairstyle under larger pose differences and preserving local hairstyle textures. The codes are available at https://github.com/Taeu/Style-Your-Hair.

**Keywords:** Hairstyle transfer · Latent optimization · Conditional image generation

## 1 Introduction

With the advance of conditional generative adversarial networks (GANs) [8,12,18], editing facial attributes has drawn great attention and shows a promising result on editing multiple attributes. Despite the success, modifying strongly correlated facial attributes is still challenging, often beyond the capacity of existing editing models. In this paper, we focus on hairstyle editing, which aims at transferring a target hairstyle to a source image, proposing high-performance neural networks to solve the problem. Hairstyle editing is similar to that of a

T. Kim, C. Chung and Y. Kim—Equal contributions.

**Supplementary Information** The online version contains supplementary material available at https://doi.org/10.1007/978-3-031-19790-1_12.

© The Author(s), under exclusive license to Springer Nature Switzerland AG 2022
S. Avidan et al. (Eds.): ECCV 2022, LNCS 13677, pp. 188–203, 2022.
https://doi.org/10.1007/978-3-031-19790-1_12

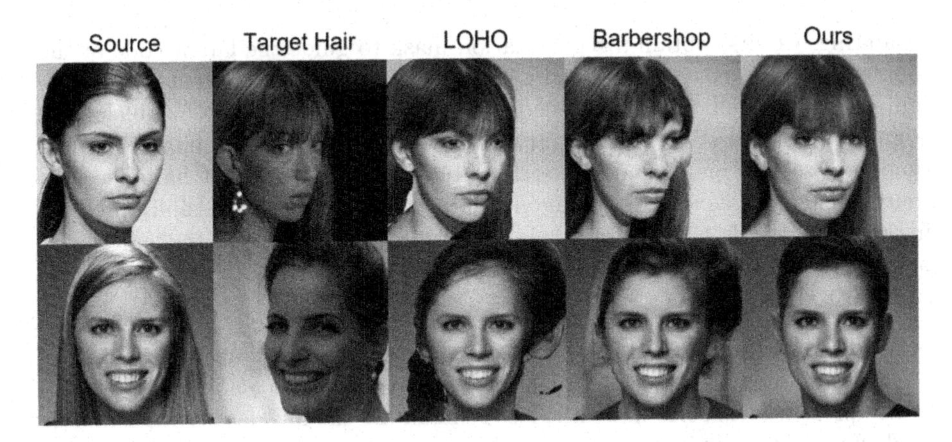


**Fig. 1.** Our model produces more realistic results compared to LOHO [20] and Barbershop [31] even with a large pose difference between a source and a target hair.

facial attribute, but it has unique, challenging aspects: (1) Due to the hairstyle's complexity and delicacy, preserving its strands given an arbitrary hairstyle is highly demanding. (2) Transferred hairstyle requires to be exactly fitted to a given source image. These challenges make the previous approaches for editing the specified facial attributes less suitable for this problem.

Recent solutions for hairstyle transfer address the problem with the power of a pre-trained image generator. For example, LOHO [20] and Barbershop [31] largely enhance the visual quality of the generated images via latent optimization based on StyleGAN2 [14]. However, these approaches produce undesirable outputs (See Fig. 1) when handling a target and source image pair with a significant pose difference.

To the best of our knowledge, HairFIT [5] is the only work to address the pose difference issue between a source and target image. HairFIT presents a pose-invariant hairstyle transfer model where a target hairstyle is aligned to a source image pose using a flow-based warping module trained on multi-view datasets such as VoxCeleb [17] and K-hairstyle dataset [15]. Although its attempt, HairFIT requires a high-quality multi-view hairstyle dataset during training, and it falls behind state-of-the-art models [20,31] in light of hair preserving capacity.

In response to these limitations, we present a framework that performs a *high-quality* pose-invariant hairstyle transfer based on latent optimization *without multi-view dataset*. Specifically, given a source and a target hair image, our model generates a hair-transferred output through embedding, hair pose alignment, inpainting, and blending step. We first take advantage of GAN-inversion algorithms [1,2,32], feeding a source and a target hair image, for the purpose of obtaining latent codes residing in the StyleGAN2 space [14], respectively. Next, we navigate the StyleGAN2 space to optimize the latent code of the target hair image to follow a source image pose. During the pose alignment, we utilize a newly-presented local-style-matching loss to penalize visually degraded hair textures by locally comparing the original target hair with the aligned one. In the inpainting, we first obtain a segmentation mask to guide the latent code of the source to fill the occluded regions by its hair. We optimize the latent code of the source image to follow the obtained segmentation mask. Lastly, we blend the aligned target hairstyle and the inpainted source image with a final optimization step. In this manner, our model is able to transfer a target hairstyle to a source image overcoming the difference in poses as well as successfully preserving the fine details of the target hair. Experiments demonstrate the superiority of our model in a quantitative and qualitative manner. Our contributions are summarized as follows:

- We propose a framework that achieves a *high-quality* pose-invariant hairstyle transfer based on latent optimization *without multi-view dataset.*
- We present local-style-matching loss to maintain the fine details of a hairstyle during pose optimization.
- Our model achieves state-of-the-art performance in quantitative and qualitative evaluations with various datasets.

## 2 Related Work

### 2.1 Latent Space Manipulation

With the understanding of the latent space in GANs, recent approaches based on latent space manipulation [9,21,24] have shown promising results in image editing. For example, GANSpace [9] and InterfaceGAN [21] modify facial attributes via manipulation in the latent space of StyleGAN [13]. While the former takes advantage of principal component analysis, the latter utilizes semantic scores to identify disentangled directions related to the target attributes. In a similar manner, Viazovetskyi et al. [24] and Zhuang et al. [33] attempt to edit the images by shifting latent vectors to semantically meaningful directions in the latent space of StyleGAN2 [14], which can easily be obtained by a pre-trained face classifier or learned transformations.

Recent hairstyle transfer approaches also actively utilize latent space manipulation to synthesize high-quality images. LOHO [20] and Barbershop [31] edit hairstyles by manipulating the extended StyleGAN2 latent space [32] via latent optimization. These methods not only significantly enhance the visual quality of the generated images but also preserve the semantic details of the target images. In particular, Barbershop introduces $FS$ space with a larger capacity than the original StyleGAN2 latent space, where the original hair structure is well-preserved. In this work, we leverage latent optimization to reach the photo-realistic image quality. Our model mainly focuses on the pose alignment of a target hairstyle to a source image without losing its detailed hair texture based on local-style-matching loss to achieve a pose-invariant hairstyle transfer.

### 2.2 Hairstyle Transfer

GAN-based facial image editing [11,12,16,19,26,27] successfully modifies the target facial attributes such as a facial expression or makeup style while maintaining other features. Common approaches for facial image editing are to utilize hand-drawn sketches [12,19,26,27] or user-edited semantic masks [16] as the conditions to precisely guide the manipulated appearance.

In spite of the remarkable progress in facial image editing, hairstyle transfer is still tricky, considering the diversity and intricacy of hairstyles. In practice, a hairstyle transfer is required to convey a wide range of target hairstyles to a given source image while preserving their subtle hair strands and color. As a prior work, MichiGAN [23] presents a hairstyle transfer framework aiming to preserve the detailed textures of a target hairstyle. Specifically, MichiGAN leverages different conditional generators responsible for decomposed hairstyle attributes (*i.e.*, hair shape, and appearance). Moreover, LOHO [20] achieves visually pleasant image quality through latent optimization and hair-related losses for reflecting a target hairstyle features. Barbershop [31] also proposes a latent optimization approach and further improves the visual quality of the outputs based on the newly presented $FS$ space. Barbershop utilizes $F$ tensor in $FS$ space to enhance the capability of preserving the overall structure of a target image, including delicate hair structure.

However, since the existing approaches have been developed to handle the images, where the head poses of a source and a target are aligned, they show limited generalization capacity for dealing with the inputs having a large pose difference. To tackle this problem, HairFIT [5] introduces a pose-invariant hairstyle transfer with flow-based target hair warping and semantic-region-aware inpainting. HairFIT leverages an optical flow estimation network and a multi-view hairstyle dataset [15] to align the target hair to the source face. Despite the aid of the high-quality multi-view dataset, the model fails to preserve the detailed features of hairstyles comprehensively. In this paper, we propose a novel latent optimization framework for pose-invariant hairstyle transfer to synthesize high-quality images regardless of the pose differences.

# 3 Method

## 3.1 Overview

Our framework consists of several optimization steps described in Fig. 2. We first find latent codes $w \in \mathbb{R}^{18\times512}$ of a source image $\mathbf{I}_{src} \in \mathbb{R}^{C\times H\times W}$ and a target hairstyle image $\mathbf{I}_{trg}$ in $W+$ space using the existing GAN-inversion algorithms [31,32]. Then, we optimize the target hair latent codes to have the pose aligned to $\mathbf{I}_{src}$. While aligning the pose, we mainly focus on preserving fine details of the target hair with a newly-presented local-style-matching loss. Local-style-matching loss allows preserving each local texture in the aligned target hair by matching the corresponding region of a similar style from the original target hair. For the next step, we inpaint the source regions occluded by its original hair by optimizing the source latent codes. Lastly, we blend the aligned target hairstyle and the inpainted source image for the final output.

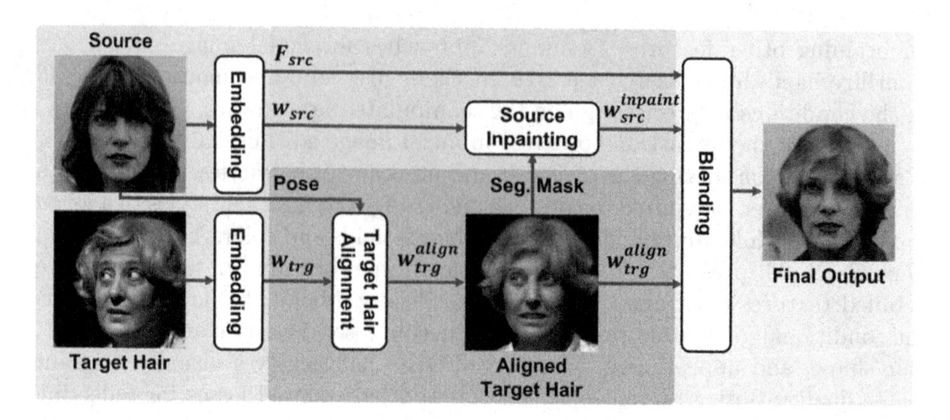


**Fig. 2.** An overview of our framework. First, we obtain $w_{src}$, $w_{trg}$, and $F_{src}$ by embedding a source and a target hair image into $W+$ and $FS$ space. Then, we optimize $w_{trg}$ to follow the source pose, resulting in $w_{trg}^{align}$. With the segmentation mask of aligned target hair, we find $w_{src}^{inpaint}$, where the source occlusions are inpainted. Finally, we blend $F_{src}$, $w_{src}^{inpaint}$, and $w_{trg}^{align}$ to generate the final output.

### 3.2 Embedding

First of all, we obtain the latent codes of each reference image (*i.e.*, source and target images) before pose alignment and blending. Given a source image $\mathbf{I}_{src}$ and a target hair image $\mathbf{I}_{trg}$, we find the source latent codes $w_{src}$ and the target latent codes $w_{trg}$ in an extended latent space of StyleGAN2 denoted as $W+$ space [1]. We employ an improved embedding algorithm [32] to enhance the reconstruction and editing quality. Moreover, we embed $\mathbf{I}_{src}$ to $FS$ space following Barbershop [31] to gain $\mathbf{F}_{src} \in \mathbb{R}^{32\times32\times512}$, which preserves the detailed structure of the source image by encoding the spatial information.

### 3.3 Target Hair Alignment

To transfer the hairstyle regardless of the pose differences, we align the target hairstyle to the source face via the latent optimization, as presented in Fig. 3. Starting from $w_{trg}$, we aim to find $w_{trg}^{align}$, where the head pose and face shape are aligned to $\mathbf{I}_{src}$, while other features, especially the hairstyle, correspond to $\mathbf{I}_{trg}$. We optimize the first $m$ style vectors among 18 style vectors of $w_{trg}$ to optimize coarse style vectors rather than fine style vectors [13]. We set $m$ as 6 in our experiments.

**Pose Align Loss.** To modify $w_{trg}$ to have a source pose, we propose a novel pose align loss $\mathcal{L}_{pose}$ based on 3D facial keypoints. Since the hairstyle significantly depends on other facial features (*e.g.*, face shape and location of eyes), the head pose alone is insufficient to fully guide the target hair alignment. Thus, we leverage 3D facial keypoints, which effectively represent the overall facial

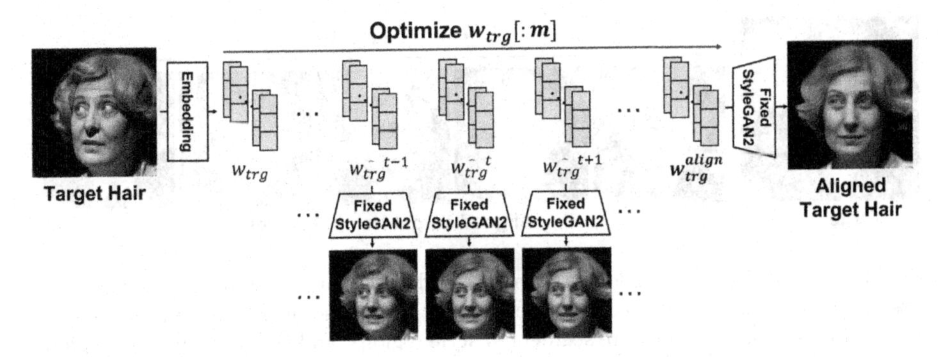


**Fig. 3.** Target hair alignment. We obtain the aligned target hair latent codes $w_{trg}^{align}$ by optimizing the first $m$ vectors of $w_{trg}$ to have a source pose with its hairstyle preserved.

features as well as the head pose. With the source 3D facial keypoints, we can provide detailed supervision of which shape and pose $w_{trg}$ should pursue.

$\mathcal{L}_{pose}$ computes the L2 distance between the 3D keypoint heatmaps of $\mathbf{I}_{src}$ and the aligned target hair image as:

$$\mathcal{L}_{pose} = \frac{1}{N_H} \|\mathbf{H}_{src} - E(G(\hat{w_{trg}}))\|_2^2. \tag{1}$$

$N_H$ indicates the number of elements in a 3D keypoint heatmap $\mathbf{H} \in \mathbb{R}^{68 \times H \times W}$ and $\mathbf{H}_{src} = E(\mathbf{I}_{src})$, where $E$ is a pre-trained keypoint extractor [4]. $G$ is a pre-trained StyleGAN2 generator and $\hat{w_{trg}}$ indicates the optimized $w_{trg}$ in progress.

**Local-Style-Matching Loss.** To preserve locally distinct hairstyles, we newly present a local-style-matching loss, which matches similar local styles between the target hair and the aligned target hair. Basically, we utilize a style loss based on the Gram matrix [7], which captures the repeated patterns (*i.e.*, texture) of given features. A style loss $\mathcal{L}_{style}$ measures the L2 distance between the Gram matrix of feature maps extracted by a VGG network [22], formulated as:

$$\mathcal{L}_{style}(\cdot, \cdot) = \frac{1}{V} \sum_{i=1}^{V} \frac{1}{N_{\mathcal{G}^i}} \|\mathcal{G}^i(\mathrm{VGG}^i(\cdot)) - \mathcal{G}^i(\mathrm{VGG}^i(\cdot))\|_2^2, \tag{2}$$

where $V$ indicates the number of VGG layers we use, which are $relu1_2$, $relu2_2$, $relu3_3$, and $relu4_3$ layer of VGG [5,20,31]. Also, $N_{\mathcal{G}^i}$ represents the number of elements in $\mathcal{G}^i$. Here, $\mathcal{G}^i$ and $\mathrm{VGG}^i$ indicate the $i$-th Gram matrix and $i$-th layer of VGG, respectively. $\mathcal{G}^i$ is calculated as $v^{i\intercal} v^i$, where $v^i \in \mathbb{R}^{H^i W^i \times N_{C^i}}$ corresponds to the activation of $\mathrm{VGG}^i$.

In local-style-matching loss $\mathcal{L}_{style}^{LSM}$, we first identify *style regions* each of which includes locally different style and apply $L_{style}$ to each style region, respectively. To identify the style regions, we leverage a simple linear iterative clustering (SLIC) [3]. The SLIC is an algorithm that conducts a K-means clustering based

**Fig. 4.** Local-style-matching loss. During target hair alignment, a local-style-matching loss is applied to style regions in the target hair and those in the aligned target hair. The white boundary regions are segmented style regions, and the red boundary regions describe an example of a consistently tracked style region. (Color figure online)

on the similarity of color and spatial distance between pixels. Since the SLIC considers both the appearance and location of neighboring pixels, it can successfully segment the target hair into proper style regions.

As presented in the first column of Fig. 4, we first find the style regions in the hair of $\mathbf{I}_{trg}$. Then, during the latent optimization, we detect the style regions of $G(\hat{w_{trg}})$ and match each region to the most similar style region of $\mathbf{I}_{trg}$, as shown in the rest columns of Fig. 4. In each step, we track the regions of similar style by setting the same label to the region of the closest centroid compared to the previous step. Figure 4 shows that an example style region marked with a red boundary is successfully tracked based on the proposed algorithm. $SLIC_{hair}(\mathbf{I}) \in \{0,1\}^{N_{style} \times H \times W}$ indicates style region masks extracted from a hair region of $\mathbf{I}$ using the SLIC algorithm. Here, $N_{style}$ indicates the number style regions. We set $N_{style}$ as 5 in our experiments. $\mathcal{L}_{style}^{LSM}$ is formulated as:

$$\mathcal{L}_{style}^{LSM} = \sum_{i=1}^{N_{style}} \mathcal{L}_{style}(SLIC_{hair}^{i}(\mathbf{I}_{trg}) \odot \mathbf{I}_{trg}, SLIC_{hair}^{i}(G(\hat{w_{trg}})) \odot G(\hat{w_{trg}})). \quad (3)$$

$SLIC_{hair}^{i}(\cdot)$ is the $i$-th channel of $SLIC_{hair}(\cdot)$ and $\odot$ indicates element-wise product. Note that a valid region of each channel, where the style region mask corresponds to 1, is cropped before calculating the style loss.

**Regularization Loss.** We add a step-wise regularization loss to keep the overall features of $\hat{w_{trg}}$, especially hairstyle, similar to the previous step. The regularization loss $\mathcal{L}_{reg}$ encourages a stable optimization via a gradual modification without a noticeable loss of the original hairstyle features. $\mathcal{L}_{reg}$ is formulated as:

$$\mathcal{L}_{reg} = \frac{1}{N_w} \|\Delta \hat{w_{trg}}\|_2^2, \quad (4)$$

where $N_w$ indicates the number of elements in $w$. $\Delta \hat{w_{trg}}$ at step $t$ is obtained by $\hat{w_{trg}}^{t} - \hat{w_{trg}}^{t-1}$, where $t$ ranges from 2 to the total number of steps.

Formally, the total objective function in the target hair alignment step is $\mathcal{L}_{pose} + \lambda_{style}^{LSM} \mathcal{L}_{style}^{LSM} + \lambda_{reg} \mathcal{L}_{reg}$, where $\lambda_{style}^{LSM}$ and $\lambda_{reg}$ denote the hyper-parameters to control relative importance between different losses.

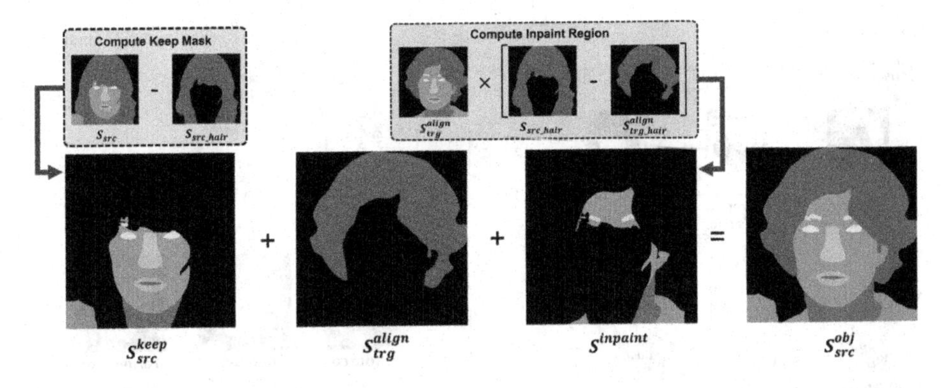


**Fig. 5.** Generation of an objective label. For source inpainting, we create an objective label $\mathbf{S}^{obj}_{src}$ to guide the occluded regions to be inpainted with proper semantics.

### 3.4 Source Inpainting

Source inpainting step aims to inpaint the regions occluded by the original source hair. As shown in Fig. 2, if we remove the source hair region from the source image, the occluded region should be filled with the proper semantics (*e.g.*, forehead, face, neck, clothes, and background) to fit the aligned target hair.

To find the inpainted source latent code $w^{inpaint}_{src}$, we generate an objective label $\mathbf{S}^{obj}_{src} \in \mathbb{Z}^{H \times W}$ to guide the occluded regions to be filled with the appropriate semantic regions. $\mathbf{S}^{obj}_{src}$ is generated by the following process, as also described in Fig. 5. First, we compute a keep label $\mathbf{S}^{keep}_{src}$, which indicates the regions that need to be maintained in the source, by removing a source hair region $\mathbf{S}_{src_hair}$ from a source semantic label $\mathbf{S}_{src}$. Here, $\mathbf{S}_{src}$ is estimated by a pre-trained segmentation network [28]. Next, we calculate a label of regions to be inpainted $\mathbf{S}^{inpaint}$ as described in Fig. 5. Finally, we obtain $\mathbf{S}^{obj}_{src}$ which indicates the inpainting regions of the source image considering the aligned target hair. Now, we optimize $w^{inpaint}_{src}$ to follow the given $\mathbf{S}^{obj}_{src}$. Here, as in the target hair alignment step, we optimize the first $m$ $w$ vectors to newly generate coarse features to fill the occlusions while preserving the fine details or the overall appearance of the source. For optimizing $w^{inpaint}_{src}$, we use a pixel-wise cross-entropy loss between the label of $\mathbf{S}^{obj}_{src}$ and a segmentation probability heatmap of the generated image, which consists of 16 semantic region categories. The heatmap is estimated by the pre-trained segmentation network.

### 3.5 Blending

The final optimization step aims to find a blending weight $w^{weight}$ that merges the optimized latent codes from the previous steps to generate the final output. First, as presented in Fig. 6(a), $w^{blend}$ is obtained by blending $w^{inpaint}_{src}$ and $w^{align}_{trg}$ with the blending weight $w^{weight}$. $w^{blend}$ is formulated as $w^{inpaint}_{src} + w^{weight} \odot w^{align}_{trg}$, where $w^{weight}$ implies how much of $w^{align}_{trg}$ needs to be reflected to synthesize the final output. Then, we prepare $\mathbf{F}$ tensors by feeding the first $m$ $w$

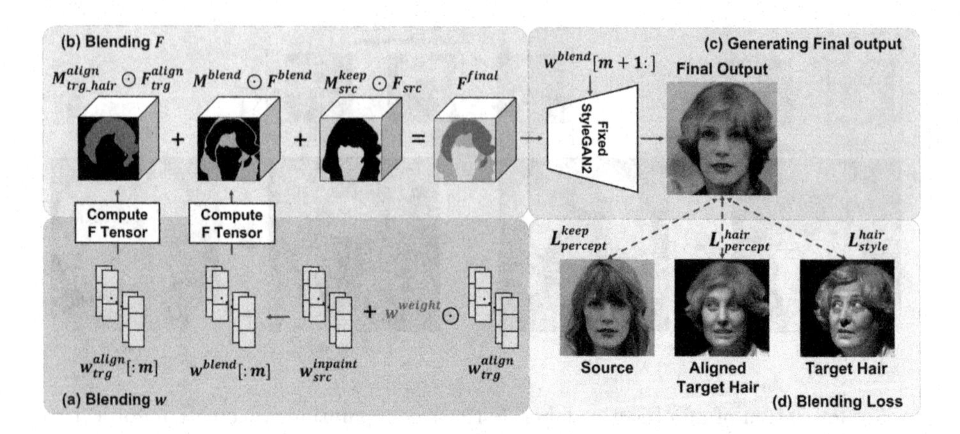


**Fig. 6.** Blending. (a) $w$ vectors from the previous steps are blended with the optimized blending weight $w^{weight}$ to obtain $w^{blend}$. (b) Next, we combine $\mathbf{F}^{align}_{trg}$, $\mathbf{F}^{blend}$, and $\mathbf{F}_{src}$ with the corresponding masks to obtain $\mathbf{F}^{final}$. (c) $\mathbf{F}^{final}$ and $w^{blend}$ are fed to the StyleGAN2 generator to synthesize the final output. (d) Blending loss consists of $\mathcal{L}^{keep}_{percept}$, $\lambda^{hair}_{percept}$, and $\mathcal{L}^{hair}_{percept}$.

vectors to the pre-trained StyleGAN2 generator, as shown in Fig. 6(b). Here, we leverage $\mathbf{F}$ tensors in $FS$ space to effectively reconstruct the detailed spatial information [31] in the further blending. We blend $\mathbf{F}^{align}_{trg}$, $\mathbf{F}^{blend}$, and $\mathbf{F}_{src}$ to gain $\mathbf{F}^{final}$ which contains detailed spatial information of the final output. $\mathbf{F}^{final}$ is calculated as follows:

$$\mathbf{F}^{final} = \mathbf{M}^{align}_{trg_hair} \odot \mathbf{F}^{align}_{trg} + \mathbf{M}^{blend} \odot \mathbf{F}^{blend} + \mathbf{M}^{keep}_{src} \odot \mathbf{F}_{src}. \tag{5}$$

$\mathbf{F}^{align}_{trg}$ and $\mathbf{F}^{blend}$ are extracted from $w^{align}_{trg}$ and $w^{blend}$, respectively, and $\mathbf{F}_{src}$ is from the embedding step. $\mathbf{M}^{align}_{trg_hair}$ is a binary mask indicating the hair region in the aligned target hair image. $\mathbf{M}^{keep}_{src}$ is also a binary mask denoting the regions which are neither the source hair nor the aligned target hair. $\mathbf{M}^{keep}_{src}$ indicates the area that needs to be preserved in the source image. Lastly, $\mathbf{M}^{blend}$ denotes the remaining regions. The final output $\hat{\mathbf{I}}$ is generated from the pre-trained StyleGAN2 generator given $w^{blend}$ and $\mathbf{F}^{final}$ as inputs. Here, vectors in $w^{blend}$ except the first $m$ vectors are fed to the generator.

**Losses.** In order to blend the previous optimized latent codes while preserving their structure and styles, we utilize the following losses.

First, to maintain the source face, clothes, background, etc., we apply the perceptual loss [29] on the valid regions in $\mathbf{S}^{keep}_{src}$ (*i.e.*, the regions to be preserved in the source image) as follows:

$$\mathcal{L}^{keep}_{percept} = \frac{1}{V}\sum_{i=1}^{V}\frac{1}{N_{\mathrm{VGG}^i}}\|\mathbf{M}^{keep}_{src} \odot (\mathrm{VGG}^i(\mathbf{I}_{src}) - \mathrm{VGG}^i(\hat{\mathbf{I}}))\|_1, \tag{6}$$

where $\text{VGG}^i$ denotes $i$-th layer of VGG16 network [22] and $N_{\text{VGG}^i}$ is the number of elements in the activation of $\text{VGG}^i$.

Also, in order to preserve the aligned target hairstyle from the aligned target latent code $w_{trg}^{align}$, we use the hair perceptual loss formulated as follows:

$$\mathcal{L}_{percept}^{hair} = \frac{1}{V}\sum_{i=1}^{V}\frac{1}{N_{\text{VGG}^i}}\|\mathbf{M}_{trg_hair}^{align}\odot(\text{VGG}^i(\mathbf{I}_{trg}^{align}) - \text{VGG}^i(\hat{\mathbf{I}}))\|_1. \quad (7)$$

Lastly, we maintain the texture of the original target hair by utilizing the hairstyle loss $\mathcal{L}_{style}^{hair}$, where the style loss $\mathcal{L}_{style}$ is applied on the hair regions of the target hair image and the final output as $\mathcal{L}_{style}(\mathbf{M}_{trg_hair}\odot\mathbf{I}_{trg}, \mathbf{M}_{\hat{I}_hair}\odot\hat{\mathbf{I}})$.

The total blending loss to optimize $w^{weight}$ is $\mathcal{L}_{percept}^{keep} + \lambda_{percept}^{hair}\mathcal{L}_{percept}^{hair} + \lambda_{style}^{hair}\mathcal{L}_{style}^{hair}$, where $\lambda_{percept}^{hair}$ and $\lambda_{style}^{hair}$ are the hyper-parameters to balance the relative importance between the losses.

# 4 Experiments

## 4.1 Experimental Setup

**Dataset.** We utilize Flickr-Faces-HQ (FFHQ) dataset [13] for hairstyle transfer and K-hairstyle [15] and VoxCeleb2 [6] for reconstruction task. For hairstyle transfer, we sample 6,000 pairs of two different identities (one for source and the other for target hairstyle) from 70,000 1,024×1,024 images in FFHQ.

For the reconstruction, we create 500 test pairs by sampling the images from the K-hairstyle dataset, which includes 500,000 high-resolution multi-view images with more than 6,400 identities. Following HairFIT [5], we filtered the images to remove the ones whose hairstyle is significantly occluded, or whose face is extremely rotated. Additionally, we sample 500 pairs of a source and a target from more than 1 million videos in VoxCeleb2. In the reconstruction task, two images in each pair have the same identity and different poses, and the source image in each pair is considered the ground truth image for the model to reconstruct. Each image is resized to 256×256 in the experiments.

**Baseline Models.** We conduct a quantitative and qualitative comparison between our model and the following baselines: LOHO [20], Barbershop [31], and HairFIT [5]. Here, we follow the official implementation code of LOHO and Barbershop. Since LOHO utilizes an external inpainting network, we use a state-of-the-art inpainting network CoModGAN [30]. Also, we implement Hair-FIT with the codes and guidelines provided by the authors of HairFIT.

## 4.2 Comparison to Baselines

**Quantitative Evaluations.** First, we compare the fréchet inception distance (FID) score [10] of LOHO, Barbershop, and our model on hairstyle transfer task. The FID score measures how similar the distributions of the synthesized images

**Table 1.** Quantitative comparison with baselines. We measure the FID scores with three different levels of pose difference and with total pairs.

| Pose difference level | Easy | Medium | Difficult | Total |
|---|---|---|---|---|
| LOHO [20] | 21.70 | 23.40 | 28.36 | 19.63 |
| Barbershop [31] | **20.75** | 21.45 | 26.30 | 18.07 |
| Ours | 20.79 | **20.56** | **22.72** | **17.06** |

**Table 2.** Quantitative comparisons with HairFIT using multi-view datasets.

| Dataset | K-hairstyle | | VoxCeleb2 | |
|---|---|---|---|---|
| Metric | SSIM$_\uparrow$ | LPIPS$_\downarrow$ | SSIM$_\uparrow$ | LPIPS$_\downarrow$ |
| HairFIT | 0.7242 | 0.2054 | 0.7520 | **0.2033** |
| Ours | **0.7424** | **0.1786** | **0.7717** | 0.2078 |

and the real images are, where the lower FID score indicates a higher similarity. We compare 6,000 pairs of real and fake images, where each image is resized to 256×256 for the evaluation. As shown in the last column of Table 1, we achieve the lowest FID score compared to the baselines.

For further analysis, we compare the FID scores on three different levels of pose difference as conducted in the previous work [5,20]. We calculate the pose difference, PD, following the protocol presented in HairFIT [5]. In particular, we use 17 facial jaw keypoints extracted by the pre-trained 3D-keypoint extraction model [4]. The pose difference is calculated as $\text{PD} = \frac{1}{17}\sum_{i=1}^{17}\|\mathbf{k}_{src}^i - \mathbf{k}_{trg}^i\|_1$, where $\mathbf{k}_{src}^i \in \mathbb{R}^3$ is a 3D coordinates of the $i$-th source keypoint and $\mathbf{k}_{trg}^i \in \mathbb{R}^3$ is a 3D coordinates of the $i$-th target keypoint. Then, we divide the 6,000 pairs of a source and a target into three categories of 2,000 pairs: Easy, Medium, and Difficult. As presented in Table 1, our model outperforms the other baselines for Medium and Difficult. Moreover, the margin between the FID scores of our model and other baselines increases as the pose difference increases from Easy to Difficult.

Additionally, we conduct a comparison with HairFIT on the reconstruction task using K-hairstyle and VoxCeleb2. As in HairFIT, we measure the structural similarity (SSIM) [25] and learned perceptual image patch similarity (LPIPS) [29] between generated images and ground truth images. Table 2 presents that our model outperforms HairFIT (except for LPIPS of VoxCeleb2) *even without* learning to reconstruct different views of a source image using a multi-view dataset.

**Qualitative Evaluations.** Figure 7 and Fig. 8 demonstrate that our model successfully transfers the target hairstyle into the source regardless of the pose differences. Especially, as presented in Fig. 7, our model shows superiority over other baselines on the Difficult level. Furthermore, although the FID score of our model is slightly higher than Barbershop on Easy level, Fig. 8 present that the

**Fig. 7.** Qualitative comparison with the baselines on difficult level of pose difference.

**Fig. 8.** Qualitative comparison on (a) easy and (b) medium level of pose difference.

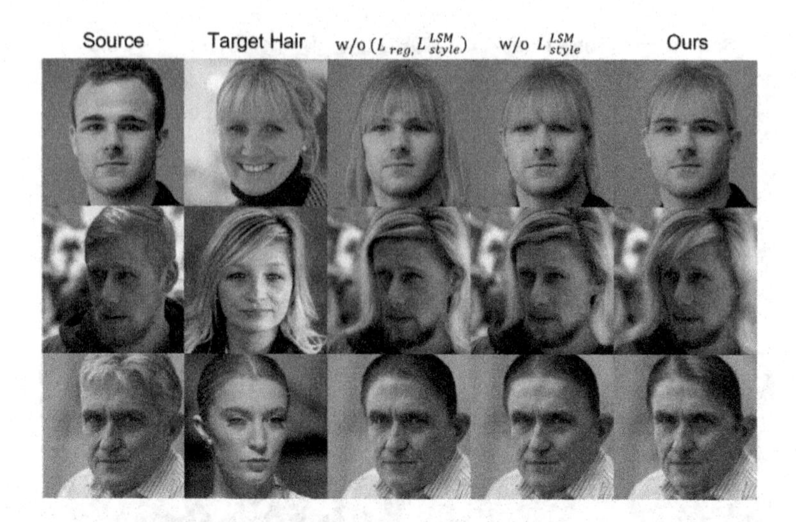


**Fig. 9.** Qualitative ablation study on the losses in target hair alignment step.

**Table 3.** Quantitative ablation study on the losses in target hair alignment step.

| Configurations | w/o ($\mathcal{L}_{reg}, \mathcal{L}_{style}^{LSM}$) | w/o $\mathcal{L}_{style}^{LSM}$ | Ours |
|---|---|---|---|
| SSIM$_{\uparrow}$ | 0.7667 | 0.7716 | **0.7717** |
| LPIPS$_{\downarrow}$ | 0.2125 | 0.2082 | **0.2078** |

quality of our model is better to reflect the target hairstyle than the baselines. The results show that our model successfully aligns the target hair to the source image, producing high-quality images of hairstyle transfer. More results of the qualitative comparison are presented in the supplementary materials.

### 4.3 Ablation Study

In the ablation study, we demonstrate the effectiveness of a local-style-matching loss and regularization loss in our target hair alignment step. We conduct a qualitative evaluation on hairstyle transfer using the FFHQ dataset and quantitative evaluation on the reconstruction task with VoxCeleb2. In Fig. 9 and Table 3, *w/o* ($\mathcal{L}_{reg}, \mathcal{L}_{style}^{LSM}$) denotes our framework without $\mathcal{L}_{reg}$ and $\mathcal{L}_{style}^{LSM}$). Also, *w/o* $\mathcal{L}_{style}^{LSM}$ indicates our framework without $\mathcal{L}_{style}^{LSM}$ and *Ours* is our full framework.

The first row of Fig. 9 indicates that the generated target hair is longer than the original target hair due to the absence of the $\mathcal{L}_{reg}$. In the second row, the direction of the front hair of the outputs without $\mathcal{L}_{hair}^{LSM}$ are different from the original target hairstyle. Moreover, in the third row, the "part" of the target hair is better reflected in the output of ours. The results present that our proposed losses effectively reflect the local style of the target hair while preserving its overall style. Additionally, as seen in Table 3, our full model outperforms

**Fig. 10.** Limitations of our proposed method.

other configurations with a gradual performance increase. Although the difference between *Ours* and *w/o* $\mathcal{L}_{style}^{LSM}$ is marginal, the qualitative results presented above clearly illustrate the high visual quality of our full model in terms of preserving delicate hair features.

## 5 Discussions

Although our model achieves a state-of-the-art performance compared to the baselines, several challenges still remain. First, since we transfer hairstyles via online latent optimization, it takes a few minutes on average for each image pair. Also, our framework cannot newly generate the occluded part of the target hair due to the extremely turned head pose. For example, the first three columns of Fig. 10 show that where the hair on the side is extremely occluded so that the final output barely has side hair. Finally, the output might contain undesired background when the hair segmentation mask is inaccurately predicted. The last three columns of Fig. 10 present undesirable background leaking.

## 6 Conclusions

This paper proposes a latent optimization framework for high-quality pose-invariant hairstyle transfer via local-style-aware hair alignment. By leveraging latent optimization, we align the target hair without a multi-view dataset, while maintaining fine details of the hairstyle. In addition, during the hair alignment, our newly-presented local-style-matching loss encourages our model to preserve the distinct structure and color of each local hair region in detail. Finally, we perform occlusion inpainting and blending via latent optimization. In this way, our model produces high-quality final output without noticeable artifacts.

**Acknowledgments.** This work was supported by the Institute of Information & communications Technology Planning & Evaluation (IITP) grant funded by the Korean government (MSIT) (No. 2019-0-00075), Artificial Intelligence Graduate School Program (KAIST) and the Ministry of Culture, Sports and Tourism and Korea Creative Content Agency (Project Number: R2021040097, Contribution Rate: 50).

## References

1. Abdal, R., Qin, Y., Wonka, P.: Image2StyleGAN: how to embed images into the StyleGAN latent space? In: Proceedings of the IEEE International Conference on Computer Vision (ICCV), pp. 4432–4441 (2019)
2. Abdal, R., Qin, Y., Wonka, P.: Image2StyleGAN++: how to edit the embedded images? In: Proceedings of the IEEE Conference on Computer Vision and Pattern Recognition (CVPR), pp. 8296–8305 (2020)
3. Achanta, R., Shaji, A., Smith, K., Lucchi, A., Fua, P., Süsstrunk, S.: SLIC superpixels compared to state-of-the-art superpixel methods. IEEE Trans. Pattern Anal. Mach. Intell. (TPAMI) **34**(11), 2274–2282 (2012)
4. Bulat, A., Tzimiropoulos, G.: How far are we from solving the 2D & 3D face alignment problem? (and a dataset of 230,000 3d facial landmarks). In: Proceedings of the IEEE International Conference on Computer Vision (ICCV) (2017)
5. Chung, C., et al.: HairFIT: pose-invariant hairstyle transfer via flow-based hair alignment and semantic-region-aware inpainting. In: Proceedings of the British Machine Vision Conference (BMVC). British Machine Vision Association (2021)
6. Chung, J.S., Nagrani, A., Zisserman, A.: VoxCeleb2: deep speaker recognition. In: Conference of the International Speech Communication Association (INTERSPEECH) (2018)
7. Gatys, L.A., Ecker, A.S., Bethge, M.: Image style transfer using convolutional neural networks. In: Proceedings of the IEEE Conference on Computer Vision and Pattern Recognition (CVPR) (2016)
8. Goodfellow, I., et al.: Generative adversarial nets. In: Proceedings of the Advances in Neural Information Processing Systems (NeurIPS) (2014)
9. Harkonen, E., Hertzmann, A., Lehtinen, J., Paris, S.: GANSpace: discovering interpretable GAN controls. In: Proceedings of the Advances in Neural Information Processing Systems (NeurIPS) (2020)
10. Heusel, M., Ramsauer, H., Unterthiner, T., Nessler, B., Hochreiter, S.: GANs trained by a two time-scale update rule converge to a local Nash equilibrium. In: Proceedings of the Advances in Neural Information Processing Systems (NeurIPS) (2017)
11. Jiang, W., et al.: PSGAN: pose and expression robust spatial-aware GAN for customizable makeup transfer. In: Proceedings of the IEEE Conference on Computer Vision and Pattern Recognition (CVPR) (2020)
12. Jo, Y., Park, J.: SC-FEGAN: face editing generative adversarial network with user's sketch and color. In: Proceedings of the IEEE Conference on Computer Vision and Pattern Recognition (CVPR) (2019)
13. Karras, T., Laine, S., Aila, T.: A style-based generator architecture for generative adversarial networks. In: Proceedings of the IEEE Conference on Computer Vision and Pattern Recognition (CVPR) (2019)
14. Karras, T., Laine, S., Aittala, M., Hellsten, J., Lehtinen, J., Aila, T.: Analyzing and improving the image quality of StyleGAN. In: Proceedings of the IEEE Conference on Computer Vision and Pattern Recognition (CVPR) (2020)
15. Kim, T., et al.: K-hairstyle: a large-scale Korean hairstyle dataset for virtual hair editing and hairstyle classification. In: Proceedings of the IEEE International Conference on Image Processing (ICIP), pp. 1299–1303. IEEE (2021)
16. Lee, C.H., Liu, Z., Wu, L., Luo, P.: MaskGAN: towards diverse and interactive facial image manipulation. In: Proceedings of the IEEE Conference on Computer Vision and Pattern Recognition (CVPR), pp. 5549–5558 (2020)

17. Nagrani, A., Chung, J.S., Zisserman, A.: VoxCeleb: a large-scale speaker identification dataset. arXiv preprint arXiv:1706.08612 (2017)
18. Odena, A., Olah, C., Shlens, J.: Conditional image synthesis with auxiliary classifier GANs. In: Proceedings of the International Conference on Learning Representations (ICLR) (2017)
19. Portenier, T., Hu, Q., Szabo, A., Bigdeli, S.A., Favaro, P., Zwicker, M.: FaceShop: deep sketch-based face image editing. arXiv preprint arXiv:1804.08972 (2018)
20. Saha, R., Duke, B., Shkurti, F., Taylor, G., Aarabi, P.: LOHO: latent optimization of hairstyles via orthogonalization. In: Proceedings of the IEEE Conference on Computer Vision and Pattern Recognition (CVPR) (2021)
21. Shen, Y., Gu, J., Tang, X., Zhou, B.: Interpreting the latent space of GANs for semantic face editing. In: Proceedings of the IEEE Conference on Computer Vision and Pattern Recognition (CVPR) (2020)
22. Simonyan, K., Zisserman, A.: Very deep convolutional networks for large-scale image recognition. In: Proceedings of the International Conference on Learning Representations (ICLR) (2015)
23. Tan, Z., et al.: MichiGAN: multi-input-conditioned hair image generation for portrait editing. ACM Trans. Graph. (TOG) **39**(4), 1–13 (2020)
24. Viazovetskyi, Y., Ivashkin, V., Kashin, E.: StyleGAN2 distillation for feed-forward image manipulation. In: Vedaldi, A., Bischof, H., Brox, T., Frahm, J.-M. (eds.) ECCV 2020. LNCS, vol. 12367, pp. 170–186. Springer, Cham (2020). https://doi.org/10.1007/978-3-030-58542-6_11
25. Wang, Z., Bovik, A.C., Sheikh, H.R., Simoncelli, E.P.: Image quality assessment: from error visibility to structural similarity. IEEE Trans. Image Process. (TIP) **13**(4), 600–612 (2004)
26. Xiao, C., Yu, D., Han, X., Zheng, Y., Fu, H.: SketchHairSalon: deep sketch-based hair image synthesis (2021)
27. Yang, S., Wang, Z., Liu, J., Guo, Z.: Deep plastic surgery: robust and controllable image editing with human-drawn sketches. In: Vedaldi, A., Bischof, H., Brox, T., Frahm, J.-M. (eds.) ECCV 2020. LNCS, vol. 12360, pp. 601–617. Springer, Cham (2020). https://doi.org/10.1007/978-3-030-58555-6_36
28. Yu, C., Wang, J., Peng, C., Gao, C., Yu, G., Sang, N.: BiSeNet: bilateral segmentation network for real-time semantic segmentation. In: Ferrari, V., Hebert, M., Sminchisescu, C., Weiss, Y. (eds.) ECCV 2018. LNCS, vol. 11217, pp. 334–349. Springer, Cham (2018). https://doi.org/10.1007/978-3-030-01261-8_20
29. Zhang, R., Isola, P., Efros, A.A., Shechtman, E., Wang, O.: The unreasonable effectiveness of deep features as a perceptual metric. In: Proceedings of the IEEE Conference on Computer Vision and Pattern Recognition (CVPR) (2018)
30. Zhao, S., et al.: Large scale image completion via co-modulated generative adversarial networks. In: Proceedings of the International Conference on Learning Representations (ICLR) (2021)
31. Zhu, P., Abdal, R., Femiani, J., Wonka, P.: Barbershop: GAN-based image compositing using segmentation masks (2021)
32. Zhu, P., Abdal, R., Qin, Y., Femiani, J., Wonka, P.: Improved StyleGAN embedding: where are the good latents? arXiv preprint arXiv:2012.09036 (2020)
33. Zhuang, P., Koyejo, O., Schwing, A.G.: Enjoy your editing: controllable GANs for image editing via latent space navigation. In: Proceedings of the International Conference on Learning Representations (ICLR) (2021)

# High-Resolution Virtual Try-On with Misalignment and Occlusion-Handled Conditions

Sangyun Lee[1], Gyojung Gu[2,3], Sunghyun Park[2], Seunghwan Choi[2], and Jaegul Choo[2(✉)]

[1] Soongsil University, Seoul, South Korea
[2] Korea Advanced Institute of Science and Technology, Daejeon, South Korea
{gyojung.gu,psh01087,shadow2496,jchoo}@kaist.ac.kr
[3] Nestyle Inc., Seoul, South Korea

**Abstract.** Image-based virtual try-on aims to synthesize an image of a person wearing a given clothing item. To solve the task, the existing methods warp the clothing item to fit the person's body and generate the segmentation map of the person wearing the item before fusing the item with the person. However, when the warping and the segmentation generation stages operate individually without information exchange, the misalignment between the warped clothes and the segmentation map occurs, which leads to the artifacts in the final image. The information disconnection also causes excessive warping near the clothing regions occluded by the body parts, so-called pixel-squeezing artifacts. To settle the issues, we propose a novel try-on condition generator as a unified module of the two stages (*i.e.*, warping and segmentation generation stages). A newly proposed feature fusion block in the condition generator implements the information exchange, and the condition generator does not create any misalignment or pixel-squeezing artifacts. We also introduce discriminator rejection that filters out the incorrect segmentation map predictions and assures the performance of virtual try-on frameworks. Experiments on a high-resolution dataset demonstrate that our model successfully handles the misalignment and occlusion, and significantly outperforms the baselines. Code is available at https://github.com/sangyun884/HR-VITON.

**Keywords:** High-resolution virtual try-on · Misalignment-free · Occlusion-handling

## 1 Introduction

As the importance of online shopping increases, a technology that allows customers to virtually try on clothes is expected to enrich the customer's experience.

S. Lee and G. Gu—Equal contributions.

**Supplementary Information** The online version contains supplementary material available at https://doi.org/10.1007/978-3-031-19790-1_13.

© The Author(s), under exclusive license to Springer Nature Switzerland AG 2022
S. Avidan et al. (Eds.): ECCV 2022, LNCS 13677, pp. 204–219, 2022.
https://doi.org/10.1007/978-3-031-19790-1_13

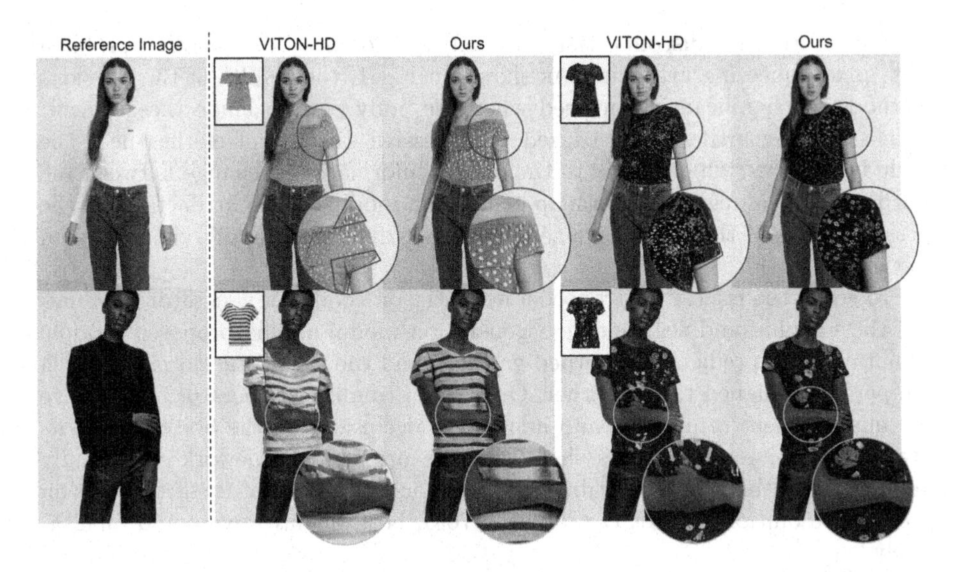


**Fig. 1.** Comparison of $1024 \times 768$ try-on synthesis results with VITON-HD [2]. (1*st* row) The red-colored areas indicate the artifact due to the misalignment between a warped clothing image and a segmentation map. (2*nd* row) The green-colored areas denote the pixel-squeezing due to the occlusion by the body parts. In contrast to the VITON-HD, our method successfully handles the misalignment and occlusion. Zoom in for the best view.

A virtual try-on task aims to change the clothing item on a person into a given clothing product. While there are 3D-based virtual try-on approaches that rely on the 3D measurement of garments [6,20,21,23], we address image-based virtual try-on [3,9,13–15,25,29,30], which only requires a garment and a person image, facilitating real-world applications.

To address this task, previous studies employ an explicit warping module that aligns the clothing image with the person's body. Moreover, predicting the segmentation map of the final image alleviates the difficulty of image generation as it guides the person's layout and separates regions to be generated and the ones to be preserved [29]. The importance of the segmentation map increases as the image resolution grows. Most image-based virtual try-on methods include these stages [2,3,9,15,25,29,30], and the outputs of the warping and segmentation map generation modules greatly influence the final try-on results.

However, the virtual try-on frameworks that consist of warping and segmentation generation modules have misaligned regions between the warped clothes and the segmentation map, so-called *misalignment*. As shown in Fig. 1, the misalignment results in the artifacts in these regions, which harm the perceptual quality of the final result significantly, especially at the high resolution. The main cause of misalignment is that the warping module and the segmentation map generator operate separately without information exchange. Although a recent study [2] tries to alleviate the artifacts in the misaligned regions, the existing methods are still not possible to solve the misalignment problem completely.

The information disconnection between two modules yields another problem (*i.e.*, pixel-squeezing artifacts). As shown in Fig. 1, the results of the previous methods are significantly impaired when the body parts occlude the garment. Pixel-squeezing artifacts are caused by excessive warping of clothes near the occluded regions, which is due to the lack of information exchange between the warping and the segmentation map generation modules. The artifacts limit the possible poses of the person images, making it difficult to apply virtual try-on to the real world.

To settle the issues, we propose a novel try-on condition generator that unifies the warping and segmentation generation modules. The proposed module simultaneously predicts the warped garment and the segmentation map, which are perfectly aligned to each other. Our try-on condition generator can remove the misalignment completely and handle the occlusions by the body parts naturally. Extensive experiments show that the proposed framework successfully handles the occlusion and misalignment, and achieves state-of-the-art results on the high-resolution dataset (*i.e.*, $1024 \times 768$), both quantitatively and qualitatively.

In addition, we introduce a discriminator rejection that filters out incorrect segmentation map predictions, which lead to unnatural final results. We demonstrate that the discriminator rejection assures the performance of virtual try-on frameworks, which is an important feature for real-world applications.

We summarize our contributions as follows:

- We propose a novel architecture that performs warping and segmentation map generation simultaneously.
- Our method is inherently *misalignment-free* and can handle the occlusion of clothes by body parts naturally.
- We adapt the discriminator rejection to filter out incorrect segmentation map predictions.
- We achieve state-of-the-art performance on a high-resolution dataset.

## 2 Related Work

### 2.1 Image-Based Virtual Try-On

An image-based virtual try-on task aims to produce a person image wearing a target clothing item given a pair of clothes and person images. Recent virtual try-on methods [2,3,9,15,25,29,30] generally consist of three separate modules: 1) segmentation map generation module, 2) clothing warping module, and 3) fusion module. The fusion module can generate the photo-realistic images by utilizing intermediate representations such as warped clothes and segmentation maps, which are produced by previous stages.

**Clothes Deformation.** To preserve the details of a clothing item, previous approaches [3,8,9,25] rely on the explicit warping module to fit the input clothing item to a given person's body. VITON [9] and CP-VTON [25] predict the parameters for thin plate spline (TPS) transformation to warp the clothing item. Since the warping modules based on the TPS transformation have a limited degree of freedom, an appearance flow is utilized to compute a pixel-wise 2D deformation field of the clothing image [3,8]. Although the warping modules have been consistently improved, the misalignment between the warped clothes and a person's body remains and results in the artifacts in the misaligned regions. Recently, VITON-HD [2] proposed a normalization technique to alleviate the issue. However, we found that the normalization method fails to naturally fill the misaligned regions with clothing texture. In this paper, we propose a method that can generate warped clothes without misaligned regions.

**Segmentation Generation for Try-On Synthesis.** To guide the try-on image synthesis, recent virtual try-on models [3,12,15,17,28,29] utilize the human segmentation maps of a person wearing the target clothes. The segmentation map disentangles the generation of appearance and shape, allowing the model to produce more spatially coherent results. In particular, the high-resolution virtual try-on methods [2,15] generally include the segmentation generation module because the importance of the segmentation map increases as the image resolution grows.

### 2.2 Rejection Sampling

There are several studies that aim to reject the low-quality generator outputs to improve the fidelity of samples. Razavi *et al.* [22] introduced rejection sampling based on the probability that the pre-trained classifier assigns to the correct class. Azadi *et al.* [1] proposed the discriminator rejection sampling, where a discriminator rejects the generated samples at test time. Under strict assumptions, this allows exact sampling from the data distribution. Although there have been several follow-up works [18,24], this technique has not been commonly used for image-conditional generation. In this paper, we utilize the discriminator to filter out the low-quality samples at test time.

## 3 Proposed Method

Given a reference image $I \in \mathbb{R}^{3\times H\times W}$ of a person and a clothing image $c \in \mathbb{R}^{3\times H\times W}$ ($H$ and $W$ denote the image height and width, respectively), our goal is to synthesize an image $\hat{I} \in \mathbb{R}^{3\times H\times W}$ of the person wearing $c$, where the pose and the body shape of $I$ are maintained. Following the training procedure of VITON [9], we train the model to reconstruct $I$ from a clothing-agnostic person representation and $c$ that the person is wearing already. The clothing-agnostic person representation eliminates any clothing information in $I$, and it allows the model to generalize at test time when an arbitrary clothing image is given.

Our framework is composed of two stages: (1) a *try-on condition generator*; (2) a *try-on image generator* (see Fig. 2). Given the clothing-agnostic person representation and $c$, our try-on condition generator deforms $c$ and produces the segmentation map simultaneously. The generator does not create any misalignment or pixel-squeezing artifacts (Sect. 3.1). Afterward, the try-on image

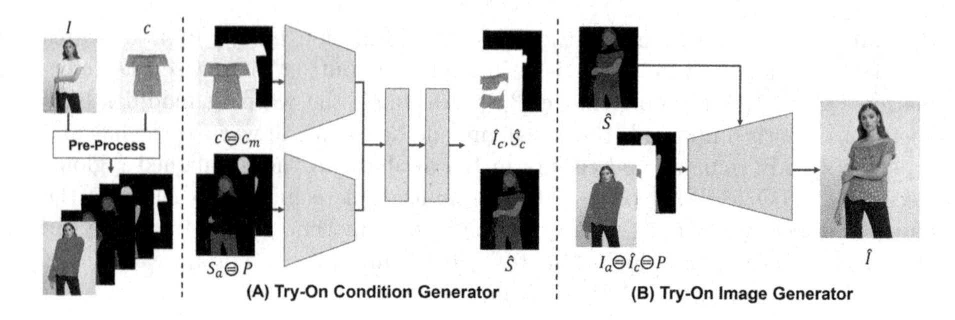


**Fig. 2.** Overview of the proposed framework (HR-VITON).

generator synthesizes the final try-on result using the outputs of the try-on condition generator (Sect. 3.2). At test time, we apply discriminator rejection that filters out incorrect segmentation map predictions (Sect. 3.3).

**Pre-processing.** In the pre-processing step, we obtain a segmentation map $S \in \mathbb{L}^{H\times W}$ of the person, a clothing mask $c_m \in \mathbb{L}^{H\times W}$, and a pose map $P \in \mathbb{R}^{3\times H\times W}$ with the off-the-shelf models [5, 7], where $\mathbb{L}$ is a set of integers indicating the semantic labels. For the pose map $P$, we utilize a dense pose [7], which maps all pixels of the person regions in the RGB image to the 3D surface of the person's body. For the clothing-agnostic person representation, we employ a clothing-agnostic person image $I_a$ and a clothing-agnostic segmentation map $S_a$ as those of VITON-HD [2].

### 3.1 Try-On Condition Generator

In this stage, we aim to generate the segmentation map $\hat{S}$ of the person wearing the target clothing item $c$ and deform $c$ to fit the body of the person. A warped clothing image $\hat{I}_c$ and a generated segmentation map $\hat{S}$ are used as the conditions for the try-on image generator. Figure 3 (A) shows the overall architecture of our try-on condition generator. Our try-on condition generator consists of two encoders (*i.e.*, a clothing encoder $E_c$ and a segmentation encoder $E_s$) and a decoder. Given $(c, c_m)$ and $(S_a, P)$, we first extract the feature pyramid $\{E_{c_k}\}_{k=0}^{4}$ and $\{E_{s_l}\}_{l=0}^{4}$ from each encoder, respectively. The extracted features are fed into the feature fusion blocks of the decoder, where the feature maps obtained from the two different feature pyramids are fused to predict the segmentation map and the appearance flow for warping the clothing image. Given the outputs of the last feature fusion block, we obtain $\hat{I}_c$, $\hat{S}_c$, and $\hat{S}$ through condition aligning.

**Feature Fusion Block.** As shown in Fig. 3 (B), there are two pathways in the feature fusion block: the *flow pathway* and the *seg pathway*. The flow and seg pathway generate the appearance flow map $F_{f_i}$ and the segmentation feature $F_{s_i}$, respectively. These two pathways exchange information with each other to estimate the appearance flow and the segmentation map jointly, which is indicated by green and blue arrows. For the green arrow, $F_{f_{i-1}}$ is used to deform the feature extracted from $c$ and $c_m$, which is then concatenated with $F_{s_{i-1}}$

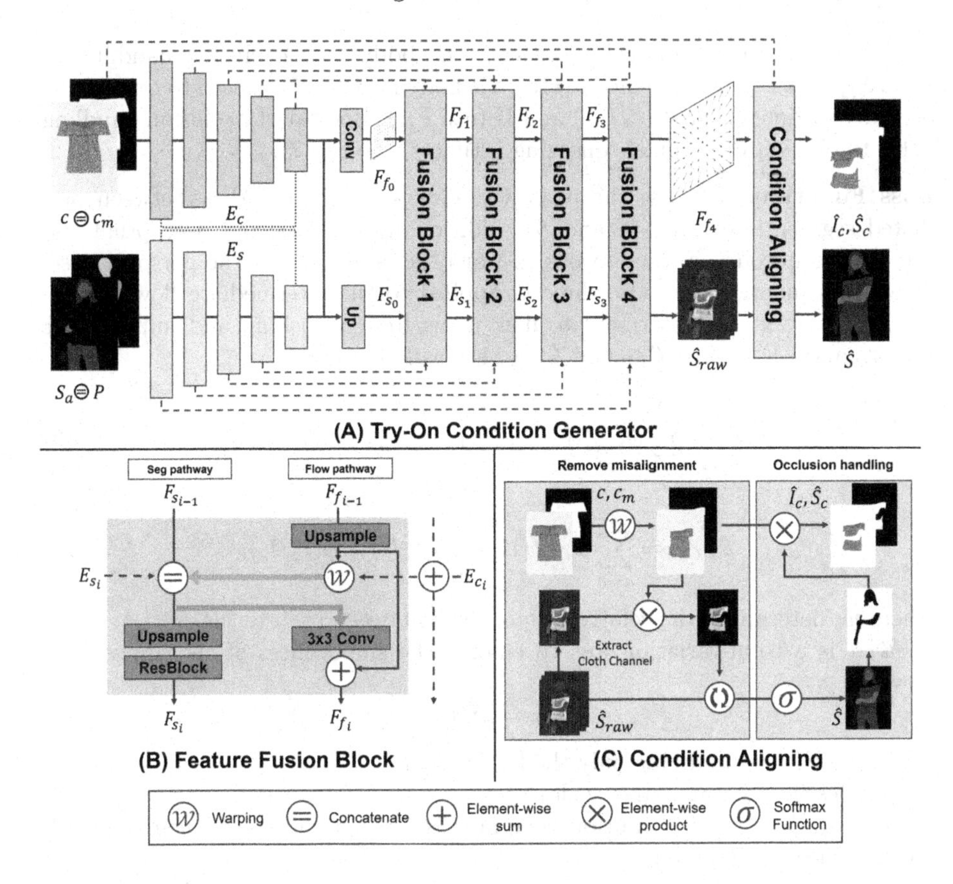


**Fig. 3.** Architecture of try-on condition generator.

and $E_{s_i}$ to generate $F_{s_i}$. For the blue arrow, $F_{s_{i-1}}$ is used to guide the flow estimation. These information exchanges are crucial in estimating the warped clothing and the segmentation map aligned each other. The feature fusion block estimates $F_{f_i}$ and $F_{s_i}$ simultaneously, which are then used to refine each other at the next block.

**Condition Aligning.** To prevent the misalignment, we obtain $\hat{S}$ by removing the non-overlapping regions of the clothing mask channel of $\hat{S}_{raw}^{k,i,j}$ with $W(c_m, F_{f_4})$:

$$\hat{S}_{logit}^{k,i,j} = \begin{cases} \hat{S}_{raw}^{k,i,j} & \text{if} \, k \neq C \\ \hat{S}_{raw}^{k,i,j} \cdot W(c_m, F_{f_4}) & \text{if} \, k = C \end{cases} \tag{1}$$

$$\hat{S} = \sigma(\hat{S}_{logit}), \tag{2}$$

where $\hat{S}_{raw}$ is equivalent to $F_{s_4}$ and $C$ denotes the index of the clothing mask channel. $i, j$, and $k$ are indices across the spatial and channel dimensions. $\sigma$ is depth-wise softmax. Note that we apply ReLU activation to assure that $\hat{S}_{raw}$ is nonnegative.

$\hat{I}_c$ and $\hat{S}_c$ are obtained by applying the body part occlusion handling to $W(c, F_{f_4})$. As Fig. 3 (C) demonstrates, the body parts of $\hat{S}$ are used to remove the occluded regions from $W(c, F_{f_4})$ and $W(c_m, F_{f_4})$. Body part occlusion handling helps to eliminate the pixel-squeezing artifacts (see Fig. 7).

**Loss Functions.** We use the pixel-wise cross-entropy loss $\mathcal{L}_{CE}$ between predicted segmentation map $\hat{S}$ and $S$. Additionally, $L1$ loss and perceptual loss are used to encourage the network to warp the clothes to fit the person's pose. These loss functions are also directly applied to the intermediate flow estimations to prevent the intermediate flow maps from vanishing and improve the performance. Formally, $\mathcal{L}_{L1}$ and $\mathcal{L}_{VGG}$ are as follows:

$$\mathcal{L}_{L1} = \sum_{i=0}^{3} w_i \cdot ||W(c_m, F_{f_i}) - S_c||_1 + ||\hat{S}_c - S_c||_1, \tag{3}$$

$$\mathcal{L}_{VGG} = \sum_{i=0}^{3} w_i \cdot \phi(W(c, F_{f_i}), I_c) + \phi(\hat{I}_c, I_c), \tag{4}$$

where $w_i$ determines the relative importance between each terms.

$\mathcal{L}_{TV}$ is a total-variation loss to enforce the smoothness of the appearance flow:

$$\mathcal{L}_{TV} = ||\nabla F_{f_4}||_1 \tag{5}$$

We found that regularizing only the last appearance flow $F_{f_4}$ is vital in learning the flow estimation at coarse scales.

Totally, our try-on condition generator is trained end-to-end using the following objective function:

$$\mathcal{L}_{TOCG} = \lambda_{CE}\mathcal{L}_{CE} + \mathcal{L}_{cGAN} + \lambda_{L1}\mathcal{L}_{L1} + \mathcal{L}_{VGG} + \lambda_{TV}\mathcal{L}_{TV}, \tag{6}$$

where $\mathcal{L}_{cGAN}$ is conditional GAN loss between $\hat{S}$ and $S$, and $\lambda_{CE}$, $\lambda_{L1}$, and $\lambda_{TV}$ denote the hyper-parameters controlling relative importance between different losses. For $\mathcal{L}_{cGAN}$, we used the least-squared GAN loss [16].

### 3.2 Try-On Image Generator

In this stage, we generate the final try-on image $\hat{I}$ by fusing the clothing-agnostic image $I_a$, the warped clothing image $\hat{I}_c$, and the pose map $P$, guided by $\hat{S}$. The try-on image generator consists of a series of residual blocks, along with upsampling layers. The residual blocks use SPADE [19] as normalization layers whose modulation parameters are inferred from $\hat{S}$. Also, the input $(I_a, \hat{I}_c, P)$ is resized and concatenated to the activation before each residual block. We train the generator with the same losses used in SPADE and pix2pixHD [26]. Details of the model architecture, hyperparameters, and the objective function are described in the supplementary.

### 3.3 Discriminator Rejection

We propose a discriminator rejection method to filter out the low-quality segmentation map generated by the try-on condition generator at the test time. In the discriminator rejection sampling [1], the acceptance probability for an input $x$ is

$$p_{accept}(x) = \frac{p_d(x)}{Lp_g(x)}, \tag{7}$$

where $p_d$ and $p_g$ are the data distribution and the implicit distribution given by the generator, and $L$ is a normalizing constant. As we use the least-squares GAN loss, the optimal discriminator is derived as follows:

$$D^*(x) = \frac{p_d(x)}{p_d(x) + p_g(x)} \tag{8}$$

Afterward, the acceptance probability can be represented using the discriminator $D(x)$:

$$p_{accept} = \frac{D(x)}{L(1 - D(x))}, \tag{9}$$

where the equality is satisfied only if $D = D^*$. $L$ is written as follows:

$$L = \max_x \frac{D(x)}{(1 - D(x))}, \tag{10}$$

which is intractable. In practice, we construct $x$ from the segmentation map and input conditions (i.e., $P, S_a, c$, and $c_m$) and obtain $L$ using the entire training dataset. Azadi *et al.* [1] sample $\psi \sim U(0,1)$ and reject $x$ if $\psi > p_{accept}(x)$. Instead, we reject $x$ if $p_{accept}(x)$ is below a certain threshold. The discriminator rejection enables us to filter out the incorrect segmentation maps faithfully.

## 4 Experiments

### 4.1 Training

For the experiments, we use a high-resolution virtual try-on dataset introduced by VITON-HD [2], which contains 13,679 frontal-view woman and top clothing image pairs. The original resolution of the images is $1024 \times 768$, and the images are bicubically downsampled to the desired resolutions when needed. We split the dataset into a training and a test set with 11,647 and 2,032 pairs, respectively. For detailed information on the model training, see our supplementary material.

### 4.2 Qualitative Results

**Comparison with Baselines.** We compare our method with several state-of-the-art baselines, including CP-VTON [25], ACGPN [29], and VITON-HD [2]. We utilize the publicly available codes for baselines. Figure 4 shows that our

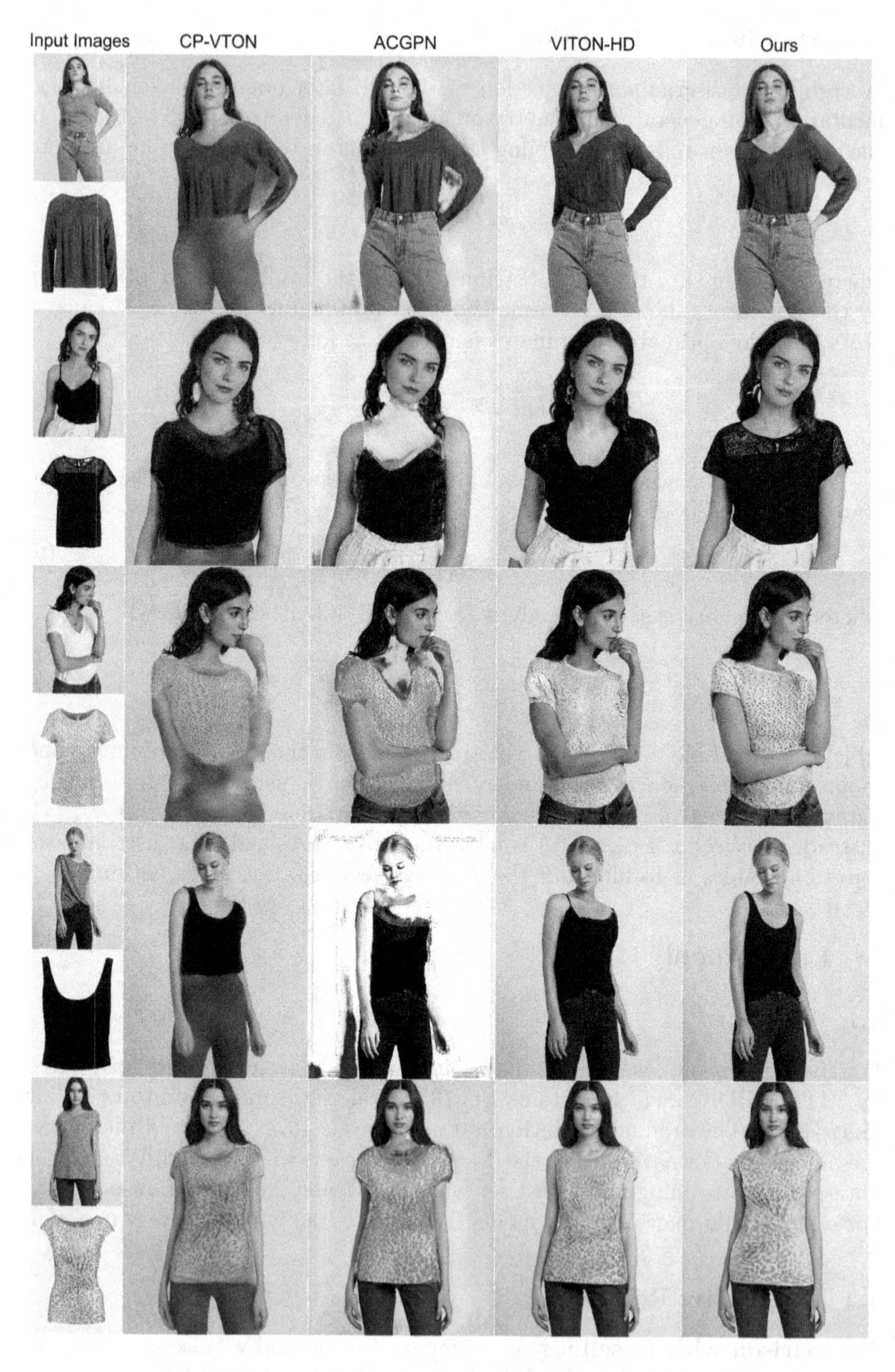


**Fig. 4.** Qualitative comparison with baselines.

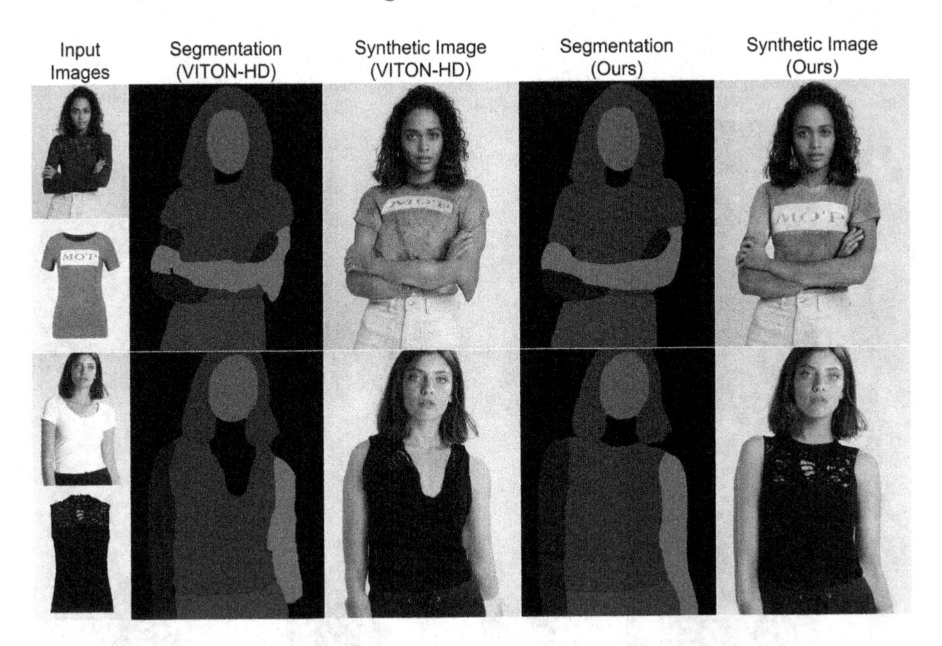


**Fig. 5.** Try-on synthesis results and corresponding segmentation maps.

method generates more photo-realistic images compared to the baselines. Specifically, we observe that our model not only preserves the details of the target clothing images but also generates the neckline naturally. As shown in Fig. 5, our try-on condition generator has the capability to produce the body shape more naturally compared to VITON-HD. These results demonstrate that the quality of the conditions for the try-on image generator is crucial in achieving perceptually convincing results. Furthermore, Fig. 6 shows that VITON-HD fails to eliminate the artifacts in the misaligned regions completely. On the other hand, since our method can produce misalignment-free segmentation maps and warped clothing images, our method solves the misalignment problem inherently. Thus, our method successfully synthesizes the high-quality images.

**Effectiveness of Occlusion Handling.** We analyze the impact of the occlusion handling process in our try-on condition generator. Figure 7 shows the effectiveness of the proposed body part occlusion handling. Without occlusion handling, the model excessively deforms the clothing image to fit the person's body shape, as shown in the $2nd$ column of Fig. 7. Due to the undesired deformation, the texture (*e.g.*, logo and stripe) of the target clothing item is squeezed, causing the missing pattern in the final results (See the $3rd$ column of Fig. 7). On the other hand, the model with occlusion handling enables to warp the clothes without the pixel-squeezing, better preserving the high-frequency details of the garment.

**Effectiveness of Discriminator Rejection.** To filter out the low-quality segmentation maps produced by our try-on condition generator, we propose a discriminator rejection method. Figure 8 shows the accepted and the rejected sam-

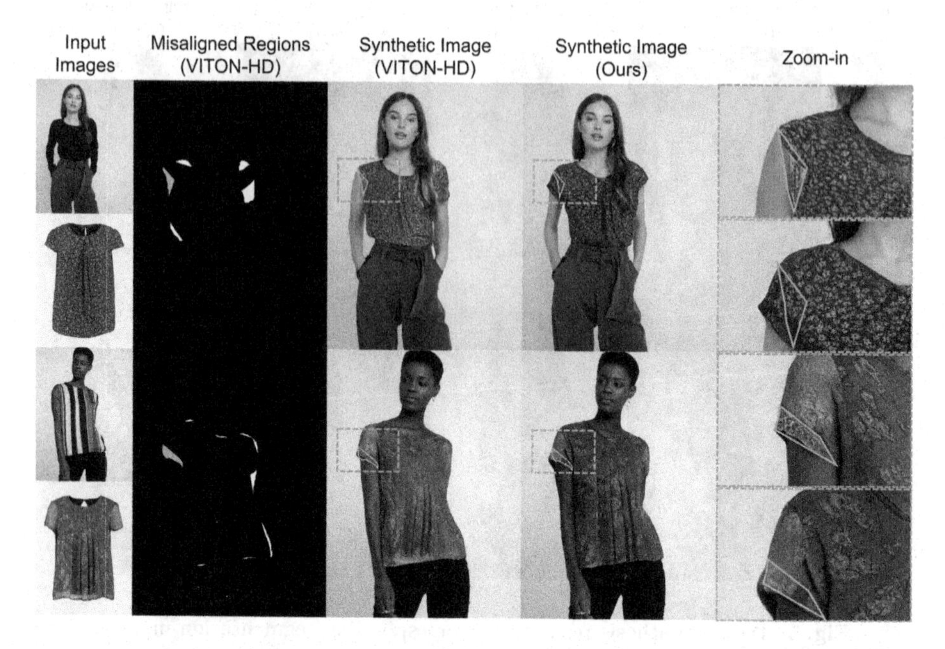


**Fig. 6.** Synthesis results and corresponding misaligned regions indicated by yellow colored areas. VITON-HD suffers from the artifacts caused by misalignment. (Color figure online)

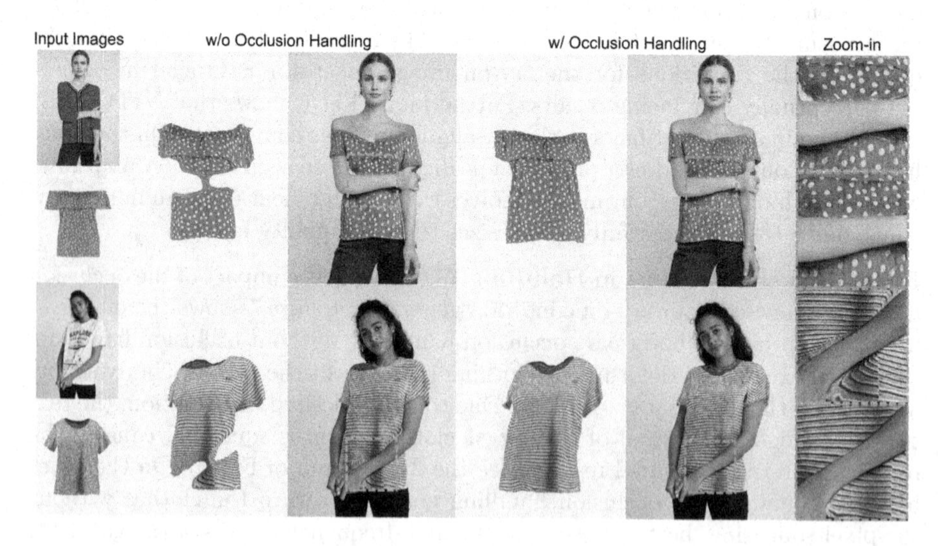


**Fig. 7.** Effects of the body part occlusion handling. The green colored areas indicate the pixel-squeezing artifacts. (Color figure online)

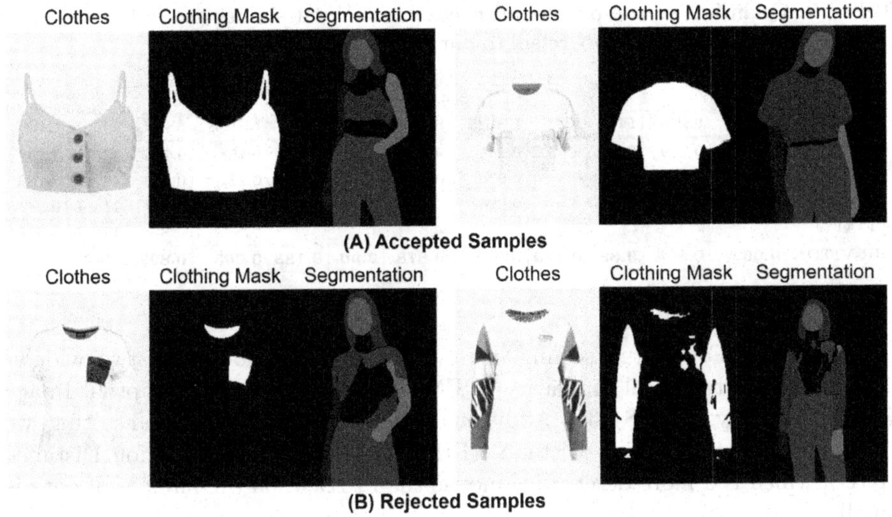


**Fig. 8.** Examples of accepted (A) and rejected (B) segmentation maps by discriminator rejection, corresponding input clothes and clothing masks.

ples of our discriminator rejection. Different from the accepted samples, the segmentation maps of the rejected samples are considerably impaired, as shown in the *2nd* row of Fig. 8. We found that the incorrect segmentation maps are caused mainly by errors in the pre-processing step, such as obtaining the clothing mask. Most virtual try-on methods rely on multiple conditions such as segmentation map and pose information obtained in the pre-processing stage and thus are prone to these errors. We believe that our discriminator rejection method can be a simple and effective solution for filtering out the low-quality outputs.

### 4.3 Quantitative Results

Following previous studies, we evaluate a paired setting and an unpaired setting, where the paired setting is to reconstruct the person image with the original clothing image, and the unpaired setting is to change the clothing item of the

**Table 1.** Ablation study in unpaired setting. We describes the KID as a value multiplied by 100. *Last row denotes that there is no information exchange.

| Method | FID$_{\downarrow}$ | KID$_{\downarrow}$ |
|---|---|---|
| HR-VITON | **10.91** | **0.179** |
| ∟ w/o Condition Aligning | 12.05 | 0.356 |
| ∟ w/o Feature Fusion Block | 12.41 | 0.381 |
| ∟ w/o Feature Fusion Block & Condition Aligning* | 12.73 | 0.415 |

**Table 2.** Quantitative comparison with baselines. We describes the KID as a value multiplied by 100. HR-VITON refers to our model.

| | 256 × 192 | | | | 512 × 384 | | | | 1024 × 768 | | | |
|---|---|---|---|---|---|---|---|---|---|---|---|---|
| | LPIPS$_\downarrow$ | SSIM$_\uparrow$ | FID$_\downarrow$ | KID$_\downarrow$ | LPIPS$_\downarrow$ | SSIM$_\uparrow$ | FID$_\downarrow$ | KID$_\downarrow$ | LPIPS$_\downarrow$ | SSIM$_\uparrow$ | FID$_\downarrow$ | KID$_\downarrow$ |
| CP-VTON | 0.159 | 0.739 | 30.11 | 2.034 | 0.141 | 0.791 | 30.25 | 4.012 | 0.158 | 0.786 | 43.28 | 3.762 |
| ACGPN | 0.074 | 0.833 | 11.33 | 0.344 | 0.076 | 0.858 | 14.43 | 0.587 | 0.112 | 0.850 | 43.29 | 3.730 |
| VITON-HD | 0.084 | 0.811 | 16.36 | 0.871 | 0.076 | 0.843 | 11.64 | 0.300 | 0.077 | 0.873 | 11.59 | 0.247 |
| PF-AFN | - | - | - | - | - | - | - | | - | - | 14.01 | 0.588 |
| HR-VITON | **0.062** | **0.864** | **9.38** | **0.153** | **0.061** | **0.878** | **9.90** | **0.188** | **0.065** | **0.892** | **10.91** | **0.179** |

person image. For paired setting, we evaluate our method using two widely-used metrics: Structural Similarity (SSIM) [27] and Learned Perceptual Image Patch Similarity (LPIPS) [31]. Additionally, to evaluate the unpaired setting, we measure Frechet Inception Distance (FID) [10] and Kernel Inception Distance (KID), which is a more descriptive metric than FID when the number of data is small.

**Ablation Study.** Table 1 shows the effectiveness of the proposed feature fusion block and condition aligning. Indeed, the benefits of fusion block and condition aligning are largely additive. Notably, the model without feature fusion block and condition aligning yields suboptimal results, demonstrating the necessity of information exchange between the warping module and the segmentation map generator.

**Comparison with Baselines.** Table 2 demonstrates that our method outperforms the baselines for all evaluation metrics, especially at the 1024 × 768 resolution. The results indicate that CP-VTON and ACGPN can not handle the high-resolution images in the unpaired setting. Furthermore, it is noteworthy that our framework surpasses VITON-HD, one of the state-of-the-art methods for high-resolution virtual try-on. Although our try-on image generator is very similar to one of VITON-HD, our framework has superior performance due to the capability to produce high-quality conditions (*i.e.*, segmentation map and warped clothing image).

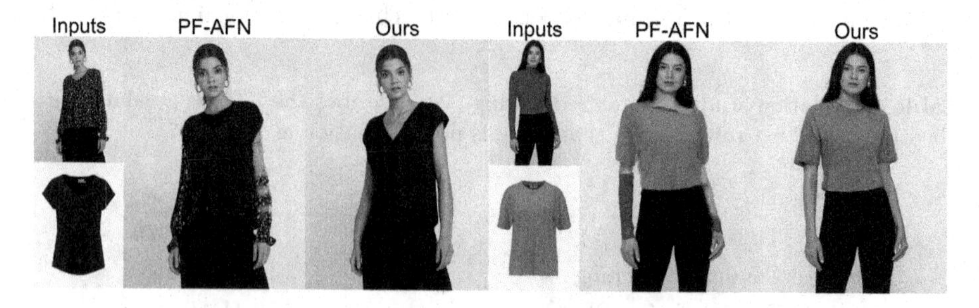


**Fig. 9.** Qualitative comparison with PF-AFN on 1024 × 768 resolution.

### 4.4 Comparison with Parser-free Virtual Try-on Methods

Recently, several approaches [4,11] propose virtual try-on models that do not rely on a predicted segmentation map. However, explicitly predicting a segmentation map helps the model distinguish the regions to be generated and the regions to be preserved, which is necessary for a high-resolution virtual try-on. To verify this, we compare our model with PF-AFN [4] on the high-resolution dataset. Figure 9 demonstrates that PF-AFN fails to remove the original clothing regions as it can not differentiate the parts to be generated and the parts to be left, resulting in significant artifacts in the outputs. Moreover, Table 2 shows that our model outperforms PF-AFN by a large margin. The results indicate that it is difficult to obtain convincing high-resolution results without predicting a segmentation map.

## 5 Discussion

**Limitation of Discriminator Rejection.** The existing image-based virtual try-on approaches assume that test data is drawn from the same distribution as the training data. However, in the real-world scenario, it is prevalent that the input images are taken at a different camera view from the training images or even do not contain humans. Since the low-quality segmentation is often predicted due to such out-of-distribution inputs, our discriminator rejection is capable of filtering out the out-of-distribution inputs. We believe that our discriminator rejection can be a solution to enhance the user experience in virtual try-on applications.

## 6 Conclusion

In this paper, we propose a novel architecture for high-resolution virtual, which performs warping clothes and segmentation generation simultaneously while exchanging information with each other. The proposed try-on condition generator completely eliminates the misaligned region and solves the pixel-squeezing problem by handling the occlusion by body parts. We also demonstrate that the discriminator of the condition generator can filter out the impaired segmentation results, which is practically helpful for real-world virtual try-on applications. Extensive experiments show that our method outperforms the existing virtual try-on methods at $1024 \times 768$ resolution.

**Acknowledgement.** This work was supported by the Institute of Information & communications Technology Planning & Evaluation (IITP) grant funded by the Korea government(MSIT) (No. 2019-0-00075, Artificial Intelligence Graduate School Program(KAIST) and No.2021-0-02068, Artificial Intelligence Innovation Hub) and the National Research Foundation of Korea (NRF) grant funded by the Korean government (MSIT) (No. NRF-2022R1A2B5B02001913).

## References

1. Azadi, S., Olsson, C., Darrell, T., Goodfellow, I., Odena, A.: Discriminator rejection sampling. arXiv preprint arXiv:1810.06758 (2018)
2. Choi, S., Park, S., Lee, M., Choo, J.: VITON-HD: high-resolution virtual try-on via misalignment-aware normalization. In: Proceedings of the IEEE/CVF Conference on Computer Vision and Pattern Recognition, pp. 14131–14140 (2021)
3. Chopra, A., Jain, R., Hemani, M., Krishnamurthy, B.: ZFlow: gated appearance flow-based virtual try-on with 3D priors. In: Proceedings of the IEEE/CVF International Conference on Computer Vision, pp. 5433–5442 (2021)
4. Ge, Y., Song, Y., Zhang, R., Ge, C., Liu, W., Luo, P.: Parser-free virtual try-on via distilling appearance flows. In: Proceedings of the IEEE/CVF Conference on Computer Vision and Pattern Recognition, pp. 8485–8493 (2021)
5. Gong, K., Liang, X., Li, Y., Chen, Y., Yang, M., Lin, L.: Instance-level human parsing via part grouping network. In: Proceedings of the European Conference on Computer Vision (ECCV), pp. 770–785 (2018)
6. Guan, P., Reiss, L., Hirshberg, D.A., Weiss, A., Black, M.J.: DRAPE: dressing any person. ACM Trans. Graph. (TOG) **31**(4), 1–10 (2012)
7. Güler, R.A., Neverova, N., Kokkinos, I.: DensePose: dense human pose estimation in the wild. In: Proceedings of the IEEE Conference on Computer Vision and Pattern Recognition (CVPR), pp. 7297–7306 (2018)
8. Han, X., Hu, X., Huang, W., Scott, M.R.: ClothFlow: a flow-based model for clothed person generation. In: Proceedings of the IEEE International Conference on Computer Vision (ICCV), pp. 10471–10480 (2019)
9. Han, X., Wu, Z., Wu, Z., Yu, R., Davis, L.S.: VITON: an image-based virtual try-on network. In: Proceedings of the IEEE Conference on Computer Vision and Pattern Recognition (CVPR), pp. 7543–7552 (2018)
10. Heusel, M., Ramsauer, H., Unterthiner, T., Nessler, B., Hochreiter, S.: GANs trained by a two time-scale update rule converge to a local Nash equilibrium. In: Proceedings of the Advances in Neural Information Processing Systems (NeurIPS) (2017)
11. Issenhuth, T., Mary, J., Calauzènes, C.: Do not mask what you do not need to mask: a parser-free virtual try-on. In: Vedaldi, A., Bischof, H., Brox, T., Frahm, J.-M. (eds.) ECCV 2020. LNCS, vol. 12365, pp. 619–635. Springer, Cham (2020). https://doi.org/10.1007/978-3-030-58565-5_37
12. Jandial, S., Chopra, A., Ayush, K., Hemani, M., Krishnamurthy, B., Halwai, A.: SieveNet: a unified framework for robust image-based virtual try-on. In: Proceedings of the IEEE/CVF Winter Conference on Applications of Computer Vision, pp. 2182–2190 (2020)
13. Jetchev, N., Bergmann, U.: The conditional analogy GAN: swapping fashion articles on people images. In: Proceedings of the IEEE International Conference on Computer Vision Workshop (ICCVW), pp. 2287–2292 (2017)
14. Lewis, K.M., Varadharajan, S., Kemelmacher-Shlizerman, I.: VOGUE: try-on by StyleGAN interpolation optimization. arXiv e-prints, pp. arXiv-2101 (2021)
15. Li, K., Chong, M.J., Zhang, J., Liu, J.: Toward accurate and realistic outfits visualization with attention to details. In: Proceedings of the IEEE/CVF Conference on Computer Vision and Pattern Recognition, pp. 15546–15555 (2021)
16. Mao, X., Li, Q., Xie, H., Lau, R.Y., Wang, Z., Paul Smolley, S.: Least squares generative adversarial networks. In: Proceedings of the IEEE International Conference on Computer Vision, pp. 2794–2802 (2017)

17. Minar, M.R., Ahn, H.: CloTH-VTON: clothing three-dimensional reconstruction for hybrid image-based virtual try-on. In: Proceedings of the Asian Conference on Computer Vision (2020)
18. Mo, S., Kim, C., Kim, S., Cho, M., Shin, J.: Mining gold samples for conditional GANs. In: Advances in Neural Information Processing Systems, vol. 32 (2019)
19. Park, T., Liu, M.Y., Wang, T.C., Zhu, J.Y.: Semantic image synthesis with spatially-adaptive normalization. In: Proceedings of the IEEE/CVF Conference on Computer Vision and Pattern Recognition, pp. 2337–2346 (2019)
20. Patel, C., Liao, Z., Pons-Moll, G.: TailorNet: predicting clothing in 3D as a function of human pose, shape and garment style. In: Proceedings of the IEEE Conference on Computer Vision and Pattern Recognition (CVPR), pp. 7365–7375 (2020)
21. Pons-Moll, G., Pujades, S., Hu, S., Black, M.J.: ClothCap: seamless 4D clothing capture and retargeting. ACM Trans. Graph. (TOG) **36**(4), 1–15 (2017)
22. Razavi, A., Van den Oord, A., Vinyals, O.: Generating diverse high-fidelity images with VQ-VAE-2. In: Advances in Neural Information Processing Systems, vol. 32 (2019)
23. Sekine, M., Sugita, K., Perbet, F., Stenger, B., Nishiyama, M.: Virtual fitting by single-shot body shape estimation. In: International Conference on 3D Body Scanning Technologies, pp. 406–413. Citeseer (2014)
24. Turner, R., Hung, J., Frank, E., Saatchi, Y., Yosinski, J.: Metropolis-hastings generative adversarial networks. In: International Conference on Machine Learning, pp. 6345–6353. PMLR (2019)
25. Wang, B., Zheng, H., Liang, X., Chen, Y., Lin, L., Yang, M.: Toward characteristic-preserving image-based virtual try-on network. In: Proceedings of the European Conference on Computer Vision (ECCV), pp. 589–604 (2018)
26. Wang, T.C., Liu, M.Y., Zhu, J.Y., Tao, A., Kautz, J., Catanzaro, B.: High-resolution image synthesis and semantic manipulation with conditional GANs. In: Proceedings of the IEEE Conference on Computer Vision and Pattern Recognition, pp. 8798–8807 (2018)
27. Wang, Z., Bovik, A.C., Sheikh, H.R., Simoncelli, E.P.: Image quality assessment: from error visibility to structural similarity. IEEE Trans. Image Process. **13**(4), 600–612 (2004)
28. Xie, Z., et al.: WAS-VTON: warping architecture search for virtual try-on network. In: Proceedings of the 29th ACM International Conference on Multimedia, pp. 3350–3359 (2021)
29. Yang, H., Zhang, R., Guo, X., Liu, W., Zuo, W., Luo, P.: Towards photo-realistic virtual try-on by adaptively generating-preserving image content. In: Proceedings of the IEEE Conference on Computer Vision and Pattern Recognition (CVPR), pp. 7850–7859 (2020)
30. Yu, R., Wang, X., Xie, X.: VTNFP: an image-based virtual try-on network with body and clothing feature preservation. In: Proceedings of the IEEE International Conference on Computer Vision (ICCV), pp. 10511–10520 (2019)
31. Zhang, R., Isola, P., Efros, A.A., Shechtman, E., Wang, O.: The unreasonable effectiveness of deep features as a perceptual metric. In: Proceedings of the IEEE Conference on Computer Vision and Pattern Recognition (CVPR) (2018)

# A Codec Information Assisted Framework for Efficient Compressed Video Super-Resolution

Hengsheng Zhang[1], Xueyi Zou[2], Jiaming Guo[2], Youliang Yan[2], Rong Xie[1], and Li Song[1,3](✉)

[1] Institute of Image Communication and Network Engineering, Shanghai Jiao Tong University, Shanghai, China
{hs_zhang,xierong,song_li}@sjtu.edu.cn
[2] Huawei Noah's Ark Lab, Shenzhen, China
{zouxueyi,guojiaming5,yanyouliang}@huawei.com
[3] MoE Key Lab of Artifical Intelligence, AI Institute, Shanghai Jiao Tong University, Shanghai, China

**Abstract.** Online processing of compressed videos to increase their resolutions attracts increasing and broad attention. Video Super-Resolution (VSR) using recurrent neural network architecture is a promising solution due to its efficient modeling of long-range temporal dependencies. However, state-of-the-art recurrent VSR models still require significant computation to obtain a good performance, mainly because of the complicated motion estimation for frame/feature alignment and the redundant processing of consecutive video frames. In this paper, considering the characteristics of compressed videos, we propose a Codec Information Assisted Framework (CIAF) to boost and accelerate recurrent VSR models for compressed videos. Firstly, the framework reuses the coded video information of Motion Vectors to model the temporal relationships between adjacent frames. Experiments demonstrate that the models with Motion Vector based alignment can significantly boost the performance with negligible additional computation, even comparable to those using more complex optical flow based alignment. Secondly, by further making use of the coded video information of Residuals, the framework can be informed to skip the computation on redundant pixels. Experiments demonstrate that the proposed framework can save up to 70% of the computation without performance drop on the REDS4 test videos encoded by H.264 when CRF is 23.

**Keywords:** Efficient video super-resolution · Compressed video · Codec information assisted · Motion vectors · Residuals

## 1 Introduction

Compressed videos are prevalent on the Internet, ranging from movies, webcasts to user-generated videos, most of which are of relatively low resolutions and

---

**Supplementary Information** The online version contains supplementary material available at https://doi.org/10.1007/978-3-031-19790-1_14.

© The Author(s), under exclusive license to Springer Nature Switzerland AG 2022
S. Avidan et al. (Eds.): ECCV 2022, LNCS 13677, pp. 220–235, 2022.
https://doi.org/10.1007/978-3-031-19790-1_14

qualities. Many terminal devices, such as smartphones, tablets, and TVs, come with a 2K/4K or even 8K definition screen. Thus, there is an urgent demand for such devices to be able to online super-resolve the low-resolution videos to the resolution of the screen definition. Video Super-Resolution (VSR) increases the video frames' resolution by exploiting redundant and complementary information along the video temporal dimension. With the wide use of neural networks in computer vision tasks, on the one hand, neural network based VSR methods outperform traditional ones. But on the other hand, they require a lot of computation and memory, which current commercial terminal devices cannot easily provide.

Most neural network based VSR models come with a lot of repeated computation or memory consumption. For example, sliding-window based VSR models [5,10,25,27] have to extract the features of adjacent frames repeatedly. Although this process can be optimized by preserving the feature maps of previous frames, it increases memory consumption. Besides, to make the most of adjacent frames' information, frame alignment is an essential part of many such models, which is usually implemented by optical flow prediction [21,24], deformable convolution [6,34], attention/correlation [16], and other complicated modules [13,32]. This frame alignment process also increases model complexity, and many of the operators are not well supported by current terminal chipsets.

Many VSR methods use recurrent neural networks to avoid repeated feature extraction and to exploit long-range dependencies. The previous frame's high-resolution information (image or features) is reused for the current frame prediction. Several information propagation schemes have been proposed, such as unidirectional propagation [8,11,23], bidirectional propagation [2,17], and the more complex grid propagation [3,31]. As expected, the more complex the propagation scheme is, the better the super-resolution performs in terms of PSNR/SSIM or visual quality. However, considering the stringent computational budget of terminal devices and the online processing requirement, most complex propagation schemes, such as bidirectional propagation and grid propagation, are not good choices. Unidirectional recurrent models seem to be good candidates, but to get better performance, frame/feature alignment is also indispensable. As mentioned above, mainstream methods for alignment are computationally heavy and not well supported by current terminal chipsets.

Compared with raw videos, compressed videos have some different characteristics. When encoding, the motion relationships of the current frame and a reference frame (e.g. the previous frame) are calculated as **Motion Vectors** (MVs). The reference frame is then warped according to MVs to get the predicted image of the current time step. The differences between the predicted image and current frame are calculated as **Residuals**. MVs and Residuals are encoded in the video streams, with MVs providing motion cues of video frames and Residuals indicating the motion-compensated differences between frames. When decoding, MVs and Residuals are extracted to rebuild the video frames sequentially based on the previous rebuilt frames.

By leveraging the characteristics of compressed videos, we propose a Codec Information Assisted Framework (CIAF) to improve the performance and the efficiency of unidirectional recurrent VSR methods. To align the features of previous frame, we reuse the MVs to model the temporal relationships between adjacent frames. The models using MV-based alignment can significantly boost the performance with negligible additional computation, even reaching a comparable performance with those using more complex optical flow based alignment. To further reduce terminal device computation burden, we apply most computation (convolutions) only to changed regions of consecutive frames. For the rest areas, we reuse features of the previous frame by warping part of the feature maps generated in the last step according to MVs. The way to determine where the change happens is based on Residuals, i.e., only pixels with Residuals not equal to zero are considered to be changed. Due to the high degree of similarity between video frames, the proposed approach can skip lots of computation. The experiments show up to 70% of computation can be saved without performance drop on the REDS4 [27] test videos encoded by H.264 when CRF is 23.

The contributions of this paper can be summarized as follows. (1) We propose to reuse the coded video information of MVs to model temporal relationships between adjacent frames for frame/feature alignment. Models with MV-based alignment can significantly boost performance with minimal additional computation, even matching the performance of optical flow based models. (2) We find that the coded information of Residuals can inform the VSR models to skip the computation on redundant pixels. The models using Residual-informed sparse processing can save lots of computation without a performance drop. (3) We disclose some of the crucial tricks to train the CIAF, and we evaluate some of the essential design considerations contributing to the efficient compressed VSR model.

## 2 Related Work

In this section, we first review the CNN-based video super-resolution work. Then, we discuss adaptive CNN acceleration techniques related to our work.

### 2.1 Video Super-Resolution

Video super-resolution (VSR) is challenging because complementary information must be aggregated across misaligned video frames for restoration. There are mainly two forms of VSR algorithms: sliding-window methods and recurrent methods.

**Sliding-Window Methods.** Sliding-window methods restore the target high-resolution frame from the current and its neighboring frames. [1,30] align the neighboring frames to the target frame with predicted optical flows between input frames. Instead of explicitly aligning frames, RBPN [10] treats each context frame as a separate source of information and employs back-projection for iterative refining of target HR features. DUF [13] utilizes generated dynamic upsampling filters to handle motions implicitly. Besides, deformable convolutions (DCNs) [6,34] are introduced to express temporal relationships. TDAN

[25] aligns neighboring frames with DCNs in the feature space. EDVR [27] uses DCNs on a multi-scale basis for more precise alignment. MuCAN [16] searches similar patches around the target position from neighboring frames instead of direct motion estimation. [5] extracts Motion Vectors from compressed video streams as motion priors for alignment and incorporates coding priors into modified SFT blocks [28] to refine the features from the input LR frames. These methods can produce pleasing results, but they are challenging to be applied in practice on the terminal devices due to repeated feature extraction or complicated motion estimation.

**Recurrent Methods.** Unlike sliding-window methods, recurrent methods take the output of the past frame processing as a prior input for the current iteration. So the recurrent networks are not only efficient but also can take account of long-range dependencies. In unidirectional recurrent methods FRVSR [23], RLSP [8] and RSDN [11], information is sequentially propagated from the first frame to the last frame, so this kind of scheme has the potential to be applied for online processing. Besides, FRVSR [23] aligns the past predicted HR frame with optical flows for the current iteration. RLSP [8] and RSDN [11] employs high-dimensional latent states to implicitly transfer temporal information between frames. Different from unidirectional recurrent networks, BasicVSR [2] proposes a bidirectional propagation scheme to better exploit temporal features. BasicVSR++ [3] redesigns BasicVSR by proposing second-order grid propagation and flow-guided deformable alignment. Similar with BasicVSR++, [31] employs complex grid propagation to boost the performance. COMISR [17] applies a bidirectional recurrent model to compressed video super-resolution and uses a CNN to predict optical flows for alignment. Although they can achieve state-of-the-art performance, the complicated information propagation scheme and complex motion estimation make them unpractical to apply to the terminal device with online processing.

### 2.2 Adaptive Inference

Most of the existing CNN methods treat all regions in the image equally. But the flat area is naturally easier to process than regions with textures. Adaptive inference can adapt the network structure according to the characteristics of the input. BlockDrop [29] proposes to dynamically pick which deep network layers to run during inference to decrease overall computation without compromising prediction accuracy. ClassSR [14] uses a "class module" to decompose the image into sub-images with different reconstruction difficulties and then applies networks with various complexity to process them separately. Liu et al. [19] establishes adaptive inference for SR by adjusting the number of convolutional layers used at various locations. Wang et al. [26] locate redundant computation by predicted spatial and channel masks and use sparse convolution to skip redundant computation. The image-based acceleration algorithms follow the internal characteristics of images, so they can only reduce spatial redundancy.

Most of the time, the changes between consecutive frames in a video are insignificant. Based on this observation, Skip-Convolutions [9] limits the computation only to the regions with significant changes between frames while skipping

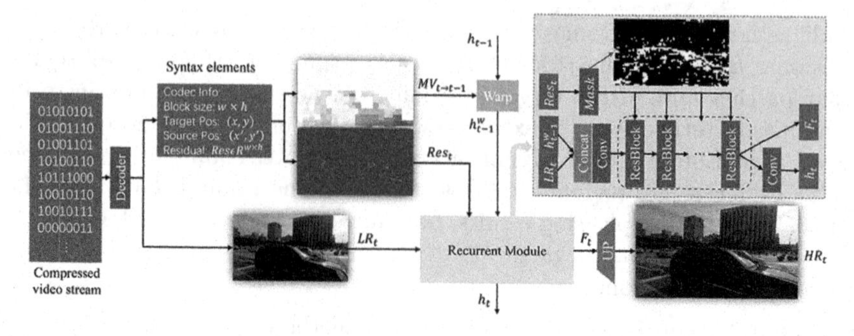


**Fig. 1.** Overview of the proposed codec information assisted framework (CIAF). The $h_{t-1}$ is the refined features from past frame $LR_{t-1}$. Motion Vector ($MV_{t \to t-1}$) and Residuals ($Res_t$) are the codec information. In our model, we utilize the Motion Vector to align the features from the past frame. Besides, the sparse processing is applied in the Resblocks only to calculate the regions with Residuals.

the others. But this model is primarily applicable to high-level tasks. FAST [33], the most similar work with ours, employs SRCNN [7] to only generate the HR image of the first frame in a group of frames. In the following iterations, the HR blocks of the last frame are transferred to the current frame according to MVs. Finally, the up-sampled Residuals are added to the transferred HR image to generate the HR output of the current frame. The operations are on the pixel level, which can easily lead to errors. Instead of directly reusing the HR pixels from past frames, we utilize MVs to conduct an efficient alignment for unidirectional recurrent VSR systems. And the Residuals are used to determine the locations of redundancy.

## 3 Codec Information Assisted Framework

In this section, we first introduce the basics of video coding related to our framework. Then we present our codec information assisted framework (CIAF, Fig. 1) consisting of two major parts, i.e., the Motion Vector (MV) based alignment and Residual informed sparse processing.

### 3.1 Video Coding Basics

The Inter-Prediction Mode (Fig. 2) of video codec inspires our framework. Generally, there is a motion relationship between the objects in each frame and its adjacent frames. The motion relationship of this kind of object constitutes the temporal redundancy between frames. In H.264 [22], temporal redundancy is reduced by motion estimation and motion compensation. As Fig. 2 shows, in motion estimation, for every current block, we can find a similar pixel block as a reference in the reference frame. The relative position between the current pixel block in the current frame and the reference block in the reference frame is represented by $(MV_x, MV_y)$, a vector of two coordinate values used to indicate this relative position, known as the **Motion Vector (MV)**. In motion compensation, we use the found reference block as a prediction of the current block. Because there are slight differences between the current and reference blocks, the encoder needs to calculate the differences as **Residual**. When decoding, we first use the decoded reference frame and MVs to generate the prediction image of the target frame. Then we add decoded Residuals to the prediction image to get the target frame. In our paper, we reuse the MVs and Residuals to increase the efficiency of unidirectional recurrent VSR models.

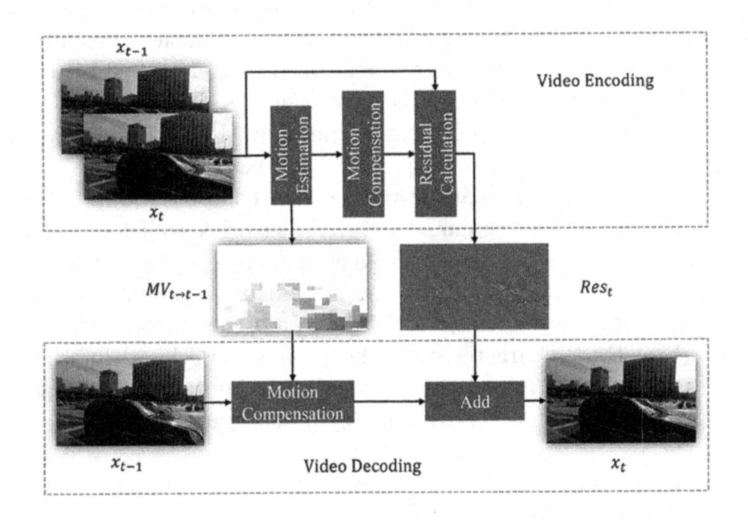


**Fig. 2.** The Inter-Prediction Mode of video codec.

### 3.2 Motion Vector Based Alignment

In VSR methods, alignment between neighboring frames is important for good performance. In this paper, for alignment, we warp the HR information of the past frame with MVs. Different from the interpolation filter used in H.264, the bilinear interpolation filter is applied to the pixels for efficiency if the MV is fractional. When there is an insufficient temporal connection between blocks, the video encoder utilizes intra-prediction. Since the intra-blocks mainly appear in the keyframe (the first frame of a video clip) and there are few intra-predicted blocks in most frames, for blocks with intra-prediction, we transfer the features of the same position in the adjacent frame. To a common format, we set $MV = (0, 0)$ for intra-blocks. We can formulate a motion field MV with size $H \times W \times 2$ like optical flow. $H$ and $W$ are the height and width of the input LR frame, respectively. The third dimension indicates the relative position in the width and height directions. So the MV is an approximate alternative to optical flow. In this way, we bypass the complicated motion estimation. The

MV-based alignment can boost the performance of existing unidirectional recurrent VSR models and even achieve comparable performance with optical flow based alignment, as demonstrated later.

### 3.3 Residual Informed Sparse Processing

As Fig. 1 shows, in the paper, we design a Residual informed sparse processing to reduce redundant computation. Residuals represent the difference between the warped frame and the current frame. The areas without Residuals indicate the current region can be directly predicted by sharing the corresponding patches from the reference frame. Therefore, Residuals can locate the areas that need to be further refined. With the guide of Residuals, we only make convolutions on the "important" pixels. The features of the rest pixels are enhanced by aggregation with the MV-warped features from the past frame. As Fig. 1 shows, to make it robust, we adopt this sparse processing to the body (Resblocks) of the network, the head and tail Conv layers are applied on all pixels.

Benifict from motion estimation and motion compensation, we can easily predict the flat regions or regular structures like brick wall for current frame according to the contents of adjacent frames without loss (Residuals). Residuals are more likely to be introduced on complex textures. Because flat regions or regular structures take up the majority of the frame, Residuals are sparse in most scenes. Based on these characteristics, the proposed Residual informed sparse processing can significantly reduce the space-time redundancy computation while maintaining the comparable performance with baseline.

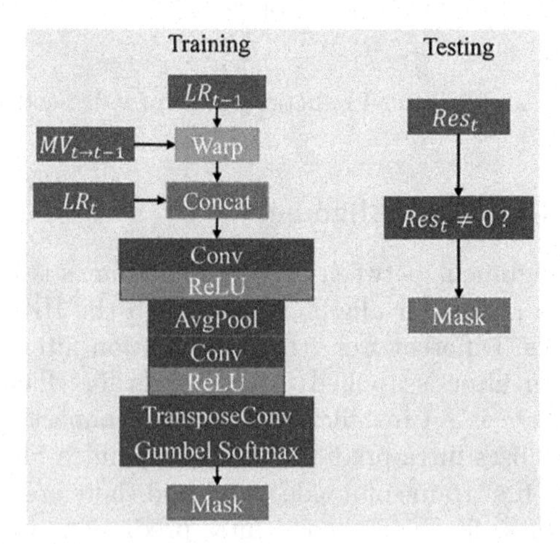


**Fig. 3.** The sparse mask generation. $Res_t$ is the Residual extracted from compressed video. When training, we use a tiny CNN to predict a spatial mask; when testing, convolutions are only applied to pixels whose Residual is not equal to 0.

Because the Residuals are sparse, only a tiny part of pixels optimize the model if we directly utilize Residuals to decide where to conduct convolutions during training. In experiments, we find it hard to converge. We design a Simulated Annealing strategy to slowly reduce the number of pixels involved in training, which is a critical trick in our sparse processing. As Fig. 3 shows, we utilize a light CNN model to identify the changed regions according to the current frame and the MV-warped past frame. Following [26], Gumbel softmax trick [12] is used to produce a spatial mask $M \in R^{H \times W}$ with the output features $F \in R^{2 \times H \times W}$.

$$M[x,y] = \frac{exp((F[1,x,y] + G[1,x,y])/\tau)}{\sum_{i=1}^{2} exp((F[i,x,y] + G[i,x,y])/\tau)} \tag{1}$$

where $x$ and $y$ are vertical and horizontal indices, $G \in R^{2 \times H \times W}$ is a Gumbel noise vector with all elements following $Gumbel(0,1)$ distribution and $\tau$ is the temperature parameter. Samples from Gumbel softmax distribution become uniform if $\tau \rightarrow \infty$. When $\tau \leftarrow 0$, samples from Gumbel softmax distribution become one-hot. The predicted mask gradually becomes sparse with training.

**Training Strategy:** During training, we utilize a sparsity regularization loss to supervise the model:

$$L_{reg} = \frac{1}{H \times W} \sum_{h,w} M[w,h] \tag{2}$$

According the Simulated Annealing strategy, we set the weight of $L_{reg}$:

$$\lambda = min(\frac{t}{T_{epoch}}, 1) \cdot \lambda_0 \tag{3}$$

where $t$ is the current number of epochs, $T_{epoch}$ is empirically set to 20, and $\lambda_0$ is set to 0.004. And the temperature parameter $\tau$ in the Gumbel softmax trick is initialized as 1 and gradually decreased to 0.5:

$$\tau = max(1 - \frac{t}{T_{temp}}, 0.5) \tag{4}$$

where $T_{temp}$ is set to 40 in this paper.

**Testing:** When testing, we directly replace the mask-prediction CNN with Residuals to select the pixels to calculate. This process is formulated as:

$$M_{test}[x,y] = (Res[x,y] \neq 0) \tag{5}$$

where $Res[x,y]$ represents the Residual value at position $[x,y]$. When Residual is equal to 0, the pixel is skipped.

# 4 Experiments

## 4.1 Implementation Details

We use dataset REDS [20] for training. REDS dataset has large motion between consecutive frames captured from a hand-held device. We evaluate the networks on the datasets REDS4 [27] and Vid4 [18]. All frames are first smoothed by a Gaussian kernel with the width of 1.5 and downsampled by 4. Because our framework is designed for compressed videos, we further encode the datasets with H.264 [22], the most common video codec, at different compression rates. The recommended CRF value in H.264 is between 18 and 28, and the default is 23. In experiments, we set CRF values to 18, 23, and 28 and use the FFmpeg codec to encode the datasets.

Our goal is to design efficient and online processing VSR systems, so we do experiments on the unidirectional recurrent VSR models. We apply our MV-based alignment to the existing models FRVSR [23], RLSP [8], and RSDN [11] to verify the effect of our MV-based alignment. In the original setting, FRVSR utilizes an optical flow to align the HR output from the past frame; RLSP and RSDN do not explicitly align the information from the previous frame. For a more comprehensive comparison, we also embed a pre-trained optical flow model SpyNet [21] into FRVSR, RLSP and RSDN to compare with our MV-based alignment. And we further fine-tune the SpyNet along with the model training. The training details follow the original works.

To evaluate the Residual informed sparse process, we first train a baseline recurrent VSR model without alignment. Then we apply MV-based alignment and Residual-based sparse processing to the baseline model to train our model. To balance model complexity and performance, the number of Resblocks for the recurrent module is set to 7. The number of feature channels is 128. We use Charbonnier loss [4] as pixel-wise loss since it better handles outliers and improves the performance over the conventional L2-loss [15]. The training details are provided in the supplementary material.

### 4.2 Effect of MV-Based Alignment

We apply our MV-based alignment approach to the FRVSR, RLSP, and RSDN. The quantitative results are summarized in Table 1. XXX+Flow means that model XXX is aligned with the SpyNet. XXX+MV represents that model XXX is aligned with MVs. Original FRVSR aligns the HR estimation from the past frame by an optical flow model trained from scratch. In FRVSR+FLow, we replace the original optical flow model with pre-trained SpyNet and further refine the SpyNet when training. From the results, we can find FRVSR+Flow outperforms the original FRVSR. Probably because SpyNet estimates the optical flow more precisely than the original model. RLSP and RSDN do not explicitly align the information from the past frame. Due to the alignment, models with MV-based alignment achieve better performance than their original counterparts, even achieving comparable performance with the models with SpyNet. And we can see that as the CRF is increased, the performance gap between optical flow-based methods and MV-based methods narrows, which makes sense since when the CRF is large, the video compression artifacts are more apparent, and the optical flow estimate mistakes are more significant. So our MV-based alignment can replace the existing optical flow estimation model in unidirectional recurrent VSR models to save computation. For RLSP and RSDN, our approach can

**Table 1. The quantitative comparison (PSNR/ SSIM/ LPIPS)** on REDS4 [27]. PSNR is calculated on Y-channel; SSIM and LPIPS are calculated on RGB-channel. Red and blue colors indicate the best and the second-best performance, respectively. 4× upsampling is performed.

| Model | Compressed Results | | | Params (M) | Runtime (ms) |
|---|---|---|---|---|---|
| | CRF18 | CRF23 | CRF28 | | |
| FRVSR [23] | 28.27/0.7367/0.3884 | 27.34/0.6965/0.4495 | 26.11/0.6492/0.5219 | 2.59 | 24 |
| FRVSR+MV | 29.01/0.7660/0.3470 | 27.77/0.7155/0.4141 | 26.32/0.6598/0.4969 | 0.84 | 20 |
| FRVSR+Flow | 29.15/0.7701/0.3393 | 27.85/0.7177/0.4076 | 26.32/0.6600/0.4928 | 2.28 | 32 |
| RLSP [8] | 28.46/0.7476/0.3614 | 27.47/0.7052/0.4243 | 26.20/0.6551/0.5015 | 4.37 | 27 |
| RLSP+MV | 29.26/0.7739/0.3309 | 27.95/0.7225/0.3973 | 26.43/0.6646/0.4815 | 4.37 | 28 |
| RLSP+Flow | 29.37/0.7769/0.3249 | 28.01/0.7242/0.3947 | 26.44/0.6651/0.4788 | 5.81 | 39 |
| RSDN [11] | 28.67/0.7575/0.3405 | 27.62/0.7144/0.3997 | 26.29/0.6642/0.4731 | 6.18 | 49 |
| RSDN+MV | 29.37/0.7804/0.3163 | 28.02/0.7294/0.3799 | 26.50/0.6724/0.4558 | 6.18 | 51 |
| RSDN+Flow | 29.59/0.7862/0.3094 | 28.13/0.7314/0.3770 | 26.51/0.6739/0.4523 | 7.62 | 62 |

achieve better performance with a tiny increase in runtime because of feature warping. It should be noted that our MV-based alignment does not increase the number of parameters. For FRVSR, because we remove its optical flow submodel, our MV-based alignment can reduce the parameters and runtime but achieve superior performance over the original version.

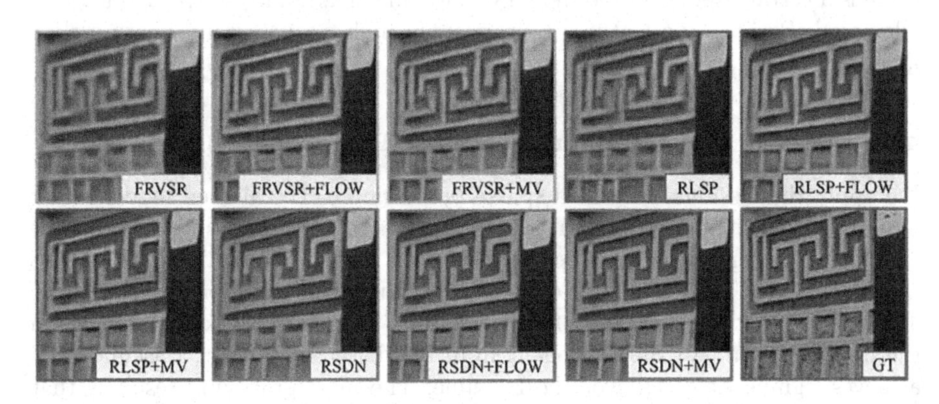


**Fig. 4.** Visual results on REDS4 [27]

Figure 4 shows the qualitative comparison. The models with our MV-based alignment restore finer details than the original FRVSR, RLSP, and RSDN. Compared with the models with optical flow estimation, our MV-aligned models achieve comparable visual results. More examples are provided in the Sect. 2.1 of supplementary material.

**Image Alignment vs Feature Alignment:** As mentioned above, spatial alignment plays an important role in the VSR systems. The existing works with alignment can be divided into two categories: image alignment and feature alignment.

**Table 2. The quantitative comparison (PSNR/ SSIM/ LPIPS)** between image alignment and feature alignment on REDS4 [27]. PSNR is calculated on Y-channel; SSIM and LPIPS are calculated on RGB-channel. The best results are highlighted in bold.

| Model | CRF18 | CRF23 | CRF28 |
|---|---|---|---|
| (a) | 28.59/0.7546/0.3420 | 27.56/0.7122/0.3999 | 26.26/0.6622/0.4719 |
| (b) | **29.32/0.7783/0.3186** | **28.00/0.7273/0.3818** | **26.47/0.6706/0.4569** |
| (c) | 29.11/0.7675/0.3294 | 27.83/0.7172/0.3957 | 26.33/0.6604/0.4787 |

We conduct experiments to analyze each of the categories and explain our design considerations about alignment. We design a recurrent baseline without alignment (Model (a)) and its MV-aligned versions. Model (b) is the MV-aligned model in feature space. And we apply MV-alignment on the HR prediction of the past frame to build a Model (c) with image alignment. The results are summarized in Table 2. The models with alignment outperform the baseline model, which further demonstrates the importance of alignment. And we find Model (b) achieves better performance than Model (c), so the alignment in feature space is more effective than in pixel level. The reason is that MV is block-wise motion estimation, the warped images inevitably suffer from information distortion. But there is a certain degree of redundancy in feature space, and this phenomenon is alleviated. Besides, the features contain more high-frequency information than images.

### 4.3 Effect of Residual Informed Sparse Processing

We apply the Residual informed sparse processing to the aligned model to get a more efficient model. The quantitative results are summarized in Table 3. The Baseline represents the baseline mentioned in Sect. 4.1; Baseline+MV means the MV-aligned model. MV+Res is the Residual-informed sparse processing. The **Sparse rate** is the ratio of pixels skipped by the network to all pixels in the image. As Table 3 shows, benefit from MV-based alignment, Baseline+MV achieves significant gains over the Baseline. The most gratifying result is that our sparse processing with MV-alignment and Residuals achieves a superior or comparable performance over Baseline with lots of computation saved. For the default CRF 23 in FFmpeg, our model can save about 70% computation on REDS4 and Vid4. CRF 18 means that the encoded video is visually lossless. So it needs more Residuals to decrease the encoding error. The sparse processing can save about 50% computation under this condition and achieve better performance than Baseline. For CRF 28, the sparse processing can save much more computation because the Residuals are sparser, and the performance is still comparable with the Baseline.

We conduct qualitative comparisons on datasets REDS4 and Vid4. The results are shown in Fig. 5. The Residual informed model achieves finer details

**Table 3. The quantitative results (PSNR/ SSIM/ Sparse rate)** of Residual informed sparse model on REDS4 [27] and Vid4 [18]. PSNR is calculated on Y-channel; SSIM is calculated on RGB-channel. The Sparse rate is the ratio of pixels skipped by the network to all pixels in the image. Red and blue colors indicate the best and the second-best performance, respectively. $4\times$ upsampling is performed.

| Model | REDS4 [27] | | | Vid4 [18] | | |
|---|---|---|---|---|---|---|
| | CRF18 | CRF23 | CRF28 | CRF18 | CRF23 | CRF28 |
| Baseline | 28.59/0.7546/0. | 27.56/0.7122/0. | 26.26/0.6622/0. | 24.61/0.6668/0. | 23.91/0.6135/0. | 22.87/0.5429/0. |
| Baseline+MV | 29.32/0.7783/0. | 28.00/0.7273/0. | 26.47/0.6706/0. | 25.13/0.6990/0. | 24.20/0.6355/0. | 23.01/0.5557/0. |
| MV+Res | 29.03/0.7639/0.56 | 27.72/0.7131/0.75 | 26.15/0.6516/0.89 | 25.02/0.6800/0.49 | 24.04/0.6132/0.72 | 22.81/0.5333/0.90 |

than the Baseline. More examples are provided in the Sect. 2.2 of supplementary material.

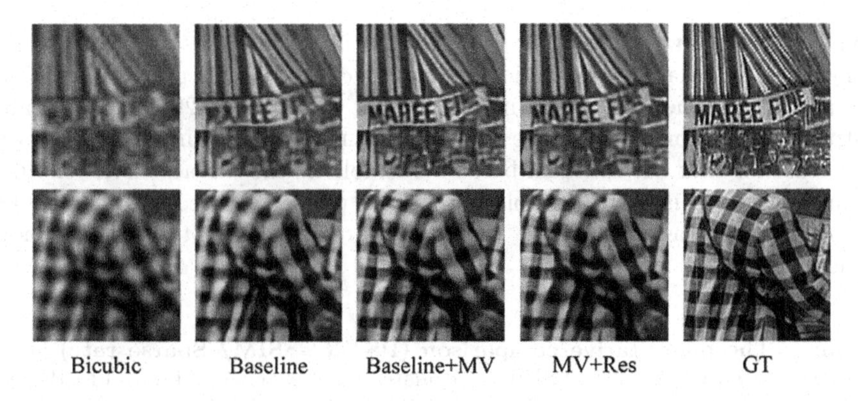


**Fig. 5.** Visual results of the Residual informed sparse process on Vid4 [18] and REDS4 [27]

**CNN-Based Mask vs Residual-Based Mask:** We use a light CNN to predict the spatial mask for our Residual informed sparse processing during training. And when testing, we directly extract the Residuals from compressed videos to generate the spatial mask. In this section, we analyze the characteristics of the CNN-predicted mask and Residual-generated mask. As Fig. 6 shows, we can quickly identify the contours of objects and locate the details and textures from CNN-based masks. The Residual-based masks focus on the errors between the recurrent frame and the MV-warped past frame. Because Residuals are more likely to appear in the areas with details, the highlights of Residual-based masks also follow the location of details. Besides, the CNN-based masks are more continuous than the Residual-based mask. We also present the performance of the models with CNN-based mask and Residual-based mask in Table 4. The results show that different from the Residual-based mask, the Sparse rate of CNN-base masks changes little with different CRF. So the CNN-based mask only highlights the main objects in the image. The Residual-based masks focus on the errors

**Fig. 6.** Visual results of the spatial mask on REDS4 [27]

about MV-based alignment. For CRF 18, the information loss is slight, so the amount of Residuals is large, and the model achieves better performance than the model with the CNN-based mask. And for CRF 23 and 28, our model also outperforms the model with the CNN-based mask with a similar Sparse rate. The reason is that our Residual-based model follows the characteristics of video compression and is more suitable for models with MV-based alignment. Our Residual-based mask locates the "important" areas that need to be refined more precisely.

**Table 4. The quantitative comparison (PSNR/ SSIM/ Sparse rate)** about spatial mask on REDS4 [27]. PSNR is calculated on Y-channel; SSIM and LPIPS are calculated on RGB-channel. The best results are highlighted in bold.

| Model | | CNN Mask | Res Mask |
|---|---|---|---|
| Compression results | CRF18 | 28.82/0.7492/**0.74** | **29.03/0.7639**/0.56 |
| | CRF23 | 27.62/0.7040/**0.76** | **27.72/0.7131**/0.75 |
| | CRF28 | 26.08/0.6456/0.79 | **26.15/0.6516/0.89** |

### 4.4 Temporal Consistency

Figure 7 shows the temporal profile of the video super-resolution results, which is produced by extracting a horizontal row of pixels at the same position from consecutive frames and stacking them vertically. The "ResSparse Model" is the model with our Residual informed sparse processing. The temporal profile produced by the model with our Residual informed sparse processing is temporally smoother, which means higher temporal consistency, and much sharper than the baseline model with about 70% computation of the baseline model saved when CRF is 23.

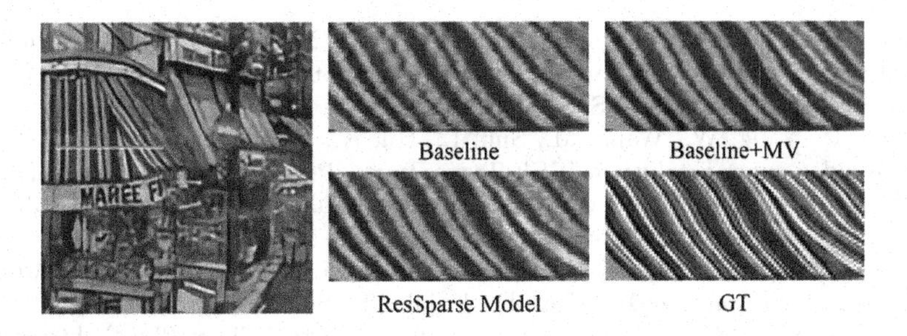


**Fig. 7.** Visualization of temporal profile for the green line on the calendar sequence with CRF 23.(Color figure online)

## 5 Conclusion

This paper proposes to reuse codec information from compressed videos to assist the video super-resolution task. We employ Motion Vector to align mismatched frames in unidirectional recurrent VSR systems efficiently. Experiments have shown that Motion Vector based alignment can significantly improve performance with negligible additional computation. It even achieves comparable performance with optical flow based alignment. To further improve the efficiency of VSR models, we extract Residuals from compressed video and design Residual informed sparse processing. Combined with Motion Vector based alignment, our Residual informed processing can precisely locate the areas needed to calculate and skip the "unimportant" regions to save computation. And the performance of our sparse model is still comparable with the baseline. Additionally, given the importance of motion information for low-level video tasks and the inherent temporal redundancy of videos, our codec information assisted framework (CIAF) has the potential to be applied to other tasks such as compressed video enhancement and denoising.

**Acknowledgement.** The authors Rong Xie and Li Song were supported by National Key R&D Project of China under Grant 2019YFB1802701, the 111 Project (B07022 and Sheitc No.150633) and the Shanghai Key Laboratory of Digital Media Processing and Transmissions.

## References

1. Caballero, J., et al.: Real-time video super-resolution with spatio-temporal networks and motion compensation. In: CVPR, pp. 2848–2857. IEEE Computer Society (2017)
2. Chan, K.C.K., Wang, X., Yu, K., Dong, C., Loy, C.C.: BasicVSR: the search for essential components in video super-resolution and beyond. In: CVPR, pp. 4947–4956. Computer Vision Foundation/IEEE (2021)
3. Chan, K.C.K., Zhou, S., Xu, X., Loy, C.C.: BasicVSR++: improving video super-resolution with enhanced propagation and alignment. CoRR abs/2104.13371 (2021)

4. Charbonnier, P., Blanc-Féraud, L., Aubert, G., Barlaud, M.: Two deterministic half-quadratic regularization algorithms for computed imaging. In: ICIP (2), pp. 168–172. IEEE Computer Society (1994)
5. Chen, P., Yang, W., Wang, M., Sun, L., Hu, K., Wang, S.: Compressed domain deep video super-resolution. IEEE Trans. Image Process. **30**, 7156–7169 (2021)
6. Dai, J., et al.: Deformable convolutional networks. In: ICCV, pp. 764–773. IEEE Computer Society (2017)
7. Dong, C., Loy, C.C., He, K., Tang, X.: Learning a deep convolutional network for image super-resolution. In: Fleet, D., Pajdla, T., Schiele, B., Tuytelaars, T. (eds.) ECCV 2014. LNCS, vol. 8692, pp. 184–199. Springer, Cham (2014). https://doi.org/10.1007/978-3-319-10593-2_13
8. Fuoli, D., Gu, S., Timofte, R.: Efficient video super-resolution through recurrent latent space propagation. In: ICCV Workshops, pp. 3476–3485. IEEE (2019)
9. Habibian, A., Abati, D., Cohen, T.S., Bejnordi, B.E.: Skip-convolutions for efficient video processing. In: CVPR, pp. 2695–2704. Computer Vision Foundation/IEEE (2021)
10. Haris, M., Shakhnarovich, G., Ukita, N.: Recurrent back-projection network for video super-resolution. In: CVPR, pp. 3897–3906. Computer Vision Foundation/IEEE (2019)
11. Isobe, T., Jia, X., Gu, S., Li, S., Wang, S., Tian, Q.: Video super-resolution with recurrent structure-detail network. In: Vedaldi, A., Bischof, H., Brox, T., Frahm, J.-M. (eds.) ECCV 2020. LNCS, vol. 12357, pp. 645–660. Springer, Cham (2020). https://doi.org/10.1007/978-3-030-58610-2_38
12. Jang, E., Gu, S., Poole, B.: Categorical reparameterization with gumbel-softmax. In: ICLR (Poster). OpenReview.net (2017)
13. Jo, Y., Oh, S.W., Kang, J., Kim, S.J.: Deep video super-resolution network using dynamic upsampling filters without explicit motion compensation. In: CVPR, pp. 3224–3232. Computer Vision Foundation/IEEE Computer Society (2018)
14. Kong, X., Zhao, H., Qiao, Y., Dong, C.: ClassSR: a general framework to accelerate super-resolution networks by data characteristic. In: CVPR, pp. 12016–12025. Computer Vision Foundation/IEEE (2021)
15. Lai, W., Huang, J., Ahuja, N., Yang, M.: Deep Laplacian pyramid networks for fast and accurate super-resolution. In: CVPR, pp. 5835–5843. IEEE Computer Society (2017)
16. Li, W., Tao, X., Guo, T., Qi, L., Lu, J., Jia, J.: MuCAN: multi-correspondence aggregation network for video super-resolution. In: Vedaldi, A., Bischof, H., Brox, T., Frahm, J.-M. (eds.) ECCV 2020. LNCS, vol. 12355, pp. 335–351. Springer, Cham (2020). https://doi.org/10.1007/978-3-030-58607-2_20
17. Li, Y., Jin, P., Yang, F., Liu, C., Yang, M., Milanfar, P.: COMISR: compression-informed video super-resolution. CoRR abs/2105.01237 (2021)
18. Liu, C., Sun, D.: A Bayesian approach to adaptive video super resolution. In: CVPR, pp. 209–216. IEEE Computer Society (2011)
19. Liu, M., Zhang, Z., Hou, L., Zuo, W., Zhang, L.: Deep adaptive inference networks for single image super-resolution. In: Bartoli, A., Fusiello, A. (eds.) ECCV 2020. LNCS, vol. 12538, pp. 131–148. Springer, Cham (2020). https://doi.org/10.1007/978-3-030-66823-5_8
20. Nah, S., et al.: NTIRE 2019 challenge on video deblurring and super-resolution: dataset and study. In: CVPR Workshops, pp. 1996–2005. Computer Vision Foundation/IEEE (2019)
21. Ranjan, A., Black, M.J.: Optical flow estimation using a spatial pyramid network. In: CVPR, pp. 2720–2729. IEEE Computer Society (2017)

22. Rec, BI: H.264, advanced video coding for generic audiovisual services (2005)
23. Sajjadi, M.S.M., Vemulapalli, R., Brown, M.: Frame-recurrent video super-resolution. In: CVPR, pp. 6626–6634. Computer Vision Foundation/IEEE Computer Society (2018)
24. Sun, D., Yang, X., Liu, M., Kautz, J.: PWC-Net: CNNs for optical flow using pyramid, warping, and cost volume. In: CVPR, pp. 8934–8943. Computer Vision Foundation/IEEE Computer Society (2018)
25. Tian, Y., Zhang, Y., Fu, Y., Xu, C.: TDAN: temporally-deformable alignment network for video super-resolution. In: CVPR, pp. 3357–3366. Computer Vision Foundation/IEEE (2020)
26. Wang, L., et al.: Learning sparse masks for efficient image super-resolution. CoRR abs/2006.09603 (2020)
27. Wang, X., Chan, K.C.K., Yu, K., Dong, C., Loy, C.C.: EDVR: video restoration with enhanced deformable convolutional networks. In: CVPR Workshops, pp. 1954–1963. Computer Vision Foundation/IEEE (2019)
28. Wang, X., Yu, K., Dong, C., Loy, C.C.: Recovering realistic texture in image super-resolution by deep spatial feature transform. In: CVPR, pp. 606–615. Computer Vision Foundation/IEEE Computer Society (2018)
29. Wu, Z., et al.: BlockDrop: dynamic inference paths in residual networks. In: CVPR, pp. 8817–8826. Computer Vision Foundation/IEEE Computer Society (2018)
30. Xue, T., Chen, B., Wu, J., Wei, D., Freeman, W.T.: Video enhancement with task-oriented flow. Int. J. Comput. Vis. **127**(8), 1106–1125 (2019)
31. Yi, P., et al.: Omniscient video super-resolution. CoRR abs/2103.15683 (2021)
32. Yi, P., Wang, Z., Jiang, K., Jiang, J., Ma, J.: Progressive fusion video super-resolution network via exploiting non-local spatio-temporal correlations. In: ICCV, pp. 3106–3115. IEEE (2019)
33. Zhang, Z., Sze, V.: FAST: a framework to accelerate super-resolution processing on compressed videos. In: CVPR Workshops, pp. 1015–1024. IEEE Computer Society (2017)
34. Zhu, X., Hu, H., Lin, S., Dai, J.: Deformable convnets V2: more deformable, better results. In: CVPR, pp. 9308–9316. Computer Vision Foundation/IEEE (2019)

# Injecting 3D Perception of Controllable NeRF-GAN into StyleGAN for Editable Portrait Image Synthesis

Jeong-gi Kwak[1(✉)], Yuanming Li[1], Dongsik Yoon[1], Donghyeon Kim[1], David Han[2], and Hanseok Ko[1]

[1] Korea University, Seoul, South Korea
kjk8557@korea.ac.kr
[2] Drexel Univiersity, Philadelphia, USA

**Abstract.** Over the years, 2D GANs have achieved great successes in photorealistic portrait generation. However, they lack 3D understanding in the generation process, thus they suffer from multi-view inconsistency problem. To alleviate the issue, many 3D-aware GANs have been proposed and shown notable results, but 3D GANs struggle with editing semantic attributes. The controllability and interpretability of 3D GANs have not been much explored. In this work, we propose two solutions to overcome these weaknesses of 2D GANs and 3D-aware GANs. We first introduce a novel 3D-aware GAN, SURF-GAN, which is capable of discovering semantic attributes during training and controlling them in an unsupervised manner. After that, we inject the prior of SURF-GAN into StyleGAN to obtain a high-fidelity 3D-controllable generator. Unlike existing latent-based methods allowing implicit pose control, the proposed 3D-controllable StyleGAN enables explicit pose control over portrait generation. This distillation allows direct compatibility between 3D control and many StyleGAN-based techniques (e.g., inversion and stylization), and also brings an advantage in terms of computational resources. Our codes are available at https://github.com/jgkwak95/SURF-GAN.

**Keywords:** 3D-aware portrait generation · Pose-disentangled GAN · Facial image editing · Novel view synthesis · Latent manipulation

## 1 Introduction

Since the advent of Generative Adversarial Networks (GANs) [15], remarkable progress has been made in the field of photorealistic image generation. The quality and diversity of images generated by 2D GANs have been improved considerably and recent models [6,25,27–29] can produce high resolution images at a level that humans cannot distinguish. Despite the expressiveness of 2D GANs, they lack 3D understanding, in that the underlying 3D geometry of an object is ignored in the

---

**Supplementary Information** The online version contains supplementary material available at https://doi.org/10.1007/978-3-031-19790-1_15.

© The Author(s), under exclusive license to Springer Nature Switzerland AG 2022
S. Avidan et al. (Eds.): ECCV 2022, LNCS 13677, pp. 236–253, 2022.
https://doi.org/10.1007/978-3-031-19790-1_15

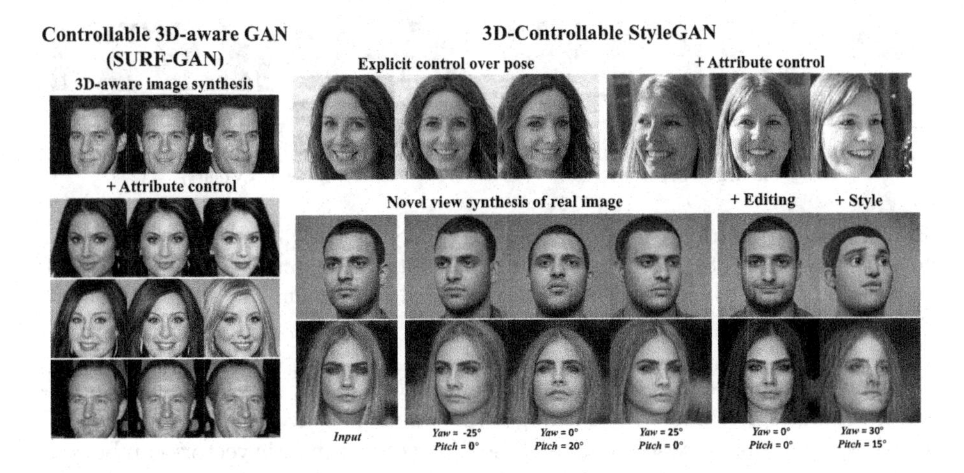


**Fig. 1.** (Left): Proposed novel 3D-aware GAN (SURF-GAN) which is capable of attribute-controllable generation as well as 3D-aware synthesis. (Right): 3D-controllable StyleGAN obtained by distilling the prior of SURF-GAN into 2D GAN.

generation process. As a result, they suffer the problem of multi-view inconsistency. To overcome the issue, many researchers have studied 3D controllable image synthesis and it has become one of the mainstream research in the community. There have been several attempts to learn 3D pose information with 2D GAN by disentangling pose in the latent space, but they require auxiliary 3D supervision such as synthetic face dataset [30] or 3DMM [12,57,58]. In addition, a few unsupervised approaches have been proposed by adopting implicit 3D feature [37,38] or differentiable renderer [42,52] in generation. However, these methods have struggled with multi-view consistency and photorealism.

Since the introduction of neural radiance fields (NeRF) by Mildenhall et al. [36] which has achieved notable success in novel view synthesis, a new paradigm has emerged in 3D-aware generation, called 3D-aware GAN. Several researchers have proposed 3D-aware generative frameworks [8,13,16,39,49,60,66] by leveraging NeRF as a 3D representation in GAN generator. NeRF-GANs learn 3D geometry from unlabelled images yet allow accurate and explicit control of 3D camera based on a volume rendering. Despite the obvious advantages, 3D GANs based on a pure NeRF network require tremendous computational resources and generate blurry images. Very recently, several approaches have alleviated the problems and have shown photorealistic output with high resolution by incorporating rear-end 2D networks [7,13,16,60,66]. However 3D GANs have difficulty with attribute-controllable generation or real image editing because their latent space has been rarely investigated for interpretable generation (Fig. 2).

In summary, these two distinct approaches have strengths and weaknesses that are complementary: 3D-aware GAN can generate novel poses but it has trouble with disentangling and manipulating attributes; 2D GAN is capable of controlling attributes but it struggles with 3D controllability. In this work, we propose novel solutions to overcome each weakness of 2D GANs and 3D GANs.

**Fig. 2.** Results of attribute manipulation when applying SeFa [51] utilized for 2D GANs to a 3D-aware GAN [8] (left). The captured attributes are entangled or not meaningful. In contrast, SURF-GAN can captures disentangled and semantic attributes (right).

First, we propose a novel 3D-aware GAN, i.e., SURF-GAN, which can discover semantic attributes by learning layer-wise **SU**bspace in INR-based Ne**RF** network in an unsupervised manner. The discovered semantic vectors can be controlled by corresponding parameters, thus this property allows us to manipulate semantic attributes (e.g., gender, hair color, etc.) as well as explicit pose.

With the proposed SURF-GAN, we take one more step to transform StyleGAN into a 3D-controllable generator. We inject the prior of 3D-aware SURF-GAN into the expressive and disentangled latent space of 2D StyleGAN. Unlike the previous methods [17,50,51] that allows implicit pose control, we make StyleGAN enable explicit control over pose. It means that the generator is capable of synthesizing accurate images based on a conditioned target view. By utilizing SURF-GAN which consists of pure NeRF layers as a generator of pseudo multi-view images, the transformed StyleGAN can learn elaborate control over 3D camera pose with latent manipulation. To this end, we proposed a method to find several orthogonal directions (not a single) related to the same pose attribute, and explicit control over the pose is accomplished by a combination of these directions. With a GAN inversion encoder, 3D controllable StyleGAN can be extended to the task of novel pose synthesis from a real image.

In addition to 3D perception, we also inject the controllability about semantic attributes that SURF-GAN finds. We can find more pose-robust latent path in the latent space of StyleGAN because SURF-GAN can manipulate a specific semantic while keeping view direction unchanged. Moreover, it allows further applications related to StyleGAN family, e.g., 3D control over stylized images generated by fine-tuned StyleGAN. It is notable that our approach neither requires 3D supervision nor exploits auxiliary off-the-shelf 3D models (e.g., 3DMM or pose detector) in both training and inference because SURF-GAN learns 3D geometry from unlabelled 2D images from scratch.

In summary, our contributions are as follow:

- We propose a novel 3D-aware GAN, called SURF-GAN, which can discover controllable semantic attributes in an unsupervised manner.
- By injecting editing directions from the low-resolution 3D-aware GAN into the high-resolution 2D StyleGAN, we achieve a 3D controllable generator which is capable of explicit control over pose and 3D consistent editing.
- Our method is directly compatible with various well-studied 2D StyleGAN-based techniques such as inversion, editing or stylization.

## 2 Related Work

**Pose-Disentangled GANs.** The remarkable advances have been achieved in photorealism by state-of-the-art GAN models [6,25,27–29]. However, pose control by image generators has been limited due to a lack of 3D understanding in the synthesizing process. Thereby, several works have attempted to disentangle the pose information from other attributes in 2D GANs. The disentanglement has been achieved by leveraging supervision such as 3DMM [12,57,58,62], landmark [21], synthetic images from 3D engine [30] or pose detector [53]. A few unsupervised approaches without 3D supervision [37,38] have been proposed by disentangling pose with implicit 3D feature projection, but they allow only implicit 3D control and show blurry results. Recently, a few methods [42,52] have incorporated a pre-trained StyleGAN with a differentiable renderer, but they struggle with photorealism, high-resolution [42] and real image editing [52].

**Interpretabilty and Controllabiltiy of GAN.** The well-trained 2D GANs, such as StyleGAN [28,29] have shown capable of disentangling the latent space. Recent works [3,17,40,43,50,51,61] have demonstrated semantic manipulation, especially for facial attributes, by analyzing the manifold and finding meaningful direction or mapping. Combining with GAN inversion [1,2,4,5,47,48,59,67], the applications of 2D GANs have been extended to real image editing. Alternatively, there have been studies [10,22,24,32] that discover and disentangle latent embeddings into interpretable dimensions during training of the generator. EigenGAN [19] that inspired our approach has demonstrated interpretable latent dimensions by designing layer-wise subspace embedding. However, both types of methods support implicit control over the discovered semantics. In the case of a pose that can be defined with camera parameters, these methods struggle to synthesize explicit novel view elaborately. Of course, the implicit methods can eventually create the desired pose through manual and iterative adjustment, but this is not an ideal situation. We can obtain a frontalized image automatically with some latent-based methods [31,47,50], but not for arbitrary target pose. Recently, Chen et al. [9] have introduced a generator allowing explicit control over pose, but it requires 3D mesh for pre-training process.

**3D-Aware GANs.** Beyond the disentanglement of pose information, many efforts have been made to obtain 3D-awareness in generation. Earlier methods have adopted several explicit 3D representations in 2D image generation such as voxel [14,20,35,68] or mesh [33,56]. However, they suffer from a lack of visual quality and limited resolution. Recently, approaches [7,8,13,16,39,41,49,60,66] based on neural fields have made significant progress in photorealism and 3D consistency. Nevertheless, these 3D-aware GANs have weakness in finding and editing semantic attribute because their latent space has been rarely investigated. Very recently, Sun et al. [55] have proposed an editable NeRF-GAN, but it does not handle diverse semantic attributes and requires semantic maps as supervision. In addition, 3D GANs struggle with novel pose generation of real image despite their capability of multi-view consistency. Recently proposed

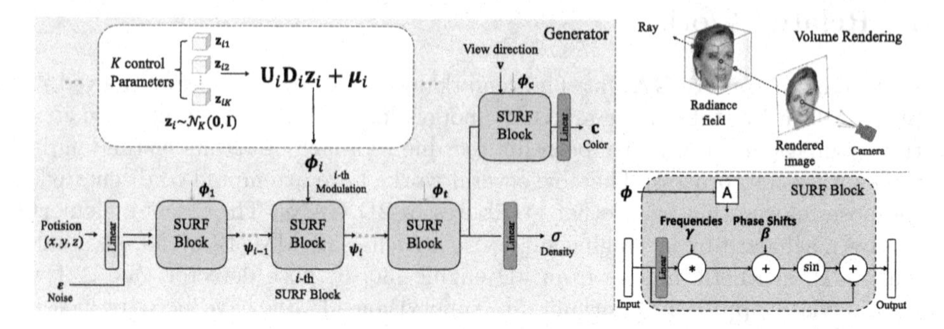


**Fig. 3.** Overview of SURF-GAN generator. Interpretable dimensions are caputured in layers with sub-modulation vectors. As like INR 3D-aware GANs, it takes position and view direction as input and predicts view dependent color ($\mathbf{c}$) and its density ($\sigma$).

EG3D [7] has shown experiments of novel view synthesis and presented outstanding results, but it requires iterative optimization for latent code and fine-tuning of the generator [48] for each target image.

# 3 Proposed Method

In this section, we describe our method, by first introducing SURF-GAN in detail and then by explaining a method to inject the prior of 3D SURF-GAN into 2D StyleGAN. Note that the word "StyleGAN" denotes StyleGAN2 [29].

## 3.1 Towards Controllable NeRF-GAN

**Preliminaries: NeRF-GANs.** Existing 2D GANs (e.g., StyleGAN [28,29]) synthesize output image directly with sampled latent vector. However, NeRF-GANs [8,16,39,66] generate a radiance field [36] before rendering 2D image. Given a position $\mathbf{x} \in \mathbb{R}^3$ and a viewing direction $\mathbf{v} \in \mathbb{S}^2$, it predicts a volume density $\sigma(\mathbf{x}) \in \mathbb{R}_+$ and the view-dependent RGB color $\mathbf{c}(\mathbf{x}, \mathbf{v}) \in \mathbb{R}^3$ of the input point. The points are sampled from rays of camera, and then an image is rendered into 2D grid with a classic volume rendering technique [23]. To produce diverse images, existing NeRF-GAN methods adopt StyleGAN-like modulation, where some components in the implicit neural network, e.g., intermediate features [8,66] or weight matrices [16] are modulated by sampled noise passing though a mapping network. Thereby, NeRF-GAN can control the pose by manipulating viewing direction $\mathbf{v}$ and change identity by injecting different noise vector. Nevertheless, it is ambiguous how to interpret the latent space and how to disentangle semantic attributes of NeRF-GAN for controllable image generation.

**Learning Layer-Wise Subspace in NeRF Network.** Inspired by EigenGAN [19], we adopt a different strategy from the existing methods [8,66] those modulation is obtained by the mapping network consisting of several

MLPs. EigenGAN learns interpretable subspaces in layers of its generator during training. However, EigenGAN is typical 2D convolution-based GAN framework, thus its concept is inapplicable to INR based NeRF-GAN. Therefore, we propose a novel framework (i.e., SURF-GAN), which captures the disentangled attributes in layers of NeRF network. Figure 3 shows the overview of SURF-GAN. The generator consists of $t+1$ SURF blocks ($t$ for shared layers and one for color layer). Following $\pi$-GAN, SURF block adopts the feature-wise linear modulation (FiLM) [44] to transform the intermediate features with frequencies $\boldsymbol{\gamma}_i$ and phase shifts $\boldsymbol{\beta}_i$, and followed SIREN activation [54]. SURF block in $i^{\text{th}}$ layer is formulated as

$$\boldsymbol{\psi}_i = \text{SURF}_i\left(\boldsymbol{\psi}_{i-1}, \boldsymbol{\phi}_i\right) = \sin\left(\boldsymbol{\gamma}_i \cdot \left(\mathbf{W}_i \boldsymbol{\psi}_{i-1} + \mathbf{b}_i\right) + \boldsymbol{\beta}_i\right) + \boldsymbol{\psi}_{i-1}, \tag{1}$$

where $\boldsymbol{\psi}_{i-1}$ and $\boldsymbol{\phi}_i$ denote input feature and modulation of $i^{th}$ layer respectively. $\mathbf{W}_i$ and $\mathbf{b}_i$ represent the weight matrix and followed bias. Unlike other NeRF-GANs, we add skip connection [18] to prevent drastic change of modulation vectors in training. In the model, a subspace embedded in each layer determines the modulation. Each subspace has orthogonal basis and it can be updated during training. The basis are learned to capture semantic modulation. Concretely, in the case of $i^{th}$ layer, a specific subspace determines the modulation of $i^{th}$ layer of NeRF network. It consists of learnable matrices, orthonormal basis $\mathbf{U}_i = [\mathbf{u}_{i1}, \ldots, \mathbf{u}_{iK}]$ and a diagonal matrix $\mathbf{D}_i = \text{diag}\left(d_{i1}, \ldots, d_{iK}\right)$. Each column of $\mathbf{U}_i$ plays a role of sub-modulation and it is updated to discover a meaningful direction that results in semantic change in image space. $d_{i1}, \ldots, d_{iK}$ serve as scaling factors of corresponding basis vectors $\mathbf{u}_{i1}, \ldots, \mathbf{u}_{iK}$. The latent $\mathbf{z}_i \in \mathbb{R}^K$ is set of $K$ scalar control parameters, i.e.,

$$\mathbf{z}_i = \left\{z_{ij} \in \mathbb{R} \mid z_{ij} \sim \mathcal{N}(0,1), j = 1, \ldots, K\right\}, \tag{2}$$

where $z_{ij}$ is a coefficient of sub-modulation $d_{ij}\mathbf{u}_{ij}$. Hence, the modulation of $i^{th}$ layer $\phi_i$ is decided by weighted summation of $K$ sub-modulations with $\mathbf{z}_i$, i.e.,

$$\boldsymbol{\phi}_i = \mathbf{U}_i \mathbf{D}_i \mathbf{z}_i + \boldsymbol{\mu}_i = \sum_{j=1}^{K} z_{ij} d_{ij} \mathbf{u}_{ij} + \boldsymbol{\mu}_i, \tag{3}$$

where the marginal vector $\boldsymbol{\mu}_i$ is employed to capture shifting bias. Finally, a simple affine transformation is applied to $\boldsymbol{\phi}_i$ for matching dimension and obtaining frequency $\boldsymbol{\gamma}_i$ and phase shift $\boldsymbol{\beta}_i$. At training phase, SURF-GAN layers learn variations of meaningful modulation controlled by randomly sampled $\mathbf{z}$. Additionally, an input noise $\epsilon$ is also injected to capture the rest variations missed by the layers. To improve the disentanglement of attributes and to prevent the basis fall into a trivial solution, we adopt the regularization loss to guarantee the column vectors of $\mathbf{U}_i$ to be orthogonal following EigenGAN, i.e.,

$$\mathcal{L}_{\text{reg}} = \mathbb{E}_i[\left\|\mathbf{U}_i^{\text{T}} \mathbf{U}_i - \mathbf{I}\right\|_1]. \tag{4}$$

Finally, output image is rendered by volume rendering technique [23]. At inference phase, we can control the discovered semantic attributes by manipulating corresponding element in $\mathbf{z}$. In addition, SURF-GAN enables explicit control over pose using viewing direction $\mathbf{v}$ as like other NeRF-based models.

### 3.2 Explicit Control over Pose with StyleGAN

In Sect. 3.2 and Sect. 3.3, we introduce a method to inject 3D perception and attribute controllability of SURF-GAN into StyleGAN.

**Leveraging 3D-Aware SURF-GAN.** The first step is to transform pre-trained StyleGAN into a 3D controllable generator. We start with a question: How can we make StyleGAN be capable of controlling over pose explicitly when given arbitrary latent code? To this end, we utilize SURF-GAN as a pseudo ground-truth generator. It provides three images, i.e., $I_s$, $I_c$, $I_t$ which denote source, canonical, and target image respectively. Here, $\mathbf{z}$ is fixed in all images but the view directions of $I_s$ and $I_t$ are randomly sampled and $I_c$ has canonical view (i.e., $\mathbf{v}=[0,0]$). Therefore, we can exploit them as multi-view supervision of the same identity. Afterwards, the images are embedded to $\mathcal{W}+$ [1] space by a GAN inversion encoder $E$, i.e., $\{\mathbf{w}_s, \mathbf{w}_c, \mathbf{w}_t\} = \{E(I_s), E(I_c), E(I_t)\}$. Here, we exploit the pre-trained pSp [47] encoder and it actually predicts the residual and adds it to the mean latent vector, but we omit the notation for simplicity.

**Mapping to a Canonical Latent Vector.** To handle arbitrary pose without employing off-the-shelf 3D models, we need to build an additional process. To this end, we propose a canonical latent mapper $T$, which converts an arbitrary code to a canonical code in the latent space of StyleGAN. Here, the canonical code implies being a canonical pose (frontal) in image space. $T$ takes $\mathbf{w}_s$ as input and predicts its frontalized version $\hat{\mathbf{w}}_c = T(\mathbf{w}_s)$ with the mapping function. In order to train $T$, we exploit latent loss to minimize the difference between the predicted $\hat{\mathbf{w}}_c$ and pseudo ground truth of canonical code $\mathbf{w}_c$, i.e.,

$$\mathcal{L}_{\mathrm{w}}^{\mathrm{c}} = \|\mathbf{w}_c - T(\mathbf{w}_s)\|_1. \tag{5}$$

To guarantee plausible translation result in image space, we also adopt pixel-level $\ell_2$-loss and LPIPS loss [63] between two decoded images, i.e.,

$$\mathcal{L}_{\mathrm{I}}^{\mathrm{c}} = \|I_c' - \hat{I}_c\|_2^2 \tag{6}$$

$$\mathcal{L}_{\mathrm{LPIPS}}^{\mathrm{c}} = \|F(I_c') - F(\hat{I}_c)\|_2^2, \tag{7}$$

where $I_c'$ and $\hat{I}_c$ represent the decoded images from $\mathbf{w}_c$ and $\hat{\mathbf{w}}_c$ respectively, and $F(\cdot)$ denotes the perceptual feature extractor. Hence, the loss for canonical view generation is formulated by

$$\mathcal{L}^{\mathrm{c}} = \lambda_1 \mathcal{L}_{\mathrm{w}}^{\mathrm{c}} + \lambda_2 \mathcal{L}_{\mathrm{I}}^{\mathrm{c}} + \lambda_3 \mathcal{L}_{\mathrm{LPIPS}}^{\mathrm{c}}. \tag{8}$$

**Target View Generation.** Next, the canonical vector is converted to a target latent vector according to given a target view $\mathbf{v}_t = [\alpha, \beta]$ as an additional input. Here, $\alpha$ and $\beta$ stand for pitch and yaw respectively. The manipulation is conducted in the latent space of StyleGAN by adding a pose vector which is obtained by a linear combination of pitch and yaw vectors ($\mathbf{p}$ and $\mathbf{y}$, respectively) with $\mathbf{v}_t$ as coefficients, i.e., $\hat{\mathbf{w}}_t = \hat{\mathbf{w}}_c + \mathbf{L}\mathbf{v}_t^T$, where $\mathbf{L} = [\mathbf{p}\ \mathbf{y}]$. Therefore, we need to

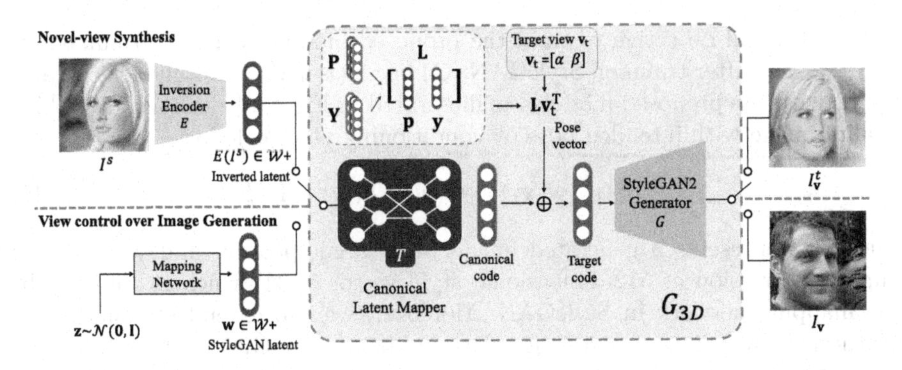


**Fig. 4.** The controllable StyleGAN allows explicit control over camera view. It can be used for novel pose synthesis (upper) and view-conditioned image generation (lower).

find optimal solution of $\mathbf{L}$ which can represent an adequate 3D control over pose. Although earlier studies [40,50] have shown successful interpolation results with the linear manipulation, unfortunately, they have found sub-optimal solutions that just control the intended pose attribute implicitly rather than explicit control over 3D camera. The interesting fact we observed is that the pose-related attribute (e.g., yaw) is not uniquely determined by a single direction. Rather, several orthogonal directions can have different effects on the same attribute. For example, two orthogonal direction A and B both can affect yaw but work differently. Based on this observation, we exploit several sub-direction vectors to compensate marginal portion that is not captured by a single direction vector. Our hypothesis is that the optimal direction that follows real geometry can be obtained by a proper combination of the sub-direction vectors. Borrowing the idea of basis in Sect. 3.1, we construct each of $N$ learnable basis to obtain final pose vectors for pitch and yaw respectively. Therefore, we optimize the matrices $\mathbf{P} = [\mathbf{d}_1^p, \ldots, \mathbf{d}_N^p]$ and $\mathbf{Y} = [\mathbf{d}_1^y, \ldots, \mathbf{d}_N^y]$. The process to obtain the target vector can be described as,

$$\hat{\mathbf{w}}_t = \hat{\mathbf{w}}_c + \sum_{i=1}^{N} (\alpha \cdot l_i^p \mathbf{d}_i^p + \beta \cdot l_i^y \mathbf{d}_i^y), \tag{9}$$

where the $l_i^p$ and $l_i^y$ represent the learnable scalining factor deciding the importance of basis $\mathbf{d}_i^p$ and $\mathbf{d}_i^y$ respectively. To penalize finding redundant directions, we add orthogonal regularization, i.e.,

$$\mathcal{L}_{\text{reg}} = \left|\left|\mathbf{P}^{\mathrm{T}}\mathbf{P} - \mathbf{I}\right|\right|_1 + \left|\left|\mathbf{Y}^{\mathrm{T}}\mathbf{Y} - \mathbf{I}\right|\right|_1 . \tag{10}$$

Similar to the canonical view generation, the model is penalized by the difference of the latent codes ($\mathbf{w}_t$ vs. $\hat{\mathbf{w}}_t$) and that of the corresponding decoded images ($I_t'$ vs. $\hat{I}_t$). In addition, we also utilize LPIPS loss. Therefore, the objective function of target view generation is described as,

$$\mathcal{L}^{\mathrm{t}} = \lambda_4 \mathcal{L}_{\mathrm{w}}^{\mathrm{t}} + \lambda_5 \mathcal{L}_{\mathrm{I}}^{\mathrm{t}} + \lambda_6 \mathcal{L}_{\mathrm{LPIPS}}^{\mathrm{t}} + \lambda_7 \mathcal{L}_{\mathrm{reg}}. \tag{11}$$

Finally, the full objective to train the proposed modules can be formulated as $\mathcal{L} = \mathcal{L}^{c} + \mathcal{L}^{t}$. After training, StyleGAN ($G$) becomes a 3D-controllable generator ($G_{3D}$) with the proposed modules as illustrated in Fig. 4. We can achieve a high quality image with intended pose by conditioning view as follow,

$$I_{\mathbf{v}} = G_{3D}(\mathbf{w}, \mathbf{v}_t) = G(\mathbf{w} + T(\mathbf{w}) + \mathbf{L}\mathbf{v}_t^T), \quad (12)$$

where $I_{\mathbf{v}}$ represents a generated image with target pose $\mathbf{v}_t$ and $\mathbf{w} \in \mathcal{W}+$ is duplicated version of 512-dimensional style vector in $\mathcal{W}$ which is obtained by the mapping network in StyleGAN. Moreover, we can extend our method to synthesize novel view of real images by combining with GAN inversion, i.e.,

$$I_{\mathbf{v}}^t = G_{3D}(E(I^s), \mathbf{v}_t), \quad (13)$$

where $I^s$ is an input source image in arbitrary view and $I_{\mathbf{v}}^t$ denotes a generated target image with target pose $\mathbf{v}_t$. Note that our method can handle arbitrary images without exploiting off-the-shelf 3D models such as pose detectors or 3D fitting models. In addition, it synthesizes output at once without an iterative optimization process for overfitting latent code into an input portrait image.

### 3.3 Finding Semantic Direction with SURF-GAN

Beyond 3D perception, we can discover semantic directions in the latent space of StyleGAN that can control facial attributes using SURF-GAN generated images. Such directions can be obtained by a simple vector arithmetic [46] with two latent codes or several interpolated samples generated by SURF-GAN. Although our approach does not overwhelm state-of-the-art methods analyzing via supervision, it would be a simple yet effective alternative that can provide pose-robust editing directions. Of course, the discovery using SURF-GAN is one of many applicable approaches and we can also utilize the existing semantic analysis methods [17,50,51] because our model is flexibly compatible with well-studied StyleGAN-based techniques.

## 4 Experimental Result

This section presents qualitative and quantitative comparisons with state-of-the-art methods and analysis of our method. Additional experiments and discussions not included in this paper can be found in the supplementary material.

### 4.1 Implementation

**SURF-GAN.** We use each of two datasets to train SURF-GAN, i.e., CelebA [34] dataset and FFHQ [28] dataset. We set the number of sub-modulations in each layer $K = 6$ (Eq. 2 and Eq. 3) and the number of modulated layers (SURF blocks) is nine ($\therefore t = 8$). The other settings are roughly the same with those of $\pi$-GAN. More details can be found in the supplementary paper.

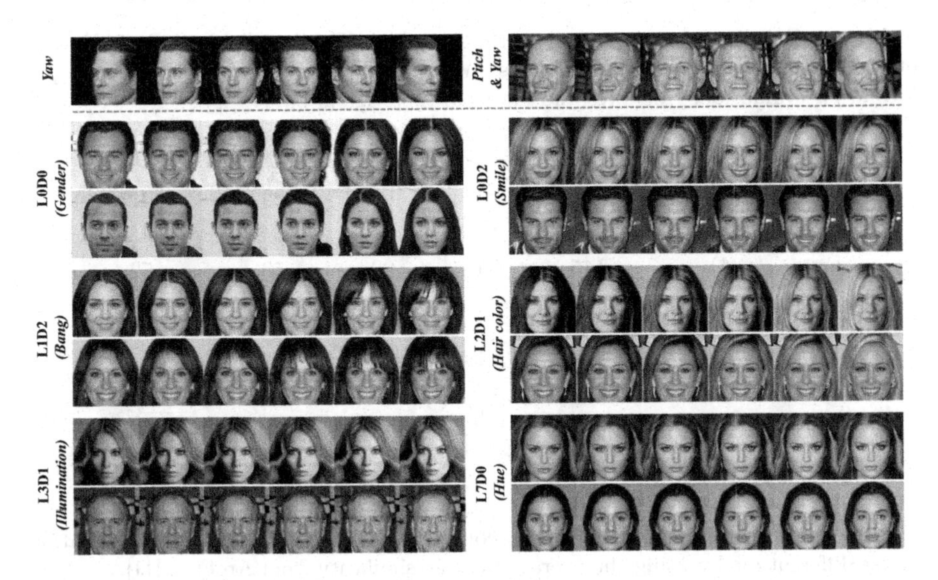


**Fig. 5.** Discovered semantic attributes at different layers in SURF-GAN. We can manipulate the attributes (e.g., hair color, gender, etc.) with the control parameters as well as explicit control over 3D camera. L$i$D$j$ denotes the $j^{th}$ basis of the layer $i^{th}$.

**3D-Controllable StyleGAN.** For training of 3D controllable StyleGAN, we exploit generated images by SURF-GAN trained with FFHQ because StyleGAN [29] and GAN inversion encoder [47] are pre-trained with FFHQ. We design the model to alter only the first four $\mathbf{w}$ vectors (i.e., $4 \times 512$) which have been known to control pose [28,64]. We set the number of sub-direction $N = 5$ (Eq. 9). The hyper-parameter of the loss function (Eq. 8 and Eq. 11) are set to $\lambda_1$, $\lambda_4$=10, $\lambda_7 = 100$, and 1.0 for the others.

## 4.2 Controllability of SURF-GAN

First, we present the attributes of CelebA discovered by SURF-GAN in Fig. 5. As like other 3D-aware GANs [7,8,16,39,49,66], it can synthesis a view-conditioned image, i.e., yaw and pitch can be controlled explicitly with input view direction (top row). In contrast to other 3D NeRF-GANs, SURF-GAN can discover semantic attributes in different layers in an unsupervised manner. Additionally, the discovered attributes can be manipulated by the corresponding control parameters. As shown in Fig. 5, different layers of SURF-GAN capture diverse attributes such as gender, hair color, illumination, etc. Interestingly, we observe the early layers capture high-level semantics (e.g., overall shape or gender) and the rear layers focus fine details or texture (e.g., illumination or hue). This property is similar to that seen in 2D GANs even though SURF-GAN consists of MLPs without convolutional layers. Additional discovered attributes, those of FFHQ and the comparison with $\pi$-GAN, which is a pure NeRF-GAN as like ours can be found in the supplementary material.

**Table 1.** Quantitative comparison of the proposed 3D controllable StyleGAN with other 3D controllable generative models. We use FID, pose accuracy, and frames per second for evaluation. † denotes quoting from the original paper.

| | ConfigNet | $\pi$-GAN | CIPS-3D | LiftedGAN | Ours |
|---|---|---|---|---|---|
| FID (↓) | 33.41† | 47.68 | 6.97† | 29.81† | **4.72** |
| Pose err.($\times 10^{-2}$) (↓) | 9.56 | **3.81** | 9.12 | 5.52 | 4.24 |
| Frames/s (↑) | **345** | 4 | 22 | 56 | 72 |

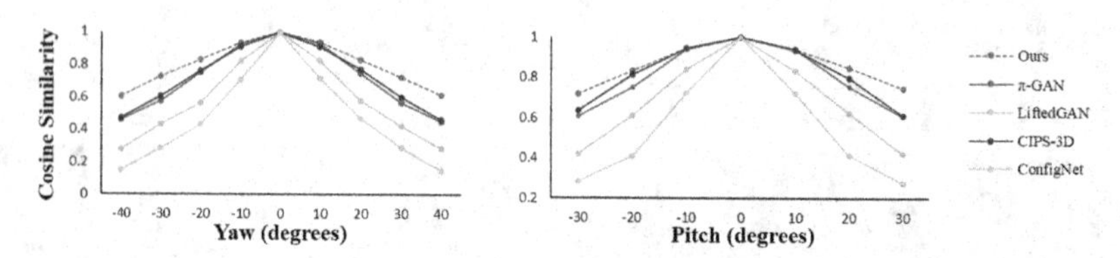


**Fig. 6.** Quantitative comparison of 3D-controllable models on identity preservation under different angles using the averaged cosine similarity from ArcFace [11].

### 4.3 Portrait Image Generation with 3D Control

To evaluate the performance of the proposed 3D-controllable StyleGAN, we report the qualitative and quantitative comparison with state-of-the-art models [8,30,52,66] whose generator allows explicit control over pose. Figure 7 shows synthesis results of each model for given target views. Here, the results are $256^2$ images generated by each method trained with FFHQ [28]. ConfigNet reveals lack of visual quality and weakness in large pose changes. $\pi$-GAN shows the accurate geometry because its generator consists of pure NeRF layers, but this property also results in some degenerated visual quality. CIPS-3D presents improved visual quality by adopting followed 2D INR network, but it suffers from 3D inaccuracy in specific poses. LiftedGAN generates reasonable outputs according to target views by utilizing differentiable renderer, but it lacks photorealism. Our method generates photorealistic images and shows plausible control over pose and multi-view consistency. We also report the quantitative comparisons of the models in Table 1 and Fig. 6. We use FID score, pose accuracy estimated by 3D model [69], frames per second, and identity similarity [11] as evaluation metrics. Compared to 3D-aware models, our method achieves a competitive score on pose accuracy and delivers superior results in efficiency, visual quality, and multi-view consistency. Although 2D-based ConfigNet shows overwhelming efficiency, it struggles with multi-view consistency and photorealism.

### 4.4 Novel View Synthesis of Real Image

By utilizing GAN inversion method, our method can perform novel view synthesis from a single portrait. Here, we use pSp [47] encoder for the inversion. To demonstrate the effectiveness of the canonical mapper, we firstly present

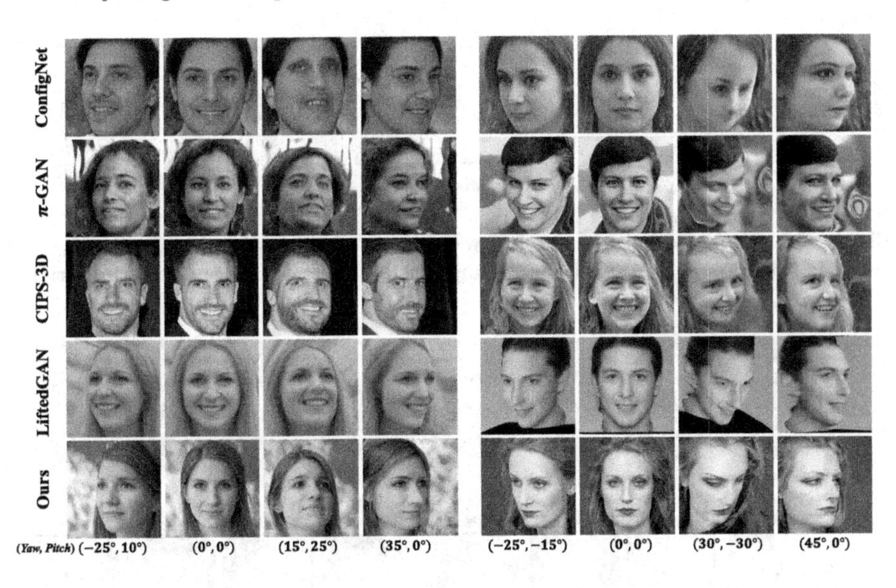


**Fig. 7.** Qualitative results of 3D-controllable generative models under target poses.

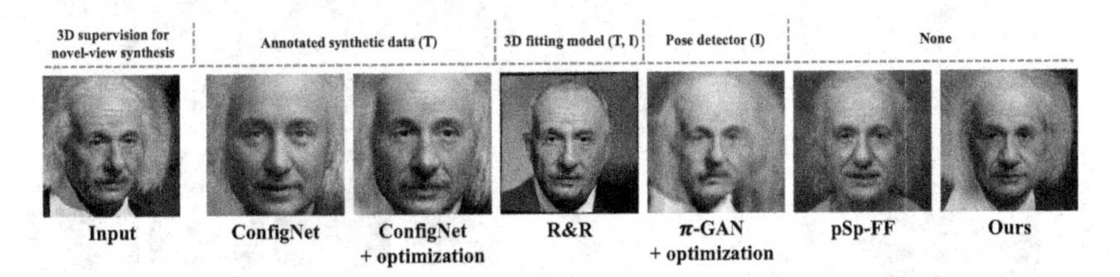


**Fig. 8.** Face frontalization results by the methods that can edit the pose of real image. Upper row denotes 3D supervision of each method for training (T) and inference (I).

the frontalization results in Fig. 8, which is a special case of novel pose synthesis. We mark the 3D supervision in training and inference for each method. Here, pSp-FF denotes the frontalization-only version of pSp [47]. Our method successfully generates a canonical view while preserving the identity. Next, we further compare the novel view synthesis results of each model. ConfigNet and $\pi$-GAN with optimizing latent code through iterative manner for overfitting to single test image show inferior results, especially in large pose variation. Rotate-and-Render (R&R) [65] presents reasonable results by exploiting off-the-shelf 3D fitting models [69] in the generation process. However, R&R loses some fine details of properties of the original, such as hair or background. Our models can edit pose successfully while preserving identity even though it does not require off-the-shelf 3D models and additional optimization for overfitting to an input. It is also demonstrated by the quantitative results in Fig. 9 which reports the averaged cosine similarity between input image and outputs at given various angles using ArcFace [11] and runtime of each method to process a single image.

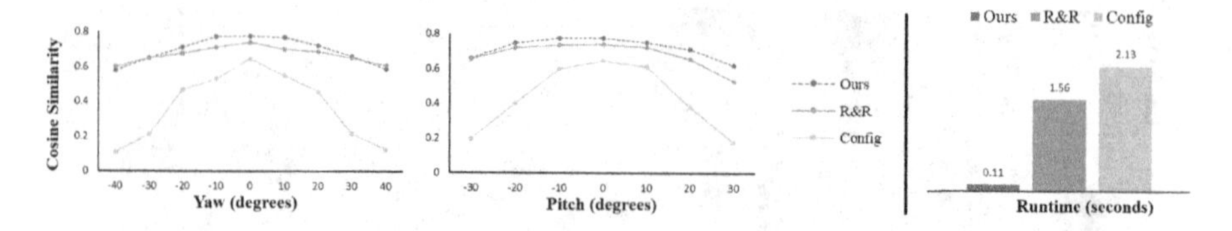


**Fig. 9.** Quantitative results of novel view synthesis models. We compute identity similarity between input and synthesized images using ArcFace (left) and runtime (right).

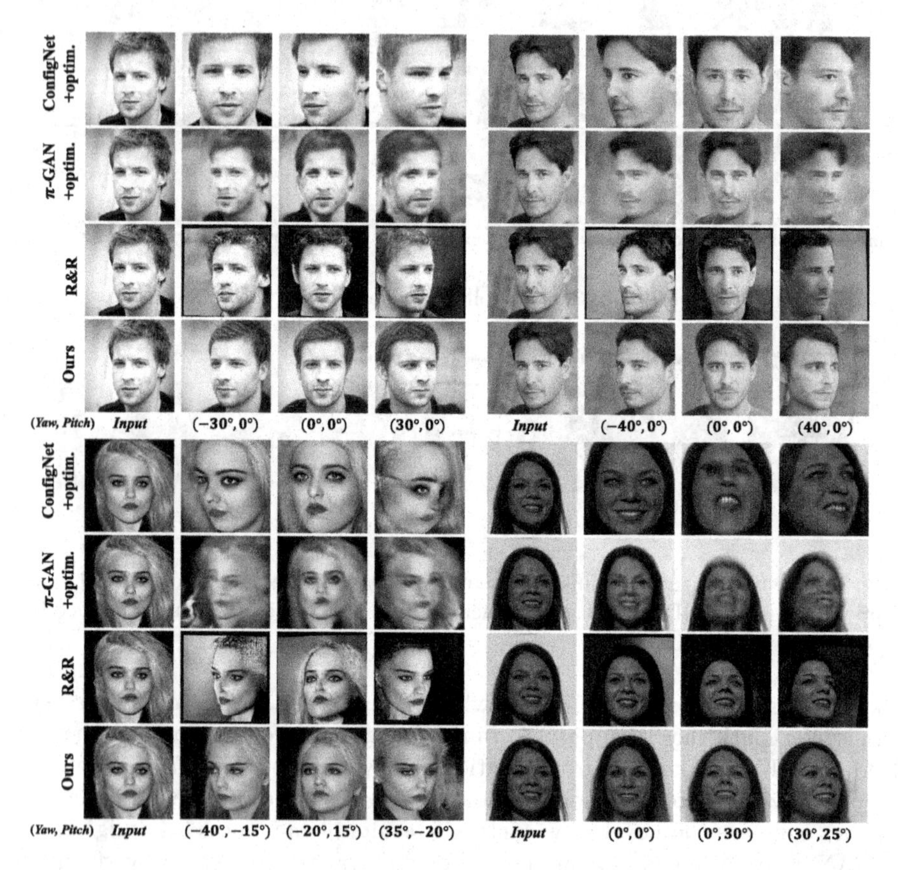


**Fig. 10.** Results of novel view synthesis under various target poses using CelebA-HQ.

### 4.5 Semantic Attribute Manipulation Under Conditioned Poses

Figure 11 presents the results of semantic attribute editing with pose control by 3D-controllable StyleGAN. The upper row stands for controllable generation (a) and the lower represents real image editing (b). The presented attributes, i.e., skin color, hue, hair color, and bangs are those discovered by SURF-GAN.

**Fig. 11.** Editing both attributes and view direction by 3D-controllable StyleGAN.

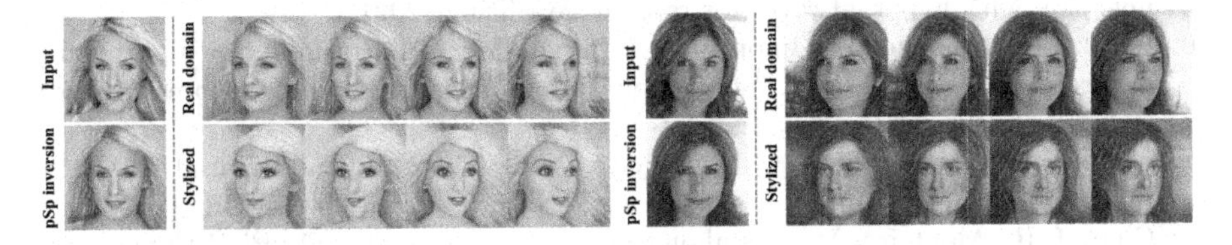


**Fig. 12.** Explicit 3D control over real and stylized images.

### 4.6 Applications

Our model can be flexibly integrated with other methods that also exploit pre-trained StyleGAN. Beyond the real image domain, we present a novel view synthesis of the stylized images such as toon or painting in Fig. 12. We use a interpolated StyleGAN proposed by Pinkney and Alder [45] for toonifying and a transferred StyleGAN trained with MetFace [26] for painting-style outputs.

## 5 Conclusion

In this paper, we solved the problems of 3D-aware GANs and 2D GANs by introducing SURF-GAN and 3D-controllable StyleGAN. Unlike other 3D-aware GANs, SURF-GAN can discover meaningful semantics and control them in an unsupervised manner. Using SURF-GAN, we convert StyleGAN to be explicitly 3D-controllable and it delivers outstanding results in both random image generation and novel view synthesis of real image. In addition, our method has the potential to be flexibly combined with other methods. We expect our work will be used practically and effectively in various tasks and hope it will open up a new direction in 3D-aware generation and editing fields.

**Acknowledgement.** This work was supported by DMLab. We also thank to Anonymous ECCV Reviewers for their constructive suggestions and discussions on our paper.

## References

1. Abdal, R., Qin, Y., Wonka, P.: Image2StyleGAN: how to embed images into the StyleGAN latent space? In: Conference on Computer Vision and Pattern Recognition (CVPR) (2019)

2. Abdal, R., Qin, Y., Wonka, P.: Image2StyleGAN++: how to edit the embedded images? In: Conference on Computer Vision and Pattern Recognition (CVPR) (2020)
3. Abdal, R., Zhu, P., Mitra, N.J., Wonka, P.: StyleFlow: Attribute-conditioned exploration of StyleGAN-generated images using conditional continuous normalizing flows. ACM Trans. Graph. **40**(3), 1–21 (2021)
4. Alaluf, Y., Patashnik, O., Cohen-Or, D.: ReStyle: a residual-based StyleGAN encoder via iterative refinement. In: International Conference on Computer Vision (ICCV) (2021)
5. Alaluf, Y., Tov, O., Mokady, R., Gal, R., Bermano, A.H.: HyperStyle: StyleGAN inversion with hypernetworks for real image editing (2021)
6. Brock, A., Donahue, J., Simonyan, K.: Large scale GAN training for high fidelity natural image synthesis. In: International Conference on Learning Representations (ICLR) (2019)
7. Chan, E.R., et al.: Efficient geometry-aware 3D generative adversarial networks. In: Conference on Computer Vision and Pattern Recognition (CVPR) (2022)
8. Chan, E.R., Monteiro, M., Kellnhofer, P., Wu, J., Wetzstein, G.: Pi-GAN: periodic implicit generative adversarial networks for 3D-aware image synthesis. In: Conference on Computer Vision and Pattern Recognition (CVPR) (2021)
9. Chen, A., Liu, R., Xie, L., Chen, Z., Su, H., Yu, J.: SofGAN: a portrait image generator with dynamic styling. ACM Trans. Graph. **41**(1), 1–26 (2022)
10. Chen, X., Duan, Y., Houthooft, R., Schulman, J., Sutskever, I., Abbeel, P.: InfoGAN: interpretable representation learning by information maximizing generative adversarial nets. In: Advances in Neural Information Processing Systems (NeurIPS) (2016)
11. Deng, J., Guo, J., Xue, N., Zafeiriou, S.: ArcFace: additive angular margin loss for deep face recognition. In: Conference on Computer Vision and Pattern Recognition (CVPR) (2019)
12. Deng, Y., Yang, J., Chen, D., Wen, F., Tong, X.: Disentangled and controllable face image generation via 3D imitative-contrastive learning. In: Conference on Computer Vision and Pattern Recognition (CVPR) (2020)
13. Deng, Y., Yang, J., Xiang, J., Tong, X.: GRAM: generative radiance manifolds for 3D-aware image generation. In: Conference on Computer Vision and Pattern Recognition (CVPR) (2022)
14. Gadelha, M., Maji, S., Wang, R.: 3D shape induction from 2D views of multiple objects. In: International Conference on 3D Vision (3DV) (2017)
15. Goodfellow, I., et al.: Generative adversarial nets. In: Advances in Neural Information Processing Systems (NeurIPS) (2014)
16. Gu, J., Liu, L., Wang, P., Theobalt, C.: StyleNeRF: a style-based 3D-aware generator for high-resolution image synthesis (2021)
17. Härkönen, E., Hertzmann, A., Lehtinen, J., Paris, S.: GANSpace: discovering interpretable GAN controls (2020)
18. He, K., Zhang, X., Ren, S., Sun, J.: Deep residual learning for image recognition. In: Conference on Computer Vision and Pattern Recognition (CVPR) (2016)
19. He, Z., Kan, M., Shan, S.: EigenGAN: layer-wise eigen-learning for GANs. In: International Conference on Computer Vision (ICCV) (2021)
20. Henzler, P., Mitra, N.J., Ritschel, T.: Escaping Plato's cave: 3D shape from adversarial rendering. In: International Conference on Computer Vision (ICCV) (2019)
21. Hu, Y., Wu, X., Yu, B., He, R., Sun, Z.: Pose-guided photorealistic face rotation. In: Conference on Computer Vision and Pattern Recognition (CVPR) (2018)

22. Jeon, I., Lee, W., Pyeon, M., Kim, G.: IB-GAN: disengangled representation learning with information bottleneck generative adversarial networks. In: AAAI Conference on Artificial Intelligence (AAAI) (2021)
23. Kajiya, J.T., Von Herzen, B.P.: Ray tracing volume densities. SIGGRAPH (1984)
24. Kaneko, T., Hiramatsu, K., Kashino, K.: Generative attribute controller with conditional filtered generative adversarial networks. In: Conference on Computer Vision and Pattern Recognition (CVPR) (2017)
25. Karras, T., Aila, T., Laine, S., Lehtinen, J.: Progressive growing of GANs for improved quality, stability, and variation. In: International Conference on Learning Representations (ICLR) (2018)
26. Karras, T., Aittala, M., Hellsten, J., Laine, S., Lehtinen, J., Aila, T.: Training generative adversarial networks with limited data. In: Advances in Neural Information Processing Systems (NeurIPS) (2020)
27. Karras, T., et al.: Alias-free generative adversarial networks. In: Advances in Neural Information Processing Systems (NeurIPS) (2021)
28. Karras, T., Laine, S., Aila, T.: A style-based generator architecture for generative adversarial networks. In: Conference on Computer Vision and Pattern Recognition (CVPR) (2019)
29. Karras, T., Laine, S., Aittala, M., Hellsten, J., Lehtinen, J., Aila, T.: Analyzing and improving the image quality of StyleGAN. In: Conference on Computer Vision and Pattern Recognition (CVPR) (2020)
30. Kowalski, M., Garbin, S.J., Estellers, V., Baltrušaitis, T., Johnson, M., Shotton, J.: CONFIG: controllable neural face image generation. In: Vedaldi, A., Bischof, H., Brox, T., Frahm, J.-M. (eds.) ECCV 2020. LNCS, vol. 12356, pp. 299–315. Springer, Cham (2020). https://doi.org/10.1007/978-3-030-58621-8_18
31. Kwak, J.G., Li, Y., Yoon, D., Han, D., Ko, H.: Generate and edit your own character in a canonical view (2022)
32. Lee, W., Kim, D., Hong, S., Lee, H.: High-fidelity synthesis with disentangled representation. In: Vedaldi, A., Bischof, H., Brox, T., Frahm, J.-M. (eds.) ECCV 2020. LNCS, vol. 12371, pp. 157–174. Springer, Cham (2020). https://doi.org/10.1007/978-3-030-58574-7_10
33. Liao, Y., Schwarz, K., Mescheder, L., Geiger, A.: Towards unsupervised learning of generative models for 3D controllable image synthesis. In: Conference on Computer Vision and Pattern Recognition (CVPR) (2020)
34. Liu, Z., Luo, P., Wang, X., Tang, X.: Deep learning face attributes in the wild. In: Conference on Computer Vision and Pattern Recognition (CVPR) (2015)
35. Lunz, S., Li, Y., Fitzgibbon, A., Kushman, N.: Inverse graphics GAN: learning to generate 3D shapes from unstructured 2D data (2020)
36. Mildenhall, B., Srinivasan, P.P., Tancik, M., Barron, J.T., Ramamoorthi, R., Ng, R.: NeRF: representing scenes as neural radiance fields for view synthesis. In: European Conference on Computer Vision (ECCV) (2020)
37. Nguyen-Phuoc, T., Li, C., Theis, L., Richardt, C., Yang, Y.L.: HoloGAN: unsupervised learning of 3D representations from natural images. In: International Conference on Computer Vision (ICCV) (2019)
38. Nguyen-Phuoc, T., Richardt, C., Mai, L., Yang, Y.L., Mitra, N.: BlockGAN: learning 3D object-aware scene representations from unlabelled images. In: Advances in Neural Information Processing Systems (NeurIPS) (2020)
39. Niemeyer, M., Geiger, A.: GIRAFFE: representing scenes as compositional generative neural feature fields. In: Conference on Computer Vision and Pattern Recognition (CVPR) (2021)

40. Nitzan, Y., Gal, R., Brenner, O., Cohen-Or, D.: LARGE: latent-based regression through GAN semantics (2021)
41. Or-El, R., Luo, X., Shan, M., Shechtman, E., Park, J.J., Kemelmacher-Shlizerman, I.: StyleSDF: high-resolution 3D-consistent image and geometry generation. In: Conference on Computer Vision and Pattern Recognition (CVPR) (2022)
42. Pan, X., Dai, B., Liu, Z., Loy, C.C., Luo, P.: Do 2D GANs know 3D shape? Unsupervised 3D shape reconstruction from 2D image GANs. In: International Conference on Learning Representations (ICLR) (2021)
43. Patashnik, O., Wu, Z., Shechtman, E., Cohen-Or, D., Lischinski, D.: StyleCLIP: text-driven manipulation of StyleGAN imagery. In: International Conference on Computer Vision (ICCV) (2021)
44. Perez, E., Strub, F., de Vries, H., Dumoulin, V., Courville, A.C.: FiLM: visual reasoning with a general conditioning layer. In: AAAI Conference on Artificial Intelligence (AAAI) (2018)
45. Pinkney, J.N., Adler, D.: Resolution dependent GAN interpolation for controllable image synthesis between domains (2020)
46. Radford, A., Metz, L., Chintala, S.: Unsupervised representation learning with deep convolutional generative adversarial networks (2015)
47. Richardson, E., et al.: Encoding in style: a StyleGAN encoder for image-to-image translation. In: Conference on Computer Vision and Pattern Recognition (CVPR) (2021)
48. Roich, D., Mokady, R., Bermano, A.H., Cohen-Or, D.: Pivotal tuning for latent-based editing of real images (2021)
49. Schwarz, K., Liao, Y., Niemeyer, M., Geiger, A.: GRAF: generative radiance fields for 3D-aware image synthesis. In: Advances in Neural Information Processing Systems (NeurIPS) (2020)
50. Shen, Y., Gu, J., Tang, X., Zhou, B.: Interpreting the latent space of GANs for semantic face editing. In: Conference on Computer Vision and Pattern Recognition (CVPR) (2020)
51. Shen, Y., Zhou, B.: Closed-form factorization of latent semantics in GANs. In: Conference on Computer Vision and Pattern Recognition (CVPR) (2021)
52. Shi, Y., Aggarwal, D., Jain, A.K.: Lifting 2D StyleGAN for 3D-aware face generation. In: Conference on Computer Vision and Pattern Recognition (CVPR) (2021)
53. Shoshan, A., Bhonker, N., Kviatkovsky, I., Medioni, G.: GAN-control: explicitly controllable GANs. In: International Conference on Computer Vision (ICCV) (2021)
54. Sitzmann, V., Martel, J., Bergman, A., Lindell, D., Wetzstein, G.: Implicit neural representations with periodic activation functions. In: Advances in Neural Information Processing Systems (NeurIPS) (2020)
55. Sun, J., et al.: FENeRF: face editing in neural radiance fields (2021)
56. Szabó, A., Meishvili, G., Favaro, P.: Unsupervised generative 3D shape learning from natural images (2019)
57. Tewari, A., et al.: PIE: portrait image embedding for semantic control. ACM Trans. Graph. **39**(6), 1–14 (2020)
58. Tewari, A., et al.: StyleRig: rigging StyleGAN for 3D control over portrait images. In: Conference on Computer Vision and Pattern Recognition (CVPR) (2020)
59. Tov, O., Alaluf, Y., Nitzan, Y., Patashnik, O., Cohen-Or, D.: Designing an encoder for StyleGAN image manipulation. ACM Trans. Graph. **40**(4), 1–14 (2021)
60. Xue, Y., Li, Y., Singh, K.K., Lee, Y.J.: GIRAFFE HD: a high-resolution 3D-aware generative model. In: Conference on Computer Vision and Pattern Recognition (CVPR) (2022)

61. Yao, X., Newson, A., Gousseau, Y., Hellier, P.: A latent transformer for disentangled face editing in images and videos. In: International Conference on Computer Vision (ICCV) (2021)
62. Yin, X., Yu, X., Sohn, K., Liu, X., Chandraker, M.: Towards large-pose face frontalization in the wild. In: International Conference on Computer Vision (ICCV) (2017)
63. Zhang, R., Isola, P., Efros, A.A., Shechtman, E., Wang, O.: The unreasonable effectiveness of deep features as a perceptual metric. In: Conference on Computer Vision and Pattern Recognition (CVPR) (2018)
64. Zhang, Y., et al.: Image GANs meet differentiable rendering for inverse graphics and interpretable 3D neural rendering (2020)
65. Zhou, H., Liu, J., Liu, Z., Liu, Y., Wang, X.: Rotate-and-render: unsupervised photorealistic face rotation from single-view images. In: Conference on Computer Vision and Pattern Recognition (CVPR) (2020)
66. Zhou, P., Xie, L., Ni, B., Tian, Q.: CIPS-3D: a 3D-aware generator of GANs based on conditionally-independent pixel synthesis (2021)
67. Zhu, J., Shen, Y., Zhao, D., Zhou, B.: In-domain GAN inversion for real image editing. In: European Conference on Computer Vision (ECCV) (2020)
68. Zhu, J.Y., et al.: Visual object networks: image generation with disentangled 3D representations. In: Advances in Neural Information Processing Systems (NeurIPS) (2018)
69. Zhu, X., Liu, X., Lei, Z., Li, S.Z.: Face alignment in full pose range: a 3D total solution. Transactions on Pattern Analysis and Machine Intelligence (TPAMI) (2017)

# AdaNeRF: Adaptive Sampling for Real-Time Rendering of Neural Radiance Fields

Andreas Kurz[1(✉)], Thomas Neff[1], Zhaoyang Lv[2], Michael Zollhöfer[2], and Markus Steinberger[1]

[1] Graz University of Technology, Graz, Austria
andreas.kurz@icg.tugraz.at
[2] Reality Labs Research, Pittsburgh, USA

**Abstract.** Novel view synthesis has recently been revolutionized by learning neural radiance fields directly from sparse observations. However, rendering images with this new paradigm is slow due to the fact that an accurate quadrature of the volume rendering equation requires a large number of samples for each ray. Previous work has mainly focused on speeding up the network evaluations that are associated with each sample point, e.g., via caching of radiance values into explicit spatial data structures, but this comes at the expense of model compactness. In this paper, we propose a novel dual-network architecture that takes an orthogonal direction by learning how to best reduce the number of required sample points. To this end, we split our network into a sampling and shading network that are jointly trained. Our training scheme employs fixed sample positions along each ray, and incrementally introduces sparsity throughout training to achieve high quality even at low sample counts. After fine-tuning with the target number of samples, the resulting compact neural representation can be rendered in real-time. Our experiments demonstrate that our approach outperforms concurrent compact neural representations in terms of quality and frame rate and performs on par with highly efficient hybrid representations. Code and supplementary material is available at https://thomasneff.github.io/adanerf.

**Keywords:** Neural rendering · Neural radiance fields · View synthesis

## 1 Introduction

The introduction of neural radiance fields [20] pushed the boundaries of modern computer graphics and vision by improving the state of the art of

A. Kurz and T. Neff—Authors contributed equally to this work.

**Supplementary Information** The online version contains supplementary material available at https://doi.org/10.1007/978-3-031-19790-1_16.

© The Author(s), under exclusive license to Springer Nature Switzerland AG 2022
S. Avidan et al. (Eds.): ECCV 2022, LNCS 13677, pp. 254–270, 2022.
https://doi.org/10.1007/978-3-031-19790-1_16

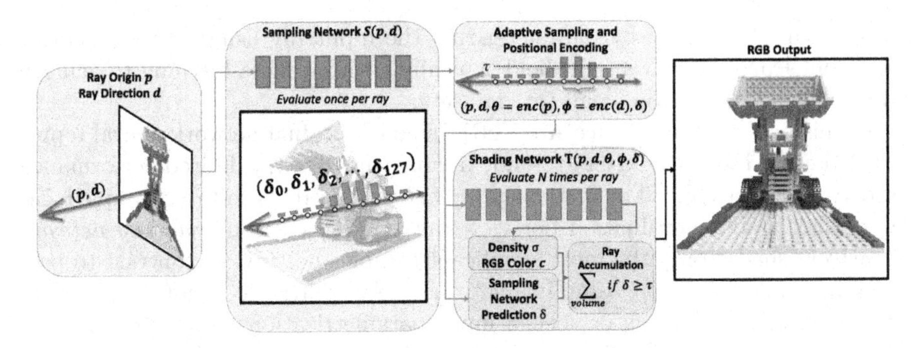


**Fig. 1.** AdaNeRF employs a single-evaluation *sampling network* and a multi-evaluation *shading network* to significantly reduce the number of required network evaluations per view ray. For each ray, the sampling network predicts a vector of estimated sample densities $\boldsymbol{\delta}$ that correspond to exactly one sample location each. We threshold the predicted $\boldsymbol{\delta}$ to cull away samples that are expected to have only minor contribution and only proceed to evaluate the shading network for this small subset of samples along the ray. Finally, when accumulating the outputs of the shading network along the ray, we additionally multiply the density $\boldsymbol{\sigma}$ predicted by the shading network with the density $\boldsymbol{\delta}$ predicted by the sampling network. This enables gradients of the RGB output loss to flow back to the sampling network, enabling end-to-end training of the full pipeline.

applications such as 3D reconstruction [35], rendering [21,22,44,45], animation [24,29] and scene relighting [18]. Furthermore, since the introduction of neural radiance fields, a significant amount of research focused on improving the resulting image quality [2], training speed [14,21], and inference performance [9,10,21,22,27,32,45]. Thus, real-time rendering of photorealistic neural radiance fields is now possible on standard consumer GPUs.

However, current real-time renderable neural radiance fields either require large amounts of memory, a restricted training data distribution or bounded training data, and it is unclear how to find an efficient compromise between those limitations. Explicit data structures have difficulties accounting for unbounded scenes, and storing large amounts of neural radiance fields (NeRFs) in sparse grids [9,10], trees [45] or hash tables [21] consumes prohibitive amounts of memory if multiple neural radiance fields need to be accessed in quick succession, such as in a streaming scenario. At a compact memory footprint, previous work relied on reducing the number of samples per view ray via dedicated *sampling networks* that estimate suitable sample locations along each view ray to improve rendering speed [14,22,27]. These sampling networks are commonly trained via supervision from depth [22] or the predicted density of a neural radiance field [27], thus requiring additional time-consuming preprocessing or pretraining steps. Alternatively, sampling networks can also be used to learn segments along each ray using an integral network [14]. Although this improves efficiency at a slight loss in quality, constructing these integral networks drastically increases the complexity and duration of training. Finally, light field networks [33] only evaluate a single sample per ray by parameterizing the input ray using Plücker coordinates, but learning such a light field typically requires meta-learning to achieve sufficient quality even on small toy-datasets.

In this paper, we introduce AdaNeRF, a compact dual-network neural representation that is optimized end-to-end, and fine-tuned to the desired performance for real-time rendering. The first *sampling network* predicts suitable sample locations using a single evaluation per view ray, while the second *shading network* adaptively shades only the most significant samples per ray. In contrast to previous methods based on sampling networks, AdaNeRF does not require any preprocessing, pretraining or special input parametrizations, lowering the overall complexity. We use fixed, discrete sample locations along each ray to set up a soft student-teacher regularization scheme by multiplying the predicted density of our sampling network with the output density of the shading network. Thus, both networks can modify the final RGB output and gradients flow throughout the whole pipeline. We ensure sparsity within our sampling network via a *4-phase training scheme*, after which we fine-tune our shading network to the desired sample counts for real-time rendering. We *adaptively sample* our shading network for each individual ray—we only evaluate the shading network for the most important samples, as predicted by the sampling network. The resulting sparse, dual-network pipeline can be rendered in real-time on consumer GPUs using our custom real-time renderer based on CUDA and TensorRT.

Our experimental results demonstrate the benefits of AdaNeRF compared to prior arts on a variety of datasets, including large, unbounded scenes. First, the adaptive sampling in AdaNeRF significantly increases the sampling efficiency of raymarching-based neural representations. Second, AdaNeRF outperforms previous sampling network based approaches in both rendering speed and quality with the same compact memory footprint. Finally, we qualitatively show that multiple AdaNeRFs can scale to complex scenes of arbitrary size.

In summary, we make the following contributions:

- A novel dual-network architecture to jointly learn sampling and shading networks for compact real-time neural radiance fields, outperforming existing sampling-network based approaches.
- An additional adjustable adaptive sampling scheme to only shade the most significant samples per ray, further improving quality and efficiency at identical average sample counts.
- A real-time rendering implementation that relies on dynamic, sparse sampling of our compact dual-network representation, targeting a sweet spot in the trade-off between performance, quality, and memory.

## 2 Related Work

Since the introduction of NeRF [20], coordinate-based neural radiance fields have improved the state-of-the-art across many domains, including dynamic scene modeling [6,8,11,13,24,25,28,37,40], animatable avatars and scenes [5,16,17,26,42], relightable objects [4,34], and object reconstruction [23,39,41,43]. The quality of object captures can be improved by incorporating different scales into the encoding, reducing aliasing and sampling artifacts when novel views are generated [2]. While this is mostly restricted to single-object captures, recent research has also investigated the reconstruction from unconstrained images [18] and large, unbounded scenes [3,46]. However, most advances of neural radiance fields focus on improving the output quality, with many of these advancements being infeasible to compute in real-time on consumer GPUs.

Our work is closely related to the advancement of neural radiance fields towards real-time rendering performance, which can be categorized into three domains: (1) Decomposed neural radiance fields. (2) Baking, caching or precomputing weights into an explicit spatial data structure, and (3) improving the sampling efficiency and reducing the total number of samples that are computed per frame.

*Decomposed Neural Radiance Fields.* By splitting a single MLP into many separate MLPs [7,30–32], subdivided scene grids [15] or primitives [17], both the quality and efficiency of rendering can be increased. Such a composition of scenes can represent scenes at city-scale [36,38]. Although these representations are useful to render single objects [32] or human avatars [17] in real-time, a real-time, scene-level representation has yet to be demonstrated.

*Baking Radiance Fields.* The radiance field can be stored inside a 3D grid [9], inside a sparse voxel octree [15,45] that does not even require any neural networks [44], or inside a sparse grid [10]. These methods run in real-time at a significantly increased memory footprint, which can be prohibitively expensive for scenarios such as streaming, where real-time swapping of neural radiance fields is desired. The concurrently introduced Instant-NGP [21] combines a hierarchical spatial hash table encoding with small MLPs to learn and render a scene representation in real-time. However, it is still unclear how it performs in demanding real-time scenarios and how to optimally tune its hash table size to still be as compact as fully neural representations.

*Improving the Sampling Efficiency of Neural Radiance Fields.* Recent work drastically improved the sampling efficiency of NeRF while keeping the same compact memory footprint. DONeRF [22] proposed a reduction in overall sample count by swapping the coarse network of the original NeRF with a *depth oracle network*, which can provide suitable sample locations for the second *shading network*, thus reducing the number of samples per ray by up to 128×. However, it cannot be trained end-to-end, and struggles without reliable depth information. Similarly to DONeRF, TermiNeRF [27] uses a *sampling network* that is conditioned on the density of a pre-trained NeRF. TermiNeRF uses the whole range of samples without the need for a depth map, which can improve quality in geometrically ambiguous scenarios. Although it can technically be trained end-to-end, TermiNeRF requires a suitable pretrained NeRF for initialization of its color network to achieve the best results. AutoInt [14] approaches sampling networks differently by predicting the lengths of segments along each ray.

By predicting ray segments instead of samples, subsequent *integral* networks can efficiently predict the density and color of each ray segment, reducing the number of network evaluations. However, the training procedure is significantly longer and more complex. Light field networks [1,33] reduce the number of network evaluations to a single sample per ray by directly mapping a view ray to the observed color. Although such an approach is advantageous in terms of memory footprint and rendering efficiency, without a meta-learned multi-view consistency prior, it fails to synthesize novel views in real-world scenes, and is thus not suitable for real-time novel view synthesis tasks for real world captures.

With AdaNeRF, we follow up on DONeRF [22], TermiNeRF [27], and AutoInt [14], and demonstrate that our approach is end-to-end trainable and more robust across a variety of training setups. Our approach achieves higher quality with faster real-time rendering performance by drastically reducing the amount of required samples, while at the same time keeping the representation compact.

## 3 Method

AdaNeRF consists of a fully end-to-end trainable pipeline that can be rendered in real-time. We replace the coarse network of the original NeRF by a *sampling network* $S$ that is only evaluated once per ray, minimizing the number of network evaluations to generate the final image. The sampling network takes the ray origin $\mathbf{p}$ and the ray direction $\mathbf{d}$ as input. The output of the sampling network is a vector of predictions $\boldsymbol{\delta}$, corresponding to the predicted importance of samples along each ray. The *shading network* $T$ takes the prediction of the sampling network, positionally encodes the samples with the largest contribution and outputs their density $\boldsymbol{\sigma}$ and color $\mathbf{c}$. By evaluating the sampling network once, a majority of samples with low contributions can be culled, increasing the overall efficiency of the pipeline. Figure 1 shows an overview of our dual-network setup.

### 3.1 End-to-End Trainable Sampling Network

We propose to multiply the predicted per-sample density $\delta_i$ of the sampling network with the predicted per-sample density $\sigma_i$ of the shading network. This formulation allows backpropagation to reach the sampling network. This is possible by using fixed sample locations along each ray, and placing exactly one sample in the center of each cell when discretizing the space along each ray. In contrast to previous work based on sampling networks, we do not require ground truth depth [22], and we avoid distinctly separated training steps [22,27].

We modify the standard ray accumulation function [20] to include an additional multiplication c

$$\hat{C}(\mathbf{r}) = \sum_{i=1}^{N} T_i(1 - \exp(-\delta_i \sigma_i t_i))\mathbf{c}_i, \text{ where } T_i = \exp\left(-\sum_{j=1}^{i-1} \delta_j \sigma_j t_j\right), \quad (1)$$

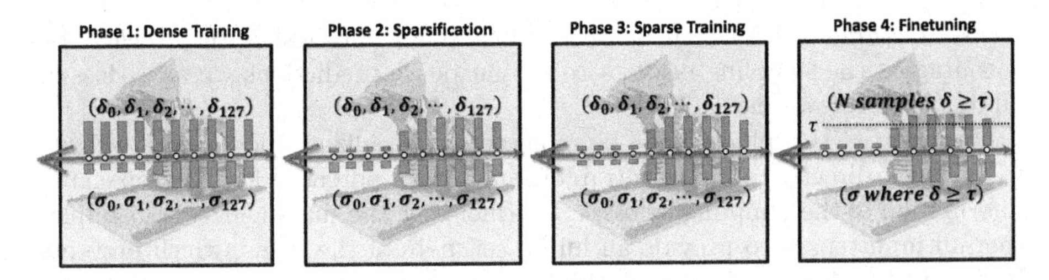


**Fig. 2.** The 4-phase training scheme of AdaNeRF. First, *dense training* forces the sampling network predictions close to $\boldsymbol{\delta} = \mathbf{1}$, enabling the shading network to densely sample the underlying scene. The *sparsification* phase then forces a majority of the sampling network predictions towards $\boldsymbol{\delta} = \mathbf{0}$, leaving only the most significant samples. The *sparse training* phase then adjusts the shading network to the newly sparsified sampling network. Finally, in the *finetuning* phase, we adaptively place samples for all sampling network predictions $\boldsymbol{\delta} \geq \tau$ up to a desired maximum number to further optimize for real-time rendering.

where $\hat{C}$ is the estimated, accumulated color, $N$ is the number of samples along the ray, $T_i$ is the accumulated transmittance along the ray, $t_i$ is the distance between adjacent samples and $\sigma_i$ is the output density of the shading network for sample $i$. The introduction of the multiplication by $\delta_i$ enables the sampling network to directly increase or decrease the importance of samples via its prediction, and to receive gradients from the MSE color reconstruction loss.

### 3.2 Sparse Adaptive Sampling Network Distillation

The modification of the ray accumulation alone does not ensure that the sampling network outputs sparse predictions—it might just as well always output the $\mathbf{1}$ vector, leading the shading network to place one sample in each cell, effectively ignoring the sampling network prediction. We disentangle $\boldsymbol{\delta}$ from the shading network density by introducing *sparsity* into the sampling network, which forces the network to select only the most important density values.

AdaNeRF trains the sampling network and the shading network end-to-end and progressively reduces the required samples per view ray. The shading network is trained via a standard MSE loss on the accumulated RGB color. The sampling network loss is composed of a sparsity loss which includes an $\ell_1$-loss that matches the output density of the shading network and an additional *density multiplication term* derived from the MSE loss of the shading network:

$$l_{sampling}(\boldsymbol{\delta}, \boldsymbol{\sigma}, \mathbf{c}) = \lambda_0 \cdot l_{mse}(\boldsymbol{\delta}, \boldsymbol{\sigma}, \mathbf{c}) + \lambda_1 \cdot l_{sparsity}(\boldsymbol{\delta}, \boldsymbol{\sigma}). \tag{2}$$

AdaNeRF uses a soft student-teacher regularization training scheme with 4 phases. We illustrate the training scheme in Fig. 2 and describe each phase in detail:

*Dense Training.* The initial dense training phase establishes the *teacher*, by encouraging the sampling network to output dense predictions via an $\ell_1$-loss of all its outputs towards $\mathbf{1}$.

In practice, this phase could be replaced by initializing the network weights to output $\mathbf{1}$ directly to increase training speed at a potential loss of propagated information to the sampling network. In either way, the shading network samples the full input space to provide an initial estimate of the scene, which prevents both networks from collapsing in the later stages.

*Sparsification.* The second phase introduces an additional $\ell_1$-loss to the sampling network that forces the majority of predictions towards $\mathbf{0}$. We linearly blend between forcing the sampling network outputs towards $\mathbf{1}$ and $\mathbf{0}$ over the course of this phase, and additionally blend in a soft student-teacher regularization loss via an $\ell_1$-term that encourages the sampling network outputs $\boldsymbol{\delta}$ to follow a similar distribution as $\boldsymbol{\sigma}$. From iteration $t_0$ over the duration of $t_d$ iterations (and the current iteration given as $t_c$), we define the sparsification loss as

$$l_{sparsity}(\boldsymbol{\delta}, \boldsymbol{\sigma}) = \lambda \cdot \frac{1}{N} \left( \sum_{i=1}^{N} |\delta_i - 0| + |\sigma_i - \delta_i| \right) + (1 - \lambda) \cdot \frac{1}{N} \sum_{i=1}^{N} |\delta_i - 1|, \quad \text{where } \lambda = \frac{t_c - t_0}{t_d}. \tag{3}$$

The $\ell_1$-term $|\sigma_i - \delta_i|$ ensures that the sampling network does not collapse to a single constant $\mathbf{0}$ or $\mathbf{1}$ vector, forcing the sampling network to follow the established scene representation of the shading network. The sparsification phase gradually increases the sparsity of the sampling network resulting in fewer significant outputs (which subsequently have zero contribution during ray accumulation).

*Sparse Training.* To allow the shading network to take advantage of the sparsification of the sampling network, we lock the sampling network's weights during sparse training. Although the shading network is still queried for all samples along each ray (as in the dense training phase) it is now free to alter the output for samples that are already dampened by the sampling network (due to the density multiplication). This enables the network to focus its capacity on those samples that actually contribute to the output.

*Fine-Tuning.* We fine-tune the shading network for a desired maximum number of samples per ray; typically 2, 4, 8 or 16. This phase is fast, as the number of samples per ray is small. Fine-tuning can increase quality as it completely removes samples that hardly contribute to the final output and allows the shading network to focus on the contributing samples only. Note that this phase results in separate shading networks for each maximum sample count, while all rely on the same sampling network.

*Real-Time Rendering with Adaptive Sampling.* We can further improve performance by enabling variable sample counts per ray. This adaptive sampling scheme exploits the fact that AdaNeRF uses fixed sample locations along the ray that can at most contain exactly one sample. First, we add an adaptive sampling threshold $\tau$ that defines the cutoff point for the sampling network's predictions $\boldsymbol{\delta}$. This enables us to save shading network evaluations in regions that do not require more than a few samples (such as a uniformly colored sky or simple geometric objects), which in turn increases the overall efficiency of our pipeline. Then, we limit the maximum number of allowed samples to $N_{max}$, and distinguish between the following cases, depending on the number of sampling network predictions $N_s$ that exceed the threshold $\tau$:

1. $N_s = 0$: If no sampling network predictions $\delta_i$ exceed $\tau$, we place one sample at the center of the ray segment corresponding to the sampling network's largest prediction.
2. $N_s \leq N_{max}$: If the number of sampling network predictions $\delta_i$ that exceed $\tau$ is at most $N_{max}$, we place samples at the center of all of their segments.
3. $N_s > N_{max}$: If the number of sampling network predictions $\delta_i$ that exceed $\tau$ is more than our maximum number of allowed samples $N_{max}$, we place one sample each at the center of the $N_{max}$ largest predictions.

This adaptive sampling scheme can be efficiently implemented on GPUs using warp communication primitives, enabling further efficiency gains compared to typical importance sampling setups that first need to generate a cumulative distribution function from a probability density function. Note that our approach is the first neural representation that relies on volume integration and (1) can go down to a single sample per ray and (2) supports variable sample counts per ray without the need for a spatial data structure.

## 4 Evaluation

*Implementation Details.* We follow the network architecture of DONeRF [22], using MLPs consisting of 8 layers with 256 units for both the sampling and shading networks. For the DONeRF dataset, we logarithmically space samples along each ray and unify rays [22]. For the LLFF dataset, we sample in normalized device coordinates [20]. We use Adam [12] with a learning rate of $5e^{-4}$ in training. We configure our 4-phase training scheme (Sect. 3.2) in the following order: 25k iterations of *dense training*, 50k iterations *sparsification*, 225k iterations *sparse training*, and 300k iterations *fine-tuning*. We vary the adaptive sampling threshold $\tau$ in comparison to other baselines at similar average sample counts. As a starting point, for the MSE loss of the sampling network (Eq. 1, Eq. 2) we use $\lambda_0 = 0.001$, and for the sparsity loss of the sampling network (Eq. 3, Eq. 2) we use $\lambda_1 = 1.0$. Please refer to the supplementary material for per-scene loss weights that were found by grid search and used in our evaluation. Finally, for the real-time performance comparison, we implemented a custom real-time renderer using CUDA and TensorRT to take advantage of our adaptive sampling strategy. All results were evaluated on a single Nvidia RTX 3090.

**Table 1.** Ablation of the 4-phase training scheme of AdaNeRF, using two maximum sample counts of $N_{max} = [2, 4]$ on the *Pavillon* scene of the DONeRF dataset.

| Method | $N_{max} = 2$ | | $N_{max} = 4$ | |
|---|---|---|---|---|
| | $N$/Ray | PSNR ↑ | $N$/Ray | PSNR ↑ |
| 1) AdaNeRF | 2.00 | **28.25** | 3.99 | **29.33** |
| 2) No dense training | 1.99 | 27.69 | 3.54 | 29.23 |
| 3) No sparsification blending | 1.66 | 27.33 | 2.75 | 28.44 |
| 4) No weight multiplication | 1.00 | 24.75 | 1.00 | 24.73 |
| 5) Only shading density supervision | 1.00 | 24.72 | 1.00 | 24.73 |
| 6) No shading density supervision | 2.00 | 27.82 | 4.00 | 28.86 |
| 7) No sparse training | 2.00 | 21.38 | 4.00 | 28.39 |

### 4.1 Ablation Studies

We provide an ablation study to validate the design of our 4-phase training scheme (Sect. 3.2) and our adaptive sampling strategy (Sect. 3.2), averaged across the *Pavillon* scene of DONeRF dataset. The number of iterations for each phase was determined in small-scale experiments and could be further optimized for training speed or image quality.

*Training Scheme.* Table 1 shows the ablation of our training scheme. Without (2) dense training or (3) sparsification, we observe a minor degradation in quality. If dense training is skipped, the shading network provides less accurate information to the sampling network; if sparsification is skipped, the sampling network is abruptly forced to be sparse by switching from "fully dense" to "fully sparse" training immediately instead of blending between them, losing potentially important samples in the process. Removing the (4) density multiplication in the ray accumulation function (Eq. 1) results in the sampling network collapsing to a constant output—the $\ell_1$-loss as the only supervision signal is insufficient to stabilize the sampling network. Similarly, using (5) $\ell_1$-loss supervision from the shading network as the sole optimization criterion (Eq. 3) leads to the sampling network collapsing towards the mean density of all rays. Removing (6) the shading density supervision $\ell_1$-term from Eq. 3 still produces reasonable sampling networks, at a quality degradation due to the lack of additional supervision. Finally, removing (7) the sparse training directly fine-tunes after sparsification. The resulting shading networks are not adapted to the sparsified sampling networks, significantly reducing quality.

*Adaptive Sampling.* We sweep the threshold $\tau$ between $[0.05, 0.40]$ and compare the resulting quality against fixed sample counts of $N = [8, 16]$, see Table 2. Compared to the fixed sample count of $N = 8$, the adaptive variant reaches similar quality between 5.07 and 6.16 samples per pixel, showing the increased efficiency even at lower sample counts. As average sample counts increase, the

**Table 2.** On the *Pavillon* scene of the DONeRF dataset, our adaptive sampling scheme manages to achieve higher quality at average sample counts of 6.16 and 7.76, compared to fixed sample counts of $N = [8, 16]$.

| | Fixed | | Adaptive | | | | | | | |
|---|---|---|---|---|---|---|---|---|---|---|
| Threshold $\tau$ | – | – | 0.05 | 0.1 | 0.15 | 0.2 | 0.25 | 0.3 | 0.35 | 0.4 |
| Samples per Ray | 16.00 | 8.00 | 12.10 | 11.89 | 11.63 | 11.03 | 9.73 | 7.76 | 6.16 | 5.07 |
| PSNR↑ | 30.89 | 30.40 | 31.66 | 31.66 | 31.68 | 31.66 | 31.62 | 31.07 | 30.64 | 30.24 |

sampling network has much more freedom in placing the samples, and thus can outperform the quality of $N = 16$ fixed samples at just 7.76 samples.

### 4.2 Results

We show a quantitative and qualitative evaluation of AdaNeRF on a variety of datasets against several baseline methods. We measure the quality of the rendered images in PSNR, and report the number of parameters required to store each method (using uncompressed 32-bit floating point) to evaluate compactness. We further present real-time rendering timings measured via TensorRT and CUDA. Please refer to the supplementary material for more visual comparisons, a discussion on training speed and a discussion on how to interleave multiple AdaNeRF for larger scenes.

*Datasets.* We evaluate our method using the following datasets.

- The **DONeRF** [22] **dataset** contains synthetic indoor and outdoor scenes of small to very large scales that are path-traced using Blender at a resolution of $800 \times 800$, with the cameras aimed at the forward hemisphere of their bounding box.
- The **LLFF** [19] **dataset** contains forward-facing real-world scenes captured using a handheld camera, which we scale to a resolution of $1008 \times 756$. We follow the convention [19] of holding out every 8th image for testing.

*Baselines.* Besides comparing to NeRF [20], we compare AdaNeRF to related work that focused on improving sampling efficiency and rendering performance:

- **DONeRF** [22]: DONeRF uses a depth oracle network trained on depth maps to improve sampling efficiency. For all experiments, we train the oracle network using depth maps extracted from a pre-trained coarse NeRF.
- **TermiNeRF*** [27]: TermiNeRF* learns a sampling network based on the density of a pre-trained NeRF. We follow the input encoding of DONeRF, and further use 128 fixed sample locations for the targets extracted from the pre-trained coarse NeRF, avoiding resampling and filtering of the targets.
- **AutoInt** [14]: AutoInt learns automatic integration via a sampling network. We compare AdaNeRF to AutoInt on a lower resolution version of the LLFF dataset, which was provided in the authors' original paper.

- **Plenoxels** [44]: Plenoxels uses a sparse grid with trilinear interpolation to directly learn a scene representation via spherical harmonics, without neural networks. For unbounded scenes, Plenoxels uses a multi-sphere-image background model in combination with its sparse foreground grid model.
- **Instant-NGP** [21]: Instant-NGP uses a hierarchical hash table to store most of its representation, with only tiny MLPs used to trilinearly interpolate the hash table entries along each ray. We show results for the default hash table size of $2^{19}$, as well as the authors' suggested smaller and faster alternative of $2^{14}$.

For both DONeRF and TermiNeRF*, we first train a coarse-only NeRF at 128 samples per ray with 8 hidden layers with 256 units each. We compare AdaNeRF to DONeRF and TermiNeRF* with fixed sample counts of $N = [2, 4, 8, 16]$. For Plenoxels we use configurations provided by the authors: *Plenoxels* uses a sparse grid resolution of $256^3$. *Plenoxels-MSI* adds a background model with 64 layers. *Plenoxels-Large* uses the authors' provided checkpoints for the LLFF dataset, which are significantly more dense. For all baselines, we use the available open source code, and the authors' suggested settings unless otherwise specified.

**Quality.** We present average output quality, memory footprint and render times for the DONeRF and LLFF datasets in Table 3, and example outputs in Fig. 3. Additional examples, per-scene data, depth reconstructions and sample placement visualizations can be found in the supplementary material.

For the DONeRF dataset, Instant-NGP-$2^{19}$ achieves the best quality, followed by all NeRF-based approaches with similar quality. Considering run-time, AdaNeRF shows the best tradeoff, allowing to choose between very fast rendering (at 3.7 samples) and competitive quality or high-quality (at 7.0 samples) and $2\times$ speed improvement over DONeRF and TermiNeRF* at the same quality. AdaNeRF only falls behind DONeRF and TermiNeRF* in image quality at extremely low sample counts while achieving greater speed improvement, suggesting that AdaNeRF operates most variably at a slightly higher sample counts. Considering memory foot footprint, AdaNeRF achieves equal or better quality than Plenoxels at a $48 - 215\times$ reduction in memory and similar run-time. Instant-NGP-$2^{19}$ is similar to AdaNeRF considering all three tradeoffs: it achieves higher quality at a higher memory and run-time cost, or similar quality with lower memory but higher run-time. For the LLFF dataset, the highest quality is achieved by NeRF, Plenoxels-Large, and AdaNeRF at 10.2 samples. Compared to other sampling-network based approaches, AdaNeRF clearly outperforms the state-of-the-art, achieving better quality than TermiNeRF* at less than half the sample count and frame time. Again, considering memory footprint and performance, both NeRF and Plenoxels show significant drawbacks compared to AdaNeRF: Plenoxels requires 3.6 GB of memory for its representation and NeRF takes 2.8 seconds to render a single frame. Interestingly, Instant-NGP ($2^{19}$ and $2^{14}$) perform worse than AdaNeRF for this data set. Compared to AutoInt, AdaNeRF achieves better quality at much faster rendering speeds. In summary, our adaptive fully neural representation shows state-of-the-art image quality at equal or better run-time performance and memory footprint, without requiring explicit data structures.

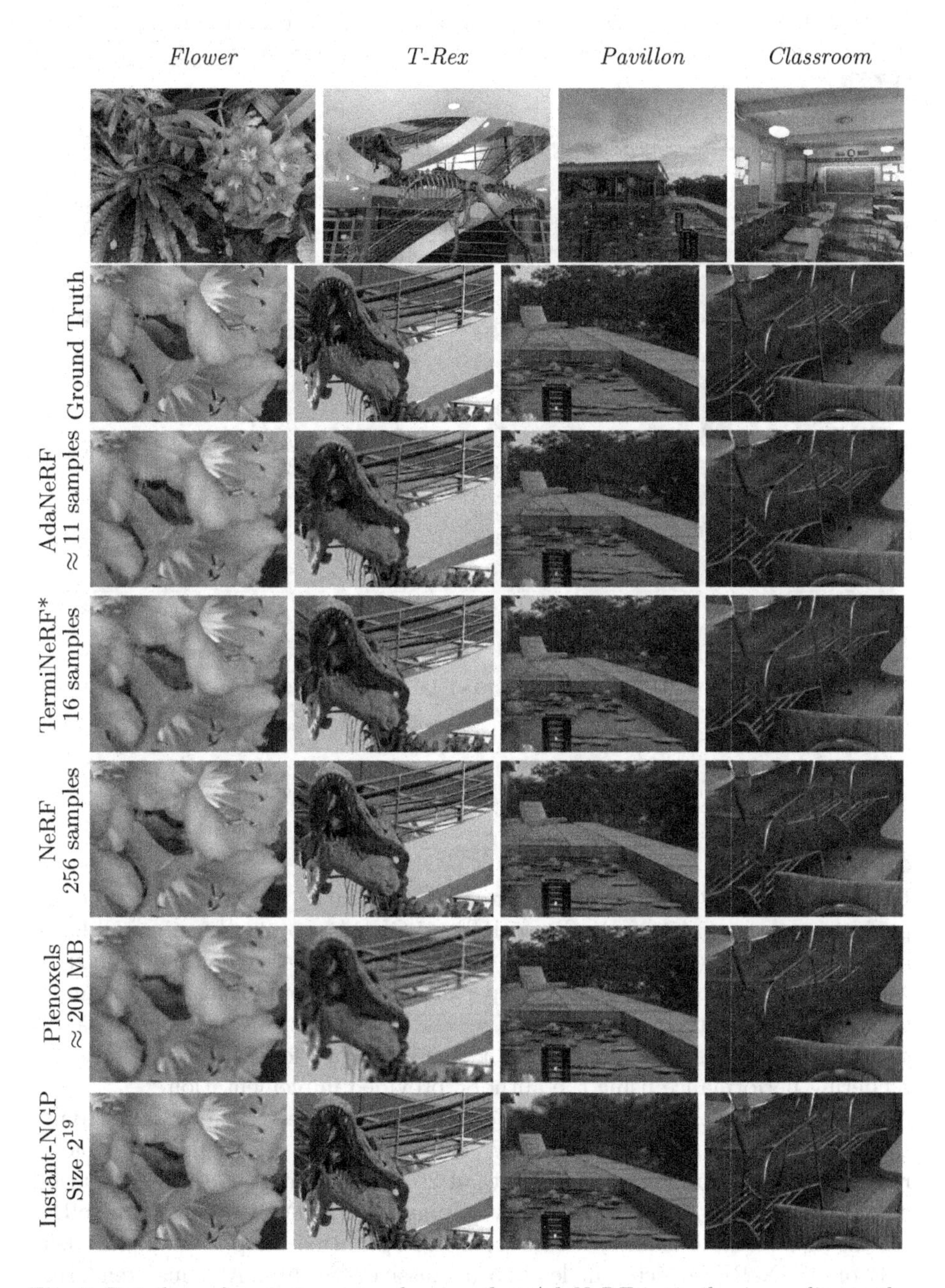


**Fig. 3.** Details on four test scenes, showing that AdaNeRF is similar in quality to the significantly slower NeRF and outperforms TermiNeRF* at lower sample counts. While using 50× more memory, Plenoxels tends to blur or leave out geometry due to lack of resolution in its grid. Instant-NGP achieves similar quality to AdaNeRF, while being slightly slower and requiring 16× more memory.

**Table 3.** Image quality, render time and memory footprint comparison on the DONeRF [22] and LLFF [19] datasets. Best results are displayed as Top 1, Top 2 and Top 3 per category.

| | | DONeRF [22] Dataset (800 × 800) | | | LLFF [19] Dataset (1008 × 756) | | | LLFF [19] Dataset (504 × 378) | | |
|---|---|---|---|---|---|---|---|---|---|---|
| Method | Memory [MB] | Samples per Ray | Time [ms]↓ | Quality PSNR↑ | Samples per Ray | Time [ms]↓ | Quality PSNR↑ | Samples per Ray | Time [ms]↓ | Quality PSNR↑ |
| AdaNeRF | 4.1 | 1.9 | 31.3 | 24.8 | 2.0 | 38.0 | 22.0 | 2.0 | 9.9 | 21.8 |
| AdaNeRF | 4.1 | 3.7 | 48.2 | 27.5 | 3.9 | 58.9 | 24.0 | 3.9 | 15.2 | 23.3 |
| AdaNeRF | 4.1 | 7.0 | 78.9 | 29.5 | 6.9 | 92.7 | 25.2 | 7.0 | 24.4 | 25.1 |
| AdaNeRF | 4.1 | 12.6 | 130.6 | 30.8 | 10.2 | 129.6 | 25.7 | 10.6 | 36.1 | 26.2 |
| DONeRF | 4.1 | 2.0 | 51.3 | 27.9 | 2.0 | 61.1 | 20.9 | - | - | - |
| DONeRF | 4.1 | 4.0 | 86.3 | 28.8 | 4.0 | 102.7 | 21.6 | - | - | - |
| DONeRF | 4.1 | 8.0 | 156.3 | 29.8 | 8.0 | 186.1 | 22.3 | - | - | - |
| DONeRF | 4.1 | 16.0 | 296.2 | 30.9 | 16.0 | 352.7 | 22.9 | - | - | - |
| TermiNeRF* | 4.1 | 2.0 | 51.3 | 27.2 | 2.0 | 61.1 | 21.7 | - | - | - |
| TermiNeRF* | 4.1 | 4.0 | 86.3 | 28.2 | 4.0 | 102.7 | 22.3 | - | - | - |
| TermiNeRF* | 4.1 | 8.0 | 156.3 | 29.2 | 8.0 | 186.1 | 23.0 | - | - | - |
| TermiNeRF* | 4.1 | 16.0 | 296.2 | 29.8 | 16.0 | 352.7 | 23.6 | - | - | - |
| NeRF | 3.8 | 256.0 | 2360.7 | 30.9 | 256.0 | 2810.9 | 26.5 | - | - | - |
| AutoInt | 4.5 | - | - | - | - | - | - | 16.0 | 44.6 | 24.1 |
| AutoInt | 4.5 | - | - | - | - | - | - | 32.0 | 88.5 | 24.9 |
| AutoInt | 4.5 | - | - | - | - | - | - | 64.0 | 176.4 | 25.5 |
| Plenoxels | 198.7 | - | 47.9 | 27.1 | - | 51.3 | 24.3 | - | - | - |
| Plenoxels-MSI | 892.9 | - | 47.5 | 29.6 | - | - | - | - | - | - |
| Plenoxels-Large | 3629.8 | - | - | - | - | 110.1 | 26.3 | - | - | - |
| Instant-NGP-$2^{14}$ | 2.0 | - | 102.1 | 29.4 | - | 100.7 | 24.8 | - | - | - |
| Instant-NGP-$2^{19}$ | 64.0 | - | 161.8 | 33.1 | - | 137.0 | 25.6 | - | - | - |

**Real-Time Rendering Performance.** We evaluate the real-time rendering performance of our AdaNeRF TensorRT and CUDA prototype against all baselines, see Table 3 (Columns "Time"). For all baselines except for Plenoxels, we evaluate their optimal rendering performance by computing the maximum throughput of identical networks in TensorRT, conservatively ignoring any additional input processing or differences in encoding. For Plenoxels, we evaluate the rendering performance using the authors' provided implementation.

In terms of real-time rendering performance vs. memory footprint, AdaNeRF at a maximum sample count of 2 achieves the best trade-off, being able to render scenes at an average frame rate of 26 frames per second at a resolution of 1008 × 756. The increased efficiency compared to DONeRF and TermiNeRF* (at identical sample counts) comes from the optimized adaptive sampling kernels of AdaNeRF, which can lead to a massive speedup. At equal rendering performance, AdaNeRF achieves significantly better quality compared to previous sampling-network based approaches, such as DONeRF, TermiNeRF* and AutoInt on both the DONeRF and LLFF datasets. At equal or improved quality, AdaNeRF outperforms DONeRF by up to 5×, TermiNeRF* by up to 6× and AutoInt by up to 7×. Compared to the densely sampled NeRF, our largest AdaNeRF shows a 20× increase in run-time, and the smallest AdaNeRF outperforms NeRF by up to 74×. Compared to the highly optimized Instant-NGP, AdaNeRF achieves a comparable trade-off between quality and run-time, achieving up to a 2× increase in run-time at equal quality. The best trade-off between quality and run-time is reached by Plenoxels, which represent their scene within a large sparse grid, with an additional optional multi-sphere image background model. Although this enables real-time rendering for single scenes, the immense memory requirements of up to multiple gigabytes prevent use-cases such as streaming or splitting complex environments into multiple representations.

The breakdown of frame time for the individual stages of AdaNeRF at a sample count of 2 is given as: 1.54 ms to generate the encoded inputs for the sampling network, 10.86 ms to evaluate the sampling network, 1.02 ms to generate the adaptively sampled inputs for the shading network, 18.16 ms to compute the shading network inference, and 0.38 ms for the final ray accumulation. Thus, our adaptive sampling kernels and overall pipeline exhibit only minor overheads compared to the inference workload, which constitutes most of the frame time. Furthermore, with higher average sample counts, the majority of additional compute load is added to the inference stages, and does not increase the overhead of input processing. Overall, AdaNeRF fills a gap in the performance-quality-memory trifecta, being extremely fast at a compact memory footprint, at a low cost in image quality for certain scenes.

## 5 Limitations and Future Work

Although AdaNeRF already achieves promising results on real-world data, our evaluation does not optimize for camera parameters, and thus can suffer from input data that is not perfectly consistent. Especially for very low sample counts (see Table 3), getting precise surface information is crucial to achieve good quality, and adding an additional optimization step for camera parameters and/or consistency in lighting could further improve results. Second, the threshold for adaptive sampling, as well as the weights of the main optimization function (maximizing the sampling network's sparsity while preventing a collapse) influences the sparsity of AdaNeRF, affecting the overall quality and real-time performance. While these parameters can be fairly robustly applied across different datasets, a grid search is recommended for best performance. In the future, these parameters could be learned from data to save the time for hyperparameter tuning.

## 6 Conclusion

We have introduced AdaNeRF, a compact real-time dual-network neural representation that can be trained fully end-to-end via a soft student-teacher optimization scheme. It is the first of its kind to adaptively place a very low amount of samples for each individual ray. We significantly outperform previous work that utilized sampling networks for very low sample count neural representations. Due to the compact nature of our neural representation, we additionally showed how multiple models can be blended in overlapping regions, which opens the door for real-time rendering of dynamically streamed neural representations of complex environments. We believe that such a fully neural real-time representation can be a useful alternative to approaches that require explicit spatial data structures.

## References


1. Attal, B., Huang, J., Zollhöfer, M., Kopf, J., Kim, C.: Learning neural light fields with ray-space embedding networks. CoRR abs/2112.01523 (2021). https://arxiv.org/abs/2112.01523
2. Barron, J.T., Mildenhall, B., Tancik, M., Hedman, P., Martin-Brualla, R., Srinivasan, P.P.: Mip-NeRF: a multiscale representation for anti-aliasing neural radiance fields. ICCV (2021)
3. Barron, J.T., Mildenhall, B., Verbin, D., Srinivasan, P.P., Hedman, P.: Mip-NeRF 360: unbounded anti-aliased neural radiance fields (2021)
4. Boss, M., Braun, R., Jampani, V., Barron, J.T., Liu, C., Lensch, H.: NeRD: neural reflectance decomposition from image collections (2020). https://arxiv.org/abs/2012.03918
5. Chen, J., et al.: Animatable neural radiance fields from monocular RGB videos (2021)
6. Du, Y., Zhang, Y., Yu, H.X., Tenenbaum, J.B., Wu, J.: Neural radiance flow for 4D view synthesis and video processing. In: Proceedings of the IEEE/CVF International Conference on Computer Vision (2021)
7. Fang, J., Xie, L., Wang, X., Zhang, X., Liu, W., Tian, Q.: NeuSample: neural sample field for efficient view synthesis. arXiv:2111.15552 (2021)
8. Gao, C., Saraf, A., Kopf, J., Huang, J.B.: Dynamic view synthesis from dynamic monocular video. In: Proceedings of the IEEE International Conference on Computer Vision (2021)
9. Garbin, S.J., Kowalski, M., Johnson, M., Shotton, J., Valentin, J.: FastNeRF: high-fidelity neural rendering at 200FPS. In: Proceedings of the IEEE/CVF International Conference on Computer Vision (ICCV), pp. 14346–14355 (2021)
10. Hedman, P., Srinivasan, P.P., Mildenhall, B., Barron, J.T., Debevec, P.: Baking neural radiance fields for real-time view synthesis (2021)
11. Jiakai, Z., et al.: Editable free-viewpoint video using a layered neural representation. In: ACM SIGGRAPH (2021)
12. Kingma, D.P., Ba, J.: Adam: a method for stochastic optimization. In: ICLR (Poster) (2015)
13. Li, T., et al.: Neural 3D video synthesis (2021)
14. Lindell, D.B., Martel, J.N., Wetzstein, G.: AutoInt: automatic integration for fast neural volume rendering (2021)
15. Liu, L., Gu, J., Zaw Lin, K., Chua, T.S., Theobalt, C.: Neural sparse voxel fields. Adv. Neural. Inf. Process. Syst. **33**, 15651–15663 (2020)
16. Liu, L., Habermann, M., Rudnev, V., Sarkar, K., Gu, J., Theobalt, C.: Neural actor: neural free-view synthesis of human actors with pose control. In: ACM SIGGRAPH Asia (2021)
17. Lombardi, S., Simon, T., Schwartz, G., Zollhoefer, M., Sheikh, Y., Saragih, J.: Mixture of volumetric primitives for efficient neural rendering (2021)

18. Martin-Brualla, R., Radwan, N., Sajjadi, M.S.M., Barron, J.T., Dosovitskiy, A., Duckworth, D.: NeRF in the wild: neural radiance fields for unconstrained photo collections. In: Proceedings of the IEEE/CVF Conference on Computer Vision and Pattern Recognition (CVPR), pp. 7210–7219 (2021)
19. Mildenhall, B., et al.: Local light field fusion: practical view synthesis with prescriptive sampling guidelines. ACM Trans. Graph. **38**(4), 1–14 (2019)
20. Mildenhall, B., Srinivasan, P.P., Tancik, M., Barron, J.T., Ramamoorthi, R., Ng, R.: NeRF: representing scenes as neural radiance fields for view synthesis. In: ECCV (2020)
21. Müller, T., Evans, A., Schied, C., Keller, A.: Instant neural graphics primitives with a multiresolution hash encoding. arXiv:2201.05989 (2022)
22. Neff, T., et al.: DONeRF: towards real-time rendering of compact neural radiance fields using depth oracle networks. Comput. Graph. Forum **40**(4), 45–59 (2021). https://doi.org/10.1111/cgf.14340
23. Oechsle, M., Peng, S., Geiger, A.: UNISURF: unifying neural implicit surfaces and radiance fields for multi-view reconstruction. In: International Conference on Computer Vision (ICCV) (2021)
24. Park, K., et al.: Nerfies: deformable neural radiance fields. In: ICCV (2021)
25. Park, K., et al.: HyperNeRF: a higher-dimensional representation for topologically varying neural radiance fields. arXiv preprint arXiv:2106.13228 (2021)
26. Peng, S., et al.: Animatable neural radiance fields for human body modeling. arXiv preprint arXiv:2105.02872 (2021)
27. Piala, M., Clark, R.: TermiNeRF: ray termination prediction for efficient neural rendering. In: 2021 International Conference on 3D Vision (3DV), pp. 1106–1114. IEEE Computer Society, Los Alamitos, CA, USA (2021). https://doi.org/10.1109/3DV53792.2021.00118
28. Pumarola, A., Corona, E., Pons-Moll, G., Moreno-Noguer, F.: D-NeRF: neural radiance fields for dynamic scenes (2020). https://arxiv.org/abs/2011.13961
29. Pumarola, A., Corona, E., Pons-Moll, G., Moreno-Noguer, F.: D-NeRF: neural radiance fields for dynamic scenes. In: Proceedings of the IEEE/CVF Conference on Computer Vision and Pattern Recognition (CVPR), pp. 10318–10327 (2021)
30. Rebain, D., Jiang, W., Yazdani, S., Li, K., Yi, K.M., Tagliasacchi, A.: DeRF: decomposed radiance fields (2020). https://arxiv.org/abs/2011.12490
31. Rebain, D., Jiang, W., Yazdani, S., Li, K., Yi, K.M., Tagliasacchi, A.: DeRF: decomposed radiance fields. In: Proceedings of the IEEE/CVF Conference on Computer Vision and Pattern Recognition (CVPR), pp. 14153–14161 (2021)
32. Reiser, C., Peng, S., Liao, Y., Geiger, A.: KiloNeRF: speeding up neural radiance fields with thousands of tiny MLPs (2021)
33. Sitzmann, V., Rezchikov, S., Freeman, W.T., Tenenbaum, J.B., Durand, F.: Light field networks: neural scene representations with single-evaluation rendering. In: Proc. NeurIPS (2021)
34. Srinivasan, P., Deng, B., Zhang, X., Tancik, M., Mildenhall, B., Barron, J.T.: NeRV: neural reflectance and visibility fields for relighting and view synthesis (2020). https://arxiv.org/abs/2012.03927
35. Takikawa, T., et al.: Neural geometric level of detail: real-time rendering with implicit 3D shapes (2021)
36. Tancik, M., et al.: Block-NeRF: scalable large scene neural view synthesis (2022)
37. Tretschk, E., Tewari, A., Golyanik, V., Zollhöfer, M., Lassner, C., Theobalt, C.: Non-rigid neural radiance fields: reconstruction and novel view synthesis of a dynamic scene from monocular video. In: IEEE International Conference on Computer Vision (ICCV). IEEE (2021)

38. Turki, H., Ramanan, D., Satyanarayanan, M.: Mega-NERF: scalable construction of large-scale NeRFs for virtual fly-throughs (2021)
39. Wang, P., Liu, L., Liu, Y., Theobalt, C., Komura, T., Wang, W.: NeuS: learning neural implicit surfaces by volume rendering for multi-view reconstruction. NeurIPS (2021)
40. Xian, W., Huang, J.B., Kopf, J., Kim, C.: Space-time neural irradiance fields for free-viewpoint video. In: Proceedings of the IEEE/CVF Conference on Computer Vision and Pattern Recognition (2021)
41. Xie, C., Park, K., Martin-Brualla, R., Brown, M.: FiG-NeRF: figure-ground neural radiance fields for 3D object category modelling (2021)
42. Yang, B., et al.: Learning object-compositional neural radiance field for editable scene rendering. In: International Conference on Computer Vision (ICCV) (2021)
43. Yariv, L., Gu, J., Kasten, Y., Lipman, Y.: Volume rendering of neural implicit surfaces. NeurIPS (2021)
44. Yu, A., Fridovich-Keil, S., Tancik, M., Chen, Q., Recht, B., Kanazawa, A.: Plenoxels: radiance fields without neural networks (2021)
45. Yu, A., Li, R., Tancik, M., Li, H., Ng, R., Kanazawa, A.: PlenOctrees for real-time rendering of neural radiance fields. In: ICCV (2021)
46. Zhang, K., Riegler, G., Snavely, N., Koltun, V.: NeRF++: analyzing and improving neural radiance fields (2020)

# Improving the Perceptual Quality of 2D Animation Interpolation

Shuhong Chen(✉) and Matthias Zwicker

University of Maryland, College Park, MD 20742, USA
{shuhong,zwicker}@cs.umd.edu

**Abstract.** Traditional 2D animation is labor-intensive, often requiring animators to manually draw twelve illustrations per second of movement. While automatic frame interpolation may ease this burden, 2D animation poses additional difficulties compared to photorealistic video. In this work, we address challenges unexplored in previous animation interpolation systems, with a focus on improving perceptual quality. Firstly, we propose SoftsplatLite (SSL), a forward-warping interpolation architecture with fewer trainable parameters and better perceptual performance. Secondly, we design a Distance Transform Module (DTM) that leverages line proximity cues to correct aberrations in difficult solid-color regions. Thirdly, we define a Restricted Relative Linear Discrepancy metric (RRLD) to automate the previously manual training data collection process. Lastly, we explore evaluation of 2D animation generation through a user study, and establish that the LPIPS perceptual metric and chamfer line distance (CD) are more appropriate measures of quality than PSNR and SSIM used in prior art.

**Keywords:** Animation · Video frame interpolation

## 1 Introduction

Traditional 2D animators typically draw each frame manually; this process is incredibly labor-intensive, requiring large production teams with expert training to sketch and color the tens of thousands of illustrations required for an animated series. With the growing global popularity of the traditional style, studios are hard-pressed to deliver high volumes of quality content. We ask whether recent advancements in computer vision and graphics may reduce the burden on animators. Specifically, we study video frame interpolation, a method of automatically generating intermediate frames in a video sequence. In the typical problem formulation, a system is expected to produce a halfway image naturally interpolating two given consecutive video frames. In the context of animation, an animator could potentially achieve the same framerate for a sequence (or "cut")

**Supplementary Information** The online version contains supplementary material available at https://doi.org/10.1007/978-3-031-19790-1_17.

© The Author(s), under exclusive license to Springer Nature Switzerland AG 2022
S. Avidan et al. (Eds.): ECCV 2022, LNCS 13677, pp. 271–287, 2022.
https://doi.org/10.1007/978-3-031-19790-1_17

by manually drawing only a fraction of the frames, and use an interpolator to generate the rest.

Though there is abundant work on video interpolation, 2D animation poses additional difficulties compared to photorealistic video. Given the high manual cost per frame, animators tend to draw at reduced framerates (e.g. "on the twos" or at 12 frames/second), increasing the pixel displacements between consecutive frames and exaggerating movement non-linearity. Unlike in natural videos with motion blur, the majority of animated frames can be viewed as stand-alone cel illustrations with crisp lines, distinct solid-color regions, and minute details. For this non-photorealistic domain with such different image and video features, even our understanding of how to evaluate generation quality is limited.

Previous animation-specific interpolation by Li et al. (AnimeInterp [37]) approached some of these challenges by improving the optical flow estimation component of a deep video interpolation system by Niklaus et al. (Softsplat [24]); in this paper, we build upon AnimeInterp by addressing some remaining challenges. Firstly, though AnimeInterp improved optical flow, it trained with an $L_1$ objective and did not modify the Softsplat feature extraction, warping, or synthesis components; this results in blurred lines/details and ghosting artifacts in supposedly solid-color regions. We alleviate these issues with architectural improvements in our proposed SoftsplatLite (SSL) model, as well as with an additional Distance Transform Module (DTM) that refines outputs using domain knowledge about line drawings. Secondly, though AnimeInterp provided a small ATD12k dataset of animation frame triplets, the construction of this dataset required intense manual filtering of evenly-spaced triplets with linear movement. We instead automate linear triplet collection from raw animation by introducing Restricted Relative Linear Discrepancy (RRLD), enabling large-scale dataset construction. Lastly, AnimeInterp only focused on PSNR/SSIM evaluation, which we show (through an exploratory user study) are less indicative of perceived quality than LPIPS [45] and chamfer line distance (CD). We summarize the contributions of this paper:

1. **SoftsplatLite (SSL)**: a forward-warping interpolation architecture with fewer trainable parameters and better perceptual performance. We tailor the feature extraction and synthesis networks to reduce overfitting, propose a simple infilling method to remove ghosting artifacts, and optimize LPIPS loss to preserve lines and details.
2. **Distance Transform Module (DTM)**: a refinement module with an auxiliary domain-specific loss that leverages line proximity cues to correct aberrations in difficult solid-color regions.
3. **Restricted Relative Linear Discrepancy (RRLD)**: a metric to quantify movement non-linearity from raw animation; this automates the previously manual training data collection process, allowing more scalable training.
4. **Perceptual user study**: we explore evaluation of 2D animation generation, establishing the LPIPS perceptual metric and chamfer line distance (CD) as more appropriate quality measures than PSNR/SSIM used in prior art.

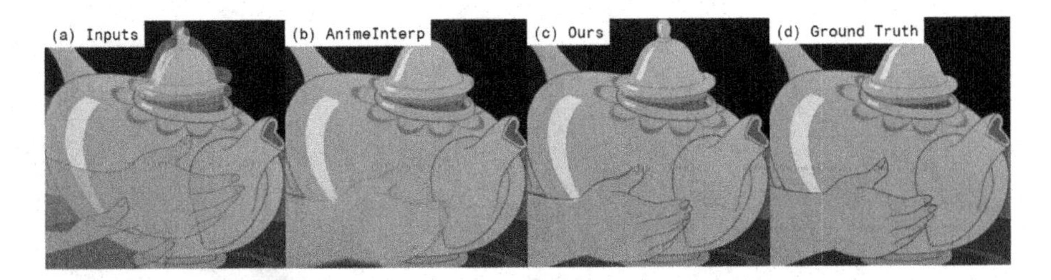


**Fig. 1.** We improve the perceptual quality of 2D animation interpolation from previous work. **(a)** Overlaid input images to interpolate; **(b)** AnimeInterp by Li et al. [37]; **(c)** Our proposed method; **(d)** Ground truth interpolation. Note the destruction of lines in (b) compared to (c), and the patchy artifacts ghosted on the teapot in (b). Our user study validates our focus on perceptual metrics and artifact removal.

## 2 Related Work

Much recent work has been published on photorealistic video interpolation. Broadly, these works fall into phase-based [21,22], kernel-based [25,26], and flow-based methods [16,24,28,43], with others using a mix of techniques [1,2,6]. The most recent state-of-the-art has seen more flow-based methods [24,28], following corresponding advancements in optical flow estimation [14,15,39,40]. Flow-based methods can be further split by forward [24], or backward [28] warping. The prior art most directly related to ours is AnimeInterp, by Li et al. [37]. While they laid the groundwork for the problem specific to the traditional 2D animation domain, their system had many shortcomings that we overcome as described in the introduction section.

Even though we focus on animations "post-production" (i.e. interpolating complete full-color sequences), there is also a body of work on automating more specific components of animation production itself. For example, sketch simplification [35,36] is a popular topic with applications to speeding up animation "tie-downs" and "cleanups". There are systems for synthesizing "in-between" line drawings from sketch keyframes in both raster [23,44] and vector [7,42] form. While the flow-based in-betweening done by Narita et al. [23] shares similarity to our work (such as the use of chamfer distance and forward warping), their system composed pretrained models without performing any form of training. Another related problem is sketch colorization, with application to both single illustrations [31] and animations [5,20,30]. These works unsurprisingly highlight the foundational role of lines and sketches in animation, and we continue the trend by introducing a Distance Transform Module to improve our generation quality.

## 3 Methodology

### 3.1 SoftsplatLite

As with AnimeInterp [37], we base our model on the state-of-the-art Softsplat [24] interpolation model, which uses bidirectional optical flow to differentiably

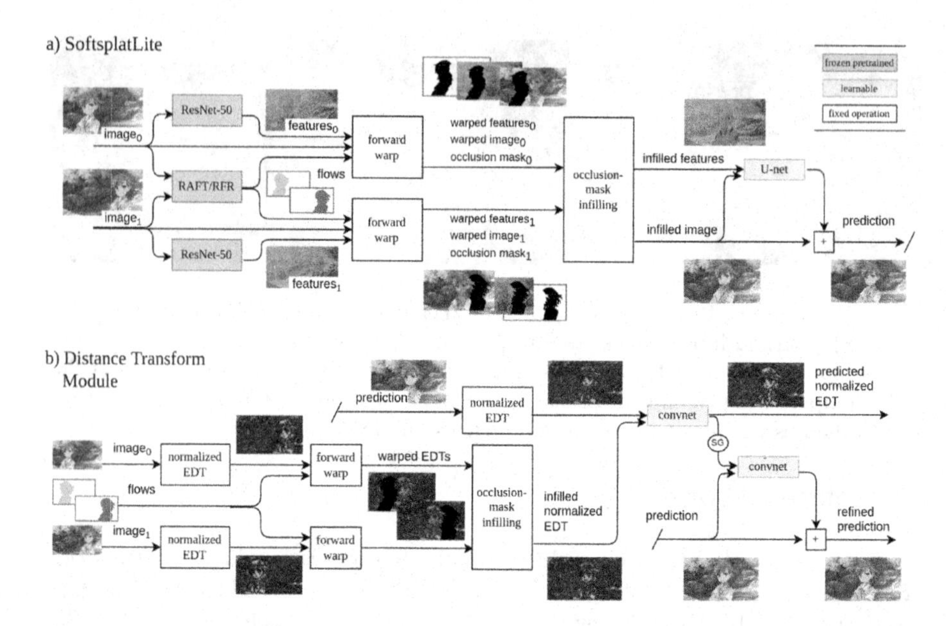


**Fig. 2.** Schematic of our proposed system. SoftsplatLite (SSL, Sect. 3.1) passes a prediction to the Distance Transform Module (DTM, Sect. 3.2) for refinement. SSL uses many fewer trainable parameters than AnimeInterp [37] to reduce overfitting, and introduces an infilling step to avoid ghosting artifacts. DTM leverages domain knowledge about line drawings to achieve more uniform solid-color regions. Artists: hariken, k.k. (hariken: https://danbooru.donmai.us/posts/5378938 k.k.: https://danbooru.donmai.us/posts/789765.)

forward-splat input image features for synthesis. Whereas AnimeInterp only focused on improving optical flow estimation, we assume a fixed flow estimator (the same RAFT [40] network from AnimeInterp, which they dub "RFR"). We instead look more closely at feature extraction, warping, and synthesis; our proposed SoftsplatLite (named similarly to PWC-Lite [19]) aims to improve convergence on LPIPS [45] while also being parameter- and training-efficient. Please see Fig. 2a for an overview of SSL.

We first note that the feature extractors in AnimeInterp [37] and Softsplat [24] are relatively shallow. The extractors must still be trained, and rely on backpropagation through the forward splatting mechanism. In practice, we found that replacing the extractor with the first four blocks of a frozen ImageNet-pretrained ResNet-50 [12] performs better; additionally, freezing the extractor contributes to reduced memory usage and compute during training, as no gradients must be backpropagated through the warping operations. Note that we also tried unfreezing the ResNet, but observed slight overfitting.

Next, we observe that forward splatting results in large empty occluded regions. If left unhandled during LPIPS training, these gaps often cause undesirable ghosting artifacts (see AnimeInterp [37] output in Fig. 3b). Additionally,

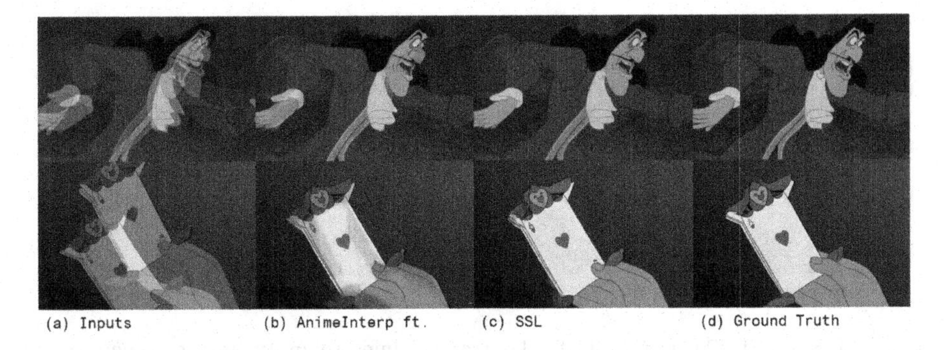


**Fig. 3.** SSL vs. AnimeInterp ft. [37]. Trained on the same ATD data [37] and LPIPS loss [45], AnimeInterp encounters many "ghosting" artifacts, which we resolve in SSL by proposing an inpainting technique.

subtle gradients at the edge of moving objects in the optical flow field may result in a spread of dots after forward warping; these later manifest as blurry patches in AnimeInterp predictions (see Fig. 1b). To remove these artifacts, we propose a simple infilling technique to generate a better warped feature stack $F$ prior to synthesis ("occlusion-mask infilling" in Fig. 2a):

$$F = \frac{1}{2}\left(M_{0\to t}W_{0\to t}(f(I_0)) + (1 - M_{0\to t})W_{1\to t}(f(I_1))\right) + \frac{1}{2}\left(M_{1\to t}W_{1\to t}(f(I_1)) + (1 - M_{1\to t})W_{0\to t}(f(I_0))\right) \quad (1)$$

$$Z_{1\to 0} = -0.1 \times ||LAB(I_1) - W'_{0\to 1}(LAB(I_0))|| \quad (2)$$

where $W_{a\to b}$ denotes forward warping from timestep $a$ to timestep $b$, $W'$ denotes backwarping, $M$ denotes the opened occlusion mask of the warp, $I$ represents either input image, and $f$ represents the feature extractor. In other words, occluded features are directly infilled with warped features from the other source image. The computation of mask $M$ involves warping an image of ones, followed by a morphological image opening with kernel $k = 5$ to remove dotted artifacts; note that though opening is non-differentiable, no gradients are needed with respect to the flow field as our flow estimator is fixed. Unlike AnimeInterp [37], we do not use average forward splatting, and instead use the more accurate soft-max weighting scheme with negative $L_2$ LAB color consistency as our Z-metric (similar as in Softsplat [24]). While it is not guaranteed that this infilling method will eliminate all holes (it is still possible for both warps to have shared occluded regions), we find that in practice the majority of image areas are covered.

Lastly, for the synthesis stage, we opt for a much more lightweight U-Net [33] instead of the GridNet [10] used in the original Softsplat [24]. We may afford this thrifty replacement by carefully placing a direct residual path from an initial warped guess to the final output. This follows the observation that

**Fig. 4.** Effect of DTM. DTM effectively leverages line proximity cues (distance transform) to refine SSL outputs. DTM not only removes minor aberrations from solid-color regions (bottom), but also corrects entire enclosures if needed (top).

directly applying our previously-described infilling method to the input RGB images produces a strong initial guess for the output; this is achieved by replacing feature extractor $f$ in Eq. 1 with the identity function. Instead of requiring a large synthesizer to reconcile two sets of warped images and features into a single final image, we employ a small network to simply refine a single good guess. Under this architecture, the additional GridNet parameters become redundant, and even contribute to overfitting.

Note that while SoftsplatLite and Softsplat have comparable parameter counts at inference (6.92M and 6.21M respectively), the frozen feature extractor and smaller synthesizer significantly reduces the number of trainable parameters compared to the original (1.28M and 2.01M respectively). We later demonstrate through ablations (Table 2) that lighter training and artifact reduction allow SSL to score better on perceptual metrics like LPIPS and chamfer distance.

### 3.2 Distance Transform Module

As seen in Fig. 4b, SoftsplatLite may struggle to choose colors for certain regions, or have trouble with large areas of flat color. These difficulties may be partly attributed to the natural texture bias of convolutional models [11]; the big monotonous regions of traditional cel animation would expectedly require convolutions with larger perceptual fields to extract meaningful features. Instead of building much deeper or wider models, we take advantage of line information inherently present in 2D animation; hypothetically, providing line proximity information to convolutions may act as a form of "stand-in" texture that helps the processing of cel-colored image data.

We thus propose a Distance Transform Module (DTM) to refine the SSL outputs by leveraging a normalized version of the Euclidean distance transform (NEDT). At a high level (see Fig. 2b), DTM first attempts to predict the ground truth NEDT of the output (middle) frame, and then uses this prediction to refine the SSL output through a residual block. To train the prediction of NEDT, we introduce an auxiliary $L_{dt}$ in addition to the $L_{lpips}$ on the final prediction, and

**Fig. 5.** RRLD filtering. RRLD quantifies whether a triplet is evenly-spaced. We show several overlaid triplets from our additional dataset ranked by RRLD; higher RRLD (bottom) indicates deviation from the halfway assumption. As RRLD is fully automatic, appropriate training data can be filtered from raw video at scale.

optimize a weighted sum of both losses end-to-end. The rest of this section provides specifics on the implementation.

The first step is to extract lines from the input images; for this, we use the simple but effective difference of gaussians (DoG) edge detector,

$$DoG(I) = \frac{1}{2} + t(G_{k\sigma}(I) - G_{\sigma}(I)) - \epsilon, \tag{3}$$

where $G_\sigma$ are Gaussian blurs after greyscale conversion, $k = 1.6$ is a factor greater than one, and $t = 2$ with $\epsilon = 0.01$ are hyperparameters. Please see Fig. 6 for examples of DoG extraction. Next, we apply the distance transform. To bound the range of values, we normalize EDT values to unit range similar to Narita et al. [23],

$$NEDT(I) = 1 - \exp\{\frac{-EDT(DoG(I) > 0.5)}{\tau d}\}, \tag{4}$$

where $\tau = 15/540$ is a steepness hyperparameter, and $d$ is the image height in pixels. Note that we thresholded DoG at 0.5 to get a binarized sketch.

This normalized EDT is extracted from both input images, and warped through the same inpainting procedure as Eq. 1; more precisely, $f$ is replaced by $NEDT$. DTM then uses this, as well as the extracted NEDT of SSL's output, to estimate the NEDT of the ground truth output frame. This prediction occurs through a small convolutional network (first yellow box in Fig. 2b), and is trained to minimize an auxiliary $L_{dt}$, the $L_1$ Laplacian pyramid loss between predicted and ground truth NEDTs. A final convolutional network (second yellow box in Fig. 2b) then incorporates the predicted NEDT to residually refine the SSL output.

Note that we detach the predicted NEDT image from the final RGB image prediction gradients ("SG" for "stop-gradient" in Fig. 2b), in order to reduce potentially competing signals from $L_{dt}$ and the final image loss. It is also important to mention that since both DoG sketch extraction and EDT are non-differentiable operations, the extraction of NEDT from the Softsplat output cannot be backpropagated. However, we found that we could still reasonably perform end-to-end training despite the required stop-gradient in this step.

Through this process, our DTM is able to predict the distance transform of the output, and utilize it in the final interpolation. Experiments show that this relatively cheap additional network is effective at improving perceptual performance (Table 2).

### 3.3 Restricted Relative Linear Discrepancy

Unlike in the natural video domain, where almost any three consecutive frames from a cut may be used as a training triplet, data collection for 2D animation is much more ambiguous. Animators often draw at variable framerates with expressive arc-like movements; when coupled with high pixel displacements, this results in a significant amount of triplets with non-linear motion or uneven spacing. However under the problem formulation, all middle frames of training triplets are assumed to be "halfway" between the inputs. While forward warping provides a way to control the interpolated $t \in [0, 1]$ at which generation occurs, it is ambiguous to label such ground truth for training. Li et al. in AnimeInterp [37] manually filter through more than 130,000 triplets to arrive at their ATD dataset with 12,000 samples, a costly manual effort with less than 10% yield.

In order to automate the training data collection process from raw animation data, we quantify the deviance of a triplet from the halfway assumption with a novel Restricted Relative Linear Discrepancy (RRLD) metric, and filter samples based on a simple threshold. In our experiments (Table 2), we demonstrate that selecting additional training data with RRLD improves generalization error, whereas training on naively-collected triplets damages performance. We additionally show that RRLD largely agrees with ATD, and that RRLD is robust to choice of flow estimator (Sect. 4.1). Please see Fig. 5 for example triplets accepted or rejected by RRLD. The rest of this section provides specifics of the filtering method. We define RRLD as follows,

$$RRLD(\omega_{0\to t}, \omega_{1\to t}) = \frac{1}{|\Omega|} \sum_{(i,j)\in\Omega} \frac{||\omega_{0\to t}[i,j] + \omega_{1\to t}[i,j]||/2}{||\omega_{0\to t}[i,j] - \omega_{1\to t}[i,j]||}, \quad (5)$$

where $\omega$ are forward flow fields extracted from consecutive frames $I_0$ and $I_t$ and $I_1$, and $\Omega$ denotes the set of $(i, j)$ pixel coordinates where both flows have norms greater than threshold 2.0 and point to pixels within the image.

RRLD takes as input flow fields from the middle frame $I_t$ to the end frames, and assumes they are correct. The numerator of Eq. 5 represents the distance from pixel $(i, j)$ to the midpoint between destination pixels, while the denominator describes the total distance between destination pixels. In other words, the interior of the summand is half the ratio between the diameters of a parallelogram formed by two flow vectors; this measures the relative distance from the

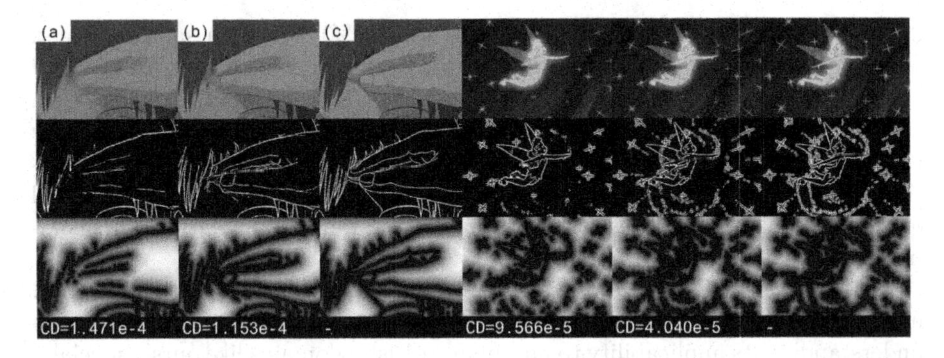


**Fig. 6.** Line and detail preservation. **(a)** AnimeInterp prediction; **(b)** our full model (SSL+DTM); **(c)** ground truth; **(middle)** extracted DoG lines; **(bottom)** normalized Euclidean distance transform. AnimeInterp blurs lines and details that are critical to animation; by focusing on perceptual metrics like LPIPS and chamfer distance (CD), we improve the generation quality.

actual to the ideal halfway point. As the estimated flows are noisy, we average over a restricted set of pixels $\Omega$. We first remove pixels with displacement close to zero, where a low denominator results in unrepresentatively high discrepancy measurement. Then, we also filter out pixels with flows pointing outside the image, which are often poor estimates. The final RRLD gives a rough measure of deviance from the halfway-frame assumption, for which we may define a cutoff (0.3 in this work).

One caveat to this method is that pans must be discarded. In some cases, a non-linear animation may be composited onto a panning background; RRLD would then include the linearly-moving background in $\Omega$, lowering the overall measurement despite having a nonlinear region of interest. We simply remove triplets with large $\Omega$, high average flow magnitude, and low flow variance. It is possible to reintroduce panning effects through data augmentation if needed, though we did not for our training.

Another important point is that even though animators may draw at framerates like 12 or 8, the final raw input videos are still at 24fps. Thus, many consecutive triplets in actuality contain two duplicates, which leads to RRLD values around 0.5; had the duplicate been removed, an adjacent frame outside the triplet may have had a qualifying RRLD. In order to maximize the data yield, we also train a simple duplicate frame detector, using linear regression over the mean and maximum $L_2$ LAB color difference between consecutive frames.

### 3.4 User Study and Quality Metrics

We perform a user study in order to evaluate our system and explore the relationship between metrics and perceived quality. To get a representative subset of the ATD test set, on which we perform all evaluations, we select 323 random samples in accordance with Fischer's sample size formula (with population

2000, margin of error 5%, and confidence level 95%). For each sample triplet, users were given a pair of animations playing back and forth at 2fps, cropped to the region-of-interest annotation provided by ATD. The middle frame of each animation was a result generated either by our best model (on LPIPS), or by the pretrained AnimeInterp [37]. Participants were asked to pick which animation had: clearer/sharper lines, more consistent shapes/colors, and better overall quality. Complete survey results, including several random animation pairs compared, are available in the supplementary.

Our main metric of interest is LPIPS [45], a general measure of perceived image quality based on deep image classification features. We are interested in understanding its applicability to non-photorealistic domains like ours, especially in comparison with PSNR/SSIM used in prior work [37].

We additionally consider the chamfer distance (CD) between lines extracted from the ground truth vs. the prediction. The chamfer metric is typically used in 3D work, where the distance between two point clouds is calculated by averaging the shortest distances from each point of one cloud to a point on the other. In the context of binary line drawings extracted from our data using DoG (Eq. 3), the 3D points are replaced by all 2D pixels that lie on lines. As chamfer distance would intuitively measure how far lines are from each other in different images, we explore the importance of this metric for our domain with images based on line drawings. Please see Fig. 6 for examples of CD evaluation. In this work, we define chamfer distance as:

$$CD(X_0, X_1) = \frac{1}{2HWD} \sum X_0 DT(X_1) + X_1 DT(X_0) \tag{6}$$

where $X$ are binary sketches with 1 on lines and 0 elsewhere, $DT$ denotes the Euclidean distance transform, the summation is pixel-wise, and $HWD$ is the product of height, width, and diameter. We normalize by both area and diameter to enforce invariance to image scale. Note that our definition is symmetric with respect to prediction and ground truth, zero if and only if they are equal, and strictly non-negative. Also observe that as neither DoG binarization nor DT is differentiable, CD cannot be optimized directly by gradient descent training; thus it is used for evaluation only.

## 4 Experiments and Discussion

We implement our system in PyTorch [29] wrapped in Lightning [8], with Kornia [32]. Our model uses the same RFR/RAFT with SGM flows as AnimeInterp for fairer comparison [37,40], and forward splatting is done with the official Softsplat [24] module. We train with the Adam [17] optimizer at learning rate $\alpha = 0.001$ for 50 epochs, and accumulate gradients for an effective batch size of 32. Our code uses the official LPIPS [45] package, with the AlexNet [18] backbone. All training minimizes the total loss $L = \lambda_{lpips} L_{lpips} + \lambda_{dt} L_{dt}$, where $\lambda_{lpips} = 30$; depending on whether DTM is trained, $\lambda_{dt}$ is either 0 or 5. Evaluations are run over the 2000-sample test set from AnimeInterp's ATD12k dataset; however we only train

**Table 1.** Comparison with baselines. Our full proposed method achieves the best perceptual performance, followed by AnimeInterp [37]. We show in our user study (Sect. 4.4) that LPIPS/CD are better indicators of quality than the PSNR/SSIM focused on in previous work; we list them here for completeness. Models from prior work are fine-tuned on LPIPS for fairer comparison. Best values are underlined, runner-ups italicized; LPIPS is scaled by 1e2, CD by 1e5.

| Model | All | | | | Eastern | | Western | |
|---|---|---|---|---|---|---|---|---|
| | LPIPS | CD | PSNR | SSIM | LPIPS | CD | LPIPS | CD |
| DAIN [1] | 4.695 | 5.288 | 28.840 | 95.28 | 5.499 | 6.537 | 4.204 | 4.524 |
| DAIN ft. [1] | 4.137 | 4.851 | 29.040 | 95.27 | 4.734 | 5.888 | 3.771 | 4.217 |
| RIFE [13] | 4.451 | 5.488 | 28.515 | 95.14 | 4.933 | 6.618 | 4.156 | 4.796 |
| RIFE ft. [13] | 4.233 | 5.411 | 27.977 | 93.70 | 4.788 | 6.643 | 3.894 | 4.658 |
| ABME [28] | 5.731 | 7.244 | 29.177 | *95.54* | 7.000 | 10.010 | 4.955 | 5.552 |
| ABME ft. [28] | 4.208 | 4.981 | 29.060 | 95.19 | 4.987 | 6.092 | 3.732 | 4.302 |
| AnimeInterp [37] | 5.059 | 5.564 | <u>29.675</u> | <u>95.84</u> | 5.824 | 7.017 | 4.590 | 4.674 |
| AnimeInterp ft. [37] | *3.757* | *4.513* | 28.962 | 95.02 | *4.113* | *5.286* | *3.540* | *4.039* |
| Ours | <u>3.494</u> | <u>4.350</u> | *29.293* | 95.15 | <u>3.826</u> | <u>4.979</u> | <u>3.291</u> | <u>3.966</u> |

on a random 9k of the remaining 10k in ATD, so that we can designate 1k for validation. Similar to Li et al. [37], we randomly perform horizontal flips and frame order reversal augmentations during training. We use single-node training with at most 4x GTX1080Ti at a time, with mixed precision where possible. All models are trained and tested at $540 \times 960$ resolution.

We wrote a custom CUDA implementation for the distance transform and chamfer distance using CuPy [27] that achieves upwards of 3000x speedup from the SciPy CPU implementation [41]; the algorithm is a simpler version of Felzenszwalb et al. [9], where we calculate the minimum of the lower envelope through brute iteration. While more efficient GPU algorithms are known [4], we found our implementation sufficient.

### 4.1 RRLD Data Collection

As RRLD was designed to replicate the manual selection of training data, we applied RRLD to AnimeInterp's ATD dataset [37] and achieved 95.3% recall (i.e. RRLD only rejected less than 5% of human-collected data); as the negative samples from the ATD collection process are not available, it is not possible to calculate RRLD's precision on ATD. Additionally we study the effect of flow estimation on RRLD, finding that filtering with FlowNet2 [14] and RFR flows [37] returns very similar results (0.877 Cohen's kappa tested over 34,128 triplets).

We use our automatic pipeline to collect additional training triplets. We source data from 14 franchises in the eastern "anime" style, with premiere dates ranging from 1989–2020, totalling 239 episodes (roughly 95hrs, 8.24M frames at 24fps); please refer to our supplementary materials for the full list of sources.

**Table 2.** Ablations of proposed methods. Firstly, each component of SSL contributes to performance (especially infilling). Secondly, new data filtered naively hurts performance, while new RRLD-filtered data helps. Lastly, DTM improvement is due to auxiliary supervision, not just increased parameter count. AnimeInterp ft. is copied from Table 1 for comparison; the last row here and in Table 1 are equivalent. Best values are underlined, runner-ups italicized; LPIPS is scaled by 1e2, CD by 1e5.

| Model | Data | All | | Eastern | | Western | |
|---|---|---|---|---|---|---|---|
| | | LPIPS | CD | LPIPS | CD | LPIPS | CD |
| AnimeInterp ft. [37] | ATD | 3.757 | 4.513 | 4.113 | 5.286 | 3.540 | 4.039 |
| SSL (no flow infill) | ATD | 3.648 | 4.496 | 4.026 | 5.160 | 3.416 | 4.089 |
| SSL (no U-net synth.) | ATD | 3.614 | 4.579 | 3.982 | 5.288 | 3.389 | 4.146 |
| SSL (no ResNet extr.) | ATD | 3.605 | 4.739 | 3.957 | 5.429 | 3.391 | 4.317 |
| SSL | ATD | 3.586 | 4.572 | 3.940 | 5.248 | 3.369 | 4.158 |
| SSL | ATD+naive | 3.702 | 4.811 | 3.997 | 5.033 | 3.521 | 4.675 |
| SSL | ATD+RRLD | 3.535 | 4.431 | 3.873 | 5.089 | 3.329 | *4.028* |
| SSL+DTM (no $L_{dt}$) | ATD+RRLD | *3.531* | *4.430* | *3.865* | *4.995* | *3.327* | 4.085 |
| SSL+DTM | ATD+RRLD | <u>3.494</u> | <u>4.350</u> | <u>3.826</u> | <u>4.979</u> | <u>3.291</u> | <u>3.966</u> |

**Table 3.** User study results. For each of the visual criteria we asked the users to judge (rows), we list the percentage of instances where users preferred the animation with a better metric score (columns). Values above 50% indicate agreement between queried criteria and metric score difference, and values under 50% indicate contradiction. "Pref. Ours" means percent of users preferring our output to AnimeInterp [37] for that criteria.

| Criteria | Prefer ours | Lower LPIPS | Lower CD | Higher PSNR | Higher SSIM |
|---|---|---|---|---|---|
| Cleaner/sharper lines | 86.01% | 86.56% | 78.20% | 18.95% | 15.48% |
| More consistent shape/color | 78.82% | 79.26% | 73.99% | 25.02% | 22.66% |
| Better overall quality | 81.11% | 81.55% | 75.67% | 22.97% | 19.88% |

Here, RRLD was calculated using FlowNet2 [14] as inference was faster than RFR [37]. While RRLD filtering presents us with 543.6k viable triplets, we only select one random triplet per cut to promote diversity; the cut detection was performed with a pretrained TransNet v2 [38]. This cuts down eligible samples to 49.7k. For the demonstrative purposes of this paper, we do not train on the full new dataset, and instead limit ourselves to doubling the ATD training set by randomly selecting 9k qualifying triplets. Please see Fig. 5 for examples of accepted and rejected triplets from franchises set aside for validation.

While we cannot release the new data collected in this work, our specific sources are listed in the supplementary and our RRLD data collection pipeline will be made public; this allows followup work to either recreate our dataset or assemble their own datasets directly from source animations.

### 4.2 Comparison with Baselines

The main focus of our work is to improve perceptual quality, namely LPIPS and chamfer distance (as validated later by our user study results). We gather four existing frame interpolation systems (ABME [28], RIFE [13], DAIN [1], and AnimeInterp [37]) for comparison to our full model incorporating all our proposed methods. For a fairer comparison, as other models may not have been trained on the same LPIPS objective or on animation data, we fine-tune their given pre-trained models with LPIPS on the ATD training set. As we can see from Table 1, our full proposed method achieves the best perceptual performance, followed by AnimeInterp. To provide more complete information on trainable parameters, our model has 1.28M (million) compared to: AnimeInterp 2.01M, RIFE 13.0M, ABME 17.5M, DAIN 24.0M. Breaking down further, our model consists of 1.266M for SSL and 0.011M for DTM.

### 4.3 Ablation Studies

We perform several ablations in Table 2. In the first group, each of the modifications to Softsplat [24] (frozen ResNet [12] feature extractor, infilling, U-net [33] replacing GridNet [10]) contributes to SSL outperforming AnimeInterp [37]. The infilling technique improves performance the most.

In the second group of Table 2, we ablate the addition of new data filtered by RRLD (Sect. 4.1). Training with RRLD-filtered data improves generalization as expected. To demonstrate the necessity of RRLD's specific filtering strategy, we train with an alternative dataset of equal size gathered from the same sources, but using a "naive" filtering approach. For simplicity, we directly follow the crude filter used in creating ATD [37]: no two frames of a triplet may contain SSIM outside [0.75, 0.95]. We see this naively-collected data actively damages model performance, validating the use of our proposed RRLD filter.

Splitting by eastern vs. western style, we clarify the distribution shift between sub-domains. Note that our new data is all anime, whereas 62.05% of ATD test set is in the western "Disney" style. From the LPIPS results, the eastern style is more difficult; adding eastern-only RRLD data has unexpectedly less of an effect on eastern testing than western. This may be because western productions tend to prioritize fluid motion (smaller displacements) over complex character designs (more details), contrary to the eastern style.

In the last group of Table 2, we train SoftsplatLite with DTM, but ablate the effect of additionally optimizing for $L_{dt}$; this way, we may see whether auxiliary supervision of NEDT improves performance under the same parameter count. Note that the upper yellow convnet of Fig. 2b receives no gradients in the ablation, effectively remaining at its random initialization. The results show that the prediction of line proximity information indeed contributes to performance.

### 4.4 User Study Results

We summarize the user study results in Table 3, and provide the full breakdown with sample animations in the supplementary. Our study had 5 participants, meaning each entry of Table 3 has support 1615 (323 compared pairs per participant). We confirm the observations made by Niklaus et al. and Blau et al. [3], that PSNR/SSIM and perceptual metrics may be at odds with one another. Despite lower PSNR/SSIM scores, users consistently preferred our outputs to those of AnimeInterp. A possible explanation is that due to animations having larger displacements, the middle ground truth frames may be quite displaced from the ideal halfway interpolation. SSIM, as noted by previous work [34,45], was not designed to assess these geometric distortions. Color metrics like PSNR and $L_1$ may penalize heavily for this perceptually minor difference, encouraging the model to reduce risk by blurring; this is consistent with behavior exhibited by the original AnimeInterp trained on $L_1$ (Fig. 6). LPIPS on the other hand has a larger perceptive field due to convolutions, and may be more forgiving of these instances. This study provides another example of the perception-distortion tradeoff [3], and establishes its transferability to 2D animation.

The user study also shows an imperfect match between LPIPS and CD. This mismatch is also reflected in Tables 1 and 2, where aggregate decreases in LPIPS do not correspond to reduced CD. This maybe because CD reflects only the line-structures of an image. However, Table 3 shows LPIPS is unexpectedly more predictive of line quality. A possible explanation is that CD is still more sensitive to offsets than LPIPS; in fact, CD grows roughly proportionally to displacement for line drawings. Thus, it may suffer the same problems as PSNR but to a lesser extent, as PSNR would penalize across an entire displaced area opposed to across a thin line.

## 5 Limitations and Conclusion

Our system still has several limitations. By design, our model can only interpolate linearly between two frames, while real animations have non-linear movements that follow arcs across long sequences. In future work, we may incorporate non-linearity from methods like QVI [43], or allow user input from an artist. Additionally, we are limited to colored frames, which are typically unavailable until the later stages of animation production; following related work [23], we can expand our scope to work on line drawings directly.

To summarize, we identify and overcome shortcomings of previous work [37] on 2D animation interpolation, and achieve state-of-the-art interpolation perceptual quality. Our contributions include an effective SoftsplatLite architecture modified to improve perceptual performance, a Distance Transform Module leveraging domain knowledge of lines to perform refinement, and a Restricted Relative Linear Discrepancy metric that allows automatic training data collection from raw animation. We validate our focus on perceptual quality through a user study, hopefully inspiring future work to maintain this emphasis for the traditional 2D animation domain.

**Acknowledgements.** The authors would like to thank Lillian Huang and Saeed Hadadan for their discussion and feedback, as well as NVIDIA for GPU support.

## References

1. Bao, W., Lai, W.S., Ma, C., Zhang, X., Gao, Z., Yang, M.H.: Depth-aware video frame interpolation. In: Proceedings of the IEEE/CVF Conference on Computer Vision and Pattern Recognition, pp. 3703–3712 (2019)
2. Bao, W., Lai, W.S., Zhang, X., Gao, Z., Yang, M.H.: MEMC-Net: motion estimation and motion compensation driven neural network for video interpolation and enhancement. IEEE Trans. Pattern Anal. Mach. Intell. **43**, 933–948 (2019)
3. Blau, Y., Michaeli, T.: The perception-distortion tradeoff. In: Proceedings of the IEEE Conference on Computer Vision and Pattern Recognition, pp. 6228–6237 (2018)
4. Cao, T.T., Tang, K., Mohamed, A., Tan, T.S.: Parallel banding algorithm to compute exact distance transform with the GPU. In: Proceedings of the 2010 ACM SIGGRAPH Symposium on Interactive 3D Graphics and Games, pp. 83–90 (2010)
5. Casey, E., Pérez, V., Li, Z.: The animation transformer: visual correspondence via segment matching. In: Proceedings of the IEEE/CVF International Conference on Computer Vision, pp. 11323–11332 (2021)
6. Choi, M., Kim, H., Han, B., Xu, N., Lee, K.M.: Channel attention is all you need for video frame interpolation. In: Proceedings of the AAAI Conference on Artificial Intelligence, vol. 34, pp. 10663–10671 (2020)
7. Dalstein, B., Ronfard, R., Van De Panne, M.: Vector graphics animation with time-varying topology. ACM Trans. Graph. (TOG) **34**(4), 1–12 (2015)
8. Falcon, W., The PyTorch Lightning team: PyTorch Lightning (2019). https://doi.org/10.5281/zenodo.3828935. https://github.com/PyTorchLightning/pytorch-lightning
9. Felzenszwalb, P.F., Huttenlocher, D.P.: Distance transforms of sampled functions. Theory Comput. **8**(1), 415–428 (2012)
10. Fourure, D., Emonet, R., Fromont, E., Muselet, D., Tremeau, A., Wolf, C.: Residual conv-deconv grid network for semantic segmentation. arXiv preprint arXiv:1707.07958 (2017)
11. Geirhos, R., Rubisch, P., Michaelis, C., Bethge, M., Wichmann, F.A., Brendel, W.: ImageNet-trained CNNs are biased towards texture; increasing shape bias improves accuracy and robustness. arXiv preprint arXiv:1811.12231 (2018)
12. He, K., Zhang, X., Ren, S., Sun, J.: Deep residual learning for image recognition. In: Proceedings of the IEEE Conference on Computer Vision and Pattern Recognition, pp. 770–778 (2016)
13. Huang, Z., Zhang, T., Heng, W., Shi, B., Zhou, S.: Rife: real-time intermediate flow estimation for video frame interpolation. arXiv preprint arXiv:2011.06294 (2020)
14. Ilg, E., Mayer, N., Saikia, T., Keuper, M., Dosovitskiy, A., Brox, T.: FlowNet 2.0: evolution of optical flow estimation with deep networks. In: Proceedings of the IEEE Conference on Computer Vision and Pattern Recognition, pp. 2462–2470 (2017)
15. Jaegle, A., et al.: Perceiver IO: a general architecture for structured inputs & outputs. arXiv preprint arXiv:2107.14795 (2021)
16. Jiang, H., Sun, D., Jampani, V., Yang, M.H., Learned-Miller, E., Kautz, J.: Super SloMo: high quality estimation of multiple intermediate frames for video interpolation. In: Proceedings of the IEEE Conference on Computer Vision and Pattern Recognition, pp. 9000–9008 (2018)
17. Kingma, D.P., Ba, J.: Adam: a method for stochastic optimization. arXiv preprint arXiv:1412.6980 (2014)

18. Krizhevsky, A., Sutskever, I., Hinton, G.E.: ImageNet classification with deep convolutional neural networks. In: Advances in Neural Information Processing Systems, vol. 25, pp. 1097–1105 (2012)
19. Liu, L., et al.: Learning by analogy: reliable supervision from transformations for unsupervised optical flow estimation. In: Proceedings of the IEEE/CVF Conference on Computer Vision and Pattern Recognition, pp. 6489–6498 (2020)
20. Maejima, A., et al.: Anime character colorization using few-shot learning. In: SIGGRAPH Asia 2021 Technical Communications, pp. 1–4 (2021)
21. Meyer, S., Djelouah, A., McWilliams, B., Sorkine-Hornung, A., Gross, M., Schroers, C.: PhaseNet for video frame interpolation. In: Proceedings of the IEEE Conference on Computer Vision and Pattern Recognition, pp. 498–507 (2018)
22. Meyer, S., Wang, O., Zimmer, H., Grosse, M., Sorkine-Hornung, A.: Phase-based frame interpolation for video. In: Proceedings of the IEEE Conference on Computer Vision and Pattern Recognition, pp. 1410–1418 (2015)
23. Narita, R., Hirakawa, K., Aizawa, K.: Optical flow based line drawing frame interpolation using distance transform to support inbetweenings. In: 2019 IEEE International Conference on Image Processing (ICIP), pp. 4200–4204. IEEE (2019)
24. Niklaus, S., Liu, F.: Softmax splatting for video frame interpolation. In: Proceedings of the IEEE/CVF Conference on Computer Vision and Pattern Recognition, pp. 5437–5446 (2020)
25. Niklaus, S., Mai, L., Liu, F.: Video frame interpolation via adaptive convolution. In: Proceedings of the IEEE Conference on Computer Vision and Pattern Recognition, pp. 670–679 (2017)
26. Niklaus, S., Mai, L., Liu, F.: Video frame interpolation via adaptive separable convolution. In: Proceedings of the IEEE International Conference on Computer Vision, pp. 261–270 (2017)
27. Okuta, R., Unno, Y., Nishino, D., Hido, S., Loomis, C.: CuPy: a NumPy-compatible library for NVIDIA GPU calculations. In: Proceedings of Workshop on Machine Learning Systems (LearningSys) in the Thirty-First Annual Conference on Neural Information Processing Systems (NIPS) (2017)
28. Park, J., Lee, C., Kim, C.S.: Asymmetric bilateral motion estimation for video frame interpolation. In: Proceedings of the IEEE/CVF International Conference on Computer Vision, pp. 14539–14548 (2021)
29. Paszke, A., et al.: Pytorch: an imperative style, high-performance deep learning library. In: Advances in Neural Information Processing Systems, vol. 32, pp. 8026–8037 (2019)
30. Qian, Z., Bo, W., Wei, W., Hai, L., Hui, L.J.: Line art correlation matching network for automatic animation colorization. arXiv e-prints, pp. arXiv-2004 (2020)
31. Ren, H., Li, J., Gao, N.: Two-stage sketch colorization with color parsing. IEEE Access **8**, 44599–44610 (2019)
32. Riba, E., Mishkin, D., Shi, J., Ponsa, D., Moreno-Noguer, F., Bradski, G.: A survey on Kornia: an open source differentiable computer vision library for Pytorch (2020)
33. Ronneberger, O., Fischer, P., Brox, T.: U-Net: convolutional networks for biomedical image segmentation. In: Navab, N., Hornegger, J., Wells, W.M., Frangi, A.F. (eds.) MICCAI 2015. LNCS, vol. 9351, pp. 234–241. Springer, Cham (2015). https://doi.org/10.1007/978-3-319-24574-4_28
34. Sampat, M.P., Wang, Z., Gupta, S., Bovik, A.C., Markey, M.K.: Complex wavelet structural similarity: a new image similarity index. IEEE Trans. Image Process. **18**(11), 2385–2401 (2009)
35. Simo-Serra, E., Iizuka, S., Ishikawa, H.: Mastering sketching: adversarial augmentation for structured prediction. ACM Trans. Graph. (TOG) **37**(1), 1–13 (2018)

36. Simo-Serra, E., Iizuka, S., Sasaki, K., Ishikawa, H.: Learning to simplify: fully convolutional networks for rough sketch cleanup. ACM Trans. Graph. (TOG) **35**(4), 1–11 (2016)
37. Siyao, L., et al.: Deep animation video interpolation in the wild. In: Proceedings of the IEEE/CVF Conference on Computer Vision and Pattern Recognition, pp. 6587–6595 (2021)
38. Souček, T., Lokoč, J.: TransNet v2: an effective deep network architecture for fast shot transition detection. arXiv preprint arXiv:2008.04838 (2020)
39. Sun, D., Yang, X., Liu, M.Y., Kautz, J.: PWC-Net: CNNs for optical flow using pyramid, warping, and cost volume. In: Proceedings of the IEEE Conference on Computer Vision and Pattern Recognition, pp. 8934–8943 (2018)
40. Teed, Z., Deng, J.: RAFT: recurrent all-pairs field transforms for optical flow. In: Vedaldi, A., Bischof, H., Brox, T., Frahm, J.-M. (eds.) ECCV 2020. LNCS, vol. 12347, pp. 402–419. Springer, Cham (2020). https://doi.org/10.1007/978-3-030-58536-5_24
41. Virtanen, P., et al.: SciPy 1.0: fundamental algorithms for scientific computing in Python. Nat. Methods **17**, 261–272 (2020). https://doi.org/10.1038/s41592-019-0686-2
42. Whited, B., Noris, G., Simmons, M., Sumner, R.W., Gross, M., Rossignac, J.: BetweenIT: an interactive tool for tight inbetweening. In: Computer Graphics Forum, vol. 29, pp. 605–614. Wiley Online Library (2010)
43. Xu, X., Siyao, L., Sun, W., Yin, Q., Yang, M.H.: Quadratic video interpolation. arXiv preprint arXiv:1911.00627 (2019)
44. Yagi, Y.: A filter based approach for inbetweening. arXiv preprint arXiv:1706.03497 (2017)
45. Zhang, R., Isola, P., Efros, A.A., Shechtman, E., Wang, O.: The unreasonable effectiveness of deep features as a perceptual metric. In: Proceedings of the IEEE Conference on Computer Vision and Pattern Recognition, pp. 586–595 (2018)

# Selective TransHDR: Transformer-Based Selective HDR Imaging Using Ghost Region Mask

Jou Won Song[1(✉)], Ye-In Park[1], Kyeongbo Kong[2], Jaeho Kwak[1], and Suk-Ju Kang[1]

[1] Department of Electronic Engineering, Sogang University, Seoul, Korea
{wn5649,yipark06,resky1111,sjkang}@sogang.ac.kr

[2] Department of Media Communication, Pukyong National University, Busan, Korea
kbkong@pknu.ac.kr

**Abstract.** The primary issue in high dynamic range (HDR) imaging is the removal of ghost artifacts afforded when merging multi-exposure low dynamic range images. In the weakly misaligned region, ghost artifacts can be suppressed using convolutional neural network (CNN)-based methods. However, in highly misaligned regions, it is necessary to extract features from the global region because the necessary information does not exist in the local region. Therefore, the CNN-based methods specialized for local features extraction cannot obtain satisfactory results. To address this issue, we propose a transformer-based selective HDR image reconstruction network that uses a ghost region mask. The proposed method separates a given image into ghost and non-ghost regions, and then, selectively applies either the CNN or the transformer. The proposed selective transformer module divides an entire image into several regions to effectively extract the features of each region for HDR image reconstruction, thereby extracting the whole information required for HDR reconstruction in the ghost regions from the entire image. Extensive experiments conducted on several benchmark datasets demonstrate the superiority of the proposed method over existing state-of-the-art methods in terms of the mitigation of ghost artifacts.

## 1 Introduction

Typical digital cameras can only capture luminance within a limited dynamic range due to sensor limitations. Therefore, low dynamic range (LDR) images with 8-bit depth obtained by these cameras have significant underexposed and overexposed regions, thereby yielding large data loss compared to the real scene. A lot of studies have been conducted to recover lost data from LDR images and to generate 10-bit or 12-bit high dynamic range (HDR) images that can provide

J. W. Song and Y.-I. Park—Equal contribution.

**Supplementary Information** The online version contains supplementary material available at https://doi.org/10.1007/978-3-031-19790-1_18.

© The Author(s), under exclusive license to Springer Nature Switzerland AG 2022
S. Avidan et al. (Eds.): ECCV 2022, LNCS 13677, pp. 288–304, 2022.
https://doi.org/10.1007/978-3-031-19790-1_18

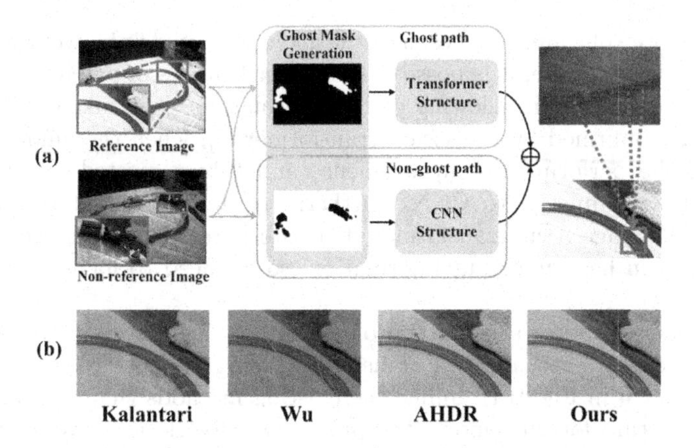


**Fig. 1.** (a) Overview of the proposed selective transformer module, and (b) sample images generated by the proposed and the state-of-the-art methods in Tursun et al.'s [32] dataset. The proposed selective transformer module separates the ghost regions from the non-ghost regions, and selectively uses one of the CNN and the transformer. (Color figure online)

a wide illuminance range. Multi-exposure image fusion is the most common HDR reconstruction method; LDR images with different exposure values are obtained by cameras with a limited dynamic range and merged into an HDR image after alignment of the LDR images [4,10,11,27,38].

However, in most cases, the aligned LDR images are difficult to obtain in the acquisition process of several LDR images with different exposure values due to the motion of a camera or an object. If an HDR image is obtained using these unaligned LDR images, ghost and blur artifacts occur in the HDR image [1,12,18,20,23,29]. To solve this problem, optical-flow methods [6,9] or patch-based methods [28] have been presented.

Although these methods can resolve some scenarios with large motions, ghost artifacts still exist. Recently, many deep neural network (DNN)-based methods have been proposed for reconstructing HDR images [7,8,19,35,37]. Typically, DNN-based methods utilize convolution neural networks (CNNs) to extract spatial features. These methods have higher performance than existing methods. However, these CNN-based methods are difficult to explicitly remove ghost artifacts due to the limitation of CNN specialized for local feature extraction from unaligned LDR images, and hence, it is difficult to consider global information required for HDR reconstruction from an entire region. As shown in Fig. 1(a), the red box regions of the reference image and the non-reference image are in the same location, but contain different information. The rail is visible in the reference image, but it is occluded by the train in the non-reference image. That is, it is impossible to extract rail information from the non-reference image due to the movement of the object. Therefore, in Fig. 1(b), Kalantari et al.'s [17], Wu et al.'s [35], and Yan et al.'s [36] yielded final images with ghost artifacts in the motion region.

To solve this limitation of CNN, the use of transformer structure in various tasks of computer vision has been studied [3,40]. In the transformer structure

[33], images are split into patches, and the attention weights between all patches are extracted. Therefore, it is possible to extract features of patches that are far from the reference patch, allowing global information to be taken into account. However, as confirmed by the vision transformer [5], when the amount of data is limited, the performance improvement cannot be expected due to lack of inductive bias compared to CNN. Furthermore, in the non-ghost region, for aligned LDR images, the transformer structure is relatively inefficient because the local region has important information, so we need to focus more on that region.

This paper proposes a novel approach that can selectively consider regions where it is effective to apply a transformer using a ghost region mask as a guide. As shown in Fig. 1(a), unlike the existing methods that utilize the same network structure for all regions, the proposed network adaptively applies the transformer and CNN structures by separating the ghost and non-ghost regions. The proposed ghost region mask-guided transformer module uses the transformer structure to extract important features for the ghost region from the global regions of the non-reference images. Figure 1(a) visualizes the attention map of the proposed selective transformer module. The proposed attention map focuses on the visible rail away from the red box where the rail is obscured by the train. The proposed method can extract meaningful information from global regions of the non-reference image, excluding unnecessary information from local regions. Therefore, as shown in Fig. 1(b), the proposed method affords significantly better reconstruction in terms of the color and details of the ghost regions. The main contributions of this paper can be summarized as follows:

- We propose a novel contents-aware ghost region detector to effectively consider both global and local features focused by the proposed model. This detector distinguishes between ghost and non-ghost regions, and the networks, suitable for each region, are selectively applied.
- We propose a transformer-based selective HDR image reconstruction network to extract the necessary features to restore the ghost region. Our method does not simply apply the transformer, but uses it for the global information analysis and selection that the transformer can be effectively used. Therefore, the proposed selective transformer module can extract important global features from the entire region of each non-reference image for HDR reconstruction of ghost regions.
- Experiments on various datasets validate the superiority of the proposed method compared to existing methods. We also demonstrate that using a transformer adaptively rather than using a single model significantly improves performance by reflecting the characteristics of images with various exposure values.

## 2 Related Work

### 2.1 Motion Detection-Based HDR Reconstruction

Motion detection-based methods are based on the assumption that the LDR images with different exposures can be globally registered in the HDR coordinate.

These methods can detect moving pixels in the images which are rejected for final weighted HDR fusion [13,16,18]. Heo et al. [14] detected motion regions with joint probability densities. Yan et al. [38] used a sparse representation to detect the object motion. However, since these methods ignore unaligned pixels and not all input regions are available, these methods heavily depend on effectiveness of motion detection and cannot expect high performance when the large motion appears.

### 2.2 Alignment-Based HDR Reconstruction

Alignment-based methods focused on aligning LDR images to a reference image, and then, merged them to reconstruct the HDR image [4,22]. For the alignment, optical flow or patch matching methods are generally used. Bogoni [2] used optical flow to estimate motion vectors. Sen et al. [28] used a patch-based energy minimization method. Hu et al. [15] aligned the images using brightness and gradient consistency in the transformed domain. However, alignment-based methods are sensitive to complex backgrounds and large motions. These methods also requires significantly high execution time.

### 2.3 CNNs-Based HDR Reconstruction

Kalantari et al. [17] introduced neural networks into the alignment-before-merging pipeline for the HDR image generation. Wu et al. [35] proposed an autoencoder that can learn to convert multiple LDR images into a ghost-free HDR image. In [39], multiple LDR images were reconstructed into HDR images using a non-local network [34]. These CNN-based HDR reconstruction methods extract features of unaligned image regions, causing geometric or color distortion. In addition, they are difficult to reconstruct regions with large motions due to the CNN specialized in the local feature extraction. Yan et al. [36] used a spatial attention mechanism to generate ghost-free HDR images. Although the attention map of the spatial attention mechanism deletes unnecessary information, it is difficult to extract features for HDR reconstruction in the global region because it still uses a CNN-based model. Prabhakar et al. [25] proposed an HDR imaging method using bilateral guided upsampler and motion compensation. This method can compensate for ghost regions in LDR images.

### 2.4 Transformer

Transformers have been actively studied in many tasks of computer vision. Carion et al. [3] used the transformer and CNN structures simultaneously for object detection. Vision transformer demonstrated that the model using a pure transformer achieved the best performance [5]. Zheng et al. [40] proposed an encoder with a transformer structure to solve the problem of the reduced sparse resolution. In a recent study related to our method, Yan et al. [39] used a non-local network to perform the HDR reconstruction. However, this method may still extract unnecessary features because the CNN structure is used in the ghost

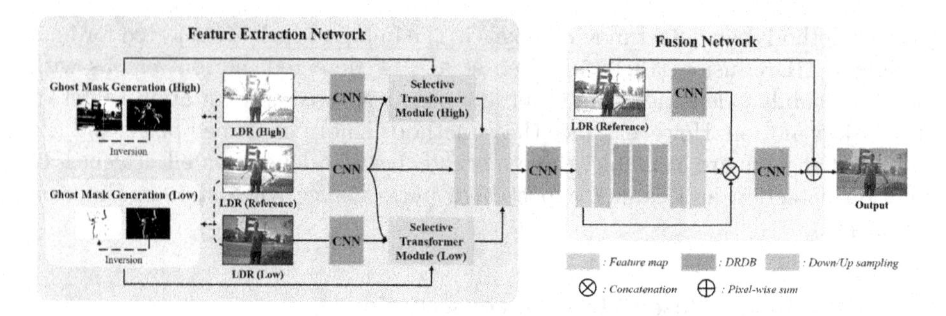


**Fig. 2.** The proposed method architecture. It consists of a feature extraction network and a fusion network. The selective transformer module selectively applies the transformer and the CNN structures to the ghost and non-ghost regions, respectively, using a proposed ghost region mask. The fusion network is constructed based on a series of dilated residual dense blocks (DRDBs) and multi-scale CNNs. The final HDR result is generated by combining the reference image with the final stage of the fusion network.

region. The proposed method selectively applies the transformer structure only to the ghost region, which is the region that requires global feature extraction. Therefore, the model can appropriately extract the necessary information for ghost and non-ghost regions.

## 3 Proposed Method

This paper proposes the novel transformer-based HDR image reconstruction and ghost mask generation to extract ghost regions in multi-exposure images. In the proposed method, the LDR image is divided into ghost and non-ghost regions, and features corresponding to these regions are extracted. Our goal is to reconstruct a ghosting-free HDR image using the given LDR images ($L_1, L_2, L_3$). We also use a middle exposure image ($L_2$) as a reference image. As used in several researches [17,26,35,39], we convert the LDR images to corresponding HDR representations through gamma correction. As confirmed in [17], LDR images are effective in detecting noise or saturated regions, and HDR images are used to measure content deviations from the reference image. We use LDR images and mapped HDR images together as inputs and pixel values are all normalized to [0, 1].

As shown in Fig. 2, the entire network comprises of a feature extraction network, which extracts the necessary features and a fusion network, which combines the extracted features to construct an HDR image. The feature extraction network consists of a ghost mask generation, which detects ghost regions to generate ghost region masks, and ghost region mask-guided transformer modules, which extract features from ghost and non-ghost regions. First, the feature extraction network extracts the feature from every LDR image through convolution layers. The extracted features are used as inputs for the ghost region mask-guided transformer module. In addition, the proposed method generates a mask that separates the regions by comparing the reference and non-reference images. In the fusion network, similar to the structure used in AHDR [36], the dilated residual dense block (DRDB) is applied and configured. In the following sub-sections, we describe the ghost region mask generation and the ghost region mask-guided transformer.

### 3.1 Feature Extraction Network

**Proposed Ghost Region Mask Generation.** This module generates ghost and non-ghost region masks in both underexposed and overexposed images. Unlike the non-ghost region, the same location of the two images has different information in the ghost region. Therefore, the information required for the ghost region in the reference image must be determined from the other regions of the non-reference image. Figure 3 shows the ghost region mask generation process between the low-exposure image ($I_{non}$) and reference images ($I_{ref}$). In the first step, the average filtering is used to blur three multi-exposure images. In the weak ghost region, the features required for HDR image reconstruction can be sufficiently extracted using the CNN structure. Therefore, the average filter is applied to select only the large motion region. The pixel value difference between blurred images decreases in the weakly misaligned regions. In the second step, the reference image is transformed into the same luminance space as the non-reference image through histogram matching. Through this process, the luminance of the reference and non-reference images becomes similar, thereby decreasing the pixel value difference for all regions other than the ghost regions. The ghost mask is determined by applying a pre-determined threshold to the pixel difference value (The experiments for changing this threshold are added in the supplemental material.). However, when histogram matching is performed, the saturation region may be falsely detected as the ghost region due to the luminance difference between the reference and non-reference images in the saturation region. Therefore, we add a process of removing the saturation region of the non-reference image from the ghost mask. Even if the saturation region includes the ghost region, the process of removing the saturation region is not a problem to extract the ghost region features because there is no information for HDR imaging in the saturation region of the non-reference image. Finally, the opening operation is performed to remove a noise region caused by weakly misaligned region. Therefore, the small noise region of the ghost mask is removed and only the strong misaligned region remains. The proposed ghost region mask ($G_i$) can be calculated by

$$G_i = |L_i - K(L_i, L_2)|, \text{if } i = 1, 3, \tag{1}$$

where $K$ is the operation for histogram matching. Finally, to compensate for the undetected ghost regions and train for various ghost masks, the kernel size of the erosion is set to 11 and the kernel size of the dilation is set to a value between 11 and 17 in the opening operation by our experiments. The kernel size of dilation is fixed to 15 during the inference process. The non-ghost region mask is generated by inverting the ghost region mask (The experiments of several kernel sizes are added in the supplemental material.).

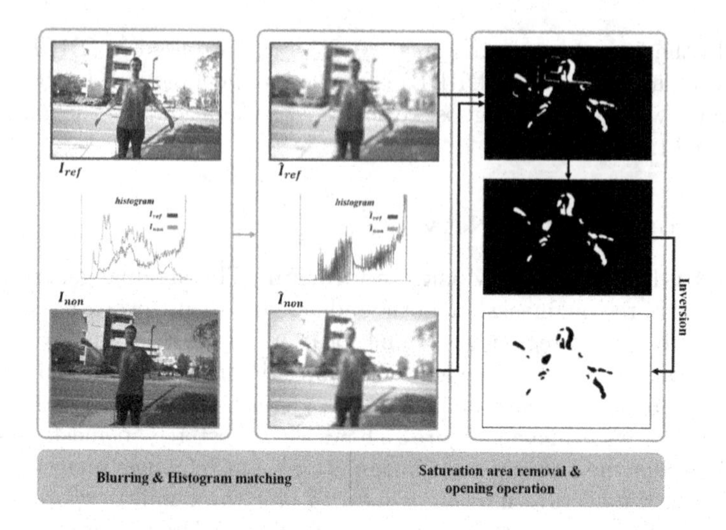


**Fig. 3.** The proposed ghost region mask generation method. First, image blurring is performed on the reference image and non-reference images to exclude weak ghost regions. After that, the luminance of the non-reference image is converted to the same as the reference image through histogram matching. The ghost mask is generated using the difference between the two images. Finally, opening operation and saturation region removal are performed to remove noise from the generated ghost mask.

**Ghost Region Mask-Guided Selective Transformer.** When an object or a camera moves, a reference image (a middle exposure image) and non-reference image (high and low exposure images) have different pixel information in the area where motion occurs. To address this problem, the proposed method employs a transformer-based module to extract important global information from entire non-reference images. As shown as Fig. 4, the selective transformer module consists of a transformer-based ghost path, which extracts the ghost region features of the reference image, and a CNN-based non-ghost path, which extracts the non-ghost region features of the reference image.

We construct the selective transformer module by applying 1 layer of cross attention. The selective transformer module first uses a CNN layer to extract the query (Q) from reference image features and key (K) and value (V) from non-reference image features. The transformer structure in this module only works on ghosted regions of the reference image. Therefore, as shown as Fig. 4, the reference image feature is multiplied by the ghost region mask to generate Q remaining only the ghost region features. In the following process, the similarity between the ghost region in a reference image and an entire non-reference image is calculated to select the best regional information. For this, all image features are unfolded into $p$-sized patches. Therefore, Q and K can be represented as $(W/p \times H/p)$ number of vectors with size $(p \times p \times D)$, denoted as $q_i(i \in [1, (W/p \times H/p)]$ and $k_j(j \in [1, (W/p \times H/p)]$, respectively. The weight $a_{ij}$ between these

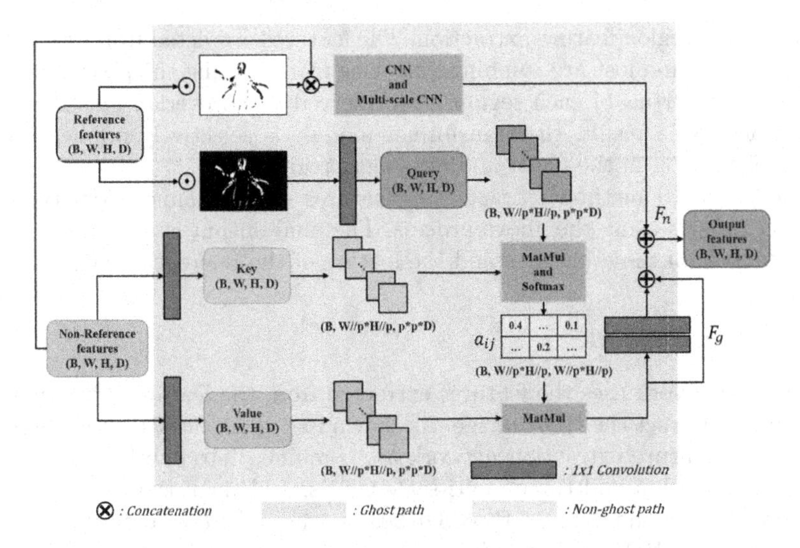


**Fig. 4.** The proposed ghost region mask-guided transformer module. $a_{ij}$ denotes the attention map calculated from K and Q. $F_g$ and $F_n$ represent the ghost and non-ghost region features, respectively. $B$ and $p$ represents batch and patch sizes, respectively, and $(H, W, D)$ is the size of the feature.

patches is calculated through the dot product of $q_i$ and $k_j$. $a_{ij}$ is defined as follows:

$$a_{ij} = softmax(\frac{q_i k_j^T}{\sqrt{p \times p \times D}}). \tag{2}$$

Also, features are extracted from the non-reference regions that are the most relevant to the reference image patch in the ghost region generated using the weight $a_{ij}$. The extracted features use two convolution layers to yield an output, and the ghost region mask is multiplied to the output so that it does not affect the other regions. The output of ghost region is as follows:

$$F_g = Conv(Relu(Conv(a_{ij}v_i))) + a_{ij}v_i, \tag{3}$$

where $Conv$ and $v_i$ denote a convolutional layer and image patches of V, respectively, and $F_g$ represent the ghost region features. Furthermore, in the non-ghost path, features from the reference and non-reference images are concatenated for feature extraction in the non-ghost region. Since non-ghost regions are aligned or weakly misaligned, the features required for HDR reconstruction can be extracted even if a conventional CNN structure is used. We use the multi-scale CNN [30,39] to extract detailed local features. The multi-scale CNN concatenates outputs of each layer by configuring CNNs with different kernel sizes in parallel. The concatenated output is transformed into same-sized features as the input channel and added to the input features. Each convolution layer uses kernel sizes of 1, 3, and 5 with the ReLU activation function. The proposed method employs a 3 × 3 convolution layer and three multi-scale convolution layers for the non-ghost region feature extraction. The features extracted from two regions of ghost and non-ghost are combined into one feature using the pixel-wise addition. The properties of each region are preserved as no overlap exists between the regions. As a result, the transformer module is selectively applied for the feature extraction in the ghost and non-ghost regions.

The proposed method extracts two features of 64 channel from two non-reference images using the above process. The final output is generated using a concatenation of these features and the feature of the reference image.

### 3.2 Fusion Network

The fusion network uses the features extracted from the feature extraction network to reconstruct the HDR image. As shown in Fig. 2, concatenated three features in the feature extraction network are combined into one feature through a $1 \times 1$ convolution and three multi-scale CNNs, and then, they are downsampled using maxpooling. Then, three DRDBs are used for the sufficient receptive field. Since the DRDB consists of dilated convolutions, the information for HDR reconstruction can be extracted using a large receptive field. Three multi-scale CNNs and a transposed CNN are used to generate features of the same size as the HDR image. Finally, as shown in Fig. 2, the HDR image is reconstructed with three features concatenated by the skip connection and reference image feature. The loss function used to reconstruct the HDR image are as follows:

$$\begin{aligned} L(H,\hat{H}) =& \parallel T(H) - T(\hat{H}) \parallel_2 \\ &+ \parallel M(T(H)) - M(T(\hat{H})) \parallel_2, \end{aligned} \tag{4}$$

where $H$ and $\hat{H}$ stands for the ground truth HDR image and the reconstructed HDR image, respectively. $M(\cdot)$ stands for the operation to extract the edge map computing the difference between adjacent pixels, and $T(\cdot)$ and $\parallel \cdot \parallel_2$ denote the tone mapping using the $\mu$-law and $l_2$ norm, respectively. Detailed information on the training and network architectures is provided in the supplementary material.

## 4 Experiments

### 4.1 Datasets

We used the Kalantari et al.'s [17] dataset to validate the performance of our method. It consists of 74 training and 15 test samples. Each sample includes three unaligned images with different exposure biases of $\{-2, 0, +2\}$ or $\{-3, 0, +3\}$. For the training, we employed randomly cropped $256 \times 256$ sized patches from the full images and applied random rotation and flip to diversify the training samples. To verify the generalization ability of the proposed method, we also used Tursun et al.'s [32] datasets used in several other papers [24,26,36]. Since these datasets do not contain ground truths of HDR images, we only displayed the tone-mapped HDR images of the proposed and conventional methods.

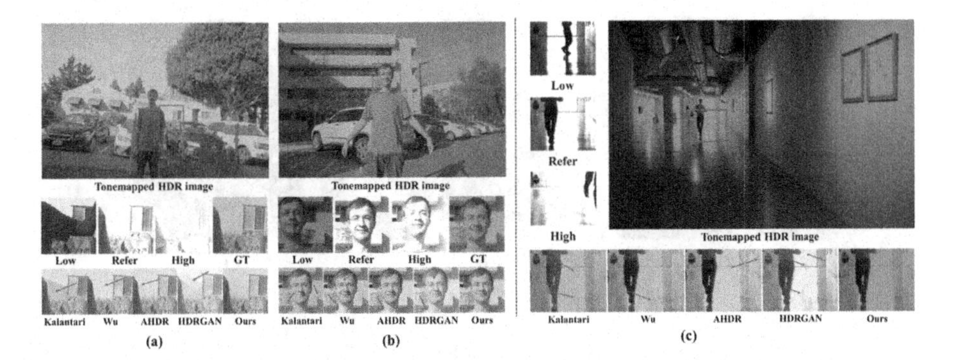


**Fig. 5.** Comparison of qualitative results of the proposed method with the state-of-the-art methods in (a), (b) the Kalantari et al.'s [18] dataset and (c) the Tursun et al.'s [31] dataset (Color figure online)

### 4.2 Evaluation Metrics

The proposed method was evaluated based on five metrics. Since HDR images may be displayed on LDR screens, the quality of the tone-mapped images needs to be checked. Therefore, we measured peak signal-to-noise ratio (PSNR) and structure similarity (SSIM) on the $\mu$-law mapped images (PSNR-$\mu$ and SSIM-$\mu$). We also measured PSNR and SSIM on the linear domain (PSNR-L and SSIM-L). Finally, we performed quantitative evaluations by calculating HDR-VDP-2 [21] designed to evaluate the HDR image quality.

### 4.3 Qualitative Results

Using the Kalantari et al.'s dataset, we compared our proposed method with several state-of-the-art methods. In Fig. 5, the first row displays the tone-mapped HDR images. The second row displays the LDR images and the ground truth; they are enlarged images of the red boxes in the images of the first row. As shown in Fig. 5(a), since the background region of the reference image (Refer) is mostly saturated, features from the low-exposure image (Low) should be extracted. However, due to the large motion of objects in the non-reference images, many details in the background are obscured, thereby making the extraction of the necessary features difficult. Therefore, the method of Kalantari et al. [17] could not completely exclude the region where the arm movement occurred in the LDR image with low-exposure value. It was confirmed that the resulting image reflected the pixel information of the corresponding region, resulting in ghost artifacts. The methods of Wu et al. [35], AHDR [36], and HDRGAN [24] succeeded in removing the regions where the arm movements occurred using LDR images with low-exposure values. However, weak ghost artifacts were observed due to the CNN limitation focusing on extracting local information. In contrast, our method used a transformer structure to extract features that are the most relevant to the reference image patch from non-reference image regions without

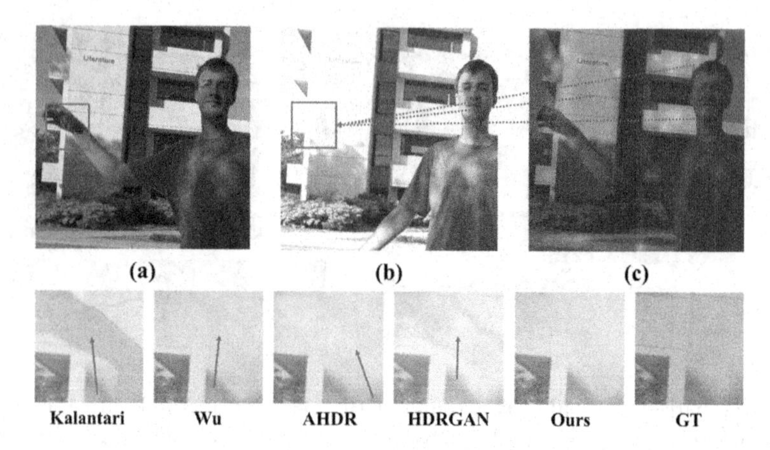


**Fig. 6.** Comparison of qualitative results of the proposed method with the state-of-the-art methods in Kalantari et al.'s [17] dataset. (a) Non-reference image (low), (b) reference image, and (c) attention region in the non-reference image. The bottom row consists of images enlarged by the red box area in the image above, in order of the state-of-the-art methods, proposed methods, and the ground truth. The red boxes shown in (a) and (b) indicate the same area, but the red box in (a) differs from that in (b). When these areas are merged, ghost artifacts appear. Using an attention map, (c) shows the area of the red box in (b) that is deeply related to (a). The closer the color is to red, the higher is the importance. (Color figure online)

focusing on the local regions where motion exists. Therefore, the corresponding ghost region is naturally reconstructed. As shown in Fig. 5(b), the object is highly saturated in the reference image. Existing methods failed to restore the color information in some regions. However, since our proposed method separately learns the ghost and non-ghost regions, it was optimized to extract the features of non-ghost regions. Therefore, the local features requiring reconstruction were well extracted from the non-reference images, thereby affording high saturation and color reconstruction performance in the corresponding regions.

Additionally, to verify the generalization ability of the proposed HDR imaging method, we evaluated its performance on the Tursun et al.'s dataset, wherein no ground truth is provided. As shown in Fig. 5(c), the methods of Kalantari et al., Wu et al., AHDR, and HDRGAN could not completely exclude the motion region information from the non-reference image. Moreover, the saturated regions were not well restored, resulting in a lot of ghost artifacts and poor color restoration. In contrast, the proposed method excluded the interference of motion region information unrelated to HDR reconstruction in the non-reference image. Therefore, our model generated a high-quality HDR image.

**Analysis of Attention Maps Generated by the Proposed Network.** In this section, we visualized the attention map of the typical motion case on the Kalantari et al.'s [17] dataset to verify the effectiveness of the attention map of the proposed transformer module. As shown in Fig. 6, the red box region

**Table 1.** Performance comparison of the proposed and the state-of-the-art methods using PSNR, SSIM, and HDR-VDP-2.

| Method | Ghost regions | | Full image | | | | |
|---|---|---|---|---|---|---|---|
| | PSNR-$\mu$ | SSIM-$\mu$ | PSNR-$\mu$ | SSIM-$\mu$ | PSNR-L | SSIM-L | HDR-VDP-2 |
| Kalantari [18] | 38.7431 | 0.9761 | 42.7423 | 0.9877 | 41.2158 | 0.9848 | 60.5088 |
| Wu [35] | 40.1244 | 0.9871 | 41.6377 | 0.9869 | 40.9082 | 0.9858 | 60.4955 |
| AHDR [36] | 40.9492 | 0.9883 | 43.6878 | 0.9902 | 41.1613 | 0.9857 | 62.0125 |
| Non [39] | - | - | 42.4143 | 0.9877 | - | - | 61.2107 |
| Robust [26] | - | - | 43.8487 | 0.9906 | 41.6452 | 0.9870 | 62.5495 |
| HDRGAN [24] | 41.2814 | 0.9887 | 43.9220 | 0.9905 | 41.5720 | 0.9865 | 63.1245 |
| Proposed | **41.6714** | **0.9890** | **44.0981** | **0.9909** | **41.7021** | **0.9872** | **63.3721** |

of the reference image (b) is saturated. Therefore, the information in the region must be extracted from the non-reference image (a) with the low exposure value. However, information cannot be extracted from non-reference images, such as Fig. 6(a), because the necessary information is deleted due to the movement of the objects. Therefore, Kalantari et al.'s [17], Wu et al.'s [35], AHDR [36], and HDRGAN [24] yielded final images with ghost artifacts in the motion region.

The proposed attention map of Fig. 6(c) visualized the importance of the pixel region to be referenced in Fig. 6(a) to restore the red box region in Fig. 6(b). The proposed attention map focused on the wall region far away from the red box, which is the saturated wall background. Therefore, compared to other state-of-the-art methods, the proposed method afforded significantly better reconstruction in terms of the color and details of the ghost regions.

### 4.4 Quantitative Results

We compared the performance of the proposed model with the state-of-the-art models using the quantitative metrics. Furthermore, to evaluate the performance of the proposed method in the ghost region, we performed evaluations on the full image and ghost regions. The ghost regions are calculated from the proposed ghost region mask. Table 1 denotes the performance on the full image; our method outperformed other methods in terms of PSNR-$\mu$, PSNR-L, SSIM-$\mu$, SSIM-L, and HDR-VDP-2 are 44.0981, 41.7021, 0.9909, 0.9872, and 63.3721, respectively. (Models with performance in bold performed best.)

Then, we evaluated the performance results for the ghost regions. The proposed method showed the best performance in all metrics among all methods compared in the region where motion occurred. The CNN-based methods, kalantari et al. [17], Wu et al. [35], and AHDR [36], were difficult to use global features. Moreover, they extracted the local features for ghost regions from non-reference images. In contrast, the proposed method generated a ghost region mask for the region where the motion occurs and applied it to the transformer-based network, so that it was possible to extract information from the relevant regions by searching all regions of the non-referenced image. Therefore, the proposed method had

**Table 2.** Performance comparison for variants of the proposed model using PSNR, SSIM, and HDR-VDP-2.

| Method | PSNR-$\mu$ | SSIM-$\mu$ | PSNR-L | SSIM-L | HDR-VDP-2 |
|---|---|---|---|---|---|
| Base model (only transformer) | 43.3244 | 0.9882 | 40.8823 | 0.9861 | 60.6124 |
| Base model (only CNN) | 43.7274 | 0.9901 | 41.2311 | 0.9866 | 62.7231 |
| Base model (no mask) | 43.8874 | 0.9903 | 41.3422 | 0.9868 | 62.9245 |
| Base model | **44.0981** | **0.9909** | **41.7021** | **0.9872** | **63.3721** |

excellent performance even with large motions. Our method effectively solved the problem of ghost artifacts in HDR images.

# 5 Ablation Studies

Ablation study demonstrates the effectiveness of selective transformer module and ghost region mask. We achieved the ablation study by comparing the performance of the following variants of the proposed model, as shown in Table 2.

- **Base model.** All modules of the proposed model are used.
- **Base model (only transformer).** In the selective transformer module of the base model, we replaced multi-scale CNNs of the non-ghost region path with a transformer structure.
- **Base model (only CNN).** In the selective transformer module of the base model, we replaced the transformer structure of the ghost region path with the multi-scale CNN.
- **Base model (no mask).** Instead of using the proposed ghost region mask, features are extracted in parallel using the transformer structure and the CNN structure from the entire input image.

## 5.1 Effectiveness of Selective Transformer Module

To verify the effectiveness of selectively applying the transformer, we designed two feature extraction networks composed of only CNNs or only transformers, respectively, and compared the performance of these models. As shown in Table 2, the base model consisting of only transformers showed the lowest performance. Due to the nature of the HDR reconstruction task, more important information exists in the local region in the case of the non-ghost region, so the performance of the base model consisting of only CNNs was higher than that of the transformer structure. However, this model failed to reconstruct high-quality HDR images in ghost regions with large motion. In contrast, the base model separated the ghost and non-ghost regions, and selectively applied the CNN and the transformer structures. Therefore, the proposed method could utilize the advantages of both structures. As a result, the base model achieved higher performance than the model using only CNN structure or only transformer structure.

### 5.2 Effectiveness of Ghost Region Mask

To confirm the effectiveness of the proposed ghost region mask, we performed an ablation experiment without using a ghost region mask on the base model. Therefore, the base model (no mask) used the transformer and the CNN structures to extract features in parallel from the entire image, and fed the combined two features to the fusion network. However, these features may contain information that is not required for each region. This problem can cause ghost artifacts or color distortion in the final HDR image. Therefore, as shown in Table 2, the model without the ghost region mask showed lower performance than the base model.

## 6 Limitation and Future Work

The proposed method significantly enhanced HDR images in strong ghost regions compared to conventional methods. However, the heuristic module, ghost mask generation, can falsely detect the ghost region if histogram matching is incorrectly performed due to severe saturation regions or camera misalignment. In this case, the proposed model will use the transformer structure in the non-ghost region and may produce a low quality HDR image as confirmed in our ablation study. We will consider these factors in our future work and design a network that outputs refined masks using the generated ghost masks. Through this future work, we will try to configure an end-to-end network including a ghost mask generation module to detect more accurate ghost regions.

## 7 Conclusion

In this paper, we proposed to selectively apply a network suitable for each region by dividing the image into ghost and non-ghost regions. In the ghost region with large motion, the proposed selective transformer module reconstructed the region well using the transformer structure. This is because the transformer can search the entire region and extract features deeply related to patches of the reference image from the global region of the non-reference image. In the non-ghost region where the LDR images are aligned, the selective transformer module used the CNN structure to effectively extract local features. In addition, through the ablation study, we found that the proposed model outperforms the CNN-only and transformer-only models. Finally, the proposed model provided ghost-free high-quality HDR images with rich details and colors compared to the state-of-the-art models.

**Acknowledgement.** This research was supported by the National Research Foundation of Korea (NRF) grant funded by the Korea government (MSIT) (No. 2021R1A2C1004208).

## References

1. An, J., Lee, S.H., Kuk, J.G., Cho, N.I.: A multi-exposure image fusion algorithm without ghost effect. In: 2011 IEEE International Conference on Acoustics, Speech and Signal Processing (ICASSP), pp. 1565–1568. IEEE (2011)
2. Bogoni, L.: Extending dynamic range of monochrome and color images through fusion. In: Proceedings 15th International Conference on Pattern Recognition, ICPR 2000, vol. 3, pp. 7–12. IEEE (2000)
3. Carion, N., Massa, F., Synnaeve, G., Usunier, N., Kirillov, A., Zagoruyko, S.: End-to-end object detection with transformers. In: Vedaldi, A., Bischof, H., Brox, T., Frahm, J.-M. (eds.) ECCV 2020. LNCS, vol. 12346, pp. 213–229. Springer, Cham (2020). https://doi.org/10.1007/978-3-030-58452-8_13
4. Debevec, P.E., Malik, J.: Recovering high dynamic range radiance maps from photographs. In: ACM SIGGRAPH 2008 Classes, pp. 1–10 (2008)
5. Dosovitskiy, A., et al.: An image is worth 16 × 16 words: transformers for image recognition at scale. In: International Conference on Learning Representations (2021)
6. Dosovitskiy, A., et al.: FlowNet: learning optical flow with convolutional networks. In: Proceedings of the IEEE International Conference on Computer Vision, pp. 2758–2766 (2015)
7. Eilertsen, G., Kronander, J., Denes, G., Mantiuk, R.K., Unger, J.: HDR image reconstruction from a single exposure using deep CNNs. ACM Trans. Graph. (TOG) **36**(6), 1–15 (2017)
8. Endo, Y., Kanamori, Y., Mitani, J.: Deep reverse tone mapping. ACM Trans. Graph. (Proceedings of SIGGRAPH ASIA 2017) **36**(6) (2017)
9. Fleet, D., Weiss, Y.: Optical flow estimation. In: Paragios, N., Chen, Y., Faugeras, O. (eds.) Handbook of Mathematical Models in Computer Vision, pp. 237–257. Springer, Boston (2006). https://doi.org/10.1007/0-387-28831-7_15
10. Granados, M., Ajdin, B., Wand, M., Theobalt, C., Seidel, H.P., Lensch, H.P.: On being 'undigital' with digital cameras: extending dynamic range by combining differently exposed pictures. In: Proceedings of IS&T, pp. 442–448 (1995)
11. Granados, M., Ajdin, B., Wand, M., Theobalt, C., Seidel, H.P., Lensch, H.P.: Optimal HDR reconstruction with linear digital cameras. In: 2010 IEEE Computer Society Conference on Computer Vision and Pattern Recognition, pp. 215–222. IEEE (2010)
12. Granados, M., Kim, K.I., Tompkin, J., Theobalt, C.: Automatic noise modeling for ghost-free HDR reconstruction. ACM Trans. Graph. (TOG) **32**(6), 1–10 (2013)
13. Grosch, T., et al.: Fast and robust high dynamic range image generation with camera and object movement. Vision, Modeling and Visualization, RWTH Aachen 277284 (2006)
14. Heo, Y.S., Lee, K.M., Lee, S.U., Moon, Y., Cha, J.: Ghost-free high dynamic range imaging. In: Kimmel, R., Klette, R., Sugimoto, A. (eds.) ACCV 2010. LNCS, vol. 6495, pp. 486–500. Springer, Heidelberg (2011). https://doi.org/10.1007/978-3-642-19282-1_39
15. Hu, J., Gallo, O., Pulli, K., Sun, X.: HDR deghosting: how to deal with saturation? In: Proceedings of the IEEE Conference on Computer Vision and Pattern Recognition, pp. 1163–1170 (2013)
16. Jinno, T., Okuda, M.: Motion blur free HDR image acquisition using multiple exposures. In: 2008 15th IEEE International Conference on Image Processing, pp. 1304–1307. IEEE (2008)

17. Kalantari, N.K., Ramamoorthi, R.: Deep high dynamic range imaging of dynamic scenes. ACM Trans. Graph. **36**(4), 144-1 (2017)
18. Khan, E.A., Akyuz, A.O., Reinhard, E.: Ghost removal in high dynamic range images. In: 2006 International Conference on Image Processing, pp. 2005–2008. IEEE (2006)
19. Khan, Z., Khanna, M., Raman, S.: FHDR: HDR image reconstruction from a single LDR image using feedback network. In: 2019 IEEE Global Conference on Signal and Information Processing (GlobalSIP), pp. 1–5 (2019). https://doi.org/10.1109/GlobalSIP45357.2019.8969167
20. Lee, C., Li, Y., Monga, V.: Ghost-free high dynamic range imaging via rank minimization. IEEE Signal Process. Lett. **21**(9), 1045–1049 (2014)
21. Mantiuk, R., Kim, K.J., Rempel, A.G., Heidrich, W.: HDR-VDP-2: a calibrated visual metric for visibility and quality predictions in all luminance conditions. ACM Trans. Graph. (TOG) **30**(4), 1–14 (2011)
22. Mitsunaga, T., Nayar, S.K.: Radiometric self calibration. In: Proceedings of the 1999 IEEE Computer Society Conference on Computer Vision and Pattern Recognition (Cat. No PR00149), vol. 1, pp. 374–380. IEEE (1999)
23. Moon, Y.S., Tai, Y.M., Cha, J.H., Lee, S.H.: A simple ghost-free exposure fusion for embedded HDR imaging. In: 2012 IEEE International Conference on Consumer Electronics (ICCE), pp. 9–10. IEEE (2012)
24. Niu, Y., Wu, J., Liu, W., Guo, W., Lau, R.W.: HDR-GAN: HDR image reconstruction from multi-exposed LDR images with large motions. IEEE Trans. Image Process. **30**, 3885–3896 (2021)
25. Prabhakar, K.R., Agrawal, S., Singh, D.K., Ashwath, B., Babu, R.V.: Towards practical and efficient high-resolution HDR deghosting with CNN. In: Vedaldi, A., Bischof, H., Brox, T., Frahm, J.-M. (eds.) ECCV 2020. LNCS, vol. 12366, pp. 497–513. Springer, Cham (2020). https://doi.org/10.1007/978-3-030-58589-1_30
26. Pu, Z., Guo, P., Asif, M.S., Ma, Z.: Robust high dynamic range (HDR) imaging with complex motion and parallax. In: Proceedings of the Asian Conference on Computer Vision (2020)
27. Reinhard, E., Heidrich, W., Debevec, P., Pattanaik, S., Ward, G., Myszkowski, K.: High Dynamic Range Imaging: Acquisition, Display, and Image-Based Lighting. Morgan Kaufmann, Burlington (2010)
28. Sen, P., Kalantari, N.K., Yaesoubi, M., Darabi, S., Goldman, D.B., Shechtman, E.: Robust patch-based HDR reconstruction of dynamic scenes. ACM Trans. Graph. **31**(6), 203-1 (2012)
29. Srikantha, A., Sidibé, D.: Ghost detection and removal for high dynamic range images: recent advances. Signal Process.: Image Commun. **27**(6), 650–662 (2012)
30. Szegedy, C., et al.: Going deeper with convolutions. In: Proceedings of the IEEE Conference on Computer Vision and Pattern Recognition, pp. 1–9 (2015)
31. Tursun, O.T., Akyüz, A.O., Erdem, A., Erdem, E.: The state of the art in HDR deghosting: a survey and evaluation. In: Computer Graphics Forum, vol. 34, pp. 683–707. Wiley Online Library (2015)
32. Tursun, O.T., Akyüz, A.O., Erdem, A., Erdem, E.: An objective deghosting quality metric for HDR images. In: Computer Graphics Forum, vol. 35, pp. 139–152. Wiley Online Library (2016)
33. Vaswani, A., et al.: Attention is all you need. In: Guyon, I., et al. (eds.) Advances in Neural Information Processing Systems, vol. 30. Curran Associates, Inc. (2017)
34. Wang, X., Girshick, R., Gupta, A., He, K.: Non-local neural networks. In: Proceedings of the IEEE Conference on Computer Vision and Pattern Recognition, pp. 7794–7803 (2018)

35. Wu, S., Xu, J., Tai, Y.W., Tang, C.K.: Deep high dynamic range imaging with large foreground motions. In: Proceedings of the European Conference on Computer Vision (ECCV), pp. 117–132 (2018)
36. Yan, Q., et al.: Attention-guided network for ghost-free high dynamic range imaging. In: Proceedings of the IEEE/CVF Conference on Computer Vision and Pattern Recognition, pp. 1751–1760 (2019)
37. Yan, Q., et al.: Multi-scale dense networks for deep high dynamic range imaging. In: 2019 IEEE Winter Conference on Applications of Computer Vision (WACV), pp. 41–50. IEEE (2019)
38. Yan, Q., Sun, J., Li, H., Zhu, Y., Zhang, Y.: High dynamic range imaging by sparse representation. Neurocomputing **269**, 160–169 (2017)
39. Yan, Q., et al.: Deep HDR imaging via a non-local network. IEEE Trans. Image Process. **29**, 4308–4322 (2020)
40. Zheng, S., et al.: Rethinking semantic segmentation from a sequence-to-sequence perspective with transformers. In: Proceedings of the IEEE/CVF Conference on Computer Vision and Pattern Recognition, pp. 6881–6890 (2021)

# Learning Series-Parallel Lookup Tables for Efficient Image Super-Resolution

Cheng Ma[1,2], Jingyi Zhang[1,2], Jie Zhou[1,2], and Jiwen Lu[1,2(✉)]

[1] Beijing National Research Center for Information Science and Technology, Beijing, China
[2] Department of Automation, Tsinghua University, Beijing, China
macheng17@tsinghua.org.cn, zhangjy20@mails.tsinghua.edu.cn, {jzhou,lujiwen}@tsinghua.edu.cn

**Abstract.** Lookup table (LUT) has shown its efficacy in low-level vision tasks due to the valuable characteristics of low computational cost and hardware independence. However, recent attempts to address the problem of single image super-resolution (SISR) with lookup tables are highly constrained by the small receptive field size. Besides, their frameworks of single-layer lookup tables limit the extension and generalization capacities of the model. In this paper, we propose a framework of series-parallel lookup tables (SPLUT) to alleviate the above issues and achieve efficient image super-resolution. On the one hand, we cascade multiple lookup tables to enlarge the receptive field of each extracted feature vector. On the other hand, we propose a parallel network which includes two branches of cascaded lookup tables which process different components of the input low-resolution images. By doing so, the two branches collaborate with each other and compensate for the precision loss of discretizing input pixels when establishing lookup tables. Compared to previous lookup table-based methods, our framework has stronger representation abilities with more flexible architectures. Furthermore, we no longer need interpolation methods which introduce redundant computations so that our method can achieve faster inference speed. Extensive experimental results on five popular benchmark datasets show that our method obtains superior SISR performance in a more efficient way. The code is available at https://github.com/zhjy2016/SPLUT.

**Keywords:** Image super-resolution · Look-up table · Series-parallel network

## 1 Introduction

As a fundamental task of computer vision, single image super-resolution (SISR) has attracted lots of research interests and plays an important role in wide

C. Ma and J. Zhang—Equal contribution.

**Supplementary Information** The online version contains supplementary material available at https://doi.org/10.1007/978-3-031-19790-1_19.

© The Author(s), under exclusive license to Springer Nature Switzerland AG 2022
S. Avidan et al. (Eds.): ECCV 2022, LNCS 13677, pp. 305–321, 2022.
https://doi.org/10.1007/978-3-031-19790-1_19

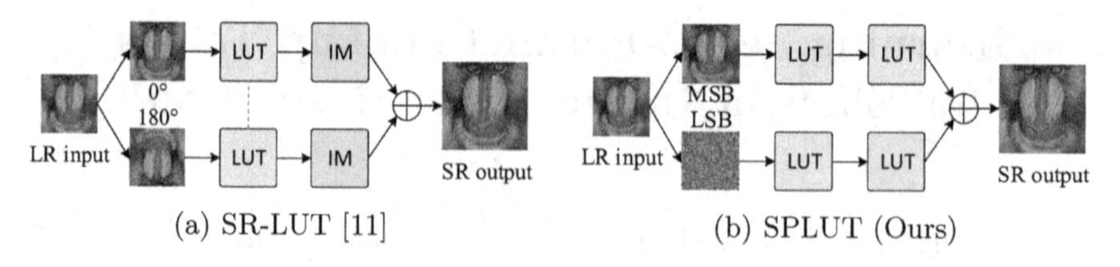


(a) SR-LUT [11]

(b) SPLUT (Ours)

**Fig. 1.** Comparison of SR-LUT and our SPLUT method. The former utilizes rotational ensemble to enlarge receptive fields from $2 \times 2$ to $3 \times 3$ and implement interpolation methods (IM) to improve recovery accuracy. The dashed line means weight sharing. For the sake of simplicity, we omit the rotations of 90° and 270°. In contrast, we stack multiple LUTs to significantly improve the receptive field size and design a new parallel architecture to compensate for the precision loss of discretizing input pixels.

applications, such as video surveillance, satellite imaging and high-definition televisions. SISR targets at recovering low-resolution (LR) images to high-resolution (HR) ones by inferring high-frequency details. Along with the rapid development of deep learning techniques, various elaborately designed frameworks [18,19,43] based on convolutional neural networks (CNNs) have achieved encouraging progress in SISR. Most of these frameworks contain a large number of parameters and are time-consuming during testing. While several methods [17,21] have been proposed to reduce computation costs, they still rely on specific high-performance computing units, for example, GPUs and CPUs. Developing practical and real-time algorithms have been a growing trend in the SISR field.

Approaches [11,33,35,40] based on lookup tables (LUTs) have emerged in low-level vision tasks, including image enhancement and image super-resolution. These methods employ LUTs to establish the mapping relation between input pixels and the desired output pixels. In the testing phase, only a small number of parameters need storing and the inference processes are liberated from heavy computational burdens by replacing time-consuming calculations with fast memory accesses. As a result, the practicality of this kind of algorithm is significantly improved on mobile devices.

However, most existing LUT-based approaches only have a single layer of LUTs, which brings some major constraints. If $n$-dimensional LUTs ($n$D LUTs) are utilized and the $n$ input entities for query all have $v$ possible values, then the LUT scale is of $v^n$, where $v$ and $n$ are the two pivotal factors. While increasing the value of $v$ and $n$ may improve the restoration accuracy, a moderate improvement may lead to a rapid increase of the LUT scale. Thus, $v$ and $n$ are usually set to small values to avoid unbearably large LUTs, which severely limits the further enhancement of recovery abilities. In fact, receptive field (RF) size is a vital factor for deep learning and super-resolution. SISR is a well-known ill-posed problem since the same LR input may correspond to various high-resolution outputs with subtle differences. When we infer the missing details, a large context area on the input image should be considered to accurately capture the semantics and structures. In this way, we can effectively reduce the ambiguity of the estimated results. Therefore, how to enlarge receptive fields without exponentially increasing the storage and computation costs of LUTs is still an open issue. Besides, due to the design of single-layer LUTs and the limited LUT scale, the extensibility and recovery capacity of existing LUT-based methods are highly constrained. Thus, a more powerful and flexible scheme is desired in order to further improve the inference ability of LUT-based methods.

To mitigate the above issues, we propose to learn series-parallel lookup tables (SPLUT) for SISR, as shown in Fig. 1. We cascade multiple lookup tables so that the query of latter LUTs are based on the outputs of former LUTs. In this way, the receptive field of each feature vector is gradually increased and the final SR results can be determined by larger local patches with more clear context information. For establishing LUTs, existing methods usually discretize the input values for reducing $v$ and apply interpolation algorithms to improve the inference accuracy. However, such operations can only be implemented for small receptive fields. In our framework with large receptive fields, interpolations are inapplicable due to the exponentially increasing computational costs. In order to compensate for the precision loss of the discretized input pixel values, we propose a new parallel network which contains two branches. The first one processes the 4 most significant bits (MSBs) of the original 8-bit pixel values while the second one processes the 4 least significant bits (LSBs). The two branches of cascaded LUTs form the framework of series-parallel lookup tables.

In each branch, we introduce three kinds of 4D LUTs whose input values for the query are from different dimensions. We further propose horizontal and vertical aggregation modules to enlarge the RF size of different dimensions. In the training procedure, we build a mapping module for each LUT and quantize the intermediate activation so that the mapping relationships of the inputs and outputs can be transferred to the corresponding LUTs. Different from previous methods [11], the whole inference procedure of our method only contains retrieval and addition operations without complex multiplications. Experimental results on benchmark datasets show that SPLUT can achieve better SR performance on the smartphone platform, which demonstrates the effectiveness and efficiency of our proposed method.

In summary, the contributions of this work are threefold:

1. To the best of our knowledge, we are the first to present cascaded LUTs for enlarging receptive fields in the SISR field.
2. We propose a new parallel network to compensate for the precision loss caused by the discretization when establishing LUTs with large RF sizes.
3. Quantitative and qualitative results show that our method can recover the missing details more precisely and efficiently. The comparison of different SPLUT models verifies the superior extensibility of our proposed method.

## 2 Related Work

**Single Image Super-Resolution.** Non-deep learning methods [6,30,31,41] and deep learning methods [2,18,25,43] have significantly promoted the development of SISR. While recent deep learning methods perform more encouragingly than non-deep learning methods, many of them have a deep neural architecture with redundant parameters, which brings heavy computing costs and makes the training and inference rely on special computing devices. Therefore, efficient super-resolution has been a prevalent research interest of the community. In this field, methods based on various techniques have been proposed to improve the efficiency of SR algorithms. Lee *et al.* [17] and Zhang *et al.* [42] take advantage of the idea of knowledge distillation to compress the original deep teacher models to small student models with strong representation abilities. Wang *et al.* [34] explore the sparsity in image super-resolution by learning sparse masks to identify important regions and unimportant regions in images. Mei *et al.* [24] propose a non-local sparse attention module to achieve efficient and robust long-range modeling. Xin *et al.* [38] develop a binary neural network for SR by proposing a bit-accumulation mechanism to improve the precision of the quantized model. Some other methods [20,21,29] accomplish efficient SR inference by designing compact neural architectures. Lee *et al.* [16] search for appropriate architectures for both the generator and the discriminator by a neural architecture search approach. However, most of these methods are still based on convolutional layers and thus lack practicability on mobile devices.

**Lookup Table.** Lookup tables (LUTs) replace complex computations by simple and fast retrieval operations so that the efficiency of algorithms can be significantly improved. LUTs are widely used in a number of applications, such as numerical computation [5,27], video coding [15,32], pedestrian detection [4], RGB-to-RGBW conversion [14], *etc.* Besides, LUT is a classic and prevalent pixel adjustment tool in camera imaging pipeline [40] and photo editing software since it can easily manipulate the appearance of an image, such as color, exposure, saturation, *etc.* Recently deep learning methods based on LUTs have also emerged in low-level vision tasks [11,33,35,40]. In the image enhancement field, Zeng *et al.* [40] first propose image-adaptive 3D lookup tables and achieve high-performance photo enhancement. On this basis, Wang *et al.* [35] consider spatial information and further propose learnable spatial-aware 3D lookup tables. Wang *et al.* [33] model local context cues and propose pixel-adaptive lookup table weights for portrait photo retouching. As for the super-resolution field, Jo *et al.* [11] has developed SR-LUT by establishing the correspondence of LR input patterns and HR output patterns. However, as mentioned above, the RF size and the extensibility of LUT-based methods are still limited.

# 3 Method

## 3.1 Network Architecture

Given an input LR image $I^{LR}$, our goal is to recover the missing details and yield the SR image $I^{SR}$ which is as similar as possible to the HR image $I^{HR}$. As shown in Fig. 2(a), we design a series-parallel lookup table (SPLUT) network which contains two parallel branches processing different components of $I^{LR}$. We treat the RGB channels equally and separate the original input pixels with 8-bit

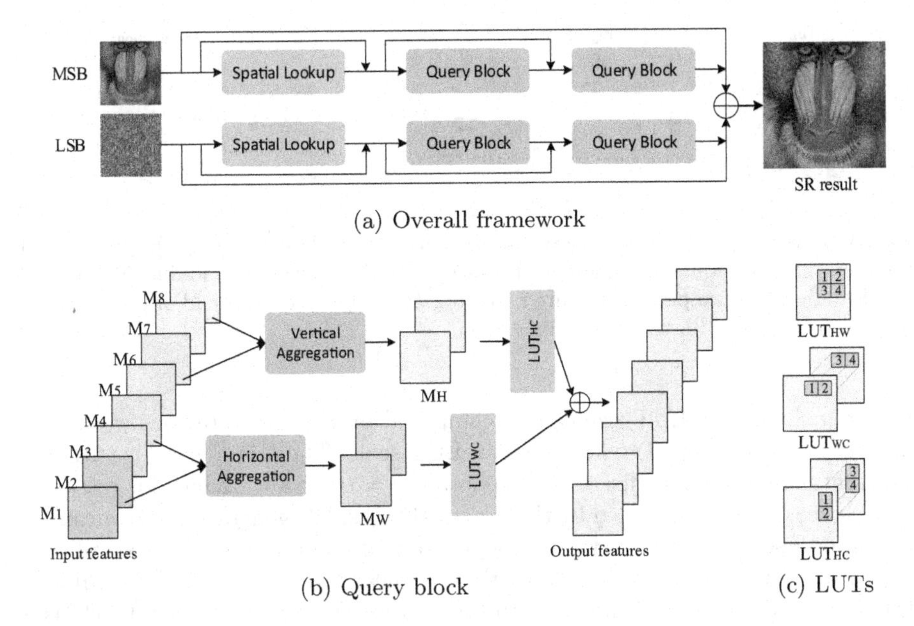


**Fig. 2.** Details of the proposed SPLUT method. (a) The overall framework of our method. The input LR images are split into $I_{MSB}$ and $I_{LSB}$, which are fed into two parallel branches, respectively. Each branch includes cascaded LUTs to extend receptive fields. (b) We take SPLUT-M with $C_f = 8$ as the example and display the details of the proposed query block, which further enlarge the RF size by aggregation modules and different kinds of LUTs. $\text{LUT}_{\text{WC}}$ and $\text{LUT}_{\text{HC}}$ can also model the correlations between different channels. (c) Illustration of different LUTs: $\text{LUT}_{\text{HW}}$, $\text{LUT}_{\text{WC}}$ and $\text{LUT}_{\text{HC}}$. Their input values for the query are in different dimensions.

values into two maps, $I_{MSB}$ with 4 most significant bits (MSBs) and $I_{LSB}$ with 4 least significant bits (LSBs). The two parallel branches take $I_{MSB}$ and $I_{LSB}$ as inputs, respectively. Then we merge the outputs of the two parallel branches to compensate for the loss of quantization when establishing LUTs. In this way, the super-resolution capacities can be significantly enhanced. In each branch, there is a spatial lookup block, query blocks and skip connections. The spatial lookup block and the query blocks increase the RF size of extracted features gradually. The query blocks include horizontal and vertical aggregation modules which enlarge the RF size by the width and height dimensions, respectively. During training, we replace each LUT with a mapping block which is built on convolutional layers. Then we establish LUTs according to the mapping relations of the inputs and outputs of these mapping blocks. During inference, we retrieve the outputs of each LUT according to the indices computed by the input patterns, which are defined as the combinations of the $n$ input entities for the query.

The LUT scale is mainly influenced by three factors, the number of pixels for retrieval $n$, the number of possible pixel values $v$, and the length of output

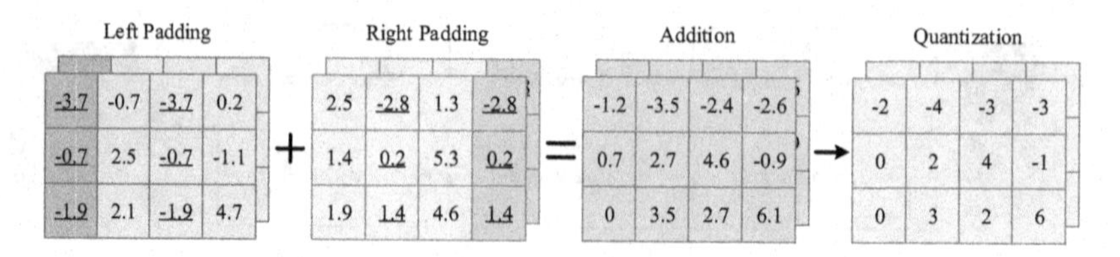


**Fig. 3.** Illustration of the proposed horizontal aggregation module. The underlined numbers in dark squares represent the results of the reflection padding operations. After adding the two padded feature maps together, the receptive field of the obtained features is enlarged on the width dimension.

vectors $c$. Then the LUT size can be computed by $v^n \cdot c$. In previous work, $n$ is usually not more than 4 to keep a small LUT scale. The original pixels with 256 different values are also discretized to obtain $v = 16$ or $v = 32$ bins for retrieval. The choice of $c$ is determined by the practical tasks. For generating the output of $\times 4$ super-resolution, $c$ is set to 16. The increase of $n$ and $v$ results in a significant increase in LUT scales while $c$ only brings a linear growth of LUT scales. In our framework, we set $n = 4$ and $v = 16$ for all the LUTs so that $v^n = 65536$ is a relatively small constant. Different from previous methods which only contain a single layer of LUTs with a small RF size and lack the model extensibility, we cascade multiple LUTs to improve the RF size and flexibly control the trade-off between efficiency and accuracy by changing the network depth and the channel number of intermediate features $C_f$. In practice, we design three models, SPLUT-S, SPLUT-M and SPLUT-L with $C_f = 4$, $C_f = 8$ and $C_f = 16$, respectively.

Since $n$ is set to 4, the indices for retrieving LUTs are computed by 4 adjacent entities. We design 3 kinds of LUTs whose input patterns are of different dimensions. For an intermediate feature, there are mainly three dimensions, W, H, and C, representing width, height, and channel, respectively. The 3 kinds of designed LUTs are $\text{LUT}_{\text{WH}}$, $\text{LUT}_{\text{HC}}$ and $\text{LUT}_{\text{WC}}$, as depicted in Fig. 2(c). The input pattern of $\text{LUT}_{\text{WH}}$ is a $2 \times 2$ area in the spatial dimensions. The $2 \times 2$ input pattern of $\text{LUT}_{\text{HC}}$ is along the height and channel dimensions while that of $\text{LUT}_{\text{WC}}$ is along the width and channel dimensions. These LUTs can capture local dependency and enlarge receptive fields of different dimensions. In our framework, they are placed in different modules for specific functions. Next, we take the model SPLUT-M as an example to describe the details of each component of our framework.

**Spatial LUT Block.** The two branches of $I_{MSB}$ and $I_{LSB}$ have similar architectures. In the beginning, spatial correlations are more important than channel correlations. Hence, we use a spatial LUT block to exploit the spatial dependency of neighboring pixels in the input images. In this block, we employ reflection padding to keep the spatial dimensions unchanged after retrievals.

**Query Blocks.** Following the spatial LUT block, there are two query blocks. The details of the query block are shown in Fig. 2(b). For the model of SPLUT-M, $C_f$ is set to 8 and thus we have 8 intermediate feature maps, named $M_1, ..., M_8$.

We split them into 4 groups, each adjacent two in one group. The first two groups are fed into the horizontal aggregation module while the last two are fed into the vertical aggregation module. The two modules enlarge the RF size by the width dimension and the height dimension, respectively. $M_W$ and $M_H$ are obtained by the two modules and they both have 2 channels. Since only exploring spatial information severely affects the representation ability of the network, we use $\text{LUT}_{\text{WC}}$ and $\text{LUT}_{\text{HC}}$ to model the correlations between different channels. We retrieve the output of a width-channel $\text{LUT}_{\text{WC}}$ by computing the query indices according to the input patterns on $M_W$. Similarly, we retrieve $\text{LUT}_{\text{HC}}$ according to $M_H$. Finally, the outputs of two kinds of LUTs are added to get the output of the query block.

**Aggregation Modules.** Here we describe the details of the horizontal and vertical aggregation modules. As shown in Fig. 3, we take the horizontal aggregation module as an example. The two input feature maps both have two channels and have the same receptive fields. First, we pad one feature map on the left by reflection padding and pad the other one on the right. Then we obtain two feature maps whose receptive fields have a shift of one pixel along the width dimension. After merging the two feature maps by addition, the obtained feature map has a larger spatial receptive field. In order to transfer the real-value responses to the query indices for the following LUTs, we need to quantize the real values to form $v = 16$ discrete values. Specifically, we set the quantization interval to 1 so that the quantization can be achieved by a simple rounding operation. By doing this, we avoid complex multiplication computations and improve the efficiency of the proposed module. The operations are similar for the vertical aggregation module. Differently, vertical aggregation enlarges the receptive field along the height dimension.

**Parallel Branches.** In prior arts [11,33,35,40], pixel values are quantized for reducing the possible values and decreasing LUT scales. However, the original continuously changing pixels become discrete, which may cause blocking effects in the SR results. Therefore, interpolation algorithms [12] are usually applied to smooth the output textures. However, they introduce additional multiplications and comparison operations. Moreover, these algorithms are only available for LUTs with a small RF size. If we use $r$ to represent the RF size of a feature, then $2^r$ nearest bounding vertices need considering for interpolating the retrieval results. In our framework with a large RF size, such a computation complexity is unacceptable. We propose a new parallel framework to alleviate this issue. The framework includes two branches with the same architecture of cascaded LUTs. One branch processes $I_{MSB}$ and mainly focuses on capturing context semantic information. The other branch processes $I_{LSB}$ and provides high-frequency details. By Merging the outputs of the two branches, we are able to compensate for the loss of quantization when establishing LUTs and hence boost SR performance.

**Skip Connections.** In order to improve the representation abilities of the network, we store low-precision real numbers in LUTs. Since the above mentioned

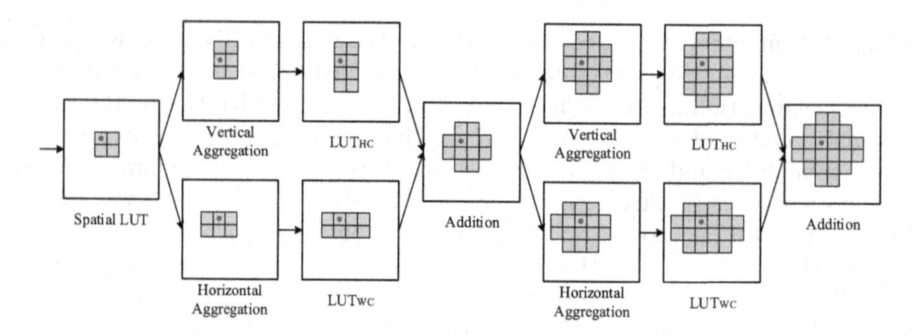


**Fig. 4.** The visualization of receptive fields with respect to the feature marked in red after each module or operation. Horizontal aggregation modules and $\mathrm{LUT}_{\mathrm{WC}}$ increase the RF size by the width dimension while vertical aggregation modules and $\mathrm{LUT}_{\mathrm{HC}}$ increase the RF size by the height dimension. The addition operations further enlarge receptive fields by fusing the two different areas of receptive fields. (Color figure online)

operations of quantization and index computation sacrifice the precision of intermediate features, we introduce skip connections to fuse the real-value inputs and the retrieval outputs to improve the precision. Besides, identity mapping [8] is a pivotal component of SR networks. Thus we adopt a skip connection between the input image and the output of the last query block to simplify optimization and enhance recovery accuracy.

### 3.2 Training Strategy

For training the network, we replace the LUTs by mapping modules, which are comprised of a convolutional layer with a kernel size of $k_h \times k_w$, GELU [9] layers, and $1 \times 1$ convolutional layers. All mapping modules output feature maps with $C_f$ channels except the last one. The last mapping module outputs $C_{sr} = s^2$ channels where $s$ is the upscaling factor. A pixel-shuffle layer [28] maps the outputs with 16 channels to the final results for $\times 4$ SR. In mapping modules of different LUTs, $k_h$ and $k_w$ are different. For the spatial LUT block, the input channel number is 1 and $k_h = k_w = 2$. For $\mathrm{LUT}_{\mathrm{HC}}$ and $\mathrm{LUT}_{\mathrm{WC}}$, the input channel number is 2. $k_h = 2$ and $k_w = 1$ for $\mathrm{LUT}_{\mathrm{HC}}$ while $k_h = 1$ and $k_w = 2$ for $\mathrm{LUT}_{\mathrm{WC}}$. The consecutive $1 \times 1$ convolutions followed by GELU layers strengthen the nonlinearity and representative abilities of the mapping modules. We jointly train the MSB and LSB branches by imposing Mean Squared Error (MSE) loss on the final SR outputs. For the quantized activations, we use the identity straight-through estimator (STE) [39] to achieve end-to-end back-propagation.

### 3.3 Analysis of SPLUT

**Receptive Field Size.** We visualize the changes of RF sizes through the whole framework in Fig. 4. After the first spatial LUT block, each feature has an RF size of $2 \times 2$. Then in the first query block, the horizontal aggregation module and

$LUT_{WC}$ enlarge the RF size along the width dimension. The vertical aggregation module and $LUT_{HC}$ enlarge the RF size along the height dimension. After fusing the two outputs by addition, we get an RF size of 12. In the second query block, we implement similar operations and further increase the RF size. Finally, we get an RF size of 24, which is much bigger than the RF size of $3 \times 3 = 9$ in SR-LUT [11] by rotational ensemble trick.

**Computational Cost.** In SR-LUT [11], rotational ensemble is proposed to extend the RF size of $2 \times 2$ to $3 \times 3$. The computational burdens of SR-LUT mainly include image rotation, retrieval of 4 input images with different orientations, and interpolation methods which contain heavy multiplication and comparison operations. While our SPLUT model has more LUTs, we do not need the rotation and interpolation operations. Therefore, our method can achieve a faster inference speed than SR-LUT.

# 4 Experiments

## 4.1 Implementation Details

**Datasets and Metrics.** We train the proposed serial-parallel lookup table (SPLUT) model on the DIV2K dataset [1] and evaluate the effectiveness of our method on 5 widely used benchmarks: Set5 [3], Set14 [41], BSD100 [22], Urban100 [10] and Manga109 [23]. We focus on the upscaling factor of $\times 4$ in our experiments. We use Peak Signal-to-Noise Ratio (PSNR) and structural similarity index (SSIM) [37] as the evaluation metrics for prediction accuracy. To compare the computation efficiency, we measure and report the runtime of super-resolving $320 \times 180$ LR images on mobile phones.

**Training Setting.** We design three SPLUT models with different model sizes, namely SPLUT-S, SPLUT-M, and SPLUT-L. The three models have the same architecture depicted in Fig. 2(a). The difference between the three models lies in the value of $C_f$, the number of lookup tables per query block and the grouping strategy of input feature maps. We have introduced the details of SPLUT-M. The details of the other two models are described in the supplementary material. We train SPLUT models with PyTorch [26] on Nvidia 2080Ti GPUs. We use Adam Optimizer [13] with $\beta_1 = 0.9, \beta_2 = 0.999$ and $\epsilon = 1 \times 10^{-8}$ to jointly train the MSB and LSB branches. The learning rate is set to $10^{-3}$. We randomly crop LR images into $48 \times 48$ patches with a mini-batch size of 32. We enhance the dataset by randomly rotating and flipping.

## 4.2 Results and Analyses

**Quantitative Comparison.** We compare our method with SR methods based on sparse coding which include NE+LLE [6], Zeyde *et al.* [41], ANR [30] and A+ [31], SR methods based on deep learning including CARN-M [2], FMEN [7] and RRDB [36], and SR method based on LUTs, SR-LUT [11]. Since the source

**Table 1.** Quantitative comparisons of different SR methods on 5 benchmark datasets. The best results among LUT-based methods are **highlighted**. Running time is measured by super-resolving 320 × 180 LR images on the mobile phone. * represents the running time is measured on computer CPUs. Size denotes the storage space or the parameter number of each model.

| Method | Time | Size | Set5 | | Set14 | | BSDS100 | | Urban100 | | Manga109 | |
|---|---|---|---|---|---|---|---|---|---|---|---|---|
| | | | PSNR | SSIM | PSNR | SSIM | PSNR | SSIM | PSNR | SSIM | PSNR | SSIM |
| NE+LLE | 7016 ms* | 1.434MB | 29.62 | 0.840 | 26.82 | 0.735 | 26.49 | 0.697 | 23.84 | 0.694 | 26.10 | 0.820 |
| Zeyde *et al.* | 8797 ms* | 1.434MB | 26.69 | 0.843 | 26.90 | 0.735 | 26.53 | 0.697 | 23.90 | 0.696 | 26.24 | 0.824 |
| ANR | 1715 ms* | 1.434MB | 29.70 | 0.842 | 26.86 | 0.737 | 26.52 | 0.699 | 23.89 | 0.696 | 26.18 | 0.821 |
| A+ | 1748 ms* | 15.17MB | 30.27 | 0.860 | 27.30 | 0.750 | 26.73 | 0.709 | 24.33 | 0.719 | 26.91 | 0.848 |
| CARN-M | 4955 ms | 1.593MB | 31.82 | 0.890 | 28.29 | 0.775 | 27.42 | 0.730 | 25.62 | 0.769 | 29.85 | 0.899 |
| FMEN | 3101 ms | 1.395MB | 32.24 | 0.896 | 28.70 | 0.784 | 27.63 | 0.738 | 26.28 | 0.791 | 30.70 | 0.911 |
| RRDB | 31717 ms | 63.83MB | 32.60 | 0.900 | 28.88 | 0.790 | 27.76 | 0.743 | 26.73 | 0.807 | 31.16 | 0.916 |
| SR-LUT | 279 ms | **1.274M** | 29.82 | 0.848 | 27.01 | 0.736 | 26.53 | 0.695 | 24.02 | 0.699 | 26.80 | 0.838 |
| SPLUT-S | **242 ms** | 5.5M | 30.01 | 0.852 | 27.20 | 0.743 | 26.68 | 0.702 | 24.13 | 0.706 | 27.00 | 0.843 |
| SPLUT-M | 265 ms | 7M | 30.23 | 0.857 | 27.32 | 0.746 | 26.74 | 0.704 | 24.21 | 0.709 | 27.20 | 0.848 |
| SPLUT-L | 545 ms | 18M | **30.52** | **0.863** | **27.54** | **0.752** | **26.87** | **0.709** | **24.46** | **0.719** | **27.70** | **0.858** |

code of SR-LUT [11] is not released, we reproduce the SR-LUT algorithm and compare our method with it under the same environment. Since the implementation of sparse coding based methods [6,30,31,41] rely on Matlab, we evaluate these methods on the CPUs of computers which may be faster than mobile phones. The quantitative comparisons are shown in Table 1. As observed, our SPLUT models achieve much faster inference than both sparse coding based methods and deep learning based methods. SPLUT-S, SPLUT-M and SPLUT-L all obtain higher PSNR and SSIM than NE+LLE, Zeyde *et al.* and ANR. The SPLUT-M is comparable to A+ and SPLUT-L is superior to A+ on all benchmarks. While deep learning based methods have the best PSNR and SSIM performance, their inference speed is much slower than our method. As a method based on LUTs, SR-LUT is much faster than the other compared methods. However, it is still slower than our SPLUT-S and SPLUT-M methods. Besides, SPLUT models all outperform SR-LUT by a large margin on PSNR and SSIM metrics. Our SPLUT-M model achieves a better trade-off between efficiency and accuracy. Compared to SR-LUT, SPLUT-M improves PSNR by 0.4 dB on Set5 and Manga109 in a faster speed. By comparing model sizes, we see that our SPLUT method only brings a linear increase in storage costs. We believe the LUT size of our method is acceptable for current mobile phones. Therefore, we regard the runtime as a more important factor for evaluating efficiency. Besides, SPLUT-L presents more powerful SR abilities than SPLUT-S and SPLUT-M. These comparisons verify that our SPLUT framework is more powerful and more flexible than the previous framework of single-layer LUTs whose scale increases exponentially with RF sizes.

**Qualitative Comparison.** Figure 5 illustrates the qualitative comparisons of Bicubic interpolation, A+, SR-LUT, our SPLUT models, and ground-truth images. We can see SR-LUT fails to present natural details for sharp edges. In

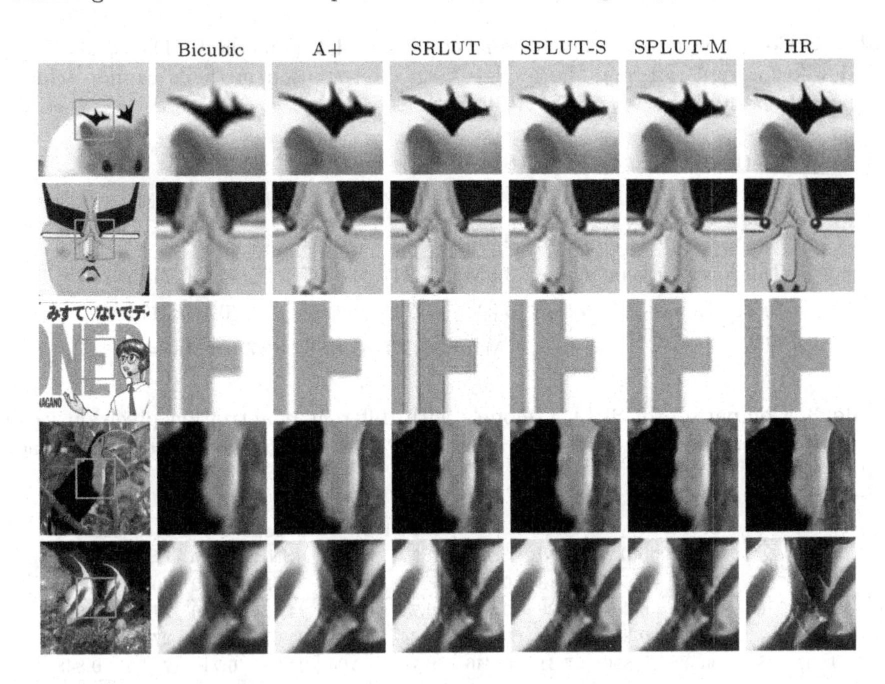


**Fig. 5.** Qualitative Comparisons of bicubic interpolation, A+ [31], SRLUT [11], our SPLUT method and HR images. The results show our method can generate sharp edges without severe artifacts.

some areas with continuously changing colors, there are often blocking artifacts. In the third row, the SR results of SR-LUT have severe ringing artifacts near edges. While A+ introduces fewer artifacts than SR-LUT, it may generate more blurry edges, as shown in the second row. On the contrary, our SPLUT models restore more natural textures. It can be seen that the expansion of the receptive field in SPLUT helps the network grasp the texture and structure information of context regions to achieve better reconstruction accuracy.

**Ablation: Parallel Network vs. Interpolation.** We take SPLUT-M as the baseline model and further investigate the effectiveness of our proposed parallel network by comparing it with interpolation algorithms. Specifically, we remain only one branch of the SPLUT model and use full-precision LR images as inputs to train this model. In the inference phase, we follow SR-LUT [11] to extract $I_{MSB}$ for retrieval and store $I_{LSB}$ for interpolation. We call the one-branch model without interpolation "OBM w/o interpolation". Since the cascaded LUTs in this model brings a large RF size of $r$, it is intractable to consider all bounding vertices and simply implement interpolations due to the computational complexity of $2^r$. To improve the SR accuracy of this model, we design two interpolation methods for the input images. For a position of $(x, y)$, SR-LUT implements 4-simplex interpolation for 4D LUTs by exploring the relation of $I_{LSB}^{(x,y)}$, $I_{LSB}^{(x+1,y)}$, $I_{LSB}^{(x,y+1)}$ and $I_{LSB}^{(x+1,y+1)}$. A weighted sum of the retrieval results for the 16 bounding

**Table 2.** Comparison of parallel network and interpolation methods. The results show OBM w/o interpolation and the models with interpolation methods cannot achieve comparable performance to our SPLUT method.

| Method | Size | Set5 | | Set14 | |
|---|---|---|---|---|---|
| | | PSNR | SSIM | PSNR | SSIM |
| OBM w/o interpolation | 3.5M | 27.24 | 0.8217 | 25.30 | 0.7175 |
| Tail-layer Interpolation | 3.5M | 27.24 | 0.8217 | 25.30 | 0.7175 |
| First-layer Interpolation | 3.5M | 27.24 | 0.8218 | 25.30 | 0.7176 |
| SPLUT | 7M | **30.23** | **0.8567** | **27.32** | **0.7460** |

**Table 3.** Comparison of SPLUT models with different quantization precision $v_f$ of intermediate features. We choose $v_f = 16$ as the optimal setting considering accuracy and efficiency.

| $v_f$ | Size | Set5 | | Set14 | | BSDS100 | | Urban100 | | Manga109 | |
|---|---|---|---|---|---|---|---|---|---|---|---|
| | | PSNR | SSIM | PSNR | SSIM | PSNR | SSIM | PSNR | SSIM | PSNR | SSIM |
| 8 | 1.375M | 29.95 | 0.849 | 27.13 | 0.738 | 26.63 | 0.698 | 24.05 | 0.702 | 26.83 | 0.839 |
| 12 | 2.898M | 30.11 | 0.854 | 27.26 | 0.743 | 26.71 | 0.702 | 24.15 | 0.706 | 27.06 | 0.844 |
| 16 | 7M | 30.23 | 0.857 | 27.32 | 0.746 | 26.74 | 0.704 | 24.21 | 0.709 | 27.20 | 0.848 |
| 20 | 15.648M | 30.22 | 0.856 | 27.31 | 0.746 | 26.74 | 0.704 | 24.22 | 0.710 | 27.25 | 0.848 |
| 24 | 31.375M | 30.25 | 0.858 | 27.34 | 0.747 | 26.75 | 0.705 | 24.24 | 0.711 | 27.24 | 0.849 |

vertices is computed as the final output since SR-LUT has only one layer of LUT. In our methods, we also utilize $I_{LSB}$ for interpolation but we cannot get the final SR output by directly fusing the retrieval results of the first layer of spatial LUT since we still have other following LUTs for retrieval. Therefore, we concatenate the 16 bounding vertices of all input pixels to form 16 index maps, which reduce the complexity of $2^r$ to $2^4$. In our first interpolation method, we take these 16 index maps as the inputs to the spatial lookup blocks and get 16 SR results through the whole network. We interpolate the 16 SR results by $I_{LSB}$ using the 4-simplex method. We call this method tail-layer interpolation. In our second method, we feed the 16 index maps to the spatial lookup blocks and get 16 intermediate feature maps. We fuse the 16 feature maps by $I_{LSB}$ to get one feature map. By feeding the feature map to the following layers, we can get a final SR output. We call this method first-layer interpolation. More implementation details are described in the supplementary. The results are shown in Table 2. Our SPLUT with the parallel network achieves an improvement of more than 2.9 dB over the two interpolation methods and OBM w/o interpolation. This proves that applying interpolation methods fails to compensate for the precision loss, which is caused by discretizing pixel values when establishing lookup tables. In contrast, our parallel network is superior to interpolation algorithms in compensating for the precision loss for large RF sizes. It can also be inferred that our parallel network is inherently robust to different receptive fields.

**Table 4.** Ablation study on LUT number. Extending the depth and width of the SPLUT network both boost the SR performance, which demonstrates the effectiveness and extensibility of the proposed method.

| Layer num | Size | Set5 | | Set14 | | BSDS100 | | Urban100 | | Manga109 | |
|---|---|---|---|---|---|---|---|---|---|---|---|
| | | PSNR | SSIM | PSNR | SSIM | PSNR | SSIM | PSNR | SSIM | PSNR | SSIM |
| $SPLUT_{1-2}$ | 5M | 29.77 | 0.846 | 27.04 | 0.737 | 26.56 | 0.696 | 23.95 | 0.697 | 26.68 | 0.836 |
| $SPLUT_{1-4}$ | 10M | 30.01 | 0.852 | 27.21 | 0.742 | 26.66 | 0.700 | 24.12 | 0.705 | 27.05 | 0.844 |
| $SPLUT_{1-2-2}$ | 7M | 30.23 | 0.857 | 27.32 | 0.746 | 26.74 | 0.704 | 24.21 | 0.709 | 27.20 | 0.848 |
| $SPLUT_{1-4-4}$ | 18M | 30.52 | 0.863 | 27.54 | 0.752 | 26.87 | 0.709 | 24.46 | 0.719 | 27.70 | 0.858 |

**Ablation: Quantization Precision.** In SPLUT, we uniformly quantize the real-value activations in the aggregation modules during training to control the size of LUTs. Since we set $n = 4$ for all the LUTs, the prediction accuracy and model scale are mainly determined by $v$. For the spatial lookup blocks, we fix the sampling interval to 16. We change the precision of quantizing the real-value intermediate features, $v_f$, and investigate the influence of it. Table 3 presents the performance of SPLUT-M with different $v_f$. The results indicate that increasing $v_f$ constantly improves the SR accuracy but the model size also increases. When $v_f$ is less than 16, the accuracy improves rapidly. However, SPLUT only has a minor improvement when $v_f$ is greater than 16. Hence we choose $v_f = 16$ as the appropriate value which presents appealing SR performance with a relatively small model size. In practice, the quantization precision can be determined by the application scenarios to achieve the flexible model design.

**Ablation: LUT Number.** We conduct ablation studies on the number of LUTs in each parallel branch to further investigate the extensibility of our SPLUT architecture. As shown in Table 4, we compare 4 models with different model depths and widths. The model is named according to the number of LUTs in each block. $SPLUT_{1-2}$ represents there are two layers of LUTs. The first layer is the spatial lookup block and the second layer is a query block which contains one $LUT_{WC}$ and one $LUT_{HC}$. For $SPLUT_{1-4}$, the second layer contains two different $LUT_{WC}$ and two different $LUT_{HC}$. In this model, the channel number of intermediate feature maps is $n_{in} = 16$. $SPLUT_{1-2-2}$ and $SPLUT_{1-4-4}$ have the similar architectures to $SPLUT_{1-2}$ and $SPLUT_{1-4}$ but have two query blocks in each branch. $SPLUT_{1-2-2}$ and $SPLUT_{1-4-4}$ are actually the same as SPLUT-M and SPLUT-L, respectively.

In Table 4, we observe a huge improvement when the network width increases by comparing the first two rows and the last two rows. This indicates that more LUTs per layer can extract more information by the limited number of channels. In this way, the model can obtain a stronger representation ability and better SR reconstruction performance. As the number of query blocks increases, we see $SPLUT_{1-2-2}$ achieves an improvement of 0.46 dB over $SPLUT_{1-2}$ while $SPLUT_{1-4-4}$ outperforms $SPLUT_{1-4}$ by about 0.51 dB on the PSNR performance of Set5. It is inferred that cascading multiple layers of LUTs is very effective in enlarging the receptive fields and boosting recovery abilities. Compar-

**Table 5.** Effects of horizontal and vertical aggregation modules. After removing the aggregation modules, the RF size gets smaller and the restoration ability is degraded.

| | Set5 | | Set14 | | BSDS100 | | Urban100 | | Manga109 | |
|---|---|---|---|---|---|---|---|---|---|---|
| | PSNR | SSIM | PSNR | SSIM | PSNR | SSIM | PSNR | SSIM | PSNR | SSIM |
| SPLUT w/o AM | 29.64 | 0.839 | 26.95 | 0.724 | 26.08 | 0.676 | 23.50 | 0.677 | 25.72 | 0.812 |
| SPLUT | **30.23** | **0.857** | **27.32** | **0.746** | **26.74** | **0.704** | **24.21** | **0.709** | **27.20** | **0.848** |

ing $\text{SPLUT}_{1-2-2}$ and $\text{SPLUT}_{1-4}$, both models have 5 lookup tables. However, $\text{SPLUT}_{1-2-2}$ gains a boost of about 0.22 dB. This indicates that the network depth is more important than network width in SPLUT. Besides, the comparisons demonstrate the effectiveness of enlarging receptive fields.

**Ablation: Aggregation Modules.** Table 5 shows the performance comparison between the original SPLUT model and the SPLUT model without aggregation modules, SPLUT w/o AM. From the table we see there is a gap of 0.59 dB between the PSNR performance of SPLUT w/o AM and SPLUT on Set5. The key insight is that we expand the receptive field of the intermediate features by fusing the feature maps padded in opposite directions in the aggregation modules. Thus the features become stronger for the subsequent lookup processes. When aggregation modules are removed, the overall receptive field of the final output is significantly reduced compared with that of the original network, leading to performance degradation correspondingly. The experimental results demonstrate the effectiveness of the proposed horizontal and vertical aggregation modules.

## 5 Conclusion

In this paper, we have proposed a series-parallel lookup table network to achieve efficient image super-resolution. On the one hand, we cascade multiple LUTs to enlarge the receptive field size progressively and enhance the representation capacity of the whole network. On the other hand, we design a parallel architecture to fuse the information of MSB inputs and LSB inputs. By doing so, we compensate for the precision loss caused by quantization when establishing LUTs and improve the prediction accuracy. Comprehensive experiments have demonstrated the effectiveness, efficiency, and flexibility of the proposed method.

**Acknowledgement.** This work was supported in part by the National Key Research and Development Program of China under Grant 2017YFA0700802, in part by the National Natural Science Foundation of China under Grant 62125603 and Grant U1813218, in part by a grant from the Beijing Academy of Artificial Intelligence (BAAI).

## References


1. Agustsson, E., Timofte, R.: NTIRE 2017 challenge on single image super-resolution: dataset and study. In: CVPR, pp. 126–135 (2017)
2. Ahn, N., Kang, B., Sohn, K.A.: Fast, accurate, and lightweight super-resolution with cascading residual network. In: ECCV, pp. 252–268 (2018)
3. Bevilacqua, M., Roumy, A., Guillemot, C., Alberi-Morel, M.L.: Low-complexity single-image super-resolution based on nonnegative neighbor embedding. In: BMVC (2012)
4. Bilal, M., Khan, A., Karim Khan, M.U., Kyung, C.M.: A low-complexity pedestrian detection framework for smart video surveillance systems. TCSVT **27**(10), 2260–2273 (2017)
5. Chang, C.C., Chou, J.S., Chen, T.S.: An efficient computation of Euclidean distances using approximated look-up table. TCSVT **10**(4), 594–599 (2000)
6. Chang, H., Yeung, D.Y., Xiong, Y.: Super-resolution through neighbor embedding. In: CVPR, vol. 1, p. I. IEEE (2004)
7. Du, Z., Liu, D., Liu, J., Tang, J., Wu, G., Fu, L.: Fast and memory-efficient network towards efficient image super-resolution. In: Proceedings of the IEEE/CVF Conference on Computer Vision and Pattern Recognition, pp. 853–862 (2022)
8. He, K., Zhang, X., Ren, S., Sun, J.: Deep residual learning for image recognition. In: CVPR, pp. 770–778 (2016)
9. Hendrycks, D., Gimpel, K.: Gaussian error linear units (GELUs). arXiv preprint arXiv:1606.08415 (2016)
10. Huang, J.B., Singh, A., Ahuja, N.: Single image super-resolution from transformed self-exemplars. In: CVPR, pp. 5197–5206 (2015)
11. Jo, Y., Kim, S.J.: Practical single-image super-resolution using look-up table. In: CVPR, pp. 691–700 (2021)
12. Kasson, J.M., Nin, S.I., Plouffe, W., Hafner, J.L.: Performing color space conversions with three-dimensional linear interpolation. J. Electron. Imaging **4**(3), 226–250 (1995)
13. Kingma, D.P., Ba, J.: Adam: a method for stochastic optimization. arXiv preprint arXiv:1412.6980 (2014)
14. Lee, C., Monga, V.: Power-constrained RGB-to-RGBW conversion for emissive displays: optimization-based approaches. TCSVT **26**(10), 1821–1834 (2016)
15. Lee, J.Y., Lee, J.J., Park, S.: New lookup tables and searching algorithms for fast H.264/AVC CAVLC decoding. TCSVT **20**(7), 1007–1017 (2010)
16. Lee, R., et al.: Journey towards tiny perceptual super-resolution. In: Vedaldi, A., Bischof, H., Brox, T., Frahm, J.-M. (eds.) ECCV 2020. LNCS, vol. 12371, pp. 85–102. Springer, Cham (2020). https://doi.org/10.1007/978-3-030-58574-7_6
17. Lee, W., Lee, J., Kim, D., Ham, B.: Learning with privileged information for efficient image super-resolution. In: Vedaldi, A., Bischof, H., Brox, T., Frahm, J.-M. (eds.) ECCV 2020. LNCS, vol. 12369, pp. 465–482. Springer, Cham (2020). https://doi.org/10.1007/978-3-030-58586-0_28
18. Li, Z., Yang, J., Liu, Z., Yang, X., Jeon, G., Wu, W.: Feedback network for image super-resolution. In: CVPR, pp. 3867–3876 (2019)
19. Lim, B., Son, S., Kim, H., Nah, S., Mu Lee, K.: Enhanced deep residual networks for single image super-resolution. In: CVPRW, pp. 136–144 (2017)

20. Liu, J., Tang, J., Wu, G.: Residual feature distillation network for lightweight image super-resolution. In: Bartoli, A., Fusiello, A. (eds.) ECCV 2020. LNCS, vol. 12537, pp. 41–55. Springer, Cham (2020). https://doi.org/10.1007/978-3-030-67070-2_2
21. Luo, X., Xie, Y., Zhang, Y., Qu, Y., Li, C., Fu, Y.: LatticeNet: towards lightweight image super-resolution with lattice block. In: Vedaldi, A., Bischof, H., Brox, T., Frahm, J.-M. (eds.) ECCV 2020. LNCS, vol. 12367, pp. 272–289. Springer, Cham (2020). https://doi.org/10.1007/978-3-030-58542-6_17
22. Martin, D.R., Fowlkes, C.C., Tal, D., Malik, J.: A database of human segmented natural images and its application to evaluating segmentation algorithms and measuring ecological statistics. In: ICCV, pp. 416–425 (2001)
23. Matsui, Y., et al.: Sketch-based manga retrieval using manga109 dataset. Multimed. Tools Appl. **76**(20), 21811–21838 (2017)
24. Mei, Y., Fan, Y., Zhou, Y.: Image super-resolution with non-local sparse attention. In: CVPR, pp. 3517–3526 (2021)
25. Park, S.J., Son, H., Cho, S., Hong, K.S., Lee, S.: SRFeat: single image super-resolution with feature discrimination. In: ECCV, pp. 439–455 (2018)
26. Paszke, A., et al.: Automatic differentiation in Pytorch. In: NIPS-W (2017)
27. Rizvi, S., Nasrabadi, N.: An efficient Euclidean distance computation for vector quantization using a truncated look-up table. TCSVT **5**(4), 370–371 (1995)
28. Shi, W., et al.: Real-time single image and video super-resolution using an efficient sub-pixel convolutional neural network. In: CVPR, pp. 1874–1883 (2016)
29. Song, D., Wang, Y., Chen, H., Xu, C., Xu, C., Tao, D.: AdderSR: towards energy efficient image super-resolution. In: CVPR, pp. 15648–15657 (2021)
30. Timofte, R., De Smet, V., Van Gool, L.: Anchored neighborhood regression for fast example-based super-resolution. In: ICCV, pp. 1920–1927 (2013)
31. Timofte, R., De Smet, V., Van Gool, L.: A+: adjusted anchored neighborhood regression for fast super-resolution. In: Cremers, D., Reid, I., Saito, H., Yang, M.-H. (eds.) ACCV 2014. LNCS, vol. 9006, pp. 111–126. Springer, Cham (2015). https://doi.org/10.1007/978-3-319-16817-3_8
32. Tsang, S.H., Chan, Y.L., Siu, W.C.: Region-based weighted prediction for coding video with local brightness variations. TCSVT **23**(3), 549–561 (2013)
33. Wang, B., Lu, C., Yan, D., Zhao, Y.: Learning pixel-adaptive weights for portrait photo retouching. arXiv preprint arXiv:2112.03536 (2021)
34. Wang, L., et al.: Exploring sparsity in image super-resolution for efficient inference. In: CVPR, pp. 4917–4926 (2021)
35. Wang, T., et al.: Real-time image enhancer via learnable spatial-aware 3D lookup tables. In: ICCV, pp. 2471–2480 (2021)
36. Wang, X., et al.: ESRGAN: enhanced super-resolution generative adversarial networks. In: ECCVW (2018)
37. Wang, Z., Bovik, A.C., Sheikh, H.R., Simoncelli, E.P., et al.: Image quality assessment: from error visibility to structural similarity. TIP **13**(4), 600–612 (2004)
38. Xin, J., Wang, N., Jiang, X., Li, J., Huang, H., Gao, X.: Binarized neural network for single image super resolution. In: Vedaldi, A., Bischof, H., Brox, T., Frahm, J.-M. (eds.) ECCV 2020. LNCS, vol. 12349, pp. 91–107. Springer, Cham (2020). https://doi.org/10.1007/978-3-030-58548-8_6
39. Yin, P., Lyu, J., Zhang, S., Osher, S., Qi, Y., Xin, J.: Understanding straight-through estimator in training activation quantized neural nets. arXiv preprint arXiv:1903.05662 (2019)
40. Zeng, H., Cai, J., Li, L., Cao, Z., Zhang, L.: Learning image-adaptive 3D lookup tables for high performance photo enhancement in real-time. TPAMI (2020)

41. Zeyde, R., Elad, M., Protter, M.: On single image scale-up using sparse-representations. In: Boissonnat, J.-D., et al. (eds.) Curves and Surfaces 2010. LNCS, vol. 6920, pp. 711–730. Springer, Heidelberg (2012). https://doi.org/10.1007/978-3-642-27413-8_47
42. Zhang, Y., Chen, H., Chen, X., Deng, Y., Xu, C., Wang, Y.: Data-free knowledge distillation for image super-resolution. In: CVPR, pp. 7852–7861 (2021)
43. Zhang, Y., Tian, Y., Kong, Y., Zhong, B., Fu, Y.: Residual dense network for image super-resolution. In: CVPR, pp. 2472–2481 (2018)

# GeoAug: Data Augmentation for Few-Shot NeRF with Geometry Constraints

Di Chen(✉), Yu Liu, Lianghua Huang, Bin Wang, and Pan Pan

Alibaba Group, Hangzhou, China
{guangpan.cd,ly103369,xuangen.hlh,ganfu.wb,panpan.pp}@alibaba-inc.com

**Abstract.** Neural Radiance Fields (NeRF) show remarkable ability to render novel views of a certain scene by learning an implicit volumetric representation with only posed RGB images. Despite its impressiveness and simplicity, NeRF usually converges to sub-optimal solutions with incorrect geometries given few training images. We hereby present GeoAug: a data augmentation method for NeRF, which enriches training data based on multi-view geometric constraint. GeoAug provides random artificial (novel pose, RGB image) pairs for training, where the RGB image is from a nearby training view. The rendering of a novel pose is warped to the nearby training view with depth map and relative pose to match the RGB image supervision. Our method reduces the risk of over-fitting by introducing more data during training, while also provides additional implicit supervision for depth maps. In experiments, our method significantly boosts the performance of neural radiance fields conditioned on few training views.

**Keywords:** Neural radiance fields · Few-shot learning · Unsupervised depth estimation

## 1 Introduction

To sense and infer our 3-dimensional world is a natural and fundamental ability of human beings. However, not until recently did we find how remarkable the ability is when creating virtual reality (VR) systems. We can memorize a scene from one perspective and imagine its appearance from another viewpoint with no effort, yet for a VR system, it is quite difficult to develop an automatic algorithm which is able to render photo-realistic images for a novel view. This task is challenging since it not only requires to understand the 3D scene geometry, but also needs to synthesize high-frequency textures with complex viewpoint-dependent effects.

Recently, real progress has been made on novel view synthesis. A representative work is Neural Radiance Fields (NeRF) [22], which learns an implicit scene representation and generate images with volume rendering. When trained on a specific scene, a Multi-layer Perceptron (MLP) is used to estimate the volume density and color for each point in the space. Volume rendering is then

© The Author(s), under exclusive license to Springer Nature Switzerland AG 2022
S. Avidan et al. (Eds.): ECCV 2022, LNCS 13677, pp. 322–337, 2022.
https://doi.org/10.1007/978-3-031-19790-1_20

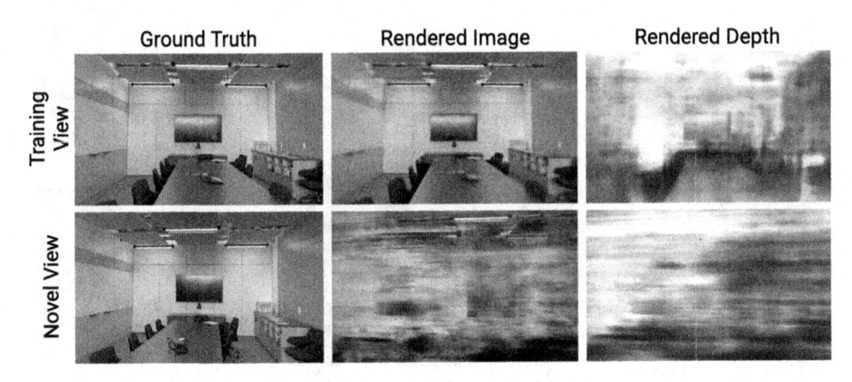


**Fig. 1. NeRF overfits to few training views.** NeRF produces favorable rendering result for training views, while fails to generalize to novel views. NeRF also struggles to learn the correct geometries with few training views.

used to generate the RGB image, supervised by the ground truth image with a photometric reconstruction loss. NeRF has shown its exceptional ability on high-quality image synthesis, while being conceptually simple and easy to train.

Typically, NeRF needs a large amount of input views for simultaneous geometry and appearance reconstruction with high fidelity. If the training views are inadequate, NeRF tends to overfit the limited training samples. Satisfactory images could only be produced at observed poses, while the renderings under novel views are polluted with many artifacts. A comparative example is shown in Fig. 1. The reason is that NeRF cannot infer the correct 3D geometry from limited views with only RGB image supervision, resulting in a sub-optimal solution which cannot generalize to novel views [41]. Figure 1 also illustrates that the depth maps for neither the training view nor the novel view are correctly inferred. Several works [6,12,33,35,40] have been proposed to address this problem, among which DSNeRF [6] shows its simplicity and effectiveness by adding sparse depth as an additional explicit supervision.

In this work, we aim to improve the performance of NeRF with few training views. A straightforward way is to increase the training samples by data augmentation. Meanwhile, DSNeRF [6] also shows that adding depth supervision is a valid strategy. To this end, we propose **GeoAug**, a data augmentation method for NeRF based on geometry constraint with implicit depth supervision. In addition to the original training views, we first generate camera poses under novel views and render the corresponding images. Since ground truth images are not available for novel views, we warp the rendered images to nearby training views based on the predicted depth maps and relative camera poses. Then we can impose a photometric loss between the warped images and training images. A comprehensive illustration is shown in Fig. 2. Since the warping operation is differentiable and involves depth information, the model is encouraged to learn depth estimation implicitly, which could be seen as a strong geometric constraint.

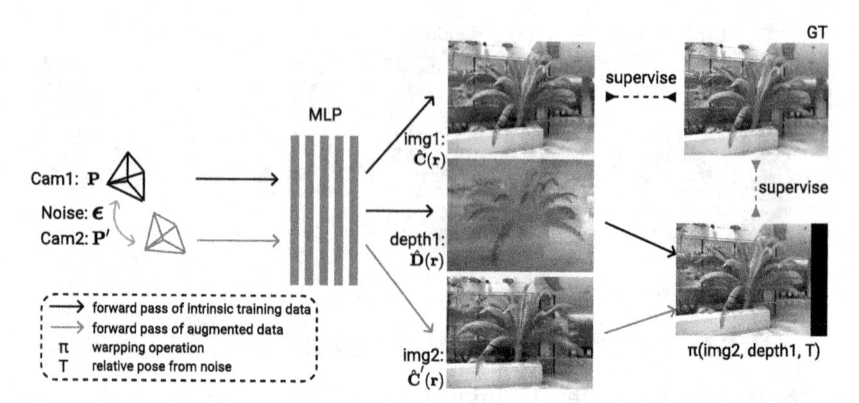


**Fig. 2. Overall pipeline of training NeRF with GeoAug.** We first sample a random camera pose (termed as Cam1) from training set, render its appearance image (img1) and depth map. A novel camera pose Cam2 is then created by adding noise to Cam1. The appearance image (img2) of Cam2 is rendered and warped to Cam1 with depth map and the relative pose between Cam1 and Cam2. Both img1 and the warped-img2 are supervised by the ground truth image of Cam1.

Our method is inspired by prior work [43] which aims for unsupervised depth estimation by means of view synthesis. In [43], view synthesis serves as a proxy task which forces the network to infer the depth map, based on the insight that a view synthesis system could only perform well across multiple views if the scene geometry is modeled correctly. Our work follows the same insight, except that we focus on view synthesis as the primary task and uses the implicit depth estimation as an extra regularizer. Compared to DSNeRF [6] which uses *explicit* and *sparse* depth supervision, our method provides *implicit* and *dense* depth signal. Both GeoAug and DSNeRF discover supervision signal for free, while being complementary to each other and easy to integrate into other NeRF based models. Empirical evaluations on NeRF Real [21,22] and DTU [13] datasets demonstrate the effectiveness of our approach on improving the synthesis quality with few training views.

## 2 Related Work

**Novel View Synthesis.** The literature of novel view synthesis could be roughly categorized into two classes: 1) *explicit* 3D reconstruction and 2) *implicit* representations. For approaches in the first category, the 3D geometry and appearance is explicitly represented by point clouds [37], voxels [20,30], meshes [11,15,29], or multi-plane images [5,7,21,34,44]. Once the 3D scene is reconstructed, it is trivial to render the 2D image under arbitrary view. These approaches are computationally efficient, while having the merit of straightforward to check and modify the 3D structure. However, these methods are typically difficult to optimize due to the discontinuous nature.

On the other hand, implicit approaches [8,22,25,31,39] directly model the appearance of 3D scene. Without the need to represent geometry explicitly, means of discretization, *i.e.* creating voxel, mesh or multi-plane, are not adopted. Therefore, images under arbitrary views could be synthesized continuously with high definition. The representative work is Neural Radiance Fields (NeRF) [22], which maps from camera pose to color and volume density of each location in the space with an MLP. RGB images are then produced with differentiable volume rendering. Due to its simplicity and exceptional rendering quality, recent works adopt NeRF for various extensions such as generative adversarial networks [4, 24,28], video synthesis [18,38], relighting [1,32], scene editing [14,19], *etc.*

**Few-Shot Neural Radiance Fields.** NeRF-based methods usually require a lot of images from different views for training. Several works have been proposed to address the data-hungry problem of NeRF by exploiting training data [35,40], meta learning [33] and additional supervision [6,12]. PixelNeRF [40] takes advantage of the training images during test time rendering, which is ignored by vanilla NeRF. Convolutional feature of the training image is projected onto the ray of novel view, which is later used as a conditional embedding for MLP inference. IBRNet [35] adopts a similar strategy and adds an additional ray transformer for better density estimation. Instead of randomly initialize the weights of MLP, MetaNeRF [33] propose to pre-train the network on a large-scale dataset before fine-tuning on each scene. DietNeRF [12] adds a pair-wise loss which regularizes multi-view consistency by pulling the cosine distance between high-level semantic features of different views. RegNeRF [23] renders image patches from unseen camera views and regularize the RGB values with a trained normalizing flow model. The density values are also regularized by a smoothness loss. DSNeRF [12] utilizes the sparse depth information generated by COLMAP [27] as an explicit supervision for the rendered depth map. Our work shares similar insight to DSNeRF, *i.e.* provide supervision for depth map, except that our approach is unsupervised, requiring no additional data nor annotations. In this paper, we mainly apply our data augmentation method upon DSNeRF, yet it is worth to notice that our method is compatible to all NeRF-based models.

**Unsupervised Depth Estimation.** Our GeoAug method is closely related to works on unsupervised depth estimation [2,3,10,43], which utilize the geometric constraints between frames as supervisory signal. During training, adjacent video frames, denoted as source/target frames, are sampled as inputs to a depth network and a pose network respectively. The output depth map and relative camera pose are used to warp the source frame to the viewpoint of target frame. The photometric error between the warped source frame and target frame is minimized during training, which represents the geometric constraint. During inference, the depth network could be used separately for depth estimation.

For NeRF models, camera poses are estimated beforehand by SfM [27], thus no pose network is needed. The depth map could be rendered in a similar way to RGB images. Therefore, the warping-based geometric constraint could be used as an additional supervision for NeRF, encouraging a better understanding of scene geometry.

## 3 Methodology

We now present our geometry-aware data augmentation method in this section. We begin by revisiting the classic NeRF method and volumetric rendering, with an important baseline: DSNeRF [6], which adds additional sparse depth supervision to NeRF. Then we introduce our data augmentation method as well as the adaptive noise module. Finally, we summarize the overall training procedure.

### 3.1 Revisiting Volume Rendering

NeRF [22] is originally proposed for the task of novel view synthesis, which aims at rendering an RGB image given a camera pose $\mathbf{P}$. This is implemented by 1) shooting rays from the center of camera pose $\mathbf{P}$, 2) predicting the radiance intensity at each point along all the rays and 3) rendering each pixel by accumulating the radiance of all points along the ray.

Specifically, given a camera center $\mathbf{o} \in \mathbb{R}^3$ and viewing direction $\mathbf{d}$, the corresponding ray is represented as $\mathbf{r}(t) = \mathbf{o} + t\mathbf{d}$, parameterized by $t$. For a specific point $\mathbf{x} \in \mathbb{R}^3$ on the ray, NeRF uses an Multi-Layer Perceptron (MLP) to predict the color $\mathbf{c} \in \mathbb{R}^3$ and volume density $\sigma \in \mathbb{R}^+$. The MLP could be seen as a function $f_\theta$ that maps from spatial location and direction to radiance field: $f_\theta(\mathbf{x}, \mathbf{d}) = (\mathbf{c}, \sigma)$.

Once the entire radiance field is available, RGB images could be rendered by integrating along rays with volume rendering:

$$\hat{\mathbf{C}}(\mathbf{r}) = \int_{t_n}^{t_f} T(t)\sigma(t)\mathbf{c}(t)\mathrm{d}t, \quad \text{where } T(t) = \exp(-\int_{t_n}^{t} \sigma(s)\mathrm{d}s) \tag{1}$$

where $t_n$ and $t_f$ represent the near and far bounds of the rays. $T(t)$ is a transmittance term which measures the probability that light could travel from $t_n$ to $t$ without being obstructed.

During training, NeRF model is supervised by a reconstruction loss:

$$\mathcal{L}_c = \frac{1}{|\mathcal{R}|} \sum_{\mathbf{r} \in \mathcal{R}} \|\hat{\mathbf{C}}(\mathbf{r}) - \mathbf{C}(\mathbf{r})\|_2^2 \tag{2}$$

where $\mathcal{R}$ is the set of rays randomly sampled during training. $\mathbf{C}$ is the ground truth RGB image under camera pose $\mathbf{P}$.

**DSNeRF: A Supervised Baseline.** Given enough training images captured under various camera poses, NeRF is effective at representing the scene implicitly and thus rendering satisfying images at novel views. However, under few-shot settings, NeRF is prone to overfit the available scenes with incorrect geometry [41], *e.g.* a plain canvas at the camera's near bound filled with the pixels from training images [6]. Adding images from more diverse views could alleviate this problem, but they are not always available in real-world applications. To this end, DSNeRF [6] is proposed to improve NeRF under few-shot settings.

Generally, DSNeRF adds a depth reconstruction loss $\mathcal{L}_d$ alongside the RGB loss $\mathcal{L}_c$:

$$\mathcal{L}_d = \frac{1}{|\mathcal{R}_d|} \sum_{\mathbf{r} \in \mathcal{R}_d} w_{\mathbf{r}} \|\hat{\mathbf{D}}(\mathbf{r}) - \mathbf{D}(\mathbf{r})\|_2^2 \tag{3}$$

where the depth ground truth $\mathbf{D}(\mathbf{r})$ and the corresponding confidence $w_{\mathbf{r}}$ are side products of camera pose estimation with structure-from-motion (SfM) [27], which is the standard preprocess for training NeRF. $\hat{\mathbf{D}}(\mathbf{r})$ is the depth map rendered in a similar way to rendering RGB images:

$$\hat{\mathbf{D}}(\mathbf{r}) = \int_{t_n}^{t_f} T(t)\sigma(t)t\mathrm{d}t \tag{4}$$

Note that SfM only produces a sparse set of points with depth, thus the sampling set $\mathcal{R}_d$ for depth supervision is different from the one for RGB supervision $\mathcal{R}$.

DSNeRF has shown it's superior performance over other NeRF variants for few-shot settings [33,35,40]. Therefore, we choose it as our baseline and test our data augmentation method upon DSNeRF.

### 3.2 Geometry-Aware Data Augmentation

Figure 2 shows the overview of our GeoAug method. During training, we first add random noise to the 6 degree-of-freedom (6-DoF) representation of a camera pose $\mathbf{P}$ in the training set:

$$\mathbf{P}' = \mathbf{P} + \epsilon, \quad \epsilon \sim \mathcal{N}(0, \delta) \tag{5}$$

where the noise vector $\epsilon$ is sampled from a Gaussian distribution with 0 mean and $\delta$ standard deviation. We then render the RGB image $\hat{\mathbf{C}}'(\mathbf{r})$ under camera view $\mathbf{P}'$ using volume rendering described in Sect. 3.1. Since there is no ground truth for $\hat{\mathbf{C}}'(\mathbf{r})$, we warp $\hat{\mathbf{C}}'(\mathbf{r})$ from $\mathbf{P}'$ to $\mathbf{P}$ and supervise the warped image with $\mathbf{C}$:

$$\mathcal{L}_a = \frac{1}{|\mathcal{R}|} \sum_{\mathbf{r} \in \mathcal{R}} \|\pi(\hat{\mathbf{C}}'(\mathbf{r}), \hat{\mathbf{D}}(\mathbf{r}), T_{\mathbf{P} \to \mathbf{P}'}) - \mathbf{C}(\mathbf{r})\| \tag{6}$$

In Eq. 6, $\pi(\cdot)$ denote the differentiable image warping function. Let $p'$ and $p$ be the homogeneous coordinates of pixels in $\hat{\mathbf{C}}'(\mathbf{r})$ and $\hat{\mathbf{C}}(\mathbf{r})$ respectively, $K$ the camera intrinsics matrix and $T_{\mathbf{P} \to \mathbf{P}'}$ the relative pose between $\mathbf{P}$ and $\mathbf{P}'$. The warping function $\pi(\cdot)$ is described as

$$p' \sim K T_{\mathbf{P} \to \mathbf{P}'} \hat{\mathbf{D}}(p) K^{-1} p \tag{7}$$

Since the projected coordinates $p'$ are continuous values, we use nearest sampling to get the pixel value from $\hat{\mathbf{C}}'(\mathbf{r})$. We also ignore the pixels if $p'$ is outside the image bound of $\hat{\mathbf{C}}'(\mathbf{r})$.

One prerequisite for successful warping between different views is that there should be no occlusion or disocclusion. However, it is not always true if the scene geometry is complex, *i.e.* a scene containing a lot of discontinuous depths. Regions with complex geometry would induce outliers in Eq. 6, corrupting the gradients and harm the training process. In our experiments, a simple L2 loss for $\mathcal{L}_a$ works just fine, which means that the outlier points are not dominating the loss. To improve the robustness of our method, we propose to designate the metric $\|\cdot\|$ in Eq. 6 as smooth L1 loss instead of the commonly used L2 loss, since smooth L1 loss is less sensitive to outliers [9]. Our experiments also show that smooth L1 loss improves the performance over L2 loss. Smooth L1 loss is defined as below:

$$\|\cdot\|_{\text{smooth L1}} = \begin{cases} 0.5 \times (\cdot)^2 & \text{if } |\cdot| < 1.0 \\ |\cdot| - 0.5 & \text{otherwise} \end{cases} \tag{8}$$

### 3.3 Adaptive Noise

One important hyper-parameter in our GeoAug method is the standard deviation $\delta$ of camera pose noise. $\delta$ controls the offset magnitude between $\mathbf{P}'$ and $\mathbf{P}$, *i.e.* the larger the $\delta$, the more $\mathbf{P}'$ is deviated from $\mathbf{P}$. It could be tedious to tune $\delta$ since we have no prior knowledge of the pose offset. If $\mathbf{P}'$ is too far away from $\mathbf{P}$, the warping operation could be highly unreliable. If the offset is otherwise too small, the efficacy of data augmentation is diminished. On the other hand, using the same noise magnitude through out the entire training process may not be optimal.

Inspired by the adaptive augmentation methods used in GANs [16], we propose to tune $\delta$ *adaptively* instead of picking the appropriate $\delta$ manually through exhaustive experiments. Our method is based on a heuristic rule regarding the discrepancy between the loss values of $\mathcal{L}_c$ and $\mathcal{L}_a$. Typically, an ideal augmentation method should keep the loss of augmented samples a little higher than the loss of intrinsic training samples. In our case, we first set an initial $\delta_0$ as a base, multiply/divide $\delta_0$ with a factor $\gamma$ ($\gamma > 1$) if $\mathcal{L}_a$ smaller/larger than a margin $m$ over $\mathcal{L}_c$:

$$\delta_t = \begin{cases} \delta_0, & \text{if } t = 0 \\ \delta_{t-1} * \gamma, & \text{if } 2\bar{\mathcal{L}}_a < \bar{\mathcal{L}}_c + m \\ \delta_{t-1}/\gamma, & \text{if } 2\bar{\mathcal{L}}_a > \bar{\mathcal{L}}_c + 2m \\ \delta_{t-1}, & \text{if } \bar{\mathcal{L}}_c + m \leq 2\bar{\mathcal{L}}_a \leq \bar{\mathcal{L}}_c + 2m \end{cases} \tag{9}$$

where $\bar{\mathcal{L}}$ is the averaged loss value of the most resent 100 training steps. In other words, the rule defined by Eq. 9 aims to keep the loss value of $2\bar{\mathcal{L}}_a$ between $\bar{\mathcal{L}}_c + m$ and $\bar{\mathcal{L}}_c + 2m$.

Although our adaptive noise rule brings additional hyper-parameters, *i.e.* $\delta_0$, $\gamma$ and $m$, we find through experiments that the choice of $\delta_0$ and $\gamma$ does not affect the performance too much, since $\delta_t$ would converge to the same range quickly.

**Algorithm 1:** Training DSNeRF with GeoAug

**Data:** Training set $\{\mathbf{P}, \mathbf{C}, \mathbf{D}\}$, initial noise standard deviation $\delta_0$, noise growing factor $\gamma$, loss margin $m$, number of augmented sample per-iteration $N$, learning rate $\eta$, loss weights $\lambda_d, \lambda_a$
**Result:** Trained radiance field function $f_\theta$
Initialize MLP parameters of $f_\theta$ ;
$t \leftarrow 0, \delta_t \leftarrow \delta_0$ ;
**for** $t \leftarrow 1$ **to** *NumIters* **do**
    Sample rays $\mathbf{r} \in \mathcal{R}$ and ground truth $\mathbf{C}(\mathbf{r})$ ;
    Render $\hat{\mathbf{C}}(\mathbf{r})$ and $\hat{\mathbf{D}}(\mathbf{r})$ under view $\mathbf{P}$;
    Calculate loss $\mathcal{L}_c$ with Eq. 2 ;
    Sample rays $\mathbf{r}_d \in \mathcal{R}_d$ and depth map $\mathbf{D}(\mathbf{r})$ ;
    Render depth value $\hat{\mathbf{D}}(\mathbf{r}_d)$ under view $\mathbf{P}$;
    Calculate loss $\mathcal{L}_d$ with Eq. 3 ;
    $\mathcal{L}_a \leftarrow 0$ ;
    **for** $n \leftarrow 1$ **to** $N$ **do**
        Draw noise vector $\epsilon \sim \mathcal{N}(0, \delta_t)$ ;
        $\mathbf{P}' \leftarrow \mathbf{P} + \epsilon$ ;
        Render $\hat{\mathbf{C}}'(\mathbf{r})$ under view $\mathbf{P}'$ ;
        Warp $\hat{\mathbf{C}}'(\mathbf{r})$ with Eq. 7 ;
        Calculate loss with Eq. 6 and add to $\mathcal{L}_a$;
    **end**
    $\mathcal{L}_a \leftarrow \mathcal{L}_a / N$ ;
    Update averaged loss value $\bar{\mathcal{L}}_c$ and $\bar{\mathcal{L}}_a$;
    Update $\delta_t$ with Eq. 9 ;
    $\mathcal{L} \leftarrow \mathcal{L}_c + \lambda_d \mathcal{L}_d + \lambda_a \mathcal{L}_a$ ;
    Update parameters $\theta \leftarrow \text{Adam}(\theta, \eta, \nabla_\theta \mathcal{L})$ ;
**end**

As for the margin $m$, it's much easier to set than setting $\delta$ directly, since the magnitude of $m$ is strongly related to the loss value of $\mathcal{L}_c$, which is intuitive to find an appropriate magnitude through experiments. Therefore, we set $\delta_0$, $\gamma$ and $m$ to 5e−5, 1.01 and 3e−3 respectively in all the experiments without further tuning. Note that $\bar{\mathcal{L}}_a$ is re-scaled by a factor of 2 in Eq. 9. It is because that we have to align the magnitude of $\bar{\mathcal{L}}_a$ and $\bar{\mathcal{L}}_c$ when compared directly, on account of the fact that smooth L1 loss [9] scales the L2 loss part by 0.5.

### 3.4 Training

During training, we can augment each sample by $N$ times and average the losses of all augmented samples as $\mathcal{L}_a$. For convenience, we set $N = 1$ throughout the paper. The final loss $\mathcal{L}$ is a linear combination of the original NeRF loss $\mathcal{L}_c$, sparse depth loss $\mathcal{L}_d$ and the loss of augmented samples $\mathcal{L}_a$:

$$\mathcal{L} = \mathcal{L}_c + \lambda_d \mathcal{L}_d + \lambda_a \mathcal{L}_a \tag{10}$$

The loss weight $\lambda_d$ and $\lambda_a$ are set to 0.01 and 0.1 throughout the paper. The complete training process is summarized in Algorithm 1, where the gray background marks the sparse depth supervision for DSNeRF. Our GeoAug procedures are marked with green.

## 4 Experiments

For experiments, we first introduce the implementation details and the datasets in Sect. 4.1. We then compare the view synthesis performance with other methods in Sect. 4.2. Finally, we conduct ablation study in Sect. 4.3, including the metric choice of $\mathcal{L}_a$, efficacy of adaptive noise and performance under multiplicity settings.

### 4.1 Settings

**Implementation Details.** Our models are implemented with PyTorch [26]. To improve computational efficiency, we made several modifications to the original NeRF model. 1) Instead of using two-stage MLPs with hierarchical sampling, we only use a single MLP with stratified sampling; 2) The network width is reduced from 256 to 128; 3) Each ray is discretized uniformly into 128 points. We keep using this configuration throughout the paper. Moreover, we also change the random pixel sampling to patch sampling [28], in order to conduct valid warping operation. We also apply patch sampling to the baseline methods for fair comparison.

During training, we sample 4096 rays under a single view for each iteration. The model is trained for 10000 epochs with a learning rate of 0.001, which is exponentially decayed with a rate of 0.9954 every 10 epochs. We use the Adam optimizer [17] with $\beta_1 = 0.9$ and $\beta_2 = 0.999$ for all models. All the experiments are conducted on a single NVIDIA Tesla V100 GPU.

**Dataset and Evaluation Protocol.** NeRF Real-world Data (NeRF Real) [21, 22] is a real-world dataset containing 8 forward-facing scenes. We use the official test split for each scene, *i.e.* test image is sampled every 8-th image. For training images, we randomly sample 2, 5 and 10 views for three different few-shot settings.

DTU MVS Dataset (DTU) [13] is a large-scale multi-view stereo dataset captured in a controlled environment. The complete dataset contains 80 scenes, from which we choose 15 scenes for testing following the configuration of [6]. For each scene, we reduce the image resolution to $400 \times 300$ and randomly sample 3, 6, 9 views for training. We use the ground truth camera poses provided by the dataset and run COLMAP [27] by initializing the camera poses with ground truths for sparse depth information.

To evaluate the synthesis quality, we report PSNR, SSIM [36] and LPIPS [42] calculated against the corresponding ground truth.

**Table 1. View synthesis results on NeRF Real dataset** [21,22]. The numbers are averaged over three few-shot settings, *i.e.* 2-view, 5-view and 10-view. Our GeoAug method effectively improves DSNeRF on all three metrics.

| Scene | PSNR ↑ | | | SSIM ↑ | | | LPIPS ↓ | | |
|---|---|---|---|---|---|---|---|---|---|
| | NeRF | DSNeRF | +GeoAug | NeRF | DSNeRF | +GeoAug | NeRF | DSNeRF | +GeoAug |
| Fern | 18.08 | 18.34 | **19.36** | 0.50 | 0.51 | **0.54** | 0.60 | 0.60 | **0.55** |
| Flower | 18.31 | 19.07 | **19.64** | 0.46 | 0.51 | **0.51** | 0.60 | 0.57 | **0.54** |
| Fortress | 20.24 | 19.86 | **20.33** | 0.54 | 0.53 | **0.57** | 0.59 | 0.59 | **0.54** |
| Horns | **16.23** | 15.42 | 15.76 | **0.44** | 0.41 | 0.41 | 0.65 | 0.67 | **0.64** |
| Leaves | **15.33** | 14.89 | 15.17 | 0.32 | 0.32 | **0.34** | 0.61 | 0.62 | **0.56** |
| Orchids | 14.05 | 14.41 | **14.62** | 0.30 | 0.31 | **0.32** | 0.62 | 0.61 | **0.60** |
| Room | 20.23 | 20.89 | **22.39** | 0.72 | 0.73 | **0.77** | 0.57 | 0.58 | **0.50** |
| Trex | 17.26 | 17.49 | **17.91** | 0.54 | 0.52 | **0.55** | 0.59 | 0.61 | **0.58** |
| Mean | 17.47 | 17.55 | **18.15** | 0.48 | 0.48 | **0.50** | 0.60 | 0.61 | **0.56** |

**Table 2. View synthesis results on DTU dataset** [13]. The numbers are averaged over 15 test scenes following the setting of DSNeRF [6].

| Method | PSNR ↑ | | | SSIM ↑ | | | LPIPS ↓ | | |
|---|---|---|---|---|---|---|---|---|---|
| | 3-view | 6-view | 9-view | 3-view | 6-view | 9-view | 3-view | 6-view | 9-view |
| NeRF [22] | 11.26 | 13.00 | 15.71 | 0.43 | 0.47 | 0.59 | 0.61 | 0.61 | 0.47 |
| DSNeRF [6] | 13.47 | 14.82 | 18.81 | 0.49 | 0.57 | **0.69** | 0.58 | 0.52 | 0.43 |
| DSNeRF +GeoAug | **14.91** | **17.12** | **19.57** | **0.52** | **0.64** | 0.68 | **0.54** | **0.47** | **0.43** |

### 4.2 Benchmark Comparison

**Comparison on NeRF Real.** In this section, we inspect the perceptual quality of novel view synthesis on NeRF Real dataset [21,22]. Three NeRF variants are compared, namely, 1) the basic NeRF described in Sect. 3.1, 2) basic NeRF with sparse depth supervision, denoted as DSNeRF [6], and 3) DSNeRF with our proposed GeoAug method. We average the experiment results under three few-shot settings and present them in Table 1. We can see from Table 1 that DSNeRF improves the mean PSNR of basic NeRF by 0.08 dB, yet did not increase SSIM or LPIPS. One explanation for the limited improvement is that the noisy depth information provided by SfM is less reliable. For instance, 'Fortress', 'Horns' and 'Leaves' are the three scenes with lowest SfM confidence for depth among all scenes. Therefore, DSNeRF performs even worse than the basic NeRF on the three scenes. In contrast, our GeoAug method does not rely on external depth information, thus won't be affected by the noise from SfM. We can see that our GeoAug boosts the performance of DSNeRF in all scenes. The three metrics are consistently better than the basic NeRF and DSNeRF.

We also present qualitative comparison in Fig. 5. Video visualizations are available at bit.ly/3wMX1Sb. When the training images are relatively abundant, *e.g.* under 10-view setting, DSNeRF already renders satisfying images.

**Table 3. Ablation study conducted on the 'fern' scene under 10-view setting. Left:** Different metric choices of augmentation loss $\mathcal{L}_a$. **Right:** GeoAug with different noise level.

| Method | PSNR ↑ | SSIM ↑ | LPIPS ↓ |
|---|---|---|---|
| DSNeRF | 21.27 | 0.61 | 0.52 |
| w. GeoAug, L2 | 22.07 | 0.64 | 0.48 |
| w. GeoAug, L1 | 21.37 | 0.61 | 0.52 |
| w. GeoAug, SmoothL1 | **22.65** | **0.66** | **0.44** |

| Method | PSNR ↑ | SSIM ↑ | LPIPS ↓ |
|---|---|---|---|
| Noise $\delta = 5\mathrm{e}{-5}$ | 22.09 | 0.64 | 0.48 |
| Noise $\delta = 2\mathrm{e}{-4}$ | 21.84 | 0.64 | 0.47 |
| Adaptive Noise | **22.65** | **0.66** | **0.44** |

Therefore, our GeoAug method only brings slight improvement w.r.t. image details like the leaves of fern and telephone wires on the desk. When the number of training images decreases, our GeoAug method shows greater improvement on preserving the overall image structure. Two representative samples are the "fern" and "room" scene in the 2-view setting, where GeoAug preserves the shape structure of fern, TV and long desk, while DSNeRF completely fail to reconstruct. For scenes with extremely complex geometries, *e.g.* stacks of petals and leaves, our method produces clearer basic structures such as flower edges and stems.

**Comparison on DTU.** We present the numerical performance on DTU in Table 2. Different to the conclusion on NeRF Real dataset [21,22] where DSNeRF only brings limited improvement over NeRF, Table 2 shows that DSNeRF consistently improves the basic NeRF on all settings by a large margin. This is mainly due to the better point cloud estimation quality, *i.e.* depth information generated by COLMAP is more reliable since the camera poses are initialized with ground truths. As a result, DSNeRF learns better geometry since the explicit depth supervision is hindered less by noise.

Our GeoAug method does not rely on SfM estimations. Therefore, the performance of GeoAug does not depend on the dataset. Moreover, our GeoAug method provides *dense* depth supervision upon the *sparse* point depth of DSNeRF. We can see in Table 2 that our GeoAug method improves the performance of DSNeRF on DTU, especially under 6 and 3 view settings. The lower block of Fig. 5 shows the qualitative comparisons. GeoAug helps to render clearer local details such as the characters on the bottle and building windows. On the other hand, NeRF and DSNeRF struggle to preserve large-scale structures, *e.g.* buildings under 6 view and 3 view settings. Our method enhances DSNeRF with the ability of inferring better structures, thanks to the additional geometry constraints provided by GeoAug.

### 4.3 Ablation Study

**Smooth L1 Loss.** As discussed in Sect. 3.2, outliers are usually inevitable during the warping process. Unsupervised depth estimation methods like [43] use an explanatory mask for each frame to cast out the outliers. However, since the augmented views in our method are randomly sampled, it is implausible to maintain a mask pool for every training view. Therefore, we choose to leave the outliers and use a robust function, *i.e.* smooth L1 loss [9] to measure the loss of augmented samples. In Table 3 (left), we compare different choices for augmented loss $\mathcal{L}_a$. For L2 loss, the re-scale factor for $\bar{\mathcal{L}}_a$ of adaptive noise is changed from 2 to 1, in order to match the L2 loss of original training samples. For L1 loss, we cannot compare its magnitude directly to L2 loss. Therefore, we remove adaptive noise and set the noise arbitrarily to 5e−5. We can see from Table 3 (left) that using L2 loss for $\mathcal{L}_a$ improves DSNeRF by 0.8 dB w.r.t. PSNR, which means that the outlier problem is not strong enough to counteract the effectiveness of our GeoAug method. Furthermore, replace L2 loss with smooth L1 loss brings an additional 0.58 dB improvement on PSNR. Therefore, smooth L1 loss is a better choice than L2 loss for handling warping outliers. Besides, we also tried to use standard L1 loss as $\mathcal{L}_a$. It only brings minor improvement to PSNR, while SSIM and LPIPS are not better. We assume it is due to its over-tolerance to large errors and the lack of adaptive noise.

**Adaptive Noise.** Our adaptive noise chooses the suitable noise standard deviation $\delta$ automatically for our GeoAug method, reduces the need for exhaustive parameter tuning. In Fig. 3, we demonstrate how adaptive noise works during the training process. As expected, we can see that the augmentation loss $2\mathcal{L}_a$ is kept above the reconstruction loss $\mathcal{L}_c$ with a reasonable margin. It ensures that the augmented samples are neither too easy nor too noisy. The noise standard deviation $\delta$ is initially set to 5e−5, which quickly converges to the range between 1.8e−4 and 2.2e−4.

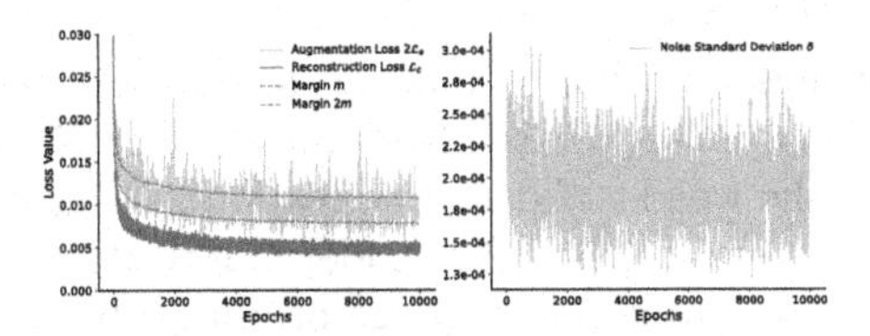


**Fig. 3. Inspection on adaptive noise. Left:** Reconstruction loss $\mathcal{L}_c$ and Augmentation loss $2\mathcal{L}_a$. **Right:** Magnitude of the noise standard deviation $\delta$. Adaptive noise is designed to ensure that $2\bar{\mathcal{L}}_a$ is roughly between $\bar{\mathcal{L}}_c + m$ and $\bar{\mathcal{L}}_c + 2m$.

We also conduct an ablative experiment to show the difference between our adaptive noise and setting a global fixed noise. The results are gathered in Table 3 (right). We can see that arbitrarily setting $\delta$ to the initial value 5e−5 or the converged mean value 2e−4 is not optimal. Both of their performance are inferior to our adaptive noise strategy. Therefore, our adaptive noise not only reduces the need for parameter tuning, but also improves the performance of GeoAug.

**Multiplicity Setting.** In this section, we investigate our data augmentation method under multiplicity settings where the training images are relatively abundant. In Table 4, we report the synthesis result on the 'fern' scene with all 17 images for training. Different from the results under few-shot settings, the basic NeRF model performs better than DSNeRF given enough training images. This is because NeRF could avoid shape-radiance-ambiguity [41] when the training views are dense, thus inferring the correct geometry and generalizing well to novel views. Under this circumstance, the depth supervision of DSNeRF only

**Table 4. View Synthesis with *all* training images on the 'fern' scene.** Given dense training view, the basic NeRF model avoids shape-radiance-ambiguity and renders high-quality novel views. DSNeRF shows inferior performance since it introduces extra noise through explicit depth supervision. Our GeoAug method harms less to the synthesis quality of NeRF.

| NeRF | PSNR↑ | SSIM ↑ | LPIPS ↓ |
|---|---|---|---|
| NeRF | **23.09** | **0.68** | *0.44* |
| NeRF + GeoAug | *23.03* | **0.68** | **0.42** |
| **DSNeRF** | **PSNR↑** | **SSIM ↑** | **LPIPS ↓** |
| $\lambda_d = 0.001$ | 22.85 | 0.66 | 0.47 |
| $\lambda_d = 0.01$ | 22.84 | 0.66 | 0.47 |
| $\lambda_d = 0.1$ | 22.60 | 0.65 | 0.47 |
| $\lambda_d = 0.01$ + GeoAug | **22.88** | **0.67** | **0.43** |

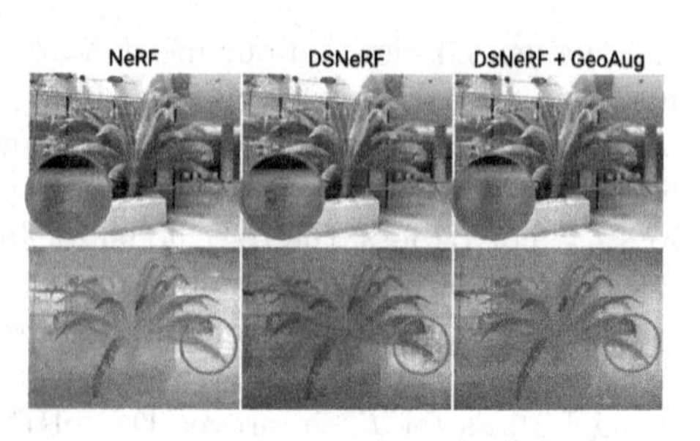


**Fig. 4. View Synthesis with *all* training images on the 'fern' scene.** The depth noise brought by DSNeRF causes stains on the rendering image. Our GeoAug method alleviates the negative effect of depth noise.

brings very limited hint on the geometry information, but instead involves additional noise resulted from imperfect SfM estimation. We can see in Table 4 that as the weight of depth regression loss $\lambda_d$ increases, the performance of DSNeRF gets worse, since the model is forced to fit the noise. Figure 4 also shows that the depth map and image produced by DSNeRF show more stain-like artifacts.

Similarly, the performance improvement of our GeoAug method is also diluted as more training views are available. However, our method does not rely on external depth information, thus free from SfM noises. When applied in companion with the basic NeRF model, GeoAug won't bring too much negative effect. The PSNR is only 0.06 dB lower, while the LPIPS is better than NeRF by 0.02. Meanwhile, GeoAug could also compensate the degradation of DSNeRF both quantitatively (Table 4) and qualitatively (Fig. 4).

## 5 Conclusion

In this paper, we present GeoAug: a data augmentation method for Neural Radiance Fields which alleviates the over-fitting problem under few-shot settings. During training, camera poses of random novel views are generated with an adaptive noise method, which are later used as inputs for the NeRF model. For each novel pose, the output rendering is warped to a nearby intrinsic training view and supervised by the corresponding ground truth image. In this way, our method enriches training data by leveraging geometry constraints through the warping operation, thereby posing implicit supervision on the rendered depth map. Experiments shows the effectiveness of GeoAug on improving the rendering quality for NeRF.

Despite good performance on standard datasets, there are challenges yet to be explored: **1)** The warping outliers caused by camera movement and scene occlusions are not handled explicitly. This is the main reason why GeoAug

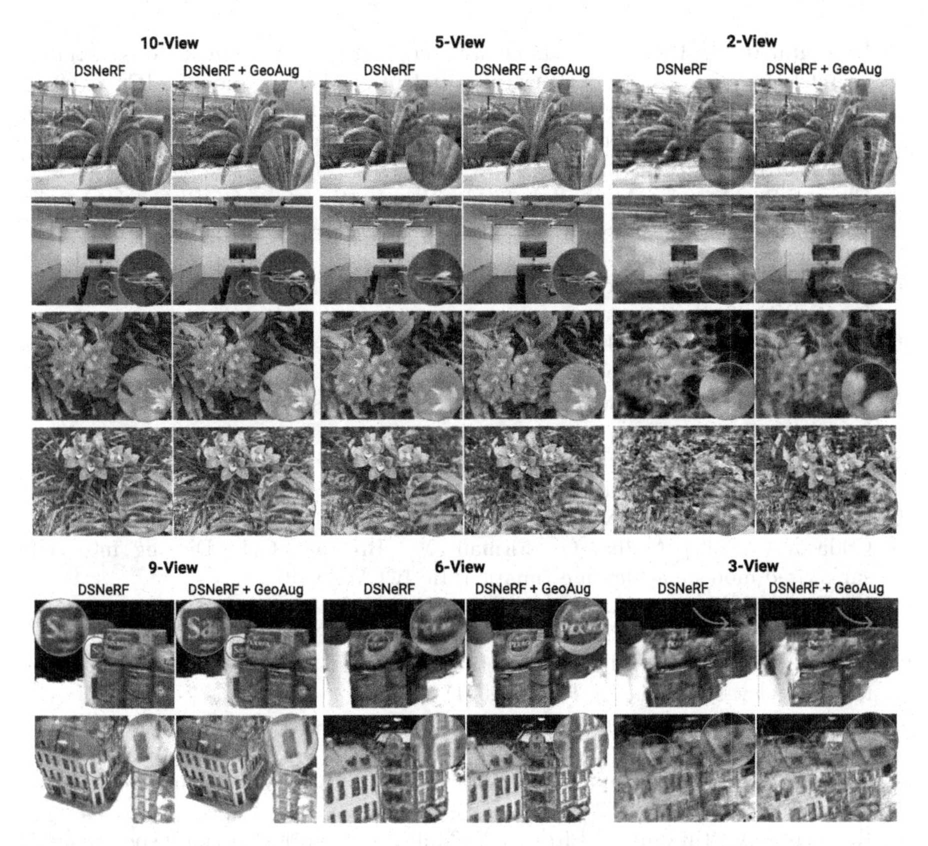


**Fig. 5. Qualitative results on NeRF Real (upper block) and DTU (lower block) datasets.** RGB images are rendered under different few-shot settings. Our GeoAug method helps to preserve image structure and retain more details.

won't improve NeRF under multiplicity settings, *i.e.* given dense training views. Although this problem is bypassed with a robust loss function, we believe that more improvement could be harvested if outliers could be managed properly. **2)** NeRF models and our GeoAug method assume that the camera pose for each training view is already known. However, under few-shot settings where the camera poses are so diverse that even SfM fails to estimate the camera pose and sparse depth, it is unlikely for NeRF models to fit training views or synthesis novel views. Therefore, extending GeoAug for NeRF models to the general purpose of multi-view stereo and structure-from-motion would be an interesting direction for future work.

## References

1. Boss, M., Braun, R., Jampani, V., Barron, J.T., Liu, C., Lensch, H.: NeRD: neural reflectance decomposition from image collections. In: ICCV (2021)

2. Bozorgtabar, B., Rad, M.S., Mahapatra, D., Thiran, J.P.: SynDeMo: synergistic deep feature alignment for joint learning of depth and ego-motion. In: ICCV (2019)
3. Casser, V., Pirk, S., Mahjourian, R., Angelova, A.: Depth prediction without the sensors: leveraging structure for unsupervised learning from monocular videos. In: AAAI (2019)
4. Chan, E.R., Monteiro, M., Kellnhofer, P., Wu, J., Wetzstein, G.: PI-GAN: periodic implicit generative adversarial networks for 3D-aware image synthesis. In: CVPR (2021)
5. Choi, I., Gallo, O., Troccoli, A., Kim, M.H., Kautz, J.: Extreme view synthesis. In: CVPR (2019)
6. Deng, K., Liu, A., Zhu, J.Y., Ramanan, D.: Depth-supervised NeRF: fewer views and faster training for free. arXiv:2107.02791 (2021)
7. Flynn, J., et al.: DeepView: high-quality view synthesis by learned gradient descent. In: CVPR (2019)
8. Flynn, J., Neulander, I., Philbin, J., Snavely, N.: DeepStereo: learning to predict new views from the world's imagery. In: CVPR (2016)
9. Girshick, R.: Fast R-CNN. In: ICCV (2015)
10. Godard, C., Mac Aodha, O., Firman, M., Brostow, G.J.: Digging into self-supervised monocular depth estimation. In: ICCV (2019)
11. Hu, R., Ravi, N., Berg, A.C., Pathak, D.: WorldSheet: wrapping the world in a 3D sheet for view synthesis from a single image. In: ICCV (2021)
12. Jain, A., Tancik, M., Abbeel, P.: Putting NeRF on a diet: semantically consistent few-shot view synthesis. In: ICCV (2021)
13. Jensen, R., Dahl, A., Vogiatzis, G., Tola, E., Aanæs, H.: Large scale multi-view stereopsis evaluation. In: CVPR (2014)
14. Jiakai, Z., et al.: Editable free-viewpoint video using a layered neural representation. In: SIGGRAPH (2021)
15. Kanazawa, A., Tulsiani, S., Efros, A.A., Malik, J.: Learning category-specific mesh reconstruction from image collections. In: ECCV (2018)
16. Karras, T., Aittala, M., Hellsten, J., Laine, S., Lehtinen, J., Aila, T.: Training generative adversarial networks with limited data. In: NeurIPS (2020)
17. Kingma, D.P., Ba, J.: Adam: a method for stochastic optimization. arXiv:1412.6980 (2014)
18. Li, Z., Niklaus, S., Snavely, N., Wang, O.: Neural scene flow fields for space-time view synthesis of dynamic scenes. In: CVPR (2021)
19. Liu, S., Zhang, X., Zhang, Z., Zhang, R., Zhu, J.Y., Russell, B.: Editing conditional radiance fields. arXiv:2105.06466 (2021)
20. Lombardi, S., Simon, T., Saragih, J., Schwartz, G., Lehrmann, A., Sheikh, Y.: Neural volumes: learning dynamic renderable volumes from images. In: SIGGRAPH (2019)
21. Mildenhall, B., et al.: Local light field fusion: practical view synthesis with prescriptive sampling guidelines. ACM TOG **38**, 1–4 (2019)
22. Mildenhall, B., Srinivasan, P.P., Tancik, M., Barron, J.T., Ramamoorthi, R., Ng, R.: NeRF: representing scenes as neural radiance fields for view synthesis. In: ECCV (2020)
23. Niemeyer, M., Barron, J.T., Mildenhall, B., Sajjadi, M.S., Geiger, A., Radwan, N.: RegNeRF: regularizing neural radiance fields for view synthesis from sparse inputs. In: CVPR (2022)
24. Niemeyer, M., Geiger, A.: GIRAFFE: representing scenes as compositional generative neural feature fields. In: CVPR (2021)

25. Niemeyer, M., Mescheder, L., Oechsle, M., Geiger, A.: Differentiable volumetric rendering: learning implicit 3D representations without 3D supervision. In: CVPR (2020)
26. Paszke, A., et al.: Pytorch: an imperative style, high-performance deep learning library. In: NeurIPS (2019)
27. Schonberger, J.L., Frahm, J.M.: Structure-from-motion revisited. In: CVPR (2016)
28. Schwarz, K., Liao, Y., Niemeyer, M., Geiger, A.: GRAF: generative radiance fields for 3D-aware image synthesis. In: NeurIPS (2020)
29. Shih, M.L., Su, S.Y., Kopf, J., Huang, J.B.: 3D photography using context-aware layered depth inpainting. In: CVPR (2020)
30. Sitzmann, V., Thies, J., Heide, F., Nießner, M., Wetzstein, G., Zollhofer, M.: DeepVoxels: learning persistent 3D feature embeddings. In: CVPR (2019)
31. Sitzmann, V., Zollhöfer, M., Wetzstein, G.: Scene representation networks: continuous 3D-structure-aware neural scene representations. In: NeurIPS (2019)
32. Srinivasan, P.P., Deng, B., Zhang, X., Tancik, M., Mildenhall, B., Barron, J.T.: NeRV: neural reflectance and visibility fields for relighting and view synthesis. In: CVPR (2021)
33. Tancik, M., et al.: Learned initializations for optimizing coordinate-based neural representations. In: CVPR (2021)
34. Tucker, R., Snavely, N.: Single-view view synthesis with multiplane images. In: CVPR (2020)
35. Wang, Q., et al.: IBRNet: learning multi-view image-based rendering. In: CVPR (2021)
36. Wang, Z., Bovik, A.C., Sheikh, H.R., Simoncelli, E.P.: Image quality assessment: from error visibility to structural similarity. IEEE TIP **13**, 600–612 (2004)
37. Wiles, O., Gkioxari, G., Szeliski, R., Johnson, J.: SynSin: end-to-end view synthesis from a single image. In: CVPR (2020)
38. Xian, W., Huang, J.B., Kopf, J., Kim, C.: Space-time neural irradiance fields for free-viewpoint video. In: CVPR (2021)
39. Yariv, L., et al.: Multiview neural surface reconstruction by disentangling geometry and appearance. In: NeurIPS (2020)
40. Yu, A., Ye, V., Tancik, M., Kanazawa, A.: pixelNeRF: neural radiance fields from one or few images. In: CVPR (2021)
41. Zhang, K., Riegler, G., Snavely, N., Koltun, V.: NeRF++: analyzing and improving neural radiance fields. arXiv:2010.07492 (2020)
42. Zhang, R., Isola, P., Efros, A.A., Shechtman, E., Wang, O.: The unreasonable effectiveness of deep features as a perceptual metric. In: CVPR (2018)
43. Zhou, T., Brown, M., Snavely, N., Lowe, D.G.: Unsupervised learning of depth and ego-motion from video. In: CVPR (2017)
44. Zhou, T., Tucker, R., Flynn, J., Fyffe, G., Snavely, N.: Stereo magnification: learning view synthesis using multiplane images. In: SIGGRAPH (2018)

# DoodleFormer: Creative Sketch Drawing with Transformers

Ankan Kumar Bhunia[1(✉)], Salman Khan[1,2], Hisham Cholakkal[1], Rao Muhammad Anwer[1,3], Fahad Shahbaz Khan[1,4], Jorma Laaksonen[3], and Michael Felsberg[4]

[1] Mohamed bin Zayed University of AI, Abu Dhabi, UAE
ankan.bhunia@mbzuai.ac.ae
[2] Australian National University, Canberra, Australia
[3] Aalto University, Espoo, Finland
[4] Linköping University, Linköping, Sweden

**Abstract.** Creative sketching or doodling is an expressive activity, where imaginative and previously unseen depictions of everyday visual objects are drawn. Creative sketch image generation is a challenging vision problem, where the task is to generate diverse, yet realistic creative sketches possessing the unseen composition of the visual-world objects. Here, we propose a novel coarse-to-fine two-stage framework, DoodleFormer, that decomposes the creative sketch generation problem into the creation of coarse sketch composition followed by the incorporation of fine-details in the sketch. We introduce graph-aware transformer encoders that effectively capture global dynamic as well as local static structural relations among different body parts. To ensure diversity of the generated creative sketches, we introduce a probabilistic coarse sketch decoder that explicitly models the variations of each sketch body part to be drawn. Experiments are performed on two creative sketch datasets: Creative Birds and Creative Creatures. Our qualitative, quantitative and human-based evaluations show that DoodleFormer outperforms the state-of-the-art on both datasets, yielding realistic and diverse creative sketches. On Creative Creatures, DoodleFormer achieves an absolute gain of 25 in Frèchet inception distance (FID) over state-of-the-art. We also demonstrate the effectiveness of DoodleFormer for related applications of text to creative sketch generation, sketch completion and house layout generation. Code is available at: https://github.com/ankanbhunia/doodleformer.

## 1 Introduction

Humans have an outstanding ability to easily communicate and express abstract ideas and emotions through sketch drawings. Generally, a sketch comprises several strokes, where each stroke can be considered as a group of points. In automatic sketch image generation, the objective is to generate recognizable sketches

**Supplementary Information** The online version contains supplementary material available at https://doi.org/10.1007/978-3-031-19790-1_21.

© The Author(s), under exclusive license to Springer Nature Switzerland AG 2022
S. Avidan et al. (Eds.): ECCV 2022, LNCS 13677, pp. 338–355, 2022.
https://doi.org/10.1007/978-3-031-19790-1_21

**Fig. 1.** Examples of visual creative sketches generated using the proposed DoodleFormer. Here, we show creative sketches generated based on (a) the random input strokes, (b) text inputs and (c) incomplete sketch images provided by the user. In all three scenarios, the generated sketches are well aligned with the user provided inputs (*e.g.*, the creative sketches generated for the text inputs "walking forward" and "fly up high" in (b)). Similarly, the diversity in terms of appearance, posture and part size can be observed within the generated creative bird sketches in (a). Furthermore, DoodleFormer accurately completes the missing bird wings, legs and beak in the bottom right example in (c). Additional examples are available in Fig. 2, Fig. 4, Fig. 6 and the supplementary.

**Fig. 2.** A visual comparison of creative sketch images generated by DoodlerGAN [10] (top row) and the proposed DoodleFormer (bottom row) for the same initial random input strokes. We show examples from both Creative Birds (a) and Creative Creatures (b) datasets. DoodlerGAN suffers from topological artefacts (*e.g.*, more than one head like region in the third bird sketch from the left), disconnected body parts (*e.g.*, the fifth sketch from the left in creatures). Further, the DoodlerGAN generated creative sketches have lesser diversity in terms of size, appearance and posture. The proposed DoodleFormer alleviates the issues of topological artefacts and disconnected body parts, generating creative sketches that are more realistic and diverse.

that are closely related to the real-world visual concepts. Here, the focus is to learn more canonical and mundane interpretations of everyday objects.

Different from the standard sketch generation problem discussed above, *creative* sketch generation [10] involves drawing more imaginative and previously unseen depictions of everyday visual concepts (see Fig. 1(a)). In this problem, creative sketches are generated according to externally provided random input strokes. Example of creative sketch generation includes doodling activity, where diverse, yet recognizable sketch images are generated through unseen composition of everyday visual concepts. Automatic generation of creative sketches can largely assist human creative process *e.g.*, inspiring further ideas by providing a possible interpretation of initial sketches by the user. However, such a creative task is more challenging compared to mimicking real-world scenes in to sketch images. This work investigates the problem of creative sketch generation.

Recently, Ge *et al.* [10] address the creative sketch image generation problem by proposing a part-based Generative Adversarial Network called DoodlerGAN. It utilizes a part-specific generator to produce each body part of the sketch. The generated body parts are then sequentially integrated with the externally provided random input, for obtaining final sketch image. Although DoodlerGAN utilizes a part-specific generator for creating each body part of the sketch, it does not comprise an explicit mechanism to ensure that each body part is placed appropriately with respect to the rest of the parts. This leads to topological artifacts and connectivity issues (see Fig. 2). Further, DoodlerGAN struggles to generate diverse sketch images, which is an especially desired property in creative sketch generation.

In this work, we argue that the aforementioned problems of topological artefacts, connectivity and diversity issues can be alleviated by imitating the natural *coarse-to-fine* creative sketch drawing process, where the artist first draws the holistic coarse structure of the sketch and then fills the fine-details to generate the final sketch. By first drawing the holistic coarse structure of the sketch aids to appropriately decide the location and the size of each sketch body part to be drawn. To imitate such a coarse-to-fine creative sketch generation process, we look into a two-stage framework where the global as well as local structural relations among different body parts can be first captured at a coarse-level followed by obtaining the fine-level sketch. The coarse-to-fine framework is expected to further improve the diversity of the creative sketch images by explicitly modeling the variations in the location and size of each sketch body part to be drawn.

## 1.1 Contributions

We propose a novel two-stage encoder-decoder framework, DoodleFormer, for creative sketch generation. DoodleFormer decomposes the creative sketch generation problem into the construction of holistic coarse sketch composition followed by injecting fine-details to generate final sketch image. To generate realistic sketch images, we introduce graph-aware transformer (GAT) encoders that effectively encode the local structural relations between different sketch body parts by integrating a static adjacency-based graph into the dynamic self-attention block. We further introduce a probabilistic coarse sketch decoder that utilizes Gaussian mixture models (GMMs) to obtain diverse locations of each body part, thereby improving the diversity of output sketches (see Fig. 2).

We evaluate the proposed DoodleFormer by conducting extensive qualitative, quantitative and human-based evaluations on the recently introduced Creative Birds and Creative Creatures datasets. Our DoodleFormer performs favorably against DoodlerGAN on all three evaluations. For instance, DoodleFormer sketches were interpreted to be drawn by a human 86%, having better strokes integration 85% and being more creative 82%, over DoodlerGAN in terms of human-based evaluation. Further, DoodleFormer outperforms DoodlerGAN with absolute gains of 25 and 23 in terms of Frèchet inception distance (FID) on Creative Creatures and Creative Birds, respectively. In addition to sketch generation based on externally provided random initial strokes, we validate the effectiveness of DoodleFormer to generate creative sketches based on text inputs, incomplete sketch images provided by user as well as generating complete house layouts given coarse-level bubble diagrams. DoodleFormer achieves impressive performance for text to sketch generation, sketch completion (see Fig. 1(b) and (c)) as well as house layout generation (see Fig. 8).

## 2 Related Work

The problem of sketch generation [5,7,12,16,20,37] has been studied extensively in literature. These methods generally aim to mimic the visual world by capturing its important aspects in the generated sketches. SketchRNN [12], utilizes sequence-to-sequence Variational Autoencoder (VAE) for conditional and unconditional generation of vector sketches. Cao *et al.* [5] propose a generative model that generates multi-class sketches. Moreover, alternative strategies such as differentiable rendering [37], attention-based architectures [27] and reinforcement learning [3,9,38] have been investigated for sketch generation. The work of [7] incorporates a convolutional encoder to capture the spatial layout of sketches, whereas [23,25,29] aim at completing the missing parts of sketches. The work of [21] targets recovering the masked parts of points in sketches. A few works [21,27] have also studied related tasks of sketch classification and retrieval.

Different from the aforementioned standard sketch generation task, creative sketch generation has been recently explored [10]. This task focuses on drawing more imaginative and previously unseen depictions of common visual concepts rather than generating canonical and mundane interpretations of visual objects. To this end, DoodlerGAN [10] introduces a part-based Generative Adversarial Network built on StyleGAN2 [17] to sequentially produce each body part of the creative sketch image. Here, the part-based GAN model needs to be trained separately for individual body parts (eye, head, beak, *etc.*) using part annotations. However, such a separate model for each body part results in a large computational overhead. During inference, these individual part-based GAN models are sequentially used to generate their respective body parts within the creative sketch. While generating recognizable creative sketches, DoodlerGANs struggles with topological artifacts, connectivity and diversity issues. In this work, we set out to overcome these issues to generate diverse, yet realistic creative sketches.

## 3 Our Approach

**Motivation:** To motivate our framework, we first distinguish two desirable properties to be considered when designing an approach for creative sketch generation.

***Holistic Sketch Part Composition:*** As discussed earlier, DoodlerGAN employs a part-specific generator to produce each body part of the sketch. However, it does not utilize any explicit mechanism to ensure that the generated part is placed in an appropriate location relative to other parts, thereby suffering from topological artifacts and connectivity issues (see Fig. 2). Here, we argue that explicitly capturing the holistic arrangement of the sketch parts is desired to generate realistic sketch images that avoid topological artifacts and connectivity issues.

***Fine-Level Diverse Sketch Generation:*** Creative sketches exhibit a large diversity in appearance, posing a major challenge when generating diverse, yet realistic fine-detailed sketch images. Existing work of DoodlerGAN struggles to generate diverse sketch images since it typically ignores the noise input in the sketch generation process [26]. Although DoodlerGAN attempts to partially address this issue by introducing heuristics in the form of randomly translating the input partial sketch, the diversity of generated sketch images is still far from satisfactory (see Fig. 2). Instead, we argue that having an explicit probabilistic modeling within the framework is expected to further improve the diversity of the generated sketch images.

**Overall Framework:** The proposed two-stage DoodleFormer framework combines the two aforementioned desired properties by decomposing the creative sketch generation problem to first capture the holistic coarse sketch composition and then injecting fine-details to generate the final sketch. The overall architecture of the proposed two-stage DoodleFormer is shown in Fig. 3. DoodleFormer comprises two stages: *Part Locator* (PL-Net) and *Part Sketcher* (PS-Net). The first stage, Part Locator (PL-Net), learns to explicitly capture the holistic arrangement of the sketch parts conditioned on the externally provided random initial stroke points $\mathcal{C}$ represented in a vector form. PL-Net comprises

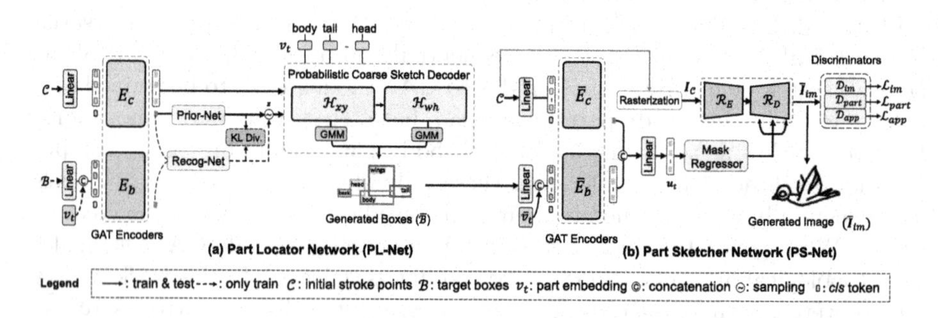


**Fig. 3.** The proposed DoodleFormer comprises two stages: Part Locator (PL-Net) and Part Sketcher (PS-Net). (a) The first stage, PL-Net, takes the initial stroke points $\mathcal{C}$ as the conditional input and learns to return the bounding boxes corresponding to each body part (coarse structure of the sketch) to be drawn. PL-Net contains two graph-aware transformer (GAT) encoders ($E_b$, $E_c$) and a probablistic coarse sketch decoder utilizing GMM modelling for the coarse box prediction. Within the decoder, the bounding box parameters are predicted by the location-predictor ($\mathcal{H}_{xy}$) and size-predictor ($\mathcal{H}_{wh}$) modules. (b) The second stage, PS-Net, then takes the predicted box locations along with $\mathcal{C}$ as inputs and generates the final sketch image $\bar{I}_{im}$. Following the design of $E_b$ and $E_c$, PS-Net also comprises GAT block-based encoders ($\bar{E}_b$, $\bar{E}_c$). Further, PS-Net contains a convolutional encoder-decoder network ($\mathcal{R}_E$, $\mathcal{R}_D$) and a mask regressor to generate rasterized high quality sketch image $\bar{I}_{im}$.

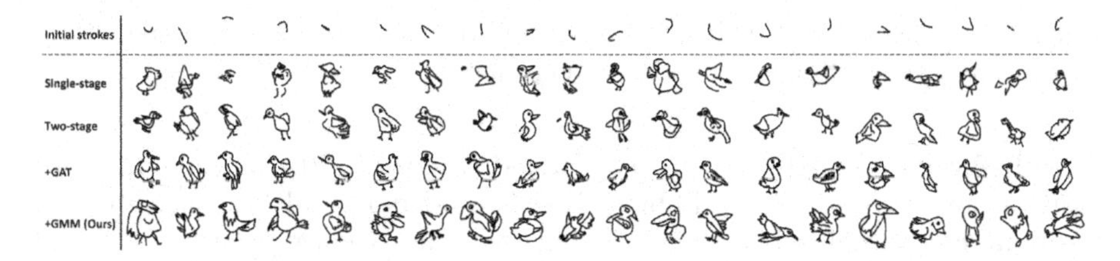


**Fig. 4.** A visual comparison in terms of progressively integrating one contribution at the time, from top to bottom, for common initial strokes. Compared to the single-stage baseline (first row), the two-stage framework (without the GAT block and probabilistic modeling in the decoder) generates sketch in a coarse-to-fine manner. As a result, the two-stage framework (second row) produces a more complete sketch where each body part is placed at an appropriate location relative to other parts. The introduction of GAT block (third row) in the encoders of the two-stage framework improves the realism of the generated sketches by capturing the structural relationship between different parts (*e.g.*, the tenth image from the left, where there is a discontinuity between the beak and the head of the bird). Further, the introduction of probabilistic modelling in the decoder of the two-stage framework (last row), improves the diversity (*e.g.*, appearance, size, orientation and posture) of generated sketch images. Our final two-stage framework (last row) produces realistic and diverse sketch images.

graph-aware transformer (GAT) block-based encoders to capture structural relationship between different regions within a sketch. To the best of our knowledge, we are the first to introduce a GAT block-based transformer encoder for the problem of creative sketch image generation. Instead of directly predicting the box parameters as deterministic points from the transformer decoder, we further introduce probabilistic coarse sketch decoders that utilize GMM modelling for box prediction. This enables our DoodleFormer to achieve diverse, yet plausible coarse structure (bounding boxes) for sketch generation. The second stage, Part Sketcher (PS-Net), creates the final sketch image with appropriate line segments based on the coarse structure obtained from PL-Net. PS-Net also comprises GAT block-based encoders, as in PL-Net, along with a convolutional encoder-decoder network to generate the final rasterized sketch image.

Our carefully designed two-stage DoodleFormer architecture possesses both desired properties (holistic sketch part composition as well as fine-level diverse sketch generation) and creates diverse, yet realistic sketch images in a coarse-to-fine manner (see Fig. 4). Next, we describe in detail PL-Net (Sect. 3.1) and PS-Net (Sect. 3.2).

### 3.1 Part Locator Network (PL-Net)

As discussed above, PL-Net takes the initial stroke points $\mathcal{C}$ as the conditional input, and learns to return a coarse structure capturing the holistic part composition of the desired sketch. The *encoders* in PL-Net contain graph-aware transformer (GAT) blocks to encode the structural relationship between different parts (holistic sketch part composition), leading to realistic sketch image generation. The *decoder* in PL-Net utilizes GMM modeling for box prediction, enabling the generations of diverse sketch images.

**Graph-Aware Transformer Block-Based Encoder.** PL-Net consists of two graph-aware transformer (GAT) block-based encoders $E_b$ and $E_c$, which are used to obtain contextualized representation of the coarse (holistic) structure $\mathcal{B}$ and the conditional input $\mathcal{C}$, respectively.

To encode the identity $t$ of each body part present in a sketch, we define $\boldsymbol{v}_t \in \mathbb{R}^d$ as a learned part embedding. We concatenate $\boldsymbol{v}_t$ with a feature representation obtained from $\boldsymbol{b}_t \in \mathcal{B}$ (box location and size information $(x_t, y_t, w_t, h_t)$ of each body part). This concatenated feature is then used as an input to the encoder $E_b$. The conditional input strokes $\mathcal{C}$ are passed through a linear layer before being input to the encoder $E_c$. We add special *cls* tokens [8] at the beginning of input sequences to the encoders ($E_b$ and $E_c$). The output of this token is considered as the contextualized representation of the whole sequence. Further, we use fixed positional encodings to the input of each attention layer to retain information regarding the sequence order. Next, we introduce our GAT block used in both encoders $(E_b, E_c)$ to encode the holistic structural composition of (sketch) body parts.

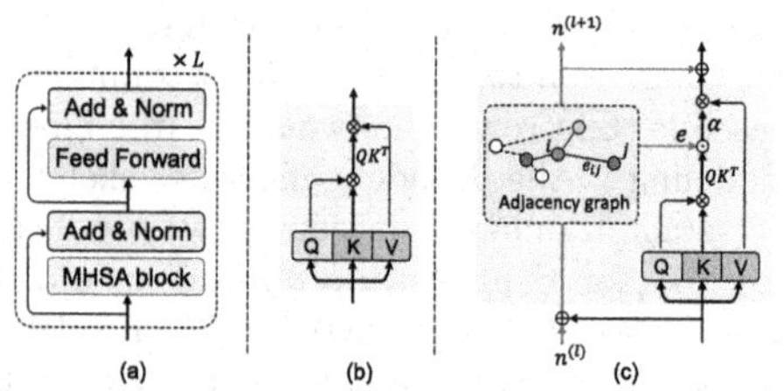


**Fig. 5.** Our proposed graph-aware transformer (GAT) block (c) replaces the standard self-attention (b) in the conventional transformer encoder layer (a). Our GAT block injects the graph structure into self-attention by learning to re-weight the attention matrix based on the pair-wise relations between the graph nodes. In this way, the proposed GAT block combines the local connectivity patterns from the learned adjacency graph with the dynamic attention from the self-attention block.

**Graph-Aware Transformer (GAT) Block:** The structure of our GAT block is shown in Fig. 5(c). Each GAT block consists of a graph-aware multi-headed self-attention (MHSA) module followed by a feed-forward network [18]. Given the queries $\boldsymbol{Q}$, keys $\boldsymbol{K}$, and values $\boldsymbol{V}$ the standard self-attention module [32] computes the attention according to the following equation (also shown in Fig. 5(b)),

$$\alpha = \text{softmax}\left(\frac{\boldsymbol{Q}\boldsymbol{K}^T}{\sqrt{d}}\right). \tag{1}$$

While the standard self-attention module is effective towards learning highly contextualized feature representation, it does not explicitly emphasize on the local structural relation. However, creative sketches are structured inputs with definite connectivity patterns between sketch parts. To model this structure, we propose to encode an adjacency based graph implemented with spectral graph convolution [19]. Our proposed GAT block combines the definite connectivity patterns from the learned adjacency graph with the dynamic attention from self-attention block. Let us consider a graph where each node $i$ is associated with a structure-aware representation $\boldsymbol{n}_i$ and corresponding neighbour set $\mathcal{N}_r(i)$. To represent the neighbor set $\mathcal{N}_r(i)$ for each node $i$, we define an adjacency matrix $\boldsymbol{A}$ where each entry represents whether two nodes $i$ and $j$ are adjacent. The edge weight $e_{ij}$ between two adjacent nodes $i$ and $j$ is given by,

$$e_{ij} = \boldsymbol{W}_b^T \mathrm{ReLU}\left(\boldsymbol{W}_a \left[\boldsymbol{n}_i, \boldsymbol{n}_j\right]\right) \ \forall j \in \mathcal{N}_r(i), \tag{2}$$

where $\boldsymbol{W}_a$ and $\boldsymbol{W}_b$ are learned parameters and $[\cdot, \cdot]$ is a concatenation operator. We set $e_{ij} = 0 \ \forall j \notin \mathcal{N}_r(i)$. For each GAT block $l$, the spectral graph convolution operation is,

$$\boldsymbol{n}_i^{(l+1)} = \mathrm{ReLU}\left(\boldsymbol{n}_i^{(l)} + \sum_{j \in \mathcal{N}_r(i)} e_{ij} \boldsymbol{W}_c \boldsymbol{n}_j^{(l)}\right), \tag{3}$$

where $\boldsymbol{W}_c$ is a learned matrix. Our main intuition is that the adjacency matrix representing the neighbourhood graph structure is static which is computed over the connected components in the graph and predetermined for each input, it is also symmetric and generally sparse. In contrast, attention learned from the self-attention layer is dynamic, can be dense and also non-symmetric. We propose to combine these two complementary representations through the following equation where we calculate the attention weight $\alpha_{ij}$ for nodes $j \in \mathcal{N}_r(i)$ as follows,

$$\alpha_{ij} = \frac{e_{ij} \exp(\varphi_{ij})}{\sum_{j \in \mathcal{N}_r(i)} e_{ij} \exp(\varphi_{ij})}, \ \text{s.t.} \ \varphi_{ij} \in \frac{\boldsymbol{Q}\boldsymbol{K}^T}{\sqrt{d}}, \tag{4}$$

where $\varphi_{ij}$ is an element of the standard attention matrix.

The special token ($cls$) output from $E_c$ is then utilized as an input to a Prior-Net for approximating the conditional prior latent distribution. Similarly, the $cls$ token outputs of both $E_b$ and $E_c$ are provided as input to a Recog-Net for approximating the variational latent distribution. Both the Prior-Net and the Recog-Net are parameterized by multi-layer perceptrons (MLPs) to approximate prior and variational latent normal distributions. During training, we sample the latent variable $\boldsymbol{z}$ from the variational distribution and provide it as input to the probabilistic coarse sketch decoder.

**Probabilistic Coarse Sketch Decoder.** The probabilistic coarse sketch decoder within our PL-Net utilizes probabilistic modelling to generate diverse coarse structure. The decoder comprises two modules: a location-predictor $\mathcal{H}_{xy}$ and a size-predictor $\mathcal{H}_{wh}$. Here, the location-predictor $\mathcal{H}_{xy}$ estimates the center coordinates $(x_t, y_t)$ of bounding boxes around body parts, while the size-predictor $\mathcal{H}_{wh}$ predicts their width and height $(w_t, h_t)$. Both these modules consist of multi-headed self- and encoder-decoder attention mechanisms [32]. The encoder-decoder attention obtains the key and value vectors from the output of the encoder $E_c$. This allows every position in the decoder to attend to all positions in the conditional input sequence. The part embedding $\boldsymbol{v}_t$ from the encoder is used as a query positional encoding to each attention layer of the decoder. Over multiple consecutive decoding layers, the decoder modules produce respective output features $\boldsymbol{f}_t^{xy} \in \mathbb{R}^d$ and $\boldsymbol{f}_t^{wh} \in \mathbb{R}^d$ that lead to the distribution parameters of bounding boxes being associated with each body part, representing the coarse structure of the final sketch to be generated.

To enhance the diversity of generated sketch images, we model the box predictions from each decoder module by Gaussian Mixture Models (GMMs) [4,11]. Different from the conventional box prediction [6,39] that directly maps the decoder output features as deterministic box parameters, our GMM-based box prediction is modeled with $M$ normal distributions $\mathcal{N}(\cdot)$ where each distribution is parameterized by $\theta_k$ and a mixture weight $\pi_k$,

$$p(\boldsymbol{b}_t|\mathcal{C}, \boldsymbol{z}) = \sum_{k=1}^{M} \pi_{k,t} \mathcal{N}(\boldsymbol{b}_t; \theta_{k,t}), \text{for} \sum_{k=1}^{M} \pi_{k,t} = 1. \tag{5}$$

The GMM parameters can be obtained by minimizing the negative log-likehood for all $P$ body parts in a sketch,

$$\mathcal{L}_b = -\frac{1}{P} \sum_{t=1}^{P} \log \left( \sum_{k=1}^{M} \pi_{k,t} \mathcal{N}(\boldsymbol{b}_t; \theta_{k,t}) \right). \tag{6}$$

Here, we simplify the quadvariate distribution of GMMs in Eq. 6 by decomposing it into two bivariate distributions as $p(\boldsymbol{b}_t|\mathcal{C}, \boldsymbol{z}) = p(x_t, y_t|\mathcal{C}, \boldsymbol{z})p(w_t, h_t|x_t, y_t, \mathcal{C}, \boldsymbol{z})$. The parameters of these bivariate GMMs are obtained by employing linear layers and appropriate normalization on the outputs $f_t^{xy}$, $f_t^{wh}$ of $\mathcal{H}_{xy}$ and $\mathcal{H}_{wh}$, respectively. In addition to GMM parameters, these linear layers also estimate the presence of a body part using an indicator variable, which is trained with a binary cross entropy loss $\mathcal{L}_c$.

**PL-Net Loss Function ($\mathcal{L}_{PL}$):** The overall loss function $\mathcal{L}_{PL}$ to train the PL-Net is the weighted sum of the reconstruction loss $\mathcal{L}_{rec}$, and the KL divergence loss $\mathcal{L}_{KL}$,

$$\mathcal{L}_{PL} = \mathcal{L}_{rec} + \lambda_{KL} \mathcal{L}_{KL}. \tag{7}$$

Here, the reconstruction loss term is $\mathcal{L}_{rec} = \mathcal{L}_b + \mathcal{L}_c$. The KL divergence loss term $\mathcal{L}_{KL}$ regularizes the variational distribution [28] from the Recog-Net to be closer to the prior distribution from the conditional Prior-Net, whereas $\lambda_{KL}$ is a scalar loss weight.

Our carefully designed PL-Net architecture, presented above, provides a coarse structure of the sketch that is used to generate a diverse, yet realistic final sketch image in the second stage (PS-Net) of the proposed two-stage Doodle-Former framework. Next, we present the PS-Net that takes the coarse structure of the sketch along with initial partial sketch $\mathcal{C}$ as inputs and generates the final sketch image.

### 3.2 Part Sketcher Network (PS-Net)

Our PS-Net comprises two graph-aware transformer (GAT) block-based encoders $\bar{E}_b$ and $\bar{E}_c$, following the design of encoders $E_b$ and $E_c$ in the PL-Net. Here, the encoder $\bar{E}_b$ produces a contextualized feature representation of bounding box $\boldsymbol{b}_t$ associated with each body part. Similarly, the encoder $\bar{E}_c$ outputs a contextualized feature representation of initial stroke points $\mathcal{C}$. Both these contextualized feature representations from $\bar{E}_b$ and $\bar{E}_c$ are then concatenated and passed through a linear layer to obtain $\boldsymbol{u}_t$.

The initial stroke points $\mathcal{C}$ is converted to its raster form $\boldsymbol{I}_\mathcal{C}$ and passed through a convolutional encoder $\mathcal{R}_E$ that outputs a spatial representation $\boldsymbol{g} = \mathcal{R}_E(\boldsymbol{I}_\mathcal{C})$. Consequently, $\boldsymbol{g}$ and $\{\boldsymbol{u}_t\}_{t=1}^P$ are provided as input to a convolutional decoder $\mathcal{R}_D$ for generating the final sketch image $\bar{\boldsymbol{I}}_{im}$,

$$\bar{\boldsymbol{I}}_{im} = \mathcal{R}_D\left(\boldsymbol{g}, \{\boldsymbol{u}_t\}_{t=1}^P\right). \tag{8}$$

The decoder network $\mathcal{R}_D$ utilizes the ResNet [13] architecture as a backbone. To introduce diversity in the generated images, a zero-mean unit-variance multivariate random noise is added with $\boldsymbol{g}$ before passing it to the decoder network. For fine-grained shape prediction, we utilize a mask regressor [30,31] having up-sampling convolutions, followed by sigmoid transformation to generate an auxiliary mask for each bounding box. The predicted masks are resized to the sizes of corresponding bounding boxes, which are then used to compute the instance-specific and structure-aware affine transformation parameters in the normalization layer of the decoder $\mathcal{R}_D$.

The training of PS-Net follows the standard GAN formulation where the PS-Net generator $\mathcal{G}$ is followed by additional discriminator networks $\mathcal{D}_{im}$, $\mathcal{D}_{part}$, and $\mathcal{D}_{app}$ to obtain image-level ($\mathcal{L}_{im}$), part-level ($\mathcal{L}_{part}$), and appearance ($\mathcal{L}_{app}$) adversarial losses [14,30], respectively. The loss function is then given by,

$$\mathcal{L}_{PS} = \mathcal{L}_{im} + \lambda_p \mathcal{L}_{part} + \lambda_a \mathcal{L}_{app}, \tag{9}$$

where $\lambda_p$ and $\lambda_a$ are the loss weight hyper-parameters.

The introduction of the GAT block in the PL-Net and PS-Net encoders contributes towards the generation of realistic sketch images, whereas the effective utilization of probabilistic modelling in the PL-Net decoder leads to improved diversity. In summary, our two-stage DoodleFormer generates diverse, yet realistic sketch images (see Fig. 4).

## 4 Experiments

**Datasets:** We perform extensive experiments on the recently introduced Creative Birds and Creative Creatures datasets [10]. The Creative Birds has 8067 sketches of birds, whereas the Creative Creatures contains 9097 sketches of various creatures. In both datasets, all sketches come with part annotations. Both datasets also contain free-form natural language phrase as a text description for each sketch.

**Table 1. Comparison of DoodleFormer with DoodlerGAN** [10], **StyleGAN2** [17] **and SketchRNN** [12] in terms of Frèchet inception distance (FID), generation diversity (GD), characteristic score (CS) and semantic diversity score (SDS). Our DoodleFormer performs favorably against existing methods on both datasets.

| Methods | Creative Birds | | | Creative Creatures | | | |
|---|---|---|---|---|---|---|---|
| | FID(↓) | GD(↑) | CS(↑) | FID(↓) | GD(↑) | CS(↑) | SDS(↑) |
| Training Data | - | 19.40 | 0.45 | - | 18.06 | 0.60 | 1.91 |
| SketchRNN [12] | 82.17 | 17.29 | 0.18 | 54.12 | 16.11 | 0.48 | 1.34 |
| StyleGAN2 [17] | 130.93 | 14.45 | 0.12 | 56.81 | 13.96 | 0.37 | 1.17 |
| DoodlerGAN [10] | 39.95 | 16.33 | **0.69** | 43.94 | 14.57 | 0.55 | 1.45 |
| **DoodleFormer (Ours)** | **16.45** | **18.33** | 0.55 | **18.71** | **16.89** | **0.56** | **1.78** |

**Implementation Details:** As discussed, both PL-Net and PS-Net utilize graph-aware transformer (GAT) block-based encoders. In each GAT-block based encoder, we define an adjacency matrix $\boldsymbol{A}$ based on the connectivity patterns of the adjacency graph. Every pair of overlapping bounding boxes on a coarse structure is connected in the adjacency graph. Similarly, for initial strokes, the corresponding adjacency graph connects adjacent points on each single stroke. Each of these encoders consist of $L = 6$ graph-aware transformer blocks. Here, each block comprises multi-headed attention having 8 heads. In the probabilistic coarse sketch decoder, the location-predictor and size-predictor utilize 3 self- and encoder-decoder attention layers. Further, we set the embedding size $d = 512$. We augment the vector sketch images by applying small affine transformations and these vector sketches are converted to raster images of size $128 \times 128$. Our DoodleFormer is trained as follows. In the first stage, for training PL-Net, we initially obtain the bounding boxes for all body parts in a sketch using the part annotations. The bounding boxes are normalized to values between 0 and 1. In the second stage, we train PS-Net using the raster sketch images and their corresponding ground-truth boxes. In both stages, the initial stroke points are provided in a vector form as a conditional input. In all experiments, we use a batch size of 32. The learning rate is set to $1e^{-4}$ and the loss weights $\lambda_{KL}$, $\lambda_p$, $\lambda_a$ are set to 1, 10 and 10.

### 4.1 Quantitative and Qualitative Comparisons

We first present a comparison (Table 1) of our DoodleFormer with state-of-the-art approaches [10,12,33] on both Creative Birds and Creative Creatures. For a fair comparison, we evaluate all methods using two widely used metrics, namely Frèchet inception distance (FID) [15] and generation diversity (GD) [5], as in DoodlerGAN [10]. Here, the FID and GD scores are computed using an Inception model trained on the QuickDraw3.8M dataset [34], that embeds the images onto a feature space [10]. Table 1 shows that DoodleFormer outperforms existing methods in terms of both FID and GD scores, on the two datasets. The higher GD score indicates the ability of DoodleFormer to generate diverse sketch images,

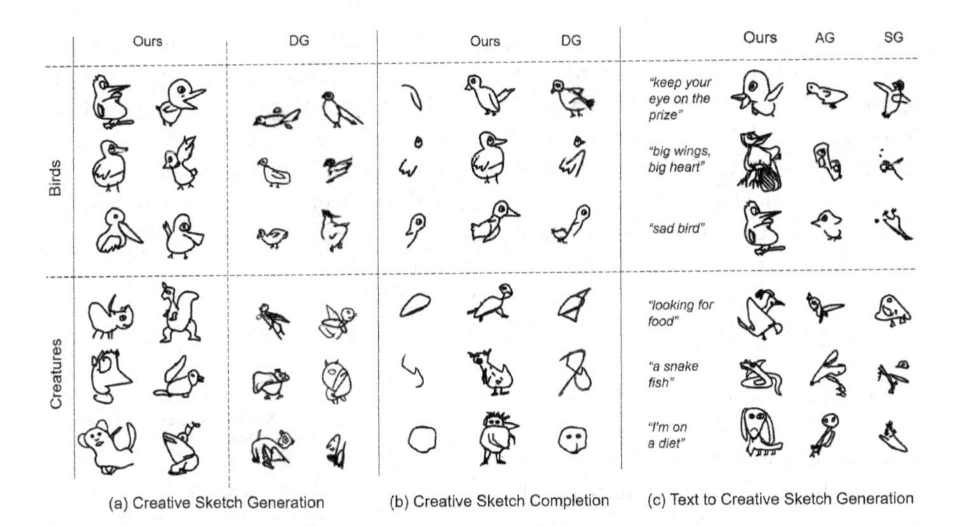


**Fig. 6.** Applications of DoodleFormer. (a) Creative sketch generation based on random input strokes. (b) Creative sketch completion: Here, DoodleFormer accurately completes missing parts (*e.g.*, beak, head and body of bird is well connected in second row and column in (b)), compared to DoodlerGAN (DG). (c) Text to creative sketch generation: We compare DoodleFormer with AttnGAN [35] (AG), StackGAN [36] (SG). DoodleFormer produces sketches that are well aligned with user provided input texts. Best viewed zoomed in.

whereas the lower FID score indicates the superior quality of its generated creative sketches.

Furthermore, similar to DoodlerGAN [10], we use two additional metrics: characteristic score (CS) and semantic diversity score (SDS). The CS metric evaluates how often a generated sketch is classified to be a bird (for Creative Birds) or creature (for Creative Creatures) by the Inception model trained on the QuickDraw3.8M dataset. The SDS metric measures the diversity of the sketches in terms of the different creature categories they represent. While the CS score can give us a basic understanding of the generation quality, it does not necessarily reflects the creative abilities of a model. For instance, if a model generates only canonical and mundane sketches of birds, then the generated sketches would more likely to be correctly classified by the trained Inception model. In that case, the CS score will still be high. In contrast, the SDS score is more reliable in measuring the diversity of the generated sketch images. Table 1 shows that DoodleFormer performs favourably against existing methods, in terms of SDS score, on both datasets. Figure 6(a) shows a visual comparison of DoodleFormer with DoodlerGAN for creative sketch generation (see Footnote 1).

### 4.2 User Study

Here, we present our user study to evaluate the human plausibility of creative sketches generated by our DoodleFormer. Specifically, we show 100 participants pairs of sketches - one generated by DoodleFormer and the other by a competing approach. For each pair of images, similar to DoodlerGAN [10], each participant is provided with 5 questions which are shown in the legend of Fig. 7(a–e). DoodleFormer performs favorably against DoodlerGAN for all five questions on both datasets. For instance, DoodleFormer sketches were interpreted to be drawn by a human 86%, having better initial strokes integration 85% and being more creative 82%, over DoodlerGAN on Creative Birds dataset. Further, for all the five questions, the DoodleFormer generated sketch images were found to be comparable with the human drawn sketches in Creative Datasets[1].

### 4.3 Ablation Study

We perform multiple ablation studies to validate the impact of proposed contributions in our framework. Table 2 shows the impact of two-stage framework, GAT blocks and GMM-based modeling on Creative Birds. Our single-stage baseline (referred as baseline*) is a standard transformer-based encoder-decoder architecture, where initial strokes are given as input to the transformer encoder. The decoder sequentially generates all body parts which are then integrated to obtain final output sketch. The generated sketches using baseline* are unrealistic and suffer from body parts misplacement. The introduction of two-stage framework leads to an absolute gain of 26.3 in terms of FID score, highlighting the importance of a coarse-to-fine framework for realistic creative sketch generation. Our two-stage framework baseline neither uses the GAT block in encoders nor employs the GMM-based modeling in decoder. Instead of GMM-based modeling, we use a deterministic L1 loss in first stage (PL-Net) of two-stage baseline. While this two-stage baseline improves realism of generated sketches, it stills suffers from topological artifacts. The introduction of GAT block in encoders of two-stage baseline improves realism of generated sketches by capturing the structural relationship between different parts. Although GAT blocks improve FID score by a margin of 3.25, the generated sketches still lack diversity as indicated by only a marginal change in GD score. The introduction of GMM-based modelling in decoder improves diversity (*e.g.*, appearance, size, orientation and posture) of generated sketches that leads to an absolute gain of 1.17 in GD (see also Fig. 4). Our final DoodleFormer (two-stage baseline + GAT blocks + GMM) achieves absolute gains of 30.0 and 3.7 in terms of FID score over baseline* and two-stage baseline, respectively.

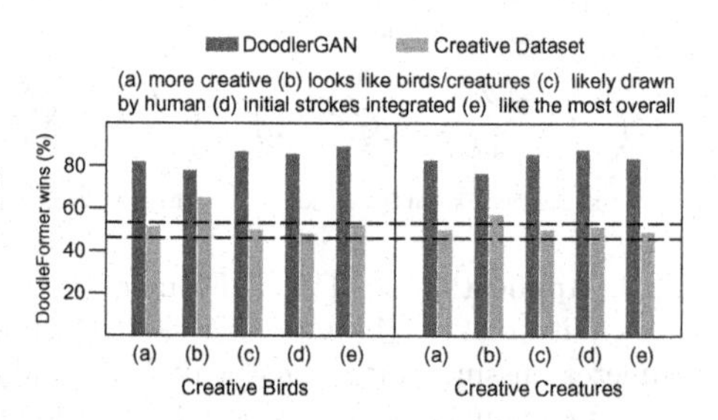


**Fig. 7.** User study results on Creative Birds (left) and Creative Creatures (right) based on the five questions (a-e) mentioned in the legend. Higher values indicate DoodleFormer is preferred more often over the compared approaches (DoodlerGAN and human drawn Creative datasets).

[1] Additional details and results are provided in supplementary material.

We also evaluate the design choices of our GAT blocks (see Table 3). First, we replace GAT blocks in encoders with simple GCN layers. The standard transformer-based encoders outperforms this GCN-based baseline by a margin of 7.31. Further, we adapt the Mesh Graphormer [22] by using their Graphormer in encoders of our framework. Mesh Graphormer stacks transformer encoder layer and GCN block together in series. In our experiments, we observe this design based on loosely connected components performs slightly worse than standard transformer-based baseline. In contrast, an integrated design like ours performs comparatively better.

**Table 2. Impact of our two-stage framework, GAT blocks and GMM-based probabilistic modelling** on Creative Birds.

| Design Choices | Methods | FID(↓) | GD(↑) |
|---|---|---|---|
| Single-stage | baseline* | 46.45 | 16.87 |
| Two-stage | baseline | 20.14 | 17.05 |
| | baseline + GAT | 16.89 | 17.16 |
| | baseline + GAT + GMM | **16.45** | **18.33** |

**Table 3. Comparison of alternative design choices for the proposed GAT blocks** on Creative Birds.

| Methods | FID(↓) | GD(↑) |
|---|---|---|
| GCN layers | 27.45 | 17.23 |
| Transformer layers | 20.14 | 17.05 |
| Mesh Graphormer [22] | 20.34 | 16.78 |
| **GAT layers (ours)** | **16.45** | **18.33** |

**Table 4. House Layout Generation:** We compare our approach with the existing methods in terms of FID and Compatibility scores (obtained by the graph edit distance). The dataset samples are split into five groups based on the room counts (1–3, 4–6, 7–9, 10–12, and 13+).

| Methods | FID (↓) | | | | | Compatibility (↓) | | | | |
|---|---|---|---|---|---|---|---|---|---|---|
| | 1–3 | 4–6 | 7–9 | 10–12 | 13+ | 1–3 | 4–6 | 7–9 | 10–12 | 13+ |
| Ashual *et al.* [12] | 64.0 | 92.2 | 87.6 | 122.8 | 149.9 | 0.2 | 2.7 | 6.2 | 19.2 | 36.0 |
| Johnson *et al.* [17] | 69.8 | 86.9 | 80.1 | 117.5 | 123.2 | 0.2 | 2.6 | 5.2 | 17.5 | 29.3 |
| House-GAN [10] | 13.6 | **9.4** | 14.4 | 11.6 | 20.1 | 0.1 | 1.1 | 2.9 | 3.9 | 10.8 |
| **Ours** | **9.6** | 10.1 | **11.2** | **9.7** | **18.2** | **0.1** | **1.0** | **2.1** | **2.4** | **8.3** |

### 4.4 Related Applications

We also analyze DoodleFormer on three related tasks: user provided text to creative sketch generation, creative sketch completion and house layout generation.

**Text to Creative Sketch Generation:** Here, the text description is given as conditional input to encoder $E_c$ in PL-Net, yielding a coarse structure of desired sketch which is fed to PS-Net to generate final sketch. We remove $\mathcal{R}_E$ from the PS-Net, and the *cls* token output from the encoder $\bar{E}_c$ is directly passed as input to $\mathcal{R}_D$. We use 80%-20% train-test split. We compare DoodleFormer with two popular text-to-image methods: StackGAN [36], AttnGAN [35] on Creative

Birds and Creative Creatures. DoodleFormer performs favorably against these methods in terms of FID and GD scores on both datasets. On Creative Birds, StackGAN, AttnGAN and DoodleFormer achieve respective FID scores of [53.1, 45.2, **18.5**], and GD scores of [16.7, 16.5, **17.3**]. Figure 6 (c) shows a qualitative comparison (see Footnote 1).

**Creative Sketch Completion:** Given an incomplete sketch as input, DoodleFormer attempts to creatively complete the rest of the sketch. First, PL-Net obtains the bounding boxes for missing parts. Then, PS-Net generates an image containing the required missing parts which is then integrated with the incomplete sketch input to obtain the final output. On both Creative Birds and Creative Creatures, DoodleFormer achieves favorable results compared to DoodlerGAN in terms of FID and GD scores. On Creative Birds, DoodlerGAN and DoodleFormer achieve respective FID scores of [44.2 and **18.3**] and GD scores of [15.1 and **17.8**]. Figure 6 (b) shows the qualitative comparison (see Footnote 1).

**House Layout Generation:** Finally, we use our proposed PL-Net architecture for the house-plan generation [24] task. The goal is to take a bubble diagram as an input, and generate a diverse set of realistic and compatible house layouts. A bubble diagram is represented by a graph where each node contains information about rooms and edges indicate their spatial adjacency. The output house layout is represented as axis-aligned bounding boxes. The Encoder $E_c$ takes the room type information as input and Probabilistic Decoder subsequently outputs boundary boxes for each room. To transform the obtained boundary box layout to a floor plan layout, we employ a floor-plan post-processing strategy. In this process, we first extract boundary lines of the generated boundary boxes. Next, we merge the adjacent line segments together and further align them to obtain a closed polygon.

We perform the house-plan generation experiments on LIFULL HOME's dataset [1]. For fair comparison, we follow the same setting used by House-

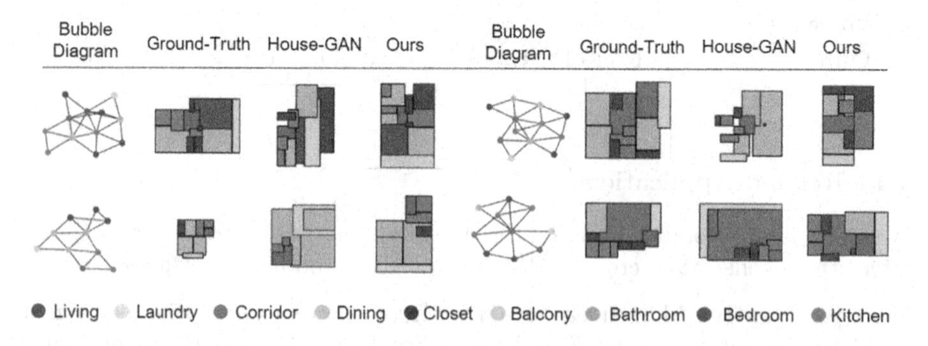


**Fig. 8.** Qualitative results of House Layout Generation. Given the input bubble diagram, We compare the house layout sample generated using our method with the House-GAN [24]. Our method produces house layouts that are well aligned with input bubble diagram texts.

GAN [24]. We divide the samples into five groups based on the number of rooms: 1–3, 4–6, 7–9, 10–12, and 13+. To test the generalization ability in each group, we train a model while excluding samples in the same group. At test time, we randomly pick a bubble input diagram from each group and generate 10 samples. Similar to House-GAN [24], we quantitatively measure the performance of our method in terms of FID and compatibility scores. The compatibility score is the graph editing distance [2] between the input bubble diagram and the bubble diagram constructed from the output layout. Table 4 shows that our method outperforms existing house-plan generation methods both in terms of FID and compatibility scores. Figure 8 shows the qualitative comparison of our house layout generation approach with House-GAN [24].

## 5 Conclusion

We proposed a novel coarse-to-fine two-stage approach, DoodleFormer, for creative sketch generation. We introduce graph-aware transformer encoders that effectively capture global dynamic as well as local static structural relations among different body parts. To ensure diversity of generated creative sketches, we introduce a probabilistic coarse sketch decoder that explicitly models variations of each sketch body part to be drawn. We show the effectiveness of DoodleFormer on two datasets by performing extensive qualitative, quantitative and human-based evaluations. In addition, we demonstrate promising results on related applications such as text to creative sketch generation, sketch completion and house layout generation.

## References


1. Lifull home's dataset. https://www.nii.ac.jp/dsc/idr/lifull. Accessed 30 Sept 2010
2. Abu-Aisheh, Z., Raveaux, R., Ramel, J.Y., Martineau, P.: An exact graph edit distance algorithm for solving pattern recognition problems. In: 4th International Conference on Pattern Recognition Applications and Methods 2015 (2015)
3. Balasubramanian, S., Balasubramanian, V.N., et al.: Teaching GANs to sketch in vector format. arXiv preprint arXiv:1904.03620 (2019)
4. Bishop, C.M.: Mixture Density Networks. Aston University (1994)
5. Cao, N., Yan, X., Shi, Y., Chen, C.: AI-sketcher: a deep generative model for producing high-quality sketches. In: AAAI (2019)
6. Carion, N., Massa, F., Synnaeve, G., Usunier, N., Kirillov, A., Zagoruyko, S.: End-to-end object detection with transformers. In: Vedaldi, A., Bischof, H., Brox, T., Frahm, J.-M. (eds.) ECCV 2020. LNCS, vol. 12346, pp. 213–229. Springer, Cham (2020). https://doi.org/10.1007/978-3-030-58452-8_13
7. Chen, Y., Tu, S., Yi, Y., Xu, L.: Sketch-pix2Seq: a model to generate sketches of multiple categories. arXiv preprint arXiv:1709.04121 (2017)
8. Devlin, J., Chang, M.W., Lee, K., Toutanova, K.: BERT: pre-training of deep bidirectional transformers for language understanding. In: NAACL (2019)
9. Ganin, Y., Kulkarni, T., Babuschkin, I., Eslami, S.A., Vinyals, O.: Synthesizing programs for images using reinforced adversarial learning. In: ICML (2018)

10. Ge, S., Goswami, V., Zitnick, C.L., Parikh, D.: Creative sketch generation. In: ICLR (2021)
11. Graves, A.: Generating sequences with recurrent neural networks. arXiv preprint arXiv:1308.0850 (2013)
12. Ha, D., Eck, D.: A neural representation of sketch drawings. In: ICLR (2018)
13. He, K., Zhang, X., Ren, S., Sun, J.: Deep residual learning for image recognition. In: CVPR (2016)
14. He, S., et al.: Context-aware layout to image generation with enhanced object appearance. In: CVPR (2021)
15. Heusel, M., Ramsauer, H., Unterthiner, T., Nessler, B., Hochreiter, S.: GANs trained by a two time-scale update rule converge to a local Nash equilibrium. In: NeurIPS (2017)
16. Hinton, G.E., Nair, V.: Inferring motor programs from images of handwritten digits. In: NeurIPS (2006)
17. Karras, T., Laine, S., Aittala, M., Hellsten, J., Lehtinen, J., Aila, T.: Analyzing and improving the image quality of StyleGAN. In: CVPR (2020)
18. Khan, S., Naseer, M., Hayat, M., Zamir, S.W., Khan, F.S., Shah, M.: Transformers in vision: a survey. arXiv preprint arXiv:2101.01169 (2021)
19. Kipf, T.N., Welling, M.: Semi-supervised classification with graph convolutional networks. In: ICLR (2017)
20. Li, Y., Song, Y.Z., Hospedales, T.M., Gong, S.: Free-hand sketch synthesis with deformable stroke models. In: IJCV (2017)
21. Lin, H., Fu, Y., Xue, X., Jiang, Y.G.: Sketch-BERT: learning sketch bidirectional encoder representation from transformers by self-supervised learning of sketch gestalt. In: CVPR (2020)
22. Lin, K., Wang, L., Liu, Z.: Mesh graphormer. In: ICCV (2021)
23. Liu, F., Deng, X., Lai, Y.K., Liu, Y.J., Ma, C., Wang, H.: SketchGAN: joint sketch completion and recognition with GAN. In: CVPR (2019)
24. Nauata, N., Chang, K.-H., Cheng, C.-Y., Mori, G., Furukawa, Y.: House-GAN: relational generative adversarial networks for graph-constrained house layout generation. In: Vedaldi, A., Bischof, H., Brox, T., Frahm, J.-M. (eds.) ECCV 2020. LNCS, vol. 12346, pp. 162–177. Springer, Cham (2020). https://doi.org/10.1007/978-3-030-58452-8_10
25. Qi, Y., Su, G., Chowdhury, P.N., Li, M., Song, Y.Z.: SketchLattice: latticed representation for sketch manipulation. In: ICCV (2021)
26. Ramasinghe, S., Farazi, M., Khan, S., Barnes, N., Gould, S.: Rethinking conditional GAN training: an approach using geometrically structured latent manifolds. In: NeurIPS (2021)
27. Ribeiro, L.S.F., Bui, T., Collomosse, J., Ponti, M.: SketchFormer: transformer-based representation for sketched structure. In: CVPR (2020)
28. Sohn, K., Lee, H., Yan, X.: Learning structured output representation using deep conditional generative models. In: NeurIPS (2015)
29. Su, G., Qi, Y., Pang, K., Yang, J., Song, Y.Z.: SketchHealer: a graph-to-sequence network for recreating partial human sketches. In: BMVC (2020)
30. Sun, W., Wu, T.: Image synthesis from reconfigurable layout and style. In: ICCV (2019)
31. Sun, W., Wu, T.: Learning layout and style reconfigurable GANs for controllable image synthesis. PAMI (2021)
32. Vaswani, A., et al.: Attention is all you need. In: NeurIPS (2017)

33. Viazovetskyi, Y., Ivashkin, V., Kashin, E.: StyleGAN2 distillation for feed-forward image manipulation. In: Vedaldi, A., Bischof, H., Brox, T., Frahm, J.-M. (eds.) ECCV 2020. LNCS, vol. 12367, pp. 170–186. Springer, Cham (2020). https://doi.org/10.1007/978-3-030-58542-6_11
34. Xu, P., Hospedales, T.M., Yin, Q., Song, Y.Z., Xiang, T., Wang, L.: Deep learning for free-hand sketch: a survey and a toolbox. arXiv preprint arXiv:2001.02600 (2020)
35. Xu, T., et al.: AttnGAN: fine-grained text to image generation with attentional generative adversarial networks. In: CVPR (2018)
36. Zhang, H., et al.: StackGAN: text to photo-realistic image synthesis with stacked generative adversarial networks. In: ICCV (2017)
37. Zheng, N., Jiang, Y., Huang, D.: StrokeNet: a neural painting environment. In: ICLR (2018)
38. Zhou, T., et al.: Learning to doodle with stroke demonstrations and deep q-networks. In: BMVC (2018)
39. Zhu, X., Su, W., Lu, L., Li, B., Wang, X., Dai, J.: Deformable DETR: deformable transformers for end-to-end object detection. In: ICLR (2021)

# Implicit Neural Representations for Variable Length Human Motion Generation

Pablo Cervantes[1], Yusuke Sekikawa[2], Ikuro Sato[1,2], and Koichi Shinoda[1](✉)

[1] Tokyo Institute of Technology, Tokyo, Japan
shinoda@c.titech.ac.jp
[2] Denso IT Laboratory Inc., Tokyo, Japan

**Abstract.** We propose an action-conditional human motion generation method using variational implicit neural representations (INR). The variational formalism enables action-conditional distributions of INRs, from which one can easily sample representations to generate novel human motion sequences. Our method offers variable-length sequence generation by construction because a part of INR is optimized for a whole sequence of arbitrary length with temporal embeddings. In contrast, previous works reported difficulties with modeling variable-length sequences. We confirm that our method with a Transformer decoder outperforms all relevant methods on HumanAct12, NTU-RGBD, and UESTC datasets in terms of realism and diversity of generated motions. Surprisingly, even our method with an MLP decoder consistently outperforms the state-of-the-art Transformer-based auto-encoder. In particular, we show that variable-length motions generated by our method are better than fixed-length motions generated by the state-of-the-art method in terms of realism and diversity. Code at https://github.com/PACerv/ImplicitMotion.

**Keywords:** Motion generation · Implicit Neural Representations

## 1 Introduction

Generative models of human motion serve as a basis for human motion prediction [2,3,5,6,12,15], human animation [36,37], and data augmentation for downstream recognition tasks [9,23,34,38]. There has been intensive research on generative models for realistic and diverse human motions [13,19,39] and in particular methods that can generate motions while controlling some semantic factors such as emotion [13], rhythm [19] or action class [10,31]. For tasks such as rare action recognition, data-efficient action-conditional motion generation has great potential, since it may provide data augmentation even for rare actions.

**Supplementary Information** The online version contains supplementary material available at https://doi.org/10.1007/978-3-031-19790-1_22.

© The Author(s), under exclusive license to Springer Nature Switzerland AG 2022
S. Avidan et al. (Eds.): ECCV 2022, LNCS 13677, pp. 356–372, 2022.
https://doi.org/10.1007/978-3-031-19790-1_22

For motion generation, the quality of generations is evaluated by their realism and diversity. Models need the ability to sample novel and rich representations to generate high-quality motions. A suitable generative model yields distributions of representations in a latent space, where a simple distance measure corresponds to semantic similarity between motions so that interpolations provide novel and high-quality motions. A common generative modeling approach is Variational Auto-Encoders (VAE) [10,11,17,31], which employ an encoder to infer a distribution from which representations of motions can be sampled and a decoder which reconstructs the data from the representation. The reconstruction loss provides strong supervision, while the variational approach results in a representation space, which allows sampling of novel data with high realism and diversity.

Since human motions naturally vary in length depending on persons or action, it is important to consider variable lengths in motion generation. For example, we would like the representations of quick (short) and slow (long) sitting motions to be different but closer to each other than the representation of a walking motion. In RNN-based VAEs [10], representations are updated each time-step; thus, it is not obvious how to sample a particular action such as quick sitting. Also, their recursive generation may accumulate error when generating long sequences. In contrast, ACTOR, a Transformer VAE [31], should conceptually provide time-independent representations and generate variable-length motions without accumulating error. Nevertheless, [31] reports directly training with variable-length motions results in almost static motions, and accordingly ACTOR requires an additional fine-tuning scheme to enable variable-length motion generation. It remains unclear what causes such issues with the Transformer architecture.

A recently proposed generative modeling approach is Implicit Neural Representations (INR), which have been shown to be highly efficient in modeling complex data such as 3D scenes [24,26,27]. INRs are representations that encode information without an explicit encoder, but through an optimization procedure as shown in Fig. 1. INRs are usually constructed with respect to a decoder that takes a target coordinate and the representation of a target sample as input and returns the signal of the target sample at the target coordinate. Such representations are optimized individually for each sample with respect to the reconstruction loss at all coordinates. For a time-series, an INR is a time-independent, optimal representation that represents one whole sequence, regardless of the sequence length. Since human motions are naturally variable-length, INRs are a very promising modeling approach. However, to the best of our knowledge, there is no INR-based motion generation method that serves as a strong baseline.

To construct distributions from which one can sample a representation to generate a novel and high-quality motion, we propose variational INRs. Compared to VAEs which infer variational distributions with an encoder, our variational INR framework models each sequence by a distribution with optimized parameters, e.g., mean and covariance, in the representation space. We further decompose INRs into an action-wise part and sequence-wise part. The action-wise part, whose distribution parameters are optimized for all sequences within the same action class, provides generalized components of an action. The sequence-wise

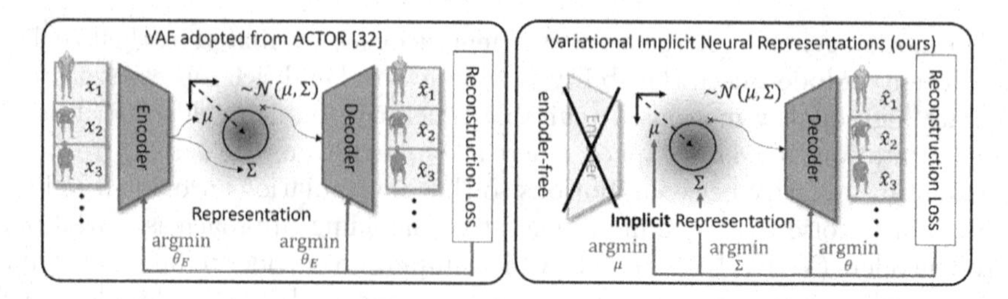


**Fig. 1.** Comparison between a Variational Auto-Encoder (VAE) baseline (top) and our variational implicit neural representation approach (bottom). In VAEs the encoder weights are optimized with respect to a full dataset and no guarantee of optimal representations for each individual sequence. In contrast, our sequence representations are directly optimized for each individual sequence and by construction offer variable-length sequence generation because a part of each INR is optimized for a whole sequence of arbitrary length. In this figure, we drop the temporal embeddings for simplicity.

part, whose distribution parameters are optimized for an individual sequence, adds fine details of a specific sequence on top of the generalized components.

The average of the sequence distributions within an action class in the representation space serves as the action-specific generative model together with an appropriately trained decoder. In our method, we further split the averaged distribution into several distributions depending on different intervals of sequence lengths. This allows sampling of novel sequences with a target action and a target length. To parameterize the action and sequence-length conditional distribution, we employ Gaussian Mixture Models (GMM). Note that existing high-performing methods are unable to control sequence length. This often results in poor motion generation with sequences ending before the action completes.

However, when fitting a GMM with a high degree of freedom to the representation space we risk simply reproducing training samples. Previous evaluation metrics such as the Fréchet Inception Distance (FID) and Diversity are not sensitive to this problematic model behavior, because they assign a high value to generated motions with a similar distribution as the training set. In this regard, we propose a novel metric, the Mean Maximum Similarity (MMS), to measure such reproducing behavior. By using this metric we confirm that our and previous studies successfully generate motions distinct from the training sequences.

We find that our proposed approach outperforms the current SOTA for action-conditional motion generation, ACTOR [31], in terms of realism and diversity. By employing an identical decoder architecture as ACTOR [31], we conduct a fair comparison between our INR-approach and a VAE-approach and find that our INR approach improves motion generation. Furthermore, since Transformer models can be difficult and expensive to train and we also explore the use of an MLP decoder and find that even such a simple, lightweight (6x fewer parameters) model can reach the SOTA performance.

Our key contributions are summarized as follows:

- We propose a variational INR framework for motion generation, which gives time-independent, optimal representations for variable-length sequences distributed such that representations for novel motions can be easily sampled.
- To improve action-conditional motion generation, we propose INRs that are decomposed into action-wise and sequence-wise INRs. The action-wise INR generalizes to features across an action-class and helps generating realistic and novel motions for a target action class.
- We show in experiments that our method outperforms SOTA (ACTOR [31]) on the HumanAct12, NTU13 and UESTC datasets in term of realism and diversity, and confirm that it generates high-quality variable-length sequences. For example on HumanAct12 we generate sequences with lengths between 8–470 time-steps and find that our motions generated with variable-length even outperform fixed-length motions generated by previous works (ACTOR [31], Action2Motion [10]) in terms of realism and diversity.

## 2 Related Works

In the following we review the context of our work first regarding human motion modeling and then regarding implicit neural representations.

*Human Motion Modeling:* The modeling of human motions is important for understanding and predicting human behavior. Most modern approaches regard a human motion as a time-series of either skeleton poses or full 3D body shapes [22,29] and previous works have proposed methods to estimate motions from videos, predicting future motions based on past motions and generating such motions conditional on signals such as emotion [13] and rhythm [19]. Our work is similar to [10,31], which generate motions conditional on the action class.

Previous works for motion generation are mostly based on Variational Auto-Encoders (VAE) [10,11,17,31], which employs an encoder to infer a variational distribution from which representations of motions can be sampled and a decoder which reconstructs data from a representation. This encoder is optimized with respect to a reconstruction loss for a whole dataset, without a guarantee that the representation for each individual sequence is optimal. The encoder may focus on the most common features in the dataset and become insensitive to rare features. In contrast, Implicit Neural Representations (INR), which optimize the representation of each sample directly, can be sensitive even to unique features.

A similarity of all sequence modeling approaches is the use of model architectures such as RNNs or Transformers. RNN are typically formulated as an auto-regressive model [2,10,13], which generates motions by recursively predicting the pose at time-step $t$ based on the prediction of the pose at time-step $t-1$. This recursive nature of RNNs means that their sequence-representations are time-dependent and representations of variable-length sequences can not be easily compared. Furthermore, the recursive generation procedure accumulates error and may result in poor performance when generating long sequences.

Petrovich et al. [31] proposes ACTOR, a Transformer VAE, which yields a single fixed-length representation for a variable-length sequence through a Transformer encoder. This representation is decoded by a Transformer decoder, which receives the representation and the temporal embedding of the target time-steps as input and generates the target sequence in one forward-pass. Since such a Transformer VAE should conceptually handle variable-length sequences well, we choose this work as our main baseline. However, [31] reports that even for ACTOR a fine-tuning scheme is needed to enable good performance for variable-length sequences.

*Implicit Neural Representations:* INR as proposed in [7,26,27] are encoder-free models which instead optimize their parameters to represent and fit a single sample. They have been popularized particularly in 3D modeling and have shown great performance on tasks such as inverse graphics [24,40], image synthesis [16,33] or scene generation [8,25]. While this work is, to the best of our knowledge, the first to explore implicit neural representations in the context of motion modeling, previous works have considered other time-series [20,26,35].

Previous works for data synthesis using INR use a GAN-like approach [1,16, 33] for image synthesis. Such approaches don't optimize the INR, but sample representations from a predetermined distribution. These representations are then used by a generator to generate images, which can fool a discriminator. However, for our task the amount of training data required by GANs is problematic.

Another approach that takes inspiration from VAEs are variational INR [4,27]. Most similar to ours is [4], however, this work doesn't optimize the INRs but rather approximates the optimal INR using an empirical Bayes. Furthermore, they predict a variational distribution from the INR and then apply a regularizing loss to this intermediate representation instead of directly regularizing the INR. Instead we directly optimize the mean and variance of the variational distribution as persistent parameters per sample. The approach in [27] optimizes point-estimates of the representations of a sample and regularizes the distribution of these estimates, but doesn't sample stochastically from this distribution during training. Our work instead optimizes a distribution for each sample and samples stochastically from it during training.

With this work we would like to show that INR are not only powerful for high dimensional data such as dense 3D point clouds, but that their flexibility is also useful for other domains.

## 3 Methodology

In this section, we will first describe how to apply INRs to model human motions and decompose the INR into sequence-wise and action-wise representations (Sect. 3.1). Then we will introduce the proposed variational INRs (Sect. 3.2), before we discuss how we fit a conditional Gaussian Mixture Model (GMM) to the representation space and how we sample novel sequence-wise representations from it (Sect. 3.3). Finally we will describe the Mean Maximum Similarity (MMS) as a measure to detect models that only reproduce training samples (Sect. 3.4).

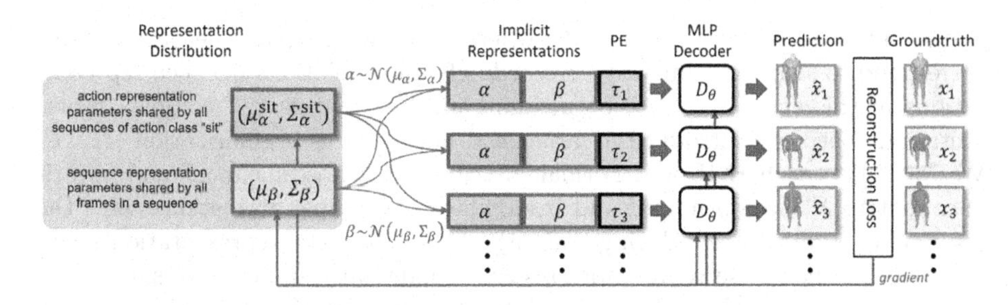


**Fig. 2.** Overview of Implicit Motion Modeling. Each representation is composed of two components, the action representation $\alpha$ and the sequence representation $\beta$. Instead of inferring these representations from an encoder, we directly optimize the parameters of a posterior normal distribution for both the action representation $(\mu_\alpha, \Sigma_\alpha)$ shared by all sequences with the same action class and sequence representation $(\mu_\beta, \Sigma_\beta)$. The representation, together with a temporal embedding (PE) $\tau_t$ of time $t$ is then input to an MLP, which predicts the pose at time $t$.

### 3.1 Implicit Neural Representations for Motion Modeling

We consider a human motion as a sequence of poses represented by a low-dimensional skeleton. Formally, we denote a skeleton pose of sequence $i$ at time $t$ as $x_t^i \in \mathbb{R}^{P \times B}$ where $P$ is the number of joints and $B$ is the dimensionality of the joint representation. The $i$-th motion (a sequence of poses) is denoted as $\mathbf{x}^i = \{x_t^i\}_{t=1}^{T^i}$ with the sequence length $T^i$ (Fig. 2).

For each sequence $i$, we construct an Implicit Neural Representation (INR) $c_i$ and a decoder $D_\theta$ (shared among all sequences) that predicts a pose $\hat{x}_t^i$ of sequence $i$ from the INR $c_i$ and a temporal embedding $\tau_t$ of time $t$

$$\hat{x}_t^i = D_\theta(c^i, \tau_t). \tag{1}$$

Note that depending on the decoder architecture, the decoder may process all time-steps of a sequence independently (MLP) or multiple time-steps simultaneously (Transformer). We obtain an INR $c_i$, shared by all time steps ($t \in \{1, 2, ..., T_i\}$), by minimizing the reconstruction loss $\mathcal{L}_{\text{rec}}^i$. Thus, INRs can represent a sequence of any sequence-length $T^i$. Also, for a given INR, the decoder can interpolate between time-steps (e.g. $t = 0.5$) or extrapolate ($t > T_i$).

To generalize INRs to all features of the same action class, we decompose the INR and introduce an action representation shared across all samples of the same action class. Formally, we divide each INR $c^i$ into a sequence-wise representation $\beta^i \in \mathbb{R}^S$ with in $i \in \mathcal{M}$ with a set of motions $\mathcal{M}$ and an action-wise representation $\alpha^z \in \mathbb{R}^A$ shared by all sequences with the same action label $z \in \mathcal{Z}$. Here $\mathcal{Z}$ is the set of action classes (e.g. $\alpha^z \in \{\alpha^{\text{sit}}, \alpha^{\text{walk}}, \alpha^{\text{run}} \dots\}$) and $S$ and $A$ denote the size of each representation respectively.

### 3.2 Variational Implicit Neural Representations

Note that each INR $c^i$ is optimized to reconstruct a single sample with an over-parameterized decoder $D_\theta$. This can make the distribution of INRs complex and result in a representation space where a simple distance measure doesn't correspond to semantic similarity. Accordingly, interpolations between representations in this space may not be meaningful. To avoid such a complex representation space, we introduce a variational approach as regularization [11,17]. We formulate each INR as a normal distribution, whose mean $\mu^i$ and covariance matrix $\Sigma^i$ are optimized and from which we sample an instance with the re-parameterization trick during training. This makes the representation space smoother so that close representations are semantically similar. We summarize the sequence-wise and action-wise variational representations as

$$\begin{aligned} c^i &\sim \mathcal{N}(\mu^i, \Sigma^i) \text{ with} \\ \mu^i &= \text{concat}(\mu^z_\alpha, \mu^i_\beta), \\ \Sigma^i &= \begin{bmatrix} \Sigma^z_\alpha & 0^{A\times S} \\ 0^{S\times A} & \Sigma^i_\beta \end{bmatrix}, \end{aligned} \tag{2}$$

where concat denotes the concatenation operation.

Furthermore, by assuming a standard normal distribution as the prior of each INR $c_i$, we further encourage a simple and compact representation space. We then use the Kullback-Leibler (KL) Divergence $\mathcal{L}^i_{\text{KL}}$ as a regularizing loss

$$\mathcal{L}^i_{\text{KL}} = \mathcal{D}_{\text{KL}}(\mathcal{N}(\mu^i, \Sigma^i) \| \mathcal{N}(0, I)). \tag{3}$$

The sequence wise training objective of our method is thus

$$\mathcal{L}^i = \mathcal{L}^i_{\text{rec}} + \lambda \mathcal{L}^i_{\text{KL}}, \tag{4}$$

where $\mathcal{L}^i_{rec}$ is the reconstruction term

$$\begin{aligned} \mathcal{L}^i_{\text{rec}} &= -\mathbb{E}_{c^i \sim \mathcal{N}(\mu^i, \Sigma^i)} \sum_{t=1}^{T^i} \log p(x^i_t | c^i, \theta), \\ \log p(x^i_t | c^i, \theta) &\propto \| x^i_t - D_\theta(\tau_t, \alpha^z, \beta^i) \|_2 + \text{const.}, \end{aligned} \tag{5}$$

and $\mathcal{L}^i_{\text{KL}}$ is the regularizing KL divergence moderated by a weight $\lambda$.

We define the optimization problem for the model parameters as:

$$\theta^\star = \underbrace{\underset{\theta}{\text{argmin}} \sum_{z=1}^{Z} \underbrace{\min_{\mu^z_\alpha, \Sigma^z_\alpha,} \sum_{i \in \mathcal{M}^z} \underbrace{\min_{\mu^i_\beta, \Sigma^i_\beta} \mathcal{L}^i}_{\text{sequence-wise minimum}}}_{\text{action-wise minimum}}}_{\text{dataset-wise minimum}}, \tag{6}$$

where $\mathcal{M}^z$ denotes a set of sequence indices within action class $z$. We optimize action-wise parameters $\mu_\alpha, \Sigma_\alpha$ for each action $z$:

$$(\mu^{z\star}_\alpha, \Sigma^{z\star}_\alpha) = \underset{\mu^z_\alpha, \Sigma^z_\alpha}{\text{argmin}} \sum_{i \in \mathcal{M}^z} \min_{\mu^i_\beta, \Sigma^i_\beta} \mathcal{L}^i. \tag{7}$$

Likewise, for the sequence-wise parameters we define the optimization problem for each sequence $i$ as:

$$(\mu_{\beta}^{i\star}, \Sigma_{\beta}^{i\star}) = \underset{\mu_{\beta}^{i}, \Sigma_{\beta}^{i}}{\operatorname{argmin}} \mathcal{L}^{i}. \tag{8}$$

### 3.3 Conditional GMM of Representation Space

To generate new sequences for a target action class, we need novel samples from the distribution of sequence-wise representations. In the distribution of sequence-wise representations obtained during training, semantic factors such as sequence/lengths and action classes may be entangled. Accordingly, the action-conditional distribution of sequence-wise representations may differ from the standard normal distribution. To control sequence-length and action class for motion generation, we fit a conditional Gaussian Mixture Model (GMM) to the sequence-wise representations $\beta^i$ sampled 50 times from the variational distributions $\beta^i \sim \mathcal{N}(\mu_{\beta}^{i\star}, \Sigma_{\beta}^{i\star})$ for each training sequence.

We fit such a conditional GMM by first constructing subsets of sequence-wise representations that have the same action class $z$ and a sequence-length within the range $[T, T + \Delta T]$. We choose the size of the sequence-length range $\Delta T$ to ensure a minimum number of samples in each subset and then fit an independent GMM to each subset of sequence-wise representations. The details for how we select such a set of sequence-length ranges are provided in Appendix A.2. Finally, we obtain the GMM of $p(\beta | z, [T, T + \Delta T])$.

To sample new sequence representations and generate corresponding novel motion sequences, we need to provide a target action class and sequence length. We sample a new sequence representation $\beta^{\text{new}}$

$$\beta^{\text{new}} \sim p(\beta | z, [T, T + \Delta T]), \tag{9}$$

by sampling from the GMM corresponding to the target action class and sequence length. With a new sequence representation we generate a new motion

$$\mathbf{x}^{\text{new}} = \{D_{\theta^\star}(\alpha^{z\star}, \beta^{\text{new}}, \tau_t)\}_{t=t_0}^{T'} \tag{10}$$

with the target action code $\alpha^{z\star}$ (obtained during the training stage) and the target sequence length $T' \in [T, T + \Delta T]$.

### 3.4 Mean Maximum Similarity

By increasing the number of components of the GMM, it can better fit the training distribution, which improves the realism of generated motions. However, we also risk fitting a GMM which only reproduces motions in the training set. Previous metrics such as the Fréchet Inception Distance (FID) or the Diversity compute the feature distribution of training and generated motions and compare these distributions. Generated motions that have a similar distribution (FID) or variance (Diversity) as real motions are considered high-quality. Generated motions identical to the training set would be considered best by such metrics.

To detect models that just reproduce training samples, we introduce the Mean Maximum Similarity (MMS) as a complementary metric. Similarly to previous metrics we extract the features from all training sample and generated motions. Then for each generated motion, we find the training sample with the smallest feature distance (most similar) to it. The mean distance over a large set of generated motions should be small for models that reproduce training samples and large for models that generate novel motions. Formally we denote the features of a motion as $f$ and the sets of generated and training motion sequences $\mathcal{M}_{\text{gen}}$ and $\mathcal{M}_{\text{train}}$ respectively, and compute the MMS as

$$\mathcal{D}_{MMS}(\mathcal{M}_{\text{gen}}, \mathcal{M}_{\text{train}}) = \frac{1}{|\mathcal{M}_{\text{gen}}|} \sum_{i \in \mathcal{M}_{\text{gen}}} \min_{j \in \mathcal{M}_{\text{train}}} (\|f_i - f_j\|_2). \tag{11}$$

We estimate the MMS of model that only reproduces motions as baseline by computing $\mathcal{D}_{MMS}(\mathcal{M}_{\text{train}}, \mathcal{M}_{\text{train}})$ of the set of training motions $\mathcal{M}_{\text{train}}$ against itself. A large gap between $\mathcal{D}_{MMS}(\mathcal{M}_{\text{gen}}, \mathcal{M}_{\text{train}})$ and $\mathcal{D}_{MMS}(\mathcal{M}_{\text{train}}, \mathcal{M}_{\text{train}})$ indicates novel generated motions distinct from the training set.

## 4 Experiments

To verify the quality of motions generated by variational INR we perform experiments with a Transformer and an MLP decoder. The Transformers is a powerful, but costly and difficult to train modeling tool, while the MLP is simple and comparatively light-weight. The comparison should highlight the efficiency of the variational INR framework independent of decoder architecture. In this section we will first explain the implementation details of our models (Sect. 4.1) and the datasets for our experiments (Sect. 4.2). Then we will describe how we quantify the realism, diversity and novelty of generated motions (Sect. 4.3). Finally we will discuss the quantitative (Sect. 4.4) and qualitative (Sect. 4.5) results.

### 4.1 Implementation

*Skeleton Representation:* We represent the human body as a kinematic tree defined by joint rotations, bone-lengths and the root joint. More specifically, we use the SMPL model [22] with pose parameters consisting of 23 joint rotations, 1 global rotation and 1 root trajectory. During training we only predict the pose parameters, which are independent of the body shape and can be used to animate any body at test time. We represent rotations with a 6D rotation parameterization as proposed by [41] which means the full body pose has 147 dimensions ($24 \times 6 + 3$). We use a reconstruction loss composed of a loss on the pose parameters (joint rotations and root joint locations) as well as the vertices of the SMLP model since [31]'s findings suggest the best performance for this configuration. On the NTU13 dataset, where at the time of writing the SMPL data was no longer available we represent the pose with a 6D rotation parameterization, but use a reconstruction loss on the joint locations (through forward-kinematics) as proposed by [10] and find similarly high performance with our method.

*Model Architecture:* We implement our MLP decoder with ELU activations and 5 hidden layers (1000, 500, 500, 200, 100). The input are temporal embeddings with 256 dimensions and sequence-wise representations/action-wise representations, which are both 128 dimensions respectively, and the decoder outputs 147 dimensional pose parameters. This results in a network with 1,399,147 parameters. Due to the larger dataset size of UESTC we also implement a larger model (2000, 2000, 1000, 1000, 200, 100) with 8,265,147 parameters which is only used on UESTC. We also implement a Transformer-decoder (same as ACTOR [31]) with 8 layers, 4 attention heads, a dropout rate of 0.1 and a feedforward network of 1024 dimensions. With temporal embeddings with 256 dimensions and the same pose parameterization this results in a network with 8,465,299 parameters (6$\times$ more than the MLP model). More details are provided in Appendix A.1.

Note that the Transformer-decoder is sensitive to the initialization of the implicit representations. If the variance parameters are initialized with a high variance the Transformer-decoder may fail to converge, while the MLP decoder is not sensitive to this phenomenon. We explore this more in Appendix B.1.

The Transformer-based decoder has an identical structure to ACTOR [31] and thus allows us a direct comparison between an auto-encoder and an implicit framework. The MLP decoder is simpler to train than the Transformer decoder and doesn't rely on self-attention. The comparison of these decoders allows us to determine if the choice of decoder architecture is critical for good performance.

### 4.2 Datasets

To evaluate the quality of action-conditional human motion generation, we used the UESTC, NTU-RGBD and HumanAct12 dataset curated by [10].[1]

*HumanAct12* [10]*:* This dataset is based on PHSPD [42] and consists of 1191 motion clips and 90099 frames in total. Action labels for 12 actions are provided with at least 47 and at most 218 samples per label. Sequence-lengths range from 8 to 470. We follow the procedure by [31] to align the poses to frontal view.

*NTU13* [21]*:* The NTU-RGBD dataset originally contains pose annotations from a MS Kinect sensor and label annotations for 120 actions. [10] re-estimated the data of a subset of 13 action, which we denote NTU13, with a state-of-the-art pose estimation method [18] to reduce noise. In this refined subset each action label has between 286–309 samples. The refined poses have 18 body joints and the sequence lengths range from 20–201.[2]

*UESTC* [14]*:* This dataset with 40 action classes, 40 subjects and 25K samples is the largest dataset we perform experiments on and the only dataset with a train/test split. We use the SMPL sequences provided by [31] and apply the

[1] We considered the CMU Mocap dataset, but manual inspection found the label annotations for some actions such as "Wash" and "Step" to be extremely noisy.

[2] Due to the release agreement of NTU RGBD, this subset can no longer be distributed. We report results to provide a complete comparison to previous studies.

same pre-proprocessing, namely we rotate all sequences to frontal view. Using the same cross-subject testing protocol we have a training split with between 225–345 samples per action class and sequence lengths between 24 and 2891 time steps (on average 300 time steps).

### 4.3 Evaluation Metrics

We use the same evaluation metrics as [10,31] (Fréchet Inception Distance (FID), action recognition accuracy, diversity and multimodality) to measure the realism and diversity of generated motions. Also, we measure the proposed **Mean Maximum Similarity** (Sect. 3.4) to detect models that reproduce training samples. We report a 95% confidence interval computed of 20 evaluations.

The features for these evaluation metrics are extracted from motions of a predetermined length (60 time-steps) by an RNN-based action recognition model (weights provided by [10]) for the NTU13 and HumanAct12 dataset and by an ST-GCN-based action recognition model (weights provided by [31]) for the UESTC dataset. However, since the real training data is variable-length, we follow [10]'s procedure for feature extraction during evaluation. This procedure adjust all sequences to a target length, by repeating the last pose of short sequences and by sampling random sub-sequence from longer sequences.

Such stochastic feature extraction means the MMS may not be zero, even for sets of identical motions. Thus we first compute a baseline MMS for identical real motions and then evaluate the MMS between real motions and generated motions. If the MMS for generated motions is larger than that of real motions only, we conclude that the generated motions are distinct from the real motions.

We find that there is a difference in the evaluation procedure of previous works in the sampling frequency of different action classes for generation. The approach by [10] generates motions uniformly for all action classes. On datasets with an action imbalance, this creates an inflated FID score. We follow [31]'s approach which generates motions according to the frequency of the action class in the training dataset, since this leads to more consistent results.

Our GMM samples novel representations conditional on the sequence-length. For model evaluation we sample sequence-lengths according to their distribution in the training dataset. We then sample corresponding representations and generate motions with the corresponding sequence-length. We perform the same feature extraction as for the variable-length real motions. More details can be found in the Appendix A.7

### 4.4 Quantitative Results

We compare our method to an RNN [10] and a Transformer [31] baseline and present some ablations for the proposed novel components of our model on HumanAct12 and NTU13 in Table 1 and UESTC in Table 2. Furthermore, we present a new state-of-the-art with the results for our Transformer-based and MLP-based models. We also investigate the contribution of variational INR by comparing them to a non-variational version and the contribution of the

**Table 1.** Comparison on HumanAct12 and NTU13 (The best in bold, the second best underlined). *Non-variational* uses action codes and *no action code* uses the variational approach. ($\pm$ indicates 95% confidence interval, $\rightarrow$ closer to real is better)

| Method | HumanAct12 | | | |
|---|---|---|---|---|
| | FID $\downarrow$ | Accuracy $\uparrow$ | Diversity $\rightarrow$ | Multimod. $\rightarrow$ |
| Real | $0.020^{\pm.010}$ | $0.997^{\pm.001}$ | $6.850^{\pm.050}$ | $2.450^{\pm.040}$ |
| Action2Motion [10] | $0.338^{\pm.015}$ | $0.917^{\pm.003}$ | $\underline{6.879^{\pm.066}}$ | $\underline{2.511^{\pm.023}}$ |
| ACTOR [31] | $0.12^{\pm.00}$ | $0.955^{\pm.008}$ | $\mathbf{6.84^{\pm.03}}$ | $2.53^{\pm.02}$ |
| INR (Transformer) | $\mathbf{0.088^{\pm.004}}$ | $\mathbf{0.973^{\pm.001}}$ | $6.881^{\pm.048}$ | $2.569^{\pm.040}$ |
| INR (MLP) | $\underline{0.114^{\pm.001}}$ | $\underline{0.970^{\pm.001}}$ | $6.786^{\pm.057}$ | $\mathbf{2.507^{\pm.034}}$ |
| - (Non-variational) | $0.551^{\pm.005}$ | $0.795^{\pm.002}$ | $6.800^{\pm.046}$ | $3.700^{\pm.032}$ |
| - (No action code) | $0.146^{\pm.003}$ | $0.955^{\pm.001}$ | $6.797^{\pm.066}$ | $2.769^{\pm.045}$ |
| | NTU13 | | | |
| Real | $0.031^{\pm.004}$ | $0.999^{\pm.001}$ | $7.108^{\pm.048}$ | $2.194^{\pm.025}$ |
| Action2Motion [10] | $0.351^{\pm.011}$ | $0.949^{\pm.001}$ | $\mathbf{7.116^{\pm.037}}$ | $\mathbf{2.186^{\pm.033}}$ |
| ACTOR [31] | $\underline{0.11^{\pm.00}}$ | $0.971^{\pm.002}$ | $\underline{7.08^{\pm.04}}$ | $2.08^{\pm.01}$ |
| INR (Transformer) | $\mathbf{0.097^{\pm..001}}$ | $\mathbf{0.977^{\pm..001}}$ | $7.060^{\pm.040}$ | $\underline{2.108^{\pm.025}}$ |
| INR (MLP) | $\underline{0.113^{\pm.001}}$ | $\underline{0.976^{\pm.001}}$ | $7.070^{\pm.052}$ | $2.070^{\pm.043}$ |
| - (Non-variational) | $0.646^{\pm.003}$ | $0.849^{\pm.001}$ | $6.905^{\pm.056}$ | $3.244^{\pm.049}$ |
| - (No action code) | $0.202^{\pm.002}$ | $0.912^{\pm.001}$ | $7.025^{\pm.050}$ | $2.648^{\pm.043}$ |

decomposed representations by comparing to a version with no action code. Note that by construction our motion generation procedure can generate high-quality motions for arbitrarily specified sequence lengths (as in Table 1) within the variation of training sequence lengths, whereas previous works reported a performance drop for variable-length generation.

The results show that our proposed method improves over both Action2Motion [10] and ACTOR [31] especially on the FID and accuracy metric. We show that our optimized INR outperforms methods with representations produced by an optimized encoder. This is most apparent when comparing our implicit Transformer model and ACTOR, since both models use the same decoder architecture.

Furthermore, we show high performance even with a simple MLP decoder. This shows that the self-attention mechanism is not necessary. We argue, that the common property of the Transformer-based models and our implicit MLP-based model, namely time-independent sequence-wise representations, are critical for motion generation performance. Such representations can avoid the error accumulation of current RNN-based models and represent variable-length sequences.

Comparing the non-variational and variational approach, we find that the realism and diversity is improved with the variational approach (see (Non-variational) in Table 1). This finding suggests that the variational approach

**Table 2.** Baseline comparison with UESTC. (± indicates 95% confidence interval, → closer to real is better

| Method | $\text{FID}_{\text{train}}$ ↓ | $\text{FID}_{\text{test}}$ ↓ | Accuracy ↑ | Diversity → | Multimod. → |
|---|---|---|---|---|---|
| Real | $2.92^{\pm.26}$ | $2.79^{\pm.29}$ | $0.988^{\pm.001}$ | $33.34^{\pm.320}$ | $14.16^{\pm.06}$ |
| ACTOR [31] | $20.49^{\pm2.31}$ | $23.43^{\pm2.20}$ | $0.911^{\pm.003}$ | $\mathbf{31.96^{\pm.33}}$ | $\mathbf{14.52^{\pm.09}}$ |
| Ours (MLP) | $\mathbf{9.55^{\pm.06}}$ | $\mathbf{15.00^{\pm.09}}$ | $\mathbf{0.941^{\pm.001}}$ | $31.59^{\pm.19}$ | $14.68^{\pm.07}$ |

**Table 3.** Mean Maximum Similarity as a sanity check to detect overfitting.

| Method | HumanAct12 | NTU13 | UESTC |
|---|---|---|---|
| Real | $0.329^{\pm.003}$ | $0.209^{\pm.002}$ | $4.925^{\pm.007}$ |
| Action2Motion [10] | $0.945^{\pm.006}$ | $0.667^{\pm.006}$ | — |
| ACTOR [31] | $0.921^{\pm.001}$ | $0.701^{\pm.001}$ | $8.645^{\pm0.008}$ |
| INR (MLP) | $0.941^{\pm.005}$ | $0.620^{\pm.001}$ | $7.113^{\pm0.006}$ |
| INR (Transformer) | $0.778^{\pm.003}$ | $0.570^{\pm.002}$ | — |

strongly regularizes the latent space and improves sampling of new motions. However, both approaches for implicit representations are able to reach similar reconstruction performance for training samples and learn effective representations that allow high quality reconstruction.

Comparing the approach with an action-wise and sequence-wise representation to an approach only with a sequence-wise representation, we find a clear advantage from using a decomposed action representation. Even the approach with only sequence-wise representations performs comparable to the RNN-baseline (See (No action code) in Table 1) However, a decomposition representation is needed to outperform the Transformer baseline.

For further ablation studies we refer to Appendix B, where we investigate various modeling choices. Among others, we investigate the effect of the number of components in the GMM and show in Appendix B.2 high performance, on-par with Action2Motion [10], even when using GMMs with a single component.

Finally we perform a sanity check to see if any model is reproducing training samples by checking the proposed Mean Maximum Similarity between real and generated sequences defined by Eq. (11) and show the results in Table 3. We interpret the gap between this baseline of real motions and all other models as an indication that no model is just reproducing training samples. Note that while the INR (Transformer) model has a lower MMS than other models, the gap to the baseline is significant. This suggests that it generates motions that are more similar to training motions than other models yet distinct from them.

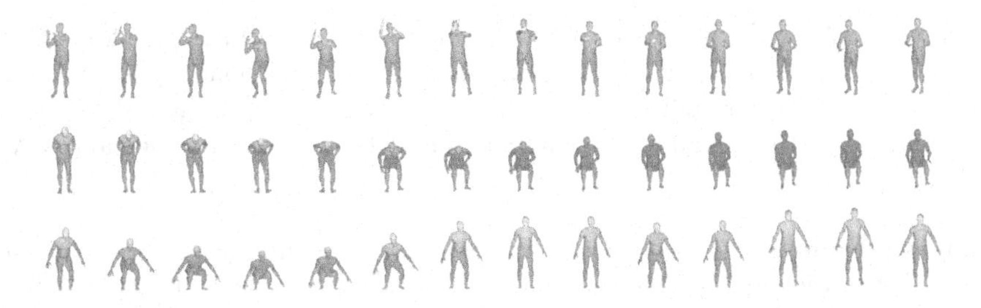

**Fig. 3.** Motions generated by our MLP model trained on HumanAct predicted with 40 time-steps (each third frame shown) for the actions *throw*, *sit* and *jump*.

### 4.5 Qualitative Results

We manually inspect the quality of the generated motions and find that our methods consistently generates high-quality motions even over a long time range. In particular we observe, that the RNN-based Action2Motion [10] shows a slow-down effect when predicting long sequences, which neither our methods nor the Transformer-based ACTOR [31] suffer from.

Also, we observe that we reliably generate complete actions, due to our model's ability to model the sequence-length. In contrast, both Action2Motion as well as ACTOR tend to generate incomplete actions, particularly when generating short sequences. As shown in Fig. 3, even for short sequences and actions with a clear start and end our generated actions are complete. For more qualitative results, we refer to supplementary videos and Appendix C.

## 5 Limitations and Future Work

While our implicit sequence representations are parameter-efficient, the effort to train sequence-wise parameters scales linearly with the size of the training dataset. Furthermore, we observe a sensitivity to the ratio of parameter updates between the sequence-wise parameters and the decoder parameters. If the decoder parameters are updated significantly more often than sequence-wise parameters, our implicit models might perform poorly.

## 6 Conclusion

We present an MLP-based model for action-conditional human motion generation. The proposed approach improves over previous RNN-based and Transformer-based baselines by employing variational implicit neural representations. We argue that the likely reason for the success of our method is implicit neural representations, which are optimized representations that can represent full variable-length sequences and we supported this hypothesis experimentally by reaching state-of-the-art performance on commonly used metrics for motion generation. While the results of our work may improve technologies for animation and action recognition, there is the potential for malicious use such a deep fakes as well. For more detailed discussions about such potential negative societal impacts and personal data of human subjects we refer to Appendices A.5 and A.6.

**Acknowledgements.** This work is an outcome of a research project, Development of Quality Foundation for Machine-Learning Applications, supported by DENSO IT LAB Recognition and Learning Algorithm Collaborative Research Chair (Tokyo Tech.). It was also supported by JST CREST JPMJCR1687.

## References

1. Anokhin, I., Demochkin, K., Khakhulin, T., Sterkin, G., Lempitsky, V., Korzhenkov, D.: Image generators with conditionally-independent pixel synthesis. In: Proceedings of the IEEE/CVF Conference on Computer Vision and Pattern Recognition (CVPR), pp. 14278–14287 (2021)
2. Barsoum, E., Kender, J., Liu, Z.: HP-GAN: probabilistic 3D human motion prediction via GAN. In: Proceedings of the IEEE Conference on Computer Vision and Pattern Recognition Workshops (CVPRW), pp. 1418–1427 (2018)
3. Battan, N., Agrawal, Y., Rao, S.S., Goel, A., Sharma, A.: GlocalNet: class-aware long-term human motion synthesis. In: Proceedings of the IEEE/CVF Winter Conference on Applications of Computer Vision (WACV), pp. 879–888 (2021)
4. Bond-Taylor, S., Willcocks, C.G.: Gradient origin networks. In: 9th International Conference on Learning Representations, ICLR 2021, Virtual Event, Austria, 3–7 May 2021. OpenReview.net (2021)
5. Butepage, J., Black, M.J., Kragic, D., Kjellstrom, H.: Deep representation learning for human motion prediction and classification. In: Proceedings of the IEEE/CVF Conference on Computer Vision and Pattern Recognition (CVPR), pp. 6158–6166 (2017)
6. Chen, Y., Liu, C., Shi, B.E., Liu, M.: CoMoGCN: Coherent motion aware trajectory prediction with graph representation. In: British Machine Vision Conference (BMVC) (2020)
7. Chen, Z., Zhang, H.: Learning implicit fields for generative shape modeling. In: Proceedings of the IEEE/CVF Conference on Computer Vision and Pattern Recognition (CVPR) (2019)
8. DeVries, T., Bautista, M.A., Srivastava, N., Taylor, G.W., Susskind, J.M.: Unconstrained scene generation with locally conditioned radiance fields. In: Proceedings of the IEEE/CVF International Conference on Computer Vision (ICCV), pp. 14304–14313 (2021)
9. Doersch, C., Zisserman, A.: Sim2real transfer learning for 3D human pose estimation: motion to the rescue. In: Advances in Neural Information Processing Systems (NeurIPS), vol. 32 (2019)
10. Guo, C., et al.: Action2Motion: conditioned generation of 3D human motions. In: Proceedings of the 28th ACM International Conference on Multimedia (MM 2020) (2020)
11. Higgins, I., et al.: $\beta$-VAE: learning basic visual concepts with a constrained variational framework. In: 5th International Conference on Learning Representations, ICLR 2017, Toulon, France, 24–26 April 2017, Conference Track Proceedings. OpenReview.net (2017)

12. Honda, Y., Kawakami, R., Naemura, T.: RNN-based motion prediction in competitive fencing considering interaction between players. In: British Machine Vision Conference (BMVC) (2020)
13. Hou, Y., Yao, H., Sun, X., Li, H.: Soul dancer: emotion-based human action generation. ACM Trans. Multimed. Comput. Commun. Appl. (TOMM) **15**(3s), 1–19 (2020)
14. Ji, Y., Xu, F., Yang, Y., Shen, F., Shen, H.T., Zheng, W.S.: A large-scale RGB-D database for arbitrary-view human action recognition. In: Proceedings of the 26th ACM International Conference on Multimedia (MM 2018) (2018)
15. Kanazawa, A., Zhang, J.Y., Felsen, P., Malik, J.: Learning 3D human dynamics from video. In: Proceedings of the IEEE/CVF Conference on Computer Vision and Pattern Recognition (CVPR) (2019)
16. Karras, T., et al.: Advances in Neural Information Processing Systems (NeurIPS), vol. 34 (2021)
17. Kingma, D.P., Welling, M.: Auto-encoding variational bayes. In: 2nd International Conference on Learning Representations, ICLR 2014, Banff, AB, Canada, 14–16 April 2014, Conference Track Proceedings (2014)
18. Kocabas, M., Athanasiou, N., Black, M.J.: Vibe: video inference for human body pose and shape estimation. In: Proceedings of the IEEE/CVF Conference on Computer Vision and Pattern Recognition (CVPR), pp. 5253–5263 (2020)
19. Li, R., Yang, S., Ross, D.A., Kanazawa, A.: AI choreographer: music conditioned 3D dance generation with AIST++. In: Proceedings of the IEEE International Conference on Computer Vision (ICCV) (2021)
20. Li, Z., Niklaus, S., Snavely, N., Wang, O.: Neural scene flow fields for space-time view synthesis of dynamic scenes. In: Proceedings of the IEEE/CVF Conference on Computer Vision and Pattern Recognition (CVPR) (2021)
21. Liu, J., Shahroudy, A., Perez, M., Wang, G., Duan, L.Y., Kot, A.C.: NTU RGB+D 120: a large-scale benchmark for 3D human activity understanding. IEEE Trans. Pattern Anal. Mach. Intell. **42**(10), 2684–2701 (2019)
22. Loper, M., Mahmood, N., Romero, J., Pons-Moll, G., Black, M.J.: SMPL: a skinned multi-person linear model. ACM Trans. Graphics (Proc. SIGGRAPH Asia) **34**(6), 248:1–248:16 (2015)
23. Meng, F., Liu, H., Liang, Y., Tu, J., Liu, M.: Sample fusion network: an end-to-end data augmentation network for skeleton-based human action recognition. IEEE Trans. Image Process. **28**(11), 5281–5295 (2019)
24. Mildenhall, B., Srinivasan, P.P., Tancik, M., Barron, J.T., Ramamoorthi, R., Ng, R.: NeRF: representing scenes as neural radiance fields for view synthesis. In: Vedaldi, A., Bischof, H., Brox, T., Frahm, J.-M. (eds.) ECCV 2020. LNCS, vol. 12346, pp. 405–421. Springer, Cham (2020). https://doi.org/10.1007/978-3-030-58452-8_24
25. Niemeyer, M., Geiger, A.: GIRAFFE representing scenes as compositional generative neural feature fields. In: Proceedings of the IEEE/CVF Conference on Computer Vision and Pattern Recognition (CVPR) (2021)
26. Niemeyer, M., Mescheder, L., Oechsle, M., Geiger, A.: Occupancy flow: 4D reconstruction by learning particle dynamics. In: Proceedings of the IEEE/CVF International Conference on Computer Vision (ICCV) (2019)
27. Park, J.J., Florence, P., Straub, J., Newcombe, R., Lovegrove, S.: DeepSDF learning continuous signed distance functions for shape representation. In: Proceedings of the IEEE/CVF Conference on Computer Vision and Pattern Recognition (CVPR), pp. 165–174 (2019)

28. Paszke, A., et al.: Pytorch: an imperative style, high-performance deep learning library. In: Advances in Neural Information Processing Systems (NeurIPS), vol. 32 (2019)
29. Pavlakos, G., et al.: Expressive body capture: 3D hands, face, and body from a single image. In: Proceedings of the IEEE/CVF Conference on Computer Vision and Pattern Recognition (CVPR) (2019)
30. Pedregosa, F., et al.: Scikit-learn: machine learning in Python. J. Mach. Learn. Res. **12**, 2825–2830 (2011)
31. Petrovich, M., Black, M.J., Varol, G.: Action-conditioned 3D human motion synthesis with Transformer VAE. In: International Conference on Computer Vision (ICCV) (2021)
32. Ravi, N., et al.: Accelerating 3D deep learning with Pytorch3D. arXiv:2007.08501 (2020)
33. Schwarz, K., Liao, Y., Niemeyer, M., Geiger, A.: GRAF generative radiance fields for 3D-aware image synthesis. In: Advances in Neural Information Processing Systems (NeurIPS) (2020)
34. Sengupta, A., Budvytis, I., Cipolla, R.: Synthetic training for accurate 3D human pose and shape estimation in the wild. In: British Machine Vision Conference (BMVC) (2020)
35. Sitzmann, V., Martel, J.N., Bergman, A.W., Lindell, D.B., Wetzstein, G.: Implicit neural representations with periodic activation functions. In: Advances in Neural Information Processing Systems (NeurIPS) (2020)
36. Starke, S., Zhao, Y., Komura, T., Zaman, K.: Local motion phases for learning multi-contact character movements. ACM Trans. Graph. **39**(4) (2020)
37. Starke, S., Zhao, Y., Zinno, F., Komura, T.: Neural animation layering for synthesizing martial arts movements. ACM Trans. Graph. **40**(4) (2021)
38. Varol, G., Laptev, I., Schmid, C., Zisserman, A.: Synthetic humans for action recognition from unseen viewpoints. Int. J. Comput. Vis. **129**(7), 2264–2287 (2021)
39. Yan, S., Li, Z., Xiong, Y., Yan, H., Lin, D.: Convolutional sequence generation for skeleton-based action synthesis. In: Proceedings of the IEEE/CVF International Conference on Computer Vision (ICCV) (2019)
40. Yariv, L., et al.: Multiview neural surface reconstruction by disentangling geometry and appearance. In: Advances in Neural Information Processing Systems (NeurIPS), vol. 33 (2020)
41. Zhou, Y., Barnes, C., Lu, J., Yang, J., Li, H.: On the continuity of rotation representations in neural networks. In: Proceedings of the IEEE/CVF Conference on Computer Vision and Pattern Recognition (CVPR), pp. 5745–5753 (2019)
42. Zou, S., et al.: 3D human shape reconstruction from a polarization image. In: Vedaldi, A., Bischof, H., Brox, T., Frahm, J.-M. (eds.) ECCV 2020. LNCS, vol. 12359, pp. 351–368. Springer, Cham (2020). https://doi.org/10.1007/978-3-030-58568-6_21

# Learning Object Placement via Dual-Path Graph Completion

Siyuan Zhou, Liu Liu, Li Niu(✉), and Liqing Zhang

MoE Key Lab of Artificial Intelligence, Shanghai Jiao Tong University, Shanghai, China
{ssluvble,Shirlley,ustcnewly}@sjtu.edu.cn
zhang-lq@cs.sjtu.edu.cn

**Abstract.** Object placement aims to place a foreground object over a background image with a suitable location and size. In this work, we treat object placement as a graph completion problem and propose a novel graph completion module (GCM). The background scene is represented by a graph with multiple nodes at different spatial locations with various receptive fields. The foreground object is encoded as a special node that should be inserted at a reasonable place in this graph. We also design a dual-path framework upon the structure of GCM to fully exploit annotated composite images. With extensive experiments on OPA dataset, our method proves to significantly outperform existing methods in generating plausible object placement without loss of diversity.

## 1 Introduction

Image composition [2,24,35] refers to the task of producing a realistic composite image based on a background image and a foreground object, which can benefit a wide range of applications of entertainment, virtual reality, and artistic creation. The main concerns of this task include both appearance compatibility (*e.g.*, shading, lighting), geometric compatibility (*e.g.*, object size, camera viewpoint), and semantic compatibility (*e.g.*, semantic context) between foreground and background [2,20]. In this work, we deal with the object placement problem [18,31,38], which is a sub-task of image composition and aims to generate reasonable locations and sizes to place foreground over background. Object placement can be applied in various conditions. For example, during artistic creation, this technique could provide designers with feedback and make recommendation for them when they are placing objects. Another application is automatic advertising, which aims to help advertisers with the product insertion in the background scene [41]. Object placement is a challenging problem, partially due to the lack of annotated composite images. Recently, the first object

**Supplementary Information** The online version contains supplementary material available at https://doi.org/10.1007/978-3-031-19790-1_23.

© The Author(s), under exclusive license to Springer Nature Switzerland AG 2022
S. Avidan et al. (Eds.): ECCV 2022, LNCS 13677, pp. 373–389, 2022.
https://doi.org/10.1007/978-3-031-19790-1_23

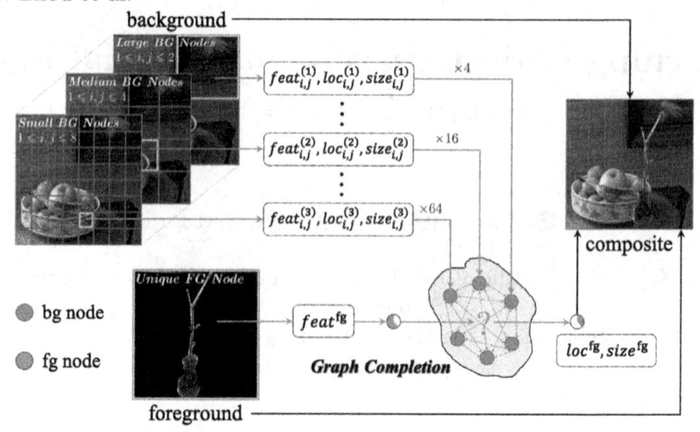


**Fig. 1.** Illustration of our graph completion module (GCM). Background nodes are extracted from different positions with different receptive fields. The unique foreground node lacks location and size information. GCM infers the missing information of the foreground node to complete the graph

placement assessment (OPA) dataset [22] was released, which contains composite images and their binary rationality labels indicating whether they are reasonable (positive sample) or not (negative sample) in terms of foreground object placement.

As far as we know, only few works focus on general object placement learning, like TERSE [31] and PlaceNet [38]. Both of them adopt adversarial training to learn the reasonable distribution from real images, and predict a set of parameters to indicate locations and sizes for placing foreground objects during inference. The drawbacks of these methods mainly come from two aspects. Firstly, they did not explicitly consider the relation between the foreground object and the background scene, which is of great importance in object placement. Secondly, they did not fully exploit the annotated composite images. Based on these considerations, we design our method to generate more plausible and diverse object placements.

To better exploit the relationship between the foreground object and the background scene, we treat object placement learning as a graph completion problem, as shown in Fig. 1. On the one hand, the background image can be considered as a graph with multiple nodes. Each node pays attention to a local region on the background feature map with a specific spatial position and receptive field. Multiple background nodes work together to form a graph, enabling the model to discover a variety of plausible solutions on the background. On the other hand, the foreground image can be seen as a special node that should be inserted into the graph with suitable location and size. Note that the background nodes have both feature information and location/size information, whereas the foreground node only has feature information. We need to infer the missing location/size information for the foreground node to complete the graph and obtain a reasonable composite layout.

To complete the graph, we propose a novel graph completion module (GCM) with two components: node extraction head (NEH) and placement seeking network (PSN). NEH aims to transform foreground and background into a node graph. We incorporate one foreground NEH to extract a foreground node and another background NEH to extract multiple background nodes from different positions and scales. PSN contains an attention layer and a regression block. The attention layer attends relevant information from different background regions for the foreground object and produces an attended feature vector, which will be transmitted to the regression block to predict transformation parameters for object placement. Note that object placement is a multi-modal problem, that is, the reasonable placement has many possible solutions given a pair of foreground and background. This guides us to incorporate a random vector in the regression block to generate diversified transformation parameters for object placement.

To take full advantage of annotated composite images, we design a dual-path framework upon the structure of GCM, including an unsupervised path and a supervised path. The whole framework follows an adversarial learning paradigm, which is composed by a GCM (functioning as a generator) and a discriminator. Transformation parameters produced by GCM are applied to predict reasonable object placements, which are then pushed to the discriminator so as to check the plausibility of the generated composite images. The distinction between two paths lies in the provided data. The unsupervised path only utilizes pairs of foreground and background as input, while the supervised path have additional annotated composite images. Recall that GCM contains a random vector in the regression block. In our implementation, two paths choose different types of random vectors. In the unsupervised path, random vectors are sampled from unit Gaussian distribution. In the supervised path, random vectors (also called latent vectors) are encoded from composite images with positive annotation via a VAE [15] encoder. We expect the generator to reconstruct the ground-truth transformation parameters of each positive composite image from its corresponding latent vector. Under this design, we establish a bijection between the latent vector and the predicted object placement. This can avoid mode collapse [43] and bring multifarious generation results. Since two paths share weights in the generator and the discriminator, we hope that the supervised path could gradually guide the unsupervised path to generate reasonable composite images. During inference, by sampling random vectors in the unsupervised path, we can obtain diverse solutions for object placement. Since the key idea of this work is **Gra**ph **co**mpletion, we name our network **GracoNet**.

In summary, the main contributions of this paper are: 1) We formulate object placement as a graph completion problem and propose a novel graph completion module (GCM). 2) We design a dual-path framework upon GCM to fully exploit annotated composite images and overcome mode collapse issue. 3) Experiments on OPA dataset demonstrate the superiority of our method in generation plausibility and diversity when compared with existing works.

# 2 Related Work

## 2.1 Image Composition

The main challenges of image composition [3,16,17,24,29,37] lie in appearance compatibility, geometric compatibility, and semantic compatibility between foreground and background. Up to now, this task has been explored from a variety of perspectives. For example, [42] refined composite images by distinguishing them from natural photographs via a simple CNN model. [14] incorporated scene graphs to explicitly learn relationships between objects and generate images from a computed scene layout. [2] introduced a new GAN architecture to explore geometric and color correction at the same time. [35] pointed out the drawback of cutting-edge methods and addressed it by a spatially-adaptive mechanism. [36,39] explored the image blending field and achieved seamless connection between foreground and background via blending boundary regions. [13,21] generated realistic shadows *w.r.t* foreground objects over background scenes. [4–6,32] proposed image harmonization to deal with color and lighting inconsistency in composite images. Additionally, object placement has been studied to realize geometric compatibility, which will be introduced next.

## 2.2 Object Placement

Learning object placement has attracted wide attention in recent years. Several early methods [10,26] attempted to design explicit rules to place foreground objects. The followers went a step further to automatically exploit reasonable placement [1,18–20,30,31,38,41]. For example, [20] employed spatial transformer networks to learn geometric corrections that warp composite images for appropriate layouts. [18] designed a two-step strategy to find where to place objects and what categories to place. [19] used VAE [15] to predict 3D locations and poses of humans. [1] achieved self-consistency in training a composition network by decomposing composite images back into individual objects. Compared with existing works, we offer a new perspective by treating object placement as a graph completion problem. Our dual-path framework could effectively boost generation plausibility and diversity by discovering placement clues from supervisions.

# 3 Methodology

Suppose we have a background image $\mathrm{I}^{\mathrm{bg}} \in \mathcal{R}^{3\times H\times W}$ and a foreground image $\mathrm{I}^{\mathrm{fg}} \in \mathcal{R}^{3\times H\times W}$ together with a binary object mask $\mathrm{M}^{\mathrm{fg}} \in \mathcal{R}^{1\times H\times W}$ delineating the foreground object, where $H$ and $W$ represent image height and width. Our objective is to output transformation parameters $\mathbf{t}$ that transform foreground and places it over background to obtain a composite image $\mathrm{I}^{\mathrm{c}} \in \mathcal{R}^{3\times H\times W}$ with a composite foreground mask $\mathrm{M}^{\mathrm{c}} \in \mathcal{R}^{1\times H\times W}$. By using $\mathcal{F}_{\mathbf{t}}$ to represent the transformation function with parameters $\mathbf{t}$, we have

$$(\mathrm{I}^{\mathrm{c}}, \mathrm{M}^{\mathrm{c}}) = \mathcal{F}_{\mathbf{t}}(\mathrm{I}^{\mathrm{bg}}, \mathrm{I}^{\mathrm{fg}}, \mathrm{M}^{\mathrm{fg}}). \tag{1}$$

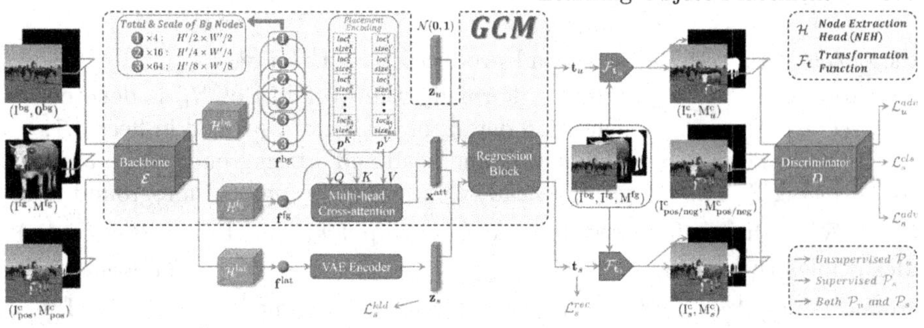


**Fig. 2.** Our GracoNet has an unsupervised path and a supervised path built upon Graph Completion Module (GCM). GCM consists of backbone network $\mathcal{E}$, node extraction head $\mathcal{H}$, and placement seeking network $\mathcal{S}$ (a multi-head cross-attention layer and a regression block). Loss functions are marked in red. More details are left to Sect. 3 (Color figure online)

The detailed definitions of $\mathbf{t}$ and $\mathcal{F}_\mathbf{t}$ are left to supplementary. In the following paragraphs, we will first introduce how GCM works to generate transformation parameters $\mathbf{t}$ in Sect. 3.1. Then, we will introduce the GCM-based dual-path framework that reasonably makes use of supervised information in Sect. 3.2.

### 3.1 Graph Completion Module (GCM)

The core module in our network is Graph Completion Module (GCM), which takes in a pair of foreground and background as well as a random vector to produce reasonable placement (transformation parameter) for the foreground object. GCM consists of three components: 1) backbone network $\mathcal{E}$, 2) node extraction head $\mathcal{H}$, and 3) placement seeking network $\mathcal{S}$. The backbone network extracts general feature maps for the input images. The node extraction head encodes graph nodes from the feature maps. After that, the placement seeking network finds the relationships between the foreground node and the background nodes, and finally outputs transformation parameters $\mathbf{t}$ that reasonably places the foreground node to complete the graph.

**Backbone Network.** Our backbone network $\mathcal{E}$ takes in the concatenation of a three-channel image and a one-channel mask to produce a feature map $\mathrm{F} \in \mathcal{R}^{C' \times H' \times W'}$. We use object masks $\mathrm{M}^{\mathrm{fg}}$ for foreground images $\mathrm{I}^{\mathrm{fg}}$ and apply all-zero masks $\mathbf{0}^{\mathrm{bg}}$ to background images $\mathrm{I}^{\mathrm{bg}}$. Formally, foreground and background features are extracted by $\mathrm{F}^{\mathrm{fg}} = \mathcal{E}(\mathrm{I}^{\mathrm{fg}}, \mathrm{M}^{\mathrm{fg}})$ and $\mathrm{F}^{\mathrm{bg}} = \mathcal{E}(\mathrm{I}^{\mathrm{bg}}, \mathbf{0}^{\mathrm{bg}})$.

**Node Extraction Head (NEH).** Given a feature map F as input and a positive integer array $\mathbf{n} = [n_1, n_2, \cdots, n_L]$ as parameter, a node extraction head $\mathcal{H}_\mathbf{n}$ consists of $L$ node extraction layers named $\mathcal{G}_1, \mathcal{G}_2, \cdots, \mathcal{G}_L$. The $l$-th layer $\mathcal{G}_l$ evenly divides the feature map F into $n_l \times n_l$ cells and sequentially encodes them into a stack of $n_l^2$ nodes with dimension $C$, denoted by $\mathbf{f}^{(l)} \in \mathcal{R}^{n_l^2 \times C}$. Each node accounts for a spatial resolution of $\frac{H'}{n_l} \times \frac{W'}{n_l}$ on the feature map. $\mathcal{H}_\mathbf{n}$ gathers the outputs from all $L$ layers, and produces $\mathbf{f} = [\mathbf{f}^{(1)}, \mathbf{f}^{(2)}, \cdots, \mathbf{f}^{(L)}] \in \mathcal{R}^{N \times C}$ with totally $N = \sum_{l=1}^{L} n_l^2$ nodes. Formally, the workflow of $\mathcal{H}_{\mathbf{n}}$ is denoted by $\mathbf{f} = \mathcal{H}(\mathrm{F}; \mathbf{n})$. More implementation details of NEH can be found in Sect. 4.1.

In GCM, we incorporate a foreground head and a background head, as illustrated in Fig. 2. The foreground head $\mathcal{H}^{\mathrm{fg}}_{\mathbf{n}=[1]}$ encodes a global foreground node $\mathbf{f}^{\mathrm{fg}} \in \mathcal{R}^{1 \times C}$ from the foreground feature map $\mathrm{F}^{\mathrm{fg}}$, *i.e.*, $\mathbf{f}^{\mathrm{fg}} = \mathcal{H}^{\mathrm{fg}}(\mathrm{F}^{\mathrm{fg}}; [1])$. Meanwhile, the background head $\mathcal{H}^{\mathrm{bg}}_{\mathbf{n}=[2,4,8]}$ produces $N = 84$ local background nodes $\mathbf{f}^{\mathrm{bg}} \in \mathcal{R}^{84 \times C}$ from three scales at different locations, *i.e.*, $\mathbf{f}^{\mathrm{bg}} = \mathcal{H}^{\mathrm{bg}}(\mathrm{F}^{\mathrm{bg}}; [2, 4, 8])$. On the whole, the unique foreground node and all the 84 background nodes work together to form a node graph, as shown in Fig. 1.

**Placement Seeking Network (PSN).** It is noteworthy that the node graph is now incomplete, because the location/size of the foreground node awaits to be determined. To address this issue, we introduce a placement seeking network $\mathcal{S}$, which consists of a multi-head cross-attention layer and a regression block.

First, we use a Transformer multi-head attention layer [33] to explore the relationship between the unique foreground node $\mathbf{f}^{\mathrm{fg}} \in \mathcal{R}^{1 \times C}$ and the 84 local background nodes $\mathbf{f}^{\mathrm{bg}} \in \mathcal{R}^{84 \times C}$ by treating $\mathbf{f}^{\mathrm{fg}}$ as query and $\mathbf{f}^{\mathrm{bg}}$ as key/value (*i.e.*, cross-attention). Inspired by Transformer that incorporates position encoding [7, 9, 23, 25, 27, 33] into the attention layer, we introduce placement encoding to encapsulate both location and size information of $\mathbf{f}^{\mathrm{bg}}$. Since different background nodes have distinct positions/scales, they should have different placement encodings. In our implementation, placement encoding includes a learnable $\boldsymbol{p}^K$ (*resp.*, $\boldsymbol{p}^V$) for key (*resp.*, value), which is based on but not exactly the same as the encoding form in [27]. In general, we denote the output of the attention layer by $\mathbf{x}^{\mathrm{att}} = \mathit{Attention}(\mathbf{f}^{\mathrm{fg}}, \mathbf{f}^{\mathrm{bg}}) \in \mathcal{R}^{1 \times C}$, and we will introduce the details as follows.

Specifically, we calculate the output $\boldsymbol{o} \in \mathcal{R}^{1 \times d_o}$ of each attention head:

$$\boldsymbol{o} = \sum_{j=1}^{84} \alpha_j (\mathbf{f}_j^{\mathrm{bg}} W^V + \boldsymbol{p}_j^V), \tag{2}$$

where coefficient $\alpha_j$ represents the edge weight between the foreground node and the $j$-th background node in the graph:

$$\alpha_j = \mathrm{Softmax}\left( \frac{(\mathbf{f}^{\mathrm{fg}} W^Q)(\mathbf{f}_j^{\mathrm{bg}} W^K + \boldsymbol{p}_j^K)^\top}{\sqrt{d_o}} \right). \tag{3}$$

In Eq. (2) and Eq. (3), $W^Q, W^K, W^V \in \mathcal{R}^{C \times d_o}$ are linear learnable weights for query, key, and value, respectively. $\boldsymbol{p}^K, \boldsymbol{p}^V \in \mathcal{R}^{84 \times d_o}$ are learnable placement encodings. Following [33], we incorporate 8 attention heads and set $d_o$ as $\frac{C}{8}$ in our implementation. $\boldsymbol{p}^K$ and $\boldsymbol{p}^V$ are not shared among different attention heads. In this way, different attention heads could potentially discover various placement information, resulting in more diversified generation results. The attention output $\boldsymbol{o}$ from different heads are concatenated and transformed with another linear layer to obtain the final output $\mathbf{x}^{\mathrm{att}} \in \mathcal{R}^{1 \times C}$ of the attention layer.

Second, we apply a regression block to predict transformation parameters from the attention output. In order to generate composite images with diversified reasonable placements, we incorporate a random vector $\mathbf{z} \in \mathcal{R}^{1\times C_z}$ with dimension $C_z$ into the block. Specifically, the regression block takes in the concatenation of $\mathbf{x}^{\text{att}}$ and $\mathbf{z}$ to predict transformation parameters $\mathbf{t} = Regression(\mathbf{x}^{\text{att}}, \mathbf{z})$. By sampling different $\mathbf{z}$ at test time, we can obtain a variety of reasonable $\mathbf{t}$ conditioned on $\mathbf{x}^{\text{att}}$. The detailed implementation for $\mathbf{z}$ is left to Sect. 3.2.

### 3.2 Dual-Path Framework

Our whole framework is designed as a Generative Adversarial Network (GAN) [11] including a generator $G$ and a discriminator $D$. The generator $G$ is comprised of a graph completion module (GCM) and a transformation function $\mathcal{F}$. As introduced in Sect. 3.1, GCM works by predicting transformation parameters $\mathbf{t}$ from a tuple of $(\text{I}^{\text{bg}}, \text{I}^{\text{fg}}, \text{M}^{\text{fg}}, \mathbf{z})$. Using $\mathbf{t}$ as parameters for $\mathcal{F}$, we could follow Eq. (1) to obtain a generated composite image $\text{I}^{\text{c}}$ with object mask $\text{M}^{\text{c}}$. Formally, the workflow of our generator $G$ is denoted by $(\text{I}^{\text{c}}, \text{M}^{\text{c}}) = G(\text{I}^{\text{bg}}, \text{I}^{\text{fg}}, \text{M}^{\text{fg}}, \mathbf{z})$. Then, the discriminator $D$ takes the concatenated $(\text{I}^{\text{c}}, \text{M}^{\text{c}})$ as input, and predicts the probability of reasonableness for the generated composite image.

To facilitate object placement learning, we adopt a dual-path adversarial training framework containing an unsupervised path $\mathcal{P}_u$ and a supervised path $\mathcal{P}_s$. In $\mathcal{P}_u$, we only have background images $\text{I}^{\text{bg}}$ and foreground images $\text{I}^{\text{fg}}$ with object masks $\text{M}^{\text{fg}}$. In $\mathcal{P}_s$, we are provided with additional annotated composite images/masks $\text{I}^{\text{c}}_{\text{pos}}/\text{M}^{\text{c}}_{\text{pos}}$ (*resp.*, $\text{I}^{\text{c}}_{\text{neg}}/\text{M}^{\text{c}}_{\text{neg}}$) with positive (*resp.*, negative) annotation in terms of object placement, as well as their corresponding original $\text{I}^{\text{bg}}/\text{I}^{\text{fg}}/\text{M}^{\text{fg}}$ that constitute them. Due to the difference of provided data, $\mathcal{P}_u$ and $\mathcal{P}_s$ adopt distinct implementations for the random vector $\mathbf{z}$. To distinguish between representations in two paths, we use notation $\mathbf{z}_u$ in $\mathcal{P}_u$ and $\mathbf{z}_s$ in $\mathcal{P}_s$, respectively. The generated composite outputs in two paths are represented by $\text{I}^{\text{c}}_u/\text{M}^{\text{c}}_u$ and $\text{I}^{\text{c}}_s/\text{M}^{\text{c}}_s$ with different subscripts correspondingly.

**Unsupervised Path ($\mathcal{P}_u$).** In unsupervised path $\mathcal{P}_u$, random vectors $\mathbf{z}_u$ are sampled from unit Gaussian distribution, *i.e.*, $\mathbf{z}_u \sim \mathcal{N}(\mathbf{0}, \mathbf{1})$. Correspondingly, the generator outputs $(\text{I}^{\text{c}}_u, \text{M}^{\text{c}}_u) = G(\text{I}^{\text{bg}}, \text{I}^{\text{fg}}, \text{M}^{\text{fg}}, \mathbf{z}_u)$. We employ an adversarial loss $\mathcal{L}^{adv}_u(G, D)$ to push the generated composite image to be undistinguishable from positive composite images.

$$\mathcal{L}^{adv}_u = \mathbb{E}_{\mathbf{z}_u \sim \mathcal{N}(\mathbf{0},\mathbf{1})}[\log(1 - D(G(\text{I}^{\text{bg}}, \text{I}^{\text{fg}}, \text{M}^{\text{fg}}, \mathbf{z}_u)))]. \tag{4}$$

**Supervised Path ($\mathcal{P}_s$).** In supervised path $\mathcal{P}_s$, random vectors (also called latent vectors) $\mathbf{z}_s$ are designed to be sampled from global features of positive composite images via an encoder network as in VAE [15]. Given a labeled positive composite image $\text{I}^{\text{c}}_{\text{pos}}$ with object mask $\text{M}^{\text{c}}_{\text{pos}}$, we first calculate its ground-truth transformation parameters $\mathbf{t}_{\text{gt}}$ (see details in supplementary), which obeys $(\text{I}^{\text{c}}_{\text{pos}}, \text{M}^{\text{c}}_{\text{pos}}) \equiv \mathcal{F}_{\mathbf{t}_{\text{gt}}}(\text{I}^{\text{bg}}, \text{I}^{\text{fg}}, \text{M}^{\text{fg}})$. Then, our idea is to employ the latent vector $\mathbf{z}_s$

to help reconstruct $\mathbf{t}_{\text{gt}}$ because $\mathbf{z}_s$ contains potential information of positive composite images. In this way, we establish a bijection between $\mathbf{z}_s$ and $(\mathrm{I}^{\mathrm{c}}_{\text{pos}}, \mathrm{M}^{\mathrm{c}}_{\text{pos}})$.

Specifically, we first use the backbone network $\mathcal{E}$ to extract a feature map $\mathrm{F}^{\mathrm{c}} = \mathcal{E}(\mathrm{I}^{\mathrm{c}}_{\text{pos}}, \mathrm{M}^{\mathrm{c}}_{\text{pos}})$. Then we add a latent head $\mathcal{H}^{\text{lat}}_{\mathbf{n}=[1]}$ to encode a global latent node $\mathbf{f}^{\text{lat}} = \mathcal{H}^{\text{lat}}(\mathrm{F}^{\mathrm{c}}; [1])$. After that, we employ the encoder network in VAE to sample a latent vector $\mathbf{z}_s$ from $\mathbf{f}^{\text{lat}}$. Following VAE, we adopt a KL divergence loss $\mathcal{L}^{kld}_s = \mathcal{D}_{\text{KL}}(\mathcal{N}(\boldsymbol{\mu}_{\mathbf{z}_s}, \boldsymbol{\sigma}^2_{\mathbf{z}_s}) \parallel \mathcal{N}(0,1))$ that forces the distribution of $\mathbf{z}_s$ to be close to $\mathbf{z}_u$. With $\mathbf{z}_s$, GCM predicts transformation parameters $\mathbf{t}_s$ from a tuple of $(\mathrm{I}^{\text{bg}}, \mathrm{I}^{\text{fg}}, \mathrm{M}^{\text{fg}}, \mathbf{z}_s)$. We utilize a reconstruction loss $\mathcal{L}^{rec}_s$ to force $\mathbf{t}_s$ to approach the ground-truth $\mathbf{t}_{\text{gt}}$, which is defined as a weighted MSE between $\mathbf{t}_s$ and $\mathbf{t}_{\text{gt}}$ (see details in supplementary). Then, transformation function $\mathcal{F}$ with parameters $\mathbf{t}_s$ produces $(\mathrm{I}^{\mathrm{c}}_s, \mathrm{M}^{\mathrm{c}}_s)$, which is finally delivered to the discriminator $D$. Similar to $\mathcal{P}_u$, we also adopt an adversarial loss in $\mathcal{P}_s$:

$$\mathcal{L}^{adv}_s = \mathbb{E}_{\mathbf{z}_s \sim \mathcal{N}(\boldsymbol{\mu}_{\mathbf{z}_s}, \boldsymbol{\sigma}^2_{\mathbf{z}_s})}[\log(1 - D(G(\mathrm{I}^{\text{bg}}, \mathrm{I}^{\text{fg}}, \mathrm{M}^{\text{fg}}, \mathbf{z}_s)))]. \tag{5}$$

Additionally, we leverage both positive and negative composite images to update the discriminator by maximizing the negative form of binary cross-entropy loss:

$$\mathcal{L}^{cls}_s = \log D(\mathrm{I}^{\mathrm{c}}_{\text{pos}}, \mathrm{M}^{\mathrm{c}}_{\text{pos}}) + \log(1 - D(\mathrm{I}^{\mathrm{c}}_{\text{neg}}, \mathrm{M}^{\mathrm{c}}_{\text{neg}})). \tag{6}$$

In summary, the loss function for $\mathcal{P}_s$ is defined by

$$\mathcal{L}_s(G, D) = \mathcal{L}^{kld}_s(G) + \lambda \mathcal{L}^{rec}_s(G) + \mathcal{L}^{adv}_s(G, D) + \mathcal{L}^{cls}_s(D), \tag{7}$$

where the hyper-parameter $\lambda$ is set as 50. Note that $\mathcal{L}^{adv}_s(G, D)$, $\mathcal{L}^{kld}_s(G)$, and $\mathcal{L}^{rec}_s(G)$ only handle positive composite images.

By using $\theta_G$ and $\theta_D$ to represent learnable weights in $G$ and $D$, our optimization objective in the whole framework is

$$\min_{\theta_G} \max_{\theta_D} \quad \mathcal{L}^{adv}_u(G, D) + \mathcal{L}_s(G, D). \tag{8}$$

Note that two paths share weights in both $G$ and $D$. Under this design, the supervised path could gradually guide the unsupervised path to generate composite images with reasonable object placement. During inference, we only use the the unsupervised path to generate composite images by sampling $\mathbf{z}_u$ from $\mathcal{N}(\mathbf{0}, \mathbf{1})$.

## 4 Experiment

### 4.1 Experimental Setting

**Dataset and Evaluation Metrics.** We perform experiments on OPA dataset [22], which provides binary rationality labels for composite images. The dataset includes 62074 (21376 positive/40698 negative) composite images for training and 11396 (3588 positive/7808 negative) composite images for testing. These

**Fig. 3.** Visualization of object placement results. Foreground is outlined in red (Color figure online)

annotated composite images in the dataset contain 1389 different background scenes and 4137 different foreground objects from 47 categories.

Since our objective is to generate composite images with reasonable object placements, we train our model on OPA *train* set and evaluate it on the 3588 positive samples of OPA *test* set. During inference, our model takes the foreground/background of each positive test sample as input, and generates 10 composite images by randomly sampling 10 different $\mathbf{z}_u$ in $\mathcal{P}_u$.

We adopt user study, accuracy, and FID [12] to evaluate generation plausibility and LPIPS [40] for generation diversity. User study will be introduced in Sect. 4.2. We extend SimOPA [22] as a binary classifier to distinguish between reasonable and unreasonable object placements. We define accuracy as the proportion of generated composite images that are classified as positive by the binary classifier. FID is calculated between the composite images generated by our method and the positive composite images in the *test* set. We compute LPIPS for all pairs of composite images among 10 generation results for each sample, and adopt the averaged LPIPS among all samples.

**Implementation Details.** Our backbone network is the beginning 34 layers (including and before the fourth MaxPool layer) of VGG16 [28] with batch normalization, except that the first Conv layer has four input channels. In NEH, each node extraction layer $\mathcal{G}_l$ with parameter $n_l$ contains three groups of ($3 \times 3$ Conv, BN, ReLU), followed by a $n_l \times n_l$ AdaptiveAvgPool. The regression block contains three fully connected layers (Fc1024, Fc1024, Fc3), followed by an activation function $(\tanh(\cdot) + 1)/2$ that normalizes transformation parameters to range $(0, 1)$. We adopt the discriminator architecture in [34]. All images are resized to $256 \times 256$ and normalized before being fed into the network (see supplementary for more details). The VGG16 backbone is pretrained on ImageNet [8]. Our model is trained with batch size 32 for 11 epochs on a single RTX 3090 GPU. We adopt Adam optimizer with $\beta_1 = 0.5$ and $\beta_2 = 0.999$ for optimization. The learning rate is initialized as $2 \times 10^{-5}$ for backbone and discriminator, and $2 \times 10^{-4}$ for the remaining parts. For hyper-parameters, we set dimension $C$ as 512 for nodes, and $C_z$ as 1024 for random vectors, which will be carefully analyzed in Sect. 4.4.

**Table 1.** Quantitative object placement results for different methods on OPA dataset

| Method | *Plausibility* | | | *Diversity* |
|---|---|---|---|---|
| | user study↑ | acc.↑ | FID↓ | LPIPS↑ |
| TERSE [31] | 0.214 | 0.679 | 46.94 | 0 |
| PlaceNet [38] | 0.249 | 0.683 | 36.69 | 0.160 |
| GracoNet | 0.537 | 0.847 | 27.75 | 0.206 |

### 4.2 Comparison with Existing Methods

We compare our GracoNet with two baselines: TERSE [31] and PlaceNet [38]. Both of them are re-implemented on OPA dataset [22]. For the first baseline, we retain the synthesizer network and the discriminator of TERSE, and remove the target network. This is because the synthesizer is enough to generate composite images that we need, and we do not need to use the target network for downstream tasks. For the second baseline, we directly use the complete PlaceNet structure to predict object placement without further adjustment. Since TERSE and PlaceNet had been proposed before OPA dataset was released, these object placement methods did not include negative samples in their method. When we conduct experiments on OPA dataset, unless otherwise stated, we fairly use positive samples and negative samples together for both the baselines and our method. Specifically, we use negative samples in baselines by introducing a binary classification loss following Eq. (6) to train the discriminator network.

Table 1 shows the quantitative object placement results for baselines and our proposed method. Among different evaluation metrics, user study, accuracy, and LPIPS are the most important ones. User study is conducted with 20 voluntary participants by comparing the composite images generated by TERSE, PlaceNet, and our method. For each sample, every participant chooses the method producing the most reasonable composite image. Then, each method is scored by the proportion of participants who choose it. The final score of each method is defined by the averaged score over all samples.

By comparison, our method significantly outperforms TERSE/PlaceNet in generation plausibility (0.537 *v.s.* 0.214/0.249 for user study, 0.847 *v.s.* 0.679/0.683 for accuracy, and 27.75 *v.s.* 46.94/36.69 for FID). Also, our method achieves better LPIPS in generation diversity than PlaceNet. Note that TERSE does not incorporate randomness, so its generation diversity is zero. Generally, our method performs satisfactorily and balances plausibility and diversity well.

Figure 3 visualizes some object placement results for different methods. As illustrated, our method works better in predicting foreground locations and sizes by comprehensively analyzing different background regions, verifying the effectiveness of our designed GCM. In the supplementary, we show more visualizations of generation plausibility and diversity from three aspects: 1) combination of an identical background scene and different foreground objects, 2) combination of an identical foreground object and different background scenes, and 3)

**Table 2.** Ablation study on background head $\mathcal{H}^{\text{bg}}$

| $\mathcal{H}^{\text{bg}}$ with parameter $\mathbf{n}$ | *Plausibility* | | *Diversity* |
|---|---|---|---|
| | acc.↑ | FID↓ | LPIPS↑ |
| $\mathbf{n} = [2]$ | 0.807 | 37.44 | 0.136 |
| $\mathbf{n} = [2, 4]$ | 0.821 | 34.23 | 0.146 |
| $\mathbf{n} = [2, 4, 8, 16]$ | 0.837 | 25.91 | 0.154 |
| $\mathbf{n} = [2, 4, 8]$ | 0.847 | 27.75 | 0.206 |

**Table 3.** Ablation study on different types of learnable placement encodings

| $\boldsymbol{p}^K$ | $\boldsymbol{p}^V$ | Shared across heads | *Plausibility* | | *Diversity* |
|---|---|---|---|---|---|
| | | | acc.↑ | FID↓ | LPIPS↑ |
| | | - | 0.793 | 28.22 | 0.120 |
| ✓ | | ✓ | 0.809 | 28.02 | 0.133 |
| | ✓ | ✓ | 0.844 | 25.14 | 0.127 |
| ✓ | ✓ | ✓ | 0.851 | 29.09 | 0.170 |
| ✓ | | | 0.836 | 26.85 | 0.156 |
| | ✓ | | 0.839 | 26.22 | 0.154 |
| ✓ | ✓ | | 0.847 | 27.75 | 0.206 |

sampling different random vectors for the same pair of foreground and background.

### 4.3 Ablation Studies

**Different Choices of Background Head $\mathcal{H}^{\text{bg}}$.** Table 2 displays four choices of parameter $\mathbf{n}$ in the background head $\mathcal{H}^{\text{bg}}$. Since the input size of images is $256 \times 256$ and the backbone $\mathcal{E}$ contains four pooling layers, feature maps F are $16 \times 16$ in size. As default, we choose $\mathbf{n} = [2, 4, 8]$ to represent large-scale, medium-scale, and small-scale background nodes, which pay attention to different regions on the feature map. In detail, each of the 4 large-scale (*resp.*, 16 medium-scale, 64 small-scale) background nodes focuses on a local receptive field of $8 \times 8$ (*resp.*, $4 \times 4$, $2 \times 2$) region on F. In Table 2, we show more choices of parameter $\mathbf{n}$ in $\mathcal{H}^{\text{bg}}$. When $\mathbf{n} = [2]$ or $\mathbf{n} = [2, 4]$, the model misses small-scale information, so the placement process lacks details in some local regions. When $\mathbf{n} = [2, 4, 8, 16]$, the model tends to learn pixel-wise knowledge, which is redundant in object placement learning and adversely affects network optimization. Overall, our choice $\mathbf{n} = [2, 4, 8]$ achieves a good balance.

**Different Types of Learnable Placement Encoding.** As discussed in Sect. 3.1, we adopt placement encoding $\boldsymbol{p}^K$ and $\boldsymbol{p}^V$ in the attention layer. In

**Table 4.** Ablation study on functionality of GCM

| Fg/Bg feature extractor | Placement finder | *Plausibility* | | *Diversity* |
|---|---|---|---|---|
| | | acc.↑ | FID↓ | LPIPS↑ |
| global vector | concat+fc | 0.793 | 33.97 | 0.082 |
| NEH | concat+fc | 0.828 | 31.36 | 0.144 |
| NEH | PSN | 0.847 | 27.75 | 0.206 |

**Table 5.** Ablation study on loss functions

| Method | *Plausibility* | | *Diversity* |
|---|---|---|---|
| | acc.↑ | FID↓ | LPIPS↑ |
| w/o $\mathcal{L}_u^{adv}$ | 0.790 | 32.64 | 0.038 |
| w/o $\mathcal{L}_s^{adv}$ | 0.800 | 34.76 | 0.057 |
| w/o $\mathcal{L}_s^{cls}$ | 0.734 | 34.62 | 0.033 |
| w/o $\mathcal{L}_s^{kld}$ | 0.844 | 29.26 | 0.199 |
| w/o $\mathcal{L}_s^{rec}$ | 0.767 | 25.53 | 0.131 |
| Ours | 0.847 | 27.75 | 0.206 |

our implementation, $\boldsymbol{p}^K$ and $\boldsymbol{p}^V$ are both learnable and not shared across different attention heads. In this paragraph, we discuss more variants of placement encoding, as shown in Table 3. Without placement encoding, the generation plausibility and diversity witness a considerable decrease, because location and size information of different background nodes are missing under this condition. With a single $\boldsymbol{p}^K$ or $\boldsymbol{p}^V$, the model performs a little better in plausibility, but the diversity is still unsatisfactory. Using $\boldsymbol{p}^K$ and $\boldsymbol{p}^V$ together works best. If placement encodings are shared across attention heads, different attention heads could not learn diversified attention regions for object placement, resulting in comparably low generation diversity. Therefore, we choose to use $\boldsymbol{p}^K$ and $\boldsymbol{p}^V$ together, and make them independent across different attention heads. Comprehensively, our choice considers both plausibility and diversity to achieve the best results.

**Functionality of GCM.** As introduced in Sect. 3.1, our graph completion module (GCM) consists of two important components: node extraction head (NEH) and placement seeking network (PSN). Table 4 provides an ablation study on GCM by replacing NEH or PSN with naive network structures. In Table 4, the first experiment uses split branches to extract global features for foreground and background respectively. Then the two global features are concatenated and regressed with several fc layers to obtain the transformation parameters for object placement. Compared with the first experiment, the second one uses NEH to extract features by considering different background locations and sizes. The background nodes are then averaged and concatenated with the foreground node to predict transformation parameters. The last experiment is our method that

**Table 6.** Choices of hyper-parameter $C_z$

| $C_z$ | *Plausibility* | | *Diversity* |
|---|---|---|---|
| | acc.↑ | FID↓ | LPIPS↑ |
| 256 | 0.807 | 36.04 | 0.151 |
| 512 | 0.838 | 35.53 | 0.175 |
| 2048 | 0.827 | 28.10 | 0.208 |
| 4096 | 0.792 | 27.39 | 0.219 |
| 1024 | 0.847 | 27.75 | 0.206 |

**Table 7.** Choices of hyper-parameter $\lambda$

| $\lambda$ | *Plausibility* | | *Diversity* |
|---|---|---|---|
| | acc.↑ | FID↓ | LPIPS↑ |
| 1 | 0.820 | 23.04 | 0.183 |
| 10 | 0.823 | 29.93 | 0.190 |
| 25 | 0.847 | 28.57 | 0.193 |
| 100 | 0.821 | 24.62 | 0.174 |
| 50 | 0.847 | 27.75 | 0.206 |

combines NEH for feature extraction and PSN for placement finding. Comparing the results of the first two experiments, we find that using NEH to explicitly encode different background locations and sizes is beneficial for object placement. By comparing the results of the last two experiments, PSN works better in discovering relationships between foreground and background, and successfully establishes a reasonable connection between foreground node and background nodes in the node graph. In summary, both NEH and PSN are crucial in our method. They work together to ensure the functionality of our proposed GCM.

**Utility of Different Loss Functions.** Table 5 gives an ablation study on different loss functions. Except for $\mathcal{L}_s^{kld}$, deleting any loss makes the performance drop sharply on plausibility or diversity, especially for $\mathcal{L}_s^{cls}$ and $\mathcal{L}_s^{rec}$. Although our model still performs passably without $\mathcal{L}_s^{kld}$, adding this loss can bring some improvement. Generally, every loss makes up an important part, and they work together to guarantee the effectiveness of our method.

### 4.4 Hyper-Parameter Analyses

**Dimension $C_z$ of Random and Latent Vectors.** As introduced in Sect. 4.1, we set the dimension of nodes as $C = 512$, and set the dimension of random vector $\mathbf{z}_u$ and latent vector $\mathbf{z}_s$ as $C_z = 1024$. The value of $C$ is chosen by considering both hardware resource occupation and model performance. Table 6 analyzes different choices of $C_z$ in the range of $[256, 4096]$. When $C_z$ increases, accuracy first increases, meets a peak at 1024, and then decreases. Meanwhile, LPIPS increases and FID decreases generally. For a balanced consideration between plausibility and diversity, we choose $C_z = 1024$ in our implementation.

**Coefficient $\lambda$ of Reconstruction Loss.** As discussed in Sect. 4.3, reconstruction loss $\mathcal{L}_s^{rec}$ plays an important role in the supervised path. It helps our model reconstruct ground-truth transformation parameters from positive composite images, enabling the supervised path to guide the unsupervised path through the unified training of two paths. In Eq. (7), reconstruction loss $\mathcal{L}_s^{rec}$ has a hyper-parameter $\lambda$ indicating the coefficient/weight of this loss during optimization.

Table 7 displays different choices of $\lambda$ in the range of $[1, 100]$. As can be analyzed, our model achieves comparably good plausibility and diversity when $\lambda = 50$, which becomes our default choice in all experiments.

### 4.5 Visualization of Multi-head Attention

In Sect. 3.1, we introduce multi-head attention with placement encoding to discover the relationship between the unique foreground node and totally 84 background nodes. Eq. (3) defines attention coefficient $\alpha_j$ $(1 \leqslant j \leqslant 84)$ in each head to represent the edge weight between the foreground node and the $j$-th background node in the node graph. Specifically, we use $j_{\max}$ to denote the index with the maximum edge weight:

$$j_{\max} = \arg\max_j \alpha_j \qquad j = 1, 2, \cdots, 84. \tag{9}$$

Since we have 8 attention heads, we use $j_{\max}^{(k)}$ with $1 \leqslant k \leqslant 8$ for differentiated representations in distinct heads. In the $k$-th attention head, $j_{\max}^{(k)}$ corresponds to a local background region with a specific location and size encoded by the $j_{\max}$-th background node. According to the definition of background nodes in Sect. 3.1, $j_{\max}^{(k)}$ corresponds to a local background region with scale $\frac{H}{2} \times \frac{H}{2}$ (*resp.*, $\frac{H}{4} \times \frac{H}{4}$, $\frac{H}{8} \times \frac{H}{8}$) when $1 \leqslant j_{\max}^{(k)} \leqslant 4$ (*resp.*, $5 \leqslant j_{\max}^{(k)} \leqslant 20$, $21 \leqslant j_{\max}^{(k)} \leqslant 84$). In Fig. 4, we visualize the local background region attended by $j_{\max}^{(k)}$ for each attention head, which represents for the region that the $k$-th head pays the most attention to on the background scene. As illustrated, the multi-head attention layer successfully discovers diversified locations and sizes of potential background regions for object placement by learning the relationship between the unique foreground node and

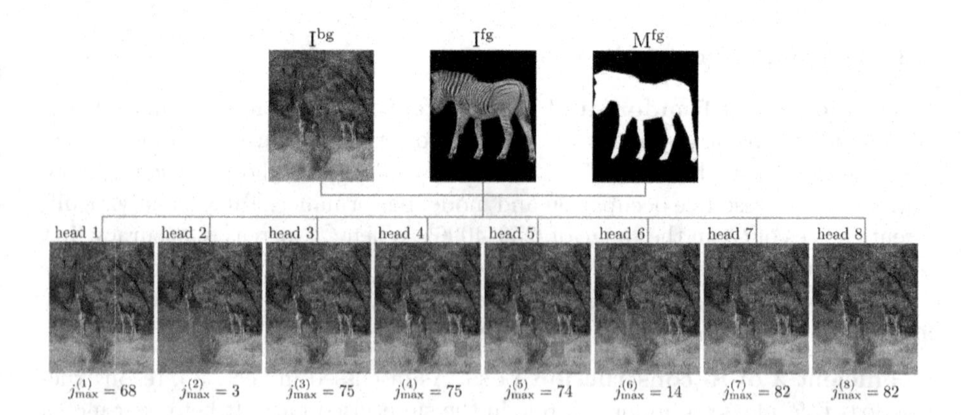


**Fig. 4.** Visualization of background regions (marked in red) that the multi-head attention layer pays the most attention to in each head. $j_{\max}^{(k)}$ is the index of background node with the maximum attention coefficient in the $k$-th head. See details in Sect. 4.5 (Color figure online)

multiple background nodes. We could also conclude from Fig. 4 that the most attended regions are reasonable enough for placing the zebra because only near-earth locations are activated, which accords with common sense.

## 5 Conclusion

In this work, we have proposed a novel graph completion module (GCM) for the object placement task to explicitly explore the relationship between the foreground object and the background image. We have also designed a dual-path framework upon the GCM structure, in which the supervised path provides additional cues for the unsupervised path so as to significantly enhance the performance. Extensive experiments on OPA dataset have demonstrated the effectiveness of our proposed method.

**Acknowledgements.** The work is supported by Shanghai Municipal Science and Technology Key Project (Grant No. 20511100300), Shanghai Municipal Science and Technology Major Project, China (2021SHZDZX0102), and National Science Foundation of China (Grant No. 61902247).

## References

1. Azadi, S., Pathak, D., Ebrahimi, S., Darrell, T.: Compositional GAN: learning image-conditional binary composition. Int. J. Comput. Vis. **128**, 2570–2585 (2020)
2. Chen, B.C., Kae, A.: Toward realistic image compositing with adversarial learning. In: CVPR (2019)
3. Chen, T., Cheng, M.M., Tan, P., Shamir, A., Hu, S.M.: Sketch2Photo: Internet image montage. ACM Trans. Graph. (TOG) **28**, 1–10 (2009)
4. Cong, W., Niu, L., Zhang, J., Liang, J., Zhang, L.: BargainNet: background-guided domain translation for image harmonization. In: ICME (2021)
5. Cong, W., et al.: High-resolution image harmonization via collaborative dual transformations. In: CVPR (2022)
6. Cong, W., et al.: DoveNet: deep image harmonization via domain verification. In: CVPR (2020)
7. Dai, Z., Yang, Z., Yang, Y., Carbonell, J., Le, Q.V., Salakhutdinov, R.: Transformer-XL: attentive language models beyond a fixed-length context (2019)
8. Deng, J., Dong, W., Socher, R., Li, L.J., Li, K., Fei-Fei, L.: ImageNet: a large-scale hierarchical image database. In: CVPR (2009)
9. Gehring, J., Auli, M., Grangier, D., Yarats, D., Dauphin, Y.N.: Convolutional sequence to sequence learning. In: ICML (2017)
10. Georgakis, G., Mousavian, A., Berg, A.C., Kosecka, J.: Synthesizing training data for object detection in indoor scenes (2017)
11. Goodfellow, I., et al.: Generative adversarial nets. NIPS (2014)
12. Heusel, M., Ramsauer, H., Unterthiner, T., Nessler, B., Hochreiter, S.: GANs trained by a two time-scale update rule converge to a local Nash equilibrium. In: NeurIPS (2017)
13. Hong, Y., Niu, L., Zhang, J.: Shadow generation for composite image in real-world scenes. In: AAAI (2022)

14. Johnson, J., Gupta, A., Fei-Fei, L.: Image generation from scene graphs. In: CVPR (2018)
15. Kingma, D.P., Welling, M.: Auto-encoding variational bayes (2014)
16. Lalonde, J.F., Efros, A.A.: Using color compatibility for assessing image realism. In: ICCV (2007)
17. Lalonde, J.F., Hoiem, D., Efros, A.A., Rother, C., Winn, J., Criminisi, A.: Photo clip art. ACM Trans. Graph. (TOG) **26**, 3-es (2007)
18. Lee, D., Liu, S., Gu, J., Liu, M.Y., Yang, M.H., Kautz, J.: Context-aware synthesis and placement of object instances (2018)
19. Li, X., Liu, S., Kim, K., Wang, X., Yang, M.H., Kautz, J.: Putting humans in a scene: learning affordance in 3D indoor environments. In: CVPR (2019)
20. Lin, C.H., Yumer, E., Wang, O., Shechtman, E., Lucey, S.: ST-GAN: spatial transformer generative adversarial networks for image compositing. In: CVPR (2018)
21. Liu, D., Long, C., Zhang, H., Yu, H., Dong, X., Xiao, C.: ARShadowGAN: shadow generative adversarial network for augmented reality in single light scenes. In: CVPR (2020)
22. Liu, L., Zhang, B., Li, J., Niu, L., Liu, Q., Zhang, L.: OPA: object placement assessment dataset. arXiv preprint arXiv:2107.01889 (2021)
23. Liu, X., Yu, H.F., Dhillon, I., Hsieh, C.J.: Learning to encode position for transformer with continuous dynamical model. In: ICML (2020)
24. Niu, L., et al.: Making images real again: a comprehensive survey on deep image composition. arXiv preprint arXiv:2106.14490 (2021)
25. Raffel, C., et al.: Exploring the limits of transfer learning with a unified text-to-text transformer (2020)
26. Schuster, M.J., Okerman, J., Nguyen, H., Rehg, J.M., Kemp, C.C.: Perceiving clutter and surfaces for object placement in indoor environments. In: ICHR (2010)
27. Shaw, P., Uszkoreit, J., Vaswani, A.: Self-attention with relative position representations (2018)
28. Simonyan, K., Zisserman, A.: Very deep convolutional networks for large-scale image recognition (2015)
29. Smith, A.R., Blinn, J.F.: Blue screen matting. In: SIGGRAPH (1996)
30. Tan, F., Bernier, C., Cohen, B., Ordonez, V., Barnes, C.: Where and who? Automatic semantic-aware person composition. In: WACV (2018)
31. Tripathi, S., Chandra, S., Agrawal, A., Tyagi, A., Rehg, J.M., Chari, V.: Learning to generate synthetic data via compositing. In: CVPR (2019)
32. Tsai, Y.H., Shen, X., Lin, Z., Sunkavalli, K., Lu, X., Yang, M.H.: Deep image harmonization. In: CVPR (2017)
33. Vaswani, A., et al.: Attention is all you need. In: NeurIPS (2017)
34. Wang, T.C., Liu, M.Y., Zhu, J.Y., Tao, A., Kautz, J., Catanzaro, B.: High-resolution image synthesis and semantic manipulation with conditional GANs. In: CVPR (2018)
35. Weng, S., Li, W., Li, D., Jin, H., Shi, B.: MISC: multi-condition injection and spatially-adaptive compositing for conditional person image synthesis. In: CVPR (2020)
36. Wu, H., Zheng, S., Zhang, J., Huang, K.: GP-GAN: towards realistic high-resolution image blending. In: ACM Multimedia (2019)
37. Xue, S., Agarwala, A., Dorsey, J., Rushmeier, H.: Understanding and improving the realism of image composites. ACM Trans. Graph. (TOG) **31**, 1–10 (2012)
38. Zhang, L., Wen, T., Min, J., Wang, J., Han, D., Shi, J.: Learning object placement by inpainting for compositional data augmentation. In: Vedaldi, A., Bischof,

H., Brox, T., Frahm, J.-M. (eds.) ECCV 2020. LNCS, vol. 12358, pp. 566–581. Springer, Cham (2020). https://doi.org/10.1007/978-3-030-58601-0_34
39. Zhang, L., Wen, T., Shi, J.: Deep image blending. In: WACV (2020)
40. Zhang, R., Isola, P., Efros, A.A., Shechtman, E., Wang, O.: The unreasonable effectiveness of deep features as a perceptual metric. In: CVPR (2018)
41. Zhang, S.-H., Zhou, Z.-P., Liu, B., Dong, X., Hall, P.: What and where: a context-based recommendation system for object insertion. Comput. Vis. Media **6**(1), 79–93 (2020). https://doi.org/10.1007/s41095-020-0158-8
42. Zhu, J.Y., Krahenbuhl, P., Shechtman, E., Efros, A.A.: Learning a discriminative model for the perception of realism in composite images. In: ICCV, pp. 3943–3951 (2015)
43. Zhu, J.Y., et al.: Multimodal image-to-image translation by enforcing bi-cycle consistency. In: NeurIPS (2017)

# Expanded Adaptive Scaling Normalization for End to End Image Compression

Chajin Shin(✉), Hyeongmin Lee, Hanbin Son, Sangjin Lee, Dogyoon Lee, and Sangyoun Lee

School of Electrical and Electronic Engineering, Yonsei University, Seoul, Korea
{chajin,minimonia,hbson,pandatimo,nemotio,syleee}@yonsei.ac.kr

**Abstract.** Recently, learning-based image compression methods that utilize convolutional neural layers have been developed rapidly. Rescaling modules such as batch normalization which are often used in convolutional neural networks do not operate adaptively for the various inputs. Therefore, Generalized Divisible Normalization (GDN) has been widely used in image compression to rescale the input features adaptively across both spatial and channel axes. However, the representation power or degree of freedom of GDN is severely limited. Additionally, GDN cannot consider the spatial correlation of an image. To handle the limitations of GDN, we construct an expanded form of the adaptive scaling module, named Expanded Adaptive Scaling Normalization (EASN). First, we exploit the swish function to increase the representation ability. Then, we increase the receptive field to make the adaptive rescaling module consider the spatial correlation. Furthermore, we introduce an input mapping function to give the module a higher degree of freedom. We demonstrate how our EASN works in an image compression network using the visualization results of the feature map, and we conduct extensive experiments to show that our EASN increases the rate-distortion performance remarkably, and even outperforms the VVC intra at a high bit rate.

**Keywords:** Image compression · Adaptive · Rescaling · End-to-end learning

## 1 Introduction

Image compression is one of the most important and fundamental tasks in image processing and computer vision. There are a countless digital images in the world, and numerous new images are generated every day. Therefore, image compression is essential to save and transmit these massive images efficiently. Many classic image compression codecs have been developed, including JPEG [29], JPEG2000 [24], HEVC [26], and VVC [22]. They use several classic methods such

**Supplementary Information** The online version contains supplementary material available at https://doi.org/10.1007/978-3-031-19790-1_24.

© The Author(s), under exclusive license to Springer Nature Switzerland AG 2022
S. Avidan et al. (Eds.): ECCV 2022, LNCS 13677, pp. 390–405, 2022.
https://doi.org/10.1007/978-3-031-19790-1_24

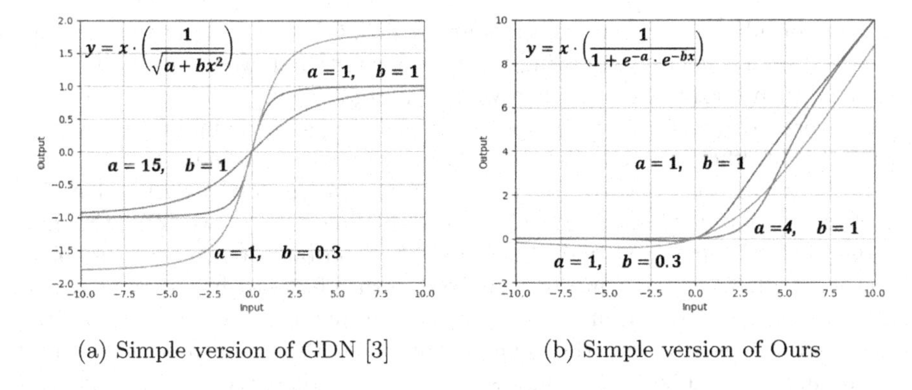


(a) Simple version of GDN [3]

(b) Simple version of Ours

**Fig. 1.** Comparison between simple version of GDN [3] and ours. $a$ and $b$ are the learnable scalar parameters. We exploit the swish [25] function (b) instead of the arithmetic sigmoidal function (a) from GDN [3].

as transformation, quantization, and entropy coding to reduce redundant information in image.

Recently, deep learning-based image processing techniques have emerged that have shown superior performance in many computer-vision tasks. There have been many attempts to apply deep learning-based methods to image compression, and convolutional Variational Autoencoder (VAE)-based architecture, which has an hourglass-shape with the encoder and decoder, is the mainstream in image compression. Ballé *et al.* [3] propose a differentiable method for both quantization and bit estimation, and entropy can be optimized directly using this method effectively. Then, HyperPrior [4], which uses additional bits to model the latent vectors as Gaussian distribution, has been proposed to further reduce redundancy in latent vectors. Since then, various models have been proposed [8,9,13,17,19,20,34], and many models now compete with VVC intra of the traditional codec.

Many methods that use a convolutional VAE structure [3,4,9,13,17,19,20,34] utilize Generalized Divisible Normalization (GDN) [3] instead of using both existing rescaling modules and activation functions, such as batch normalization [14] and ReLU [2] function. The reason is that existing rescaling modules and activation functions cannot operate adaptively for the various inputs since they apply the same value or manner to all spatial locations equally. In contrast, GDN controls the input value scale of intermediate features adaptively and non-linearly across spatial and channel axes.

However, GDN has several limitations. First, the representation power of GDN is severely limited. The reason is that GDN can only have non-negative learnable parameters, and the input features of GDN should be squared since they are included within the square root of GDN. We empirically find that increasing the receptive field of GDN or adding more layers cannot increase the performance due to the square root term in the equation of GDN. Second, GDN has only a $1 \times 1$ receptive field, which cannot deal with the spatial correlation of images. Natural images have strong redundancy between adjacent pixels, thus spatial correlation must be considered for image compression. Third, GDN has a limitation in non-linearity because it has a only single $1 \times 1$ convolutional layer. Finally, GDN is unstable at the training stage when we use a larger convolutional kernel or add more layers. Although GDN is initialized as a scaled identity matrix for convergence, it is insufficient to stabilize the adaptive rescaling module.

In this paper, we propose Expanded Adaptive Scaling Normalization (EASN), which is an expanded form of the adaptive scaling module, to overcome the limitations of GDN. First, we exploit the swish [25] function instead of the sigmoidal function of GDN, as shown in Fig. 1. Since the swish function has no square root, both the non-negative and negative learnable parameters are available, and the input feature does not need to be squared. This allows the scaling module to utilize the full range of parameters and inputs. Second, we increase the receptive field and add more layers, which allows the scaling module to consider the spatial correlation of the features and approximate more complex functions, and utilize the skip connection to stabilize the training. Additionally, we add an input mapping function to the scaling module to transform the input features to increase the degree of freedom of modules. Furthermore, we use the features before the spatial resolution is reduced by downsampling for a scaling function, and obtain a better performance. Moreover, we show that simply increasing the layers of EASN does not increase the performance by ablation study, and we propose a structure that makes the EASN deeper effectively to further improve performance. Finally, we visualize the output feature map of the scaling function of low and high bit rate models and reveal that our EASN can adjust and scale the highfrequency components in accordance with the bit rate. We evaluate our model on the Kodak dataset [16] and CLIC2021 validation dataset [10], and our EASN achieves the rate-distortion performance dramatically, even outperforms VVC intra at a high bit rate.

## 2 Related Works

**Traditional Codec.** There are various traditional hand-crafted image compression methods. JPEG [29], JPEG2000 [24], HEVC [26] and VVC [22] are very popular image compression standard methods. To compress and reduce spatial redundancy of the image effectively, encoder modules divide the image into multiple blocks, and convert spatial domain of the image to the frequency domain with traditional transforms such as discrete cosine transform (DCT). After transforming, quantization and entropy coding, like Huffman coding, are conducted. Moreover, HEVC or VVC have many modes of each module and they check every case to get best rate-distortion performance.

**Learning-Based.** Recently, deep learning-based image processing methods have emerged and shown superior performance in various computer vision tasks, and there have been many efforts to utilize deep learning-based methods for image compression. In the first stage, some works [27,28] utilize recurrent neural networks for image compression. These methods can have variable bit rates using the recurrent scheme. However, entropy of the image is not optimized directly since the constraint of entropy is not in the loss function, thus these methods show lower performance than JPEG2000.

The second stage, which is convolutional VAE-based architectures, has become the mainstream in image compression with optimizing entropy directly through loss function. From Ballé *et al.* [3], minimizing the expectation of Kullback-Leibler divergence is equal to minimizing distortion and entropy at the same time using a variational autoencoder. Furthermore, they [3] proposes a differentiable method for both quantization and bit estimation to consider the bit rate constraint at the training step. They add uniform random noise in the range of $[-0.5, 0.5]$ to the latent representation $y$, which is the output of encoder $g_a$. By adding the noise, they can approximate the probability mass function (PMF) of the quantized latent representation $\hat{y}$ with integrating the probability density function (PDF) of latent representation $y$. Using approximated PMF, they [3] directly optimize the entropy of the image with the following total loss function.

$$\mathcal{L} = -\mathbb{E}[log_2 P] + \lambda \cdot D(x, \hat{x}) \tag{1}$$

where $P$ is the estimated PMF of the latent representation, and $D$ is the distortion between the original image $x$ and the reconstruction $\hat{x}$. Thereafter, Hyperprior [4] introduces an auxiliary convolutional autoencoder to utilize side information to model the latent representation $y$ as a Gaussian distribution to further reduce the spatial redundancy in the latent representation $y$. Through these, the performance of convolutional VAE-based image compression has greatly improved. However they [3,4] are still transform-based models, and there are no spatial or context prediction modules. Some works [17,19] predict the context of an image by using an autoregressive context prediction module with latent representation $y$ and Hyperprior. Another work [13] proposes a parallelizable context model to accelerate the sequential process of the autoregressive context prediction module. Further works [8,9] consider the latent representation $y$ as a more generalized distribution such as the asymmetric Gaussian or Gaussian mixture distribution.

## 3 Preliminary

Existing rescaling modules and activation functions such as batch normalization [14] or ReLU [2] function are not adaptive since they operate the same way to all spatial location equally. To deal with this problem, Ballé *et al.* [3] propose GDN, which rescale input features adaptively and non-linearly across spatial and channel axes. GDN is used in image compression neural network instead of batch normalization [14] or ReLU [2] functions. GDN of normal version $g_i$ is used in the analysis transform, which is encoder, and inverse version $g_i^{inv}$ is used in the synthesis transform, which is decoder.

$$g_i(m,n) = x_i(m,n) \cdot \frac{1}{\sqrt{\beta_i + \sum_j \gamma_{ij}(x_j(m,n))^2}} \tag{2}$$

$$g_i^{inv}(m,n) = x_i(m,n) \cdot \sqrt{\beta_i + \sum_j \gamma_{ij}(x_j(m,n))^2)} \tag{3}$$

where $i$ is output channel index, and $j$ is input channel index. We can interpret $\gamma$ as the $1 \times 1$ convolutional kernel, and $\beta$ as the bias. $(m, n)$ are coordinates of spatial height and width axis. If we focus on the normal version of GDN, $g_i$, it can be simplified as follows.

$$g = \frac{x}{\sqrt{a + bx^2}} \tag{4}$$

where $a$ and $b$ are non-negative scalar learnable parameters. Output $g$ is adaptively changed according to input $x$, since scaling factor function, which is rescaling part $s(x) = 1/\sqrt{a + bx^2}$, is various with respect to input features $x$. Therefore, we can consider GDN as the adaptive rescaling module. Furthermore, Fig. 1a shows graph of output $y$ with respect to input $x$ with different values of $a$ and $b$. We can find that the network can learn non-linear sigmoidal shape using $a$ and $b$, and can use it as a learnable sigmoidal shape activation function. GDN uses multivariate parameters for $a$ and $b$ instead of the scalar parameter, thus GDN is a multivariate sigmoidal function.

# 4 Method

In this section, we introduce the limitations of GDN [3] and our proposals to cope with the limitations. Furthermore, we propose more deeper scaling module architecture to obtain higher performance.

The scaling module of GDN can be expressed as follows.

$$g(x) = x \cdot s(x) \tag{5}$$

where $x$ is an input feature and $s(\cdot)$ is a scaling factor function. We only consider the normal version of GDN $g_i$ in Eq. 2 for scaling factor function $s(x)$ in this section. We replace the inverse version of GDN $g_i^{inv}$ in Eq. 3 with the normal version $g_i$ to consider only a single case when we modify the scaling factor function $s(x)$ in GDN. We empirically confirm that the network shows the same rate-distortion performance when we only use the normal version of GDN. Figure 2a represents the result that Joint Autoregressive [19] model with only normal version of GDN shows the same performance as the base Joint Autoregressive model. Therefore, we only use the normal version for simplicity.

## 4.1 Swish Function

In this section, we describe that GDN [3] has limited representation power or degree of freedom, and we show that using a swish [25] function for an adaptive rescaling module allows it to cope with the problem of GDN.

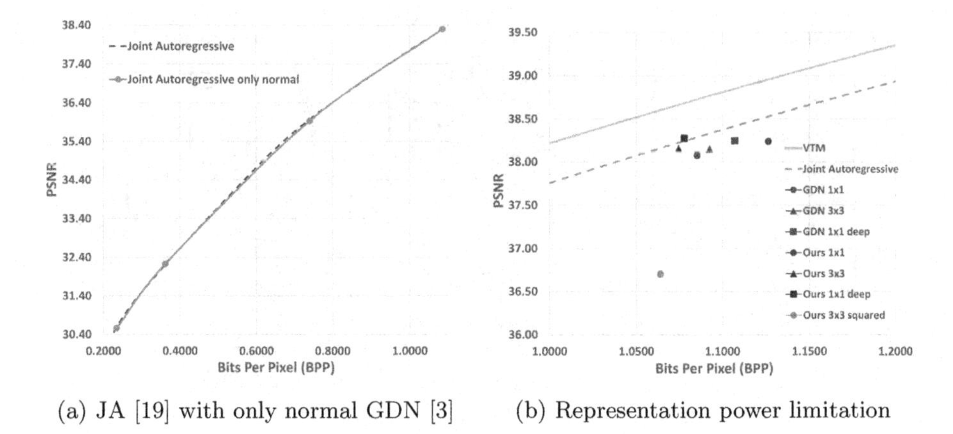


(a) JA [19] with only normal GDN [3]

(b) Representation power limitation

**Fig. 2.** (a) There is no difference in performance even if only the normal version of GDN [3] is used. (b) Red points denote Eq. 6 of GDN, and blue points indicate Eq. 8 of ours. The performance increases steadily with Eq. 8 of ours. By contrast, Eq. 6 of GDN does not show a performance increase owing to limitations of representation power or degree of freedom. (Color figure online)

Considering the scaling factor function $s(x)$ of GDN,

$$s(x) = \frac{1}{\sqrt{\beta_i + \sum_j \gamma_{ij}(x_j(m,n))^2}} \tag{6}$$

we can notice that $\beta_i + \sum_j \gamma_{ij}(x_j(m,n))^2$ should be non-negative. First, to keep it non-negative, GDN set $\beta_i$ and $\gamma_{ij}$ as non-negative learnable parameters, which limits the degree of freedom of the rescaling module. Second, the input features $x$ should be squared to be non-negative. This leads to information loss because the two different values that have the same magnitude but opposite sign attain the same value after the square operation. Finally, the scaling factor function of Eq. 6 is even symmetric, and it equally scales for inputs that have the same magnitude but opposite sign. These characteristics significantly limit the representation power and degree of freedom of the rescaling module. Therefore, we modify the scaling factor function $s(\cdot)$.

In Eq. 6, $\beta_i$ is a vector with output channel axis $i$, and the convolution operation $\sum_j \gamma_{ij} x_j^2$ is calculated along the input channel axis $j$. Thus, we can factorize Eq. 6 by $\frac{1}{\sqrt{\beta_i}}$ with $\delta_{ij} = \frac{\gamma_{ij}}{\beta_i}$. Then, $\frac{1}{\sqrt{\beta_i}}$ can be considered a constant scaling factor along the input channel axis of the next convolutional layer. This means that the convolution kernel of next layer can learn a constant scaling factor $\frac{1}{\sqrt{\beta_i}}$, thus we can ignore this term. Therefore, we can consider the scaling factor function as below.

$$\bar{s}(x) = \frac{1}{\sqrt{1 + \sum_j \delta_{ij}(x_j(m,n))^2}} \tag{7}$$

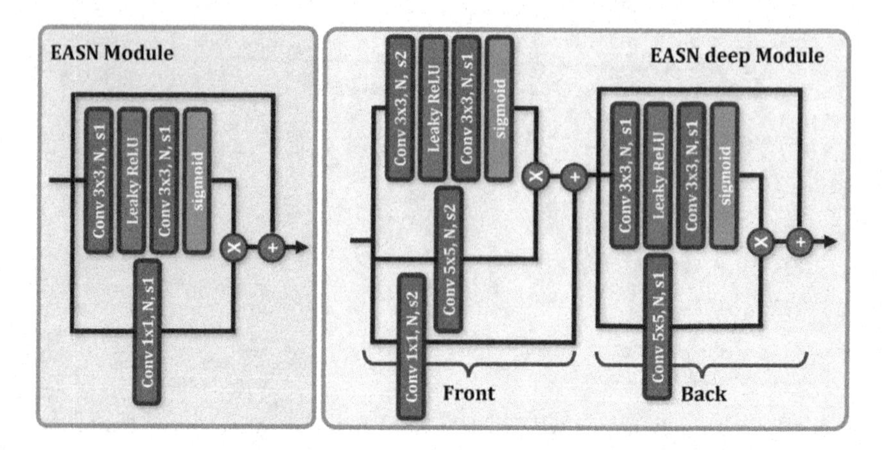


**Fig. 3.** Our EASN and EASN-deep. $N$ is output channel, and $s1$, $s2$ represent stride 1 and 2, respectively.

We replace the even symmetric function of Eq. 7 with a sigmoid function that has the same output range of $[0, 1]$, but is a bijective function as follows.

$$\hat{s}_i(x) = \frac{1}{1 + e^{\beta_i} \cdot e^{[\mathcal{F}(x)]_i}} \tag{8}$$

where $\beta_i$ represents one-dimensional learnable parameters along the output channel axis and $\mathcal{F}(\cdot)$ represents an arbitrary convolutional neural block. Using Eq. 8, all learnable parameters can have both negative and non-negative values, which have a higher degree of freedom than Eq. 7. Moreover, input feature $x$ does not need to be squared, and the scaling factor function of Eq. 8 can rescale the different inputs that have the same magnitude but opposite sign with different scale values.

We directly compare the scaling factor function of Eq. 6 from GDN and Eq. 8 of ours in Fig. 2b. All points in Fig. 2b are based on Joint Autoregressive [19] models with only the normal version of GDN and the skip connection for stability. Red points represent Eq. 6 from GDN, and blue points denote Eq. 8. The circle represents the models with only one $1 \times 1$ convolution, and the triangle is the model in which the $1 \times 1$ convolution is replaced by a $3 \times 3$ convolution. The rectangle represents the models with an additional $1 \times 1$ convolutional layers. We use the ReLU [2] activation function between the $1 \times 1$ convolutional layers of Eq. 6 from GDN to maintain the non-negative values, and we use a Leaky ReLU activation function for Eq. 8. As we can see, in the case of Eq. 6 from GDN, which is red points, even if the receptive field is expanded or more layers are added, the performance does not increase since the square root term limits the representation power or degree of freedom of the scaling module. In contrast, Eq. 8, which is blue points, shows steady performance improvements as the scaling module expands. Additionally, the green circle point in Fig. 2b represents Eq. 8 with a $3 \times 3$ convolution and the squared input. We can confirm that the squared input limits the representation

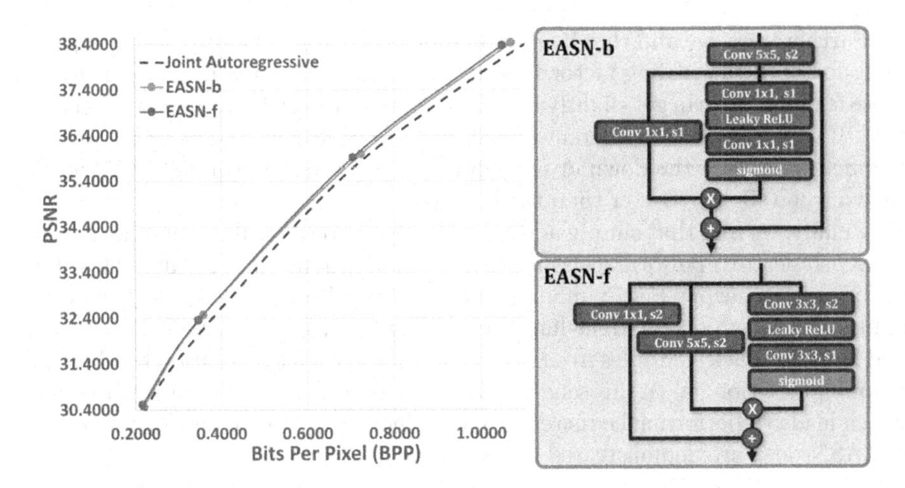


**Fig. 4.** Comparison results with feature location. EASN-f uses features before down or upsampling for scaling factor function and input mapping function. Both models have a $5 \times 5$ receptive field, including down or upsampling layer.

power or degree of freedom of the network and decreases the performance. Therefore, we can confirm that replacing Eq. 6 with Eq. 8 allows networks to overcome the limitations of representation power or degree of freedom.

### 4.2 EASN

GDN [3] has a single $1 \times 1$ convolution layer. Thus, GDN cannot deal with the spatial correlation, which is an important key in compression to reduce the spatial redundancy, and cannot approximate a more complex function. Since we can now utilize full representation power with a swish [25] function from Sect. 4.1, the scaling module can be expanded to consider the spatial correlation or obtain a higher degree of freedom.

We use two $3 \times 3$ convolutions with an intermediate Leaky ReLU activation function for the scaling factor function $\hat{s}(x)$ to increase the receptive field and make a function $\hat{s}(x)$ to be more complex. For these expansions, we add a skip connection to stabilize the training. Without a skip connection, such expansions make training unstable and training loss diverges very early. In many works [7,11, 12,21,23,31,33], a skip connection is used to ensure stability when two different features are multiplied in the neural network. Therefore, we use a skip connection. Furthermore, we introduce another function, input mapping function $m(x)$, to provide the scaling module with the option of transforming the input features to increase the degree of freedom. We call this rescaling module, Expanded Adaptive Scaling Normalization (EASN), and the final equation for EASN is as follows.

$$EASN(x) = m(x) \cdot \hat{s}(x) + x \tag{9}$$

Furthermore, we find that it is useful to utilize features before down or upsampling for both the scaling factor function and input mapping function. If we use these features, we can get slightly better performance even with the same receptive field. If we compare two different models in Fig. 4 that have the same $5 \times 5$ receptive field including the down or upsampling layer, the performance of EASN-f is shown to be slightly better than the EASN-b model.

Finally, we find that simply adding more layers to the scaling module does not efficiently lead to performance increases from ablation study results. Therefore, we propose a deeper EASN module called EASN-deep to obtain higher performance. As shown on the right-hand side of Fig. 3, we cascade the EASN-f(front) and EASN-e(back) modules from Fig. 5 that have a $5 \times 5$ convolution for the input mapping function. With this scheme, EASN-deep rescales the input feature twice, which leads to performance increases more efficiently. More experimental details of EASN-deep are demonstrated in Sect. 5.4.

## 5 Experiments

### 5.1 Implementation Details

We use MSE loss or MS-SSIM [30] loss to measure distortion for each PSNR or MS-SSIM performance comparison. Total loss is given as follows.

$$L_{total} = -\mathbb{E}[log_2 P] + \lambda \cdot D(x, \hat{x}) \tag{10}$$

where $P$ is estimated PMF, $x$ is the original image, and $\hat{x}$ is the reconstructed image. In case of MSE loss for $D$, we use $D(x, \hat{x}) = 255^2 \cdot MSE(x, \hat{x})$, and for MS-SSIM loss, we use $D(x, \hat{x}) = (1 - MSSSIM(x, \hat{x}))$. We set Hyperprior [4] and Joint Autoregressive [19] model as the baseline. In case of EASN, we replace normal and inverse GDN [3] of baseline with ours. For EASN-deep, we replace both down or upsampling convolution and GDN with ours, because the down or upsampling process is included in the EASN-deep module. Rate distortion trade-off parameter $\lambda$ is set to [0.005, 0.010, 0.020, 0.035, 0.080, 0.180] for MSE distortion loss, and to [7, 15, 30, 48, 110, 220] for MS-SSIM distortion loss. $N$ is the base channel number, and $M$ is the output channel number of the latent representation $y$. For the Hyperprior baseline, we select $N = 128$, $M = 192$ for the front two $\lambda$ values. For the other $\lambda$ values, we set $N = 192$, and $M = 320$. For the Joint Autoregressive baseline, we set $N = 192$, $M = 192$ for the front two $\lambda$ values and we select $N = 192$, $M = 320$ for the other $\lambda$ values.

### 5.2 Training

Basically, we follow the training process of CompressAI [5] framework. We use Vimeo90K [32] dataset for training. We randomly crop training images into $256 \times 256$ size, and randomly flip them horizontally. We use Adam optimizer [15] with batch size of 16, and learning rate is set to $1e^{-4}$ initially. We evaluate every epoch

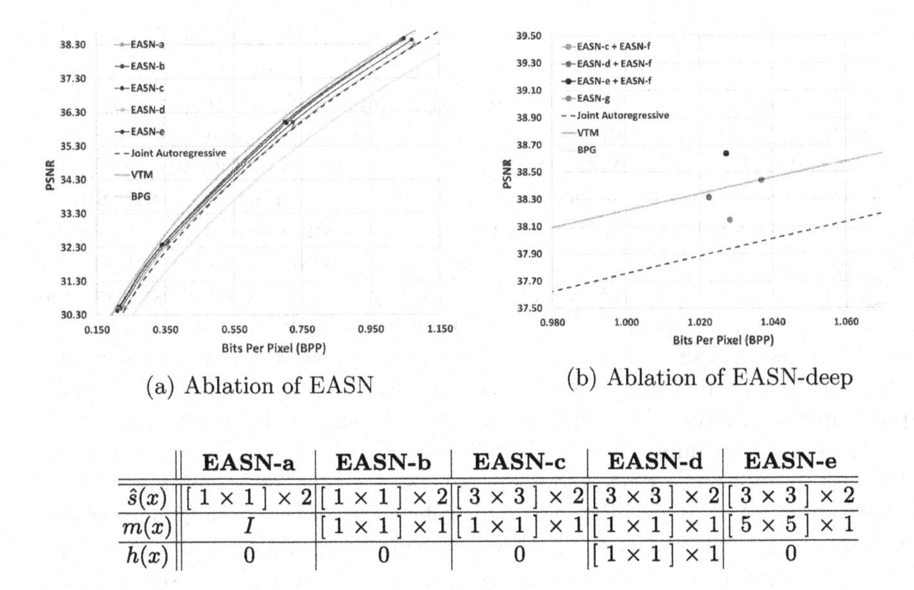


(a) Ablation of EASN

(b) Ablation of EASN-deep

| | **EASN-a** | **EASN-b** | **EASN-c** | **EASN-d** | **EASN-e** |
|---|---|---|---|---|---|
| $\hat{s}(x)$ | $[1 \times 1] \times 2$ | $[1 \times 1] \times 2$ | $[3 \times 3] \times 2$ | $[3 \times 3] \times 2$ | $[3 \times 3] \times 2$ |
| $m(x)$ | $I$ | $[1 \times 1] \times 1$ | $[1 \times 1] \times 1$ | $[1 \times 1] \times 1$ | $[5 \times 5] \times 1$ |
| $h(x)$ | 0 | 0 | 0 | $[1 \times 1] \times 1$ | 0 |

**Fig. 5.** Ablation study result. [k × k] × L means using L number of k × k convolution. $I$ is identity function. 0 means multiplying zero.

using validation set of COCO [18] dataset to get the total loss of validation dataset. We crop the COCO validation dataset at the center with a size of 256 × 256. We reduce learning rate of factor 0.5 if the validation loss does not improve during 10 epochs. We stop training when the learning rate decrease 4 times. In case of MS-SSIM loss, we fine-tune the pretrained model with MSE loss using initial learning rate of $0.5e^{-4}$ and stop training when the learning rate decreases 3 times.

### 5.3 Evaluation

We use Kodak dataset [16] for evaluation for both PSNR and MS-SSIM metrics. Moreover, we use CLIC2021 validation dataset [10], which consists of 41 high resolution images for confirming robustness for more high resolution images. To evaluate rate-distortion performance, we measure bits per pixel (bpp). We save the bitstreams to a hard disk drive to get a physical file size and divide the size with the total pixels number of the image to get bpp. We draw rate-distortion (RD) curves to check the compression performance.

### 5.4 Ablation Study

Figure 5a represents the ablation study results of EASN, and Fig. 5b shows the results of EASN-deep. We expand existing GDN [3] to the following equation.

$$EASN(x) = m(x) \cdot \hat{s}(x) + h(x) + x \tag{11}$$

where $h(x)$ is shift function. The table from Fig. 5 shows module structure of each modules. $[k \times k] \times L$ represents that it has L number of $k \times k$ convolutions. $I$ is identity function, and 0 means that those modules do not use a corresponding function. GDN [3] and EASN modules with skip connection have a worse performance than original GDN of Joint Autoregressive [19]. However, if we look at the Fig. 5a, we can make up slightly poor performance using only one more $1 \times 1$ convolution (EASN-a). EASN-a from Fig. 5 shows the same performance as the GDN-based Joint Autoregressive [19] model. As we add $1 \times 1$ convolution to input mapping function $m(x)$ (EASN-b), and replace $1 \times 1$ convolution of scaling factor function $\hat{s}(x)$ with $3 \times 3$ convolution (EASN-c), performance of the EASN modules increase steadily. In case of EASN-d, we use shift function with $1 \times 1$ convolution. However, we find that the performance does not increase. EASN-e has $5 \times 5$ convolution for input mapping function. Although the performance slightly increases, but considering the parameter numbers, we select EASN-c for the final EASN.

Figure 5b shows the performance comparison results of combining EASN-f with EASN-c, EASN-d, and EASN-e, which show the highest performance within EASN ablation results. In case of EASN-g, we simply add two more $3 \times 3$ convolution layers to scaling factor function $\hat{s}(x)$ of EASN-f and one more $5 \times 5$ convolution layer to input mapping function $m(x)$ of EASN-f to make same receptive field as the EASN-e + EASN-f module. As shown in Fig. 5b, we can confirm that simply adding more layers decreases the performance. Therefore, for constructing a deeper adaptive rescaling module effectively, we cascade the EASN-f module with other EASN modules. We find that the receptive field of input mapping function is important in terms of cascading two modules. Using $5 \times 5$ convolution for input mapping function (EASN-e) shows significant performance improvement. Therefore, we choose the combination of EASN-f(front) and EASN-e(back) modules for EASN-deep version.

### 5.5 Rate Distortion Performance

For comparison, we use traditional codecs of JPEG [29], JPEG2000 [24], BPG [6] which is image codec based on HEVC [26], and VTM [1] which is the official test model of VVC [22]. For learned image compression method, we use Hyper-Prior [4], Joint Autoregressive [19] and GMM [8]. For GMM, we use two different version, Anchor and Attention. The only difference between them is existence of the attention module. We adapt our EASN and EASN-deep to HyperPrior and Joint Autoregressive models which have a GDN-based structure. We plot two separate figures optimized by MSE or MS-SSIM [30], respectively. In case of MS-SSIM, we use log scale for visualization.

Figure 6a shows the rate-distortion performance with PSNR metric on both dataset. HP and JA represent HyperPrior and Joint Autoregressive, respectively. As we can see, HP + EASN outperforms HyperPrior that has a similar performance with traditional BPG, and HP + EASN-deep model shows higher performance than the HP + EASN model. The JA + EASN model outperforms the Joint Autoregressive model, and even shows similar performance with GMM Anchor model. At high bit rate, Our JA + EASN model reach the rate-distortion

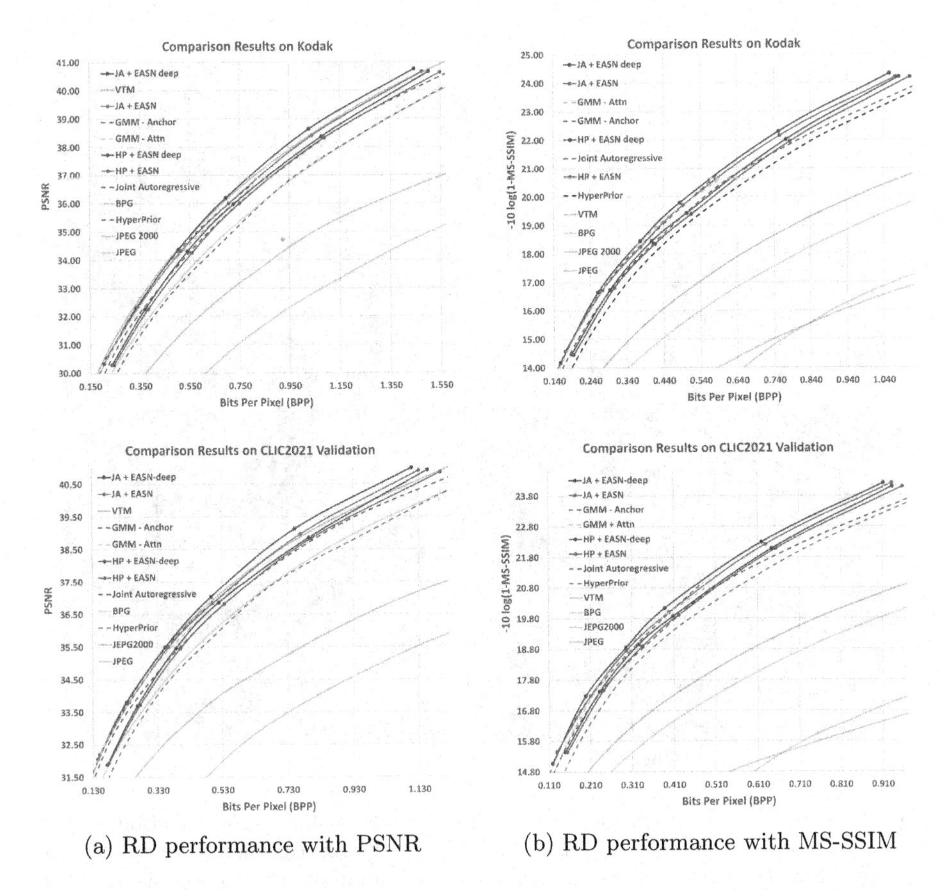


(a) RD performance with PSNR (b) RD performance with MS-SSIM

**Fig. 6.** Rate-distortion Performance comparison results on Kodak [16] and CLIC2021 validation dataset [10].

performance of traditional codec of VTM. In case of JA + EASN-deep model, although its performance is similar with JA + EASN model at low bit rate, our model outperforms all other learning-based and traditional codecs at high bit rate on both datasets. Figure 6b shows the performance comparison results with MS-SSIM metric. They show a similar tendency to PSNR results. HP + EASN and JA + EASN models outperform the baselines of HyperPrior and Joint Autoregressive, respectively. The HP + EASN-deep model has a higher performance than our HP + EASN models, and our JA + EASN-deep model outperforms all other learning-based models on both datasets.

### 5.6 Scale Feature Map

In this section, we demonstrate how the EASN module works along the bit rates with the visualization results of feature maps of the scaling factor function $\hat{s}(x)$. A low bit rate model discards many high frequency information to obtain a high

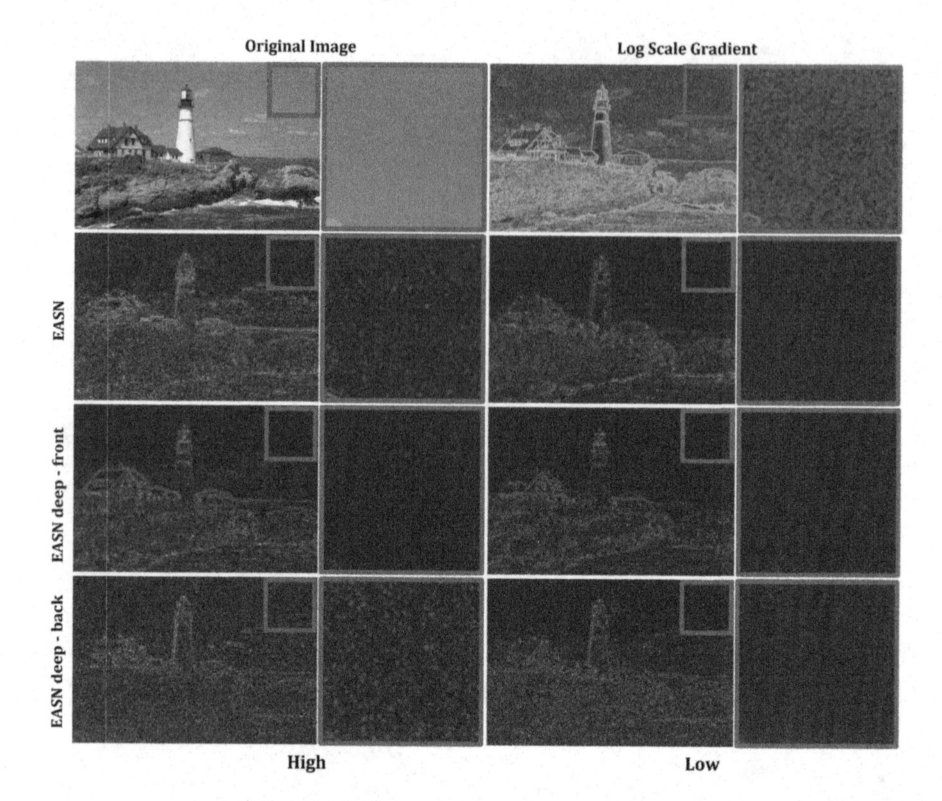


**Fig. 7.** Visualization of high frequency components of scaling factor function output $\hat{s}(x)$ with kodim21 image from Kodak dataset [16]. The top-right image represents the log scale gradient result of original image. The left column images denotes high bit rate results, and right column images represents low bit rate results.

compression rate. Whereas, a high bit rate model should generate reconstructed images with low distortion that comprise many fine details. To confirm the difference between various bit rate models, we remove low-frequency components to focus on high frequency details. Intuitively, the two models with different bit rates may rescale the blue color pixels with different values, such as 0.1 and 0.8, respectively. Therefore, the exact scale values are not important, and we should focus on the variety of scale values in accordance with the pixel variety of input images. We remove the low-frequency components using the following equation.

$$x_{avg}^{hf} = \frac{1}{N} \sum_{n=0}^{N} x_n - (x_n * k_{3\times3}) \tag{12}$$

where $k_{3\times3}$ is the mean filter with a kernel size of $3 \times 3$, and $1/9$ value for all components. Symbol $*$ is a convolution operator, $n$ indicates channel axis, and $N$ is the channel number. $x$ is the feature map from the scaling factor function $\hat{s}(x)$ of the first EASN module in the encoder.

Figure 7 is the visualization results of the high frequency components of the feature $x$ of the scaling factor function $\hat{s}(x)$ with the kodim21 image from the Kodak dataset [16]. The top-left image is the original image, and the top-right image is the gradient of the original image with log scale. The vertical axis represents each module, the left column images represent high bit rate models, and the right column images represent low bit rate models. For EASN-deep model, there are two scaling factor functions. EASN-deep front is the first rescaling part, which is the EASN-f module, and EASN-deep back is the second rescaling part, which is the EASN-e module represented in Fig. 3.

The log scale gradient of the original image shows high frequency components in the sky of the image, which is a flat region. Unlike the textures or edges, these details are not clearly visible to the human eye. The high bit rate module results of the EASN show that they catch these details in the red boxes of the sky region. In the case of GDN-deep front, this does not show a difference for the sky region but the EASN-deep back shows high frequency components in the sky. In contrast, there are no high frequency components in the sky region for all models trained for low bit rate. This means that models trained with a high bit rate catch more fine details in images. From these results, we can confirm that the scaling factor function $\hat{s}(x)$ in our EASN can adjust and rescale high frequency components of input features depending on the bit rates. We can also interpret these results as the scaling factor function $\hat{s}(x)$ determines how many details to remove to save bits.

## 6 Conclusions

We propose Expanded Adaptive Scaling Normalization (EASN), which is an expanded structure of existing GDN. For constructing EASN, first we exploit the swish function for the scaling factor function to make the module to utilize representation power fully. Second, we increase receptive field and make the scaling factor function deeper to consider spatial correlation and approximate more complex function. Additionally, we add input mapping function to increase degree of freedom, and we propose more EASN-deep module to make the module more deeper effectively. Furthermore, we reveal the process of how our EASN works along the bit rates within an image compression network using the visualization results of feature map. We conduct extensive experiments to show that each of the proposed methods is effective through ablation study, and our EASN shows dramatic increase of performance, and even outperforms other image compression methods.

**Acknowledgement.** This work was supported by Institute of Information & communications Technology Planning & Evaluation (IITP) grant funded by the Korea government (MSIT) (No. 2021-0-02068, Artificial Intelligence Innovation Hub).

## References


1. VVC VTM reference software. https://vcgit.hhi.fraunhofer.de/jvet/VVCSoftware_VTM
2. Agarap, A.F.: Deep learning using rectified linear units (RELU). arXiv preprint arXiv:1803.08375 (2018)
3. Ballé, J., Laparra, V., Simoncelli, E.P.: End-to-end optimized image compression. arXiv preprint arXiv:1611.01704 (2016)
4. Ballé, J., Minnen, D., Singh, S., Hwang, S.J., Johnston, N.: Variational image compression with a scale hyperprior. arXiv preprint arXiv:1802.01436 (2018)
5. Bégaint, J., Racapé, F., Feltman, S., Pushparaja, A.: CompressAI: a PyTorch library and evaluation platform for end-to-end compression research. arXiv preprint arXiv:2011.03029 (2020)
6. Bellard, F.: BPG image format (2015). Signalprocessing: Imagecommunication
7. Chen, H., Gu, J., Zhang, Z.: Attention in attention network for image super-resolution. arXiv preprint arXiv:2104.09497 (2021)
8. Cheng, Z., Sun, H., Takeuchi, M., Katto, J.: Learned image compression with discretized Gaussian mixture likelihoods and attention modules. In: Proceedings of the IEEE/CVF Conference on Computer Vision and Pattern Recognition, pp. 7939–7948 (2020)
9. Cui, Z., Wang, J., Gao, S., Guo, T., Feng, Y., Bai, B.: Asymmetric gained deep image compression with continuous rate adaptation. In: Proceedings of the IEEE/CVF Conference on Computer Vision and Pattern Recognition, pp. 10532–10541 (2021)
10. CVPR2021: Workshop and challenge on learned image compression (2021). http://clic.compression.cc/2021/tasks/index.html
11. Dai, T., Cai, J., Zhang, Y., Xia, S.T., Zhang, L.: Second-order attention network for single image super-resolution. In: Proceedings of the IEEE/CVF Conference on Computer Vision and Pattern Recognition, pp. 11065–11074 (2019)
12. Feichtenhofer, C., Pinz, A., Wildes, R.P.: Spatiotemporal multiplier networks for video action recognition. In: Proceedings of the IEEE Conference on Computer Vision and Pattern Recognition, pp. 4768–4777 (2017)
13. He, D., Zheng, Y., Sun, B., Wang, Y., Qin, H.: Checkerboard context model for efficient learned image compression. In: Proceedings of the IEEE/CVF Conference on Computer Vision and Pattern Recognition, pp. 14771–14780 (2021)
14. Ioffe, S., Szegedy, C.: Batch normalization: accelerating deep network training by reducing internal covariate shift. In: International Conference on Machine Learning, pp. 448–456. PMLR (2015)
15. Kingma, D.P., Ba, J.: Adam: a method for stochastic optimization. arXiv preprint arXiv:1412.6980 (2014)
16. Kodak, E.: Kodak lossless true color image suite (PhotoCD PCD0992). http://r0k.us/graphics/kodak/
17. Lee, J., Cho, S., Beack, S.K.: Context-adaptive entropy model for end-to-end optimized image compression. arXiv preprint arXiv:1809.10452 (2018)
18. Lin, T.-Y., et al.: Microsoft COCO: common objects in context. In: Fleet, D., Pajdla, T., Schiele, B., Tuytelaars, T. (eds.) ECCV 2014. LNCS, vol. 8693, pp. 740–755. Springer, Cham (2014). https://doi.org/10.1007/978-3-319-10602-1_48
19. Minnen, D., Ballé, J., Toderici, G.D.: Joint autoregressive and hierarchical priors for learned image compression. In: Advances in Neural Information Processing Systems, vol. 31 (2018)

20. Minnen, D., Singh, S.: Channel-wise autoregressive entropy models for learned image compression. In: 2020 IEEE International Conference on Image Processing (ICIP), pp. 3339–3343. IEEE (2020)
21. Niu, B., et al.: Single image super-resolution via a holistic attention network. In: Vedaldi, A., Bischof, H., Brox, T., Frahm, J.-M. (eds.) ECCV 2020. LNCS, vol. 12357, pp. 191–207. Springer, Cham (2020). https://doi.org/10.1007/978-3-030-58610-2_12
22. Ohm, J.R., Sullivan, G.J.: Versatile video coding-towards the next generation of video compression. In: Picture Coding Symposium (2018)
23. Park, J., Woo, S., Lee, J.Y., Kweon, I.S.: BAM: bottleneck attention module. arXiv preprint arXiv:1807.06514 (2018)
24. Rabbani, M., Joshi, R.: An overview of the JPEG 2000 still image compression standard. Signal Process.: Image Commun. **17**(1), 3–48 (2002)
25. Ramachandran, P., Zoph, B., Le, Q.V.: Searching for activation functions. arXiv preprint arXiv:1710.05941 (2017)
26. Sullivan, G.J., Ohm, J.R., Han, W.J., Wiegand, T.: Overview of the high efficiency video coding (HEVC) standard. IEEE Trans. Circuits Syst. Video Technol. **22**(12), 1649–1668 (2012). https://doi.org/10.1109/TCSVT.2012.2221191
27. Toderici, G., et al.: Variable rate image compression with recurrent neural networks. arXiv preprint arXiv:1511.06085 (2015)
28. Toderici, G., et al.: Full resolution image compression with recurrent neural networks. In: Proceedings of the IEEE Conference on Computer Vision and Pattern Recognition, pp. 5306–5314 (2017)
29. Wallace, G.K.: The JPEG still picture compression standard. IEEE Trans. Consum. Electron. **38**(1), xviii–xxxiv (1992)
30. Wang, Z., Simoncelli, E., Bovik, A.: Multiscale structural similarity for image quality assessment. In: The Thirty-Seventh Asilomar Conference on Signals, Systems Computers, vol. 2, pp. 1398–1402 (2003). https://doi.org/10.1109/ACSSC.2003.1292216
31. Woo, S., Park, J., Lee, J.Y., Kweon, I.S.: CBAM: convolutional block attention module. In: Proceedings of the European Conference on Computer Vision (ECCV), pp. 3–19 (2018)
32. Xue, T., Chen, B., Wu, J., Wei, D., Freeman, W.T.: Video enhancement with task-oriented flow. Int. J. Comput. Vis. (IJCV) **127**(8), 1106–1125 (2019)
33. Zhang, Y., Li, K., Li, K., Wang, L., Zhong, B., Fu, Y.: Image super-resolution using very deep residual channel attention networks. In: Proceedings of the European Conference on Computer Vision (ECCV), pp. 286–301 (2018)
34. Zhou, L., Sun, Z., Wu, X., Wu, J.: End-to-end optimized image compression with attention mechanism. In: CVPR Workshops (2019)

# Generator Knows What Discriminator Should Learn in Unconditional GANs

Gayoung Lee[1], Hyunsu Kim[1], Junho Kim[1], Seonghyeon Kim[2], Jung-Woo Ha[1,2], and Yunjey Choi[1](✉)

[1] NAVER AI Lab, Seongnam-si, Republic of Korea
{gayoung.lee,hyunsu1125.kim,jhkim.ai,kim.seonghyeon, jungwoo.ha,yunjey.choi}@navercorp.com

[2] NAVER CLOVA, Seongnam-si, Republic of Korea
kim.seonghyeon@navercorp.com

**Abstract.** Recent methods for conditional image generation benefit from dense supervision such as segmentation label maps to achieve high-fidelity. However, it is rarely explored to employ dense supervision for unconditional image generation. Here we explore the efficacy of dense supervision in unconditional generation and find generator feature maps can be an alternative of cost-expensive semantic label maps. From our empirical evidences, we propose a new *generator-guided discriminator regularization* (*GGDR*) in which the generator feature maps supervise the discriminator to have rich semantic representations in unconditional generation. In specific, we employ an U-Net architecture for discriminator, which is trained to predict the generator feature maps given fake images as inputs. Extensive experiments on mulitple datasets show that our GGDR consistently improves the performance of baseline methods in terms of quantitative and qualitative aspects. Code is available at https://github.com/naver-ai/GGDR.

**Keywords:** Generative adversarial networks · Unconditional image generation · Discriminator regularization · Generator feature maps

## 1 Introduction

Generative adversarial networks(GANs) have achieved promising results in various computer vision tasks including image [26–28] or video generation [52,58,59,64], translation [7,20,31,33,72], manipulation [3,15,21,30,35,49], and cross-domain translation [18,34] for the past several years. In GANs, building an effective discriminator is one of the key components for generation quality since the generator is trained by the feedback from the discriminator. Existing studies proposed various methods to make the discriminator learn better representations

---

**Supplementary Information** The online version contains supplementary material available at https://doi.org/10.1007/978-3-031-19790-1_25.

© The Author(s), under exclusive license to Springer Nature Switzerland AG 2022
S. Avidan et al. (Eds.): ECCV 2022, LNCS 13677, pp. 406–422, 2022.
https://doi.org/10.1007/978-3-031-19790-1_25

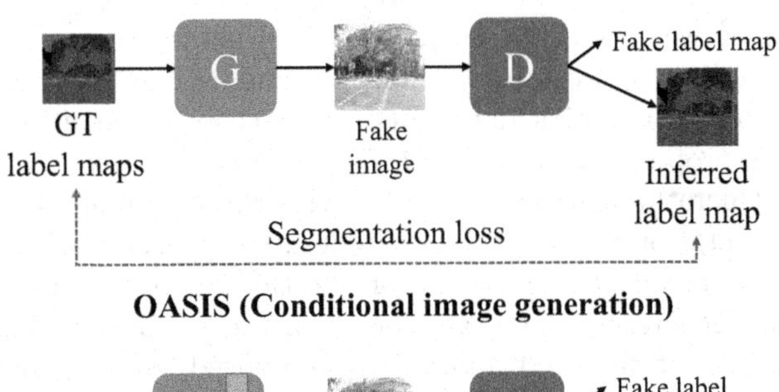


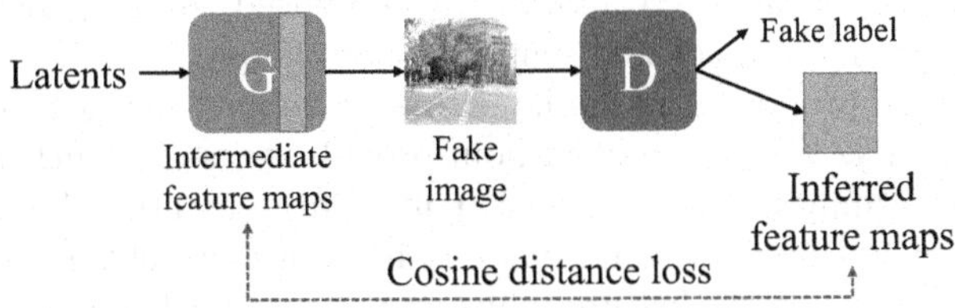


**Fig. 1.** Comparison of how to provide semantic information between OASIS and our method. OASIS enhance the discriminator with the ground truth label maps in conditional image generation setting. GGDR, on the other hand, aims at unconditional image synthesis, and uses the generator feature maps instead of human-annotating label maps.

by data augmentation [25,67,69,70], gradient penalty [41,42,45], and carefully designed architectures [24,48].

One simple yet effective way to improve the discriminator is to provide available additional annotations such as class labels [5,43], pose descriptors [51], normal maps [61], and semantic label maps [20,40,44,55]. Among these annotations, semantic label maps contain rich and dense descriptions about images, and have been frequently used in conditional scene generation. To provide dense semantic information to the discriminator, Pix2pix [20] and SPADE [44] concatenate the label maps with input images, and CC-FPSE [40] uses projection instead of the concatenation to inject the embedding of label maps. OASIS [55] further enhances the discriminator by providing strong supervision using auxiliary semantic segmentation task and achieves better performance.

Despite the success of dense semantic supervision in conditional generation, it has been rarely explored in an unconditional setting. Dense semantic supervision can be useful here as well, as GAN models often struggle when the data has varied and complex layout images. However, in unconditional generation, most large datasets do not have pairs of images and semantic label maps, since collecting them has a significant human annotation cost. Therefore, unlike the conditional setting, which requires a dense label map for the generator input, unconditional image generation assumes no dense map, and most studies use discriminators that learn only from images.

In this paper, we show that guiding a discriminator using dense and rich semantic information is also useful in unconditional image generation, and propose the method that avoids data annotation costs while utilizing semantic supervision. We propose *generator-guided discriminator regularization* (*GGDR*) in which the generator feature maps supervise the discriminator to have rich semantic representations. Specifically, we redesign the discriminator architecture in U-Net style, and train the discriminator to estimate the generator feature map when input is a generated image. As shown in Fig. 1, GGDR differs from the previous work in that the discriminator is supervised by the generator feature maps instead of human-annotated semantic label maps.

To justify our proposed method, we first compare the generation performance of StyleGAN2 [28] with and without providing ground-truth segmentation maps to the discriminator, and show that utilizing semantic label maps indeed improves the generation performance in an unconditional setting (Sect. 2.1). We then visualize the generator feature maps and show that they contain semantic information rich enough to guide the discriminator, replacing the ground-truth label maps (Sect. 2.2). Utilizing the generator feature maps, GGDR improves the discriminator representation, which is the key component to enhance the generation performance (Sect. 3). We provide thorough comparisons to demonstrate that GGDR consistently improves the baseline models on a variety of data. Our method can be easily attached to any setting without burdensome cost; only 3.7% of the network parameter increased. Our contributions can be summarized as follows:

1. We investigate the effectiveness of dense semantic supervision on unconditional image generation.
2. We show that generator feature maps can be used as an effective alternative of human-annotated semantic label maps.
3. We propose generator-guided discriminator regularization (GGDR), which encourages the discriminator to have rich semantic representation by utilizing generator feature maps.
4. We demonstrate that GGDR consistently improves the state-of-the-art methods on multiple datasets, especially in terms of generation diversity.

## 2 Dense Semantic Supervision in Unconditional GANs

We first conduct a preliminary experiment using ground-truth segmentation maps to show the efficacy of providing dense semantic supervision for the discriminator (Sect. 2.1). Then, we study whether the generator feature maps can be used as a guide instead of using human annotating ground-truth label maps to avoid expensive manual annotations. We visualize the internal feature maps of the generator and show that they have semantic information rich enough to be used as pseudo-semantic labels (Sect. 2.2).

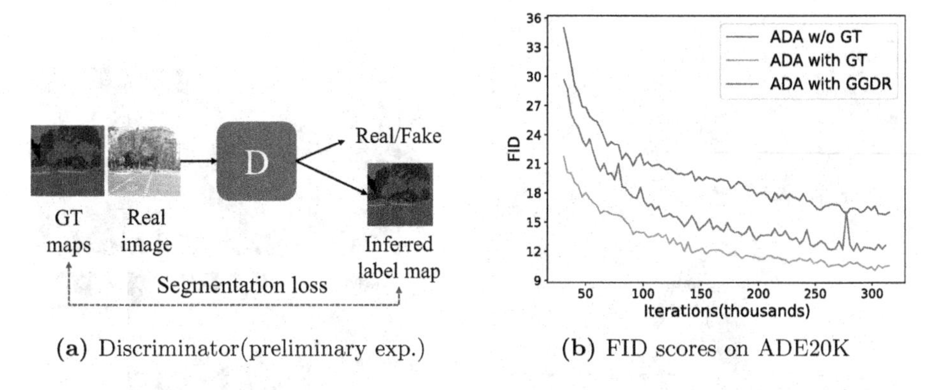


(a) Discriminator(preliminary exp.)

(b) FID scores on ADE20K

**Fig. 2.** Discriminator architecture and FID scores of the preliminary experiment to show the effects of the semantic label maps for unconditional image generation. (a) Discriminator with the auxiliary segmentation loss using semantic label maps (b) FID scores on ADE20K with and without dense semantic supervision.

### 2.1 Utilization of Semantic Label Maps for Discriminator

Although it is natural to utilize semantic label maps in conditional image generation [40,44,55], it has been still underexplored whether label maps are beneficial for unconditional image generation [25,27,28]. We conduct a preliminary experiment to validate the effects of the semantic label maps. We use ADE20K scene parsing benchmark dataset [71] consisting of 20,210 paired images and semantic label map annotations with 150 class labels, which is frequently used to evaluate conditional generation models. We choose StyleGAN2 [28] as our baseline, which is a standard model for unconditional image generation and apply adaptive discriminator augmentation [25]. To provide semantic supervision for the network, we redesign the discriminator to perform additional segmentation task similar to OASIS [55]. The modified task for the discriminator is described in Fig. 2 (a). The detailed architecture is similar to Fig. 4, except that it upsamples the decoder output until the image size is reached. The decoder in a U-Net style is attached to the discriminator and the segmentation loss is applied to the last layer of the decoder to provide dense supervision. The segmentation loss is the usual cross-entropy loss. Since ground-truth label maps are not available for generated images, we activate the segmentation loss for real images only.

As shown in Fig. 2 (b), the model with the discriminator leveraging the semantic supervision(ADA with GT) outperforms the baseline(ADA w/o GT). As argued in OASIS, the stronger semantic supervision seems to help discriminator learn more semantically and spatially-aware representations and give the generator more meaningful feedback. Our experiment supports that providing additional semantic guide for the discriminator can improve the model performance in unconditional image synthesis. However, dense label maps are rare in datasets for unconditional image synthesis, and it is time-consuming to collect them manually. In the next section, we analyze the feature maps from the generator as an effective alternative for the ground-truth label maps.

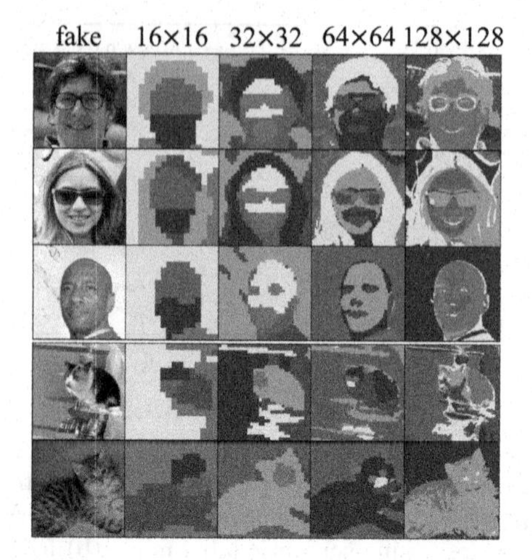


(a) Generator feature map visualization

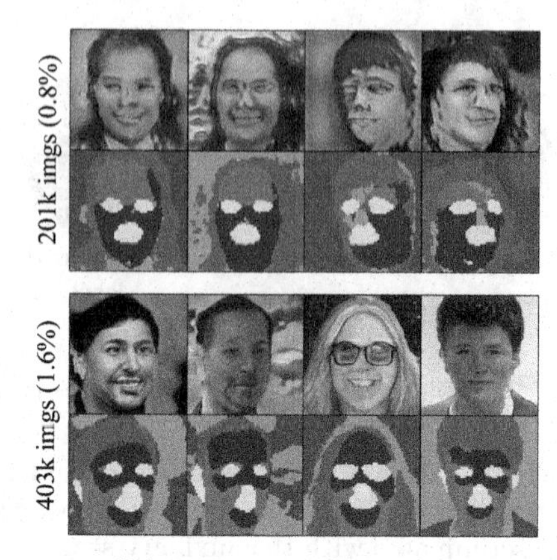


(b) Early training phase

**Fig. 3.** Visualization of the generator feature maps using $k$-means ($k = 6$) clustering. (a) StyleGAN2 generator feature map. The visualized feature maps reveal semantically consistent and meaningful regions such as ears in cats. (b) Generated images and their $32 \times 32$ feature maps in the early training phase.

### 2.2 Analysis of Generator Feature Maps

Recent studies have reported that the feature maps of the trained generator of GANs contain rich and dense semantic information [9,11,62]. Collins et al. [9] showed that applying $k$-means clustering to the feature maps of the generator reveals semantics and parts of objects, and used the clusters to edit images. We notice that these feature maps are rich semantic descriptors of the generated images and can be the substitute for the ground truth label maps. To visualize what information is captured in each feature map, we run $k$-means algorithm on each layer using the batch of generated images. We set $k = 6$ in this experiment. As shown in Fig. 3 (a), the pixels are clustered by the semantic information instead of the low-level features except the last feature map. For example, the hairs of the people have different colors, but are clustered in to the same cluster. The early feature maps show coarse object location, and those from the latter layers contain detailed object parts. The visualized feature maps look like pseudo-semantic label maps and might be regarded as rich descriptions including spatial and semantic information about the images. Therefore, we choose the feature maps of the generator as the substitute for the semantic label maps to guide the discriminator using semantic supervision. The generator feature maps are useful in our case. First, we do not need perfect semantic segmentation maps because our goal is image generation not semantic segmentation. Second, the feature maps are intermediate by-products essential for the generation, so acquiring them is free and does not require additional human annotations.

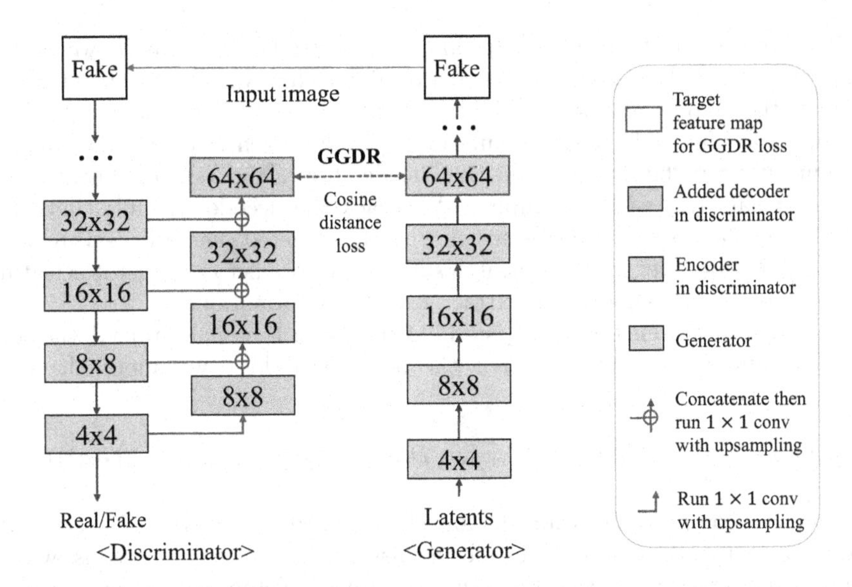


**Fig. 4.** Visualization of our framework. Our method can be applied to a GAN model by adding a decoder and the cosine distance loss with the reference generator feature maps to a discriminator. The generator feature maps guide the discriminator to learn more semantically-aware representations.

Dissimilar to previous works [9,11,62] that utilize the generator feature map for separate tasks, our method utilizes them during the training to enhance the generation performance itself. Therefore, it is essential to check whether the feature maps from the generator in the middle of training are still semantically meaningful for the guidance. In Fig. 3 (b), we visualize the feature maps of the generator during training to check how early the feature maps become semantically meaningful. Surprisingly, thanks to the powerful modern GANs, we can observe that even in the early stage, the feature maps and the corresponding generated images capture coarse shapes and location of objects. Therefore, we utilize the feature maps from the beginning of training, but for more complex data where the generator needs more iterations to produce meaningful semantics, one may choose when to attach our objective function.

## 3 Generator-Guided Discriminator Regularization

Based on our observations, we propose *generator-guided discriminator regularization* (GGDR) in which the generator feature maps supervise the discriminator to have rich semantic representations. The overall framework is shown in Fig. 4.

The design of our discriminator $D$ is inspired by that of OASIS [55] where the U-Net encoder-decoder structure is adopted and the last layer predicts semantic label maps. However, unlike OASIS, we leverage feature maps of the generator instead of ground-truth label maps. Thus, there are several differences in the design. First, since the feature maps are not discrete labels anymore, we cannot simply add real/fake class to the decoder output as done in OASIS. Therefore, we separate the decoder and adversarial loss. Next, we use more compact and lighter modules to reduce additional calculation costs. For each layer, we concatenate the output from the decoder and the encoder layer, and run one linear $1 \times 1$ convolutional layer with upsampling. We stack the decoder modules until the decoder output has the same resolution with the targeted generator feature map. Although the decoder is compact, it is sufficient to predict the generator feature map as the shared encoder can extract semantic information.

Meanwhile, the encoder part is still shared, and thus it is trained via both semantic and adversarial loss. For the adversarial loss, we adopt the non-saturating adversarial loss [13]:

$$\min_G \max_D \mathcal{L}_{adv}(G, D) = \mathbb{E}_{\boldsymbol{x}\sim p_{data}(\boldsymbol{x})}[\log D(\boldsymbol{x})] + \mathbb{E}_{\boldsymbol{z}\sim p(\boldsymbol{z})}[\log(1 - D(G(\boldsymbol{z})))]. \quad (1)$$

Next, we compute the cosine distance loss between the output of the decoder and the target feature map. We use cosine distance loss since it gives a loss within a specified range even between denormalized feature vectors, so it is convenient to scale according to the adversarial loss. Here, we denote $l \in \{1, 2, ..., L\}$ as a layer index, where $L$ is the number of decoding layers. Our discriminator $D$ contains an U-Net-style decoder $F$ ($F \subset D$) and the output of each layer denoted as $F^l$ has the same resolution with the corresponding generator feature map $G(\boldsymbol{z})^l$. Let us denote the target layer index for guidance as $t$. Our generator-guided discriminator regularization (GGDR) is defined as:

$$\max_D \mathcal{L}_{ggdr}(G, D) = -\mathbb{E}_{\boldsymbol{z}\sim p(\boldsymbol{z})}\left[1 - \frac{F^t(G(z)) \cdot G(z)^t}{\|F^t(G(z))\|_2 \cdot \|G(z)^t\|_2}\right], \quad (2)$$

Our full objective functions can be summarized as follows:

$$\mathcal{L}_{total} = \mathcal{L}_{adv} + \lambda_{reg}\mathcal{L}_{ggdr}, \quad (3)$$

where $\lambda_{reg}$ is a hyperparameter for relative strength compared to the adversarial term. Note that $k$-means clustering used in Sect. 2.2 is only for visualization purpose and we directly compare raw feature maps without any clustering. We expect the $\mathcal{L}_{ggdr}$ term to enhance the semantic representation of $D$. While the generator feature maps participate the regularization loss, we do not update the generator with $\mathcal{L}_{ggdr}$ to prevent a feature collapse which is a trivial solution making the cosine distance to zeros.

Our framework is simple and easy to apply existing GAN models, and does not require any additional annotation. Acquiring the intermediate feature maps from a generator is free because the generator already produces them in order to generate fake images. Despite its simplicity, in the next section, we show the effectiveness of our method in unconditional image generation for various datasets.

**Table 1.** FID scores of ours and comparison methods on FFHQ(left) and CIFAR-10(right) datasets. We run three training for each data and show their means and standard deviations. The numbers are largely brought from ADA [25] and we follow their evaluation protocol. We brought the numbers of diffusion models from [53]. The bold numbers indicate the best FID for each baseline.

| FFHQ | 2k | 10k | 140k |
|---|---|---|---|
| PA-GAN | 56.49 ±7.28 | 27.71 ±2.77 | 3.78 ±0.06 |
| WGAN-GP | 79.19 ±6.30 | 35.68 ±1.27 | 6.43 ±0.37 |
| zCR | 71.61 ±9.64 | 23.02 ±2.09 | 3.45 ±0.19 |
| AR | 66.64 ±3.64 | 25.37 ±1.45 | 4.16 ±0.05 |
| StyleGAN2 | 78.80 ±2.31 | 30.73 ±0.48 | 3.66 ±0.10 |
| +GGDR | 70.59 ±5.16 | 24.44 ±0.63 | **3.14±0.03** |
| ADA | **16.49±0.65** | 8.29 ±0.31 | 3.88 ±0.13 |
| +GGDR | 18.28 ±0.77 | **6.11±0.15** | 3.57 ±0.10 |

| CIFAR-10 | FID | IS |
|---|---|---|
| ProGAN | 15.52 | 8.56 ±0.06 |
| AutoGAN | 12.42 | 8.55 ±0.10 |
| StyleGAN2 | 8.32 ±0.09 | 9.21 ±0.09 |
| ADA | 2.92 ±0.05 | 9.83 ±0.04 |
| FSMR | 2.90 | 9.68 |
| DDPM | 3.17 ±0.05 | 9.46 ±0.11 |
| NCSN++ | 2.2 | 9.89 |
| ADA+GGDR | **2.15±0.02** | **10.02±0.06** |

## 4 Experiments

We validate the efficacy of our GGDR on various datasets including CIFAR-10 [36], FFHQ [27], LSUN cat, horse, church [63], AFHQ [8] and Landscapes [2]. CIFAR-10 consists of 50,000 tiny color images in 10 classes. FFHQ contains 70,000 face images, and AFHQ includes approximately 5,000 images per cat, dog and wild animal faces. LSUN cat, horse and church consist of scenes with cat, horse and church respectively, and we have used 200,000 images per each dataset. Landscapes contains photographs 4,320 landscape images collected from Flickr [1]. Following StyleGAN2 and ADA [25], we have applied horizontal flips for FFHQ and small datasets. All images are resized to $256 \times 256$ except AFHQ ($512 \times 512$) and CIFAR-10 ($32 \times 32$). For GGDR loss, we select the $64 \times 64$ feature map of the generator as the guidance map, except CIFAR-10 where we select the $8 \times 8$ feature map. We set $\lambda_{reg} = 10$ for the weight of the proposed regularization in all experiments and let other hyperparameters unmodified. We apply $R_1$ regularization [41] for StyleGAN2 and ADA models. In the case of the ADA, we apply the augmentation to the generator feature maps to make them consistent with corresponding fake images. We use only geometric operations and skip color transformation for the feature map augmentation.

For evaluation metrics, we have used Fréchet Inception Distance(FID) [16] and Precision & Recall [38]. FID measures the distance between the real images and the generate samples in feature space, and Precision & Recall scores indicate sample quality and variety. We compare 50,000 generated images and all training images following previous works [25]. For CIFAR-10, we also use Inception Score (IS) [46] following the previous works [25,53].

**Table 2.** Comparision on FFHQ, LSUN Cat, LSUN Horse and LSUN Church. Our method improves StyleGAN2 [28] in large datasets in terms of FID and recall. P and R denote precision and recall. Lower FID and higher precision and recall mean better performance. The bold numbers indicate the best FID, P, R for each dataset.

| Method | FFHQ | | | LSUN Cat | | | LSUN Horse | | | LSUN Church | | |
|---|---|---|---|---|---|---|---|---|---|---|---|---|
| | FID↓ | P↑ | R↑ | FID | P | R | FID | P | R | FID | P | R |
| UT [4] | 6.11 | **0.73** | 0.48 | – | – | – | – | – | – | 4.07 | **0.71** | 0.45 |
| Polarity [19] | – | – | – | 6.39 | **0.64** | 0.32 | – | – | – | 3.92 | 0.61 | 0.39 |
| StyleGAN2 | 3.71 | 0.69 | 0.44 | 7.98 | 0.60 | 0.27 | 3.62 | 0.63 | 0.36 | 3.97 | 0.59 | 0.39 |
| +GGDR | **3.14** | 0.69 | **0.50** | **5.28** | 0.58 | **0.38** | **2.50** | **0.64** | **0.43** | **3.15** | 0.61 | **0.46** |

**Table 3.** Comparision on AFHQ Cat, Dog, Wild and Landscape. Our method improves ADA [25] in small datasets in terms of FID and recall. P and R denote precision and recall. The bold numbers indicate the best FID, P, and R of the models.

| Method | AFHQ Cat | | | AFHQ Dog | | | AFHQ Wild | | | Landscape | | |
|---|---|---|---|---|---|---|---|---|---|---|---|---|
| | FID↓ | P↑ | R↑ | FID | P | R | FID | P | R | FID | P | R |
| FastGAN [39] | 4.69 | **0.78** | 0.31 | 13.09 | 0.75 | 0.38 | 3.14 | 0.76 | 0.20 | 16.44 | **0.77** | 0.16 |
| ContraD [22] | 3.82 | – | – | 7.16 | – | – | 2.54 | – | – | – | – | – |
| ADA | 3.55 | 0.77 | 0.41 | 7.40 | 0.76 | 0.48 | 3.05 | 0.76 | 0.13 | 13.87 | 0.72 | 0.20 |
| +GGDR | **2.76** | 0.74 | **0.52** | **4.59** | **0.79** | **0.53** | **2.06** | **0.80** | **0.27** | **10.38** | 0.69 | **0.29** |

### 4.1 Comparison with Baselines

We apply GGDR to StyleGAN2 [28] which is one of the standard models for unconditional image generation. Instead of the original StyleGAN2 setting, we use the baseline setting used in ADA [25] which has less parameters and shorter training iterations but shows comparable performance. For small datasets, we apply the adaptive discriminator augmentations (ADA) [25] that prevents overfitting of the discriminator and shows the superior performance in small datasets.

Table 1 shows that the proposed GGDR improves the performance of the baselines in terms of FID scores which indicate the overall quality of the synthesized images. Following ADA [25], we run the experiment multiple times on FFHQ with varying numbers (2k, 10k and 140k) of training images. We largely borrowed the reported scores from ADA [25] and NCSN++ [53]. In Table 1 left, we compare our method in a varied number of training images with various regularizing methods WGAN-GP [14], PA-GAN [66], zCR [70], AR [6] and ADA [25]. In the table, StyleGAN2 with our GGDR achieves the best score on FFHQ with full training images. With a sufficient number of training images, regularization methods based on data augmentation show limited improvement or degradation, whereas in our method the discriminator learns from a more accurate semantic map provided by a better quality generator. In most cases, GGDR improves the baseline performance significantly except the FFHQ 2k setting. We conjecture the dataset is too small to learn semantically meaningful feature maps in the generator. Since the quality of generator feature maps directly affects the discriminator in our method, our method is more effective when there is suffi-

**Fig. 5.** Selective samples generated by our method. For FFHQ and LSUN datasets, we show the results of StyleGAN2 with GGDR. For AFHQ and Landscape, we show the results of ADA with GGDR.

cient number of training data. However, as shown in Table 3, our method shows effectiveness on the datasets with approximately 5,000 images which are not too many to collect. Table 1 right shows that our GGDR significantly improves ADA performance in terms of both FID and IS scores, and makes it superior to various models including ProGAN [24], AutoGAN [12], FSMR [32] and DDPM [17], and comparable to the NCSN++ [53].

In Tables 2 and 3, we conduct extensive experiments to validate the performance improvement using GGDR on various datasets. For FFHQ and LSUN datasets, we report the scores of UT [4], and Polarity [19] which are the state-of-the-arts models that show the improvement on these datasets. For AFHQ and Landscape datasets, we report the score of ContraD [22] and FastGAN [39] that show significant improvements on image synthesis with small size datasets. We brought the numbers from their papers except FastGAN whose scores are brought from ProjGAN [47]. GGDR consistently improves the baseline in terms of FID scores with large gap. In terms of precision and recall metrics, GGDR improves the recall with significant margins compared to the baselines, which indirectly indicates where the advantages of our method come from. Better recall scores mean that our model generates more diverse images and less prone to the mode collapse. As it is known that incorporating image-level labels to the discriminator enhances coverage of classes in the data, utilizing pseudo

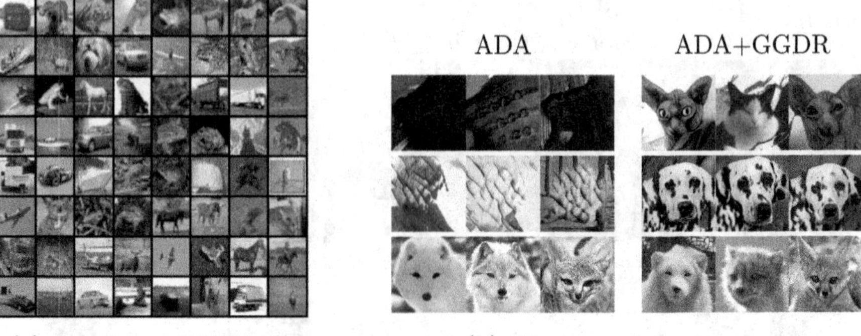


(a) Samples on CIFAR-10

(b) Worst samples comparison

**Fig. 6.** (a) Random samples by ADA with GGDR on CIFAR-10 dataset. (b) Qualitative comparison of worst-sample images. (top) AFHQ-Cat (middle) AFHQ-Dog (bottom) AFHQ-Wild.

dense semantic information could facilitate semantic diversities in the generated images.

In Fig. 5 and Fig. 6 (a), we show some selected results of our method on the evaluated datasets. We visualize StyleGAN2 with GGDR for FFHQ and LSUN, and ADA with GGDR for other datasets. More uncurated images are shown in the supplementary material. Since our method tends to improve the recall than the precision, it is hard to show visual improvement with the limited numbers of samples. Instead, we compare the worst samples in Fig. 6 (b). We follow the method of [37] to sort the samples, which uses the Inception [56] model to fit a gaussian model and sorts by the log-likelihood using it. We can see the worst samples of our method still contain objects unlike those of ADA. It is interesting that [37] reports similar improvements on the worst samples by utilizing pretrained models for the discriminator. We conjecture that the feature maps of the generator plays similar role with the pretrained models in their works [37,47].

### 4.2 Analysis and Ablation Study

The proposed GGDR supervises a discriminator using the intermediate feature maps of a generator in a GAN model, so it depends on the quality of the generator feature maps. In Sect. 2.2, we show that the feature maps of the generator contain valid semantic information even in the early stage. In addition to the visualization, we measure the FIDs in early iterations and show in Fig. 7 (a). To check the effect of initializations, we run multiple experiments on LSUN Cat. In early iterations, GGDR can interfere the performance by using less trained feature maps and bad initialization. However, after only several thousand iterations, StyleGAN2 with GGDR starts to converge faster and shows better scores.

In Table 4, we conduct ablation studies to investigate the effects of the decoder architecture and the feature map resolution. In our experiments, we

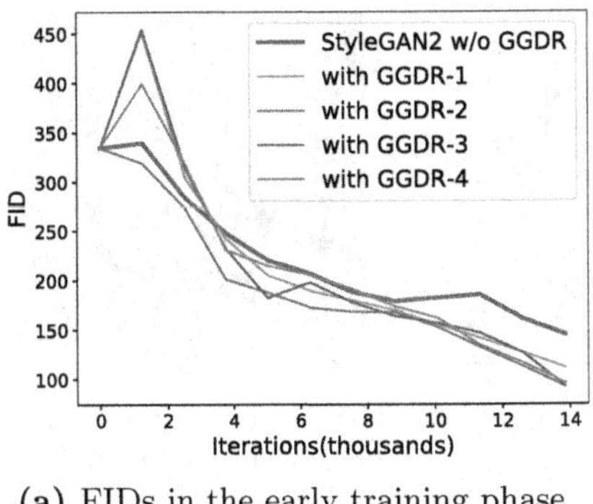


(a) FIDs in the early training phase

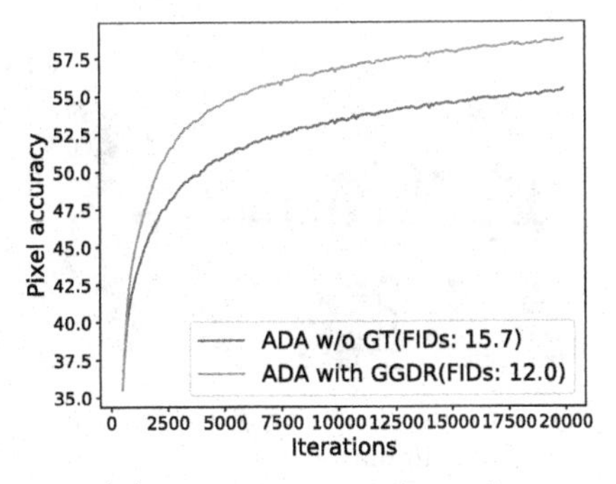


(b) Downstream task result

**Fig. 7.** (a) FID score graph of multiple experiments with GGDR in the early training phase on LSUN Cat. (b) Validation pixel accuracy on the ADE20K segmentation task with the frozen discriminator trained with and without GGDR.

**Table 4.** Ablation studies and calculation costs on LSUN Cat with eight V100 GPUs. Ablation study on (a) the target feature map size and (b) the decoder design. (c) Calculation costs with and without GGDR.

| Target | FID |
|---|---|
| None | 7.98 |
| $8 \times 8$ | 7.57 |
| $16 \times 16$ | 6.56 |
| $32 \times 32$ | 5.98 |
| $64 \times 64$ | **5.28** |

(a) Target size

| Activation | FID |
|---|---|
| Linear | **5.28** |
| leaky ReLU | 5.43 |
| Kernel size | FID |
| $1 \times 1$ | 5.28 |
| $3 \times 3$ | **5.25** |

(b) Decoder design

| Method | # params |
|---|---|
| Baseline | 4.87M |
| + GGDR | 5.05M (+3.7%) |
| Method | time(s) |
| Baseline | 5.60 |
| + GGDR | 6.05 (+8.0%) |

(c) Calculation costs

select $64 \times 64$ feature maps as the guidance. One may curious the performance differences if we use different sizes of the guidance feature maps. As shown in Table 4 (a), utilizing the large and dense feature maps achieves the best FID scores. Meanwhile, we design a compact decoder with $1 \times 1$ convolutional filters and linear activation for fast training and convergence. In Table 4 (b), we show that changing decoder activation and kernel size affects only negligible performance difference.

To analyze the effects of GGDR, we visualize and compare the original discriminator part in Fig. 8 (a). We run $k$-means clustering as done in Sect. 2.2. With GGDR loss, the shared encoder part prefers to learn high-level features which are useful for both tasks, so its feature maps reveal more semantically meaningful clusters. By guiding to learn semantic features, our approach can help discriminators to focus on salient parts of the image instead of meaningless features. To further analyze, we conduct a downstream experiment, training a

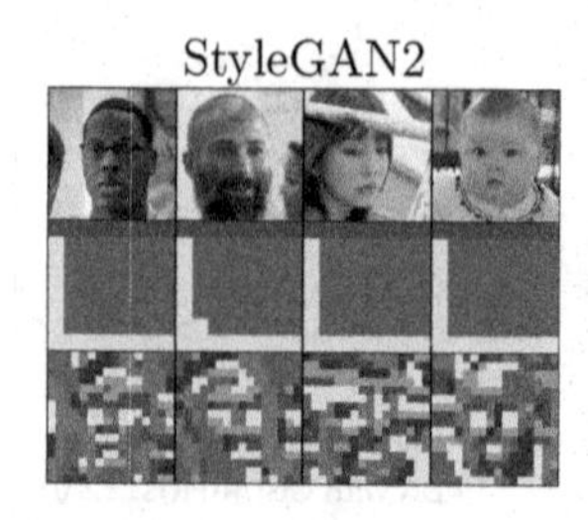

(a) $K$-means clustering of the feature maps of the encoder in the discriminators.

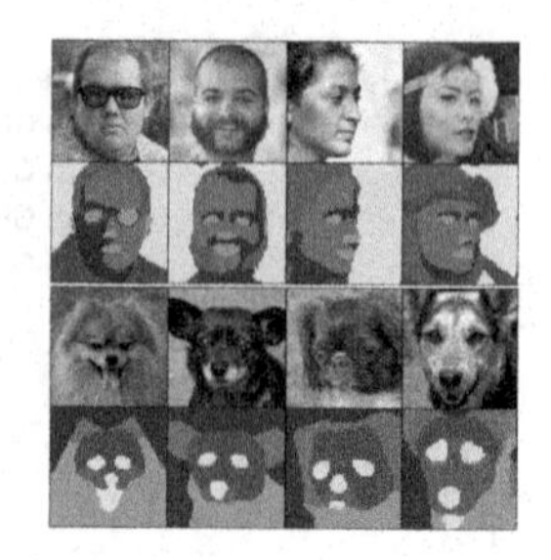

(b) Visualization of the decoder feature map

**Fig. 8.** $k$-means clustering of the feature maps of (a) the encoder in the discriminator with and without GGDR loss. From top to bottom, real images, feature maps that are 8 ($k = 3$) and 16 ($k = 6$) pixels wide. (b) the decoder in our discriminator on real images ($k = 6$).

shallow segmentation network using the extracted features by the discriminator with or without GGDR. As shown in Fig. 7 (b), the accuracy on validation data shows the discriminator with GGDR has more representation power on semantic information. Meanwhile, since we trained on fake images only, it may be curious if the guidance by fake images is still valid for real images. In Fig. 8 (b), we visualize the outputs of the decoder of our discriminator on real image. While the decoder of our method learned using fake images, we can see that its features well capture the semantically meaningful regions of the real images.

In Table 4 (c), we show the additional calculation costs when use GGDR. We can see the additional costs are marginal where the parameters increase 3.7% and the time increases 8.0%. For these measurement, we run the StyleGAN2 on the 256 × 256 dataset with eight V100 GPUs.

## 5 Related Work

**Conditional Image Synthesis Utilizing Semantic Label Map.** For the controllability of the generated images, it is common to exploit semantic layout-level information for the conditional image generation [20,40,60]. SPADE [44] utilizes Spatilally-Adaptive Denormalization which preserves semantic informations, and OASIS [55] have shown that it is able to train the conditional GANs using the discriminator that predicts pixel-level semantic labels, without incorporating semantic maps as additional conditions. Also, instead of using explicit semantic ground-truths, it is possible to use the features from the deep networks for the semantic guidance of the generator as shown in [10,50]. Unlike these works, our method aims at unconditional image generation.

**Regularization for GANs.** Several works interest to stabilizing the GAN trainings, especially by regularizing the discriminators [32,41,42]. Recently, utilizing augmentations for the discriminator gained a lot of interests, which was proven successful in general vision tasks [65,68]. In consistency regularization (CR) [67,70], in addition to using regular GAN losses, a discriminator is penalized by the differences in the outputs between augmented and non-augmented images. APA [23] regularizes the discriminator by utilizing fake images as psuedo-real data adaptively. DiffAugment [69] and ADA [25] use non-leaking augmentations for both generator and discriminator losses. Recently, several papers use pretrained models to help discriminator for fast and stable training. ProjGAN [47] uses EfficientNet [57] as a feature extractor for the discriminator, and Vision-aided GAN [37] provides automatic selection from model bank of pretrained networks to get optimal features for real and fake discrimination.

**Utilization of Generator Features.** Recent studies have shown that the generators contain rich and disentangled semantic structures in the features. Collins et al. [9] show that by applying $k$-means clustering on feature activations of the generator is possible to extract semantic objects and object parts in the generated images. Xu et al. [62] have trained linear mapping between feature maps in the generator and semantic maps, and Endo et al. [11] used the nearest neighbor matching between feature maps and representative vectors by averaging the feature vectors corresponding to the ground-truth semantic labels of inverted images. StyleMapGAN [30] has used spatial dimensions in the latent codes and grouping of the channels to further disentangle spatial semantic features.

## 6 Conclusion and Limitation

In this paper, we present the efficacy of the dense semantic label maps for unconditional image generation. Inspired by this observation, we propose a new regularization method to leverage the feature maps of the generator instead of human annotating ground-truth semantic annotations to allow the discriminator to learn richer semantic representation. With negligible additional parameters and no ground-truth semantic segmentation map, the proposed GGDR consistently outperforms strong baselines. Since our method depends on the performance of the generator, if the generator cannot learn meaningful representations due to the extremely limited number of data or initial training collapse, GGDR will fail to improve the performance. However, thanks to modern GANs and training techniques, we believe our method can be easily applied in various situations.

**Acknowledgement.** The experiments in the paper were conducted on NAVER Smart Machine Learning (NSML) platform [29,54]. We thank to Jun-Yan Zhu, Jaehoon Yoo, NAVER AI LAB researchers and the reviewers for their helpful comments and discussion.

## References

1. Flickr. https://www.flickr.com/
2. Kaggle landscapes dataset. https://www.kaggle.com/arnaud58/landscape-pictures (2019)

3. Bau, D., et al.: Gan dissection: visualizing and understanding generative adversarial networks. In: ICLR (2019)
4. Bond-Taylor, S., Hessey, P., Sasaki, H., Breckon, T.P., Willcocks, C.G.: Unleashing transformers: parallel token prediction with discrete absorbing diffusion for fast high-resolution image generation from vector-quantized codes. arXiv preprint. arXiv:2111.12701 (2021)
5. Brock, A., Donahue, J., Simonyan, K.: Large scale GAN training for high fidelity natural image synthesis. In: ICLR (2019)
6. Chen, T., Zhai, X., Ritter, M., Lucic, M., Houlsby, N.: Self-supervised gans via auxiliary rotation loss. In: CVPR (2019)
7. Choi, Y., Choi, M., Kim, M., Ha, J.W., Kim, S., Choo, J.: Stargan: unified generative adversarial networks for multi-domain image-to-image translation. In: CVPR (2018)
8. Choi, Y., Uh, Y., Yoo, J., Ha, J.W.: Stargan v2: diverse image synthesis for multiple domains. In: CVPR (2020)
9. Collins, E., Bala, R., Price, B., Susstrunk, S.: Editing in style: uncovering the local semantics of gans. In: CVPR (2020)
10. Ditria, L., Meyer, B.J., Drummond, T.: Opengan: open set generative adversarial networks. In: ACCV (2020)
11. Endo, Y., Kanamori, Y.: Few-shot semantic image synthesis using stylegan prior. CoRR abs/2103.14877 (2021)
12. Gong, X., Chang, S., Jiang, Y., Wang, Z.: Autogan: neural architecture search for generative adversarial networks. In: ICCV (2019)
13. Goodfellow, I., et al.: Generative adversarial networks. In: NeurIPS (2014)
14. Gulrajani, I., Ahmed, F., Arjovsky, M., Dumoulin, V., Courville, A.C.: Improved training of wasserstein gans. In: NeurIPS (2017)
15. Härkönen, E., Hertzmann, A., Lehtinen, J., Paris, S.: Ganspace: discovering interpretable gan controls. NeurIPS (2020)
16. Heusel, M., Ramsauer, H., Unterthiner, T., Nessler, B., Hochreiter, S.: Gans trained by a two time-scale update rule converge to a local nash equilibrium. In: NeurIPS (2017)
17. Ho, J., Jain, A., Abbeel, P.: Denoising diffusion probabilistic models. In: NeurIPS (2020)
18. Huang, S.W., Lin, C.T., Chen, S.P., Wu, Y.Y., Hsu, P.H., Lai, S.H.: Auggan: cross domain adaptation with gan-based data augmentation. In: ECCV (2018)
19. Humayun, A.I., Balestriero, R., Baraniuk, R.: Polarity sampling: quality and diversity control of pre-trained generative networks via singular values. In: CVPR (2022)
20. Isola, P., Zhu, J.Y., Zhou, T., Efros, A.A.: Image-to-image translation with conditional adversarial networks. In: CVPR (2017)
21. Jahanian, A., Chai, L., Isola, P.: On the "steerability" of generative adversarial networks. In: ICLR (2020)
22. Jeong, J., Shin, J.: Training gans with stronger augmentations via contrastive discriminator. In: ICLR (2021)
23. Jiang, L., Dai, B., Wu, W., Loy, C.C.: Deceive d: adaptive pseudo augmentation for gan training with limited data. In: NeurIPS (2021)
24. Karras, T., Aila, T., Laine, S., Lehtinen, J.: Progressive growing of gans for improved quality, stability, and variation. In: ICLR (2018)
25. Karras, T., Aittala, M., Hellsten, J., Laine, S., Lehtinen, J., Aila, T.: Training generative adversarial networks with limited data. In: NeurIPS (2020)
26. Karras, T., et al.: Alias-free generative adversarial networks. In: NeurIPS (2021)

27. Karras, T., Laine, S., Aila, T.: A style-based generator architecture for generative adversarial networks. In: CVPR (2019)
28. Karras, T., Laine, S., Aittala, M., Hellsten, J., Lehtinen, J., Aila, T.: Analyzing and improving the image quality of stylegan. In: CVPR (2020)
29. Kim, H., et al.: Nsml: meet the mlaas platform with a real-world case study. arXiv preprint. arXiv:1810.09957 (2018)
30. Kim, H., Choi, Y., Kim, J., Yoo, S., Uh, Y.: Exploiting spatial dimensions of latent in gan for real-time image editing. In: CVPR (2021)
31. Kim, H., Jhoo, H.Y., Park, E., Yoo, S.: Tag2pix: line art colorization using text tag with secat and changing loss. In: ICCV (2019)
32. Kim, J., Choi, Y., Uh, Y.: Feature statistics mixing regularization for generative adversarial networks. In: CVPR (2022)
33. Kim, J., Kim, M., Kang, H., Lee, K.H.: U-gat-it: unsupervised generative attentional networks with adaptive layer-instance normalization for image-to-image translation. In: ICLR (2020)
34. Kim, T., Cha, M., Kim, H., Lee, J.K., Kim, J.: Learning to discover cross-domain relations with generative adversarial networks. In: ICML (2017)
35. Kim, Y., Ha, J.W.: Contrastive fine-grained class clustering via generative adversarial networks. ICLR (2022)
36. Krizhevsky, A., Hinton, G., et al.: Learning multiple layers of features from tiny images. Department of Computer Science, University of Toronto, Technical report (2009)
37. Kumari, N., Zhang, R., Shechtman, E., Zhu, J.Y.: Ensembling off-the-shelf models for gan training. In: CVPR (2022)
38. Kynkäänniemi, T., Karras, T., Laine, S., Lehtinen, J., Aila, T.: Improved precision and recall metric for assessing generative models. NeurIPS (2019)
39. Liu, B., Zhu, Y., Song, K., Elgammal, A.: Towards faster and stabilized gan training for high-fidelity few-shot image synthesis. In: ICLR (2020)
40. Liu, X., Yin, G., Shao, J., Wang, X., Li, H.: Learning to predict layout-to-image conditional convolutions for semantic image synthesis. In: NeurIPS (2019)
41. Mescheder, L., Nowozin, S., Geiger, A.: Which training methods for gans do actually converge? In: ICML (2018)
42. Miyato, T., Kataoka, T., Koyama, M., Yoshida, Y.: Spectral normalization for generative adversarial networks. ICLR (2018)
43. Odena, A., Olah, C., Shlens, J.: Conditional image synthesis with auxiliary classifier gans. In: ICML (2017)
44. Park, T., Liu, M.Y., Wang, T.C., Zhu, J.Y.: Semantic image synthesis with spatially-adaptive normalization. In: CVPR (2019)
45. Roth, K., Lucchi, A., Nowozin, S., Hofmann, T.: Stabilizing training of generative adversarial networks through regularization. NeurIPS (2017)
46. Salimans, T., Goodfellow, I., Zaremba, W., Cheung, V., Radford, A., Chen, X.: Improved techniques for training gans. In: NeurIPS (2016)
47. Sauer, A., Chitta, K., Müller, J., Geiger, A.: Projected gans converge faster. NeurIPS (2021)
48. Schonfeld, E., Schiele, B., Khoreva, A.: A u-net based discriminator for generative adversarial networks. In: CVPR (2020)
49. Shen, Y., Gu, J., Tang, X., Zhou, B.: Interpreting the latent space of gans for semantic face editing. In: CVPR (2020)
50. Shocher, A., et al.: Semantic pyramid for image generation. In: CVPR (2020)
51. Siarohin, A., Sangineto, E., Lathuiliere, S., Sebe, N.: Deformable gans for pose-based human image generation. In: CVPR (2018)

52. Skorokhodov, I., Tulyakov, S., Elhoseiny, M.: Stylegan-v: A continuous video generator with the price, image quality and perks of stylegan2. In: CVPR (2022)
53. Song, Y., Sohl-Dickstein, J., Kingma, D.P., Kumar, A., Ermon, S., Poole, B.: Score-based generative modeling through stochastic differential equations. In: ICLR (2021)
54. Sung, N., et al.: Nsml: a machine learning platform that enables you to focus on your models. arXiv preprint. arXiv:1712.05902 (2017)
55. Sushko, V., Schönfeld, E., Zhang, D., Gall, J., Schiele, B., Khoreva, A.: You only need adversarial supervision for semantic image synthesis. In: ICLR (2021)
56. Szegedy, C., Vanhoucke, V., Ioffe, S., Shlens, J., Wojna, Z.: Rethinking the inception architecture for computer vision. In: CVPR (2016)
57. Tan, M., Le, Q.: Efficientnet: rethinking model scaling for convolutional neural networks. In: ICML (2019)
58. Tian, Y., et al.: A good image generator is what you need for high-resolution video synthesis. In: ICLR (2021)
59. Tulyakov, S., Liu, M.Y., Yang, X., Kautz, J.: Mocogan: decomposing motion and content for video generation. In: CVPR (2018)
60. Wang, T.C., Liu, M.Y., Zhu, J.Y., Tao, A., Kautz, J., Catanzaro, B.: High-resolution image synthesis and semantic manipulation with conditional gans. In: CVPR (2018)
61. Wang, X., Gupta, A.: Generative image modeling using style and structure adversarial networks. In: Leibe, B., Matas, J., Sebe, N., Welling, M. (eds.) ECCV 2016. LNCS, vol. 9908, pp. 318–335. Springer, Cham (2016). https://doi.org/10.1007/978-3-319-46493-0_20
62. Xu, J., Zheng, C.: Linear semantics in generative adversarial networks. In: CVPR (2021)
63. Yu, F., Seff, A., Zhang, Y., Song, S., Funkhouser, T., Xiao, J.: Lsun: construction of a large-scale image dataset using deep learning with humans in the loop. arXiv preprint. arXiv:1506.03365 (2015)
64. Yu, S., et al.: Generating videos with dynamics-aware implicit generative adversarial networks. In: ICLR (2022)
65. Yun, S., Han, D., Oh, S.J., Chun, S., Choe, J., Yoo, Y.: Cutmix: regularization strategy to train strong classifiers with localizable features. In: ICCV (2019)
66. Zhang, D., Khoreva, A.: PA-GAN: Improving gan training by progressive augmentation. In: NeurIPS (2019)
67. Zhang, H., Zhang, Z., Odena, A., Lee, H.: Consistency regularization for generative adversarial networks. In: ICLR (2020)
68. Zhang, H., Cisse, M., Dauphin, Y.N., Lopez-Paz, D.: mixup: beyond empirical risk minimization. arXiv preprint .arXiv:1710.09412 (2017)
69. Zhao, S., Liu, Z., Lin, J., Zhu, J.Y., Han, S.: Differentiable augmentation for data-efficient gan training. In: NeurIPS (2020)
70. Zhao, Z., Singh, S., Lee, H., Zhang, Z., Odena, A., Zhang, H.: Improved consistency regularization for gans. In: AAAI (2021)
71. Zhou, B., Zhao, H., Puig, X., Fidler, S., Barriuso, A., Torralba, A.: Scene parsing through ade20k dataset. In: CVPR (2017)
72. Zhu, J.Y., Park, T., Isola, P., Efros, A.A.: Unpaired image-to-image translation using cycle-consistent adversarial networks. In: ICCV (2017)

# Compositional Visual Generation with Composable Diffusion Models

Nan Liu[1], Shuang Li[2(✉)], Yilun Du[2(✉)], Antonio Torralba[2], and Joshua B. Tenenbaum[2]

[1] University of Illinois Urbana-Champaign, Champaign, USA
nanliu4@illinois.edu
[2] Massachusetts Institute of Technology, Cambridge, USA
{lishuang,yilundu,torralba,jbt}@mit.edu
https://energy-based-model.github.io/Compositional-Visual-Generation-with-Composable-Diffusion-Models/

**Abstract.** Large text-guided diffusion models, such as DALLE-2, are able to generate stunning photorealistic images given natural language descriptions. While such models are highly flexible, they struggle to understand the composition of certain concepts, such as confusing the attributes of different objects or relations between objects. In this paper, we propose an alternative structured approach for compositional generation using diffusion models. An image is generated by composing a set of diffusion models, with each of them modeling a certain component of the image. To do this, we interpret diffusion models as energy-based models in which the data distributions defined by the energy functions may be explicitly combined. The proposed method can generate scenes at test time that are substantially more complex than those seen in training, composing sentence descriptions, object relations, human facial attributes, and even generalizing to new combinations that are rarely seen in the real world. We further illustrate how our approach may be used to compose pre-trained text-guided diffusion models and generate photorealistic images containing all the details described in the input descriptions, including the binding of certain object attributes that have been shown difficult for DALLE-2. These results point to the effectiveness of the proposed method in promoting structured generalization for visual generation.

**Keywords:** Compositionality · Diffusion models · Energy-based models · Visual generation

## 1 Introduction

Our understanding of the world is highly compositional in nature. We are able to rapidly understand new objects from their components, or compose words into

N. Liu, S. Li and Y. Du—indicates equal contribution.

**Supplementary Information** The online version contains supplementary material available at https://doi.org/10.1007/978-3-031-19790-1_26.

© The Author(s), under exclusive license to Springer Nature Switzerland AG 2022
S. Avidan et al. (Eds.): ECCV 2022, LNCS 13677, pp. 423–439, 2022.
https://doi.org/10.1007/978-3-031-19790-1_26

**Fig. 1.** Our method allows compositional visual generation across a variety of domains, such as language descriptions, objects, object relations, and human attributes.

complex sentences to describe the world states we encounter [21]. We are able to make 'infinite use of finite means' [4], *i.e.*, repeatedly reuse and recombine concepts we have acquired in a potentially infinite manner. We are interested in constructing machine learning systems to have such compositional capabilities, particularly in the context of generative modeling.

Existing text-conditioned diffusion models such as DALLE-2 [30] have recently made remarkable strides towards compositional generation, and are capable in generating photorealistic images given textual descriptions. However, such systems are not fully compositional in nature and generate incorrect images when given more complex descriptions [24,39]. An underlying difficulty may be that such models encode text descriptions as fixed-size latent vectors. However, as textual descriptions become more complex, more information needs to be squeezed into the fixed-size vector. Thus it is impossible to encode arbitrarily complex textual descriptions.

In this work, we propose to factorize the compositional generation problem, using different diffusion models to capture different subsets of a compositional specification. These diffusion models are then explicitly composed together to jointly generate an image. By explicitly factorizing the compositional generative modeling problem, our method is able to generalize to significantly more complex combinations that are unseen during training.

Such an explicit form of compositionality has been explored before under the context of Energy-Based Models (EBMs) [7,8,23]. However, directly training EBMs has been proved to be unstable and hard to scale. We show that diffusion models can be interpreted as implicitly parameterized EBMs, which can be further composed for image generation, significantly improving training stability and image quality.

Our proposed method enables zero-shot compositional generation across different domains as shown in Fig. 1. First, we illustrate how our approach may be applied to large pre-trained diffusion models, such as GLIDE [26], to compose multiple text descriptions. Next, we illustrate how our approach can be applied to compose objects and object relations, enabling zero-shot generalization to a larger number of objects. Finally, we illustrate how our framework can compose different facial attributes to generate human faces.

**Contributions:** In this paper, we introduce an approach towards compositional visual generation using diffusion models. First, we show that diffusion models can be composed by interpreting them as energy-based models and drawing on this connection, show how we may compose diffusion models together.

Second, we propose two compositional operators, conjunction and negation, on top of diffusion models that allow us to compose concepts in different domains during inference without any additional training. We show that the proposed method enables effective zero-shot combinatorial generalization. Finally, we evaluate our method on composing language descriptions, objects, object relations, and human facial attributes. Our method can generate high-quality images containing all the concepts and outperforms baselines by a large margin. For example, the accuracy of our method is 24.02% higher than the best baseline for composing three objects in the specified positions on the CLEVR dataset.

## 2 Related Work

**Controllable Image Generation.** Our work is related to existing work on controllable image generation. One type of approach towards controllable image generation specifies the underlying content of an image utilizing text through either GANs [2,43,44], VQ-VAEs [31], or diffusion models [26]. An alternative type of approach towards controllable image generation manipulates the underlying attributes in an image [35,42,46]. In contrast, we are interested in *compositionally controlling* the underlying content of an image at test time, generating images that exhibit compositions of multiple different types of image content. Thus, most relevant to our work, existing work has utilized EBMs to compose different factors describing a scene [7,8,23,27]. We illustrate how we may implement such probabilistic composition on diffusion models, achieving better performance.

**Diffusion Models.** Diffusion models have emerged as a promising class of generative models that formulates the data-generating process as an iterative denoising procedure [15,36]. The denoising procedure can be seen as parameterizing the gradients of the data distribution [38], connecting diffusion models to EBMs [10–12,22,28]. Diffusion models have recently shown great promise in image generation [6], enabling effective image editing [20,25], text conditioning [13,26,32], and image inpainting [33]. The iterative, gradient-based sampling of diffusion models lends itself towards flexible conditioning [6], enabling us to compose factors across different images. While diffusion models have been developed for image generation [37], they have further proven successful in the generation of waveforms [3], 3D shapes [45], decision making [16] and text [1], suggesting that our proposed composition operators may further be applied in such domains.

# 3 Background

## 3.1 Denoising Diffusion Models

Denoising Diffusion Probabilistic Models (DDPMs) are a class of generative models where generation is modeled as a denoising process. Starting from sampled noise, the diffusion model performs $T$ denoising steps until a sharp image is formed. In particular, the denoising process produces a series of intermediate images with decreasing levels of noise, denoted as $\boldsymbol{x}_T, \boldsymbol{x}_{T-1}, ..., \boldsymbol{x}_0$, where $\boldsymbol{x}_T$ is sampled from a Gaussian prior and $\boldsymbol{x}_0$ is the final output image.

DDPMs construct a forward diffusion process by gradually adding Gaussian noise to the ground truth image. A diffusion model then learns to revert this noise corruption process. Both the *forward processes* $q(\boldsymbol{x}_t|\boldsymbol{x}_{t-1})$ and the *reverse process* $q(\boldsymbol{x}_{t-1}|\boldsymbol{x}_t)$ are modeled as the products of Markov transition probabilities:

$$q(\boldsymbol{x}_{0:T}) = q(\boldsymbol{x}_0) \prod_{t=1}^{T} q(\boldsymbol{x}_t|\boldsymbol{x}_{t-1}), \quad p_\theta(\boldsymbol{x}_{T:0}) = p(\boldsymbol{x}_T) \prod_{t=T}^{1} p_\theta(\boldsymbol{x}_{t-1}|\boldsymbol{x}_t), \tag{1}$$

where $q(\boldsymbol{x}_0)$ is the real data distribution and $p(\boldsymbol{x}_T)$ is a standard Gaussian prior.

A *generative process* $p_\theta(\boldsymbol{x}_{t-1}|\boldsymbol{x}_t)$ is trained to generate realistic images by approximating the reverse process through variational inference. Each step of the *generative process* is a Gaussian distribution with learned mean and covariance:

$$p_\theta(\boldsymbol{x}_{t-1}|\boldsymbol{x}_t) := \mathcal{N}(\mu_\theta(\boldsymbol{x}_t, t), \sigma_t^2) = \mathcal{N}(\boldsymbol{x}_t + \epsilon_\theta(\boldsymbol{x}_t, t), \sigma_t^2), \tag{2}$$

where $\boldsymbol{x}_{t-1}$ is parameterized by a mean $\mu_\theta(\boldsymbol{x}_t, t)$ represented by a perturbation $\epsilon_\theta(\boldsymbol{x}_t, t)$ to a noisy image $\boldsymbol{x}_t$. The goal is to remove the noise gradually by predicting a less noisy image at timestep $\boldsymbol{x}_{t-1}$ given a noisy image $\boldsymbol{x}_t$. To generate real images, we sample $\boldsymbol{x}_{t-1}$ from $t = T$ to $t = 1$ using the parameterized marginal distribution $p_\theta(\boldsymbol{x}_{t-1}|\boldsymbol{x}_t)$, with an individual step corresponding to:

$$\boldsymbol{x}_{t-1} = \boldsymbol{x}_t + \epsilon_\theta(\boldsymbol{x}_t, t) + \mathcal{N}(0, \sigma_t^2). \tag{3}$$

The generated images become more realistic over multiple iterations.

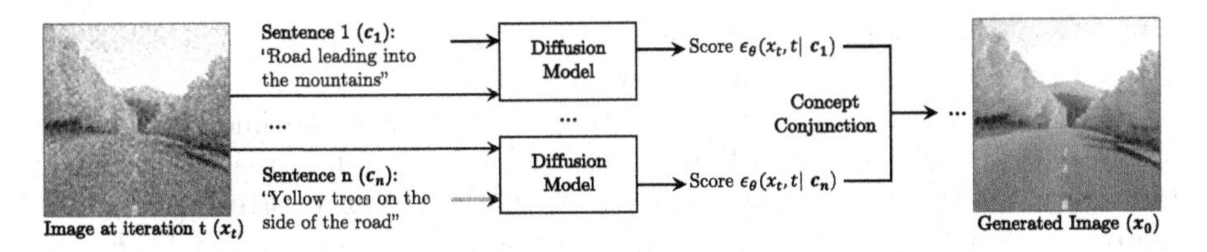


**Fig. 2. Compositional generation.** Our method can compose multiple concepts during inference and generate images containing all the concepts without further training. We first send an image from iteration $t$ and each of the concept to the diffusion model to generate a set of scores $\{\epsilon_\theta(\boldsymbol{x}_t, t|\boldsymbol{c}_1), \ldots, \epsilon_\theta(\boldsymbol{x}_t, t|\boldsymbol{c}_n)\}$. We then compose different concepts using the proposed compositional operators, such as conjunction, to denoise the generated images. The final image is obtained after $T$ iterations.

### 3.2 Energy Based Models

Energy-Based Models (EBMs) [9,10,12,28] are a class of generative models where the data distribution is modeled using an unnormalized probability density. Given an image $\boldsymbol{x} \in \mathbb{R}^D$, the probability density of image $\boldsymbol{x}$ is defined as:

$$p_\theta(\boldsymbol{x}) \propto e^{-E_\theta(\boldsymbol{x})}, \tag{4}$$

where the energy function $E_\theta(\boldsymbol{x}) : \mathbb{R}^D \to \mathbb{R}$ is a learnable neural network.

A gradient based MCMC procedure, Langevin dynamics [10], is then used to sample from an unnormalized probability distribution to iteratively refine the generated image $\boldsymbol{x}$:

$$\boldsymbol{x}_t = \boldsymbol{x}_{t-1} - \frac{\lambda}{2} \nabla_{\boldsymbol{x}} E_\theta(\boldsymbol{x}_{t-1}) + \mathcal{N}(0, \sigma^2). \tag{5}$$

The procedure for sampling from diffusion models in Eq. (3) is functionally similar to the sampling procedure used by EBMs in Eq. (5). In both settings, images are iteratively refined starting from Gaussian noise, with a small amount of additional Gaussian noise added at each iterative step.

## 4 Our Approach

In this section, we first introduce how we may interpret diffusion models as energy-based models in Sect. 4.1 and then introduce how we compose diffusion models for visual generation in Sect. 4.2.

### 4.1 Diffusion Models as Energy Based Models

The sampling procedure of diffusion models in Eq. (3) and EBMs in Eq. (5) are functionally similar. At a timestep $t$, in diffusion models, images are updated using a learned denoising network $\epsilon_\theta(\boldsymbol{x}_t, t)$ while in EBMs, images are updated using the gradient of the energy function $\nabla_{\boldsymbol{x}} E_\theta(\boldsymbol{x}_t) \propto \nabla_{\boldsymbol{x}} \log p_\theta(\boldsymbol{x})$, which is the score of the estimated probability distribution $p_\theta(\boldsymbol{x})$.

The denoising network $\epsilon_\theta(\boldsymbol{x}_t, t)$ is trained to predict the underlying score of the data distribution [37,41] when the number of diffusion steps increases to infinity. Similarly, an EBM is trained so that $\nabla_{\boldsymbol{x}} E_\theta(\boldsymbol{x}_t)$ corresponds to the score of the data distribution as well. In this sense, $\epsilon_\theta(\boldsymbol{x}_t, t)$ and $\nabla_{\boldsymbol{x}} E_\theta(\boldsymbol{x}_t)$ are fuctionally the same, and the underlying sampling procedure in Eq. (3) and Eq. (5) are equivalent. We may view a trained diffusion model $\epsilon_\theta(\boldsymbol{x}_t, t)$ as implicitly parameterizing an EBM by defining its data gradient $\nabla_{\boldsymbol{x}} E_\theta(\boldsymbol{x}_t)$ at each data point, and we will subsequently refer to $\epsilon_\theta(\boldsymbol{x}_t, t)$ as the score function. Such a parameterization enables us to leverage past work towards composing EBMs and apply it to diffusion models.

**Composing EBMs.** Previous EBMs [7,14] have shown good compositionality ability for visual generation. Given $n$ independent EBMs, $E_\theta^1(\boldsymbol{x}), \cdots, E_\theta^n(\boldsymbol{x})$, the functional form of EBMs in Eq. (4) enable us to compose multiple separate EBMs together to obtain a new EBM. The composed distribution can be represented as:

$$p_{\text{compose}}(\boldsymbol{x}) \propto p_\theta^1(\boldsymbol{x}) \cdots p_\theta^n(\boldsymbol{x}) \propto e^{-\sum_i E_\theta^i(\boldsymbol{x})} = e^{-E_\theta(\boldsymbol{x})}, \tag{6}$$

where $p_\theta^i \propto e^{-E_\theta^i(\boldsymbol{x})}$ is the probability density of image $\boldsymbol{x}$ (Eq. (4)). Langevin dynamics is used to iteratively refine the generated image $\boldsymbol{x}$.

$$\boldsymbol{x}_t = \boldsymbol{x}_{t-1} - \frac{\lambda}{2}\nabla_{\boldsymbol{x}}(\sum_i E_\theta^i(\boldsymbol{x}_{t-1})) + \mathcal{N}(0, \sigma^2). \tag{7}$$

**Composing Diffusion Models.** By leveraging the interpretation that diffusion models are functionally similar to EBMs, we may compose diffusion models in a similar way. The *generative process* and the score function of a diffusion model can be represented as $p_\theta^i(\boldsymbol{x}_{t-1}|\boldsymbol{x}_t)$ and $\epsilon_\theta^i(\boldsymbol{x}, t)$, respectively. If we treat the individual score function in diffusion models as the learned gradient of energy functions in EBMs, the composition of diffusion models has a score function of $\sum_i \epsilon_\theta^i(\boldsymbol{x}, t)$. Thus the *generative process* of composing multiple diffusion models becomes:

$$p_{\text{compose}}(\boldsymbol{x}_{t-1}|\boldsymbol{x}_t) = \mathcal{N}(\boldsymbol{x}_t + \sum_i \epsilon_\theta^i(\boldsymbol{x}_t, t), \sigma_t^2). \tag{8}$$

A complication when parameterizing of a gradient field of EBM $\nabla_{\boldsymbol{x}} E_\theta(\boldsymbol{x}_t)$ with a learned score function $\epsilon_\theta(\boldsymbol{x}, t)$, is that the gradient field may not be conservative, and thus does not lead to a valid probability density. However, as discussed in [34], explicitly parameterizing the learned function $\epsilon_\theta(\mathbf{x}, t)$ as the gradient of EBM achieves similar performance as the non-conservative parameterization of diffusion models, suggesting this is not problematic.

### 4.2 Compositional Generation Through Diffusion Models

Next, we discuss how we compose diffusion models for image generation. We aim to generate images conditioned on a set of concepts $\{\boldsymbol{c}_1, \boldsymbol{c}_2, \ldots, \boldsymbol{c}_n\}$. To do this, we represent each concept $\boldsymbol{c}_i$ as an individual diffusion model, which are composed to generate images. Inspired by EBMs [7,23], we define two compositional operators, **conjunction (AND)** and **negation (NOT)**, to compose diffusion models. We learn a set of diffusion models representing the conditional image generation $p(\boldsymbol{x}|\boldsymbol{c}_i)$ given factor $\boldsymbol{c}_i$ and an unconditional image generation $p(\boldsymbol{x})$.

**Concept Conjunction (AND).** We aim to generate images containing certain attributes. Following [7], the conditional probability can be factorized as

$$p(\boldsymbol{x}|\boldsymbol{c}_1, \ldots, \boldsymbol{c}_n) \propto p(\boldsymbol{x}, \boldsymbol{c}_1, \ldots, \boldsymbol{c}_n) = p(\boldsymbol{x}) \prod_i p(\boldsymbol{c}_i|\boldsymbol{x}). \tag{9}$$

We can represent $p(\boldsymbol{c}_i|\boldsymbol{x})$ using a combination of a conditional distribution $p(\boldsymbol{x}|\boldsymbol{c}_i)$ and an unconditional distribution $p(\boldsymbol{x})$, with both of them are parameterized as diffusion models $p(\boldsymbol{c}_i|\boldsymbol{x}) \propto \frac{p(\boldsymbol{x}|\boldsymbol{c}_i)}{p(\boldsymbol{x})}$. The expression of $p(\boldsymbol{c}_i|\boldsymbol{x})$ corresponds to the implicit classifier that represents the likelihood of $\boldsymbol{x}$ exhibiting factor $\boldsymbol{c}_i$. Substituting $p(\boldsymbol{c}_i|\boldsymbol{x})$ in Eq. 9, we can rewrite Eq. 9 as the probability distribution:

$$p(\boldsymbol{x}|\boldsymbol{c}_1, \ldots, \boldsymbol{c}_n) \propto p(\boldsymbol{x}) \prod_i \frac{p(\boldsymbol{x}|\boldsymbol{c}_i)}{p(\boldsymbol{x})}. \tag{10}$$

We sample from this resultant distribution using Eq. (8), with a new composed score function $\epsilon^*(\boldsymbol{x}_t, t)$:

$$\epsilon^*(\boldsymbol{x}_t, t) = \epsilon_\theta(\boldsymbol{x}_t, t) + \alpha \sum_i (\epsilon_\theta(\boldsymbol{x}_t, t|\boldsymbol{c}_i) - \epsilon_\theta(\boldsymbol{x}_t, t)), \tag{11}$$

where the constant $\alpha$ corresponds to a temperature scaling on $\frac{p(\boldsymbol{x}|\boldsymbol{c}_i)}{p(\boldsymbol{x})}$. We may then generate the composed sample using the following *generative process*:

$$p^*(\boldsymbol{x}_{t-1}|\boldsymbol{x}_t) := \mathcal{N}(\boldsymbol{x}_t + \epsilon^*(\boldsymbol{x}_t, t), \sigma_t^2). \tag{12}$$

In the setting in which image generation is conditioned on a single concept, the above sampling procedure reduces to classifier-free guidance.

**Concept Negation (NOT).** In concept negation, we aim to generate image with the absence of a certain factor $\tilde{\boldsymbol{c}}_j$. We also need to generate images that look realistic. One easy way to do this is to make the generated images contain another factor $\boldsymbol{c}_i$. Following [7], concept negation can be represented as the composed probability distribution $p(\boldsymbol{x}|\text{not } \tilde{\boldsymbol{c}}_j, \boldsymbol{c}_1, \boldsymbol{c}_2, \ldots, \boldsymbol{c}_n) = \frac{\prod_i p(\boldsymbol{x}|\boldsymbol{c}_i)}{p(\boldsymbol{x}|\tilde{\boldsymbol{c}}_j)^\alpha}$. Following [7], we refactorize the joint probability distribution as:

$$p(\boldsymbol{x}|\text{not } \tilde{\boldsymbol{c}}_j, \boldsymbol{c}_1, \boldsymbol{c}_2, \ldots, \boldsymbol{c}_n) \propto p(\boldsymbol{x}, \text{not } \tilde{\boldsymbol{c}}_j, \boldsymbol{c}_1, \boldsymbol{c}_2, \ldots, \boldsymbol{c}_n) = p(\boldsymbol{x}) \frac{\prod_i p(\boldsymbol{c}_i|\boldsymbol{x})}{p(\tilde{\boldsymbol{c}}_j|\boldsymbol{x})^\beta}. \tag{13}$$

Using the implicit classifier factorization $p(\boldsymbol{c}_i|\boldsymbol{x}) \propto \frac{p(\boldsymbol{x}|\boldsymbol{c}_i)}{p(\boldsymbol{x})}$, we can rewrite the above expression as:

$$p(\boldsymbol{x}|\text{not } \tilde{\boldsymbol{c}}_j, \boldsymbol{c}_1, \boldsymbol{c}_2, \ldots, \boldsymbol{c}_n) \propto p(\boldsymbol{x}) \frac{p(\boldsymbol{x})^\beta}{p(\boldsymbol{x}|\tilde{\boldsymbol{c}}_j)^\beta} \prod_i \frac{p(\boldsymbol{x}|\boldsymbol{c}_i)}{p(\boldsymbol{x})}. \tag{14}$$

**Algorithm 1.** Code for Composing Diffusion Models

1: **Require** Diffusion model $\epsilon_\theta(\boldsymbol{x}, t|\boldsymbol{c})$, scale $\alpha$, negation factor $\beta$, noises $\sigma_t$
2: // Code for conjunction
3: Initialize sample $\boldsymbol{x}_T \sim \mathcal{N}(\boldsymbol{0}, \boldsymbol{I})$
4: **for** $t = T, \ldots, 1$ **do**
5: $\quad \epsilon_i \leftarrow \epsilon_\theta(\boldsymbol{x}_t, t|\boldsymbol{c}_i)$ // compute conditional scores for each factor $\boldsymbol{c}_i$
6: $\quad \epsilon \leftarrow \epsilon_\theta(\boldsymbol{x}_t, t)$ // compute unconditional score
7: $\quad \boldsymbol{x}_{t-1} \sim \mathcal{N}(\boldsymbol{x}_t + \epsilon + \alpha \sum_i (\epsilon_i - \epsilon), \sigma_t^2)$ // sampling
8: **end for**
9:
10: // Code for negation
11: Initialize sample $\boldsymbol{x}_T \sim \mathcal{N}(\boldsymbol{0}, \boldsymbol{I})$
12: **for** $t = T, \ldots, 1$ **do**
13: $\quad \tilde{\epsilon}_j \leftarrow \epsilon_\theta(\boldsymbol{x}_t, t|\tilde{\boldsymbol{c}}_j)$ // compute conditional scores for negated factor $\tilde{\boldsymbol{c}}_j$
14: $\quad \epsilon_i \leftarrow \epsilon_\theta(\boldsymbol{x}_t, t|\boldsymbol{c}_i)$ // compute conditional scores for each factor $\boldsymbol{c}_i$
15: $\quad \epsilon \leftarrow \epsilon_\theta(\boldsymbol{x}_t, t)$ // compute unconditional score
16: $\quad \boldsymbol{x}_{t-1} \sim \mathcal{N}(\boldsymbol{x}_t + \epsilon + \alpha\{-\beta(\tilde{\epsilon}_j - \epsilon) + \sum_i (\epsilon_i - \epsilon)\}, \sigma_t^2)$ // sampling
17: **end for**

Similarly, we may construct a new learned score $\epsilon^*(\boldsymbol{x}_t, t)$ using Eq. (8) to sample from the *generative process* to represent this negated probability distribution at each timestep:

$$\epsilon^*(\boldsymbol{x}_t, t) = \epsilon_\theta(x_t, t) + \alpha\{-\beta(\epsilon_\theta(\boldsymbol{x}_t, t|\tilde{\boldsymbol{c}}_j) - \epsilon_\theta(\boldsymbol{x}_t, t)) + \sum_i (\epsilon_\theta(\boldsymbol{x}_t, t|\boldsymbol{c}_i) - \epsilon_\theta(\boldsymbol{x}_t, t))\}, \quad (15)$$

where the constant $\alpha$ corresponds to a temperature scaling on each implicit classifier $\frac{p(\boldsymbol{x}|\boldsymbol{c}_i)}{p(\boldsymbol{x})}$. We may then generate samples from this modified learned score using Eq. 12.

Algorithm 1 provides the pseudo-code for composing diffusion models using concept conjunction and negation. Our method can compose pre-trained diffusion models during inference time without any additional training.

# 5 Experiment Setup

## 5.1 Datasets

**CLEVR.** CLEVR [17] is a synthetic dataset containing objects with different shapes, colors, and sizes. The training set consists of 30,000 images at $128 \times 128$ resolution. Each image contains $1 - 5$ objects and a 2D coordinate $(x, y)$ label indicating that the image contains an object at $(x, y)$. In our experiments, the 2D coordinate label is the coordinate of one random object in the image.

**Relational CLEVR.** Relational CLEVR [23] contains relational descriptions between objects in the image, such as "a red cube to the left of a blue cylinder". The training dataset contains $50,000$ images at $128 \times 128$ resolution. Each training image contains $1 - 5$ objects, and one label describing a relation between two objects. If there is only one object in the image, the second object in the relational description is null.

**FFHQ.** FFHQ [18] is a real world human face dataset. The original FFHQ dataset consists of 70,000 human face images without labels. [5] annotates three binary attributes, including *smile*, *gender*, and *glasses*, for the images using pre-trained classifiers. As a result, there are 51,067 images labeled by the classifiers.

### 5.2 Evaluation Metrics

**Binary Classification Accuracy.** During testing, we evaluate the performance of the proposed method and baselines on three different settings. The first test setting, **1 Component**, generates images conditioned on a single concept (matching the training distribution). The second and third test settings, **2 Components** and **3 Components**, generate images by composing two and three concepts respectively using the *conjunction* and *negation* operators. They are used to evaluate the models' generalization ability to new combinations.

For each task, we use the training data (real images) to train a binary classifier that takes an image and a concept, *e.g.*'smiling', as input, and predicts whether the image contains or represents the concept. We then apply this classifier to a generated image, checking whether it faithfully captures each of the concepts. In each test setting, each method generates 5,000 images for evaluation. The accuracy of the method is the percentage of generated images capturing all the concepts (See Appendix B).

**Fréchet Inception Distance (FID)** is a commonly used metric for evaluating the quality of generated images. It uses a pretrained inception model [40] to extract features for the generated images and real images and measure their feature similarity. Specifically, we use Clean-FID [29] to evaluate the generated images. FID is usually computed on 50,000 generated images, but we use 5,000 images in our experiments, thus causing our FID scores to be higher than usual.

## 6 Experiments

We compare the proposed method and baselines (Sect. 6.1) on compositional generation on different domains. We show results of composing natural language descriptions (Sect. 6.2), objects (Sect. 6.3), object relational descriptions (Sect. 6.4), and human facial attributes (Appendix A). Results analysis are shown in Sect. 6.5.

### 6.1 Baselines

We compare our method with baselines for compositional visual generation.

**Energy-Based Models (EBM)** [7] is the first paper using EBMs for compositional visual generation. They propose three compositional operators for composing different concepts. Our works is inspired by [7], but we compose diffusion models and achieve better results.

**Fig. 3. Composing Language Descriptions.** We develop *Composed GLIDE (Ours)*, a version of *GLIDE* [26] that utilizes our compositional operators to combine textual descriptions, without further training. We compare it to the original *GLIDE*, which directly encodes the descriptions as a single long sentence. Our approach more accurately captures text details, such as the "overwater bungalows" in the third example.

**StyleGAN2** [19] is one of the state-of-the-art GAN methods for unconditional image generation. To enable compositional image generation, We optimize the latent code by decreasing the loss between a trained binary classifier and the given labels. We use the final latent code to generate images.

**LACE** [27] uses pre-trained classifiers to generate energy scores in the latent space of the pre-trained StyleGAN2 model. To enable compositional image synthesis, LACE uses compositional operators [7].

**GLIDE** [26] is a recent state-of-the-art text-conditioned diffusion model. For composing language descriptions, we use the classifier-free model released by OpenAI for comparison. For the rest tasks, we train the GLIDE model using the same data as our method.

### 6.2 Composing Language Descriptions

We first validate that our approach can compose natural language descriptions. We use the pre-trained text conditional diffusion models from *GLIDE* [26]. The image generation results of the released *GLIDE* model (a small model) is shown in Fig. 3. We develop *Composed GLIDE (Ours)*, a version of GLIDE [26] that utilizes our compositional operators to combine textual descriptions, without further training. We compare this model to the original GLIDE model, which directly encodes the descriptions as a single long sentence.

In Fig. 3, *GLIDE* takes a single long sentence as input, for example "A pink sky in the horizon, a sailboat at the sea, and overwater bungalows". In contrast,

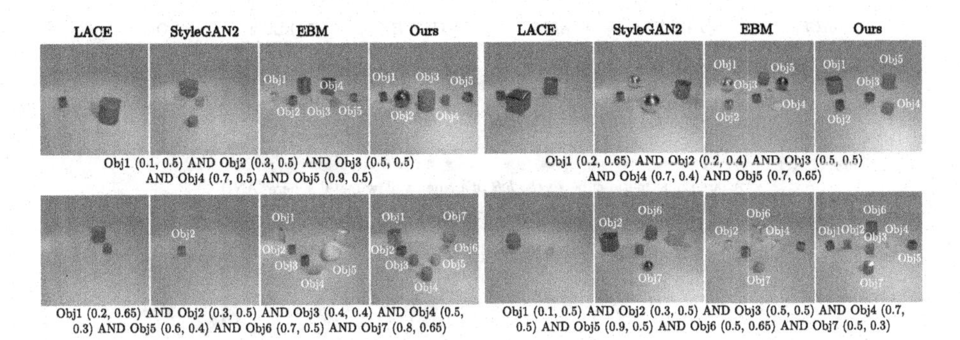


**Fig. 4. Composing Objects.** Our method can compose multiple objects while baselines either miss or generate more objects.

**Table 1.** Quantitative evaluation of 128 × 128 image generation results on CLEVR. The binary classification accuracy (Acc) and FID scores are reported. Our method outperforms baselines on all the three test settings.

| Models | 1 Component | | 2 Components | | 3 Components | |
|---|---|---|---|---|---|---|
| | Acc (%) ↑ | FID ↓ | Acc (%) ↑ | FID ↓ | Acc (%) ↑ | FID ↓ |
| EBM [7] | 70.54 | 78.63 | 28.22 | 65.45 | 7.34 | 58.33 |
| StyleGAN2 [19] | 1.04 | 51.37 | 0.04 | 23.29 | 0.00 | 19.01 |
| LACE [27] | 0.70 | 50.92 | 0.00 | 22.83 | 0.00 | 19.62 |
| GLIDE [26] | 0.86 | 61.68 | 0.06 | 38.26 | 0.00 | 37.18 |
| **Ours** | **86.42** | **29.29** | **59.20** | **15.94** | **31.36** | **10.51** |

*Composed GLIDE (Ours)* composes several short sentences using the concept conjunction operator, *e.g.*"A pink sky in the horizon" AND "A sailboat at the sea" AND "Overwater bungalows". While both *GLIDE* and *Composed GLIDE (Ours)* can generate reasonable images containing objects described in the text prompt, our approach with the compositional operators can more accurately capture text details, such as the presence of "a polar bear" in the first example and the "overwater bungalows" in the third example.

### 6.3 Composing Objects

Given a set of 2D object positions, we aim to generate images containing objects at those positions.

**Qualitative Results.** We compare the proposed method and baselines on composing objects in Fig. 4. We only show the concept conjunction here because the object positions are not binary values, and thus negation of object positions is not interpretable. Given a set of object position labels, we compose them to generate images. Our model can generate images of objects at certain locations while the baseline methods either miss objects or generate incorrect objects.

**Fig. 5. Composing Visual Relations.** Image generation results on the Relational CLEVR dataset. Our model is trained to generate images conditioned on a single object relation, but during inference, our model can compose multiple object relations, generating better results than baselines.

**Table 2.** Quantitative evaluation of 128 × 128 image generation results on the Relational CLEVR dataset. The binary classification accuracy (Acc) and FID score on three test settings are reported. Although *EBM* performs well on the binary classification accuracy, its FID score is much lower than other methods. Our method achieves comparable or better results than baselines.

| Models | 1 Component | | 2 Components | | 3 Components | |
|---|---|---|---|---|---|---|
| | Acc (%) ↑ | FID ↓ | Acc (%) ↑ | FID ↓ | Acc (%) ↑ | FID ↓ |
| EBM [23] | **78.14** | 44.41 | **24.16** | 55.89 | **4.26** | 58.66 |
| StyleGAN2 [19] | 20.18 | **22.29** | 1.66 | 30.58 | 0.16 | 31.30 |
| LACE [27] | 1.10 | 40.54 | 0.10 | 40.61 | 0.04 | 40.60 |
| GLIDE [26] | 32.68 | 57.48 | 7.48 | 59.47 | 2.14 | 61.52 |
| **Ours** | 60.40 | 29.06 | 21.84 | **29.82** | 2.80 | **26.11** |

**Quantitative Results.** As shown in Table 1, our method outperforms baselines by a large margin. The binary classification accuracy of our method is 15.88% higher than the best baseline, EBM, in the *1 component* test setting and is 24.02 higher than EBM on the more challenging *3 Components* setting. Our method is more effective in zero-shot compositional generalization. In addition, our method can generate images with lower FID scores (more similar to the real images).

### 6.4 Composing Object Relations

**Qualitative Results.** We further compare the proposed approach and baselines on composing object relational descriptions in Fig. 5. Our model is trained to generate images conditioned on a single object relation, but it can compose multiple object relations during inference without additional training. Both *LACE*

**Fig. 6. Qualitative results.** Successful examples (a) and failure examples (b-d) generated by the proposed method. There are three main types of failures: (b) The pre-trained diffusion model does not understand certain concepts, such as "person". (c) The pre-trained diffusion model confuses objects' attributes. (d) The composition fails. This usually happens when the objects are in the center of the images.

and *StyleGAN2* fail to capture object relations in the input sentences, but *EBM* and our method can correctly compose multiple object relations. Our method generates higher-quality images compared with *EBM*, *e.g.*the object boundaries are sharper in our results than *EBM*. Surprisingly, *DALLE-2* and *GLIDE* can generate high-quality images, but they fail to understand object relations.

**Quantitative Results.** Same as experiments in Sect. 6.3, we evaluate the proposed method and baselines on three test settings in Table 2. We train a binary classifier to evaluate whether an image contains objects that satisfy the input relational description. For binary classification accuracy, our method outperforms *StyleGAN2 (CLIP)*, *LACE*, and *GLIDE* on all three test settings. *EBMs* perform well on composing relational descriptions, but their FID scores are much worse than other methods, *i.e.*their generated images are not realistic.

### 6.5 Results Analysis

We show the image generation results conditioned on each individual sentence description, and our composition results in Fig. 6. We provide four successful compositional examples, where the generated image contains all the concepts mentioned in the input sentences.

**Failure Cases.** We observed three main failure cases of the proposed method. The first one is the pre-trained diffusion models do not understand certain concepts, such as "person" in (b). We used the pre-trained diffusion model, *GLIDE* [26], which is trained to avoid generating human images. The second type of failure is because the diffusion models confuse the objects' attributes. In (c), the generated image contains "a red bear" while the input is "a bear in a red forest". The third type of failure is because the composition does not work, *e.g.*the "bird-shape and flower-color object" and the "dog-fur and sofa-shape object" in (d). Such failures usually happen when the objects are in the center of the images.

## 7 Conclusion

In this paper, we compose diffusion models for image generation. By interpreting diffusion models as energy-based models, we may explicitly compose them and generate images with significantly more complex combinations that are never seen during training. We propose two compositional operators, concept conjunction and negation, allowing us to compose diffusion models during the inference time without any additional training. The proposed composable diffusion models can generate images conditioned on sentence descriptions, objects, object relations, human facial attributes, and even generalize to new combinations that are rarely seen in the real world. These results demonstrate the effectiveness of the proposed method for compositional visual generation.

A limitation of our current approach is that while we are able to compose multiple diffusion models together, they are instances of the same model. We found limited success when composing diffusion models trained on different datasets. In contrast, compositional generation with EBMs [7] can successfully compose multiple separately trained models. Incorporating additional structures into diffusion models from EBMs [10], such as a conservative score field, may be a promising direction towards enabling compositions of separately trained diffusion models.

**Acknowledgments.** Shuang Li is supported by Raytheon BBN Technologies Corp. under the project Symbiant (reg. no. 030256-00001 90113), Mitsubishi Electric Research Laboratory (MERL) under the project Generative Models For Annotated Video, and CMU and Army Research Laboratory under the project Contrastive dissection to visualize the differences between synthetic and real trained representations (reg. no. 1130233-442111). Yilun Du is supported by a NSF Graduate Fellowship.

## References


1. Austin, J., Johnson, D.D., Ho, J., Tarlow, D., van den Berg, R.: Structured denoising diffusion models in discrete state-spaces. In: Advances in Neural Information Processing Systems (2021)
2. Bau, D., et al.: Paint by word. arXiv preprint. arXiv:2103.10951 (2021)
3. Chen, N., Zhang, Y., Zen, H., Weiss, R.J., Norouzi, M., Chan, W.: Wavegrad: estimating gradients for waveform generation. arXiv preprint. arXiv:2009.00713 (2020)
4. Chomsky, N.: Aspects of the Theory of Syntax. The MIT Press, Cambridge (1965). http://www.amazon.com/Aspects-Theory-Syntax-Noam-Chomsky/dp/0262530074
5. DCGM: Gender, age, and emotions extracted for flickr-faces-hq dataset (ffhq) (2020). https://github.com/DCGM/ffhq-features-dataset
6. Dhariwal, P., Nichol, A.: Diffusion models beat gans on image synthesis. In: Advances in Neural Information Processing Systems, vol. 34 (2021)
7. Du, Y., Li, S., Mordatch, I.: Compositional visual generation with energy based models. In: Advances in Neural Information Processing Systems, vol. 33, pp. 6637–6647 (2020)
8. Du, Y., Li, S., Sharma, Y., Tenenbaum, J., Mordatch, I.: Unsupervised learning of compositional energy concepts. In: Advances in Neural Information Processing Systems, vol. 34 (2021)
9. Du, Y., Li, S., Tenenbaum, J., Mordatch, I.: Improved contrastive divergence training of energy based models. arXiv preprint. arXiv:2012.01316 (2020)
10. Du, Y., Mordatch, I.: Implicit generation and generalization in energy-based models. arXiv preprint. arXiv:1903.08689 (2019)
11. Gao, R., Song, Y., Poole, B., Wu, Y.N., Kingma, D.P.: Learning energy-based models by diffusion recovery likelihood. In: International Conference on Learning Representations (2021). https://openreview.net/forum?id=v_1Soh8QUNc
12. Grathwohl, W., Wang, K.C., Jacobsen, J.H., Duvenaud, D., Zemel, R.: Learning the stein discrepancy for training and evaluating energy-based models without sampling. In: International Conference on Machine Learning (2020)
13. Gu, S., et al.: Vector quantized diffusion model for text-to-image synthesis. In: Proceedings of the IEEE/CVF Conference on Computer Vision and Pattern Recognition, pp. 10696–10706 (2022)
14. Hinton, G.E.: Training products of experts by minimizing contrastive divergence. Neural Comput. **14**(8), 1771–1800 (2002)
15. Ho, J., Jain, A., Abbeel, P.: Denoising diffusion probabilistic models. In: Advances in Neural Information Processing Systems, vol. 33, pp. 6840–6851 (2020)
16. Janner, M., Du, Y., Tenenbaum, J., Levine, S.: Planning with diffusion for flexible behavior synthesis. In: International Conference on Machine Learning (2022)
17. Johnson, J., Hariharan, B., Van Der Maaten, L., Fei-Fei, L., Lawrence Zitnick, C., Girshick, R.: Clevr: a diagnostic dataset for compositional language and elementary visual reasoning. In: Proceedings of the IEEE Conference on Computer Vision and Pattern Recognition, pp. 2901–2910 (2017)
18. Karras, T., Laine, S., Aila, T.: A style-based generator architecture for generative adversarial networks. In: Proceedings of the IEEE/CVF Conference on Computer Vision and Pattern Recognition, pp. 4401–4410 (2019)
19. Karras, T., Laine, S., Aittala, M., Hellsten, J., Lehtinen, J., Aila, T.: Analyzing and improving the image quality of stylegan. In: Proceedings of the IEEE/CVF Conference on Computer Vision and Pattern Recognition, pp. 8110–8119 (2020)

20. Kim, G., Ye, J.C.: Diffusionclip: text-guided image manipulation using diffusion models (2021)
21. Lake, B.M., Salakhutdinov, R., Tenenbaum, J.B.: Human-level concept learning through probabilistic program induction. Science **350**(6266), 1332–1338 (2015). https://doi.org/10.1126/science.aab3050
22. LeCun, Y., Chopra, S., Hadsell, R., Ranzato, M., Huang, F.: A tutorial on energy-based learning. Predicting Struct. Data **1**(0) (2006)
23. Liu, N., Li, S., Du, Y., Tenenbaum, J., Torralba, A.: Learning to compose visual relations. In: Advances in Neural Information Processing Systems, vol. 34 (2021)
24. Marcus, G., Davis, E., Aaronson, S.: A very preliminary analysis of dall-e 2. arXiv preprint. arXiv:2204.13807 (2022)
25. Meng, C., et al.: Sdedit: guided image synthesis and editing with stochastic differential equations. In: International Conference on Learning Representations (2021)
26. Nichol, A., et al.: Glide: Towards photorealistic image generation and editing with text-guided diffusion models. arXiv preprint. arXiv:2112.10741 (2021)
27. Nie, W., Vahdat, A., Anandkumar, A.: Controllable and compositional generation with latent-space energy-based models. In: Advances in Neural Information Processing Systems, vol. 34 (2021)
28. Nijkamp, E., Hill, M., Han, T., Zhu, S.C., Wu, Y.N.: On the anatomy of mcmc-based maximum likelihood learning of energy-based models. arXiv preprint. arXiv:1903.12370 (2019)
29. Parmar, G., Zhang, R., Zhu, J.Y.: On aliased resizing and surprising subtleties in gan evaluation. In: CVPR (2022)
30. Ramesh, A., Dhariwal, P., Nichol, A., Chu, C., Chen, M.: Hierarchical text-conditional image generation with clip latents. arXiv preprint. arXiv:2204.06125 (2022)
31. Ramesh, A., et al.: Zero-shot text-to-image generation. In: International Conference on Machine Learning, pp. 8821–8831. PMLR (2021)
32. Rombach, R., Blattmann, A., Lorenz, D., Esser, P., Ommer, B.: High-resolution image synthesis with latent diffusion models. In: Proceedings of the IEEE/CVF Conference on Computer Vision and Pattern Recognition, pp. 10684–10695 (2022)
33. Saharia, C., et al.: Palette: Image-to-image diffusion models. arXiv preprint. arXiv:2111.05826 (2021)
34. Salimans, T., Ho, J.: Should EBMs model the energy or the score? In: Energy Based Models Workshop-ICLR 2021 (2021)
35. Shoshan, A., Bhonker, N., Kviatkovsky, I., Medioni, G.: Gan-control: explicitly controllable gans. In: Proceedings of the IEEE/CVF International Conference on Computer Vision, pp. 14083–14093 (2021)
36. Sohl-Dickstein, J., Weiss, E., Maheswaranathan, N., Ganguli, S.: Deep unsupervised learning using nonequilibrium thermodynamics. In: International Conference on Machine Learning, pp. 2256–2265. PMLR (2015)
37. Song, J., Meng, C., Ermon, S.: Denoising diffusion implicit models. In: International Conference on Learning Representations (2021)
38. Song, Y., Sohl-Dickstein, J., Kingma, D.P., Kumar, A., Ermon, S., Poole, B.: Score-based generative modeling through stochastic differential equations. arXiv preprint. arXiv:2011.13456 (2020)
39. Swimmer963: What dall-e 2 can and cannot do (2022). https://www.lesswrong.com/posts/uKp6tBFStnsvrot5t/what-dall-e-2-can-and-cannot-do
40. Szegedy, C., Vanhoucke, V., Ioffe, S., Shlens, J., Wojna, Z.: Rethinking the inception architecture for computer vision. In: Proceedings of the IEEE Conference on Computer Vision and Pattern Recognition, pp. 2818–2826 (2016)

41. Vincent, P.: A connection between score matching and denoising autoencoders. Neural Comput. **23**(7), 1661–1674 (2011)
42. Xiao, T., Hong, J., Ma, J.: Elegant: exchanging latent encodings with gan for transferring multiple face attributes. In: Proceedings of the European conference on computer vision (ECCV), pp. 168–184 (2018)
43. Xu, T., Zhang, P., Huang, Q., Zhang, H., Gan, Z., Huang, X., He, X.: Attngan: fine-grained text to image generation with attentional generative adversarial networks. In: Proceedings of the IEEE Conference on Computer Vision and Pattern Recognition, pp. 1316–1324 (2018)
44. Zhang, H., et al.: Stackgan: text to photo-realistic image synthesis with stacked generative adversarial networks. In: Proceedings of the IEEE International Conference on Computer Vision, pp. 5907–5915 (2017)
45. Zhou, L., Du, Y., Wu, J.: 3D shape generation and completion through point-voxel diffusion. In: International Conference on Computer Vision (2021)
46. Zhu, J., Shen, Y., Zhao, D., Zhou, B.: In-domain gan inversion for real image editing. In: Vedaldi, A., Bischof, H., Brox, T., Frahm, J.-M. (eds.) ECCV 2020. LNCS, vol. 12362, pp. 592–608. Springer, Cham (2020). https://doi.org/10.1007/978-3-030-58520-4_35

# ManiFest: Manifold Deformation for Few-Shot Image Translation

Fabio Pizzati[1,2](✉), Jean-François Lalonde[3], and Raoul de Charette[1]

[1] Inria, Paris, France
{fabio.pizzati,raoul.de-charette}@inria.fr
[2] VisLab, Parma, Italy
[3] Université Laval, Québec, Canada
jflalonde@gel.ulaval.ca

**Abstract.** Most image-to-image translation methods require a large number of training images, which restricts their applicability. We instead propose ManiFest: a framework for few-shot image translation that learns a context-aware representation of a target domain from a few images only. To enforce feature consistency, our framework learns a style manifold between source and additional anchor domains (assumed to be composed of large numbers of images). The learned manifold is interpolated and deformed towards the few-shot target domain via patch-based adversarial and feature statistics alignment losses. All of these components are trained simultaneously during a single end-to-end loop. In addition to the general few-shot translation task, our approach can alternatively be conditioned on a single exemplar image to reproduce its specific style. Extensive experiments demonstrate the efficacy of ManiFest on multiple tasks, outperforming the state-of-the-art on all metrics. Our code is avaliable at https://github.com/cv-rits/ManiFest.

**Keywords:** Image-to-image translation · Few-shot learning · Generative networks · Night generation · Adverse weather

## 1 Introduction

Image-to-image translation (i2i) frameworks are gaining traction on multiple applications such as autonomous driving [24,45] as well as photo editing [36,37]. Those methods rely on the availability of large-scale datasets, and as such they are restricted to applications where large quantities of images are available.

Unfortunately, it is unrealistic to impose significant data collection constraints every time a new i2i scenario is pursued. In addition to the complex logistics involved in acquiring large quantities of images, some scenarios may be rare (*e.g.*, auroras) or dangerous (*e.g.*, erupting volcanoes) thereby preventing even the capture of sufficient training data. Existing methods have been proposed

**Supplementary Information** The online version contains supplementary material available at https://doi.org/10.1007/978-3-031-19790-1_27.

© The Author(s), under exclusive license to Springer Nature Switzerland AG 2022
S. Avidan et al. (Eds.): ECCV 2022, LNCS 13677, pp. 440–456, 2022.
https://doi.org/10.1007/978-3-031-19790-1_27

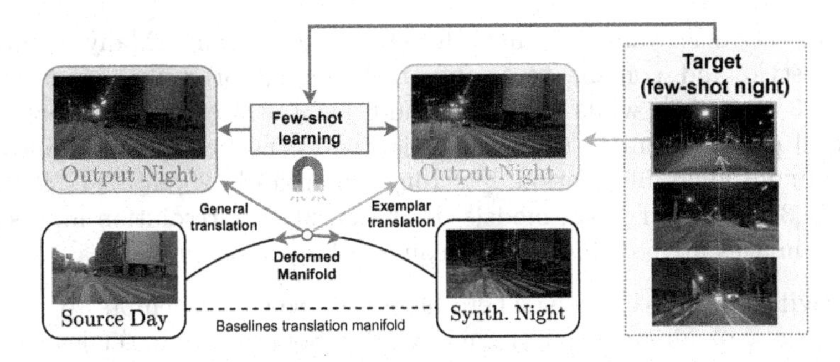


**Fig. 1.** Overview of ManiFest, which translates images from a source domain (here, Day) to a few-shot target (Night). Our framework learns a manifold between *anchor* domains, in the example spanning the translation between Day and Synthetic Nighttime. Our system deforms the manifold by injecting the few-shot domain information between anchor style representations, and further departs from the deformed manifold by learning to approximate the target domain *general* appearance, or to reproduce the style of a particular *exemplar*.

to alleviate the requirement for large datasets, but they mostly show realistic results in highly structured environments such as face translation [19,29,44]. In this context, we propose ManiFest, a framework for few-shot image-to-image translation which is shown to be robust to highly unstructured transformations such as adverse weather generation or night rendering. Our approach, illustrated in Fig. 1, starts from the observation that features consistency (*i.e.* which image parts should be translated together) is crucial for unstructured i2i [31] and that the few-shot domain offers little cues to train efficiently without overfitting [34]. Indeed, rather than directly addressing few-shot i2i, ManiFest exploits features learned on a stable manifold for the few-shot domain transformation. To do so, it leverages techniques inspired from style transfer and patch-based training. We learn either to translate to some *general* style approximating the entire few-shot set, or to reproduce a specific *exemplar* from it. In short, our contributions are:

- ManiFest, a few-shot image translation framework using feature consistency by weighted manifold interpolation (WMI) and local-global few-shot loss (LGFS).
- We introduce GERM, a novel residual correction mechanism for enabling general and exemplar translation, that also boosts performances.
- Our framework outperforms previous work on adverse weather and low-light few-shot image translation tasks. We also present qualitative evaluations on rare (auroras) and dangerous (volcanoes) events.

We discuss related works in Sect. 2 and present our method in Sect. 3. The latter is thoroughly evaluated in Sect. 4 and we show several extensions in Sect. 5.

## 2 Related Work

**Image-to-Image Translation (i2i).** Although the early i2i translation methods required paired data [14,64], cycle-consistency [28,63] or recent alternatives with contrastive learning [35] have lifted such constraint. Many approaches separate style and content to enable multi-modal or multi-target translations [5,13,16,18,57], while others use additional strategies to increase scene contextual preservation [15,62]. Translation networks can be conditioned on a variety of additional information, including semantics [4,20,25,31,42,48,65,66], instances [33], geometry [55], models [11,39–41,50], low-resolution inputs [1] or exemplar images [31,58,59]. Still, all require a large amount of data.

**GANs with Limited Data.** There have been several attempts to overcome the large data requirement for training GANs. Some use transfer learning [49] to adapt previously-trained networks to new few-shot tasks [23,54]. In particular, [34] uses a patch-based discriminator to generalize to few-shot domains. However, these methods are designed for generative networks and do not immediately apply to i2i. Another line of work focuses on the limited data scenario [2,17,38,61], but usually performs poorly when very few (10–15) images are used for training. Others exploit additional knowledge to enable few-shot or zero-shot learning, such as pose-appearance decomposition [53], image conditioning [8] or textual inputs [27]. FUNIT [29] and COCO-FUNIT [44] use few-shot style encoders to adapt the network behavior at inference time. Some use meta-learning to adapt quickly to newly seen domains [26]. Those methods show limited performance on highly unstructured scenarios. [7] exploits geometry for patch-based few-shot training, but only on limited domains with specific characteristics.

**Neural Style Transfer.** Style transfer could be seen as an instance of few-shot i2i, where the goal is combining content and style of two images [9]. This may result in distortions, which some work tried to mitigate [30]. The first examples of style transfer with arbitrary input styles are in [12,21]. Others try to transfer styles in a photo-realistic manner by using a smoothing step [22] or with wavelet transforms [56]. These methods provide good results in some controlled scenarios, but they may fail to understand contextual mappings between source and style elements (*e.g.* sky, buildings, etc.) which we learn accurately.

## 3 ManiFest

The few-shot i2i task consists in learning a $\mathcal{S} \mapsto \mathcal{T}$ mapping between images of a source domain $\mathcal{S}$ and a target domain $\mathcal{T}$ containing few training samples (*e.g.*, $|\mathcal{T}| \leq 25$). Figure 2 presents an overview of our approach. We learn a style manifold in a standard multi-target GAN fashion (Sect. 3.1) from a set of domains which contain large amounts of training data. We call these domains *anchors*, and denote them $\mathbb{A}$. The idea of ManiFest is to simultaneously 1) learn a stable manifold using anchor domains and 2) perform few-shot training by enforcing the target style appearance to lie within the learned manifold. This allows to exploit additional knowledge, like feature consistency (*i.e.*, image parts to be translated together), learned on anchors. To this end, Weighted Manifold Interpolation (WMI, Sect. 3.2) exploits style interpolation to benefit from the learned feature consistency on anchors. We allow to further depart from the interpolated

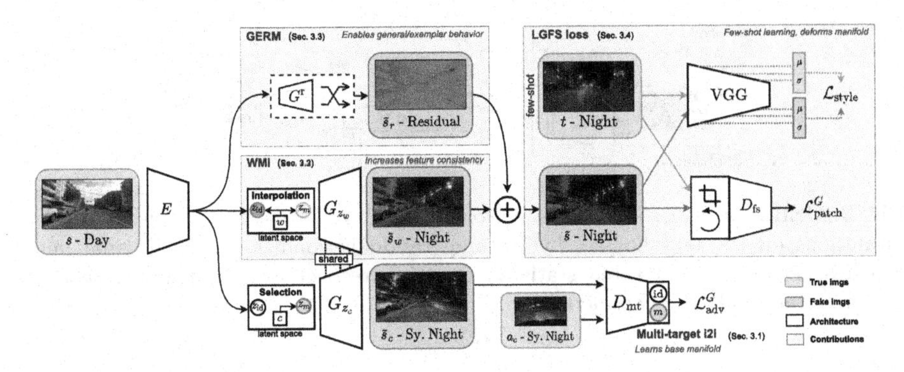


**Fig. 2.** ManiFest architecture, here translating Day $\mapsto$ Night using few real night images and a synthetic night anchor domain. The encoded image representation $E(s)$ is separated into content and style codes, and translated to the few-shot domain by injecting $\mathcal{T}$ on a manifold learned on anchor domains in a multi-target i2i setting (bottom). We correct the output by using residuals estimated by the GERM (top). The LGFS loss (top-right), based on statistics alignment and patch-based adversarial learning, deforms the manifold and injects $\mathcal{T}$ in it. The reconstruction cycle with a style encoder is omitted for simplicity and follows [13].

manifold with the General-Exemplar Residual Module (GERM, Sect. 3.3) which learns a residual image refining the overall appearance and thus enabling style transfer to the *general* few-shot style (approximating the entire set $\mathcal{T}$), or to a single *exemplar* in $\mathcal{T}$ as in [31]. We learn the appearance of $\mathcal{T}$ and inject it in the manifold with the Local-Global Few-Shot loss (LGFS, Sect. 3.4). In the following, *real* images are $s \in \mathcal{S}, t \in \mathcal{T}$, and *fake* ones $\tilde{s} \in \mathcal{T}$ where $\tilde{s}$ is our output.

### 3.1 Multi-target i2i

Instead of learning $\mathcal{S} \mapsto \mathcal{T}$ directly, we assume the availability of a set of two *anchor* domains, $\mathbb{A} = \{\mathcal{A}_{\text{id}}, \mathcal{A}_m\}$, with abundant data (equivalent to the "base" categories in few-shot classification, *e.g.*, [3]). By construction, one anchor is always the identity domain ($\mathcal{A}_{\text{id}} = \mathcal{S}$), while the other ($\mathcal{A}_m$) contains images easier to collect than $\mathcal{T}$, for example synthetic images or images from existing datasets. We formalize the multi-target image translation problem as learning the $\mathcal{S} \mapsto \mathbb{A}$ mapping. At training time, we disentangle image content and appearance by using content and style encoders $E(\cdot)$ and $Z(\cdot)$ respectively. We use $Z$ for reconstruction and translation as in [13], which we refer for details. We reconstruct $s = G_{Z(s)}(E(s))$, where $G_{Z(s)}$ is the style injection of $Z(s)$ into $G$ as in [13]. This effectively learns latent style distributions as in [13], namely here for each anchor $\{z_{\text{id}}, z_{\text{m}}\}$. A multi-target mechanism (following [5]) is employed in $Z$ since we have two anchors. We translate to a randomly selected domain $c \in \{\text{id}, m\}$ with

$$z_c = [\![c = \text{id}]\!] z_{\text{id}} + [\![c = m]\!] z_m \,, \quad \tilde{s}_c = G_{z_c}(E(s)) \,, \tag{1}$$

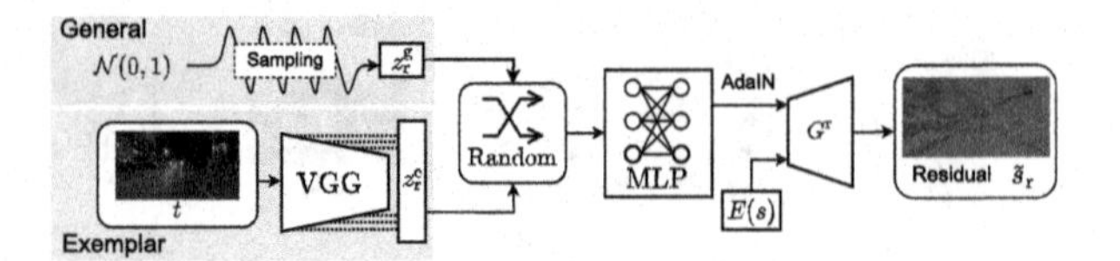


**Fig. 3.** GERM-based residuals. We perform either *exemplar*- or *general*-based transformations on the few-shot set by learning residuals conditioned on original image features ($E(s)$) and extracted statistics or noise, respectively. At training time, we alternate randomly the two modalities.

where $[\![\cdot]\!]$ are the Iverson brackets. The translation to a given anchor style is depicted in Fig. 2 as "**selection**". The multi-target discriminator $D_{\text{mt}}$ employs adversarial losses $\mathcal{L}_{\text{adv}}^{G}$ and $\mathcal{L}_{\text{adv}}^{D}$ to force fake images $\tilde{s}_c$ to resemble $a_c \in \mathcal{A}_c$. Additional training details are in the supp. material.

### 3.2 Weighted Manifold Interpolation (WMI)

Our intuition is that encoding $\mathcal{T}$ between the linearly interpolated style representations of $\mathbb{A}$ should enforce feature consistency in $\mathcal{T}$. For instance, assuming $\mathcal{S}$ = *day*, $\mathcal{T}$ = *night*, $\mathcal{A}_m$ = *synthetic night*, the network will be provided with the information that all sky pixels should be darkened together.

In practice, we learn weights $w = \{w_{\text{id}}, w_m\}$ which sum to 1 and encode an image $\tilde{s}_w$ with feature consistency by interpolating the anchors style representations:

$$z_w = w_{\text{id}} z_{\text{id}} + w_m z_m, \quad \tilde{s}_w = G_{z_w}(E(s)) \,. \tag{2}$$

This is visualized in Fig. 2 as **"interpolation"**. Learning $w$ allows us to determine the point in the $\mathbb{A}$ manifold which is most consistent with $\mathcal{T}$. This point is learned with the LGFS loss (Sect. 3.4).

### 3.3 General-Exemplar Residual Module (GERM)

Our GERM seeks to further increase realism by learning a residual in image space. Moreover, our design enables distinguishing between *general* and *exemplar* translations. The idea is to allow deviations from the $\mathbb{A}$ manifold by learning a residual image $\tilde{s}_r$ which helps encode missing characteristics from $\mathcal{T}$. This is done by processing the input image features $E(s)$ with a generator $G^{\text{r}}$ such that

$$\tilde{s}_{\text{r}} = G^{\text{r}}_{z_{\text{r}}}(E(s)) \,, \quad \text{and} \quad \tilde{s} = \tilde{s}_w + \tilde{s}_{\text{r}} \,, \tag{3}$$

where $z_{\text{r}}$ is a vector controlling general- or exemplar-based modalities. In both cases, we draw inspiration from AdaIN style injection [13] and condition the injected parameters on different vectors, as illustrated in Fig. 3.

For the *exemplar* residual, the style of a specific image $t \in \mathcal{T}$ as in [31] is reproduced by conditioning the residual on $t$. In this case,

$$z^{\text{e}}_{\text{r}} = (\mu_k(t), \sigma_k(t))|_{k=1}^{K} \,, \quad \tilde{s}_{\text{r}} = G^{\text{r}}_{z^{\text{e}}_{\text{r}}}(E(s)) \,, \tag{4}$$

where $\mu_k(\cdot) = \mu(\phi_k(\cdot))$ and $\sigma_k(\cdot) = \sigma(\phi_k(\cdot))$ are mean and variance of the $k$-th out of $K$ layer outputs $\phi_k$ of a pretrained VGG network [12], and $|$ is the concatenation operator. Since the LGFS loss exploits VGG statistics (Sect. 3.4), $G^{\mathrm{r}}$ will be driven to exploit the additional information provided by the input statistics vector, effectively making the generated image more similar to $t$.

We learn a *general* residual by removing the conditioning on $t$ and by injecting random noise instead:

$$z_{\mathrm{r}}^{\mathrm{g}} \sim \mathcal{N}(0,1)\,, \quad \tilde{s}_{\mathrm{r}} = G^{\mathrm{r}}_{z_{\mathrm{r}}^{\mathrm{g}}}(E(s))\,. \tag{5}$$

### 3.4 Local-Global Few-Shot Loss (LGFS)

To guide the learning, the resulting image $\tilde{s}$ is compared against the few-shot training set $\mathcal{T}$ with a combination of two loss functions. First, we take inspiration from the state-of-the-art of image style transfer where one image is enough for transferring the *global* appearance of the style scene [12]. Our intuition is that feature statistics alignment, widely used in style transfer, could be less prone to overfitting with respect to adversarial training. Therefore, we align features between $\tilde{s}$ and a target image $t \in \mathcal{T}$ using style loss $\mathcal{L}_{\text{style}}$ as in [12]

$$\mathcal{L}_{\text{style}} = \sum_{k=1}^{K} ||\mu_k(\tilde{s}) - \mu_k(t)||_2 + ||\sigma_k(\tilde{s}) - \sigma_k(t)||_2\,, \tag{6}$$

where $(\mu_k, \sigma_k)$ are the same as in Sect. 3.3. While this is effective in modifying the general image appearance, aligning statistics alone is insufficient to produce realistic outputs. Thus, to provide *local* guidance, *i.e.*, on more fine-grained characteristics, we employ an additional discriminator $D_{\text{fs}}$ which is trained to distinguish between rotated patches sampled from $\tilde{s}$ and $t$. We define the adversarial losses [32]:

$$\begin{aligned} \mathcal{L}^{G}_{\text{patch}} &= ||D_{\text{fs}}(p(\tilde{s})) - 1||_2\,, \\ \mathcal{L}^{D}_{\text{patch}} &= ||D_{\text{fs}}(p(\tilde{s}))||_2 + ||D_{\text{fs}}(p(t)) - 1||_2\,, \end{aligned} \tag{7}$$

where $p$ is a random cropping and rotation function. Note how the *exemplar* residual (from Sect. 3.3) is conditioned on the *same* feature statistics used here—this is what enables the exemplar-based behavior of the network. Also note the interaction between components: backpropagating the LGFS loss *deforms* the manifold learned by multi-target i2i, at the point identified by WMI, thereby injecting $\mathcal{T}$ "between" $\{\mathcal{A}_{\text{id}}, \mathcal{A}_m\}$. We provide additional visualizations of the deformed manifold in the supp. video and material.

### 3.5 Training Strategy

Our framework is fully trained end-to-end and optimizes

$$\min_{\Theta(E,G,G^{r},Z),w} \mathcal{L}_{\text{style}} + \mathcal{L}^{G}_{\text{patch}} + \mathcal{L}^{G}_{\text{adv}} \quad \text{and} \quad \min_{\Theta(D_{\text{fs}},D_{\text{mt}})} \mathcal{L}^{D}_{\text{patch}} + \mathcal{L}^{D}_{\text{adv}}\,, \tag{8}$$

where $\Theta(\cdot)$ refers to the network parameters. We train GERM (Sect. 3.3) by randomly selecting one of the exemplar or general mode at each training iteration. For the multi-target settings, we adapt the discriminator and the style encoder of our backbone in a multi-target setup following [5].

## 4 Experiments

We leverage 4 datasets [6,43,45,46] and 3 translation tasks (Sect. 4.1) and evaluate performances against recent baselines [13,29,31,44,56] (Sect. 4.2). We further demonstrate the benefit of our few-shot translation on a downstream segmentation task (Sect. 4.3), and rare few-shot scenarios (Sect. 4.4), and finally ablate our contributions (Sect. 4.5). In all, we use MUNIT [13] as our backbone.

### 4.1 Training Setup

**Datasets.** We use four datasets for our experiments.

**ACDC.** We use ACDC [46] for most of our experiments, using the night/rain/snow/fog conditions with 400/100/500 images for train/val/test respectively, following official splits. For any individual condition, ACDC also includes geolocalized weakly-paired clear weather day images of same splits.

**Dark Zurich.** Similar to ACDC, Dark Zurich (DZ) [45] has daytime images paired with nighttime/twilight conditions. Here, we focus on twilight conditions exclusively and use training images from the GOPRO348 sequence only since it exhibits a distinctive twilight appearance. We split the total 819 image pairs into 25/794 for train/test, respectively.

**Cityscapes.** Cityscapes [6] is used to evaluate ManiFest for training segmentation networks robust to nighttime[1]. It includes 2975/500/1525 annotated images for train/val/test.

**VIPER.** As anchors, we employ synthetic images from the VIPER dataset [43], using the condition metadata to define splits. 4137/3090/1305/2018/2817 images are extracted from the VIPER training set for day/night/rain/snow/sunset conditions, respectively.

**Tasks and Evaluation.** We train our framework on three main tasks:
**Day $\mapsto$ Night** on ACDC daytime ($\mathcal{S}$) and nighttime ($\mathcal{T}$).
**Clear** $\mapsto$ Fog on ACDC daytime ($\mathcal{S}$) and fog ($\mathcal{T}$).
**Day $\mapsto$ Twilight** on DZ daytime ($\mathcal{S}$) and twilight ($\mathcal{T}$).
Unless mentioned otherwise, the (synthetic) anchor domains from VIPER are "night" for Day $\mapsto$ Night and Day $\mapsto$ Twilight, and "day" for Clear $\mapsto$ Fog. We evaluate with the FID [10] and LPIPS [60] metrics. While FID compares feature distance globally, LPIPS compares translated source images and the geolocalized paired image in the target dataset. This is beneficial for evaluating our exemplar modality. For all, we train on downsampled x4 images.

[1] ACDC does not provide annotated daytime clear weather sequences.

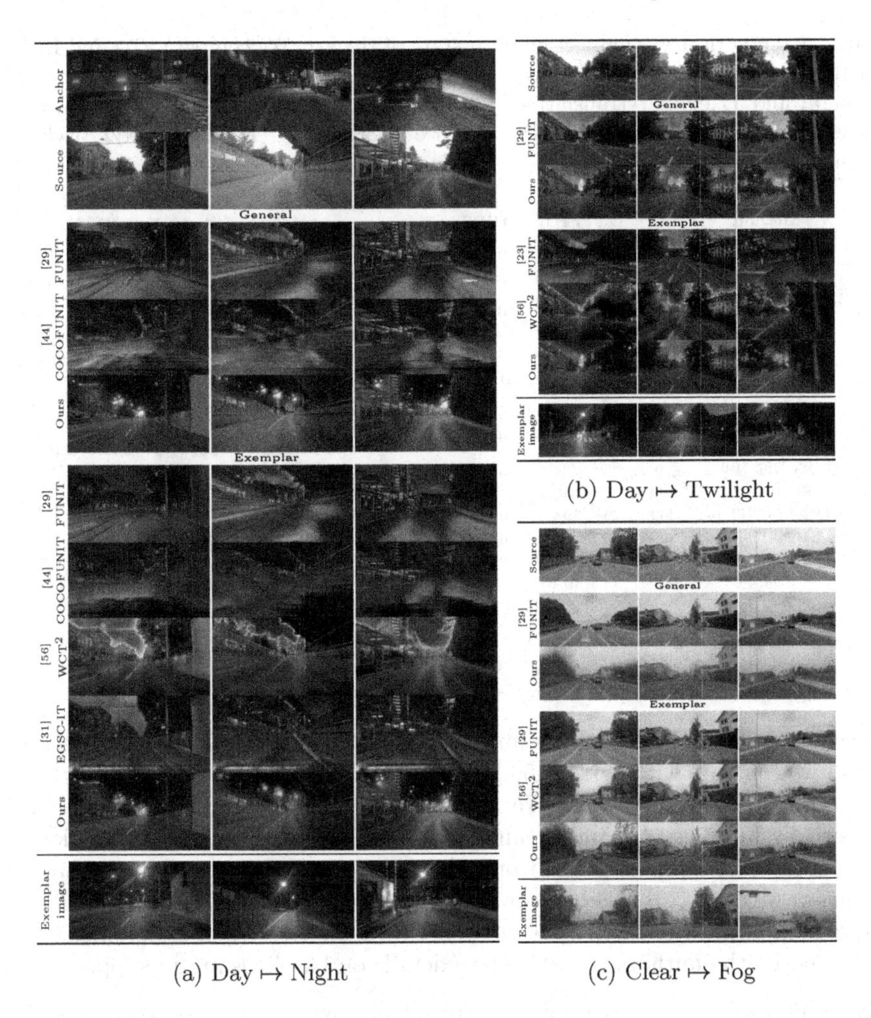


(a) Day $\mapsto$ Night

(b) Day $\mapsto$ Twilight

(c) Clear $\mapsto$ Fog

**Fig. 4.** Qualitative evaluation and comparison with the state of the art. We evaluate the a Day $\mapsto$ Night, b Day $\mapsto$ Twilight, and c Clear $\mapsto$ Fog tasks. In all cases, our approach extracts a *general* realistic representation of the few-shot target, and correctly reproduces the style of paired *exemplar* target images. In comparison, existing baselines either has unnecessary similarity with anchors (e.g. FUNIT, COCO-FUNIT, EGSC-IT) or unrealistic artifacts (e.g. WCT$^2$).

## 4.2 Comparison with the State-of-the-Art

**Baselines.** We compare with four baselines for few-shot image translation with $|\mathcal{T}| = 25$. We extensively evaluate on the most challenging Day $\mapsto$ Night task, and provide insights and comparison for the two others tasks. We evaluate the impact of the few-shot image selection and of $|\mathcal{T}|$ in Sect. 4.5. We compare against the recent FUNIT [29] and COCO-FUNIT [44], trained on $\mathcal{S} \mapsto \mathcal{A}_m$ and adapted following [29,44] to the few-shot $\mathcal{T}$ (general) or to a single reference

**Table 1.** Quantitative comparison with state of the art. We compare FID and LPIPS on the (a) Day ↦ Night, (b) Day ↦ Twilight and (c) Clear ↦ Fog tasks, for both **G**eneral and **E**xemplar translations. Our approach outperforms all baselines on all tasks, while also being on par (**G**) or even outperforming (**E**) the MUNIT backbone trained on the full dataset for Day ↦ Night in a.

| | Method | $\|\mathcal{A}_m\|$ | $\|\mathcal{T}\|$ | FID$_\downarrow$ | LPIPS$_\downarrow$ |
|---|---|---|---|---|---|
| G | MUNIT [13] | 0 | 400 | 79.20 | 0.529 |
| G | MUNIT [13] | 3090 | 0 | 132.72 | 0.613 |
| G | MUNIT [13] | 0 | 25 | 91.61 | 0.553 |
| G | FUNIT [29] | 3090 | 25 | 156.97 | 0.573 |
| G | COCO-FUNIT [44] | 3090 | 25 | 201.67 | 0.644 |
| G | Ours | 3090 | 25 | **81.01** | **0.535** |
| E | MUNIT [13] | 0 | 400 | 87.71 | 0.522 |
| E | MUNIT [13] | 3090 | 0 | 142.04 | 0.559 |
| E | MUNIT [13] | 0 | 25 | 128.73 | 0.562 |
| E | FUNIT [29] | 3090 | 25 | 136.2 | 0.572 |
| E | COCO-FUNIT [44] | 3090 | 25 | 193.4 | 0.646 |
| E | EGSC-IT [31] | 3090 | 25 | 106.68 | 0.574 |
| E | WCT$^2$ [56] | - | - | 105.58 | 0.580 |
| E | Ours | 3090 | 25 | **80.57** | **0.525** |

(a) Day ↦ Night

| | Method | $\|\mathcal{A}_m\|$ | $\|\mathcal{T}\|$ | FID$_\downarrow$ | LPIPS$_\downarrow$ |
|---|---|---|---|---|---|
| G | FUNIT [29] | 3090 | 25 | 69.53 | 0.511 |
| G | Ours | 3090 | 25 | **63.15** | **0.510** |
| E | FUNIT [29] | 3090 | 25 | 69.97 | 0.501 |
| E | WCT$^2$ [56] | - | - | 71.77 | 0.536 |
| E | Ours | 3090 | 25 | **58.07** | **0.483** |

(b) Day ↦ Twilight

| | Method | $\|\mathcal{A}_m\|$ | $\|\mathcal{T}\|$ | FID$_\downarrow$ | LPIPS$_\downarrow$ |
|---|---|---|---|---|---|
| G | FUNIT [29] | 3090 | 25 | 152.90 | 0.580 |
| G | Ours | 3090 | 25 | **89.57** | **0.520** |
| E | FUNIT [29] | 3090 | 25 | 137.7 | 0.568 |
| E | WCT$^2$ [56] | - | - | 120.9 | 0.591 |
| E | Ours | 3090 | 25 | **89.89** | **0.521** |

(c) Clear ↦ Fog

image (exemplar). For exemplar image translation, we also add specific baselines. First, we compare with WCT$^2$ [56], used to transfer the style of the paired target condition to the source one. We also evaluate EGSC-IT [31]. The method is trained by merging $\mathcal{A}_m$ and $\mathcal{T}$ since it should be able to identify inter-domain variability, separating $\mathcal{T}$ styles from $\mathcal{A}_m$ [31]. To define metrics bounds, we also train our MUNIT [13] backbone on $\mathcal{A}_m$, on the full $\mathcal{T}$ set and on $\mathcal{T}$ with $|\mathcal{T}| = 25$. More comparisons with the backbone are in Sect. 4.5. We use the official code provided by the authors for all[2]. More details on baselines are in supp.

**Evaluation.** We compare qualitative results in Fig. 4. In Day ↦ Night (Fig. 4a), even if the appearance of images in $\mathcal{T}$ is partially transferred on translated images (e.g. road color, darker sky), FUNIT and COCO-FUNIT still retain some characteristics of $\mathbb{A}$ (note, for example, how the street is similar to the GTA one) which worsens the overall image realism.

The same can be observed with EGSC-IT, where the hood of the ego-vehicle in anchor images (first column) is retained and significantly impacts visual results. While WCT$^2$ exhibits sharp results, it does not correctly map the image context, and it is limited to appearance alignment which leads to artifacts (e.g. yellow sky with white halos). Our method generates significantly better results than the baselines in both the *general* and *exemplar* modalities, with visible differences in all three tasks: the *general* appearance is consistent across test samples, and each result adapts to its *exemplar*. For example, observe how the

[2] For FUNIT [29] and COCO-FUNIT [44], we modify hyperparameters per authors suggestions to adapt to the ACDC and Dark Zurich datasets.

| Model | mIoU % ↑ | Acc. % ↑ |
|---|---|---|
| Baseline (*CS day*) | 12.93 | 45.15 |
| MUNIT [13] | 21.22 | 56.65 |
| Ours | **24.31** | **60.50** |
| Oracle (*ACDC night*) | 49.23 | 88.47 |

(a) Quantitative evaluation

Input GT MUNIT Ours

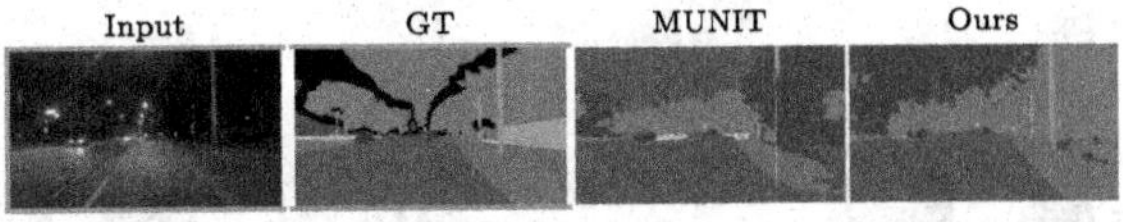

(b) Qualitative evaluation

**Fig. 5.** Segmentation on ACDC-night, for few-shot Day↦Night translations ($|\mathcal{T}| = 25$) (a). We outperform the baseline with noticeably better segmentation in (b) due to the increased quality of our translation.

overall sky colors (Day ↦ Twilight, Fig. 4b) match the exemplar. Here, the exemplars were unseen in training (not part of the few-shot set $\mathcal{T}$), thus GERM generalizes the few-shot learned exemplar behavior. The quantitative evaluation in Table 1 is coherent with the qualitative results, as we always outperform baselines. We perform on par (*general*), or even better (*exemplar*) than the backbone trained on the entire set of 400 training images on Day ↦ Night (Table 1a). This result shows that GERM (Sect. 3.3) improves modeling of the exemplar style over AdaIN exemplar style injection [13]. The exemplar behavior may force artifacts following subtle characteristics of the scene (as trees in Fig. 4c), for which the general translation may be advisable.

### 4.3 Segmentation Downstream Task

We exploit semantic segmentation to evaluate ManiFest for increasing robustness in challenging scenarios. In Fig. 5 we train HRNet [51] on nighttime versions of Cityscapes [6] obtained by translating the dataset with ManiFest or MUNIT, and evaluating on the ACDC-night validation set labels. We choose the best MUNIT and ManiFest configurations following nighttime realism in Table 1a with $|\mathcal{T}| = 25$. As lower and upper bounds we train HRNet either on original Cityscapes (baseline) or on ACDC-night training set (oracle). Figure 5 shows we outperform the MUNIT backbone (+3.09 mIoU) thanks to our better target domain modeling. Additional results on other domains are in supplementary.

### 4.4 Rare Few-Shot Scenarios

Few-shot plays its full role with conditions that are rare by nature, difficult or even dangerous to photograph, such as auroras or erupting volcanoes. Figure 6 shows the capability of ManiFest to learn Mountain↦Volcano or Day↦Aurora, by taking as source and anchor the summer and winter Yosemite dataset [63] splits respectively. Each task uses only 4 images from Google Images as $\mathcal{T}$. We generate realistic erupting volcanoes or auroras starting from mountain images, with contextual understanding (Fig. 6, cols 1–4), where only one mountain is mapped to a volcano and auroras only partially cover the sky. Figure 6 (cols 5–6) also demonstrate how exemplar characteristics are preserved.

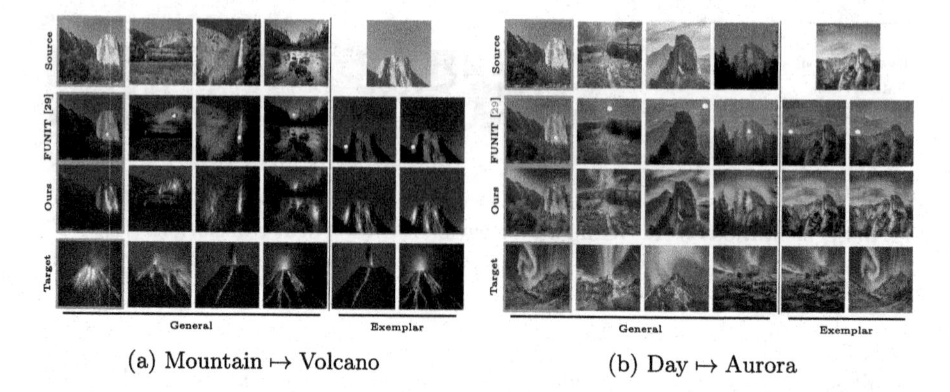


(a) Mountain $\mapsto$ Volcano

(b) Day $\mapsto$ Aurora

**Fig. 6.** Qualitative results for the Mountain $\mapsto$ Volcano (a) and Day $\mapsto$ Aurora (b) tasks. We retain contextual information by only partially mapping mountains to volcanoes and sky to auroras. In the green box we process the same image for ease of comparison. *Exemplar* results show how Ours conforms to Target, effectively reproducing the exemplar image style (cols 5–6). For space reason, we show WCT$^2$ [56] outputs in Supp.

| Component | FID$_\downarrow$ | LPIPS$_\downarrow$ |
|---|---|---|
| w/o $\mathcal{L}_{\text{style}}$ | 143.66 | 0.614 |
| w/o $\mathcal{L}_{\text{patch}}$ | 93.42 | 0.566 |
| w/o GERM | 85.62 | 0.544 |
| w/o WMI | 101.57 | 0.589 |
| LGFS-only | 84.29 | 0.558 |
| Ours | **81.01** | **0.535** |

(a) Quantitative evaluation

w/o $\mathcal{L}_{\text{style}}$ w/o $\mathcal{L}_{\text{patch}}$ w/o GERM

w/o WMI LGFS-only Ours

(b) Qualitative evaluation

**Fig. 7.** Ablation study for architectural components. a Removing each component individually lowers quantitative performances, which maps to b decreased visual quality in the generated images.

## 4.5 Ablation Studies

**Architectural Components.** We evaluate the contribution of each component in ManiFest (*c.f.* Fig. 2, Sect. 3) using the Day $\mapsto$ Night task in the *general* scenario, and report results in Fig. 7a. The impact of LGFS is studied by removing $\mathcal{L}_{\text{style}}$ or $\mathcal{L}_{\text{patch}}$, showing that both local *and* global guidance are improving translations. Removing the GERM from the training pipeline simultaneously precludes the *exemplar* behavior and worsens the performance, demonstrating the effectiveness of encoding complementary characteristics outside of the manifold spanned by $\mathbb{A}$. The benefit of WMI is evaluated in two experiments. First, the "w/o WMI" setting applies the residual directly on the fake anchor images $\tilde{s}_c$, instead of the interpolated $\tilde{s}_w$ as in Eq. (3). The worse performance relate to synthetic characteristics present in $\tilde{s}_c$ (*e.g.* road texture in Fig. 7a). Second, "LGFS-only" directly uses the LGFS losses in substitution to $\mathcal{L}_{\text{adv}}$, without WMI and GERM components. While it only slightly worsens metrics, the impact

**Table 2.** Study of the impact of anchor domains $\mathbb{A}$ on the $\mathcal{S} \mapsto \mathcal{T}$ translations for intra-dataset (a) and cross-dataset (b) tasks. The stable performance across all tested anchors demonstrates the robustness of our method. For all, we test a multi-anchor setup by using all anchors ("All"). In (b), * means LPIPS cannot be computed due to lack of pairs of matched images (Sect. 4.1).

Day $\mapsto$ Night

| $\mathcal{S}$ | $\mathcal{T}$ | $\mathcal{A}_m$ | FID$_\downarrow$ | LPIPS$_\downarrow$ |
|---|---|---|---|---|
| ACDC-Day | ACDC-Night | Day | 85.73 | 0.553 |
| | | Night | **81.01** | **0.535** |
| | | Rain | 81.38 | 0.549 |
| | | Snow | 86.74 | 0.554 |
| | | Sunset | 83.83 | 0.571 |
| | | All | 83.71 | 0.547 |

Day $\mapsto$ Twilight

| $\mathcal{S}$ | $\mathcal{T}$ | $\mathcal{A}_m$ | FID$_\downarrow$ | LPIPS$_\downarrow$ |
|---|---|---|---|---|
| DZ-Day | DZ-Twilight | Day | 64.19 | 0.505 |
| | | Night | 63.15 | 0.510 |
| | | Rain | 65.33 | 0.501 |
| | | Snow | 64.09 | 0.513 |
| | | Sunset | 63.78 | 0.504 |
| | | All | **60.98** | **0.469** |

Clear $\mapsto$ Fog

| $\mathcal{S}$ | $\mathcal{T}$ | $\mathcal{A}_m$ | FID$_\downarrow$ | LPIPS$_\downarrow$ |
|---|---|---|---|---|
| ACDC-Clear | ACDC-Fog | Day | **89.57** | **0.520** |
| | | Night | 91.79 | **0.520** |
| | | Rain | 93.15 | 0.522 |
| | | Snow | 90.28 | 0.524 |
| | | Sunset | 90.11 | 0.525 |
| | | All | 92.19 | **0.520** |

(a) Intra-dataset

Day $\mapsto$ Twilight

| $\mathcal{S}$ | $\mathcal{T}$ | $\mathcal{A}_m$ | FID$_\downarrow$ | LPIPS$_\downarrow$ |
|---|---|---|---|---|
| ACDC-Day | DZ-Twilight | Day | 89.61 | * |
| | | Night | 90.48 | * |
| | | Rain | 89.47 | * |
| | | Snow | 91.49 | * |
| | | Sunset | 91.77 | * |
| | | All | **85.15** | * |

(b) Cross-dataset

on feature consistency is dramatic as shown in Fig. 7b, where the sky presents obvious artifacts and road trivially darkens.

**Anchor Selection.** We ablate the choice of anchor domain $\mathbb{A}$ by selecting different conditions from the VIPER dataset, namely {Day, Night, Rain, Snow, Sunset}. In particular, we experiment on previous *intra-dataset* ($\mathcal{S}$ and $\mathcal{T}$ taken from the same dataset) tasks, as well as on a *cross-dataset* task in which $\mathcal{S}$ = ACDC-Day and $\mathcal{T}$ = DZ-Twilight. Results in Table 2 show how performance remains relatively stable across most anchors. This may seem counter-intuitive since one could, for example, expect that the "Rain" anchor would be a poor choice for the Day $\mapsto$ Night task since rainy and night scenes look different. The results instead show that the WMI only encodes consistency in the transformation, and is thus robust to the choice of anchors. We also test a multi-anchor setup ("All" in Table 2), where $\mathbb{A} = \{\mathcal{A}_{\text{id}}$, Day, Night, Rain, Snow, Sunset}. In general, more anchor domains improve performances, ranking either first or second in all cases for at least one metric, due to the additional information available for shaping the manifold in WMI. We hypothesize that multiple anchors helps identifying correspondences between $\mathcal{S}$ and $\mathcal{T}$, benefiting especially the cross-dataset tasks.

**Number of Images and Variability.** First we compare our Day $\mapsto$ Night translations against MUNIT [13] for $|\mathcal{T}| = \{25, 20, 15, 10, 5, 1\}$, to understand the effects of few-shot training on the backbone network. Some qualitative *general* outputs are shown in Fig. 8a. While MUNIT overfits and creates unrealistic appearance (25–10 images) or collapse (5, 1 image), we output realistic transformations in all cases, even retaining the image context in the extreme one-shot scenario. This is confirmed by the FID and LPIPS in Figs. 8b and 8c for the *general* and *exemplar* scenarios respectively.

In Table 3 we also study variability, evaluating FID and LPIPS for the *general* and *exemplar* cases for $|\mathcal{T}| = \{25, 15, 5, 1\}$ images reporting the results of 7 runs. Overall, the performance remains relatively constant with the exception of the one-shot setup, where despite realistic transfer, the metrics are penalized since the target image itself might not accurately represent the style distribution of the test set.

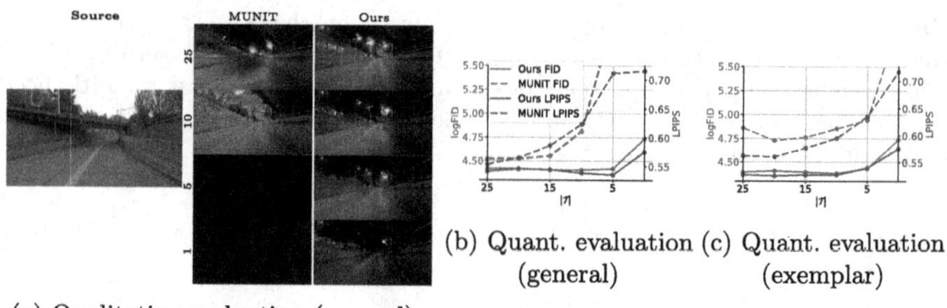


(b) Quant. evaluation (general)

(c) Quant. evaluation (exemplar)

(a) Qualitative evaluation (general).

**Fig. 8.** Comparison against MUNIT for varying $|\mathcal{T}|$: a qualitatively for the *general* scenario; as well as quantitatively for b *general* and c *exemplar*, always outperforming it.

**Table 3.** Day $\mapsto$ Night ablation on variability by training on 4 few-shot configurations with 7 runs each on general (a) and exemplar (b). $|\mathcal{T}|$ does not impact performance much except for the extreme one-shot scenario, where the network overfits to the seen style. The exemplar behavior performs better due to the style conditioning mechanism.

| $\vert\mathcal{T}\vert$ | **FID**$_\downarrow$ | **LPIPS**$_\downarrow$ |
|---|---|---|
| 25 | 82.95 ±2.95 | 0.541 ±1.85e-2 |
| 15 | 82.21 ±3.09 | 0.544 ±2.35e-2 |
| 5 | 83.11 ±2.49 | 0.535 ±2.24e-2 |
| 1 | 114.5 ±34.2 | 0.575 ±2.37e-2 |

(a) General

| $\vert\mathcal{T}\vert$ | **FID**$_\downarrow$ | **LPIPS**$_\downarrow$ |
|---|---|---|
| 25 | 80.78 ±2.91 | 0.527 ±0.64e-2 |
| 15 | 80.55 ±2.85 | 0.527 ±1.07e-2 |
| 5 | 84.40 ±1.88 | 0.540 ±1.88e-2 |
| 1 | 114.3 ±33.5 | 0.575 ±2.40e-2 |

(b) Exemplar

# 5 Extensions

## 5.1 Few-Shot Continuous Manifolds

We investigate the use of ManiFest for performing continuous image translation as in CoMoGAN [40], thus learning the transformation from $\mathcal{S} = day$ to $\mathcal{A}_m = night$ on the Waymo [47] dataset by generating realistically intermediate frontal sun/twilight conditions where we have only few images. Here, we consider *two* few-shot sets ($|\mathcal{T}| = 10$), each one associated to one set of learned weights $(w^1, w^2)$ between identity and night anchors. Results are in Fig. 9, where we also perform comparably to DNI-MUNIT [52] and CoMoGAN which are trained with significantly more intermediate data (4721 vs 20). Please note that estimating $w^1$ and $w^2$ reorganizes the transformation realistically (*i.e.* Day $\mapsto$ Frontal sun $\mapsto$ Twilight $\mapsto$ Night) without prior knowledge on the order of few-shot set in the manifold. We evaluate mean rolling FID (mrFID) as in [40] and perform on par or better than baselines (for Model/StarGAN V2 / DNI - CycleGAN/DNI - MUNIT/CoMoGAN/<u>Ours</u> we get 195/177/155/144/145/<u>145</u>). Only CoMoGAN (mrFID 137) outperforms thanks to its physical guidance.

**Fig. 9.** By enforcing multiple few-shot manifold deformations, we discover a realistic timelapse appearance comparable with CoMoGAN [40] and DNI-MUNIT [52] (all non few-shot).

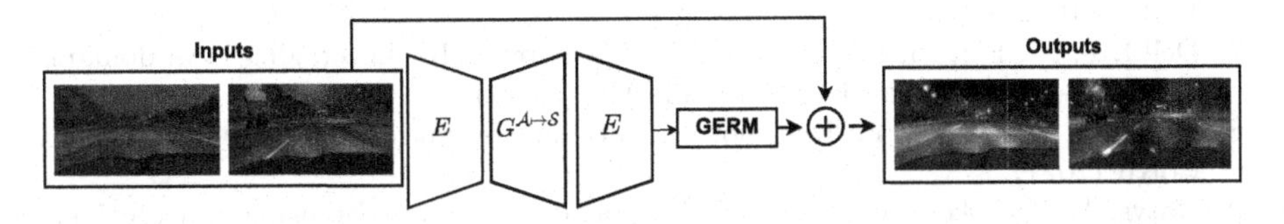


**Fig. 10.** In the reconstruction cycle $\mathcal{A}_m \mapsto \mathcal{S} \mapsto \mathcal{A}_m$, we can inject the extracted residual with GERM on anchor images to perform an alternative $\mathcal{A}_m \mapsto \mathcal{T}$ transformation.

### 5.2 Anchor-Based Translation

The GERM extracts residual information from encoded source images. We investigate the application of residuals on the anchor images themselves, by first translating from $\mathcal{A}_m \mapsto \mathcal{S}$ using our backbone cycle consistency [13], and afterwards re-encoding the fake image in a $\mathcal{S} \mapsto \mathcal{A}_m$ reconstruction *without retraining* (see Fig. 10). This shows how ManiFest simultaneously learns $\mathcal{S} \mapsto \mathcal{T}$ and acceptable $\mathcal{A}_m \mapsto \mathcal{T}$ transformations. The FID w.r.t. ACDC-Night improves from 142 to 130 when applying the residual on the synthetic anchors, thus confirming their shift towards $\mathcal{T}$.

## 6 Conclusion

In this paper we presented ManiFest, a framework for few-shot i2i which enables translating images to a single *general* style approximating the entire few-shot set (*e.g.*, for photo editing), or reproducing any specific *exemplar* from the set for more variability (*e.g.*, for domain adaptation). We demonstrated its effectiveness outperforming the state-of-the-art on many tasks, ablated its components and provided extensions to the framework.

**Acknowledgements.** This work was partly funded by Vislab Ambarella, the French project SIGHT (ANR-20-CE23-0016), and received support from Service de coopération et d'action culturelle du Consulat général de France à Québec. It used HPC resources from GENCI-IDRIS (Grant 2021-AD011012808).

## References

1. Abid, M.A., Hedhli, I., Lalonde, J.F., Gagne, C.: Image-to-image translation with low resolution conditioning. arXiv (2021)
2. Cao, J., Hou, L., Yang, M.H., He, R., Sun, Z.: Remix: towards image-to-image translation with limited data. In: CVPR (2021)
3. Chen, W.Y., Liu, Y.C., Kira, Z., Wang, Y.C.F., Huang, J.B.: A closer look at few-shot classification. In: ICLR (2019)
4. Cherian, A., Sullivan, A.: SEM-GAN: semantically-consistent image-to-image translation. In: WACV (2019)
5. Choi, Y., Uh, Y., Yoo, J., Ha, J.W.: Stargan v2: diverse image synthesis for multiple domains. In: CVPR (2020)
6. Cordts, M., et al.: The cityscapes dataset for semantic urban scene understanding. In: CVPR (2016)
7. Dell'Eva, A., Pizzati, F., Bertozzi, M., de Charette, R.: Leveraging local domains for image-to-image translation. In: VISAPP (2022)
8. Endo, Y., Kanamori, Y.: Few-shot semantic image synthesis using stylegan prior. CoRR (2021)
9. Gatys, L.A., Ecker, A.S., Bethge, M.: Image style transfer using convolutional neural networks. In: CVPR (2016)
10. Heusel, M., Ramsauer, H., Unterthiner, T., Nessler, B., Hochreiter, S.: Gans trained by a two time-scale update rule converge to a local nash equilibrium. In: NeurIPS (2017)
11. Hu, H., Wang, W., Zhou, W., Zhao, W., Li, H.: Model-aware gesture-to-gesture translation. In: CVPR (2021)
12. Huang, X., Belongie, S.: Arbitrary style transfer in real-time with adaptive instance normalization. In: ICCV (2017)
13. Huang, X., Liu, M.Y., Belongie, S., Kautz, J.: Multimodal unsupervised image-to-image translation. In: ECCV (2018)
14. Isola, P., Zhu, J.Y., Zhou, T., Efros, A.A.: Image-to-image translation with conditional adversarial networks. In: CVPR (2017)
15. Jia, Z., et al.: Semantically robust unpaired image translation for data with unmatched semantics statistics. In: ICCV (2021)
16. Jiang, L., Zhang, C., Huang, M., Liu, C., Shi, J., Loy, C.C.: TSIT: a simple and versatile framework for image-to-image translation. In: Vedaldi, A., Bischof, H., Brox, T., Frahm, J.-M. (eds.) ECCV 2020. LNCS, vol. 12348, pp. 206–222. Springer, Cham (2020). https://doi.org/10.1007/978-3-030-58580-8_13
17. Karras, T., Aittala, M., Hellsten, J., Laine, S., Lehtinen, J., Aila, T.: Training generative adversarial networks with limited data. In: NeurIPS (2020)
18. Lee, H.-Y.: DRIT++: diverse image-to-image translation via disentangled representations. Int. J. Comput. Vis. **128**(10), 2402–2417 (2020). https://doi.org/10.1007/s11263-019-01284-z
19. Li, P., Yu, X., Yang, Y.: Super-resolving cross-domain face miniatures by peeking at one-shot exemplar. In: ICCV (2021)
20. Li, P., Liang, X., Jia, D., Xing, E.P.: Semantic-aware grad-gan for virtual-to-real urban scene adaption. In: BMVC (2018)
21. Li, Y., Fang, C., Yang, J., Wang, Z., Lu, X., Yang, M.H.: Universal style transfer via feature transforms. In: NeurIPS (2017)
22. Li, Y., Liu, M.Y., Li, X., Yang, M.H., Kautz, J.: A closed-form solution to photorealistic image stylization. In: ECCV (2018)

23. Li, Y., Zhang, R., Lu, J., Shechtman, E.: Few-shot image generation with elastic weight consolidation. In: NeurIPS (2020)
24. Li, Y., Yuan, L., Vasconcelos, N.: Bidirectional learning for domain adaptation of semantic segmentation. In: CVPR (2019)
25. Lin, C.T., Wu, Y.Y., Hsu, P.H., Lai, S.H.: Multimodal structure-consistent image-to-image translation. In: AAAI (2020)
26. Lin, J., Wang, Y., He, T., Chen, Z.: Learning to transfer: Unsupervised meta domain translation. In: AAAI (2020)
27. Lin, J., Xia, Y., Liu, S., Zhao, S., Chen, Z.: Zstgan: an adversarial approach for unsupervised zero-shot image-to-image translation. Neurocomputing **461**, 327–335 (2021)
28. Liu, M.Y., Breuel, T., Kautz, J.: Unsupervised image-to-image translation networks. In: NeurIPS (2017)
29. Liu, M.Y., et al.: Few-shot unsupervised image-to-image translation. In: ICCV (2019)
30. Luan, F., Paris, S., Shechtman, E., Bala, K.: Deep photo style transfer. In: CVPR (2017)
31. Ma, L., Jia, X., Georgoulis, S., Tuytelaars, T., Van Gool, L.: Exemplar guided unsupervised image-to-image translation with semantic consistency. In: ICLR (2019)
32. Mao, X., Li, Q., Xie, H., Lau, R.Y., Wang, Z., Paul Smolley, S.: Least squares generative adversarial networks. In: ICCV (2017)
33. Mo, S., Cho, M., Shin, J.: Instagan: instance-aware image-to-image translation. In: ICLR (2019)
34. Ojha, U., et al.: Few-shot image generation via cross-domain correspondence. In: CVPR (2021)
35. Park, T., Efros, A.A., Zhang, R., Zhu, J.-Y.: Contrastive learning for unpaired image-to-image translation. In: Vedaldi, A., Bischof, H., Brox, T., Frahm, J.-M. (eds.) ECCV 2020. LNCS, vol. 12354, pp. 319–345. Springer, Cham (2020). https://doi.org/10.1007/978-3-030-58545-7_19
36. Park, T., Liu, M.Y., Wang, T.C., Zhu, J.Y.: Semantic image synthesis with spatially-adaptive normalization. In: CVPR (2019)
37. Park, T., et al.: Swapping autoencoder for deep image manipulation. In: NeurIPS (2020)
38. Patashnik, O., Danon, D., Zhang, H., Cohen-Or, D.: Balagan: cross-modal image translation between imbalanced domains. In: CVPR Workshops (2021)
39. Pizzati, F., Cerri, P., de Charette, R.: Model-based occlusion disentanglement for image-to-image translation. In: Vedaldi, A., Bischof, H., Brox, T., Frahm, J.-M. (eds.) ECCV 2020. LNCS, vol. 12365, pp. 447–463. Springer, Cham (2020). https://doi.org/10.1007/978-3-030-58565-5_27
40. Pizzati, F., Cerri, P., de Charette, R.: CoMoGAN: continuous model-guided image-to-image translation. In: CVPR (2021)
41. Pizzati, F., Cerri, P., de Charette, R.: Guided disentanglement in generative networks. arXiv (2021)
42. Ramirez, P.Z., Tonioni, A., Di Stefano, L.: Exploiting semantics in adversarial training for image-level domain adaptation. In: IPAS (2018)
43. Richter, S.R., Hayder, Z., Koltun, V.: Playing for benchmarks. In: ICCV (2017)
44. Saito, K., Saenko, K., Liu, M.-Y.: COCO-FUNIT: few-shot unsupervised image translation with a content conditioned style encoder. In: Vedaldi, A., Bischof, H., Brox, T., Frahm, J.-M. (eds.) ECCV 2020. LNCS, vol. 12348, pp. 382–398. Springer, Cham (2020). https://doi.org/10.1007/978-3-030-58580-8_23

45. Sakaridis, C., Dai, D., Van Gool, L.: Map-guided curriculum domain adaptation and uncertainty-aware evaluation for semantic nighttime image segmentation. In: T-PAMI (2020)
46. Sakaridis, C., Dai, D., Van Gool, L.: ACDC: the adverse conditions dataset with correspondences for semantic driving scene understanding. In: ICCV (2021)
47. Sun, P., et al.: Scalability in perception for autonomous driving: Waymo open dataset. In: CVPR (2020)
48. Tang, H., Xu, D., Yan, Y., Corso, J.J., Torr, P.H., Sebe, N.: Multi-channel attention selection gans for guided image-to-image translation. In: CVPR (2019)
49. Torrey, L., Shavlik, J.: Transfer learning. In: Handbook of Research on Machine Learning Applications and Trends: Algorithms, Methods, and Techniques. IGI Global (2010)
50. Tremblay, M., Halder, S.S., de Charette, R., Lalonde, J.-F.: Rain rendering for evaluating and improving robustness to bad weather. Int. J. Comput. Vis. **129**, 1–20 (2020). https://doi.org/10.1007/s11263-020-01366-3
51. Wang, J., et al.: Deep high-resolution representation learning for visual recognition. In: T-PAMI (2019)
52. Wang, X., Yu, K., Dong, C., Tang, X., Loy, C.C.: Deep network interpolation for continuous imagery effect transition. In: CVPR (2019)
53. Wang, Y., Khan, S., Gonzalez-Garcia, A., Weijer, J.V.D., Khan, F.S.: Semi-supervised learning for few-shot image-to-image translation. In: CVPR (2020)
54. Wang, Y., Mantecon, H.L., Lopez-Fuentes, J.V.D.W., Raducanu, B.: Transferi2i: transfer learning for image-to-image translation from small datasets. In: ICCV (2021)
55. Wu, W., Cao, K., Li, C., Qian, C., Loy, C.C.: Transgaga: geometry-aware unsupervised image-to-image translation. In: CVPR (2019)
56. Yoo, J., Uh, Y., Chun, S., Kang, B., Ha, J.W.: Photorealistic style transfer via wavelet transforms. In: ICCV (2019)
57. Yu, X., Chen, Y., Liu, S., Li, T., Li, G.: Multi-mapping image-to-image translation via learning disentanglement. In: NeurIPS (2019)
58. Zhan, F., et al.: Unbalanced feature transport for exemplar-based image translation. In: CVPR (2021)
59. Zhang, P., Zhang, B., Chen, D., Yuan, L., Wen, F.: Cross-domain correspondence learning for exemplar-based image translation. In: CVPR (2020)
60. Zhang, R., Isola, P., Efros, A.A., Shechtman, E., Wang, O.: The unreasonable effectiveness of deep features as a perceptual metric. In: CVPR (2018)
61. Zhao, S., Liu, Z., Lin, J., Zhu, J.Y., Han, S.: Differentiable augmentation for data-efficient gan training. In: NeurIPS (2020)
62. Zheng, C., Cham, T.J., Cai, J.: The spatially-correlative loss for various image translation tasks. In: CVPR (2021)
63. Zhu, J.Y., Park, T., Isola, P., Efros, A.A.: Unpaired image-to-image translation using cycle-consistent adversarial networks. In: CVPR (2017)
64. Zhu, J.Y., et al.: Toward multimodal image-to-image translation. In: NeurIPS (2017)
65. Zhu, P., Abdal, R., Qin, Y., Wonka, P.: Sean: image synthesis with semantic region-adaptive normalization. In: CVPR (2020)
66. Zhu, Z., Xu, Z., You, A., Bai, X.: Semantically multi-modal image synthesis. In: CVPR (2020)

# Supervised Attribute Information Removal and Reconstruction for Image Manipulation

Nannan Li(✉) and Bryan A. Plummer

Boston University, Boston, USA
{nnli,bplum}@bu.edu

**Abstract.** The goal of attribute manipulation is to control specified attribute(s) in given images. Prior work approaches this problem by learning disentangled representations for each attribute that enables it to manipulate the encoded source attributes to the target attributes. However, encoded attributes are often correlated with relevant image content. Thus, the source attribute information can often be hidden in the disentangled features, leading to unwanted image editing effects. In this paper, we propose an Attribute Information Removal and Reconstruction (AIRR) network that prevents such information hiding by learning how to remove the attribute information entirely, creating attribute excluded features, and then learns to directly inject the desired attributes in a reconstructed image. We evaluate our approach on four diverse datasets with a variety of attributes including DeepFashion Synthesis, DeepFashion Fine-grained Attribute, CelebA and CelebA-HQ, where our model improves attribute manipulation accuracy and top-k retrieval rate by 10% on average over prior work. A user study also reports that AIRR manipulated images are preferred over prior work in up to 76% of cases (Code and models are available at https://github.com/NannanLi999/AIRR).

## 1 Introduction

Attribute manipulation translates images based on desired attributes, which has applications to face editing [26,31,32], image retrieval [12,27,34], and image synthesis [3,16], among others. In these tasks, the goal is to be able to control a specified attribute without affecting other information in the source image. While Generative Adversarial Networks (GANs) have achieved impressive performance on attribute manipulation, a major challenge is that the generator tends to take a shortcut by utilizing the preserved source attribute information instead of the target attribute for manipulation [13,18,29], thus causing improper image editing effects in manipulated images. Prior work has tried to address this by adding random noise during the reconstruction [4,30] or learning disentangled attribute

**Supplementary Information** The online version contains supplementary material available at https://doi.org/10.1007/978-3-031-19790-1_28.

© The Author(s), under exclusive license to Springer Nature Switzerland AG 2022
S. Avidan et al. (Eds.): ECCV 2022, LNCS 13677, pp. 457–473, 2022.
https://doi.org/10.1007/978-3-031-19790-1_28

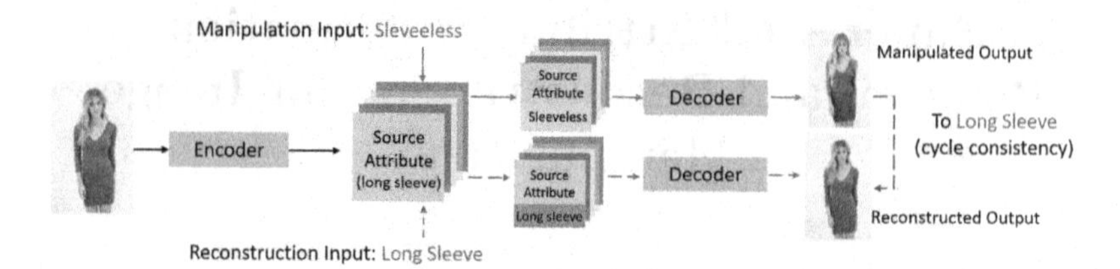


(a) Framework of previous methods [3,12,34,35,11]. Dashed arrows mean two different ways of obtaining the reconstructed image

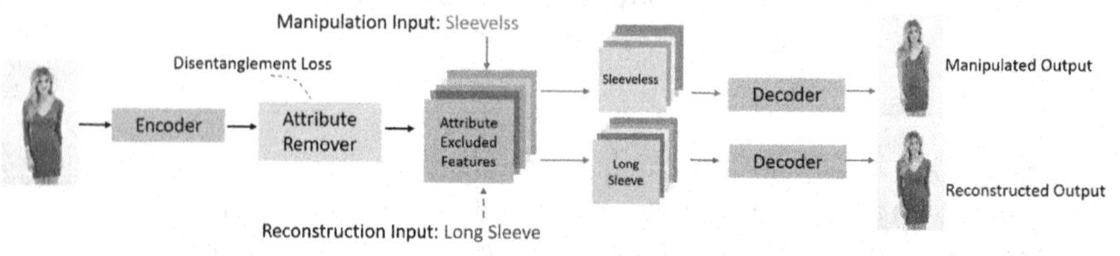


(b) Pipeline of the proposed method

**Fig. 1.** In (a), the generator like those used by [3,11,12,34,35] incorrectly utilizes the hidden source attribute information instead of the target attribute *sleeveless* for image manipulation. As a result, the manipulated image still contains the source attribute *long sleeve*, causing improper image editing effects. To avoid this issue, the proposed method in (b) erases the source attribute information in the encoded features through an attribute remover with a disentanglement loss, conditioning the manipulated output only on the input target attribute *sleeveless* and the attribute excluded features.

representations which are used to manipulate the images [26,33]. However, low-magnitude random noise is not targeted to the source attribute, which could be intentionally ignored by the model or inadvertently suppress key source features. On the other hand, even with a disentangled image representation, the correlation between attributes and relevant image content could cause source attribute information to be hidden in the rest of the image features. For example, attribute *formal* in a dress is often correlated with the dress's *long length*.

To address these issues, we propose a supervised Attribute Information Removal and Reconstruction (AIRR) model that learns an attribute excluded representation and reconstructs the image with desired attributes. The key challenge is in identifying the preserved source attribute information and decorrelating it from the image representation [15,26,31,33]. Prior research on feature disentanglement either doesn't consider the decorrelation [15,31,35], or has limitations on the number of attributes it can decorrelate in a forward pass [26,33], unlike our approach which can disentangle any number of attributes. In addition, as illustrated in Fig. 1a, these methods often rely on the full image information for both manipulation and reconstruction. As mentioned earlier, this can lead to information hiding in the manipulated image. In contrast, as shown in Fig. 1b, we use our remover to erase attribute information to obtain attribute excluded features, which are then used to directly generate both the reconstructed and manipulated images. Since this should eliminate the information that could potentially be hidden in the disentangled representation, we avoid the information hiding issues in prior work.

One challenge in our approach is our reliance on being able to identify and remove attribute information in real images. For example, although the color *white* appears in the background of the input images in Fig. 1, a good attribute classifier would predict that the clothing item is *gray* and not *white*. This means that the background information could mislead the attribute recognition. To address this issue, we segment the object of interest (*e.g.*, using [19,36]) to split the image encoder into two branches for the object of interest and the background, respectively. This helps AIRR to concentrate the manipulation on the object of interest without influencing the background information.

Our main contributions are:

- We propose the Attribute Information Removal and Reconstruction method (AIRR), a controllable disentangled attribute manipulation framework that produces high quality images. The key insight in AIRR is the attribute information removal and reconstruction module that produces an attribute excluded representation, eliminating sources of information hiding that degrades performance in prior work.
- Extensive experiments across DeepFashion Synthesis [21], DeepFashion Fine-grained Attribute [21], CelebA [22] and CelebA-HQ [17] report that AIRR improves the attribute manipulation accuracy and top-k retrieval rate by 10% on average over the state-of-the-art. Moreover, we show that AIRR can effectively control attribute strength as well as efficiently manipulating multiple attributes in a single forward pass.
- A user study further validates the effectiveness of our approach, where our methods are shown to produce high quality images that more accurately achieve the target attribute manipulation by up to 76% over prior work.

## 2 Related Work

Early research in attribute manipulation [7,11] combined the target attribute label directly with the image or image features, and decoded them into manipulated output. However, the decoder could incorrectly use the preserved source attribute information for image manipulation. Thus, more recent work (including this paper), has focused on learning disentangled attribute representations, which we will discuss in more detailed below.

**Unsupervised Disentanglement.** Several studies explored disentanglement in the latent space of GANs in an unsupervised manner [9,23,25,28,32]. These methods aim to manipulate the attributes on synthetic data, where the image content is randomly generated. In [9], the authors found that the principle components of features on pretrained GANs represent high-level semantic concepts. In [32], the authors introduced channel-wise disentanglement of StyleGAN [14]. Shoshan *et al.* [28] utilized contrastive learning to disentangle the latent space, achieving explicit control over synthetic facial images. However, without manual examinations on the feature space, it's difficult to locate the exact attribute representation that we want to manipulate, especially for attributes with high-level semantics. Thus, in our work we focus on cases where attributes we wish to manipulate are known, enabling us to directly target our feature learning.

**Supervised Disentanglement.** Supervised disentanglement methods edit real images based on attribute annotations. Prior work on this task can be categorized in two types: spatial disentanglement and feature disentanglement. In methods that focus on spatial disentanglement [15,31], attributes are located spatially and thus disentangled in the feature map. These methods can find attribute-specific features by an attention map, whereas attribute-relevant information is implicitly kept and thus influences the image manipulation. On the other hand, feature disentanglement identifies certain features corresponding to the manipulated attribute. [35] presents a method that learns a linear transformation function that maps StyleGAN's latent code. Although StyleGAN's latent space is disentangled [1], without orthogonal constraints, such linear combination could result in correlation between different image attributes and content. To address this, Yang *et al.* [33] learned attribute relevant and irrelevant features, but each manipulated attribute requires training its own model, which is computationally costly. Instead, Shen *et al.* [26] manipulated attributes with a conditional subspace projection via Support Vector Machines (SVM), whereas the manipulation accuracy depends on the capability of SVM and each forward pass can control only a single attribute. In contrast, our proposed approach can manipulate multiple source attributes in a single forward pass by utilizing the injected attribute embedding.

**Attribute Manipulation in Fashion.** Apart from the above mentioned methods that are mainly applied to facial attribute editing, attribute manipulation in fashion images has also gained a lot of attention. Recent work on this topic mainly aim to improve the image retrieval accuracy for item recommendation. For example, researches have leveraged spatial information when manipulating attributes [2,3], learned a dictionary of attribute transformations [27], or used the attribute probability distribution as an disentangled representation for image retrieval [12]. Kwon *et al.* [16] predicted changes to an item's shape as a result of changing an attribute, enabling them to make more significant alterations to the clothing in images. However, many of these methods also suffered from issues with disentangling attributes, often due to misinformation hiding, which our work minimizes.

## 3 Attribute Information Removal and Reconstruction

Given image $I$ and its attributes $\mathbf{A} = \{\mathbf{a_1}, \mathbf{a_2}, ..., \mathbf{a_n}\}$, where $a_i$ denotes the $i$th attribute, we aim to manipulate any number of attributes in $\mathbf{A}$. To achieve this goal, the generator first takes a real image as input, and uses our attribute remover to decorrelate the image attributes from the image features. The resulting attribute excluded representation is then combined with

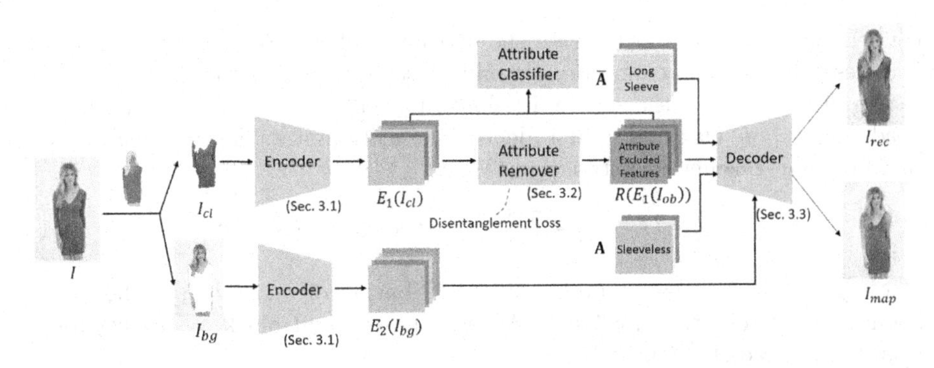


**Fig. 2. AIRR framework.** In the AIRR generator, a given image is parsed into an object of interest $I_{cl}$ and background $I_{bg}$ through an offline parser [19,36]. $I_{cl}$ and $I_{bg}$ are encoded in separate branches (Sect. 3.1). In the $I_{cl}$ branch, the source attribute information in the encoded features are erased by an attribute remover using a disentanglement loss (Sect. 3.2). Subsequently, the source attributes in $\mathbf{A}$ and the target attributes in $\bar{\mathbf{A}}$ are embedded into the attribute excluded features for image reconstruction and image manipulation, respectively (Sect. 3.3).

the target attribute embeddings to produce the manipulated output $I_{map}$. In the following, we introduce the four components of our model: image encoder (Sect. 3.1), attribute remover (Sect. 3.2), decoder (Sect. 3.3), and learning objectives (Sect. 3.4). Figure 2 shows an overview of our approach.

### 3.1 Image Encoder

To concentrate the manipulation on the object that we want to manipulate, the image encoder in AIRR is split into two branches for the object of interest $I_{cl}$ and the background $I_{bg}$, respectively. Prior work often achieves the segmentation of $I_{cl}$ by learning an attention map [2,3,15], while we found empirically that using an offline parser [19,36] is more accurate in segmenting instances. This segmentation is especially helpful for images with multiple objects, *e.g.*, an image of a fashion model wearing top, leggings and boots. As shown in Fig. 2, after obtaining $I_{cl}$ and $I_{bg}$, the image encoder encodes $I_{cl}$ into $E_1(I_{cl})$ in the first branch, and $I_{bg}$ into $E_2(I_{bg})$ in the second branch. Later on, AIRR only manipulates $E_1(I_{cl})$ without influencing the background information $E_2(I_{bg})$.

### 3.2 Attribute Remover

While prior work directly used the image features $E_1(I_{cl})$ to generate the image [3,11,12,34,35], in AIRR, image features $E_1(I_{cl})$ from our base encoder are fed into an attribute remover to learn an attribute-excluded representation $R(E_1(I_{cl}))$. The attribute remover is an $n$-layer convolutional block that is used to decorrelate the source attribute information from the image representation. A design requirement for the attribute remover is that it does not have skip connections since we aim to erase the attribute information from the encoded features, whereas skip connections would preserve this information.

To disentangle all the source attribute information from $E_1(I_{cl})$, we would need an attribute classifier to first identify these attributes. This can be achieved by Maximum Likelihood Estimation (MLE):

$$L_d(E_1(I_{cl})) = -\sum\nolimits_{\mathbf{a_i} \in \mathbf{A}} \mathbf{y_i}^T \log p_c(\mathbf{a_i}|E_1(I_{cl})) \quad (1)$$

where $p_c(\mathbf{a_i}|E_1(I_{cl}))$ is the probability distribution of $\mathbf{a_i}$, and $\mathbf{y_i}$ is the corresponding one-hot attribute label. We use one residual block as the attribute classifier to predict $p_c(\cdot)$.

After identifying the source attributes, we can then eliminate the attribute information in $R(E_1(I_{cl}))$ by minimizing their mutual information. Alternatively, it's easier to minimize the upper bound of this mutual information, which is the maximum log probability in the attribute class distribution added by a constant $c$ (See the Supplementary for proof of this upper bound):

$$\mathrm{MI}(\mathbf{a}_i, R(E_1(I_{cl}))) \leq c + \max\nolimits_{\mathbf{a}_i} \log p_c(\mathbf{a}_i|R(E_1(I_{cl}))) \quad (2)$$

Intuitively, minimizing this upper bound gives a uniform distribution over the attributes, meaning that the uncertainty for a specific attribute is maximized. Therefore, the generated features would have little knowledge of what the original attributes are. This loss function is thus defined as a margin loss:

$$L_d(R(E_1(I_{cl}))) = \sum_{\mathbf{a_i} \in \mathbf{A}} \max\{\max\nolimits_{\mathbf{a}_i} \log p_c(\mathbf{a_i}|R(E_1(I_{cl}))) - \log \frac{1}{|\mathbf{a_i}|}, c\}, \quad (3)$$

where $|\mathbf{a_i}|$ is the number of attribute values in $\mathbf{a_i}$, and $c$ indicates the proximity to a uniform distribution, which is set to 0.01 in our experiments as we found it performed well. We refer to $L_d(R(E_1(I_{cl})))$ as a Mutual Information Minimization (MIM) loss. Note that our attribute classifier and remover are trained end-to-end with our other generator and decoder components.

### 3.3 Decoder

After the attribute remover, a new learned attribute representation is integrated with the disentangled features $R(E_1(I_{cl}))$ to generate the output image. Assuming a scale embedding vector $\boldsymbol{\beta}_{a_i}$ and a bias embedding vector $\gamma_{a_i}$ for attribute $\mathbf{a_i}$, the original image $I$ thus can be reconstructed by combining $R(E_1(I_{cl}))$, $E_2(I_{bg})$ and the attribute embeddings, *i.e.*,

$$I_{rec} = G\Big(\mathrm{concat}\big[\sum\nolimits_{\mathbf{a_i} \in \mathbf{A}} \boldsymbol{\beta}_{\mathbf{a_i}} \cdot R(E_1(I_{cl})) + \gamma_{\mathbf{a_i}}, E_2(I_{bg})\big]\Big), \quad (4)$$

where $G$ is the decoder. Similarly, given target attributes of our desired output $\bar{\mathbf{A}}$, $R(E_1(I_{cl}))$ produces the manipulated image $I_{map}$ by

$$I_{map} = G\Big(\mathrm{concat}\big[\sum\nolimits_{\bar{\mathbf{a}}_i \in \bar{\mathbf{A}}} \boldsymbol{\beta}_{\bar{\mathbf{a}}_i} \cdot R(E_1(I_{cl})) + \gamma_{\bar{\mathbf{a}}_i}, E_2(I_{bg})\big]\Big) \quad (5)$$

In contrast to some prior work [3,7,15], where the attribute embedding vector and the encoded features are combined by concatenation, we multiply these two features such that linear interpolation between different attribute embeddings can better control the strength of these attributes in the output [8,33]. Further discussion can be found in Sect. 4.5.

### 3.4 Learning Objectives

**Disentanglement Loss.** The disentanglement loss combines the MLE loss in Eq. (1) and the MIM loss in Eq. (3) as

$$L_d = L_d\big(E_1(I_{cl})\big) + L_d\big(R\big(E_1(I_{cl})\big)\big) \tag{6}$$

By first identifying the source attributes in $E_1(I_{cl})$ and then minimizing the attribute information in $R(E_1(I_{cl}))$, this disentanglement loss enables the decoder to condition the output on the new attribute that is injected to the generator. In Sect. 4.5, we also show empirically that with the disentanglement loss, the attribute remover indeed gets rid of all the source attribute information.

**Reconstruction Loss.** The reconstructed image $I_{rec}$ is evaluated by its $l_1$ distance to the original image $I$:

$$L_{rec} = ||I_{rec} - I||_1 \tag{7}$$

**Adversarial Loss.** The manipulated image $I_{map}$ doesn't have a paired ground truth of how it should look like, for which its plausibility is evaluated by the discriminator $D$. Using LSGAN [24], the adversarial loss of the generator can be written as

$$L^g_{adv} = (1 - D(I_{map}))^2 + (1 - D(I_{rec}))^2 \tag{8}$$

In the discriminator, this adversarial loss includes both the reconstructed image $I_{rec}$ and the manipulated image $I_{map}$ since they are both fake samples:

$$L^d_{adv} = (1 - D(I))^2 + \frac{1}{2}\big((D(I_{map})^2 + D(I_{rec})^2\big) \tag{9}$$

**Image Attribute Classification Loss** [6]. This loss maximizes the mutual information between the injected attribute embedding and the generated image:

$$L^g_{attr} = -\sum\nolimits_{\mathbf{a_i}\in\mathbf{A}} \mathbf{y_i}^T \log p_d(\mathbf{a_i}|I_{rec}) - \sum\nolimits_{\bar{\mathbf{a}}_i\in\bar{\mathbf{A}}} \bar{\mathbf{y}}_i^T \log p_d(\bar{\mathbf{a}}_i|I_{map}) \tag{10}$$

where $p_d(\cdot)$ is the probability distribution of attributes predicted by a classification branch in the discriminator. For the discriminator, this loss is defined on the real image $I$.

$$L^d_{attr} = L^d_{attr}(I) = -\sum\nolimits_{\mathbf{a_i}\in\mathbf{A}} \mathbf{y_i}^T \log p_d(\mathbf{a_i}|I) \tag{11}$$

**Table 1.** Statistics of the datasets used in our experiments in Sect. 4.

| Dataset | Image size | #training images | #test images | #attributes | #attribute values |
|---|---|---|---|---|---|
| DeepFashion Synthesis [21] | 128 × 128 | 76,979 | 2000 | 2 | 21 |
| DeepFashion Fine-grained Attribute [21] | 256 × 256 | 19,000 | 1000 | 6 | 26 |
| CelebA [22] | 128 × 128 | 200,599 | 2000 | 8 | 21 |
| CelebA-HQ [17] | 1024 × 1024 | 29,000 | 1000 | 8 | 21 |

**Perceptual Loss.** To further improve the quality of the generated images, a perceptual loss is introduced in the generator as in [3]. It is based on the distance of paired real and fake images in the CNN feature space

$$L_p = ||\text{CNN}(I) - \text{CNN}(I_{rec})||_1 + ||\text{CNN}(I_{ref}) - \text{CNN}(I_{map})||_1 \tag{12}$$

where $I_{ref}$ is selected from the real images in the dataset to have exactly the same attributes $\bar{\mathbf{A}}$ as $I_{map}$.

**Full Objective.** Including all the above loss functions, the full objectives for the generator and discriminator are

$$L_{gen} = L^g_{adv} + \lambda_1 L_d + \lambda_2 L^g_{attr} + \lambda_3 L_{rec} + \lambda_4 L_p \tag{13}$$

$$L_{dis} = L^d_{adv} + 2\lambda_2 L^d_{attr} \tag{14}$$

where $\lambda_1, \lambda_2, \lambda_3$ are trade-off parameters. As in prior work [5,20], these parameters must be set carefully to control the degree of disentanglement. Note that except for $L_{rec}$, these loss functions are all symmetrical with respect to $\mathbf{A}$ and $\bar{\mathbf{A}}$ to enforce that the manipulated image is a plausible reconstructed image.

## 4 Experiments

To prove the efficiency of the proposed method, we evaluate our model on four publicly available datasets: DeepFashion Synthesis [21], DeepFashion Fine-grained Attribute [21], CelebA [22] and CelebA-HQ [17]. On CelebA and CelebA-HQ, we group the attributes into 8 attribute categories and 21 attribute values following [2]. See Table 1 for detailed statics of each dataset.

### 4.1 Implementation Details

We use the model architecture in [37] as our backbone on DeepFashion and CelebA datasets. Following [35], on CelebA-HQ we adopt another backbone: StyleGAN2 [14], which is better suited to high resolution images. To improve training stability, we froze the weights of StyleGAN2's encoder and generator. Except for CelebA-HQ, the CNN used in Eq. (12) is a ResNet-50 model [10] pretrained on image attribute classification task. On CelebA-HQ, we use StyleGAN2's encoder as the CNN in order to match the identity loss defined in [35].

For DeepFashion the target attributes are uniformly and randomly sampled from items in the same clothing category, *e.g.*, dress and leggings. Here we did not sample the target attributes from the whole dataset because some annotated attributes can only appear in certain clothing categories. For example, leggings can't have V-neckline, and skirts can't have long sleeve. On CelebA-HQ, we traverse the values of the 8 attributes for each image during test for fair comparison with existing methods that use binary attribute values [26,33,35].

### 4.2 Experimental Settings

**Baselines.** We compare our model with related approaches on attribute manipulation: StarGAN [7], AMNet [3], FLAM [27], VPTNet [16], AttGAN [11], Student [18], FSNet-v2 [2], CAFE-GAN [15], InterfaceGAN [26], L2M-GAN [33] and LatentTransformer [35]. Among these approaches, FLAM, Student, InterfaceGAN and L2M-GAN also aim to achieve feature-level disentanglement, while AMNet, FSNet-v2 and CAFE-GAN aim to learn spatially disentangled representations. For StarGAN, AttGAN, Student, InterfaceGAN, L2M-GAN and LatentTransformer, we used the official implementations at author-provided links. AMNet and FLAM are reproduced by us following the configurations provided in the corresponding papers. Results of FashionSearchNet-v2, VPTNet and CAFE-GAN are directly copied from the original papers. For fair comparison on CelebA-HQ, we used StyleGAN2 encoded image features in InterfaceGAN.

**Evaluation Metrics.** Following [2,3], we use human evaluation and two standard metrics to evaluate the model's performance on attribute manipulation: attribute manipulation accuracy and top-k retrieval. Attribute manipulation accuracy, which is the classification accuracy of the target attribute on the manipulated images, measures the extent to which a model can modify the target attribute. We use a ResNet-50 model [10] pretrained on attribute classification to evaluate the attribute manipulation accuracy on DeepFashion and CelebA. On CelebA-HQ, the accuracy is computed using the same facial attribute classifier as [35] for a fair comparison. Top-k retrieval, on the other hand, evaluates both the attribute changing and preservation capability. It is defined as the number of hits divided by the total number of queries. A query is called a hit if any of the manipulated image's top-k matches has exactly the target attributes in $\bar{\mathbf{A}}$. The top-k retrieval rate is averaged across all attributes. In all experiments, we use the deep features in the last fully-connected layer of the attribute classifier for image retrieval. All retrieval galleries have 20,000 images.

### 4.3 Quantitative Results

As shown in Table 2, 3, 4 and 5, AIRR outperforms the state-of-the-art by a significant margin on most evaluation metrics. For example, Table 5 shows the

**Table 2.** Results on DeepFahison Synthesis. In our models, $\lambda_1 = 0.25, \lambda_2 = 0.125, \lambda_3 = 1.0, \lambda_4 = 1.0$.

| Method | Manipulation accuracy | | | Top-K retrieval | |
|---|---|---|---|---|---|
| | Color | Sleeve | Avg. | R@5 | R@20 |
| StarGAN [7] | 70.4 | 77.2 | 73.8 | 71.1 | 82.9 |
| AMNet [3] | 74.4 | 82.1 | 78.3 | 85.1 | 90.5 |
| AttGAN [11] | 80.2 | 91.0 | 85.6 | 90.6 | 95.2 |
| FLAM [27] | – | – | – | 26.7 | 41.3 |
| VPTNet [16] | – | 85.7 | – | – | – |
| AIRR (w/o mask) | 88.8 | 92.2 | 90.5 | 94.4 | 97.1 |
| AIRR (w/o $L_d$) | 93.9 | 89.5 | 91.7 | 95.0 | 97.3 |
| AIRR (w $L_h$) | 89.4 | 90.5 | 90.0 | 90.8 | 92.4 |
| AIRR | **94.1** | **96.5** | **95.3** | **97.6** | **98.8** |

**Table 3.** Results on DeepFahison Fine-grained Attributes. In our models, $\lambda_1 = 0.05, \lambda_2 = 0.125, \lambda_3 = 2.0, \lambda_4 = 1.0$.

| Method | Manipulation accuracy | | | | | | | Top-K retrieval | |
|---|---|---|---|---|---|---|---|---|---|
| | Pattern | Sleeve | Length | Neckline | Material | Style | Avg. | R@5 | R@20 |
| StarGAN [7] | 54.0 | 38.7 | 22.4 | 44.3 | 47.2 | 24.6 | 38.5 | 25.1 | 39.3 |
| AttGAN [11] | 47.4 | 31.5 | 19.0 | 33.7 | 40.3 | 23.5 | 32.6 | 28.3 | 39.1 |
| AMNet [3] | 53.6 | 56.5 | 24.4 | 68.3 | 47.9 | 27.4 | 46.4 | 44.1 | 47.5 |
| FLAM [27] | – | – | – | – | — | – | – | 17.6 | 29.8 |
| AIRR (w/o mask) | 70.7 | 44.3 | 19.0 | 57.3 | 55.0 | 25.9 | 45.4 | 38.2 | 52.4 |
| AIRR (w/o $L_d$) | **89.1** | 60.6 | 31.8 | 73.5 | 74.9 | 34.2 | 60.8 | 56.4 | 68.7 |
| AIRR (w $L_h$) | 85.1 | 59.2 | 25.7 | 72.4 | 72.6 | 34.7 | 58.2 | 51.1 | 60.2 |
| AIRR | 87.8 | **65.9** | **32.2** | **74.5** | **76.3** | **35.8** | **62.1** | **57.7** | **70.3** |

average attribute manipulation accuracy and top-k retrieval rates on CelebA-HQ, which boost performance by more than 20% compared to existing methods. The improvements are more obvious on additive attributes, such as *wearing hat*, which reports gains over prior work by more than 40% in Table 4 and 5. Further discussion on ablations of our model can be found in Sect. 4.5.

**Attribute Preservation Analysis.** To analyze what influence that changing a specific attribute has on preserving others, we gradually increase the ratio of manipulated images (*i.e.*, the number of manipulated images divided by the number of all test images), and observe the ratio of successfully preserved attributes. Figure 3 provides the attribute changing rate (*i.e.*, attribute manipulation accuracy) vs. attribute preservation rate for each attribute in the DeepFashion dataset. In each graph, the preservation rates are averaged over all attributes excluding the target attribute. In Fig. 3, our method achieves the

**Table 4.** Results on CelebA. In our models, $\lambda_1 = 0.5, \lambda_2 = 0.5, \lambda_3 = 1.0, \lambda_4 = 1.0$.

| Method | Manipulation accuracy | | | | | | | | | Top-K retrieval | |
|---|---|---|---|---|---|---|---|---|---|---|---|
| | Hair Color | Beard | Hair Type | Smiling | Eyeglasses | Gender | Hat | Age | Avg. | R@5 | R@20 |
| StarGAN [7] | 55.2 | 51.6 | 35.6 | 64.0 | 86.1 | 36.5 | 6.4 | 44.9 | 38.8 | 39.9 | 55.2 |
| Student [18] | 47.7 | 43.5 | 37.3 | 60.0 | 12.1 | 42.7 | 11.5 | 39.8 | 36.8 | 38.2 | 53.9 |
| AMNet [3] | 58.6 | 34.4 | 26.7 | 43.7 | 10.0 | 21.0 | 13.0 | 22.4 | 28.7 | 33.1 | 46.0 |
| AttGAN [11] | 72.6 | 88.5 | 48.1 | 79.8 | 94.7 | 89.7 | 21.7 | 60.6 | 69.5 | 72.4 | 86.5 |
| CAFE-GAN [15] | 83.6 | 40.1 | – | – | – | **95.2** | – | **88.6** | – | – | – |
| FSNet-v2 [2] | – | – | – | – | – | – | – | – | – | 68.0 | 77.5 |
| AIRR(w/o mask) | **86.1** | 96.0 | **58.5** | 92.7 | 98.9 | 91.3 | 75.6 | 64.9 | 83.0 | 86.5 | 94.3 |
| AIRR (w/o $L_d$) | 74.3 | 93.9 | 51.4 | 92.4 | 99.4 | 89.5 | 83.4 | 69.7 | 81.8 | 87.1 | 94.9 |
| AIRR | 75.9 | **96.4** | 58.4 | **94.8** | **99.1** | 91.6 | **93.1** | 80.6 | **86.2** | **89.1** | **95.6** |

**Table 5.** Results on CelebA-HQ. In our models, $\lambda_1 = 0.25, \lambda_2 = 0.125, \lambda_3 = 20.0, \lambda_4 = 10.0$.

| Method | Manipulation accuracy | | | | | | | | | Top-K retrieval | |
|---|---|---|---|---|---|---|---|---|---|---|---|
| | Hair Color | Beard | Hair Type | Smiling | Eyeglasses | Gender | Hat | Age | Avg. | R@5 | R@20 |
| InterfaceGAN [26] | 38.4 | **80.8** | 36.4 | **97.7** | 29.7 | 55.8 | 2.9 | 42.0 | 48.0 | 19.1 | 38.8 |
| LatentTrans [35] | 37.0 | 78.8 | 48.9 | 85.3 | 49.6 | 62.2 | 5.6 | 47.4 | 51.9 | 20.7 | 41.8 |
| L2M-GAN [33] | – | – | – | 89.7 | — | – | – | – | – | - | – |
| AIRR(w/o mask+$L_d$) | 40.8 | 73.1 | 50.7 | 93.2 | 63.1 | 71.4 | 40.1 | 83.3 | 64.5 | 51.3 | 67.9 |
| AIRR(w/o mask) | **54.8** | 76.9 | **58.4** | 95.4 | **88.2** | **79.0** | **49.2** | **88.0** | **73.7** | **60.9** | **75.5** |

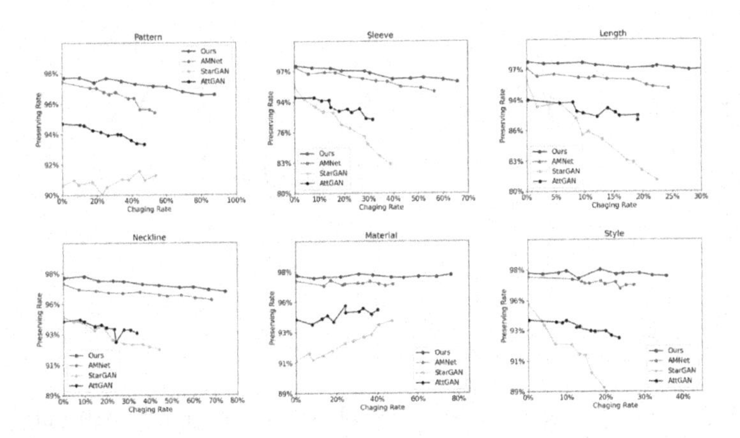


**Fig. 3.** Attribute changing rate vs. attribute preservation rate. The interval of $y$ axis is made to be unequal for better visualization purposes.

highest preservation rate under the same attribute changing rate, proving its capability of controllable attribute manipulation as well as preservation.

**User Study.** We also conducted human evaluation experiments on DeepFashion Fine-grained Attribute and CelebA-HQ using Amazon Mechanical Turk service to verify the quality of manipulated images. We tested on 50 images in each dataset, and different 5 worker were assigned per image. Each worker was presented 3 pictures: the original image, the manipulated image produced by AIRR,

**Table 6.** A/B user judgements for attribute manipulation correctness on the DeepFashion Fine-grained Attribute dataset.

| AIRR/StarGAN | AIRR/AttGAN | AIRR/AMNet |
|---|---|---|
| 65%/35% | 61%/39% | 54%/46% |

**Table 7.** A/B user judgements for attribute manipulation correctness on the CelebA-HQ dataset.

| AIRR/InterfaceGAN | AIRR/LatentTrans |
|---|---|
| 76%/24% | 70%/30% |

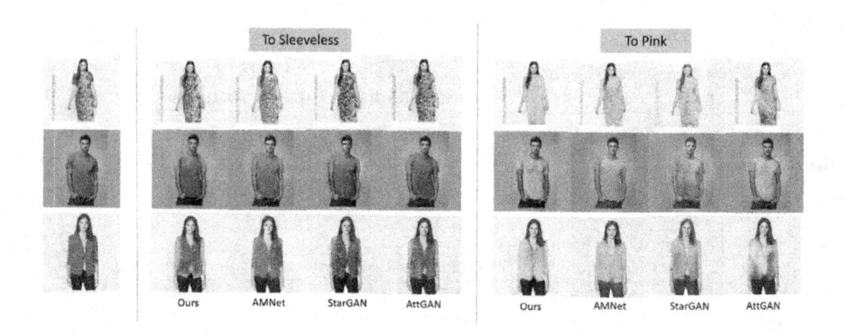


**Fig. 4.** Qualitative examples on DeepFashion Synthesis. The first column shows original images.

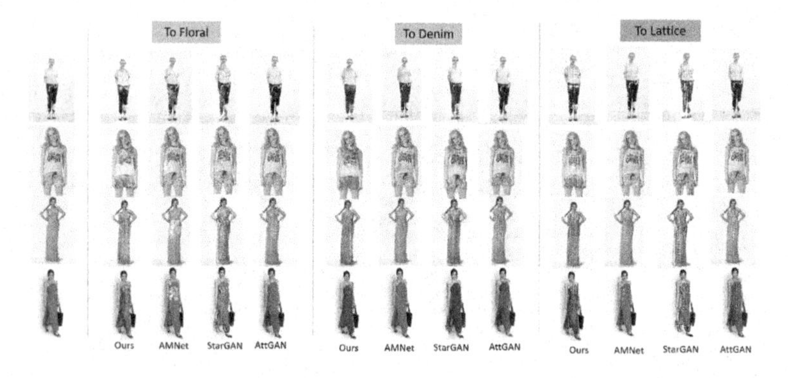


**Fig. 5.** Qualitative examples on DeepFashion Fine-grained Attribute.

and the manipulated image generated by a randomly chosen baseline approaches in Table 6 or 7. The worker was asked to pick an image that better converts the specified attribute in the given image. Table 6 shows that 54–65% of workers think our method achieves better attribute manipulation on the DeepFashion Fine-grained Attribute dataset. On CelebA-HQ, reported in Table 7, 70–76% workers voted for AIRR, verifying its improved capability of manipulating facial attributes compared to prior work.

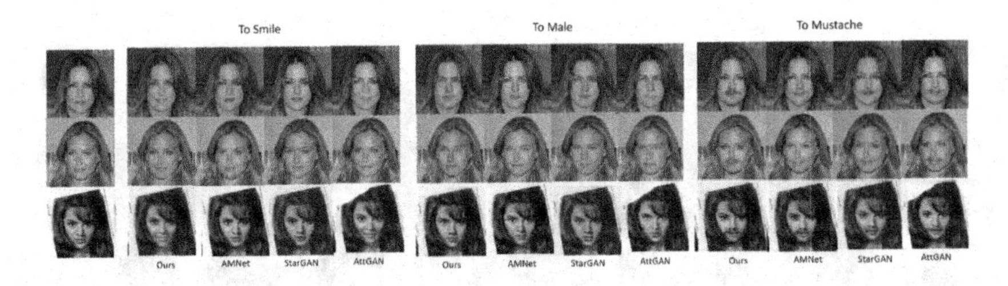


**Fig. 6.** Qualitative examples on CelebA.

## 4.4 Qualitative Results

Figure 4, 5, 6 and 7 presents some qualitative examples on each dataset that we used. For attributes that are relatively shallow and easy to learn, such as *color* in Fig. 5, all methods perform well in transforming the source attribute into the target attribute. Whereas for attributes with relatively more complicated semantics, *e.g.*, *lattice* in Fig. 5, our method better represents the target attribute in the generated images. In addition, it can also be observed that due to the information hiding problem, several failures of previous methods exhibit as visually not altering the original image, such as AMGAN's *To Denim* in Fig. 5 and LatentTrans's *To Hat* in Fig. 7. On the other hand, although our method avoids using the source attribute information for manipulation, it's failures can be more significant due to excessive manipulation. For example, the last two rows of *To Floral* in Fig. 5 alter the dresses' appearance completely.

## 4.5 Model Analysis

**Ablations of Model Components.** To demonstrate the contribution of the components of our proposed framework, we evaluate the performance of our model without the disentanglement loss and the parsing mask. Tables 2, 3, 4 and 5 report quantitative results of ablations of our model. AIRR (w/o $L_d$), which disables the attribute remover, causes losses on average attribute manipulation accuracy and top-5 retrieval on all four datasets. Especially in CelebA-HQ, AIRR (w/o $L_d$) losses 9% accuracy compared to AIRR. This suggests that disentangling attribute information by decorrelation is effective in image manipulation. We also explored what the generated images would look like when injecting no attribute, *i.e.*, setting the target attribute's scale embedding vector to be $\mathbf{1}$ and the bias embedding vector to be $\mathbf{0}$. This way we can check if the model is hiding source attribute information in the encoded features. If it suffers from information hiding, then the source attributes should been seen in generated images. In Fig. 8a, without the proposed disentanglement loss, most source attributes, including material and pattern, indeed appear in the generated images. With the proposed attribute excluded representation, all these attribute information is successfully removed, avoiding the information hiding problem suffered by prior work.

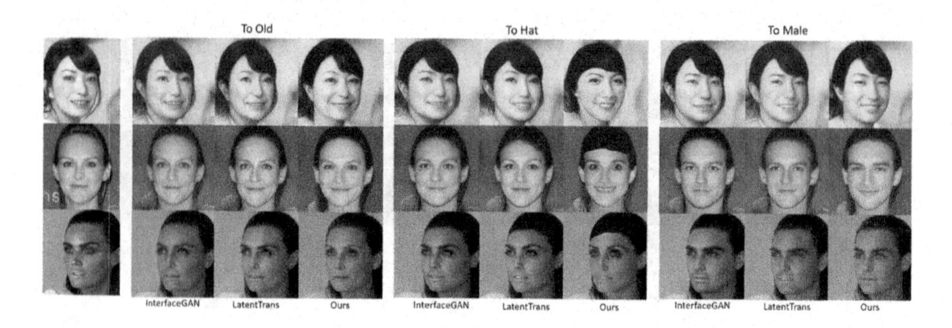


**Fig. 7.** Qualitative examples on CelebA-HQ.

In the meanwhile, AIRR (w/o mask), which removes the parsing mask along with the second encoder, also degrades the evaluation metrics as seen in Tables 2, 3, 4 and 5. This indicates the importance of concentrating manipulation on the object of interest. However, we note that even without the parsing mask, our approach still outperforms prior work on most metrics. Note that in the two ablations of CelebA-HQ, we didn't add the parsing mask in order to reduce the computational costs for generating high resolution images.

We also tried replacing the proposed disentanglement loss with the honesty loss in [4], which was introduced to avoid the general information hidden problem when using cycle consistency. In Table 2 and 3, AIRR still outperforms AIRR (w $L_h$) that adopts the honesty loss, suggesting that the proposed disentanglement loss is more targeted to the attribute manipulation task. See the Supplementary for more ablation results for hyperparameters used by our model.

**Interpolation of Attribute Values.** Linear interpolating between different attribute embedding vectors $\boldsymbol{\beta}_{a_i}$ and $\gamma_{a_i}$ corresponds to an interpolation between different values of the target attribute. Take "smile" for example, Fig. 8b gives the outputs of interpolating from the *not smiling* embedding vector to the *smiling* embedding vector. Let $c$ be the weight (*i.e.*, interpolation coefficient) of the target attribute *smiling*. The smile in generated images gradually builds up as $c$ increases, showing a continuous control over the attribute strength.

**Controlling Multiple Attributes in One Forward Pass.** In some prior work, *e.g.*, [31,33,35], multi-attribute editing is often accomplished by sequential manipulation, *i.e.*, edit one attribute at a time. In contrast, AIRR is capable of changing multiple attributes in a single forward-pass by directly specifying the input target attributes in $\bar{\mathbf{A}}$. Figure 8c gives some examples on CelebA-HQ. Even manipulating 3 or 4 attributes at the same time, our model is able to edit only the specified attributes without influencing other information in the image.

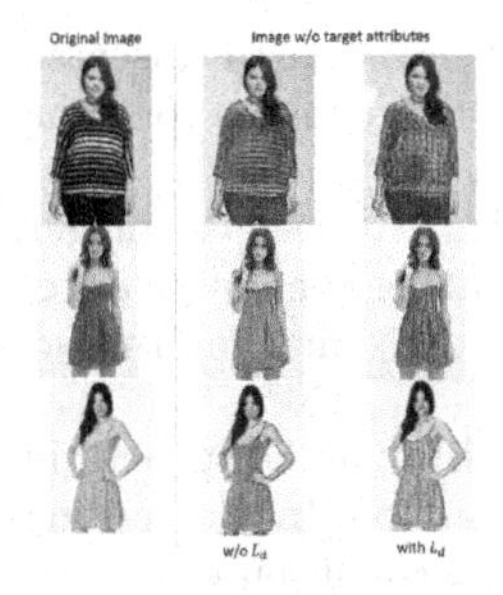

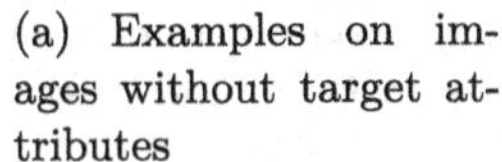

(a) Examples on images without target attributes

(b) Examples on controlling attribute strength

(c) Examples on controlling multiple attributes

**Fig. 8.** Visualizations used for model analysis in Sect. 4.5.

# 5 Conclusion

In this paper, we propose Attribute Information Removal and Reconstruction (AIRR) network for image editing. The attribute information removal and reconstruction module in AIRR produces an attribute excluded representation, eliminating sources of information hiding suffered by prior work. Results on four diverse datasets including DeepFashion Synthesis, DeepFashion Fine-grained Attribute, CelebA and CelebA-HQ, report that our model improves attribute manipulation accuracy and top-k retrieval rate by 10% on average over prior work. A user study also demonstrates that images with attributes manipulated with our approach are preferred in up to 76% of cases. One direction for future work is to explore controllable attribute manipulation in unsupervised setting.

**Acknowledgements.** This material is based upon work supported, in part, by DARPA under agreement number HR00112020054 and the National Science Foundation under Grant No. DBI-2134696. Any opinions, findings, and conclusions or recommendations expressed in this material are those of the author(s) and do not necessarily reflect the views of the supporting agencies.

# References

1. Abdal, R., Qin, Y., Wonka, P.: Image2stylegan++: how to edit the embedded images? In: Proceedings of the IEEE Conference on Computer Vision and Pattern Recognition (2020)
2. Ak, K.E., Lim, J.H., Sun, Y., Tham, J.Y., Kassim, A.A.: Fashionsearchnet-v2: learning attribute representations with localization for image retrieval with attribute manipulation. arXiv preprint. arXiv:2111.14145 (2021)
3. Ak, K.E., Lim, J.H., Tham, J.Y., Kassim, A.A.: Attribute manipulation generative adversarial networks for fashion images. In: Proceedings of the IEEE International Conference on Computer Vision (2019)
4. Bashkirova, D., Usman, B., Saenko, K.: Adversarial self-defense for cycle-consistent GANs. arXiv preprint. arXiv:1908.01517 (2019)

5. Burns, A., Sarna, A., Krishnan, D., Maschinot, A.: Unsupervised disentanglement without autoencoding: Pitfalls and future directions. arXiv preprint. arXiv:2108.06613 (2021)
6. Chen, X., Duan, Y., Houthooft, R., Schulman, J., Sutskever, I., Abbeel, P.: Infogan: interpretable representation learning by information maximizing generative adversarial nets. In: Proceedings of the International Conference on Neural Information Processing Systems (2016)
7. Choi, Y., Choi, M., Kim, M., Ha, J., Kim, S., Choo, J.: StarGAN: unified generative adversarial networks for multi-domain image-to-image translation. In: Proceedings of the IEEE Conference on Computer Vision and Pattern Recognition (2018)
8. Choi, Y., Uh, Y., Yoo, J., Ha, J.W.: StarGAN v2: diverse image synthesis for multiple domains. In: Proceedings of the IEEE Conference on Computer Vision and Pattern Recognition (2020)
9. Härkönen, E., Hertzman, A., Lehtinen, J., Paris, S.: GANSpace: discovering interpretable GAN controls. In: Proceedings of the IEEE Conference on Neural Information Processing Systems (2020)
10. He, K., Zhang, X., Ren, S., Sun, J.: Deep residual learning for image recognition. In: Proceedings of the IEEE Conference on Computer Vision and Pattern Recognition (2016)
11. He, Z., Zuo, W., Kan, M., Shan, S., Chen, X.: AttGAN: facial attribute editing by only changing what you want. IEEE Trans. Image Process. **28**(11), 5464–5478 (2019)
12. Hou, Y., Vig, E., Donoser, M., Bazzani, L.: Learning attribute-driven disentangled representations for interactive fashion retrieval. In: Proceedings of the IEEE International Conference on Computer Vision (2021)
13. Hu, Q., Szabó, A., Portenier, T., Favaro, P., Zwicker, M.: Disentangling factors of variation by mixing them. In: Proceedings of the IEEE Conference on Computer Vision and Pattern Recognition (2018)
14. Karras, T., Laine, S., Aila, T.: A style-based generator architecture for generative adversarial networks. In: Proceedings of the IEEE Conference on Computer Vision and Pattern Recognition (2019)
15. Kwak, J., Han, D.K., Ko, H.: CAFE-GAN: arbitrary face attribute editing with complementary attention feature. In: Vedaldi, A., Bischof, H., Brox, T., Frahm, J.-M. (eds.) ECCV 2020. LNCS, vol. 12359, pp. 524–540. Springer, Cham (2020). https://doi.org/10.1007/978-3-030-58568-6_31
16. Kwon, Y., Petrangeli, S., Kim, D., Wang, H., Swaminathan, V., Fuchs, H.: Tailor me: ln editing network for fashion attribute shape manipulation. In: Proceedings of the IEEE Winter Conference on Applications of Computer Vision (2022)
17. Lee, C.H., Liu, Z., Wu, L., Luo, P.: MaskGAN: towards diverse and interactive facial image manipulation. In: Proceedings of the IEEE Conference on Computer Vision and Pattern Recognition (2020)
18. Lezama, J.: Overcoming the disentanglement vs reconstruction trade-off via Jacobian supervision. In: International Conference on Learning Representations (2018)
19. Li, P., Xu, Y., Wei, Y., Yang, Y.: Self-correction for human parsing. IEEE Trans. Pattern Anal. Mach. Intell. (2020). https://doi.org/10.1109/TPAMI.2020.3048039
20. Liu, X., Thermos, S., Valvano, G., Chartsias, A., O'Neil, A., Tsaftaris, S.A.: Measuring the biases and effectiveness of content-style disentanglement. In: Proceedings of the British Machine Vison Conference (2021)
21. Liu, Z., Luo, P., Qiu, S., Wang, X., Tang, X.: DeepFashion: powering robust clothes recognition and retrieval with rich annotations. In: Proceedings of IEEE Conference on Computer Vision and Pattern Recognition (2016)

22. Liu, Z., Luo, P., Wang, X., Tang, X.: Deep learning face attributes in the wild. In: Proceedings of International Conference on Computer Vision (2015)
23. Locatello, F., et al.: Challenging common assumptions in the unsupervised learning of disentangled representations. In: International Conference on Machine Learning (2019)
24. Mao, X., Li, Q., Xie, H., Lau, R.Y., Wang, Z., Paul Smolley, S.: Least squares generative adversarial networks. In: Proceedings of the IEEE International Conference on Computer Vision (2017)
25. Ramesh, A., Choi, Y., LeCun, Y.: A spectral regularizer for unsupervised disentanglement. In: International Conference on Machine Learning (2018)
26. Shen, Y., Gu, J., Tang, X., Zhou, B.: Interpreting the latent space of GANs for semantic face editing. In: Proceedings of the IEEE Conference on Computer Vision and Pattern Recognition (2020)
27. Shin, M., Park, S., Kim, T.: Semi-supervised feature-level attribute manipulation for fashion image retrieval. In: Proceedings of the British Machine Vison Conference (2019)
28. Shoshan, A., Bhonker, N., Kviatkovsky, I., Medioni, G.: GAN-control: explicitly controllable GANs. In: Proceedings of the IEEE International Conference on Computer Vision (2021)
29. Szabo, A., Hu, Q., Portenier, T., Zwicker, M., Favaro, P.: Understanding degeneracies and ambiguities in attribute transfer. In: Proceedings of the European Conference on Computer Vision (2018)
30. Usman, B., Bashkirova, D., Saenko, K.: Disentangled unsupervised image translation via restricted information flow (2021)
31. Wang, R., et al.: Attribute-specific control units in StyleGAN for fine-grained image manipulation. In: Proceedings of the ACM International Conference on Multimedia (2021)
32. Wu, Z., Lischinski, D., Shechtman, E.: StyleSpace analysis: disentangled controls for StyleGAN image generation. In: Proceedings of the IEEE Conference on Computer Vision and Pattern Recognition (2021)
33. Yang, G., Fei, N., Ding, M., Liu, G., Lu, Z., Xiang, T.: L2M-GAN: learning to manipulate latent space semantics for facial attribute editing. In: Proceedings of the IEEE Conference on Computer Vision and Pattern Recognition (2021)
34. Yang, X., Song, X., Han, X., Wen, H., Nie, J., Nie, L.: Generative attribute manipulation scheme for flexible fashion search. In: Proceedings of the International ACM SIGIR Conference on Research and Development in Information Retrieval (2020)
35. Yao, X., Newson, A., Gousseau, Y., Hellier, P.: A latent transformer for disentangled face editing in images and videos. In: Proceedings of the IEEE International Conference on Computer Vision (2021)
36. Yu, C., Wang, J., Peng, C., Gao, C., Yu, G., Sang, N.: Bisenet: bilateral segmentation network for real-time semantic segmentation. In: Proceedings of the European Conference on Computer Vision (2018)
37. Zhu, J.Y., Park, T., Isola, P., Efros, A.A.: Unpaired image-to-image translation using cycle-consistent adversarial networks. In: Proceedings of the IEEE International Conference on Computer Vision (2017)

# BLT: Bidirectional Layout Transformer for Controllable Layout Generation

Xiang Kong[2(✉)], Lu Jiang[1], Huiwen Chang[1], Han Zhang[1], Yuan Hao[1], Haifeng Gong[1], and Irfan Essa[1,3]

[1] Google, Mountain View, USA
lujiang@google.com
[2] LTI, Carnegie Mellon University, Pittsburgh, USA
xiangk@cs.cmu.edu
[3] Georgia Institute of Technology, Atlanta, USA

**Abstract.** Creating visual layouts is a critical step in graphic design. Automatic generation of such layouts is essential for scalable and diverse visual designs. To advance conditional layout generation, we introduce BLT, a bidirectional layout transformer. BLT differs from previous work on transformers in adopting non-autoregressive transformers. In training, BLT learns to predict the masked attributes by attending to surrounding attributes in two directions. During inference, BLT first generates a draft layout from the input and then iteratively refines it into a high-quality layout by masking out low-confident attributes. The masks generated in both training and inference are controlled by a new hierarchical sampling policy. We verify the proposed model on six benchmarks of diverse design tasks. Experimental results demonstrate two benefits compared to the state-of-the-art layout transformer models. First, our model empowers layout transformers to fulfill controllable layout generation. Second, it achieves up to 10x speedup in generating a layout at inference time than the layout transformer baseline. Code is released at https://shawnkx.github.io/blt.

**Keywords:** Design · Layout creation · Transformer · Non-autoregressive

## 1 Introduction

Graphic layout dictates the placement and sizing of graphic components, playing a central role in how viewers interact with the information provided [24]. Layout generation is emerging as a new research area with a focus of generating realistic and diverse layouts to facilitate design tasks. Recent works show promising progress for various applications such as graphic user interfaces [2,18],

X. Kong—Work done during their research internship at Google.

**Supplementary Information** The online version contains supplementary material available at https://doi.org/10.1007/978-3-031-19790-1_29.

© The Author(s), under exclusive license to Springer Nature Switzerland AG 2022
S. Avidan et al. (Eds.): ECCV 2022, LNCS 13677, pp. 474–490, 2022.
https://doi.org/10.1007/978-3-031-19790-1_29

presentation slides [13], magazines [43,45], scientific publications [1], commercial advertisements [14,24,36], computer-aided design [41], indoor scenes [4], layout representations [28,42], *etc.*.

Previous work explores neural models for layout generation using Generative Adversarial Networks (GANs) [10,25] or Variational Autoencoder (VAEs) [19, 22,24,34]. Currently, layout transformers hold the state-of-the-art performance for layout generation [1,15]. These transformers represent a layout as a sequence of objects and an object as a (sub)sequence of attributes (See Fig. 1a). Layout transformers predict the attribute sequentially based on previously generated output (*i.e.* autoregressive decoding). Like other vision tasks, by virtue of the powerful self-attention [38], transformer models yield superior quality and diversity than GAN or VAE models for layout generation [1,15].

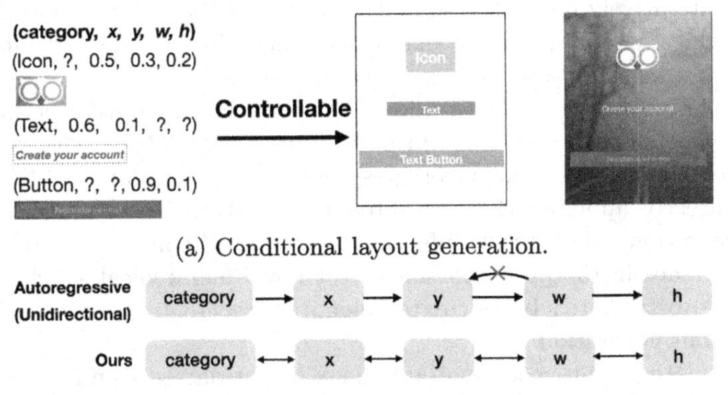


(a) Conditional layout generation.

(b) Unidirectional autoregressive (top) and non-autoregressive (bottom) decoding.

**Fig. 1. (a) Conditional layout generation**. Each object is modeled by 5 attributes 'category', '$x$', '$y$', '$w$' (width) and '$h$' (height). In conditional generation, attributes are partially given by the user and the goal is to generate the unknown attributes, *e.g.* putting the icon or button on the canvas. **(b) Illustration of immutable dependency chain in autoregressive decoding**.

Unlike Layout VAE (or GAN) models that are capable of generating layouts considering user requirements, layout transformers, however, have difficulties in conditional generation as a result of an acknowledged limitation discussed in [15] (*c.f.* order of primitives). Figure 1a illustrates a scenario in which a designer has objects with partially known attributes and hopes to generate the missing attributes. Specifically, each object is modeled by five attributes 'category', '$x$', '$y$', '$w$' (width) and '$h$' (height). The designer wants the layout model to 1) place the "icon" and "button" with known sizes onto the canvas (*i.e.* generating $x$, $y$ from $w$, $h$, and 'category'), and 2) determines the size of the centered "text object" (*i.e.* generating $w$, $h$ from $x$, $y$, and 'category').

Such functionality is currently missing in the layout transformers [1,15] due to *immutable dependency chain.* This is because autoregressive transformers follow a pre-defined generation order of object attributes. As shown in Fig. 1b, attributes must be generated starting from the category $c$, then $x$ and $y$, followed by $w$ and $h$. The dependency chain is immutable *i.e.* it cannot be changed at decoding time. Therefore, autoregressive transformers fail to perform conditional layout generation when the condition disagrees with the pre-defined dependency, *e.g.* generating position $y$ from the known width $w$ in Fig. 1b.

In this work, we introduce Bidirectional Layout Transformer (or BLT) for controllable layout generation. Different from the traditional transformer models [1,15], BLT enables controllable layout generation where every attribute in the layout can be modified, with high flexibility, based on the user inputs (*c.f.* Fig. 1a). During training, BLT learns to predict the masked attributes by attending to attributes in two directions (*c.f.* Fig. 2a). At inference time, BLT adopts a non-autoregressive decoding algorithm to refine the low-confident attributes iteratively into a high-quality layout (*c.f.* Fig. 2b). We propose a simple hierarchical sampling policy that is used both in training and inference to guide the mask generation over attribute groups.

BLT eliminates a critical limitation in the prior layout transformer models [1, 15] that prevents transformers from performing controllable layout generation. Our model is inspired by the autoregressive work in NLP [5,9,11,12]. However, we find directly applying the non-autoregressive translation models [9,23] to layout generation only leads to inferior results than the autoregressive baseline. Our novelty lies in the proposed simple yet novel hierarchical sampling policy, which, as substantiated by our experiments in Sect. 5.4, is essential for high-quality layout generation.

We evaluate the proposed method on six layout datasets under various metrics. These datasets cover representative design applications for graphic user interface [2], magazines [45] and publications [46], commercial ads [24], natural scenes [27] and home decoration [8]. Experiments demonstrate two benefits to several strong baseline models [1,15,19,19]. First, our model empowers transformers to fulfill controllable layout generation and thereby outperforms the previous conditional models based on VAE (*i.e.*, LayoutVAE [19] and NDN [24]). Even though our model is not designed for unconditional layout generation, it achieves quality on-par with the state-of-the-art. Second, our new method reduces the time complexity in [1,15] while achieving 4x–10x speedups in layout generation.

To summarize, we make the following contributions:

1. We address a critical limitation in state-of-the-art layout transformers [1,15] and hence empower transformers to fulfill controllable layout generation.
2. Though our idea is inspired by the autoregressive work in NLP [5,9,11,12], a novel hierarchical mask sampling policy is introduced in training and decoding, which is essential for high-quality layout generation.
3. Extensive experiments validate that our method performs favorably against state-of-the-art models in terms of realism, alignment, and semantic relevance on six diverse layout benchmarks.

## 2 Related Work

*Layout Synthesis:* Recently, automatic generation of high-quality and realistic layouts has fueled increasing interest. Unlike early work [6,7,29,31–33,35,39,40,44], recent data-driven methods rely on deep generative models such as GANs [10] and VAE [22]. For example, LayoutGAN [25] uses a GANs-based framework to synthesize semantic and geometric properties for scene elements. During inference time, LayoutGAN generates layouts from the Gaussian noise. Afterwards, LayoutGAN is extended to attribute-conditioned design tasks [26]. LayoutVAE [19] introduces two conditional VAEs. The first aims to learn the distribution of category counts which will be used during layout generation. The second produces layouts conditioning on the number and category of objects generated from the first VAE or ground-truth data. Recently, various VAE models are proposed [20,24,34]. Among them, Neural Design Networks (NDN) [24] is a competitive VAEs-based model for conditional layout generation, which focuses on modeling the asset relations and constraints by graph convolution. Our work is different from LayoutVAE and NDN in modeling layout and user inputs by the transformer, which, as shown in Table 1, perform more favorably thanks to the transformer architecture. Our finding is consistent with [1] where Arroyo *et al.* find VAEs underperforming transformers for unconditional layout generation [1].

Currently, the state-of-the-art for layout generation is held by the transformer models [38]. In particular, [15] employs the standard autoregressive Transformer decoder with unidirectional attention. They find out that self-attention is able to explicitly learn relationships between objects in the layout, resulting in superior quality compared to prior work. Furthermore, to increase the diversity of generated layout, [1] incorporates the standard autoregressive Transformer decoder into a VAE framework and [30] employs multi-choice prediction and winner-takes-all loss. Despite the superior performance, this work addresses a critical limitation acknowledged in [15] that prevents transformers from performing controllable layout generation. Following LayoutGAN [20,25] proposes a Transformer based layout GAN model, LayoutGAN++. In this framework, the input is a set of asset labels and randomly generated code and the output is the location and size of these asset. Different from the LayoutGAN++, the input to our proposed model is more flexible and can support unconditional generation and various types of conditional generation tasks.

*Bidirectional Transformer and Non-autoregressive Decoding:* The classic Transformer [38] decoder uses the unidirectional self-attention mechanism to generate the sequence token-by-token from left to right, leaving the right-to-left contexts unexploited. Several NLP works [9,23,37] are proposed to investigate language generation tasks by non-autoregressive generation with bidirectional Transformers, which allow representations to attend in both directions [3]. However, non-autoregressive decoding process leads to an apparent performance drop compared to the autoregressive decoding algorithm [9,11,12]. In this work, we finds that applying the non-autoregressive NLP model [9] to layout generation also leads to inferior results than the autoregressive baseline. To this end, we propose a simple yet effective hierarchical sampling policy which is essential for high-quality layout generation.

## 3 Problem Formulation

Following [15], we use 5 attributes to describe an object, *i.e.*, $(c, x, y, w, h)$, in which the first element $c \in C$ is the object category such as the logo or button, and the remainder details the bounding box information *i.e.* the center location $(x, y) \in \mathbb{R}^2$ and the width and height $(w, h) \in \mathbb{R}^2$. Furthermore, float values in bounding box information is discretized using 8-bit uniform quantization. For instance, the $x$-coordinate after the quantization becomes $\{x | x \in \mathbb{Z}, 0 \leq x \leq 31\}$. A layout $l$ of $K$ assets is hence denoted as a flattened sequence of integer indices:

$$l = [\langle \text{bos} \rangle, c_1, x_1, y_1, w_1, h_1, c_2 \cdots, h_K, \langle \text{eos} \rangle] \tag{1}$$

where $\langle \text{bos} \rangle$ and $\langle \text{eos} \rangle$ are special tokens to denote the start and the end of sequence. We use a shared vocabulary and represent each element in $l$ as an integer index or equivalently as a one-hot vector with the same length. It is trivial to extend the attribute dimension to model more complex layouts.

*Issues.* To train the model, prior work [1,15] estimates the joint likelihood of observing a layout as $p(l) = \prod_{i=1}^{|l|} p(l_i | l_{1:i})$.

During training, an autoregressive Transformer model is learned to maximize the likelihood using ground-truth attribute as input (*i.e.* teacher forcing). At inference time, the transformer model predicts the attribute sequentially based on previously generated output (*i.e.* autoregressive decoding), starting from the begin-of-sequence or $\langle \text{bos} \rangle$ token until yielding the end-of-sequence token $\langle \text{eos} \rangle$. The generation must follow a fixed conditional dependency. For example, Eq. (1) defines an immutable generation order $x \rightarrow y \rightarrow w \rightarrow h$. And in order to generate the height $h$ for an object, one must know its $x$-$y$ coordinates and width $w$.

There are two issues with autoregressive decoding for the conditional generation. First, it is infeasible to process user conditions that differ from the dependency order used in training. For instance, the model using Eq. (1) is not able to generate $x$-$y$ coordinates from width and height, which corresponds to a practical example of placing an object with given size. This issue is exacerbated by complex layouts that require more attributes to represent an object. Second, the autoregressive inference is not parallelizable, rendering it inefficient for the dense layout with a large number of objects or attributes.

## 4 Approach

Our goal is to design a transformer model for controllable layout generation. We propose a method to learn non-autoregressive transformers. Unlike existing layout transformers [1,15], the new layout transformer is bidirectional and can generate all attributes simultaneously in parallel, which allows not only for flexible conditional generation but also more efficient inference. In this section, we first discuss the model and training objective; then detail a novel hierarchical sampling policy for training and parallel decoding.

### 4.1 Model and Training

The BLT backbone is the multi-layer bidirectional Transformer encoder [38] as shown in Fig. 2. We use the identical architecture as in the existing autoregressive layout transformers [1,15] but a bidirectional attention mechanism.

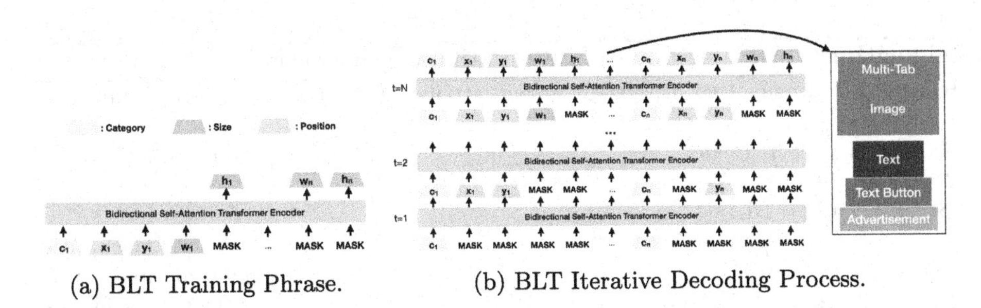


(a) BLT Training Phrase.

(b) BLT Iterative Decoding Process.

**Fig. 2.** The training (left) and decoding (right) stages of the proposed Bidirectional Layout Transformer (BLT).

Inspired by BERT [3], during training, we randomly select a subset of attributes in the input sequence, replace them with a special "[MASK]" token, and optimize the model to predict the masked attributes. For a layout sequence $l$, let $\mathcal{M}$ denote a set of masked positions. Replacing attributes in $l$ with "[MASK]" at $\mathcal{M}$ yields the masked sequence $l^{\mathcal{M}}$.

Given a layout set $\mathcal{D}$, the training objective is to minimize the negative log-likelihood of the masked attributes:

$$\mathcal{L}_{mask} = - \mathop{\mathbb{E}}_{l \in \mathcal{D}} \Big[ \sum_{i \in \mathcal{M}} \log p(l_i | l^{\mathcal{M}}) \Big], \tag{2}$$

The masking strategy greatly affects the quality of the masked language model [3]. BERT [3] applies random masking with a fixed ratio where a constant 15% masks are randomly generated for each input. Similarly, we find masking strategy is important for layout generation, but the random masking used in BERT does not work well. We propose to use a new sampling policy. Specifically, we divide the attributes of an object into semantic groups, *e.g.* Fig. 2 showing 3 groups: category, position, and size. First, we randomly select a semantic group. Next, we dynamically sample the number of masked tokens from a uniform distribution between one and the number of attributes belonging to the chosen group, and then randomly mask that number of tokens in the selected group. As such, it is guaranteed that the model only predicts attributes of the same semantic meaning each time. Therefore, given the hierarchical relations between these groups, we call this method as the hierarchical sampling. We will discuss how to apply the hierarchical sampling policy to decoding in the next subsection.

### 4.2 Parallel Decoding by Iterative Refinement

In BLT, all attributes in the layout are generated simultaneously in parallel. Since generating layouts in a single pass is challenging [9], we employ a parallel language model. The core idea is to generate a layout iteratively in a small number of steps where parallel decoding is applied at each.

**Algorithm 1.** Decoding by Iterative Attribute Refinement

**Require:** Sequence $l$ with partially-known attributes. Constant $T$ for the number of iterations.

1: **for** $g$ in $[C, S, P]$ **do** ▷ Loop over semantic group
2: **for** $i \leftarrow 1$ to $T/3$ **do**
3: $p, l^i = \text{BLT}(l)$
4: $\gamma_i = \frac{T-3i}{T}$ ▷ Compute mask ratio
5: $n_i = \lfloor \gamma_i \times |g| \rfloor$ ▷ $|g|$: # attributes in $g$
6: $\mathcal{M} = \arg_{k=n_i} \text{top-k}(-p)$ ▷ Get mask indices
7: Obtain $l$ by masking $l^i$ with respect to $\mathcal{M}$
8: **end for**
9: **end for**
10: **return** $l$

Algorithm 1 presents the non-autoregressive decoding algorithm. The procedure is also illustrated in Fig. 2b. The input to the decoding algorithm is a mixture sequence of known and unknown attributes, where the known attributes are given by the user inputs, and the model aims at generating the unknown attributes denoted by the [MASK] token. Like in training, we employ the hierarchical sampling policy to generate attributes of three semantic groups: category ($C$), size ($S$), and position ($P$). For each iteration, one group of attributes is sampled. In Step 3 of Algorithm 1, the model makes parallel predictions for all unknown attributes, where $p$ denotes the prediction scores. Step 6 samples the attributes that belong to the selected group and have the lowest prediction scores. Finally, it masks low-confident attributes on which the model has doubts. The prediction probabilities from the softmax layer are used as the confidence scores. These masked attributes will be re-predicted in the next iteration of decoding conditioning on all other ascertained attributes so far. The masking ratio calculated in Step 4 decreases with the number of iterations. This process will repeat $T$ times until all attributes of all objects are generated (*c.f.* Fig. 2b).

Our model is inspired by the autoregressive models in NLP [9,23]. It is noteworthy that Algorithm 1 differs from the non-autoregressive NLP models [9,23] in the proposed hierarchical sampling. This paper finds applying [9,23] to layout generation only leads to inferior results than our autoregressive baseline. We hypothesize that it is because layout attributes, unlike natural language, have apparent structures, and the non-autoregressive models designed for word sequences [9,23] might not sufficiently capture the complex correlation between layout attributes. We empirically demonstrate that Algorithm 1 outperforms our non-autoregressive NLP baselines in Sect. 5.4.

Algorithm 1 can be extended to unconditional generation. In this case, the input is a layout sequence of only "[MASK]" tokens, and the same algorithm is used to generate all attributes in the layout. Unlike conditional generation, we need to know the sequence length in advance, *i.e.* the number of objects to be generated. Here, we can use the prior distribution obtained on the training dataset. During decoding, we obtain the number of objects through sampling from this prior distribution.

## 5 Experimental Results

This section verifies the proposed method on six diverse layout benchmarks under various metrics to examine realism, alignment, and semantic relevance. The results show our model performs favorably against the strong baselines and achieves a 4x–10x speedup than autoregressive decoding in layout generation.

### 5.1 Setups

*Datasets.* We employ six datasets that cover representative graphic design applications. *RICO* [2] is a dataset of user interface designs for mobile applications. It contains 91K entries with 27 object categories (button, toolbar, list item, *etc.*). *PubLayNet* [46] contains 330K examples of machine annotated scientific documents crawled from the Internet. Its objects come from 5 categories: text, title, figure, list, and table. *Magazine* [45] contains 4K images of magazine pages and six categories (texts, images, headlines, over-image texts, over-image headlines, backgrounds). *Image Ads* [24] is the commercial ads dataset with layout annotation detailed in [24]. *COCO* [27] contains ∼100K images of natural scenes. We follow [1] to use the Stuff variant, which contains 80 things and 91 stuff categories, after removing small bounding boxes ($\leq 2\%$ image area), and instances tagged as "iscrowd". *3D-FRONT* [8] is a repository of professionally designed indoor layouts. It contains around 7K room layouts with objects belonging to 37 categories, *e.g.*, the table and bed. Different from previous datasets, objects in 3D-FRONT are represented by 3D bounding boxes. The maximum number of objects in our experiments is 25 in the RICO dataset and 22 in the PubLayNet dataset.

*Evaluation Metrics.* We employ five common metrics in the literature as well as a user study to validate the proposed method's effectiveness. Specifically, *IOU* measures the intersection over the union between the generated bounding boxes. We use an improved perceptual IOU (see more discussions in the Appendix). *Overlap* [25] measures the total overlapping area between any pair of bounding boxes inside the layout. *Alignment* [24] computes an alignment loss with the intuition that objects in graphic design are often aligned either by center or edge. *FID* [16] measures the distributional distance of the generated layout to the real layout. Following [24], we compute FID using a binary layout classifier to discriminate real layouts. We employ a 2-layer Transformer to train the classifier. Notice that the lower, the better for all IOU, Overlap, Alignment, and FID.

The above metrics ignore the input condition. For conditional generation, we employ a metric called *Similarity* [34] and a user study, where the former compares the generated layout with the ground-truth layout under the same input. Following [34], *DocSim* is used to calculate the similarity between two layouts. The user study is used to further evaluate human's perception about the conditionally-generated layouts.

*Generation Settings.* We examine three layout generation scenarios (2 conditional and 1 unconditional).

- Conditional on **Category**: only object categories are given by users. The model needs to predict the size and position of each object.
- Conditional on **Category + Size**: the object category and size are specified. The model needs to predict the positions, *i.e.* placing objects on the canvas.
- **Unconditional** Generation: no information is provided by users. Prior layout transformer work focuses on this setting.

In unconditional generation, the model generates 1K samples from the random seed. The test split of each dataset is used for conditional generation.

*Implementation Details.* The model is trained for five trials with random initialization and the averaged metrics with standard deviations are reported. All models including ours have the same configuration, *i.e.*, 4 layers, 8 attention heads, 512 embedding dimensions and 2,048 hidden dimensions. Adam optimizer [21] with $\beta_1 = 0.9$ and $\beta_1 = 0.98$ is used. Models are trained on 2×2 TPU devices with batch size 64. For conditional generation, we randomly shuffle objects in the layout. For unconditional generation, to improve diversity, we use the nucleus sampling [17] with $p = 0.9$ for the baseline Transformers and the top-k sampling ($k = 5$) for our model. Greedy decoding method is used for conditional generation. Please refer to the Appendix for more detailed hyper-parameter configurations.

### 5.2 Quantitative Comparison

*Conditional Generation.* The results are shown in Table 1 and Table 2. State-of-the-art layout transformers are compared *i.e.* **LayoutTransformer (Trans.)** [15] and **Variational Transformer Network (VTN)** [1]. In addition, two representative VAEs for conditional generation: **LayoutVAE (L-VAE)** [19] and **Neural Design Network (NDN)** [24] are also compared on the large datasets of RICO and PubLayNet. Two conditional generation tasks are examined *i.e.* Conditioned on Category and Conditioned on Category + Size (Column "+ Size"). The same model is used for both conditional cases and "-" indicates the baseline models fail to process the condition "Category + Size". The results are aggregated on independently trained models, where the mean and standard deviation over five trails are reported.

**Table 1.** Conditional layout generation on two settings (Category and Category+ Size) on the large datasets of RICO and PubLayNet.

| RICO | Conditioned on Category | | | | | + Size | |
|---|---|---|---|---|---|---|---|
| Model | IOU↓ | Overlap↓ | Alignment↓ | FID↓ | Sim.↑ | Sim.↑ | FID↓ |
| L-VAE [19] | 0.41±1.5% | 0.39±2.3% | 0.38±1.9% | 122±19 | 0.13±1.5% | 0.19 | 76 |
| NDN [24] | 0.37±1.7% | 0.36±1.9% | 0.41±1.6% | 97±21 | 0.15±2.3% | 0.21 | 63 |
| Trans. [15] | 0.31±0.2% | 0.33±0.8% | 0.30±0.8% | 76±24 | 0.20±0.1% | – | – |
| VTN [1] | **0.30**±0.1% | 0.30±0.3% | 0.32±0.9% | 82±23 | 0.20±0.1% | – | – |
| Ours | **0.30**±0.4% | **0.23**±0.2% | **0.20**±1.1% | **70**±29 | **0.21**±0.2% | **0.30** | **26** |
| PubLayNet | Conditioned on Category | | | | | + Size | |
| Model | IOU↓ | Overlap↓ | Alignment↓ | FID↓ | Sim.↑ | Sim.↑ | FID↓ |
| L-VAE | 0.45±1.3% | 0.15±0.9% | 0.37±0.7% | 513±26 | 0.07±0.3% | 0.09 | 239 |
| NDN | 0.34±1.8% | 0.12±0.8% | 0.39±0.4% | 425±37 | 0.06±0.3% | 0.09 | 178 |
| Trans. | **0.19**±0.3% | 0.06±0.3% | 0.33±0.3% | **127**±29 | **0.11**±0.1% | – | – |
| VTN | 0.21±0.6% | 0.06±0.2% | 0.33±0.4% | 159±21 | 0.10±0.1% | – | – |
| Ours | **0.19**±0.2% | **0.04**±0.1% | **0.25**±0.7% | 134±24 | **0.11**±0.2% | **0.18** | **87** |

**Table 2.** Category (+ Size) conditional layout generation on four datasets.

| COCO | Conditioned on Category | | + Size |
|---|---|---|---|
| Model | IOU↓ | Sim.↑ | Sim.↑ |
| Trans. [15] | 0.60±0.4% | 0.20±0.2% | – |
| VTN [1] | 0.63±0.4% | 0.22±0.1% | – |
| Ours | **0.43**±0.5% | **0.24**±0.1% | **0.44** |
| Ads | Conditioned on Category | | + Size |
| Model | IOU↓ | Sim.↑ | Sim.↑ |
| Trans. [15] | 0.19±0.1% | 0.30±0.1% | – |
| VTN [1] | 0.18±0.2% | 0.30±0.1% | – |
| Ours | **0.10**±0.4% | **0.31**±0.1% | **0.41** |

| Magazine | Conditioned on Category | | + Size |
|---|---|---|---|
| Model | IOU↓ | Sim.↑ | Sim.↑ |
| Trans. | 0.20±0.8% | 0.15±0.3% | – |
| VTN | **0.18**±1.8% | 0.15±0.9% | – |
| Ours | **0.18**±0.6% | **0.18**±0.4% | **0.27** |

| 3D-FRONT | Conditioned on Category | + size |
|---|---|---|
| Model | Sim.↑ | Sim.↑ |
| Trans. | 0.04±0.7% | – |
| VTN | 0.04±0.4% | – |
| Ours | **0.06**±0.7% | **0.10** |

Because of the non-autoregressive decoding, our model is able to conduct conditional generation on category + size while the baseline transformer models (Trans. [15] and VTN [1]) fail. Our model also outperforms VAE-based conditional layout models (L-VAE [19] and NDN [24]) across all metrics in Table 1 by statistically significant margins. This result is consistent with the prior finding in [1] that transformers outperform VAEs for unconditional layout generation.

*Unconditional Generation.* Although our model is not designed for this task, we compare it to the models [1,15,19] on unconditional layout generation. From

**Table 3.** Unconditional layout generation comparison to the state-of-the-art on three benchmarks. Results of baselines are cited from [1] and our scores are calculated following the same method described in [1].

| Methods | RICO | | | PubLayNet | | | COCO | | |
|---|---|---|---|---|---|---|---|---|---|
| | IOU↓ | Overlap↓ | Alignment↓ | IOU↓ | Overlap↓ | Alignment↓ | IOU↓ | Overlap↓ | Alignment↓ |
| LayoutVAE [19] | 0.193 | 0.400 | 0.416 | 0.171 | 0.321 | 0.472 | 0.325 | 2.819 | 0.246 |
| Trans. [15] | 0.086 | 0.145 | 0.366 | 0.039 | 0.006 | 0.361 | 0.194 | 1.709 | 0.334 |
| VTN [1] | 0.115 | 0.165 | 0.373 | 0.031 | 0.017 | 0.347 | 0.197 | 2.384 | 0.330 |
| Ours | 0.127 | 0.102 | 0.342 | 0.048 | 0.012 | 0.337 | 0.227 | 1.452 | 0.311 |

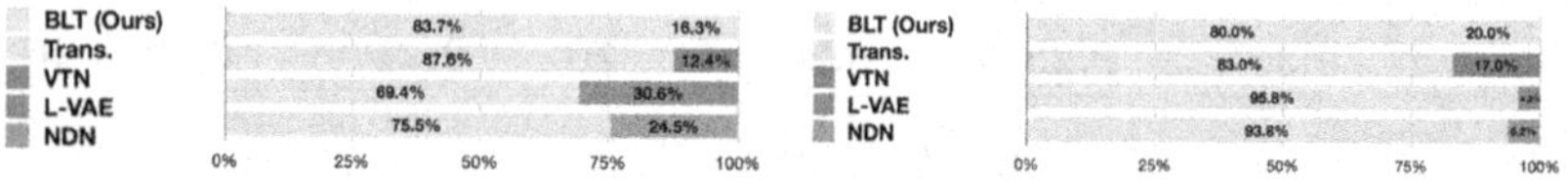

**Fig. 3.** We conduct a user study to compare the quality of generated samples from our model and baseline models on RICO (left) and PubLayNet (Right).

Table 3, our model outperforms LayoutVAE [19] and achieves comparable performance with two autoregressive transformers (Trans. [15] and VTN [1]).

*User Study.* We conduct user studies on RICO and PubLayNet to assess generated layouts for conditional generation. We randomly select 50 generated layouts under both conditional settings specified in Sect. 5.1 and collect their golden layouts. For each trial, we present Amazon Mechanical Turk workers two layouts generated by different methods along with the golden layout for reference, and ask "which layout is more similar to the true reference layout?". There are 75 unique workers participating in the study. Qualitative comparison is shown in the Appendix. The results, which are plotted in Fig. 3, verify that the proposed model outperforms all baseline models for conditional layout generation.

### 5.3 Qualitative Result

We show some generated layouts, along with the rendered examples for visualization, in Fig. 4. The setting is conditional generation on category and size for three design applications, including the mobile UI interface, scientific paper, and magazine. We observe that our method yields reasonable layouts, which facilitates generating high-quality outputs by rendering.

Next, we explore the home design task on the 3D-Front dataset [8]. The goal is to place the furniture with the user-given category and length, height, and width information. Examples are shown in Fig. 5. Unlike previous tasks, the model needs to predict the position of the 3D bounding box. The result suggests the feasibility of our method extending to 3D object attributes. The low similarity score on this dataset indicates that housing design layout is still a challenging task that needs future research.

To further understand what relationships between attributes BLT has learned, we visualize the patterns in how our model's attention heads behave.

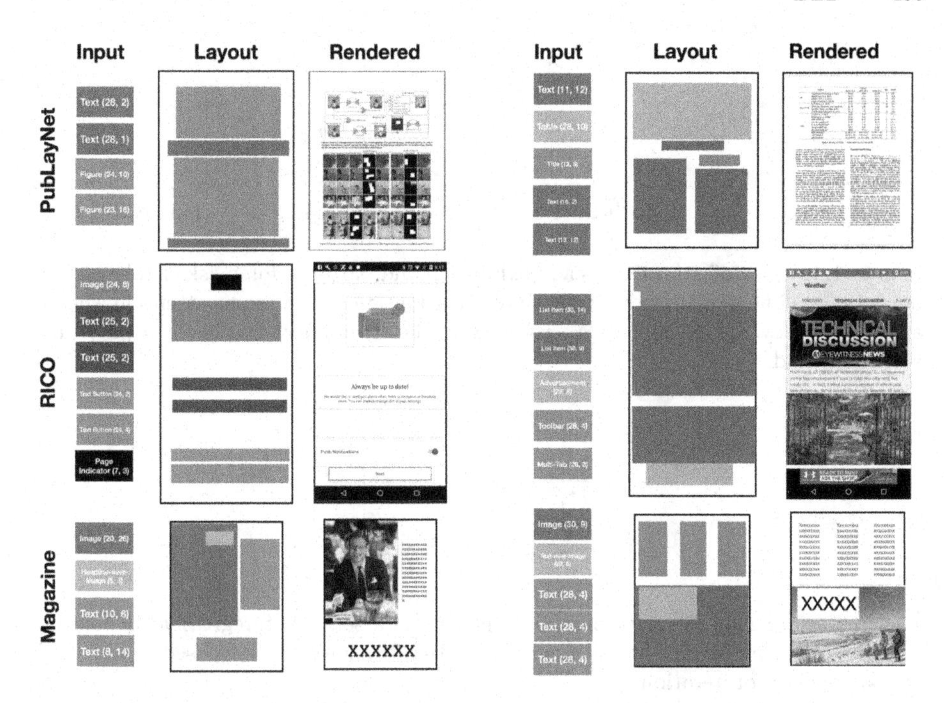


**Fig. 4.** Conditional layout generation for scientific papers, user interface, and magazine. The user inputs are the object category and their size (width, height). We present the rendered examples constructed based on the generated layouts.

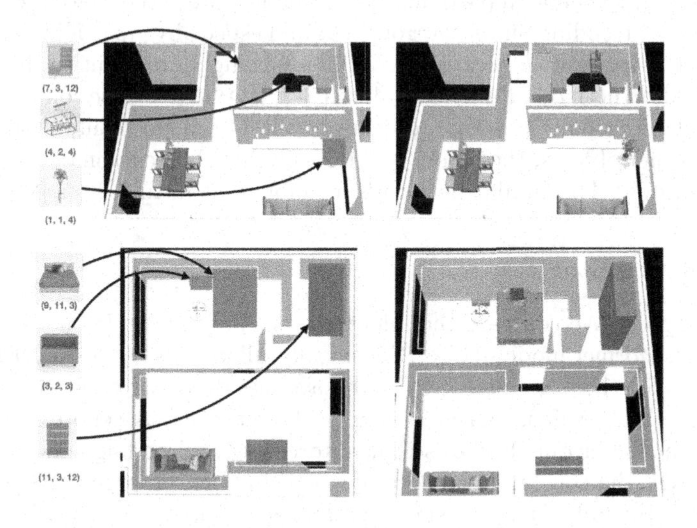


**Fig. 5.** 3D-FRONT sample layouts.

We choose a simple layout with two objects and mask their positions $(x, y)$. The model needs to predict these masked attributes from other known attributes.

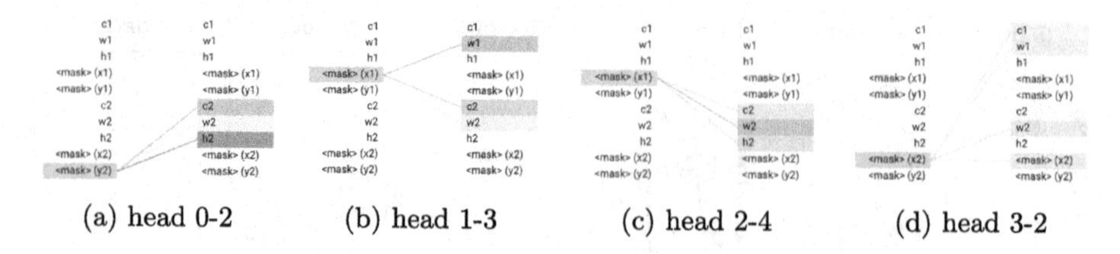


**Fig. 6.** Examples of attention heads exhibiting the patterns for masked tokens. The darkness of a line indicates the strength of the attention weight (some attention weights are so low they are invisible). We use ⟨layer⟩-⟨head number⟩ to denote a particular attention head.

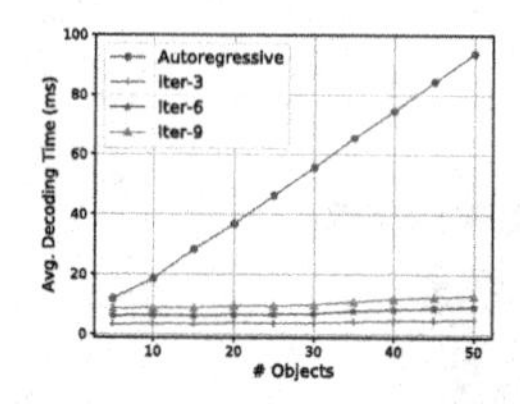


**Fig. 7.** Decoding speed versus number of generated assets. 'Autoregressive' denote the autogressive Transformer-based model [15]. 'Iter-*' shows the proposed model with various number of iterations.

Examples of heads exhibiting these patterns are shown in Fig. 6. We use ⟨layer⟩-⟨head number⟩ to denote a particular attention head. For the head 0–2, $[\text{MASK}]_{y_2}$ specializes to attending on its category (c2) and especially, its height information (h2), which is reasonable because $y$-coordinate is highly relevant to the height of the object. Furthermore, for heads 2-4 and 3-2, $[\text{MASK}]_{x_1}$ focuses on the width of not only the first but the second object as well. Given this contextual information from other objects, the model is able to predict the position of these objects more accurately. The similar pattern is also found at head 3-2 for $[\text{MASK}]_{x_2}$.

### 5.4 Ablation Study

*Decoding Speed.* We compare the inference speed of our model and the autoregressive transformer models [1,15]. Specifically, all models generate 1,000 layouts with batch size 1 on a single GPU. The average decoding time in millisecond is reported. The result is shown in Fig. 7, where the $x$-axis denotes the number of objects in the layout. It shows that autoregressive decoding time grows with #objects. On the contrary, the decoding speed of the proposed model appears not affected by #objects. The speed advantage becomes evident when producing dense layouts. For example, our fastest model obtains a 4× speedup when generating around 10 objects and a 10× speedup for 20 objects.

*Hierarchical Sampling.* This experiment investigates the effectiveness of the hierarchical sampling strategy used in training (Sect. 4.1) and non-autoregressive

**Table 4.** Comparison with the non-autoregressive method [9] in NLP on the RICO and PubLayNet datasets. Autoregressive results are included for reference. HSP denotes hierarchical sampling policy proposed in this work.

| RICO | IoU ↓ | Overlap↓ | Align.↓ | FID↓ | Sim. ↑ |
|---|---|---|---|---|---|
| Autoregressive [15] | **0.30** | 0.33 | 0.30 | 76 | 0.20 |
| Non-autoregressive [9] | 0.37 | 0.33 | 0.24 | 104 | 0.17 |
| Non-autoregressive + HSP (Ours) | **0.30** | **0.23** | **0.20** | **70** | **0.21** |
| PubLayNet | IoU ↓ | Overlap↓ | Align.↓ | FID↓ | Sim. ↑ |
| Autoregressive [15] | **0.19** | 0.06 | 0.33 | **127** | **0.11** |
| Non-autoregressive [9] | 0.16 | 0.12 | 0.32 | 217 | 0.09 |
| Non-autoregressive + HSP (Ours) | **0.19** | **0.04** | **0.25** | 134 | **0.11** |

**Table 5.** Layout generation results with different iteration group orders on the RICO dataset. C, S, and P denote category, size, and position attribute groups, respectively.

| Order | IoU ↓ | Overlap↓ | Alignment↓ |
|---|---|---|---|
| C→S→P | **0.127** | **0.102** | **0.342** |
| C→P→S | 0.129 | 0.107 | 0.344 |
| S→C→P | 0.147 | 0.109 | 0.351 |
| S→P→C | 0.162 | 0.121 | 0.357 |

decoding (Sect. 4.2). Specifically, we compare with the non-autoregressive method [9] in NLP on the large datasets of RICO and PubLayNet in Table 4. Autoregressive transformer results [15] are also included for reference but notice that autoregressive methods [15] have difficulties with conditional generation.

The results in Table 4 show that the non-autoregressive baseline yields inferior results than the autoregressive one. We hypothesize that it is because the non-autoregressive models designed for word sequences [9,23] might not sufficiently capture the apparently-structural correlation between layout attributes. The proposed method with hierarchical sampling significantly outperforms the non-autoregressive NLP baseline, which suggests the necessity of the proposed hierarchical sampling strategy. We also explore the effect of hierarchical sampling order. In Algorithm 1, we prespecify an order of attribute groups, *i.e.*, Category (C) → Size (S) → Position (P). Here, more orders are explored in Table 5. It seems better to first generate the category and afterward determine either location or size.

## 6 Conclusion and Future Work

We present BLT, a bidirectional layout transformer capable of empowering the transformer-based models to carry out conditional and controllable layout generation. Moreover, we propose a hierarchical sampling policy during BLT training and inference processes which has been shown to be essential for producing high-quality layouts. Thanks to the high computation parallelism, BLT achieves 4–10 times speedup compared to the autoregressive transformer baselines during inference. Experiments on six benchmarks show the effectiveness and flexibility of BLT. A limitation of our work is content-agnostic generation. We leave this out to have a fair and lateral comparison to our baselines which do not use visual information either. In the future, we will explore using rich visual information.

**Acknowledgement.** The authors would like to thank all anonymous reviewers and area chairs for helpful comments.

## References


1. Arroyo, D.M., Postels, J., Tombari, F.: Variational transformer networks for layout generation. In: CVPR (2021)
2. Deka, B., et al.: Rico: A mobile app dataset for building data-driven design applications. In: UIST (2017)
3. Devlin, J., Chang, M.W., Lee, K., Toutanova, K.: Bert: Pre-training of deep bidirectional transformers for language understanding. In: NAACL (2019)
4. Di, X., Yu, P.: Multi-agent reinforcement learning of 3d furniture layout simulation in indoor graphics scenes. arXiv preprint. arXiv:2102.09137 (2021)
5. Donahue, J., Dieleman, S., Binkowski, M., Elsen, E., Simonyan, K.: End-to-end adversarial text-to-speech. In: ICLR (2020)
6. Fan, H., Su, H., Guibas, L.J.: A point set generation network for 3d object reconstruction from a single image. In: Proceedings of the IEEE Conference on Computer Vision and Pattern Recognition, pp. 605–613 (2017)
7. Fisher, M., Ritchie, D., Savva, M., Funkhouser, T., Hanrahan, P.: Example-based synthesis of 3d object arrangements. ACM Trans. Graph. (TOG) **31**(6), 1–11 (2012)
8. Fu, H., et al.: 3d-front: 3d furnished rooms with layouts and semantics. arXiv preprint. arXiv:2011.09127 (2020)
9. Ghazvininejad, M., Levy, O., Liu, Y., Zettlemoyer, L.: Mask-predict: parallel decoding of conditional masked language models. In: EMNLP-IJCNLP (2019)
10. Goodfellow, I., et al.: Generative adversarial nets. NeurIPS, vol. 27 (2014)
11. Gu, J., Bradbury, J., Xiong, C., Li, V.O., Socher, R.: Non-autoregressive neural machine translation. In: ICLR (2018)
12. Gu, J., Kong, X.: Fully non-autoregressive neural machine translation: tricks of the trade. In: Findings of ACL-IJCNLP (2021)
13. Guo, M., Huang, D., Xie, X.: The layout generation algorithm of graphic design based on transformer-cvae. arXiv preprint. arXiv:2110.06794 (2021)
14. Guo, S., Jin, Z., Sun, F., Li, J., Li, Z., Shi, Y., Cao, N.: Vinci: an intelligent graphic design system for generating advertising posters. In: Proceedings of the 2021 CHI Conference on Human Factors in Computing Systems, pp. 1–17 (2021)
15. Gupta, K., Lazarow, J., Achille, A., Davis, L.S., Mahadevan, V., Shrivastava, A.: Layouttransformer: layout generation and completion with self-attention. In: ICCV (2021)
16. Heusel, M., Ramsauer, H., Unterthiner, T., Nessler, B., Hochreiter, S.: Gans trained by a two time-scale update rule converge to a local nash equilibrium. NeurIPS (2017)

17. Holtzman, A., Buys, J., Du, L., Forbes, M., Choi, Y.: The curious case of neural text degeneration. In: ICLR (2019)
18. Jiang, Z., Sun, S., Zhu, J., Lou, J.G., Zhang, D.: Coarse-to-fine generative modeling for graphic layouts (2022)
19. Jyothi, A.A., Durand, T., He, J., Sigal, L., Mori, G.: Layoutvae: stochastic scene layout generation from a label set. In: CVPR (2019)
20. Kikuchi, K., Simo-Serra, E., Otani, M., Yamaguchi, K.: Constrained graphic layout generation via latent optimization. In: MM (2021)
21. Kingma, D.P., Ba, J.: Adam: a method for stochastic optimization. arXiv preprint. arXiv:1412.6980 (2014)
22. Kingma, D.P., Welling, M.: Auto-encoding variational bayes. arXiv preprint arXiv:1312.6114 (2013)
23. Kong, X., Zhang, Z., Hovy, E.: Incorporating a local translation mechanism into non-autoregressive translation. arXiv preprint. arXiv:2011.06132 (2020)
24. Lee, H.Y., et al.: Neural design network: graphic layout generation with constraints. In: Vedaldi, A., Bischof, H., Brox, T., Frahm, J.-M. (eds.) ECCV 2020. LNCS, vol. 12348, pp. 491–506. Springer, Cham (2020). https://doi.org/10.1007/978-3-030-58580-8_29
25. Li, J., Yang, J., Hertzmann, A., Zhang, J., Xu, T.: Layoutgan: generating graphic layouts with wireframe discriminators. In: ICLR (2018)
26. Li, J., Yang, J., Zhang, J., Liu, C., Wang, C., Xu, T.: Attribute-conditioned layout gan for automatic graphic design. IEEE Trans. Visual Comput. Graphics **27**(10), 4039–4048 (2020)
27. Lin, T.Y., et al.: Microsoft COCO: common objects in context. In: Fleet, D., Pajdla, T., Schiele, B., Tuytelaars, T. (eds.) ECCV 2014. LNCS, vol. 8693, pp. 740–755. Springer, Cham (2014). https://doi.org/10.1007/978-3-319-10602-1_48
28. Manandhar, D., Ruta, D., Collomosse, J.: Learning structural similarity of user interface layouts using graph networks. In: Vedaldi, A., Bischof, H., Brox, T., Frahm, J.-M. (eds.) ECCV 2020. LNCS, vol. 12367, pp. 730–746. Springer, Cham (2020). https://doi.org/10.1007/978-3-030-58542-6_44
29. Merrell, P., Schkufza, E., Li, Z., Agrawala, M., Koltun, V.: Interactive furniture layout using interior design guidelines. ACM Trans. Graph. (TOG) **30**(4), 1–10 (2011)
30. Nguyen, D.D., Nepal, S., Kanhere, S.S.: Diverse multimedia layout generation with multi choice learning. In: MM (2021)
31. O'Donovan, P., Lībeks, J., Agarwala, A., Hertzmann, A.: Exploratory font selection using crowdsourced attributes. ACM Trans. Graph. (TOG) **33**(4), 1–9 (2014)
32. O'Donovan, P., Agarwala, A., Hertzmann, A.: Learning layouts for single-pagegraphic designs. IEEE Trans. Visual Comput. Graphics **20**(8), 1200–1213 (2014)
33. Pang, X., Cao, Y., Lau, R.W., Chan, A.B.: Directing user attention via visual flow on web designs. ACM Trans. Graph. (TOG) **35**(6), 1–11 (2016)
34. Patil, A.G., Ben-Eliezer, O., Perel, O., Averbuch-Elor, H.: Read: recursive autoencoders for document layout generation. In: CVPR Workshops (2020)
35. Qi, S., Zhu, Y., Huang, S., Jiang, C., Zhu, S.C.: Human-centric indoor scene synthesis using stochastic grammar. In: Proceedings of the IEEE Conference on Computer Vision and Pattern Recognition, pp. 5899–5908 (2018)
36. Qian, C., Sun, S., Cui, W., Lou, J.G., Zhang, H., Zhang, D.: Retrieve-then-adapt: example-based automatic generation for proportion-related infographics. IEEE TVCG **27**(2), 443–452 (2020)

37. Stern, M., Chan, W., Kiros, J., Uszkoreit, J.: Insertion transformer: flexible sequence generation via insertion operations. In: ICML, pp. 5976–5985. PMLR (2019)
38. Vaswani, A., et al.: Attention is all you need. In: NeurIPS (2017)
39. Wang, K., Lin, Y.A., Weissmann, B., Savva, M., Chang, A.X., Ritchie, D.: Planit: Planning and instantiating indoor scenes with relation graph and spatial prior networks. ACM Trans. Graph. (TOG) **38**(4), 1–15 (2019)
40. Wang, K., Savva, M., Chang, A.X., Ritchie, D.: Deep convolutional priors for indoor scene synthesis. ACM Trans. Graph. (TOG) **37**(4), 1–14 (2018)
41. Willis, K.D., Jayaraman, P.K., Lambourne, J.G., Chu, H., Pu, Y.: Engineering sketch generation for computer-aided design. In: CVPR (2021)
42. Xie, Y., Huang, D., Wang, J., Lin, C.Y.: Canvasemb: learning layout representation with large-scale pre-training for graphic design. In: Proceedings of the 29th ACM International Conference on Multimedia, pp. 4100–4108 (2021)
43. Yamaguchi, K.: Canvasvae: learning to generate vector graphic documents. In: Proceedings of the IEEE/CVF International Conference on Computer Vision, pp. 5481–5489 (2021)
44. Yu, L.F., Yeung, S.K., Tang, C.K., Terzopoulos, D., Chan, T.F., Osher, S.J.: Make it home: automatic optimization of furniture arrangement. In: ACM Transactions on Graphics (TOG)-Proceedings of ACM SIGGRAPH 2011, vol. 30, no. 4 (2011). Article no. 86
45. Zheng, X., Qiao, X., Cao, Y., Lau, R.W.: Content-aware generative modeling of graphic design layouts. TOG **38**(4), 1–15 (2019)
46. Zhong, X., Tang, J., Yepes, A.J.: Publaynet: largest dataset ever for document layout analysis. In: IEEE ICDAR (2019)

# Diverse Generation from a Single Video Made Possible

Niv Haim(✉), Ben Feinstein, Niv Granot, Assaf Shocher, Shai Bagon, Tali Dekel, and Michal Irani

Weizmann Institute of Science, Rehovot, Israel
niv.haim@weizmann.ac.il
https://nivha.github.io/vgpnn

**Abstract.** GANs are able to perform generation and manipulation tasks, trained on a single video. However, these single video GANs require unreasonable amount of time to train on a single video, rendering them almost impractical. In this paper we question the necessity of a GAN for generation from a single video, and introduce a non-parametric baseline for a variety of generation and manipulation tasks. We revive classical space-time patches-nearest-neighbors approaches and adapt them to a scalable unconditional generative model, without any learning. This simple baseline surprisingly outperforms single-video GANs in visual quality and realism (confirmed by quantitative and qualitative evaluations), and is disproportionately faster (runtime reduced from several days to seconds). Other than diverse video generation, we demonstrate other applications using the same framework, including video analogies and spatio-temporal retargeting. Our proposed approach is easily scaled to Full-HD videos. These observations show that the classical approaches, if adapted correctly, significantly outperform heavy deep learning machinery for these tasks. This sets a new baseline for single-video generation and manipulation tasks, and no less important – makes diverse generation from a single video practically possible for the first time.

## 1 Introduction

Generation and editing of natural videos remain challenging, mainly due to their large dimensionality and the enormous space of motion they span. Most modern frameworks train generative models on a large collection of videos, producing high quality results for only a limited class of videos. These include extensions of GANs [23] to video data [2,36,48,58,62,66], video to video translation [8,16, 40,63–65,71] and autoregressive sequence prediction [3,6,7,17,22,59–61].

While externally-trained generative models produce impressive results, they are restricted to the types of video dynamics in their training set. On the other side of the spectrum are *single-video GANs*. These video generative models train on a *single* input video, learn its distribution of space-time patches, and are then

N. Haim and B. Feinstein—Equal contribution.

© The Author(s), under exclusive license to Springer Nature Switzerland AG 2022
S. Avidan et al. (Eds.): ECCV 2022, LNCS 13677, pp. 491–509, 2022.
https://doi.org/10.1007/978-3-031-19790-1_30

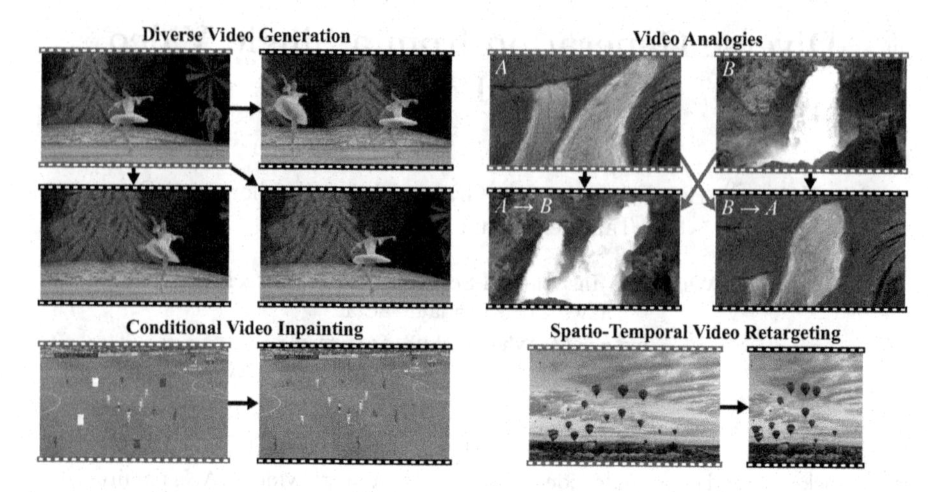


**Fig. 1.** We adapt classical patch-based approaches as a better, much faster non-parametric alternative to single video GANs, for a variety of video generation and manipulation tasks.

able to generate a diversity of new videos with the same patch distribution [5,25]. However, these take very long time to train for each input video, making them applicable to only small spatial resolutions and to very short videos (typically, very few small frames). Furthermore, their output oftentimes shows poor visual quality and noticeable visual artifacts. These shortcomings render existing single-video GANs impractical and unscalable.

Video synthesis and manipulation of a single video sequence based on its distribution of space-time patches dates back to classical pre-deep learning methods. These classical methods demonstrated impressive results for various applications, such as video retargeting [30,47,55,70], video completion [27,34,39,41,42,69], video texture synthesis [14,21,28,31–33] and more. With the rise of deep-learning, these methods gradually, perhaps unjustifiably, became less popular. Recently, Granot et al. [24] revived classical patch-based approaches for image synthesis, and was shown to significantly outperform *single-image* GANs in both run-time and visual quality.

In light of the above-mentioned deficiencies of single-video GANs, and inspired by [24], we propose a fast and practical method for video generation from a single video that we term VGPNN (*Video Generative Patch Nearest Neighbors*). In order to handle the huge amounts of space-time patches in a single video sequence, we use the classical fast approximate nearest neighbor search method PatchMatch by Barnes et al. [10]. By adding stochastic noise to the process, our approach can generate a large diversity of random different video outputs from a single input video in an unconditional manner.

Like single-video GANs, our approach enables the diverse and random generation of videos. However, in contrast to existing single-video GANs, we can generate *high resolution* videos, while reducing runtime by many orders of magnitude, thus making diverse unconditional video generation from a single video realistically possible for the first time.

In addition to diverse generation from a single video, by employing robust optical-flow based descriptors we use our framework to transfer the dynamics and motions between two videos with different appearance (which we call "video analogies"). We also show the applicability of our framework to spatio-temporal video retargeting and to conditional video inpainting.

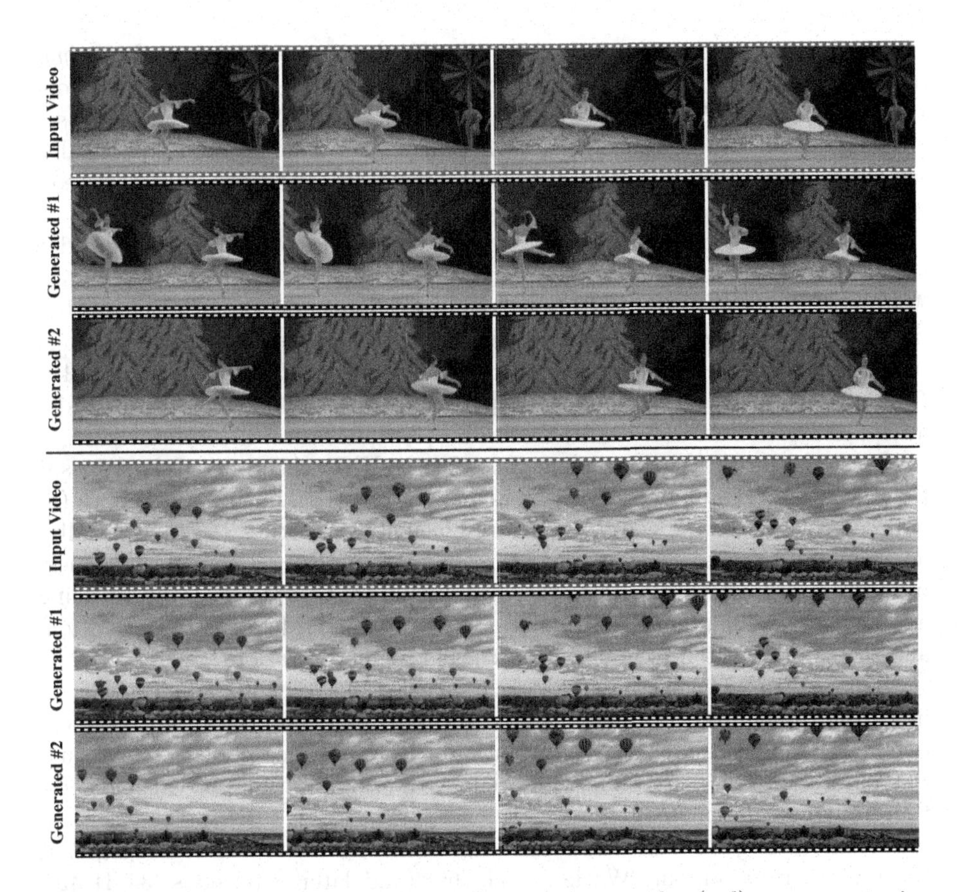


**Fig. 2.** Diverse Single Video Generation: Given an input video (red), we generate similarly looking videos (black) capturing both appearance of objects as well as their dynamics. Note the high quality of our generated videos. *Please watch the full resolution videos in the supplementary material.* (Color figure online)

To summarize, our contributions are as follows:

- We show that our space-time patch nearest-neighbors approach, despite its simplicity, outperforms single-video GANs by a large margin, both in runtime and in quality.

- Our approach is the first to generate diverse high resolution videos (spatial or temporal) from a single video.
- We demonstrate the applicability of our framework to other applications: video analogies, sketch-to-video, spatio-temporal video retargeting and conditional video inpainting.

## 2 Related Work

Classical video generation methods, many of whom inspired by similar *image* methods [19,20,67], include video texture synthesis [31–33], MRF-based controllable synthesis [51], flow-guided synthesis [14,28,33,43,49,50] and more (see surveys by [9,68]). While some used a generative model to model patch distribution, none of them considered unconditional generation of natural videos, beyond dynamic textures.

Classical methods typically involve comparing and matching of image/video patches. Efficient computation for such matching is therefore critical. *PatchMatch* [10] proposed a fast method for finding an approximate nearest-neighbor field (NNF) between patches of two images $A, B$. Namely, for each patch in image $A$, find its nearest neighbor in $B$. While solving NNF exhaustively takes $O(N^2)$ time ($N$ being the number of patches to match), PatchMatch provides an approximation in $O(N \log N)$ time.

PatchMatch starts by a random guess for the NNF, then iteratively refines it for each patch. The main observation is that since natural images are smooth (as opposed to e.g., noise), w.h.p, the nearest neighbour patches of two (spatially) adjacent patches are also adjacent. Therefore, each patch can refine its own guess by either "peeking" at its neighbor's guess, keeping the current guess or sampling a new guess. At random guess, w.h.p at least one patch has a correct solution, and this is propagated to adjacent patches in further iterations.

Being very efficient, PatchMatch allowed for many applications of image/video manipulation [9]. It was also extended for k-nearest neighbors search [11], faster search [12] and being differentiably learnable [18]. We use PatchMatch in order to dramatically reduce the running times of video generation from a single video.

The tasks of generation and inpainting are closely related. Both are required to "invent" new content. Wexler et al. [69] (and later extensions [34,41,42]) proposed a patch-based method for video completion. Missing space-time patches are replaced by their nearest neighbours from the rest of the video. This is also done in a multi-scale manner by computing a spatio-temporal pyramid. The task is first solved in the coarsest level, and the upsampled result is the initial guess for the next level in the pyramid. Our approach uses a different metric for patch similarity and much deeper spatio-temporal pyramids (higher down-scale ratio). More importantly, we focus on video generation and video analogies and their relation to recent deep learning methods.

# 3 Method

Our main task is to generate diverse video samples based on a single input video, such that the generated outputs have similar appearance and motions as the original input video, but are also visually different from one another. We want our model to operate on natural input videos that can vary in their appearance and dynamics. In order to capture both spatial and temporal information of a single video, we start by building a spatio-temporal pyramid and operate coarse-to-fine to capture the internal statistics of the input video at multiple scales (Fig. 3). This multi-scale approach is extensively used in classical image synthesis methods as well as in modern GANs [eg., [29], [52], [25]). At each scale we employ a Video-Patch-Nearest-Neighbor module (*VPNN*); VGPNN is in fact a sequence of VPNN layers. The inputs to each layer depend on the application, where we first focus on our main application of diverse video generation (see Sect. 5 for the specific details of the other applications).

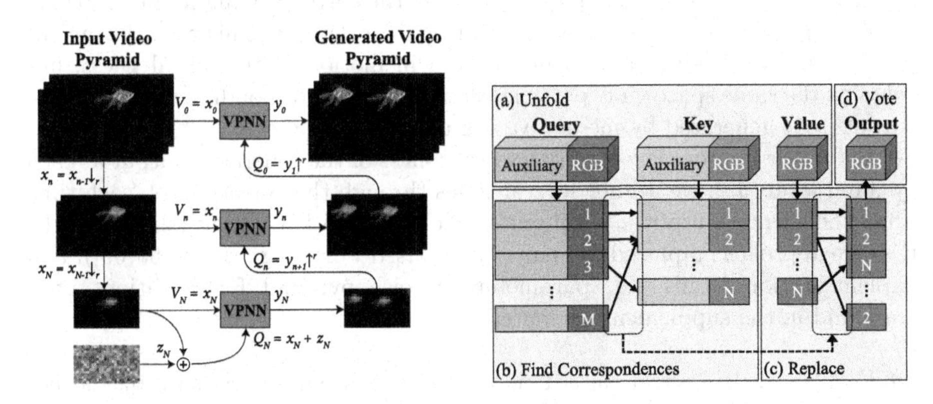


**Fig. 3.** VGPNN Architecture *Left*: given a single input video $x_0$, a spatio-temporal pyramid is constructed and an output video $y_0$ is generated coarse-to-fine. At each scale, VPNN module (*right*) is applied to transfer an initial guess $Q_n$ to the output $y_n$ which shares the same space-time patch distribution as the input $x_n$. At the coarsest scale, noise is injected to induce spatial and temporal randomness. *Right*: VPNN module gets as input query, key and value RGB videos (QKV respectively) and outputs an RGB video. Q and K can be concatenated to additional auxiliary channels. (a) Inputs are unfolded to patches (each position holds a concatenation of neighboring positions); (b) Each patch in Q finds its nearest neighbor patch in K. This is achieved by solving the NNF using PatchMatch [10]; (c) Each patch in Q is replaced with a patch from V, according to the correspondences found in stage (b); (d) Resulting patches are "folded" back to an RGB video output (using the *median* of all suggested votes).

*Multi-scale Approach:* Given an input video $x$, we construct a spatio-temporal pyramid $\{x_0 \ldots, x_N\}$, where $x_0 = x$, and $x_n = x_{n-1}{\downarrow_r}$ is a bicubically down-scaled version of $x_{n-1}$ by factor $r$ ($r = (r_H, r_W, r_T)$, where $r_H = r_W$ are the spatial factors and $r_T$ is the temporal factor, which can be different).

At the coarsest level, the input to the first VPNN layer is an initial coarse guess of the output video. This is created by adding random Gaussian noise $z_N$ to $x_N$. The noise $z_N$ promotes high diversity in the generated output samples from the single input. The global structure (e.g., a head is above the body) and global motion (e.g., humans walk forward), is prompted by $x_N$, where such structure and motion can be captured by *small space-time* patches. The std of the noise $z_n$ is much larger than that of $x_n$ and is related to the typical distance between neighbouring patches in the video (for full details see our supplementary).

Each space-time patch of the initial coarse guess ($x_N + z_N$) is then replaced with its nearest neighbor patch from the corresponding coarse input $x_N$. The coarsest-level output $y_N$ is generated by choosing at each space-time position the median of all suggestions from neighboring patches (known as "voting" or "folding").

At each subsequent level, the input to the VPNN layer is the bicubically-upscaled output of the previous layer ($y_{n+1} \uparrow^r$). Each space-time patch is replaced with its nearest neighbor patch from the corresponding input $x_n$ (using the same patch-size as before, now capturing finer details). This way, the output $y_n$ in each level is similar in structure and in motion to the initial guess, but contains the same space-time patch statistics of the corresponding input $x_n$. The output $y_n$ is generated by median voting as described above.

To further improve the quality and sharpness of the generated output at each pyramid level ($y_n$), we iterate several times through the current level, each time using the current output $y_n$ as input to the current VPNN layer (similar to the EM-like approach employed in many patch-based works e.g., [10,24,55,69]). Full implementation details (e.g., parameters of noise, pyramid, EM-iterations, etc.) are found in the supplementary material.

*QKV Scheme:* In several cases it is necessary to compare patches in another search space than the original RGB input space. To this end we adopt a QKV scheme (query, key and value, respectively) as used by [24]. We denote $V = x_n$ (the corresponding level from the pyramid of the original video) and $Q = y_{n+1} \uparrow$ (the upscaled output of previous layer). Note that since $Q$ is an *upscaled* version of previous output, its patches are blurry. Seeking their nearest neighbors in $V$ (whose patches are sharp) often results in improper matches. This is mitigated by setting $K = x_{n+1}\uparrow^r$ (in the first iteration of each level), which has a similar degree of blur/degradation as $Q$. After finding its match in $K_j$, each patch $Q_i$ is then replaced with a patch $V_j$ (where $i, j$ are spatio-temporal positions. Also note that $K$ and $V$ are of the same shape). The QKV scheme is especially important in our video analogies application where it is used to include additional temporal information in the queries and the keys. We discuss it in detail in Sect. 5.1.

*Completeness Score:* In the applications of video analogies, spatio-temporal video retargeting and conditional video inpainting we use the normalized similarity score [24] that encourages visual completeness. The score between a query patch $Q_i$ and a key patch $K_j$ is defined as:

$$S(Q_i, K_j) := \frac{1}{\alpha + \min_\ell D(Q_\ell, K_j)} D(Q_i, K_j) \tag{1}$$

where $D$ is mean square error, and $\alpha$ controls the degree of completeness (smaller $\alpha$ encourages more completeness). $S$ is essentially a weighted version of $D$, whose weights depend *globally* on $K$ and $Q$.

*Finding Correspondences:* We find the nearest neighbors between $Q$ and $K$ (Fig. 3right-b) using PatchMatch (Barnes et al. [10]). To cope with the completeness score, we apply PatchMatch twice. First we find a "rareness" score for the keys - for each *key* we find its closest *query*. Then, for each *query* we find its closest *key* while factoring in the rareness of the keys as weights in the PatchMatch search. Namely, we solve for:

$$\text{NNF}(\mathbf{p}) = \arg\min_{\mathbf{v}} W(\mathbf{p} + \mathbf{v}) \cdot D(Q(\mathbf{p}), K(\mathbf{p} + \mathbf{v})) \tag{2}$$

where $D$ is a distance function, $W$ are per-patch weights, $\mathbf{p} = (t, x, y)$ a position in $Q$ and $\mathbf{v} = (t', x', y')$ are possible NNF candidates (such as the NNF at the current position $\text{NNF}(t, x, y)$ or at a neighbor position $\text{NNF}(t, x - 1, y)$ in the propagation step).

This requires a slight modification of PatchMatch to support per-key weights. This additional support makes it possible to approximately solve Eq. 1 with two passes of PatchMatch. Even though this gives an approximation of Eq. 1, we do not suffer loss in quality or lack of completeness, as apparent from our results.

The algorithm is implemented on GPU using PyTorch [44], with time complexity $O(n \times d)$ and $O(n)$ additional memory (where $n$ is the video size and $d$ is the patch size; also see Fig. 5).

*Temporal Diversity and Consistency:* A simple but effective trick to enhance the temporal diversity of our samples is to generate outputs with less frames than in the input video. Generating samples with similar number of frames as in the input video result in outputs that are "in sync". Intuitively, the motion of the input video is the only motion that is coherent for this amount of frames. By generating shorter videos allow for shorter motions from different times in the input video, to occur simultaneously in the generated outputs (see for example how the generated dancers in Fig. 2 are not "synced"). We also found that the temporal consistency is best preserved in the generated output when the initial noise $z_N$ is randomized for each spatial position, but is the same (replicated) in the temporal dimension.

## 4 Experimental Results

In this section we evaluate and compare the performance of our main application – diverse video generation from a single input video. Figures 1 and 2 illustrate diverse videos generated from a single input video, all sharing the same space-time patch distribution. The diversity is both spatially (e.g., number of dancers and their positions are different from the input video) and temporally (generated dancers are not synced). **Please refer to the supplementary material** to view the full resolution videos and many more examples.

*Evaluation of Video Generation from a Single Video:* We compare our results to recently published methods for diverse video generation from single video: HP-VAE-GAN [25] and SinGAN-GIF [5]. We show that our results are both qualitatively and quantitatively superior while reducing the runtime by a factor of $3 \times 10^4$ (from 8 d training on one video to 18 s for new generated video). Since SinGAN-GIF did not make their code available, and the training time of HP-VAE-GAN for a single video is roughly 8 days, we are only able to compare to the videos published by these methods (we use all published videos).

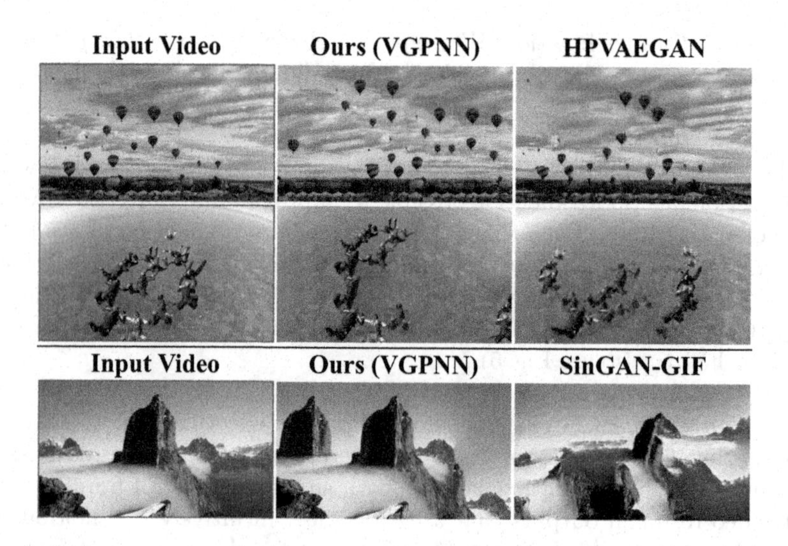


**Fig. 4.** Comparing Visual Quality between our generated frames and those of HP-VAE-GAN [25] and SinGAN-GIF [5] (please **zoom in** on the frames). Note that our generated frames are sharper and also exhibit more coherent and plausible arrangements of the scene. For details see Sect. 4. See supplementary for full videos and more comparisons.

*Evaluation Set:* "HP-VAE-GAN dataset" comprises of 10 input videos with 13 frames each, and of spatial resolution of 144×256 pixels. "SinGAN-GIF dataset" has 5 input videos with maximal resolution of 168×298 pixels and 8–16 frames.

*Qualitative Comparison:* In Fig. 4 we show a side-by-side comparison of representative generated frames of our method to frames generated by HP-VAE-GAN [25] and SinGAN-GIF [5]. Note that while [5,25] are limited to generated outputs of small resolution (144×256), we can generate outputs in the same resolution of the input video (full-HD 1280×1920, shown in the figure). The full videos (as well as a comparison to our generated outputs of similar low resolution) can be viewed in the supplementary material. As can be seen, our generated samples (in low and high resolution) are more spatially and temporally coherent, as well as having higher visual quality. It is evident that generating videos using the space-time patches of the original input video, rather than regressing output RGB values, gives rise to high quality outputs.

*Quantitative Comparison:* In Table 1 we report the Single-Video-FID (SVFID) [5] of our generated samples, compared to those generated by HP-VAE-GAN [25] and SinGAN-GIF [5][1]. SVFID was proposed by [25] to measure the patch statistics similarity between the input video and a generated video. It computes the Fréchet distance between the statistics of the input video and the generated video using pre-computed C3D [57] features (Lower SVFID is better). As can be seen in Table 1, our generated samples bear more substantial similarity to the input videos (indicated by lower SVFID). [52] proposed a diversity index to make sure that generated outputs are indeed different (and not simply "copying" the input). We adapt the index for videos. The index is zero if all generated outputs are the same, and higher otherwise. While our and HP-VAE-GAN generated samples have similar index (0.45/0.41 respectively), those of SinGAN-GIF have higher index (0.86 vs. our 0.6). Such high diversity is not an advantage, when paired with SVFID about twice worse than ours. It stems from low quality appearance with out-of-distribution patches. All inputs and generated videos can be found in the supplementary material.

*User Study:* We conducted a user study evaluation using Amazon Mechanical Turk (AMT). For each dataset, 100 subjects were shown multiple pairs of videos, each consisting of a video generated by our method, and a video generated by the other method (both were generated from the same input video). The subjects were asked to judge which sample is better in terms of sharpness, natural look and coherence. In Table 1 we report the percentage of users who favored our method over the other. Compared to videos generated from HP-VAE-GAN dataset, there is a clear preference in favor of our patch-based method. The results on the SinGAN-GIF dataset are not that clear-cut, this might be due to the somewhat restricted nature of the videos in that particular dataset (as mentioned above, it was not possible to check SinGAN-GIF on other samples, since the authors did not publish their code, nor stated the amount of time it took to generate their samples).

*Reducing Running Times:* In Fig. 5 we show a comparison of the runtime taken to generate random video samples using our method, compared to a naïve extension

[1] All quantitative comparisons were done on generated samples of the same resolution and video length as that of the other method.

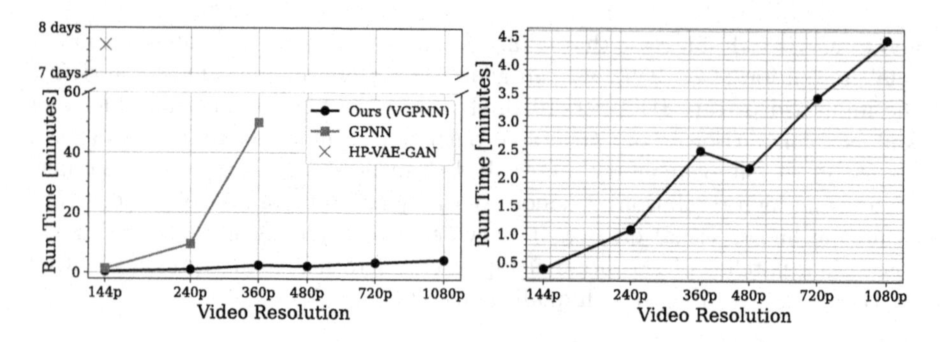


**Fig. 5.** Generation Runtime. **Left:** Comparing between our approach (VGPNN), a naïve extension of GPNN [24] from 2D to 3D and HP-VAE-GAN [25]. We compared the generation time of 13-frames videos with different spatial resolutions (X-axis). All videos have 16:9 aspect ratio (e.g., 144p is $144 \times 256$ and 1080p is $1080 \times 1920$ – full-HD). **Right:** Close-up of our generation run-time (black line in left). Our approach takes seconds/minutes to generate low-res/high-res video outputs. The drop in 480p is the result of decreasing the patch-size in finer-levels of high-res videos. See Sect. 4 for details, and supplementary for implementation details.

of GPNN [24] (from 2D to 3D patches) and compared to the training time of HP-VAE-GAN [25]. As discussed in Sect. 3, the use of efficient PatchMatch algorithm for nearest neighbors search, as opposed to the exhaustive search done in GPNN, dramatically reduces both run time and memory footprint used for video generation, making it possible to generate high-resolution videos (including Full-HD 1080p). All experiments were conducted on Quadro RTX 8000 GPU.

**Table 1.** Quantitative Evaluation: A comparison of our generated video samples to that of HP-VAE-GAN [25] and SinGAN-GIF [5], conducted on input videos provided in their papers. Our diverse samples have more resemblance to the input videos (indicated by lower SVFID). In a user study, users scored in favor of our method (see Sect. 4 for details).

| Method | SVFID [25] ↓ | Head-on comparison (User study)[%]↑ | Runtime ↓ |
|---|---|---|---|
| HP-VAE-GAN [25] | 0.0081 | **67.84** ± 1.77 | 7.625 d |
| **VGPNN** (Ours) | **0.0072** | | **18** s |
| SinGAN-GIF [5] | 0.0119 | **50.57** ± 3.27 | Unpublished |
| **VGPNN** (Ours) | **0.0058** | | **10** s |

## 5 Applications

Other than unconditional diverse generation, we demonstrate the utility of our framework on several other video manipulation applications.

### 5.1 Video Analogies

Video to video translation methods typically train on large datasets and are either conditioned on human poses or keypoint detection e.g. [15,40,63–65], or require knowledge of a human/animal model e.g. [1,16,37,46,53,54,71]. We show that when videos' dynamics are similar in both their motion and semantic context within their video, one can use our framework to transfer the motion and appearance between the two (see Fig. 6). We term this task "video analogies" (inspired by *image analogies* [13,26,38]). More formally, we generate a new video whose spatio-temporal layout is taken from a content video $C$, and overall appearance and dynamics from a style video $S$.

Our goal is to find a mapping of dynamic elements (3D patches) between the two videos, which can be very different in their appearance (RGB space). This is achieved by using the magnitude of the optical flow (extracted via RAFT [56]), quantized into few bins (using k-means)[2]. We term this the *dynamic structure* of the video. By concatenating the dynamic structure to the RGB values of the video (along the channels axis), each patch can now be compared using its RGB values and its dynamic values. This provides a good mapping between the dynamic elements of the two input videos.

We compute spatio-temporal pyramids from the style video $S$, as well as from the dynamic structure of the content video $\text{dyn}(C)$, and the dynamic structure the style video $\text{dyn}(S)$. The output video is generated by setting $Q, K, V$ at each level as follows:

| Level | $Q$ | $K$ | $V$ |
|---|---|---|---|
| N (coarsest) | $\text{dyn}(C)_N$ | $\text{dyn}(S)_N$ | $S_N$ |
| n (any other) | $\text{dyn}(C)_n \| Q_{n+1} \uparrow$ | $\text{dyn}(S)_n \| S_n$ | $S_n$ |

where $\|$ denotes concatenation along the channels axis, and $n$ denote the current level in the pyramid. Note that in the coarsest level, the two videos are only compared by their dynamic structure. In finer levels, the dynamic structure of $C$ (the content video) is used to "guide" the output to the desired spatio-temporal layout.

In Fig. 1 we show a snapshot of the analogies between a waterfall and a lava stream, and in Fig. 6 we show snapshots of the analogies of all possible pairs between three videos (the lava stream, a waterfall and a meat grinder). *The full videos are in the supplementary material.*

We can use the above mentioned mechanism for "sketch-to-video" transfer, where the dynamic structure is given by a sketch video instead of an actual video. See Fig. 6 for a few snapshots of transfering the motion of morphed MNIST [35] digits to a video of marching soldiers, and *please see the full videos and many more results in the supplementary material.*

Flow-based appearance transfer of fluids has been studied by [14,28,33,43, 49,50]. Most similar to us is [28] that uses a patch nearest neighbor approach

[2] Each cluster has an integer cluster index. We divide each index by the total number of clusters/bins to be in $[0, 1]$.

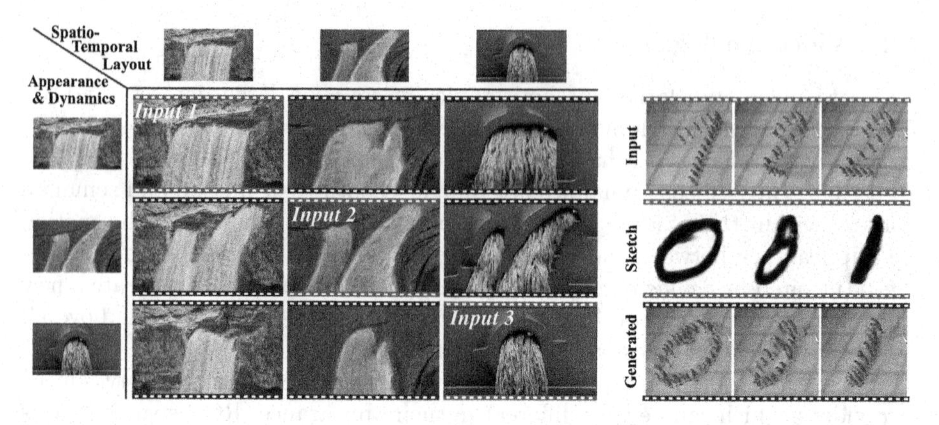


**Fig. 6.** Video Analogies: *Left:* an example of video analogies between all pairs of three input videos (red). Each generated video (black) takes the spatio-temporal layout from the input video in its row, and the appearance and dynamics of the input video from its column. *Right:* an example of sketch-to-video – the generated video (bottom) takes its spatio-temporal layout from the sketch video of morphed MNIST digits (middle) and its appearance and dynamics from the input video of parading soldiers (top). *Please find full videos and additional examples in the supplementary material.* (Color figure online)

to transfer the appearance of a fluid exemplar (a still image) into a video given a human annotated flow+alpha mask. Our method differs in how we model the flow guidance and in the mapping we have between two flows of two videos (instead of a still image exemplar). Also similar to us is Recycle-GAN by Bansal et al. [8] that pose unsupervised video-to-video translation as a domain transfer problem (each video is a domain). They train convolutional encoders to map between the two videos using adversarial loss with cyclic constraints.

In Fig. 7 we compare our results to [8]. As can be seen, [8] results generally fail to converge to the visual quality of the original inputs (partially due to the difficulty of training a parameteric model on small amounts of data), and in many cases converge to the input style video (probably due to the instability of training a GAN).

In Fig. 7 we show ablations for the main parameters used for video analogies. Figure 7(i) we show the role of using the completeness term $\alpha$ (see Eq. 1). No completeness term (equivalent to $\alpha \rightarrow \infty$) results in over smoothed outputs due to many patches in $Q$ being mapped to a similar patch in $K$. On the other hand, too "strong" completeness term results in many undesired visual details. We found $\alpha = 1$ to be a good balance for our results. In Fig. 7(ii) we ablate the use of quantized optical flow in the auxiliary channel. Without employing temporal features (RGB only) or using optical flow without quantizing to $[0-1]$, the resulting mapping fails to match similar dynamical elements to those with similar semantics in the context of their video. *Please find full videos in the supplementary material.*

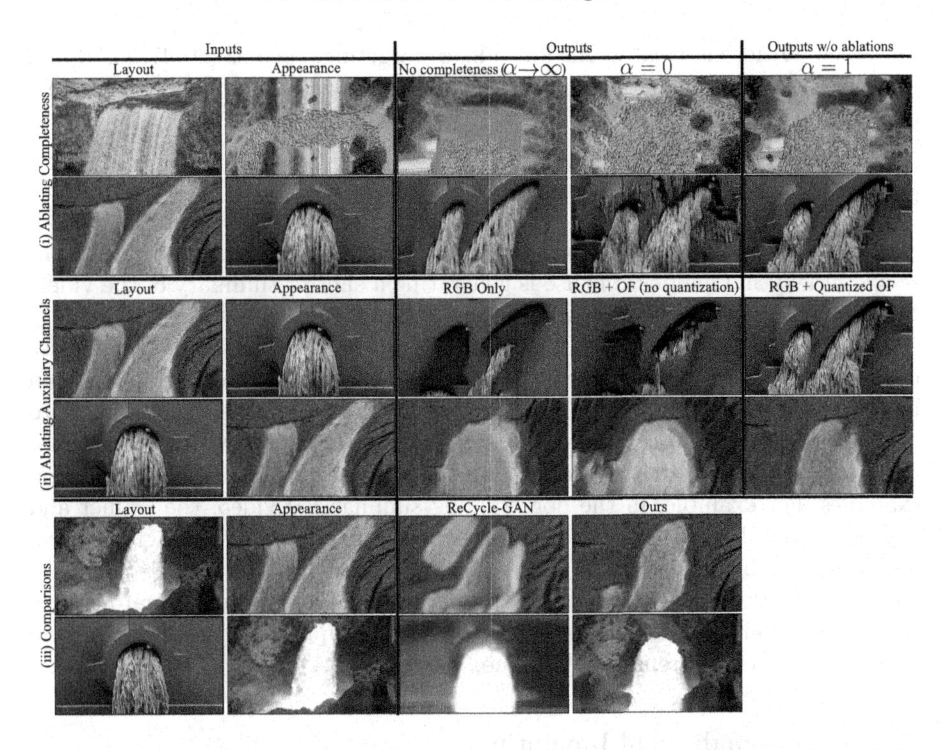


**Fig. 7.** Video Analogies: (i) Ablations for the choice of $\alpha$ (completeness score); (ii) Ablations of choice of auxiliary channel; (iii) Comparisons our results to that of ReCycle-GAN [8].

## 5.2 Spatial Retargeting

The goal of video retargeting is to change the dimensions of a video without distorting its visual contents (e.g., fit a portrait video to a wide screen display). It can be performed in a very similar manner to our video generation described in Sect.3. Given a target shape, we first resize (bicubically) the input video to the target shape, then compute two pyramids (for the input and resized videos) with the same depth and downscale factor. The initial guess at the coarsest level $Q_N$ would be the coarsest level of the resized pyramid (without any additional noise). We then compute the rest of the output video in the same manner as in Sect.3. Note that at each level, $V_n$ are unchanged, hence no distortion is introduced to the patches reconstructing the retargeted video.

As can be seen in Fig. 1 and in the supplementary material, the results preserve the original size and aspect ratio of objects from the input videos while keeping the overall appearance coherent even though the aspect ratio is significantly altered. The dynamics and motions in the videos are also preserved. For instance, the balloons are not "squashed" but rather packed more compactly in the sky and more members were added to the choir instead of stretching them. Nevertheless, the motion of the balloons or the sway of the choir members are preserved. Other classical works for video retargeting, such as [30,47,55,70] did not make their implementation available, therefore we were unable to provide a comparison.

### 5.3 Temporal Retargeting

Similar to spatial retargeting, one can generate a realistic video with a different *temporal* length. One possible use is generating a shorter summary of the video. While most deep video summarization techniques are achieved by selecting a subset of frames (see survey [4]), classical methods have demonstrated summaries that consist of *novel frames* in which dynamics that are originally sequential can be parallelized or vice versa [45,55]. By applying the retargeting approach to the temporal dimenstion, we are able to generate summaries with novel frames. The temporal retargeting section in the supplementary material shows several examples. For example, in the dog training summarized video, the trainer and dog turn around simultaneously as opposed to sequentially in the original video. Moreover, we can, in a similar manner, extend the temporal duration of a video creating longer dynamics while preserving the speed of the individual actions. In the ballet dancer video for example, the choreography is longer, but the pace of the dance motions remains the same.

### 5.4 Video Conditional Inpainting

In this task we are given an input video with some occluded space-time volume, where the missing parts should be completed based on crude color cues placed by the user in the occluded space (similar to conditional image inpainting [24]). Here we set the number of levels in the pyramid such that the occluded part in the coarsest level is roughly in the size of a single patch. The masked part is then coherently reconstructed using other space-time patches of similar colors to that of the cue. In finer levels, details and dynamic elements are added correctly. The conditional inpainting section in the supplementary material shows how different cues are completed with different elements from the non-occluded parts. See for instance, how a blue cue will be replaced by a player from Barcelona while a white cue by a player from Real Madrid. See more examples in the supplementary material.

## 6 Limitations

Generation of local patches lacks high-level semantic or global geometric understanding of the scene. For example, This is apparent when scenes with significant depth variations are introduced with large camera motion. While each frame is plausible, different patches are not being transformed consistently, resulting in non-rigid deformations to entities that are realistically rigid. See the generated videos of mountains in the supplementary.

## 7 Conclusion

We demonstrated that random diverse video generation from a single video can be efficiently done by simple patch-based methods. We also demonstrated how small modifications to our framework give rise to other tasks such as video analogies and spatio-temporal retargeting. We showed that our non-parametric approach outperforms existing single-video GANs in the visual quality of the generated outputs, while being orders of magnitude faster. The low run time required for generating videos using our approach makes it a good baseline for future works in the field.

**Acknowledgements.** This project received funding from the European Research Council (ERC) under the European Union's Horizon 2020 research and innovation programme (grant agreement No 788535), from the D. Dan and Betty Kahn Foundation, and from the Israel Science Foundation (grant 2303/20). Dr. Bagon is a Robin Chemers Neustein AI Fellow.

## References

1. Aberman, K., Weng, Y., Lischinski, D., Cohen-Or, D., Chen, B.: Unpaired motion style transfer from video to animation. ACM Trans. Graph. (TOG) **39**(4), 1–64 (2020)
2. Aigner, S., Körner, M.: Futuregan: anticipating the future frames of video sequences using spatio-temporal 3d convolutions in progressively growing gans. arXiv preprint arXiv:1810.01325 (2018)
3. Aksan, E., Hilliges, O.: Stcn: stochastic temporal convolutional networks. arXiv preprint arXiv:1902.06568 (2019)
4. Apostolidis, E., Adamantidou, E., Metsai, A.I., Mezaris, V., Patras, I.: Video summarization using deep neural networks: a survey. arXiv preprint arXiv:2101.06072 (2021)
5. Arora, R., Lee, Y.J.: Singan-gif: learning a generative video model from a single gif. In: Proceedings of the IEEE/CVF Winter Conference on Applications of Computer Vision, pp. 1310–1319 (2021)
6. Babaeizadeh, M., Finn, C., Erhan, D., Campbell, R.H., Levine, S.: Stochastic variational video prediction. arXiv preprint arXiv:1710.11252 (2017)
7. Ballas, N., Yao, L., Pal, C., Courville, A.: Delving deeper into convolutional networks for learning video representations. arXiv preprint arXiv:1511.06432 (2015)
8. Bansal, A., Ma, S., Ramanan, D., Sheikh, Y.: Recycle-gan: unsupervised video retargeting. In: Proceedings of the European Conference on Computer Vision (ECCV), pp. 119–135 (2018)
9. Barnes, C., Zhang, F.-L.: A survey of the state-of-the-art in patch-based synthesis. Comput. Vis. Media **3**(1), 3–20 (2017). https://doi.org/10.1007/s41095-016-0064-2
10. Barnes, C., Shechtman, E., Finkelstein, A., Goldman, D.B.: Patchmatch: a randomized correspondence algorithm for structural image editing. ACM Trans. Graph. **28**(3), 24 (2009)

11. Barnes, C., Shechtman, E., Goldman, D.B., Finkelstein, A.: The generalized patch-match correspondence algorithm. In: Daniilidis, K., Maragos, P., Paragios, N. (eds.) ECCV 2010. LNCS, vol. 6313, pp. 29–43. Springer, Heidelberg (2010). https://doi.org/10.1007/978-3-642-15558-1_3
12. Barnes, C., Zhang, F.-L., Lou, L., Xian, W., Shi-Min, H.: Patchtable: efficient patch queries for large datasets and applications. ACM Trans. Graph. (ToG) **34**(4), 1–10 (2015)
13. Benaim, S., Mokady, R., Bermano, A., Wolf, L.: Structural analogy from a single image pair. In: Computer Graphics Forum, vol. 40, pp. 249–265. Wiley Online Library (2021)
14. Bhat, K.S., Seitz, S.M., Hodgins, J.K., Khosla, P.K.: Flow-based video synthesis and editing. In: ACM SIGGRAPH 2004 Papers, pp. 360–363 (2004)
15. Blattmann, A., Milbich, T., Dorkenwald, M., Ommer, B.: ipoke: poking a still image for controlled stochastic video synthesis. In: Proceedings of the IEEE/CVF International Conference on Computer Vision, pp. 14707–14717 (2021)
16. Chan, C., Ginosar, S., Zhou, T., Efros, A.A.: Everybody dance now. In: Proceedings of the IEEE/CVF International Conference on Computer Vision, pp. 5933–5942 (2019)
17. Denton, E., Fergus, R.: Stochastic video generation with a learned prior. In: International Conference on Machine Learning, pp. 1174–1183. PMLR (2018)
18. Duggal, S., Wang, S., Ma, W. C., Hu, R., Urtasun, R.: Deeppruner: learning efficient stereo matching via differentiable patchmatch. In: Proceedings of the IEEE/CVF International Conference on Computer Vision, pp. 4384–4393 (2019)
19. Efros, A.A., Freeman, W.T.: Image quilting for texture synthesis and transfer. In: Proceedings of the 28th Annual Conference on Computer Graphics and Interactive Techniques, pp. 341–346 (2001)
20. Efros, A.A., Leung, T.K.: Texture synthesis by non-parametric sampling. In: Proceedings of the Seventh IEEE International Conference on Computer Vision, vol. 2, pp. 1033–1038. IEEE (1999)
21. Fišer, J., et al.: Stylit: illumination-guided example-based stylization of 3d renderings. ACM Trans. Graph. (TOG) **35**(4), 1–11 (2016)
22. Franceschi, J.-Y., Delasalles, E., Chen, M., Lamprier, S., Gallinari, P.: Stochastic latent residual video prediction. In: International Conference on Machine Learning, pp. 3233–3246. PMLR (2020)
23. Goodfellow, I.J., et al.: Generative adversarial networks. arXiv preprint arXiv:1406.2661 (2014)
24. Granot, N., Feinstein, B., Shocher, A., Bagon, S., Irani, M.: Drop the gan: in defense of patches nearest neighbors as single image generative models. In: Proceedings of the IEEE/CVF Conference on Computer Vision and Pattern Recognition, pp. 13460–13469 (2022)
25. Gur, S., Benaim, S., Wolf, L.: Hierarchical patch vae-gan: generating diverse videos from a single sample. arXiv preprint arXiv:2006.12226 (2020)
26. Hertzmann, A., Jacobs, C.E., Oliver, N., Curless, B., Salesin, D.H.: Image analogies. In: Proceedings of the 28th Annual Conference on Computer Graphics and Interactive Techniques, pp. 327–340 (2001)
27. Huang, J.-B., Kang, S.B., Ahuja, N., Kopf, J.: Temporally coherent completion of dynamic video. ACM Trans. Graph. (TOG) **35**(6), 1–11 (2016)
28. Jamriška, O., Fišer, J., Asente, P., Jingwan, L., Shechtman, E., Sỳkora, D.: Lazyfluids: appearance transfer for fluid animations. ACM Trans. Graph. (TOG) **34**(4), 1–10 (2015)

29. Karras, T., Aila, T., Laine, S., Lehtinen, J.: Progressive growing of gans for improved quality, stability, and variation. arXiv preprint arXiv:1710.10196, 2017
30. Krähenbühl, P., Lang, M., Hornung, A., Gross, M.: A system for retargeting of streaming video. In: ACM SIGGRAPH Asia 2009 papers (2009)
31. Kwatra, V., Schödl, A., Essa, I., Turk, G., Bobick, A.: Graphcut textures: Image and video synthesis using graph cuts. Acm Trans. Graph. (ToG) **22**(3), 277–286 (2003)
32. Kwatra, V., Essa, I., Bobick, A., Kwatra, N.: Texture optimization for example-based synthesis. In: ACM SIGGRAPH 2005 Papers, pp. 795–802 (2005)
33. Kwatra, V., Adalsteinsson, D., Kim, T., Kwatra, N., Carlson, M., Lin, M.: Texturing fluids. IEEE Trans. Visual Comput. Graph. **13**(5), 939–952 (2007)
34. Le, T.T., Almansa, A., Gousseau, Y., Masnou, S.: Motion-consistent video inpainting. In: 2017 IEEE International Conference on Image Processing (ICIP), pp. 2094–2098. IEEE (2017)
35. LeCun, Y.: The mnist database of handwritten digits. http://yann.lecun.com/exdb/mnist/ (1998)
36. Lee, A.X., Zhang, R., Ebert, F., Abbeel, P., Finn, C., Levine, S.: Stochastic adversarial video prediction. arXiv preprint arXiv:1804.01523 (2018)
37. Lee, J., Ramanan, D., Girdhar, R.: Metapix: few-shot video retargeting. arXiv preprint arXiv:1910.04742 (2019)
38. Liao, J., Yao, Y., Yuan, L., Hua, G., Kang, S.B.: Visual attribute transfer through deep image analogy. arXiv preprint arXiv:1705.01088 (2017)
39. Liu, M., Chen, S., Liu, J., Tang, X.: Video completion via motion guided spatial-temporal global optimization. In: Proceedings of the 17th ACM International Conference on Multimedia, pp. 537–540 (2009)
40. Mallya, A., Wang, T.-C., Sapra, K., Liu, M.-Y.: World-consistent video-to-video synthesis. arXiv preprint arXiv:2007.08509 (2020)
41. Newson, A., Almansa, A., Fradet, M., Gousseau, Y., Pérez, P.: Towards fast, generic video inpainting. In: Proceedings of the 10th European Conference on Visual Media Production, pp. 1–8 (2013)
42. Newson, A., Almansa, A., Fradet, M., Gousseau, Y., Pérez, P.: Video inpainting of complex scenes. SIAM J. Imag. Sci. **7**(4), 1993–2019 (2014)
43. Okabe, M., Anjyo, K., Igarashi, T., Seidel, H.P.: Animating pictures of fluid using video examples. In: Computer Graphics Forum, vol. 28, pp. 677–686. Wiley Online Library (2009)
44. Paszke, A., et al.: Pytorch: an imperative style, high-performance deep learning library. In: Wallach, H., Larochelle, H., Beygelzimer, A., d'Alché-Buc, F., Fox, E., Garnett, R. (eds.) Advances in Neural Information Processing Systems 32, pp. 8024–8035. Curran Associates Inc (2019). http://papers.neurips.cc/paper/9015-pytorch-an-imperative-style-high-performance-deep-learning-library.pdf
45. Rav-Acha, A., Pritch, Y., Peleg, S.: Making a long video short: dynamic video synopsis. In: 2006 IEEE Computer Society Conference on Computer Vision and Pattern Recognition (CVPR'06), vol. 1, pp. 435–441. IEEE (2006)
46. Ren, J., Chai, M., Tulyakov, S., Fang, C., Shen, X., Yang, J.: Human motion transfer from poses in the wild. In: Bartoli, A., Fusiello, A. (eds.) ECCV 2020. LNCS, vol. 12537, pp. 262–279. Springer, Cham (2020). https://doi.org/10.1007/978-3-030-67070-2_16
47. Rubinstein, M., Shamir, A., Avidan, S.: Improved seam carving for video retargeting. ACM Trans. Graph. (TOG) **27**(3), 1–9 (2008)

48. Saito, M., Matsumoto, E., Saito, S.: Temporal generative adversarial nets with singular value clipping. In: Proceedings of the IEEE International Conference on Computer Vision, pp. 2830–2839 (2017)
49. Sato, S., Dobashi, Y., Kim, T., Nishita, T.: Example-based turbulence style transfer. ACM Trans. Graph. (TOG) **37**(4), 1–9 (2018)
50. Sato, S., Dobashi, Y., Nishita, T.: Editing fluid animation using flow interpolation. ACM Trans. Graph. (TOG) **37**(5), 1–12 (2018)
51. Schödl, A., Szeliski, R., Salesin, D.H., Essa, I.: Video textures. In: Proceedings of the 27th Annual Conference on Computer Graphics and Interactive Techniques, pp. 489–498 (2000)
52. Shaham, T.R., Dekel, T., Michaeli, T.: Singan: learning a generative model from a single natural image. In: Proceedings of the IEEE/CVF International Conference on Computer Vision, pp. 4570–4580 (2019)
53. Siarohin, A., Lathuilière, S., Tulyakov, S., Ricci, E., Sebe, N.: Animating arbitrary objects via deep motion transfer. In: Proceedings of the IEEE/CVF Conference on Computer Vision and Pattern Recognition, pp. 2377–2386 (2019)
54. Siarohin, A., Lathuilière, S., Tulyakov, S., Ricci, E., Sebe, N.: First order motion model for image animation. Adv. Neural. Inf. Process. Syst. **32**, 7137–7147 (2019)
55. Simakov, D., Caspi, Y., Shechtman, E., Irani, M.: Summarizing visual data using bidirectional similarity. In: 2008 IEEE Conference on Computer Vision and Pattern Recognition, pp. 1–8. IEEE (2008)
56. Teed, Z., Deng, J.: RAFT: recurrent all-pairs field transforms for optical flow. In: Vedaldi, A., Bischof, H., Brox, T., Frahm, J.-M. (eds.) ECCV 2020. LNCS, vol. 12347, pp. 402–419. Springer, Cham (2020). https://doi.org/10.1007/978-3-030-58536-5_24
57. Tran, D., Bourdev, L., Fergus, R., Torresani, L., Paluri, M.: Learning spatiotemporal features with 3d convolutional networks. In: Proceedings of the IEEE International Conference on Computer Vision, pp. 4489–4497 (2015)
58. Tulyakov, S., Liu, M. Y., Yang, X., Kautz, J.: Mocogan: decomposing motion and content for video generation. In: Proceedings of the IEEE Conference on Computer Vision and Pattern Recognition, pp. 1526–1535 (2018)
59. Villegas, R., Yang, J., Hong, S., Lin, X., Lee, H.: Decomposing motion and content for natural video sequence prediction. arXiv preprint arXiv:1706.08033 (2017)
60. Villegas, R., Erhan, D., Lee, H., et al.: Hierarchical long-term video prediction without supervision. In: International Conference on Machine Learning, pp. 6038–6046. PMLR (2018)
61. Villegas, R., Pathak, A., Kannan, H., Erhan, D., Le, Q.V., Lee, H.: High fidelity video prediction with large stochastic recurrent neural networks. arXiv preprint arXiv:1911.01655 (2019)
62. Vondrick, C., Pirsiavash, H., Torralba, A.: Generating videos with scene dynamics. arXiv preprint arXiv:1609.02612 (2016)
63. Wang, T.-C., et al.: Video-to-video synthesis. arXiv preprint arXiv:1808.06601 (2018)
64. Wang, T.-C., Liu, M.-Y., Tao, A., Liu, G., Kautz, J., Catanzaro, B.: Few-shot video-to-video synthesis. arXiv preprint arXiv:1910.12713 (2019)
65. Wang, Y., Bilinski, P., Bremond, F., Dantcheva, A: Imaginator: conditional spatio-temporal gan for video generation. In: Proceedings of the IEEE/CVF Winter Conference on Applications of Computer Vision, pp. 1160–1169 (2020)
66. Wang, Y., Bremond, F., Dantcheva, A.: Inmodegan: interpretable motion decomposition generative adversarial network for video generation. arXiv preprint arXiv:2101.03049 (2021)

67. Wei, L.-Y., Levoy, M.: Fast texture synthesis using tree-structured vector quantization. In: Proceedings of the 27th Annual Conference on Computer Graphics and Interactive Techniques, pp. 479–488 (2000)
68. Wei, L. Y., Lefebvre, S., Kwatra, V., Turk, G.: State of the art in example-based texture synthesis. In: Eurographics 2009, State of the Art Report, EG-STAR, pp. 93–117. Eurographics Association (2009)
69. Wexler, Y., Shechtman, E., Irani, M.: Space-time video completion. In: Proceedings of the 2004 IEEE Computer Society Conference on Computer Vision and Pattern Recognition, 2004. CVPR 2004, vol. 1, pp. I-I. IEEE (2004)
70. Wolf, L., Guttmann, M., Cohen-Or, D.: Non-homogeneous content-driven video-retargeting. In: Proceedings of the Eleventh IEEE International Conference on Computer Vision (ICCV) (2007)
71. Yang, Z., et al.: Transmomo: invariance-driven unsupervised video motion retargeting. In: Proceedings of the IEEE/CVF Conference on Computer Vision and Pattern Recognition, pp. 5306–5315 (2020)

# Rayleigh EigenDirections (REDs): Nonlinear GAN Latent Space Traversals for Multidimensional Features

Guha Balakrishnan[1,2(✉)], Raghudeep Gadde[2], Aleix Martinez[2], and Pietro Perona[2]

[1] Rice University, Houston, TX 77005, USA
guha@rice.edu
[2] Amazon, Seattle, WA 98109, USA

**Abstract.** We present a method for finding paths in a deep generative model's latent space that can maximally vary one set of image features while holding others constant. Crucially, unlike past traversal approaches, ours can manipulate arbitrary multidimensional features of an image such as facial identity and pixels within a specified region. Our method is principled and conceptually simple: optimal traversal directions are chosen by maximizing differential changes to one feature set such that changes to another set are negligible. We show that this problem is nearly equivalent to one of Rayleigh quotient maximization, and provide a closed-form solution to it based on solving a generalized eigenvalue equation. We use repeated computations of the corresponding optimal directions, which we call Rayleigh EigenDirections (REDs), to generate appropriately curved paths in latent space. We empirically evaluate our method using StyleGAN2 and BigGAN on the following image domains: faces, living rooms and ImageNet. We show that our method is capable of controlling various multidimensional features: face identity, geometric and semantic attributes, spatial frequency bands, pixels within a region, and the appearance and position of an object. Our work suggests that a wealth of opportunities lies in the local analysis of the geometry and semantics of latent spaces.

## 1 Introduction

Latent spaces of deep generative networks like generative adversarial networks (GANs) [13,17,18,29] and variational autoencoders (VAEs) [19] are known to organize semantic attributes into disentangled subspaces without supervision [14,16,29,37,39]. This property is the basis of several latent space *traversal* algorithms that can modify specific image attributes while holding others constant by moving along carefully-chosen latent space directions [4,12,28,31,43].

**Supplementary Information** The online version contains supplementary material available at https://doi.org/10.1007/978-3-031-19790-1_31.

© The Author(s), under exclusive license to Springer Nature Switzerland AG 2022
S. Avidan et al. (Eds.): ECCV 2022, LNCS 13677, pp. 510–526, 2022.
https://doi.org/10.1007/978-3-031-19790-1_31

Traversal methods have many potential applications including dataset creation/augmentation, image editing, entertainment and graphic design.

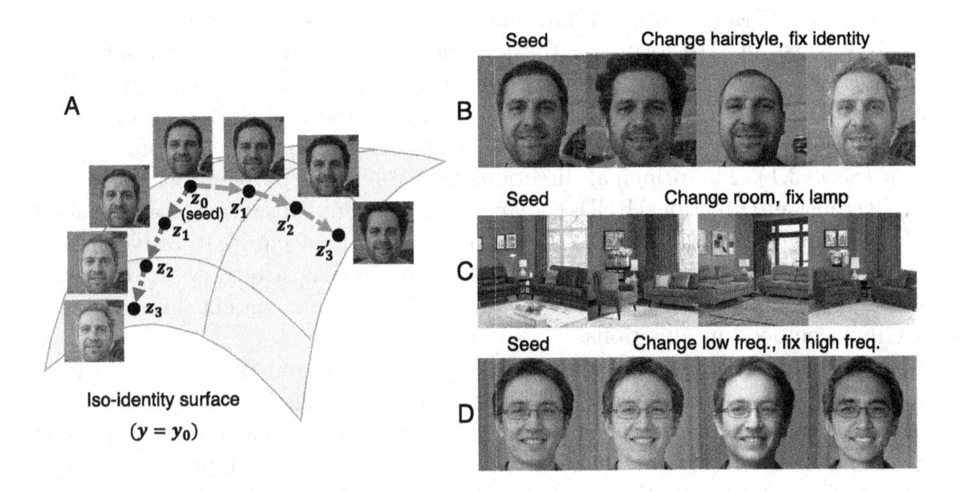


**Fig. 1.** Method and examples. (A) Our method traverses the local latent space around a seed point $\mathbf{z}_0$ along optimally chosen paths to synthesize images that share the same high-dimensional attribute value $\mathbf{y}_0$ (e.g., identity), and vary as much as possible across other image attributes (e.g., hairstyle). Also shown are samples from our method modifying (B) hairstyle while preserving identity (C) a living room with a fixed object (lamp in red box of seed image) and (D) low spatial frequencies while modifying high ones. These results are produced by our method using StyleGAN2 generators.

Virtually all existing traversal methods assume *scalar* attributes of interest that may be modeled well with global linear functions, e.g., a linear regressor or a support vector machine, in the latent space. This approach works well for attributes like gender, hair color and smile of faces [4,31] and image transformations like translation, color change and camera movements [16,28]. But these approaches cannot be easily extended to work with attributes like 'style of a couch' and 'face identity' which are best described with high-dimensional vectors[1]. For example, to find a latent space traversal that preserves identity in our experiments, we need a representation that can compute the similarity between two 512-dimensional embeddings returned by a face recognition model [10]. In addition, faces with the same identity or rooms with the same furniture layout (see Fig. 1C) tend to be tightly clustered in latent space, requiring methods tuned to local latent space geometry unlike the common global linear models used for scalar attributes.

[1] There is no physical 'identity' ground truth behind a GAN-generated portrait. However, human observers or face recognition algorithms can respond to the question "Is this the same person?" and can produce consistent judgments. Therefore 'identity' here denotes 'perceptual identity'.

We propose a method to tackle this broader class of traversal problems. Given a point in latent space, we aim to generate many traversals, or sequences of images, such that we vary one multidimensional feature ($\mathbf{x}$) in as many ways as possible subject to other multidimensional features ($\mathbf{y}$) being held approximately constant. We formalize the task of finding local latent directions that fulfill these criteria as a constrained optimization problem. By using differential approximations of the feature functions, we recast the problem into an instance of Rayleigh quotient maximization, which has a well-known closed-form solution (Sect. 3.1). The principal directions that solve this problem, which we call Rayleigh EigenDirections(REDs), span the local latent subspace containing good paths. Using REDs, we propose a fast linear and more accurate iterative nonlinear projection traversal algorithms (Sect. 3.3) to produce arbitrary-length paths. Our approach is agnostic to network architecture, scene content, and choice of attribute embedding functions.

We evaluate our method using StyleGAN2 [17,18] and BigGAN [7] generators. We consider a number of challenging applications outside the scope of previous GAN traversal algorithms: face traversals that preserve identity (Fig. 3) while changing hairstyle and facial geometries, face traversals that preserve/change content from specific spatial frequency bands (Fig. 5), and living room traversals that preserve the appearance and location of selected pieces of furniture (Fig. 6). We provide a number of qualitative results demonstrating the perceptual quality of our generated image sequences, and quantitatively demonstrate the necessity for nonlinear traversal strategies in these applications. Finally, we also compare our method against well-known global linear model baselines [4,31] for scalar attributes and perform comparably, though with some failure cases that we discuss in Sect. 5.

Our main contributions are: (a) REDs, a *local* method for synthesizing a diverse set of images that share a chosen set of multidimensional attribute. The method is principled, simple, and versatile – applicable to pretrained generators, any image type, and to both low-level and semantically meaningful features. (b) A nonlinear technique for long-distance traversals in latent space; (c) Qualitative and quantitative validation experiments on a number of challenging synthesis tasks in different image domains (faces, livingrooms and ImageNet) using two different models (StyleGAN2 and BigGAN).

## 2 Related Work

Several studies focus on finding interpretable directions in GAN latent spaces for editing and synthesizing images. Most propose finding global linear directions correlated with scalar attributes of interest [4,12,14,28,31,38,43]. Unfortunately, multidimensional features like face identity and image regions lie on complex latent space manifolds rather than on simple linear ones, meaning that a single global direction is not appropriate to model them. Recently, a method called LowRankGAN [46] was proposed for manipulating image regions by finding low-rank subspaces around a latent point. As in our work, they compute a local Jacobian matrix to discover steerable latent subspaces that change one feature while fixing another. Our work is similar in spirit to LowRankGAN with the following differences: we cast the traversal problem as a constrained optimization related to a generalized eigenvalue problem, we propose and demonstrate the superiority of nonlinear traversals for multidimensional features (see *Projection* algorithm in Sect. 3.3), and our method is applicable to general multidimensional features. Another work [41] introduces a new disentangled intermediate latent space where linear traversals along a single dimension provides control over specific visual properties by computing local gradients. Overall, our nonlinear traversals produce samples that satisfy the enforced constraints for longer traversal lengths compared to [41] and [46].

A few nonlinear traversal strategies for scalar features also exist. Several are based on training deep neural networks to map latent codes to features [16,36,44]. Our method is complementary to these – ours requires no additional training, but also does not leverage global latent space structure as theirs presumably can. Finally, our focus on a local rather than global view of the latent space may also complement various theoretical studies on understanding GAN latent space structure [3,5,8,21,30,39].

A more explicit way to control GAN outputs is to train the generator using attribute values as inputs. Many of these so-called "conditional GANs" have been proposed, particularly for altering face attributes [2,6,9,15,20,22–26,34,35,42,45], controlling face identity [6,32–34], and conditioning on semantic maps [27,40]. Our approach is complementary to all of these in that offers the benefit of not needing to design and train a GAN from scratch with apriori-known attribute controls. Working with a general-purpose black-box GAN has the advantage of keeping all control objectives open and not committing to a specific goal, e.g. preserving identity, from the beginning.

## 3 Method

Given a point $\mathbf{z}_0 \in \mathcal{R}^d$ in latent space defining an image, we want to generate a set of images that holds fixed the multidimensional features $\mathbf{y}_0 \in \mathcal{R}^n$ while maximally changing the features $\mathbf{x}_0 \in \mathcal{R}^m$. For ease of explanation, we assume $\mathbf{y}_0$ and $\mathbf{x}_0$ each define a single multidimensional feature like facial identity or hairstyle, though our method easily handles features from multiple semantic attributes as explained in Sect. 3.2.

We denote the function that computes the *fixed* features $f(\cdot) : \mathbf{z} \rightarrow \mathbf{y} \in \mathcal{R}^n$, and the function that computes the *changing* features $c(\cdot) : \mathbf{z} \rightarrow \mathbf{x} \in \mathcal{R}^m$. For example, in one of our experiments with faces, $f(\cdot)$ is the concatenation of two functions: the GAN generator on the input latent vector, and a face recognition embedding model on the synthesized face. $c(\cdot)$ may be the generator itself (i.e., $\mathbf{x}$ are the raw pixels of the image) or the concatenation of the generator with learning models computing various image attributes.

Starting at $\mathbf{z}_0$, our method traverses different paths in latent space to generate latent code sequences. For each such trajectory $t$ of length $L$, $\mathbf{z}_0, \mathbf{z}_1^t \cdots, \mathbf{z}_L^t$, we want $\mathbf{y}_i^t \approx \mathbf{y}_0$ for all $i$ and $\mathbf{x}_0, \mathbf{x}_1^t, \cdots, \mathbf{x}_L^t$ to progressively change such that $\|\mathbf{x}_i^t - \mathbf{x}_{i+1}^t\| < \|\mathbf{x}_i^t - \mathbf{x}_{i+2}^t\|$, where $\|\cdot\|$ is a norm. We return all points from all sequences.

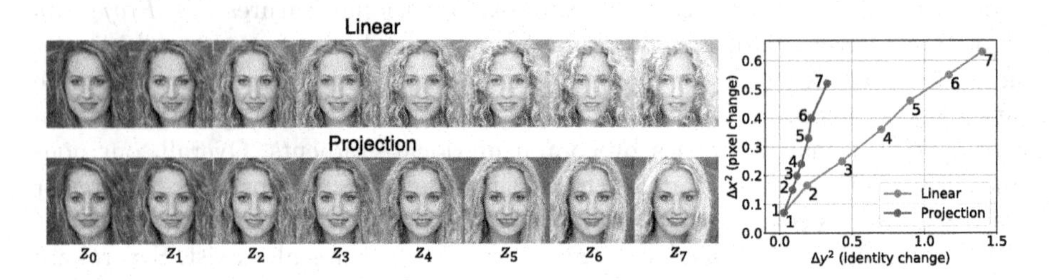


**Fig. 2.** Comparison of *Linear* and *Projection* traversal. We show a *Linear* and *Projection* traversal originating from the same latent seed code (left-most face), and top RED vector at the seed. $f(\cdot)$ measures identity and $c(\cdot)$ measures raw face pixel values. We also plot squared pixel distance versus squared identity distance. *Projection* and *Linear* change pixels by roughly the same amount, but *Projection* is better at preserving identity (lower distance values). (Color figure online)

The key intuition behind our approach is that there exists a manifold on which $\mathbf{y}$ does not change around $\mathbf{z}_0$ (see Fig. 1). This is true whenever $d > n$ (and thus the iso-$\mathbf{y}$ manifold has dimension $n - d$) and the generator function is continuous (which, by inspection, it is, apart from a zero-size set). When $d \leq n$, our approach naturally transitions to a "soft" constraint $\mathbf{y}_i \approx \mathbf{y}_0$ as will become clear below. We find directions, which we call Rayleigh EigenDirections (REDs), that maximally change $\mathbf{x}$ within this subspace. This procedure is described in Sect. 3.1. We propose two traversal strategies using REDs in Sect. 3.3: a linear method which simply extrapolates the local REDs throughout the latent space, and a nonlinear method (*Projection*) which updates traversal directions based on local latent space geometry.

### 3.1 Rayleigh EigenDirections (REDs)

Let $\mathbf{z}$ be a generic point in the generator's latent space with fixed and changing features $\mathbf{y} = f(\mathbf{z})$ and $\mathbf{x} = c(\mathbf{z})$. Given a displacement $\delta\mathbf{z}$, the displacements to $\mathbf{y}$ and $\mathbf{x}$ are:

$$\delta\mathbf{y} = f(\mathbf{z} + \delta\mathbf{z}) - f(\mathbf{z}) \tag{1}$$

$$\delta\mathbf{x} = c(\mathbf{z} + \delta\mathbf{z}) - c(\mathbf{z}). \tag{2}$$

We aim to find the displacement $\delta\mathbf{z}^*$ that maximizes $\delta\mathbf{x}$ with insignificant changes to $\delta\mathbf{y}$:

$$\delta\mathbf{z}^* = \underset{\delta\mathbf{z}:\|\delta\mathbf{z}\|=\epsilon}{\operatorname{argmax}} \|\delta\mathbf{x}(\mathbf{z}, \delta\mathbf{z})\|^2 \quad \text{s.t. } \|\delta\mathbf{y}(\mathbf{z}, \delta\mathbf{z})\|^2 \approx 0, \tag{3}$$

**Algorithm 1:** Compute local REDs (solves optimization (4))

**Input**: $A_f$, $A_c$, $\beta_f$
**Output**: $R$ (REDs matrix with directions as columns, from best to worst)
$A_f, A_c \leftarrow A_f/\|A_f\|_2, A_c/\|A_c\|_2$
$\mathbf{u}_f, V_f \leftarrow \text{eig}(A_f)$
$\rho \leftarrow$ smallest $k$ s.t. $\sum_{i=0}^{k} \mathbf{u}_f^2(i) \geq \beta_f \|\mathbf{u}_f\|^2$
$\text{null}(A_f) \leftarrow$ columns $\rho$ to $d$ of $V_f$
$\tilde{\mathbf{u}}_c, \tilde{V}_c \leftarrow \text{eig}(\text{null}(A_f)^T A_c \, \text{null}(A_f))$
$R \leftarrow \text{null}(A_f)\tilde{V}_c$

where we write $\delta\mathbf{x}$ and $\delta\mathbf{y}$ as functions of $\mathbf{z}$ and $\delta\mathbf{z}$, and $\epsilon$ is a small, fixed constant. For sufficiently small $\epsilon$, we can approximate $\delta\mathbf{y}$ and $\delta\mathbf{x}$ with local linear expansions: $\delta\mathbf{y} \approx J_f(\mathbf{z})\delta\mathbf{z}$ and $\delta\mathbf{x} \approx J_c(\mathbf{z})\delta\mathbf{z}$, where $J_f \in \mathcal{R}^{n\times d}$ and $J_c \in \mathcal{R}^{m\times d}$ are Jacobian matrices. Letting $A_f(\mathbf{z}) = J_f^T(\mathbf{z})J_f(\mathbf{z})$ and $A_c(\mathbf{z}) = J_c^T(\mathbf{z})J_c(\mathbf{z})$:

$$\delta\mathbf{z}^* = \underset{\delta\mathbf{z}:\|\delta\mathbf{z}\|=\epsilon}{\text{argmax}} \; \delta\mathbf{z}^T A_c(\mathbf{z})\delta\mathbf{z} \quad \text{s.t. } \delta\mathbf{z}^T A_f(\mathbf{z})\delta\mathbf{z} \approx 0 \tag{4}$$

This optimization is similar to one of finding the $\delta\mathbf{z}$ that maximizes the Rayleigh quotient $\left(\delta\mathbf{z}^T A_c(\mathbf{z})\delta\mathbf{z}\right) / \left(\delta\mathbf{z}^T A_f(\mathbf{z})\delta\mathbf{z}\right)$, known to be the solution of the generalized eigenvalue problem $A_c\delta\mathbf{x} = \lambda A_f\delta\mathbf{x}$, or the principal eigenvector of $A_f^{-1}A_c$ (see Supplementary). The main point of difference is that in our applications $A_f$ is often singular ($n < d$) and therefore not invertible. Put another way, $f(\cdot)$ is constant in a subspace $\text{null}(A_f)$ around $\mathbf{z}$ and any $\delta\mathbf{z}$ in that subspace will exactly satisfy the constraint in (4). We instead first project $A_c$ onto $\text{null}(A_f)$, and then find the principal eigenvectors of the resulting matrix (Algorithm 1) [11]. We return the eigenvectors (REDs) in matrix $R \in \mathcal{R}^{d\times d}$, ordered from best to worst.

For some high-dimensional features, the rank of $\text{null}(A_f)$ may be too small (or even 0 when $d < n$), yielding little to no diversity of $\mathbf{x}$ in the generated trajectories. To address this, we introduce hyperparameter $\beta_f$ in Algorithm 1 that lets users smoothly control the approximation of $A_f$'s rank based on explained variance.

The main computational cost of finding REDs is in calculating the Jacobian matrices $J_f$ and $J_c$. We compute them using one-sided finite difference approximations with step size $\epsilon$, which requires $d + 1$ forward evaluations of $f(\cdot)$ and $c(\cdot)$. See Sect. 4.6 for more on this topic.

### 3.2 Handling Multiple Attributes

To fix multiple attributes $\mathbf{y}^1, \cdots, \mathbf{y}^{n_f}$, we replace the constraint in (4) with multiple constraints: $\delta\mathbf{z}^T A_f^i(\mathbf{z})\delta\mathbf{z} \approx 0, i = 1\cdots n_f$, and introduce a separate $\beta_f^i$ for computing the rank of each $A_f^i$. We compute REDs by projecting $A_c$ onto $\cap_{i=1}^{n_f}\text{null}(A_f^i)$ – the intersection of the fixed attribute nullspaces – and returning the eigenvectors of the resulting matrix as before. To change multiple attributes, we compute REDs separately for each changing attribute, and return all vectors formed by summing together one RED chosen from each set.

### 3.3 Traversal Algorithms

We propose two traversal algorithms using REDs. The first is a simple *Linear* traversal (see Supplementary for algorithm). We randomly select a direction in the span of $R_0$ (the REDs of $\mathbf{z}_0$), and generate a sequence of latent codes $\mathbf{z}_1, \cdots, \mathbf{z}_K$ by moving in that direction starting from $\mathbf{z}_0$ with step size $s$. In the likely case that the constant-$\mathbf{y}$ manifold is curved, the linear traversal is expected to diverge quadratically from $||\delta\mathbf{y}|| = 0$ as a function of $||\delta\mathbf{z}||$.

Our second algorithm, *Projection* (see Supplementary for algorithm), addresses this shortcoming by recomputing the space of local REDs along the traversal path. We again start by selecting a random direction in $R_0$. However, at each step $i$ (of length $s$), we project the previous direction, $\delta\mathbf{z}_{i-1}$, onto $R_i$. This results in a path that more faithfully adheres to the local geometries of $f(\cdot)$ and $c(\cdot)$ in latent space.

A visual example of a *Linear* and *Projection* traversal for the same initial latent code is shown in Fig. 2, where $f(\cdot)$ measures identity and $c(\cdot)$ measures raw face pixels. *Projection* is better than *Linear* at preserving identity for long trajectories (right plot), while achieving similar levels of image change (left plot).

## 4 Experiments

We focus our evaluations on two image domains: faces and living rooms, modeled with StyleGAN2 [18]. For faces, we use the public *config-f* model from NVIDIA trained on the Flickr Faces HQ (FFHQ) dataset at 1024×1024 resolution. For living rooms, we train a StyleGAN2 generator from scratch on an in-house dataset of 100K 1024 × 1024 living room scenes from the web. For both domains, we use the "style" space, $\mathbf{w} \in \mathcal{R}^{512}$, as our latent space. We also demonstrate results on BigGAN [7] trained on ImageNet, where we use the input noise space (in $\mathcal{R}^{128}$) as our latent space.

We set $\beta_f$ to 0.95 or 0.99 for our experiments, depending on the perceptual characteristics of the generated samples that the user prefers ($\beta_f = 0.95$ results in more diverse samples at the expense of letting the fixed features change to a larger degree). See Supplementary for further analysis and figures.

### 4.1 Identity, Hairstyle and Geometry Traversals for Faces

We first evaluate our method on controlling three multidimensional facial features: identity, hairstyle, and geometry (quantified by 3D facial landmark positions). We use ArcFace [10], a popular open-source face identification model, to encode identity with a 512-dimensional vector. To encode hairstyle, we run a public face segmentation model[2] on each image, set pixels outside of the hair

[2] https://github.com/zllrunning/face-parsing.PyTorch.

region to 0, and flatten all pixels into a $256 \times 256 \times 3 = 196,608$-dimensional vector. We encode 3D geometry using the MediaPipe mesh model [1], which predicts 468 landmarks around the face. This results in a $468 \times 3 = 1404$-dimensional vector. We set $\beta_f$ to 0.95.

We performed four experiments: changing hairstyle while keeping landmarks fixed, changing hairstyle while keeping landmarks and identity fixed, changing landmarks while keeping hairstyle fixed, and changing landmarks while keeping identity and hairstyle fixed. Fig. 4 presents sample results for three test seed points using REDs and *Projection*. We set both the Jacobian finite difference step and path step $s$ to 1, and the path length $L = 4$. Along with changing the

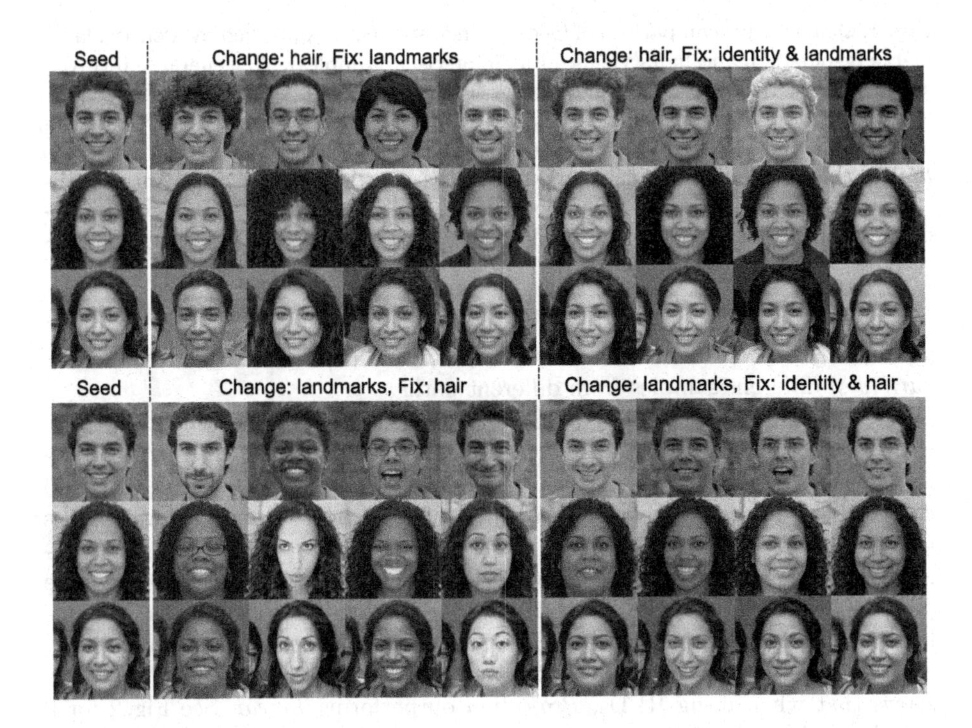


**Fig. 3.** Results for face traversals controlled by hairstyle, landmarks (geometry) and identity. We generated results using REDs with *Projection*. For each seed face and experiment, we selected four illustrative samples from the first few principal trajectories. (Top) Changing hairstyle while fixing facial landmarks (columns 2–5) and changing hairstyle while fixing identity and facial landmarks (columns 6–9). (Bottom) Changing landmarks while fixing hairstyle (columns 2–5) and changing landmarks while fixing identity and hairstyle (columns 6–9). Our method is able to generate a perceptually diverse set of faces while adhering to the fixed attribute constraints. Explicitly fixing identity greatly helps preserve the identity in the seed images. See Fig. 4 for quantitative analysis.

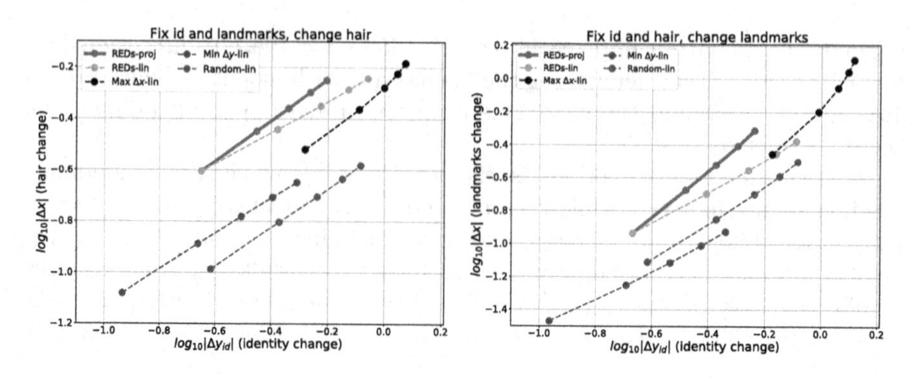


**Fig. 4.** Quantitative comparison of face traversal strategies controlled by identity, landmarks (geometry), and hairstyle (see Fig. 3 for visual samples). We generated 6 traversals with $L = 4$ steps for each method for 50 random seeds. We plot changes to hair (left) and landmarks (right) versus changes to identity in log-log scale, where each dot in the plot is the average value for each traversal step over all examples. ***Leftward and higher values are better***. Our method using *Linear* traversal (REDs-lin) outperforms the baselines also using *Linear*. Our method with *Projection* traversal (REDs-proj) outperforms REDs-lin by reducing identity changes with no impact to hair or landmark changes.

input images along the intended features, our method is able to produce a *wide variety* of different samples from different paths.

We quantitatively evaluated REDs against three baseline direction-finding approaches: choosing directions at random (**Random**), choosing the most significant eigenvectors of $A_c$, thereby maximizing changes to $\mathbf{x}$ (**Max-**$\boldsymbol{\Delta}\mathbf{x}$), and choosing the least significant eigenvectors of $A_f$, thereby minimizing changes to $\mathbf{y}$ (Min-$\Delta\mathbf{y}$). The plots in Fig. 4 present our results for two of the experiments. When using *Linear* traversal, REDs outperforms the three baseline direction-finding approaches. Max-$\boldsymbol{\Delta}\mathbf{x}$ finds directions that significantly change hairstyle/landmarks and identity, Min-$\boldsymbol{\Delta}\mathbf{y}$ preserves identity but also minimally changes hairstyle/landmarks, and Random performs worst of all. The figure also shows that when using REDs, *Projection* outperforms *Linear*. See Fig. 2 for a visual sample of this comparison and Supplementary for complete traversals and more plots.

## 4.2 Frequency Band Traversals

Our method can handle arbitrary low-level image representations. We demonstrate this by controlling specific spatial frequency bands for StyleGAN2 and BigGAN in Fig. 5. We let $f(\cdot)$ and $c(\cdot)$ encode the raw pixels of low-pass and high-pass filtered versions of the input image (and vice versa). We set $\beta_f$ to 0.99. High-pass modifications change fine details like face physiognomies and expressions, while low-pass modifications mainly change colors, lighting and shading.

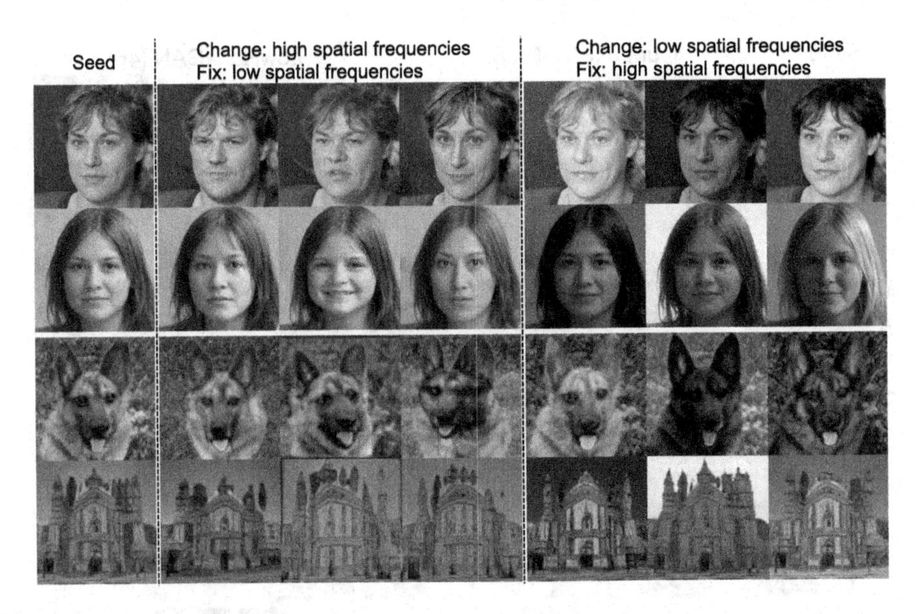


**Fig. 5.** Samples from traversals controlled by spatial frequency bands. The first two rows are generated by StyleGAN2, and the last two rows are generated by BigGAN. (Columns 2–4) The embedding function $f(\cdot)$ returns the raw pixels of the low-pass filtered image and $c(\cdot)$ the high-pass one. High-pass modifications change fine details. (Columns 5–7) $f(\cdot)$ and $c(\cdot)$ are inverted. Low-pass modifications change colors, lighting and shading.

### 4.3 Object-Preserving Living Room Traversals

We next apply our method to living room scenes. We aim to keep selected furniture fixed while changing other parts of the scene. We generated furniture bounding boxes with an object detector. We let $f(\cdot)$ encode the raw pixels within the bounding box, and let $c(\cdot)$ encode all remaining pixels in the scene. We set the Jacobian finite difference step to 0.75, path step $s = 0.25$, and a path length $L = 10$. We set $\beta_f$ to 0.99.

Figure 6 shows several sample sequences from REDs with *Projection* traversal and LowRankGAN [46]. Samples from LowRankGAN deviate significantly from the desired constraint of preserving the object in the bounding box, due mostly to the linear traversal strategy used in that method. See caption of Fig. 6 for a detailed description. In Supplementary, we show sample strips of full traversals and also compare against StyleSpace [41]. We observe two notable degradations in all methods the farther we move away from the seed image. First, the 'fixed' object often moves slightly at each step. Second, artifacts become more prominent because we rapidly advance to low-probability regions of the latent space.

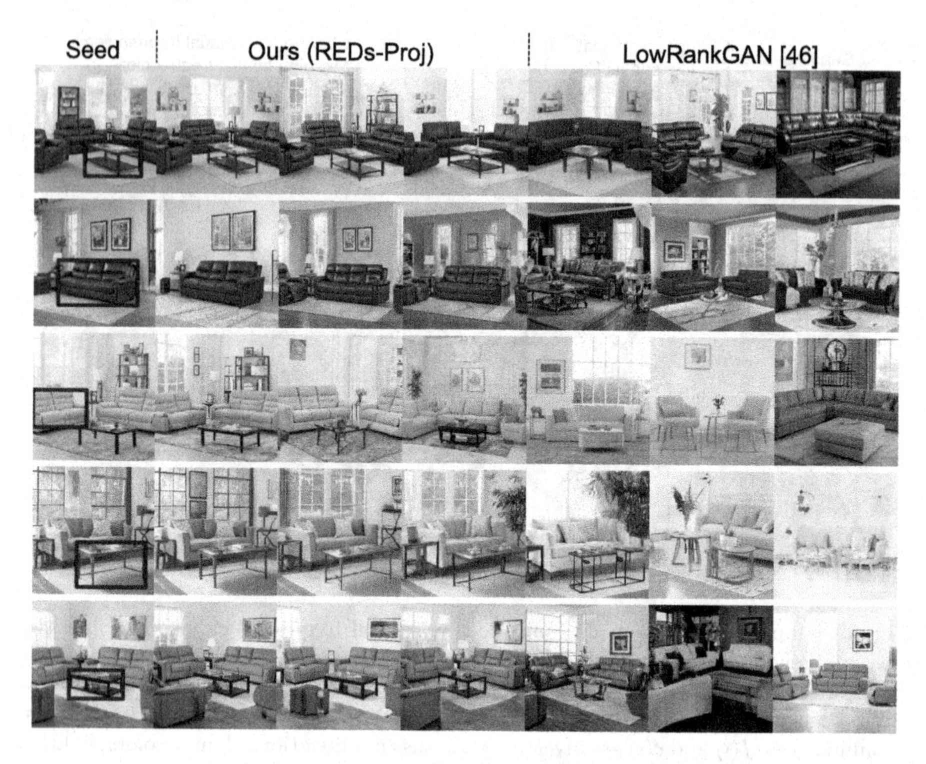


**Fig. 6.** Object-preserving living room traversals. We used REDs with *Projection* traversal, with $f(\cdot)$ and $c(\cdot)$ encoding raw pixel values inside and outside a bounding box on a piece of furniture (red box on seed image at left). The object within the box often stays fixed, but can undergo stylistic changes and movements (examples in rows 3, 5) due to feature correlations in latent space. There are diverse changes to the rooms outside of the boxes, including new furniture (rows 2, 3, 5), wall and window properties/decorations (all rows), and house plants (rows 3, 4). Samples in the right three columns are edits following the approach in [46] using the same path step. Clearly REDs-Proj is better at preserving objects. (Color figure online)

### 4.4 Spatial Region Traversals for Faces

We next demonstrate our method's effectiveness at manipulating facial regions (mouth and eyes). We use the same fixed bounding boxes for the mouth and eyes used in the LowRankGAN study [46]. The changing features are the pixels within the box, and the fixed features are pixels outside the box. Figure 7 presents sample visual results using REDs with *Projection* traversal. Our method is better than LowRankGAN and StyleSpace [41] at generating a variety of changes within the bounding boxes while roughly adhering to the constraints (the degree to fixing these constraints can be adjusted with $\beta_f$).

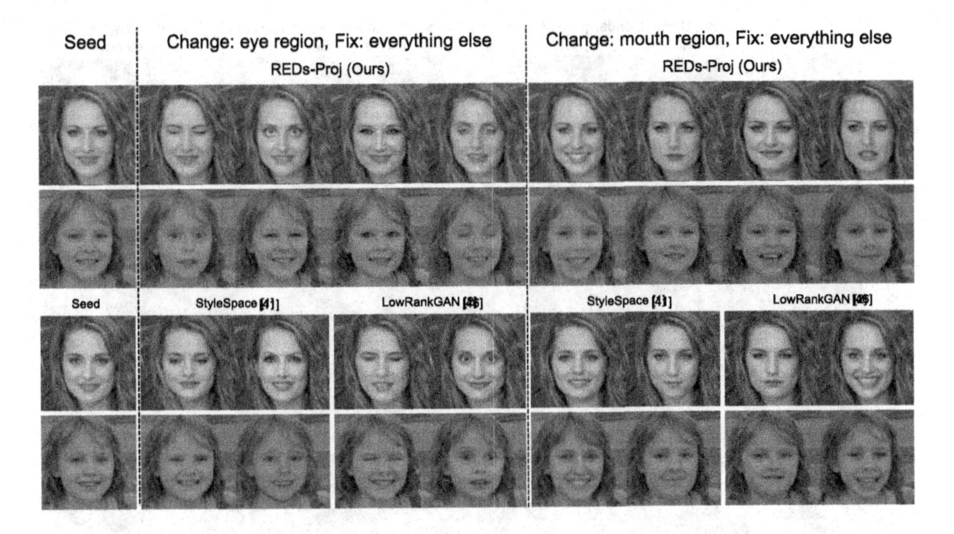


**Fig. 7.** Samples from traversals controlled by spatial image regions. We let the changing features be the pixels inside a bounding box (green boxes overlaid on images for visualization), and the fixed features be pixels outside the box. (Top) For each seed, we show several output samples of our method changing the eyes (columns 2–5) and mouth (columns 6–9). (Bottom) Results for the same task using StyleSpace [41] and LowRankGAN [46]. Notice the change in other attributes like identity and landmarks which REDS-proj better preserves. (Color figure online)

### 4.5 Scalar Attribute Traversals

We finally compare our method against a popular technique for modifying *scalar* attributes with global linear directions [4,31]: train a linear model (regressor for a continuous attribute or an SVM for a binary attribute) to predict an attribute value from the latent code, and change the attribute by moving along the hyperplane's normal direction. To fix attributes, we orthogonalize the changing attribute's direction with respect to the fixed attribute directions.

Figure 8 presents our results for four face attributes: age, pose, smile, and gender. Overall, REDS-proj achieves similar qualitative performance to the baseline for most samples, but also has more failures cases when changing an attribute like gender, which often does not have a large local gradient in latent space. We discuss this more in Sect. 5.

### 4.6 Computation Time

Virtually all the computation time of our method is spent on computing the Jacobian matrices in each traversal step, which involves generating $d+1$ images and evaluating $d+1$ feature functions. For the livingroom traversals shown in Sect. 4.3 at $1024 \times 1024$ resolution, evaluating one of the $d$ dimensions (assuming no parallelization) on an NVIDIA A100 GPU required approximately 25

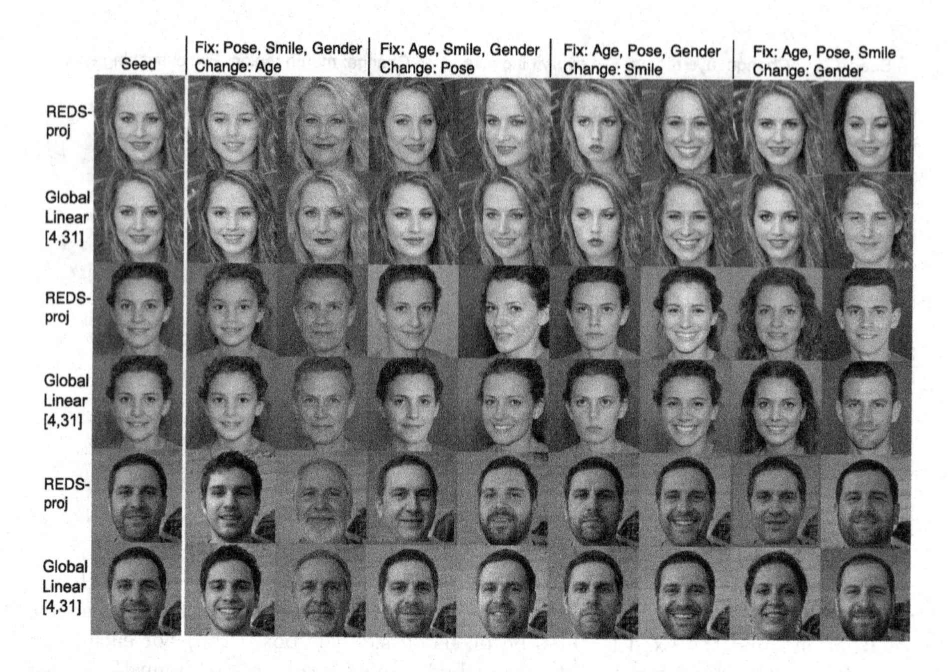


**Fig. 8.** Samples from traversals controlled by scalar semantic attributes. On scalar attributes one may compare our method to a baseline of using a global linear model (SVM or ridge regressor) in latent space [4,31] (the global method is not defined and cannot handle multi-dimensional attributes). We change one attribute (age, pose, smile, gender) at a time while fixing the other three. Both methods are comparable for many cases. REDs sometimes fails (red-boxed images), particularly for gender (see Sect. 5 for further discussion). (Color figure online)

milliseconds with StyleGAN2, translating to 15 s for all $d$ dimensions. For faces, the time for one Jacobian computation ranged from 15–50 s depending on the features being extracted. This time may be reduced dramatically if operations are parallelized in batches and across multiple GPUs.

## 5 Discussion

Our experiments demonstrate the effectiveness of REDs at finding locally optimal orientations. By contrast, selecting random traversal directions or local directions that prioritize only one of the objective or constraint in Eq. (4) do not work well due to the high dimensionality of the latent space (see Fig. 4).

The superiority of *Projection* over *Linear* traversal (Figs. 4, 7) also demonstrates the need for localized approximations of latent space geometry for complex image features. This is in contrast to past traversal studies [4,14,16,28,31] that found global linear directions to suffice for simple scalar attributes.

A consideration in all image synthesis works is the balance between perceptual quality based on human judgment, and quantitative optimization and analysis. In the application of faces, the user may have his/her own internal tradeoff curve between identity preservation and image diversity. Our method offers a principled way to explore different points on this curve by tuning the $\beta$ parameters (see Supplementary). Image perception also factors into the embedding functions used to measure image changes.

GAN latent spaces are not all alike, and each requires different considerations. Faces are easier to model than living rooms, because the latter are a composition of many discrete objects interacting with one another. As a result, we found the face latent space and traversals to be smoother. Our living room traversals often exhibit large perceptual "jumps" due to discontinuities in latent space (see Supplementary). The complexity of a distribution also affects the degree of correlation between attributes. As Fig. 6 shows, it is not always possible to exactly fix a particular region of a living room while obtaining enough diversity elsewhere due to entangled features. Different regions of the latent space are also not alike. We found that high-likelihood regions produce the most realistic images and diverse traversals. Thus, the biases of the generative model have a direct effect on how well our method performs for a given image.

Our method takes a local view of the latent space to identify good traversal directions. However, as our results in Sect. 4.5 suggest, there are benefits to taking a global view. Global linear models are likely better for attributes that are discrete, such as 'wearing eyeglasses,' or approximately discrete for a large majority of samples like gender. For such attributes, local gradients in latent space can be near zero and swamped by noise. Another limitation of a local view is that gradients are undefined near sharp discontinuities in the latent space. We did not find this to be a decisive issue for faces, but did notice perceptual 'jumps' in the living room scenes during traversals (see Supplementary for traversal strips). However, we note that our framework could be extended to use both global and local directions per traversal step, which we leave for future work.

### 5.1 Ethics

**Fairness**: As in past work [4] we observed bias in StyleGAN's face distribution: Caucasian faces are most likely to be generated. This bias also affects trajectory quality, with light-skinned seed faces producing more diverse trajectories than dark-skinned ones. Biases in fixed and changing functions that use learning models also affect results. One example are face recognition models, like the one we used in our experiments to fix identity, which are known to have gender and ethnicity biases. To reduce bias one will want to train GANs and any learned models on rich and diverse datasets. **Fake portrayals**: GANs could be used to generate fake images of individuals under different conditions. This could include the case where the image of the face of a real person is projected onto the GAN latent space and then manipulated.

## 6 Conclusion

We presented a simple, principled and versatile method designed to explore a generative model's latent space to produce sets of synthetic samples where one group of multidimensional features is held constant while another is varied as much as possible. We demonstrated traversal results on several features that previous works are not capable of handling: landmark locations, pixels within regions, frequency information, and facial identity as measured by a deep neural network. Our experiments show the need for modeling local geometry of latent spaces for high-dimensional features. Understanding the complex nature and geometry of the latent space of image generators is a fascinating question which we have only started to explore.

## References

1. Mediapipe. https://github.com/google/mediapipe
2. Antipov, G., Baccouche, M., Dugelay, J.L.: Face aging with conditional generative adversarial networks. In: 2017 IEEE International Conference on Image Processing (ICIP), pp. 2089–2093. IEEE (2017)
3. Arvanitidis, G., Hansen, L.K., Hauberg, S.: Latent space oddity: on the curvature of deep generative models. arXiv preprint arXiv:1710.11379 (2017)
4. Balakrishnan, G., Xiong, Y., Xia, W., Perona, P.: Towards causal benchmarking of bias in face analysis algorithms. In: European Conference on Computer Vision, pp. 547–563. Springer (2020). https://doi.org/10.1007/978-3-030-74697-1_15
5. Balestriero, R., Paris, S., Baraniuk, R.: Max-affine spline insights into deep generative networks. arXiv preprint arXiv:2002.11912 (2020)
6. Bao, J., Chen, D., Wen, F., Li, H., Hua, G.: Towards open-set identity preserving face synthesis. In: Proceedings of the IEEE Conference on Computer Vision and Pattern Recognition, pp. 6713–6722 (2018)
7. Brock, A., Donahue, J., Simonyan, K.: Large scale GAN training for high fidelity natural image synthesis. In: International Conference on Learning Representations (2019)
8. Chen, N., Klushyn, A., Kurle, R., Jiang, X., Bayer, J., Smagt, P.: Metrics for deep generative models. In: International Conference on Artificial Intelligence and Statistics, pp. 1540–1550. PMLR (2018)
9. Choi, Y., Choi, M., Kim, M., Ha, J.W., Kim, S., Choo, J.: Stargan: Unified generative adversarial networks for multi-domain image-to-image translation. In: Proceedings of the IEEE Conference on Computer Vision and Pattern Recognition, pp. 8789–8797 (2018)
10. Deng, J., Guo, J., Xue, N., Zafeiriou, S.: Arcface: additive angular margin loss for deep face recognition. In: Proceedings of the IEEE/CVF Conference on Computer Vision and Pattern Recognition, pp. 4690–4699 (2019)
11. Ghojogh, B., Karray, F., Crowley, M.: Eigenvalue and generalized eigenvalue problems: Tutorial. arXiv preprint arXiv:1903.11240 (2019)
12. Goetschalckx, L., Andonian, A., Oliva, A., Isola, P.: Ganalyze: toward visual definitions of cognitive image properties. In: Proceedings of the IEEE/CVF International Conference on Computer Vision, pp. 5744–5753 (2019)

13. Goodfellow, I., et al: Generative adversarial nets. In: Advances in Neural Information Processing Systems, vol. 27 (2014)
14. Härkönen, E., Hertzmann, A., Lehtinen, J., Paris, S.: Ganspace: discovering interpretable gan controls. In: Advances in Neural Information Processing Systems, pp. 9841–9850 (2020)
15. He, Z., Zuo, W., Kan, M., Shan, S., Chen, X.: Attgan: facial attribute editing by only changing what you want. IEEE Trans. Image Process. **28**(11), 5464–5478 (2019)
16. Jahanian*, A., Chai*, L., Isola, P.: On the "steerability" of generative adversarial networks. In: International Conference on Learning Representations (2020)
17. Karras, T., Laine, S., Aila, T.: A style-based generator architecture for generative adversarial networks. In: Proceedings of the IEEE/CVF Conference on Computer Vision and Pattern Recognition, pp. 4401–4410 (2019)
18. Karras, T., Laine, S., Aittala, M., Hellsten, J., Lehtinen, J., Aila, T.: Analyzing and improving the image quality of stylegan. arXiv preprint arXiv:1912.04958 (2019)
19. Kingma, D.P., Welling, M.: Auto-encoding variational bayes. In: 2nd International Conference on Learning Representations, ICLR 2014, Banff, AB, Canada, 14–16 April 2014, Conference Track Proceedings (2014)
20. Kocaoglu, M., Snyder, C., Dimakis, A.G., Vishwanath, S.: Causalgan: learning causal implicit generative models with adversarial training. In: International Conference on Learning Representations (2018)
21. Kuhnel, L., Fletcher, T., Joshi, S., Sommer, S.: Latent space non-linear statistics. arXiv preprint arXiv:1805.07632 (2018)
22. Lample, G., Zeghidour, N., Usunier, N., Bordes, A., Denoyer, L., Ranzato, M.: Fader networks: Manipulating images by sliding attributes. In: 31st Conference on Neural Information Processing Systems (NIPS 2017), pp. 5969–5978 (2017)
23. Liu, M., et al.: Stgan: a unified selective transfer network for arbitrary image attribute editing. In: Proceedings of the IEEE/CVF Conference on Computer Vision and Pattern Recognition, pp. 3673–3682 (2019)
24. Mirza, M., Osindero, S.: Conditional generative adversarial nets. arXiv preprint arXiv:1411.1784 (2014)
25. Odena, A., Olah, C., Shlens, J.: Conditional image synthesis with auxiliary classifier gans. In: International Conference on Machine Learning, pp. 2642–2651. PMLR (2017)
26. Or-El, R., Sengupta, S., Fried, O., Shechtman, E., Kemelmacher-Shlizerman, I.: Lifespan age transformation synthesis. In: Vedaldi, A., Bischof, H., Brox, T., Frahm, J.-M. (eds.) ECCV 2020. LNCS, vol. 12351, pp. 739–755. Springer, Cham (2020). https://doi.org/10.1007/978-3-030-58539-6_44
27. Park, T., Liu, M.Y., Wang, T.C., Zhu, J.Y.: Semantic image synthesis with spatially-adaptive normalization. In: Proceedings of the IEEE/CVF Conference on Computer Vision and Pattern Recognition, pp. 2337–2346 (2019)
28. Plumerault, A., Borgne, H.L., Hudelot, C.: Controlling generative models with continuous factors of variations. In: International Conference on Learning Representations (2020)
29. Radford, A., Metz, L., Chintala, S.: Unsupervised representation learning with deep convolutional generative adversarial networks. arXiv preprint arXiv:1511.06434 (2015)
30. Shao, H., Kumar, A., Thomas Fletcher, P.: The riemannian geometry of deep generative models. In: Proceedings of the IEEE Conference on Computer Vision and Pattern Recognition Workshops, pp. 315–323 (2018)

31. Shen, Y., Gu, J., Tang, X., Zhou, B.: Interpreting the latent space of gans for semantic face editing. In: Proceedings of the IEEE/CVF Conference on Computer Vision and Pattern Recognition, pp. 9243–9252 (2020)
32. Shen, Y., Luo, P., Yan, J., Wang, X., Tang, X.: Faceid-gan: Learning a symmetry three-player gan for identity-preserving face synthesis. In: Proceedings of the IEEE Conference on Computer Vision and Pattern Recognition, pp. 821–830 (2018)
33. Shen, Y., Zhou, B., Luo, P., Tang, X.: Facefeat-gan: a two-stage approach for identity-preserving face synthesis. arXiv preprint arXiv:1812.01288 (2018)
34. Shoshan, A., Bhonker, N., Kviatkovsky, I., Medioni, G.: Gan-control: Explicitly controllable gans. arXiv preprint arXiv:2101.02477 (2021)
35. Tran, L., Yin, X., Liu, X.: Disentangled representation learning gan for pose-invariant face recognition. In: Proceedings of the IEEE Conference on Computer Vision and Pattern Recognition, pp. 1415–1424 (2017)
36. Tzelepis, C., Tzimiropoulos, G., Patras, I.: Warpedganspace: finding non-linear rbf paths in gan latent space. In: Proceedings of the IEEE/CVF International Conference on Computer Vision, pp. 6393–6402 (2021)
37. Upchurch, P., et al.: Deep feature interpolation for image content changes. In: Proceedings of the IEEE Conference on Computer Vision and Pattern Recognition, pp. 7064–7073 (2017)
38. Voynov, A., Babenko, A.: Unsupervised discovery of interpretable directions in the gan latent space. In: International Conference on Machine Learning, pp. 9786–9796. PMLR (2020)
39. Wang, B., Ponce, C.R.: A geometric analysis of deep generative image models and its applications. In: International Conference on Learning Representations (2021)
40. Wang, T.C., Liu, M.Y., Zhu, J.Y., Tao, A., Kautz, J., Catanzaro, B.: High-resolution image synthesis and semantic manipulation with conditional gans. In: Proceedings of the IEEE Conference on Computer Vision and Pattern Recognition, pp. 8798–8807 (2018)
41. Wu, Z., Lischinski, D., Shechtman, E.: Stylespace analysis: disentangled controls for stylegan image generation. In: Proceedings of the IEEE/CVF Conference on Computer Vision and Pattern Recognition, pp. 12863–12872 (2021)
42. Xiao, T., Hong, J., Ma, J.: Elegant: Exchanging latent encodings with gan for transferring multiple face attributes. In: Proceedings of the European conference on computer vision (ECCV), pp. 168–184 (2018)
43. Yang, C., Shen, Y., Zhou, B.: Semantic hierarchy emerges in deep generative representations for scene synthesis. Int. J. Comput. Vis. **129**(5), 1451–1466 (2021). https://doi.org/10.1007/s11263-020-01429-5
44. Yang, H., Chai, L., Wen, Q., Zhao, S., Sun, Z., He, S.: Discovering interpretable latent space directions of gans beyond binary attributes. In: Proceedings of the IEEE Conference on Computer Vision and Pattern Recognition, pp. 12177–12185 (2021)
45. Yin, X., Yu, X., Sohn, K., Liu, X., Chandraker, M.: Towards large-pose face frontalization in the wild. In: Proceedings of the IEEE International Conference on Computer Vision, pp. 3990–3999 (2017)
46. Zhu, J., et al.: Low-rank subspaces in gans. In: Advances in Neural Information Processing Systems 34 (2021)

# Bridging the Domain Gap Towards Generalization in Automatic Colorization

Hyejin Lee[1], Daehee Kim[1,2], Daeun Lee[3], Jinkyu Kim[4(✉)], and Jaekoo Lee[1(✉)]

[1] Department of Computer Science, Kookmin University, Seoul, South Korea
jaekoo@kookmin.ac.kr
[2] Clova AI Research, NAVER Corp., Seongnam-si, South Korea
[3] Department of Statistics, Korea University, Seoul, South Korea
[4] Department of Computer Science and Engineering, Korea University, Seoul, South Korea
jinkyukim@korea.ac.kr

**Abstract.** We propose a novel automatic colorization technique that learns domain-invariance across multiple source domains and is able to leverage such invariance to colorize grayscale images in unseen target domains. This would be particularly useful for colorizing sketches, line arts, or line drawings, which are generally difficult to colorize due to a lack of data. To address this issue, we first apply existing domain generalization (DG) techniques, which, however, produce less compelling desaturated images due to the network's over-emphasis on learning domain-invariant contents (or shapes). Thus, we propose a new domain generalizable colorization model, which consists of two modules: (i) a domain-invariant content-biased feature encoder and (ii) a source-domain-specific color generator. To mitigate the issue of insufficient source domain-specific color information in domain-invariant features, we propose a skip connection that can transfer content feature statistics via adaptive instance normalization. Our experiments with publicly available PACS and Office-Home DG benchmarks confirm that our model is indeed able to produce perceptually reasonable colorized images. Further, we conduct a user study where human evaluators are asked to (1) answer whether the generated image looks naturally colored and to (2) choose the best-generated images against alternatives. Our model significantly outperforms the alternatives, confirming the effectiveness of the proposed method. The code is available at https://github.com/Lhyejin/DG-Colorization.

**Keywords:** Automatic colorization · Domain generalization · Generative adversarial networks

**Supplementary Information** The online version contains supplementary material available at https://doi.org/10.1007/978-3-031-19790-1_32.

© The Author(s), under exclusive license to Springer Nature Switzerland AG 2022
S. Avidan et al. (Eds.): ECCV 2022, LNCS 13677, pp. 527–543, 2022.
https://doi.org/10.1007/978-3-031-19790-1_32

## 1 Introduction

Recent successes in applying deep learning to computer vision tasks suggest that a ConvNet-based model can learn a fully automated data-driven colorization of grayscale images. This model can be trained to predict the $a$ and $b$ color channels in the CIE-$Lab$ color space using semantic information and the surface texture of given grayscale image [1,6,15,43]. In practice, data-driven colorization is used in applications such as coloring assistance of cartoonist or legacy photo restoration. However, current learning-based methods often fail to generalize colorizing performance in out-of-distribution. While successfully producing plausible and visually compelling colorization within the same known test domain, such models would not intrinsically generalize well to novel domains outside the training distribution because different domains involve different textures [11,18,30]. For example, a model trained with photo and cartoon images would fail to colorize art painting images (see Fig. 1). To address this issue, we first analyze the effect of domain shift in the colorization task and propose a novel domain generalization technique that enables a colorization model to generalize well to novel domains outside the training distribution. This model would be particularly useful for colorizing sketches, line arts, or line drawings, which are generally difficult to colorize due to a lack of data. Also, we would emphasize that, to our best knowledge, improving the model's ability to generalize across multiple domains for colorization has not yet been explored.

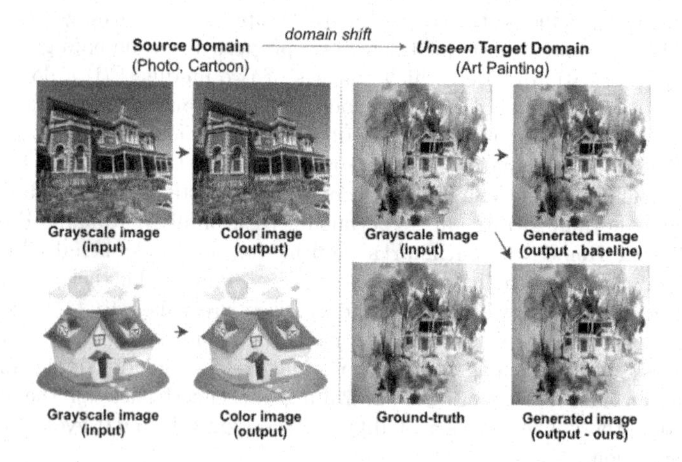


**Fig. 1.** We propose a data-driven fully automated colorization technique that can learn a mapping function from the grayscale image to color channels by regression onto continuous color space. With a novel domain generalization technique, our model can generalize well to unseen target domains, whereas existing colorization approaches and conventional DG techniques often fail to generate.

In literature, various domain generalization techniques have been introduced to make models generalize well to out-of-training-distribution. These techniques primarily focused on generalizable object class classification models by learning the domain-invariances across multiple source domains so that the classifier can robustly leverage such invariances in unseen target domains [10,30,33,35]. Examples from target domains are not available during training, thus these approaches differ from domain adaptation, semi-supervised domain adaptation, and unsupervised domain generations.

In our experiment, applying existing DG techniques to the colorization models was not successful. We have found that these models tend to over-emphasize generating domain-invariant contents and fail to leverage source domain-specific color information. Interestingly, such domain-specific information is particularly useful for colorization, as it can provide a domain-sensitive and essential prior for synthesizing colorized images of better quality.

To mitigate this issue, we propose the domain generalizable color generation network, which is divided into two components: (i) domain-invariant content-biased encoder and (ii) source domain-specific color generator. The former is trained to match content information across multiple domains to generalize well across domains by learning a domain-invariant texture (i.e., shapes), whereas the latter captures source domain-specific color information. Because the extracted features delivered through the skip connection between the encoder and decoder have little source domain-specific color information, it is necessary to adjust the color information to the source domain. Thus, we propose to transfer domain-invariant content feature statistics from encoder to decoder via skip connections followed by adaptive instance normalization (AdaIN) [14]. We observe such a connection significantly improves the quality of colored images.

We use two publicly available domain generalization benchmark datasets (PACS [23] and Office-Home [37]) to evaluate the effectiveness of the proposed method. We also analyze and compare with existing state-of-the-art domain generalization techniques, and we observe that our model generally outperforms these alternatives. In addition, we conduct a user study where human evaluators are asked to answer the following two questions: (i) Naturalness: "Do you think the provided image looks naturally colored" and (ii) Perceptual Realism: "Which of the images are the best". Our work significantly outperforms others, which confirms the effectiveness of the proposed method. We summarize our contribution as follows:

- We propose a novel fully automated colorization method that can generalize well to unseen target domains.
- Our model utilizes domain-invariant contents-biased encoder and source domain-specific color generator where a skip connection is used to transfer domain-invariant content feature statistics between the encoder and the decoder for a better quality colorization.
- We effectively show the benefit of our proposed methods on two large domain generalization benchmark datasets: PACS and Office-home. Our proposed method quantitatively and qualitatively outperforms alternative colorization approaches and domain generalization techniques.

- We conduct a user study where human participants evaluate the quality in terms of naturalness and the perceptual realism. Our method significantly outperforms other alternative domain generational approaches.

## 2 Related Work

### 2.1 Image Colorization

Colorization algorithms use various approaches, which can be broadly categorized into the following three types: (i) scribble-based colorization techniques, (ii) example-based colorization, and (iii) fully automatic colorization. Given an input grayscale image, the scribble-based colorization propagates scribbles (provided by the user) to the whole image [21]. Example-based colorization techniques, however, exploit user-provided [5,12,16,40], or automatically retrieved [7,27] reference images to match the luminance and texture information between a reference image and the input image (i.e., transferring color onto the input grayscale image from analogous regions of the reference image). These scribble-based and example-based approaches, though promising, depend primarily on user input, which can be time-consuming and expensive for achieving an acceptable result.

Recent work suggests that a data-driven, fully automated approach can successfully learn a mapping function from the lightness channel to color channels by regressing onto continuous color space or classifying them into quantized color values [3,6,15,17,31,34,38,43]. These approaches have developed similar systems, such as (i) convolutional neural networks, which are trained end-to-end to predict color channels of the image from the lightness channel using large-scale data, (ii) conditional GANs, which have a sharpening effect in the spectral dimension and make images more colorful, have recently become a key architectural component for colorization, (iii) Isola et al. [17] used a U-Net-based architecture [32] for the generator and a convolutional PatchGAN classifier [22] for the discriminator, and they got promising colorization results. Our method is also based on (iii).

Most of these fully automatic colorization techniques produce promising results in the same known test domains, but their generation quality tends to become sub-optimal in unseen different test domains. In previous work [34,38] because the pretrained ImageNet and COCO models were also used, domain shift occurred when training domains other than photo. Improving such ability to generalize across multiple domains has not yet been investigated (though important) in the community to the best of our knowledge. Here, we explore the effect of domain shift in the colorization task, and propose a fully automatic colorization model that can generalize well to novel domains outside the training distribution.

### 2.2 Domain Generalization

Domain generalization (DG) techniques focus on generating domain-invariant latent representations so that the model is to better generalize to the unseen target domains outside the training distribution. Vapnik et al. [36] introduces Empirical Risk Minimization (ERM) that minimizes the sum of errors across the domains of landmark work. Notable variants have been proposed to learn domain-invariant features by matching distributions across different domains. Ganin et al. [10] use an adversarial network for such distribution matching, while Li et al. [26] match the conditional distributions across the domains. Minimizing maximum mean discrepancy [25], transformed feature distribution distance [29], or covariances [35] is frequently used to optimize such a shared feature space. In this study, we also follow this workstream; however, we focus on the benefits of these techniques for the image colorization task, specifically how the model can learn domain-invariant content information across different domains and generalize effectively to unknown target domains.

Inter-domain mixup [39,41,42] techniques are used on linearly interpolated examples from random pairs across domains to perform an ERM. JiGen [4] improves generalization using self-supervised clues obtained by solving a jigsaw puzzle as a secondary task. Meta-learning frameworks [24] are also investigated for DG to meta-learn how to generalize across domains by leveraging MAML [9]. Low-rank parameterization [23], style-agnostic network [30], and domain-specific aggregation modules [8] have recently been used to partition the model into domain-invariant and domain-variant components.

In stylized ImageNet [11] and Kim et al. [18], texture and content are defined respectively as a domain-specific feature and domain-invariant feature. They applied a random representation mixture to training data with AdaIN to make the extracted feature robust. Here, we also use such disentangled image representations (i.e., content vs. texture). However, we advocate learning not only a content-biased network (i.e., reducing domain-specific texture information), but also source domain-specific color information for the colorization task. As the latter can provide domain-sensitive content information about source domain-specific color features, which can serve as an essential prior to generate high-quality colorized images.

# 3 Method

## 3.1 Conditional GANs for Colorization

Our model is built upon the pix2pix architecture [17]. This model uses GANs in the conditional setting, which is suitable for image-to-image mappings as they can generate a corresponding image conditioned on an input image. As shown in Fig. 2, our model consists of two main components: (i) a generator $G(E(x))$ and (ii) a discriminator $D$.

For (i), we use a U-Net-based architecture, a typical encoder ($E$) - decoder ($G$) architecture with skip connections, as a generator and a PatchGAN classifier for the discriminator. Given an input lightness channel $x \in \mathrm{R}^{h \times w \times 1}$, our U-Net-based generator is trained to predict associated $a$ and $b$ color channels (in the CIE $Lab$ color space) $y_{ab} \in \mathrm{R}^{h \times w \times 2}$, where $h$ and $w$ are image dimensions. For (ii), we use a convolutional classifier that is trained to classify whether the

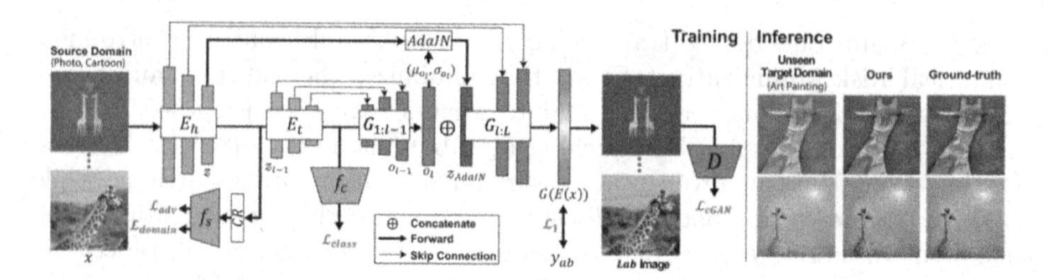


**Fig. 2.** An overview of our proposed generalizable colorization model. Our model is built upon the existing pix2pix [17] model that utilizes U-Net-based encoder as a generator and a PatchGAN classifier as a discriminator. Our model takes a grayscale image and predicts its color channels. During training, to improve the model's ability to generalize well to unseen target domains, we propose to use the following three regularizations: (i) Object Class Classifier $f_c$, (ii) Texture-biased Domain Classifier $f_s$, and (iii) Content Feature Statistics Transfer via AdaIN (adaptive instance normalization).

output image is real (i.e. ground-truth images) or fake (i.e. generated images). Following the work by Mao et al. [28], we adopt the least squares loss function for the discriminator instead of using the sigmoid cross entropy loss function, which may lead to the vanishing gradients problem during the learning process. Our loss functions for the generator and the discriminator are as follows:

$$\begin{aligned}\mathcal{L}_D(x, y_{ab}) &= \frac{1}{2}\mathbb{E}_{x,y_{ab}\sim\mathbb{D}}\left[(D(x, y_{ab}) - a)^2\right] \\ &\quad + \frac{1}{2}\mathbb{E}_{x\sim\mathbb{D}}\left[(D(x, G(E(x))) - b)^2\right] \\ \mathcal{L}_G(x) &= \mathbb{E}_{x\sim\mathbb{D}}\left[(D(x, G(E(x))) - c)^2\right]\end{aligned} \tag{1}$$

where we use the $a$-$b$ coding scheme for the discriminator and $a$ and $b$ are the labels for real data and fake data, respectively. We use $\mathbb{D}$ to denote the training data distribution. We use $c$ to denote the value that $G$ wants $D$ to believe for fake data. In our implementation, we use $a = 1$, $b = 0$, and $c = 1$. In literature, mixing the GAN objective with a traditional loss function, such as $\mathcal{L}_2$ distance, is advantageous for being stable during learning process. Thus, we add the following $\mathcal{L}_1$ distance rather than $\mathcal{L}_2$, which often makes the output blurred.

$$\mathcal{L}_1(x, y_{ab}) = \mathbb{E}_{x,y_{ab}\sim\mathbb{D}}\left[||y_{ab} - G(E(x))||_1\right] \tag{2}$$

Concretely, our loss function for training the conditional GANs is as follows:

$$\mathcal{L}_{cGAN} = \mathcal{L}_D(x, y_{ab}) + \mathcal{L}_G(x) + \lambda_{\text{color}}\mathcal{L}_1(x, y_{ab}) \tag{3}$$

where we use a hyperparameter $\lambda_{\text{color}}$ to control the strength of the traditional loss function.

### 3.2 Learning Domain-Invariant Contents Features

Unlike the human vision system that largely focuses on contents (i.e. shapes) for object recognition, recent studies show that ConvNets have a strong inductive bias towards image texture [2,11,13,30]. Moreover, models trained for the colorization task inherently exhibit a strong bias towards texture, which is an essential prior to a high-quality image generation. Such texture information often varies across different domains than the content, and thus makes models more sensitive to domain shift. We empirically observe this domain shift effect in colorization as shown in Fig. 3, where all conventional colorization models generally fail to generalize to unseen target domains.

From recent work [30], we constrain the head encoder $E_h$ from learning domain-invariant information (i.e., more content-biased and texture-invariant representation) via an adversarial framework. We use a content randomization (CR) module that interpolates contents feature statistics between different examples (regardless of their object class), i.e. it replaces the content of the input with the randomized content through AdaIN.

Formally, given an input image $x_i$ and a randomly-chosen $x'$ in same batch, we first obtain corresponding latent representations $z$ and $z'$ from the head encoder $E_h$, respectively. Given the channel-wise means $\mu(z)$ and standard deviations $\sigma(z)$ as style representation, we apply AdaIN to the content of $z'$ with the texture of $z$.

$$CR(z, z') = \sigma(z) \cdot \left( \frac{z' - \mu(z')}{\sigma(z')} \right) + \mu(z) \tag{4}$$

Such content-randomized representation is then fed into a domain classifier $f_s$, which needs to correctly predict its source domain given texture-biased representations. Thus, this classifier $f_s$ is trained by minimizing a domain classification loss $\mathcal{L}_{\text{domain}}$:

$$\mathcal{L}_{\text{domain}} = -\mathbb{E}_{x,y_s \sim \mathbb{D}} \left[ \sum_{s=1}^{S} y_s \log f_s(CR(z, z'))_s \right] \tag{5}$$

where $S$ is the number of source domains and $y_i \in \{0, 1\}^S$ is the one-hot domain label.

Our head encoder $E_h$ is then trained by minimizing the following adversarial loss $\mathcal{L}_{\text{adv}} = -\lambda_{\text{adv}} \mathcal{L}_{\text{domain}}$ where $\lambda_{\text{adv}}$ is a hyperparameter to control the strength of $\mathcal{L}_{adv}$.

**Emphasizing Semantic Information.** We also add an object class predictor $f_c$ that consumes image representations from the tail encoder $E_t$ and performs the object recognition task to regularize the model to learn semantic features useful for our generator $G$ to generate better quality images. Formally, we add the cross entropy loss $\mathcal{L}_{class}$ given an image $x$ and its class label $y_c$.

$$\mathcal{L}_{\text{class}}(x, y_c) = -\mathbb{E}_{x,y_c \sim \mathbb{D}} \left[ \sum_{c=1}^{C} y_c \log f_c(E_t(E_h(x)))_c \right] \tag{6}$$

where $\mathbb{D}$ is the training data distribution and $C$ is the number of class categories.

### 3.3 Transferring Domain-Invariant Features

In the previous section, we discussed how we train our encoder to learn content-biased representations by applying a content randomization (CR) module as well as a domain classifier. We observe, however, our generator $G$ exhibits color bias towards source domain during training as the network is forced to learn "source-domain style-sensitive color information". This is achieved using skip connections between encoder and decoder (see Fig. 2). In particular, statistical color adjustment is required to generate source-domain style-sensitive color for realistic colorization from the encoder's content-biased feature. So, Instead of using a simple concatenation via skip connection, we use a style transfer technique using an AdaIN – i.e. given content information from $z$, we transfer style feature statistics of $o_l$ from the intermediate layer of $G$. We empirically observe that such a "content" skip connection allows the generator to use such content information directly from the encoder during synthesizing output images. We also observe such a skip connection is critical to improving the colorization performance in the unseen target domains. We summarize our results in Experiment section.

Formally, we first obtain a representation $o_l$ by concatenating the output $z$ from $E_h$ and the latent representation $o_l$ from the $l$-th layer of $G$: i.e. $o_l = G_l(o_{l-1} \oplus z_{l-1})$ where $G_l$ is the $l$-th layer of $G$. Given the representations $z$ and $o_l$, we apply AdaIN to the content of $z$ with the style of $o_l$:

$$z_{\text{AdaIN}} = \sigma(o_l) \cdot \left( \frac{z - \mu(z)}{\sigma(z)} \right) + \mu(o_l) \tag{7}$$

where $z_{\text{AdaIN}}$ can be interpreted as a style-transferred representation of $z$. We concatenate $z_{\text{AdaIN}}$ with $o_l$ and feed into the next layer of $G$: $o_{l+1} = G_{l+1}(o_l \oplus z_{\text{AdaIN}})$. In our experiment, we set $l = 5$. Details of the relevant experiments are provided in the supplementary material.

## 4 Experiments

### 4.1 Implementation and Evaluation Details

**Implements Details.** Following [17], we use the same architectural choices for the generator and (color) discriminator. For our $G$, we use a U-Net-based architecture as a backbone, and for our (color) $D$, we use a convolutional PatchGAN classifier. The model architectures for the $f_s$ and the $f_c$ are based on the same architecture as that of our $E_t$. Note that domain labels are finally computed followed by an average pooling layer and three fully connected layers. We train our model using an Adam optimizer [19] for approximately 50 epochs. The batch size is set to 128 and the learning rate to 0.001. Our implementation is based on PyTorch.

**Dataset.** We evaluate the effectiveness of the proposed method on the publicly available PACS [23] and Office-Home [37] benchmark datasets. The PACS dataset contains over 10k images from four diverse domains: Photo, Art Painting, Cartoon, and Sketch. This dataset is particularly useful in domain generalization research as it provides a bigger domain shift than existing photo-only benchmarks. As the Sketch domain does not provide color information, we exclude it from the experiment. This PACS dataset provides seven object categories: dog, elephant, giraffe, guitar, horse, house, and person. We split examples from training domains in the ratio 8:2 (training:validation) and test on the entire held-out domain. Note that we use the best-performed model on validation for testing. We also use the Office-Home dataset, which contains over 15 k images from four domains: Art, Clipart, Product, and Real-World. We exclude the Product domain from our experiment owing to its lack of color information. This dataset provides 65 object categories.

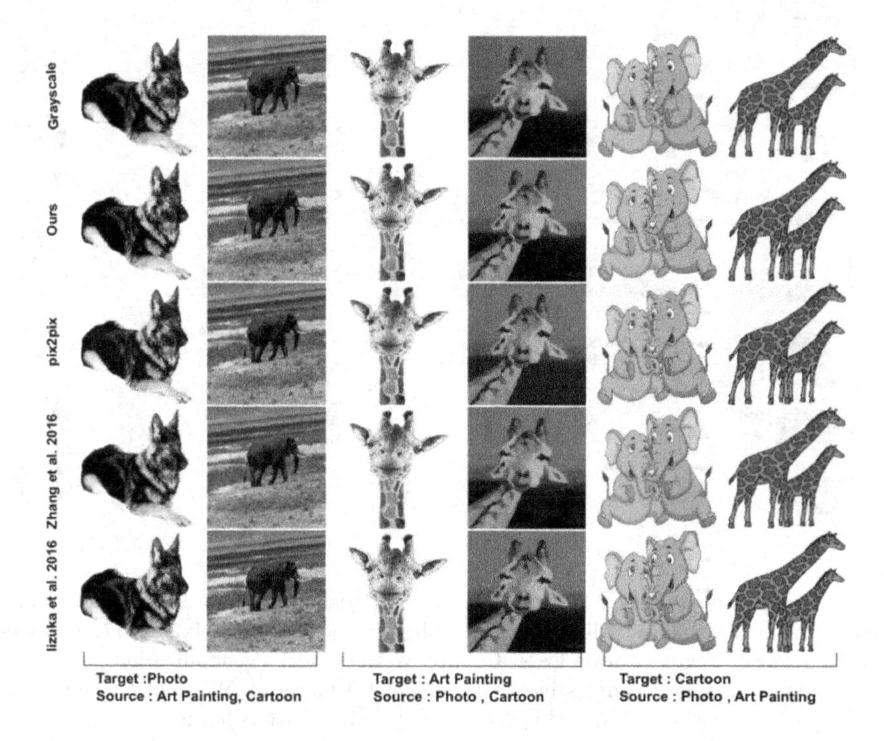


**Fig. 3.** Colorization performance comparison with conventional colorization approaches. Target and source domains are listed in the bottom row. Data: PACS.

**Evaluation Metrics.** Evaluation of the quality of colorized images is known to be challenging. We first use the following four widely-used quantitative metrics: peak signal-to-noise ratio (PSNR), structural similarity index measure (SSIM), image quality metrics (IQM), and Frechet Inception distance (FID). The first two metrics compute the pixel-wise distances between the ground-truth and synthesized images–thus quantifying the similarity of colors. Unlike PSNR and SSIM, IQM only uses the $a$ and $b$ color channels of the image in the CIE *Lab* color space to quantify the image quality in terms of colorfulness, sharpness, and contrast. Thus, IQM is a suitable metric for the colorization task [1]. FID measures the distance between latent image representations for the ground-truth and synthesized images. We use an ImageNet-pretrained Inception v3 model to extract such feature vectors for FID. Note that we also finetuned the Inception v3 model with the corresponding domains to remove the negative effect of domain shift when computing FID.

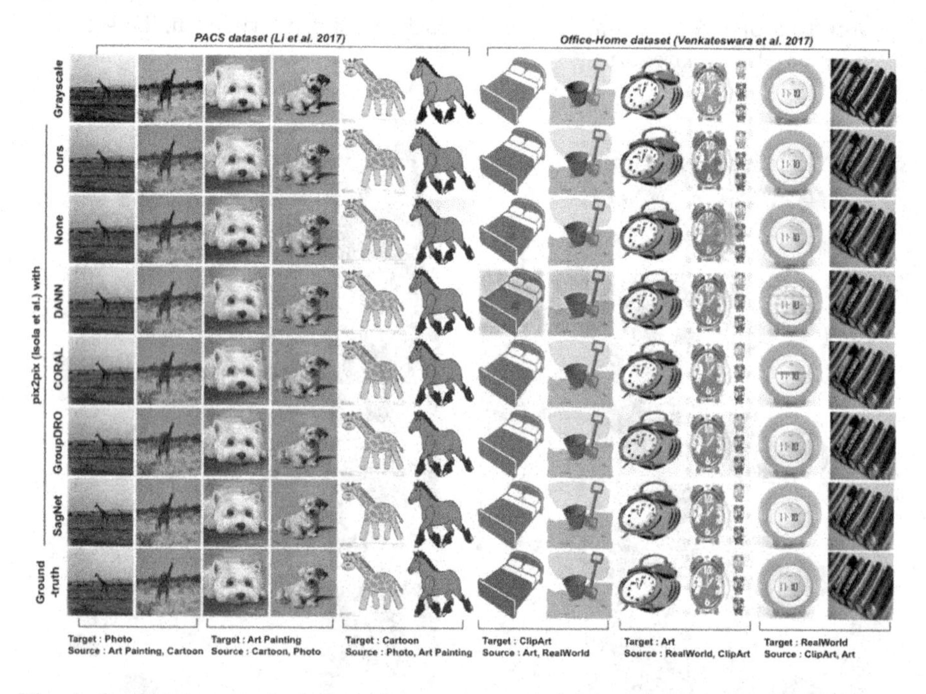


**Fig. 4.** Qualitative colorization performance comparison with four alternative domain generalization techniques. All models are built upon our baseline pix2pix [17] architecture and we add regularization losses to improve the model's generalization power. We provide more diverse examples in the supplemental material. We used two datasets: PACS (see 1st–6th columns) and Office-Home (see 7th–12th columns).

However, these metrics often fail to capture visual realism. Although the aforementioned four quantitative metrics are frequently used in colorization tasks, further discussions of clear quantitative performance indicators of colorization are still ongoing [1,3,20,44,45]. To address the shortcomings of any of the above individual evaluations, we further evaluate our model using a user study.

### 4.2 Effect of Domain Shift in Colorization

We first investigate the effect of domain shift in the colorization task with the following three landmark colorization models: Zhang et al. [43], Iizuka et al. [15], and pix2pix [17]. We use the leave-one-out setting, i.e. a pre-selected single domain is used as a test domain and the others as training domains. We use the PACS and Office-Home datasets for this experiment. As shown in Fig. 3, we observe that all models generally fail to generate successful colorized images; in fact, they often show a failure to capture long-range color consistency, a sepia-tone on complex scenes, and confusion between red and blue information. This is further confirmed by our quantitative analysis in terms of the four image quality metrics: PSNR, SSIM, IQM, and FID. In Table 1, a large degradation with these metrics is observed for all models. This clearly indicates that the current colorization models do not generalize well to novel target domains and additional treatment is needed to deal with such a domain shift. We provide more detailed numerical values and failed examples for each domain in the supplementary material.

**Table 1.** Colorization performance comparison in terms of four image quality evaluation metrics. An average value across domains is reported. To observe any performance degradation in the domain generalization setting, we also compare each model with the non-domain generalization (non-DG) setting, i.e. models are trained using the same target domain. Data: PACS [23] and Office-Home [37].

| Models | PACS [23] | | | | Office-home [37] | | | |
|---|---|---|---|---|---|---|---|---|
| | PSNR ↑ | SSIM ↑ | IQM ↑ | FID ↓ | PSNR ↑ | SSIM ↑ | IQM ↑ | FID ↓ |
| A. Zhang et al. [43] | 30.56 | 0.65 | 1.89 | 30.78 | 32.25 | 0.74 | 1.63 | 17.77 |
| B. A w/non-DG setting | 33.48 | 0.79 | 1.84 | 12.96 | – | – | – | – |
| C. Iizuka et al. [15] | 30.81 | 0.80 | 1.91 | 23.52 | 32.67 | 0.79 | 1.48 | 17.25 |
| D. C w/non-DG setting | 33.49 | 0.82 | 1.84 | 14.18 | – | – | – | – |
| E. pix2pix [17] | 30.38 | 0.63 | 1.80 | 25.28 | 32.69 | 0.74 | 1.63 | 19.41 |
| F. E w/ non-DG setting | 31.56 | 0.71 | 1.89 | 21.28 | – | – | – | – |
| G. E + DANN [10] | 30.12 | 0.59 | 1.68 | 25.39 | 31.07 | 0.67 | 1.53 | 22.53 |
| H. E + CORAL [35] | 30.59 | 0.65 | 1.84 | 24.63 | 32.15 | 0.73 | 1.57 | 16.09 |
| I. E + GroupDRO [33] | 30.51 | 0.65 | 1.82 | 25.93 | 31.98 | 0.75 | 1.47 | 18.35 |
| J. E + SagNet [30] | 30.67 | 0.68 | 1.81 | 27.46 | 31.92 | 0.67 | 1.61 | 18.69 |
| K. E + Ours | 30.71 | 0.66 | 1.88 | 24.92 | 32.29 | 0.73 | 1.52 | 14.92 |

### 4.3 Effect of Domain Generalization Techniques

To improve the generalization power of the colorization models, we explore the effect of applying some existing domain generalization techniques. Note that these models originally focused on the object recognition task, not on the colorization task. Thus, for a fair comparison, all models are based on the pix2pix model that generally shows better colorization performance than alternatives.

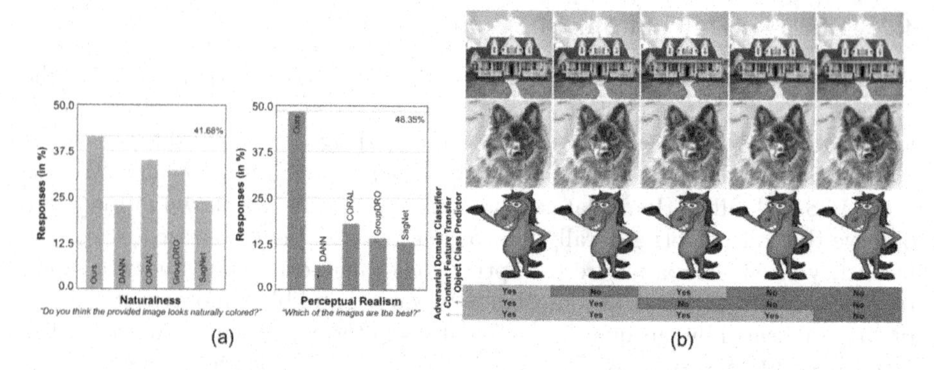


**Fig. 5.** (a) Evaluation of perceptual realism by a user study. Participants were asked to answer two questions for evaluating naturalness (left) and perceptual realism (right). (b) Ablation study results between variants of our models with and without the following three modules: Adversarial Domain Classifier, Content Feature Transfer, and Object Class Predictor.

We then implement the core idea of each domain generalization technique from DANN [10], CORAL [35], GroupDRO [33], and SagNet [30]. Note that other techniques may also be applicable, but we leave them as future work. As summarized in Table 1 (in the 7th–10th rows for alternatives and in the 11th row for ours), all domain generalization techniques (except DANN [10]) generally provide better results than our baseline pix2pix model (5th row) in terms of four image quality evaluation metrics. In particular, ours generally outperforms others and shows better capability in generalizing to unseen target domains. This indicates that considering both domain-invariant contents and source domain-specific color information is suitable for the generalized colorization task.

In Fig. 4, we provide some random examples of the generated colorized images. Models are trained on the PACS (1st–6th columns) and Office-Home (7th–12th columns) datasets. For a better comparison, we provide the corresponding ground-truth images in the 8th row as well as the results from our baseline model (i.e. pix2pix) in the 3rd row. Not surprisingly, the baseline model generates lower-quality images; they often generate gray-tone images on the Office-Home dataset and fail to capture long-range color consistency. Applying existing domain generation techniques (as observed by comparing the 3nd vs. 4–7th rows) generally provides better-quality colorized images than our baseline, but their generated images still show limitations as they show confusion between blue and red channels. This failure is more apparent in the Office-Home dataset, which is more challenging for colorizing unseen-domain images.

Ours, however, generally shows comparable or better performance against alternative domain generalization techniques. Our quantitative analysis, as presented in Table 1, confirms that our proposed method outperforms others in terms of four image quality evaluation metrics. Our qualitative analysis in Fig. 4 further confirms that exploiting domain-invariant contents and source domain-specific color features enables the proper colorization of images (see 2nd row vs. others). In Office-Home benchmark, neither baseline nor existing domain generalization techniques succeeded in colorizing the object in novel domains, whereas our model could realistically color these objects.

**Table 2.** Ablation study results between variants of our models. An average value is reported. Data: PACS.

| Models | $\lambda_{\text{adv}}$ | PSNR ↑ | SSIM ↑ | IQM ↑ | FID ↓ | Acc. |
|---|---|---|---|---|---|---|
| A. Ours | 1.0 | 30.71 | 0.66 | 1.88 | 24.92 | 38.5 |
| B. Ours ($\lambda_{\text{adv}} = 0.5$) | 0.5 | 30.61 | 0.66 | 1.84 | 25.48 | 36.8 |
| C. Ours ($\lambda_{\text{adv}} = 0.1$) | 0.1 | 30.63 | 0.65 | 1.93 | 27.06 | 36.4 |
| D. A - Object class predictor $f_c$ | 1.0 | 30.67 | 0.67 | 1.84 | 25.23 | – |
| E. A - Content feature transfer | 1.0 | 29.77 | 0.62 | 1.76 | 28.25 | – |
| F. A - Content feature transfer - Object class predictor $f_c$ | 1.0 | 30.42 | 0.63 | 1.82 | 26.10 | – |
| G. F - Adversarial domain classifier $f_s$ (baseline) | – | 30.38 | 0.63 | 1.80 | 25.28 | – |
| H. Ours w/ style feature transfer | 1.0 | 30.52 | 0.67 | 1.77 | 30.10 | – |

### 4.4 Evaluation of Perceptual Realism by User Study

Evaluating colorized images is generally difficult using a set of automatic metrics, such as PSNR. As our main goal is to make the colorized images that are more compelling to human observers, we set up a user study, in which we show participants synthesized colors for an image, and ask them to answer the following two questions: (i) Naturalness: do you think the provided image looks naturally colored? (ii) Perceptual Realism: which of the images are the best? Images were randomly sampled from each domain on the PACS and Office-Home datasets. For this user study, 34 participants were recruited and each participant answers overall 360 questions and submitted 12, 240 votes. A detailed explanation is provided in the supplementary material.

As shown in Fig. 5(a), ours significantly outperforms alternatives (including existing domain generalization approaches) with a large gap in both questions. 41.68% (2,551 out of 6,120 votes) of colorized images by ours were perceived as naturally colored. For evaluating perceptual realism, 49.26% (2,959 out of 6,120 votes) of colorized images by ours was chosen as the best-colorized images among five approaches: DANN, CORAL, GroupDRO, SagNet, and ours. This number is significantly higher than all compared approaches, and these results validate the effectiveness of the proposed method.

### 4.5 Ablation Study

**Effect of Object Class Predictor.** Recall that our model uses an object class predictor that takes domain-invariant content features as an input and performs object recognition tasks to encourage the model to learn semantic features necessary to predict their classes. We observe in Fig. 5(b) that this predictor is particularly useful to generate more saturated images (compare A vs. E columns). As shown in Table 2, our quantitative analysis also confirms that performance is generally degraded without the use of object class predictor as a regularization (compare Model A vs. D).

**Effect of Transferring Domain Feature Statistics.** Recall that we transfer content feature statistics from encoder to decoder as a skip connection followed by AdaIN, which is widely used in the style transfer task. To see its effect, we conduct an ablation study and we observe in Table 2 that colorization performance is generally degraded as we turn the content feature transfer off (see Model A vs. E), which is probably because our generator is easily biased towards source domain-specific color information. This is more apparent in our qualitative analysis as shown in Fig. 5(b) where the color for source domain is better recovered (see 1st vs. 3rd columns).

Further, to verify our motivation behind transferring content feature statistics, we evaluate a variant of ours where we apply AdaIN to the content of $o_l$ with the style of $z$ (instead of using the content of $z$ with the style of $o_l$). As expected, constraining the generator towards using content-biased features generally degrades the overall performance in colorization (see model A vs. H in Table 2). This may confirm that the generator needs to be source domain color-biased network, while the encoder needs to be content-biased network. We provide examples in the supplementary material.

**Effect of $\lambda_{\text{adv}}$.** We observe in Table 2 that decreasing $\lambda_{\text{adv}}$ generally degrades the colorization evaluation scores (see models A–C), which may confirm that learning content-biased representations is beneficial for the model to generalize well to unseen domains. This trend is more apparent in object class classification performance (see the rightmost column in the table). Ours outperforms other alternative domain generalization approaches (Acc. of SagNet: 36.5).

## 5 Conclusion

We proposed an innovative fully automatic colorization algorithm that can generalize well to unseen target domains. Built upon a conditional GAN-based colorization deep neural network architecture, we proposed three modules to learn domain-invariant content-biased encoder and source domain-specific color generator. A skip connection between them transfers rich information about content feature statistics. Our extensive experiments demonstrate ours generally outperforms alternative domain generalization techniques, and our user study further confirmed this. To the best of our knowledge, we are the first to explore the effect of domain shift in the colorization task.

**Acknowledgement.** This work was supported by Institute of Information & communications Technology Planning & Evaluation (IITP) grant funded by the Korea government (MSIT) (No. 2021-0-00994, Sustainable and robust autonomous driving AI education/development integrated platform). J. Kim was supported by the MSIT (Ministry of Science and ICT), Korea, under the ICT Creative Consilience program (IITP-2022-2020-0-01819) supervised by the IITP (Institute for Information & communications Technology Planning & Evaluation)

## References

1. Anwar, S., Tahir, M., Li, C., Mian, A., Khan, F.S., Muzaffar, A.W.: Image colorization: a survey and dataset. arXiv preprint arXiv:2008.10774 (2020)
2. Baker, N., Lu, H., Erlikhman, G., Kellman, P.J.: Deep convolutional networks do not classify based on global object shape. PLoS Comput. Biol. **14**(12), e1006613 (2018)
3. Cao, Y., Zhou, Z., Zhang, W., Yu, Y.: Unsupervised diverse colorization via generative adversarial networks. In: Ceci, M., Hollmén, J., Todorovski, L., Vens, C., Džeroski, S. (eds.) ECML PKDD 2017. LNCS (LNAI), vol. 10534, pp. 151–166. Springer, Cham (2017). https://doi.org/10.1007/978-3-319-71249-9_10
4. Carlucci, F.M., D'Innocente, A., Bucci, S., Caputo, B., Tommasi, T.: Domain generalization by solving jigsaw puzzles. In: Proceedings of the IEEE Conference on Computer Vision and Pattern Recognition (CVPR), pp. 2229–2238 (2019)
5. Charpiat, G., Hofmann, M., Schölkopf, B.: Automatic image colorization via multimodal predictions. In: Forsyth, D., Torr, P., Zisserman, A. (eds.) ECCV 2008. LNCS, vol. 5304, pp. 126–139. Springer, Heidelberg (2008). https://doi.org/10.1007/978-3-540-88690-7_10
6. Cheng, Z., Yang, Q., Sheng, B.: Deep colorization. In: Proceedings of the IEEE International Conference on Computer Vision, pp. 415–423 (2015)
7. Chia, A.Y.S., Zhuo, S., Gupta, R.K., Tai, Y.W., Cho, S.Y., Tan, P., Lin, S.: Semantic colorization with internet images. ACM Trans. Graph. (TOG) **30**(6), 1–8 (2011)
8. D'Innocente, A., Caputo, B.: Domain generalization with domain-specific aggregation modules. In: Brox, T., Bruhn, A., Fritz, M. (eds.) GCPR 2018. LNCS, vol. 11269, pp. 187–198. Springer, Cham (2019). https://doi.org/10.1007/978-3-030-12939-2_14
9. Finn, C., Abbeel, P., Levine, S.: Model-agnostic meta-learning for fast adaptation of deep networks. In: Proceedings of the International Conference on Machine Learning (ICML), pp. 1126–1135. PMLR (2017)
10. Ganin, Y., Ustinova, E., Ajakan, H., Germain, P., Larochelle, H., Laviolette, F., Marchand, M., Lempitsky, V.: Domain-adversarial training of neural networks. J. Mach. Learn. Res. **17**(1), 2030–2096 (2016)
11. Geirhos, R., Rubisch, P., Michaelis, C., Bethge, M., Wichmann, F.A., Brendel, W.: Imagenet-trained cnns are biased towards texture; increasing shape bias improves accuracy and robustness. arXiv preprint arXiv:1811.12231 (2018)
12. Gupta, R.K., Chia, A.Y.S., Rajan, D., Ng, E.S., Zhiyong, H.: Image colorization using similar images. In: Proceedings of the 20th ACM International Conference on Multimedia, pp. 369–378 (2012)
13. Hermann, K., Chen, T., Kornblith, S.: The origins and prevalence of texture bias in convolutional neural networks. Adv. Neural. Inf. Process. Syst. **33**, 19000–19015 (2020)
14. Huang, X., Belongie, S.: Arbitrary style transfer in real-time with adaptive instance normalization. In: Proceedings of the IEEE International Conference on Computer Vision, pp. 1501–1510 (2017)

15. Iizuka, S., Simo-Serra, E., Ishikawa, H.: Let there be color! joint end-to-end learning of global and local image priors for automatic image colorization with simultaneous classification. ACM Trans. Graph. (ToG) **35**(4), 1–11 (2016)
16. Ironi, R., Cohen-Or, D., Lischinski, D.: Colorization by example. Rendering Techn. **29**, 201–210 (2005)
17. Isola, P., Zhu, J.Y., Zhou, T., Efros, A.A.: Image-to-image translation with conditional adversarial networks. In: Proceedings of the IEEE Conference on Computer Vision and Pattern Recognition, pp. 1125–1134 (2017)
18. Kim, M., Byun, H.: Learning texture invariant representation for domain adaptation of semantic segmentation. In: Proceedings of the IEEE/CVF Conference on Computer Vision and Pattern Recognition, pp. 12975–12984 (2020)
19. Kingma, D.P., Ba, J.: Adam: a method for stochastic optimization. arXiv preprint arXiv:1412.6980 (2014)
20. Lei, C., Chen, Q.: Fully automatic video colorization with self-regularization and diversity. In: Proceedings of the IEEE/CVF Conference on Computer Vision and Pattern Recognition, pp. 3753–3761 (2019)
21. Levin, A., Lischinski, D., Weiss, Y.: Colorization using optimization. In: ACM SIGGRAPH 2004 Papers, pp. 689–694 (2004)
22. Li, C., Wand, M.: Precomputed real-time texture synthesis with markovian generative adversarial networks. In: Leibe, B., Matas, J., Sebe, N., Welling, M. (eds.) ECCV 2016. LNCS, vol. 9907, pp. 702–716. Springer, Cham (2016). https://doi.org/10.1007/978-3-319-46487-9_43
23. Li, D., Yang, Y., Song, Y.Z., Hospedales, T.M.: Deeper, broader and artier domain generalization. In: Proceedings of the IEEE International Conference on Computer Vision (ICCV) (2017)
24. Li, D., Yang, Y., Song, Y.Z., Hospedales, T.M.: Learning to generalize: meta-learning for domain generalization. In: Proceedings of the AAAI Conference on Artificial Intelligence, vol. 32 (2018)
25. Li, H., Pan, S.J., Wang, S., Kot, A.C.: Domain generalization with adversarial feature learning. In: Proceedings of the IEEE Conference on Computer Vision and Pattern Recognition, pp. 5400–5409 (2018)
26. Li, Y., et al.: Deep domain generalization via conditional invariant adversarial networks. In: Proceedings of the European Conference on Computer Vision (ECCV), pp. 624–639 (2018)
27. Liu, X., Wan, L., Qu, Y., Wong, T.T., Lin, S., Leung, C.S., Heng, P.A.: Intrinsic colorization. In: ACM SIGGRAPH Asia 2008 papers, pp. 1–9 (2008)
28. Mao, X., Li, Q., Xie, H., Lau, R.Y., Wang, Z., Paul Smolley, S.: Least squares generative adversarial networks. In: Proceedings of the IEEE International Conference on Computer Vision, pp. 2794–2802 (2017)
29. Muandet, K., Balduzzi, D., Schölkopf, B.: Domain generalization via invariant feature representation. In: Proceedings of the International Conference on Machine Learning (ICML), pp. 10–18. PMLR (2013)
30. Nam, H., Lee, H., Park, J., Yoon, W., Yoo, D.: Reducing domain gap by reducing style bias. In: Proceedings of the IEEE/CVF Conference on Computer Vision and Pattern Recognition, pp. 8690–8699 (2021)
31. Nazeri, K., Ng, E., Ebrahimi, M.: Image colorization using generative adversarial networks. In: Perales, F.J., Kittler, J. (eds.) AMDO 2018. LNCS, vol. 10945, pp. 85–94. Springer, Cham (2018). https://doi.org/10.1007/978-3-319-94544-6_9
32. Ronneberger, O., Fischer, P., Brox, T.: U-Net: convolutional Networks for Biomedical Image Segmentation. In: Navab, N., Hornegger, J., Wells, W.M., Frangi,

A.F. (eds.) MICCAI 2015. LNCS, vol. 9351, pp. 234–241. Springer, Cham (2015). https://doi.org/10.1007/978-3-319-24574-4_28
33. Sagawa, S., Koh, P.W., Hashimoto, T.B., Liang, P.: Distributionally robust neural networks for group shifts: on the importance of regularization for worst-case generalization. arXiv preprint arXiv:1911.08731 (2019)
34. Su, J.W., Chu, H.K., Huang, J.B.: Instance-aware image colorization. In: Proceedings of the IEEE/CVF Conference on Computer Vision and Pattern Recognition, pp. 7968–7977 (2020)
35. Sun, B., Saenko, K.: Deep CORAL: correlation alignment for deep domain adaptation. In: Hua, G., Jégou, H. (eds.) ECCV 2016. LNCS, vol. 9915, pp. 443–450. Springer, Cham (2016). https://doi.org/10.1007/978-3-319-49409-8_35
36. Vapnik, V.: Statistical Learning Theory. Wiley, New York (1998)
37. Venkateswara, H., Eusebio, J., Chakraborty, S., Panchanathan, S.: Deep hashing network for unsupervised domain adaptation. In: Proceedings of the IEEE Conference on Computer Vision and Pattern Recognition, pp. 5018–5027 (2017)
38. Vitoria, P., Raad, L., Ballester, C.: Chromagan: adversarial picture colorization with semantic class distribution. In: The IEEE Winter Conference on Applications of Computer Vision, pp. 2445–2454 (2020)
39. Wang, Y., Li, H., Kot, A.C.: Heterogeneous domain generalization via domain mixup. In: ICASSP 2020–2020 IEEE International Conference on Acoustics, Speech and Signal Processing (ICASSP), pp. 3622–3626. IEEE (2020)
40. Welsh, T., Ashikhmin, M., Mueller, K.: Transferring color to greyscale images. In: Proceedings of the 29th Annual Conference on Computer Graphics and Interactive Techniques, pp. 277–280 (2002)
41. Xu, M., et al.: Adversarial domain adaptation with domain mixup. In: Proceedings of the AAAI Conference on Artificial Intelligence, vol. 34, pp. 6502–6509 (2020)
42. Yan, S., Song, H., Li, N., Zou, L., Ren, L.: Improve unsupervised domain adaptation with mixup training. arXiv preprint arXiv:2001.00677 (2020)
43. Zhang, R., Isola, P., Efros, A.A.: Colorful image colorization. In: Leibe, B., Matas, J., Sebe, N., Welling, M. (eds.) ECCV 2016. LNCS, vol. 9907, pp. 649–666. Springer, Cham (2016). https://doi.org/10.1007/978-3-319-46487-9_40
44. Zhang, R., et al.: Real-time user-guided image colorization with learned deep priors. arXiv preprint arXiv:1705.02999 (2017)
45. Zhao, J., Han, J., Shao, L., Snoek, C.G.M.: Pixelated semantic colorization. Int. J. Comput. Vis. **128**(4), 818–834 (2019). https://doi.org/10.1007/s11263-019-01271-4

# Generating Natural Images with Direct Patch Distributions Matching

Ariel Elnekave(✉) and Yair Weiss

The Hebrew University of Jerusalem, Jerusalem, Israel
{Ariel.Elnekave,Yair.Weiss}@mail.huji.ac.il

**Abstract.** Many traditional computer vision algorithms generate realistic images by requiring that each patch in the generated image be similar to a patch in a training image and vice versa. Recently, this classical approach has been replaced by adversarial training with a patch discriminator. The adversarial approach avoids the computational burden of finding nearest neighbors of patches but often requires very long training times and may fail to match the distribution of patches.

In this paper we leverage the Sliced Wasserstein Distance to develop an algorithm that explicitly and efficiently minimizes the distance between patch distributions in two images. Our method is conceptually simple, requires no training and can be implemented in a few lines of codes. On a number of image generation tasks we show that our results are often superior to single-image-GANs, and can generate high quality images in a few seconds. Our implementation is publicly available at https://github.com/ariel415el/GPDM.

## 1 Introduction

In a wide range of computer vision problems (e.g. image retargeting, super-resolution, novel view synthesis) an algorithm needs to generate a realistic image as an output. A classical approach to ensuring that the output image appears realistic is based on *local patches* [1–3]: if each patch in the output image is similar to a patch in a training image, we can assume that the generated image will be realistic. Similarly, if each patch in the generated image is similar to a patch in a Van-Gogh painting, we can assume that the generated image captures the "style" of Van-Gogh. This insight led to a large number of papers over the past two decades that use patch nearest neighbors to ensure high quality image outputs [3–6].

One observation shared by successful methods based on patches is that the similarity between patches should be *bidirectional*: it is not enough to require that each patch in the generated image be similar to a patch in the training image. Consider a generated image that consists of many repetitions of a single

**Supplementary Information** The online version contains supplementary material available at https://doi.org/10.1007/978-3-031-19790-1_33.

© The Author(s), under exclusive license to Springer Nature Switzerland AG 2022
S. Avidan et al. (Eds.): ECCV 2022, LNCS 13677, pp. 544–560, 2022.
https://doi.org/10.1007/978-3-031-19790-1_33

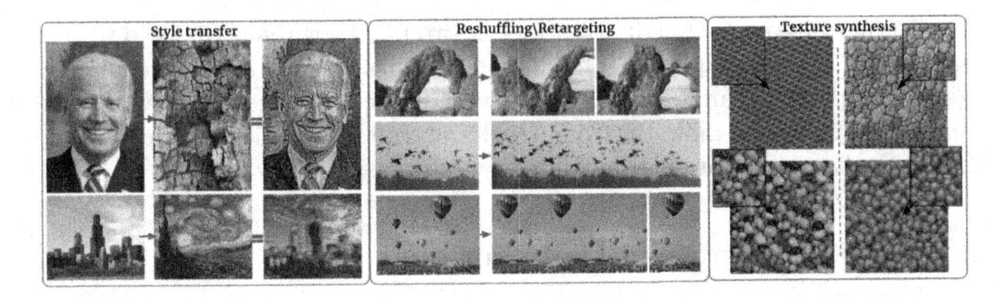


**Fig. 1.** By efficiently matching the distribution of patches between images we can solve a broad spectrum of single-image generative tasks without training a per-image GAN or computing patch nearest neighbors.

patch from a Van-Gogh image: even though each patch in the generated image is similar to a patch in the target image, no one would consider such a generated image to capture the "style" of Van-Gogh. In order to rule out such solutions, the bidirectional similarity (BDS) method used in [1,2,7] also requires that each patch in the training image be similar to a patch in the generated image. As we show in Sect. 2.1, while bidirectional similarity indeed helps push the distribution of patches in the generated image towards that of the target image, it still falls short of matching the distributions. Furthermore, optimizing the BDS loss function requires finding nearest neighbors of patches and this is both memory and computation intensive (given M patches in each image, BDS is based on an $M^2$ matrix of similarities between all pairs of patches).

In recent years, these classical, patch-based approaches have been overtaken by Generative Adversarial Networks (GANs) and related methods [8–13]. In the adversarial approach, a discriminator is trained to classify patches at different scales as "real" or "fake" and it can be shown that under certain conditions training with an adversarial loss and a patch discriminator is equivalent to minimizing the distance between the distribution of patches in the generated image and the training image [14,15]. The requirement that the generated image has the same *distribution* over patches as the target image addresses the limitation of early patch-based approaches that simply required that each patch in the generated image be similar to a patch in the training image. Indeed Single-Image GANs [9,10] have yielded impressive results in generating novel images that have approximately the same distribution over patches as the target image.

Despite the considerable success of GAN-based methods, they have some notable disadvantages. While there are theoretical guarantees that globally optimizing the adversarial objective is equivalent to optimizing the distance between distributions, in practice GAN training often suffers from "mode collapse" [16] and the generated image may contain only few types of possible patches. Furthermore, GAN training is computationally intensive and a separate generator needs to be trained for different image tasks.

In this paper we leverage the previously proposed Sliced Wasserstein Distance [17–19] to develop an algorithm that explicitly and efficiently minimizes the distance between patch distributions in two images without the need to compute patch nearest neighbors. Our method is conceptually simple, requires no training and can be implemented in a few lines of codes. On a number of image generation tasks we show that our results are often superior to single-image-GANs, and can generate high quality images in a few seconds.

## 2 Distances Between Distributions

Given $M$ patches in two images, how do we compute the distance between the distribution of patches in the two images? The Wasserstein (or Earth Movers) distance between two distributions $P, Q$ is defined as:

$$W(P,Q) = \inf_{\gamma \in \Pi(P,Q)} E_{x,y \sim \gamma} \|x - y\| \tag{1}$$

where $\Pi(P,Q)$ denotes the set of joint distributions whose marginal probabilities are $P, Q$. Intuitively, $\Pi$ can be thought of as a soft correspondence between samples in $P$ and $Q$ and so the Wasserstein distance is the average distance between corresponding samples with the optimal correspondence. Calculating this optimal correspondence is computationally intensive ($O(M^{2.5})$ [17]) making it unsuitable for use as a loss function that we wish to optimize for many iterations.

The sliced Wasserstein distance (SWD) makes use of the fact that for one dimensional data, the optimal correspondence can be solved by simply sorting the samples and so the distance between two samples of size $M$ can be computed in $O(M \log M)$. For a projection vector $w$ define $P^w$ as the distribution of samples from $P$ projected in direction $w$, the Sliced Wasserstein Distance is defined as:

$$SWD(P,Q) = E_w W(P^w, Q^w) \tag{2}$$

where the expectation is over random unit norm vectors $\omega$.

### 2.1 Properties and Comparisons

As mentioned in the introduction, many classical approaches to comparing patch probability distributions are based on bidirectional similarity. Suppose we are given a set of samples $\{p_i\}, \{q_j\}$ from two distributions $P, Q$ the Bidirectional Similarity (BDS) is defined as:

$$BDS(P,Q) = \frac{1}{M} \sum_i \min_j \|p_i - q_j\| + \frac{1}{M} \sum_j \min_i \|q_j - p_i\|$$

The first term ("coherence") measures the average distance between a patch in $\{p_i\}$ and its closest patch in $\{q_j\}$ and the second term ("completeness") measures the average distance between a patch in $\{q_i\}$ and its closest patch in $\{p_j\}$. Thus two images are judged to be similar if each patch in one image has a close match in the second image and vice versa.

Kolkin et al., [20] used a closely related measure which they called the "Relaxed Earth Movers Distance" (REMD):

$$REMD(P,Q) = \max(\frac{1}{M}\sum_i \min_j \|p_i - q_j\|, \frac{1}{M}\sum_j \min_i \|q_j - p_i\|)$$

Again, two images are judged to be similar if each patch in one image has a close match in the second image and vice versa.

Thus $SWD(P,Q)$, $BDS(P,Q)$ and $REMD(P,Q)$ are all methods to measure the similarity between the patch distributions. Why should one method be preferred over the others? The following theorem shows that neither BDS nor REMD can be considered as distance metric between distributions, while SWD can.

**Theorem:** $SWD(P,Q) = 0$ if and only if $P = Q$. On the other hand for both BDS and REMD, there exist an infinite number of pairs of distributions $P, Q$ that are arbitrarily different (i.e. $W(P,Q)$ is arbitrarily large) and yet $BDS(P,Q) = 0$ and $REMD(P,Q) = 0$.

**Proof:** The fact that $SWD(P,Q) = 0$ if and only if $P = Q$ follows from the fact that the Wasserstein distance is a metric [17]. To see that neither BDS nor REMD are metrics, note that any two discrete distributions that have the same support will satisfy $BDS(P,Q) = 0$ and $REMD(P,Q) = 0$ regardless of the densities on the support. This is because a single sample in one distribution can serve as an exact match for an arbitrarily large number of samples in the other distribution. For example, suppose $P$ and $Q$ are both distributions over the set $\{0, a\}$ for some constant $a$ and $P(0) = \epsilon, Q(0) = 1 - \epsilon$. For any $\epsilon$, $BDS(P,Q) = REMD(P,Q) = 0$ (since all samples from $P$ will have an exact matching sample in $Q$ and vice-versa) even though as $\epsilon \to 0$ $W(P,Q) \to a$. By increasing $a$ we can increase $W(P,Q)$ arbitrarily. ■

To illustrate the difference between SWD, BDS and REMD in the context of image patches, consider three images of sky and grass, each with 1000 patches. Images A and B, have 999 sky patches and one grass patch, while image C has 1 sky patch and 999 grass patches. Note that when comparing A and B, and when comparing B and C *both coherence and completeness losses will be zero.* This means that BDS and REMD will consider A and B (which have the same patch distribution) to be as similar as B and C (which have very different distributions). In contrast, the SWD will be zero if and only if the two distributions are identical, and $SWD(A,B)$ will be much lower than $SWD(B,C)$.

## 2.2 SWD for Image Patch Distributions

SWD has been previously suggested for use as a training loss for different image processing methods [17,19,21,22]. Here we point out that using SWD to measure the similarity of patch distributions in two images allows us to efficiently optimize the patch distribution similarity.

As pointed out in [18], while exact SWD requires an integral over all unit norm filters $w$, an *approximate* SWD can be obtained by considering a set of $k$ random unit vectors, $\{w_i\}$:

$$\tilde{SWD}(P,Q) = \frac{1}{k}\sum_i W(P^{w_i}, Q^{w_i})$$

While for any fixed set of $k$ vectors the approximate SWD is not the same as the exact SWD, by taking a gradient of the approximate SWD we obtain an *unbiased* estimate of the gradient of the exact SWD. The fact that the estimate is unbiased means that by changing the set of $k$ random vectors at each iterations, we can efficiently optimize the exact SWD, just as is done in training of neural networks with stochastic gradient descent.

Our additional observation is that for a single projection $w$, the calculation of $\tilde{SWD}$ requires convolving the two images with $w$, sorting the two convolved images with respect to their value and then calculating the L1 distance between the sorted vectors. Similarly, the derivative of the $\tilde{SWD}$ with respect to an image requires a second convolution of the thresholded and sorted difference image with a flipped version of $w$. Combining this observation with the result of [18] means that a stochastic gradient update with respect to the SWD of the patch distribution in two images can be performed with two convolutions and a sorting operation.

Our full algorithm, which we call Generative Patch Distribution Matching (GPDM) is given in Algorithm 1. Note that it allows us to optimize the difference between patch distributions in two images *without finding patch nearest neighbors* and at a complexity of $O(M \log M)$.

**Algorithm 1.** A pseudo-code of the GPDM module where SWD over sets of patches in two images is computed and differentiated through. The "flat/unflat" operators reshapes a tensor into a vector and vice-versa.

**Input**: Target image x, initial guess $\hat{y}$,
learning rate $\beta$
**Output**: Optimized image y

1: $y \leftarrow \hat{y}$
2: **while** not converged **do**
3: $\quad L \leftarrow 0$
4: $\quad$ **for** i=1,k **do**
5: $\quad\quad \omega \sim N(0, \sigma I)$
6: $\quad\quad \omega \leftarrow unflat(\frac{\omega}{\|\omega\|})$
7: $\quad\quad p \leftarrow flat(conv2d(x, \omega))$
8: $\quad\quad q \leftarrow flat(conv2d(y, \omega))$
9: $\quad\quad L \leftarrow L + \frac{1}{k}|sort(p) - sort(q)|$
10: $\quad$ **end for**
11: $\quad y \leftarrow y - \beta\nabla_y L$
12: **end while**

## 3 Method

Our method uses the same multi scale structure as previous works [5,7,9]. At each level an initial guess is transformed into an output image which is either used as an initial guess for the next level or is the final output. More formally, given a target image $x$ we build an image pyramid out of it $(x_0, x_1, ....x_n)$ specified with a downscale ratio $r < 1$ and a minimal height for the coarsest level. We start with an initial guess $\hat{y_n}$ of the same size as $x_n$ and at each level $i$ we optimize the initial guess using Algorithm 1 to minimize its patch-SWD with $x_i$. The optimization output $y_i$ is a final output or up-scaled by $\frac{1}{r}$ to serve as an initial guess for the next level, $i+1$, optimization. The first initial guess can be a blurred version of the target, a color map or simple pixel noise.

The same learning rate and number of Adam steps is used to optimize SWD at all scales. Images are normalized to $[-1, 1]$ and the optimized image values are clipped to this interval at the end of each level optimization.

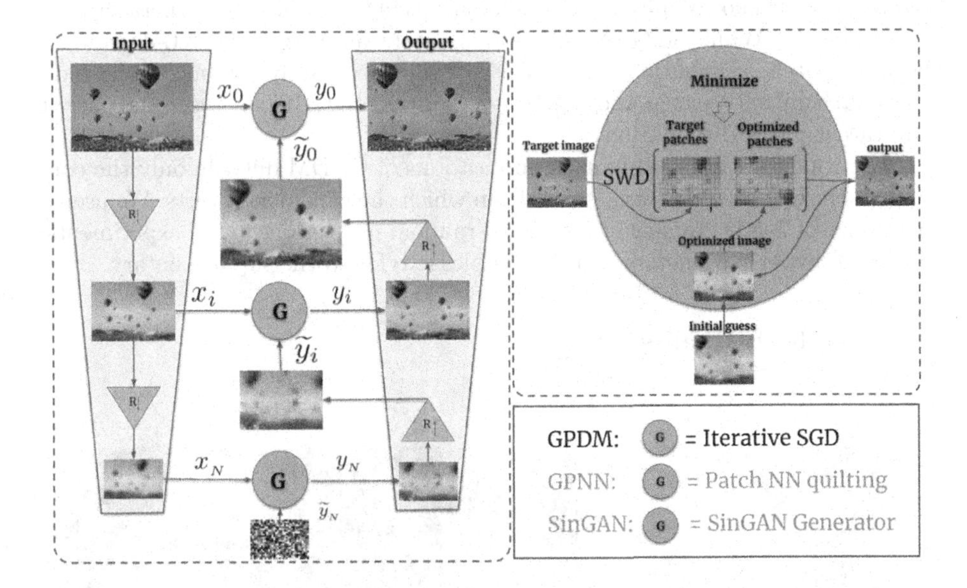


**Fig. 2. Left:** GPDM's multi-scale architecture: At each scale, $i$, an image is optimized to have similar patch distribution as the target $x_i$. **Right:** The generation module $G$ is an optimization process of the differentiable patch-distribution metric $SWD(x, y)$. Triangles marked with **R** stand for up/downscale.

Figure 2 (based on a similar figure in [7]) shows the overall coarse to fine structure and the minimization process at each scale.

## 4 Experiments

In this section we conduct experiments to compare GPDM to other methods. We first compare the synthesis quality on a number of different single image generation tasks and later compare the algorithms in terms of their run-time efficiency.

The methods we compare to are SinGAN [9] as well as a recent method, GPNN [7] which approximately optimizes the bidirectional similarity between the generated image and the target image. GPNN generates a new image by copying patches from the training image in a coarse to fine manner: at each iteration it searches for patches in the training image that are closest to the current estimated image and then aggregates these patches to form a new image. By construction, this method achieves high coherence (since all patches in the new image are copied from the training image) and completeness is encouraged by transforming distances into similarities using a free parameter $\alpha$. When $\alpha \rightarrow \infty$ the patch with maximal similarity is the patch in the second image with minimal L2 distance, but for small $\alpha$ patches in the second image that have already been used will receive low similarity.

Our method, SINGAN and GPNN all use exactly the same coarse-to-fine strategy but as shown in Fig. 2, the difference is in the form of the generator used at each scale. While SinGAN [9] approximately minimizes the KL-divergence between patch distributions and GPNN [7] approximately minimizes the bidirectional similarity, our method directly minimizes the SWD between the output and target patch distribution.

The configuration used in each applications of GPDM differ in only the content and size of the first initial guess from which the generation starts. We used a patch size of 7, 300 gradient steps and 64 random projections for all experiments. For more hyper-parameters and details please refer to the supplementary.

### 4.1 Synthesis Quality

**Fig. 3.** Image reshuffling on images from the Places50 and SIGD16 selected by [7].

**Image Reshuffling:** We first compare the three algorithms on the image reshuffling task. Given a natural image the task is to generate more images from the same scene which are different from the original one but show the same scene and are visually coherent. For this task, we start the optimization from a small noise image. Figure 3 visually compares our method to [7,9] on the same images from Fig. 4 in [7]. It can be seen that our method provides comparable visual quality to that of GPNN, and both methods generate more realistic and artifact-free images compared to SinGAN. Some of the images in Fig. 3 generated by GPDM do show more artifacts compare to GPNN but they are also much less similar to the reference image.

**Table 1.** Qualitative comparison image-reshuffling.

| Dataset | Method | SFID↓ | Diversity↑ | Dataset | Method | SFID↓ | Diversity↑ |
|---|---|---|---|---|---|---|---|
| Places5 | Ours | 0.068 | 0.56 | SIGD16 | Ours | 0.069 | 0.67 |
| | GPNN | 0.065 | 0.5 | | GPNN | 0.122 | 0.52 |
| | SINGAN | 0.082 | 0.5 | | SINGAN | 0.172 | 0.49 |

Table 1 shows a quantitative comparison between our method and [7,9] using the SFID metric (a full reference image quality metric described in [9]) and also using diversity, i.e. the normalized per-pixel standard deviation over 50 generated images [9]. Good results should have high diversity (i.e. the generated images are not all identical to the target image) and low SFID (i.e. the generated images are statistically similar to the training image).

We used the Places50 and SIGD16 datasets from [7,9]. We recomputed the SIFID scores for SinGAN and GPNN using the published generated images sets from each paper's supplementary. Our scores differ slightly from those reported in [7], presumably due to different implementations. The numerical scores are consistent with the visual inspection: our method is comparable to GPNN in terms of image quality and diversity and both methods outperform SinGAN.

**Retargeting:** Figure 4 shows the results of our method in retargeting images into various aspect ratios. We start the optimization from a stretched version of the target's smallest pyramid level that matches the desired aspect ratio.

In such tasks, where the number of patches in the output and target images differ, we duplicate randomly selected patches from the smaller image so that we compute SWD on two equaly large sets of patches.

**Style-Transfer:** Figure 5 shows our results in style transfer: triplets of content, style and the mixed output generated by our method. These results are surprising due to the simplicity of our method. In this task we start the optimization from a rather big image or even use a single-scale configuration. We start the optimization from the content image and match the patch distribution to the style image.

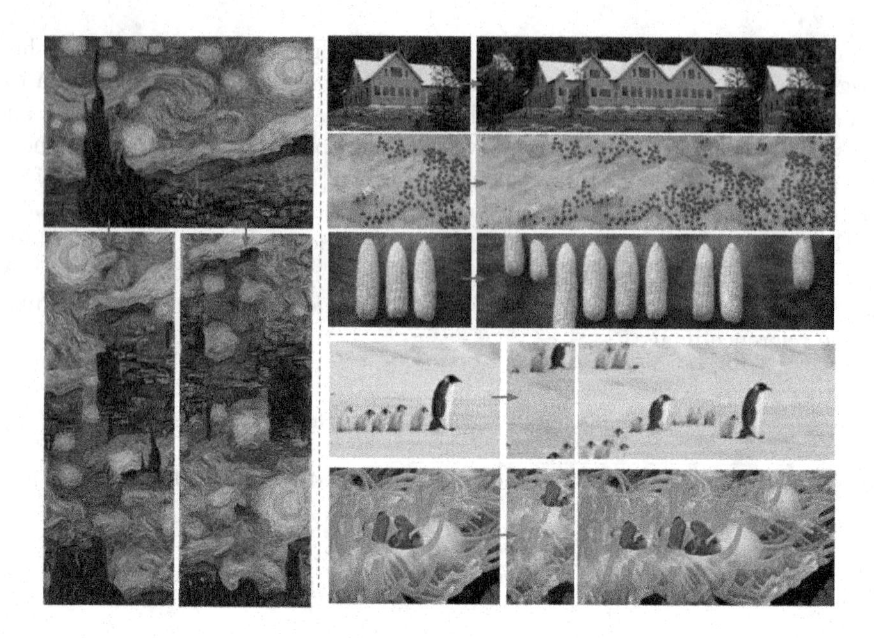

**Fig. 4.** Image retargeting results.

**Additional Tasks:** We also applied our method to generating texture images from texture samples. This is done similarly to retargeting but the initial condition is a noise image. Image editing can be performed by first crudely editing the image and then using our algorithm to harmonize its fine details so that it looks real. Figure 1 shows examples of texture synthesis and image editing is discussed in Sect. 6.

Results for additional tasks and more results of reshuffling, retargeting and style-transfer are available in the supplementary material.

### 4.2 Efficiency

A major disadvantage of GAN based methods is the need to train a generator and discriminator for every image. As reported in [7] this means that generating a new image of size $180 \times 250$ using SinGAN will take about one hour while GPNN (and other methods based on bidirectional similarity) take about two seconds. Our method also does not require any new training and the run times are similar to those of GPNN when the images are small. However, as the size of the image increases, the fact that bidirectional similarity requires computing $M^2$ distances (where $M$ is the number of patches in each image) means that GPNNs run time grows approximately quadratically and generating a novel image of size $1024 \times 678$ with GPNN takes more than half an hour on a GPU. In contrast, our method has complexity of $O(M \log M)$ and takes less than a minute to generate an image of the same size on the same GPU (see Fig. 6).

**Fig. 5.** Style/texture transfer: Each image triplet shows a content image, a style/texture image and the results of combining them with our synthesis method.

The runtime of methods based on bidirectional similarity can be improved by using approximate nearest neighbor search, rather than exact search. Table 2 shows an analysis of SWD compute time compared to an exact nearest neighbor search and its approximated counterpart with inverted index [23]. As can be seen, the use of an inverted index speeds up the search substantially.

While in some of our experiments we found that the approximation performs reasonably well, a major disadvantage of using approximate nearest neighbor in the context of GPNN is that it is impractical to modify the similarity metric during the image generation process. Recall that GPNN uses a parameter $\alpha$ to change the similarity of patches based on which patches have already been used but when an inverted index is used, this would require recomputing the inverted index at each iteration and the approximate search would become slower than exact search.

Figure 6 demonstrates this with an example of high resolution style transfer (see Fig. 5 for the two input images): while using approximate nearest neighbor (left) speeds up the computation, it requires using $\alpha \rightarrow \infty$ in GPDM and therefore creates lower quality images in which the same patches are reused many times. In contrast, GPNN with exact nearest neighbor (middle) can use a small $\alpha$ parameter and avoid the reuse of patches. Our method (right) is very fast and since it explicitly optimizes the distance between patch distributions, does a better job of capturing the style It should be noted that using GPNN with exact nearest neighbor search and with $\alpha \rightarrow \infty$ produces practically the same results as those generated with an approximate search, indicating that the drop in quality is due to the $\alpha$ parameter and not to the inaccuracy of the approximation. We refer the reader to our supplementary material where one can find further details about GPNN's $\alpha$ parameter and a quantitative evaluation of the accuracy of nearest neighbor approximation methods.

**Quality-Efficiency Tradeoff.** GPDM is an iterative algorithm and there is a natural trade-off between the number of iterations and the quality of the generated images. More random projections at each optimization step and more optimization steps help the patch-distributions to match more closely, ensuring more realistic outputs. We refer the interested reader to our supplementary material where we visualize this tradeoff by comparing SIFID [9] scores and average running times on the SIGD16 dataset compared for different number of SWD random projections.

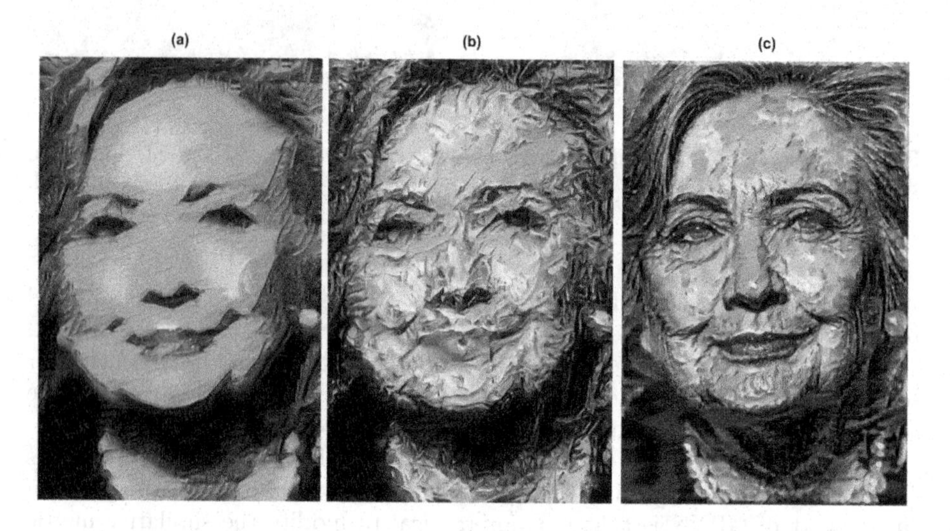


**Fig. 6.** High resolution style transfer images generated (a) GPNN (no-$\alpha$) with approximate nearest neighbor (~70 s, GPU), (b) GPNN ($\alpha = 0.005$) with exact nearest neighbor (~1900 s, GPU). (c) Our result (~60 s, GPU). $\alpha$ is the completeness enforcing parameter and no-$\alpha$ means no constraint.

**Table 2.** Compute time (seconds) of SWD (64 projections), exact nearest neighbor (FaissFlat) and approximated nearest neighbor (Inverted Index with $\sqrt{M}$ bins where M is the number of patches) for different image sizes. All computation are done on a NVIDIA TITAN X GPU.

| Image-size | $64^2$ | $128^2$ | $256^2$ | $512^2$ | $1024^2$ |
|---|---|---|---|---|---|
| SWD(64) | 0.002 | 0.006 | 0.021 | 0.086 | 0.335 |
| NN-exact | 0.086 | 0.110 | 0.336 | 4.144 | 73.23 |
| NN-IVF | 0.101 | 0.137 | 0.382 | 1.082 | 5.530 |

## 5 Related Work

The idea of synthesizing realistic images using patches from a target image goes back to Efros and Leung [24]. Efros and Freeman [3] and Hertzmann et al. [5] extended this idea to style transfer and other tasks. These classical, nonparametric approaches required finding patch nearest neighbors in high dimensions and a great deal of subsequent work attempted to make the search for nearest neighbors more efficient. The PatchMatch algorithm [2] pointed out that the coherence of patches in natural images can be used to greatly speed up the search for nearest neighbors and this enabled the use of these nonparametric techniques in real-time image editing. Our work is very much inspired by these classical papers but we use the SWD to directly optimize the similarity of patch distributions without computing nearest neighbors. Furthermore, in our approach patches in the synthesized image are not constrained to be direct copies of patches in the training image.

As mentioned in the introduction, a key insight behind successful patch-based methods is the use of some form of bidirectional similarity [1]: it is not enough to require that patches in the generated image be similar to patches in the training image. While [1] optimized the BDS directly, the GPNN approach [7] rewards bidirectional similarity indirectly by modifying the similarity measure to penalize patches that have already been used. In a parallel line of work [25–28] attempts are made to make a uniform use of all patches in the source image through patch histogram matching. Our approach optimizes a well-understood similarity (the Sliced Wasserstein Distance) that is guaranteed to be zero only if the two distributions are equal and importantly it optimizes this similarity very efficiently.

SINGAN and InGAN [9,10] are both variants of GANs that are trained on a single image with a patch based discriminator. Thus they can be seen as approximately matching the distribution of patches in the generated image and the target image. Our approach directly optimizes the similarity between patch distributions in the two images, requires no training in the traditional sense, and provides superior quality results. More recently [12], [13] were able to push the performance of SinGAN to match that of our method and that of GPNN but the train time are still orders of magnitude slower than our method.

The Sliced Wasserstein Distance was used in a number of image generation tasks but it is most often used to estimate the distance between distributions of full images [29]. Thus [22] train a generative model by replacing the GAN objective with an SWD objective. In contrast, here our focus is on estimating the distance between two patch distributions and we have shown that an unbiased estimate of the distance between patch distributions in the two images can be estimated using a single convolution.

Although not immediately apparent, neural style transfer [30] can be seen as a single image generative model that preserves patch distribution. [31] showed that the Gram loss in neural style transfer is equivalent to MMD [32] over features of intermediate layers of VGG hence the style transfer objective is to minimize distribution of patches (in size of the layers' receptive fields) between the optimized image and the style target. [33] also recognized this nature of the Gram loss and suggested replacing it with SWD between the VGG features. Similarly [34–36] use the closed form of the Wasserstein distance between Gaussians to push the spatial distribution in VGG feature maps of a content image into that of a style image and then decode it as a mix image.

The works of [4,37] closely relate to the neural patch distribution for style transfer. They both suggest objectives that focus on the coherence of neural patches in the synthesized image. Similar to [7], they enforce NN similarity of neural-patches but they compute similarity in a pre-trained neural network representations rather than in pixel space. Similar to the classic algorithms, both of these work require explicit computation of patch nearest neighbors. In contrast, our use of SWD allows avoiding the computation of patch nearest neighbors which make our algorithm much faster.

The work of [18] also use SWD for texture synthesis. They minimize SWD on wavelet coefficients with SGD to synthesize new samples from a given texture image. Our paper differs from theirs in that we work in multiple scales, on pixel-level and compute SWD in a more efficient way. We are able thus to produce much better results and to apply our method to more complicated generative tasks.

## 6 Limitations and Extensions

**Limitations:** As mentioned previously, approaches based on bidirectional similarity do not explicitly optimize the similarity of patch distributions while our method does. In some applications, the fact that our method attempts to reproduce the relative frequencies of different patches is a limitation. It is often easier to find an image that is coherent (i.e. all patches in the synthesized image come from the target image) than to find one that preserves the relative frequencies. Furthermore, in some tasks such as image editing, the desired image should have a different distribution over patches (e.g. if the user wishes to add a new object to the image). Finally, since our method is based on optimization, it is not guaranteed to find a good local minima and so the final results sometimes show artifacts as can be seen in Fig. 3.

**Extensions:** We propose here two extensions to our method. Both extensions are designed to soften the patch frequency constraint on the output allowing the generation of more diverse outputs.

The first extension of our method allows the user to explicitly manipulate the distribution over patches that is matched by our algorithm. Figure 7a illustrates this. Here the user first manipulates the target distribution of patches by indicating that certain patches should be increased in frequency and the algorithm then generates an image to match the augmented target distribution. As can be seen, this simple modification allows us to generate images with different number of objects based on the user's preference.

The second extension is to use an ensemble of images so that the patch distribution in the ensemble matches that of the target image. Formally, instead

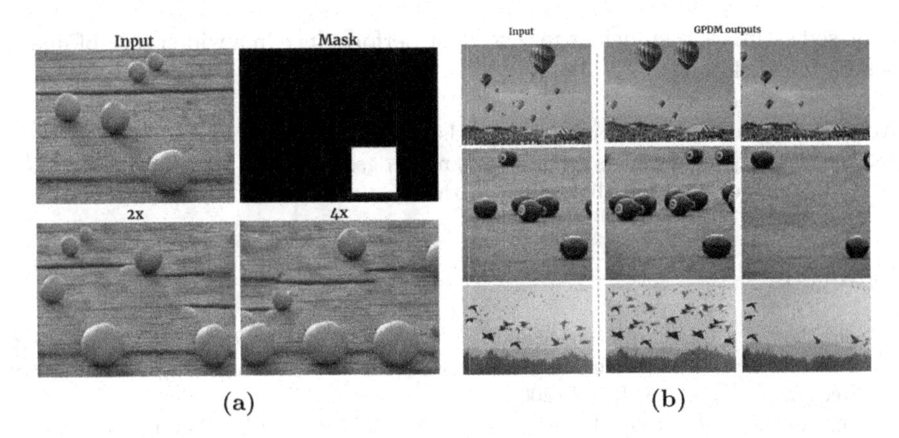


**Fig. 7.** Two Extensions of our method. (a): A mask that specifies target patches whose frequency should be increased is given as an additional input. This allows the user to manipulate the number of objects in the generated images. (b): Batched GPDM. Each row shows the two most diverse outputs out of 10 image batch.

of matching the patch distribution of a single output image to a single target we generate $N$ output images and match their total patch distribution to that of all the patches in $N$ copies of the target image. This way, while the total patch distribution is still strictly preserved, the distribution within each generated image can diverge from that of the original target image. For example when reshuffling 2 images to match the distribution of 2 copies of an image with black balls on grass, one output may contain no balls at all and the other will have twice as many balls while still preserving the frequency of patches in the two images together. Figure 7b shows the results from this algorithm which we call "batched-GPD": For each input image we generated 10 images in parallel and showed the two most different results.

This extension works surprisingly well and yields cleaner and more diverse results. Moreover since the memory footprint of GPDM is linear in the number of patches (in-place sorting) we were able to run in large batches reducing the generation time for a single image.

## 7 Conclusion

Classical approaches for generating realistic images are based on the insight that if we can match the distribution of patches between the generated image and the target image then we will have a realistic image. In this paper we have used the same insight but with a novel twist. We use the Sliced Wasserstein Distance which was recently used as a method to compare distributions of full images when training generative models and we showed that when applied to distributions of patches, an unbiased estimate of the SWD can be computed with a single convolution of the two images. Our experiments shows that this unbiased estimate is sufficient for excellent performance in a wide range of image generation tasks.

**Acknowledgements.** Support from the Israeli Ministry of Science and Technology and the Gatsby Foundation is gratefully acknowledged. We also thank the authors of [7] for answering our question about their method.

## References

1. Simakov, D., Caspi, Y., Shechtman, E., Irani, M.: Summarizing visual data using bidirectional similarity. In: 2008 IEEE Conference on Computer Vision and Pattern Recognition, pp. 1–8. IEEE (2008)
2. Barnes, C., Shechtman, E., Finkelstein, A., Goldman, D.B.: PatchMatch: a randomized correspondence algorithm for structural image editing. ACM Trans. Graph. **28**(3), 24 (2009)
3. Efros, A.A., Freeman, W.T.: Image quilting for texture synthesis and transfer. In: Proceedings of the 28th Annual Conference on Computer Graphics and Interactive Techniques, pp. 341–346 (2001)
4. Mechrez, R., Talmi, I., Zelnik-Manor, L.: The contextual loss for image transformation with non-aligned data. In: Proceedings of the European Conference on Computer Vision (ECCV), pp. 768–783 (2018)
5. Hertzmann, A., Jacobs, C.E., Oliver, N., Curless, B., Salesin, D.H.: Image analogies. In: Proceedings of the 28th Annual Conference on Computer Graphics and Interactive Techniques, pp. 327–340 (2001)
6. Michaeli, T., Irani, M.: Blind deblurring using internal patch recurrence. In: Fleet, D., Pajdla, T., Schiele, B., Tuytelaars, T. (eds.) ECCV 2014. LNCS, vol. 8691, pp. 783–798. Springer, Cham (2014). https://doi.org/10.1007/978-3-319-10578-9_51
7. Granot, N., Feinstein, B., Shocher, A., Bagon, S., Irani, M.: Drop the GAN: in defense of patches nearest neighbors as single image generative models. In: Proceedings of the IEEE/CVF Conference on Computer Vision and Pattern Recognition, pp. 13460–13469 (2022)
8. Iizuka, S., Simo-Serra, E., Ishikawa, H.: Globally and locally consistent image completion. ACM Trans. Graph. (ToG) **36**(4), 1–14 (2017)
9. Shaham, T.R., Dekel, T., Michaeli, T.: SinGAN: learning a generative model from a single natural image. In: Proceedings of the IEEE/CVF International Conference on Computer Vision, pp. 4570–4580 (2019)
10. Shocher, A., Bagon, S., Isola, P., Irani, M.: InGAN: capturing and retargeting the "DNA" of a natural image. In: Proceedings of the IEEE/CVF International Conference on Computer Vision, pp. 4492–4501 (2019)
11. Wang, T.-C., Liu, M.-Y., Zhu, J.-Y., Tao, A., Kautz, J., Catanzaro, B.: High-resolution image synthesis and semantic manipulation with conditional GANs. In: Proceedings of the IEEE Conference on Computer Vision and Pattern Recognition (CVPR) (2018)
12. Hinz, T., Fisher, M., Wang, O., Wermter, S.: Improved techniques for training single-image GANs. In: Proceedings of the IEEE/CVF Winter Conference on Applications of Computer Vision, pp. 1300–1309 (2021)
13. Sushko, V., Zhang, D., Gall, J., Khoreva, A.: Generating novel scene compositions from single images and videos. arXiv preprint arXiv:2103.13389 (2021)

14. Arjovsky, M., Chintala, S., Bottou, L.: Wasserstein generative adversarial networks. In: International Conference on Machine Learning, pp. 214–223. PMLR (2017)
15. Goodfellow, I., et al.: Generative adversarial nets. In: Advances in Neural Information Processing Systems, vol. 27 (2014)
16. Srivastava, A., Valkov, L., Russell, C., Gutmann, M.U., Sutton, C.: VEEGAN: reducing mode collapse in GANs using implicit variational learning. In: Proceedings of the 31st International Conference on Neural Information Processing Systems, pp. 3310–3320 (2017)
17. Pitie, F., Kokaram, A.C., Dahyot, R.: N-dimensional probability density function transfer and its application to color transfer. In: Tenth IEEE International Conference on Computer Vision (ICCV 2005), vol. 1, 2, pp. 1434–1439. IEEE (2005)
18. Rabin, J., Peyré, G., Delon, J., Bernot, M.: Wasserstein barycenter and its application to texture mixing. In: Bruckstein, A.M., ter Haar Romeny, B.M., Bronstein, A.M., Bronstein, M.M. (eds.) SSVM 2011. LNCS, vol. 6667, pp. 435–446. Springer, Heidelberg (2012). https://doi.org/10.1007/978-3-642-24785-9_37
19. Bonneel, N., Rabin, J., Peyré, G., Pfister, H.: Sliced and radon Wasserstein barycenters of measures. J. Math. Imaging Vis. **51**(1), 22–45 (2015)
20. Kolkin, N., Salavon, J., Shakhnarovich, G.: Style transfer by relaxed optimal transport and self-similarity. In: Proceedings of the IEEE/CVF Conference on Computer Vision and Pattern Recognition, pp. 10051–10060 (2019)
21. Kolouri, S., Pope, P.E., Martin, C.E., Rohde, G.K.: Sliced-Wasserstein autoencoder: an embarrassingly simple generative model. arXiv preprint arXiv:1804.01947 (2018)
22. Deshpande, I., Zhang, Z., Schwing, A.G.: Generative modeling using the sliced Wasserstein distance. In: Proceedings of the IEEE Conference on Computer Vision and Pattern Recognition, pp. 3483–3491 (2018)
23. Johnson, J., Douze, M., Jégou, H.: Billion-scale similarity search with GPUs. IEEE Trans. Big Data **7**(3), 535–547 (2019)
24. Efros, A.A., Leung, T.K.: Texture synthesis by non-parametric sampling. In: Proceedings of the Seventh IEEE International Conference on Computer Vision, vol. 2, pp. 1033–1038. IEEE (1999)
25. Kopf, J., Fu, C.-W., Cohen-Or, D., Deussen, O., Lischinski, D., Wong, T.-T.: Solid texture synthesis from 2D exemplars. In: ACM SIGGRAPH 2007 Papers, p. 2-es (2007)
26. Kaspar, A., Neubert, B., Lischinski, D., Pauly, M., Kopf, J.: Self tuning texture optimization. In: Computer Graphics Forum, vol. 34, pp. 349–359. Wiley Online Library (2015)
27. Jamriška, O., Fišer, J., Asente, P., Jingwan, L., Shechtman, E., Sỳkora, D.: LazyFluids: appearance transfer for fluid animations. ACM Trans. Graph. (TOG) **34**(4), 1–10 (2015)
28. Fišer, J., Jamriška, O., Lukáč, M., Shechtman, E., Asente, P., Jingwan, L., Sỳkora, D.: StyLit: illumination-guided example-based stylization of 3D renderings. ACM Trans. Graph. (TOG) **35**(4), 1–11 (2016)
29. Karras, T., Aila, T., Laine, S., Lehtinen, J.: Progressive growing of GANs for improved quality, stability, and variation. arXiv preprint arXiv:1710.10196 (2017)
30. Gatys, L.A., Ecker, A.S., Bethge, M.: Image style transfer using convolutional neural networks. In: Proceedings of the IEEE Conference on Computer Vision and Pattern Recognition, pp. 2414–2423 (2016)
31. Li, Y., Wang, N., Liu, J., Hou, X.: Demystifying neural style transfer. arXiv preprint arXiv:1701.01036 (2017)

32. Gretton, A., Borgwardt, K., Rasch, M., Schölkopf, B., Smola, A.: A kernel method for the two-sample-problem. In: Advances in Neural Information Processing Systems, vol. 19, pp. 513–520 (2006)
33. Heitz, E., Vanhoey, K., Chambon, T., Belcour, L.: A sliced Wasserstein loss for neural texture synthesis. In: Proceedings of the IEEE/CVF Conference on Computer Vision and Pattern Recognition, pp. 9412–9420 (2021)
34. Li, Y., Fang, C., Yang, J., Wang, Z., Lu, X., Yang, M.-H.: Universal style transfer via feature transforms. In: Advances in Neural Information Processing Systems, vol. 30 (2017)
35. Mroueh, Y.: Wasserstein style transfer. arXiv preprint arXiv:1905.12828 (2019)
36. Li, P., Zhao, L., Xu, D., Lu, D.: Optimal transport of deep feature for image style transfer. In: Proceedings of the 2019 4th International Conference on Multimedia Systems and Signal Processing, pp. 167–171 (2019)
37. Li, C., Wand, M.: Combining Markov random fields and convolutional neural networks for image synthesis. In: Proceedings of the IEEE Conference on Computer Vision and Pattern Recognition, pp. 2479–2486 (2016)

# Context-Consistent Semantic Image Editing with Style-Preserved Modulation

Wuyang Luo[1], Su Yang[1(✉)], Hong Wang[1], Bo Long[1], and Weishan Zhang[2]

[1] Shanghai Key Laboratory of Intelligent Information Processing, School of Computer Science, Fudan University, Shanghai, China
{wyluo18,suyang}@fudan.edu.cn

[2] School of Computer Science and Technology, China University of Petroleum, Qingdao, China

https://github.com/WuyangLuo/SPMPGAN

**Abstract.** Semantic image editing utilizes local semantic label maps to generate the desired content in the edited region. A recent work borrows SPADE block to achieve semantic image editing. However, it cannot produce pleasing results due to style discrepancy between the edited region and surrounding pixels. We attribute this to the fact that SPADE only uses an image-independent local semantic layout but ignores the image-specific styles included in the known pixels. To address this issue, we propose a style-preserved modulation (SPM) comprising two modulations processes: The first modulation incorporates the contextual style and semantic layout, and then generates two fused modulation parameters. The second modulation employs the fused parameters to modulate feature maps. By using such two modulations, SPM can inject the given semantic layout while preserving the image-specific context style. Moreover, we design a progressive architecture for generating the edited content in a coarse-to-fine manner. The proposed method can obtain context-consistent results and significantly alleviate the unpleasant boundary between the generated regions and the known pixels.

**Keywords:** Semantic image editing · Style-preserved modulation

## 1 Introduction

Image editing aims to generate the desired content in a specific region under users' control. This task attracts a lot of research enthusiasm due to its wide application in social media, image and video re-creation, and virtual human-object interaction. The well-known commercial software Photoshop has achieved success in this field. However, the use of such software requires many professional skills and much manual effort.

Most image editing methods fall into a few categories. The first category is low-level-guided editing methods [3,6,18,28]. They introduce low-level information such as lines and color. These methods can deal with editing simple contours

© The Author(s), under exclusive license to Springer Nature Switzerland AG 2022
S. Avidan et al. (Eds.): ECCV 2022, LNCS 13677, pp. 561–578, 2022.
https://doi.org/10.1007/978-3-031-19790-1_34

**Fig. 1.** Applications of the proposed method. Our image editing system is flexible in responding to a wide variety of editing requirements.

or shapes but only provide very limited editing control and cannot manipulate the high-level semantics of the image. The second category is classification-based methods [9,12]. They utilize an auxiliary classifier to guide synthesis and edit images. These methods can only control discrete attributes and cannot provide spatial control. The third category methods employ GAN inversion technique [1,4,27,39], which relies on a pre-trained GAN and dissects GANs' latent spaces, finding disentangled latent codes suitable for editing. They require a powerful well-trained StyleGAN, which is impossible in many cases because training a strong StyleGAN [20,22] model is not easy, especially for complex scenes. Further, such methods lack flexibility, and the editing of each attribute may require independent training. The fourth category methods [11,32] utilize pixel-level semantic label maps, which define the class labels of pixels in edited regions to control edited content. This task is also known as Semantic Image Editing. Following this line of work, our approach can provide users with greater editing flexibility than the other three categories of methods. Our method includes the following editing capabilities: (1) Our method can be applied to complex scene editing. (2) Users can flexibly edit the image via manipulating semantic layout, such as modifying the shape of objects, adding or removing objects. (3) Edited regions can be selected at arbitrary positions, even beyond the original image boundaries. The Fig. 1 demonstrates the versatility of our approach.

Semantic image editing is a non-trivial task. Its challenge lies in keeping context style consistent between edited and known regions. Here, "context" refers to the non-edited region of the input image, and "style" is the features of "context" such as color/texture. The previous state-of-the-art method SESAME [32] leverages SPADE block [33] to build their generator. SPADE is remarkably effective in conditional image synthesis. Conditional image synthesis learns a mapping from the semantic map domain to the real image domain, synthesizing the entire image according to the given semantic label map. Therefore, the generator may synthesize simple textures to get visually plausible results. However, since known pixels

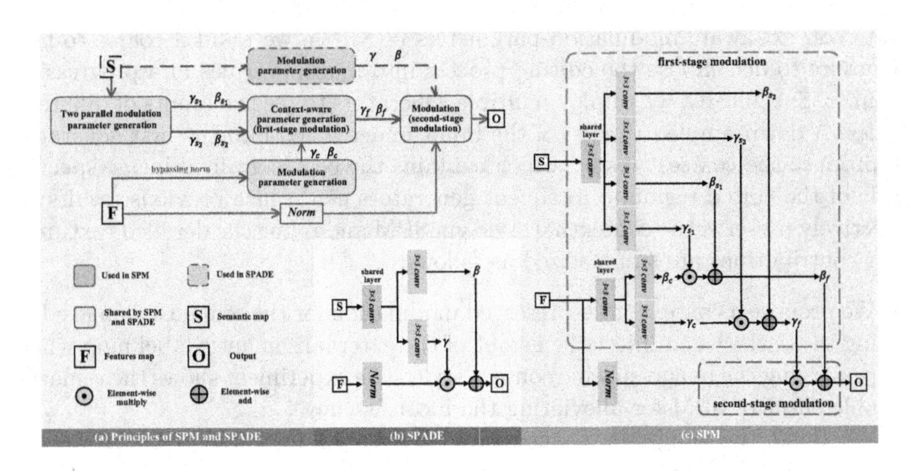


**Fig. 2.** (a) Principle difference between SPM and SPADE; (b) The structure of SPADE; (c) The structure of the proposed SPM.

and fake pixels coexist for the image editing task, our task becomes tougher in that the requirement is synthesizing realistic textures and retaining consistency to the context style. Aside from that, image synthesis requires a full semantic label map, but semantic image editing can only see the semantic layout of the edited region. Thus, if SPADE is employed directly on the editing task, only meaningless modulation parameters would be generated in the known region. Previous work [32] often causes significant style inconsistency and unpleasant boundaries for the above reasons.

To address such limitations of the existing works, we propose a style-preserved modulation module (SPM). Compared with SPADE, which only utilizes one modulation operation, SPM consists of a two-stage modulation process. Inspired by the style transfer [14], which show that non-normalized feature maps contain high-level "style" information, we use non-normalized feature maps for context preserving via"bypassing norm". The principle difference between SPM and SPADE is illustrated in Fig. 2(a) and their details are described in the Sect. 3. Specifically, we first generate two parallel pairs of modulation parameters from semantic maps and a pair of modulation parameters from feature maps. Then we fuse them through the first modulation operation to generate two context-aware modulation parameters. The second stage modulation uses the context-aware modulation parameters to modulate feature maps. Through two-stage modulation, SPM can effectively integrate external semantic maps while preserving the image-specific context style.

SPM involves feature maps into the modulation process for preserving contextual style. For image editing tasks, the input is empty in the edited region. The contextual information of the known region is gradually transferred to the edited region through the enlargement of the receptive field of the generator. In order to make the edited region more effectively perceive the contextual style to generate context-aware modulation parameters of SPMs, we build a coarse-to-fine structure to decompose the editing process into multiple scales in a progressive manner. Specifically, we employ multiple generators to receive inputs of different scales. A downsampled version of the input image is fed into the first generator to produce the coarsest result, which contains the coarse-grained image-specific style of the edited region. Subsequent generators can utilize previous results to effectively preserve the contextual style via SPM and refine the detailed textures. Our contributions are summarized as follows:

- We propose a context style-preserved modulation for the semantic image editing task, which can inject the layout of the external semantic label map while preserving the image-specific context style. The experiment shows the remarkable effect of SPM for alleviating the inconsistency.
- We build the progressive generative adversarial networks with SPMs for coarse-to-fine generation of edited regions.
- Extensive qualitative and quantitative experiments conducted on several benchmark datasets indicate that our model outperforms the state-of-the-art methods, especially in the sense of contextual style consistency.

## 2 Related Works

### 2.1 Image-to-Image Translation

Image translation attempts to learn a mapping from a source domain to a target domain. It can be applied to various tasks, such as image synthesis [46–48], image editing [3,11,18], style transfer [7,14], image inpainting [34,44,45], image extension [38,43], and image super-resolution [24,25]. Existing works utilize different conditional inputs as source domains such as semantic label maps, scene layouts, key points, and edge maps. Among them, the most relevant subtask is semantic image synthesis, which aims at generating photo-realistic images conditioned on semantic label maps.

Semantic image synthesis has achieved remarkable progress benefitting from GAN [8]. Pix2pix [17] is the seminal work based on cGAN framework [30]. The following work Pix2pixHD [41] is devoted to generating high-resolution images. SPADE [33] proposes a spatially-adaptive normalization that learns transformation parameters from the semantic layout to modulate the activations in normalization layers. CLADE [37] proposes a lightweight class-adaptive normalization to improve the efficiency of SPADE. Semantic image synthesis has been applied to different downstream tasks in recent works, such as semantic image editing [32], semantic view synthesis [13], and portrait editing [26,53].

### 2.2 Semantic Image Editing

Semantic image editing refers to users providing semantic label maps as a clue to edit the local region of a given image at pixel level. Semantic concepts are more intuitive and fundamental image features than colors, edges, key points, and textures. By manipulating the semantic label map, users can easily edit the image content in many ways, including re-painting, adding, removing, and out-painting semantic objects. Semantic image editing has not been fully developed because it is challenging. Semantic image editing requires that the edited content not only has high fidelity but also must be consistent with the style of the remaining region. HIM [11] is the earliest attempt at this task. HIM can only operate on one foreground target each time. Furthermore, HIM requires a full semantic label map of the entire image as input, which is inconvenient for users. SESAME [32] only inputs the semantic label map of the edited region, making the image editing tool more practical. SESAME builds its generator with SPADE and uses a new discriminator to process the semantic and image information in separate streams. Although the previous methods can synthesize plausible results, they ignore the consistency of the context between the edited region and the known region. In contrast, our work is dedicated to reducing this inconsistency.

### 2.3 Modulation Technique

Modulation, also called denormalization, is an effective way to inject external control information. Unlike the unconditional normalization technique, such as BN [16], IN [40], and GN [42], modulation techniques require external data and follow a similar operating flow. First, feature maps are normalized to zero mean and unit deviation using an unconditional normalization layer. Then the normalized feature maps are modulated with scaling and shifting parameters learned from external data. Modulation techniques were initially applied to style transfer tasks, such as AdaIN [14] and later adopted in various vision tasks [15,21,35]. AdaIN only learns global style representation. [33] proposes SPADE for semantic image synthesis to handle external data with spatial dimensions. However, the previous methods only consider the external conditional input and ignore the internal contextual information, which is a fatal disadvantage for our task. This paper proposes a new modulation scheme that can aggregate internal context style and external semantic layout. The experimental results show that the proposed method can effectively preserve the context style and improve consistency for semantic image editing.

## 3 Approach

We describe our approach from bottom to top. We first analyze the limitations of SPADE for semantic image editing and introduce SPM proposed in this paper. Then, we introduce how to build a progressive architecture based on SPM.

### 3.1 Rethinking SPADE for Semantic Image Editing

SPADE is a state-of-the-art modulation technology remarkably successful in semantic image synthesis, as shown in Fig. 2(b). $F^i \in \mathbb{R}^{N\times C\times H\times W}$ is the input feature maps of the i-th layers. $N$ is the number of samples in one batch. $C$ is the number of channels. $H$ and $W$ represent the height and width, respectively. SPADE learns two modulation parameters, scaling parameters $\gamma$ and shifting parameters $\beta$, via two convolutional layers from the given semantic label map $S$. First, $F^i$ is normalized in a channel-wise manner:

$$\bar{F^i} = \frac{F^i - \mu^i}{\sigma^i} \tag{1}$$

where $\mu^i \in \mathbb{R}^{N \times C \times 1 \times 1}$ and $\sigma^i \in \mathbb{R}^{N \times C \times 1 \times 1}$ are the channel-wise means and standard deviations of $F^i$. Then, we perform the modulation operation:

$$\widetilde{F^i} = (\mathbf{1} + \gamma) \odot \bar{F^i} + \beta \tag{2}$$

Previous work [32] applies SPADE for semantic image editing. However, SPADE is ill-fitted for semantic image editing for the following two reasons: First, SPADE can only generate image-independent modulation parameters from the given external semantic label map. Thus, if two edited images are given the same semantic label map, SPADE will generate the same modulation parameters. This is unreasonable because SPADE ignores image-specific style. Second, for semantic image editing, the generator can only see the semantic layout of the edited region, and the semantic labels of the rest known regions are set to a fixed value. Therefore, SPADE cannot learn effective parameters on the known region. If we naively transfer SPADE to semantic image editing, the above two limitations will cause style inconsistency and unpleasant boundaries.

### 3.2 Style-Preserved Modulation

To solve the issues mentioned above, we propose a two-stage modulation mechanism for style preserving, as shown in Fig. 2(c). The first stage of modulation aims to integrate the context style and the external semantic layout. The second stage of modulation is to inject the fused information into feature maps.

In the first modulation, we generate two kinds of parameters: Four semantic modulation parameters and two context modulation parameters. Semantic modulation parameters include two groups: $(\gamma_{s_1}, \beta_{s_1})$ and $(\gamma_{s_2}, \beta_{s_2})$. The context modulation parameters $(\gamma_c, \beta_c)$ are generated from the original feature maps without passing through the normalization layer. The previous style transfer works [14] revealed that the style of the image could be washed away by normalization layers. The non-normalized feature maps can retain the context style more. So, we use the original feature maps to generate two context modulation parameters. Finally, we perform the first modulation to generate the fused modulation parameters $\gamma_f$ and $\beta_f$:

$$\gamma_f = (\mathbf{1} + \gamma_{s_2}) \odot \gamma_c + \beta_{s_2} \tag{3}$$

$$\beta_f = (\mathbf{1} + \gamma_{s_1}) \odot \beta_c + \beta_{s_1} \tag{4}$$

where $\odot$ denotes element-wise multiplication. All modulation parameters have the same shape as the feature maps $F^i$.

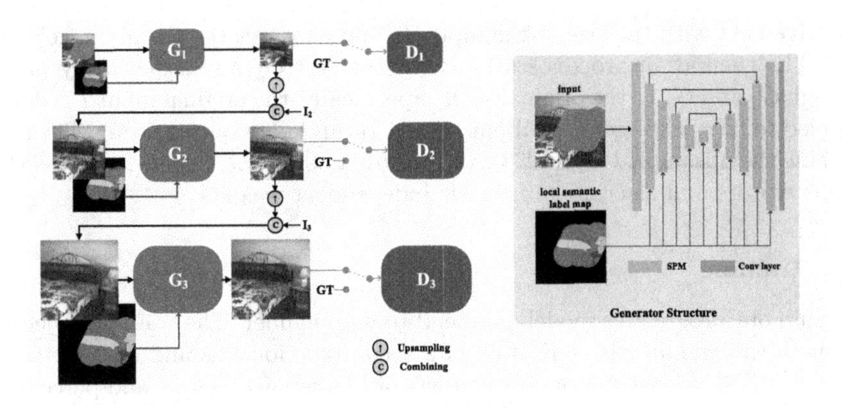


**Fig. 3.** Overview of the progressive architecture.

In the second modulation, we use fused modulation parameters to modulate the normalized feature maps $\bar{F^i}$.

$$\widetilde{F^i} = (\mathbf{1} + \gamma_f) \odot \bar{F^i} + \beta_f \tag{5}$$

Through two-stage modulation process, SPM overcomes the two shortcomings of SPADE: First, the fused modulation parameters integrate the external semantic layout and retain the internal context style. Second, the fused modulation parameters can generate meaningful modulation parameters for known regions.

### 3.3 Progressive Editing Architecture

We propose a progressive architecture for image editing based on SPM, called *SPMPGAN*. Our model has three inputs: (1) The input image $I \in \mathbb{R}^{256\times256\times3}$ which contains only known pixels with masked edited region; (2) the local semantic map $S$ providing the semantic layouts of the edited region; and (3) the corresponding mask map $M$ whose value is 0 in the non-edited region and 1 in the edited region. Our progressive architecture consists of a pyramid of generators $\{G_1, G_2, G_3\}$ and discriminators $\{D_1, D_2, D_3\}$ with an image pyramid of $I$: $\{I_1, I_2, I_3\}$, where $I_n$ is a downsampled version of $I$ by a factor $2^{3-n}$, mask pyramid of $M$: $\{M_1, M_2, M_3\}$, and semantic map pyramid of $S$: $\{S_1, S_2, S_3\}$. Each generator $G_n$ is trained with an associated discriminator $D_n$. $G_n$ learns to generate realistic new content in the edited region and try to fool the corresponding discriminator. $D_n$ attempts to distinguish the edited result and the real image. We adopt an encoder-decoder architecture with skip connections [36] for all generators, as shown in Fig. 3. Each generator adds a down-sampling layer in the encoder and an up-sampling layer in the decoder on the previous generator. Inspired by [45], the discriminators are composed of several convolutional layers with $5 \times 5$ convolution kernel and spectral normalization [31]. The number of layers of $D_1$, $D_2$, and $D_3$ are 4, 5, and 6, respectively. Thus, each $D_n$ has the receptive field with the size of the input $I_n$ and captures the entire image's feature. The generation process starts at the coarsest $G_1$ and sequentially passes through $G_2$ and $G_3$ to the original scale. Specifically, the original input $I$ is downsampled to $64 \times 64$ to get $G_1$'s input: $I_{G1} = I_1$, and $G_1$'s output is $O_1$. Then, we combine the upsampled $O_1$ with $I_2$ as $G_2$'s input: $I_{G2} = O_1 \odot M_2 + I_2 \odot (1 - M_2)$. All generators and discriminators have independent weights.

### 3.4 Training

We train our progressive model in an end-to-end manner. The training objective for the n-th generator is comprised of a reconstruction loss and an adversarial loss $\mathcal{L}_{\text{adv}}$. The reconstruction loss consists of L1 distance loss $\mathcal{L}_1$ and perceptual loss $\mathcal{L}_{\text{p}}$ [19]. We employ the hinge version adversarial loss [2,29]. The overall loss can be written as:

$$\mathcal{L} = \mathcal{L}_1 + 10.0\mathcal{L}_{\text{p}} + \mathcal{L}_{adv} \tag{6}$$

## 4 Experiments

### 4.1 Datasets

**ADE20K-Room.** ADE20K [51] has over 20,000 images together with detailed semantic labels of 150 classes. We select a subset of the ADE20K comprised of `Bedroom`, `Hotel Room`, and `Living Room`. This subset is called ADE20K-room. We resize all the images with their longer sides no more than 384 and their shorter sides no less than 256. We crop them to $256 \times 256$ when training. This dataset has 2246 images for training and 255 for testing.

**ADE20K-Landscape.** We also selected the landscape subclass from ADE20K and use the same preprocessing approach. The difference is that this dataset has only background and no foreground objects. The training set and the testing set contain 1689 images and 155 images, respectively.

**Cityscapes.** [5] The dataset collects streetscapes of 50 German cities, which contains 33 semantic categories. The training and testing set has 2975 and 500 images, respectively, with a resolution of $2048 \times 1024$. We downsample all images to $512 \times 256$ and crop them to $256 \times 256$ patches.

### 4.2 Baselines

**Semantic Image Editing Methods.** We employ two existing works [11,32] as baselines. HIM [11] introduces a two-stage method for image editing. They first predict semantic layout from object bounding boxes. Then, they generate new content according to the predicted semantic layout. Because in our setting, the ground truth semantic layout of the edited region is known, we directly input the ground truth layout to the second stage of HIM to get the results. SESAME [32] has similar settings with our work.

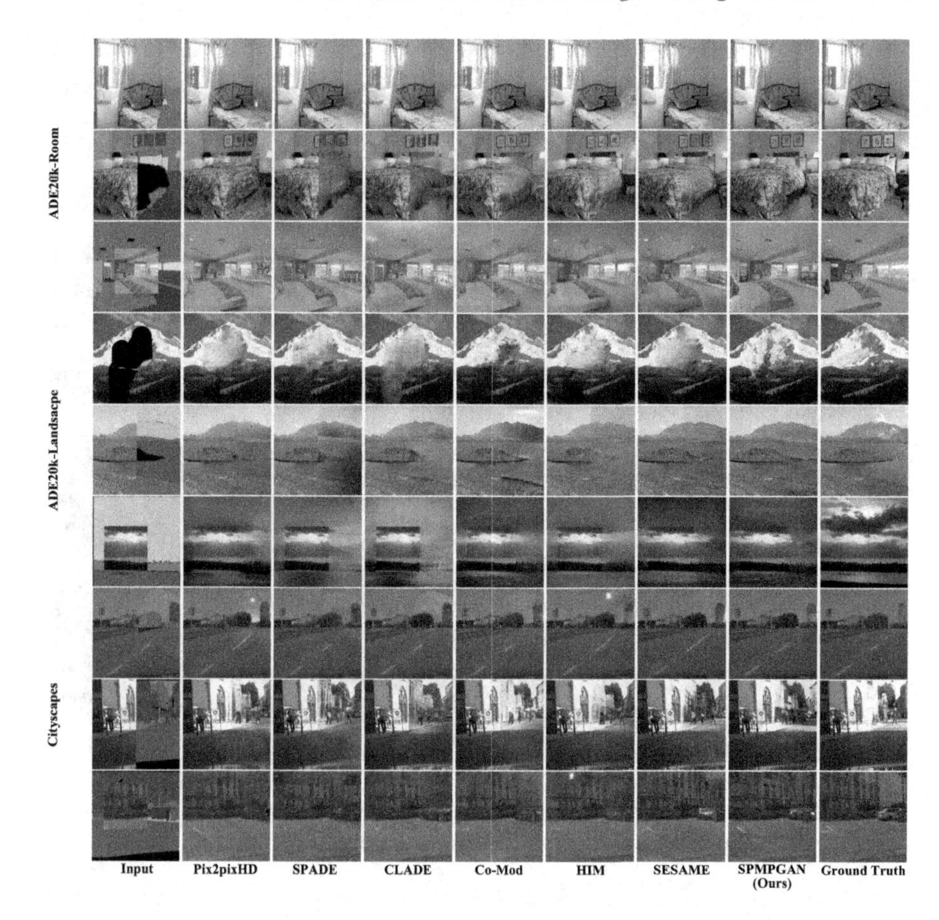


**Fig. 4.** Visual comparison with other methods.

**Image Synthesis Methods.** Our experiments also include several image generation methods for comparison. These recent works [33,37,41,50] can be directly transferred to our task via only modifying their generators' input. It is worth mentioning that some recent works cannot be simply adapted for our task. For example, SEAN [53] requires a full segmentation map to calculate their style codes. CoCosNetv2 [52] requires a full segmentation map to perform their domain alignment. However, our task can only see local semantic label maps.

### 4.3 Implementation Details

To obtain a more flexible model, we employ five types of masks for training: Free-form mask, extension mask, outpainting mask, instance mask, and class mask. The extension mask is the right half of the input. For the outpainting mask, we randomly retain a $128 \times 128$ patch as the known region. The instance mask contains only a single foreground target, and the class mask drops all the

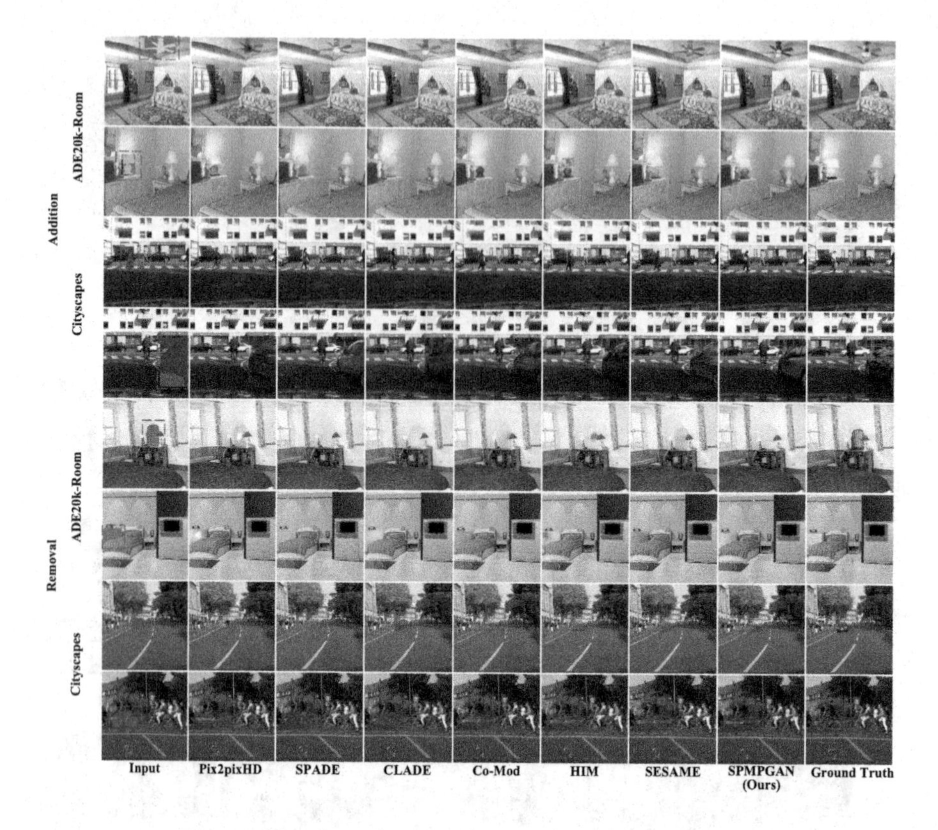


**Fig. 5.** Visual results of addition and removal objects.

pixels belonging to a semantic class. During training, each mask is randomly selected and sent to the network at each iteration. We use Adam optimizers [23] for both the generator and the discriminators with momentum $\beta_1 = 0.5$ and $\beta_2 = 0.999$. The learning rates for the generator and the discriminators are set to 0.0001 and 0.0004, respectively. All models are trained for 500 epochs on all datasets. The batch size is set to the maximum value to fit the memory size of a single NVIDIA RTX 3090 GPU.

### 4.4 Semantic Image Editing

We compare our results with state-of-the-art methods using free-form masks, extension masks, and outpainting masks on the three benchmarks. Figure 4 provides some visual comparisons. Pix2pixHD [41] and HIM [11] only use semantic label maps as conditions in the input layer, and they often generate artifacts. SPADE [33], CLADE [37], and SESAME [32] can synthesize reasonable structures and realistic textures, but they severely suffer from style inconsistencies leading to unpleasant boundaries. Because they only use the image-independent external semantic map when injecting the semantic label map and completely

**Table 1.** Quantitative comparison with different mask types (↑: Higher is better; ↓: Lower is better). In the leftmost column, M, F, E, and O represent Mask Type, Free-Form Mask, Extension Mask, and Outpainting Mask, respectively.

| M | Method | ADE20k-Room | | | ADE20k-Landscape | | | Cityscapes | | |
|---|---|---|---|---|---|---|---|---|---|---|
| | | FID↓ | LPIPS↓ | mIoU↑ | FID↓ | LPIPS↓ | mIoU↑ | FID↓ | LPIPS↓ | mIoU↑ |
| F | pix2pixHD | 23.72 | 0.107 | 27.49 | 33.90 | 0.120 | 28.30 | 15.28 | 0.090 | 58.69 |
| | SPADE | 27.65 | 0.124 | 27.47 | 41.92 | 0.134 | 28.41 | 15.83 | 0.099 | **59.10** |
| | CLADE | 30.77 | 0.126 | 25.91 | 46.59 | 0.139 | 26.39 | 17.06 | 0.103 | 57.72 |
| | Co-Mod | 27.37 | 0.111 | 27.52 | 32.35 | 0.124 | 28.60 | 15.88 | 0.097 | 56.50 |
| | HIM | 28.64 | 0.133 | 28.04 | 35.89 | 0.116 | 28.43 | 15.58 | 0.093 | 58.99 |
| | SESAME | 21.73 | 0.101 | 27.50 | 30.30 | 0.116 | 28.28 | 12.89 | **0.082** | 58.88 |
| | SPMPGAN | **18.83** | **0.090** | **28.22** | **23.11** | **0.105** | **28.73** | **11.90** | 0.084 | 58.80 |
| E | pix2pixHD | 38.08 | 0.223 | 27.32 | 56.15 | 0.242 | 28.10 | 26.14 | 0.176 | 58.55 |
| | SPADE | 36.43 | 0.211 | 27.62 | 68.96 | 0.277 | 28.44 | 25.78 | 0.194 | 59.01 |
| | CLADE | 41.77 | 0.242 | 25.67 | 65.33 | 0.267 | 26.39 | 25.29 | 0.195 | 58.09 |
| | Co-Mod | 38.61 | 0.231 | 27.13 | 53.96 | 0.249 | 28.09 | 29.27 | 0.188 | 56.44 |
| | HIM | 40.69 | 0.239 | 27.61 | 52.14 | 0.234 | 28.42 | 25.20 | 0.180 | 58.91 |
| | SESAME | 36.43 | 0.211 | 27.62 | 48.16 | 0.232 | 28.31 | 20.30 | 0.168 | 59.08 |
| | SPMPGAN | **32.61** | **0.199** | **27.73** | **45.10** | **0.217** | **28.48** | **19.46** | **0.167** | **59.10** |
| O | pix2pixHD | 52.14 | 0.323 | 27.49 | 82.56 | 0.360 | 28.30 | 39.50 | 0.253 | 58.72 |
| | SPADE | 47.72 | 0.305 | 27.40 | 88.79 | 0.389 | 28.30 | 33.97 | 0.268 | **59.07** |
| | CLADE | 52.45 | 0.346 | 25.47 | 86.77 | 0.388 | 24.49 | 34.19 | 0.276 | 57.49 |
| | Co-Mod | 51.45 | 0.325 | 26.54 | 79.77 | 0.360 | 26.70 | 50.29 | 0.264 | 55.39 |
| | HIM | 54.51 | 0.337 | **28.19** | 77.18 | 0.352 | **28.57** | 36.27 | 0.252 | 58.99 |
| | SESAME | 47.72 | 0.305 | 27.40 | 72.28 | 0.344 | 28.13 | 28.27 | 0.237 | 58.75 |
| | SPMPGAN | **41.52** | **0.288** | 27.85 | **63.32** | **0.328** | 27.56 | **27.63** | **0.233** | 58.53 |

ignoring the context information. Co-Mod [50] also has the apparent texture inconsistency as it lacked a specific design for the image editing tasks. The proposed method can effectively integrate the contextual style and the semantic layout to produce realistic textures while preserving the contextual style. Table 1 also shows the quantitative comparison results. FID [10] has been widely demonstrated that it is consistent with human visual perception. A lower FID value indicates that results have higher fidelity. LPIPS[49] evaluates the similarity between the generated image and the corresponding ground truth in a pairwise manner. A lower LPIPS indicates that the generated image is closer to the ground truth. mIoU is employed in the semantic synthesis task [33] to evaluate the alignment between the semantic label map and the generated result. Our method outperforms the other methods in most evaluation metrics.

### 4.5 Addition and Removal of Objects

Our work is capable of adding or removing individual objects by modifying the semantic label maps. Visual results are demonstrated in Fig. 5. For the object addition, we randomly select an instance of input and extract the boundary boxes to generate its local semantic label map. For object removal, we delete a

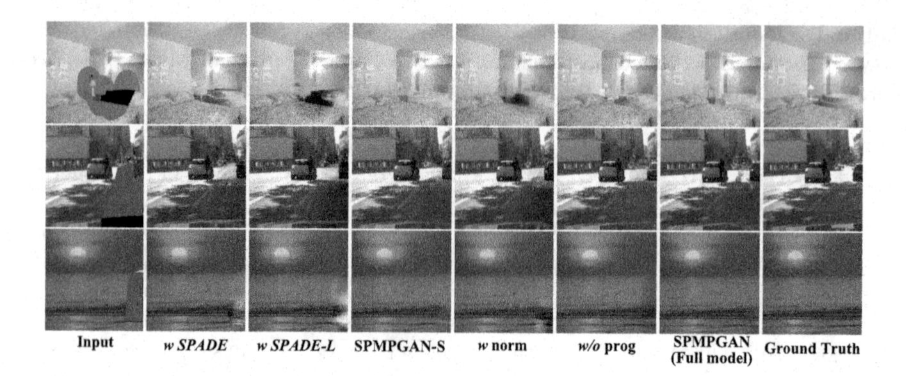


**Fig. 6.** Visual comparison of ablation studies.

**Table 2.** Addition and removal results for Cityscapes and ADE20k-Room.

| Manipulation | Method | ADE20k-Room | | | Cityscapes | | |
|---|---|---|---|---|---|---|---|
| | | FID↓ | LPIPS↓ | mIoU↑ | FID↓ | LPIPS↓ | mIoU↑ |
| Addition | pix2pixHD | 6.29 | 0.027 | 27.09 | 11.77 | 0.030 | 58.28 |
| | SPADE | 5.66 | 0.027 | 27.17 | 10.48 | 0.031 | 58.66 |
| | CLADE | 6.21 | 0.028 | 27.16 | 11.03 | 0.031 | 57.55 |
| | Co-Mod | 5.75 | 0.026 | 27.23 | 11.28 | 0.031 | 56.40 |
| | HIM | 9.80 | 0.046 | 27.22 | 11.41 | 0.030 | **58.75** |
| | SESAME | 5.50 | 0.024 | 27.14 | 9.70 | 0.027 | 58.56 |
| | SPMPGAN | **5.14** | **0.022** | **27.43** | **9.04** | **0.026** | 58.68 |
| Removal | pix2pixHD | 4.52 | 0.019 | 28.32 | 15.01 | 0.039 | 55.02 |
| | SPADE | 3.96 | 0.019 | 28.35 | 15.48 | 0.040 | 55.04 |
| | CLADE | 4.12 | 0.019 | 28.34 | 16.18 | 0.040 | 54.22 |
| | Co-Mod | 4.03 | 0.019 | 28.33 | 15.05 | 0.041 | 55.10 |
| | HIM | 7.44 | 0.035 | 28.33 | 15.10 | 0.040 | **55.11** |
| | SESAME | 4.02 | 0.018 | 28.34 | 15.52 | 0.041 | 55.08 |
| | SPMPGAN | **3.68** | **0.016** | **28.35** | **14.63** | **0.039** | 55.01 |

instance and fill it with nearby background semantic class. Quantitative results shown in Table 2 indicate that our method achieves the best results in style preservation and fidelity.

### 4.6 Controllable Panorama Generation

A well-trained model can be used recursively to obtain panoramas. Specifically, we employ the generated region of the previous step as the known region of the next step in a sliding window manner. Thus, the input is extended to the right by 128 pixels in each step so that images with arbitrary width can be controllable synthesized. Figure 1 shows a recursive generated result.

**Table 3.** Ablation study with different mask types.

| M | Method | ADE20k-Room | | | ADE20k-Landscape | | | Cityscapes | | |
|---|---|---|---|---|---|---|---|---|---|---|
| | | FID↓ | LPIPS↓ | mIoU↑ | FID↓ | LPIPS↓ | mIoU↑ | FID↓ | LPIPS↓ | mIoU↑ |
| F | *w SPADE* | 23.27 | 0.098 | 27.60 | 34.71 | 0.118 | 28.39 | 14.20 | 0.091 | 58.80 |
| | *w norm* | 20.51 | 0.098 | 27.58 | 29.87 | 0.111 | 28.31 | 12.64 | 0.085 | 58.78 |
| | *w/o prog* | 20.47 | 0.096 | 27.42 | 25.87 | 0.109 | 28.42 | 13.07 | 0.089 | 58.73 |
| | *w SPADE-L* | 24.11 | 0.098 | 27.61 | 34.68 | 0.116 | 28.43 | 14.40 | 0.090 | 58.79 |
| | *SPMPGAN-S* | 18.93 | 0.090 | **28.24** | 23.21 | 0.106 | 28.70 | **11.89** | **0.084** | **58.82** |
| | *SPMPGAN* | **18.83** | **0.090** | 28.22 | **23.11** | **0.105** | **28.73** | 11.90 | 0.084 | 58.80 |
| E | *w SPADE* | 36.84 | 0.220 | 27.51 | 53.02 | 0.239 | 28.92 | 21.99 | 0.173 | 58.88 |
| | *w norm* | 32.76 | 0.205 | 27.56 | 48.43 | 0.228 | 28.92 | 20.50 | 0.176 | 59.01 |
| | *w/o prog* | 33.87 | 0.205 | 27.44 | 45.96 | 0.222 | 28.86 | 21.00 | 0.170 | 59.09 |
| | *w SPADE-L* | 36.14 | 0.218 | 27.48 | 53.13 | 0.240 | **28.93** | 21.86 | 0.174 | 58.81 |
| | *SPMPGAN-S* | **31.92** | 0.200 | **27.74** | 45.17 | 0.218 | 28.47 | **19.12** | **0.167** | **59.12** |
| | *SPMPGAN* | 32.61 | **0.199** | 27.73 | **45.10** | **0.217** | 28.48 | 19.46 | 0.167 | 59.10 |
| O | *w SPADE* | 47.37 | 0.321 | 28.38 | 71.52 | 0.357 | **28.84** | 31.33 | 0.244 | 58.95 |
| | *w norm* | 42.31 | 0.300 | **28.52** | 66.52 | 0.337 | 28.82 | 27.74 | 0.235 | 57.98 |
| | *w/o prog* | 43.98 | 0.297 | 28.05 | 66.32 | 0.329 | 27.39 | 29.54 | 0.238 | 58.53 |
| | *w SPADE-L* | 47.16 | 0.318 | 28.39 | 70.33 | 0.354 | 28.83 | 31.43 | 0.243 | **58.95** |
| | *SPMPGAN-S* | **41.49** | 0.289 | 27.80 | **62.43** | 0.330 | 27.63 | **27.39** | **0.228** | 58.59 |
| | *SPMPGAN* | 41.52 | **0.288** | 27.85 | 63.32 | **0.328** | 27.56 | 27.63 | 0.233 | 58.53 |

**Table 4.** Comparison of the number of parameters.

| | *w SPADE* | *w SPADE-L* | *SPMPGAN* | *SPMPGAN-S* |
|---|---|---|---|---|
| ADE20k-Room | 63.4 M | 90.0 M | 118.4 M | 76.9 M |
| Cityscapes | 57.8 M | 81.5 M | 112.7 M | 74.0 M |

## 4.7 Ablation Study

**Style-Preserved Modulation**

We study the importance of SPM for style preserving. We replace all SPMs with SPADE blocks ("*w SPADE*"). The visual results are shown in Fig. 6. It can be observed that SPADE leads to unpleasant boundaries. This is because SPADE completely ignores the image-specific context style and only uses local semantic label maps to modulate feature maps. As a comparison, SPM can relieve the inconsistency. The two-stage modulation can integrate the context style and the external semantic label map. In addition, SPM can also help the generator to synthesize more realistic texture details. We also study the influence of "bypassing norm" for style preserving. Specifically, for the generation of $\gamma_c$ and $\beta_c$ in SPM, we replace original feature maps by normalized feature maps ("*w norm*"). The experimental results show that the style preserving is significantly weakened. It proves that the normalization operation washes away context style. Therefore, we use the original feature maps without normalization in SPM. Quantitative results are also demonstrated in Table 3.

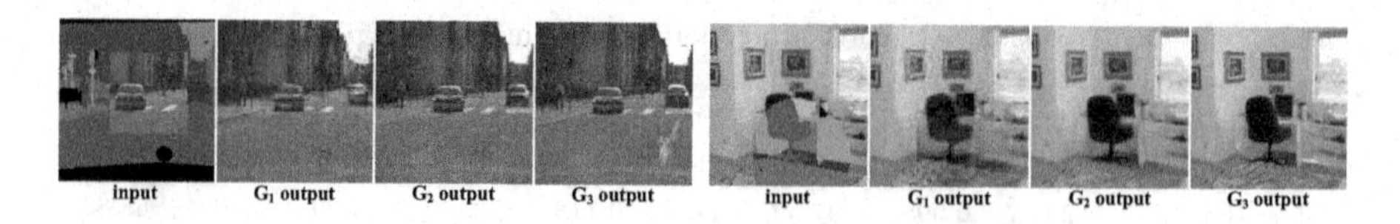


**Fig. 7.** Outputs of all generators.

**Effectiveness of Progressive Architecture**
We conduct an ablation study to demonstrate the effectiveness of the progressive design for synthesizing high-quality results. We only use the last level generator as the baseline ("*w/o prog*"). Figure 6(c) shows that without the progressive generation, the model will produce style inconsistency and unrealistic textures. The outputs of the generators of all scales are shown in the Figure 7. It can be seen, $G_1$ synthesizes the global structure, and $G_2$ and $G_3$ produce the sharper detail. Quantitative results are given in Table 3, which indicates that progressive architecture contributes to performance improvement.

### 4.8 Study of Model Scale

This study demonstrates that our performance improvement stems from the novel design of SPM rather than increasing parameters. As shown in Table 4, our model follows SPADE to set the number of output channels $C^h$ of the shared layer to 128. We reduce $C^h$ of all SPMs to 64 and keep the structure unchanged ("*SPMGAN-S*"). We do not observe the performance drop. In addition, we insert more SPADE blocks into "*w SPADE*" to obtain a new baseline "*w SPADE-L*" . The experimental results are shown in the Table 3, "*w SPADE-L*" does not obtain performance gain by simply increasing the network scale and computational consumption. The performance of "*SPMGAN-S*" still significantly outperforms "*w SPADE-L*" with fewer parameters.

## 5 Conclusion

This paper is dedicated to solving style inconsistency for the semantic editing task. We propose a style-preserved modulation and a progressive architecture that effectively injects the structure from semantic label maps while preserving the context style. The key of SPM lies in effectively integrating contextual information and semantic label maps. We also demonstrate the ability of our method for various applications.

**Acknowledgement.** This work is supported by State Grid Corporation of China (Grant No. 5500-202011091A-0-0-00).

## References

1. Alaluf, Y., Tov, O., Mokady, R., Gal, R., Bermano, A.H.: Hyperstyle: stylegan inversion with hypernetworks for real image editing. arXiv preprint arXiv:2111.15666 (2021)
2. Brock, A., Donahue, J., Simonyan, K.: Large scale GAN training for high fidelity natural image synthesis. arXiv preprint arXiv:1809.11096 (2018)
3. Chen, S.Y., et al.: Deepfaceediting: deep face generation and editing with disentangled geometry and appearance control. arXiv preprint arXiv:2105.08935 (2021)
4. Chong, M.J., Lee, H.Y., Forsyth, D.: Stylegan of all trades: image manipulation with only pretrained stylegan. arXiv preprint arXiv:2111.01619 (2021)
5. Cordts, M., et al.: The cityscapes dataset for semantic urban scene understanding. In: Proceedings of the IEEE Conference on Computer Vision and Pattern Recognition, pp. 3213–3223 (2016)
6. Dong, H., et al.: Fashion editing with adversarial parsing learning. In: Proceedings of the IEEE/CVF Conference on Computer Vision and Pattern Recognition, pp. 8120–8128 (2020)
7. Gatys, L.A., Ecker, A.S., Bethge, M.: Image style transfer using convolutional neural networks. In: Proceedings of the IEEE Conference on Computer Vision and Pattern Recognition, pp. 2414–2423 (2016)
8. Goodfellow, I., et al.: Generative adversarial nets. In: Advances in Neural Information Processing Systems, pp. 2672–2680 (2014)
9. He, Z., Zuo, W., Kan, M., Shan, S., Chen, X.: Attgan: facial attribute editing by only changing what you want. IEEE Trans. Image Process. **28**(11), 5464–5478 (2019)
10. Heusel, M., Ramsauer, H., Unterthiner, T., Nessler, B., Hochreiter, S.: Gans trained by a two time-scale update rule converge to a local nash equilibrium. In: Advances in Neural Information Processing Systems, pp. 6626–6637 (2017)
11. Hong, S., Yan, X., Huang, T., Lee, H.: Learning hierarchical semantic image manipulation through structured representations. arXiv preprint arXiv:1808.07535 (2018)
12. Hou, X., Zhang, X., Liang, H., Shen, L., Lai, Z., Wan, J.: Guidedstyle: attribute knowledge guided style manipulation for semantic face editing. Neural Netw. **145**, 209–220 (2022)
13. Huang, H.-P., Tseng, H.-Y., Lee, H.-Y., Huang, J.-B.: Semantic view synthesis. In: Vedaldi, A., Bischof, H., Brox, T., Frahm, J.-M. (eds.) ECCV 2020. LNCS, vol. 12357, pp. 592–608. Springer, Cham (2020). https://doi.org/10.1007/978-3-030-58610-2_35
14. Huang, X., Belongie, S.: Arbitrary style transfer in real-time with adaptive instance normalization. In: Proceedings of the IEEE International Conference on Computer Vision, pp. 1501–1510 (2017)
15. Huang, X., Liu, M.Y., Belongie, S., Kautz, J.: Multimodal unsupervised image-to-image translation. In: Proceedings of the European Conference on Computer Vision (ECCV), pp. 172–189 (2018)
16. Ioffe, S., Szegedy, C.: Batch normalization: accelerating deep network training by reducing internal covariate shift. In: International Conference on Machine Learning, pp. 448–456. PMLR (2015)
17. Isola, P., Zhu, J.Y., Zhou, T., Efros, A.A.: Image-to-image translation with conditional adversarial networks. In: Proceedings of the IEEE Conference on Computer Vision and Pattern Recognition, pp. 1125–1134 (2017)

18. Jo, Y., Park, J.: Sc-fegan: face editing generative adversarial network with user's sketch and color. In: Proceedings of the IEEE/CVF International Conference on Computer Vision, pp. 1745–1753 (2019)
19. Johnson, J., Alahi, A., Fei-Fei, L.: Perceptual losses for real-time style transfer and super-resolution. In: Leibe, B., Matas, J., Sebe, N., Welling, M. (eds.) ECCV 2016. LNCS, vol. 9906, pp. 694–711. Springer, Cham (2016). https://doi.org/10.1007/978-3-319-46475-6_43
20. Karras, T., Laine, S., Aila, T.: A style-based generator architecture for generative adversarial networks. In: Proceedings of the IEEE/CVF Conference on Computer Vision and Pattern Recognition, pp. 4401–4410 (2019)
21. Karras, T., Laine, S., Aila, T.: A style-based generator architecture for generative adversarial networks. In: Proceedings of the IEEE Conference on Computer Vision and Pattern Recognition, pp. 4401–4410 (2019)
22. Karras, T., Laine, S., Aittala, M., Hellsten, J., Lehtinen, J., Aila, T.: Analyzing and improving the image quality of stylegan. In: Proceedings of the IEEE/CVF Conference on Computer Vision and Pattern Recognition, pp. 8110–8119 (2020)
23. Kingma, D.P., Ba, J.: Adam: a method for stochastic optimization. arXiv preprint arXiv:1412.6980 (2014)
24. Lai, W.S., Huang, J.B., Ahuja, N., Yang, M.H.: Deep laplacian pyramid networks for fast and accurate super-resolution. In: Proceedings of the IEEE Conference on Computer Vision and Pattern Recognition, pp. 624–632 (2017)
25. Ledig, C., et al.: Photo-realistic single image super-resolution using a generative adversarial network. In: Proceedings of the IEEE Conference on Computer Vision and Pattern Recognition, pp. 4681–4690 (2017)
26. Lee, C.H., Liu, Z., Wu, L., Luo, P.: Maskgan: towards diverse and interactive facial image manipulation. In: Proceedings of the IEEE/CVF Conference on Computer Vision and Pattern Recognition, pp. 5549–5558 (2020)
27. Ling, H., Kreis, K., Li, D., Kim, S.W., Torralba, A., Fidler, S.: Editgan: high-precision semantic image editing. Adv. Neural. Inf. Process. Syst. **34**, 16331–16345 (2021)
28. Liu, H., et al.: Deflocnet: deep image editing via flexible low-level controls. In: Proceedings of the IEEE/CVF Conference on Computer Vision and Pattern Recognition, pp. 10765–10774 (2021)
29. Liu, M.Y., Huang, X., Mallya, A., Karras, T., Aila, T., Lehtinen, J., Kautz, J.: Few-shot unsupervised image-to-image translation. In: Proceedings of the IEEE/CVF International Conference on Computer Vision, pp. 10551–10560 (2019)
30. Mirza, M., Osindero, S.: Conditional generative adversarial nets. arXiv preprint arXiv:1411.1784 (2014)
31. Miyato, T., Kataoka, T., Koyama, M., Yoshida, Y.: Spectral normalization for generative adversarial networks. arXiv preprint arXiv:1802.05957 (2018)
32. Ntavelis, E., Romero, A., Kastanis, I., Van Gool, L., Timofte, R.: SESAME: semantic editing of scenes by adding, manipulating or erasing objects. In: Vedaldi, A., Bischof, H., Brox, T., Frahm, J.-M. (eds.) ECCV 2020. LNCS, vol. 12367, pp. 394–411. Springer, Cham (2020). https://doi.org/10.1007/978-3-030-58542-6_24
33. Park, T., Liu, M.Y., Wang, T.C., Zhu, J.Y.: Semantic image synthesis with spatially-adaptive normalization. In: Proceedings of the IEEE Conference on Computer Vision and Pattern Recognition, pp. 2337–2346 (2019)
34. Pathak, D., Krahenbuhl, P., Donahue, J., Darrell, T., Efros, A.A.: Context encoders: feature learning by inpainting. In: Proceedings of the IEEE Conference on Computer Vision and Pattern Recognition, pp. 2536–2544 (2016)

35. Perez, E., Strub, F., De Vries, H., Dumoulin, V., Courville, A.: Film: visual reasoning with a general conditioning layer. In: Proceedings of the AAAI Conference on Artificial Intelligence, vol. 32 (2018)
36. Ronneberger, O., Fischer, P., Brox, T.: U-Net: convolutional networks for biomedical image segmentation. In: Navab, N., Hornegger, J., Wells, W.M., Frangi, A.F. (eds.) MICCAI 2015. LNCS, vol. 9351, pp. 234–241. Springer, Cham (2015). https://doi.org/10.1007/978-3-319-24574-4_28
37. Tan, Z., et al.: Efficient semantic image synthesis via class-adaptive normalization. IEEE Trans. Pattern Anal. Mach. Intell. (2021)
38. Teterwak, P., et al.: Boundless: generative adversarial networks for image extension. In: Proceedings of the IEEE International Conference on Computer Vision, pp. 10521–10530 (2019)
39. Tov, O., Alaluf, Y., Nitzan, Y., Patashnik, O., Cohen-Or, D.: Designing an encoder for stylegan image manipulation. ACM Trans. Graph. (TOG) **40**(4), 1–14 (2021)
40. Ulyanov, D., Vedaldi, A., Lempitsky, V.: Instance normalization: the missing ingredient for fast stylization. arXiv preprint arXiv:1607.08022 (2016)
41. Wang, T.C., Liu, M.Y., Zhu, J.Y., Tao, A., Kautz, J., Catanzaro, B.: High-resolution image synthesis and semantic manipulation with conditional gans. In: Proceedings of the IEEE Conference on Computer Vision and Pattern Recognition, pp. 8798–8807 (2018)
42. Wu, Y., He, K.: Group normalization. In: Proceedings of the European Conference on Computer Vision (ECCV), pp. 3–19 (2018)
43. Yang, Z., Dong, J., Liu, P., Yang, Y., Yan, S.: Very long natural scenery image prediction by outpainting. In: Proceedings of the IEEE International Conference on Computer Vision, pp. 10561–10570 (2019)
44. Yu, J., Lin, Z., Yang, J., Shen, X., Lu, X., Huang, T.S.: Generative image inpainting with contextual attention. In: Proceedings of the IEEE Conference on Computer Vision and Pattern Recognition, pp. 5505–5514 (2018)
45. Yu, J., Lin, Z., Yang, J., Shen, X., Lu, X., Huang, T.S.: Free-form image inpainting with gated convolution. In: Proceedings of the IEEE International Conference on Computer Vision, pp. 4471–4480 (2019)
46. Zhan, F., Lu, S.: Esir: end-to-end scene text recognition via iterative image rectification. In: Proceedings of the IEEE/CVF Conference on Computer Vision and Pattern Recognition, pp. 2059–2068 (2019)
47. Zhan, F., Lu, S., Xue, C.: Verisimilar image synthesis for accurate detection and recognition of texts in scenes. In: Proceedings of the European Conference on Computer Vision (ECCV), pp. 249–266 (2018)
48. Zhan, F., Zhu, H., Lu, S.: Spatial fusion GAN for image synthesis. In: Proceedings of the IEEE/CVF Conference on Computer Vision and Pattern Recognition, pp. 3653–3662 (2019)
49. Zhang, R., Isola, P., Efros, A.A., Shechtman, E., Wang, O.: The unreasonable effectiveness of deep features as a perceptual metric. In: Proceedings of the IEEE Conference On Computer Vision And Pattern Recognition, pp. 586–595 (2018)
50. Zhao, S., Cui, J., Sheng, Y., Dong, Y., Liang, X., Eric, I., Chang, C., Xu, Y.: Large scale image completion via co-modulated generative adversarial networks. In: International Conference on Learning Representations (2020)
51. Zhou, B., Zhao, H., Puig, X., Fidler, S., Barriuso, A., Torralba, A.: Scene parsing through ade20k dataset. In: Proceedings of the IEEE Conference on Computer Vision and Pattern Recognition, pp. 633–641 (2017)

52. Zhou, X., et al.: Cocosnet v2: full-resolution correspondence learning for image translation. In: Proceedings of the IEEE/CVF Conference on Computer Vision and Pattern Recognition, pp. 11465–11475 (2021)
53. Zhu, P., Abdal, R., Qin, Y., Wonka, P.: Sean: image synthesis with semantic region-adaptive normalization. In: Proceedings of the IEEE/CVF Conference on Computer Vision and Pattern Recognition, pp. 5104–5113 (2020)

# Eliminating Gradient Conflict in Reference-based Line-Art Colorization

Zekun Li[1], Zhengyang Geng[2], Zhao Kang[1(✉)], Wenyu Chen[1], and Yibo Yang[3]

[1] University of Electronic Science and Technology of China, Chengdu, China
kunkun0w0@std.uestc.edu.cn, {zkang,cwy}@uestc.edu.cn
[2] School of AI, Peking University, Beijing, China
[3] JD Explore Academy, Beijing, China
ibo@pku.edu.cn

**Abstract.** Reference-based line-art colorization is a challenging task in computer vision. The color, texture, and shading are rendered based on an abstract sketch, which heavily relies on the precise long-range dependency modeling between the sketch and reference. Popular techniques to bridge the cross-modal information and model the long-range dependency employ the attention mechanism. However, in the context of reference-based line-art colorization, several techniques would intensify the existing training difficulty of attention, for instance, self-supervised training protocol and GAN-based losses. To understand the instability in training, we detect the gradient flow of attention and observe gradient conflict among attention branches. This phenomenon motivates us to alleviate the gradient issue by preserving the dominant gradient branch while removing the conflict ones. We propose a novel attention mechanism using this training strategy, Stop-Gradient Attention (SGA), outperforming the attention baseline by a large margin with better training stability. Compared with state-of-the-art modules in line-art colorization, our approach demonstrates significant improvements in Fréchet Inception Distance (FID, up to 27.21%) and structural similarity index measure (SSIM, up to 25.67%) on several benchmarks. The code of SGA is available at https://github.com/kunkun0w0/SGA.

**Keywords:** GAN · Attention mechanism · Stop-gradient

## 1 Introduction

Reference-based line-art colorization has achieved impressive performance in generating a realistic color image from a line-art image [32,59]. This technique is in high demand in comics, animation, and other content creation applications [2,55]. Different from painting with other conditions such as color strokes [12,53], palette [56], or text [25], using a style reference image as condition input

**Supplementary Information** The online version contains supplementary material available at https://doi.org/10.1007/978-3-031-19790-1_35.

© The Author(s), under exclusive license to Springer Nature Switzerland AG 2022
S. Avidan et al. (Eds.): ECCV 2022, LNCS 13677, pp. 579–596, 2022.
https://doi.org/10.1007/978-3-031-19790-1_35

not only provides richer semantic information for the model but also eliminates the requirements of precise color information and the geometric hints provided by users for every step. Nevertheless, due to the huge information discrepancy between the sketch and reference, it is challenging to correctly transfer colors from reference to the same semantic region in the sketch.

Several methods attempt to tackle the reference-based colorization by fusing the style latent code of reference into the sketch [31,36,38]. Inspired by the success of the attention mechanism [41,46], researchers adopt attention modules to establish the semantic correspondence and inject colors by mapping the reference to the sketch [28,54,55]. However, as shown in Fig. 1, the images generated by these methods often contain color bleeding or semantic mismatching, indicating considerable room for improving attention methods in line-art colorization.

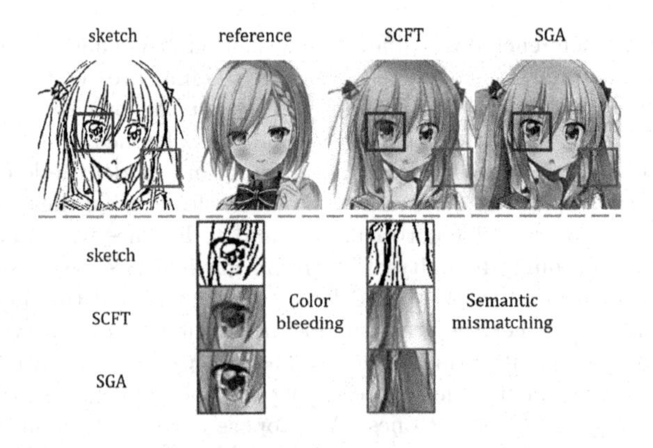


**Fig. 1.** The comparison between the images produced by SCFT [28] and SGA (Ours). SCFT subjects to color bleeding (orange box) and semantic mismatching (green box). (Color figure online)

There are many possible reasons for the deficiency of line-art colorization using attention: model pipeline, module architecture, or training. Motivated by recent works [5,13] concerning the training issues of attention models, we are particularly interested in the training stability of attention modules in line-art colorization. It is even more challenging to train attention models in line-art colorization because state-of-the-art models [28] deploy multiple losses using a GAN-style training pipeline, which can double the training instability. Therefore, we carefully analyze the training dynamics of attention in terms of its gradient flow in the context of line-art colorization. We observe the gradient conflict phenomenon, namely, a gradient branch contains a negative cosine similarity with the summed gradient.

To eliminate the gradient conflict, we detach the conflict one while preserving the dominant gradient, which ensures that the inexact gradient has a positive cosine similarity with the exact gradient and meet theory requirements [14,50]. This training strategy visibly boosts the training stability and performance compared with the baseline attention colorization models. Combined with architecture design, this paper introduces **S**top-**G**radient **A**ttention, **SGA**, whose training strategy eliminates the gradient conflict and helps the model learn better colorization correspondence. SGA properly transfers the style of reference images to the sketches, establishing accurate semantic correspondence between sketch-reference pairs. Our experiment results on several image domains show clear improvements over previous methods, *i.e.,* up to 27.21% and 25.67% regarding FID and SSIM, respectively.

Our contributions are summarized as follows:

- We reveal the gradient conflict in attention mechanism for line-art colorization, *i.e.,* a gradient branch contains a negative cosine similarity with the summed gradient.
- We propose a novel attention mechanism with gradient and design two attention blocks based on SGA, *i.e.,* cross-SGA and self-SGA.
- Both quantitative and qualitative results verify that our method outperforms state-of-the-art modules on several image datasets.

## 2 Related Work

**Reference-based Line-Art Colorization.** The reference-based line-art colorization is a user-friendly approach to assist designers in painting the sketch with their desired color [2,28,31,55]. Early studies attempt to get the style latent code of reference and directly mix it with sketch feature maps to generate the color image [31,38]. To make better use of reference images, some studies propose spatial-adaptive normalization methods [36,60].

Different from the aforementioned methods that adopt latent vectors for style control, [28,54,55] learn dense semantic correspondences between sketch-reference pairs. These approaches utilize the dot-product attention [41,46] to model the semantic mapping between sketch-reference pairs and inject color into sketch correctly. Although traditional non-local attention is excellent in feature alignment and integration between different modalities, the model cannot learn robust representation due to the gradient conflict in attention's optimization. Thus, our work proposes the stop-gradient operation for attention to eliminate the gradient conflict problem in line-art colorization.

**Attention Mechanism.** The attention mechanism [41,49] is proposed to capture long-range dependencies and align signals from different sources. It is widely applied in vision [46,52], language [10,41], and graph [42] areas. Due to the quadratic memory complexity of standard dot-product attention, many researchers from the vision [6,7,13,29,57] and language [24,37,44] communities endeavor to reduce the memory consumption to linear complexity. Recently, vision transformer [11] starts a new era for modeling visual data through the attention mechanism. The booming researches using transformer substantially change the trend in image [30,39,45,48], point cloud [16,58], gauge [19], and video [1,34] processing.

Unlike existing works concerning the architectures of attention mechanism, we focus on the training of attention modules regarding its gradient flow. Although some strategies have been developed to improve the training efficiency [3,39] for vision transformer, they mainly modify the objective function to impose additional supervision. From another perspective, our work investigates the gradient issue in the attention mechanism.

**Stop-Gradient Operation.** Backpropagation is the foundation for training deep neural networks. Recently some researchers have paid attention to the gradient flow in the deep models. Hamburger [13] proposes the one-step gradient to tackle the gradient conditioning and gradient norm issues in the implicit global context module, which helps obtain stable learning and performance. SimSiam [4] adopts the one-side stop-gradient operation to implicitly introduce an extra set of variables to implement Expectation-Maximization (EM) like algorithm in contrastive learning. VQ-VAE [35] also encourages discrete codebook learning by the stop-gradient supervision. All of these works indicate the indispensability of the gradient manipulation, which demonstrates that the neural network performance is related to both the advanced architecture and the appropriate training strategy.

Inspired by prior arts, our work investigates the gradient conflict issue for training non-local attention. The stop-gradient operation clips the conflict gradient branches while preserving correction direction for model updates.

# 3 Proposed Method

## 3.1 Overall Workflow

As illustrated in Fig. 2, we adopt a self-supervised training process similar to [28]. Given a color image $I$, we first use XDoG [47] to convert it into a line-art image $I_s$. Then, the expected coloring result $I_{gt}$ is obtained by adding a random color jittering on $I$ . Additionally, we generate a style reference image $I_r$ through applying the thin plate splines transformation on $I_{gt}$.

In the training process, utilizing $I_r$ as the reference to color the sketch $I_s$, our model first uses encoder $E_s$ and $E_r$ to extract sketch feature $f_s \in \mathbb{R}^{c \times h \times w}$ and reference feature $f_r \in \mathbb{R}^{c \times h \times w}$. In order to leverage multi-level representation simultaneously for feature alignment and integration, we concatenate the feature maps of all convolution layers outputs after using 2D adaptive average pooling function to down-sample them into the same spatial size.

To integrate the content in sketch and the style in reference, we employ our SGA blocks. There are two types of SGA blocks in our module: cross-SGA integrates the features from different domains and self-SGA models the global context of input features. Then several residual blocks and a U-net decoder $Dec$ with skip connections to sketch encoder $E_s$ are adopted to generate the image $I_{gen}$ by the mixed feature map $f_{gen}$. In the end, we add an adversarial loss [15] by using a discriminator $\boldsymbol{D}$ to distinguish the output $I_{gen}$ and the ground truth $I_{gt}$.

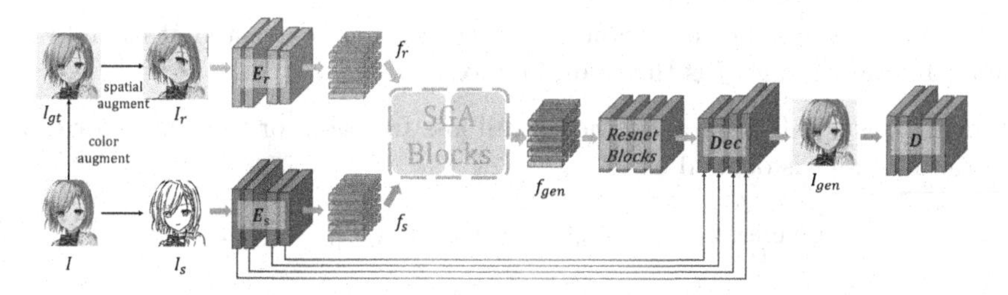


**Fig. 2.** The overview of our reference-based line-art colorization framework with a discriminator $\boldsymbol{D}$: Given the sketch $I_s$, the target image $I_{gt}$ and the reference $I_r$ obtained through original image $I$, we input $I_s$ and $I_r$ into encoder $E_s$ and $E_r$ to extract feature maps $f_s$ and $f_r$. The SGA blocks, which contain cross-SGA and self-SGA, integrate $f_s$ and $f_r$ into the mixed feature map $f_{gen}$. Then $f_{gen}$ is passed through several residual blocks and a U-net decoder $Dec$ with skip connection to generate the image $I_{gen}$. The $I_{gen}$ is supposed to be similar to $I_{gt}$.

### 3.2 Loss Function

**Image Reconstruction Loss.** According to the Sect. 3.1, both generated images $I_{gen}$ and ground truth images $I_{gt}$ should keep style consistency with reference $I_r$ and outline preservation with sketch $I_s$. Thus, we adopt $L_1$ regularization to measure the difference between $I_{gen}$ and $I_{gt}$, which ensures that the model colors correctly and distinctly:

$$\mathcal{L}_{\text{rec}} = \mathbb{E}_{I_s,I_r,I_{gt}}\left[\left\|\boldsymbol{G}\left(I_s, I_r\right) - I_{gt}\right\|_1\right] \tag{1}$$

where $\boldsymbol{G}\left(I_s, I_r\right)$ means coloring the sketch $I_s$ with the reference $I_r$.

**Adversarial Loss.** In order to generate a realistic image with the same outline as the prior sketch $I_s$, we leverage a conditional discriminator $\boldsymbol{D}$ to distinguish the generated images from real ones [21]. The least square adversarial loss [33] for optimizing our GAN-based model is formulated as:

$$\mathcal{L}_{adv} = \mathbb{E}_{I_{gt},I_s}\left[\left\|\boldsymbol{D}\left(I_{gt}, I_s\right)\right\|_2^2\right] + \mathbb{E}_{I_s,I_r}\left[\left\|(\mathbf{1} - \boldsymbol{D}\left(\boldsymbol{G}\left(I_s, I_r\right), I_s\right))\right\|_2^2\right] \tag{2}$$

**Style and Perceptual Loss.** As shown in previous works [22,28], perceptual loss and style loss encourage a network to produce a perceptually plausible output. Leveraging the ImageNet pretrained network, we reduce the gaps in multi-layer activation outputs between the target image $I_{gt}$ and generated image $I_{gen}$ by minimizing the following losses:

$$\mathcal{L}_{perc} = \mathbb{E}_{I_{gt},I_{gen}}\left[\sum_l \left\|\phi_l(I_{gt}) - \phi_l\left(I_{gen}\right)\right\|_1\right] \tag{3}$$

$$\mathcal{L}_{\text{style}} = \mathbb{E}_{I_{gt},I_{gen}}\left[\left\|\mathcal{G}\left(\phi_l(I_{gt})\right) - \mathcal{G}\left(\phi_l\left(I_{gen}\right)\right)\right\|_1\right] \tag{4}$$

where $\phi_l$ represents the activation map of the $l_{th}$ layer extracted at the relu from VGG19 network, and $\mathcal{G}$ is the gram matrix.

**Overall Loss.** In summary, the overall loss function for the generator $\boldsymbol{G}$ and discriminator $\boldsymbol{D}$ is defined as:

$$\min_{G} \max_{D} \mathcal{L}_{\text{total}} = \mathcal{L}_{adv} + \lambda_1 \mathcal{L}_{\text{rec}} + \lambda_2 \mathcal{L}_{\text{perc}} + \lambda_3 \mathcal{L}_{\text{style}} \tag{5}$$

### 3.3 Gradient Issue in Attention

In this section, we use SCFT [28], a classic attention-based method in colorization, as an example to study the gradient issue in attention. $\boldsymbol{Q} \in \mathbb{R}^{n \times d}$ is the feature projection transformed by $\mathbf{W}_\text{q}$ from the input $\boldsymbol{X} \in \mathbb{R}^{n \times d}$. The feature projections $\boldsymbol{K}, \boldsymbol{V} \in \mathbb{R}^{n \times d}$ from input $\boldsymbol{Y} \in \mathbb{R}^{n \times d}$ are transformed by $\mathbf{W}_\text{k}$ and $\mathbf{W}_\text{v}$ . Given the attention map $\boldsymbol{A} \in \mathbb{R}^{n \times n}$, the classic dot-product attention mechanism can be formulated as follows:

$$\boldsymbol{Z} = \text{softmax}(\frac{\boldsymbol{Q}\boldsymbol{K}^\top}{\sqrt{d}})\boldsymbol{V} + \boldsymbol{X} = \boldsymbol{A}\boldsymbol{V} + \boldsymbol{X} \tag{6}$$

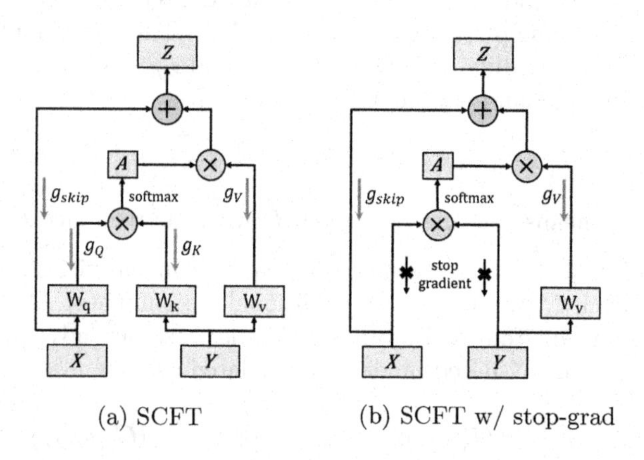


**Fig. 3.** Stop-gradient in attention module. The $g_{skip}$, $g_Q$, $g_K$ and $g_V$ separately represent the gradient along their branches. The stop-gradient operation (**stop-grad**) truncates the backpropagation of conflict gradients existing in attention map calculation.

Previous works [3,5,13,39] present the training difficulty of vision attention: instability, worse generalization, *etc* . For line-art colorization, it is even more challenging to train the attention models, as the training involves GAN-style loss and reconstruction loss, which are understood to lead to mode collapse [15] or trivial solutions. Given a training schedule, the loss of colorization network can shake during training and finally deteriorate.

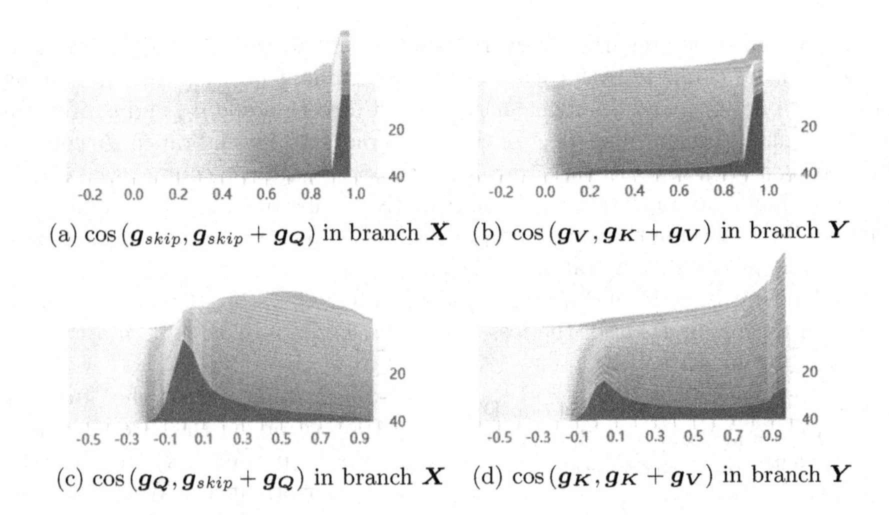


(a) $\cos(\boldsymbol{g}_{skip}, \boldsymbol{g}_{skip} + \boldsymbol{g}_{\boldsymbol{Q}})$ in branch $\boldsymbol{X}$ (b) $\cos(\boldsymbol{g}_{\boldsymbol{V}}, \boldsymbol{g}_{\boldsymbol{K}} + \boldsymbol{g}_{\boldsymbol{V}})$ in branch $\boldsymbol{Y}$

(c) $\cos(\boldsymbol{g}_{\boldsymbol{Q}}, \boldsymbol{g}_{skip} + \boldsymbol{g}_{\boldsymbol{Q}})$ in branch $\boldsymbol{X}$ (d) $\cos(\boldsymbol{g}_{\boldsymbol{K}}, \boldsymbol{g}_{\boldsymbol{K}} + \boldsymbol{g}_{\boldsymbol{V}})$ in branch $\boldsymbol{Y}$

**Fig. 4.** The histograms of the gradient cosine value distribution in 40 epochs. A large cosine value means that the network mainly uses this branch of gradient to optimize the loss function.

To better understand reasons behind the training difficulty of attention in colorization, we analyze the gradient issue through the classic SCFT model [28]. We visualize the gradient flow back through the attention module in terms of each gradient branch and the summed gradient.

Figure 4 offers the cosine value between different gradient branches and the total gradient. We separately calculate $\cos(\boldsymbol{g}_{skip}, \boldsymbol{g}_{skip} + \boldsymbol{g}_{\boldsymbol{Q}})$ and $\cos(\boldsymbol{g}_{\boldsymbol{Q}}, \boldsymbol{g}_{skip} + \boldsymbol{g}_{\boldsymbol{Q}})$ for each pixel in branch $\boldsymbol{X}$ (means gradient in sketch feature maps $f_s$), $\cos(\boldsymbol{g}_{\boldsymbol{V}}, \boldsymbol{g}_{\boldsymbol{K}} + \boldsymbol{g}_{\boldsymbol{V}})$ and $\cos(\boldsymbol{g}_{\boldsymbol{K}}, \boldsymbol{g}_{\boldsymbol{K}} + \boldsymbol{g}_{\boldsymbol{V}})$ in branch $\boldsymbol{Y}$ (means gradient in reference feature maps $f_r$) to explore the gradient flow of the network during learning.

Note that first order optimization methods usually require the surrogate gradient $\tilde{\boldsymbol{g}}$ for update to be ascent, *i.e.*, $\cos(\tilde{\boldsymbol{g}}, \boldsymbol{g}) > 0$, where $\boldsymbol{g}$ is the exact gradient. Then the update direction based on the surrogate gradient can be descent direction. The visualization in Fig. 4 implies that the gradient $\boldsymbol{g}_{skip}$ from the skip connection for the branch $\boldsymbol{X}$ and the gradient $\boldsymbol{g}_V$ from $\boldsymbol{V}$ for the branch $\boldsymbol{Y}$ has already become an ascent direction for optimization, denoting that $\boldsymbol{g}_Q$ and $\boldsymbol{g}_K$ from the attention map construct the "conflict gradient" $\not{\boldsymbol{g}}$ in respect of the total gradient $\boldsymbol{g}$, *i.e.*, $\cos(\not{\boldsymbol{g}}, \boldsymbol{g}) < 0$.

Figures 4a and 4b show that $\boldsymbol{g}_{skip}$ and $\boldsymbol{g}_V$ are usually highly correlated with the total gradient, where over **78.09%** and **52.39%** of the cosine values are greater than **0.935** in the 40th epoch, respectively. Moreover, these percentages increase during training, indicating the significance of the representative gradient. On the other hand, nearly **30.57%** of $\boldsymbol{g}_q$ in Fig. 4c and **10.77%** of $\boldsymbol{g}_K$ in Fig. 4d have negative cosine values in the 40th epoch. These proportions are **22.81%** and **5.32%** in the 20th epoch, respectively, gradually increasing during training.

The visualization regarding the gradient flows demonstrates that the two gradient branches compete with each other for a dominant position during training process, while $\boldsymbol{g}_{skip}$ and $\boldsymbol{g}_V$ construct an ascent direction and $\boldsymbol{g}_Q$ and $\boldsymbol{g}_K$ remain as the conflict gradient in respect of the total gradient in each branch. According to existing works in multi-task learning [50], large gradient conflict ratios may result in significant performance drop. It motivates us to detach the conflict gradient while preserving the dominant gradient as inexact gradient to approximate the original gradient, illustrated in Fig. 3.

Verified by Figs. 4a and 4b, the gradient after the stop-gradient operation forms an ascent direction of the loss landscape, *i.e.*, $\cos(\tilde{\boldsymbol{g}}, \boldsymbol{g}) > 0$, and thus be valid for optimization [14].

**Table 1.** Test the Fréchet Inception Distance (FID) and SSIM with different settings of SCFT on anime dataset. ↑ means the higher the better, while ↓ indicates the lower the better.

| SCFT Setting | | FID↓ | SSIM↑ |
|---|---|---|---|
| **Stop-grad** | $\mathbf{W}_q$ & $\mathbf{W}_k$ | | |
| ✗ | ✓ | 44.65 | 0.788 |
| ✗ | ✗ | 48.04 | 0.799 |
| ✓ | ✓ | 38.20 | 0.835 |
| ✓ | ✗ | **36.78** | **0.841** |

Table 1 shows that the gradient clipping through the stop-gradient operation can effectively improve the model performance. We can also remove $\mathbf{W}_k$ and $\mathbf{W}_q$ since there is no gradient propagating in them and they will not be updated in the training process. The lower FID and higher SSIM mean that model can generate more realistic images with higher outline preservation during colorization after the stop-gradient clipping.

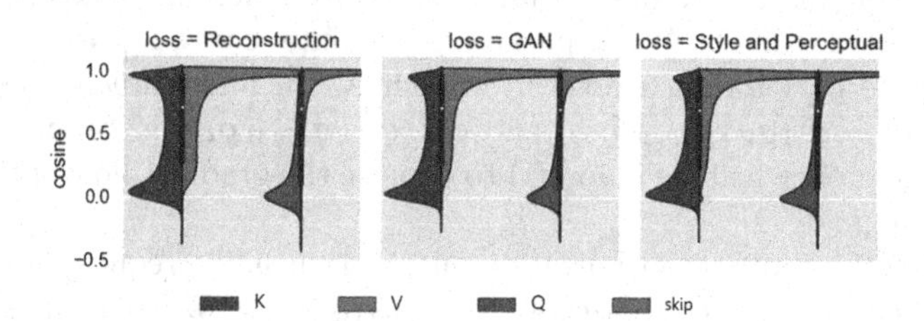


**Fig. 5.** Visualizations of the gradient cosine distribution when using a single loss on the pretrained SCFT model.

In order to investigate the reliability of gradient conflicts, we test the gradient cosine distributions when using a certain loss to confirm the trigger to gradient issue is the dot-product attention. We use the SCFT model to compute the gradients cosine distribution of each loss to investigate whether loss functions or architectures cause the conflict. Figure 5 shows that all loss terms cause similar conflicts, implying that the attention architecture leads to gradient conflicts.

### 3.4 Stop-Gradient Attention

Combining with the training strategy, we propose the **S**top-**G**radient **A**ttention (SGA). As Fig. 6a illustrates, in addition to the stop-gradient operation, we also design a new feature integration and normalization strategy for SGA. Treating stop-gradient attention map $\boldsymbol{A}$ as a prior deep graph structure input, inspired by [27,43], features can be effectively aggregated from adjacency nodes and the node itself:

$$\boldsymbol{Z} = \sigma(\boldsymbol{X}\mathbf{W}_{\mathrm{x}}) + \widehat{\boldsymbol{A}}\sigma(\boldsymbol{Y}\mathbf{W}_{\mathrm{y}}) \tag{7}$$

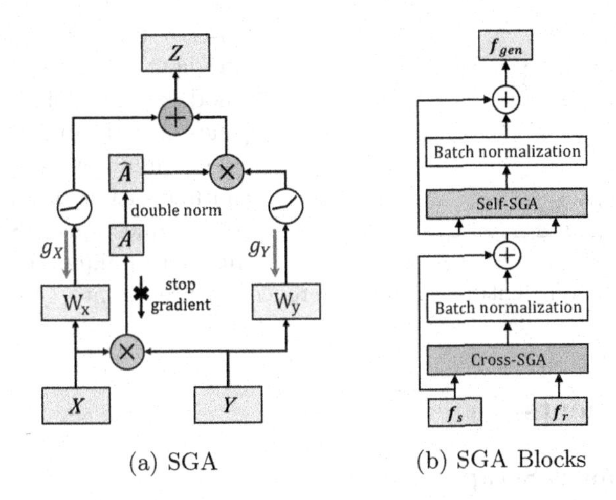


(a) SGA (b) SGA Blocks

**Fig. 6.** SGA computes the attention map with stop-gradient, which truncates the gradient propagation of $\boldsymbol{g}_{att}$ and adopts a double normalization technique in addition. In our colorization network, we stack two types of SGA to integrate features: cross-SGA (yellow box) and self-SGA (green box). (Color figure online)

where $\sigma$ is the leaky relu activate function and $\widehat{\boldsymbol{A}}$ is the attention map normalized by double normalization method analogous to Sinkhorn algorithm [9]. Different from softmax employed in classic non-local attention, the double normalization makes the attention map insensitive to the scale of input features [17]. The normalized attention map $\widehat{\boldsymbol{A}}$ can be formulated as follows:

$$\boldsymbol{A} = \boldsymbol{X}\boldsymbol{Y}^{\top} \tag{8}$$

$$\tilde{\boldsymbol{A}}_{ij} = \exp(\boldsymbol{A}_{ij}) / \sum_k \exp(\boldsymbol{A}_{ik}) \tag{9}$$

$$\widehat{\boldsymbol{A}}_{ij} = \tilde{\boldsymbol{A}}_{ij} / \sum_k \tilde{\boldsymbol{A}}_{kj} \tag{10}$$

where $\widehat{\boldsymbol{A}}_{ij}$ means correlation between $i_{th}$ feature vector in $\boldsymbol{X}$ and $j_{th}$ feature vector in $\boldsymbol{Y}$. The pseudo-code of SGA is summarized in Algorithm 1.

**Algorithm 1** SGA Pseudocode pytorch

```
# input:
# X: feature maps -> tensor(b, wh, c)
# Y: feature maps -> tensor(b, wh, c)

# output:
# Z: feature maps -> tensor(b, wh, c)

# other objects:
# Wx, Wy: embedding matrix -> nn.Linear(c,c)
# A: attention map -> tensor(b, wh, wh)
# leaky_relu: leaky relu activation function

with torch.no_grad():
    A = X.bmm(Y.permute(0, 2, 1))
    A = softmax(A, dim=-1)
    A = normalize(A, p=1, dim=-2)

X = leaky_relu(Wx(X))
Y = leaky_relu(Wy(Y))

Z = torch.bmm(A,Y) + X
```

Furthermore, we design two types of SGA, called cross-SGA and self-SGA. Both of their calculation are based on Algorithm 1. As shown in Fig. 6b, the only difference between them is whether the inputs are the same or not. Cross-SGA calculates pixel correlation between features from different image domains and integrates features under a stop-gradient attention map. Self-SGA models the global context and fine-tunes the integration. For stable training, we also adopt batch normalization layer and shortcut connections [18]. Combining above techniques, our SGA blocks integrate the sketch feature $f_s$ and reference feature $f_r$ into generated feature $f_{gen}$ effectively.

# 4 Experiments

## 4.1 Experiment Setup

**Dataset.** We test our method on popular anime portraits [40] and Animal FacesHQ (AFHQ) [8] dataset. The anime portraits dataset contains 33323 anime faces for training and 1000 for evaluation. AFHQ is a dataset of animal faces consisting of 15,000 high-quality images at 512 × 512 resolution, which contains three categories of pictures, *i.e.*, cat, dog, and wildlife. Each class in AFHQ provides 5000 images for training and 500 for evaluation. To simulate the line-art drawn by artists, we use XDoG [47] to extract sketch inputs and set the parameters of XDoG algorithm with $\phi = 1 \times 10^9$ to keep a step transition at the border of sketch lines. We randomly set $\sigma$ to be 0.3/0.4/0.5 to get different levels of line thickness, which generalizes the network on various line widths to avoid overfitting. And we set $p = 19, k = 4.5, \epsilon = 0.01$ by default in XDoG.

**Implementation Details.** We implement our model with the size of input image fixed at 256×256 for each dataset. For training, we set the coefficients for each loss terms as follows: $\lambda_1 = 30, \lambda_2 = 0.01$, and $\lambda_3 = 50$. We use Adam solver [26] for optimization with $\beta_1 = 0.5, \beta_2 = 0.999$. The learning rate of generator and discriminator are initially set to 0.0001 and 0.0002, respectively. The training lasts 40 epochs on each dataset.

**Evaluation Metrics.** In evaluation process, we randomly select reference images and sketch images for colorization as Fig. 7 shows. The popular Fréchet

Inception Distance (FID) [20] is used to assess the perceptual quality of generated images by comparing the distance between distributions of generated and real images in a deep feature embedding. Besides measuring the perceptual credibility, we also adopt the structural similarity index measure (SSIM) to quantify the outline preservation during colorization, by calculating the SSIM between generated image and original color image of sketch.

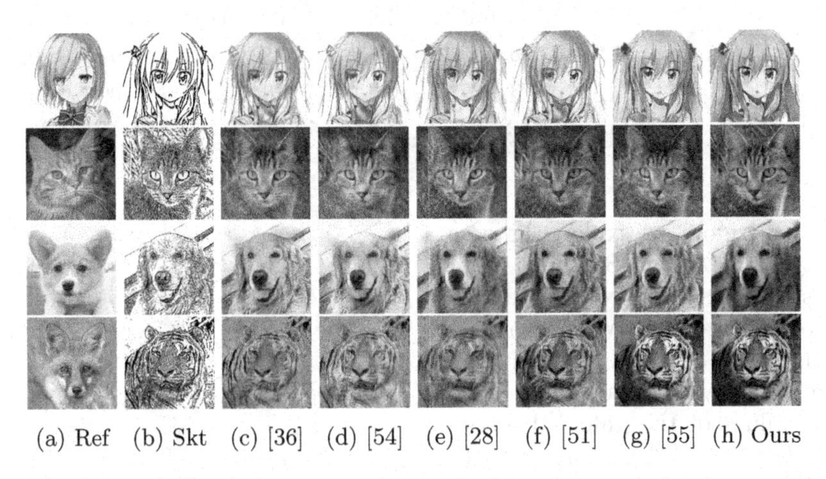

(a) Ref (b) Skt (c) [36] (d) [54] (e) [28] (f) [51] (g) [55] (h) Ours

**Fig. 7.** Visualization of colorization results. "Ref" stands for "reference". "Skt" indicates "sketch". Compared with other methods, SGA shows correct correspondence between the sketch and reference images.

### 4.2 Comparison Results

We compare our method with existing state-of-the-art modules include not only reference-based line-art colorization [28] but also image-to-image translation, *i.e.*, SPADE [36], CoCosNet [54], UNITE [51] and CMFT [55]. For fairness, in our experiments, all networks use the same encoders, decoder, residual blocks and discriminator implemented in SCFT [28] with aforementioned train losses. Table 2 shows that SGA outperforms other techniques by a large margin. With respect to our main competitor SCFT, SGA improves by 27.21% and 25.67% on average for FID and SSIM, respectively. This clear-cut improvement means that SGA produces a more realistic image with high outline preservation compared with previous methods. According to Fig. 7, the images generated by SGA have less color-bleeding and higher color consistency in perceptual.

**Table 2.** Quantitative comparison with different methods. Boldface represents the best value. Underline stands for the second score.

| Method | Anime | | Cat | | Dog | | Wild | |
|---|---|---|---|---|---|---|---|---|
| | FID↓ | SSIM↑ | FID↓ | SSIM↑ | FID↓ | SSIM↑ | FID↓ | SSIM↑ |
| SPADE [36] | 57.55 | 0.681 | 36.11 | 0.526 | 76.57 | 0.631 | 24.56 | 0.573 |
| CoCosNet [54] | 52.06 | 0.672 | 35.02 | 0.511 | <u>68.69</u> | 0.603 | <u>23.10</u> | 0.554 |
| SCFT [28] | 44.65 | 0.788 | 36.33 | 0.636 | 79.08 | 0.683 | 24.93 | 0.633 |
| UNITE [51] | 52.19 | 0.676 | **33.26** | 0.636 | 72.38 | 0.677 | 23.97 | 0.592 |
| CMFT [55] | <u>38.94</u> | <u>0.873</u> | 37.78 | <u>0.813</u> | 73.18 | <u>0.809</u> | 23.90 | <u>0.822</u> |
| SGA | **29.65** | **0.912** | <u>34.35</u> | **0.843** | **54.76** | **0.841** | **15.19** | **0.831** |

Furthermore, we explore the superiority of SGA over SCFT in terms of rescaling spectrum concentration of the representations. We compare the accumulative ratios of squared top $r$ singular values over total squared singular values of the unfolded feature maps (*i.e.*, $\mathbb{R}^{C\times HW}$) before and after passing through the attention module, illustrated in Fig. 8. The sum of singular values is the nuclear norm, *i.e.*, the convex relaxation for matrix rank that measures how compact the representations are, which is widely applied in machine learning [23]. The accumulative ratios are obviously lifted after going through SCFT and SGA, which facilitates the model to focus more on critical global information [13]. However, our effective SGA can not only further denoise feature maps but also enforce the encoder before attention module to learn energy-concentrated representations, *i.e.*, under the effect of SGA, the CNN encoder can also learn to focus on the global information.

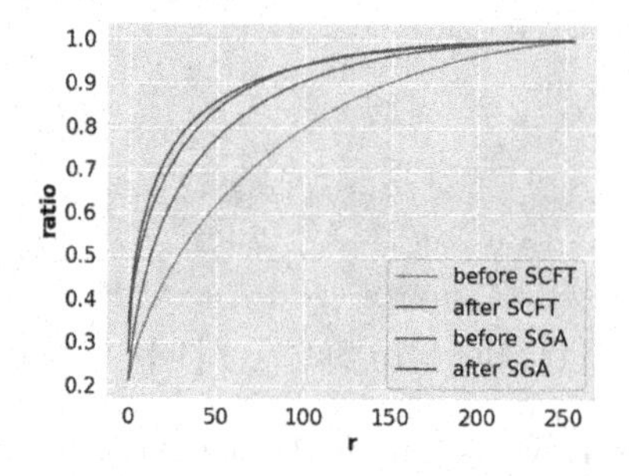


**Fig. 8.** Accumulative ratio of the squared top $r$ singular values over total squared singular values in feature maps. The ratios of feature maps before and after the attention module in SCFT and SGA are displayed.

### 4.3 Ablation Study

We perform several ablation experiments to verify the effectiveness of SGA blocks in our framework, *i.e.*, stop-gradient operation, attention map normalization, and self-SGA. The quantitative results are reported in Table 3, showing the superiority of our SGA blocks.

Specifically, to evaluate the necessity of stop-gradient in non-local attention, we design a variant SGA without stop-gradient. In Table 3, it obtains inferior performance, which verifies the benefit of eliminating gradient conflict through stop-gradient.

**Table 3.** Ablation study result with different settings. Boldface represents the best value. Underline stands for the second best.

| Setting | Anime | | cat | | Dog | | Wilds | |
|---|---|---|---|---|---|---|---|---|
| | FID↓ | SSIM↑ | FID↓ | SSIM↑ | FID↓ | SSIM↑ | FID↓ | SSIM↑ |
| SGA | **29.65** | 0.912 | 34.35 | **0.843** | **54.76** | **0.841** | **15.19** | **0.831** |
| SGA w/o stop-gradient | 36.34 | 0.876 | 40.73 | 0.796 | 72.34 | 0.808 | 19.90 | 0.791 |
| SGA w/o double-norm | 33.42 | 0.861 | 34.42 | 0.811 | 55.08 | 0.828 | 15.95 | 0.809 |
| SGA w/o self-SGA | 31.56 | **0.917** | **34.26** | 0.842 | 55.69 | 0.839 | 16.36 | 0.821 |

Furthermore, we conduct an ablation study on the attention map normalization to validate the advantage of double normalization in our framework. Table 3 demonstrates that SGA with double normalization outperforms that with classic softmax function. Although classic softmax can generate realistic images, it suffers a low outline preservation, *i.e.*, the SSIM measure.

Based on the framework with stop-gradient and double normalization, we make an ablation study on the improvement of self-SGA additionally. Although our model has achieved excellent performance without self-SGA, there is still a clear-cut enhancement on most datasets after employing the self-SGA according to Table 3. The stacks of SGA can help model not only integrate feature effectively, but also fine-tune a better representation with global awareness for coloring.

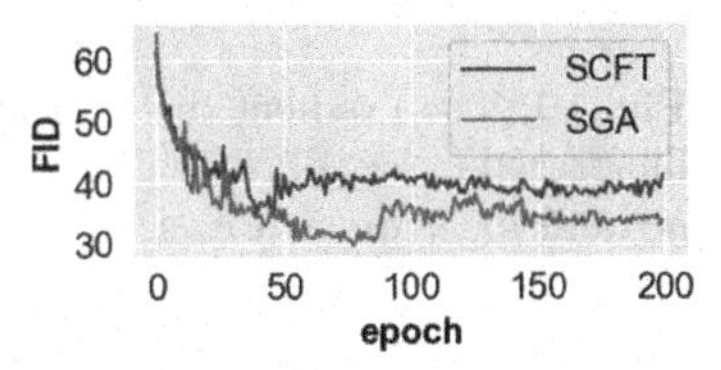


**Fig. 9.** Cat's FID during 200 epochs training.

Extending the training schedule to 200 epochs, Fig. 9 shows that SGA can still perform better with more epochs (29.71 in the 78th epoch) and collapse later than SCFT [28], demonstrating the training stability for attention models in line-art colorization.

Additionally, to be more rigorous, we visualize the gradient distributions in the "SGA w/o stop-gradient". Figure 10 implies the existing of gradient conflicts is a general phenomena in dot-product attention mechanism.

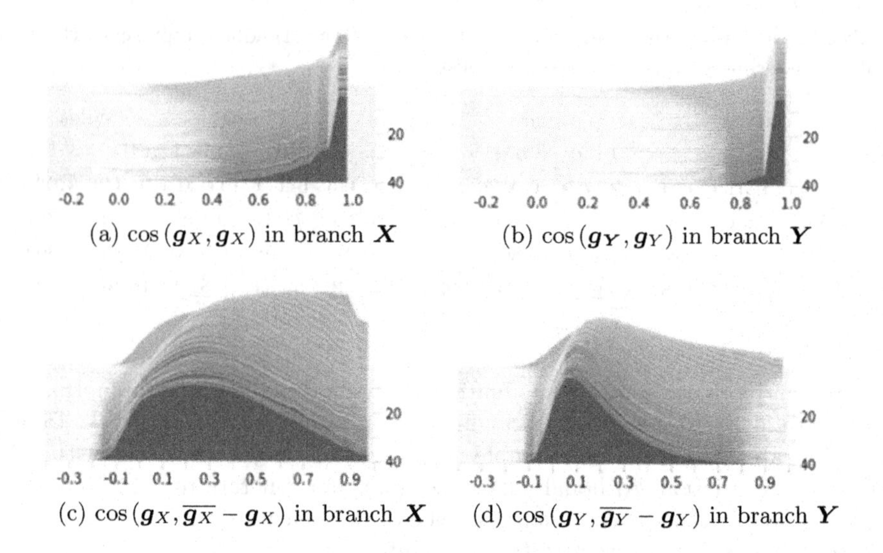


(a) $\cos(\boldsymbol{g}_X, \boldsymbol{g}_X)$ in branch $\boldsymbol{X}$ (b) $\cos(\boldsymbol{g}_Y, \boldsymbol{g}_Y)$ in branch $\boldsymbol{Y}$

(c) $\cos(\boldsymbol{g}_X, \overline{\boldsymbol{g}_X} - \boldsymbol{g}_X)$ in branch $\boldsymbol{X}$ (d) $\cos(\boldsymbol{g}_Y, \overline{\boldsymbol{g}_Y} - \boldsymbol{g}_Y)$ in branch $\boldsymbol{Y}$

**Fig. 10.** The gradient distribution of "SGA w/o stop-gradient". The $\boldsymbol{g}_X$ and $\boldsymbol{g}_Y$ are illustrated in Fig. 6a. The $\overline{\boldsymbol{g}_X}$ and $\overline{\boldsymbol{g}_Y}$ represent the total gradient, similar to the $\boldsymbol{g}_{skip} + \boldsymbol{g}_Q$ and $\boldsymbol{g}_K + \boldsymbol{g}_V$ in Fig. 4.

## 5 Conclusion

In this paper, we investigate the gradient conflict phenomenon in classic attention networks for line-art colorization. To eliminate the gradient conflict issue, we present a novel cross-modal attention mechanism, **S**top-**G**radient **A**ttention (**SGA**) by clipping the conflict gradient through the stop-gradient operation. The stop-gradient operation can unleash the potential of attention mechanism for reference-based line-art colorization. Extensive experiments on several image domains demonstrate that our simple technique significantly improves the reference-based colorization performance with better the training stability.

**Acknowledgment.** This research was funded in part by the Sichuan Science and Technology Program (Nos. 2021YFG0018, 2022YFG0038).

## References


1. Arnab, A., Dehghani, M., Heigold, G., Sun, C., Lučić, M., Schmid, C.: Vivit: a video vision transformer. In: Proceedings of the IEEE/CVF International Conference on Computer Vision (ICCV), pp. 6836–6846 (2021)
2. Casey, E., Perez, V., Li, Z.: The animation transformer: visual correspondence via segment matching. In: Proceedings of the IEEE/CVF International Conference on Computer Vision (ICCV), pp. 11323–11332 (2021)

3. Chen, X., Hsieh, C.J., Gong, B.: When vision transformers outperform resnets without pretraining or strong data augmentations. arXiv preprint arXiv:2106.01548 (2021)
4. Chen, X., He, K.: Exploring simple siamese representation learning. In: Proceedings of the IEEE/CVF Conference on Computer Vision and Pattern Recognition (CVPR), pp. 15750–15758 (2021)
5. Chen, X., Xie, S., He, K.: An empirical study of training self-supervised vision transformers. In: Proceedings of the IEEE/CVF International Conference on Computer Vision, pp. 9640–9649 (2021)
6. Chen, Y., Kalantidis, Y., Li, J., Yan, S., Feng, J.: A $\wedge$ 2-nets: double attention networks. In: Advances in Neural Information Processing Systems, vol. 31, pp. 352–361 (2018)
7. Chen, Y., Rohrbach, M., Yan, Z., Shuicheng, Y., Feng, J., Kalantidis, Y.: Graph-based global reasoning networks. In: IEEE Conference on Computer Vision and Pattern Recognition (CVPR) (2019)
8. Choi, Y., Uh, Y., Yoo, J., Ha, J.W.: Stargan v2: diverse image synthesis for multiple domains. In: IEEE/CVF Conference on Computer Vision and Pattern Recognition (CVPR) (2020)
9. Cuturi, M.: Sinkhorn distances: Lightspeed computation of optimal transport. Adv. Neural. Inf. Process. Syst. **26**, 2292–2300 (2013)
10. Dai, Z., Yang, Z., Yang, Y., Carbonell, J.G., Le, Q., Salakhutdinov, R.: Transformer-xl: attentive language models beyond a fixed-length context. In: Annual Meeting of the Association for Computational Linguistics (ACL), pp. 2978–2988 (2019)
11. Dosovitskiy, A., et al.: An image is worth 16 $\times$ 16 words: transformers for image recognition at scale. In: International Conference on Learning Representations (2021)
12. Dou, Z., Wang, N., Li, B., Wang, Z., Li, H., Liu, B.: Dual color space guided sketch colorization. IEEE Trans. Image Process. **30**, 7292–7304 (2021)
13. Geng, Z., Guo, M.H., Chen, H., Li, X., Wei, K., Lin, Z.: Is attention better than matrix decomposition? In: International Conference on Learning Representations (2021)
14. Geng, Z., Zhang, X.Y., Bai, S., Wang, Y., Lin, Z.: On training implicit models. In: Thirty-Fifth Conference on Neural Information Processing Systems (2021)
15. Goodfellow, I., et al.: Generative adversarial nets. Adv. Neural Inf. Process. Syst. **27** (2014)
16. Guo, M.H., Cai, J.X., Liu, Z.N., Mu, T.J., Martin, R.R., Hu, S.M.: PCT: point cloud transformer. Comput. Visual Media **7**(2), 187–199 (2021). https://doi.org/10.1007/s41095-021-0229-5
17. Guo, M.H., Liu, Z.N., Mu, T.J., Hu, S.M.: Beyond self-attention: external attention using two linear layers for visual tasks. arXiv preprint arXiv:2105.02358 (2021)
18. He, K., Zhang, X., Ren, S., Sun, J.: Identity mappings in deep residual networks. In: Leibe, B., Matas, J., Sebe, N., Welling, M. (eds.) ECCV 2016. LNCS, vol. 9908, pp. 630–645. Springer, Cham (2016). https://doi.org/10.1007/978-3-319-46493-0_38
19. He, L., Dong, Y., Wang, Y., Tao, D., Lin, Z.: Gauge equivariant transformer. Adv. Neural. Inf. Process. Syst. **34**, 27331–27343 (2021)
20. Heusel, M., Ramsauer, H., Unterthiner, T., Nessler, B., Hochreiter, S.: Gans trained by a two time-scale update rule converge to a local nash equilibrium. Adv. Neural Inf. Process. Syst. **30** (2017)

21. Isola, P., Zhu, J.Y., Zhou, T., Efros, A.A.: Image-to-image translation with conditional adversarial networks. In: Proceedings of the IEEE Conference on Computer Vision and Pattern Recognition, pp. 1125–1134 (2017)
22. Johnson, J., Alahi, A., Fei-Fei, L.: Perceptual losses for real-time style transfer and super-resolution. In: European Conference on Computer Vision (2016)
23. Kang, Z., Peng, C., Cheng, J., Cheng, Q.: Logdet rank minimization with application to subspace clustering. Comput. Intell. Neurosci. **2015** (2015)
24. Katharopoulos, A., Vyas, A., Pappas, N., Fleuret, F.: Transformers are RNNs: fast autoregressive transformers with linear attention. In: International Conference on Machine Learning, pp. 5156–5165. PMLR (2020)
25. Kim, H., Jhoo, H.Y., Park, E., Yoo, S.: Tag2pix: line art colorization using text tag with secat and changing loss. In: Proceedings of the IEEE/CVF International Conference on Computer Vision, pp. 9056–9065 (2019)
26. Kingma, D.P., Ba, J.: Adam: a method for stochastic optimization. In: International Conference on Learning Representations (ICLR) (2015)
27. Kipf, T.N., Welling, M.: Semi-supervised classification with graph convolutional networks. In: Proceedings of the 5th International Conference on Learning Representations. ICLR 2017 (2017)
28. Lee, J., Kim, E., Lee, Y., Kim, D., Chang, J., Choo, J.: Reference-based sketch image colorization using augmented-self reference and dense semantic correspondence. In: The IEEE/CVF Conference on Computer Vision and Pattern Recognition (CVPR) (2020)
29. Li, X., Zhong, Z., Wu, J., Yang, Y., Lin, Z., Liu, H.: Expectation-maximization attention networks for semantic segmentation. In: International Conference on Computer Vision (2019)
30. Liu, Z., et al.: Swin transformer: hierarchical vision transformer using shifted windows. In: International Conference on Computer Vision (ICCV) (2021)
31. LvMin Zhang, Y.J., Liu, C.: Style transfer for anime sketches with enhanced residual u-net and auxiliary classifier GAN. In: Asian Conference on Pattern Recognition (ACPR) (2017)
32. Maejima, A., Kubo, H., Funatomi, T., Yotsukura, T., Nakamura, S., Mukaigawa, Y.: Graph matching based anime colorization with multiple references. In: ACM SIGGRAPH 2019 (2019)
33. Mao, X., Li, Q., Xie, H., Lau, R.Y., Wang, Z., Paul Smolley, S.: Least squares generative adversarial networks. In: Proceedings of the IEEE International Conference on Computer Vision (ICCV) (2017)
34. Neimark, D., Bar, O., Zohar, M., Asselmann, D.: Video transformer network. In: Proceedings of the IEEE/CVF International Conference on Computer Vision (ICCV) Workshops, pp. 3163–3172 (2021)
35. van den Oord, A., Vinyals, O., Kavukcuoglu, K.: Neural discrete representation learning. In: NIPS (2017)
36. Park, T., Liu, M.Y., Wang, T.C., Zhu, J.Y.: Semantic image synthesis with spatially-adaptive normalization. In: Proceedings of the IEEE/CVF Conference on Computer Vision and Pattern Recognition (CVPR) (2019)
37. Roy, A., Saffar, M.T., Vaswani, A., Grangier, D.: Efficient content-based sparse attention with routing transformers. Trans. Assoc. Comput. Linguist. **9**, 53–68 (2021)
38. Sun, T.H., Lai, C.H., Wong, S.K., Wang, Y.S.: Adversarial colorization of icons based on contour and color conditions. In: Proceedings of the 27th ACM International Conference on Multimedia, pp. 683–691 (2019)

39. Touvron, H., Cord, M., Douze, M., Massa, F., Sablayrolles, A., Jégou, H.: Training data-efficient image transformers & distillation through attention. In: International Conference on Machine Learning, pp. 10347–10357. PMLR (2021)
40. Tseng, H.-Y., Fisher, M., Lu, J., Li, Y., Kim, V., Yang, M.-H.: Modeling artistic workflows for image generation and editing. In: Vedaldi, A., Bischof, H., Brox, T., Frahm, J.-M. (eds.) ECCV 2020. LNCS, vol. 12363, pp. 158–174. Springer, Cham (2020). https://doi.org/10.1007/978-3-030-58523-5_10
41. Vaswani, A., et al.: Attention is all you need. In: Advances in Neural Information Processing Systems, pp. 5998–6008 (2017)
42. Veličković, P., Cucurull, G., Casanova, A., Romero, A., Liò, P., Bengio, Y.: Graph attention networks. In: International Conference on Learning Representations (ICLR) (2018)
43. Wang, R., Yan, J., Yang, X.: Learning combinatorial embedding networks for deep graph matching. In: Proceedings of the IEEE/CVF International Conference on Computer Vision, pp. 3056–3065 (2019)
44. Wang, S., Li, B.Z., Khabsa, M., Fang, H., Ma, H.: Linformer: self-attention with linear complexity. ArXiv abs/2006.04768 (2020)
45. Wang, W., et al.: Pyramid vision transformer: a versatile backbone for dense prediction without convolutions. In: IEEE ICCV (2021)
46. Wang, X., Girshick, R., Gupta, A., He, K.: Non-local neural networks. In: IEEE Conference on Computer Vision and Pattern Recognition (CVPR), pp. 7794–7803 (2018)
47. Winnemöller, H., Kyprianidis, J.E., Olsen, S.C.: XDoG: an extended difference-of-gaussians compendium including advanced image stylization. Comput. Graph. **36**(6), 740–753 (2012)
48. Wu, H., et al.: Cvt: introducing convolutions to vision transformers. arXiv preprint arXiv:2103.15808 (2021)
49. Xu, K., et al.: Show, attend and tell: neural image caption generation with visual attention. In: International Conference on Machine Learning (ICML), pp. 2048–2057 (2015)
50. Yu, T., Kumar, S., Gupta, A., Levine, S., Hausman, K., Finn, C.: Gradient surgery for multi-task learning. Adv. Neural. Inf. Process. Syst. **33**, 5824–5836 (2020)
51. Zhan, F., et al.: Unbalanced feature transport for exemplar-based image translation. In: Proceedings of the IEEE/CVF Conference on Computer Vision and Pattern Recognition (CVPR), pp. 15028–15038 (2021)
52. Zhang, H., Goodfellow, I., Metaxas, D., Odena, A.: Self-attention generative adversarial networks. In: ICML (2019)
53. Zhang, L., Li, C., Simo-Serra, E., Ji, Y., Wong, T.T., Liu, C.: User-guided line art flat filling with split filling mechanism. In: IEEE/CVF Conference on Computer Vision and Pattern Recognition (CVPR) (2021)
54. Zhang, P., Zhang, B., Chen, D., Yuan, L., Wen, F.: Cross-domain correspondence learning for exemplar-based image translation. In: Proceedings of the IEEE/CVF Conference on Computer Vision and Pattern Recognition, pp. 5143–5153 (2020)
55. Zhang, Q., Wang, B., Wen, W., Li, H., Liu, J.: Line art correlation matching feature transfer network for automatic animation colorization. In: Proceedings of the IEEE/CVF Winter Conference on Applications of Computer Vision (WACV), pp. 3872–3881 (2021)
56. Zhang, R.Y., et al.: Real-time user-guided image colorization with learned deep priors. ACM Trans. Graph. **36**(4), 119 (2017)

57. Zhang, S., Yan, S., He, X.: LatentGNN: learning efficient non-local relations for visual recognition. In: International Conference on Machine Learning (ICML). Proceedings of Machine Learning Research, vol. 97, pp. 7374–7383. PMLR (2019)
58. Zhao, H., Jiang, L., Jia, J., Torr, P.H., Koltun, V.: Point transformer. In: Proceedings of the IEEE/CVF International Conference on Computer Vision, pp. 16259–16268 (2021)
59. Zhou, X., et al.: Cocosnet v2: full-resolution correspondence learning for image translation. In: Proceedings of the IEEE/CVF Conference on Computer Vision and Pattern Recognition (CVPR), pp. 11465–11475 (2021)
60. Zhu, P., Abdal, R., Qin, Y., Wonka, P.: Sean: image synthesis with semantic region-adaptive normalization. In: IEEE/CVF Conference on Computer Vision and Pattern Recognition (CVPR) (2020)

# Unsupervised Learning of Efficient Geometry-Aware Neural Articulated Representations

Atsuhiro Noguchi[1(✉)], Xiao Sun[2], Stephen Lin[2], and Tatsuya Harada[1,3]

[1] The University of Tokyo, Tokyo, Japan
noguchi@mi.t.u-tokyo.ac.jp
[2] Microsoft Research Asia, Beijing, China
[3] RIKEN, Tokyo, Japan

**Abstract.** We propose an unsupervised method for 3D geometry-aware representation learning of articulated objects, in which no image-pose pairs or foreground masks are used for training. Though photorealistic images of articulated objects can be rendered with explicit pose control through existing 3D neural representations, these methods require ground truth 3D pose and foreground masks for training, which are expensive to obtain. We obviate this need by learning the representations with GAN training. The generator is trained to produce realistic images of articulated objects from random poses and latent vectors by adversarial training. To avoid a high computational cost for GAN training, we propose an efficient neural representation for articulated objects based on tri-planes and then present a GAN-based framework for its unsupervised training. Experiments demonstrate the efficiency of our method and show that GAN-based training enables the learning of controllable 3D representations without paired supervision.

**Keywords:** Image synthesis · Articulated objects · Neural radiance fields · Unsupervised learning

## 1 Introduction

3D models that allow free control over the pose and appearance of articulated objects are essential in various applications, including computer games, media content creation, and augmented/virtual reality. In early work, articulated objects were typically represented by explicit models such as skinned meshes. More recently, the success of learned implicit representations such as neural radiance fields (NeRF) [40] for rendering static 3D scenes has led to extensions for modeling dynamic scenes and articulated objects. Much of this attention has focused on the photorealistic rendering of humans, from novel viewpoints and with controllable poses, by learning from images and videos.

---

**Supplementary Information** The online version contains supplementary material available at https://doi.org/10.1007/978-3-031-19790-1_36.

© The Author(s), under exclusive license to Springer Nature Switzerland AG 2022
S. Avidan et al. (Eds.): ECCV 2022, LNCS 13677, pp. 597–614, 2022.
https://doi.org/10.1007/978-3-031-19790-1_36

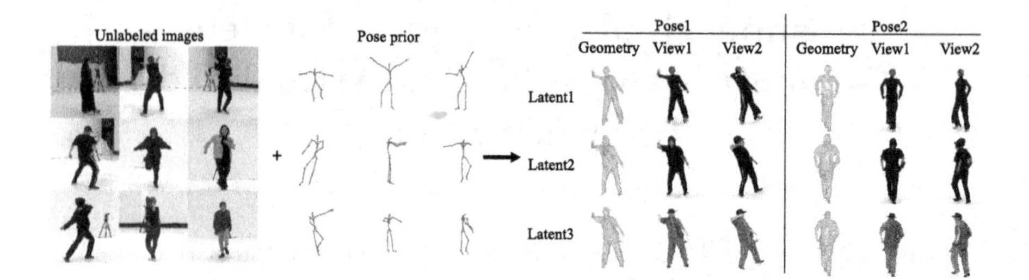


**Fig. 1.** Our ENARF-GAN is a geometry-aware, 3D-consistent image generation model that allows independent control of viewpoint, object pose, and appearance information. It is learned from unlabeled images and a prior distribution on object pose.

Existing methods for learning explicitly pose-controllable articulated representations, however, require much supervision, such as videos with 3D pose/mesh annotation and a mask for each frame. Preparing such data involves tremendous annotation costs; thus, reducing annotation is very important. In this paper, we propose a novel unsupervised learning framework for 3D pose-aware generative models of articulated objects, which are learned only from unlabeled images of objects sharing the same structure and a pose prior distribution of the objects.

We exploit recent advances in 3D-aware GAN [8,9,19,44,56] for unsupervised learning of the articulated representations. They learn 3D-aware image generation models from images without supervision, such as viewpoints or 3D shapes. The generator is based on NeRF [40] and is optimized with a GAN objective to generate realistic images from randomly sampled viewpoints and latent vectors from a prior distribution defined before training. As a result, the generator learns to generate 3D-consistent images without any supervision. We employ the idea for articulated objects by defining a pose prior distribution for the target object and optimizing the GAN objective on randomly generated images from random poses and latent variables. It becomes possible to learn a generative model with free control of poses. We demonstrate this approach by modeling the pose prior as a skeletal distribution [47,59], while noting that other models like meshes [52,53] may bring potential performance benefits.

However, the direct application of existing neural articulated representations to GANs is not computationally practical. While NeRF can produce high-quality images, its processing is expensive because it requires network inference for every point in space. Some methods [19,44] reduce computational cost by volume rendering at low resolution followed by 2D CNN based upsampling. Although this technique achieves high-resolution images with real-time inference speed, it is not geometry-aware (i.e., the surface mesh cannot be extracted). Recently, a method which we call Efficient NeRF [8] overcomes the problem. The method is based on an efficient tri-plane based neural representation and GAN training on it. Thanks to the computational efficiency, it can produce relatively high-resolution ($128 \times 128$) images with volumetric rendering. We extend the tri-plane representation to articulated objects for efficient GAN training. An overview of the method is visualized in Fig. 1. The contributions of this work are as follows:

- We propose a novel efficient neural representation for articulated objects based on an efficient tri-plane representation.
- We propose an efficient implementation of deformation fields using tri-planes for dynamic scene training, achieving 4 times faster rendering than NARF [47] with comparable or better performance.
- We propose a novel GAN framework to learn articulated representations without using any 3D pose or mask annotation for each image. The controllable 3D representation can be learned from real unlabeled images.

## 2 Related Work

**Articulated 3D Representations.** The traditional approach for modeling pose-controllable 3D representations of articulated objects is by skinned mesh [25,26,31], where each vertex of the mesh is deformed according to the skeletal pose. Several parametric skinned mesh models have been developed specifically for humans and animals [23,37,48,51,73]. For humans, the skinned multi-person linear model (SMPL) [37] is commonly used. However, these representations can only handle tight surfaces with no clothing and cannot handle non-rigid or topology-changing objects such as clothing or hair. Some work alleviates the problem by deforming the mesh surface or using a detailed 3D body scan [1,20]. Recently, implicit 3D shape representations have achieved state-of-the-art performance in pose-conditioned shape reconstruction. These methods learn neural occupancy/indicator functions [11,39] or signed distance functions [49] of articulated objects [6,10,13,61]. Photorealistic rendering of articulated objects, especially for humans, is also achieved with 3D implicit representations [2,35,47,52,59,65]. However, all these models require ground truth 3D pose and/or object shape for training. Very recently, methods have been proposed to reproduce the 3D shape and motion of objects from video data without using 3D shape and pose annotation [46,67,68]. However, they either do not allow free control of poses or are limited to optimizing for a single object. An SMPL mesh-based generative model [18] and image-to-image translation methods [7,14,36,57] can learn pose controllable image synthesis models for humans. However, the rendering process is completely in 2D and thus is not geometry-aware.

**Implicit 3D Representations.** Implicit 3D representations are memory efficient, continuous, and topology free. They have achieved the state-of-the art in learning 3D shape [11,39,49], static [4,40,58] and dynamic scenes [33,50,54], articulated objects [2,6,10,13,35,47,52,53,59,61,65], and image synthesis [9,56]. Although early works rely on ground truth 3D geometry for training [11,39,49], developments in differentiable rendering have enabled learning of networks from only photometric reconstruction losses [40,58,69]. In particular, neural radiance fields (NeRF) [40] applied volumetric rendering on implicit color and density fields, achieving photorealistic novel view synthesis of complex static scenes using multi-view posed images. Dynamic NeRF [33,50,54] extends NeRF to dynamic scenes, but these methods just reenact the motion in the scene and cannot repose objects based on their structure. Recently, articulated representations based on NeRF have been proposed [35,47,52,53,59,65,66]. These methods can render images conditioned on pose configurations. However, all of them require ground truth skeletal poses, SMPL meshes, or foreground masks for training, which makes them unsuitable for in-the-wild images.

Another NeRF improvement is the reduction of computational complexity: NeRF requires forward computation of MLPs to compute color and density for every point in 3D. Thus, the cost of rendering is very high. Fast NeRF algorithms [16,41,55,70] reduce the computational complexity of neural networks by creating caches or using explicit representations. However, these methods can only be trained on a single static scene. Very recently, a hybrid explicit and implicit representation was proposed [8]. In this representation, the feature field is constructed using a memory-efficient explicit representation called a tri-plane, and color and density are decoded using a lightweight MLP. This method can render images at low cost and is well suited for image generation models.

In this work, we propose an unsupervised learning framework for articulated objects. We extend tri-planes to articulated objects for efficient training.

**Generative 3D-Aware Image Synthesis.** Advances in generative adversarial networks (GANs) [17] have made it possible to generate high-resolution, photorealistic images [28–30]. In recent years, many 3D-aware image generation models have been proposed by combining GANs with 3D generators that use meshes [60], voxels [15,22,42,43,64,72], depth [45], or implicit representations [8,9,19,44,56]. These methods can learn 3D-aware generators without 3D supervision. Among these, image generation methods using implicit functions, thanks to their continuous and topology-free properties, have been successful in producing 3D-consistent and high-quality images. However, fully implicit models [9,56] are computationally expensive, making the training of GANs inefficient. Therefore, several innovations have been proposed to reduce the rendering cost of generators. Neural rendering-based methods [19,44] reduce computation by performing volumetric rendering at low resolution and upsampling the rendered feature images using a 2D CNN. Though this enables the generation of high-resolution images at a faster rate, 2D upsampling does not consider 3D consistency and cannot generate detailed 3D geometry. Very recently, a hybrid of explicit and implicit methods [8] has been developed for 3d geometry-aware image generation. Instead of using a coordinate-based implicit representation, this method uses tri-planes, which are explicit 3D feature representations, to reduce the number of forward computations of the network and to achieve volumetric rendering at high resolution.

The existing research is specific to scenes of static objects or objects that exist independently of each other, and do not allow free control of the skeletal pose of the generated object. Therefore, we propose a novel GAN framework for articulated objects.

# 3 Method

Recent advances in implicit neural rendering [47,59] have made it possible to generate 3D-aware pose-controllable images of articulated objects from images with accurate 3D pose and foreground mask annotations. However, training such models from only in-the-wild images remains challenging since accurate 3D pose annotations are generally difficult to obtain for them. In the following, we first briefly review Neural Articulated Radiance Field (NARF) [47], then propose an adversarial-based framework, named ENARF-GAN, to efficiently train the NARF model without any paired image-pose and foreground mask annotations.

## 3.1 Neural Articulated Radiance Field Revisited

NARF is an implicit 3D representation for articulated objects. It takes a kinematic 3D pose configuration of an articulated object $o = \{l_k, R_k, \mathbf{t}_k\}_{k=1:K}$ as input and predicts the color and the density of any 3D location $\mathbf{x}$, where $l_k$ is the length of the $k^{\text{th}}$ part, and $R_k$ and $\mathbf{t}_k$ are its rotation and translation matrices, respectively. Given the pose configuration $o$, NARF first transforms a global 3D position $\mathbf{x}$ into several local coordinate systems defined by the rigid parts of the articulated object. Specifically, the transformed local location $\mathbf{x}^l_k$ for the $k^{\text{th}}$ part is computed as $\mathbf{x}^l_k = (R^k)^{-1}(\mathbf{x} - \mathbf{t}^k)$ for $k \in \{1, ..., K\}$.

NARF first trains an extra lightweight selector $S$ in the local space to decide which part a global 3D location $\mathbf{x}$ belongs to. Specifically, it outputs the probability $p^k$ of $\mathbf{x}$ belonging to the $k^{th}$ part. Then NARF computes color $c$ and density $\sigma$ at the location $\mathbf{x}$ from a concatenation of local locations masked by the corresponding part probability $p^k$.

$$c, \sigma = G(\text{Cat}(\{\gamma(\mathbf{x}^l_k) * p^k\}_{k=1:K})), \tag{1}$$

where $\gamma$ is a positional encoding [40], Cat is the concatenation operation, and $G$ is an MLP network. The RGB color $\mathbf{C}$ and foreground mask value $\mathbf{M}$ for each pixel are generated by volumetric rendering [40]. The network is trained with a reconstruction loss between the generated and ground truth color $\hat{\mathbf{C}}$ and mask $\hat{\mathbf{M}}$,

$$\mathcal{L}_{\text{supervised}} = \sum_{r \in \mathcal{R}} \left( ||\mathbf{C} - \hat{\mathbf{C}}||_2^2 + ||\mathbf{M} - \hat{\mathbf{M}}||_2^2 \right), \tag{2}$$

where $\mathcal{R}$ is the set of rays in each batch. Please refer to the original NARF paper [47] for more details.

## 3.2 Unsupervised Learning by Adversarial Training

In this work, we propose a method for efficient and unsupervised training of the NARF model from unposed image collections. Without loss of generality, we consider humans as the articulated objects here. To this end, we first define a human pose distribution $\mathcal{O}$. For one training iteration, our NARF based generator $G$ takes a latent vector $\mathbf{z}$ and a sampled human pose instance $o$ from

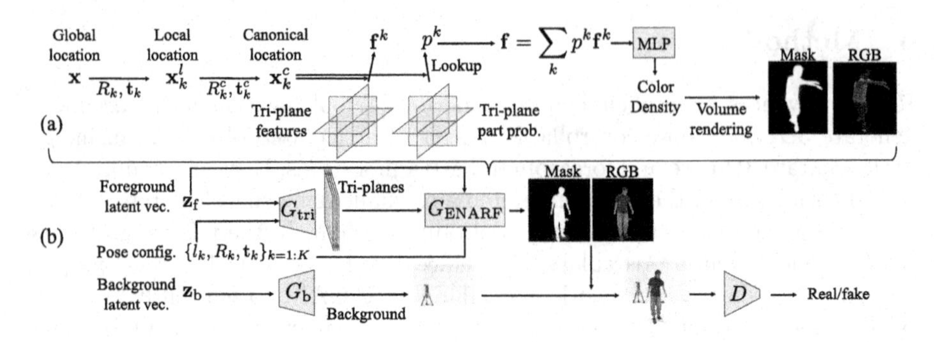


**Fig. 2.** Overview of (a) Efficient NARF (ENARF) and (b) GAN training.

$\mathcal{O}$ as input and predicts a synthesized image $\mathbf{C}$. Following standard adversarial training of GANs, a discriminator $\mathcal{D}$ is used to distinguish the synthesized image $\mathbf{C}$ from real ones $\tilde{\mathbf{C}}$. Formally, the training objectives of the generator $\mathcal{L}_{\text{adv}}^{G}$ and discriminator $\mathcal{L}_{\text{adv}}^{D}$ are defined as follows,

$$\mathcal{L}_{\text{adv}}^{G} = -\mathbb{E}\left[\log(D(G(\mathbf{z}, o)))\right], \mathcal{L}_{\text{adv}}^{D} = -\mathbb{E}\left[\log(D(\tilde{\mathbf{C}})) + \log(1 - D(G(\mathbf{z}, o)))\right]. \tag{3}$$

An overview of this method is illustrated in Fig. 2(b).

However, this training would be computationally expensive. The rendering cost of NARF is heavy because computation is performed for many 3D locations in the viewed space. Even though in supervised training, the time and memory cost of computing the reconstruction loss could be reduced by evaluating it over just a small proportion of the pixels [47], the adversarial loss in Eq. 3 requires the generation of the full image for evaluation. As a result, the amount of computation becomes impractical.

In the following, we propose a series of changes in feature computation and the selector to address this issue. Note that these changes not only enable the GAN training but also greatly improve the efficiency of the original NARF.

### 3.3 Efficiency Improvements on NARF

Recently, Chan et al. [8] proposed a hybrid explicit-implicit 3D-aware network that uses a memory-efficient tri-plane representation to explicitly store features on axis-aligned planes. With this representation, the efficiency of feature extraction for a 3D location is greatly improved. Instead of forwarding all the sampled 3D points through the network, the intermediate features of arbitrary 3D points can be obtained via simple lookups on the tri-planes. The tri-plane representation can be more efficiently generated with Convolutional Neural Networks (CNNs) instead of MLPs. The intermediate features of 3D points are then transformed into the color and density using a lightweight decoder MLP. The decoder significantly increases the non-linearity between features at different positions, thus greatly enhancing the expressiveness of the model.

Here, we adapt the tri-plane representation to NARF for more efficient training. Similar to [8], we first divide the original NARF network $G$ into an intermediate feature generator (the first linear layer of $G$) $W$ and a decoder network $G_{\text{dec}}$. Then, Eq. 1 is rewritten as follows.

$$c, \sigma = G_{\text{dec}}(\mathbf{f}), \text{ where } \mathbf{f} = W(\text{Cat}(\{\gamma(\mathbf{x}_k^l) * p^k | k \in \{1, ..., K\}\})), \quad (4)$$

where $\mathbf{f}$ is an intermediate feature vector of input 3D location $\mathbf{x}$. However, re-implementing the feature generator $W$ to produce the tri-plane representation is not straightforward because its input, a weighted concatenation of $\mathbf{x}_k^l$, does not form a valid location in a specific 3D space. We make two important changes to address this issue. First, we decompose $W$ into $K$ sub-matrices $\{W_k\}$, one for each part, where each one takes the corresponding local position $\mathbf{x}_k^l$ as input and outputs an intermediate feature for the $k^{\text{th}}$ part. Then, the intermediate feature in Eq. 4 can be equivalently rewritten as follows.

$$\mathbf{f} = \sum_{k=1}^{K} p^k * \mathbf{f}^k, \text{ where } \mathbf{f}^k = W_k(\gamma(\mathbf{x}_k^l)), \quad (5)$$

where $\mathbf{f}^k$ is a feature generated in the local coordinate system of the $k^{\text{th}}$ part. Now, $W_k(\gamma(\mathbf{x}_k^l))$ can be directly re-implemented by tri-planes $F$. However, the computational complexity of this implementation is still proportional to $K$. In order to train a single tri-plane for all parts, the second change is to further transform the local coordinates $\mathbf{x}_k^l$ into a canonical space defined by a canonical pose $o^c$, similar to Animatable NeRF [52].

$$\mathbf{x}_k^c = R_k^c \mathbf{x}_k^l + \mathbf{t}_k^c = R_k^c (R^k)^{-1}(\mathbf{x} - \mathbf{t}^k) + \mathbf{t}_k^c, \quad (6)$$

where $R_k^c$ and $\mathbf{t}_k^c$ are the rotation and translation matrices of the canonical pose. Intuitively, $\mathbf{x}_k^c$ is the corresponding point location of $\mathbf{x}$ transformed into the canonical space when $\mathbf{x}$ is considered to belong to the $k^{\text{th}}$ part. Finally, the tri-plane feature $F$ is learned in the canonical space. The feature extraction for location $\mathbf{x}$ is achieved by retrieving the 32-dimensional feature vector $\mathbf{f}^k$ on the tri-plane $F$ at $\mathbf{x}_k^c$ for all parts, then taking a weighted sum of those features as in Eq. 5,

$$\mathbf{f} = \sum_{k=1}^{K} p^k * \mathbf{f}^k, \text{ where } \mathbf{f}^k = \sum_{ij \in \{xy, yz, xz\}} F_{ij}(\mathbf{x}_k^c). \quad (7)$$

$F_{**}(\mathbf{a})$ is a retrieved feature vector from each axis-aligned plane at location $\mathbf{a}$.

We estimate the RGB color $c$ and density $\sigma$ from $\mathbf{f}$ using a lightweight decoder network $G_{\text{dec}}$ consisting of three FC layers with a hidden dimension of 64 and output dimension of 4. We apply volume rendering [40] on the color and density to output an RGB image $\mathbf{C}$ and a foreground mask $\mathbf{M}$.

Although we efficiently parameterize the intermediate features, the probability $p^k$ needs to be computed for every 3D point and every part. In the original NARF, though lightweight MLPs are used to estimate the probabilities, they are still computationally infeasible.

Therefore, we propose an efficient selector network using tri-planes. Since $p^k$ is used to mask out features of irrelevant parts, this probability can be a rough approximation of the shape of each part. Thus the tri-plane representation is expressive enough to model the probability. We use $K$ separate 1-channel tri-planes to represent $P^k$, the part probability projected to each axis-aligned plane. We retrieve the three probability values $(p^k_{xy}, p^k_{xz}, p^k_{yz}) = (P^k_{xy}(\mathbf{x}^c_k), P^k_{xz}(\mathbf{x}^c_k), P^k_{yz}(\mathbf{x}^c_k))$ of the $k^{\text{th}}$ part by querying the 3D location in the canonical space $\mathbf{x}^c_k$. The probability $p^k$ that $\mathbf{x}^c_k$ belongs to the $k^{\text{th}}$ part is approximated as $p^k = p^k_{xy}p^k_{xz}p^k_{yz}$.

In this way, the features $F$ and part probabilities $P$ are modeled efficiently with a single tri-plane representation. The tri-plane is represented by a $(32 + K) \times 3$ channel image. The first 96 channels represent the tri-plane features in the canonical space. The remaining $3K$ channels represent the tri-plane probability maps for each of the $K$ parts. We call this approach Efficient NARF, or ENARF.

### 3.4 GAN

To condition the generator on latent vectors, we utilize a StyleGAN2 [30] based generator to produce tri-plane features. We condition each layer of the proposed ENARF by the latent vector with a modulated convolution [30]. Since the proposed tri-plane based generator can only represent the foreground object, we use an additional StyleGAN2 based generator for the background.

We randomly sample latent vectors for our tri-plane generator and background generator: $\mathbf{z} = (\mathbf{z}_{\text{tri}}, \mathbf{z}_{\text{ENARF}}, \mathbf{z}_b) \sim \mathcal{N}(0, I)$, where $\mathbf{z}_{\text{tri}}$, $\mathbf{z}_{\text{ENARF}}$, and $\mathbf{z}_b$ are latent vectors for the tri-plane generator, ENARF, and background generator, respectively. The tri-plane generator $G_{\text{tri}}$ generates tri-plane feature $F$ and part probability $P$ from randomly sampled $\mathbf{z}_{\text{tri}}$ and bone length $\{l_k\}_{k=1:K}$. $G_{\text{tri}}$ takes $l_k$ as inputs to account for the diversity of bone lengths.

$$F, P = G_{\text{tri}}(\mathbf{z}_{\text{tri}}, \{l_k\}_{k=1:K}) \tag{8}$$

The ENARF based foreground generator $G_{\text{ENARF}}$ generates the foreground RGB image $\mathbf{C}_f$ and mask $\mathbf{M}_f$ from the generated tri-planes, and the background generator $G_{\text{b}}$ generates background RGB image $\mathbf{C}_b$.

$$\mathbf{C}_f, \mathbf{M}_f = G_{\text{ENARF}}(\mathbf{z}_{\text{ENARF}}, \{l_k, R_k, \mathbf{t}_k\}_{k=1:K}), \mathbf{C}_b = G_{\text{b}}(\mathbf{z}_b) \tag{9}$$

The final output RGB image is $\mathbf{C} = \mathbf{C}_f + \mathbf{C}_b * (1 - \mathbf{M}_f)$, which is a composite of $\mathbf{C}_f$ and $\mathbf{C}_b$. To handle the diversity of bone lengths, we replace Eq. 6 with one normalized by the length of the bone: $\mathbf{x}^c_k = \frac{l^c_k}{l_k} R^c_k \mathbf{x}_k + \mathbf{t}^c_k$, where $l^c_k$ is the bone length of the $k^{\text{th}}$ part in the canonical space.

We optimize these generator networks with GAN training. We use a bone loss in addition to an adversarial loss on images, R1 regularization on the discriminator [38], and L2 regularization on the tri-planes. The bone loss ensures that an object is generated in the foreground. Based on the input pose of the object, a skeletal image $B$ is created, where pixels with skeletons are 1 and others are 0, and the generated mask $M$ at pixels with skeletons is made close to 1: $\mathcal{L}_{\text{bone}} = \frac{\sum_{r \in \mathcal{R}} (1-M)^2 B}{\sum_{r \in \mathcal{R}} B}$. Additional details are provided in the supplement. The final loss is the linear combination of these losses.

### 3.5 Dynamic Scene Overfitting

Since ENARF is an improved version of NARF, we can directly use it for single dynamic scene overfitting. For training, we use the ground truth 3D pose and foreground mask of each frame and optimize the reconstruction loss in Eq. 2.

If the object shape is strictly determined by the poses that comprise the kinematic motion, we can use the same tri-plane features for the entire sequence and directly optimize them. However, real-world objects have time or pose-dependent non-rigid deformation such as clothing and facial expression change in a single sequence. Therefore, the tri-plane features should change depending on time and pose. We use a technique based on deformation fields [50,54] proposed in time-dependent NeRF, also known as Dynamic NeRF. Deformation field based methods learn a mapping network from observation space to canonical space and learn the NeRF in the canonical frame. Since learning the deformation field with an MLP is expensive, we also approximate it with tri-planes. We approximate the deformation in 3D space by independent 2D deformations in each tri-plane. First, a StyleGAN2 [30] generator takes positionally encoded time $t$ and a rotation matrix of each part $R_k$ and generates 6-channel images representing the relative 2D deformation from the canonical space of each tri-plane feature. We deform the tri-plane feature based on the generated deformation. Please refer to the supplement for more details. We use constant tri-plane probabilities $P$ for all frames since the object shares the same coarse part shape throughout the entire sequence. The remaining networks are the same. We refer to this method as D-ENARF.

## 4 Experiments

Our experimental results are presented in two parts. First, in Sect. 4.1, we compare the proposed Efficient NARF (ENARF) with the state-of-the-art methods [47,52] in terms of both efficiency and effectiveness, and we conduct ablation studies on the deformation modeling and the design choices for the selector. Second, in Sect. 4.2, we present our results of using adversarial training on ENARF, namely, ENARF-GAN, and compare it with baselines. Then, we discuss the effectiveness of the pose prior and the generalization ability of ENARF-GAN.

### 4.1 Training on a Dynamic Scene

Following the training setting in Animatable NeRF [52], we train our ENARF model on synchronized multi-view videos of a single moving articulated object. The ZJU mocap dataset [53] consisting of three subjects (313, 315, 386) is used for training. We use the same pre-processed data provided by the official implementation of Animatable NeRF. All images are resized to 512 × 512. We use 4 views and the first 80% of the frames for training, and the remaining views or frames for testing. In this setting, the ground truth camera and articulated object poses, as well as the ground truth foreground mask, are given for each frame. More implementation details can be found in the supplement.

**Table 1.** Quantitative comparison on dynamic scenes.

| | Cost | | | Novel view | | | Novel pose | | |
|---|---|---|---|---|---|---|---|---|---|
| | #Memory | #FLOPS | Time(s) | PSNR↑ | SSIM↑ | LPIPS↓ | PSNR↑ | SSIM↑ | LPIPS↓ |
| Animatable NeRF [52] | - | - | **0.42** | 28.28 | 0.9484 | 0.05818 | 29.09 | 0.9507 | 0.05706 |
| NARF [47]. | 283.9 GB | 15.9T | 2.17 | 30.62 | 0.9625 | 0.05228 | 29.51 | **0.959** | 0.05208 |
| ENARF | 27.0 GB | 71.7G | 0.47 | 31.94 | 0.9655 | 0.04792 | 29.66 | 0.953 | 0.05702 |
| D-ENARF | 27.6 GB | 354G | 0.49 | **32.93** | **0.9713** | **0.03718** | **30.06** | 0.9396 | **0.05205** |
| ENARF w/o selector | **23.0 GB** | **70.2G** | 0.43 | 29.16 | 0.9493 | 0.07234 | 27.9 | 0.9377 | 0.08316 |
| ENARF w/MLP selector | 83.2 GB | 337G | 1.13 | 32.27 | 0.9684 | 0.04633 | 29.74 | 0.9573 | 0.05228 |

First, we compare our method with the state-of-the-art supervised methods NARF [47] and Animatable NeRF [52]. Our comparison with Neural Actor [35] is provided in the supplement. Note that our method and NARF [47] take ground truth kinematic pose parameters (joint angles and bone lengths) as inputs, while Animatable NeRF needs the ground truth SMPL mesh parameters. In addition, Animatable NeRF requires additional training on novel poses to render novel-pose images, which is not necessary for our model.

Table 1 shows the quantitative results. To compare the efficiency between models, we examine the GPU memory, FLOPS, and the running time used to render an entire image of resolution 512 × 512 on a single A100 GPU as evaluation metrics. To compare the quality of synthesized images under novel view and novel pose settings, PSNR, SSIM [63], and LPIPS [71] are used as evaluation metrics. Table 1 shows that the proposed Efficient NARF achieves competitive or even better performance compared to existing methods with far fewer FLOPS and 4.6 times the speed of the original NARF. Although the runtime of ENARF is a bit slower than Animatable NeRF (0.05 s), its performance is superior under both novel view and novel pose settings. In addition, it does not need extra training on novel poses. Our dynamic model D-ENARF further improves the performance of ENARF with little increased overhead in inference time, and outperforms the state-of-the-arts Animatable NeRF and NARF by a large margin.

Qualitative results for novel view and pose synthesis are shown in Fig. 3. ENARF produces much sharper images than NARF due to the more efficient explicit-implicit tri-plane representation. D-ENARF, which utilizes deformation fields, further improves the rendering quality. In summary, the proposed D-ENARF method achieves better performance in both image quality and computational efficiency.

**Ablation Study.** To evaluate the effectiveness of the tri-plane based selector, we compare our method against models using an MLP selector or without a selector. A quantitative comparison is provided in Table 1, and a qualitative comparison is provided in the supplement. Although an MLP based selector improves the metrics a bit, it results in a significant increase in testing time. In contrast, the model without a selector is unable to learn clean/sharp part shapes and textures, because a 3D location will inappropriately be affected by all parts without a selector. These results indicate that our tri-plane based selector is efficient and effective.

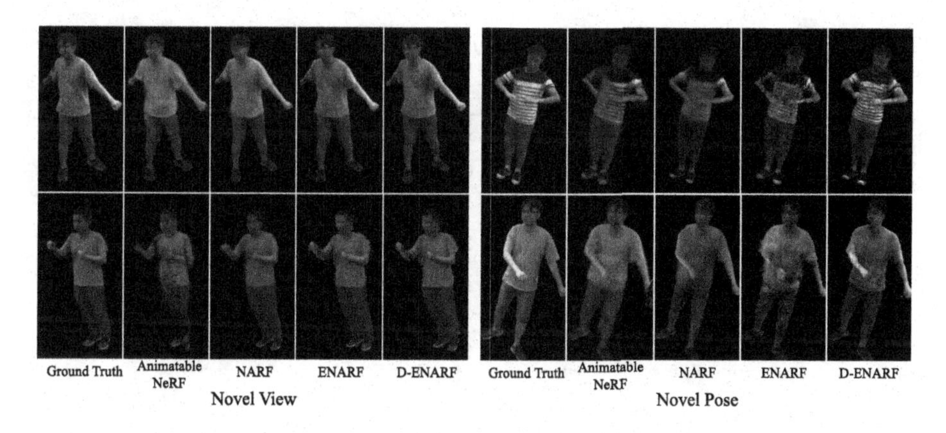


**Fig. 3.** Qualitative comparison of novel view and pose synthesis.

### 4.2 Unsupervised Learning with GAN

In this section, we train the proposed efficient NARF using GAN objectives without any image-pose pairs or mask annotations.

**Comparison with Baselines.** Since this is the first work to learn an articulated representation without image-pose pairs or mesh shape priors, no exact competitor exists. We thus compare our method against two baselines. The first is a *supervised* baseline called ENARF-VAE, inspired by the original NARF [47]. Here, a ResNet50 [21] based encoder estimates the latent vectors $\mathbf{z}$ from images, and the efficient NARF based decoder decodes the original images from estimated latent vectors $\mathbf{z}$ and ground truth pose configurations $o$. These networks are trained with the reconstruction loss defined in Eq. 2 and the KL divergence loss on the latent vector $\mathbf{z}$. Following [47], ENARF-VAE is trained with images with a black background. The second model is called StyleNARF, which is a combination of the original NARF and the state-of-the-art high-resolution 3D-aware image generation model called StyleNeRF [19]. To reduce the computational cost, the original NARF first generates low-resolution features using volumetric rendering. Subsequently, a 2D CNN-based network upsamples them into final images. Additional details are provided in the supplement. Please note that ENARF-VAE is a *supervised* method and cannot handle the background, and StyleNARF loses 3D consistency and thus cannot generate high-resolution geometry.

We use the SURREAL dataset [62] for comparison. It is a synthetic human image dataset with a resolution of $128 \times 128$. Dataset details are given in the supplement. For the pose prior, we use the ground truth pose distribution of the training dataset, where we randomly sample poses from the entire dataset. Please note that we do not use image-pose pairs for these unsupervised methods.

**Table 2.** Quantitative comparison on generative models. * indicates that the methods are modified from the cited papers.

| | FID↓ | FG-FID↓ | Depth↓ | PCKh@0.5 ↑ |
|---|---|---|---|---|
| ENARF-VAE [47]*. | – | 63.0 | **3.2** | **0.984** |
| StyleNARF [19]*. | **20.8** | – | 16.5 | 0.924 |
| ENARF-GAN | 22.6 | **21.3** | 8.8 | 0.947 |
| ENARF-GAN CMU pri. | 24.2 | 25.6 | 12.8 | 0.915 |
| ENARF-GAN rand. pri. | 25.9 | 37.0 | 13.8 | 0.887 |
| w/ trunc. $\psi = 0.4$ | 26.5 | 24.1 | 8.0 | 0.884 |

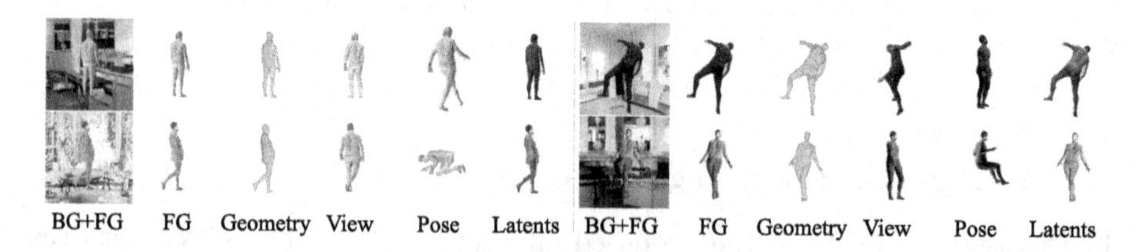


**Fig. 4.** Learned geometry and disentangled representations on the SURREAL dataset by ENARF-GAN. For each of the generated results, the leftmost three columns show the generated images with background, foreground, and corresponding geometry. The rightmost three images show the results of changing only the viewpoint, object pose, and latent variables for the same results, respectively.

Quantitative results are shown in Table 2. We measure image quality with the Fréchet Inception Distance (FID) [24]. To better evaluate the quality of foreground, we use an extra metric called FG-FID that replaces the background with black color, using the generated or ground truth mask. We measure depth plausibility by comparing the real and generated depth map. Although there is no ground truth depth for the generated images, the depth generated from a pose would have a similar depth to the real depth that arises from the same pose. We compare the L2 norm between the inverse depth generated from poses sampled from the dataset and the real inverse depth of them. Finally, we measure the correspondence between pose and appearance following the contemporary work named GNARF [5]. We apply an off-the-shelf 2D human keypoint estimator [12] to both generated and real images with the same poses and compute the Percentage of Correct Keypoints (PCK) between them, which is commonly used for evaluating 2D pose estimators. We report the averaged PCKh@0.5 metric [3] for all keypoints. Details are provided in the supplement. Qualitative results are shown in Fig. 5. Not surprisingly, ENARF-VAE produces the most plausible depth/geometry and learns the most accurate pose conditioning among the three since it uses image-pose pairs for supervised training. However, compared to styleNARF, its FID is worse and the images lack photorealism. styleNARF achieves the best FID among the three, thanks to the effective CNN renderer. However, it cannot explicitly render the foreground only or generate accurate geometry of the generated images. In contrast, our method performs volumetric rendering at the output resolution, and the generated geometry perfectly matches the generated foreground image (Fig. 5).

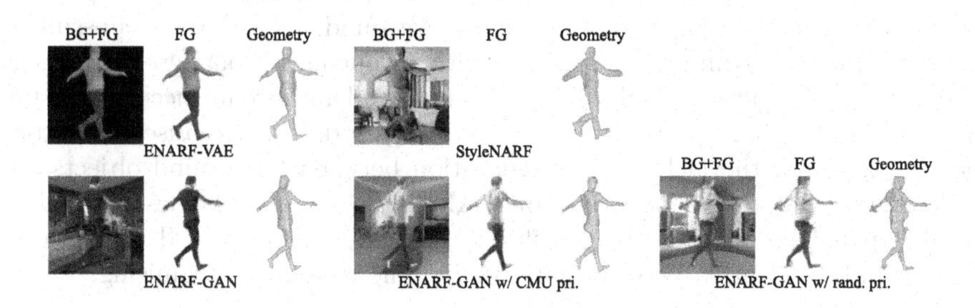


**Fig. 5.** Qualitative comparison on generative models.

**Using Different Pose Distribution.** Obtaining a ground truth pose distribution of the training images is not feasible for in-the-wild images or new categories. Thus, we train our model with a pose distribution different from the training images. Here, we consider two pose prior distributions. The first uses poses from CMU Panoptic [27] as a prior, which we call the CMU prior. During training, we randomly sample poses from the entire dataset. In addition, to show that our method works without collecting actual human motion capture data, we also create a much simpler pose prior. We fit a multi-variate Gaussian distribution on each joint angle of the CMU Panoptic dataset, and we use randomly sampled poses from the distribution for training. Each Gaussian distribution only defines the rotation angle of each part, which can be easily constructed for novel objects. We call this the random prior.

Quantitative and qualitative results are shown in Table 2 and Fig. 5. We can confirm that even when using the CMU prior, our model learns pose-controllable 3D representations with just a slight sacrifice in image quality. When using the random prior, the plausibility of the generated images and the quality of the generated geometry are worse. This may be because the distribution of the random prior is so far from the distribution of poses in the dataset that the learned space of latent vectors too often falls outside the distribution of the actual data. Therefore, we used the truncation trick [29] to restrict the diversity of the latent space, and the results are shown in the bottom row of Table 2. By using the truncation trick, even with a simple prior, we can eliminate latent variables outside the distribution and improve the quality of the generated images and geometry. Further experimental results on truncation are given in the supplement.

**Additional Results on Real Images.** To show the generalization ability of the proposed framework, we train our model on two real image datasets, namely AIST++ [32] and MSCOCO [34]. AIST++ is a dataset of dancing persons with relatively simple backgrounds. We use the ground truth pose distribution for training. MSCOCO is a large scale in-the-wild image dataset. We choose images capturing roughly the whole human body and crop them around the persons. Since 3D pose annotations are not available for MSCOCO, we use poses in CMU Panoptic as the pose prior. Note that we do not use any image-pose or mask supervisions for training. Qualitative results are shown in Fig. 6. Experimental results with AIST++, which has a simple background, show that it is possible to generate detailed geometry and images with independent control of viewpoint, pose, and appearance. For MSCOCO, two successful and two unsuccessful results are shown in Fig. 6. MSCOCO is a very challenging dataset because of the complex background, the lack of clear separation between foreground objects and background, and the many occlusions. Although our model does not always produce plausible results, it is possible to generate geometry and control each element independently. As an initial attempt, the results are promising.

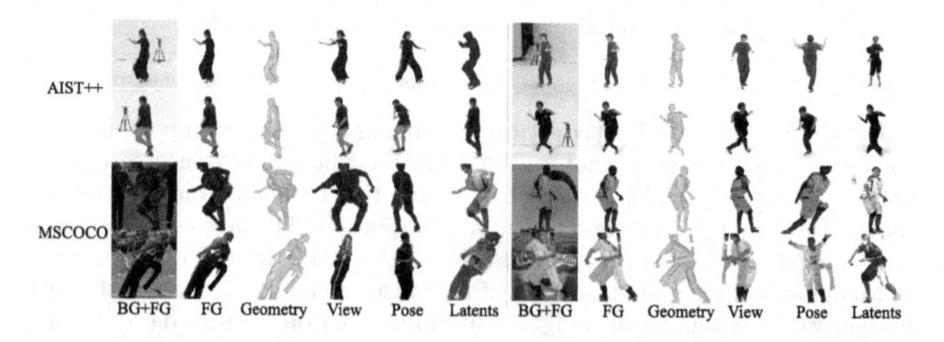


**Fig. 6.** Qualitative results on AIST++ and MSCOCO.

## 5 Conclusion

In this work, we propose a novel unsupervised learning framework for 3D geometry-aware articulated representations. We showed that our framework is able to learn representations with controllable viewpoint and pose. We first propose a computationally efficient neural 3D representation for articulated objects by adapting the tri-plane representation to NARF, then show it can be trained with GAN objectives without using ground truth image-pose pairs or mask supervision. However, the resolution and the quality of the generated images are still limited compared to recent NeRF-based GAN methods; meanwhile, we assume that a prior distribution of the object's pose is available, which may not be easily obtained for other object categories. Future work includes incorporating the neural rendering techniques proposed in 3D-aware GANs to generate photorealistic high-quality images while preserving the 3D consistency and estimating the pose prior distribution directly from the training data.

**Acknowledgements.** This work was supported by D-CORE Grant from Microsoft Research Asia and partially supported by JST AIP Acceleration Research JPMJCR20U3, Moonshot R&D Grant Number JPMJPS2011, CREST Grant Number JPMJCR2015, JSPS KAKENHI Grant Number JP19H01115, and JP20H05556 and Basic Research Grant (Super AI) of Institute for AI and Beyond of the University of Tokyo. We would like to thank Haruo Fujiwara, Lin Gu, Yuki Kawana, and the authors of [5] for helpful discussions.

## References

1. Alldieck, T., Magnor, M., Xu, W., Theobalt, C., Pons-Moll, G.: Detailed human avatars from monocular video. In: 2018 International Conference on 3D Vision (3DV) (2018)
2. Alldieck, T., Xu, H., Sminchisescu, C.: imghum: implicit generative models of 3D human shape and articulated pose. In: ICCV (2021)
3. Andriluka, M., Pishchulin, L., Gehler, P., Schiele, B.: 2D human pose estimation: new benchmark and state of the art analysis. In: CVPR (2014)
4. Barron, J.T., Mildenhall, B., Tancik, M., Hedman, P., Martin-Brualla, R., Srinivasan, P.P.: Mip-NeRF: a multiscale representation for anti-aliasing neural radiance fields. In: ICCV (2021)
5. Bergman, A.W., Kellnhofer, P., Wang, Y., Chan, E.R., Lindell, D.B., Wetzstein, G.: Generative neural articulated radiance fields. arXiv preprint arXiv:2206.14314 (2022)
6. Bozic, A., Palafox, P., Zollhofer, M., Thies, J., Dai, A., Nießner, M.: Neural deformation graphs for globally-consistent non-rigid reconstruction. In: CVPR (2021)
7. Chan, C., Ginosar, S., Zhou, T., Efros, A.A.: Everybody dance now. In: ICCV (2019)
8. Chan, E.R., et al.: Efficient geometry-aware 3D generative adversarial networks. In: CVPR (2022)
9. Chan, E.R., Monteiro, M., Kellnhofer, P., Wu, J., Wetzstein, G.: pi-GAN: periodic implicit generative adversarial networks for 3D-aware image synthesis. In: CVPR (2021)

10. Chen, X., Zheng, Y., Black, M.J., Hilliges, O., Geiger, A.: SNARF: differentiable forward skinning for animating non-rigid neural implicit shapes. In: ICCV (2021)
11. Chen, Z., Zhang, H.: Learning implicit fields for generative shape modeling. In: CVPR (2019)
12. Contributors, M.: Openmmlab pose estimation toolbox and benchmark (2020). https://github.com/open-mmlab/mmpose
13. Deng, B., et al.: NASA neural articulated shape approximation. In: ECCV (2020)
14. Esser, P., Sutter, E., Ommer, B.: A variational u-net for conditional appearance and shape generation. In: CVPR (2018)
15. Gadelha, M., Maji, S., Wang, R.: 3D shape induction from 2D views of multiple objects. In: 2017 International Conference on 3D Vision (3DV) (2017)
16. Garbin, S.J., Kowalski, M., Johnson, M., Shotton, J., Valentin, J.: Fastnerf: high-fidelity neural rendering at 200fps. In: ICCV (2021)
17. Goodfellow, I., et al.: Generative adversarial nets. In: NeurIPS (2014)
18. Grigorev, A., et al.: Stylepeople: a generative model of fullbody human avatars. In: CVPR (2021)
19. Gu, J., Liu, L., Wang, P., Theobalt, C.: StyleNeRF: a style-based 3D aware generator for high-resolution image synthesis. In: ICLR (2022)
20. Habermann, M., Liu, L., Xu, W., Zollhoefer, M., Pons-Moll, G., Theobalt, C.: Real-time deep dynamic characters. ACM Trans. Graph. (TOG) **40**(4), 1–16 (2021)
21. He, K., Zhang, X., Ren, S., Sun, J.: Deep residual learning for image recognition. In: CVPR (2016)
22. Henzler, P., Mitra, N.J., Ritschel, T.: Escaping Plato's cave: 3D shape from adversarial rendering. In: ICCV (2019)
23. Hesse, N., Pujades, S., Black, M.J., Arens, M., Hofmann, U.G., Schroeder, A.S.: Learning and tracking the 3D body shape of freely moving infants from RGB-D sequences. IEEE Trans. Pattern Anal. Mach. Intell. **42**(10), 2540–2551 (2019)
24. Heusel, M., Ramsauer, H., Unterthiner, T., Nessler, B., Hochreiter, S.: Gans trained by a two time-scale update rule converge to a local nash equilibrium. In: NeurIPS (2017)
25. Jacobson, A., Deng, Z., Kavan, L., Lewis, J.P.: Skinning: real-time shape deformation (full text not available). In: ACM SIGGRAPH 2014 Courses (2014)
26. James, D.L., Twigg, C.D.: Skinning mesh animations. ACM Trans. Graph. (TOG) **24**(3), 399–407 (2005)
27. Joo, H., et al.: Panoptic studio: a massively multiview system for social motion capture. In: ICCV (2015)
28. Karras, T., et al.: Alias-free generative adversarial networks. In: NeurIPS (2021)
29. Karras, T., Laine, S., Aila, T.: A style-based generator architecture for generative adversarial networks. In: CVPR (2019)
30. Karras, T., Laine, S., Aittala, M., Hellsten, J., Lehtinen, J., Aila, T.: Analyzing and improving the image quality of stylegan. In: CVPR (2020)
31. Lewis, J.P., Cordner, M., Fong, N.: Pose space deformation: a unified approach to shape interpolation and skeleton-driven deformation. In: Proceedings of the 27th Annual Conference on Computer Graphics and Interactive Techniques (2000)
32. Li, R., Yang, S., Ross, D.A., Kanazawa, A.: Ai choreographer: music conditioned 3D dance generation with aist++. In: ICCV (2021)
33. Li, Z., Niklaus, S., Snavely, N., Wang, O.: Neural scene flow fields for space-time view synthesis of dynamic scenes. In: CVPR (2021)
34. Lin, T.Y., et al.: Microsoft coco: common objects in context. In: ECCV (2014)

35. Liu, L., Habermann, M., Rudnev, V., Sarkar, K., Gu, J., Theobalt, C.: Neural actor: neural free-view synthesis of human actors with pose control. In: ACM SIGGRAPH Asia (2021)
36. Liu, L., et al.: Neural human video rendering by learning dynamic textures and rendering-to-video translation. IEEE Trans. Visual. Comput. Graph. (2020)
37. Loper, M., Mahmood, N., Romero, J., Pons-Moll, G., Black, M.J.: SMPL: a skinned multi-person linear model. ACM Trans. Graph. (TOG) **34**(6), 1–16 (2015)
38. Mescheder, L., Geiger, A., Nowozin, S.: Which training methods for gans do actually converge? In: ICML (2018)
39. Mescheder, L., Oechsle, M., Niemeyer, M., Nowozin, S., Geiger, A.: Occupancy networks: learning 3D reconstruction in function space. In: CVPR (2019)
40. Mildenhall, B., Srinivasan, P.P., Tancik, M., Barron, J.T., Ramamoorthi, R., Ng, R.: NeRF: representing scenes as neural radiance fields for view synthesis. In: ECCV (2020)
41. Müller, T., Evans, A., Schied, C., Keller, A.: Instant neural graphics primitives with a multiresolution hash encoding. ACM Trans. Graph. **41**(4), 102:1-102:15 (2022)
42. Nguyen-Phuoc, T., Li, C., Theis, L., Richardt, C., Yang, Y.L.: Hologan: unsupervised learning of 3D representations from natural images. In: ICCV (2019)
43. Nguyen-Phuoc, T.H., Richardt, C., Mai, L., Yang, Y., Mitra, N.: Blockgan: learning 3D object-aware scene representations from unlabelled images. In: NeurIPS (2020)
44. Niemeyer, M., Geiger, A.: Giraffe: representing scenes as compositional generative neural feature fields. In: CVPR (2021)
45. Noguchi, A., Harada, T.: RGBD-GAN: unsupervised 3D representation learning from natural image datasets via RGBD image synthesis. In: ICLR (2020)
46. Noguchi, A., Iqbal, U., Tremblay, J., Harada, T., Gallo, O.: Watch it move: unsupervised discovery of 3D joints for re-posing of articulated objects. In: CVPR (2022)
47. Noguchi, A., Sun, X., Lin, S., Harada, T.: Neural articulated radiance field. In: ICCV (2021)
48. Osman, A.A.A., Bolkart, T., Black, M.J.: STAR: sparse trained articulated human body regressor. In: Vedaldi, A., Bischof, H., Brox, T., Frahm, J.-M. (eds.) ECCV 2020. LNCS, vol. 12351, pp. 598–613. Springer, Cham (2020). https://doi.org/10.1007/978-3-030-58539-6_36
49. Park, J.J., Florence, P., Straub, J., Newcombe, R., Lovegrove, S.: Deepsdf: learning continuous signed distance functions for shape representation. In: CVPR (2019)
50. Park, K., et al.: Nerfies: deformable neural radiance fields. In: ICCV (2021)
51. Pavlakos, G., et al.: Expressive body capture: 3D hands, face, and body from a single image. In: CVPR (2019)
52. Peng, S., et al.: Animatable neural radiance fields for modeling dynamic human bodies. In: ICCV (2021)
53. Peng, S., et al.: Neural body: implicit neural representations with structured latent codes for novel view synthesis of dynamic humans. In: CVPR (2021)
54. Pumarola, A., Corona, E., Pons-Moll, G., Moreno-Noguer, F.: D-nerf: neural radiance fields for dynamic scenes. In: CVPR (2021)
55. Fridovich-Keil, S., Yu, A., Tancik, M., Chen, Q., Recht, B., Kanazawa, A.: Plenoxels: radiance fields without neural networks. In: CVPR (2022)
56. Schwarz, K., Liao, Y., Niemeyer, M., Geiger, A.: GRAF: generative radiance fields for 3D-aware image synthesis. In: NeurIPS (2020)
57. Shysheya, A., et al.: Textured neural avatars. In: CVPR (2019)

58. Sitzmann, V., Zollhöfer, M., Wetzstein, G.: Scene representation networks: continuous 3D-structure-aware neural scene representations. In: NeurIPS (2019)
59. Su, S.Y., Yu, F., Zollhöfer, M., Rhodin, H.: A-NeRF: articulated neural radiance fields for learning human shape, appearance, and pose. In: NeurIPS (2021)
60. Szabó, A., Meishvili, G., Favaro, P.: Unsupervised generative 3D shape learning from natural images. arXiv preprint arXiv:1910.00287 (2019)
61. Tiwari, G., Sarafianos, N., Tung, T., Pons-Moll, G.: Neural-GIF: neural generalized implicit functions for animating people in clothing. In: ICCV (2021)
62. Varol, G., Romero, J., Martin, X., Mahmood, N., Black, M.J., Laptev, I., Schmid, C.: Learning from synthetic humans. In: CVPR (2017)
63. Wang, Z., Bovik, A.C., Sheikh, H.R., Simoncelli, E.P.: Image quality assessment: from error visibility to structural similarity. IEEE Trans. Image Process. **13**(4), 600–612 (2004)
64. Wu, J., Zhang, C., Xue, T., Freeman, B., Tenenbaum, J.: Learning a probabilistic latent space of object shapes via 3D generative-adversarial modeling. In: NeurIPS (2016)
65. Xu, H., Alldieck, T., Sminchisescu, C.: H-NeRF: neural radiance fields for rendering and temporal reconstruction of humans in motion. In: NeurIPS (2021)
66. Xu, T., Fujita, Y., Matsumoto, E.: Surface-aligned neural radiance fields for controllable 3D human synthesis. In: CVPR (2022)
67. Yang, G., Sun, D., Jampani, V., Vlasic, D., Cole, F., Liu, C., Ramanan, D.: ViSER: video-specific surface embeddings for articulated 3D shape reconstruction. In: NeurIPS (2021)
68. Yang, G., Vo, M., Natalia, N., Ramanan, D., Andrea, V., Hanbyul, J.: BANMo: building animatable 3D neural models from many casual videos. In: CVPR (2022)
69. Yariv, L., et al.: Multiview neural surface reconstruction by disentangling geometry and appearance. In: NeurIPS (2020)
70. Yu, A., Li, R., Tancik, M., Li, H., Ng, R., Kanazawa, A.: PlenOctrees for real-time rendering of neural radiance fields. In: ICCV (2021)
71. Zhang, R., Isola, P., Efros, A.A., Shechtman, E., Wang, O.: The unreasonable effectiveness of deep features as a perceptual metric. In: CVPR (2018)
72. Zhu, J.Y., et al.: Visual object networks: Image generation with disentangled 3D representations. In: NeurIPS (2018)
73. Zuffi, S., Kanazawa, A., Jacobs, D.W., Black, M.J.: 3D menagerie: modeling the 3D shape and pose of animals. In: CVPR (2017)

# JPEG Artifacts Removal via Contrastive Representation Learning

Xi Wang, Xueyang Fu(✉), Yurui Zhu, and Zheng-Jun Zha

University of Science and Technology of China, Hefei, China
{wangxxi,zyr}@mail.ustc.edu.cn, {xyfu,zhazj}@ustc.edu.cn
https://github.com/wang-xi-1/JPEG

**Abstract.** To meet the needs of practical applications, current deep learning-based methods focus on using a single model to handle JPEG images with different compression qualities, while few of them consider the auxiliary effects of the compression quality information. Recently, several methods estimate quality factors in a supervised learning manner to guide their network to remove JPEG artifacts. However, they may fail to estimate unseen compression types, affecting the subsequent restoration performance. To remedy this issue, we propose an unsupervised compression quality representation learning strategy for the blind JPEG artifacts removal. Specifically, we utilize contrastive learning to obtain discriminative compression quality representations in the latent feature space. Then, to fully exploit the learned representations, we design a compression-guided blind JPEG artifacts removal network, which integrates the discriminative compression quality representations in an information lossless way. In this way, our single network can flexibly handle various JPEG compression images. Experiments demonstrate that our method can adapt to different compression qualities to obtain discriminative representations and outperform state-of-art methods.

**Keywords:** JPEG Artifacts Removal · Unsupervised representation learning · Contrastive learning · Image restoration

## 1 Introduction

Due to the explosive growth of images and videos on the website, lossy compression has become a widely adopted strategy to save transmission bandwidth and storage. JPEG compression [1], which uses discrete cosine transform (DCT), is a popular compression standard due to the ease and speed of its application. First, the JPEG compression divides the image into $8 \times 8$ blocks. Then a discrete cosine transform is implemented to obtain DCT coefficients. After the critical lossy step of quantizing and rounding the coefficients on each block, the information is lost, and complex artifacts inevitably appear in the compressed images.

**Supplementary Information** The online version contains supplementary material available at https://doi.org/10.1007/978-3-031-19790-1_37.

© The Author(s), under exclusive license to Springer Nature Switzerland AG 2022
S. Avidan et al. (Eds.): ECCV 2022, LNCS 13677, pp. 615–631, 2022.
https://doi.org/10.1007/978-3-031-19790-1_37

These artifacts not only cause visual discomfort but also lead to the performance degradation of subsequent computer vision tasks.

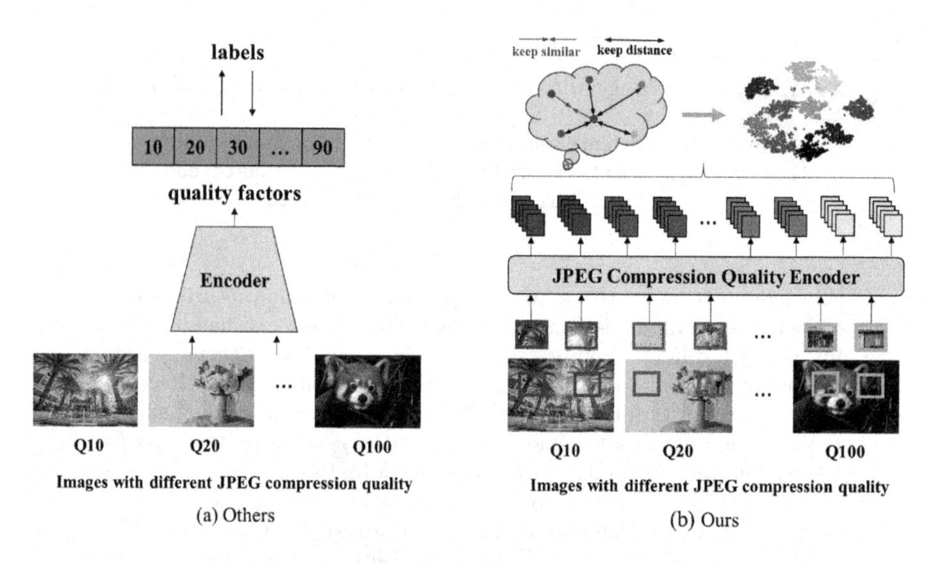


**Fig. 1.** An illustration of the other **supervised** method with our **unsupervised** compression quality representation learning method. In our method, the same color represents the same compression quality. We use the t-SNE [3] approach to cluster the output compression quality representations. (Color figure online)

To mitigate the impact of JPEG compression artifacts, many methods have been proposed. Generally, these methods can be roughly divided into model-based methods and deep learning (DL)-based methods. Model-based methods are primarily based on the filter design [2], they are usually limited to solving certain artifacts (*e.g.*, blocking and ringing artifacts). In recent years, thanks to the rapid development of deep learning network, which has powerful non-linear mapping capabilities, DL-based methods achieve better performance and dominate the field of JPEG artifacts removal.

However, most of the existing DL-based methods [4–7] train a specific network for each compression quality, which significantly limits the practicability of the network. Several blind JPEG compression artifacts removal methods [8,9] employ a single model to handle different compression qualities. However, these methods ignore the compression quality, and thus cannot explicitly reflect the degradation degree. Parts of the methods take into account compression qualities, but they all have certain shortcomings. For example, DCT-based methods [10,11] use a quantization table to guide the restoration of the image, but when the image is compressed multiple times, the quantization table information is incomplete. Wang *et al.* [12] utilize the ranker of image compression qualities but treat them to design loss functions instead of adding them into the JPEG artifacts removal network, which cannot fully exploit the distinguishable compression quality information. Jiang *et al.* [13] predicte quality factors in a supervised way which requires label quality factors. But the supervised manner is difficult to generalize to the unseen compression quality. When the prediction deviates from the accurate compression, the recovery performance will drop.

Unlike previous approaches, we manage to obtain the discriminative compression quality representation rather than predict the exact quality factor. Motivated by the success of contrastive learning [14–16], we propose an unsupervised contrastive learning strategy to obtain compression quality representations, which can fully mine the discrepancy between different compression qualities. Specifically, as shown in Fig. 1, we learn discriminative compression quality representations in the latent feature space by utilizing the variations of different JPEG compression images. In order to fully exploit this information, the learned representations are integrated into the JPEG artifacts removal network in an information lossless way to guide the network training. In this way, our network is able to flexibly process JPEG images with different compression qualities. Compared with directly predicting quality factors in a supervised way, our method does not require ground truth information of specific quality factors, which is accomplished in an unsupervised manner. Therefore, our method has a better generalization ability so that it is more applicable to unseen compression qualities, *e.g.*, real-world scenes (Fig. 2). Not only seen images but also unseen compressed images can be well recovered using our method.

The main contributions of our paper are as follows:

1. We propose a new framework for blind JPEG artifacts removal By taking advantage of the potential compression quality information in JPEG compressed images, our model can work well with all compressed quality JPEG images.
2. We propose an unsupervised manner to extract the discriminative compression quality representation hidden in the JPEG images, then integrate these learned representations into the compression quality-guided JPEG artifacts removal network in an information lossless way to guide the restoration of images with different compression qualities.
3. Experiments demonstrate that our network can flexibly handle various compression qualities and achieve state-of-the-art performance both in seen and unseen JPEG images, *e.g.*, improving 0.3dB in terms of PSNR on the widely used BSDS500 dataset of RGB channels.

## 2 Related Work

### 2.1 JPEG Artifacts Removal

There are mainly model-based and DL-based methods for JPEG artifacts removal. The earlier methods perform filtering operations to achieve compression artifacts removal. Foi *et al.* [2], based on shape-adaptive transformations

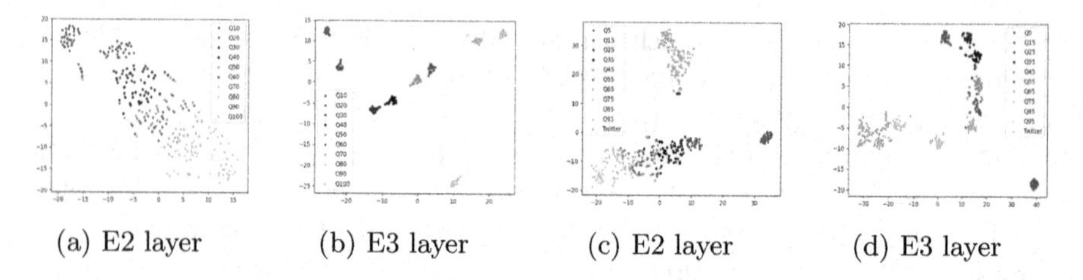

(a) E2 layer (b) E3 layer (c) E2 layer (d) E3 layer

**Fig. 2.** (a)(b) Visualization of different compression quality representations for LIVE1 with quality in **[10, 100]** in steps of 10. (c)(d) **Generalization Capabilities** Visualization of **unseen** compression quality representations for LIVE1 with quality in **[5, 95]** in steps of 10 and the **real-world** dataset (Twitter).

provide image filtering algorithms, clean edges are reconstructed, and no introduce unpleasant ring artifacts. Because it has a natural ill-posed characteristic, Probabilistic-Prior Methods play an important role. Many effective priors, *e.g.*, non-local similarity [17], low-rank [18,19], sparse coding [20], and adaptive DCT transformations [2], are explored. In recent years, DL-based methods have made significant progress in JPEG artifacts removal due to the powerful nonlinear mapping capability. ARCNN [4], proposed by Dong *et al.*, is a pioneering work that uses only four layers of CNN. Wang *et al.* [21] introduce a DCT domain prior to facilitating the JPEG artifacts removal. Mao *et al.* [22] use a deep encoding-decoding structure to exploit the rich dependencies of deep features. Some work also embeds traditional priors into deep networks, *e.g.*, multi-scale constraints [5] and wavelet signal structures [7]. Zhang *et al.* [8] achieve blind JPEG artifact removal using BN [23] and residual learning [24]. Since GANs [25] can be used to generate realistic textures, Galteri *et al.* [26] demonstrate that the GAN is able to produce more realistic details than MSE or SSIM based networks. Ehrlish *et al.* [10] also utilize the GAN loss to generate significantly more visually pleasing results. Zhang *et al.* [27] achieve effective image restoration by super-imposing local and non-local attention blocks to construct a residual non-local attention network. Zini *et al.* [9] exploit RRDB to remove JPEG artifacts in a blind way. Recently, there have been some attempts to use compressed quality information. Kim *et al.* [28] utilize the estimated quality factor for JPEG artifacts removal. AGARNet [29] estimates the pixel-wise quality factor in achieving using a single network to cover a wide range of quality factors. Wang *et al.* [12] propose compression quality ranker-guided networks. Jiang *et al.* [13] use a supervised way to predict the compression quality factor directly and embed the predicted quality factor into the subsequent network to guide the JPEG artifacts removal.

Although some methods take the compression qualities into account, they do not fully exploit this information or are limited by the supervised learning method. We propose an unsupervised compression quality representation learning strategy and make adequate exploitation of the learned representations to achieve restoration of all compressed quality JPEG images.

### 2.2 Contrastive Learning

Unsupervised learning [30–35] is a popular learning technique that does not rely on the label. Unsupervised contrastive learning [14,33,36] is the most popular method for generating discriminative representation via distinguishing positive and negative samples in an unsupervised way. In computer vision tasks, there are many flexible choices of positive and negative samples, which allows for the great application of contrastive learning. Although contrastive learning has been widely used in high-level tasks, it has not been widely applied in low-level tasks, especially in the field of JPEG artifacts removal. In this paper, we utilize contrast learning to obtain discriminative compression quality representations to guide our single model in processing JPEG images of all compression qualities.

## 3 Proposed Method

We propose a blind JPEG artifacts removal network, which consists of two parts: Unsupervised Compression Quality Encoder and Compression Quality-guided JPEG Artifacts Removal Network, as shown in Fig. 3. Our network is trained in two stages. First, we train a compression quality encoder in an unsupervised way to generate discriminative representations for different compression qualities. Second, based on the learned compression quality representations, we design the compression quality-guided JPEG artifacts removal network. In the next section, we describe the network structure and training strategy in detail.

### 3.1 Unsupervised Compression Quality Encoder

The goal of unsupervised representation learning [14,35] is to learn an encoder that converts input data to general-purpose representations. Unsupervised contrastive learning [30,33,36], which is trained by positive and negative samples, intends to generate similar representations for similar data and to make the representations of different data as different as possible. In order to achieve this goal, the InfoNCE loss [32] is often used to measure the similarity of representations, which uses the dot product measure of similarity:

$$L_q = -\log \frac{\exp(q \cdot k^+/\tau)}{\sum_{i=1}^{N_{neg}} \exp(q \cdot k_i^-/\tau)}, \tag{1}$$

where $k^+$ denotes a positive sample similar to q, $k^-$ denotes a negative sample not similar to $q$, $\cdot$ represents the dot product, $N_{neg}$ is the total number of negative samples and $\tau$ is a temperature hyper-parameter.

In this paper, we use the unsupervised contrast learning method to extract discriminative compression quality representations of JPEG images. To achieve this goal, we set the patch on the same image to have the same JPEG compression quality and the patch from the different images to have different JPEG compression qualities. Multiple patches can be cropped from each image, where

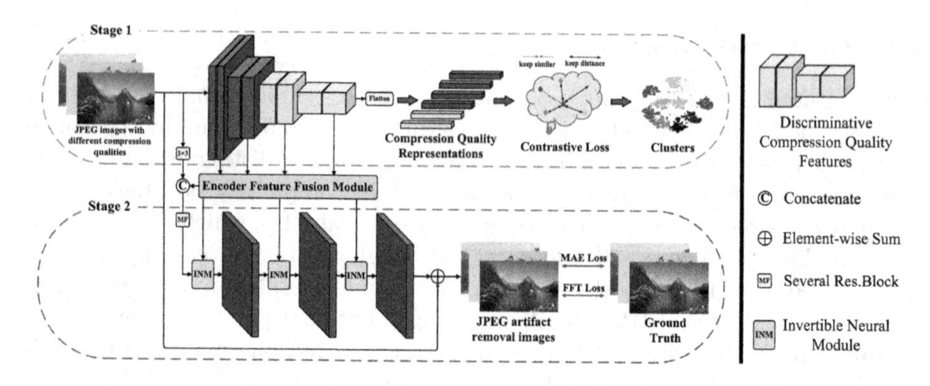


**Fig. 3.** The architecture of the proposed framework for blind JPEG artifacts removal. The training is divided into two stages. **First**, we train the compression quality encoder and generate discriminative compression quality representations. **Second**, we integrate the learned discriminative compression quality representations into the JPEG artifacts removal network in an information lossless way to handle various JPEG compression images flexibly.

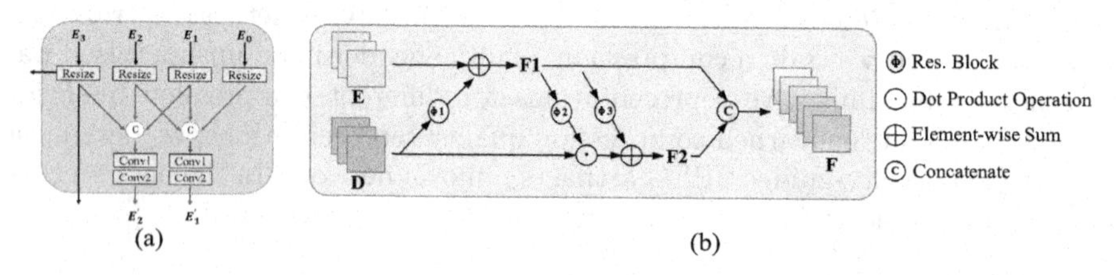


**Fig. 4.** Multi-scale Information Lossless Fusion Module. It consists of two parts: (a) Encoder Feature Fusion Module, (b) Invertible Neural Module.

patches from the same image can be used as positive samples, while patches from different images can be used as negative samples.

In the training phase of the compression quality encoder, we randomly select a mini-batch consisting of $B$ images with different compression qualities. Then, two patches are randomly cropped from each JPEG compression image, denoted as $p_i^1$ and $p_i^2$, where $p_i$ indicates that the patch is from the $i^{th}$ JPEG compression image. Then they are fed into the compression quality encoder to get compression quality representations $c_i^1$ and $c_i^2$. For each image, we set $c_i^1$ as a query and $c_i^2$ as a positive sample, and the compression representation $c_j^1$ and $c_j^2$ (i $\neq$ j) of patches from other JPEG compressed quality images as negative samples. $c_i^1$ should be as similar to $c_i^2$ as possible and as different from $c_j^1$ and $c_j^2$ as possible. Recent studies have shown that a large number of negative samples are crucial in unsupervised contrast representation learning. Following MoCo [16], we utilize a queue to store negative samples. The queue stores multiple representations of recent training images, and is dynamically, constantly updating, with representations of the latest images entering the queue and representations of the oldest images leaving the queue. The loss function of the compression quality encoder is:

$$L_{CQE} = \sum_{i=1}^{B} -\log \frac{\exp(c_i^1 \cdot c_i^2/\tau)}{\sum_{j=1}^{N_{neg}} \exp(c_i^1 \cdot c_j^{1,2}/\tau)}, \tag{2}$$

where the numerator represents the dot product of query and positive sample, and the denominator represents the dot product of query and all negative samples in the queue, where the dot product is used to measure the relative distance.

We use a multi-scale feature extraction network as an encoder network, as shown in Fig. 3. The output of each scale of the compression quality encoder is denoted as $E_0$, $E_1$, $E_2$ and $E_3$, respectively. We provide the detailed network structure in the supplementary material. To demonstrate that our compression quality encoder learns discriminative representations, we visualize the features in the compression quality encoder network using the t-SNE [3] method, as shown in Fig. 2. Our network can obtain discriminative representations on various compression qualities, including unseen degradation types.

### 3.2 Compression Quality-Guided JPEG Artifact Removal Network

After obtaining compression quality representations, to fully exploit this information, we design a compression quality-guided JPEG artifacts removal network, which integrates the learned compression quality representations in an information lossless way. The network contains multi-scale information lossless feature fusion module and restoration decoder, as shown in Fig. 3.

**Multi-scale Information Lossless Fusion Module.** In order to better integrate the discriminative representations learned by the compression quality encoder into the subsequent JPEG artifacts removal network, we use a multi-scale information lossless fusion module for this operation. First, since the features at different scales of the network are closely related, we try to fuse feature maps from multi-scales in the comparison quality encoder as much as possible. Specifically, we feed the output of each scale into the Encoder Feature Fusion Module (EFFM) and resize them to the same scales. $E_1$ and $E_2$ incorporate feature maps from nearby scales. If these feature maps were to be directly concatenated into the network afterward, this would result in a large number of operations, so to reduce the computational effort we introduce two convolution operations to fuse them, the output feature maps are denoted $E_1^{'}$ and $E_2^{'}$. The formulations are as:

$$E_1^{'} = EFFM[E_0, E_1, E_2], \tag{3}$$

$$E_2^{'} = EFFM[E_1, E_2, E_3], \tag{4}$$

where $EFFM$ includes convolution and resizes operations, $[\cdot]$ represents concatenation along the channel dimension, as shown in Fig. 4(a).

In order to fully exploit the learned discriminative compression quality feature representations, we use invertible fusion modules designed based on invert-

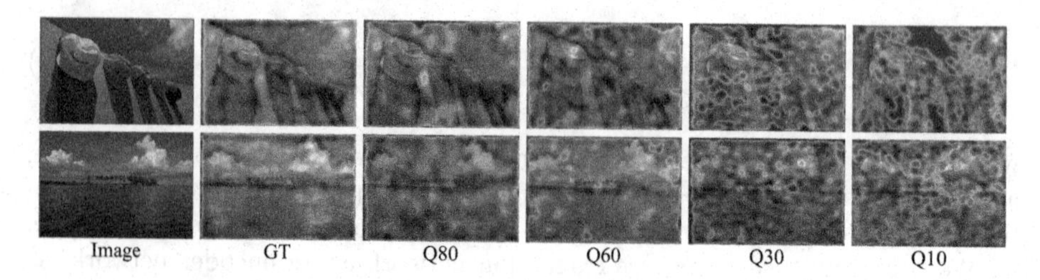


**Fig. 5.** Visualization of the intermediate feature maps of our Compression Quality Encoder at JPEG images with different compression qualities.

ible neural architecture [37,38] to preserve all information about the input features. Compared to simple concatenation operations, the invertible neural network [39–41] is information lossless in the processes of the transformation. In our work, a total of three invertible fusion modules are used, corresponding to the outputs of the last three scales in the compression quality encoder. Invertible networks require the input to be divided into two parts, we set the feature maps from the EFFM and subsequent restoration decoder as inputs, noted as E and D, respectively. Take one module as an example, it performs the following operations:

$$F_1 = E + \phi_1(D), \tag{5}$$

$$F_2 = D \odot \exp(\phi_2(F_1)) + \phi_3(F_1), \tag{6}$$

$$F = Concat(F_1, F_2), \tag{7}$$

where $\exp(\cdot)$ and $\odot$ indicate exponential function and dot product operation, respectively. As shown in Fig. 4(b), we choose residual blocks to perform $\phi_1$, $\phi_2$ and $\phi_3$, each residual block is composed of two $3 \times 3$ convolutions layers with the LeakyReLU activation function [42] in the middle.

**Restoration Decoder.** The outputs of each invertible neural module are fed into the local recovery module referenced from the RNAN [27]. The network then applies a $1 \times 1$ convolution layer to restore the feature maps to the original image channel. Finally, we use global residual learning to connect the input and output images to achieve faster training.

### 3.3 Loss Function

**MAE Loss.** We adopt the Mean Absolute Error (MAE) loss to reduce the distance between the predicted image $I_{pre}$ and the ground truth $I_{gt}$, which is defined as:

$$L_{MAE} = \frac{1}{N} \sum_{i=1}^{N} \| I_{pre}^i - I_{gt}^i \|_1, \tag{8}$$

where $N$ is the number of training samples within a mini-batch.

**Table 1.** Quantitative comparisons of different methods on **grayscale** JPEG images. PSNR/SSIM/PSNR-B format. The best and the second best results are **boldfaced** and underlined, respectively.

| Dataset | Quality | JPEG | ARCNN [4] | DnCNN [8] | MWCNN [7] | DCSC [44] |
|---|---|---|---|---|---|---|
| *Classic5* | 10 | 27.82/0.760/25.21 | 29.03/0.793/28.76 | 29.40/0.803/29.13 | 30.01/0.820/29.59 | 29.62/0.810/29.30 |
| | 20 | 30.12/0.834/27.50 | 31.15/0.852/30.59 | 31.63/0.861/31.19 | 32.16/0.870/31.52 | 31.81/0.864/31.34 |
| | 30 | 31.48/0.867/28.94 | 32.51/0.881/31.98 | 32.91/0.886/32.38 | 33.43/0.893/32.62 | 33.06/0.888/32.49 |
| | 40 | 32.43/0.885/29.92 | 33.32/0.895/32.79 | 33.77/0.900/33.23 | 34.27/0.906/33.35 | 33.87/0.902/33.30 |
| *LIVE1* | 10 | 27.77/0.773/25.33 | 28.96/0.808/28.68 | 29.19/0.812/28.90 | 29.69/0.825/29.32 | 29.34/0.818/29.01 |
| | 20 | 30.07/0.851/27.57 | 31.29/0.873/30.76 | 31.59/0.880/31.07 | 32.04/0.889/31.51 | 31.70/0.883/31.18 |
| | 30 | 31.41/0.885/28.92 | 32.67/0.904/32.14 | 32.98/0.909/32.34 | 33.45/0.915/32.80 | 33.07/0.911/32.43 |
| | 40 | 32.35/0.904/29.96 | 33.61/0.920/33.11 | 33.96/0.925/33.28 | 34.45/0.930/33.78 | 34.02/0.926/33.36 |
| *BSDS500* | 10 | 27.80/0.768/25.10 | 29.10/0.804/28.73 | 29.21/0.809/28.80 | 29.61/0.820/29.14 | 29.32/0.813/28.91 |
| | 20 | 30.05/0.849/27.22 | 31.28/0.870/30.55 | 31.53/0.878/30.79 | 31.92/0.885/31.15 | 31.63/0.880/30.92 |
| | 30 | 31.37/0.884/28.53 | 32.67/0.902/31.94 | 32.90/0.907/31.97 | 33.30/0.912/32.34 | 32.99/0.908/32.08 |
| | 40 | 32.30/0.903/29.49 | 33.55/0.918/32.78 | 33.85/0.923/32.80 | 34.27/0.928/33.19 | 33.92/0.924/32.92 |

| Dataset | Quality | RNAN [27] | RDN [45] | QGAC [10] | FBCNN [13] | Ours |
|---|---|---|---|---|---|---|
| *Classic5* | 10 | 29.96/0.819/29.42 | 30.03/0.819/29.59 | 29.84/0.812/29.43 | 30.12/**0.822**/29.80 | **30.16/0.822/29.85** |
| | 20 | 32.11/0.869/31.26 | 32.19/0.870/31.53 | 31.98/0.869/31.37 | 32.31/0.872/31.74 | **32.37/0.873/31.84** |
| | 30 | 33.38/0.892/32.35 | 33.46/0.893/32.59 | 33.22/0.892/32.42 | 33.54/0.894/32.78 | **33.60/0.895/32.89** |
| | 40 | 34.27/0.906/33.40 | - | 34.05/0.905/33.12 | 34.35/0.907/33.48 | **34.43/0.908/33.58** |
| *LIVE1* | 10 | 29.63/0.824/29.13 | 29.70/0.825/29.37 | 29.51/0.825/29.13 | 29.75/0.827/29.40 | **29.80/0.827/29.44** |
| | 20 | 32.03/0.888/31.12 | 32.10/0.889/31.29 | 31.83/0.888/31.25 | 32.13/0.889/31.57 | **32.19/0.890/31.63** |
| | 30 | 33.45/0.915/32.22 | 33.54/0.916/32.62 | 33.20/0.914/32.47 | 33.54/0.916/32.83 | **33.62/0.918/32.91** |
| | 40 | 34.47/0.930/33.66 | - | 34.16/0.929/33.36 | 34.53/0.931/33.74 | **34.62/0.931/33.84** |
| *BSDS500* | 10 | 29.08/0.805/28.48 | 29.24/0.808/28.71 | 29.46/0.821/28.97 | 29.67/0.821/29.22 | **29.70/0.822/29.27** |
| | 20 | 31.25/0.875/30.27 | 31.48/0.879/30.45 | 31.73/0.884/30.93 | 32.00/0.885/31.19 | **32.06/0.886/31.27** |
| | 30 | 32.70/0.907/31.33 | 32.83/0.908/31.60 | 33.07/0.912/32.04 | 33.37/0.913/32.32 | **33.45/0.914/32.41** |
| | 40 | 33.47/0.923/32.27 | - | 34.01/0.927/32.81 | 34.33/0.928/33.10 | **34.42/0.929/33.22** |

**FFT loss.** Since the quantization operation of JPEG compression results in the loss of high-frequency information in the image, we further employ the difference between the predicted image and the ground truth in the frequency domain [43] to optimize our network. The frequency loss is defined as:

$$L_{\text{FFT}} = \frac{1}{N}\sum_{i=1}^{N} \|\text{FFT}(I^i_{pre}) - \text{FFT}(I^i_{gt})\|_1, \tag{9}$$

where FFT stands for fast Fourier transform, which converts an image to the frequency domain. The total loss function is defined as:

$$L_{total} = L_{MAE} + \lambda L_{\text{FFT}}. \tag{10}$$

In our experiment, we set $\lambda$ equal to 0.1.

# 4 Experiments

## 4.1 Experimental Datasets and Implementation Details

**Datasets.** In our experiments, we use a total of six datasets: DIV2K [46], BSDS500 [47], LIVE1 [48], Classic5 [49], ICB [10] and Twitter [4]. 900 images

**Table 2.** Perceptual metrics results of LIPIS↓ / FID↓.

| Dataset | Classic5 | | | | LIVE1 | | | |
|---|---|---|---|---|---|---|---|---|
| Quality | Q10 | Q20 | Q30 | Q40 | Q10 | Q20 | Q30 | Q40 |
| FBCNN | **0.1543**/103.85 | 0.1072/46.98 | 0.0817/31.53 | 0.0665/23.66 | **0.1603**/69.47 | 0.0924/32.78 | 0.0637/21.87 | 0.0479/15.50 |
| Ours | 0.1545/**101.68** | **0.1062/43.08** | **0.0806/30.19** | **0.0651/22.44** | 0.1633/**65.92** | **0.0923/32.03** | **0.0630/20.79** | **0.0469/14.72** |

from the training and validation sets of DIV2K and 200 images from the training sets of BSDS500 are used for training. The test set of BSDS500, Classic5, LIVE1, ICB and Twitter are used for testing. We used the Y channel of YCbCr space for grayscale image recovery and the RGB channel for color image recovery.

**Training Settings.** The compression quality of the training images is set to [Q10, Q90] at step 10 and we randomly crop 256 $\times$ 256 patches from the images. Note that our model is trained in two stages. For the first stage, when we train the compression quality encoder, the learning rate is set to 0.001, and the number of training epochs is set to 200, then we freeze the model weights. For the second stage of training the JPEG artifacts removal network, the initial learning rate is set to 0.0001 and decayed by a cosine annealing algorithm with T = 600. For the optimization model, we set the epochs for 600 with a batch size of 8 and choose the Adam optimizer [50] with $\beta_1 = 0.9$ and $\beta_2 = 0.999$. In addition, our single model can handle multiple JPEG compression qualities. We train our model on two NVIDIA GeForce GTX 3090 GPUs by using PyTorch.

**Testing Settings.** For grayscale images, we evaluate the performance of our model on Classic5 [49], LIVE1 [48], Twitter [4] and the test set of BSDS500 [47]. During the standard testing phase, all test datasets are all applied JPEG compression with compression quality factors of Q10, Q20, Q30 and Q40. During the testing phase of the model generalizability capability, these test datasets are compressed into Q15, Q25, Q35 and Q45. For color images, we do not use the Classic5 [49] but the ICB [10] instead.

**Evaluation Metrics.** We use PSNR, SSIM(structural similarity) [51], and PSNR-B(specially designed for JPEG artifacts removal) [52] to quantitatively assess the performance of our JPEG artifacts removal model.

### 4.2 Experiments on Synthetic Datasets

**Feature Maps Visualisation for Compression Quality Encoder.** To demonstrate the ability of our compression quality encoder to distinguish different quality factors, we perform Grad-CAM [53] to visualize the learned feature maps in Fig. 5. It is clear that our compression quality encoder generates different feature maps for different compression qualities, which can provide discriminative information to guide subsequent JPEG artifacts removal.

**Table 3.** Quantitative comparisons of different methods on **color** JPEG images. PSNR/SSIM/PSNR-B format. The best and the second best results are **boldfaced** and underlined, respectively.

| Dataset | Quality | JPEG | QGAC [10] | FBCNN [13] | Ours |
|---|---|---|---|---|---|
| *LIVE1* | 10 | 25.69/0.743/24.20 | 27.62/0.804/27.43 | 27.77/0.803/27.51 | **27.80/0.805/27.57** |
| | 20 | 28.06/0.826/26.49 | 29.88/0.868/29.56 | 30.11/0.868/29.70 | **30.23/0.872/29.85** |
| | 30 | 29.37/0.861/27.84 | 31.17/0.896/30.77 | 31.43/0.897/30.92 | **31.58/0.900/31.13** |
| | 40 | 30.28/0.882/28.84 | 32.05/0.912/31.61 | 32.34/0.913/31.80 | **32.53/0.916/32.04** |
| *BSDS500* | 10 | 25.84/0.741/24.13 | 27.74/0.802/27.47 | 27.85/0.799/27.52 | **27.91/0.803/27.59** |
| | 20 | 28.21/0.827/26.37 | 30.01/0.869/29.53 | 30.14/0.867/29.56 | **30.31/0.872/29.74** |
| | 30 | 29.57/0.865/27.72 | 31.33/0.898/30.70 | 31.45/0.897/30.72 | **31.69/0.901/30.96** |
| | 40 | 30.52/0.887/28.69 | 32.25/0.915/31.50 | 32.36/0.913/31.52 | **32.66/0.918/31.82** |
| *ICB* | 10 | 29.44/0.757/28.53 | 32.06/**0.816**/32.04 | **32.18**/0.815/**32.15** | 32.05/0.813/32.04 |
| | 20 | 32.01/0.806/31.11 | 34.13/0.843/34.10 | **34.38/0.844/34.34** | 34.32/0.842/34.31 |
| | 30 | 33.20/0.831/32.35 | 35.07/**0.857**/35.02 | **35.41/0.857/35.35** | 35.37/0.856/**35.35** |
| | 40 | 33.95/0.840/33.14 | 32.25/**0.915**/31.50 | **36.02**/0.866/35.95 | 35.99/0.860/**35.97** |

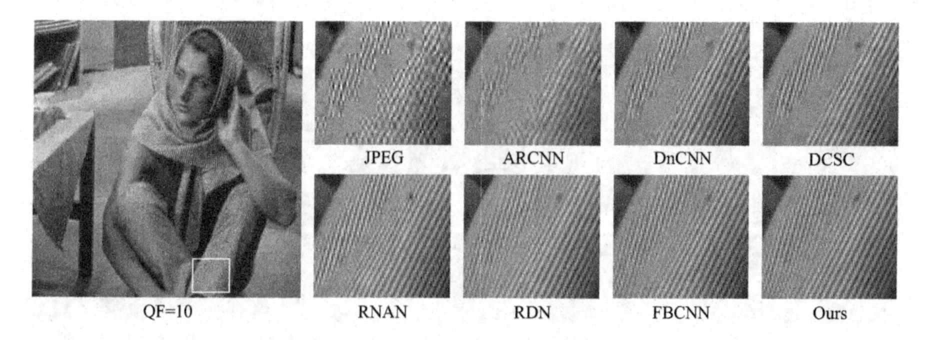


**Fig. 6.** Visual comparisons of JPEG image "Classic5: barbara" with QF = 10.

**Y Channel JPEG Artifacts Removal.** We first evaluate the effect of our model on the Y-channel JPEG compressed images. For LIVE1 [48], Classic5 [49], BSDS500 [47], we compared our model with a series of JPEG artifact removal network: *i.e.*, ARCNN [4], DnCNN [8], MWCNN [7], DCSC [44], RNAN [27], RDN [45], QGAC [10] and FBCNN [13]. For quantitative evaluation, we use PSNR, SSIM and PSNR-B, the results of them are presented in Table 1. As can be seen that our proposed model outperforms all previous methods. This proves the validity of our proposed model. Note that we use a single model for all compression qualities, this allows for greater flexibility in our models, and our method outperforms all those methods that train one model for one compression quality. We show some visual results of the Classic5 recovery image in Fig. 6, demonstrating the more pleasing visual effect of our method. Moreover, we utilize LIPIS [54] and FID [55] to evaluate the perceptual performance in Table 2.

**Table 4.** Quantitative comparisons of **Generalization Capabilities**. PSNR/SSIM/PSNR-B format. The best results are **boldfaced**. Our model has not seen the compression quality of the test phase during the training phase.

| Dataset | Training Quality (step) | Testing Quality | JPEG | FBCNN [13] | Ours |
|---|---|---|---|---|---|
| *Classic5* | $Q10-Q90(10)$ | 15 | 29.17/0.807/26.53 | **31.42/0.854/30.97** | 31.37/**0.854**/30.86 |
| | | 25 | 30.87/0.853/28.30 | 33.02/**0.885**/32.34 | **33.04/0.885/32.41** |
| | | 35 | 32.01/0.877/29.50 | 33.99/0.9015/33.20 | **34.05/0.902/33.30** |
| | | 45 | 32.84/0.892/30.37 | 34.72/0.9122/33.83 | **34.79/0.913/33.92** |
| *LIVE1* | $Q10-Q90(10)$ | 15 | 29.13/0.822/26.65 | **31.15**/0.866/**30.69** | 31.12/**0.867**/30.53 |
| | | 25 | 30.81/0.871/28.29 | 32.91/**0.905/32.26** | **32.95/0.905/32.26** |
| | | 35 | 31.93/0.896/29.48 | 34.09/**0.925**/33.33 | **34.15/0.925/33.40** |
| | | 45 | 32.778/0.912/30.43 | 34.96/0.936/34.12 | **35.04/0.937/34.22** |
| *BSDS500* | $Q10-Q90(10)$ | 15 | 29.13/0.819/26.34 | **31.04/0.862/30.39** | 30.99/**0.862**/30.22 |
| | | 25 | 30.77/0.869/27.93 | 32.75/0.901/31.81 | **32.79/0.902/31.83** |
| | | 35 | 31.88/0.895/29.05 | 33.91/**0.922**/32.76 | **33.97/0.922/32.85** |
| | | 45 | 32.73/0.911/29.94 | 34.76/0.934/33.46 | **34.85/0.935/33.58** |

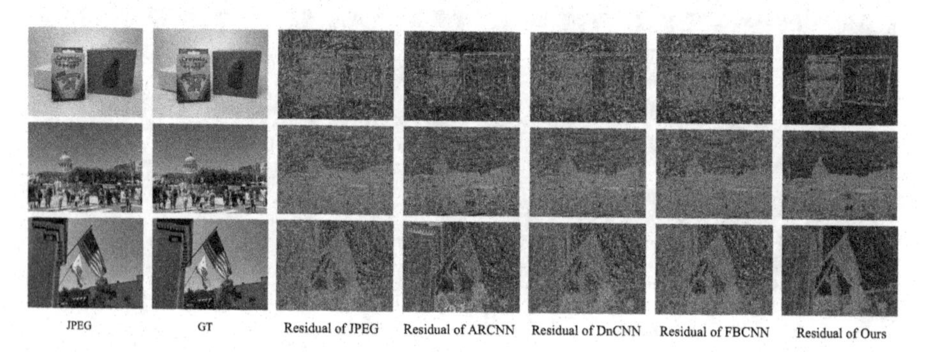


**Fig. 7.** Visual comparisons on **real-world** images from "Twitter" dataset.

**RGB Channels JPEG Artifacts Removal.** To evaluate the effectiveness of our model on color images, we also trained our model on color images. We set the number of input and output channels to 3, while the other model settings remain unchanged. The test data set is selected LIVE1 [48] and test sets of BSDS500 [47]. Quantitative results are shown in Table 3. It can be seen that our method achieves better JPEG artifacts removal results on color images as well.

**Study of Generalization Capabilities.** Both our compression quality encoder and JPEG artifacts removal network are trained only on training data with compression quality set to [Q10, Q90] at step 10. To explore whether our model can perform well on unseen JPEG compressed quality images, we choose images with compression qualities of Q15, Q25, Q35 and Q45. As shown in Table 4, our single model consistently performs well on unseen compression qualities. All these processes are performed on the Y-channel of LIVE1 [48], Classic5 [49] and the test set of BSDS500 [47].

**Table 5.** Quantitative **Ablation Analysis** on PSNR/SSIM/PSNR-B Values. The dataset used in this experiment is Classic5.

| Models | CQE | CQE(pre-trained) | MILFM | | FFT loss | Compression quality | | | |
|---|---|---|---|---|---|---|---|---|---|
| | | | INM | EFFM | | Q10 | Q20 | Q30 | Q40 |
| model-1 | | | | | ✓ | 29.97/0.818/29.63 | 32.21/0.871/31.68 | 33.48/0.893/32.76 | 34.31/0.906/33.46 |
| model-2 | ✓ | | ✓ | ✓ | ✓ | 30.04/0.820/29.73 | 32.26/0.871/31.78 | 33.52/0.894/32.85 | 34.36/0.907/33.57 |
| model-3(a) | ✓ | ✓ | | ✓ | ✓ | 30.11/0.821/29.77 | 32.30/0.872/31.74 | 33.55/0.894/32.81 | 34.39/0.907/33.52 |
| model-3(b) | ✓ | ✓ | | | ✓ | 30.09/0.821/29.72 | 32.30/0.872/31.72 | 33.54/0.894/32.78 | 34.37/0.907/33.48 |
| model-4 | ✓ | ✓ | ✓ | ✓ | | 30.03/0.819/29.66 | 32.21/0.871/31.69 | 33.50/0.893/32.77 | 34.32/0.906/33.51 |
| Ours | ✓ | ✓ | ✓ | ✓ | ✓ | 30.16/0.822/29.85 | 32.37/0.873/31.84 | 33.60/0.895/32.89 | 34.43/0.908/33.58 |

### 4.3 Experiments on Real-World Compression Qualities

To avoid taking up too much storage and transmission resources, social platforms such as Twitter often compress uploaded images, which inevitably reduces visual feelings of users. To test the performance of our model on real data, we use the Twitter dataset to test the real image directly using the model we trained on the synthetic datasets. Since the real images were too large in resolution, we first crop images and then feed them into the network. We show in Fig. 7 the visual residual maps of other methods and ours to increase the distinction of the visualization. Note that the residual map means the difference between the estimated result and its ground truth. It is clear that our method achieves the better visual result. This result shows that our method works better than other methods on unseen JPEG compressed quality images.

### 4.4 Ablation Analysis

We remove some parts of the network that we designed and report their effect. We choose Q10 of the Classic5 dataset to report the results. For all ablation experiments, quantitative results are presented in Table 5.

**Effect of Compression Quality Encoder (CQE).** The compression quality encoder generates discriminative representations that provide recovery guidance for subsequent JPEG artifacts removal networks. To demonstrate the effectiveness of compression quality encoder, we compare two-stage joint training and two-stage separate training strategies: (1) remove the entire compression quality encoder, denoted as model-1, (2) train the entire network directly without removing the compression quality encoder, but without pre-training, denoted as model-2. The results of the quantitative evaluation show that networks that remove the compression quality representation learning encoder would cause the network performance to drop. Moreover, we show the second option in which the compression quality encoder visualizes the clustered feature maps with and without the pre-trained weights in the supplementary material.

**Effect of Multi-scale Information Lossless Fusion Module (MILFM).** With the compression quality encoder and the extracted compression quality feature representations, we utilize the multi-scale information lossless feature fusion module to integrate them with the subsequent JPEG artifacts removal network. To demonstrate the effectiveness of this fusion module, we replace INM with the concatenation operation and convolution layers, denoted as model-3(a). Moreover, we replace INM and EFFM with the concatenation operation and convolution layers to achieve feature fusion, denoted as model-3(b). In this way, the network does not make much difference in terms of the number of parameters. As can be seen from the PSNR values taken, the performance of the model will drop if the concatenation operation and convolution layers are used as the fusion module. On the contrary, better JPEG artifacts removal results can be achieved by using the multiscale information lossless fusion module.

**Effect of FFT loss.** In order to better recover the lost high-frequency information, we introduced the FFT Loss. To test the capability of this loss function, we removed this Loss without changing the other parts of the model, which was noted as model-4. It is seen from the experimental results that the recovery of the model decreased due to the disappearance of the FFT loss.

## 5 Conclusions

In this paper, we propose an unsupervised JPEG compression quality representation learning to guide the blind JPEG artifacts removal. Rather than directly predicting the exact quality factor, our approach focuses on mining the discrepancy in compression quality of various compressed images. Moreover, to fully exploit the learned representations, we design a compression-guided blind JPEG artifacts removal network, which specially integrates the learned discriminative compression quality representations in an information lossless way. Experiments demonstrate that our unsupervised compression quality learning strategy could extract discriminative representations, and our network achieves state-of-the-art performances for various types of JPEG compressed quality images.

**Acknowledgement.** This work was supported by the National Key R&D Program of China under Grant 2020AAA0105702, the National Natural Science Foundation of China (NSFC) under Grants U19B2038 and 61901433, the University Synergy Innovation Program of Anhui Province under Grants GXXT-2019-025, the Fundamental Research Funds for the Central Universities under Grant WK2100000024, and the USTC Research Funds of the Double First-Class Initiative under Grant YD2100002003.

## References

1. Wallace, G.K.: The jpeg still picture compression standard. IEEE Trans. Consum. Electr. **38**(1), xviii–xxxiv (1992)
2. Foi, A., Katkovnik, V., Egiazarian, K.: Pointwise shape-adaptive DCT for high-quality denoising and deblocking of grayscale and color images. IEEE Trans. Image Process. **16**(5), 1395–1411 (2007)
3. Van der Maaten, L., Hinton, G.: Visualizing data using t-SNE. J. Mach. Learn. Res. **9**(11) (2008)

4. Dong, C., Deng, Y., Loy, C. C., Tang, X.: Compression artifacts reduction by a deep convolutional network. In: Proceedings of the IEEE International Conference on Computer Vision, pp. 576–584 (2015)
5. Cavigelli, L., Hager, P., Benini, L.: CAS-CNN: a deep convolutional neural network for image compression artifact suppression. In: 2017 International Joint Conference on Neural Networks (IJCNN). IEEE, pp. 752–759 (2017)
6. Chen, Y., Pock, T.: Trainable nonlinear reaction diffusion: a flexible framework for fast and effective image restoration. IEEE Trans. Pattern Anal. Mach. Intell. **39**(6), 1256–1272 (2016)
7. Liu, P., Zhang, H., Zhang, K., Lin, L., Zuo, W.: Multi-level wavelet-CNN for image restoration. In: Proceedings of the IEEE Conference on Computer Vision and Pattern Recognition Workshops, pp. 773–782 (2018)
8. Zhang, K., Zuo, W., Chen, Y., Meng, D., Zhang, L.: Beyond a gaussian denoiser: residual learning of deep CNN for image denoising. IEEE Trans. Image Process. **26**(7), 3142–3155 (2017)
9. Zini, S., Bianco, S., Schettini, R.: Deep residual autoencoder for blind universal jpeg restoration. IEEE Access **8**, 63283–63294 (2020)
10. Ehrlich, M., Davis, L., Lim, S.-N., Shrivastava, A.: Quantization guided JPEG artifact correction. In: Vedaldi, A., Bischof, H., Brox, T., Frahm, J.-M. (eds.) ECCV 2020. LNCS, vol. 12353, pp. 293–309. Springer, Cham (2020). https://doi.org/10.1007/978-3-030-58598-3_18
11. Guo, J., Chao, H.: Building dual-domain representations for compression artifacts reduction. In: Leibe, B., Matas, J., Sebe, N., Welling, M. (eds.) ECCV 2016. LNCS, vol. 9905, pp. 628–644. Springer, Cham (2016). https://doi.org/10.1007/978-3-319-46448-0_38
12. Wang, M., Fu, X., Sun, Z., Zha, Z.J.: JPEG artifacts removal via compression quality ranker-guided networks. In: Proceedings of the Twenty-Ninth International Conference on International Joint Conferences on Artificial Intelligence, pp. 566–572 (2021)
13. Jiang, J., Zhang, K., Timofte, R.: Towards flexible blind jpeg artifacts removal. In: Proceedings of the IEEE/CVF International Conference on Computer Vision, pp. 4997–5006 (2021)
14. Chen, T., Kornblith, S., Norouzi, M., Hinton, G.: A simple framework for contrastive learning of visual representations. In: International Conference on Machine Learning, PMLR, pp. 1597–1607 (2020)
15. Dosovitskiy, A., Springenberg, J.T., Riedmiller, M., Brox, T.: Discriminative unsupervised feature learning with convolutional neural networks. Adv. Neural Inf. Process. Syst. **27** (2014)
16. He, K., Fan, H., Wu, Y., Xie, S., Girshick, R.: Momentum contrast for unsupervised visual representation learning. In: Proceedings of the IEEE/CVF Conference on Computer Vision and Pattern Recognition, pp. 9729–9738 (2020)
17. Zhang, X., Xiong, R., Ma, S., Gao, W.: Reducing blocking artifacts in compressed images via transform-domain non-local coefficients estimation. In: 2012 IEEE International Conference on Multimedia and Expo, pp. 836–841. IEEE (2012)
18. Zhang, X., Xiong, R., Fan, X., Ma, S., Gao, W.: Compression artifact reduction by overlapped-block transform coefficient estimation with block similarity. IEEE Trans. Image Process. **22**(12), 4613–4626 (2013)
19. Ren, J., Liu, J., Li, M., Bai, W., Guo, Z.: Image blocking artifacts reduction via patch clustering and low-rank minimization. In: Data Compression Conference, 516–516. IEEE (2013)

20. Chang, H., Ng, M.K., Zeng, T.: Reducing artifacts in jpeg decompression via a learned dictionary. IEEE Trans. Signal Process. **62**(3), 718–728 (2013)
21. Wang, Z., Liu, D., Chang, S., Ling, Q., Yang, Y., Huang, T.S.: D3: deep dual-domain based fast restoration of JPEG-compressed images. In: Proceedings of the IEEE Conference on Computer Vision and Pattern Recognition, pp. 2764–2772 (2016)
22. Mao, X.J., Shen, C., Yang, Y.B.: Image restoration using very deep convolutional encoder-decoder networks with symmetric skip connections. IEEE Trans. Image Process. **15**(13), 3142–3155 (2017)
23. Ioffe, S., Szegedy, C.: Batch normalization: accelerating deep network training by reducing internal covariate shift. In: International Conference on Machine Learning, PMLR, pp. 448–456 (2015)
24. He, K., Zhang, X., Ren, S., Sun, J.: Deep residual learning for image recognition. In: Proceedings of the IEEE Conference on Computer Vision and Pattern Recognition, pp. 770–778 (2016)
25. Goodfellow, I., et al.: Generative adversarial nets. Adv. Neural Inf. Process. Syst. **27** (2014)
26. Galteri, L., Seidenari, L., Bertini, M., Del Bimbo, A.: Deep generative adversarial compression artifact removal. In: Proceedings of the IEEE International Conference on Computer Vision, pp. 4826–4835 (2017)
27. Zhang, Y., Li, K., Li, K., Zhong, B., Fu, Y.: Residual non-local attention networks for image restoration. arXiv preprint arXiv:1903.10082 (2019)
28. Kim, Y., et al.: A pseudo-blind convolutional neural network for the reduction of compression artifacts. IEEE Trans. Circuits Syst. Video Technol. **30**(4), 1121–1135 (2019)
29. Kim, Y., Soh, J.W., Cho, N.I.: AGARNet: adaptively gated jpeg compression artifacts removal network for a wide range quality factor. IEEE Access **8**, 20160–20170 (2020)
30. Wu, Z., Xiong, Y., Yu, S.X., Lin, D.: Unsupervised feature learning via non-parametric instance discrimination. In: Proceedings of the IEEE Conference on Computer Vision and Pattern Recognition, pp. 3733–3742 (2018)
31. Zhuang, C., Zhai, A.L., Yamins, D.: Local aggregation for unsupervised learning of visual embeddings. In: Proceedings of the IEEE/CVF International Conference on Computer Vision, pp. 6002–6012 (2019)
32. Oord, A.V.D., Li, Y., Vinyals, O.: Representation learning with contrastive predictive coding. arXiv e-prints, pp. arXiv-1807 (2018)
33. Hjelm, R. D., et al.: Learning deep representations by mutual information estimation and maximization. arXiv preprint arXiv:1808.06670 (2018)
34. Henaff, O.: Data-efficient image recognition with contrastive predictive coding. In: International Conference on Machine Learning. PMLR, pp. 4182–4192 (2020)
35. Bachman, P., Hjelm, R.D., Buchwalter, W.: Learning representations by maximizing mutual information across views. Adv. Neural Inf. Process. Syst. **32** (2019)
36. Chen, X., Fan, H., Girshick, R., He, K.: Improved baselines with momentum contrastive learning. arXiv preprint arXiv:2003.04297 (2020)
37. Dinh, L., Krueger, D., Bengio, Y.: Nice: non-linear independent components estimation. arXiv preprint arXiv:1410.8516 (2014)
38. Dinh, L., Sohl-Dickstein, J., Bengio, S.: Density estimation using real nvp. arXiv preprint arXiv:1605.08803 (2016)
39. Liu, Y., et al.: Invertible denoising network: a light solution for real noise removal. In: Proceedings of the IEEE/CVF Conference on Computer Vision and Pattern Recognition, pp. 13365–13374 (2021)

40. Zhang, S., Zhang, C., Kang, N., Li, Z.: ivpf: numerical invertible volume preserving flow for efficient lossless compression. In: Proceedings of the IEEE/CVF Conference on Computer Vision and Pattern Recognition, pp. 620–629 (2021)
41. Xing, Y., Qian, Z., Chen, Q.: Invertible image signal processing. In: Proceedings of the IEEE/CVF Conference on Computer Vision and Pattern Recognition, pp. 6287–6296 (2021)
42. Xu, B., Wang, N., Chen, T., Li, M.: Empirical evaluation of rectified activations in convolutional network. arXiv preprint arXiv:1505.00853 (2015)
43. Cho, S.J., Ji, S.W., Hong, J.P., Jung, S.W., Ko, S.J.: Rethinking coarse-to-fine approach in single image deblurring. In: Proceedings of the IEEE/CVF International Conference on Computer Vision, pp. 4641–4650 (2021)
44. Fu, X., Zha, Z.J., Wu, F., Ding, X., Paisley, J.: JPEG artifacts reduction via deep convolutional sparse coding. In: 2019 IEEE/CVF International Conference on Computer Vision (ICCV) (2019)
45. Zhang, Y., Tian, Y., Kong, Y., Zhong, B., Fu, Y.: Residual dense network for image restoration. IEEE Trans. Pattern Anal. Mach. Intell. **43**(7), 2480–2495 (2020)
46. Agustsson, E., Timofte, R.: Ntire: challenge on single image super-resolution: dataset and study. In: Proceedings of the IEEE Conference on Computer Vision and Pattern Recognition Workshops, vol. 2017, pp. 126–135 (2017)
47. Arbelaez, P., Maire, M., Fowlkes, C., Malik, J.: Contour detection and hierarchical image segmentation. IEEE Trans. Pattern Anal. Mach. Intell. **33**(5), 898–916 (2011)
48. Sheikh, H.: Live image quality assessment database release 2 (2005). http://live.ece.utexas.edu/research/quality
49. Zeyde, R., Elad, M., Protter, M.: On single image scale-up using sparse-representations. In: Boissonnat, J.-D., et al. (eds.) Curves and Surfaces 2010. LNCS, vol. 6920, pp. 711–730. Springer, Heidelberg (2012). https://doi.org/10.1007/978-3-642-27413-8_47
50. Kingma, D.P., Ba, J.: Adam: a method for stochastic optimization. arXiv preprint arXiv:1412.6980 (2014)
51. Wang, Z., Bovik, A.C., Sheikh, H.R., Simoncelli, E.P.: Image quality assessment: from error visibility to structural similarity. IEEE Trans. Image Process. **13**(4), 600–612 (2004)
52. Tadala, T., Narayana, S.E.V.: A novel PSNR-B approach for evaluating the quality of de-blocked images (2012)
53. Selvaraju, R.R., Cogswell, M., Das, A., Vedantam, R., Parikh, D., Batra, D.: Grad-cam: visual explanations from deep networks via gradient-based localization. In: Proceedings of the IEEE International Conference on Computer Vision, pp. 618–626 (2017)
54. Zhang, R., Isola, P., Efros, A.A., Shechtman, E., Wang, O.: The unreasonable effectiveness of deep features as a perceptual metric. In: Proceedings of the IEEE Conference on Computer Vision and Pattern Recognition, pp. 586–595 (2018)
55. Heusel, M., Ramsauer, H., Unterthiner, T., Nessler, B., Hochreiter, S.: Gans trained by a two time-scale update rule converge to a local nash equilibrium. Adv. Neural Inf. Process. Syst. **30** (2017)

# Unpaired Deep Image Dehazing Using Contrastive Disentanglement Learning

Xiang Chen[1,2], Zhentao Fan[1], Pengpeng Li[3], Longgang Dai[1], Caihua Kong[1], Zhuoran Zheng[2], Yufeng Huang[1], and Yufeng Li[1(✉)]

[1] Gaofen Lab, Shenyang Aerospace University, Shenyang, China
gaofenlab@foxmail.com
[2] CSE, Nanjing University of Science and Technology, Nanjing, China
[3] ISE, Dalian Polytechnic University, Dalian, China

**Abstract.** We offer a practical unpaired learning based image dehazing network from an unpaired set of clear and hazy images. This paper provides a new perspective to treat image dehazing as a two-class separated factor disentanglement task, *i.e.*, the task-relevant factor of clear image reconstruction and the task-irrelevant factor of haze-relevant distribution. To achieve the disentanglement of these two-class factors in deep feature space, contrastive learning is introduced into a CycleGAN framework to learn disentangled representations by guiding the generated images to be associated with latent factors. With such formulation, the proposed contrastive disentangled dehazing method (CDD-GAN) employs negative generators to cooperate with the encoder network to update alternately, so as to produce a queue of challenging negative adversaries. Then these negative adversaries are trained end-to-end together with the backbone representation network to enhance the discriminative information and promote factor disentanglement performance by maximizing the adversarial contrastive loss. During the training, we further show that hard negative examples can suppress the task-irrelevant factors and unpaired clear exemples can enhance the task-relevant factors, in order to better facilitate haze removal and help image restoration. Extensive experiments on both synthetic and real-world datasets demonstrate that our method performs favorably against existing unpaired dehazing baselines.

**Keywords:** Single image dehazing · Haze removal · Contrastive learning · Factor disentanglement · Unpaired data · CycleGAN

## 1 Introduction

Single image dehazing (SID) is a typical low-level vision problem emerging in recent years, whose aim is to predict the haze-free image from the observed hazy image. Most existing SID methods are immersed in learning supervised models

---

X. Chen and Z. Fan—Contributed equally to this work.

© The Author(s), under exclusive license to Springer Nature Switzerland AG 2022
S. Avidan et al. (Eds.): ECCV 2022, LNCS 13677, pp. 632–648, 2022.
https://doi.org/10.1007/978-3-031-19790-1_38

from paired synthetic data [19], which inevitably limits their generalization capability in real-world applications. Therefore, learning the practical SID network from an unpaired set of clear and hazy images is significant as obtaining paired real-world data is almost prohibitively expensive and time-consuming [7,54].

How to learn a SID network when paired data is not available? To solve this issue, some recent studies [11,12,17,30] attempt to explore different unpaired dehazing solutions that mainly divided into two trends. The first one is semi/unsupervised transfer learning [11,12,23], where they either utilize the circulatory structure of CycleGAN [55] or design domain adaption paradigms [3,40] to boost the generalization abilities of the algorithm themselves. The above transfer learning based approaches regard SID as an image-to-image translation case, which are performed by making use of the limited labeled data and adding the auxiliary optimization terms. Due to the fact that the domain knowledge of the hazy and haze-free images is asymmetrical, it is laborious for these CycleGAN-based strategies to capture accurate mapping between two different domains using only weak constraints. Furthermore, these methods ignore the potential association in the latent space [7] and do not fully mine the useful feature information for SID, resulting in sub-optimal performance.

In consideration of the hazy input as the entanglement of several simple layers (*i.e.*, the scene radiance layer, the transmission map layer, and the atmospheric light layer), another popular way can be seen as a problem of the physical-based disentanglement. With this idea, several works [20,21,30,49] fully consider the physical model of haze process, and employ three joint subnetworks to disentangle the given hazy image into these three component parts, so as to estimate the haze and recover the clear image. Although learning disentangled representations has certain natural advantages, it is not easy to disentangle into three hidden factors from the hazy input. Furthermore, since the model is only a rough approximation of the real world, relying on a physics-based model to design the SID network would not make the method robust, especially under non-uniform haze conditions.

Following the above two lines of thinking, we rethink hazy image formation by simplifying the entanglement model itself. Motivated by the similar intuition in [28,42], we make a simple and elegant assumption of factor disentanglement which views an hazy image as an entanglement of two separable parts, a task-relevant factor (*e.g.*, the color, texture, and semantic information of the clear background image) and a task-irrelevant factor (*e.g.*, the distribution of the haze component). In this work, our key insight is that a good dehazing model is formulated by enhancing task-relevant factors, while suppressing task-irrelevant factors in the latent space. In other words, it could be helpful to reconstruct a clear image from the learned unambiguous embeddings by clustering these factors with the same value together and isolating other factors with the different value. The intuitive fact is that the same factor values produce the similar image features related to that factor [33], and vise versa. Therefore, this encourages us to introduce recent successful contrastive learning into the frequently-used unpaired adversarial framework, CycleGAN, to guide the generated images to be associated with latent factors, so that we can facilitate the learned representation to fulfill factor disentanglement and help image restoration.

In this paper, a contrastive disentangled dehazing method (CDD-GAN) is formulated without using paired training information. Specifically, we introduce a bidirectional disentangled translation network as the backbone of the proposed CDD-GAN. Different from the conventional contrastive loss [34] in GANs, we employ negative generators to perform adversarial contrastive mechanism [44] on the image generator encoder, so as to produce a series of challenging negative adversaries. With these hard negative adversaries, the image encoder on the backbone representation network will learn more distinguishing representation of the latent factors, so that we can disentangle the discrete variation of these factors during the bidirectional translation process. When the above-mentioned two-class factors are well separated, the image decoder will better isolate those task-irrelevant factors and obtain a more accurate representation for achieving high-quality outputs. To summarize, we offer the following contributions:

- We rethink the image dehazing task and propose an effective unpaired learning framework CDD-GAN, which first attempts to leverage disentangled factor representations to facilitate haze removal in the latent space.
- We introduce adversarial contrastive loss into CDD-GAN to fulfill factor disentanglement, where hard negative examples can suppress the task-irrelevant factors and unpaired clear exemples can enhance the task-relevant factors.
- Extensive experiments are carried out on both synthesis and real-world datasets, and demonstrate that our method is superior to existing unpaired dehazing networks and achieves encouraging performance.

## 2 Related Work

### 2.1 Single Image Dehazing

**For the paired dehazing aspect**, many classical methods [2,18,29,37,52] continuously comply with atmospheric scattering model and restore haze-free image through the estimation of the global atmospheric light and transmission map. Nevertheless, these algorithms tend to fail drastically when the corresponding parameter estimation is not accurate enough, thereby resulting in sub-optimal performance. To remedy this, numerous end-to-end dehazing networks [4,10,24,27,35,38] are recently developed for directly outputting dehazed images from hazy inputs without estimating atmospheric lights and transmission maps. However, those paired supervised models in dealing with real-world images will rapidly drop due to the inter-domain and intra-domain gap [51] between the training and test data.

**For the unpaired dehazing field**, inspired by the popular CycleGAN [55], previous works pursue directly learning the translation relationship from hazy domain to haze-free domain without using paired training information. In [1,11,12,25], several dehazing methods based on improved CycleGAN structure are proposed by utilizing unpaired adversarial learning strategy. Due to the domain knowledge between hazy and clear images is asymmetrical [3], it is not effective to restore high-quality results only relying on limited cycle-consistency constraints. Afterwards, Li et al. [23] first explore a semi-supervised dehazing framework, which can promote the learning ability by using unlabeled real hazy images and synthetic images. Recently, the idea of physical-based disentanglement [17,22,54] has emerged to further increase the unpaired dehazing performance. For instance, Yang et al. [49] design disentangled dehazing network (DisentGAN) to estimate the scene radiance, the medium transmission, and global atmosphere light by exploiting different generators jointly. Similarly, numerous novel unsupervised disentangled network architectures have been developed, such as you only look yourself (YOLY) [20], zero-shot image dehazing (ZID) [21], disentangled-consistency mean-teacher network (DMT-Net) [30]. Unlike these methods based on complex multilayer disentanglement, our assumption is simpler, that is, the latent space can be further divided into two separated parts, including the task-relevant factors and the task-irrelevant factors.

### 2.2 Contrastive Learning

Contrastive loss has demonstrated its effectiveness in self-supervised and unsupervised representation learning [5]. Recent researches have employed contrastive learning into low-level vision tasks and obtained improved performance, such as haze removal [47], rain removal [7], image super-resolution [43] and image-to-image translation [14,34]. The most critical design in contrastive learning is how to select the negatives. Different with previous methods sampling negative examples from patches at different positions in the source image, we actively train a set of negative examples as a whole in an adversarial manner. The closest thing to our method is [44], but the difference is that our method performs the contrastive operation in the CycleGAN framework, which benefits from mining the attributes of unpaired clear exemples in the backward cycle.

### 2.3 Disentanglement in GANs

Disentanglement methods in GANs [6,13] have been proposed and used to decompose and recombine the representations of individual factors from hidden representations. Most disentanglement frameworks attempt to learn representations which capture different factors of variation in the latent space. Recently, Ye et al. [50] decomposed the rainy image into the rain-free background and the rain layer in disentangle image translation framework. Inspired by [33], we flexibly embed the contrastive learning into the disentangle translation network to enable the end-to-end training, which could be beneficial to image disentanglement.

## 3 Proposed Method

### 3.1 Problem Formulation

Let $\mathcal{D}_{unpair} = \left\{\left(I_H^i, I_N^{\pi_i}\right)\right\}_{i=1}^{l}$ be a training dataset for unpaired SID, where the permutation $\pi$ indicates that each pair of the hazy image $I_H^i$ and the clear image

$I_N^{\pi_i}$ does not have any content correspondence. The goal of unpaired SID is to learn a deep model to explore the intrinsic connection based on the unpaired dataset $D_{unpair}$ without the supervision of the ground truth labels to estimate the haze-free images. To achieve the goal, most of the existing disentanglement-based unpaired SID methods empirically construct three joint disentanglement subnetworks under the assumption of atmospheric scattering model. Formally,

$$I_H(x) = I_N(x)t(x) + A(1 - t(x)), \tag{1}$$

where $A$ represents the atmospheric light, and $t(x)$ describes the transmission map on each pixel coordinates. Different from these methods that guide the layer disentanglement by describing the hazing process in image space, we rethink hazy image formation by simplifying the entanglement model itself in feature space. From the perspective of feature distribution learning, it can be formulated as

$$p(I_H) = p(I_N, I_h) = p(I_N)p(I_h \mid I_N), \tag{2}$$

where the distribution of the hazy image $p(I_H)$ is a joint distribution of the clear image $p(I_N)$ (contains task-relevant factor) and haze component $p(I_h)$ (contains task-irrelevant factor). The clean representation can be achieved if we can disentangle task-relevant factor $c_r$ and task-irrelevant factor $c_{ir}$ from $p(I_H)$. Then, the clear images can be recovered with the disentangled task-relevant factor.

To achieve factor disentanglement, the recent contrastive representation learning may open a door for guiding the learning of an unambiguous embedding. Due to the intuitive fact is that the same factor values produce the similar image features related to that factor, we propose to compare the features of the generated images to disentangle the discrete variation of these two-class factors. The details of our proposed framework are described below.

### 3.2 Framework Architecture

Based on above analysis, we formulate contrastive disentanglement in a GAN framework to achieve better unpaired SID performance. Figure 1 shows the overall architecture of our developed contrastive disentangled dehazing method (CDD-GAN). Since the natural advantages of CycleGAN can fully excavates the useful feature properties of unpaired clear images for SID, we introduce a bidirectional disentangled translation network as the backbone of the proposed CDD-GAN. Intuitively, the first half of the generators are presented as encoders while the second half are decoders, and defined as $G_{enc}$ and $F_{enc}$ followed by $G_{dec}$ and $F_{dec}$ respectively. In our framework, two alternately updated paths (*i.e.*, contrastive path and adversarial contrastive path) are playing a minimax game to achieve factor disentanglement in the latent space. Ideally, we note that enhancing task-relevant factors as well as isolating task-irrelevant factors will be a double benefit for building a better SID framework [42]. We will illustrate it with feature visualization in Sect. 4.4. Thus, the advantage of such a contrastive disentanglement design is twofold. First, the isolation of the task-irrelevant factors can reduce the ambiguity of the encoder network representation to guide the

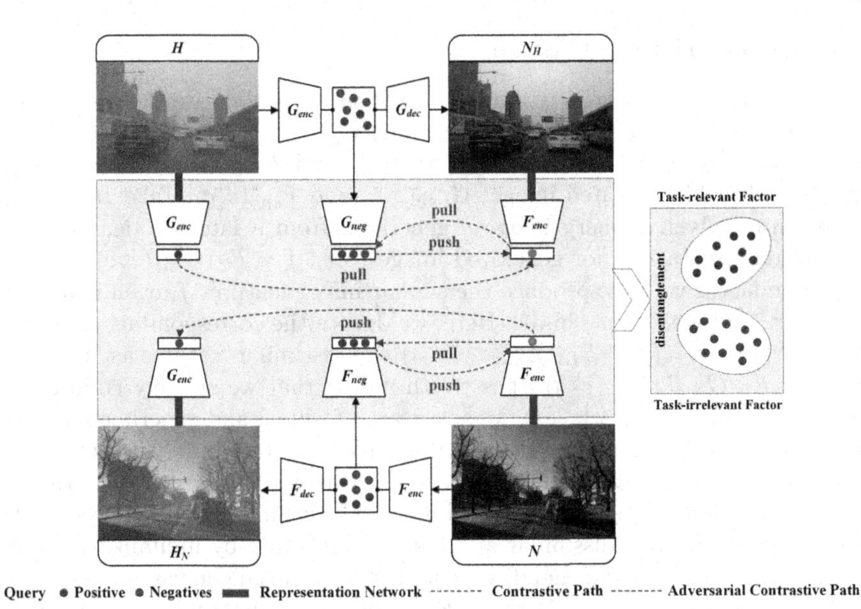


**Fig. 1.** The overview of the proposed contrastive disentangled dehazing method (CDD-GAN). In our framework, two alternately updated paths are playing a minimax game to achieve factor disentanglement in the latent feature space. On one path, the backbone representation network is trained in the conventional contrastive learning. On the other path, the negative generators enforce adversarial contrastive learning to pull negatives to closely track the positive query. Here, we omit two discriminators.

learning of more unambiguous embedding. On the other hand, the enhancement of the task-relevant factors can encourage the capability of the decoder network representation to guide the learning of more accurate mapping.

To capture variability between $p(I_H)$ and $p(I_N)$ in feature space, similar to the dual learning setting in [14], we first extract features of images from the $L$ layers of $G_{enc}$ and $F_{enc}$, and then send them to a two-layer multi-layer perceptron (MLP) representation network $\mathcal{R}$. Compared with the previous method [7,14,34] of randomly sampling negative exemples, we introduce negative generators $G_{neg}$ and $F_{neg}$ to produce more challenging negatives $\mathcal{N}$ based on the embedded features of the image in an adversarial manner, thereby allowing them to closely track the change of representations during the optimization [44]. With the help of the adversarial contrastive loss, these hard negative adversaries are trained end-to-end together with the image encoder network (*i.e.*, the image generator and representation network) to guide the generated images to be associated with latent factors, so that we can facilitate the learned representation to fulfill factor disentanglement during the bidirectional translation process. Finally, the image decoder of CDD-GAN will remove those task-irrelevant factors that are not used to generate the recovered image for generating high-quality dehazed results. The details of the adversarial contrastive loss are illustrated below.

### 3.3 Adversarial Contrastive Loss

To disentangle the discrete variation of two-class factors, we define the latent code consists of two parts: category value $c \in \mathcal{Y} = \{c_r, c_{ir}\}$ and distribution value $z \sim \mathcal{Z}(0,1)$. The generators $G_{enc}$ and $F_{enc}$ take both $c$ and $z$ as inputs and yield generated images $G_{enc}(z,c)$ and $F_{enc}(z,c)$. Take $H \rightarrow N_H$ as an example, given a query image $q$ generated from a latent code, we extract feature representations for generated images, *i.e.*, $f = E_L(G_{enc}(z,c))$. We wish the same factor values $c$ produce the similar image features $f$, even match with various $z$, and vice versa [26,33]. Here, we denote the corresponding similar feature as "positive" $f^+ = E_L(G_{enc}(z^+, c^+))$ and dissimilar features as "negatives" $f^-_{adv,i} = E_L(G_{enc}(z^-_{adv,i}, c^-_i))$. It is worth noting that we actively train a set of negative examples as a whole in an adversarial fashion, and experiments demonstrate our strategy can promote disentanglement performance (see Sect. 4.4).

To be specific, conventional contrastive learning is performed to learn a representation for training network backbone, which aims to pull similar feature distribution and push dissimilar apart in feature space by minimizing the contrastive loss. On the other hand, we conduct adversarial contrastive learning on the $G_{enc}$ and $F_{enc}$ to cooperate with $\mathcal{R}$ to update alternately by maximizing the contrastive loss. With these negative adversaries produced by the $G_{neg}$ and $F_{neg}$, the encoder network will learn more distinguishing representation of the latent factors, which in turn causes the negative exemples to closely track the positive query. In a word, this leads to a minimax problem, that is, training two mutually interacted players (*i.e.*, $\mathcal{R}$ and $\mathcal{N}$) jointly with the adversarial contrastive loss $\mathcal{L}_{ac}$ for CDD-GAN. Mathematically, it takes the form:

$$\mathcal{R}^\star, \mathcal{N}^\star = \arg\min_{\mathcal{R}} \max_{\mathcal{N}} \mathcal{L}_{ac}(\mathcal{R}, \mathcal{N}), \tag{3}$$

where $\mathcal{R}$ and $\mathcal{N}$ will reach an equilibrium by alternate training. Generally, a pair of gradient descent and ascent are applied to alternately update the network parameters $\theta_R$ and $\theta_N$, which are formulated as follow,

$$\theta_R \leftarrow \theta_R - \eta_R \frac{\partial \mathcal{L}_{ac}(\mathcal{R}, \mathcal{N})}{\partial \theta_R}, \tag{4}$$

$$\theta_N \leftarrow \theta_N + \eta_N \frac{\partial \mathcal{L}_{ac}(\mathcal{R}, \mathcal{N})}{\partial \theta_N}, \tag{5}$$

where $\eta_R$ and $\eta_N$ are the positive learning rates for updating the network and negative adversaries. By constraining the contrastive distribution learning with $\mathcal{L}_{ac}$, these representations become well distinguished and can be formulated as

$$\mathcal{L}_{ac} = \mathbb{E}_{\mathcal{D}_{unpair}} \left[ -\log \frac{\text{sim}\,(f, f^+)}{\text{sim}\,(f, f^+) + \sum_{i=1}^{N} \text{sim}\left(f, f^-_{adv,i}\right)} \right], \tag{6}$$

where $\tau$ is the scalar temperature parameter, and $\text{sim}(u,v) = \exp\left(\frac{u^T v}{\|u\|\|v\|\tau}\right)$ is the similarity between the two normalized feature vectors.

### 3.4 Other Objectives

Since ground truths are not available, it is essential to constrain CDD-GAN with several effective loss functions. As well as the adversarial contrastive loss mentioned above, we introduce other objectives to regularize the network training process.

**Diversity Loss.** To encourage the generation of diverse hard negative exemples $\mathcal{N}_i = \{N_0, N_1, \cdots, N_l\}$ in $\mathcal{L}_{ac}$, similar to [44], we introduce the diversity loss by combining different input noises, which is formulated as follows,

$$\mathcal{L}_{div} = -\left\|\mathcal{N}_i\left(\overline{\mathcal{R}}, v_1\right) - \mathcal{N}_i\left(\overline{\mathcal{R}}, v_2\right)\right\|_1, \tag{7}$$

where $\overline{\mathcal{R}}$ denotes the spatially-average features from $\mathcal{R}$, and $v_i$ is noise vector randomly sampled from standard Gaussian distribution.

**Total Variation Loss.** To remove the artifacts in the restored images, we apply the total variation to $N_H$:

$$\mathcal{L}_{tv} = \left\|\partial_h N_H\right\|_1 + \left\|\partial_v N_H\right\|_1, \tag{8}$$

where $\partial_h$ and $\partial_v$ represent the horizontal and vertical gradient operators, respectively.

**Dark Channel Loss.** Inspired by [15,23,40], we also take advantage of the dark channel of clear images, which is written as:

$$D(I) = \min_{y \in N(x)} \left[\min_{c \in \{r,g,b\}} I^c(y)\right], \tag{9}$$

where $x$ and $y$ are pixel coordinates, $N(x)$ is an image patch centered at $x$, and $I^c$ denotes c-th color channel. Thus, we impose dark channel loss to further constrain the sparsity of the dark channel of the dehazed images:

$$\mathcal{L}_{dc} = \left\|D\left(N_H\right)\right\|_1. \tag{10}$$

**Full Objective.** The full objective function for the negative generator and encoder network are as follows:

$$\mathcal{L}_{neg} = -\mathcal{L}_{ac} + \lambda_1 \mathcal{L}_{div}, \tag{11}$$

$$\mathcal{L}_{enc} = \mathcal{L}_{ac} + \lambda_2 \mathcal{L}_{adv} + \lambda_3 \mathcal{L}_{cycle} + \lambda_4 \mathcal{L}_{tv} + \lambda_5 \mathcal{L}_{dc}, \tag{12}$$

where $\lambda_i$ is balance weight, $\mathcal{L}_{adv}$ and $\mathcal{L}_{cycle}$ are the generative adversarial loss and the cycle-consistency loss. Here, we empirically set $\lambda_1 = \lambda_2 = 1$, $\lambda_3 = 10^{-1}$, $\lambda_4 = 10^{-3}$, and $\lambda_5 = 10^{-2}$.

# 4 Experimental Results

## 4.1 Datasets Setup

**SOTS and HSTS.** We conduct experiments on a large-scale benchmark dataset, named REalistic Single Image DEhazing (RESIDE) [19], which consists of two testing sets, SOTS and HSTS. In detail, SOTS has 500 indoor and outdoor hazy images generated using the physical model with manual parameters. HSTS provides a synthetic set and a real-world set, each containing 10 hazy images.

**Foggy Cityscapes.** Sakaridis et al. [39] apply fog simulation on the Cityscapes dataset [9] and generate Foggy Cityscapes with 20,550 images. Here, we select elaborately 4,000 high-quality synthetic hazy-clear images, which contains 3,600 hazy images for training and the remaining 400 ones for evaluation.

## 4.2 Training Details

The developed CDD-GAN is based on CycleGAN [55], a Resnet-based generator with nine residual blocks and a PatchGAN [16] discriminator. The whole framework is implemented using the PyTorch with two Tesla V100 GPUs. We perform the adversarial contrastive learning on the 1-st, 5-th, 9-th, 13-th, 17-th layers of $G_{enc}$ and $F_{enc}$. The number of negative exemples $\mathcal{N}_i$ for contrastive learning is set to 256. The temperature parameter $\tau$ is set to 0.07. We apply the Adam optimizer and the batch size is set to 1 and the models are trained for total 400 epochs. Initially, the proposed network is trained with 0.0001 learning rate for 200 epochs, followed by another 200 epochs with linearly decaying learning rate. $256 \times 256$ patches are randomly cropped from all training images in an unpaired learning procedure.

## 4.3 Comparison Results

We compare our method with those of two prior-based approaches (i.e., DCP [15] and CAP [56]), three paired learning-based models (i.e., MSCNN [37], AODNet [18], and GFN [38]), four unpaired learning-based networks (i.e., CycleGAN [55], DisentGAN [49], SSID [23], and RefineDNet [54]). With the help of the corresponding labels in synthetic datasets, we adopt two evaluation criteria: PSNR and SSIM [45]. To compare real-world hazy cases that lack ground truth, we use the no-reference quality metric NIQE [32].

**Results on Synthetic Datasets.** Table 1 summarizes quantitative values of different approaches on synthetic datasets including SOTS, HSTS, and Foggy Cityscapes. We can notice that our method remarkably outperforms all existing unpaired dehazing nets and achieves state-of-the-art performance. Despite the unsupervised characteristics of our proposed CDD-GAN, it can also deliver comparable results against several paired supervised models, clearly demonstrating that the potential advantages of our proposed contrastive disentanglement

**Table 1.** Comparison of quantitative results on three synthetic datasets. Bold and underline indicate the best and second-best results.

| Datasets | | SOTS | | HSTS | | Cityscapes | |
|---|---|---|---|---|---|---|---|
| Metrics | | PSNR | SSIM | PSNR | SSIM | PSNR | SSIM |
| Prior-based methods | DCP [15] | 16.62 | 0.817 | 14.84 | 0.761 | 15.09 | 0.795 |
| | CAP [56] | 19.05 | 0.836 | 21.53 | 0.872 | 17.34 | 0.844 |
| Paired/supervised methods | MSCNN [37] | 17.57 | 0.810 | 18.64 | 0.817 | 17.98 | 0.828 |
| | AOD-Net [18] | 19.06 | 0.850 | 20.55 | 0.897 | 18.51 | 0.836 |
| | GFN [38] | 22.30 | 0.884 | 21.87 | 0.893 | 19.69 | 0.857 |
| Unpaired/without paired supervised methods | CycleGAN [55] | 17.78 | 0.725 | 18.52 | 0.831 | 17.82 | 0.812 |
| | DisentGAN [49] | 22.12 | 0.899 | 19.68 | 0.866 | 18.66 | 0.837 |
| | SSID [23] | 24.44 | 0.896 | 21.83 | 0.882 | 19.50 | 0.841 |
| | RefineDNet [54] | 24.39 | 0.912 | 21.69 | 0.904 | 20.24 | 0.866 |
| | **Ours** | **24.61** | **0.918** | **22.16** | **0.911** | **20.93** | **0.874** |

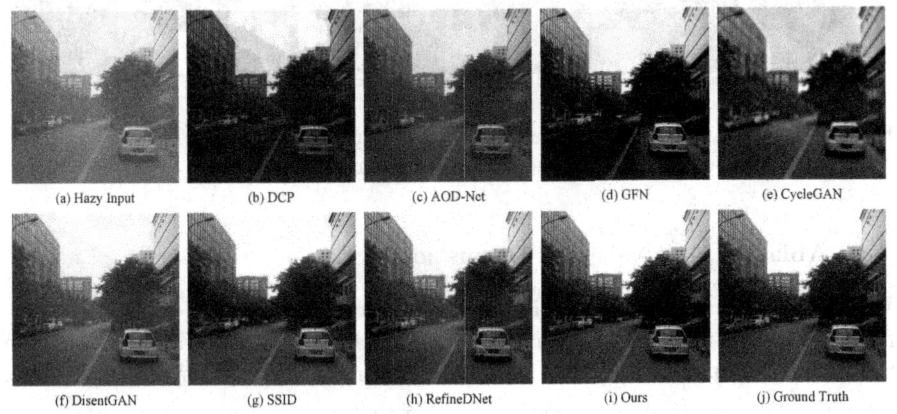


**Fig. 2.** Comparison of qualitative results on the SOTS synthetic dataset.

framework. Besides the quantitative results, we further present visual observation comparisons in Fig. 2 and Fig. 3. All the competitive methods contain more haze residue and obtain unsatisfactory results in detail restoration, which keep consistent with the above quantitative scores. In contrast, the proposed method generates much clearer results that are visually close to the ground truth.

**Results on Real-World Datasets.** To demonstrate the effectiveness of our dehazing model on real hazy images, we conduct comparisons against other algorithms on the HSTS real-world image set and present results in Fig. 4. According the values of NIQE under the images, the proposed method obtains the lowest score, which indicates a high-quality dehazed result with better fidelity and higher naturalness. This benefits from the fact that the decoder of CDD-GAN can reconstruct high-quality outputs with the help of factor disentanglement.

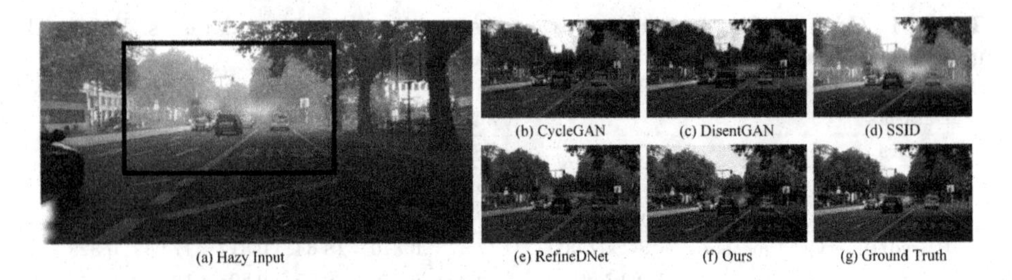


**Fig. 3.** Comparison of qualitative results on the Foggy Cityscapes synthetic dataset.

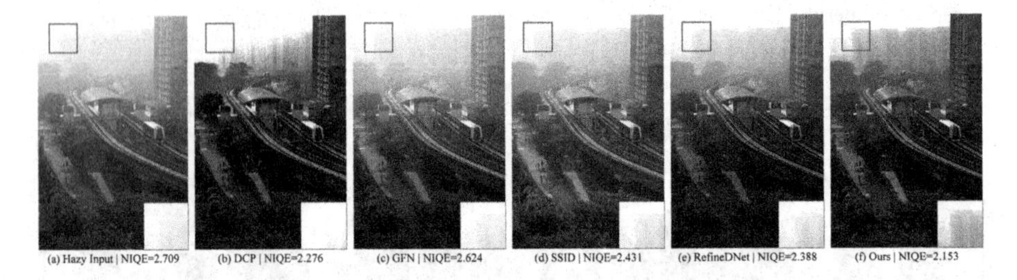


**Fig. 4.** Comparison of qualitative and quantitative results on the HSTS real-world set. Note that lower values of NIQE indicate better image quality.

## 4.4 Ablation Analysis and Discussion

We study the main component impacts and parameter choices on the final performance. To ensure the fair comparison, all the ablation studies are performed in the same environment and training settings using the Foggy Cityscapes dataset.

**Effectiveness of Negative Generator.** To investigate the impact of the proposed negative generator, we consider two variants of our framework, including (a) without contrastive loss, and (b) the developed adversarial contrastive loss $\mathcal{L}_{ac}$ is replaced by conventional contrastive loss $\mathcal{L}_{con}$ in [14,34]. Table 2 reports the quantitative results of different models. Obviously, contrastive learning can bring great performance gain to the baseline model (a), which shows its potential in unsupervised vision tasks. By comparing model (b) and model (c) in Table 2, it reveals that the design of negative generator is more effective than the previous strategy of generating negatives by randomly sampling from the images. To better understand the influence of the negative adversaries in $\mathcal{L}_{ac}$, we further use t-SNE [31] to visualize learned features in Fig. 5. As can be seen, conventional contrastive method is not enough to fulfill factor disentanglement, because their negatives are not effective to push the positives close to the query examples. By contrast, our strategy can produce more discriminative representations, thanks to the challenging negative adversaries provided by negative generators. In such case, the task-irrelevant factors are suppressed by generating hard negative exemples to guide the dehazing process and clear image reconstruction.

**Table 2.** Ablation study for different components and designs. PSNR and SSIM results among different models of CDD-GAN on the Foggy Cityscapes dataset. Note that $\mathcal{L}_{con}$ indicates general contrastive loss [34] and $S_{dual}$ indicates dual learning setting.

| Models | $\mathcal{L}_{ac}$ | $\mathcal{L}_{con}$ | $\mathcal{L}_{div}$ | $\mathcal{L}_{tv}$ | $\mathcal{L}_{dc}$ | $\mathcal{L}_{adv}$ | $\mathcal{L}_{cycle}$ | $S_{dual}$ | PSNR/SSIM |
|---|---|---|---|---|---|---|---|---|---|
| (a) | × | × | × | × | × | ✓ | ✓ | × | 18.64/0.830 |
| (b) | × | ✓ | × | × | × | ✓ | ✓ | ✓ | 19.88/0.846 |
| (c) | ✓ | × | × | × | × | ✓ | ✓ | ✓ | 20.35/0.843 |
| (d) | ✓ | × | ✓ | × | × | ✓ | ✓ | ✓ | 20.71/0.865 |
| (e) | ✓ | × | ✓ | ✓ | × | ✓ | ✓ | ✓ | 20.79/0.871 |
| (f) | ✓ | × | ✓ | ✓ | ✓ | ✓ | ✓ | ✓ | **20.93/0.874** |
| (g) | ✓ | × | ✓ | ✓ | ✓ | ✓ | ✓ | × | 20.72/0.863 |

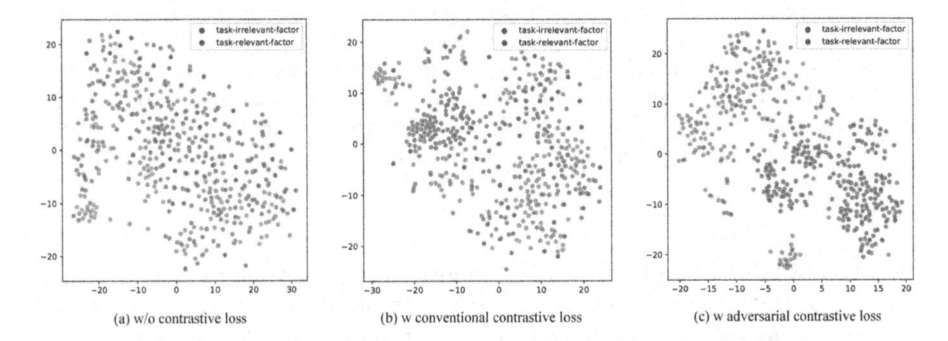


**Fig. 5.** The t-SNE visualization of features learned for task-relevant (blue round point) and task-irrelevant (red round point) factors. With the adversarial contrastive loss, the same factors are pulled closer and gathered together in the latent space. Hence, isolating the task-irrelevant factors by disentanglement is able to generate clear images. (Color figure online)

**Effectiveness of Other Objectives.** To better demonstrate the effectiveness of other objective function, we also conduct an ablation study by considering the combinations of the diversity loss $\mathcal{L}_{div}$, total variation loss $\mathcal{L}_{tv}$, and dark channel loss $\mathcal{L}_{dc}$. Since our framework is based on CycleGAN, the generative adversarial loss $\mathcal{L}_{adv}$ and cycle-consistency loss $\mathcal{L}_{cycle}$ are the default common items, which will not be discussed here. Correspondingly, we regularly add one component to each configuration at one time. By comparing model (c) and model (d) in Table 2, there is a sharp decline in performance without $\mathcal{L}_{div}$, especially on the metric of PSNR. This is because the lack of $\mathcal{L}_{div}$ will cause the negative exemples to maintain less diversity in the training stage. Under this case, the negative generator fails to produce challenging negative exemples, resulting in suboptimal performance. With the combination of all objectives, our model (f) can achieve the best performance, which also demonstrates that each loss term contributes in its own way during dehazing process.

**Table 3.** Ablation study for different number of negative exemples. PSNR and SSIM results among different settings of CDD-GAN on the Foggy Cityscapes dataset.

| Number of $\mathcal{N}_i$ | $N = 64$ | $N = 128$ | $N = 256$ (default) | $N = 512$ |
|---|---|---|---|---|
| PSNR/SSIM | 19.84/0.859 | 20.52/0.868 | **20.93**/0.874 | 20.85/**0.876** |

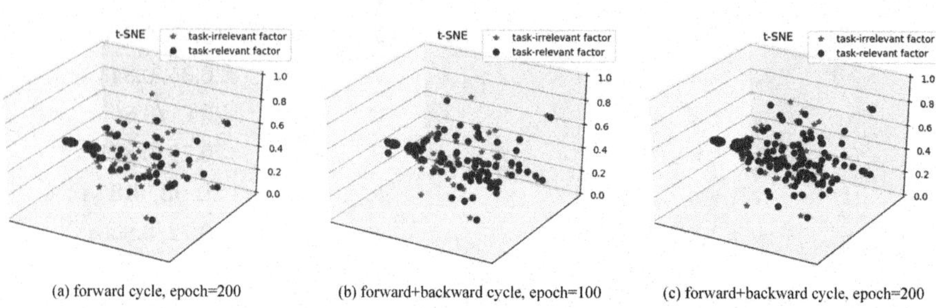


(a) forward cycle, epoch=200 (b) forward+backward cycle, epoch=100 (c) forward+backward cycle, epoch=200

**Fig. 6.** The t-SNE visualization of features learned for task-relevant (blue round point) and task-irrelevant (red pentagram) factors. With the forward and backward dual-path cycle, more task-relevant factors are produced during the optimization. Thus, removing haze can be facilitated by utilizing the features from the unpaired clear exemplars. (Color figure online)

**Effectiveness of Dual Setting.** We remove the dual setting $S_{dual}$ for comparison, see model (f) and model (g) in Table 2. It can be seen that $S_{dual}$ achieves dehazing performance improvement, due to its ability to stabilize the training and learn better embeddings for different domains.

**Influences of Unpaired Clear Exemples.** Unlike [34,44], we extend unidirectional mapping to bidirectional mapping, which is more suitable for SID task because it can take advantage of the characteristics of backward cycle. We visualize the corresponding features using t-SNE [31] in Fig. 6. It can be observed that the features from unpaired clear exemples generated by the backward cycle can continuously enhance the task-relevant factors during the optimization.

**Number of Negative Exemples.** We study the different number influences of negative exemples in Table 3. For one thing, too few negatives may weaken the ability to pull the positives closer to the query. For another, too many negatives may increase the computation cost and produce unnecessary interference. To balance the model performances and memory, we choose $N = 256$ as the default.

### 4.5 Other Applications

**Generality to Other Low-Level Vision Tasks.** It is a general assumption to regard the degraded image as the entanglement of task-relevant factor and task-irrelevant factor, so our method can be easily applied to similar vision tasks, such as image denoising and deraining. Here, we provide one deraining example in Fig. 7 for comparison. Surprisingly, our model even outperforms recent unsupervised deraining method [46] with more than 5 dB in PSNR. This is because, previous approaches learn complex mapping in the high-dimensional image space, while our method learns latent restoration in the low-dimensional feature space.

**Table 4.** Comparison of object detection quantitative results on the RTTS dataset.

| | Hazy input | CycleGAN | DisentGAN | SSID | RefineDNet | Ours |
|---|---|---|---|---|---|---|
| mAP(%) | 61.35 | 53.82 | 63.59 | 61.74 | 65.22 | **66.04** |
| Gain | - | −7.53 | +2.24 | +0.39 | +3.87 | **+4.69** |

**Fig. 7.** Comparison of qualitative and quantitative results on the Rain800 synthetic dataset [53].

**Preprocessing for High-Level Vision Tasks.** As suggested in [8,41,48], SID has become a frequently-used preprocessing step, so we further examine whether our method bring benefits to downstream high-level vision tasks. Here, we randomly select 100 real hazy images from RTTS [19] with bounding boxes and object categories. We adopt YOLOv3 [36] to evaluate object detection performance and then calculate the mean Average Precision (mAP). As can be seen from Table 4, the quantitative gain of CycleGAN is negative. This is because it is not completely developed for SID task, which leads to the destruction of the semantic information of the original image. Compared with other unsupervised models, our dehazed results bring higher recognition accuracy for object detection, which further demonstrates the effectiveness of our designed CDD-GAN.

### 4.6 Limitations

In this study, we have three limitations: 1) For $\mathcal{L}_{ac}$, no theory can guarantee convergence to the saddle point, so our method only achieves an approximate equilibrium by updating $\mathcal{R}$ and $\mathcal{N}$ alternately. 2) Due to the common inference of GAN and contrastive learning, the network training is not stable. 3) Our method fails to deal with heavy fog scenes.

## 5 Conclusions

This paper provides a new contrastive disentangled dehazing method (CDD-GAN) to address the challenging unpaired SID problem. To fulfill factor disentanglement in latent feature space, negative generators are introduced into the CycleGAN framework to isolate task-irrelevant factors with the help of adversarial contrastive loss. On the other hand, the features from unpaired clear exemples are utilized for enhancing the learning of task-relevant factors. Extensive experiments considerably show that the effectiveness and scalability of our model. In future work, we plan to explore the possibility of applying contrastive learning for more complex factor disentangling problem in the field of low-level vision.

## References


1. Anvari, Z., Athitsos, V.: Dehaze-GLCGAN: unpaired single image de-hazing via adversarial training. arXiv preprint arXiv:2008.06632 (2020)
2. Cai, B., Xu, X., Jia, K., Qing, C., Tao, D.: DehazeNet: an end-to-end system for single image haze removal. IEEE TIP **25**(11), 5187–5198 (2016)
3. Chang, C.M., Sung, C.S., Lin, T.N.: DAMix: density-aware data augmentation for unsupervised domain adaptation on single image dehazing. arXiv preprint arXiv:2109.12544 (2021)
4. Chen, D., et al.: Gated context aggregation network for image dehazing and deraining. In: WACV, pp. 1375–1383. IEEE (2019)
5. Chen, T., Kornblith, S., Norouzi, M., Hinton, G.: A simple framework for contrastive learning of visual representations. In: ICML, pp. 1597–1607. PMLR (2020)
6. Chen, X., Duan, Y., Houthooft, R., Schulman, J., Sutskever, I., Abbeel, P.: Info-GAN: interpretable representation learning by information maximizing generative adversarial nets. In: NIPS, pp. 2180–2188 (2016)
7. Chen, X., et al.: Unpaired deep image deraining using dual contrastive learning. In: CVPR, pp. 2017–2026 (2022)
8. Chen, Z., Wang, Y., Yang, Y., Liu, D.: PSD: principled synthetic-to-real dehazing guided by physical priors. In: CVPR, pp. 7180–7189 (2021)
9. Cordts, M., et al.: The cityscapes dataset for semantic urban scene understanding. In: CVPR, pp. 3213–3223 (2016)
10. Dong, H., et al.: Multi-scale boosted dehazing network with dense feature fusion. In: CVPR, pp. 2157–2167 (2020)
11. Dudhane, A., Murala, S.: CDNet: single image de-hazing using unpaired adversarial training. In: WACV, pp. 1147–1155. IEEE (2019)
12. Engin, D., Genç, A., Kemal Ekenel, H.: Cycle-dehaze: enhanced CycleGAN for single image dehazing. In: CVPRW, pp. 825–833 (2018)
13. Esser, P., Rombach, R., Ommer, B.: A disentangling invertible interpretation network for explaining latent representations. In: CVPR, pp. 9223–9232 (2020)
14. Han, J., Shoeiby, M., Petersson, L., Armin, M.A.: Dual contrastive learning for unsupervised image-to-image translation. In: CVPR, pp. 746–755 (2021)
15. He, K., Sun, J., Tang, X.: Single image haze removal using dark channel prior. IEEE TPAMI **33**(12), 2341–2353 (2010)
16. Isola, P., Zhu, J.Y., Zhou, T., Efros, A.A.: Image-to-image translation with conditional adversarial networks. In: CVPR, pp. 1125–1134 (2017)

17. Jin, Y., Gao, G., Liu, Q., Wang, Y.: Unsupervised conditional disentangle network for image dehazing. In: ICIP, pp. 963–967. IEEE (2020)
18. Li, B., Peng, X., Wang, Z., Xu, J., Feng, D.: AOD-Net: all-in-one dehazing network. In: ICCV, pp. 4770–4778 (2017)
19. Li, B., et al.: Benchmarking single-image dehazing and beyond. IEEE TIP **28**(1), 492–505 (2018)
20. Li, B., Gou, Y., Gu, S., Liu, J.Z., Zhou, J.T., Peng, X.: You only look yourself: unsupervised and untrained single image dehazing neural network. IJCV **129**(5), 1754–1767 (2021)
21. Li, B., Gou, Y., Liu, J.Z., Zhu, H., Zhou, J.T., Peng, X.: Zero-shot image dehazing. IEEE TIP **29**, 8457–8466 (2020)
22. Li, B., Lin, Y., Liu, X., Hu, P., Lv, J., Peng, X.: Unsupervised neural rendering for image hazing. arXiv preprint arXiv:2107.06681 (2021)
23. Li, L., et al.: Semi-supervised image dehazing. IEEE TIP **29**, 2766–2779 (2019)
24. Li, R., Pan, J., Li, Z., Tang, J.: Single image dehazing via conditional generative adversarial network. In: CVPR, pp. 8202–8211 (2018)
25. Liu, C., Fan, J., Yin, G.: Efficient unpaired image dehazing with cyclic perceptual-depth supervision. arXiv preprint arXiv:2007.05220 (2020)
26. Liu, R., Ge, Y., Choi, C.L., Wang, X., Li, H.: DivCo: diverse conditional image synthesis via contrastive generative adversarial network. In: CVPR, pp. 16377–16386 (2021)
27. Liu, X., Ma, Y., Shi, Z., Chen, J.: GridDehazeNet: attention-based multi-scale network for image dehazing. In: ICCV, pp. 7314–7323 (2019)
28. Liu, Y., Anwar, S., Qin, Z., Ji, P., Caldwell, S., Gedeon, T.: Disentangling noise from images: a flow-based image denoising neural network. arXiv preprint arXiv:2105.04746 (2021)
29. Liu, Y., Pan, J., Ren, J., Su, Z.: Learning deep priors for image dehazing. In: ICCV, pp. 2492–2500 (2019)
30. Liu, Y., et al.: From synthetic to real: image dehazing collaborating with unlabeled real data. arXiv preprint arXiv:2108.02934 (2021)
31. Van der Maaten, L., Hinton, G.: Visualizing data using t-SNE. JMLR **9**(11), 2579–2605 (2008)
32. Mittal, A., Moorthy, A.K., Bovik, A.C.: No-reference image quality assessment in the spatial domain. IEEE TIP **21**(12), 4695–4708 (2012)
33. Pan, L., Tang, P., Chen, Z., Xu, Z.: Contrastive disentanglement in generative adversarial networks. arXiv preprint arXiv:2103.03636 (2021)
34. Park, T., Efros, A.A., Zhang, R., Zhu, J.-Y.: Contrastive learning for unpaired image-to-image translation. In: Vedaldi, A., Bischof, H., Brox, T., Frahm, J.-M. (eds.) ECCV 2020. LNCS, vol. 12354, pp. 319–345. Springer, Cham (2020). https://doi.org/10.1007/978-3-030-58545-7_19
35. Qin, X., Wang, Z., Bai, Y., Xie, X., Jia, H.: FFA-Net: feature fusion attention network for single image dehazing. In: AAAI, vol. 34, pp. 11908–11915 (2020)
36. Redmon, J., Farhadi, A.: YOLOv3: an incremental improvement. arXiv preprint arXiv:1804.02767 (2018)
37. Ren, W., Liu, S., Zhang, H., Pan, J., Cao, X., Yang, M.-H.: Single image dehazing via multi-scale convolutional neural networks. In: Leibe, B., Matas, J., Sebe, N., Welling, M. (eds.) ECCV 2016. LNCS, vol. 9906, pp. 154–169. Springer, Cham (2016). https://doi.org/10.1007/978-3-319-46475-6_10
38. Ren, W., et al.: Gated fusion network for single image dehazing. In: CVPR, pp. 3253–3261 (2018)

39. Sakaridis, C., Dai, D., Van Gool, L.: Semantic foggy scene understanding with synthetic data. IJCV **126**(9), 973–992 (2018). Sep
40. Shao, Y., Li, L., Ren, W., Gao, C., Sang, N.: Domain adaptation for image dehazing. In: CVPR, pp. 2808–2817 (2020)
41. VidalMata, R.G., Banerjee, S., RichardWebster, B., Albright, M., Davalos, P., McCloskey, S., Miller, B., Tambo, A., Ghosh, S., Nagesh, S., et al.: Bridging the gap between computational photography and visual recognition. IEEE TPAMI **43**(12), 4272–4290 (2020)
42. Wang, G., Sun, C., Xu, X., Li, J., Wang, Z., Ma, Z.: Disentangled representation learning and enhancement network for single image de-raining. In: ACM MM, pp. 3015–3023 (2021)
43. Wang, L., Wang, Y., Dong, X., Xu, Q., Yang, J., An, W., Guo, Y.: Unsupervised degradation representation learning for blind super-resolution. In: CVPR, pp. 10581–10590 (2021)
44. Wang, W., Zhou, W., Bao, J., Chen, D., Li, H.: Instance-wise hard negative example generation for contrastive learning in unpaired image-to-image translation. In: ICCV, pp. 14020–14029 (2021)
45. Wang, Z., Bovik, A.C., Sheikh, H.R., Simoncelli, E.P.: Image quality assessment: from error visibility to structural similarity. IEEE TIP **13**(4), 600–612 (2004)
46. Wei, Y., et al.: DerainCycleGAN: rain attentive CycleGAN for single image deraining and rainmaking. IEEE TIP **30**, 4788–4801 (2021)
47. Wu, H., et al.: Contrastive learning for compact single image dehazing. In: CVPR, pp. 10551–10560 (2021)
48. Yang, W., et al.: Advancing image understanding in poor visibility environments: a collective benchmark study. IEEE TIP **29**, 5737–5752 (2020)
49. Yang, X., Xu, Z., Luo, J.: Towards perceptual image dehazing by physics-based disentanglement and adversarial training. In: AAAI, vol. 32 (2018)
50. Ye, Y., Chang, Y., Zhou, H., Yan, L.: Closing the loop: joint rain generation and removal via disentangled image translation. In: CVPR, pp. 2053–2062 (2021)
51. Yi, X., Ma, B., Zhang, Y., Liu, L., Wu, J.: Two-step image dehazing with intra-domain and inter-domain adaptation. arXiv preprint arXiv:2102.03501 (2021)
52. Zhang, H., Patel, V.M.: Densely connected pyramid dehazing network. In: CVPR, pp. 3194–3203 (2018)
53. Zhang, H., Sindagi, V., Patel, V.M.: Image de-raining using a conditional generative adversarial network. IEEE TCSVT **30**(11), 3943–3956 (2019)
54. Zhao, S., Zhang, L., Shen, Y., Zhou, Y.: RefinedNet: a weakly supervised refinement framework for single image dehazing. IEEE TIP **30**, 3391–3404 (2021)
55. Zhu, J.Y., Park, T., Isola, P., Efros, A.A.: Unpaired image-to-image translation using cycle-consistent adversarial networks. In: ICCV, pp. 2223–2232 (2017)
56. Zhu, Q., Mai, J., Shao, L.: A fast single image haze removal algorithm using color attenuation prior. IEEE TIP **24**(11), 3522–3533 (2015)

# Efficient Long-Range Attention Network for Image Super-Resolution

Xindong Zhang[1,2], Hui Zeng[2], Shi Guo[1], and Lei Zhang[1,2](✉)

[1] Department of Computing, The Hong Kong Polytechnic University, Hung Hom, Hong Kong
{csxdzhang,csshiguo,cslzhang}@comp.polyu.edu.hk
[2] OPPO Research, Chengdu, China

**Abstract.** Recently, transformer-based methods have demonstrated impressive results in various vision tasks, including image super-resolution (SR), by exploiting the self-attention (SA) for feature extraction. However, the computation of SA in most existing transformer based models is very expensive, while some employed operations may be redundant for the SR task. This limits the range of SA computation and consequently limits the SR performance. In this work, we propose an efficient long-range attention network (ELAN) for image SR. Specifically, we first employ shift convolution (shift-conv) to effectively extract the image local structural information while maintaining the same level of complexity as $1 \times 1$ convolution, then propose a group-wise multi-scale self-attention (GMSA) module, which calculates SA on non-overlapped groups of features using different window sizes to exploit the long-range image dependency. A highly efficient long-range attention block (ELAB) is then built by simply cascading two shift-conv with a GMSA module, which is further accelerated by using a shared attention mechanism. Without bells and whistles, our ELAN follows a fairly simple design by sequentially cascading the ELABs. Extensive experiments demonstrate that ELAN obtains even better results against the transformer-based SR models but with significantly less complexity. The source codes of ELAN can be found at https://github.com/xindongzhang/ELAN.

**Keywords:** Super-resolution · Long-range attention · Transformer

## 1 Introduction

Singe image super-resolution (SR) aims at reproducing a high-resolution (HR) output from its degraded low-resolution (LR) counterpart. In recent years, deep convolutional neural network (CNN) [24] based SR models [6,11,12,30,47,48,51,

X. Zhang and H. Zeng—Have contributed equally.

This work is supported by the Hong Kong RGC RIF grant (R5001-18) and the PolyU-OPPO Joint Innovation Lab.

**Supplementary Information** The online version contains supplementary material available at https://doi.org/10.1007/978-3-031-19790-1_39.

© The Author(s), under exclusive license to Springer Nature Switzerland AG 2022
S. Avidan et al. (Eds.): ECCV 2022, LNCS 13677, pp. 649–667, 2022.
https://doi.org/10.1007/978-3-031-19790-1_39

55,65] have become prevalent for their strong capability in recovering or generating [25,56] image high-frequency details, showing high practical value in image and video restoration, transition and display. However, many CNN-based methods extract local features with spatially invariant kernels, which are inflexible to adaptively model the relations among pixels. In addition, to enlarge the receptive field, the CNN-based SR models tend to employ very deep and complicated network topology [10,41,42,65,69] to recover more details, resulting in much computational resource consumption.

Recently, the transformer based methods have shown impressive performance on natural language processing tasks [4,14,34,44] for their strong ability on modeling the self-attention (SA) of input data. The great success of transformers on NLP has inspired researchers to explore their application to computer vision tasks, and interesting results have been shown on several high-level vision tasks [13,27,35,36,45,54]. Generally speaking, the input image is first divided into non-overlapped patches with different scales as tokens, and then the local feature extraction and SA modules are applied to the collection of patches. Though SA has proven to be effective for high-level vision tasks, its complexity grows quadratically with the input feature size, limiting its utilization in low-level vision tasks such as SR, where the feature size is usually very large.

A few attempts have been made to reduce the computational cost of SA in the application of SR. Mei *et al.* [41] divided an image into non-overlapped patches for modeling local feature and SA independently, which however may introduce border effect and deteriorate the visual quality of restored images. SwinIR [29] follows the design of Swin Transformer [36], where the local feature is first extracted by two cascaded $1 \times 1$ convolutions and then SA is calculated within a small sized window (i.e., $8 \times 8$) with a shifting mechanism to build connections with other windows. However, local features extracted by $1 \times 1$ convolutions with small receptive field may produce weak feature representations to calculate the SA for long-range modeling. Furthermore, calculating SA with a small window size restricts the ability of modeling long-range dependency among image pixels. Restormer [61] calculates the SA with dependency on the channel space which remains applicable to large images. However, modeling dependency on channel space may sacrifice some useful spatial information of textures and structures which is important for reproducing high quality HR images.

In this paper, we aim to develop an effective and efficient way to exploit image long-range attention for image SR with a simple network architecture. The existing transformer-based models such as SwinIR [29] have many redundant and fragmented components which are not cost-effective for the SR task, such as relative position bias, masking mechanism, layer normalization, and several sub-branches created with residual shortcut. We therefore aim to build a highly neat and efficient SR model, where the LR to HR image mapping is simply built by stacking local feature extraction operations and SA sequentially. Two successive shift convolution (shift-conv) operations are used to efficiently extract local structural information, while a shift-conv has larger receptive field but shares the same arithmetic complexity as a $1 \times 1$ convolution. For SA calculation, we propose a group-wise multi-scale self-attention (GMSA) operator, which divides features into different groups of different window sizes and calculates SA separately. This strategy provides a larger receptive field than a small fixed-size window for long-range attention modeling, while being more flexible than a large fixed-size window for reducing the computation resource cost. Furthermore, a shared attention mechanism is proposed to accelerate the calculation for successive GMSA modules. Arming with the shift-conv and GMSA with shared attention, an efficient long-range attention network, namely ELAN, is readily obtained for SR. Our contributions are summarized as follows:

1) We propose an efficient long-range attention block to model image long-range dependency, which is important for improving the SR performance.
2) We present a fairly simple yet powerful network, namely ELAN, which records new state-of-the arts for image SR with significantly less complexity than existing vision transformer based SR methods.

## 2 Related Work

Numerous deep learning based image SR methods have been developed in the past decade. Here we briefly discuss the related work from the perspective of CNN-based methods and transformer-based methods.

### 2.1 CNN-Based SR Methods

CNN-based methods have demonstrated impressive performance in the SR task. SRCNN [11] makes the first attempt to employ CNN for image SR by learning a non-linear mapping from the bicubically upsampled LR image to the HR output with only three convolution layers. Kim *et al.* [21] deepened the network with VGG-19 and residual learning and achieved much better performance. Since the pre-upsampling strategy increases the amount of input data to CNN and causes large computational cost, FSRCNN [12] adopts a post-upsampling strategy to accelerate the CNN model. An enhanced residual block was proposed in [30] to train the deep model without batch normalization, and the developed EDSR network won the first prize of the NTIRE2017 challenge [1].

To build more effective models for SR, the recently developed methods tend to employ deeper and more complicated architectures as well as the attention techniques. Zhang *et al.* proposed a residual-in-residual structure coupled with channel attention to train a very deep network over 400 layers. Other works like MemNet [50] and RDN [67] are designed by employing the dense blocks [17] to utilize the intermediate features from all layers. In addition to increasing the depth of network, some other works, such as SAN [10], NLRN [31], HAN [42] and NLSA [41], excavate the feature correlations along the spatial or channel dimension to boost the SR performance. Our proposed ELAN takes the advantage of fast local feature extraction, while models the long-range dependency of features via efficient group-wise multi-scale self-attention.

### 2.2 Transformer-Based SR Methods

The breakthrough of transformer networks in natural language processing (NLP) inspired of use of self-attention (SA) in computer vision tasks. The SA mechanism in transformers can effectively model the dependency across data, and it has achieved impressive results on several high-level vision tasks, such as image classification [13,27,35,36,45,54], image detection [8,33,36,54], and segmentation [7,36,59,68]. Very recently, transformer has also been applied to low-level vision tasks [9,29,57,61]. IPT [9] is an extremely large pre-trained model for various low-level vision tasks based on the standard vision transformer. It computes both the local feature and SA on non-overlapped patches, which however may lose some useful information for reproducing image details. SwinIR [29] hence adapts the Swin Transformer [36] to image restoration, which combines the advantages of both CNNs and transformers. Though SwinIR has achieved impressive results for image SR, its network structure is mostly borrowed from the Swin Transformer, which is designed for high-level vision tasks. In particular, the network design of SwinIR is redundant for the SR problem, and it calculates SA on small fixed-size windows, preventing it from exploiting long-range feature dependency. Our proposed ELAN is not only much more efficient than SwinIR but also able to compute the SA in larger windows.

## 3 Methodology

In this section we first present the pipeline of our efficient long-range attention network (ELAN) for SR tasks, and then discuss in detail its key component, the efficient long-range attention block (ELAB).

### 3.1 Overall Pipeline of ELAN

The overall pipeline of ELAN is shown in Fig. 1(a), which consists of three modules: shallow feature extraction, ELAB based deep feature extraction, and HR image reconstruction. The network follows a fairly simple topology with a global shortcut connection from the shallow feature extraction module to the output of deep feature extraction module before fed into the HR reconstruction module. In specific, given a degraded LR image $X_l \in \mathbb{R}^{3\times H\times W}$, where $H$ and $W$ are the height and width of the LR image, respectively, we first apply the shallow feature extraction module, denoted by $H_{SF}(\cdot)$, which consists of only a single $3\times 3$ convolution, to extract the local feature $X_s \in \mathbb{R}^{C\times H\times W}$:

$$X_s = H_{SF}(X_l) \tag{1}$$

where $C$ is the channel number of the intermediate feature.

$X_s$ then goes to the deep feature extraction module, denoted by $H_{DF}(\cdot)$, which is composed of $M$ cascaded ELABs. That is:

$$X_d = H_{DF}(X_s), \tag{2}$$

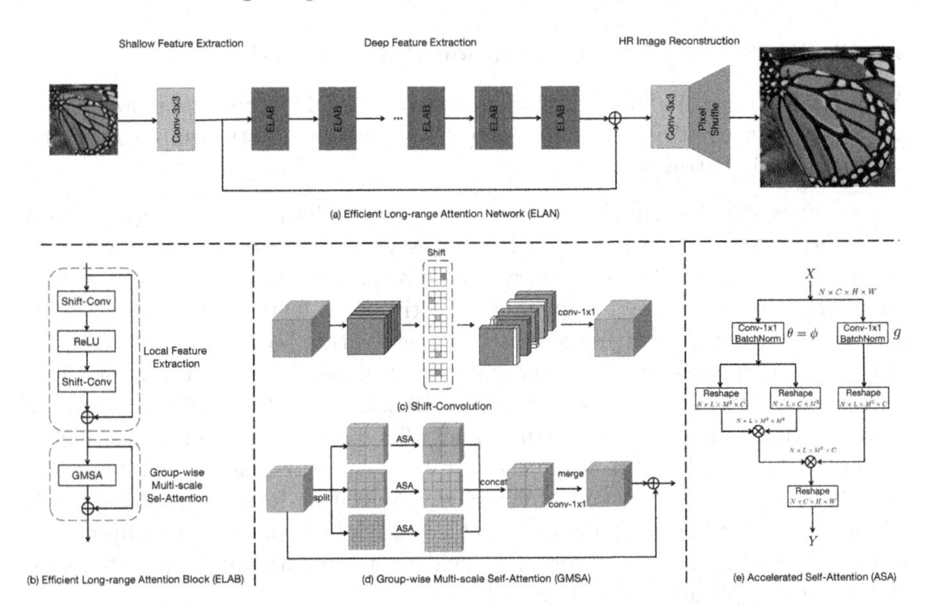


**Fig. 1.** Illustration of the proposed efficient long-range attention network (ELAN). (a) The overall pipeline of ELAN, which contains several ELABs, two $3 \times 3$ convolutions and one pixel shuffle operator. (b) The architecture of the efficient long-range attention block (ELAB). (c) Illustration of shift-conv, which is composed of a shift operation followed by one $1 \times 1$ convolution. (d) Illustration of the computation of group-wise multi-scale self-attention (GMSA). (e) Illustration of our accelerated self-attention (ASA) computation.

where $X_d \in \mathbb{R}^{C \times H \times W}$ denotes the output. By taking $X_d$ and $X_s$ as inputs, the HR image $X_h$ is reconstructed as:

$$X_h = H_{RC}(X_s + X_d), \tag{3}$$

where $H_{RC}$ is the reconstruction module. There are some choices for the design of the reconstruction module [11,12,30,48]. To achieve high efficiency, we build it simply with a single $3 \times 3$ convolution and a pixel shuffle operation.

The ELAN can be optimized with the commonly used loss functions for SR, such as $L_2$ [11,21,49,50], $L_1$ [23,30,67] and perceptual losses [17,46]. For simplicity, given a number of $N$ ground-truth HR images $\{X_{t,i}\}_{i=1}^{N}$, we optimize the parameters of ELAN by minimizing the pixel-wise $L_1$ loss:

$$\mathcal{L} = \frac{1}{N} \sum_{i=1}^{N} ||X_{h,i} - X_{t,i}||_1. \tag{4}$$

The Adam optimizer [22] is employed to optimize our ELAN for its good performance in low-level vision tasks.

### 3.2 Efficient Long-Range Attention Block (ELAB)

As shown in Fig. 1(b), our ELAB is composed of a local feature extraction module and a group-wise multi-scale attention (GMSA) module, both equipped with the residual learning strategy.

**Local Feature Extraction.** Given the intermediate features $X$, previous researches mostly extract the local features via multi-layer perception or two cascaded $1 \times 1$ convolutions [9,35], which however have only $1 \times 1$ receptive field. To enlarge the receptive field for more effective local feature extraction, we utilize two shift-conv [58] with a simple ReLU activation between them. As shown in Fig. 1(c), the shift-conv is composed of a set of shift operations and one $1 \times 1$ convolution. Specifically, we split the input feature equally into five groups, and move the first four groups of feature along different spatial dimensions, including left, right, top, bottom, while the last group remains unchanged. Therefore, the followed $1 \times 1$ convolution can leverage information from neighboring pixels. Without introducing additional learnable parameters and much computation, shift-conv can provide larger receptive fields while maintaining almost the same arithmetic complexity as $1 \times 1$ convolution.

**Group-Wise Multi-scale Self-attention (GMSA).** Given a feature map of $C \times H \times W$, the computational complexity of the window-based self-attention [29,36] using $M \times M$ non-overlapped windows is $2M^2HWC$. The window size $M$ determines the range of SA calculation, and a larger $M$ contributes to exploit more self-similarity information. However, directly enlarging $M$ will quadratically increase the computational cost and resources. To more efficiently calculate the long-range SA, we propose the GMSA module, which is illustrated in Fig. 1(d). We first split the input feature $X$ into $K$ groups, denoted by $\{X_k\}_{k=1}^{K}$, then calculate SA on the $k$-th group of features using window size $M_k$. In this way, we can flexibly control the computational cost by setting the ratio of different window size. For example, supposing the $K$ groups of features are equally split with $\frac{C}{K}$ channels, the computational cost of $K$ groups of SA is $\frac{2}{K}(\sum_k M_k^2)HWC$. The SA calculated on different groups are then concatenated and merged via a $1 \times 1$ convolution.

**Accelerated Self-attention (ASA).** The calculation of SA is computation and memory-intensive in existing transformer models [7,29,36,57]. We make several modifications to accelerate the calculation of SA, especially in the inference stage. First, we discard the layer normalization (LN), which is widely employed in previous transformer models [7,29,36,57], because the LN fragments the calculation of SA into many element-wise operations, which are not friendly for efficient inference. Instead, we utilize batch normalization (BN) [20] to stabilize the training process. It is worth mentioning that the BN can be merged into the convolution operation, which does not cause additional computation cost in the inference stage. Second, the SA in SwinIR [29] is calculated on the embedded Gaussian space, where three independent $1 \times 1$ convolutions, denoted by $\theta$, $\phi$ and $g$, are employed to map the input feature $X$ into three different feature maps. We set $\theta = \phi$ and calculate the SA in the symmetric embedded Gaussian

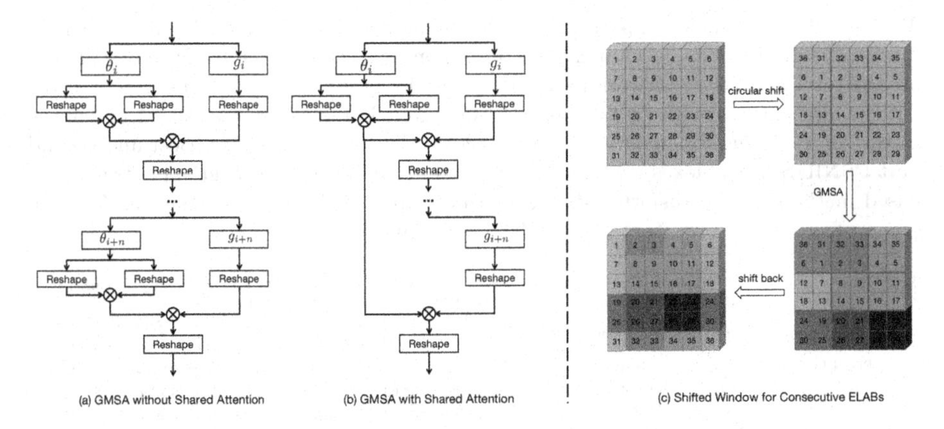


**Fig. 2.** (a) and (b) illustrate the calculation of GMSA without and with the shared attention mechanism, respectively. (c) Illustration of the shifted window mechanism used in consecutive ELABs.

space [5,31], which can save one $1 \times 1$ convolution in each SA. This modification further alleviates the computation and memory cost of SA without sacrificing the SR performance. Our ASA is shown in Fig. 1(e).

**Shared Attention.** Despite the above acceleration, one single forward pass of SA still consists of two $1 \times 1$ convolutions and four IO-intensive reshape operations. Although the reshape operation is FLOPs-free, it is time-consuming due to large feature size in the SR task. To further accelerate the SA computation of the entire network, we propose to share the attention scores among adjacent SA modules. As shown in Fig. 2(b), the calculated attention scores in the $i$-th SA module is directly re-used by the following $n$ SA modules on the same scale. In this way, we can avoid $2n$ reshape and $n$ $1 \times 1$ convolution operations for the following $n$ SAs. We found that the proposed shared attention mechanism only leads to slight drop in SR performance by using a number of $n$ (e.g., 1 or 2), while it saves much computation resources during inference.

**Shifted Window.** The calculated SA on the group-wise multi-scale windows still lacks connection across local windows within the same scale. We improve the shifted window mechanism of SwinIR [29] to reach a simple yet effective shifting scheme for the SR task. The whole process is visualized in Fig. 2(c). We first employ circular shift to the feature along the diagonal direction and calculate GMSA on the shifted feature. Then we shift the result back via inverse circular shift. The circular shift with half window size leads to a new partition of feature map and introduces connections among neighboring non-overlapped windows in the previous GMSA module. Although some pixels on the border are shifted to distant areas via circular shift, we found it has negligible impact on the final SR performance since such pixels only occupy a small portion of the entire feature map in the SR task. Benefiting from the circular shift mechanism, we remove the masking strategy and relative positional encoding adopted in SwinIR [29], making our network neater and more efficient.

**Table 1.** Performance comparison of different light-weight SR models on five benchmarks. PSNR/SSIM on Y channel are reported on each dataset. #Params and #FLOPs are the total number of network parameters and floating-point operations, respectively. All the efficiency proxies (#Params, #FLOPs and Latency) are measured under the setting of upscaling SR images to $1280 \times 720$ resolution on all scales. Best and second best PSNR/SSIM indexes are marked in red and blue colors, respectively. The CNN-based methods and transformer-based methods are separated via a dash line for each scaling factor. '–' means that the result is not available.

| Scale | Model | #Params (K) | #FLOPs (G) | Latency (ms) | Set5 [3] PSNR/SSIM | Set14 [62] PSNR/SSIM | B100 [39] PSNR/SSIM | Urban100 [18] PSNR/SSIM | Manga109 [40] PSNR/SSIM |
|---|---|---|---|---|---|---|---|---|---|
| ×2 | CARN [2] | 1,592 | 222.8 | 72 | 37.76/0.9590 | 33.52/0.9166 | 32.09/0.8978 | 31.92/0.9256 | 38.36/0.9765 |
| | EDSR-baseline [30] | 1370 | 316.3 | 71 | 37.99/0.9604 | 33.57/0.9175 | 32.16/0.8994 | 31.98/0.9272 | 38.54/0.9769 |
| | IMDN [19] | 694 | 158.8 | 54 | 38.00/0.9605 | 33.63/0.9177 | 32.19/0.8996 | 32.17/0.9283 | 38.88/0.9774 |
| | LAPAR-A [26] | 548 | 171.0 | 73 | 38.01/0.9605 | 33.62/0.9183 | 32.19/0.8999 | 32.10/0.9283 | 38.67/0.9772 |
| | LatticeNet [38] | 756 | 169.5 | 66 | 38.06/0.9607 | 33.70/0.9187 | 32.20/0.8999 | 32.25/0.9288 | —/— |
| | ESRT [37] | 677 | — | — | 38.03/0.9600 | 33.75/0.9184 | 32.25/0.9001 | 32.58/0.9318 | 39.12/0.9774 |
| | SwinIR-light [29] | 878 | 195.6 | 1007 | 38.14/0.9611 | 33.86/0.9206 | 32.31/0.9012 | 32.76/0.9340 | 39.12/0.9783 |
| | ELAN-light (ours) | 582 | 168.4 | 230 | 38.17/0.9611 | 33.94/0.9207 | 32.30/0.9012 | 32.76/0.9340 | 39.11/0.9782 |
| ×3 | CARN [2] | 1,592 | 118.8 | 39 | 34.29/0.9255 | 30.29/0.8407 | 29.06/0.8034 | 28.06/0.8493 | 33.50/0.9440 |
| | EDSR-baseline [30] | 1555 | 160.2 | 37 | 34.37/0.9270 | 30.28/0.8417 | 29.09/0.8052 | 28.15/0.8527 | 33.45/0.9439 |
| | IMDN [19] | 703 | 71.5 | 27 | 34.36/0.9270 | 30.32/0.8417 | 29.09/0.8046 | 28.17/0.8519 | 33.61/0.9445 |
| | LAPAR-A [26] | 544 | 114.0 | 55 | 34.36/0.9267 | 30.34/0.8421 | 29.11/0.8054 | 28.15/0.8523 | 33.51/0.9441 |
| | LatticeNet [38] | 765 | 76.3 | 33 | 34.40/0.9272 | 30.32/0.8416 | 29.10/0.8049 | 28.19/0.8513 | —/— |
| | ESRT [37] | 770 | — | — | 34.42/0.9268 | 30.43/0.8433 | 29.15/0.8063 | 28.46/0.8574 | 33.95/0.9455 |
| | SwinIR-light [29] | 886 | 87.2 | 445 | 34.62/0.9289 | 30.54/0.8463 | 29.20/0.8082 | 28.66/0.8624 | 33.98/0.9478 |
| | ELAN-light (ours) | 590 | 75.7 | 105 | 34.61/0.9288 | 30.55/0.8463 | 29.21/0.8081 | 28.69/0.8624 | 34.00/0.9478 |
| ×4 | CARN [2] | 1,592 | 90.9 | 30 | 32.13/0.8937 | 28.60/0.7806 | 27.58/0.7349 | 26.07/0.7837 | 30.47/0.9084 |
| | EDSR-baseline [30] | 1518 | 114.0 | 28 | 32.09/0.8938 | 28.58/0.7813 | 27.57/0.7357 | 26.04/0.7849 | 30.35/0.9067 |
| | IMDN [19] | 715 | 40.9 | 19 | 32.21/0.8948 | 28.58/0.7811 | 27.56/0.7353 | 26.04/0.7838 | 30.45/0.9075 |
| | LAPAR-A [26] | 659 | 94.0 | 47 | 32.15/0.8944 | 28.61/0.7818 | 27.61/0.7366 | 26.14/0.7871 | 30.42/0.9074 |
| | LatticeNet [38] | 777 | 43.6 | 23 | 32.18/0.8943 | 28.61/0.7812 | 27.57/0.7355 | 26.14/0.7844 | —/— |
| | ESRT [37] | 751 | — | — | 32.19/0.8947 | 28.69/0.7833 | 27.69/0.7379 | 26.39/0.7962 | 30.75/0.9100 |
| | SwinIR-light [29] | 897 | 49.6 | 271 | 32.44/0.8976 | 28.77/0.7858 | 27.69/0.7406 | 26.47/0.7980 | 30.92/0.9151 |
| | ELAN-light (ours) | 601 | 43.2 | 62 | 32.43/0.8975 | 28.78/0.7858 | 27.69/0.7406 | 26.54/0.7982 | 30.92/0.9150 |

# 4 Experiments

In this section, we conduct extensive experiments to quantitatively and qualitatively validate the superior performance of our ELAN for light-weight and classic SR tasks on five SR benchmark datasets. We also present comprehensive ablation studies to evaluate the design of our proposed ELAN.

## 4.1 Experimental Setup

**Dataset and Evaluation Metrics.** We employ the DIV2K dataset [53] with 800 training images to train our ELAN model, and use the five benchmark datasets, including Set5 [3], Set14 [62], BSD100 [39], Urban100 [18] and Manga109 [40], for performance comparison. PSNR and SSIM are used as the evaluation metrics, which are calculated on the Y channel after converting RGB to YCbCr format. For efficiency comparison, we report the latency evaluated on a single NVIDIA 2080Ti GPU in the inference stage. We also report the number of network parameters and FLOPs as reference, although they may not be able to faithfully reflect the network complexity and efficiency.

Note that since some competing methods do not release the source codes, we can only copy their PSNR/SSIM results from the original papers, but cannot report their results of latency and FLOPS.

**Training Details.** Following SwinIR [29], we train two versions of ELAN with different complexity. The light-weight version, i.e., ELAN-light consists of 24 ELABs with 60 channels, while the normal version of ELAN has 36 ELABs with 180 channels. We calculate GMSA on three equally split scales with window size: $4 \times 4$, $8 \times 8$ and $16 \times 16$. By default, we set $n = 1$ for the shared attention mechanism. Bicubic downsampling is used to generate training image pairs. We randomly crop 64 patches of size $64 \times 64$, and 32 patches of size $48 \times 48$ from the LR images as training mini-batch for the light-weight and normal ELAN models, respectively. We employ randomly rotating 90°, 180°, 270° and horizontal flip for data augmentation. Both models are trained using the ADAM optimizer with $\beta_1 = 0.9$, $\beta_2 = 0.999$, and $\epsilon = 10^{-8}$ for 500 epochs. The learning rate is initialized as $2 \times 10^{-4}$ and multiplied with 0.5 after $\{250, 400, 425, 450, 475\}$-th epoch. We train the model by Pytorch [43] using 4 NVIDIA 2080Ti GPUs. It takes about 42 and 63 h to train ELAN-light and ELAN models from scratch, respectively.

### 4.2 Comparison with Light-Weight SR Models

We first compare our ELAN-light with state-of-the-art light-weight SR models, including CNN-based models CARN [2], IMDN [19], LAPAR-A [26], LatticeNet [38], and transformer-based models ESRT [37] and SwinIR-light [29].

**Quantitative Comparison.** The quantitative indexes of different methods are reported in Table 1. Several observations can be made from the table. First, with similar #Params and #FLOPs, the transformer-based methods especially SwinIR-light outperform much CNN-based methods on PSNR/SSIM indexes, by exploiting the self-similarity of images. However, the latency time of SwinIR-light is more than ×10 slower than the CNN-based methods, because the calculation of self-attention in SwinIR is a heavy burden for inference. Benefiting from our efficient long-range attention design, our ELAN-light model not only obtains the best or second best PSNR/SSIM indexes on all the five datasets and on all the three zooming scales, but also is about ×4.5 faster than SwinIR-light. Its #Params and #FLOPs are also smaller compared with SwinIR-light.

**Qualitative Comparison.** We then qualitatively compare the SR quality of different light-weight models. The ×4 SR results on three example images are shown in Fig. 3. One can see that all the CNN-based models result in very blurry and distorted edges on the three images. The transformer based SwinIR-light can recover the main structure in the first image and half edges in the second one. Our ELAN-light is the only method that succeeds in recovering the main structures on all the three images with clear and sharp edges. It is worth mentioning that the advantage of ELAN-light over SwinIR-light is achieved using much less computational cost, as we have validated in Table 1. More visual examples can be found in the **supplementary file**.

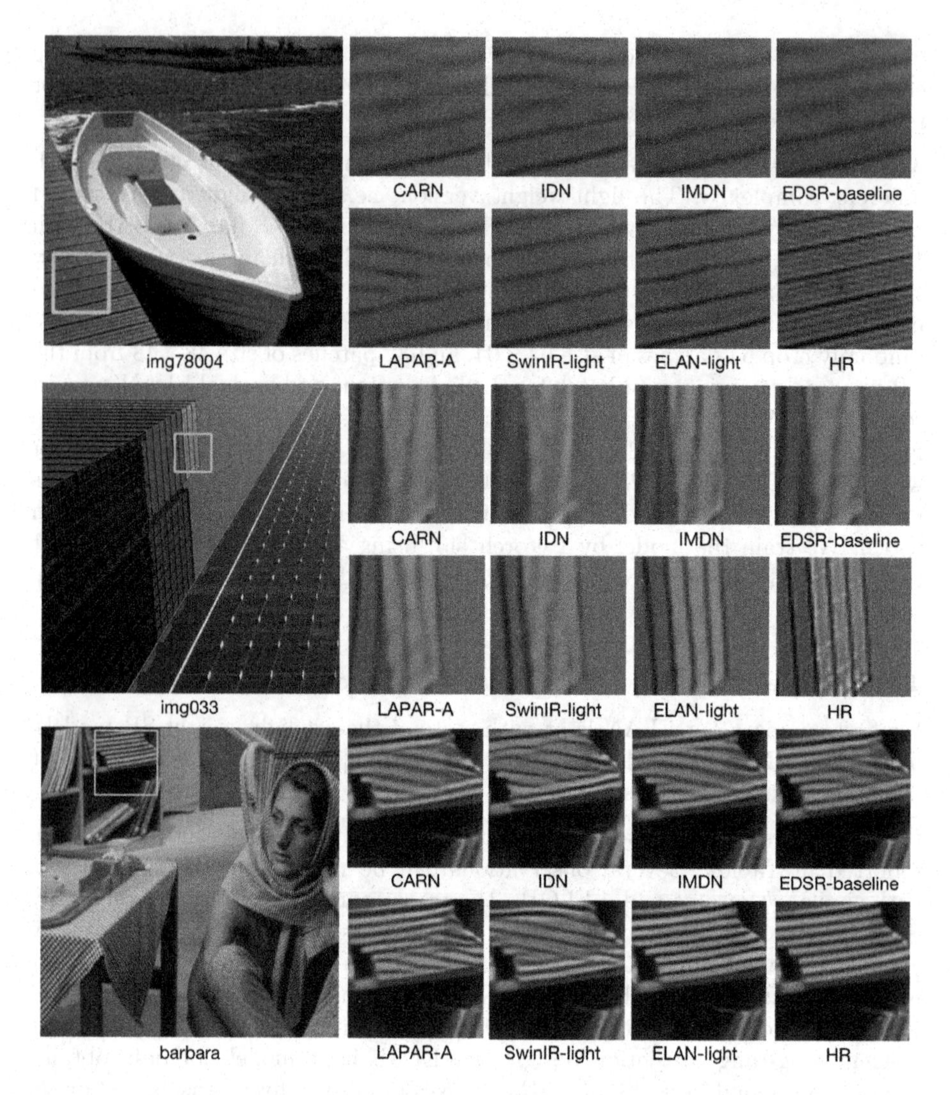


**Fig. 3.** Qualitative comparison of state-of-the-art light-weight SR models for $\times 4$ upscaling. ELAN-light can restore more accurate and clear structures than other models.

## 4.3 Comparison with Classic SR Models

To validate the scalability of ELAN, we further compare the normal version of ELAN with state-of-the-art classic performance-oriented SR models, including EDSR [30], SRFBN [28], RNAN [66], RDN [67], RCAN [32], SAN [10], IGNN [69], HAN [42], NLSA [41] and SwinIR [29]. Note that these models employ either very deep network topology with large channel number or complicated self-attention and non-local strategies.

**Table 2.** Performance comparison of different classic performance-oriented SR models on five benchmarks. PSNR/SSIM on Y channel are reported on each dataset. #Params and #FLOPs are the total number of network parameters and floating-point operations, respectively. Noted that all the efficiency proxies (#Params, #FLOPs and Latency) are measured under the setting of upscaling SR images to 1280 × 720 resolution on all scales. Best and second best PSNR/SSIM indexes are marked in red and blue colors, respectively. The CNN-based methods and transformer-based methods are separated by a dash line. 'NaN' indicates that corresponding models are too heavy to run on a single NVIDIA 2080Ti GPU. '–' means that the result is not available.

| Scale | Model | #Params (K) | #FLOPs (G) | Latency (ms) | Set5 [3] PSNR/SSIM | Set14 [62] PSNR/SSIM | B100 [39] PSNR/SSIM | Urban100 [18] PSNR/SSIM | Manga109 [40] PSNR/SSIM |
|---|---|---|---|---|---|---|---|---|---|
| × 2 | EDSR [30] | 40730 | 9387 | 1143 | 38.11/0.9602 | 33.92/0.9195 | 32.32/0.9013 | 32.93/0.9351 | 39.10/0.9773 |
| | SRFBN [28] | 2140 | 5044 | 920 | 38.11/0.9609 | 33.82/0.9196 | 32.29/0.9010 | 32.62/0.9328 | 39.08/0.9779 |
| | RNAN [66] | 9107 | NaN | NaN | 38.17/0.9611 | 33.87/0.9207 | 32.32/0.9014 | 32.73/0.9340 | 39.23/0.9785 |
| | RDN [67] | 22123 | 5098 | 846 | 38.24/0.9614 | 34.01/0.9212 | 32.34/0.9017 | 32.89/0.9353 | 39.18/0.9780 |
| | OISR [16] | 41910 | 9657 | — | 38.21/0.9612 | 33.94/0.9206 | 32.36/0.9019 | 33.03/0.9365 | — |
| | RCAN [67] | 15445 | 3530 | 743 | 38.27/0.9614 | 34.12/0.9216 | 32.41/0.9027 | 33.34 0.9384 | 39.44/0.9786 |
| | SAN [10] | 15861 | 3050 | NaN | 38.31/0.9620 | 34.07/0.9213 | 32.42/0.9028 | 33.10/0.9370 | 39.32/0.9792 |
| | IGNN [69] | 49513 | — | — | 38.24/0.9613 | 34.07/0.9217 | 32.41/0.9025 | 33.23/0.9383 | 39.35/0.9786 |
| | HAN [42] | 63608 | 14551 | 2278 | 38.27/0.9614 | 34.16/0.9217 | 32.41/0.9027 | 33.35/0.9385 | 39.46/0.9785 |
| | NLSA [41] | 41796 | 9632 | NaN | 38.34/0.9618 | 34.08/0.9231 | 32.43/0.9027 | 33.42/0.9394 | 39.59/0.9789 |
| | SwinIR [29] | 11752 | 2301 | 2913 | 38.35/0.9620 | 34.14/0.9227 | 32.44/0.9030 | 33.40/0.9393 | 39.60/0.9792 |
| | ELAN (ours) | 8254 | 1965 | 1244 | 38.36/0.9620 | 34.20/0.9228 | 32.45/0.9030 | 33.44/0.9391 | 39.62/0.9793 |
| × 3 | EDSR [30] | 43680 | 4470 | 573 | 34.65/0.9280 | 30.52/0.8462 | 29.25/0.8093 | 28.80/0.8653 | 34.17/0.9476 |
| | SRFBN [28] | 2833 | 6024 | 672 | 34.70/0.9292 | 30.51/0.8461 | 29.24/0.8084 | 28.73/0.8641 | 34.18/0.9481 |
| | RNAN [66] | 9292 | 809 | NaN | 34.66/0.9290 | 30.52/0.8462 | 29.26/0.8090 | 28.75/0.8646 | 34.25/0.9483 |
| | RDN [67] | 22308 | 2282 | 406 | 34.71/0.9296 | 30.57/0.8468 | 29.26/0.8093 | 28.80/0.8653 | 34.13/0.9484 |
| | OISR [16] | 44860 | 4590 | — | 34.72/0.9297 | 30.57/0.8470 | 29.29/0.8103 | 28.95/0.8680 | — |
| | RCAN [67] | 15629 | 1586 | 367 | 34.74/0.9299 | 30.65/0.8482 | 29.32/0.8111 | 29.09/0.8702 | 34.44/0.9499 |
| | SAN [10] | 15897 | 1620 | NaN | 34.75/0.9300 | 30.59/0.8476 | 29.33/0.8112 | 28.93/0.8671 | 34.30/0.9494 |
| | IGNN [69] | 49512 | — | — | 34.72/0.9298 | 30.66/0.8484 | 29.31/0.8105 | 29.03/0.8696 | 34.39/0.9496 |
| | HAN [42] | 64346 | 6534 | 1014 | 34.75/0.9299 | 30.67/0.8483 | 29.32/0.8110 | 29.10/0.8705 | 34.48/0.9500 |
| | NLSA [41] | 44747 | 4579 | 840 | 34.85/0.9306 | 30.70/0.8485 | 29.34/0.8117 | 29.25/0.8726 | 34.57 0.9508 |
| | SwinIR [29] | 11937 | 1026 | 1238 | 34.89/0.9312 | 30.77/0.8503 | 29.37/0.8124 | 29.29/0.8744 | 34.74/0.9518 |
| | ELAN (ours) | 8278 | 874 | 530 | 34.90/0.9313 | 30.80/0.8504 | 29.38/0.8124 | 29.32/0.8745 | 34.73/0.9517 |
| × 4 | EDSR [30] | 43090 | 2895 | 360 | 32.46/0.8968 | 28.80/0.7876 | 27.71/0.7420 | 26.64/0.8033 | 31.02/0.9148 |
| | SRFBN [28] | 3631 | 7466 | 551 | 32.47/0.8983 | 28.81/0.7868 | 27.72/0.7409 | 26.60/0.8015 | 31.15/0.9160 |
| | RNAN [66] | 9255 | 480 | NaN | 32.49/0.8982 | 28.83/0.7878 | 27.72/0.7421 | 26.61/0.8023 | 31.09/0.9149 |
| | RDN [67] | 22271 | 1310 | 243 | 32.47/0.8990 | 28.81/0.7871 | 27.72/0.7419 | 26.61/0.8028 | 31.00 0.9151 |
| | OISR [16] | 44270 | 2963 | — | 32.53/0.8992 | 28.86/0.7878 | 27.75/0.7428 | 26.79/0.8068 | — |
| | RCAN [67] | 15592 | 918 | 223 | 32.63/0.9002 | 28.87/0.7889 | 27.77/0.7436 | 26.82/0.8087 | 31.22/0.9173 |
| | SAN [10] | 15861 | 937 | NaN | 32.64/0.9003 | 28.92/0.7888 | 27.78/0.7436 | 26.79/0.8068 | 31.18/0.9169 |
| | IGNN [69] | 49513 | — | — | 32.57/0.8998 | 28.85/0.7891 | 27.77/0.7434 | 26.84/0.8090 | 31.28 0.9182 |
| | HAN [42] | 64199 | 3776 | 628 | 32.64/0.9002 | 28.90/0.7890 | 27.80/0.7442 | 26.85/0.8094 | 31.42/0.9177 |
| | NLSA [41] | 44157 | 2956 | 502 | 32.59/0.9000 | 28.87/0.7891 | 27.78/0.7444 | 26.96/0.8109 | 31.27/0.9184 |
| | SwinIR [29] | 11900 | 584 | 645 | 32.72/0.9021 | 28.94/0.7914 | 27.83/0.7459 | 27.07/0.8164 | 31.67/0.9226 |
| | ELAN (ours) | 8312 | 494 | 298 | 32.75/0.9022 | 28.96/0.7914 | 27.83/0.7459 | 27.13/0.8167 | 31.68/0.9226 |

**Quantitative Comparison.** The quantitative results are shown in Table 2. As can be seen, our ELAN achieves the best results on almost all benchmarks and all upscaling factors. In particular, compared with SwinIR, our ELAN obtains better PSNR and SSIM indexes on almost all settings with less number of parameters and FLOPs, and more than ×2 faster inference speed. These performance gains show that the proposed ELAB block succeeds in capturing useful infor-

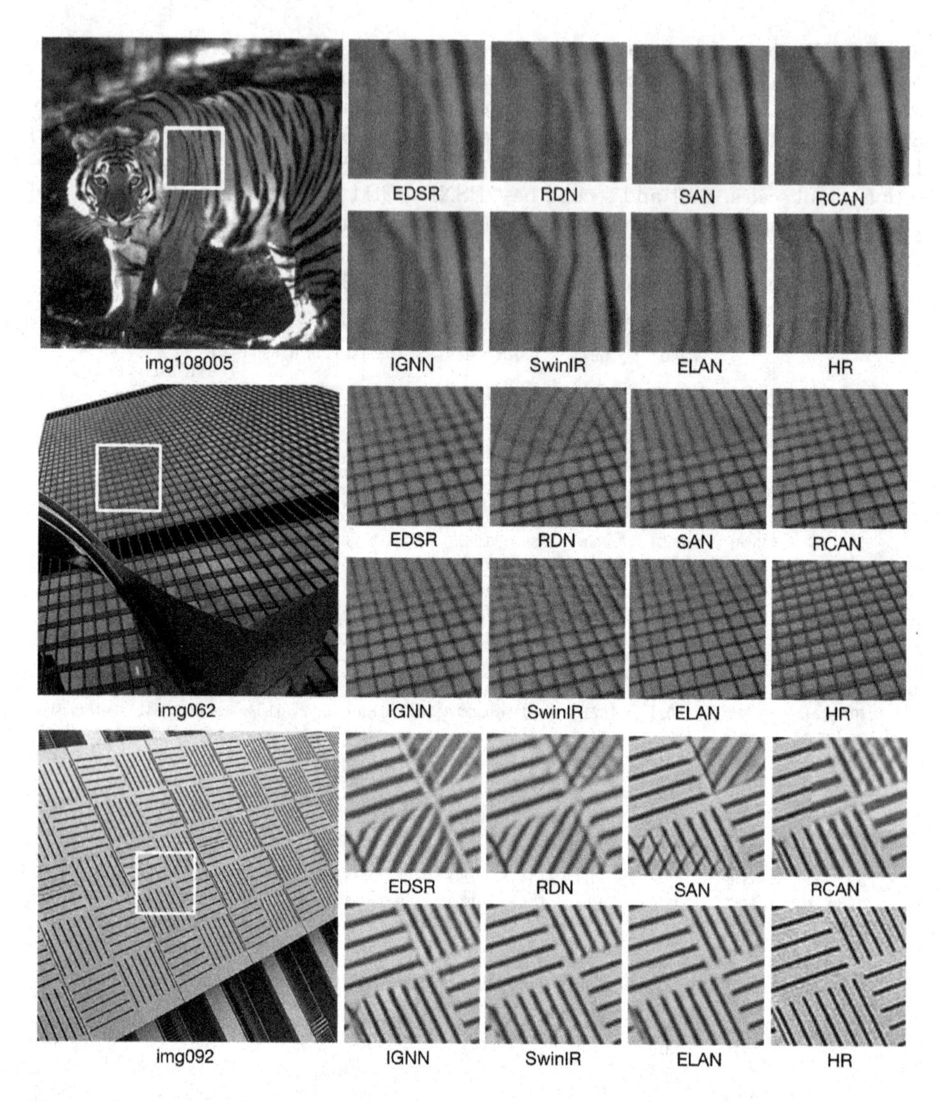


**Fig. 4.** Qualitative comparison of state-of-the-art classic SR models for ×4 upscaling task. The ELAN can restore more accurate and sharper details than the other models.

mation (such as long-range dependency) to conduct super-resolution in a more efficient and effective manner. Although the classical CNN-based SR models have huge amount of parameters and FLOPs, their performance is less competitive than the transformer-based models. Comparing with most CNN-based methods, our ELAN shows significant advantages in reconstruction performance metrics, benefiting from its larger receptive field and capability of modeling long-range spatial feature correlation. Some methods such as HAN and NLSA can also obtain competitive performance by exploiting dedicated attention mechanisms and very deep network, but their computation and memory cost are very expensive. For example, the NLSA model on ×2 upscaling task is too heavy to execute on a single NVIDIA 2080Ti GPU. Nevertheless, our ELAN obtains better performance than these complicated CNN-based models at much less cost.

**Qualitative Comparison.** Because of the space limitation, we compare the visual quality of SR results by our ELAN and six representative models, including EDSR [30], RDN [67], SAN [10], RCAN [67], IGNN [28] and SwinIR [29], on ×4 upscaling task. The results on three example images are shown in Fig. 4. Regarding the first example, all the compared methods yield either blurry or inaccurate fur textures on the body of tiger, while our ELAN can restore more accurate and sharper edges. The advantage of ELAN is more obvious on the second and third images with repetitive patterns. Most of the compared methods fail to reconstruct the correct structures, and even generate undesired artifacts. In contrast, our ELAN can accurately recover the structure with distinct edges in both cases. Such obvious advantages of ELAN validate the importance of modeling long-range self-attention to the SR task. More visual comparison examples can be found in the **supplementary file**.

### 4.4 Ablation Studies and Discussions

To better understand how ELAN works, we present comprehensive ablation studies to evaluate the roles of different components of ELAN, the depth selection for the shared attention mechanism, and the window size of GMSA. We also discuss the generalization capability of ELAN to other low-level vision tasks such as image denoising.

**The Components of ELAN.** We first utilize the architecture and setting of ELAN-light to ablate the roles of different components of ELAB, and observe the change in performance and efficiency. The PSNR/SSIM indexes on five benchmarks and inference time of variant models are reported in Table 3. The indexes of SwinIR-light is also listed for reference. Specifically, we start with a naive baseline by removing the redundant sub-branches from SwinIR-light, where the deep feature extraction module is composed of 24 sequential swin transformer block. As expected, this simplification leads to slight performance drop compared with the SwinIR-light while reducing the latency from 271 ms to 247 ms. By utilizing our shifted windows mechanism, the performance remain almost unchanged while the inference speed is decreased from 247 ms to 177 ms.

We then replace the SA calculation with our proposed ASA, the inference latency is greatly reduced from 177 ms to 66 ms without losing PSNR/SSIM performance. By employing the GMSA which can efficiently model long-range dependency, the PSNR and SSIM indexes get significantly improved on all five datasets. Specifically, the PSNR increases by 0.21 dB and 0.19 dB on the Urban100 and Manga109 datasets, respectively, while the latency is only slightly increased by 9 ms. This indicates the effectiveness of GMSA over the SA in SwinIR with small fixed window size. Finally, by employing the proposed shared attention mechanism, we can further speed up the inference time of ELAN-light with little performance drop. Combining all the improvements, the final version

**Table 3.** Ablation study on network design for ELAN.

| Scale | Model | Different components | | | | #Params (K) | #FLOPs (G) | Latency(ms) | Set5 [3] | Set14 [62] | B100 [39] | U100 [18] | Manga109 [40] |
|---|---|---|---|---|---|---|---|---|---|---|---|---|---|
| | | Shifted window | ASA | GMSA | Shared attention | | | | PSNR/SSIM | PSNR/SSIM | PSNR/SSIM | PSNR/SSIM | PSNR/SSIM |
| × 4 | SwinIR-light [29] | | | | | 897 | 49.6 | 271 | 32.44/0.8976 | 28.77/0.7858 | 27.69/0.7406 | 26.47/0.7980 | 30.92/0.9151 |
| | ELAN-light | | | | | 767 | 44.6 | 247 | 32.38/0.8971 | 28.68/0.7832 | 27.62/0.7368 | 26.40/0.7973 | 30.78/0.9142 |
| | ELAN-light | ✓ | | | | 765 | 44.4 | 177 | 32.38/0.8971 | 28.69/0.7833 | 27.61/0.7367 | 26.42/0.7973 | 30.78/0.9141 |
| | ELAN-light | ✓ | ✓ | | | 641 | 43.8 | 66 | 32.39/0.8970 | 28.67/0.7831 | 27.62/0.7368 | 26.39/0.7972 | 30.76/0.9139 |
| | ELAN-light | ✓ | ✓ | ✓ | | 641 | 45.5 | 75 | 32.47/0.8977 | 28.79/0.7858 | 27.71/0.7405 | 26.60/0.7985 | 30.95/0.9151 |
| | ELAN-light | ✓ | ✓ | ✓ | ✓ | 601 | 43.2 | 62 | 32.43/0.8975 | 28.78/0.7858 | 27.69/0.7406 | 26.54/0.7982 | 30.92/0.9150 |

of ELAN-light achieves about ×4.5 acceleration while maintaining comparable performance to SwinIR-light.

**Depth of Shared Attention.** We further conduct a detailed ablation study on the depth of shared attention blocks (i.e., $n$ in Fig. 2(b)) in sequential layers. ELAN-light is employed for this study and the depth of shared attention blocks varies in {0, 1, 3, 5}. Note that $n = 0$ means that each ELAB calculates self-attention independently. The results of ELAN-light using different $n$ are reported in Table 4. We also provide one representative CNN-based method EDSR-baseline for intuitive comparison. One can observe that using larger $n$ can effectively reduce the number of network parameters, FLOPs and latency time in the inference stage, at the cost of performance drop. By choosing an appropriate $n$, our ELAN can achieve a good trade-off between efficiency and performance. It is worth mentioning that even using $n = 5$, the PSNR of ELAN-light on the challenging Urban100 and Manga109 still outperform EDSR-baseline by a large margin (up to 0.31 dB and 0.28 dB, respectively), validating the advantage of modeling long-range attention in our model.

**Window Size of GMSA.** To further validate the efficacy of our design, we ablate the setting of window size of ELAN-light on ×4 upscaling task and report the results in the Table 5. One can see that the PSNR/SSIM indices are steadily increased with the increase of window size. In particular, ELAN-light with [8, 16,32] window size outperforms SwinIR-light by 0.27 dB on Urban100 yet with more than 2× faster the inference speed.

**Generalization of ELAN.** To verify the generalization capability of ELAN to other tasks, we apply it to color image denoising with noise level 25, following the settings of SwinIR [29]. In Table 6, we see that although ELAN is designed for SR, it achieves slightly better PSNR result than SwinIR and surpasses other state-of-the-art CNN-based denoising methods. What's more, as for the inference speed, ELAN runs 2.3× faster than SwinIR given 480p image as input.

**Table 4.** Ablation study on depth selection of shared attention.

| Scale | Model | Depth($n$) | #Params (K) | #FLOPs (G) | Latency (ms) | Set5 [3] PSNR/SSIM | Set14 [62] PSNR/SSIM | B100 [39] PSNR/SSIM | U100 [18] PSNR/SSIM | Manga109 [40] PSNR/SSIM |
|---|---|---|---|---|---|---|---|---|---|---|
| × 4 | ELAN-light | 0 | 641 | 45.5 | 75 | 32.47/0.8977 | 28.79/0.7858 | 27.71/0.7405 | 26.60/0.7985 | 30.95/0.9151 |
| | ELAN-light | 1 | 601 | 43.2 | 62 | 32.43/0.8976 | 28.78/0.7858 | 27.69/0.7406 | 26.54/0.7982 | 30.92/0.9150 |
| | ELAN-light | 3 | 575 | 40.4 | 51 | 32.35/0.8970 | 28.71/0.7852 | 27.65/0.7401 | 26.40/0.7071 | 30.74/0.9133 |
| | ELAN-light | 5 | 571 | 39.3 | 48 | 32.31/0.8967 | 28.66/0.7846 | 27.62/0.7400 | 26.35/0.7967 | 30.63/0.9120 |
| | EDSR-baseline [30] | — | 1518 | 114.0 | 28 | 32.09/0.8938 | 28.58/0.7813 | 27.57/0.7357 | 26.04/0.7849 | 30.35/0.9067 |

**Table 5.** Ablation study on window size of GMSA.

| Scale | Model | Latency | Set5 [3] | Set14 [62] | B100 [39] | Urban100 [18] | Manga109 [40] |
|---|---|---|---|---|---|---|---|
| × 4 | [8, 16, 32] | 109 | 32.47/0.8985 | 28.83/0.7873 | 27.73/0.7419 | 26.74/0.8042 | 31.06/0.9161 |
| | [4, 8, 16] | 62 | 32.43/0.8975 | 28.78/0.7858 | 27.69/0.7406 | 26.54/0.7982 | 30.92/0.9150 |
| | [8, 8, 8] | 58 | 32.38/0.8971 | 28.68/0.7832 | 27.62/0.7368 | 26.40/0.7973 | 30.78/0.9142 |
| | SwinIR-light | 271 | 32.44/0.8976 | 28.77/0.7858 | 27.69/0.7406 | 26.47/0.7980 | 30.92/0.9151 |

**Table 6.** Quantitative comparison with SOTAs for color image denoising.

| Dataset | RPCNN [60] | BRDNet [52] | DRUNet [63] | SwinIR [29] | ELAN |
|---|---|---|---|---|---|
| CBSD68 [39] | 31.24 | 31.43 | 31.69 | 31.78 | 31.82 |
| Kodak24 [15] | 32.34 | 32.41 | 32.89 | 32.89 | 32.89 |
| McMaster [64] | 32.33 | 32.75 | 33.14 | 33.20 | 33.21 |
| Urban100 [18] | 31.81 | 31.99 | 32.60 | 32.90 | 32.94 |

## 5 Conclusion

In this paper, we proposed an efficient long-range attention network (ELAN) for single image super resolution. ELAN had a neat topology with sequentially cascaded efficient long-range attention blocks (ELAB). Each ELAB was composed of a local feature extraction module with two sequential shift-conv and a group-wise multi-scale self-attention (GMSA) module to gradually increase the receptive field of self-attention (SA). Benefiting from our accelerated SA calculation and shared attention mechanism, ELAB can effectively capture the local structure and long-range dependency in a very efficient manner. Extensive experiments show that ELAN can obtain highly competitive performance than previous state-of-the-art SR models on both light-weight and performance-oriented settings, while being much more economical than previous transformer-based SR methods. Though ELAN achieved significant speedup than SwinIR, the calculation of SA was still computation- and memory-intensive compared to those light-weight CNN-based models. In the future, we will explore more efficient implementations or approximations of SA for low-level vision tasks.

## References


1. Agustsson, E., Timofte, R.: NTIRE 2017 challenge on single image super-resolution: dataset and study. In: Proceedings of the IEEE Conference on Computer Vision and Pattern Recognition Workshops, pp. 126–135 (2017)
2. Ahn, N., Kang, B., Sohn, K.-A.: Fast, accurate, and lightweight super-resolution with cascading residual network. In: Ferrari, V., Hebert, M., Sminchisescu, C., Weiss, Y. (eds.) ECCV 2018. LNCS, vol. 11214, pp. 256–272. Springer, Cham (2018). https://doi.org/10.1007/978-3-030-01249-6_16
3. Bevilacqua, M., Roumy, A., Guillemot, C., Alberi-Morel, M.L.: Low-complexity single-image super-resolution based on nonnegative neighbor embedding (2012)
4. Brown, T., et al.: Language models are few-shot learners. In: Advances in Neural Information Processing Systems 33, pp. 1877–1901 (2020)
5. Buades, A., Coll, B., Morel, J.M.: A non-local algorithm for image denoising. In: 2005 IEEE Computer Society Conference on Computer Vision and Pattern Recognition (CVPR 2005), vol. 2, pp. 60–65. IEEE (2005)
6. Caballero, J., et al.: Real-time video super-resolution with spatio-temporal networks and motion compensation. In: Proceedings of the IEEE Conference on Computer Vision and Pattern Recognition, pp. 4778–4787 (2017)
7. Cao, H., et al.: Swin-Unet: Unet-like pure transformer for medical image segmentation. arXiv preprint arXiv:2105.05537 (2021)
8. Carion, N., Massa, F., Synnaeve, G., Usunier, N., Kirillov, A., Zagoruyko, S.: End-to-end object detection with transformers. In: Vedaldi, A., Bischof, H., Brox, T., Frahm, J.-M. (eds.) ECCV 2020. LNCS, vol. 12346, pp. 213–229. Springer, Cham (2020). https://doi.org/10.1007/978-3-030-58452-8_13
9. Chen, H., et al.: Pre-trained image processing transformer. In: Proceedings of the IEEE/CVF Conference on Computer Vision and Pattern Recognition, pp. 12299–12310 (2021)
10. Dai, T., Cai, J., Zhang, Y., Xia, S.T., Zhang, L.: Second-order attention network for single image super-resolution. In: Proceedings of the IEEE/CVF Conference on Computer Vision and Pattern Recognition, pp. 11065–11074 (2019)
11. Dong, C., Loy, C.C., He, K., Tang, X.: Image super-resolution using deep convolutional networks. IEEE Trans. Pattern Anal. Mach. Intell. **38**(2), 295–307 (2015)
12. Dong, C., Loy, C.C., Tang, X.: Accelerating the super-resolution convolutional neural network. In: Leibe, B., Matas, J., Sebe, N., Welling, M. (eds.) ECCV 2016. LNCS, vol. 9906, pp. 391–407. Springer, Cham (2016). https://doi.org/10.1007/978-3-319-46475-6_25
13. Dosovitskiy, A., et al.: An image is worth 16x16 words: transformers for image recognition at scale. arXiv preprint arXiv:2010.11929 (2020)
14. Fedus, W., Zoph, B., Shazeer, N.: Switch transformers: scaling to trillion parameter models with simple and efficient sparsity. arXiv preprint arXiv:2101.03961 (2021)
15. Franzen, R.: Kodak lossless true color image suite (1998). http://r0k.us/graphics/kodak/
16. He, X., Mo, Z., Wang, P., Liu, Y., Yang, M., Cheng, J.: ODE-inspired network design for single image super-resolution. In: Proceedings of the IEEE/CVF Conference on Computer Vision and Pattern Recognition, pp. 1732–1741 (2019)
17. Huang, G., Liu, Z., Van Der Maaten, L., Weinberger, K.Q.: Densely connected convolutional networks. In: Proceedings of the IEEE Conference on Computer Vision and Pattern Recognition, pp. 4700–4708 (2017)

18. Huang, J.B., Singh, A., Ahuja, N.: Single image super-resolution from transformed self-exemplars. In: Proceedings of the IEEE Conference on Computer Vision and Pattern Recognition, pp. 5197–5206 (2015)
19. Hui, Z., Gao, X., Yang, Y., Wang, X.: Lightweight image super-resolution with information multi-distillation network. In: Proceedings of the 27th ACM International Conference on Multimedia, pp. 2024–2032 (2019)
20. Ioffe, S., Szegedy, C.: Batch normalization: accelerating deep network training by reducing internal covariate shift. In: International Conference on Machine Learning, pp. 448–456. PMLR (2015)
21. Kim, J., Kwon Lee, J., Mu Lee, K.: Accurate image super-resolution using very deep convolutional networks. In: Proceedings of the IEEE Conference on Computer Vision and Pattern Recognition, pp. 1646–1654 (2016)
22. Kingma, D.P., Ba, J.: Adam: a method for stochastic optimization. arXiv preprint arXiv:1412.6980 (2014)
23. Lai, W.S., Huang, J.B., Ahuja, N., Yang, M.H.: Deep Laplacian pyramid networks for fast and accurate super-resolution. In: Proceedings of the IEEE Conference on Computer Vision and Pattern Recognition, pp. 624–632 (2017)
24. LeCun, Y., Bengio, Y., Hinton, G.: Deep learning. Nature **521**(7553), 436–444 (2015)
25. Ledig, C., et al.: Photo-realistic single image super-resolution using a generative adversarial network. In: Proceedings of the IEEE Conference on Computer Vision and Pattern Recognition, pp. 4681–4690 (2017)
26. Li, W., Zhou, K., Qi, L., Jiang, N., Lu, J., Jia, J.: LAPAR: linearly-assembled pixel-adaptive regression network for single image super-resolution and beyond. In: Advances in Neural Information Processing Systems 33, pp. 20343–20355 (2020)
27. Li, Y., Zhang, K., Cao, J., Timofte, R., Van Gool, L.: LocalViT: bringing locality to vision transformers. arXiv preprint arXiv:2104.05707 (2021)
28. Li, Z., Yang, J., Liu, Z., Yang, X., Jeon, G., Wu, W.: Feedback network for image super-resolution. In: Proceedings of the IEEE/CVF Conference on Computer Vision and Pattern Recognition, pp. 3867–3876 (2019)
29. Liang, J., Cao, J., Sun, G., Zhang, K., Van Gool, L., Timofte, R.: SwinIR: image restoration using swin transformer. In: Proceedings of the IEEE/CVF International Conference on Computer Vision, pp. 1833–1844 (2021)
30. Lim, B., Son, S., Kim, H., Nah, S., Mu Lee, K.: Enhanced deep residual networks for single image super-resolution. In: Proceedings of the IEEE Conference on Computer Vision and Pattern Recognition Workshops, pp. 136–144 (2017)
31. Liu, D., Wen, B., Fan, Y., Loy, C.C., Huang, T.S.: Non-local recurrent network for image restoration. In: Advances in Neural Information Processing Systems 31 (2018)
32. Liu, J., Tang, J., Wu, G.: Residual feature distillation network for lightweight image super-resolution. arXiv preprint arXiv:2009.11551 (2020)
33. Liu, L., et al.: Deep learning for generic object detection: a survey. Int. J. Comput. Vis. **128**(2), 261–318 (2020)
34. Liu, Y., et al.: RoBERTa: a robustly optimized BERT pretraining approach. arXiv preprint arXiv:1907.11692 (2019)
35. Liu, Y., Sun, G., Qiu, Y., Zhang, L., Chhatkuli, A., Van Gool, L.: Transformer in convolutional neural networks. arXiv preprint arXiv:2106.03180 (2021)
36. Liu, Z., et al.: Swin transformer: hierarchical vision transformer using shifted windows. In: Proceedings of the IEEE/CVF International Conference on Computer Vision, pp. 10012–10022 (2021)

37. Lu, Z., Liu, H., Li, J., Zhang, L.: Efficient transformer for single image super-resolution. arXiv preprint arXiv:2108.11084 (2021)
38. Luo, X., Xie, Y., Zhang, Y., Qu, Y., Li, C., Fu, Y.: LatticeNet: towards lightweight image super-resolution with lattice block. In: Vedaldi, A., Bischof, H., Brox, T., Frahm, J.-M. (eds.) ECCV 2020. LNCS, vol. 12367, pp. 272–289. Springer, Cham (2020). https://doi.org/10.1007/978-3-030-58542-6_17
39. Martin, D., Fowlkes, C., Tal, D., Malik, J.: A database of human segmented natural images and its application to evaluating segmentation algorithms and measuring ecological statistics. In: Proceedings Eighth IEEE International Conference on Computer Vision, ICCV 2001, vol. 2, pp. 416–423. IEEE (2001)
40. Matsui, Y., et al.: Sketch-based manga retrieval using Manga109 dataset. Multimed. Tools Appl. **76**(20), 21811–21838 (2017)
41. Mei, Y., Fan, Y., Zhou, Y.: Image super-resolution with non-local sparse attention. In: Proceedings of the IEEE/CVF Conference on Computer Vision and Pattern Recognition, pp. 3517–3526 (2021)
42. Niu, B., et al.: Single image super-resolution via a holistic attention network. In: Vedaldi, A., Bischof, H., Brox, T., Frahm, J.-M. (eds.) ECCV 2020. LNCS, vol. 12357, pp. 191–207. Springer, Cham (2020). https://doi.org/10.1007/978-3-030-58610-2_12
43. Paszke, A., et al.: Pytorch: an imperative style, high-performance deep learning library. In: Advances in Neural Information Processing Systems 32, pp. 8026–8037 (2019)
44. Radford, A., Narasimhan, K., Salimans, T., Sutskever, I.: Improving language understanding by generative pre-training (2018)
45. Ramachandran, P., Parmar, N., Vaswani, A., Bello, I., Levskaya, A., Shlens, J.: Stand-alone self-attention in vision models. In: Advances in Neural Information Processing Systems 32 (2019)
46. Sajjadi, M.S., Scholkopf, B., Hirsch, M.: EnhanceNet: single image super-resolution through automated texture synthesis. In: Proceedings of the IEEE International Conference on Computer Vision, pp. 4491–4500 (2017)
47. Sajjadi, M.S., Vemulapalli, R., Brown, M.: Frame-recurrent video super-resolution. In: Proceedings of the IEEE Conference on Computer Vision and Pattern Recognition, pp. 6626–6634 (2018)
48. Shi, W., et al.: Real-time single image and video super-resolution using an efficient sub-pixel convolutional neural network. In: Proceedings of the IEEE Conference on Computer Vision and Pattern Recognition, pp. 1874–1883 (2016)
49. Tai, Y., Yang, J., Liu, X.: Image super-resolution via deep recursive residual network. In: Proceedings of the IEEE Conference on Computer Vision and Pattern Recognition, pp. 3147–3155 (2017)
50. Tai, Y., Yang, J., Liu, X., Xu, C.: MemNet: a persistent memory network for image restoration. In: Proceedings of the IEEE International Conference on Computer Vision, pp. 4539–4547 (2017)
51. Tao, X., Gao, H., Liao, R., Wang, J., Jia, J.: Detail-revealing deep video super-resolution. In: Proceedings of the IEEE International Conference on Computer Vision, pp. 4472–4480 (2017)
52. Tian, C., Xu, Y., Zuo, W.: Image denoising using deep CNN with batch renormalization. Neural Netw. **121**, 461–473 (2020)
53. Timofte, R., Agustsson, E., Van Gool, L., Yang, M.H., Zhang, L.: NTIRE 2017 challenge on single image super-resolution: methods and results. In: Proceedings of the IEEE Conference on Computer Vision and Pattern Recognition Workshops, pp. 114–125 (2017)

54. Touvron, H., Cord, M., Douze, M., Massa, F., Sablayrolles, A., Jégou, H.: Training data-efficient image transformers & distillation through attention. In: International Conference on Machine Learning, pp. 10347–10357. PMLR (2021)
55. Wang, X., Chan, K.C., Yu, K., Dong, C., Change Loy, C.: EDVR: video restoration with enhanced deformable convolutional networks. In: Proceedings of the IEEE/CVF Conference on Computer Vision and Pattern Recognition Workshops (2019)
56. Wang, X., et al.: ESRGAN: enhanced super-resolution generative adversarial networks. In: Leal-Taixé, L., Roth, S. (eds.) ECCV 2018. LNCS, vol. 11133, pp. 63–79. Springer, Cham (2019). https://doi.org/10.1007/978-3-030-11021-5_5
57. Wang, Z., Cun, X., Bao, J., Liu, J.: Uformer: a general U-shaped transformer for image restoration. arXiv preprint arXiv:2106.03106 (2021)
58. Wu, B., et al.: Shift: a zero flop, zero parameter alternative to spatial convolutions. In: Proceedings of the IEEE Conference on Computer Vision and Pattern Recognition, pp. 9127–9135 (2018)
59. Wu, B., et al.: Visual transformers: token-based image representation and processing for computer vision. arXiv preprint arXiv:2006.03677 (2020)
60. Xia, Z., Chakrabarti, A.: Identifying recurring patterns with deep neural networks for natural image denoising. In: Proceedings of the IEEE/CVF Winter Conference on Applications of Computer Vision, pp. 2426–2434 (2020)
61. Zamir, S.W., Arora, A., Khan, S., Hayat, M., Khan, F.S., Yang, M.H.: Restormer: efficient transformer for high-resolution image restoration. arXiv preprint arXiv:2111.09881 (2021)
62. Zeyde, R., Elad, M., Protter, M.: On single image scale-up using sparse-representations. In: Boissonnat, J.-D., et al. (eds.) Curves and Surfaces 2010. LNCS, vol. 6920, pp. 711–730. Springer, Heidelberg (2012). https://doi.org/10.1007/978-3-642-27413-8_47
63. Zhang, K., Li, Y., Zuo, W., Zhang, L., Van Gool, L., Timofte, R.: Plug-and-play image restoration with deep denoiser prior. IEEE Trans. Pattern Anal. Mach. Intell. **44**, 6360–6376 (2021)
64. Zhang, L., Wu, X., Buades, A., Li, X.: Color demosaicking by local directional interpolation and nonlocal adaptive thresholding. J. Electron. Imaging **20**(2), 023016 (2011)
65. Zhang, Y., Li, K., Li, K., Wang, L., Zhong, B., Fu, Y.: Image super-resolution using very deep residual channel attention networks. In: Ferrari, V., Hebert, M., Sminchisescu, C., Weiss, Y. (eds.) ECCV 2018. LNCS, vol. 11211, pp. 294–310. Springer, Cham (2018). https://doi.org/10.1007/978-3-030-01234-2_18
66. Zhang, Y., Li, K., Li, K., Zhong, B., Fu, Y.: Residual non-local attention networks for image restoration. arXiv preprint arXiv:1903.10082 (2019)
67. Zhang, Y., Tian, Y., Kong, Y., Zhong, B., Fu, Y.: Residual dense network for image super-resolution. In: Proceedings of the IEEE Conference on Computer Vision and Pattern Recognition, pp. 2472–2481 (2018)
68. Zheng, S., et al.: Rethinking semantic segmentation from a sequence-to-sequence perspective with transformers. In: Proceedings of the IEEE/CVF Conference on Computer Vision and Pattern Recognition, pp. 6881–6890 (2021)
69. Zhou, S., Zhang, J., Zuo, W., Loy, C.C.: Cross-scale internal graph neural network for image super-resolution. In: Advances in Neural Information Processing Systems 33, pp. 3499–3509 (2020)

# FlowFormer: A Transformer Architecture for Optical Flow

Zhaoyang Huang[1,3], Xiaoyu Shi[1,3], Chao Zhang[2], Qiang Wang[2], Ka Chun Cheung[3], Hongwei Qin[4], Jifeng Dai[4], and Hongsheng Li[1](✉)

[1] Multimedia Laboratory, The Chinese University of Hong Kong, Shatin, Hong Kong
{drinkingcoder,xiaoyushi}@link.cuhk.edu.hk, hsli@ee.cuhk.edu.hk
[2] Samsung Telecommunication Research, Suwon, South Korea
[3] NVIDIA AI Technology Center, Shanghai, China
[4] SenseTime Research, Shanghai, China

**Abstract.** We introduce optical Flow transFormer, dubbed as FlowFormer, a transformer-based neural network architecture for learning optical flow. FlowFormer tokenizes the 4D cost volume built from an image pair, encodes the cost tokens into a cost memory with alternate-group transformer (AGT) layers in a novel latent space, and decodes the cost memory via a recurrent transformer decoder with dynamic positional cost queries. On the Sintel benchmark, FlowFormer achieves 1.144 and 2.183 average end-ponit-error (AEPE) on the clean and final pass, a 17.6% and 11.6% error reduction from the best published result (1.388 and 2.47). Besides, FlowFormer also achieves strong generalization performance. Without being trained on Sintel, FlowFormer achieves 0.95 AEPE on the Sintel training set clean pass, outperforming the best published result (1.29) by 26.9%.

**Keywords:** Optical flow · Cost volume · Transformer · RAFT

## 1 Introduction

Optical flow targets at estimating per-pixel correspondences between a source image and a target image, in the form of a 2D displacement field. In many downstream video tasks, such as action recognition [36,45,60], video inpainting [13, 28,49], video super-resolution [5,30,38], and frame interpolation [20,33,50], optical flow serves as a fundamental component providing dense correspondences as valuable clues for prediction.

A general assumption adopted in optical flow estimation is that the appearance of corresponding locations in the two images induced from optical flows remains unchanged. Traditionally, optical flow is modeled as an optimization

Z. Huang and X. Shi—Assert equal contributions.

**Supplementary Information** The online version contains supplementary material available at https://doi.org/10.1007/978-3-031-19790-1_40.

© The Author(s), under exclusive license to Springer Nature Switzerland AG 2022
S. Avidan et al. (Eds.): ECCV 2022, LNCS 13677, pp. 668–685, 2022.
https://doi.org/10.1007/978-3-031-19790-1_40

problem that maximizes visual similarities between cross-image corresponding locations with regularization terms. With the rapid development of deep learning and emerging training data, this field has been significantly advanced by deep convolutional neural network-based methods. The recent methods compute costs (i.e. visual similarities) between feature pairs, upon which flows are regressed. Most successful architecture designs in optical flow are achieved via better designs of cost encoding and decoding. PWC-Net [42] and RAFT [46] are two recent representative deep learning-based methods. PWC-Net [42] builds hierarchical local cost volumes with warped features and progressively estimates flows from such local costs. RAFT [46] forms an $H \times W \times H \times W$ 4D cost volume that measures similarities between all pairs of pixels of the $H \times W$ image pair and iteratively retrieves local costs within local windows for regressing flow residuals.

Recently, transformers have attracted much attention for their ability of modeling long-range relations, which can benefit optical flow estimation. Perceiver IO [24] is the pioneering work that learns optical flow regression with a transformer-based architecture. However, it directly operates on pixels of image pairs and ignores the well-established domain knowledge of encoding visual similarities to costs for flow estimation. It thus requires a large number of parameters and $\sim$80$\times$ training examples to capture the desired input-output mapping. We therefore raise a question: can we enjoy both advantages of transformers and the cost volume from the previous milestones? Such a question calls for designing novel transformer architectures for optical flow estimation that can effectively aggregate information from the cost volume. In this paper, we introduce the novel optical Flow TransFormer (FlowFormer) to address this challenging problem.

FlowFormer adopts an encoder-decoder architecture for cost volume encoding and decoding. After building a 4D cost volume, FlowFormer consists of two main components: 1) a cost volume encoder that embeds the 4D cost volume into a latent cost space and fully encodes the cost information in such a space, and 2) a recurrent cost decoder that estimates flows from the encoded latent cost features. Compared with previous works, the main characteristic of our FlowFormer is to adapt the transformer architectures to effectively process cost volumes, which are compact yet rich representations widely explored in optical flow estimation communities, for estimating accurate optical flows.

A naive strategy to transform the 4D cost volume with transformers is directly tokenizing the 4D cost volume and applying transformers. However, such a strategy needs to use thousands of tokens, which is computationally unbearable. To tackle this challenge, we propose two key designs in our cost encoder. We propose a two-step tokenization: 1) converting each of the 2D cost maps, which records visual similarities between one source pixel and all target pixels, from the 4D cost volume into patches as commonly done in transformer networks, and 2) further projecting cost-map patches of each cost map into $K$ latent cost tokens. In this way, the $H \times W \times H \times W$ 4D cost volume can be transformed into $H \times W \times K$ tokens. Secondly, instead of performing self-attention among all tokens, we alternatively conduct attention over tokens within the same cost map and tokens across different cost maps. In other words, an interweaving stack of aggregations of latent cost tokens belonging to the same source pixel and those across different source pixels. Combining these two designs, FlowFormer encodes the cost volume into compact and globally aware latent cost tokens, dubbed as the *cost memory*.

Classical transformer architectures, such as DETR [4], decodes information from the encoded memory via stacked cross-attention layers. In contrast to them, inspired by RAFT, our cost decoder adopts only a recurrent attention layer that formulates the cost decoding as a recurrent query process with dynamic positional cost queries: based on current estimated flows, we query the cost memory for regressing the flow residuals. In each iteration, we compute the corresponding positions in the target image for all source pixels according to current flows and then dynamically update positional cost queries with such positions. Then, we fetch cost features from the cost memory via cross-attention and use a shared gated recurrent unit (GRU) head for residual flow regression. Moreover, RAFT only utilizes a shallow CNN as the image feature encoder. We find that our FlowFormer can be benefited from using an ImageNet-pretrained transformer backbone.

Our contributions can be summarized as fourfold. 1) We propose a novel transformer-based neural network architecture, FlowFormer, for optical flow estimation, which achieves state-of-the-art flow estimation performance. 2) We design a novel cost volume encoder, effectively aggregating cost information into compact latent cost tokens. 3) We propose a recurrent cost decoder that recurrently decodes cost features with dynamic positional cost queries to iteratively refine the estimated optical flows. 4) To the best of our knowledge, we validate for the first time that an ImageNet-pretrained transformer can benefit the estimation of optical flow.

## 2 Related Work

**Optical Flow.** Traditionally, optical flow was modeled as an optimization problem that maximizes visual similarity between image pairs with regularizations [1,2,17,40]. Major improvements in this era came from better designs of similarity and regularization terms. The rise of deep neural networks significantly advanced this field. FlowNet [12] was the first end-to-end convolutional network for optical flow estimation. Its successive work, FlowNet2.0 [23], adopted a stacked architecture with warping operation, performing on par with state-of-the-art (SOTA) methods. Then a series of works, represented by SpyNet [37], PWC-Net [42,43], LiteFlowNet [21,22] and VCN [53], employed coarse-to-fine and iterative estimation methodology. These models inherently suffered from missing small fast-motion objects in coarse stage. To remedy this issue, Teed and Deng [46] proposed RAFT [46], which performs optical flow estimation in a coarse-and-fine (i.e. multi-scale search window in each iteration) and recurrent manner. Based on RAFT architecture, many works [16,25,26,48,57] were proposed to either reduce the computational costs or improve the flow accuracy. Recently, optical flow was extended to more challenging settings, such as low-light [61], foggy [52], and lighting variations [18].

Among these explorations, visual similarity is computed by the correlation of high dimensional features encoded by a convolutional neural network, and the cost volume that contains visual similarity of pixels pairs acts as a core component supporting optical flow estimation. However, their cost information utilization lacks effectiveness. We propose FlowFormer that aggregates the cost volume in a latent space with transformers [47]. Perceiver IO [24] pioneered the use of transformers [4,11,47] that is able to establish long-range relationship in optical flow and achieved state-of-the-art performance. It ignored the cost volume, showing the strong expressive capacity of transformer architecture at the cost of $\sim 80\times$ training examples. In contrast, we propose to keep cost volume as a compact similarity representation and push search space to the extreme by globally aggregating similarity information via a transformer architecture. Such global encoding operation is especially beneficial in the hard cases of large displacement and occlusion.

**Transformers for Computer Vision.** Transformers achieved great success in Natural Language Processing [9,10,47], which inspired the development of self-attention for image classification [8,11,34]. Since then, transformer-based architectures has been introduced into many other vision tasks, such as detection [4], point cloud processing [15,58], image restoration [6,31], video inpainting [32,56], visual grounding [54], etc., and achieves state-of-the-art in most tasks. The appealing performance is generally attributed to the long-range modeling capacity, which is also a desired property in optical flow estimation. One of the challenges that vision transformers are faced with is the large number of visual tokens because the computational cost quadratically increases along with the token number. Twins [8] proposed a spatially separable self-attention (SS Self-Attention) layer that propagates information over tokens arranged in a 2D plane. We also adopt the SS Self-Attention in the cost volume encoder to propagate information inter-cost-maps. Perceiver IO [24] proposed a general transformer backbone, which although requires a large amount of parameters, achieves state-of-the-art optical flow performance. Visual correspondence tasks [7,19,27,44,51] is a main stream in computer vision. Recently, transformers also lead a trend in such tasks [7,27,39,44], which is more related to ours.

## 3 Method

The task of optical flow estimation requires to output a per-pixel displacement field $\mathbf{f} : \mathbb{R}^2 \rightarrow \mathbb{R}^2$ that maps every 2D location $\mathbf{x} \in \mathbb{R}^2$ of a source image $\mathbf{I}_s$ to its corresponding 2D location $\mathbf{p} = \mathbf{x} + \mathbf{f}(\mathbf{x})$ of a target image $\mathbf{I}_t$. To take advantage of the recent vision transformer architectures as well as the 4D cost volumes widely utilized by previous CNN-based optical flow estimation methods, we propose FlowFormer, a transformer-based architecture that encodes and decodes the 4D cost volume to achieve accurate optical flow estimation. In Fig. 1, we show the overview architecture of FlowFormer, which processes the 4D cost volumes from siamese features with two main components: 1) a cost volume encoder that encodes the 4D cost volume into a latent space to form cost memory, and 2) a cost memory decoder for predicting a per-pixel displacement field based on the encoded cost memory and contextual features.

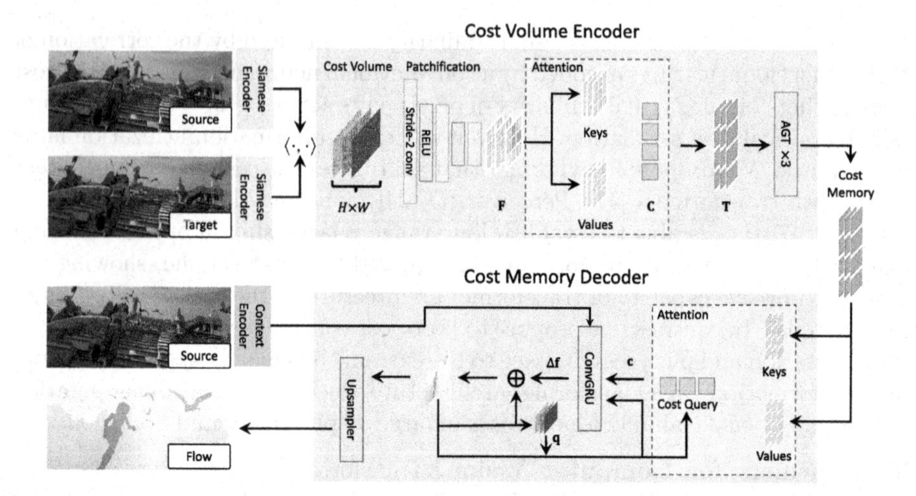


**Fig. 1.** Architecture of FlowFormer. FlowFormer estimates optical flow in three steps: 1) building a 4D cost volume from image features. 2) A cost volume encoder that encodes the cost volume into the cost memory. 3) A recurrent transformer decoder that decodes the cost memory with the source image context features into flows.

### 3.1 Building the 4D Cost Volume

A backbone vision network is used to extract an $H \times W \times D_f$ feature map from an input $H_I \times W_I \times 3$ RGB image, where typically we set $(H, W) = (H_I/8, W_I/8)$. After extracting the feature maps of the source image and the target image, we construct an $H \times W \times H \times W$ 4D cost volume by computing the dot-product similarities between all pixel pairs between the source and target feature maps.

### 3.2 Cost Volume Encoder

To estimate optical flows, the corresponding positions in the target image of source pixels need to be identified based on source-target visual similarities encoded in the 4D cost volume. The built 4D cost volume can be viewed as a series of 2D cost maps of size $H \times W$, each of which measures visual similarities between a single source pixel and all target pixels. We denote source pixel $\mathbf{x}$'s cost map as $\mathbf{M}_\mathbf{x} \in \mathbb{R}^{H \times W}$. Finding corresponding positions in such cost maps is generally challenging, as there might exist repeated patterns and non-discriminative regions in the two images. The task becomes even more challenging when only considering costs from a local window of the map, as previous CNN-based optical flow estimation methods do. Even for estimating a single source pixel's accurate displacement, it is beneficial to take its contextual source pixels' cost maps into consideration.

To tackle this challenging problem, we propose a transformer-based cost volume encoder that encodes the whole cost volume into a *cost memory*. Our cost volume encoder consists of three steps: 1) cost map patchification, 2) cost patch token embedding, and 3) cost memory encoding. We elaborate the details of the three steps as follows.

**Cost Map Patchification.** Following existing vision transformers, we patchify the cost map $\mathbf{M_x} \in \mathbb{R}^{H \times W}$ of each source pixel $\mathbf{x}$ with strided convolutions to obtain a sequence of cost patch embeddings. Specifically, given an $H \times W$ cost map, we first pad zeros at its right and bottom sides to make its width and height multiples of 8. The padded cost map is then transformed by a stack of three stride-2 convolutions followed by ReLU into a feature map $\mathbf{F_x} \in \mathbb{R}^{\lceil H/8 \rceil \times \lceil W/8 \rceil \times D_p}$. Each feature in the feature map stands for an $8 \times 8$ patch in the input cost map. The three convolutions have output channels of $D_p/4$, $D_p/2$, $D_p$, respectively.

**Patch Feature Tokenization via Latent Summarization.** Although the patchification results in a sequence of cost patch feature vectors for each source pixel, the number of such patch features is still large and hinders the efficiency of information propagation among different source pixels. Actually, a cost map is highly redundant because only a few high costs are most informative. To obtain more compact cost features, we further summarize the patch features $\mathbf{F_x}$ of each source pixel $\mathbf{x}$ via $K$ latent codewords $\mathbf{C} \in \mathbb{R}^{K \times D}$. Specifically, the latent codewords query each source pixel's cost-patch features to further summarize each cost map into $K$ latent vectors of $D$ dimensions via the dot-product attention mechanism. The latent codewords $\mathbf{C} \in \mathbb{R}^{K \times D}$ are randomly initialized, updated via back-propagation, and shared across all source pixels. The latent representations $\mathbf{T_x}$ for summarizing $\mathbf{F_x}$ are obtained as

$$\begin{aligned} \mathbf{K_x} &= \text{Conv}_{1\times1}\left(\text{Concat}(\mathbf{F_x}, \text{PE})\right), \\ \mathbf{V_x} &= \text{Conv}_{1\times1}\left(\text{Concat}(\mathbf{F_x}, \text{PE})\right), \\ \mathbf{T_x} &= \text{Attention}(\mathbf{C}, \mathbf{K_x}, \mathbf{V_x}). \end{aligned} \tag{1}$$

Before projecting the cost-patch features $\mathbf{F_x}$ to obtain keys $\mathbf{K_x}$ and values $\mathbf{V_x}$, the patch features are concatenated with a sequence of positional embeddings $\text{PE} \in \mathbb{R}^{\lceil H/8 \rceil \times \lceil W/8 \rceil \times D_p}$. Given a 2D position $\mathbf{p}$, we encode it into a positional embedding of length $D_p$ following COTR [27]. Finally, the cost map of the source pixel $\mathbf{x}$ can be summarized into $K$ latent representations $\mathbf{T_x} \in \mathbb{R}^{K \times D}$ by conducting multi-head dot-product attention with the queries, keys, and values. Generally, $K \times D \ll H \times W$ and the latent summarizations $\mathbf{T_x}$ therefore provides more compact representations than each $H \times W$ cost map for each source pixel $\mathbf{x}$. For all source pixels in the image, there are a total of $(H \times W)$ 2D cost maps. Their summarized representations can consequently be converted into a latent 4D cost volume $\mathbf{T} \in \mathbb{R}^{H \times W \times K \times D}$.

**Attention in the Latent Cost Space.** The aforementioned two stages transform the original 4D cost volume into a latent and compact 4D cost volume $\mathbf{T}$. However, it is still too expensive to directly apply self-attention over all the vectors in the 4D volume because the computational cost quadratically increases with the number of tokens. As shown in Fig. 2, we propose an alternate-group

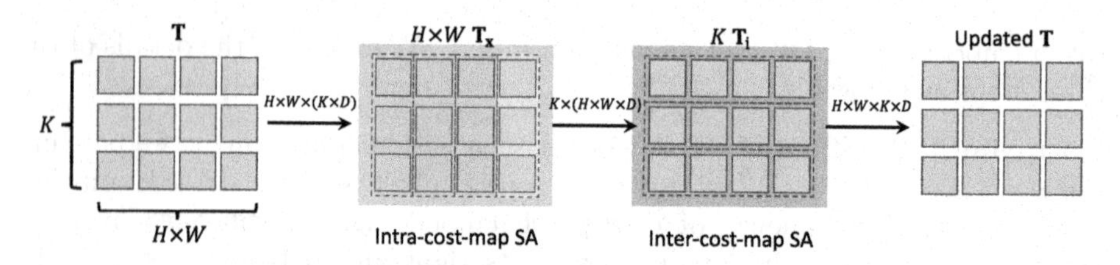


**Fig. 2.** Alternate-Group Transformer Layer. The alternate-group transformer layer (AGT) alternatively groups tokens in $\mathbf{T}$ into $H \times W$ groups that contains $K$ tokens ($\mathbf{T}_{\mathbf{x}}$) and $K$ groups that contains $H \times W$ tokens ($\mathbf{T}_i$), and encode tokens inside groups via self-attention and ss self-attention [8] respectively.

transformer layer (AGT) that groups the tokens in two mutually orthogonal manners and apply attentions in the two groups alternatively, which reduces the cost of attention while still being able to propagate information among all tokens.

The first grouping is conducted for each source pixel, i.e., each $\mathbf{T}_{\mathbf{x}} \in \mathbb{R}^{K \times D}$ forms a group and the self-attention is conducted within each group.

$$\mathbf{T}_{\mathbf{x}} = \text{FFN}(\text{Self-Attention}(\mathbf{T}_{\mathbf{x}}(1), \dots, \mathbf{T}_{\mathbf{x}}(K)) \quad \text{for all } \mathbf{x} \text{ in } \mathbf{I}_s, \tag{2}$$

where $\mathbf{T}_{\mathbf{x}}(i)$ denotes the $i$-th latent representation for encoding the source pixel $\mathbf{x}$'s cost map. After the self-attention is conducted between all $K$ latent tokens for each source pixel $\mathbf{x}$, updated $\mathbf{T}_{\mathbf{x}}$ are further transformed by a feed-forward network (FFN) and then re-organized back to form the updated 4D cost volume $\mathbf{T}$. Both the self-attention and FFN sub-layers adopt the common designs of residual connection and layer normalization of transformers. This self-attention operation propagates the information within each cost map and we name it as intra-cost-map self-attention.

The second way groups all the latent cost tokens $\mathbf{T} \in \mathbb{R}^{H \times W \times K \times D}$ into $K$ groups according to the $K$ different latent representations. Each group would therefore have $(H \times W)$ tokens of dimension $D$ for information propagation in the spatial domain via the spatially separable self-attention (SS-SelfAttention) proposed in Twins [8],

$$\mathbf{T}_i = \text{FFN}(\text{SS-SelfAttention}(\mathbf{T}_i)) \quad \text{for } i = 1, 2, \dots, K, \tag{3}$$

where we slightly abuse the notation and denote $\mathbf{T}_i \in \mathbb{R}^{(H \times W) \times D}$ as the $i$-th group. The updated $\mathbf{T}_i$'s are then re-organized back to obtain the updated 4D latent cost volume $\mathbf{T}$. Moreover, visually similar source pixels should have coherent flows, which has been validated by previous methods [7,25]. Thus, we integrate appearance affinities between different source pixels into SS-SelfAttention via concatenating the source image's context features $\mathbf{t}$ with the cost tokens when generating queries and keys. We call this layer inter-cost-map self-attention layer as it propagates information of cost volume across different source pixels. Note that these two operations are different from CATs [7], which augmented correlations 'intra' a level of cost map and 'inter' multi-level correlation layers.

The above self-attention operations' parameters are shared across different groups and they are sequentially operated to form the proposed alternate-group attention layer. By stacking the alternate-group transformer layer multiple times, the latent cost tokens can effectively exchange information across source pixels and across latent representations to better encode the 4D cost volume. In this way, our cost volume encoder transforms the $H \times W \times H \times W$ 4D cost volume to $H \times W \times K$ latent tokens of length $D$. We call the final $H \times W \times K$ tokens as the *cost memory*, which is to be decoded for optical flow estimation.

### 3.3 Cost Memory Decoder for Flow Estimation

Given the cost memory encoded by the cost volume encoder, we propose a cost memory decoder to predict optical flows. Since the original resolution of the input image is $H_I \times W_I$, we estimate optical flow at the $H \times W$ resolution and then upsample the predicted flows to the original resolution with a learnable convex upsampler [46]. However, in contrast to previous vision transformers that seek abstract semantic features, optical flow estimation requires recovering dense correspondences from the cost memory. Inspired by RAFT [46], we propose to use cost queries to retrieve cost features from the cost memory and iteratively refine flow predictions with a recurrent attention decoder layer.

**Cost Memory Aggregation.** For predicting the flows of the $H \times W$ source pixels, we generate a sequence of $(H \times W)$ cost queries, each of which is responsible for estimating the flow of a single source pixel via co-attention on the cost memory. To generate the cost query $\mathbf{Q_x}$ for a source pixel $\mathbf{x}$, we first compute its corresponding location in the target image given its current estimated flow $\mathbf{f}(\mathbf{x})$ as $\mathbf{p} = \mathbf{x} + \mathbf{f}(\mathbf{x})$. We then retrieve a local $9 \times 9$ cost-map patch $\mathbf{q_x} = \mathrm{Crop}_{9\times 9}(\mathbf{M_x}, \mathbf{p})$ by cropping costs inside the $9 \times 9$ local window centered at $\mathbf{p}$ on the cost map $\mathbf{M_x}$. The cost query $\mathbf{Q_x}$ is then formulated based on the features $\mathrm{FFN}(\mathbf{q_x})$ that encoded from the local costs $\mathbf{q_x}$ and $\mathbf{p}$'s positional embedding $\mathrm{PE}(\mathbf{p})$, which can aggregate information from source pixel $\mathbf{x}$'s cost memory $\mathbf{T_x}$ via cross-attention,

$$\begin{aligned} \mathbf{Q_x} &= \mathrm{FFN}\left(\mathrm{FFN}(\mathbf{q_x}) + \mathrm{PE}(\mathbf{p})\right), \\ \mathbf{K_x} &= \mathrm{FFN}\left(\mathbf{T_x}\right), \quad \mathbf{V_x} = \mathrm{FFN}\left(\mathbf{T_x}\right), \\ \mathbf{c_x} &= \mathrm{Attention}(\mathbf{Q_x}, \mathbf{K_x}, \mathbf{V_x}). \end{aligned} \tag{4}$$

The cross-attention summarizes information from the cost memory for each source pixel to predict its flow. As $\mathbf{Q_x}$ is dynamically updated in terms of the fed position at each iteration, we call it as dynamic positional cost query. We note that keys and values can be generated at the beginning and re-used in subsequent iterations, which saves computation as a benefit of our recurrent decoder.

**Recurrent Flow Prediction.** Our cost decoder iteratively regresses flow residuals $\Delta\mathbf{f}(\mathbf{x})$ to refine the flow of each source pixel $\mathbf{x}$ as $\mathbf{f}(\mathbf{x}) \leftarrow \mathbf{f}(\mathbf{x}) + \Delta\mathbf{f}(\mathbf{x})$. We adopt a ConvGRU module and follow the similar design to that in GMA-RAFT [25] for flow refinement. However, the key difference of our recurrent module is the use of cost queries to adaptively aggregate information from the cost memory for more accurate flow estimation. Specifically, at each iteration, the ConvGRU unit takes as input the concatenation of retrieved cost features and cost-map patch Concat($\mathbf{c}_\mathbf{x}, \mathbf{q}_\mathbf{x}$), the source-image context feature $\mathbf{t}_\mathbf{x}$ from the context network, and the current estimated flow $\mathbf{f}$, and outputs the predicted flow residuals as follows,

$$\Delta\mathbf{f}(\mathbf{x}) = \text{ConvGRU}(\text{Concat}(\mathbf{c}_\mathbf{x}, \mathbf{q}_\mathbf{x}), \mathbf{t}_\mathbf{x}, \mathbf{f}(\mathbf{x})). \quad (5)$$

The flows generated at each iteration are unsampled to the size of the source image via a convex upsampler following [46] and supervised by ground-truth flows at all recurrent iterations with increasing weights.

# 4 Experiment

We evaluate our FlowFormer on the Sintel [3] and the KITTI-2015 [14] benchmarks. Following previous works, we train FlowFormer on FlyingChairs [12] and FlyingThings [35], and then respectively finetune it for Sintel and KITTI benchmark. Flowformer achieves state-of-the-art performance on both benchmarks.

**Experimental Setup.** We use the average end-point-error (AEPE) and F1-All (%) metric for evaluation. The AEPE computes mean flow error over all valid pixels. The F1-all, which refers to the percentage of pixels whose flow error is larger than 3 pixels or over 5% of length of ground truth flows. The Sintel dataset is rendered from the same model but in two passes, i.e. clean pass and final pass. The clean pass is rendered with smooth shading and specular reflections. The final pass uses full rendering settings including motion blur, camera depth-of-field blur, and atmospheric effects.

**Implementation Details.** The image feature encoder of our final FlowFormer is chosen as the first two stages of ImageNet-pretrained Twins-SVT [8], which encodes an image into $D_f = 256$-channel feature map of 1/8 image size. The cost volume encoder patchifies each cost map to a $D_p = 64$-channel feature map and further summarizes the feature map to $N = 8$ cost tokens of $K = 128$ dimensions. Then, the cost volume encoder encodes the cost tokens with 3 AGT layers. Following previous optical flow training procedure [25], we pre-train FlowFormer on FlyingChairs [12] for 120k iterations with a batch size of 8, and on FlyingThings [35] for 120k iterations with a batch size of 6 (denoted as 'C+T'). After pre-training, we finetune FlowFormer on the data combined from FlyingThings, Sintel, KITTI-2015, and HD1K [29] (denoted as 'C+T+S+K+H') for 120k iterations with a batch size of 6. To achieve the best performance on the KITTI benchmark, we also further finetune FlowFormer on the KITTI-2015 for 50k iterations with a batch size of 6. We use the one-cycle learning rate scheduler. The highest learning rate is set as $2.5 \times 10^{-4}$ on FlyingChairs and $1.25 \times 10^{-4}$ on the other training sets. As positional encodings used in transformers are sensitive to image size, we crop the image pairs for flow estimation and tile them to obtain complete flows following Perceiver IO [24]. We use fixed Gaussian weights for tile, which will be detailed in the supplementary materials.

**Table 1.** Experiments on Sintel [3] and KITTI [14] datasets. * denotes that the methods use the warm-start strategy [46], which relies on previous image frames in a video. 'A' denotes the autoflow dataset. 'C + T' denotes training only on the FlyingChairs and FlyingThings datasets. '+ S + K + H' denotes finetuning on the combination of Sintel, KITTI, and HD1K training sets. Our FlowFormer achieves best generalization performance (C+T) and ranks 1st on the Sintel benchmark (C+T+S+K+H).

| Training data | Method | Sintel (train) | | KITTI-15 (train) | | Sintel (test) | | KITTI-15 (test) |
|---|---|---|---|---|---|---|---|---|
| | | Clean | Final | F1-epe | F1-all | Clean | Final | F1-all |
| A+S+K+H | Perceiver IO [24] | - | - | - | - | 1.81 | 2.42 | 4.98 |
| | PWC-Net [42] | - | - | - | - | 2.17 | 2.91 | 5.76 |
| | RAFT [46] | - | - | - | - | 1.95 | 2.57 | 4.23 |
| C+T | HD3 [55] | 3.84 | 8.77 | 13.17 | 24.0 | - | - | - |
| | LiteFlowNet [21] | 2.48 | 4.04 | 10.39 | 28.5 | - | - | - |
| | PWC-Net [42] | 2.55 | 3.93 | 10.35 | 33.7 | - | - | - |
| | LiteFlowNet2 [22] | 2.24 | 3.78 | 8.97 | 25.9 | - | - | - |
| | S-Flow [57] | 1.30 | 2.59 | 4.60 | 15.9 | | | |
| | RAFT [46] | 1.43 | 2.71 | 5.04 | 17.4 | - | - | - |
| | FM-RAFT [26] | 1.29 | 2.95 | 6.80 | 19.3 | - | - | - |
| | GMA [25] | 1.30 | 2.74 | 4.69 | 17.1 | - | - | - |
| | Ours | **0.95** | **2.35** | **4.09** | **14.72** | - | - | - |
| C+T+S+K+H | LiteFlowNet2 [22] | (1.30) | (1.62) | (1.47) | (4.8) | 3.48 | 4.69 | 7.74 |
| | PWC-Net+ [43] | (1.71) | (2.34) | (1.50) | (5.3) | 3.45 | 4.60 | 7.72 |
| | VCN [53] | (1.66) | (2.24) | (1.16) | (4.1) | 2.81 | 4.40 | 6.30 |
| | MaskFlowNet [59] | - | - | - | - | 2.52 | 4.17 | 6.10 |
| | S-Flow [57] | (0.69) | (1.10) | (0.69) | (1.60) | 1.50 | 2.67 | **4.64** |
| | RAFT [46] | (0.76) | (1.22) | (0.63) | (1.5) | 1.94 | 3.18 | 5.10 |
| | FM-RAFT [26] | (0.79) | (1.70) | (0.75) | (2.1) | 1.72 | 3.60 | 6.17 |
| | GMA [25] | - | - | - | - | 1.40 | 2.88 | 5.15 |
| | Ours | (0.48) | (0.74) | (0.53) | (1.11) | **1.14** | **2.18** | 4.68 |
| | RAFT* [46] | (0.77) | (1.27) | - | - | 1.61 | 2.86 | - |
| | GMA* [25] | (0.62) | (1.06) | (0.57) | (1.2) | 1.39 | 2.47 | - |

### 4.1 Quantitative Experiment

We evaluate FlowFormer on the well-known Sintel and KITTI benchmarks as shown in Table 1. GMA [25], an improved version of RAFT [46], is the most competitive flow estimation method at present. After being trained on FlyingChairs and FlyingThings, we evaluate the generalization performance of FlowFormer on the training set of Sintel and KITTI-2015. By further finetuning FlowFormer on the combination of HD1K, Sintel and KITTI training sets, we compare the dataset-specific accuracy of optical flow models. Autoflow [41] is a dataset that provides training data covering various challenging visual disturbance, but its training code is not released yet.

**Generalization Performance.** We train FlowFormer on the FlyingChairs and FlyingThings (C+T), and evaluate it on the training set of Sintel and KITTI-2015. This settings evaluates the generalization performance of optical flow models. FlowFormer ranks 1st among all compared methods on both benchmarks. FlowFormer achieves 0.95 and 2.35 on the clean and final pass of Sintel. On the KITTI-2015 training set, FlowFormer achieves 4.09 F1-epe and 14.72 F1-all.

Compared to GMA, FlowFormer reduces 26.9% and 14.2% errors on Sintel clean and final, and 13.9% errors on KITTI-2015 F1-all, which shows its extraordinary generalization performance.

**Sintel Benchmark.** We finetune the pretrained FlowFormer on the combination of training data of FlyingThings, HD1K, Sintel and KITTI-2015, and then evaluate it on the Sintel test set. FlowFormer achieves 1.14 and 2.18 on the Sintel clean and final, 17.6% and 11.6% lower error compared to GMA*, which ranks both 1st on the Sintel benchmark. It is noteworthy that RAFT* and GMA* use the warm-start strategy that requires image sequences while FlowFormer does not. Compared with GMA, which also does not use the warm-start, FlowFormer obtains 18.6% and 24.3% error reduction. RAFT trained on the autoflow dataset (A+S+K+H) significantly outperforms RAFT trained on the C+T+S+K+H on final pass because autoflow provides training image pairs that are more challenging. We believe training FlowFormer with autoflow can achieve better accuracy but it is not released yet.

**KITTI-2015 Benchmark.** We further finetune the FlowFormer on the KITTI-2015 training set after the Sintel finetuning stage and evaluate it on the KITTI test set. FlowFormer achieves 4.68, ranking 2nd on the KITTI-2015 benchmark. S-Flow [57] obtains slightly smaller error than FlowFormer on KITTI (−0.85%), which, however, is significantly worse on Sintel (31.6% and 22.5% larger error on clean and final pass). S-Flow finds corresponding points by computing the coordinate expectation weighted by refined cost maps. Images in the KITTI dataset are captured in urban traffic scenes, which contains objects that are mostly rigid. Flows on rigid objects are rather simple, which is easier for cost-based coordinate expectation, but the assumption can be easily violated in non-rigid scenarios such as Sintel.

### 4.2 Qualitative Experiment

We visualize flows that estimated by our FlowFormer and GMA of three examples in Fig. 3 to qualitatively show how FlowFormer outperforms GMA. As transformers can encode the cost information at a large perceptive field, FlowFormer can distinguish overlapping objects via contextual information and thus reduce the leakage of flows over boundaries. Compared with GMA, the flows that are estimated by FlowFormer on boundaries of the bamboo and the human body are more precise and clear. Besides, FlowFormer can also recover motion details that are ignored by GMA, such as the hair and the holes on the box.

### 4.3 Ablation Study

We conduct a series of ablation experiments in Table 2. We start from RAFT as the baseline, which directly regresses residual flows with the multi-level cost retrieval (MCR) decoder, and gradually replace its components with our proposed components. We first replace RAFT's MCR decoder with the latent cost tokenization (LCT) part of our encoder and the cost memory decoder (CMD)

**Fig. 3.** Qualitative comparison on the Sintel test set. FlowFormer greatly reduces the flow leakage around object boundaries (pointed by red arrows) and clearer details (pointed by blue arrows). (Color figure online)

(denoted as 'MCR→LCT+CMD'). Note that our cost memory decoder cannot be used alone on top of the 4D cost volume of RAFT because of the too large number of tokens. It must be combined with our latent cost tokens ($\mathbf{T_x}$ from Eq. (1)). Encoding $K = 8$ latent tokens of $D = 128$ dimensions for each source pixel achieves the best performance. Based on LCT+CMD with $K = 8$ and $D = 128$, we replace RAFT's CNN image feature encoder with Twins-SVT (denoted as 'CNN→Twins'). We then further add attention layers of the proposed cost volume encoder to encode and update latent cost tokens. The proposed Alternate-Group Transformer (AGT) layer consists of two types of attention, i.e., intra-cost-map attention and inter-cost-map attention. We first add a single intra-cost-map attention layer (denoted as '+Intra.'), and then add the inter-cost-map attention (denoted as 'AGT×1 (+Intra.+Inter.)', which is equivalent to adding a single AGT layer. We then test on increasing the number of AGT layers to 2 and 3. Following RAFT, all models are trained on FlyingChairs [12] with 100k iterations and FlyingThings [35] with 60k iterations, and then evaluated on the training set of Sintel [3] and KITTI-2015 [14].

**MCR → LCT+MCD.** The number of latent tokens $K$ and token dimension $D$ determine how much cost volume information the cost tokens can encode. From $K = 4, D = 32$ to $K = 8, D = 128$, the AEPE decreases because the cost tokens summarizes more cost map information and benefits the residual flow regression. The latent cost tokens are capable of summarizing whole-image information and our MCD can absorb interested information from them through co-attention, while the MCR decoder of RAFT only retrieves multi-level costs inside flow-guided local windows. Therefore, even without our AGT layers in our encoder, LCT+MCD still shows better performance than MCR decoder of RAFT.

**Table 2.** Ablation study. We gradually change one component of the RAFT at a time to obtain our FlowFormer model. MCR→LCT+CMD: replacing RAFT's decoder with OUR latent cost tokens + cost memory decoder. CNN→Twins: replacing RAFT's CNN encoder with Twins-SVT transformer. Cost Encoding: adding intra-cost-map and inter-cost-map to form an Alternate-Group Transformer layer in the encoder. 3 AGT layers are used in our final model.

| Experiment | Method | Sintel (train) | | KITTI-15 (train) | | Params. |
|---|---|---|---|---|---|---|
| | | Clean | Final | F1-epe | F1-all | |
| Baseline | RAFT | 1.53 | 2.99 | 5.73 | 18.29 | 5.3M |
| MCR→LCT+CMD | $K = 4$, $D = 32$ | 1.66 | 2.93 | 5.60 | 19.67 | 5.5M |
| | $K = 8$, $D = 32$ | 1.58 | 2.90 | 5.50 | 18.71 | 5.5M |
| | $K - 8$, $D = 128$ | 1.44 | 2.80 | 5.22 | 17.64 | 5.6M |
| CNN → Twins | CNN | 1.44 | 2.80 | 5.22 | 17.64 | 5.6M |
| | Twins from Scratch | 1.44 | 2.86 | 5.38 | 17.58 | 14.0M |
| | Pretrained Twins | 1.29 | 2.72 | 4.82 | 16.16 | 14.0M |
| Cost Encoding | None | 1.29 | 2.72 | 4.82 | 16.16 | 14.0M |
| | +Intra | 1.29 | 2.89 | 4.74 | 15.71 | 14.1M |
| | AGT×1 (+Intra.+Inter.) | 1.20 | 2.85 | 4.57 | 15.46 | 15.2M |
| | AGT×2 | 1.16 | 2.66 | 4.70 | 16.01 | 16.4M |
| | AGT×3 | 1.10 | 2.57 | 4.45 | 15.15 | 17.6M |

**CNN vs. Transformer Image Encoder.** In the CNN→Twins experiment, the AEPE of Twins trained from scratch is marginally worse than CNN, but the ImageNet-pretraining is beneficial, because Twins is a transformer architecture with larger receptive field and model capacity, which requires more training examples for sufficient training.

**Cost Encoding.** In the cost volume encoder, we encode and update the latent cost tokens with an intra-cost-map attention operation and an inter-cost-map attention operation. The two operations form an Alternate-Group Transformer (AGT) layer. Then we gradually increase the number of AGT layers to 3. From no attention layer to AGT×3, the errors gradually decrease, which demonstrates that encoding latent cost tokens with our AGT layers benefits flow estimation.

**FlowFormer vs. GMA.** We train all the models with the settings of GMA. The full version of FlowFormer has 18.2M parameters, which is larger than GMA. One of the causes is that FlowFormer uses the first two stages of ImageNet-pretrained Twins-SVT as the image feature encoder while GMA uses a CNN. We present an experiment to compare FlowFormer and GMA with aligned settings in Table 3. We first provide a small version of FlowFormer using GMA's CNN image encoder and also set $K = 4$, $D = 32$, and AGT×1. Although the smaller version of FlowFormer (denoted as 'Ours (small)') has a significant performance drop compared to the full version of FlowFormer, it still outperforms GMA in terms of all metrics. We also design two enhanced GMA models and compare them with the full version of FlowFormer to show that the performance improvements

**Table 3.** FlowFormer v.s. GMA. Ours (small) is a small version of FlowFormer and uses the CNN image feature encoder of GMA. GMA-L is a large version of GMA. GMA-Twins replace its CNN image feature encoder with pre-trained Twins. (+x%) indicates that this model obtains x% larger error than ours.

| Method | Sintel (train) | | KITTI-15 (train) | | Parameters |
|---|---|---|---|---|---|
| | Clean | Final | F1-epe | F1-all | |
| GMA [25] | 1.30 (+30%) | 2.74 (+12%) | 4.69 (+15%) | 17.1 (+16%) | 5.9M |
| Ours (small) | 1.20 (+20%) | 2.64 (+8%) | 4.57 (+12%) | 16.62 (+13%) | 6.2M |
| GMA-L [25] | 1.33 (+33%) | 2.56 (+4%) | 4.40 (+8%) | 15.93 (+8%) | 17.0M |
| GMA-Twins [25] | 1.15 (+15%) | 2.73 (+11%) | 4.98 (+22%) | 16.82 (+14%) | 14.2M |
| Ours | **1.00** | **2.45** | **4.09** | **14.72** | 18.2M |

are not simply derived from adding more parameters. The first one is denoted as 'GMA-L', a large version of GMA and the second one is denoted as 'GMA-Twins' which also adopts the pretrained Twins as the image encoder. In this experiment, we train all models on FlyingChairs with 120k iterations and FlyingThings with 120k iterations. Similar to reducing RAFT to RAFT (small) [46], GMA-L enlarges GMA by doubling feature channels, which has 17M parameters, comparable to FlowFormer. However, its performance degrades in Sintel clean, a 33% larger error than FlowFormer. GMA-Twins replaces the CNN image encoder with the shallow Image-Net pre-trained Twins-SVT as FlowFormer does. The largest improvement of GMA-Twins upon GMA is on the Sintel clean, but it still has a 15% larger error than FlowFormer. GMA-Twins does not lead to significant error reduction on other metrics and is even worse on the KITTI-15. In conclusion, the performance improvement of FlowFormer is not derived from more parameters but the novel design of the architecture.

## 5 Conclusion

We have proposed FlowFormer, a Transformer-based architecture for optical flow estimation. FlowFormer summarizes the $H \times W \times H \times W$ 4D cost volume built from a pair of images as $H \times W \times K$ tokens of length $D$, and then efficiently and effectively encodes the cost tokens via the alternate-group transformer (AGT). Thanks to such design, the generated cost memory is able to grasp essential information over the cost volume and obtain compact cost features. Finally, the cost memory decoder absorbs cost information from the cost memory with dynamic positional cost queries, which gets rid of the limitation of local windows, for residual flow regression. To our best knowledge, FlowFormer is the first method that deeply integrates transformers with cost volumes for optical flow estimation. Thanks to the compact cost tokens and long-range relation modeling ability of transformers, FlowFormer achieves state-of-the-art accuracy and shows strong cross-dataset generalization.

**Acknowledgements.** Hongsheng Li is also a Principal Investigator of Centre for Perceptual and Interactive Intelligence Limited (CPII). This work is supported in part by CPII, in part by the General Research Fund through the Research Grants Council of Hong Kong under Grants (Nos. 14204021, 14207319), in part by CUHK Strategic Fund.

## References

1. Black, M.J., Anandan, P.: A framework for the robust estimation of optical flow. In: 1993 (4th) International Conference on Computer Vision, pp. 231–236. IEEE (1993)
2. Bruhn, A., Weickert, J., Schnörr, C.: Lucas/Kanade meets Horn/Schunck: combining local and global optic flow methods. Int. J. Comput. Vis. **61**(3), 211–231 (2005)
3. Butler, D.J., Wulff, J., Stanley, G.B., Black, M.J.: A naturalistic open source movie for optical flow evaluation. In: Fitzgibbon, A., Lazebnik, S., Perona, P., Sato, Y., Schmid, C. (eds.) ECCV 2012. LNCS, vol. 7577, pp. 611–625. Springer, Heidelberg (2012). https://doi.org/10.1007/978-3-642-33783-3_44
4. Carion, N., Massa, F., Synnaeve, G., Usunier, N., Kirillov, A., Zagoruyko, S.: End-to-end object detection with transformers. In: Vedaldi, A., Bischof, H., Brox, T., Frahm, J.-M. (eds.) ECCV 2020. LNCS, vol. 12346, pp. 213–229. Springer, Cham (2020). https://doi.org/10.1007/978-3-030-58452-8_13
5. Chan, K.C., Wang, X., Yu, K., Dong, C., Loy, C.C.: BasicVSR: the search for essential components in video super-resolution and beyond. In: Proceedings of the IEEE/CVF Conference on Computer Vision and Pattern Recognition, pp. 4947–4956 (2021)
6. Chen, H., et al.: Pre-trained image processing transformer. In: Proceedings of the IEEE/CVF Conference on Computer Vision and Pattern Recognition, pp. 12299–12310 (2021)
7. Cho, S., Hong, S., Jeon, S., Lee, Y., Sohn, K., Kim, S.: CATs: cost aggregation transformers for visual correspondence. In: Advances in Neural Information Processing Systems, vol. 34 (2021)
8. Chu, X., et al.: Twins: revisiting spatial attention design in vision transformers. arXiv preprint arXiv:2104.13840 (2021)
9. Dai, Z., Yang, Z., Yang, Y., Carbonell, J., Le, Q.V., Salakhutdinov, R.: Transformer-XL: attentive language models beyond a fixed-length context. arXiv preprint arXiv:1901.02860 (2019)
10. Devlin, J., Chang, M.W., Lee, K., Toutanova, K.: BERT: pre-training of deep bidirectional transformers for language understanding. arXiv preprint arXiv:1810.04805 (2018)
11. Dosovitskiy, A., et al.: An image is worth $16 \times 16$ words: transformers for image recognition at scale. arXiv preprint arXiv:2010.11929 (2020)
12. Dosovitskiy, A., et al.: FlowNet: learning optical flow with convolutional networks. In: Proceedings of the IEEE International Conference on Computer Vision, pp. 2758–2766 (2015)
13. Gao, C., Saraf, A., Huang, J.-B., Kopf, J.: Flow-edge guided video completion. In: Vedaldi, A., Bischof, H., Brox, T., Frahm, J.-M. (eds.) ECCV 2020. LNCS, vol. 12357, pp. 713–729. Springer, Cham (2020). https://doi.org/10.1007/978-3-030-58610-2_42

14. Geiger, A., Lenz, P., Stiller, C., Urtasun, R.: Vision meets robotics: the KITTI dataset. Int. J. Robot. Res. **32**(11), 1231–1237 (2013)
15. Guo, M.H., Cai, J.X., Liu, Z.N., Mu, T.J., Martin, R.R., Hu, S.M.: PCT: point cloud transformer. Comput. Vis. Media **7**(2), 187–199 (2021)
16. Hofinger, M., Bulò, S.R., Porzi, L., Knapitsch, A., Pock, T., Kontschieder, P.: Improving optical flow on a pyramid level. In: Vedaldi, A., Bischof, H., Brox, T., Frahm, J.-M. (eds.) ECCV 2020. LNCS, vol. 12373, pp. 770–786. Springer, Cham (2020). https://doi.org/10.1007/978-3-030-58604-1_46
17. Horn, B.K., Schunck, B.G.: Determining optical flow. Artif. Intell. **17**(1–3), 185–203 (1981)
18. Huang, Z., et al.: Life: lighting invariant flow estimation. arXiv preprint arXiv:2104.03097 (2021)
19. Huang, Z., et al.: VS-Net: voting with segmentation for visual localization. In: Proceedings of the IEEE/CVF Conference on Computer Vision and Pattern Recognition, pp. 6101–6111 (2021)
20. Huang, Z., Zhang, T., Heng, W., Shi, B., Zhou, S.: RIFE: real-time intermediate flow estimation for video frame interpolation. arXiv preprint arXiv:2011.06294 (2020)
21. Hui, T.W., Tang, X., Loy, C.C.: LiteFlowNet: a lightweight convolutional neural network for optical flow estimation. In: Proceedings of the IEEE Conference on Computer Vision and Pattern Recognition, pp. 8981–8989 (2018)
22. Hui, T.W., Tang, X., Loy, C.C.: A lightweight optical flow CNN-revisiting data fidelity and regularization. IEEE Trans. Pattern Anal. Mach. Intell. **43**(8), 2555–2569 (2020)
23. Ilg, E., Mayer, N., Saikia, T., Keuper, M., Dosovitskiy, A., Brox, T.: FlowNet 2.0: evolution of optical flow estimation with deep networks. In: Proceedings of the IEEE Conference on Computer Vision and Pattern Recognition, pp. 2462–2470 (2017)
24. Jaegle, A., et al.: Perceiver IO: a general architecture for structured inputs & outputs. arXiv preprint arXiv:2107.14795 (2021)
25. Jiang, S., Campbell, D., Lu, Y., Li, H., Hartley, R.: Learning to estimate hidden motions with global motion aggregation. arXiv preprint arXiv:2104.02409 (2021)
26. Jiang, S., Lu, Y., Li, H., Hartley, R.: Learning optical flow from a few matches. In: Proceedings of the IEEE/CVF Conference on Computer Vision and Pattern Recognition, pp. 16592–16600 (2021)
27. Jiang, W., Trulls, E., Hosang, J., Tagliasacchi, A., Yi, K.M.: COTR: correspondence transformer for matching across images. arXiv preprint arXiv:2103.14167 (2021)
28. Kim, D., Woo, S., Lee, J.Y., Kweon, I.S.: Deep video inpainting. In: Proceedings of the IEEE/CVF Conference on Computer Vision and Pattern Recognition, pp. 5792–5801 (2019)
29. Kondermann, D., et al.: The HCI benchmark suite: stereo and flow ground truth with uncertainties for urban autonomous driving. In: Proceedings of the IEEE Conference on Computer Vision and Pattern Recognition Workshops, pp. 19–28 (2016)
30. Lai, W.S., Huang, J.B., Ahuja, N., Yang, M.H.: Deep Laplacian pyramid networks for fast and accurate super-resolution. In: Proceedings of the IEEE Conference on Computer Vision and Pattern Recognition, pp. 624–632 (2017)
31. Liang, J., Cao, J., Sun, G., Zhang, K., Van Gool, L., Timofte, R.: SwinIR: image restoration using Swin transformer. In: Proceedings of the IEEE/CVF International Conference on Computer Vision, pp. 1833–1844 (2021)

32. Liu, R., et al.: FuseFormer: fusing fine-grained information in transformers for video inpainting. In: Proceedings of the IEEE/CVF International Conference on Computer Vision, pp. 14040–14049 (2021)
33. Liu, X., Liu, H., Lin, Y.: Video frame interpolation via optical flow estimation with image inpainting. Int. J. Intell. Syst. **35**(12), 2087–2102 (2020)
34. Liu, Z., et al.: Swin transformer: hierarchical vision transformer using shifted windows. arXiv preprint arXiv:2103.14030 (2021)
35. Mayer, N., et al.: A large dataset to train convolutional networks for disparity, optical flow, and scene flow estimation. In: Proceedings of the IEEE Conference on Computer Vision and Pattern Recognition, pp. 4040–4048 (2016)
36. Piergiovanni, A., Ryoo, M.S.: Representation flow for action recognition. In: Proceedings of the IEEE/CVF Conference on Computer Vision and Pattern Recognition, pp. 9945–9953 (2019)
37. Ranjan, A., Black, M.J.: Optical flow estimation using a spatial pyramid network. In: Proceedings of the IEEE Conference on Computer Vision and Pattern Recognition, pp. 4161–4170 (2017)
38. Sajjadi, M.S., Vemulapalli, R., Brown, M.: Frame-recurrent video super-resolution. In: Proceedings of the IEEE Conference on Computer Vision and Pattern Recognition, pp. 6626–6634 (2018)
39. Sarlin, P.E., DeTone, D., Malisiewicz, T., Rabinovich, A.: SuperGlue: learning feature matching with graph neural networks. In: Proceedings of the IEEE/CVF Conference on Computer Vision and Pattern Recognition, pp. 4938–4947 (2020)
40. Sun, D., Roth, S., Black, M.J.: A quantitative analysis of current practices in optical flow estimation and the principles behind them. Int. J. Comput. Vis. **106**(2), 115–137 (2014)
41. Sun, D., et al.: AutoFlow: learning a better training set for optical flow. In: Proceedings of the IEEE/CVF Conference on Computer Vision and Pattern Recognition, pp. 10093–10102 (2021)
42. Sun, D., Yang, X., Liu, M.Y., Kautz, J.: PWC-Net: CNNs for optical flow using pyramid, warping, and cost volume. In: Proceedings of the IEEE Conference on Computer Vision and Pattern Recognition, pp. 8934–8943 (2018)
43. Sun, D., Yang, X., Liu, M.Y., Kautz, J.: Models matter, so does training: an empirical study of CNNs for optical flow estimation. IEEE Trans. Pattern Anal. Mach. Intell. **42**(6), 1408–1423 (2019)
44. Sun, J., Shen, Z., Wang, Y., Bao, H., Zhou, X.: LoFTR: detector-free local feature matching with transformers. In: Proceedings of the IEEE/CVF Conference on Computer Vision and Pattern Recognition, pp. 8922–8931 (2021)
45. Sun, S., Kuang, Z., Sheng, L., Ouyang, W., Zhang, W.: Optical flow guided feature: a fast and robust motion representation for video action recognition. In: Proceedings of the IEEE Conference on Computer Vision and Pattern Recognition, pp. 1390–1399 (2018)
46. Teed, Z., Deng, J.: RAFT: recurrent all-pairs field transforms for optical flow. In: Vedaldi, A., Bischof, H., Brox, T., Frahm, J.-M. (eds.) ECCV 2020. LNCS, vol. 12347, pp. 402–419. Springer, Cham (2020). https://doi.org/10.1007/978-3-030-58536-5_24
47. Vaswani, A., et al.: Attention is all you need. In: Advances in Neural Information Processing Systems, pp. 5998–6008 (2017)
48. Xu, H., Yang, J., Cai, J., Zhang, J., Tong, X.: High-resolution optical flow from 1D attention and correlation. In: Proceedings of the IEEE/CVF International Conference on Computer Vision, pp. 10498–10507 (2021)

49. Xu, R., Li, X., Zhou, B., Loy, C.C.: Deep flow-guided video inpainting. In: Proceedings of the IEEE/CVF Conference on Computer Vision and Pattern Recognition, pp. 3723–3732 (2019)
50. Xu, X., Siyao, L., Sun, W., Yin, Q., Yang, M.H.: Quadratic video interpolation. In: Advances in Neural Information Processing Systems, vol. 32 (2019)
51. Xu, Y., Lin, K.Y., Zhang, G., Wang, X., Li, H.: RNNPose: recurrent 6-DoF object pose refinement with robust correspondence field estimation and pose optimization (2022)
52. Yan, W., Sharma, A., Tan, R.T.: Optical flow in dense foggy scenes using semi-supervised learning. In: Proceedings of the IEEE/CVF Conference on Computer Vision and Pattern Recognition, pp. 13259–13268 (2020)
53. Yang, G., Ramanan, D.: Volumetric correspondence networks for optical flow. In: Advances in Neural Information Processing Systems, vol. 32, pp. 794–805 (2019)
54. Yang, L., Xu, Y., Yuan, C., Liu, W., Li, B., Hu, W.: Improving visual grounding with visual-linguistic verification and iterative reasoning. In: Proceedings of the IEEE/CVF Conference on Computer Vision and Pattern Recognition, pp. 9499–9508 (2022)
55. Yin, Z., Darrell, T., Yu, F.: Hierarchical discrete distribution decomposition for match density estimation. In: Proceedings of the IEEE/CVF Conference on Computer Vision and Pattern Recognition, pp. 6044–6053 (2019)
56. Zeng, Y., Fu, J., Chao, H.: Learning joint spatial-temporal transformations for video inpainting. In: Vedaldi, A., Bischof, H., Brox, T., Frahm, J.-M. (eds.) ECCV 2020. LNCS, vol. 12361, pp. 528–543. Springer, Cham (2020). https://doi.org/10.1007/978-3-030-58517-4_31
57. Zhang, F., Woodford, O.J., Prisacariu, V.A., Torr, P.H.: Separable flow: learning motion cost volumes for optical flow estimation. In: Proceedings of the IEEE/CVF International Conference on Computer Vision, pp. 10807–10817 (2021)
58. Zhao, H., Jiang, L., Jia, J., Torr, P.H., Koltun, V.: Point transformer. In: Proceedings of the IEEE/CVF International Conference on Computer Vision, pp. 16259–16268 (2021)
59. Zhao, S., Sheng, Y., Dong, Y., Chang, E.I., Xu, Y., et al.: MaskFlowNet: asymmetric feature matching with learnable occlusion mask. In: Proceedings of the IEEE/CVF Conference on Computer Vision and Pattern Recognition, pp. 6278–6287 (2020)
60. Zhao, Y., Man, K.L., Smith, J., Siddique, K., Guan, S.U.: Improved two-stream model for human action recognition. EURASIP J. Image Video Process. **2020**(1), 1–9 (2020)
61. Zheng, Y., Zhang, M., Lu, F.: Optical flow in the dark. In: Proceedings of the IEEE/CVF Conference on Computer Vision and Pattern Recognition, pp. 6749–6757 (2020)

# Coarse-to-Fine Sparse Transformer for Hyperspectral Image Reconstruction

Yuanhao Cai[1,2], Jing Lin[1,2], Xiaowan Hu[1,2], Haoqian Wang[1,2(✉)], Xin Yuan[3], Yulun Zhang[4], Radu Timofte[4,5], and Luc Van Gool[4]

[1] Shenzhen International Graduate School, Tsinghua University, Shenzhen, China
wanghaoqian@tsinghua.edu.cn
[2] Shenzhen Institute of Future Media Technology, Shenzhen, China
[3] Westlake University, Hangzhou, China
[4] ETH Zürich, Zürich, Switzerland
[5] University of Würzburg, Würzburg, Germany

**Abstract.** Many learning-based algorithms have been developed to solve the inverse problem of coded aperture snapshot spectral imaging (CASSI). However, CNN-based methods show limitations in capturing long-range dependencies. Previous Transformer-based methods densely sample tokens, some of which are uninformative, and calculate multi-head self-attention (MSA) between some tokens that are unrelated in content. In this paper, we propose a novel Transformer-based method, coarse-to-fine sparse Transformer (CST), firstly embedding HSI sparsity into deep learning for HSI reconstruction. In particular, CST uses our proposed spectra-aware screening mechanism (SASM) for *coarse patch selecting*. Then the selected patches are fed into our customized spectra-aggregation hashing multi-head self-attention (SAH-MSA) for *fine pixel clustering* and self-similarity capturing. Comprehensive experiments show that our CST significantly outperforms state-of-the-art methods while requiring cheaper computational costs. https://github.com/caiyuanhao1998/MST

**Keywords:** Compressive imaging · Transformer · Image restoration

## 1 Introduction

Hyperspectral images (HSIs), which contain multiple continuous and narrow spectral bands, can provide more detailed information of the captured scene than normal RGB images. Based on the inherently rich and detailed spectral signatures, HSIs have been widely applied to many computer vision tasks and graphical applications, *e.g.*, image classification [26,58,102], object tracking [30,40,76,77], remote sensing [4,62,74,96], medical imaging [1,54,65], *etc.*

To collect HSI cubes, traditional imaging systems scan the scenes with multiple exposures using 1D or 2D sensors. This imaging process is time-consuming

Y. Cai and J. Lin—Equal Contribution.

© The Author(s), under exclusive license to Springer Nature Switzerland AG 2022
S. Avidan et al. (Eds.): ECCV 2022, LNCS 13677, pp. 686–704, 2022.
https://doi.org/10.1007/978-3-031-19790-1_41

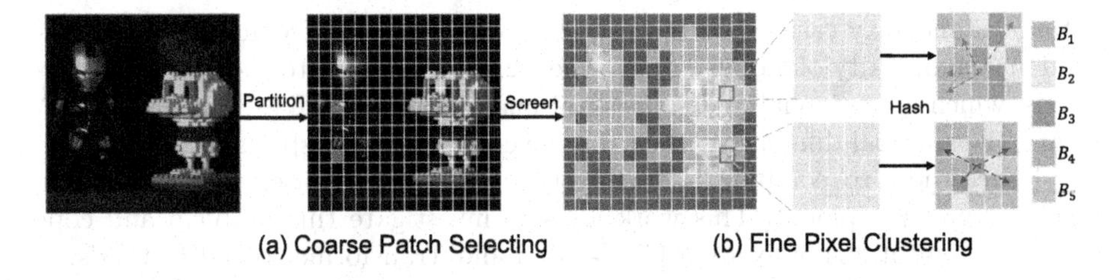


**Fig. 1.** Diagram of our coarse-to-fine learning scheme. (a) The image is firstly partitioned into patches. Then the informative patches (yellow) are screened out. (b) Tokens with correlated content are clustered into the same *bucket* ($B_1$~$B_5$). (Color figure online)

and limited to static objects [38]. Thus, conventional imaging systems cannot capture dynamic scenes. Recently, researchers have developed several snapshot compressive imaging (SCI) systems to capture HSIs, where the 3D HSI cube is compressed into a single 2D measurement [10,11,59,79]. Among these SCI systems, coded aperture snapshot spectral imaging (CASSI) stands out as a promising solution and has become an active research direction [31,37,64,79]. CASSI systems modulate HSI signals at different wavelengths by a coded aperture (physical mask) and then vary the modulation by a disperser, *i.e.*, to shift the modulated images at different wavelengths to different spatial locations on the detector plane. Subsequently, a reconstruction algorithm is used to restore the 3D HSI cube from the 2D compressive image, which is a core task in CASSI.

To solve this ill-posed inverse problem, traditional methods [51,83,94] mainly depend on hand-crafted priors and assumptions. The main drawbacks of these model-based methods are that they need to tweak parameters manually, leading to poor generality and slow reconstruction speed. In recent years, deep learning methods have shown the potential to speed up the reconstruction and improve restoration quality for natural images [5,34–36,71,98–101,103,104]. Thus, convolutional neural networks (CNNs) have been used to learn the underlying mapping function from the measurement to the HSI signal. Nonetheless, these CNN-based methods show limitations in capturing long-range dependencies.

In the past few years, the natural language processing (NLP) model Transformer [78] has achieved great success in computer vision. Transformer provides a powerful model that excels at exploring global inter-dependence between different regions to alleviate the constraints of CNN-based methods. Yet, directly applying vision Transformers to HSI reconstruction encounters two main issues. **Firstly**, HSI signals exhibit high spatial sparsity as shown in Fig. 1 (a). Some dark regions are almost uninformative. However, previous local [52] or global [22] Transformers process all spatial pixel vectors inside non-overlapping windows or global images into tokens without screening and then feed the tokens into the multi-head self-attention (MSA) mechanism. Many regions with limited information are sampled, which degrades the model efficiency and limits the reconstruction performance. **Secondly**, previous Transformers linearly project all the tokens into *query*, *key*, and *value*, and then perform matrix multiplication for calculating MSA without clustering. Yet, some of the tokens are not related in content. Attending to all these tokens at once lowers down the cost-effectiveness of model and may easily lead to over-smooth results [45]. **Besides**, the computational complexity of global Transformer [22] is quadratic to the spatial dimensions, which is nontrivial and sometimes unaffordable. MST [6] calculates MSA along the spectral dimension, thus circumventing the HSI spatial sparsity.

Hence, how to combine HSI sparsity with learning-based algorithms still remains under-explored. This work aims to investigate this problem and cope with the limitations of existing CNN-based and Transformer-based methods.

In this paper, we propose a novel method, coarse-to-fine sparse Transformer (CST), for HSI reconstruction. Our CST composes two key techniques. **Firstly**, due to the large variation in HSI informativeness of spatial regions, we propose a spectra-aware screening mechanism (SASM) for *coarse patch selecting*. To be specific, in Fig. 1 (a), our SASM partitions the image into non-overlapping patches and then detects the patches that are informative of HSI representations. Subsequently, only the detected patches (yellow) are fed into the self-attention mechanisms to decrease the inefficient calculation of uninformative regions (green) and promote the model cost-effectiveness. **Secondly**, instead of using all projected tokens at once like previous Transformers, we aim to calculate self-attention of tokens that are closely related in content. Toward this end, we customize spectra-aggregation hashing multi-head self-attention (SAH-MSA) for *fine pixel clustering* as shown in Fig. 1 (b). SAH-MSA learns to cluster tokens into different groups (termed *buckets* in this paper) by searching similar elements that produce the max inner product. Tokens inside each *bucket* are considered closely related in content. Then the MSA operation is applied within each *bucket*. **Finally**, with the proposed techniques, we enable a coarse-to-fine learning scheme that embeds the HSI spatial sparsity into learning-based methods. We establish a series of small-to-large CST families that outperform state-of-the-art (SOTA) methods while requiring much cheaper computational costs.

The main contributions of this work can be summarized as follows:

- We propose a novel Transformer-based method, CST, for HSI reconstruction. To the best of our knowledge, it is the first attempt to embed the HSI spatial sparsity nature into learning-based algorithms for this task.
- We present SASM to locate informative regions with HSI signals.
- We customize SAH-MSA to capture interactions of closely related patterns.
- Our CST with much lower computational complexity significantly surpasses SOTA algorithms on all scenes in simulation. Moreover, our CST yields more visually pleasant results than existing methods in real HSI restoration.

## 2 Related Work

### 2.1 Hyperspectral Image Reconstruction

Conventional HSI reconstruction methods [3,27,50,51,83,94] rely on hand-crafted image priors. Nonetheless, these traditional model-based methods suffer from low reconstruction speed and poor generalization ability. Recently, CNNs have been used to solve the inverse problem of spectral SCI. These

CNN-based algorithms can be divided into three categories, *i.e.*, end-to-end (E2E) methods, deep unfolding methods, and plug-and-play (PnP) methods. E2E algorithms [6,29,33,64,67,89] apply a deep CNN as a powerful model to learn the E2E mapping function of HSI restoration. Deep unfolding methods [7,28,37,57,63,82] employ multi-stage CNNs trained to map the measurements into the desired signal. Each stage contains two parts, *i.e.*, linear projection and passing the signal through a CNN functioning as a denoiser. PnP methods [13,72,95] plug pre-trained CNN denoisers into model-based methods to solve the HSI reconstruction problem. Nevertheless, these CNN-based algorithms show limitations in capturing long-range spatial dependencies and modeling the non-local self-similarity. Besides, the sparsity property of HSI representations is not well addressed, posing a low-efficiency problem to HSI reconstruction models.

### 2.2 Vision Transformer

Transformer [78] is proposed for machine translation in NLP. Recently, it has gained much popularity in computer vision because of its superiority in modeling long-range interactions between spatial regions. Vision Transformer has been widely applied in image classification [2,14,22,23,32,43,73,86,87], object detection [12,19,20,25,68,69,93,108], semantic segmentation [17,52,55, 75,88,92,97,107], human pose estimation [8,39,44,46,49,56,60,106], and so on. Besides high-level vision, Transformer has also been used in image restoration [6,9,15,21,47,48,80,84]. For example, Cai *et al.* [6] propose the first Transformer-based model MST for HSI reconstruction. MST treats spectral maps as tokens and calculates the self-attention along the spectral dimension. However, existing Transformers densely sample tokens, some of which corresponding to the regions with limited information, and calculate MSA between some tokens that are unrelated in content. How to embed HSI spatial sparsity into Transformer to boost the model efficiency still remains under-studied. Our work aims to fill this research gap.

## 3 Mathematical Model of CASSI

The input HSI is denoted as $\mathbf{F} \in \mathbb{R}^{H\times W\times N_\lambda}$, where $H$, $W$, and $N_\lambda$ refer to the HSI's height, width, and number of wavelengths, respectively. Firstly, a coded aperture $\mathbf{M}^* \in \mathbb{R}^{H\times W}$ is used to modulate $\mathbf{F}$ along the channel dimension:

$$\mathbf{F}'(:,:,n_\lambda) = \mathbf{F}(:,:,n_\lambda) \odot \mathbf{M}^*, \tag{1}$$

where $\mathbf{F}' \in \mathbb{R}^{H\times W\times N_\lambda}$ indicates the modulated signals, $n_\lambda \in [1,\ldots,N_\lambda]$ indexes the spectral wavelengths, and $\odot$ represents the element-wise product. After undergoing the disperser, $\mathbf{F}'$ becomes tilted and could be treated as sheared along the $y$-axis. We denote this tilted data cube as $\mathbf{F}'' \in \mathbb{R}^{H\times(W+d(N_\lambda-1))\times N_\lambda}$, where $d$ refers to the step of spatial shifting. Suppose $\lambda_c$ is the reference wavelength, which means that $\mathbf{F}''(:,:,n_{\lambda_c})$ works like an anchor image that is not sheared along the $y$-axis. Then the dispersion can be formulated as

$$\mathbf{F}''(u,v,n_\lambda) = \mathbf{F}'(x, y + d(\lambda_n - \lambda_c), n_\lambda), \tag{2}$$

where $(u, v)$ locates the coordinate on the sensoring detector, $\lambda_n$ represents the wavelength of the $n_\lambda$-th channel, and $d(\lambda_n - \lambda_c)$ refers to the spatial shifting offset of the $n_\lambda$-th channel on $\mathbf{F}''$. Eventually, the data cube is compressed into a 2D measurement $\mathbf{Y} \in \mathbb{R}^{H\times(W+d(N_\lambda-1))}$ by integrating all the channels as

$$\mathbf{Y} = \sum_{n_\lambda=1}^{N_\lambda} \mathbf{F}''(:,:,n_\lambda) + \mathbf{G}, \tag{3}$$

where $\mathbf{G} \in \mathbb{R}^{H\times(W+d(N_\lambda-1))}$ is the random noise generated during the imaging process. Given the 2D measurement $\mathbf{Y}$ captured by CASSI, the core task of HSI reconstruction is to restore the 3D HSI data cube $\mathbf{F}$ as mentioned in Eq. (1).

# 4 Method

As shown in Fig. 2. CST consists of two key components, *i.e.*, spectra-aware screening mechanism (SASM) for *coarse patch selecting* and spectra-aggregation hashing multi-head self-attention (SAH-MSA) for *fine pixel clustering*. Figure 2 (a) depicts SASM and the network architecture of CST. Figure 2 (b) shows the basic unit of CST, spectra-aware hashing attention block (SAHAB). Figure 2 (c) illustrates our SAH-MSA, which is the most important component of SAHAB.

## 4.1 Network Architecture

Given a 2D measurement $\mathbf{Y} \in \mathbb{R}^{H\times(W+d(N_\lambda-1))}$, we reverse the dispersion in Eq. (2) and shift back $\mathbf{Y}$ to obtain an initialized input signal $\mathbf{H} \in \mathbb{R}^{H\times W\times N_\lambda}$ as

$$\mathbf{H}(x,y,n_\lambda) = \mathbf{Y}(x, y - d(\lambda_n - \lambda_c)).$$

Then $\mathbf{H}$ concatenated with the 3D physical mask $\mathbf{M} \in \mathbb{R}^{H\times W\times N_\lambda}$ (copy the physical mask $\mathbf{M}^*$ $N_\lambda$ times) passes through a *conv*1×1 (convolutional layer with kernel size = 1×1) to generate the initialized feature $\mathbf{X} \in \mathbb{R}^{H\times W\times N_\lambda}$.

**Firstly**, a sparsity estimator is developed to process $\mathbf{X}$ into a sparsity mask $\mathbf{M}_s \in \mathbb{R}^{H\times W}$ and shallow feature $\mathbf{X}_0 \in \mathbb{R}^{H\times W\times C}$. The sparsity estimator is detailed in Sect. 4.2. **Secondly**, the shallow feature $\mathbf{X}_0$ passes through a three-stage symmetric encoder-decoder and is embedded into deep feature $\mathbf{X}_d \in \mathbb{R}^{H\times W\times C}$. The $i$-th stage of encoder or decoder contains $N_i$ SAHABs. As shown in Fig. 2 (b), SAHAB consists of two layer normalization (LN), an SAH-MSA, and a Feed-Forward Network (FFN). The encoder features are aggregated with the decoder features via the identity connection. **Finally**, a *conv*3 × 3 is applied to $\mathbf{X}_d$ to produce the residual HSIs $\mathbf{R} \in \mathbb{R}^{H\times W\times N_\lambda}$. Then the reconstructed HSIs $\mathbf{X}'$ can be obtained by the sum of $\mathbf{R}$ and $\mathbf{X}$ , *i.e.*, $\mathbf{X}' = \mathbf{X} + \mathbf{R}$.

In our implementation, we change the combination ($N_1$, $N_2$, $N_3$) in Fig. 2 (a) to establish CST families with small, medium, and large model sizes and computational costs. They are CST-S (1, 1, 2), CST-M (2, 2, 2), and CST-L (2, 4, 6).

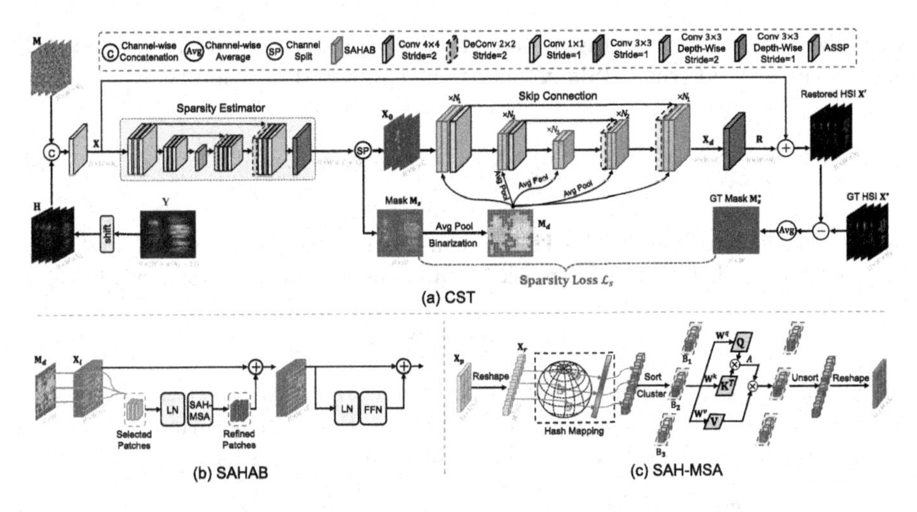


**Fig. 2.** Framework of CST. (a) Spectra-aware screening mechanism (SASM) and the architecture of CST. (b) The components of spectra-aware hashing attention block (SAHAB), which is the basic unit of CST. (c) Spectra-aggregation hashing multi-head self-attention (SAH-MSA) is the key component of SAHAB.

### 4.2 Spectra-Aware Screening Mechanism

The original global Transformer [22] samples all tokens on the feature map while the window-based local Transformer [52] samples all tokens inside every non-overlapping window. These Transformers sample many uninformative regions to calculate MSA, which degrades the model efficiency. To cope with this problem, we propose SASM for *coarse patch selecting*, *i.e.*, screening out regions with dense HSI information to produce tokens. In this section, we introduce SASM in three parts, *i.e.*, sparsity estimator, sparsity loss, and patch selection.

**Sparsity Estimator.** As shown in Fig. 2 (a), the sparsity estimator adopts a U-shaped structure including a two-stage encoder, an ASSP module [16], and a two-stage decoder. Each stage of the encoder consists of two $conv1{\times}1$ and a strided depth-wise $conv3{\times}3$. Each stage of the decoder contains a strided $deconv2{\times}2$, two $conv1{\times}1$, and a depth-wise $conv3{\times}3$. The sparsity estimator takes the initialized feature $\mathbf{X}$ as the input to produce shallow feature $\mathbf{X}_0$ and sparsity mask $\mathbf{M}_s$ that localizes and screens out informative spatial regions with HSI representations. We achieve this by minimizing our proposed sparsity loss.

**Sparsity Loss.** To supervise $\mathbf{M}_s$, we need a reference that can tell where the spatially sparse HSI information on the HSI is. Since the background is dark and uninformative, the regions with HSI representations are roughly equivalent to the regions that are hard to reconstruct. This statement can be verified by the visual analysis of sparsity mask in Sect. 5.4. Therefore, we design our reference signal $\mathbf{M}_s^* \in \mathbb{R}^{H \times W}$ by averaging the differences between the reconstructed HSIs $\mathbf{X}'$ and the ground-truth HSIs $\mathbf{X}^*$ along the spectral dimension to avoid bias as

$$\mathbf{M}_s^* = \frac{1}{N_\lambda} \sum_{n_\lambda=1}^{N_\lambda} |\mathbf{X}'(:,:,n_\lambda) - \mathbf{X}^*(:,:,n_\lambda)|. \tag{4}$$

Subsequently, our sparsity loss $\mathcal{L}_s$ is constructed as the mean squared error between the predicted sparsity mask $\mathbf{M}_s$ and the reference sparsity mask $\mathbf{M}_s^*$ as

$$\mathcal{L}_s = ||\mathbf{M}_s - \mathbf{M}_s^*||_2. \tag{5}$$

By minimizing $\mathcal{L}_s$, the sparsity estimator is encouraged to detect the foreground hard-to-reconstruct regions with HSI representations. In addition, the overall training objective $\mathcal{L}$ is the weighted sum of $\mathcal{L}_s$ and $\mathcal{L}_2$ loss as

$$\mathcal{L} = \mathcal{L}_2 + \lambda \cdot \mathcal{L}_s = ||\mathbf{X}' - \mathbf{X}^*||_2 + \lambda \cdot ||\mathbf{M}_s - \mathbf{M}_s^*||_2, \tag{6}$$

where $\mathbf{X}^*$ represents the ground-truth HSIs and $\lambda$ refers to the hyperparameter that controls the importance balance between $\mathcal{L}_2$ and $\mathcal{L}_s$.

**Patch Selection.** Our SASM partitions the feature map into non-overlapping patches at the size of $M \times M$. Then the patches with HSI representations are screened out by the predicted sparsity mask $\mathbf{M}_s$ and fed into SAH-MSA as shown in Fig. 2 (b). To be specific, $\mathbf{M}_s$ is firstly downsampled by average pooling and then binarized into $\mathbf{M}_d \in \mathbb{R}^{\frac{H}{M} \times \frac{W}{M}}$. We use a hyperparameter, sparsity ratio $\sigma$, to control the binarization. More specifically, we select the top $k$ patches with the highest values on the downsampled sparsity mask. $k$ is controlled by $\sigma$ that $k = \lfloor (1-\sigma)\frac{HW}{M^2} \rfloor$. Each pixel on $\mathbf{M}_d$ corresponds to an $M \times M$ patch on the feature map and its 0–1 value classifies whether this patch is screened out. Then $\mathbf{M}_d$ is applied to the SAH-MSA of each SAHAB. When $\mathbf{M}_d$ is used in the $i$-th stage ($i > 1$), an average pooling operation is exploited to downsample $\mathbf{M}_d$ into $\frac{1}{2^{i-1}}$ size to match the spatial resolution of the feature map of the $i$-th stage.

### 4.3 Spectra-Aggregation Hashing Multi-head Self-attention

Previous Transformers calculate MSA between all the sampled tokens, some of which are even unrelated in content. This may lead to inefficient computation that lowers down the model cost-effectiveness and easily hamper convergence [108]. The sparse coding methods [24,61,90,91,105] assume that image signals can be represented by a sparse linear combination over dictionary signals. Inspired by this, we propose SAH-MSA for *fine pixel clustering*. SAH-MSA enforces a sparsity constraint on the MSA mechanism. In particular, SAH-MSA only calculates self-attention between tokens that are closely correlated in content, which addresses the limitation of previous Transformers.

Our SAH-MSA learns to cluster tokens into different *buckets* by searching elements that produce the max inner product. As shown in Fig. 2 (c), We denote a patch feature map as $\mathbf{X}_p \in \mathbb{R}^{M \times M \times C}$ that is screened out by the sparsity mask. We reshape $\mathbf{X}_p$ into $\mathbf{X}_r \in \mathbb{R}^{N \times C}$, where $N = M \times M$ is the number of elements. Subsequently, we use a hash function to aggregate the information in spectral wise and map a $C$-dimensional element (pixel vector) $\boldsymbol{x} \in \mathbb{R}^C$ into an integer hash code. We formulate this hash mapping $h : \mathbb{R}^C \to \mathbb{Z}$ as

$$h(\boldsymbol{x}) = \lfloor \frac{\boldsymbol{a} \cdot \boldsymbol{x} + b}{r} \rfloor, \tag{7}$$

where $r \in \mathbb{R}$ is a constant, $\boldsymbol{a} \in \mathbb{R}^C$ and $b \in \mathbb{R}$ are random variables satisfying $\boldsymbol{a} = (a_1, a_2, ..., a_C)$ with $a_i \sim \mathcal{N}(0,1)$ and $b \sim \mathcal{U}(0,r)$ follows a uniform distribution. Then we sort the elements in $\mathbf{X}_r$ according to their hash codes. The $i$-th sorted element is denoted as $\boldsymbol{x}_i \in \mathbb{R}^C$. Then we split the elements into *buckets* as

$$\mathbf{B}_i = \{\boldsymbol{x}_j : im + 1 \leq j \leq (i+1)m\}, \tag{8}$$

where $\mathbf{B}_i$ represents the $i$-th *bucket*. Each *bucket* has $m$ elements. There are $\frac{M \times M}{m}$ *buckets* in total. With our hash clustering scheme, the closely content-correlated tokens are grouped into the same *bucket*. Therefore, the model can reduce the computational burden between content-unrelated elements by only applying the MSA operation to the tokens within the same *bucket*. More specifically, for a *query* element $\boldsymbol{q} \in \mathbf{B}_i$, our SAH-MSA can be formulated as

$$\text{SAH-MSA}(\boldsymbol{q}, \mathbf{B}_i) = \sum_{n=1}^{N} \mathbf{W}_n \ \text{head}_n(\boldsymbol{q}, \mathbf{B}_i), \tag{9}$$

where $N$ is the number of attention heads. $\mathbf{W}_n \in \mathbb{R}^{C \times d}$ and $\mathbf{W'}_n \in \mathbb{R}^{d \times C}$ are learnable parameters, where $d = \frac{C}{N}$ denotes the dimension of each head. $A_{nqk}$ and $\text{head}_n$ refer to the attention and output of the $n$-th head, formulated as

$$\mathbf{A}_{nqk} = \underset{\boldsymbol{k} \in \mathbf{B}_i}{\text{softmax}}(\frac{\boldsymbol{q}^T \mathbf{U}_n^T \mathbf{V}_n \boldsymbol{k}}{\sqrt{d}}), \quad \text{head}_n(\boldsymbol{q}, \mathbf{B}_i) = \sum_{\boldsymbol{k} \in \mathbf{B}_i} \mathbf{A}_{nqk} \mathbf{W'}_n \boldsymbol{k}, \tag{10}$$

where $\mathbf{U}_n$ and $\mathbf{V}_n \in \mathbb{R}^{d \times C}$ are learnable parameters. With our hashing scheme, the similar elements are at small possibility to fall into different *buckets*. This probability can be further reduced by conducting multiple rounds of hashing in parallel [42]. $\mathbf{B}_i^r$ denotes the $i$-th *bucket* of the $r$-th round. Then for each head, the multi-round output is the weighted sum of each single-round output, *i.e.*,

$$\text{head}_n(\boldsymbol{q}, \mathbf{B}_i) = \sum_{r=1}^{R} w_n^r \ \text{head}_n(\boldsymbol{q}, \mathbf{B}_i^r), \tag{11}$$

where $R$ refers to the round number and $w_n^r$ represents the weight importance of the $r$-th round in the $n$-th head, which scores the similarity between the *query* element $\boldsymbol{q}$ and the elements belonging to *bucket* $\mathbf{B}_i^r$. $w_n^r$ can be obtained by

**Table 1.** Comparisons of Params, FLOPS, PSNR (upper entry in each cell), and SSIM (lower entry in each cell) of different methods on 10 simulation scenes (S1∼S10). Best results are in bold. * denotes setting the sparsity ratio to 0.

| Algorithms | Params | GFLOPS | S1 | S2 | S3 | S4 | S5 | S6 | S7 | S8 | S9 | S10 | Avg |
|---|---|---|---|---|---|---|---|---|---|---|---|---|---|
| TwIST [3] | - | - | 25.16<br>0.700 | 23.02<br>0.604 | 21.40<br>0.711 | 30.19<br>0.851 | 21.41<br>0.635 | 20.95<br>0.644 | 22.20<br>0.643 | 21.82<br>0.650 | 22.42<br>0.690 | 22.67<br>0.569 | 23.12<br>0.669 |
| GAP-TV [94] | - | - | 26.82<br>0.754 | 22.89<br>0.610 | 26.31<br>0.802 | 30.65<br>0.852 | 23.64<br>0.703 | 21.85<br>0.663 | 23.76<br>0.688 | 21.98<br>0.655 | 22.63<br>0.682 | 23.10<br>0.584 | 24.36<br>0.669 |
| DeSCI [50] | - | - | 27.13<br>0.748 | 23.04<br>0.620 | 26.62<br>0.818 | 34.96<br>0.897 | 23.94<br>0.706 | 22.38<br>0.683 | 24.45<br>0.743 | 22.03<br>0.673 | 24.56<br>0.732 | 23.59<br>0.587 | 25.27<br>0.721 |
| λ-net [67] | 62.64M | 117.98 | 30.10<br>0.849 | 28.49<br>0.805 | 27.73<br>0.870 | 37.01<br>0.934 | 26.19<br>0.817 | 28.64<br>0.853 | 26.47<br>0.806 | 26.09<br>0.831 | 27.50<br>0.826 | 27.13<br>0.816 | 28.53<br>0.841 |
| HSSP [81] | - | - | 31.48<br>0.858 | 31.09<br>0.842 | 28.96<br>0.823 | 34.56<br>0.902 | 28.53<br>0.808 | 30.83<br>0.877 | 28.71<br>0.824 | 30.09<br>0.881 | 30.43<br>0.868 | 28.78<br>0.842 | 30.35<br>0.852 |
| DNU [82] | 1.19M | 163.48 | 31.72<br>0.863 | 31.13<br>0.846 | 29.99<br>0.845 | 35.34<br>0.908 | 29.03<br>0.833 | 30.87<br>0.887 | 28.99<br>0.839 | 30.13<br>0.885 | 31.03<br>0.876 | 29.14<br>0.849 | 30.74<br>0.863 |
| DIP-HSI [66] | 33.85M | 64.42 | 32.68<br>0.890 | 27.26<br>0.833 | 31.30<br>0.914 | 40.54<br>0.962 | 29.79<br>0.900 | 30.39<br>0.877 | 28.18<br>0.913 | 29.44<br>0.874 | 34.51<br>0.927 | 28.51<br>0.851 | 31.26<br>0.894 |
| TSA-Net [64] | 44.25M | 110.06 | 32.03<br>0.892 | 31.00<br>0.858 | 32.25<br>0.915 | 39.19<br>0.953 | 29.39<br>0.884 | 31.44<br>0.908 | 30.32<br>0.878 | 29.35<br>0.888 | 30.01<br>0.890 | 29.59<br>0.874 | 31.46<br>0.894 |
| DGSMP [37] | 3.76M | 646.65 | 33.26<br>0.915 | 32.09<br>0.898 | 33.06<br>0.925 | 40.54<br>0.964 | 28.86<br>0.882 | 33.08<br>0.937 | 30.74<br>0.886 | 31.55<br>0.923 | 31.66<br>0.911 | 31.44<br>0.925 | 32.63<br>0.917 |
| HDNet [33] | 2.37M | 154.76 | 35.14<br>0.935 | 35.67<br>0.940 | 36.03<br>0.943 | 42.30<br>0.969 | 32.69<br>0.946 | 34.46<br>0.952 | 33.67<br>0.926 | 32.48<br>0.941 | 34.89<br>0.942 | 32.38<br>0.937 | 34.97<br>0.943 |
| MST-S [6] | **0.93M** | 12.96 | 34.71<br>0.930 | 34.45<br>0.925 | 35.32<br>0.943 | 41.50<br>0.967 | 31.90<br>0.933 | 33.85<br>0.943 | 32.69<br>0.911 | 31.69<br>0.933 | 34.67<br>0.939 | 31.82<br>0.926 | 34.26<br>0.935 |
| MST-M [6] | 1.50M | 18.07 | 35.15<br>0.937 | 35.19<br>0.935 | 36.26<br>0.950 | **42.48**<br>0.973 | 32.49<br>0.943 | 34.28<br>0.948 | 33.29<br>0.921 | 32.40<br>0.943 | 35.35<br>0.942 | 32.53<br>0.935 | 34.94<br>0.943 |
| MST-L [6] | 2.03M | 28.15 | 35.40<br>0.941 | 35.87<br>0.944 | 36.51<br>0.953 | 42.27<br>0.973 | 32.77<br>0.947 | 34.80<br>0.955 | 33.66<br>0.925 | 32.67<br>0.948 | 35.39<br>0.949 | 32.50<br>0.941 | 35.18<br>0.948 |
| **CST-S** | 1.20M | **11.67** | 34.78<br>0.930 | 34.81<br>0.931 | 35.42<br>0.944 | 41.84<br>0.967 | 32.29<br>0.939 | 34.49<br>0.949 | 33.47<br>0.922 | 32.89<br>0.945 | 34.96<br>0.944 | 32.14<br>0.932 | 34.71<br>0.940 |
| **CST-M** | 1.36M | 16.91 | 35.16<br>0.938 | 35.60<br>0.942 | 36.57<br>0.953 | 42.29<br>0.972 | 32.82<br>0.948 | 35.15<br>0.956 | 33.85<br>0.927 | 33.52<br>0.952 | 35.28<br>0.946 | 32.84<br>0.940 | 35.31<br>0.947 |
| **CST-L** | 3.00M | 27.81 | 35.82<br>0.947 | 36.54<br>0.952 | 37.39<br>0.959 | 42.28<br>0.972 | 33.40<br>0.953 | 35.52<br>0.962 | 34.44<br>0.937 | 33.83<br>0.959 | 35.92<br>0.951 | **33.36**<br>**0.948** | 35.85<br>0.954 |
| **CST-L*** | 3.00M | 40.10 | **35.96**<br>**0.949** | **36.84**<br>**0.955** | **38.16**<br>**0.962** | 42.44<br>**0.975** | **33.25**<br>**0.955** | **35.72**<br>**0.963** | **34.86**<br>**0.944** | **34.34**<br>**0.961** | **36.51**<br>**0.957** | 33.09<br>0.945 | **36.12**<br>**0.957** |

$$w_n^r = \frac{\sum_{k \in \mathbf{B}_i^r} A_{nqk}}{\sum_{\hat{r}=1}^{R} \sum_{k \in \mathbf{B}_i^{\hat{r}}} A_{nqk}}. \tag{12}$$

# 5 Experiment

## 5.1 Experiment Settings

The same with TSA-Net [64], 28 wavelengths from 450 nm to 650 nm are derived by spectral interpolation manipulation for simulation and real experiments.

**Synthetic Data.** Two HSI datasets, CAVE [70] and KAIST [18], are adopted for simulation experiments. CAVE contains 32 HSIs with spatial size 512×512.

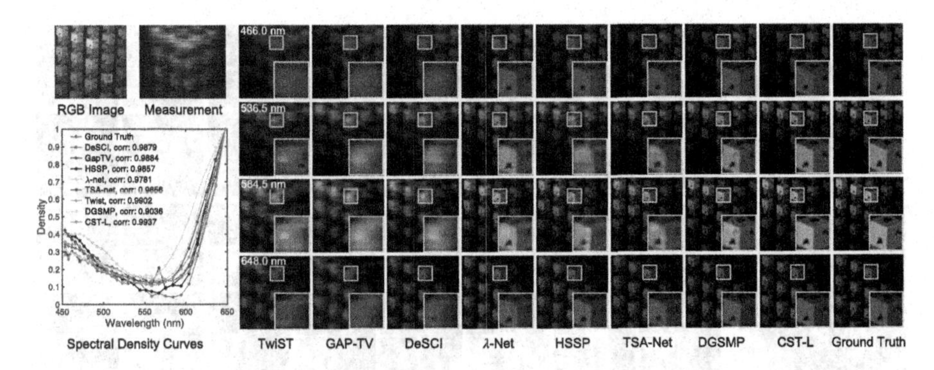


**Fig. 3.** Reconstructed simulation HSI comparisons of *Scene* 2 with 4 out of 28 spectral channels. 7 SOTA methods and CST-L are included. Please zoom in.

KAIST is composed of 30 HSIs with spatial size 2704×3376. Similar to [6,37,64], CAVE is used for training and 10 scenes from KAIST are selected for testing.

**Real Data.** We adopt the real HSI dataset collected by TSA-Net [64].

**Evaluation Metrics.** We use peak signal-to-noise ratio (PSNR) and structural similarity (SSIM) [85] as metrics to evaluate HSI reconstruction methods.

**Implementation Details.** Our CST models are implemented by Pytorch. They are trained with Adam [41] optimizer ($\beta_1 = 0.9$ and $\beta_2 = 0.999$) using Cosine Annealing scheme [53] for 500 epochs. The learning rate is initially set to $4\times10^{-4}$. In simulation experiments, patches at the spatial size of 256×256 are randomly cropped from the 3D HSI cubes with 28 channels as training samples. For real HSI reconstruction, we set the spatial size of patches to 660×660 with the same size of the real physical mask. We set the shifting step $d$ in the dispersion to 2. The batch size is set to 5. $r$ and $m$ in Eq. (7) and (8) are set to 1 and 64. The training data is augmented with random rotation and flipping.

### 5.2 Quantitative Results

We compare our CST with SOTA methods, including three model-based methods (TwIST [3], GAP-TV [94], and DeSCI [50]), six CNN-based methods ($\lambda$-net [67], HSSP [81], DNU [82], PnP-DIP-HSI [66], TSA-Net [64], DGSMP [37]), and a recent Transformer-based method (MST [6]). For fairness, we test all these algorithms with the same settings as [6,37]. The results on 10 simulation scenes are reported in Table 1. As can be seen: **(i)** When we set the sparsity ratio to 0, our best model CST-L* achieves very impressive results, *i.e.*, 36.12 dB in PSNR and 0.957 in SSIM, showing the effectiveness of our method. **(ii)** Our CST families significantly outperform other SOTA algorithms while requiring cheaper computational costs. Particularly, when compared to the recent best Transformer-based method MST, our CST-S, CST-M, and CST-L achieve 0.45, 0.37, and 0.67 dB improvements while costing 1.29G, 1.16G, and 0.34G less

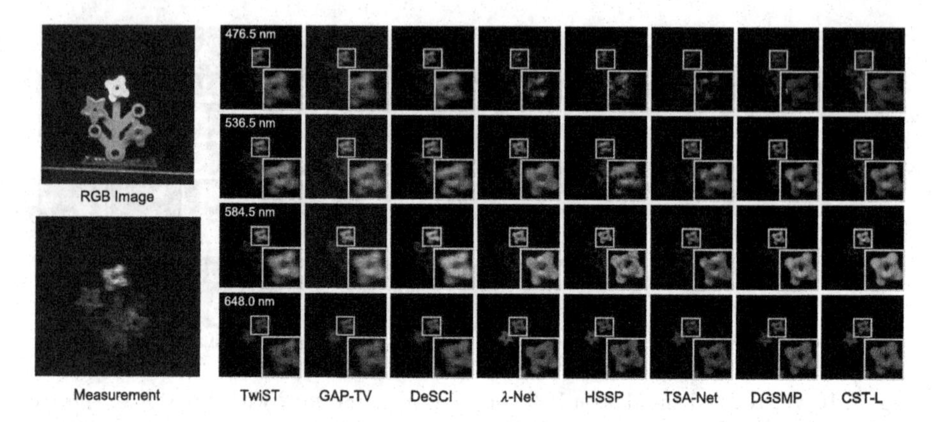


**Fig. 4.** Reconstructed real HSI comparisons of *Scene* 1 with 4 out of 28 spectral channels. 7 SOTA methods and CST-L are included. Zoom in for a better view.

FLOPS than MST-S, MST-M, and MST-L. When compared to CNN-based methods, our CST exhibits extreme efficiency advantages. For instance, CST-L outperforms DGSMP, TSA-Net, and $\lambda$-Net by 3.22, 4.39, and 7.32 dB while costing 79.8% (3.00/3.76), 6.8%, 4.8% Params and 4.3% (27.81/646.65), 25.3%, 23.6% FLOPS. Surprisingly, even our smallest model CST-S surpasses DGSMP, TSA-Net, and $\lambda$-Net by 2.08, 3.25, and 6.18 dB while requiring 31.9%, 2.7%, 1.9% Params and 1.8%, 10.6%, 9.9% FLOPS. These results demonstrate the cost-effectiveness superiority of our CST. This is mainly because CST embeds the HSI sparsity into the learning-based model, which reduces the inefficient computation of less informative dark regions and self-attention between content-unrelated tokens.

### 5.3 Qualitative Results

**Simulation HSI Restoration.** Figure 3 compares the restored simulation HSIs of our CST-L and seven SOTA algorithms on *Scene* 2 with 4 out of 28 spectral channels. It can be observed from the reconstructed HSIs (right) and the zoomed-in patches in the yellow boxes that CST is effective in producing perceptually pleasant images with more sharp edge details while maintaining the spatial smoothness of the homogeneous regions without introducing artifacts. In contrast, other methods fail to restore fine-grained details. They either achieve over-smooth results sacrificing structural contents and high-frequency details, or generate blotchy textures and chromatic artifacts. Besides, Fig. 3 depicts the spectral density curves (bottom-left) corresponding to the selected region of the green box in the RGB image (top-left). CST achieves the highest correlation coefficient with the ground-truth. This evidence demonstrates the spectral-dimension consistency reconstruction effectiveness of our proposed CST.

**Table 2.** Ablations. Models are trained on CAVE and tested on KAIST.

(a) Break-down ablation study.

| Method | Baseline | + SAH-MSA | + SASM |
|---|---|---|---|
| PSNR | 32.57 | **35.53** | 35.31 (↓ 0.60 %) |
| SSIM | 0.906 | **0.948** | 0.947 (↓ 0.10 %) |
| Params (M) | 0.51 | 1.36 | 1.36 (↓ 0.00 %) |
| FLOPS (G) | 6.40 | 24.60 | 16.91 (↓ 31.3 %) |

(b) Ablation study of sparse mechanisms.

| Method | Baseline | Random Sparsity | Uniform Sparsity | **SASM** |
|---|---|---|---|---|
| PSNR | 32.57 | 34.37 | 34.33 | **35.31** |
| SSIM | 0.906 | 0.937 | 0.936 | **0.947** |
| Params (M) | 0.51 | 1.36 | 1.36 | 1.36 |
| FLOPS (G) | 6.40 | 16.89 | 16.89 | 16.91 |

(c) Ablation study of self-attention mechanisms.

| Method | Baseline | G-MSA | W-MSA | Swin-MSA | S-MSA | **SAH-MSA** |
|---|---|---|---|---|---|---|
| PSNR | 32.57 | 35.04 | 35.02 | 35.12 | 35.21 | **35.53** |
| SSIM | 0.906 | 0.944 | 0.943 | 0.945 | 0.946 | **0.948** |
| Params (M) | 0.51 | 1.85 | 1.85 | 1.85 | 1.66 | **1.36** |
| FLOPS (G) | 6.40 | 35.58 | 24.98 | 24.98 | 24.74 | **24.60** |

(d) Study of clustering scope.

| Method | Baseline | Global | **Local** |
|---|---|---|---|
| PSNR | 32.57 | 35.33 | **35.53** |
| SSIM | 0.906 | 0.946 | **0.948** |
| Params (M) | 0.51 | 1.36 | 1.36 |
| FLOPS (G) | 6.40 | 24.60 | 24.60 |

**Real HSI Restoration.** We also evaluate our CST in real HSI reconstruction. Following the setting of [6,37,64], we re-train our CST-L with all samples of the KAIST and CAVE datasets. To simulate real CASSI, 11-bit shot noise is injected into the measurement during the training procedure. The reconstructed HSI comparisons are depicted in Fig. 4. Our CST-L shows significant advantages in fine-grained content restoration and real noise removal.

### 5.4 Ablation Study

We adopt the simulation HSI datasets [18,70] to conduct ablation studies. The baseline model is derived by removing our SAH-MSA and SASM from CST-M.

**Break-Down Ablation.** We firstly perform a break-down ablation to investigate the effect of each component and their interactions. The results are listed in Table 2a. The baseline model yields 32.57 dB in PSNR and 0.906 in SSIM. When SAH-MSA is applied, the performance gains by 2.96 dB in PSNR and 0.042 in SSIM, showing its significant contribution. When we continue to exploit SASM, the computational cost dramatically declines by 31.3% (7.69/24.60) while the performance only degrades by 0.6 % in PSNR and 0.1% in SSIM. This evidence suggests that our SASM can reduce the computational burden while sacrificing minimal reconstruction performance, thus increasing the model efficiency.

**Sparsity Scheme Comparison.** We conduct ablation to study the effects of sparsity schemes including: (i) random sparsity, *i.e.*, the patches to be calculated are randomly selected, (ii) uniform sparsity, *i.e.*, the patches to be calculated are uniformly distributed, and (iii) our SASM. The results are listed in Table 2b. Our SASM yields the best results and drastically outperforms other schemes (over 0.9 dB). Additionally, we conduct visual analysis of the sparsity mask generated by the three sparsity schemes. As depicted in Fig. 5, the sparsity mask produced by our SASM generates more complete and accurate responses to the informative regions with HSI information. In contrast, both random and uniform sparsity schemes are not aware of HSI signals and rigidly pick the preset positions. These

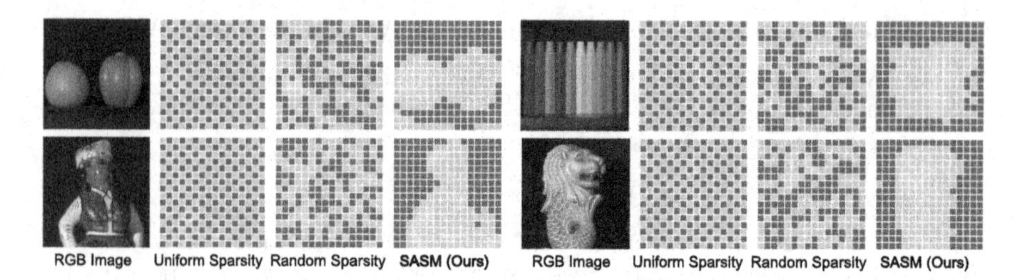


**Fig. 5.** Visual analysis of uniform sparsity scheme, random sparsity scheme, and our SASM. We visualize the sparsity masks produced by different sparsity schemes. Yellow indicates the patch is selected while green means vice versa. (Color figure online)

results demonstrate the superiority of our SASM in perceiving spatially sparse HSI signals and locating regions with dense HSI representations.

**Self-Attention Mechanism Comparison.** We compare our SAH-MSA with other self-attention mechanisms. The results are reported in Table 2c. The baseline yields 32.57 dB with 0.51 M Params and 6.40 G FLOPS. We respectively apply global MSA (G-MSA) [22], local window-based MSA (W-MSA) [52], Swin-MSA [52], spectral-wise MSA (S-MSA) [6], and SAH-MSA. Our SAH-MSA yields the most significant improvement but requires the cheapest FLOPS and Params. Please note that we downscale the input feature of G-MSA into $\frac{1}{4}$ size to avoid memory bottlenecks. This evidence shows the cost-effectiveness advantage of SAH-MSA, which is mainly because SAH-MSA applies MSA calculation between tokens that are closely related in content within each *bucket* while cutting down the burden of computation between content-uncorrelated elements.

**Clustering Scope.** We study the effect of the scope of clustering, *i.e.*, local *vs.* global. Local means constraining the hash clustering operation inside each $M \times M$ patch while global indicates applying the hash clustering to the whole image. In the beginning, we thought that expanding the receptive field would improve the performance. However, the experimental results in Table 2d point out the opposite. The model with local clustering scope performs better. We now analyze the reason for this observation. The hash clustering is essentially a linear dimension reduction ($h : \mathbb{R}^C \rightarrow \mathbb{Z}$) suffering from limited discriminative ability. It is suitable for simple, linearly separable situations with a small number of samples. When the clustering scope is enlarged from the local patch to the global image, the number of tokens increases dramatically ($M \times M \rightarrow H \times W$). As a result, the situation becomes more complex and may be linearly inseparable. Thus, the hash clustering performance degrades. Then the elements clustered into the same *bucket* are less content-related and the MSA calculation of each *bucket* becomes less effective, leading to the degradation of HSI restoration.

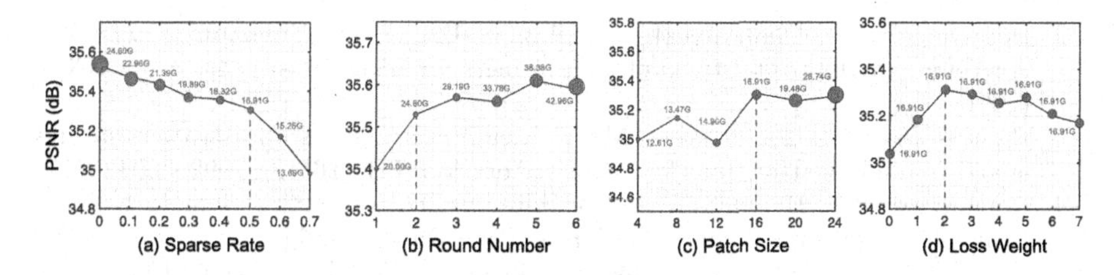


**Fig. 6.** Parameter analysis of sparsity ratio $\sigma$, round number $R$, patch size $M$, and loss weight $\lambda$. The vertical axis is PNSR. The circle radius is FLOPS.

**Parameter Analysis.** We adopt CST-M to conduct parameter analysis of sparsity rate $\sigma$, round number $R$ in Eq. (11), patch size $M$, and loss weight $\lambda$ in Eq. (6) as shown in Fig. 6, where the vertical axis is PSNR and the circle radius is FLOPS. As can be observed: **(i)** When increasing $\sigma$, the computational cost declines but the performance is sacrificed. When $\sigma$ is larger than 50%, the performance degrades dramatically. **(ii)** When changing $R$ from 1 to 6, the reconstruction quality increases. Nonetheless, when $R \geq 2$, further increasing $R$ does not lead to a significant improvement. **(iii)** The two maximums are achieved when $M = 16$ and $\lambda = 2$, respectively, without costing too much FLOPS. Since our goal is not to pursue the best results with heavy computational burden sacrificing the model efficiency but to yield a better trade-off between performance and computational cost, we finally set $\sigma = 0.5$, $R = 2$, $M = 16$, and $\lambda = 2$.

## 6 Conclusion

In this paper, we investigate a critical problem in HSI reconstruction, *i.e.*, how to embed HSI sparsity into learning-based algorithms. To this end, we propose a novel Transformer-based method, named CST, for HSI restoration. CST firstly exploits SASM to detect informative regions with HSI representations. Then the detected patches are fed into our SAH-MSA to cluster spatially scattered tokens with closely correlated contents for calculating MSA. Extensive quantitative and qualitative experiments demonstrate that our CST significantly outperforms other SOTA methods while requiring cheaper computational costs. Additionally, our CST yields more visually pleasing results with more fine-grained details and structural contents than existing algorithms in real-world HSI reconstruction.

**Acknowledgements.** This work is partially supported by the NSFC fund (61831 014), the Shenzhen Science and Technology Project under Grant (JSGG20 210802153150005, CJGJZD20200617102601004), and the Westlake Foundation (2021B1 501-2). Xin Yuan would like to thank the funding from Lochn Optics.

## References

1. Backman, V., et al.: Detection of preinvasive cancer cells. Nature **406**(6791), 35–36 (2000)

2. Bhojanapalli, S., Chakrabarti, A., Glasner, D., Li, D., Unterthiner, T., Veit, A.: Understanding robustness of transformers for image classification. In: ICCV (2021)
3. Bioucas-Dias, J., Figueiredo., M.: A new twist: Two-step iterative shrinkage/thresholding algorithms for image restoration. TIP, **16**(12), 2992–3004 (2007)
4. Borengasser, M., Hungate, W.S., Watkins, R.: Hyperspectral Remote Sensing: Principles and Applications. CRC Press, Boca Raton (2007)
5. Cai, Y., Hu, X., Wang, H., Zhang, Y., Pfister, H., Wei, D.: Learning to generate realistic noisy images via pixel-level noise-aware adversarial training. In: NeurIPS (2021)
6. Cai, Y., et al.: Mask-guided spectral-wise transformer for efficient hyperspectral image reconstruction. In: CVPR (2022)
7. Cai, Y., et al.: Degradation-aware unfolding half-shuffle transformer for spectral compressive imaging. arXiv preprint arXiv:2205.10102 (2022)
8. Cai, Y., et al.: Learning delicate local representations for multi-person pose estimation. arXiv preprint arXiv:2003.04030 (2020)
9. Cao, J., Li, Y., Zhang, K., Van Gool, L.: Video super-resolution transformer. arXiv preprint arXiv:2106.06847 (2021)
10. Cao, X., Du, H., Tong, X., Dai, Q., Lin, S.: A prism-mask system for multispectral video acquisition. TPAMI **33**(12), 2423–2435 (2011)
11. Cao, X., et al.: Computational snapshot multispectral cameras: toward dynamic capture of the spectral world. Signal Process. Mag. **33**(5), 95–108 (2016)
12. Carion, N., Massa, F., Synnaeve, G., Usunier, N., Kirillov, A., Zagoruyko, S.: End-to-end object detection with transformers. In: Vedaldi, A., Bischof, H., Brox, T., Frahm, J.-M. (eds.) ECCV 2020. LNCS, vol. 12346, pp. 213–229. Springer, Cham (2020). https://doi.org/10.1007/978-3-030-58452-8_13
13. Chan, S.H., Wang, X., Elgendy, O.A.: Plug-and-play ADMM for image restoration: fixed-point convergence and applications. Transactions on Computational Imaging **3**(1), 84–98 (2016)
14. Chen, C.F.R., Fan, Q., Panda, R.: CrossVIT: cross-attention multi-scale vision transformer for image classification. In: ICCV (2021)
15. Chen, H., et al.: Pre-trained image processing transformer. In: CVPR (2021)
16. Chen, L.C., Papandreou, G., Schroff, F., Adam, H.: Rethinking Atrous convolution for semantic image segmentation. arXiv preprint arXiv:1706.05587 (2017)
17. Cheng, B., Schwing, A., Kirillov, A.: Per-pixel classification is not all you need for semantic segmentation. In: NeurIPS (2021)
18. Choi, I., Kim, M., Gutierrez, D., Jeon, D., Nam, G.: High-quality hyperspectral reconstruction using a spectral prior. In: Technical report (2017)
19. Dai, X., Chen, Y., Yang, J., Zhang, P., Yuan, L., Zhang, L.: Dynamic DETR: end-to-end object detection with dynamic attention. In: ICCV (2021)
20. Dai, Z., Cai, B., Lin, Y., Chen, J.: UP-DETR: unsupervised pre-training for object detection with transformers. In: CVPR (2021)
21. Deng, Z., et al.: RFormer: transformer-based generative adversarial network for real fundus image restoration on a new clinical benchmark. arXiv preprint arXiv:2201.00466 (2022)
22. Dosovitskiy, A., et al.: An image is worth 16x16 words: transformers for image recognition at scale. In: ICLR (2021)
23. El-Nouby, A., et al.: XCiT: Cross-covariance image transformers. arXiv preprint arXiv:2106.09681 (2021)
24. Elad, M., Aharon, M.: Image denoising via learned dictionaries and sparse representation. In: CVPR (2006)

25. Fang, Y., et al.: You only look at one sequence: rethinking transformer in vision through object detection. In: NeurIPS (2021)
26. Fauvel, M., Tarabalka, Y., Benediktsson, J.A., Chanussot, J., Tilton, J.C.: Advances in spectral-spatial classification of hyperspectral images. Proc. IEEE **101**(3), 652–675 (2012)
27. Figueiredo, M.A., Nowak, R.D., Wright, S.J.: Gradient projection for sparse reconstruction: application to compressed sensing and other inverse problems. IEEE J. Sel. Top. Sign. Process. **1**(4), 586–597 (2007)
28. Fu, Y., Liang, Z., You, S.: Bidirectional 3D quasi-recurrent neural network for hyperspectral image super-resolution. J. Sel. Top. Appl. Earth Obs. Remote Sens. **14**, 2674–2688 (2021)
29. Fu, Y., Zhang, T., Wang, L., Huang, H.: Coded hyperspectral image reconstruction using deep external and internal learning. TPAMI, **44**(7) (2021)
30. Fu, Y., Zheng, Y., Sato, I., Sato, Y.: Exploiting spectral-spatial correlation for coded hyperspectral image restoration. In: CVPR (2016)
31. Gehm, M.E., John, R., Brady, D.J., Willett, R.M., Schulz, T.J.: Single-shot compressive spectral imaging with a dual-disperser architecture. Opt. Express **15**(21), 14013–14027 (2007)
32. Han, K., Xiao, A., Wu, E., Guo, J., Xu, C., Wang, Y.: Transformer in transformer. In: NeurIPS (2021)
33. Hu, X., et al.: HDNET: high-resolution dual-domain learning for spectral compressive imaging. In: CVPR (2022)
34. Hu, X., Cai, Y., Liu, Z., Wang, H., Zhang, Y.: Multi-scale selective feedback network with dual loss for real image denoising. In: IJCAI (2021)
35. Hu, X., et al.: Pseudo 3D auto-correlation network for real image denoising. In: CVPR (2021)
36. Hu, X., Wang, H., Cai, Y., Zhao, X., Zhang, Y.: Pyramid orthogonal attention network based on dual self-similarity for accurate MR image super-resolution. In: ICME (2021)
37. Huang, T., Dong, W., Yuan, X., Wu, J., Shi, G.: Deep gaussian scale mixture prior for spectral compressive imaging. In: CVPR (2021)
38. James, J.: Spectrograph Design Fundamentals. Cambridge University Press, Cambridge (2007)
39. Jiang, T., Camgoz, N.C., Bowden, R.: Skeletor: skeletal transformers for robust body-pose estimation. In: CVPR (2021)
40. Kim, M.H., et al.: 3D imaging spectroscopy for measuring hyperspectral patterns on solid objects. ACM Trans. Graph. **31**(4), 1–11 (2012)
41. Kingma, D.P., Ba, J.L.: Adam: a method for stochastic optimization. In: ICLR (2015)
42. Kitaev, N., Kaiser, Ł., Levskaya, A.: Reformer: the efficient transformer. arXiv preprint arXiv:2001.04451 (2020)
43. Lanchantin, J., Wang, T., Ordonez, V., Qi, Y.: General multi-label image classification with transformers. In: CVPR (2021)
44. Li, W., Liu, H., Ding, R., Liu, M., Wang, P.: Lifting transformer for 3D human pose estimation in video. arXiv preprint arXiv:2103.14304 (2021)
45. Li, X., Zhang, L., You, A., Yang, M., Yang, K., Tong, Y.: Global aggregation then local distribution in fully convolutional networks. In: BMVC (2019)
46. Li, Y., Hao, M., Di, Z., Gundavarapu, N.B., Wang, X.: Test-time personalization with a transformer for human pose estimation. In: NeurIPS (2021)
47. Liang, J., Cao, J., Sun, G., Zhang, K., Van Gool, L., Timofte, R.: SwinIR: image restoration using swin transformer. In: ICCVW (2021)

48. Lin, J., et al.: Flow-guided sparse transformer for video deblurring. arXiv preprint arXiv:2201.01893 (2022)
49. Lin, K., Wang, L., Liu, Z.: End-to-end human pose and mesh reconstruction with transformers. In: CVPR (2021)
50. Liu, Y., Yuan, X., Suo, J., Brady, D., Dai, Q.: Rank minimization for snapshot compressive imaging. TPAMI (2019)
51. Liu, Y., Yuan, X., Suo, J., Brady, D.J., Dai, Q.: Rank minimization for snapshot compressive imaging. TPAMI **41**(12), 2990–3006 (2018)
52. Liu, Z., et al.: Swin transformer: hierarchical vision transformer using shifted windows. In: ICCV (2021)
53. Loshchilov, I., Hutter, F.: SGDR: stochastic gradient descent with warm restarts. arXiv preprint arXiv:1608.03983 (2016)
54. Lu, G., Fei, B.: Medical hyperspectral imaging: a review. J. Biomed. Opt. **19**(1), 010901 (2014)
55. Lu, Z., He, S., Zhu, X., Zhang, L., Song, Y.Z., Xiang, T.: Simpler is better: few-shot semantic segmentation with classifier weight transformer. In: ICCV (2021)
56. Ludwig, K., Harzig, P., Lienhart, R.: Detecting arbitrary intermediate keypoints for human pose estimation with vision transformers. In: WACV (2022)
57. Ma, J., Liu, X.Y., Shou, Z., Yuan, X.: Deep tensor admm-net for snapshot compressive imaging. In: ICCV (2019)
58. Maggiori, E., Charpiat, G., Tarabalka, Y., Alliez, P.: Recurrent neural networks to correct satellite image classification maps. Trans. Geosci. Remote Sens. **55**(9), 4962–4971 (2017)
59. Manakov, A., et al.: A reconfigurable camera add-on for high dynamic range, multispectral, polarization, and light-field imaging. Trans. Graph. (2013)
60. Mao, W., Ge, Y., Shen, C., Tian, Z., Wang, X., Wang, Z.: TFPOSE: direct human pose estimation with transformers. arXiv preprint arXiv:2103.15320 (2021)
61. Mei, Y., Fan, Y., Zhou, Y.: Image super-resolution with non-local sparse attention. In: CVPR (2021)
62. Melgani, F., Bruzzone, L.: Classification of hyperspectral remote sensing images with support vector machines. Trans. Geosci. Remote Sens. **42**(8), 1778–1790 (2004)
63. Meng, Z., Jalali, S., Yuan, X.: Gap-net for snapshot compressive imaging. arXiv preprint arXiv:2012.08364 (2020)
64. Meng, Z., Ma, J., Yuan, X.: End-to-end low cost compressive spectral imaging with spatial-spectral self-attention. In: Vedaldi, A., Bischof, H., Brox, T., Frahm, J.-M. (eds.) ECCV 2020. LNCS, vol. 12368, pp. 187–204. Springer, Cham (2020). https://doi.org/10.1007/978-3-030-58592-1_12
65. Meng, Z., Qiao, M., Ma, J., Yu, Z., Xu, K., Yuan, X.: Snapshot multispectral endomicroscopy. Opt. Lett. **45**(14), 3897–3900 (2020)
66. Meng, Z., Yu, Z., Xu, K., Yuan, X.: Self-supervised neural networks for spectral snapshot compressive imaging. In: ICCV (2021)
67. Miao, X., Yuan, X., Pu, Y., Athitsos, V.: l-net: reconstruct hyperspectral images from a snapshot measurement. In: ICCV (2019)
68. Misra, I., Girdhar, R., Joulin, A.: An end-to-end transformer model for 3D object detection. In: ICCV (2021)
69. Pan, X., Xia, Z., Song, S., Li, L.E., Huang, G.: 3D object detection with point-former. In: CVPR (2021)
70. Park, J.I., Lee, M.H., Grossberg, M.D., Nayar, S.K.: Multispectral imaging using multiplexed illumination. In: ICCV (2007)

71. Patrick, W., Hirsch, M., Scholkopf, B., Lensch, H.P.A.: Learning blind motion deblurring. In: ICCV (2017)
72. Qiao, M., Liu, X., Yuan, X.: Snapshot spatial-temporal compressive imaging. Opt. Lett. **45**(7), 1659–1662 (2020)
73. Ramachandran, P., Parmar, N., Vaswani, A., Bello, I., Levskaya, A., Shlens, J.: Stand-alone self-attention in vision models. In: NeurIPS (2019)
74. Solomon, J., Rock, B.: Imaging spectrometry for earth remote sensing. Science **228**(4704), 1147–1153 (1985)
75. Strudel, R., Garcia, R., Laptev, I., Schmid, C.: Segmenter: transformer for semantic segmentation. In: ICCV (2021)
76. Uzkent, B., Hoffman, M.J., Vodacek, A.: Real-time vehicle tracking in aerial video using hyperspectral features. In: CVPRW (2016)
77. Uzkent, B., Rangnekar, A., Hoffman, M.: Aerial vehicle tracking by adaptive fusion of hyperspectral likelihood maps. In: CVPRW (2017)
78. Vaswani, A., et al.: Attention is all you need. In: NeurIPS (2017)
79. Wagadarikar, A., John, R., Willett, R., Brady, D.: Single disperser design for coded aperture snapshot spectral imaging. Appl. Opt. **47**(10), B44–B51 (2008)
80. Wang, L., Wu, Z., Zhong, Y., Yuan, X.: Spectral compressive imaging reconstruction using convolution and spectral contextual transformer. arXiv preprint arXiv:2201.05768 (2022)
81. Wang, L., Sun, C., Fu, Y., Kim, M.H., Huang, H.: Hyperspectral image reconstruction using a deep spatial-spectral prior. In: CVPR (2019)
82. Wang, L., Sun, C., Zhang, M., Fu, Y., Huang, H.: DNU: deep non-local unrolling for computational spectral imaging. In: CVPR (2020)
83. Wang, L., Xiong, Z., Gao, D., Shi, G., Wu, F.: Dual-camera design for coded aperture snapshot spectral imaging. Appl. Opt. **54**(4), 848–858 (2015)
84. Wang, Z., Cun, X., Bao, J., Liu, J.: Uformer: a general u-shaped transformer for image restoration. arXiv preprint 2106.03106 (2021)
85. Wang, Z., Bovik, A.C., Sheikh, H.R., Simoncell, E.P.: Image quality assessment: from error visibility to structural similarity. TIP **13**(4), 600–612 (2004)
86. Wu, B., et al.: Visual transformers: token-based image representation and processing for computer vision. arXiv preprint arXiv:2006.03677 (2020)
87. Wu, K., Peng, H., Chen, M., Fu, J., Chao, H.: Rethinking and improving relative position encoding for vision transformer. In: ICCV (2021)
88. Xie, E., Wang, W., Yu, Z., Anandkumar, A., Alvarez, J.M., Luo, P.: SegFormer: simple and efficient design for semantic segmentation with transformers. In: NeurIPS (2021)
89. Xiong, Z., Shi, Z., Li, H., Wang, L., Liu, D., Wu, F.: HSCNN: CNN-based hyperspectral image recovery from spectrally undersampled projections. In: ICCVW (2017)
90. Yang, J., Wang, Z., Lin, Z., Cohen, S., Huang, T.: Coupled dictionary training for image super-resolution. TIP **21**(8), 3467–3478 (2012)
91. Yang, J., Wright, J., Huang, T.S., Ma, Y.: Image super-resolution via sparse representation. TIP **19**(11), 2861–2873 (2010)
92. Yang, J., et al.: Focal self-attention for local-global interactions in vision transformers. arXiv preprint arXiv:2107.00641 (2021)
93. Yang, Z., Wei, Y., Yang, Y.: Associating objects with transformers for video object segmentation. In: NeurIPS (2021)
94. Yuan, X.: Generalized alternating projection based total variation minimization for compressive sensing. In: ICIP (2016)

95. Yuan, X., Liu, Y., Suo, J., Dai, Q.: Plug-and-play algorithms for large-scale snapshot compressive imaging. In: CVPR (2020)
96. Yuan, Y., Zheng, X., Lu, X.: Hyperspectral image superresolution by transfer learning. J. Sel. Top. Appl. Earth Obs. Remote Sens. **10**(5), 1963–1974 (2017)
97. Yuan, Y., Fu, R., Huang, L., Lin, W., Zhang, C., Chen, X., Wang, J.: HRFormer: high-resolution transformer for dense prediction. In: NeurIPS (2021)
98. Zamir, S.W., Arora, A., Khan, S., Hayat, M., Khan, F.S., Yang, M.H.: Restormer: efficient transformer for high-resolution image restoration. ArXiv 2111.09881 (2021)
99. Zamir, S.W., et al.: CycleISP: real image restoration via improved data synthesis. In: CVPR (2020)
100. Zamir, S.W., et al.: Learning enriched features for real image restoration and enhancement. In: Vedaldi, A., Bischof, H., Brox, T., Frahm, J.-M. (eds.) ECCV 2020. LNCS, vol. 12370, pp. 492–511. Springer, Cham (2020). https://doi.org/10.1007/978-3-030-58595-2_30
101. Zamir, S.W., et al.: Multi-stage progressive image restoration. In: CVPR (2021)
102. Zhang, F., Du, B., Zhang, L.: Scene classification via a gradient boosting random convolutional network framework. Trans. Geosci. Remote Sens. **54**(3), 1793–1802 (2015)
103. Zhang, Y., Li, K., Li, K., Wang, L., Zhong, B., Fu, Y.: Image super-resolution using very deep residual channel attention networks. In: ECCV (2018)
104. Zhang, Y., Li, K., Li, K., Zhong, B., Fu, Y.: Residual non-local attention networks for image restoration. In: ICLR (2019)
105. Zhao, C., Zhang, J., Ma, S., Fan, X., Zhang, Y., Gao, W.: Reducing image compression artifacts by structural sparse representation and quantization constraint prior. TCSVT **27**(10), 2057–2071 (2016)
106. Zheng, C., Zhu, S., Mendieta, M., Yang, T., Chen, C., Ding, Z.: 3D human pose estimation with spatial and temporal transformers. In: ICCV (2021)
107. Zheng, S., et al.: Rethinking semantic segmentation from a sequence-to-sequence perspective with transformers. In: CVPR (2021)
108. Zhu, X., Su, W., Lu, L., Li, B., Wang, X., Dai, J.: Deformable DETR: deformable transformers for end-to-end object detection. In: ICLR (2021)

# Learning Shadow Correspondence for Video Shadow Detection

Xinpeng Ding[1], Jingwen Yang[1], Xiaowei Hu[2], and Xiaomeng Li[1,3(✉)]

[1] The Hong Kong University of Science and Technology, Kowloon, Hong Kong
{xdingaf,jyangbv}@connect.ust.hk, eexmli@ust.hk
[2] Shanghai AI Laboratory, Shanghai, China
huxiaowei@pjlab.org.cn
[3] The Hong Kong University of Science and Technology Shenzhen Research Institute, Kowloon, Hong Kong

**Abstract.** Video shadow detection aims to generate consistent shadow predictions among video frames. However, the current approaches suffer from inconsistent shadow predictions across frames, especially when the illumination and background textures change in a video. We make an observation that the inconsistent predictions are caused by the shadow feature inconsistency, *i.e.*, the features of the same shadow regions show dissimilar proprieties among the nearby frames. In this paper, we present a novel **S**hadow-**C**onsistent **Cor**respondence method (**SC-Cor**) to enhance pixel-wise similarity of the specific shadow regions across frames for video shadow detection. Our proposed SC-Cor has three main advantages. Firstly, without requiring the dense pixel-to-pixel correspondence labels, SC-Cor can learn the pixel-wise correspondence across frames in a weakly-supervised manner. Secondly, SC-Cor considers intra-shadow separability, which is robust to the variant textures and illuminations in videos. Finally, SC-Cor is a plug-and-play module that can be easily integrated into existing shadow detectors with no extra computational cost. We further design a new evaluation metric to evaluate the temporal stability of the video shadow detection results. Experimental results show that SC-Cor outperforms the prior state-of-the-art method, by 6.51% on IoU and 3.35% on the newly introduced temporal stability metric. Code is available at https://github.com/xmed-lab/SC-Cor.

**Keywords:** Shadow detection · Video understanding · Correspondence learning

## 1 Introduction

Shadows in natural images or videos present different colors and brightness. Known where the shadow is, we can infer light source directions [26,36], scene

**Supplementary Information** The online version contains supplementary material available at https://doi.org/10.1007/978-3-031-19790-1_42.

© The Author(s), under exclusive license to Springer Nature Switzerland AG 2022
S. Avidan et al. (Eds.): ECCV 2022, LNCS 13677, pp. 705–722, 2022.
https://doi.org/10.1007/978-3-031-19790-1_42

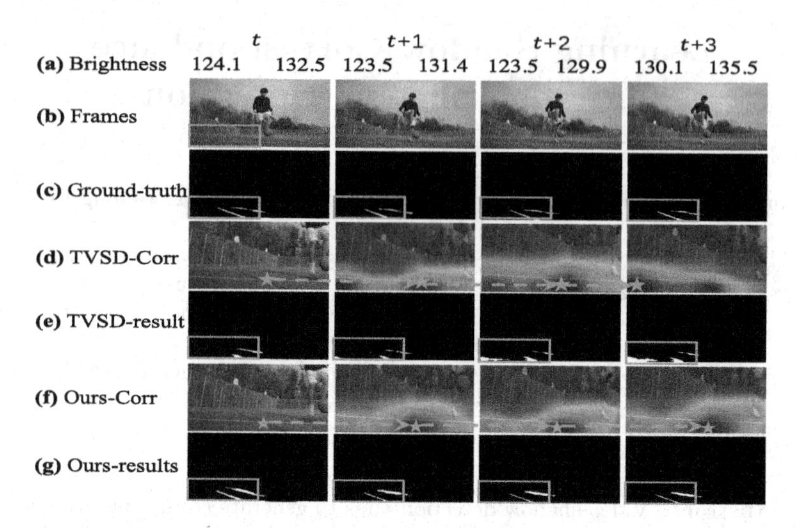


**Fig. 1.** Comparison of correspondences and results of TVSD [7] and Ours. We compute the brightness of four selected frames. Green and blue values indicate the non-shadow and shadow regions respectively. Given a query shadow region in the $t$-th frame, *i.e.*, the orange pentagram, we find the most similar features in nearby frames for the query, and regard the found features as its correspondences. "TVSD-Corr" and "Ours-Corr" indicate the correspondences found by TVSD [7] and our method respectively. "TVSD-result" and "Our-result" refer to the results predicted by TVSD and our method. It is clear that the found correspondences in the $(t + 3)$-th frame are in non-shadow regions (dark areas). This shadow features inconsistency, *i.e.*, features of the same shadow region may be dissimilar across frames, would result inconsistent prediction (red boxes). Our method can address the shadow feature inconsistency, and generate the contiguous results. (Color figure online)

geometry [21,22,35], and camera locations or parameters [21]. Therefore, shadow detection has attracted a lot of attention and achieved remarkable progress. However, most of the recent methods [8,16,23,33,45,55,56] detect shadows from single images while shadow detection over dynamic scenes, *i.e.*, in videos, is less explored.

To explore the powerful representation capability of deep learning for video shadow detection (VSD), Chen *et al.* [7] collect a large-scale video shadow detection (ViSha) dataset covering various scenarios. Then, a global contrastive objective is applied on the frame-level, which enhances the similarity between frames in the same video and push away the representations of frames from different videos. However, video shadow detection is a fine-grained pixel-level detection task, this frame-level semantic constraint may ignore shadow details, i.e., the same shadow regions across frames show dissimilar, resulting in inconsistent predictions; see red boxes in Fig. 1(d). Video data has the inherent property of *temporal consistency*, where the nearby frames are expected to contain similar shadow regions. Hence, we aim to explore both frame-level accuracy and temporal-level consistency for video shadow detection. In this paper, we make

**Fig. 2.** (a) Supervised contrastive learning pulls close all pixels in shadow regions, which is too strict to generate complete shadow detection; see Fig. 6 and Table 2 for details. (b) We aims to leverage correspondence learning to consider intra-shadow separability. Unlike existing correspondence learning [20,31] that require *pixel-to-pixel* labels among video frames, for each pixel in the shadow, we only know its corresponding pixel is within a shadow region in another frame, denoted as *pixel-to-set* correspondence learning.

a critical observation that this inconsistent prediction is caused by the shadow feature inconsistency, *i.e.*, the features of the same shadow regions show dissimilar proprieties among the nearby frames. For example, due to the illumination change (see Fig. 1(a)), the extracted features in a specific shadow region may show higher similarity with dark non-shadow regions in the nearby frames; see orange pentagram in the $(t+3)$-th frame in Fig. 1(c).

To address this problem, we aim to enhance temporal pixel-wise similarity for the specific shadow regions across frames, thus improving the detection accuracy and consistency in shadow videos; see Figs. 1(e) and (f). The supervised contrastive learning aims to increase intra-class compactness and inter-class separability, and has been used for image classification [24] or semantic segmentation [51,54]. An intuitive way is to use the supervised contrastive learning to pull close pixels in shadow regions across frames, and push away shadowed pixels and non-shadowed pixels. However, as objects move and illumination changes, the same shadow region may appear on backgrounds with different textures across frames. Simply adopting the supervised contrastive learning for video, *i.e.*, pulling close all pixels in shadows, leads to incomplete shadow regions; see Fig. 6 and Table 2 for the comparison results.

Hence, in this paper, we leverage the correspondence learning to learn a more fine-grained pixel-wise similarity. *i.e.*, only encouraging a pixel to be similar with its corresponding pixels in nearby frames. However, unlike the existing correspondence learning [11,20,31] that requires *pixel-to-pixel* correspondence labels across video frames, we do not need the pixel-wise correspondence labels. To this end, we present a novel **S**hadow-**C**onsistent **Cor**respondence method, namely **SC-Cor**, to learn the dense shadow correspondence in a *pixel-to-set* way, based on a key prior knowledge that a corresponding *(pixel)* of shadow is within a shadow region *(set)* in another frame; see Fig. 2(a). Different from the supervised contrastive learning [24,51,54], our proposed SC-Cor keeps the pixel most similar to the anchor in the shadow region and considers inter-shadow separability, which is robust to the variant textures and illuminations in videos. Note that our SC-Cor is a plug-and-play module and is only used in the training process. Therefore, SC-Cor can be easily applied to any deep-learning-based video/image shadow detection method without additional computational cost in testing.

Finally, existing metrics only evaluate the performance of VSD in frame-level, *e.g.*, frame-level Balance Error Rate (BER), and ignores the temporal consistency of shadow predictions. To this end, we introduce a new evaluation metric, temporal stability (TS), which computes the intersection over union score between the adjacent frames, thus helping to evaluate the temporal consistency of shadow predictions in videos. Below, we summarize the major contributions of this work:

- We present a novel and plug-and-play shadow-consistent correspondence (SC-Cor) method for video shadow detection. Compared with the existing *pixel-to-pixel* learning, our proposed SC-Cor is learned in a *pixel-to-set* way, without requiring pixel-wise correspondence labels.
- To fairly evaluate the temporal consistency of different shadow detection approaches, we introduce a new evaluation metric, which evaluates the flow-warped IoU between the adjacent video frames.
- We evaluate our SC-Cor on the benchmark dataset for video shadow detection and the experimental results show that our method clearly outperforms various state-of-the-art approaches in terms of both frame-level and temporal-level evaluation metrics.

## 2 Related Work

**Image Shadow Detection.** Early traditional methods are based on the hand-crafted shadow features, *e.g.*, intensity, chromaticity, physical properties, geometry, and textures [37]. Recently, deep-learning-based methods become the mainstream algorithms for shadow detection [9,16,18,23,27,33,40,45,55,56]. Khan *et al.* [23] build the first method based on deep neural network, which is a seven-layer CNN that learns from super-pixel level features and object boundaries. Hu *et al.* [17] present a fast shadow detection network by designing a detail enhancement module to refine shadow details. In the most recent work, Zhu *et al.* [57] design a feature decomposition and re-weighting scheme, which leverages intensity-variant and intensity-invariant features via self-supervision to mitigate the susceptibility of the intensity cue. Except the general shadow detection, Wang *et al.* [47,48] detecte the shadow regions associated with the objects simultaneously.

**Video Shadow Detection.** Early traditional video shadow detection (VSD) methods adopt the hand-crafted spectral and spatial features [1,19,32] to detect the shadow regions. To exploit the capability of deep-learning-based methods on this task, Chen *et al.* [7] collect the first large-scale VSD dataset ViSha. To detect the shadows in videos, they design a deep-learning-based method that contains a dual gated co-attention module and an auxiliary similarity loss to mine frame-level consistency information between different videos. Hu *et al.* [15] capture the temporal consistency by an optical-flow-based warping module to align and combine features between video frames. However, due to lack of the temporal pixel-level relation, these methods would suffer from shadow feature inconsistency and generate temporal-inconsistent results. Unlike existing methods, this paper presents a novel solution to learn pixel-wise consistency by formulating the dense shadow correspondence objective. Our method is flexible and can be easily integrated into many existing methods designed for both single-image and video shadow detection methods.

**Correspondence Learning.** Finding correspondences between pairs of images is a fundamental task in computer vision [3,20,31,39,42,43]. However, these methods require pixel-level correspondence labels and can hardly be obtained in videos. Hence, numerous works aim to learn temporal correspondence in the unsupervised way [49,50,52]. These methods perform unsupervised correspondence learning on videos and show obvious improvement on the obvious foreground objects. However, shadows are usually less obvious than the foreground, and may show different appearances and deformation due to illumination and texture changes. Our SC-Cor can address the above problems in a weakly supervised way, which is proved by experiments (see Fig. 6 and Table 2). In this paper, different from all of these methods, we aim to learn pixel-wise similarity in a *pixel-to-set* way.

**Contrastive Learning.** Contrastive learning pulls close an anchor and a positive sample, and pushes the anchor away from many negative samples, which has show great success in self-supervised learning [4–6,10–13]. Recently, the supervised contrastive learning aims to increase intra-class compactness and inter-class separability to improve image classification [24,28] or semantic segmentation [51,54]. However, as objects move and illumination changes, the same shadow region may appear on backgrounds with different textures across frames. The supervised contrastive learning, *i.e.*, simply pulling close all pixels in shadow regions is too strict, resulting in generating incomplete shadow regions; see Fig. 6 and Table 2 for details. Differently, our proposed SC-Cor aims to keep the pixel most similar to the anchor in the shadow regions, which considers inter-class separability due to the varying shadows in the videos.

## 3 Methodology

Figure 3(a) shows the training process of the overall SC-Cor framework, which can generate temporal-consistent and accurate shadow detection results. Formally, we denote a video sequence and the corresponding ground-truth (GT) masks as $\{V_t\}_{t=1}^{T}$ and $\{\mathbf{Y}_t\}_{t=1}^{T}$, respectively, where $T$ is the frame number of this video sequence. Given two video frames, which are denoted as $V_t$ and $V_{t+\delta}$ and $\delta$ is the time interval, we feed them into two branches of the framework; see Fig. 3(a). Each branch contains a feature extractor, which is used to capture

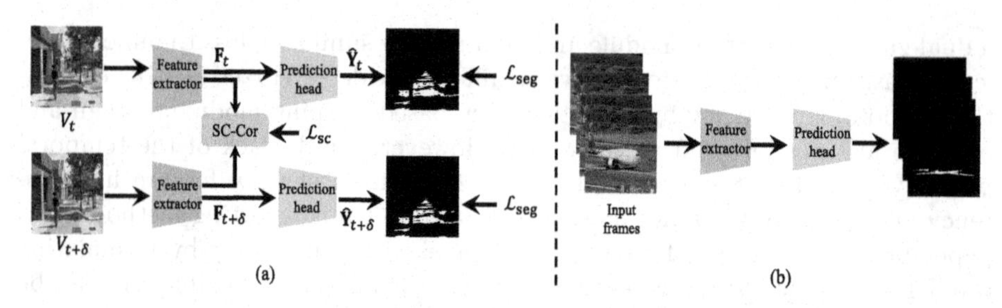


**Fig. 3.** (a) Illustration of the training process integrated with our shadow-consistent correspondence (SC-Cor) learning objective. Given two frames from one video, besides using the segmentation loss $\mathcal{L}_{seg}$ to supervise their frame-level predictions individually, we also enhance their temporal consistency by SC-Cor, described in Sect. 3.1. (b) Illustration of the inference phase. The proposed SC-Cor is only applied during training. We can improve the temporal consistency as well as the frame-level accuracy without any extra parameters or computation cost during inference.

the spatial features, *i.e.*, $\mathbf{F}_t$ and $\mathbf{F}_{t+\delta}$ of the input frames. Note that the weights in these two feature extractors are shared. Then, we adopt shadow-consistent correspondence (Fig. 4) to extract the temporal information of $\mathbf{F}_t$ and $\mathbf{F}_{t+\delta}$; see details in Sect. 3.1 and 3.2. Next, we send $\mathbf{F}_t$ and $\mathbf{F}_{t+\delta}$ into the shared prediction head to obtain the shadow detection results $\hat{\mathbf{Y}}_t$ and $\hat{\mathbf{Y}}_{t+\delta}$, which are supervised by the ground-truth masks $\mathbf{Y}_t$ and $\mathbf{Y}_{t+\delta}$. Note that the SC-Cor module is flexible and is only used in the training stage without any extra parameters introduced in the test stage, as shown in Fig. 3(b). Therefore, it can serve as a plug-and-play component and can be used in many single-image or video shadow detection methods.

### 3.1 Shadow-Consistent Correspondence

To explore the temporal consistency for VSD, we aim to learn shadow correspondence to capture the pixel-wise relations between shadows across frames in the video, which acts as a regularizer to optimize the framework. As discussed in Sect. 1, instead of the dense *pixel-to-pixel* labels [3,39], in this paper, we only obtain the *pixel-to-set* labels. To learn dense shadow correspondence, we introduce a novel shadow consistent correspondence method. The proposed shadow-consistent correspondence contains three modules: (a) a shadow guidance module, (b) a cross-frame correspondence module, and (c) a consistency regularization, as shown in Fig. 4.

**(a) Shadow guidance module.** The shadow guidance module aims to obtain a feature map that only contains feature vector on the shadow regions. Let $\mathbf{F}_t \in \mathbb{R}^{H\times W\times D}$ be the feature map of the frame $V_t$, where $D$, $H$ and $W$ denote the dimension, height, and width of the feature map, respectively. Here, we define the ground-truth shadow mask as $\mathbf{Y}_t \in \mathbb{R}^{H\times W\times D}$ and $\mathbf{Y}_t = \{0,1\}$, where $\mathbf{Y}_t(h,w) = 0$ indicates that the position $(h,w)$ in $\mathbf{Y}_t$ is in the non-shadow regions

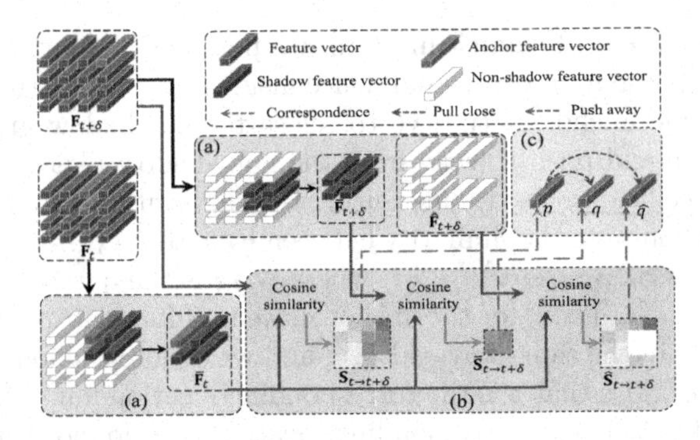


**Fig. 4.** Procedure of shadow-consistent correspondence from frame $t$ to $t+\delta$. The proposed method consists of three modules: (a) the shadow guidance module (Sect. 3.1(a)), (b) the cross-frame correspondence module (Sect. 3.1(b)), and (c) the consistency regularization (Sect. 3.1(c)).

and $\mathbf{Y}_t(h,w)=1$ represents the position $(h,w)$ is in the shadow regions. Then, we can obtain the set of shadow indexes $\mathcal{O}$:

$$\mathcal{O} = \{(h,w) \mid \mathbf{Y}_t(h,w)=1\}\ . \tag{1}$$

$$\overline{\mathbf{F}}_t = \{\mathbf{F}_t(h,w) \mid (h,w) \in \mathcal{O}\}, \tag{2}$$

where $\overline{\mathbf{F}}_t \in \mathbb{R}^{N_t \times D}$ and $N_t$ indicates the number of shadow feature vectors on $\mathbf{F}_t$, *i.e.*, $|\mathcal{O}| = N_t$. We define the operation of the shadow guidance module as $\overline{\mathbf{F}}_t = \mathrm{SG}(\mathbf{F}_t)$.

**(b) Cross-frame correspondence module.** For each shadow feature vector of a frame, cross-frame correspondence module aims to find its correspondence feature vector, *i.e.*, most relevant one, from another frame. Formally, we define the features of two frames from a video as $\mathbf{F}_t$ and $\mathbf{F}_{t+\delta}$, where $\delta > 0$ is the time interval. In the following, we will illustrate how to find the correspondence from $\mathbf{F}_t$ to $\mathbf{F}_{t+\delta}$, as well as in the other way around, *i.e.*, from $\mathbf{F}_{t+\delta}$ to $\mathbf{F}_t$.

To find the most correlated feature vector from $\mathbf{F}_{t+\delta}$, we first obtain the shadow feature map of $\mathbf{F}_t$ by $\overline{\mathbf{F}}_t = \mathrm{SG}(\mathbf{F}_t)$. Then, we measure the similarity between $\overline{\mathbf{F}}_t$ and $\mathbf{F}_{t+\delta}$ by the cosine similarity:

$$\mathbf{S}_{t\to t+\delta} = \frac{\overline{\mathbf{F}}_t \cdot \mathbf{F}_{t+\delta}}{||\overline{\mathbf{F}}_t||\,||\mathbf{F}_{t+\delta}||}\ , \tag{3}$$

where $\mathbf{S}_{t\to t+\delta} \in \mathbb{R}^{N_t \times L}$ and $L = H \times W$. Here, we take the $n$-th ($n \in N_t$) feature vector on $\overline{\mathbf{F}}_t$ as the anchor vector, and compute its most relevant feature vector on $\mathbf{F}_{t+\delta}$ based on the similarity map $\mathbf{S}_{t\to t+\delta}$:

$$p = \max_m \mathbf{S}_{t\to t+\delta}(n,m), m \in [1, L]\ , \tag{4}$$

where $p$ is the index of the corresponding location on $\mathbf{F}_{t+\delta}$. Note that we perform the same operation to find the corresponding $p$ for each feature vector on $\overline{\mathbf{F}}_t$.

**(c) Consistency regularization.** Here, we perform the consistency regularization to enforce the found correspondence inside the ground-truth sets. Specifically, we pull the anchor feature vector close to the shadow ground-truth on the second frame and push it away from the non-shadow regions on the second frame. More specifically, besides measuring the similarity between $\overline{\mathbf{F}}_t$ and $\mathbf{F}_{t+\delta}$, we additionally compute the similarity between $\overline{\mathbf{F}}_t$ and $\overline{\mathbf{F}}_{t+\delta}$ through Eq. 3, which can be defined as $\overline{\mathbf{S}}_{t\to t+\delta} \in \mathbb{R}^{N_t \times N_{t+\delta}}$, where $N_{t+\delta}$ is the number of shadow feature vectors on $\mathbf{F}_{t+\delta}$. Then, for the anchor $n$, we find its correspondence on $\overline{\mathbf{F}}_{t+\delta}$ based on $\overline{\mathbf{S}}_{t\to t+\delta}$ in the same way as Eq. 4, and we denote the found corresponding location as $q$. Note that $p$ is the corresponding location found from the whole feature map while $q$ is the corresponding location only found from the shadow set (indicated by the ground-truth mask). To pull $p$ and $q$ together, we minimize the discrepancy in two feature similarities by Eq. 5.

$$\mathcal{L}_{\text{shadow}}^{t\to t+\delta} = \frac{1}{N_t}\sum_{n=1}^{N_t}\left(\mathbf{S}_{t\to t+\delta}(n,p) - \overline{\mathbf{S}}_{t\to t+\delta}(n,q)\right)^2 . \tag{5}$$

To push the anchor $p$ away from the non-shadow regions, we first compute the set of non-shadow indexes $\hat{\mathcal{O}}$:

$$\hat{\mathcal{O}} = \{(h,w) \mid \mathbf{Y}_{t+\delta}(h,w) = 0\} . \tag{6}$$

Based on $\hat{\mathcal{O}}$, we obtain the non-shadow feature $\hat{\mathbf{F}}_{t+\delta}$. Then, we compute the similarity between $\overline{\mathbf{F}}_t$ and $\hat{\mathbf{F}}_{t+\delta}$ in the same way as Eq. 3 to obtain $\hat{\mathbf{S}}_{t\to t+\delta} \in \mathbb{R}^{N_t \times M_{t+\delta}}$, where $M_{t+\delta}$ is the number of non-shadow features on $\mathbf{F}_{t+\delta}$, *i.e.*, $|\hat{\mathcal{O}}| = M_{t+\delta}$. For the anchor $n$, we find its correspondence on $\hat{\mathbf{F}}_{t+\delta}$ based on $\hat{\mathbf{S}}_{t\to t+\delta}$ in the same way as Eq. 4, which is denoted as $\hat{q}$. To push away $p$ and $\hat{q}$, we maximize the margin in two feature similarities by the following loss function:

$$\mathcal{L}_{\text{n-shadow}}^{t\to t+\delta} = \frac{1}{N_t}\sum_{n=1}^{N_t}\max\left(0, \beta - |\mathbf{S}_{t\to t+\delta}(n,p) - \hat{\mathbf{S}}_{t\to t+\delta}(n,\hat{q})|\right) , \tag{7}$$

where $\beta$ controls the margin between $\mathbf{S}_{t\to t+\delta}(n,p)$ and $\hat{\mathbf{S}}_{t\to t+\delta}(n,\hat{q})$. In the same way, we can obtain the consistency regularization in the other way around, *i.e.*, from $\mathbf{F}_{t+\delta}$ to $\mathbf{F}_t$, and we define the loss functions as $\mathcal{L}_{\text{shadow}}^{t+\delta\to t}$ and $\mathcal{L}_{\text{n-shadow}}^{t+\delta\to t}$. Finally, shadow-consistent correspondence learning can be formulated as follows:

$$\mathcal{L}_{\text{sc}} = \mathcal{L}_{\text{shadow}}^{t\to t+\delta} + \mathcal{L}_{\text{n-shadow}}^{t\to t+\delta} + \mathcal{L}_{\text{shadow}}^{t+\delta\to t} + \mathcal{L}_{\text{n-shadow}}^{t+\delta\to t} . \tag{8}$$

### 3.2 Brightness-Invariant Correspondence

Due to the changes of brightness in a video, the shadow and non-shadow regions in different frames may present similar appearance. To learn brightness-invariant correspondence, we randomly shift the brightness of one frame and learn the brightness-invariant shadow consistency between the shifted frame and another frame in the same video [57]. Formally, for two frames in a video, *i.e.*, $V_t$ and $V_{t+\delta}$, we randomly shift the intensity of $V_{t+\delta}$ to produce the shifted frame $V'_{t+\delta} = V_{t+\delta} + \gamma$, where $\gamma \in [-\Delta, \Delta]$ is a randomly generated shift parameter and $\Delta$ is a hyper-parameter to control the shift range. Next, we use the shadow-consistent learning to learn the cross-frame correspondence between $V_t$ and $V'_{t+\delta}$, as introduced in Sect. 3.1.

### 3.3 Overall Objective

The overall objective of our framework is defined as:

$$\mathcal{L} = \mathcal{L}_{\text{seg}} + \lambda \mathcal{L}_{\text{sc}} , \tag{9}$$

where $\lambda$ is a hyper-parameter to control the trade-off between these two losses and $\mathcal{L}_{\text{seg}}$ is the segmentation loss to supervise the pixel-wise prediction. The segmentation loss is different for different shadow detection works. For example, [7,55,57] adopt the binary cross entropy (BCE) loss as the segmentation loss:

$$\mathcal{L}_{\text{seg}} = -\frac{1}{T}\sum_{t=1}^{T} \mathbf{Y}_t \cdot \log\left(\hat{\mathbf{Y}}_t\right) + (1 - \mathbf{Y}_t) \cdot \log\left(1 - \hat{\mathbf{Y}}_t\right), \tag{10}$$

where $\mathbf{Y}_t$ and $\hat{\mathbf{Y}}_t$ are ground-truth and predicted shadow masks, respectively. TVSD-Net [7] uses BCE combined with a lovász-hinge loss [2]. In the experiments, to highlight the effectiveness of the proposed shadow-consistent correspondence, the segmentation loss keeps consistent with existing papers [7,55].

## 4 Experimental Results

### 4.1 Evaluation Metrics and Datasets

**Temporal Stability.** Compared with the previous works that only evaluates the performance on each single image (frame-level), in this paper, we introduce a new evaluation metric to evaluate the temporal stability across the video frames, motivated by [25,44]. In detail, different from [25,44] that compute the optical flow between RGB frames, we calculate the optical flow between the ground-truth labels of two adjacent frames, *i.e.*, $\mathbf{Y}_t$ and $\mathbf{Y}_{t+1}$ through ARFlow [30], since the motions of shadows are hard to be captured on the RGB frames. For instance, the optical flows generated by RGB are focus on objects, which can not capture shadows since the motions of shadows are hard to be captured on the RGB frames; see supplementary materials for more details. Then, assume $I_{t\rightarrow t+1}$ as the optical flow between $\mathbf{Y}_t$ and $\mathbf{Y}_{t+1}$, and we define the reconstructed result that warps $\hat{\mathbf{Y}}_{t+1}$ by the optical flow $I_{t\rightarrow t+1}$ as $\mathbb{Y}_t$. Next, we measure the temporal stability of VSD for a video based on the flow warping IoU between the adjacent frames as:

$$\text{TS} = \frac{1}{T-1}\sum_{t=1}^{T-1} IoU(\hat{\mathbf{Y}}_t, \mathbb{Y}_t) . \tag{11}$$

**Table 1. Comparison with the state-of-the-art methods.** "↓" indicates the lower the scores, the better the results, while "↑" indicates the higher the scores, the better the results. "AVG" is the average score of IoU and TS, which presents the frame-level and temporal-level IoUs. "ISD" and "VSD" stand for the single-image shadow detection and video shadow detection, respectively. "SOD" stands for salient object detection. "VOS" stands for video object segmentation. "S-BER" and "N-BER" stand for BER of shadow regions and non-shadow regions, respectively.

| Task | Method | Frame-level | | | | | | Temporal-level | AVG ↑ |
|---|---|---|---|---|---|---|---|---|---|
| | | MAE ↓ | $F_\beta$ ↑ | BER ↓ | S-BER ↓ | N-BER ↓ | IoU [%] ↑ | TS [%] ↑ | |
| Scene Parsing | FPN [29] | 0.044 | 0.707 | 19.49 | 36.59 | 2.40 | 51.28 | 74.27 | 62.78 |
| | PSPNet [53] | 0.052 | 0.642 | 19.75 | 36.44 | 3.07 | 47.65 | 76.63 | 62.14 |
| SOD | DSS [14] | 0.045 | 0.697 | 19.78 | 36.96 | 2.59 | 50.28 | 75.02 | 62.65 |
| VOS | PDBM [41] | 0.066 | 0.623 | 19.74 | 34.32 | 5.16 | 46.65 | 80.00 | 63.33 |
| | FEELVOS [46] | 0.043 | 0.710 | 19.76 | 37.27 | 2.26 | 51.20 | 74.89 | 63.05 |
| | STM [34] | 0.064 | 0.639 | 23.77 | 43.88 | 3.65 | 44.69 | 75.30 | 60.00 |
| ISD | BDRAR [56] | 0.050 | 0.695 | 21.30 | 40.28 | 2.31 | 48.39 | 72.63 | 60.51 |
| | MTMT [8] | 0.043 | 0.729 | 20.29 | 38.71 | 1.86 | 51.69 | 74.44 | 63.07 |
| | FSD [17] | 0.057 | 0.671 | 20.57 | 38.06 | 3.06 | 48.56 | 74.88 | 61.72 |
| | DSD [55] | 0.044 | 0.702 | 19.89 | 37.89 | 1.88 | 51.89 | 74.68 | 63.29 |
| | DSD + ours | 0.039 | 0.730 | 15.15 | 27.78 | 2.52 | 58.40 | 78.03 | 68.22 |
| | - | **+0.05** | **+0.028** | **+4.65** | **+10.11** | −0.64 | **+6.51** | **+3.35** | **+4.93** |
| VSD | Hu *et al.* [15] | 0.078 | 0.683 | 17.03 | 30.13 | 3.93 | 51.03 | 83.67 | 67.35 |
| | TVSD [7] | 0.033 | 0.757 | 17.70 | 33.97 | 1.45 | 56.57 | 78.25 | 67.41 |
| | TVSD + ours | 0.042 | 0.762 | 13.61 | 24.31 | 2.91 | 61.50 | 81.44 | 71.47 |
| | - | −0.09 | **+0.005** | **+4.09** | **+9.66** | −0.46 | **+4.93** | **+3.19** | **+4.06** |

**Frame-Level Accuracy.** Except using the proposed evaluation metric to measure temporal stability, we follow the previous works [7,55,57] and adopt four common evaluation metrics that have been widely used in image/video shadow detection to evaluate the detection accuracy in frame-level. Specifically, they are Mean Absolute Error (MAE) [7], F-measure ($F_\beta$) [7,17], Intersection over Union (IoU) [7], and Balance Error Rate (BER) [18,57].

**Evaluation Dataset.** We conduct our experiments on the ViSha dataset [7] to evaluate the performance. ViSha consists of 11,685 image frames and 390$s$ duration, which is adjusted to 30 fps for all video sequences. This dataset is split into 50 videos for training and 70 videos for testing.

### 4.2 Implementation Details

Since our framework is a plug-and-play module that can be used in any shadow detectors, we insert our framework into two state-of-the-art methods on single-image shadow detection and video shadow detection, *i.e.*, DSD [55] and TVSD [7], for evaluation. During the training process of our shadow-consistent correspondence module, $\lambda$ in Eq. 9 and $\beta$ in Eq. 7 are set to 10 and 0.5, respectively; please refer to Sect. 4.4 for the analysis of $\lambda$. $\Delta$ is used to control the range of the brightness shift and it is set to 0.3 following [57]. Note, shifting brightness

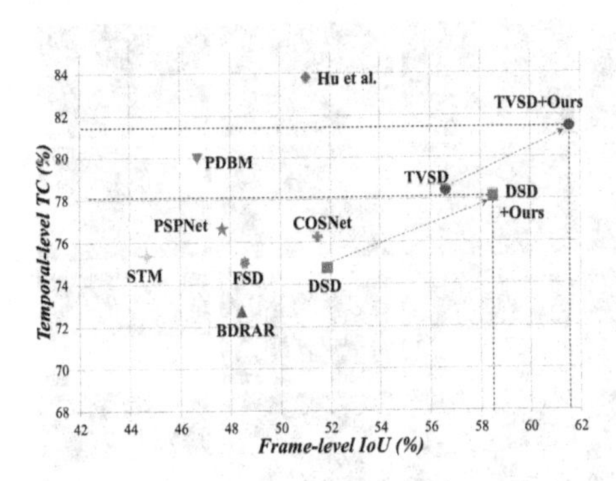


**Fig. 5.** Trade-off between temporal and frame-level accuracy.

**Table 2. Comparison of the supervised contrastive learning and ours.** Baseline is DSD [55]. "SCon" indicates the supervised contrastive learning. "Ratio" indicates the ratio of the found correspondence in the shadow regions. "AVG" refers to the average score of IoU and TS.

| Method | Ratio ↑ | AVG ↑ |
|---|---|---|
| Baseline | 36.71 | 63.29 |
| Baseline + [50] | 82.33 | 66.17 |
| Baseline + SCon | 85.30 | 65.13 |
| Baseline + Ours | 85.12 | 68.22 |

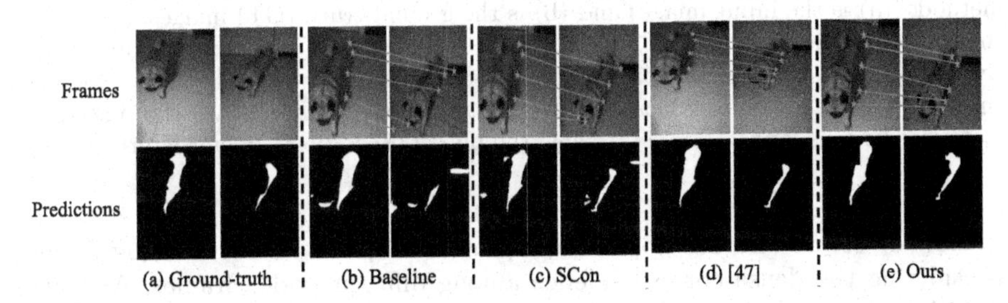


**Fig. 6.** Visualization of the correspondence found by different models. Note that Baseline is DSD [55]. We sample five pixels in one frame and find their correspondence in the other one.

of frames will change the distribution of the images, resulting in degrading the detection performance [57]. Hence, we only shift the brightness of the frames after 2,000 training iterations and freeze the batch normalization [38]; see supplementary materials for more details.

### 4.3 Comparison with the State-of-the-Art Methods

We conduct the experiments on ViSha [7] to compare with the state-of-the-art methods designed for scene parsing, salient object detection, video object segmentation, single-image shadow detection, and video shadow detection; please see the compared methods in Table 1. We obtain the results of these methods by retraining them on the ViSha dataset for video shadow detection with the recommendation training parameters or by downloading their results directly from Internet.

Table 1 provides the comparison results, which clearly shows that our proposed method can largely improve the performance of both single-image and video shadow detection approaches, *i.e.*, DSD [55] and TVSD [7], in terms of both frame-level and temporal-level accuracy. DSD is designed for single-image

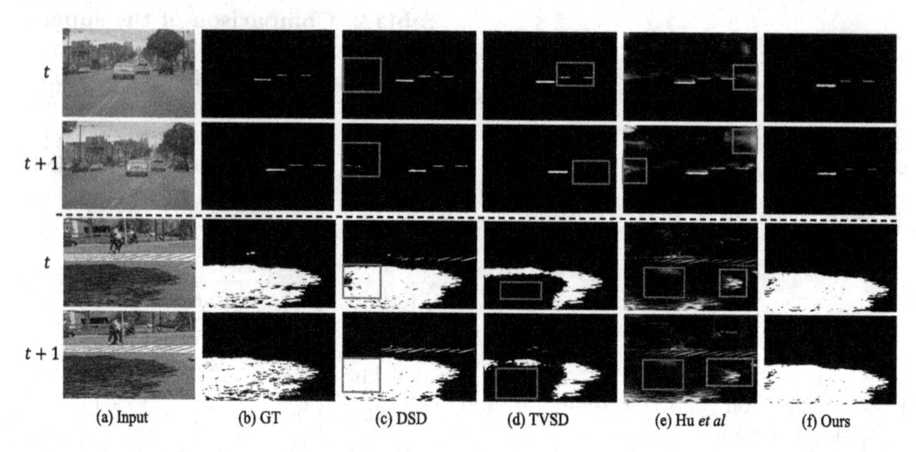


**Fig. 7.** Visual comparison of video shadow detection results produced by different methods. (a) is the input images and (b) is the ground-truth (GT) images. (c)–(f) are the results predicted by DSD [55], TVSD [7], Hu *et al.* [15], and our method, respectively. Our method takes the DSD as the basic network. Note that red boxes indicate the inconsistent predictions across video frames, blue boxes indicate the inaccurate static predictions, and green boxes show the blurry predictions. (Color figure online)

shadow detection and our approach improves the performance a lot by further considering the dense correspondence among different video frames. Although TVSD is designed for video shadow detection and has explored the temporal consistency in videos, our method further explores the shadow-region correspondence and learns the brightness-invariant features. Besides, our method achieves the best trade-off on both temporal-level and frame-level accuracy, as shown in Fig. 5.

Figure 7 illustrates the visual comparison of the shadow masks produced by DSD [55], TVSD [7], Hu *et al.* [15], and ours. From the results, we can see that our method provides more accurate and consistent shadow detection results across different videos frames than others. More examples and failure cases please refer to supplementary materials.

**Comparison of the Supervised Contrastive Learning, Unsupervised Correspondence Learning and our SC-Cor.** Specifically, for the sampled pixel of one frame, we find the most correlated pixel in the other frame and denote the found pixel as its correspondence. It is clear that the correspondences found by the baseline would be in dark non-shadow regions. Those found by the supervised contrastive learning would only focus on the shadow regions with similar textures. For unsupervised correspondence learning, we select [50], a recently SOTA, for comparison. It is clear that ours outperforms [47]. In order to further evaluate the effect of our method, we report the ratio of the found correspondence in ground truth masks and the average performance in Table 2. The results show that our method can largely improve the accuracy of found correspondence, *e.g.*, over 43.41% on DSD [55]. Although the ratio of correspondences

**Table 3. Ablation on the effectiveness of SC-Cor and BS.** "SC-Cor" indicates the shadow-consistent correspondence and "BS" indicates the brightness shift operation.

| Method | SC | BS | Frame-level | | | Temporal-level | AVG ↑ |
|---|---|---|---|---|---|---|---|
| | | | MAE ↓ | BER ↓ | IoU [%] ↑ | TS [%] ↑ | |
| Baseline | × | × | 0.044 | 19.89 | 51.89 | 74.68 | 63.29 |
| Ours (SC-Cor) | √ | × | 0.040 | 15.67 | 56.89 | 77.09 | 67.00 |
| Ours (BS) | × | √ | 0.043 | 16.82 | 53.65 | 74.92 | 64.29 |
| Ours (Full) | √ | √ | **0.039** | **14.89** | **58.40** | **78.03** | **68.22** |

**Table 4.** Ablation Study on Shadow-consistent correspondence.

(a) **Bidirectional correspondence**. "Bi-D" refers to the bidirectional correspondence, which has been defined by $t \rightarrow t+\delta$ and $t+\delta \rightarrow t$ in Eq. 8.

| Bi-D | *Frame-level* | | *Temporal-level* | AVG ↑ |
|---|---|---|---|---|
| | BER ↓ | IoU [%] ↑ | TS [%] ↑ | |
| × | 15.54 | 57.25 | 77.08 | 67.17 |
| ✓ | **14.89** | **58.40** | **78.03** | **68.22** |

(b) **Consistency regularization.** "Shadow" and "N-Shadow" indicate the regularization method described in Eq. 5 and Eq. 7.

| | *Frame-level* | | *Temporal-level* | AVG ↑ |
|---|---|---|---|---|
| | BER ↓ | IoU [%] ↑ | TS [%] ↑ | |
| Shadow | 15.67 | 57.18 | 76.89 | 67.04 |
| N-shadow | 15.79 | 57.09 | 76.72 | 66.91 |
| Full | **14.89** | **58.40** | **78.03** | **68.22** |

found by supervised contrastive learning in ground-truth is high, the generated shadow mask may be incomplete; see Fig. 6(c).

## 4.4 Ablation Study

We conduct ablation experiments to show how each module in our framework design contributes to video shadow detection. We regard DSD [55] as our baseline module in this section. All the detection results are reported on the testing set of the ViSha dataset [7].

**Effectiveness of SC-Cor and BS.** Table 3 reports the effectiveness of the shadow-consistent correspondence (SC-Cor) and the brightness shift (BS) operation. Training with SC-Cor, we can see a clear improvement in terms of both frame-level accuracy and temporal accuracy, *i.e.*, 4.22 on BER and 2.40% on TS. It is worth noting that only adopting with BS cannot obtain the clear improvement on the temporal stability, *i.e.*, 74.92% vs. 74.68%, due to the lack of exploring temporal information. By combining with both SC-Cor and BS, the model achieves the best performance.

**Bidirectional Correspondence and Consistency Regularization.** Table 4a reports the results of bidirectional correspondence in Eq. 8 and shows the effectiveness of the designed bidirectional correspondence. Furthermore, we

**Table 5.** Ablation Study on different frame setting.

(a) **Multiple frames**. "Frame number" denotes the number of sampled frames.

| Frame number | Frame-level | | Temporal-level | |
|---|---|---|---|---|
| | BER ↓ | IoU [%] ↑ | TS [%] ↑ | AVG ↑ |
| 2 | 15.91 | 58.06 | 77.96 | 68.01 |
| 3 | 15.86 | 58.28 | 78.46 | 68.37 |
| 4 | **15.82** | **58.26** | **78.51** | **68.39** |

(b) **Frame interval** $\delta$

| $\delta$ | Frame-level | | Temporal-level | |
|---|---|---|---|---|
| | BER ↓ | IoU [%] ↑ | TS [%] ↑ | AVG ↑ |
| 1 | **14.51** | **58.62** | 76.94 | 67.78 |
| 3 | 14.73 | 58.55 | 77.23 | 67.89 |
| 5 | 14.89 | 58.40 | 78.03 | **68.22** |
| 7 | 15.64 | 57.12 | **78.23** | 67.68 |

perform the ablation study on the shadow and non-shadow consistency regularization in Table 4b, showing that the combination of them achieves the best results.

**Multiple Frames and Frame Interval.** We integrate our SC-Cor with multiple pairs of frames in a video and analyze the effectiveness in Table 5a. We observe that training with more frames brings a slight improvement on both frame-level and temporal-level accuracy. Considering the training efficiency, we choose two pairs of frames. Furthermore, we study the frame sampling strategy and report the detection results in Table 5b. It is clear that the longer time interval achieves the higher temporal stability while the short one performs better in frame-level accuracy. For instances, $\delta = 1$ achieves the best BER, *i.e.*, 14.51, and the lowest TS, *i.e.*, 76.94%. On the contrary, $\delta = 7$ obtains the best TS performance 78.23%. In this paper, we set $\delta$ as five to balance the temporal-level accuracy and the frame-level accuracy.

## 5 Conclusion

In this paper, we present a novel and plug-and-play shadow-consistent correspondence (SC-Cor) method for video shadow detection (VSD). A shadow-consistent correspondence is formulated to enforce the network to learn temporal-consistent shadows. A brightness shifting operation is employed to further regularize the network to be brightness-invariant. Considering current metrics only evaluate the frame-level accuracy, we introduce a new temporal stability metric, namely TS, for VSD. Experimental results on the benchmark dataset prove that our SC-Cor outperforms various shadow detection methods.

**Acknowledgement.** This work was supported by a research grant from HKUST-BICI Exploratory Fund under HCIC-004 and a research grant from Foshan HKUST Projects under FSUST21-HKUST11E.

## References

1. Benedek, C., Szirányi, T.: Bayesian foreground and shadow detection in uncertain frame rate surveillance videos. IEEE Trans. Image Process. **17**(4), 608–621 (2008)

2. Berman, M., Triki, A.R., Blaschko, M.B.: The lovász-softmax loss: a tractable surrogate for the optimization of the intersection-over-union measure in neural networks. In: Proceedings of the IEEE Conference on Computer Vision and Pattern Recognition, pp. 4413–4421 (2018)
3. Bian, J., Lin, W.Y., Matsushita, Y., Yeung, S.K., Nguyen, T.D., Cheng, M.M.: GMS: grid-based motion statistics for fast, ultra-robust feature correspondence. In: Proceedings of the IEEE Conference on Computer Vision and Pattern Recognition, pp. 4181–4190 (2017)
4. Caron, M., Misra, I., Mairal, J., Goyal, P., Bojanowski, P., Joulin, A.: Unsupervised learning of visual features by contrasting cluster assignments. Adv. Neural. Inf. Process. Syst. **33**, 9912–9924 (2020)
5. Chen, X., Fan, H., Girshick, R., He, K.: Improved baselines with momentum contrastive learning. arXiv preprint arXiv:2003.04297 (2020)
6. Chen, X., He, K.: Exploring simple siamese representation learning. In: Proceedings of the IEEE/CVF Conference on Computer Vision and Pattern Recognition, pp. 15750–15758 (2021)
7. Chen, Z., et al.: Triple-cooperative video shadow detection. In: Proceedings of the IEEE/CVF Conference on Computer Vision and Pattern Recognition, pp. 2715–2724 (2021)
8. Chen, Z., Zhu, L., Wan, L., Wang, S., Feng, W., Heng, P.A.: A multi-task mean teacher for semi-supervised shadow detection. In: Proceedings of the IEEE/CVF Conference on Computer Vision and Pattern Recognition, pp. 5611–5620 (2020)
9. Ding, B., Long, C., Zhang, L., Xiao, C.: Argan: attentive recurrent generative adversarial network for shadow detection and removal. In: Proceedings of the IEEE/CVF International Conference on Computer Vision (ICCV) (2019)
10. Ding, X., Liu, Z., Li, X.: Free lunch for surgical video understanding by distilling self-supervisions. arXiv preprint arXiv:2205.09292 (2022)
11. Ding, X., et al.: Support-set based cross-supervision for video grounding. In: Proceedings of the IEEE/CVF International Conference on Computer Vision, pp. 11573–11582 (2021)
12. Ding, X., et al.: Exploring language hierarchy for video grounding. IEEE Trans. Image Process. **31**, 4693–4706 (2022)
13. He, K., Fan, H., Wu, Y., Xie, S., Girshick, R.: Momentum contrast for unsupervised visual representation learning. In: Proceedings of the IEEE/CVF Conference on Computer Vision and Pattern Recognition, pp. 9729–9738 (2020)
14. Hou, Q., Cheng, M.M., Hu, X., Borji, A., Tu, Z., Torr, P.H.S.: Deeply supervised salient object detection with short connections. IEEE Trans. Pattern Anal. Mach. Intell. **41**(4), 815–828 (2019). https://doi.org/10.1109/TPAMI.2018.2815688
15. Hu, S., Le, H., Samaras, D.: Temporal feature warping for video shadow detection. arXiv preprint arXiv:2107.14287 (2021)
16. Hu, X., Fu, C.W., Zhu, L., Qin, J., Heng, P.A.: Direction-aware spatial context features for shadow detection and removal. IEEE Trans. Pattern Anal. Mach. Intell. **42**(11), 2795–2808 (2020)
17. Hu, X., Wang, T., Fu, C.W., Jiang, Y., Wang, Q., Heng, P.A.: Revisiting shadow detection: a new benchmark dataset for complex world. IEEE Trans. Image Process. **30**, 1925–1934 (2021). https://doi.org/10.1109/TIP.2021.3049331
18. Hu, X., Zhu, L., Fu, C.W., Qin, J., Heng, P.A.: Direction-aware spatial context features for shadow detection. In: Proceedings of the IEEE Conference on Computer Vision and Pattern Recognition, pp. 7454–7462 (2018)

19. Jacques, J.C.S., Jung, C.R., Musse, S.R.: Background subtraction and shadow detection in grayscale video sequences. In: XVIII Brazilian Symposium on Computer Graphics and Image Processing (SIBGRAPI 2005), pp. 189–196. IEEE (2005)
20. Jiang, W., Trulls, E., Hosang, J., Tagliasacchi, A., Yi, K.M.: COTR: correspondence transformer for matching across images. arXiv preprint arXiv:2103.14167 (2021)
21. Junejo, I.N., Foroosh, H.: Estimating geo-temporal location of stationary cameras using shadow trajectories. In: Forsyth, D., Torr, P., Zisserman, A. (eds.) ECCV 2008. LNCS, vol. 5302, pp. 318–331. Springer, Heidelberg (2008). https://doi.org/10.1007/978-3-540-88682-2_25
22. Karsch, K., Hedau, V., Forsyth, D., Hoiem, D.: Rendering synthetic objects into legacy photographs. ACM Trans. Graph. (TOG) **30**(6), 1–12 (2011)
23. Khan, S.H., Bennamoun, M., Sohel, F., Togneri, R.: Automatic feature learning for robust shadow detection. In: 2014 IEEE Conference on Computer Vision and Pattern Recognition, pp. 1939–1946. IEEE (2014)
24. Khosla, P., et al.: Supervised contrastive learning. Adv. Neural. Inf. Process. Syst. **33**, 18661–18673 (2020)
25. Lai, W.S., Huang, J.B., Wang, O., Shechtman, E., Yumer, E., Yang, M.H.: Learning blind video temporal consistency. In: Proceedings of the European Conference on Computer Vision (ECCV), pp. 170–185 (2018)
26. Lalonde, J.F., Efros, A.A., Narasimhan, S.G.: Estimating natural illumination from a single outdoor image. In: 2009 IEEE 12th International Conference on Computer Vision, pp. 183–190. IEEE (2009)
27. Le, H., Vicente, T.F.Y., Nguyen, V., Hoai, M., Samaras, D.: A+D Net: training a shadow detector with adversarial shadow attenuation. In: Proceedings of the European Conference on Computer Vision (ECCV), September 2018
28. Li, H., Wang, N., Ding, X., Yang, X., Gao, X.: Adaptively learning facial expression representation via CF labels and distillation. IEEE Trans. Image Process. **30**, 2016–2028 (2021)
29. Lin, T., Dollar, P., Girshick, R., He, K., Hariharan, B., Belongie, S.: Feature pyramid networks for object detection. In: 2017 IEEE Conference on Computer Vision and Pattern Recognition (CVPR), Los Alamitos, CA, USA, pp. 936–944. IEEE Computer Society, July 2017. https://doi.org/10.1109/CVPR.2017.106
30. Liu, L., et al.: Learning by analogy: reliable supervision from transformations for unsupervised optical flow estimation. In: Proceedings of the IEEE/CVF Conference on Computer Vision and Pattern Recognition, pp. 6489–6498 (2020)
31. Melekhov, I., Tiulpin, A., Sattler, T., Pollefeys, M., Rahtu, E., Kannala, J.: DGC-Net: dense geometric correspondence network. In: 2019 IEEE Winter Conference on Applications of Computer Vision (WACV), pp. 1034–1042. IEEE (2019)
32. Nadimi, S., Bhanu, B.: Physical models for moving shadow and object detection in video. IEEE Trans. Pattern Anal. Mach. Intell. **26**(8), 1079–1087 (2004)
33. Nguyen, V., Vicente, T.F.Y., Zhao, M., Hoai, M., Samaras, D.: Shadow detection with conditional generative adversarial networks. In: Proceedings of the IEEE International Conference on Computer Vision, pp. 4510–4518 (2017)
34. Oh, S.W., Lee, J.Y., Xu, N., Kim, S.J.: Video object segmentation using space-time memory networks. In: Proceedings of the IEEE/CVF International Conference on Computer Vision (ICCV), October 2019
35. Okabe, T., Sato, I., Sato, Y.: Attached shadow coding: estimating surface normals from shadows under unknown reflectance and lighting conditions. In: 2009 IEEE 12th International Conference on Computer Vision, pp. 1693–1700. IEEE (2009)

36. Panagopoulos, A., Samaras, D., Paragios, N.: Robust shadow and illumination estimation using a mixture model. In: 2009 IEEE Conference on Computer Vision and Pattern Recognition, pp. 651–658. IEEE (2009)
37. Sanin, A., Sanderson, C., Lovell, B.C.: Shadow detection: a survey and comparative evaluation of recent methods. Pattern Recogn. **45**(4), 1684–1695 (2012)
38. Santurkar, S., Tsipras, D., Ilyas, A., Madry, A.: How does batch normalization help optimization? In: Proceedings of the 32nd International Conference on Neural Information Processing Systems, pp. 2488–2498 (2018)
39. Sarlin, P.E., DeTone, D., Malisiewicz, T., Rabinovich, A.: Superglue: learning feature matching with graph neural networks. In: Proceedings of the IEEE/CVF Conference on Computer Vision and Pattern Recognition, pp. 4938–4947 (2020)
40. Shen, L., Chua, T.W., Leman, K.: Shadow optimization from structured deep edge detection. In: 2015 IEEE Conference on Computer Vision and Pattern Recognition (CVPR), pp. 2067–2074 (2015). https://doi.org/10.1109/CVPR.2015.7298818
41. Song, H., Wang, W., Zhao, S., Shen, J., Lam, K.M.: Pyramid dilated deeper ConvLSTM for video salient object detection. In: Proceedings of the European Conference on Computer Vision (ECCV), September 2018
42. Truong, P., Danelljan, M., Timofte, R.: GLU-Net: global-local universal network for dense flow and correspondences. In: Proceedings of the IEEE/CVF Conference on Computer Vision and Pattern Recognition, pp. 6258–6268 (2020)
43. Tyszkiewicz, M.J., Fua, P., Trulls, E.: Disk: learning local features with policy gradient. arXiv preprint arXiv:2006.13566 (2020)
44. Varghese, S., et al.: Unsupervised temporal consistency metric for video segmentation in highly-automated driving. In: Proceedings of the IEEE/CVF Conference on Computer Vision and Pattern Recognition Workshops, pp. 336–337 (2020)
45. Vicente, T.F.Y., Hou, L., Yu, C.-P., Hoai, M., Samaras, D.: Large-scale training of shadow detectors with noisily-annotated shadow examples. In: Leibe, B., Matas, J., Sebe, N., Welling, M. (eds.) ECCV 2016. LNCS, vol. 9910, pp. 816–832. Springer, Cham (2016). https://doi.org/10.1007/978-3-319-46466-4_49
46. Voigtlaender, P., Chai, Y., Schroff, F., Adam, H., Leibe, B., Chen, L.: FEELVOS: fast end-to-end embedding learning for video object segmentation. CoRR abs/1902.09513 (2019). http://arxiv.org/abs/1902.09513
47. Wang, T., Hu, X., Fu, C.W., Heng, P.A.: Single-stage instance shadow detection with bidirectional relation learning. In: Proceedings of the IEEE/CVF Conference on Computer Vision and Pattern Recognition, pp. 1–11 (2021)
48. Wang, T., Hu, X., Wang, Q., Heng, P.A., Fu, C.W.: Instance shadow detection. In: Proceedings of the IEEE/CVF Conference on Computer Vision and Pattern Recognition, pp. 1880–1889 (2020)
49. Wang, X., Jabri, A., Efros, A.A.: Learning correspondence from the cycle-consistency of time. In: Proceedings of the IEEE/CVF Conference on Computer Vision and Pattern Recognition, pp. 2566–2576 (2019)
50. Xu, J., Wang, X.: Rethinking self-supervised correspondence learning: a video frame-level similarity perspective. arXiv preprint arXiv:2103.17263 (2021)
51. Zhang, F., Torr, P., Ranftl, R., Richter, S.: Looking beyond single images for contrastive semantic segmentation learning. In: Advances in Neural Information Processing Systems, vol. 34 (2021)
52. Zhang, Q., Xiao, T., Efros, A.A., Pinto, L., Wang, X.: Learning cross-domain correspondence for control with dynamics cycle-consistency. arXiv preprint arXiv:2012.09811 (2020)
53. Zhao, H., Shi, J., Qi, X., Wang, X., Jia, J.: Pyramid scene parsing network. In: CVPR (2017)

54. Zhao, X., et al.: Contrastive learning for label efficient semantic segmentation. In: Proceedings of the IEEE/CVF International Conference on Computer Vision, pp. 10623–10633 (2021)
55. Zheng, Q., Qiao, X., Cao, Y., Lau, R.W.: Distraction-aware shadow detection. In: Proceedings of the IEEE/CVF Conference on Computer Vision and Pattern Recognition, pp. 5167–5176 (2019)
56. Zhu, L., et al.: Bidirectional feature pyramid network with recurrent attention residual modules for shadow detection. In: Proceedings of the European Conference on Computer Vision (ECCV), pp. 121–136 (2018)
57. Zhu, L., Xu, K., Ke, Z., Lau, R.W.: Mitigating intensity bias in shadow detection via feature decomposition and reweighting. In: Proceedings of the IEEE/CVF International Conference on Computer Vision (ICCV), pp. 4702–4711 (2021)

# Metric Learning Based Interactive Modulation for Real-World Super-Resolution

Chong Mou[1,3], Yanze Wu[3], Xintao Wang[3], Chao Dong[4,5], Jian Zhang[1,2(✉)], and Ying Shan[3]

[1] Peking University Shenzhen Graduate School, Shenzhen, China
eechongm@gmail.com,zhangjian.sz@pku.edu.cn
[2] Peng Cheng Laboratory, Shenzhen, China
[3] ARC Lab, Tencent PCG,Shenzhen, China
{yanzewu,xintaowang,yingsshan}@tencent.com
[4] Shenzhen Institutes of Advanced Technology, Chinese Academy of Sciences, Beijing, China
chao.dong@siat.ac.cn
[5] Shanghai AI Laboratory, Shanghai, China

**Abstract.** Interactive image restoration aims to restore images by adjusting several controlling coefficients, which determine the restoration strength. Existing methods are restricted in learning the controllable functions under the supervision of known degradation types and levels. They usually suffer from a severe performance drop when the real degradation is different from their assumptions. Such a limitation is due to the complexity of real-world degradations, which can not provide explicit supervision to the interactive modulation during training. However, how to realize the interactive modulation in real-world super-resolution has not yet been studied. In this work, we present a Metric Learning based Interactive Modulation for **Real**-World **S**uper-**R**esolution (**MM-RealSR**). Specifically, we propose an unsupervised degradation estimation strategy to estimate the degradation level in real-world scenarios. Instead of using known degradation levels as explicit supervision to the interactive mechanism, we propose a metric learning strategy to map the unquantifiable degradation levels in real-world scenarios to a metric space, which is trained in an unsupervised manner. Moreover, we introduce an anchor point strategy in the metric learning process to normalize the distribution of metric space. Extensive experiments demonstrate that the proposed MM-RealSR achieves excellent modulation and restoration performance in real-world super-resolution. Codes are available at https://github.com/TencentARC/MM-RealSR.

X. Wang—Project lead.
Chong Mou is an intern in ARC Lab, Tencent PCG.

**Supplementary Information** The online version contains supplementary material available at https://doi.org/10.1007/978-3-031-19790-1_43.

© The Author(s), under exclusive license to Springer Nature Switzerland AG 2022
S. Avidan et al. (Eds.): ECCV 2022, LNCS 13677, pp. 723–740, 2022.
https://doi.org/10.1007/978-3-031-19790-1_43

**Keywords:** Metric learning · Real-world super-resolution · Interactive modulation · Generative adversarial network

## 1 Introduction

Image super-resolution (SR) is the task of recovering details of a high-resolution (HR) image from its low-resolution (LR) counterpart. Most SR methods [7–9,24,46] assume an ideal bicubic downsampling kernel, which is different from real-world degradations. This degradation mismatch makes those approaches unpractical in real-world scenarios. Recently, some attempts [2,27,35,43] are proposed to address real-world super-resolution (RWSR). Nevertheless, existing RWSR methods can only perform restoration with fixed one-to-one mapping. In other words, they lack the flexibility, *i.e.*, interactive modulation [12,13], to alter the outputs by adjusting different restoration strength levels.

On the other hand, several modulation-based methods [3,12,13,40] are proposed to enable controllable image restoration according to users' flavours. These modulation-based methods generally take the degradation level as a part of the network inputs, and then construct mapping between the reconstruction result and the degradation level during training. During inference, users can simply adjust the value of the input degradation level, and the network will generate reconstruction results according to the corresponding restoration strength. However, existing controllable image restoration methods can only be trained on datasets with *simple* degradation processes and *known* degradation types/levels.

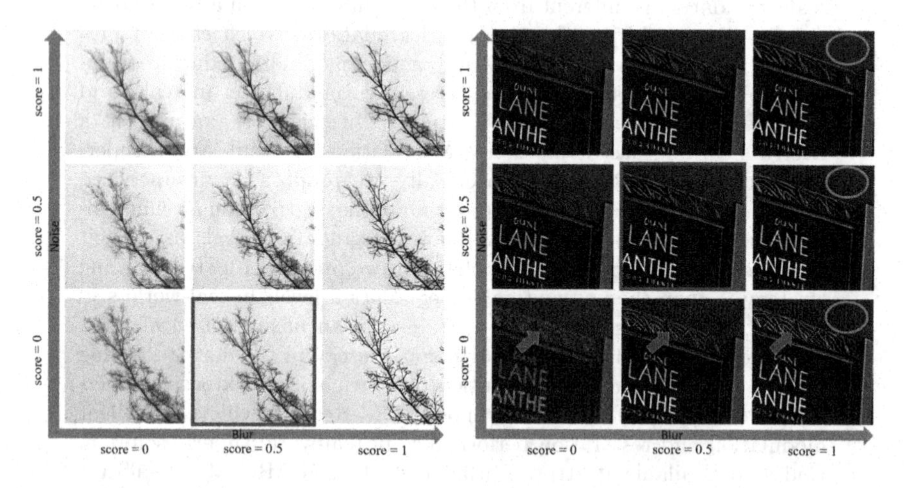


**Fig. 1.** Visual results of our interactive modulation method for real-world super-resolution (×4). Best results are labeled with red boxes. **(Zoom in for best view)** (Color figure online)

In real-world scenarios, corrupted images usually contain mixture and complex degradations (*e.g.*, blur [29], noise [42] and JPEG compression). Such a complex real-world degradation process can be simulated by a random shuffle order [43] or a high-order degradation process [35]. Due to the mixtures of different degradation levels and complicated degradation types in high-order degradations, the explicit degradation levels can not reflect the actual degradation effects in the corrupted images. For example, a corrupted image is degraded by a sequence of Gaussian blur with $\sigma_1$, Gaussian noise, and Gaussian blur with $\sigma_2$. We cannot know the final equivalent blur level on this degraded image. This case becomes more complicated when the real-world degradation process involves more degradation types and more complex degradation combinations. Therefore, existing modulation-based methods trained with known degradation types and levels can not get effective supervision in real-world settings. How to extend the interactive modulation to RWSR is worth investigating and has not yet been studied.

In this work, we present a **M**etric Learning based Interactive **M**odulation method for **Real**-World **S**uper-**R**esolution (MM-RealSR). Specifically, we propose a metric learning scheme to map the unquantifiable degradation level to a ranking score (named *degradation score* in this work) in a metric space. The degradation score generated from the metric space can reflect the relative strength of the degradation level and provide pseudo supervision to the interactive restoration mechanism. A brief illustration and comparison to the existing interactive modulation strategy for image restoration are presented in Fig. 2. To restrict the learned degradation scores to a reasonable range and normalize the distribution of metric space, we further introduce an anchor point strategy in the metric learning process. Specifically, when the degradation score is zero, the network almost learns an identity mapping, while the network has the strongest restoration ability when the degradation score is one. Equipped with such pseudo estimations for real-world degradations, our MM-RealSR can learn the controllable restoration by mapping the degradation scores to different restoration strengths. Concretely, a condition network is used to generate controllable condition vectors to adjust the restoration quality. A base network is used to perform RWSR based on the input LR image and condition vectors.

We summarize our contributions as follows. **1)** We propose a metric learning strategy to map unquantifiable degradation levels in real-world scenarios to a metric space in an unsupervised manner. We further introduce an anchor point strategy to normalize the distribution of metric space. **2)** Our proposed MM-RealSR is the first work to investigate the interactive modulation for RWSR. **3)** Extensive experiments show that the proposed MM-RealSR achieves excellent modulation (*e.g.*, Fig. 1) and restoration performance in RWSR.

## 2 Related Work

### 2.1 Image Super-Resolution

As a pioneer work, a three-convolutional-layer network is used in SRCNN [7] to learn the LR-HR mapping for single image SR. Since then, motivated by the

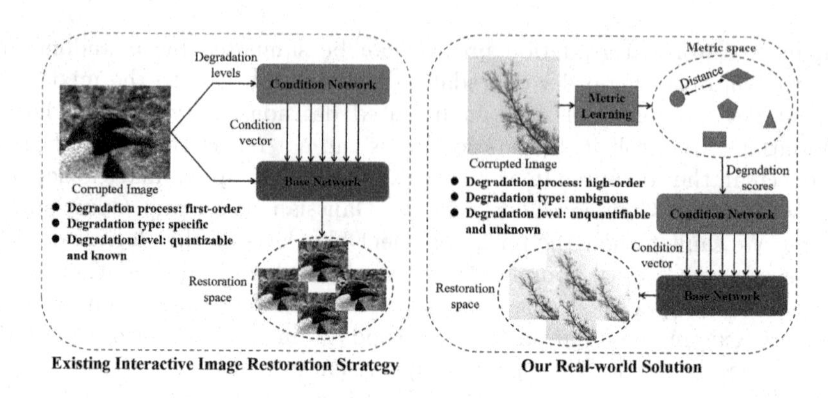


**Fig. 2.** Illustration of the existing interactive modulation strategy and our solution. Existing methods are confined to the sample with simple (*e.g.*, the first-order) and known degradations. Our method is an unsupervised solution to solve the more challenging real-world problem.

promising performance of CNN, several CNN-based SR methods have been proposed. Realizing the importance of the network depth, [17] designed a 20-layer network with residual blocks. [49] then combined residual learning and dense connection to extend the network depth to 100 layers. To enlarge the receptive field, [23,30,48] proposed to utilize non-local self-similarity in SR. Recently, channel attention and second-order channel attention were further introduced by RCAN [47] and SAN [5] to exploit feature correlation for improved performance. Some GAN-based methods [38] were proposed to produce higher perceptual quality. Among existing methods, setting downsampling operation as a known prior is the most popular choice. However, they have poor generalization ability due to the domain gap between real-world LR images and clean-LR images used for training (typically clean bicubically downsampled images).

There have been several attempts in blind SR [15,22] or real-world SR. [2,44] used a separate degradation estimation network to guide the blind SR process. Such methods usually consider simple synthetic degradations, and the accuracy of degradation representation largely affects the restoration quality. [4,39] capture real-world training pairs through specific cameras followed by tedious alignments. However, it is expensive to get such real-world training pairs, and it only works with specific imaging devices. [27,41] directly learned from unpaired data with cycle consistency loss. Nevertheless, learning fine-grained degradations with unpaired data is challenging and usually unsatisfactory. [14,35,43,50] generated training pairs as close to real data as possible, including various degradation factors and random/high-order degradation processes.

### 2.2 Interactive Modulation for Image Restoration

Most existing image restoration methods can only restore a specific image to a fixed result. Controllable image restoration methods aim to solve this problem by allowing users to adjust the restoration strength. Some pioneer works (*e.g.*, DNI [37] and AdaFM [12]) find that the learned filters of restoration models trained with different restoration levels are pretty similar at visual patterns. Based on this observation, DNI directly interpolated kernels to attain a smooth control of diverse imagery effects. AdaFM adopts a more efficient way by inserting AdaFM layers to change the statistics of filters. The work in CFSNet [34] adaptively learns the interpolation coefficients and uses them to couple intermediate features from the main branch and tuning branch. Recently, CResMD [13] distinguished different degradation types in the modulation process and proposed a framework that accepts both corrupted images and their degradation information as input to realize multi-dimension modulation. Following this strategy, [3] proposed a GAN-based image restoration framework to produce higher perceptual quality with interactive modulation. However, existing modulation strategies can only be trained in a supervised manner to deal with simple and first-order degradations, which have limitations in real-world applications.

### 2.3 Metric Learning

Metric learning [21] aims to measure the distance among samples while using an optimal distance metric for learning tasks. Instead of providing an explicit label, metric learning compares the distance among inputs to construct a metric space. In the last few years, deep learning and metric learning have been brought together to introduce the concept of deep metric learning [25]. Most of the existing deep metric learning methods can be roughly categorized based on loss functions: 1) contrastive loss [11]; 2) triplet loss [32]; 3) margin ranking loss [19,20]. Contrastive loss and triplet loss have a similar formulation to close the distance between semantic-similar samples and enlarge the distance between dissimilar samples. Similar to the hinge loss in support vector machine (SVM) [31], margin ranking loss is designed for capturing the ranking relationship between input training samples. In this paper, we utilize the margin ranking loss to construct the metric space of degradation levels, which drives our model to predict degradation levels in an unsupervised manner.

## 3 Methodology

### 3.1 Overview

The overview of our proposed method (MM-RealSR) is presented in Fig. 3. It is composed of a base network, a condition network, and an unsupervised degradation estimation module. The condition network takes the degradation scores as inputs to generate a condition vector. During inference, we can simply change the degradation scores of each degradation factor to adjust the restoration strength. The base network takes condition vectors and degraded images as inputs to restore a clean result controlled by the condition vectors. The unsupervised degradation estimation module is used to map the unquantifiable degradation,

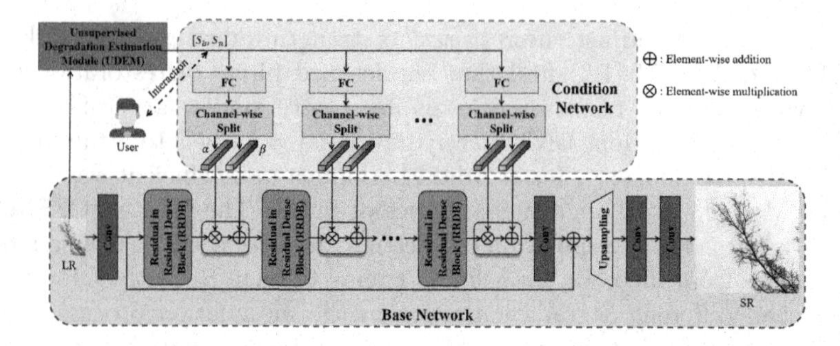


**Fig. 3.** Overview of our proposed metric learning based interactive modulation method for real-world super-resolution (MM-RealSR). It is composed of a base network, a condition network, and an unsupervised degradation estimation module.

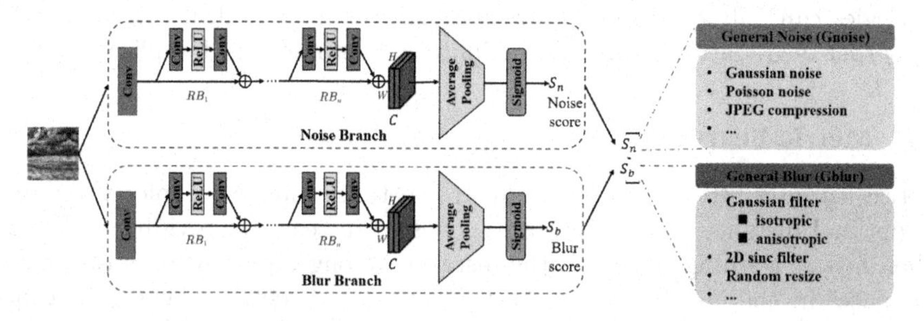


**Fig. 4.** The architecture of our proposed unsupervised degradation estimation module (UDEM). It is composed of two branches to estimate the degradation score of the general noise (Gnoise) and general blur (Gblur), respectively.

which involve complex and real-world degradations, into a metric space. Each degraded image can find a degradation score in this metric space by the unsupervised degradation estimation module. Note that the degradation scores reflect the relative strength in the metric space instead of the absolute values of the degradation level. The estimated degradation scores can further provide pseudo supervision to train the condition network.

### 3.2 Unsupervised Degradation Estimation Module (UDEM)

**Generalized Degradation Factors for Controllable Dimensions.** Before we delve into the details of the unsupervised degradation estimation module (UDEM), we first present the controllable dimensions for real-world interactive modulation. Recall that we want to investigate the interactive modulation under real-world degradations. We adopt the high-order degradation process [35,43] to simulate real-world degradations. In this setting, there are different degradation types (*e.g.*, Gaussian noise, Poisson noise, anisotropic Gaussian blur) and thus we cannot simply determine the controllable dimensions according to their types. Furthermore, due to the mixture of different degradation levels in high-order

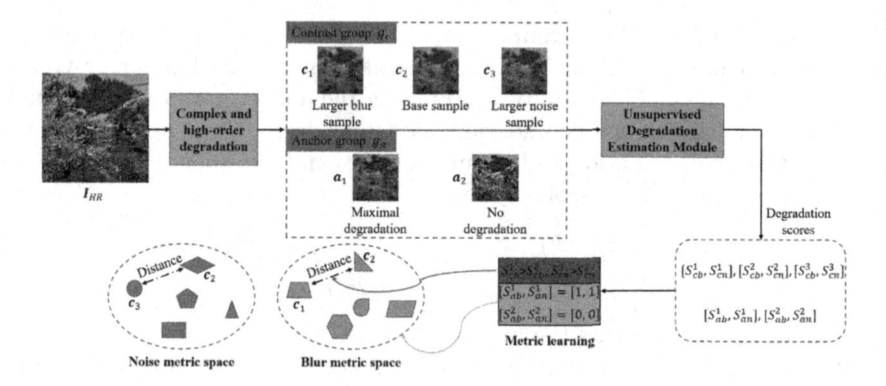


**Fig. 5.** The illustration of the optimization pipeline of our proposed unsupervised degradation estimation module. We utilize metric learning to construct a metric space for each degradation. We also apply two anchor points to restrict the learned degradation scores to a reasonable range and normalize the distribution of metric space.

degradations, the explicit degradation levels can not reflect the actual degradation effects in the corrupted images. For example, a corrupted image is degraded by a sequence of Gaussian blur with $\sigma_1$, Gaussian noise, and Gaussian blur with $\sigma_2$. We cannot know the final equivalent blur levels on this degraded image.

Therefore, we need to re-define the controllable dimensions for real-world interactive modulation. In this paper, we propose to adopt the *general noise* (Gnoise) and *general blur* (Gblur) as the two controllable dimensions, as users usually focus on the visual effects brought by those two factors. Specifically, general noise includes Gaussian noise, Poisson noise, and JPEG compression, as they all introduce unpleasant "artifacts" into images. Note that we classify JPEG compression artifacts as Gnoise, as the JPEG blocking artifacts are visually more like noise. The general blur includes Gaussian filter (isotropic/anisotropic), 2D sinc filter [35], and random resize operation, as they all weaken image details.

**Module Architecture.** Our proposed UDEM is used to construct two metric spaces, corresponding to Gblur and Gnoise, respectively. Each metric space can convert the unquantifiable degradation level to a degradation score. A brief illustration of our UDEM is presented in Fig. 4. Specifically, there are two prediction branches to predict the Gnoise score $S_n$ and Gblur score $S_b$, respectively. Each branch comprises several residual blocks end with an average pooling operation.

**Metric Learning Pipeline.** To construct the metric space with UDEM, we propose a metric learning pipeline, as illustrated in Fig. 5. Given the HR image $\mathbf{I}_{HR}$, we first corrupt it with complex and high-order degradations as [35] to simulate the real-world degraded images. To train the metric space, we generate two groups of samples – the contrast group $\mathbf{g}_c$ and anchor group $\mathbf{g}_a$. $\mathbf{g}_c$ is used to model the distance in the metric space by comparing the sample distance among this group, and $\mathbf{g}_a$ is to normalize the distribution of the metric space.

Concretely, there are three samples in the contrast group $\mathbf{g}_c = \{\mathbf{c}_1, \mathbf{c}_2, \mathbf{c}_3\}$, in which $\mathbf{c}_1$ has larger Gblur degradation than $\mathbf{c}_2$, and $\mathbf{c}_3$ has larger Gnoise degradation than $\mathbf{c}_2$. There are two samples in the anchor group $\mathbf{g}_a = \{\mathbf{a}_1, \mathbf{a}_2\}$. $\mathbf{a}_1$ has the maximal Gnoise degradation and Gblur degradation, and $\mathbf{a}_2$ does not have any Gnoise degradation and Gblur degradation. Correspondingly, there are two groups of degradation scores generated as:

$$\{[S_{cn}^1, S_{cb}^1], [S_{cn}^2, S_{cb}^2], [S_{cn}^3, S_{cb}^3]\} = \mathcal{F}_{UDEM}(\{\mathbf{c}_1, \mathbf{c}_2, \mathbf{c}_3\}) \tag{1}$$

$$\{[S_{an}^1, S_{ab}^1], [S_{an}^2, S_{ab}^2]\} = \mathcal{F}_{UDEM}(\{\mathbf{a}_1, \mathbf{a}_2\}), \tag{2}$$

where $\mathcal{F}_{UDEM}(\cdot)$ represents the function of UDEM.

We then adopt the margin ranking loss [19,20] to construct the Gnoise metric space and Gblur metric space. The margin ranking loss is defined as:

$$\mathcal{L}_{ML} = \frac{1}{2N} \sum_{i,j}^{N} \mathtt{max}(0, \gamma - \delta(\mathbf{s}_i, \mathbf{s}_j) \cdot (\hat{\mathbf{s}}_i - \hat{\mathbf{s}}_j)) \tag{3}$$

$$\delta(\mathbf{s}_i, \mathbf{s}_j) = \begin{cases} 1, & \mathbf{s}_i \geq \mathbf{s}_j \\ -1, & \mathbf{s}_i < \mathbf{s}_j, \end{cases} \tag{4}$$

where $\mathbf{s}_i$ and $\mathbf{s}_j$ are the ground truth scores. $\hat{\mathbf{s}}_i$ and $\hat{\mathbf{s}}_j$ are predicted scores. $\gamma$ is the margin parameter to constrain the distance between two samples. $N$ refers to the number of training samples. Note that during optimization, we do not need to know the explicit values of $\mathbf{s}_i$ and $\mathbf{s}_j$ but their relative values. In our cases, two margin ranking losses are applied:

$$\mathcal{L}_{ML}^{n} = \frac{1}{N} \sum_{i}^{N} \mathtt{max}(0, \gamma - (S_{cn}^{1\circledast i} - S_{cn}^{2\circledast i})) \tag{5}$$

$$\mathcal{L}_{ML}^{b} = \frac{1}{N} \sum_{i}^{N} \mathtt{max}(0, \gamma - (S_{cb}^{3\circledast i} - S_{cb}^{2\circledast i})), \tag{6}$$

where $\circledast i$ represents the *i-th* training sample. Note that since we choose the larger Gblur and Gnoise samples for $\mathbf{c}_1$ and $\mathbf{c}_3$, we already know their relative values, thus omitting the $\delta(\cdot,\cdot)$ function.

In order to further restrict the learned degradation scores to a reasonable range and normalize the distribution of metric space, we introduce an anchor point strategy in the metric learning process. Specifically, when the degradation score is zero, the network is enforced to learns an identity mapping, while the network has the strongest restoration ability when the degradation score is one. The introduced anchor loss ($\mathcal{L}_{AC}$) is defined as:

$$\mathcal{L}_{AC} = \mathcal{L}_{AC}^{u} + \mathcal{L}_{AC}^{l} \tag{7}$$

$$\begin{cases} \mathcal{L}_{AC}^{u} = \frac{1}{N} \sum_{i}^{N} (||S_{an}^{1\circledast i} - 1||_2^2 + ||S_{ab}^{1\circledast i} - 1||_2^2) \\ \mathcal{L}_{AC}^{l} = \frac{1}{N} \sum_{i}^{N} (||S_{an}^{2\circledast i}||_2^2 + ||S_{ab}^{2\circledast i}||_2^2), \end{cases} \tag{8}$$

where $||\cdot||_2^2$ represents the $\ell_2$ norm. $\mathcal{L}_{AC}^{u}$ and $\mathcal{L}_{AC}^{l}$ constraint the upper bound and lower bound of degradation scores.

### 3.3 Base Network

Our proposed MM-RealSR is a GAN-based solution for interactive RWSR. Thus, the base network consists of a generator $G$ and a discriminator $D$. The generator $G$ accepts the input image and outputs the restored result, while the discriminator $D$ aims to discriminate the restored result and the ground truth.

We adopt the same generator as ESRGAN [38] and Real-ESRGAN [35]. Specifically, it comprises several residual-in-residual dense blocks (RRDB), as shown in Fig. 3. At the end of the network, a pixel-shuffle upsampling and two convolution layers are applied for reconstruction. For discriminator, we apply the same architecture as Real-ESRGAN, *i.e.*, a U-Net design to better consider the realness values for each pixel. Note that the restoration performance of Real-ESRGAN is the theoretical upper bound of our proposed MM-RealSR, as MM-RealSR needs to learn another objective for interactive modulation.

### 3.4 Condition Network

As shown in Fig. 3, the condition network is composed of several fully-connected (FC) layers, placed at the end of each RRDB. Each FC takes degradation scores $[S_n, S_b]$ as input and generates a condition vector $Z_i \in \mathbb{R}^{B\times 2C}$, where $B$ and $C$ represent the batch size and the number of channels. $i$ represents the modulation operation in the *i-th* RRDB. Then, a channel-wise splitting operation divides $Z_i$ into two affine transformation parameters $\alpha_i$, $\beta_i \in \mathbb{R}^{B\times C}$. The modulation operation is performed through an affine transformation with $\alpha_i$ and $\beta_i$:

$$\widetilde{\mathbf{F}}_i = \mathbf{F}_i \cdot \alpha_i + \beta_i, \tag{9}$$

where $\mathbf{F}_i$ and $\widetilde{\mathbf{F}}_i$ refer to the output feature in the *i-th* RRDB and the transformation results, respectively.

### 3.5 Loss Function

The proposed MM-RealSR constructs a conditional mapping in real-world super-resolution. The training objective is to drive our MM-RealSR to match this mapping relation, leading to two specific goals, *i.e.*, interactive modulation mechanism and GAN-based image restoration. The interactive modulation mechanism is optimized with the metric learning strategy, as illustrated in Sect. 3.2, including two margin ranking losses ($\mathcal{L}^n_{ML}$, $\mathcal{L}^b_{ML}$) and two anchor losses ($\mathcal{L}^u_{AC}$, $\mathcal{L}^l_{AC}$). The GAN-based image restoration is optimized with $\mathcal{L}_1$ loss, perceptual loss $\mathcal{L}_{per}$ [16], and GAN loss $\mathcal{L}_{GAN}$ [10]. The total training objective is defined as:

$$\mathcal{L} = \lambda_1\mathcal{L}_{GAN} + \lambda_2\mathcal{L}_{per} + \lambda_3\mathcal{L}_1 + \lambda_4(\mathcal{L}^n_{ML} + \mathcal{L}^b_{ML} + \mathcal{L}^u_{AC} + \mathcal{L}^l_{AC}), \tag{10}$$

where $\{\lambda_1,\ \lambda_2,\ \lambda_3,\ \lambda_4\}$ are hyper parameters to balance these loss items. In our implementation, we set these four weights as $\{\lambda_1,\ \lambda_2,\ \lambda_3,\ \lambda_4\} = \{0.1,\ 1,\ 1,\ 0.05\}$.

# 4 Experiment

## 4.1 Experimental Settings

**Datasets.** Similar to ESRGAN [38] and Real-ESRGAN [35], we adopt DIV2K [1], Flickr2K [33] and OutdoorSceneTraining [36] datasets for training. For evaluation, we use the dataset provided in the challenge of Real-World Super-Resolution: AIM19 [26], and we also use a test set: RealSRSet [4] modeling DLSR camera corruptions. In all cases, we use $\times 4$ upsampling.

**Training Details.** Our training process consists of two stages. First, we pre-train the PSNR-oriented modulation model through Eq. 10, without perceptual loss $\mathcal{L}_{per}$ and GAN loss $\mathcal{L}_{GAN}$. In this stage, We train our system for $1000K$ iterations with the learning rate $2 \times 10^{-4}$. The well-trained model is served as the starting point for the next GAN training. In the second stage, our system is optimized with Eq. 10 for $400K$ iterations. The learning rate is fixed as $1 \times 10^{-4}$. For optimization, we use Adam [18] with $\beta_1 = 0.9$ and $\beta_2 = 0.99$. In both two stages, we set the batch size to 48, with the input patch size being 64.

**Evaluation Framework.** The main contribution of this paper is extending the interactive modulation mechanism to RWSR with the help of metric learning. The evaluation includes three aspects. **First**, we evaluate the estimation ability of our UDEM to present that the proposed metric learning strategy can well perform degradation estimation in an unsupervised manner. **Second**, we evaluate the restoration performance without manually adjusting the degradation score estimated by UDEM. This aims to present that our MM-RealSR can achieve state-of-the-art RWSR performance. **Third**, we compare our MM-RealSR with the recent modulation SR methods to present that existing modulation methods can not deal with real-world scenarios. Since AIM19 and RealSRSet datasets provide a paired validation set, we compute the learned perceptual image patch similarity (LPIPS) [45] and deep image structure and texture similarity (DISTS) [6] as the evaluation items. We also adopt a non-reference image quality assessment (IQA) metric (*i.e.*, NIQE [28]) for quantitative evaluation.

## 4.2 Effectiveness of the Unsupervised Degradation Estimation

As mentioned above, our degradation estimation module is optimized in an unsupervised manner to adapt the real-world degradation settings. To verify the effectiveness of this module, we first evaluate it on the synthesized data, corrupted by noise, blur, and JPEG compression degradations, respectively. For the noise degradation, we choose the Gaussian noise with the noise level range [1, 30]. For the blur degradation, we choose the Gaussian blur, with the $\sigma$ range $[0.2, 3]$. For the JPEG compression, we set the quality factor ranging from 30 to 95. We divide the degradation range of these three kinds of degradations equally into 20 points and add them to natural images, respectively. The evaluation results are presented in Fig. 6, including the estimation performance of our proposed MM-RealSR and the same model trained without anchor loss (w/o anchor loss). One

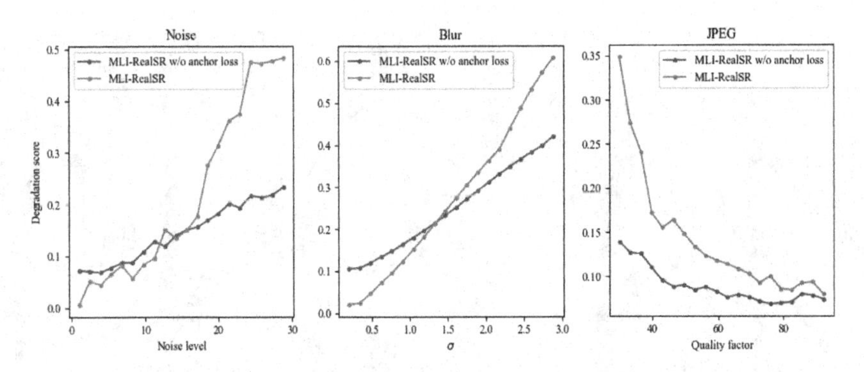


**Fig. 6.** The evaluation of our UDEM, including the model trained with and without the anchor loss. It shows that metric learning can generate good metric spaces to rank the degradation levels for various degradation types. The anchor loss can further normalize the space distribution and better distinguish degradation levels in a larger score range.

**Fig. 7.** Visualization of the degradation scoring ability of our unsupervised degradation estimation module (UDEM) on real-world data.

can see that with the increase of degradation level, the estimated degradation score increases monotonously. Note that for JPEG compression, the degradation artifact is weakened with the increase of the quality factor.

Additionally, the anchor loss allows the estimated score to have a larger dynamic range. Therefore, our unsupervised degradation estimation module can accurately estimate the relative value from the corresponding metric space. The anchor loss (Eq. 8) can constrain the distribution of points in metric space. In Fig. 6, there are some minor fluctuations for estimating noise and JPEG artifacts, while the curve is pretty smooth for the blur degradation. We conjecture that estimating noise is more difficult than estimating blur, as distinguishing the textures on bricks, wood, and noisy artifacts are challenging.

Apart from the quantitative evaluation, we also visualize the scoring performance on the real-world data, as presented in Fig. 7. We can find that the scoring results on the real-world data are reasonable and satisfactory. The first sample is noisy, and the second sample is blurry. We can observe that they obtain large scores on noise and blur, respectively, while they have a very low score on the other degradations. The fifth sample has large noise with some blur, and the

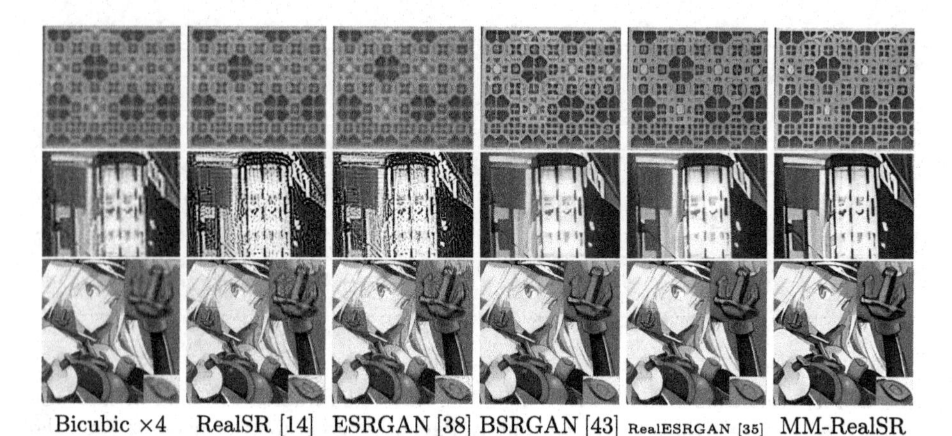


**Fig. 8.** Visual comparison between our proposed MM-RealSR and several recent methods on real-world super-resolution. The degradation scores are estimated by our UDEM without specific adjustment during inference.

predicted scores can also reflect such an observation. Note that the scoring for noise and blur is achieved in two separate metric spaces. Therefore, the noise score and blur score are not suitable for strict corresponding, and there might be a slight difference with visual comparison.

### 4.3 Evaluation for Restoration Performance

**Comparison Details.** In this part, we fully compare our proposed MM-RealSR with recent modulation (*i.e.*, CUGAN [3]) and non-modulation methods. The non-modulation methods include several well-known SR methods (*e.g.*, RealSR [14] and ESRGAN [38]) and some recent top-performing methods (*e.g.*, BSR-GAN [43] and Real-ESRGAN [35]). Real-ESRGAN is also the theoretical upper bound of our proposed MM-RealSR, as MM-RealSR needs to learn another objective for interactive modulation. Note that CUGAN must be provided with the degradation levels ($S_b$, $S_n \in [0,\ 1]$), which are not available in real-world test sets, *i.e.*, RealSRSet and AIM19. Thus, we divide the noise and blur degradation levels into 11 points uniformly distributed between 0 and 1, and then traversed all cases (121 cases in total). We select the best result (lowest LPIPS) for each image. Unlike the above two settings, our MM-RealSR can be used either with the degradation scores estimated by our UDEM module or with the user input scores. In this part, to make a fair comparison, the input degradation score of our MM-RealSR is provided by UDEM without manual adjustment.

**Comparison Results.** The quantitative comparison is presented in Table 1. One can see that compared with other methods, our proposed MM-RealSR achieves excellent performance even with interactive modulation ability. Especially, our performance is comparable or even better than the upper-bound model Real-ESRGAN. Even selecting the best result for each image, the performance

**Table 1.** Quantitative comparisons (NIQE/LPIPS/DISTS) on RealSRSet [4] and AIM19 [26] test sets. Best results are **highlighted**. We compare our MM-RealSR with both modulation (Mod.) and non-modulation (Non-mod.) methods. Real-ESRGAN [35] is the upper-bound model of our MM-RealSR.

| Setting | | RealSR [14] | ESRGAN [38] | CUGAN [3] | BSRGAN [43] | Upper bound [35] | MM-RealSR |
|---|---|---|---|---|---|---|---|
| | | Non-mod. | Non-mod. | Mod. | Non-mod. | Non-mod. | For both |
| Real SRSet | NIQE ↓ | 8.31 | 8.23 | 6.17 | 4.62 | **4.53** | 4.62 |
| | LPIPS ↓ | 0.4047 | 0.4054 | 0.3757 | 0.3647 | 0.3640 | **0.3606** |
| | DISTS ↓ | 0.2201 | 0.2190 | 0.1764 | **0.1535** | 0.1545 | 0.1664 |
| AIM19 | NIQE ↓ | **2.86** | 3.60 | 4.77 | 3.92 | 3.60 | 3.61 |
| | LPIPS ↓ | 0.5566 | 0.5558 | 0.4635 | 0.4048 | 0.3957 | **0.3948** |
| | DISTS ↓ | 0.2325 | 0.2336 | 0.1944 | 0.1596 | 0.1545 | **0.1507** |

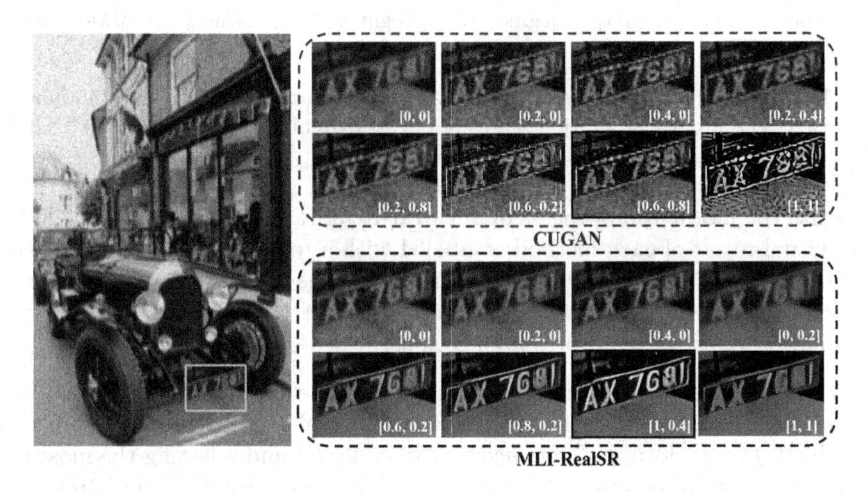


**Fig. 9.** The evaluation of the modulation performance on real-world super-resolution (×4). The degradation scores (*i.e.*, $[S_b,\ S_n]$) are manually selected for CUGAN [3] and our proposed MM-RealSR to present the visual quality and adjusting ability. More comparison videos are in the appendix.

of CUGAN is still inferior to our proposed MM-RealSR. It demonstrates the incapability of existing modulation methods in real-world scenarios. Though RealSR and ESRGAN achieve better performance on NIQE on the AIM19 test set, the reconstruction quality of these two methods is unsatisfactory, as shown in Fig. 8. From the visual comparison in Fig. 8, we can find that our MM-RealSR can achieve better restoration quality with vivid details, demonstrating that our approach can estimate the degradation scores to guide the restoration process in real-world scenarios.

## 4.4 Real-World Modulation Performance

**Comparison with CUGAN.** In this part, We compare our method with the most recent modulation-based SR method CUGAN [3] on real-world data. We

**Table 2.** The quantitative comparison between our MM-RealSR and the segmented modulation approach. The quantitative results are evaluated on RealSRSet test set [4].

| | Segment-1 | Segment-2 | Segment-3 | Segment-4 | Best-LPIPS | MM-RealSR |
|---|---|---|---|---|---|---|
| LPIPS ↓ | 0.3957 | 0.3910 | 0.3950 | 0.4125 | 0.3905 | **0.3662** |
| DISTS ↓ | 0.1764 | 0.1820 | 0.1910 | 0.1942 | 0.1760 | **0.1632** |

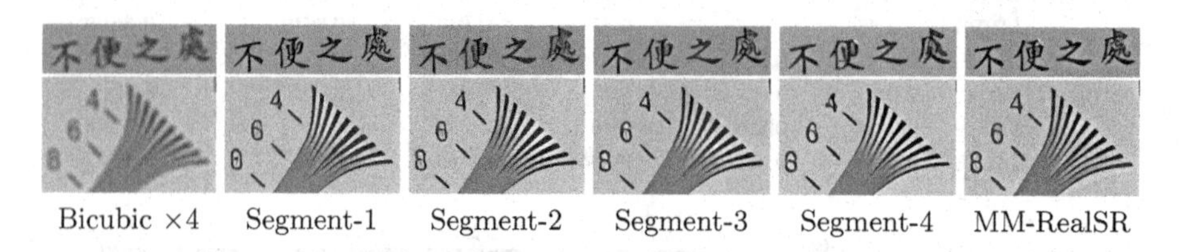


**Fig. 10.** Visual comparison between our MM-RealSR and the segmented modulation approach. The input degradation scores of our MM-RealSR are estimated by UDEM.

visualize the modulation process of our MM-RealSR and CUGAN in Fig. 9. The best visual result of each method is labeled with a red box We can find that the CUGAN trained with simple degradations can not deal with real-world scenarios. On the contrary, our MM-RealSR has better reconstruction quality with a larger modulation range in real-world applications.

**Further Discussions.** A direct interactive modulation method for RWSR is to train several models with different degradation levels. The modulation can then be achieved by performing inference on those models and selecting the most satisfactory result. To compare our method with this case, we uniformly divide the degradations in training Real-ESRGAN [35] into four segments, corresponding to four models. In the evaluation, we select the best result (lowest LPIPS) among the four results for each sample. We experiment on the RealSRSet [4] and show the results in Table 2 and Fig. 10. Obviously, the performance of such a simple approach is inferior to our method both quantitatively and qualitatively. Besides, this simple method only has four points for modulation.

## 5 Conclusion

In this paper, we present the first attempt to study the interactive modulation in real-world super-resolution through a novel metric learning based strategy. Specifically, we utilize metric learning to map the unquantifiable degradation level in real-world scenarios to a metric space and train it in an unsupervised manner. Equipped with the metric space, we can get a relative ranking score of the unquantifiable real-world degradation to guide the restoration and modulation processes. We also propose an anchor point strategy to constrain the distribution in the metric space. To adapt to the complex degradations in real-world scenarios, we also carefully design controllable dimensions in which each degradation factor has a more general degradation meaning. Extensive experiments demonstrate that our proposed MM-RealSR achieves excellent modulation and restoration performance in real-world super-resolution.

**Limitations.** Our MM-RealSR adopts the complex and high-order degradation synthesis [35,43] to simulate the real-world degradations. However, it is still hard to directly apply our method to train on real samples. There are also some failure cases (*e.g.*, local jitter in Fig. 6, and our UDEM regards some art effects such as bokeh as degradation). Our work makes the first attempt towards interactive modulation in RWSR, and there is still a long way with challenges.

**Acknowledgement.** This work was partially supported by the Shenzhen Fundamental Research Program (No. GXWD20201231165807007-20200807164903001), National Natural Science Foundation of China (61906184, U1913210), and the Shanghai Committee of Science and Technology, China (Grant No. 21DZ1100100).

## References

1. Agustsson, E., Timofte, R.: NTIRE 2017 challenge on single image super-resolution: dataset and study. In: Proceedings of the IEEE Conference on Computer Vision and Pattern Recognition Workshops (CVPRW), pp. 126–135 (2017)
2. Bell-Kligler, S., Shocher, A., Irani, M.: Blind super-resolution kernel estimation using an internal-GAN. In: Proceedings of the Advances in Neural Information Processing Systems (NeurIPS) (2019)
3. Cai, H., He, J., Qiao, Y., Dong, C.: Toward interactive modulation for photo-realistic image restoration. In: Proceedings of the IEEE Conference on Computer Vision and Pattern Recognition (CVPR), pp. 294–303 (2021)
4. Cai, J., Zeng, H., Yong, H., Cao, Z., Zhang, L.: Toward real-world single image super-resolution: a new benchmark and a new model. In: Proceedings of the IEEE International Conference on Computer Vision (ICCV), pp. 3086–3095 (2019)
5. Dai, T., Cai, J., Zhang, Y., Xia, S.T., Zhang, L.: Second-order attention network for single image super-resolution. In: Proceedings of the IEEE Conference on Computer Vision and Pattern Recognition (CVPR), pp. 11065–11074 (2019)
6. Ding, K., Ma, K., Wang, S., Simoncelli, E.P.: Image quality assessment: unifying structure and texture similarity. IEEE Trans. Pattern Anal. Mach. Intell. **44**(05), 2567–2581 (2022)
7. Dong, C., Loy, C.C., He, K., Tang, X.: Learning a deep convolutional network for image super-resolution. In: Fleet, D., Pajdla, T., Schiele, B., Tuytelaars, T. (eds.) ECCV 2014. LNCS, vol. 8692, pp. 184–199. Springer, Cham (2014). https://doi.org/10.1007/978-3-319-10593-2_13
8. Dong, C., Loy, C.C., He, K., Tang, X.: Image super-resolution using deep convolutional networks. IEEE Trans. Pattern Anal. Mach. Intell. **38**(2), 295–307 (2015)
9. Glasner, D., Bagon, S., Irani, M.: Super-resolution from a single image. In: Proceedings of the International Conference on Computer Vision (ICCV), pp. 349–356 (2009)
10. Goodfellow, I., et al.: Generative adversarial nets. In: Advances in Neural Information Processing Systems (NeurIPS), vol. 27 (2014)

11. Hadsell, R., Chopra, S., LeCun, Y.: Dimensionality reduction by learning an invariant mapping. In: Proceedings of the IEEE Conference on Computer Vision and Pattern Recognition (CVPR), pp. 1735–1742 (2006)
12. He, J., Dong, C., Qiao, Y.: Modulating image restoration with continual levels via adaptive feature modification layers. In: Proceedings of the IEEE Conference on Computer Vision and Pattern Recognition (CVPR), pp. 11056–11064 (2019)
13. He, J., Dong, C., Qiao, Yu.: Interactive multi-dimension modulation with dynamic controllable residual learning for image restoration. In: Vedaldi, A., Bischof, H., Brox, T., Frahm, J.-M. (eds.) ECCV 2020. LNCS, vol. 12365, pp. 53–68. Springer, Cham (2020). https://doi.org/10.1007/978-3-030-58565-5_4
14. Ji, X., Cao, Y., Tai, Y., Wang, C., Li, J., Huang, F.: Real-world super-resolution via kernel estimation and noise injection. In: Proceedings of the IEEE Conference on Computer Vision and Pattern Recognition Workshops (CVPRW), pp. 466–467 (2020)
15. Jiang, J., Zhang, K., Timofte, R.: Towards flexible blind JPEG artifacts removal. In: Proceedings of the IEEE International Conference on Computer Vision (ICCV), pp. 4997–5006 (2021)
16. Johnson, J., Alahi, A., Fei-Fei, L.: Perceptual losses for real-time style transfer and super-resolution. In: Leibe, B., Matas, J., Sebe, N., Welling, M. (eds.) ECCV 2016. LNCS, vol. 9906, pp. 694–711. Springer, Cham (2016). https://doi.org/10.1007/978-3-319-46475-6_43
17. Kim, J., Lee, J.K., Lee, K.M.: Accurate image super-resolution using very deep convolutional networks. In: Proceedings of the IEEE Conference on Computer Vision and Pattern Recognition (CVPR), pp. 1646–1654 (2016)
18. Kingma, D.P., Ba, J.: Adam: a method for stochastic optimization. In: Proceedings of the International Conference on Learning Representations (ICLR) (2015)
19. Koltchinskii, V., Panchenko, D.: Empirical margin distributions and bounding the generalization error of combined classifiers. Ann. Stat. **30**(1), 1–50 (2002)
20. Kong, S., Shen, X., Lin, Z., Mech, R., Fowlkes, C.: Photo aesthetics ranking network with attributes and content adaptation. In: Leibe, B., Matas, J., Sebe, N., Welling, M. (eds.) ECCV 2016. LNCS, vol. 9905, pp. 662–679. Springer, Cham (2016). https://doi.org/10.1007/978-3-319-46448-0_40
21. Kulis, B., et al.: Metric learning: a survey. Found. Trends Mach. Learn. **5**(4), 287–364 (2012)
22. Liu, A., Liu, Y., Gu, J., Qiao, Y., Dong, C.: Blind image super-resolution: a survey and beyond. arXiv preprint arXiv:2107.03055 (2021)
23. Liu, D., Wen, B., Fan, Y., Loy, C.C., Huang, T.S.: Non-local recurrent network for image restoration. In: Proceedings of the Advances in Neural Information Processing Systems (NeurIPS), pp. 1673–1682 (2018)
24. Liu, Y., Wang, S., Zhang, J., Wang, S., Ma, S., Gao, W.: Iterative network for image super-resolution. IEEE Trans. Multimed. **24**, 2259–2272 (2021)
25. Lu, J., Hu, J., Zhou, J.: Deep metric learning for visual understanding: an overview of recent advances. IEEE Signal Process. Mag. **34**(6), 76–84 (2017)
26. Lugmayr, A., et al.: AIM 2019 challenge on real-world image super-resolution: methods and results. In: Proceedings of the IEEE International Conference on Computer Vision Workshop (ICCVW), pp. 3575–3583 (2019)
27. Maeda, S.: Unpaired image super-resolution using pseudo-supervision. In: Proceedings of the IEEE Conference on Computer Vision and Pattern Recognition (CVPR), pp. 291–300 (2020)
28. Mittal, A., Soundararajan, R., Bovik, A.C.: Making a "completely blind" image quality analyzer. IEEE Signal Process. Lett. **20**(3), 209–212 (2012)

29. Mou, C., Wang, Q., Zhang, J.: Deep generalized unfolding networks for image restoration. In: Proceedings of the IEEE Conference on Computer Vision and Pattern Recognition (CVPR), pp. 17399–17410 (2022)
30. Mou, C., Zhang, J., Wu, Z.: Dynamic attentive graph learning for image restoration. In: Proceedings of the IEEE International Conference on Computer Vision (ICCV), pp. 4328–4337 (2021)
31. Platt, J.: Sequential minimal optimization: a fast algorithm for training support vector machines (1998)
32. Schroff, F., Kalenichenko, D., Philbin, J.: FaceNet: a unified embedding for face recognition and clustering. In: Proceedings of the IEEE Conference on Computer Vision and Pattern Recognition (CVPR), pp. 815–823 (2015)
33. Timofte, R., Agustsson, E., Van Gool, L., Yang, M.H., Zhang, L.: NTIRE 2017 challenge on single image super-resolution: methods and results. In: Proceedings of the IEEE Conference on Computer Vision and Pattern Recognition Workshops (CVPRW), pp. 114–125 (2017)
34. Wang, W., Guo, R., Tian, Y., Yang, W.: CFSNet: toward a controllable feature space for image restoration. In: Proceedings of the IEEE International Conference on Computer Vision (ICCV), pp. 4140–4149 (2019)
35. Wang, X., Xie, L., Dong, C., Shan, Y.: Real-ESRGAN: training real-world blind super-resolution with pure synthetic data. In: Proceedings of the International Conference on Computer Vision Workshops (ICCVW), pp. 1905–1914 (2021)
36. Wang, X., Yu, K., Dong, C., Loy, C.C.: Recovering realistic texture in image super-resolution by deep spatial feature transform. In: Proceedings of the IEEE Conference on Computer Vision and Pattern Recognition (CVPR), pp. 606–615 (2018)
37. Wang, X., Yu, K., Dong, C., Tang, X., Loy, C.C.: Deep network interpolation for continuous imagery effect transition. In: Proceedings of the IEEE Conference on Computer Vision and Pattern Recognition (CVPR), pp. 1692–1701 (2019)
38. Wang, X., et al.: ESRGAN: enhanced super-resolution generative adversarial networks. In: Proceedings of the European Conference on Computer Vision Workshops (ECCVW) (2018)
39. Wei, P., et al.: Component divide-and-conquer for real-world image super-resolution. In: Vedaldi, A., Bischof, H., Brox, T., Frahm, J.-M. (eds.) ECCV 2020. LNCS, vol. 12353, pp. 101–117. Springer, Cham (2020). https://doi.org/10.1007/978-3-030-58598-3_7
40. You, D., Zhang, J., Xie, J., Chen, B., Ma, S.: Coast: controllable arbitrary-sampling network for compressive sensing. IEEE Trans. Image Process. **30**, 6066–6080 (2021)
41. Yuan, Y., Liu, S., Zhang, J., Zhang, Y., Dong, C., Lin, L.: Unsupervised image super-resolution using cycle-in-cycle generative adversarial networks. In: Proceedings of the IEEE Conference on Computer Vision and Pattern Recognition Workshops (CVPRW), pp. 701–710 (2018)
42. Zhang, J., Xiong, R., Zhao, C., Ma, S., Zhao, D.: Exploiting image local and nonlocal consistency for mixed gaussian-impulse noise removal. In: Proceedings of the IEEE International Conference on Multimedia and Expo, pp. 592–597 (2012)
43. Zhang, K., Liang, J., Van Gool, L., Timofte, R.: Designing a practical degradation model for deep blind image super-resolution. In: Proceedings of the IEEE International Conference on Computer Vision (ICCV), pp. 4791–4800 (2021)
44. Zhang, K., Zuo, W., Zhang, L.: Learning a single convolutional super-resolution network for multiple degradations. In: Proceedings of the IEEE Conference on Computer Vision and Pattern Recognition (CVPR), pp. 3262–3271 (2018)

45. Zhang, R., Isola, P., Efros, A.A., Shechtman, E., Wang, O.: The unreasonable effectiveness of deep features as a perceptual metric. In: Proceedings of the IEEE Conference on Computer Vision and Pattern Recognition (CVPR), pp. 586–595 (2018)
46. Zhang, Y., et al.: Collaborative representation cascade for single-image super-resolution. IEEE Trans. Syst. Man Cybern. Syst. **49**, 845–860 (2017)
47. Zhang, Y., Li, K., Li, K., Wang, L., Zhong, B., Fu, Y.: Image super-resolution using very deep residual channel attention networks. In: Proceedings of the European Conference on Computer Vision (ECCV), pp. 286–301 (2018)
48. Zhang, Y., Li, K., Li, K., Zhong, B., Fu, Y.: Residual non-local attention networks for image restoration. In: International Conference on Learning Representations (2019)
49. Zhang, Y., Tian, Y., Kong, Y., Zhong, B., Fu, Y.: Residual dense network for image super-resolution. In: Proceedings of the IEEE Conference on Computer Vision and Pattern Recognition (CVPR), pp. 2472–2481 (2018)
50. Zhou, R., Susstrunk, S.: Kernel modeling super-resolution on real low-resolution images. In: Proceedings of the IEEE International Conference on Computer Vision (ICCV), pp. 2433–2443 (2019)

# Author Index

Printed in the United States
by Baker & Taylor Publisher Services